P9-CJX-276

Classical and Modern Fourier Analysis

Loukas Grafakos

University of Missouri

PEARSON
Prentice Hall

Pearson Education, Inc.
Upper Saddle River, New Jersey 07458

Library of Congress Cataloging-in-Publication Data

Loukas Grafakos
Classical and modern Fourier analysis / Loukas Grafakos
p. cm.
Includes bibliographical references and index.
ISBN 0-13-035399-X

1. Fourier analysis. I. Title.

QA403.5.G73 2004
515'.2433—dc21 2003051280

Editor in Chief: *Sally Yagan*
Acquisitions Editor: *George Lobell*
Production Editor: *Lynn Savino Wendel*
Vice President/Director of Production and Manufacturing: *David W. Riccardi*
Senior Managing Editor: *Linda Mihatov Behrens*
Assistant Managing Editor: *Bayani Mendoza DeLeon*
Executive Managing Editor: *Kathleen Schiaparelli*
Assistant Manufacturing Manager/Buyer: *Michael Bell*
Manufacturing Manager: *Trudy Pisciotti*
Marketing Manager: *Halee Dinsey*
Marketing Assistant: *Rachel Beckman*
Art Director: *Jayne Conte*
Editorial Assistant: *Jennifer Brady*
Cover Designer: *Bruce Kenselaar*
Cover Photo Credits: *Stephen Montgomery-Smith*

Pearson Education, Inc., Upper Saddle River, New Jersey 07458

Printed in the United States of America
10 9 8 7 6 5 4 3 2 1

ISBN 0-13-035399-X

Pearson Education LTD., London
Pearson Education Australia PTY, Limited, Sydney
Pearson Education Singapore, Pte. Ltd
Pearson Education North Asia Ltd, Hong Kong
Pearson Education Canada, Ltd., Toronto
Pearson Educación de Mexico, S.A. de C.V.
Pearson Education—Japan, Tokyo
Pearson Education Malaysia, Pte. Ltd

Mathematics compares the most diverse phenomena and discovers the secret analogies that unite them.

Jean Baptiste Joseph Fourier (1768–1830)

About The Author

Loukas Grafakos is a native of Athens, Greece. He earned his doctoral degree at UCLA and is currently a Professor of Mathematics at the University of Missouri. He has taught at Yale University and Washington University in St. Louis and he has also held visiting positions at the Mathematical Sciences Research Institute in Berkeley and the University of Pittsburgh. He has been named a Kemper Fellow for Excellence in Teaching and he has authored or co-authored over forty research articles in Fourier analysis. An avid traveller, he has visited over one hundred countries and has given many international lectures.

CONTENTS

PREFACE

Prologue

The word *analysis* comes from the Greek $\alpha\nu\acute{\alpha}\lambda\upsilon\sigma\iota s$, which means "dissolving into pieces." This is usually the first step of a process that leads to a careful study and understanding of an object or phenomenon. The antithetical process, called $\sigma\acute{\upsilon}\nu\theta\varepsilon\sigma\iota s$, is equally significant as it assembles the analyzed pieces after they have been individually examined. This procedure is the heart of Fourier analysis. Through its aorta, this heart disseminates information to a variety of applications. Fourier analysis is therefore a prism that diffracts ideas into a rainbow of uses and applications, making the subject one of the richest and most far-reaching in mathematics.

The primary goal of this text is to present the theoretical foundation of the field of Fourier analysis. This book is mainly addressed to graduate students in mathematics and is designed to serve for a three-course sequence on the subject. The only prerequisite for understanding the text is satisfactory completion of a course in measure theory, Lebesgue integration, and complex variables. This book is intended to present the selected topics in some depth and stimulate further study. Although the emphasis falls on real variable methods in Euclidean spaces, a chapter is devoted to the fundamentals of analysis on the torus. This material is included for historical reasons, as the genesis of Fourier analysis can be found in trigonometric expansions of periodic functions in several variables.

The choice of the material in the text reflects a measure of personal taste; however, a certain effort has been made to include a variety of topics of general interest. Much attention is given to details, which are designed to facilitate the understanding of first-time readers. Based on my personal experience, I felt a need to include details related to topics that articles often omit, leaving beginners to struggle through without explanation. Although it will behoove many readers to skim through the more technical aspects of the presentation and concentrate on the flow of ideas, the mere fact that details are here for reference will be comforting to some. I hope that students will profit from this comprehensive presentation and learn how to do mathematics rigorously. Unfortunately, including so many details has led to the large size of the book. But as one's maturity and familiarity with the subject increases, topics slowly become natural and reading is significantly accelerated.

The exercises that follow each section enrich the material of the corresponding section and provide an opportunity to develop additional intuition and deeper comprehension. Some of them are rather rudimentary and require minimal skill, while others are more interesting and challenging. Only a few exercises are considered difficult, but these are given with hints. A special effort has been made to prepare the exercises, which unfortunately did not double, but almost tripled, the amount of time and effort it took to complete this text. I hope that the reader will find this extra effort beneficial.

The historical notes given at the end of each chapter are intended to provide an accurate account of past research but also to suggest directions for further investigation. This book was partly written with the purpose of attracting students to research. Many of the topics in Chapter 10 lead to open problems that have bewildered mathematicians for decades. It is hoped that many students will be fascinated by the easy statements, yet the delicate complexity of some of these problems, and pursue a deeper understanding.

The text is completely self-contained as the appendix includes the miscellaneous material needed throughout. Certain user-friendly conventions have been adopted to facilitate searching. For instance, theorems, propositions, definitions, lemmas, remarks, and examples are numbered according to the order in which they appear in each section. Exercises are numbered similarly and can be easily located.

As this book is intended for a three-course sequence on the subject, I would like to suggest a slowly paced initial breakdown of the material, flexible enough to accommodate adjustments: Semester I: Chapters 1, 2, 3, and 4. Semester II: Chapters 5, 6, 7, and 9. Semester III: Chapters 8, 10, and other topics. Sections or subsections marked by a star would normally be omitted in a yearly course.

I am solely responsible for any misprints, mistakes, and historical omissions in this book. Please contact me directly (loukas@math.missouri.edu) if you have any comments, suggestions, improvements, or corrections. Instructors are also welcome to contact me to obtain further hints on the existing exercises in the text. Suggestions for other exercises are also welcome. A list of current errata with acknowledgements will be kept at the following URL:

http://math.missouri.edu/~loukas/Fourier-Analysis

Acknowledgments

I gratefully acknowledge the following individuals, who have assisted me in making this book a reality:

All the anonymous and eponymous reviewers who provided me with an abundance of meaningful remarks, corrections, and suggestions for improvements.

All Prentice Hall staff who dealt with the production of this book, especially George Lobell, Lynn Savino Wendel, and Dennis Kletzing.

For their valuable suggestions, corrections, and other important assistance at different stages in my preparation of this monograph, I would like to offer my deep appreciation and sincere thanks to Carmen Chicone, David Cramer, Brenda Frazier, Derrick Hart, Mark Hoffmann, Helge Holden, Brian Hollenbeck, Petr Honzík, Tunde Jakab, Gregory Jones, Douglas Kurtz, Antonios Melas, Keith Mersman, Krzysztof Oleszkiewicz, Cristina Pereyra, Daniel Redmond, Jorge Rivera-Noriega, Dmitry Ryabogin, Shih-Chi Shen, Elias Stein, Christoph Thiele, Deanie Tourville, Don Vaught, Brett Wick, and James Wright.

Many thanks to Georges Alexopoulos, who used my manuscript to teach a course at the Université Paris-Sud at Orsay in the spring 2002 and contributed in identifying a variety of mistakes.

I am grateful to Nakhlé Asmar for making me believe in the project, for his constant encouragement and support, and for bringing me in contact with the editorial staff of Prentice Hall. Several of his suggestions have significantly improved the text.

Many thanks to Steven Hofmann, who explained thoroughly to me the material in Section 8.7 and provided me with the proof of Theorem 8.6.3.

My deep appreciation to Alexander Iosevich, who provided me with the simple proof of Lemma 5.5.3, resulting in the slick exposition of Section 5.5.

I am indebted to Nigel Kalton, who assisted me with the exposition of Sections 1.4, 4.5, and 4.6 and provided me with the proof of Lemma 1.4.20. I have learned a lot of mathematics and I have immensely profited from my numerous conversations with him. I have been very fortunate to interact with him and I thank him sincerely for everything he has taught me.

I am also indebted to Steven Krantz, who sent me several pages of corrections and suggestions that have significantly improved my presentation. His contribution added some fine touches to this book.

I can't find words to thank José María Martell, who assisted me in the exposition in Chapter 9 and provided me with the proof of Lemma 9.5.4. I have greatly profited from my mathematical discussions with him.

I would like to express my deep gratitude to Stephen Montgomery-Smith, who provided me with the proof of Theorem 6.3.6 and helped me with a variety of technical issues related to the presentation. The picture on the cover of the book is courtesy of him as well.

I especially appreciate Andrea Nahmod's encouragement and consistent praise of this work. My sincere thanks goes to her for using my notes to teach a graduate course at the University of Massachusetts at Amherst and for giving me important feedback from her experience.

I am also thankful to Carlos Pérez who provided me with the material that shaped the exposition in section 7.5.

Many thanks to Terence Tao for providing me with the proof of Lemma 3.3.4 and for his instrumental collaboration in the proof of Theorem 10.6.1.

I am especially grateful to Rodolfo Torres, who assisted me with the exposition in Chapter 6, fueled my patience during the last eight years, and encouraged me so strongly during the seemingly endless writing of the manuscript.

Many thanks to Igor Verbitsky for his elegant improvement of the proof of Theorem 4.5.1 and for his crucial comments and remarks.

I would like to acknowledge the careful proofreading of Bruno Calado, Jakub Duda, Emmanouil Katsoprinakis, Nguyen Cong Phuc, Roman Shvidkoy, and Nikolaos Tzirakis and express my deep gratitude to them. They have all preserved the mathematical integrity of this book, and I am truly indebted to them for their effort.

I would like to express my thanks to my graduate students Atanas Stefanov, Xiaochun Li, Erin Terwilleger, Dmytro Bilyk, Geoffrey Diestel, Linqiao Zhao, and Christopher Sansing, who read various parts of the manuscript and enriched it with their corrections and suggestions; but most important I would like to thank them for making my world of mathematics more meaningful.

Finally, none of this work would have been possible without the constant love and support of my wife Suzanne. I am taking this opportunity to express my deep gratitude to her. The recent arrival of our beautiful children Joanna and Constandina has changed the meaning of my life. It is to Suzanne, Joanna, and the newborn Constandina that this book is dedicated.

Loukas Grafakos
Columbia, Missouri
loukas@math.missouri.edu

April 2003

CHAPTER 1

L^p Spaces and Interpolation

The primary focus of this monograph is the study of Fourier series and integrals of functions. Many of their properties are quantitatively expressed in terms of the integrability of the function. For this reason it is desirable to acquire a good understanding of spaces of functions whose modulus to a power p is integrable. These are called Lebesgue spaces and are denoted by L^p. Although an in-depth study of Lebesgue spaces falls outside the scope of this book, it seems appropriate to devote a chapter to reviewing some of their fundamental properties.

The emphasis of our review will be basic interpolation between Lebesgue spaces. Many problems in Fourier analysis concern boundedness of operators on Lebesgue spaces and interpolation provides a framework that often simplifies their study. For instance, in order to show that a linear operator maps L^p into itself for all $1 < p < \infty$, it is sufficient to show that it maps the (smaller) Lorentz space $L^{p,1}$ into the (larger) Lorentz space $L^{p,\infty}$ for the same range of p's. Moreover, some further reductions can be made in terms of the Lorentz space $L^{p,1}$. This and other considerations indicate that interpolation is a powerful tool in the study of boundedness of operators.

Although we will be mainly concerned with L^p subspaces of the Euclidean space $\mathbf{R}^n$, we discuss in this chapter L^p spaces of arbitrary measure spaces, as they often present a useful general setting. Moreover, many proofs in the text go through when Lebesgue measure is replaced by a more general measure.

1.1. L^p and Weak L^p

Let X be a measure space and let μ be a positive, not necessarily finite, measure on X. For $0 < p < \infty$, $L^p(X, \mu)$ will denote the set of all complex-valued μ-measurable functions on X whose modulus to the pth power is integrable. $L^\infty(X, \mu)$ will be the set of all complex-valued μ-measurable functions f on X such that for some $B > 0$, the set $\{x : |f(x)| > B\}$ has μ-measure zero. Two functions in $L^p(X, \mu)$ will be considered equal if they are equal μ-almost everywhere. The notation $L^p(\mathbf{R}^n)$ will be reserved for the space $L^p(\mathbf{R}^n, |\cdot|)$, where $|\cdot|$ denotes n-dimensional Lebesgue measure. Lebesgue measure on $\mathbf{R}^n$ will also be denoted by dx. Within context and in the lack of ambiguity, $L^p(X,\mu)$ will simply be L^p. The space $L^p(\mathbf{Z})$ equipped with counting measure will be denoted by $\ell^p(\mathbf{Z})$ or simply ℓ^p.

For $0 < p < \infty$, we define the L^p quasi-norm of a function f by

$$\left\|f\right\|_{L^p(X,\mu)} = \left(\int_X |f(x)|^p \, d\mu(x)\right)^{\frac{1}{p}} \tag{1.1.1}$$

and for $p = \infty$ by

$$\left\|f\right\|_{L^\infty(X,\mu)} = \inf\big\{B > 0 \,:\, \mu(\{x : |f(x)| > B\}) = 0\big\}. \tag{1.1.2}$$

It is well known that Minkowski's (or the triangle) inequality

$$\left\|f+g\right\|_{L^p(X,\mu)} \le \left\|f\right\|_{L^p(X,\mu)} + \left\|g\right\|_{L^p(X,\mu)} \tag{1.1.3}$$

holds for all f, g in $L^p = L^p(X,\mu)$, whenever $1 \le p \le \infty$. Since in addition $\left\|f\right\|_{L^p(X,\mu)} = 0$ implies that $f = 0$ (μ-a.e.), the L^p spaces are normed linear spaces for $1 \le p \le \infty$. For $0 < p < 1$, inequality (1.1.3) is reversed when $f, g \ge 0$. However, the following substitute of (1.1.3) holds:

$$\left\|f+g\right\|_{L^p(X,\mu)} \le 2^{(1-p)/p}\big(\left\|f\right\|_{L^p(X,\mu)} + \left\|g\right\|_{L^p(X,\mu)}\big) \tag{1.1.4}$$

and thus the spaces $L^p(X,\mu)$ are quasi-normed linear spaces. See also Exercise 1.1.5. For all $0 < p \le \infty$, it can be shown that every Cauchy sequence in $L^p(X,\mu)$ is convergent, and hence the spaces $L^p(X,\mu)$ are complete. For the case $0 < p < 1$ we refer to Exercise 1.1.8. Therefore, the L^p spaces are Banach spaces for $1 \le p \le \infty$ and quasi-Banach spaces for $0 < p < 1$. For any $p \in (0,\infty) \setminus \{1\}$ we will use the notation $p' = \frac{p}{p-1}$. Moreover we set $1' = \infty$ and $\infty' = 1$ so that $p'' = p$ for all $p \in (0,\infty]$. Hölder's inequality says that for all $p \in [1,\infty]$ and all measurable functions f, g on (X,μ) we have

$$\left\|fg\right\|_{L^1} \le \left\|f\right\|_{L^p}\left\|g\right\|_{L^{p'}}.$$

It is a well-known fact that the dual $(L^p)^*$ of L^p is isometric to $L^{p'}$ for all $1 \le p < \infty$. Furthermore, the L^p norm of a function can be obtained via duality when $1 \le p \le \infty$ as follows:

$$\left\|f\right\|_{L^p} = \sup_{\left\|g\right\|_{L^{p'}}=1} \left|\int_X f\,g\,d\mu\right|.$$

For the endpoint cases $p = 1$, $p = \infty$, see Exercise 1.4.12(a), (b).

1.1.a. The Distribution Function

Definition 1.1.1. For f a measurable function on X, the *distribution function* of f is the function d_f defined on $[0,\infty)$ as follows:

$$d_f(\alpha) = \mu(\{x \in X : |f(x)| > \alpha\}). \tag{1.1.5}$$

The distribution function d_f provides information about the size of f but not about the behavior of f itself near any given point. For instance, a function on $\mathbf{R}^n$ and any of its translates have the same distribution function. It follows from Definition 1.1.1 that d_f is a decreasing function of α (not necessarily strictly).

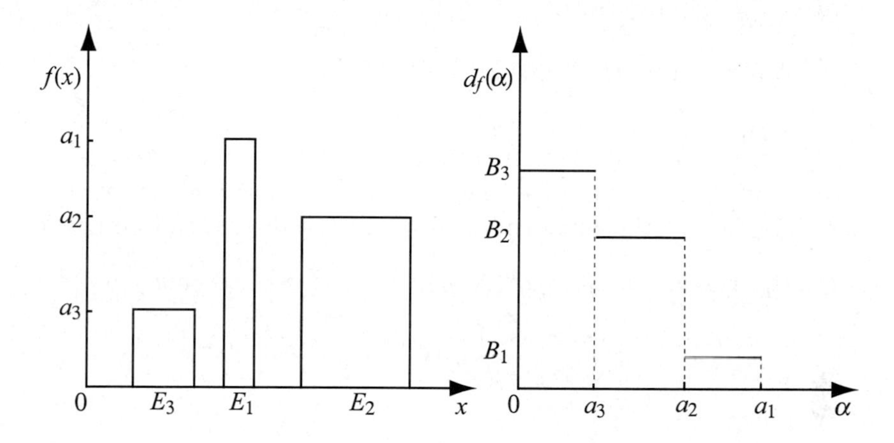

FIGURE 1.1. The graph of a simple function $f = \sum_{k=1}^{3} a_k \chi_{E_k}$ and its distribution function $d_f(\alpha)$. Here $B_j = \sum_{k=1}^{j} \mu(E_k)$.

Example 1.1.2. Recall that simple functions are finite linear combinations of characteristic functions of sets of finite measure. For pedagogical reasons we compute the distribution function d_f of a nonnegative simple function

$$f(x) = \sum_{j=1}^{N} a_j \chi_{E_j}(x),$$

where the sets E_j are pairwise disjoint and $a_1 > \cdots > a_N > 0$. If $\alpha \geq a_1$, then clearly $d_f(\alpha) = 0$. However, if $a_2 \leq \alpha < a_1$ then $|f(x)| > \alpha$ precisely when $x \in E_1$ and, in general, if $a_{j+1} \leq \alpha < a_j$, then $|f(x)| > \alpha$ precisely when $x \in E_1 \cup \cdots \cup E_j$. Setting

$$B_j = \sum_{k=1}^{j} \mu(E_k)$$

we have

$$d_f(\alpha) = \sum_{j=1}^{N} B_j \chi_{[a_{j+1}, a_j)}(\alpha),$$

where $a_{N+1} = 0$. Figure 1.1 illustrates this example when $N = 3$.

We now state a few simple facts about the distribution function d_f. We have

Proposition 1.1.3. *Let f and g be measurable functions on (X, μ). Then for all $\alpha, \beta > 0$ we have*

1. $|g| \le |f|$ μ-*a.e. implies that* $d_g \le d_f$
2. $d_{cf}(\alpha) = d_f(\alpha/|c|)$, *for all* $c \in \mathbf{C} \setminus \{0\}$
3. $d_{f+g}(\alpha+\beta) \le d_f(\alpha) + d_g(\beta)$
4. $d_{fg}(\alpha\beta) \le d_f(\alpha) + d_g(\beta)$.

Proof. The simple proofs are left to the reader. □

Knowledge of the distribution function d_f provides sufficient information to evaluate precisely the L^p norm of a function f. We state and prove the following important description of the L^p norm in terms of the distribution function.

Proposition 1.1.4. *For f in $L^p(X,\mu)$, $0 < p < \infty$, we have*

$$\|f\|_{L^p}^p = p\int_0^\infty \alpha^{p-1} d_f(\alpha)\, d\alpha. \tag{1.1.6}$$

Proof. We have

$$\begin{aligned} p\int_0^\infty \alpha^{p-1} d_f(\alpha)\, d\alpha &= p\int_0^\infty \alpha^{p-1} \int_X \chi_{\{x:\, |f(x)|>\alpha\}}\, d\mu(x)\, d\alpha \\ &= \int_X \int_0^{|f(x)|} p\alpha^{p-1}\, d\alpha\, d\mu(x) \\ &= \int_X |f(x)|^p\, d\mu(x) \\ &= \|f\|_{L^p}^p, \end{aligned}$$

where we used Fubini's theorem to obtain the second equality. This proves (1.1.6). □

Notice that the same argument yields the more general fact that for any increasing continuously differentiable function φ on $[0,\infty)$ we have

$$\int_X \varphi(|f|)\, d\mu = \int_0^\infty \varphi'(\alpha) d_f(\alpha)\, d\alpha\,. \tag{1.1.7}$$

Definition 1.1.5. For $0 < p < \infty$, the space *weak* $L^p(X,\mu)$ is defined as the set of all μ-measurable functions f such that

$$\|f\|_{L^{p,\infty}} = \inf\{C > 0 : d_f(\alpha) \le \frac{C^p}{\alpha^p} \quad \text{for all} \quad \alpha > 0\} \tag{1.1.8}$$

$$= \sup\{\gamma\, d_f(\gamma)^{\frac{1}{p}} : \gamma > 0\} \tag{1.1.9}$$

is finite. The space *weak*-$L^\infty(X,\mu)$ is by definition $L^\infty(X,\mu)$.

The reader should check that (1.1.9) and (1.1.8) are in fact equal. The weak L^p spaces will also be denoted by $L^{p,\infty}(X,\mu)$. Two functions in $L^{p,\infty}(X,\mu)$ will be considered equal if they are equal μ-a.e. The notation $L^{p,\infty}(\mathbf{R}^n)$ is reserved for $L^{p,\infty}(\mathbf{R}^n, |\cdot|)$. Using Proposition 1.1.3 (2), we can easily show that

$$\big\|kf\big\|_{L^{p,\infty}} = |k|\big\|f\big\|_{L^{p,\infty}}, \tag{1.1.10}$$

for any complex nonzero constant k. The analogue of (1.1.3) is

$$\big\|f+g\big\|_{L^{p,\infty}} \le c_p\big(\big\|f\big\|_{L^{p,\infty}} + \big\|g\big\|_{L^{p,\infty}}\big), \tag{1.1.11}$$

where $c_p = \max(2, 2^{1/p})$ a fact that follows from Proposition 1.1.3 (3) with $\alpha_1 = \alpha_2 = \alpha/2$. We also have that

$$\big\|f\big\|_{L^{p,\infty}(X,\mu)} = 0 \Rightarrow f = 0 \qquad \mu\text{-a.e.} \tag{1.1.12}$$

In view of (1.1.10), (1.1.11), and (1.1.12), $L^{p,\infty}$ is a quasi-normed linear space for $0<p<\infty$.

The weak L^p spaces are larger than the usual L^p spaces. We have the following:

Proposition 1.1.6. *For any $0 < p < \infty$, and any f in $L^p(X,\mu)$ we have $\big\|f\big\|_{L^{p,\infty}} \le \big\|f\big\|_{L^p}$; hence $L^p(X,\mu) \subseteq L^{p,\infty}(X,\mu)$.*

Proof. This is just a trivial consequence of Chebychev's inequality:

$$\alpha^p d_f(\alpha) \le \int_{\{x:\,|f(x)|>\alpha\}} |f(x)|^p \, d\mu(x)\,. \tag{1.1.13}$$

As the integral in (1.1.13) is at most $\big\|f\big\|_{L^p}^p$, using (1.1.9) we obtain that $\big\|f\big\|_{L^{p,\infty}} \le \big\|f\big\|_{L^p}$.

□

The inclusion $L^p \subseteq L^{p,\infty}$ is strict. For example, on $\mathbf{R}^n$ with the usual Lebesgue measure, let $h(x) = |x|^{-\frac{n}{p}}$. Obviously, h is not in $L^p(\mathbf{R}^n)$ but h is in $L^{p,\infty}(\mathbf{R}^n)$ and we may check easily that $\big\|h\big\|_{L^{p,\infty}(\mathbf{R}^n)} = v_n$, where v_n is the measure of the unit ball of $\mathbf{R}^n$.

It is not immediate from their definition that the weak L^p spaces are complete with respect to the quasi-norm $\big\|\cdot\big\|_{L^{p,\infty}}$. The completeness of these spaces is proved in Theorem 1.4.11, but it is also a consequence of Theorem 1.1.13 proved in this section.

1.1.b. Convergence in Measure

Next we discuss some convergence notions. The following notion is of importance in probability theory.

Definition 1.1.7. Let f, f_n, $n = 1, 2, \dots$ be measurable functions on the measure space (X,μ). The sequence f_n is said to *converge in measure* to f if for all $\varepsilon > 0$ there exists an $n_0 \in \mathbf{Z}^+$ such that

$$n > n_0 \implies \mu(\{x \in X : \ |f_n(x) - f(x)| > \varepsilon\}) < \varepsilon. \tag{1.1.14}$$

Remark 1.1.8. The preceding definition is equivalent to the following statement:

$$\text{For all } \varepsilon > 0 \quad \lim_{n\to\infty} \mu(\{x \in X : |f_n(x) - f(x)| > \varepsilon\}) = 0. \tag{1.1.15}$$

Clearly (1.1.15) implies (1.1.14). To see the converse given $\varepsilon > 0$, pick $0 < \delta < \varepsilon$ and apply (1.1.14) for this δ. There exists an $n_0 \in \mathbf{Z}^+$ such that

$$\mu(\{x \in X : |f_n(x) - f(x)| > \delta\}) < \delta$$

holds for $n > n_0$. Since

$$\mu(\{x \in X : |f_n(x) - f(x)| > \varepsilon\}) \le \mu(\{x \in X : |f_n(x) - f(x)| > \delta\}),$$

we conclude that

$$\mu(\{x \in X : |f_n(x) - f(x)| > \varepsilon\}) < \delta$$

for all $n > n_0$. Let $n \to \infty$ to deduce that

$$\limsup_{n\to\infty} \mu(\{x \in X : |f_n(x) - f(x)| > \varepsilon\}) \le \delta. \tag{1.1.16}$$

Since (1.1.16) holds for all $0 < \delta < \varepsilon$, (1.1.15) follows by letting $\delta \to 0$.

Convergence in measure is a more general property than convergence in either L^p or $L^{p,\infty}$, $0 < p \le \infty$, as the following proposition indicates:

Proposition 1.1.9. *Let $0 < p \le \infty$ and f_n, f be in $L^{p,\infty}(X,\mu)$.*

1. *If f_n, f are in L^p and $f_n \to f$ in L^p, then $f_n \to f$ in $L^{p,\infty}$.*
2. *If $f_n \to f$ in $L^{p,\infty}$, then f_n converges to f in measure.*

Proof. Fix $0 < p < \infty$. Proposition 1.1.6 gives that for all $\varepsilon > 0$ we have

$$\mu(\{x \in X : |f_n(x) - f(x)| > \varepsilon\}) \le \frac{1}{\varepsilon^p} \int_X |f_n - f|^p \, d\mu \,.$$

This shows that convergence in L^p implies convergence in weak L^p. The case $p = \infty$ is tautological.

Given $\varepsilon > 0$ find an n_0 such that for $n > n_0$, we have

$$\big\|f_n - f\big\|_{L^{p,\infty}} = \sup_{\alpha>0} \alpha\mu(\{x \in X : |f_n(x) - f(x)| > \alpha\})^{\frac{1}{p}} < \varepsilon^{\frac{1}{p}+1}.$$

Taking $\alpha = \varepsilon$, we conclude that convergence in $L^{p,\infty}$ implies convergence in measure. □

Example 1.1.10. Fix $0 < p < \infty$. On $[0,1]$ define the functions

$$f_{k,j} = k^{1/p} \chi_{(\frac{j-1}{k}, \frac{j}{k})}, \qquad k \ge 1,\ 1 \le j \le k.$$

Consider the sequence $\{f_{1,1}, f_{2,1}, f_{2,2}, f_{3,1}, f_{3,2}, f_{3,3}, \dots\}$. Observe that

$$|\{x : f_{k,j}(x) > 0\}| = 1/k \,.$$

Therefore, $f_{k,j}$ converges to 0 in measure. Likewise, observe that

$$\big\|f_{k,j}\big\|_{L^{p,\infty}} = \sup_{\alpha>0} \alpha|\{x : \ f_{k,j}(x) > \alpha\}|^{1/p} \geq \sup_{k\geq 1} \frac{(k-1/k)^{1/p}}{k^{1/p}} = 1,$$

which implies that $f_{k,j}$ does not converge to 0 in $L^{p,\infty}$.

It turns out that every sequence convergent in $L^p(X,\mu)$ or in $L^{p,\infty}(X,\mu)$ has a subsequence that converges a.e. to the same limit.

Theorem 1.1.11. *Let f_n and f be complex-valued measurable functions on a measure space (X,μ) and suppose that f_n converges to f in measure. Then some subsequence of f_n converges to f μ-a.e.*

Proof. For all $k = 1, 2, \dots$ choose inductively n_k such that

$$\mu(\{x \in X : \ |f_{n_k}(x) - f(x)| > 2^{-k}\}) < 2^{-k} \tag{1.1.17}$$

and such that $n_1 < n_2 < \dots < n_k < \dots$. Define the sets

$$A_k = \{x \in X : \ |f_{n_k}(x) - f(x)| > 2^{-k}\}.$$

Equation (1.1.17) implies that

$$\mu\Big(\bigcup_{k=m}^{\infty} A_k\Big) \leq \sum_{k=m}^{\infty} \mu(A_k) \leq \sum_{k=m}^{\infty} 2^{-k} = 2^{1-m} \tag{1.1.18}$$

for all $m = 1, 2, 3, \dots$. It follows from (1.1.18) that

$$\mu\Big(\bigcup_{k=1}^{\infty} A_k\Big) \leq 1 < \infty. \tag{1.1.19}$$

Using (1.1.18) and (1.1.19), we conclude that the sequence of the measures of the sets $\{\bigcup_{k=m}^{\infty} A_k\}_{m=1}^{\infty}$ converges as $m \to \infty$ to

$$\mu\Big(\bigcap_{m=1}^{\infty}\bigcup_{k=m}^{\infty} A_k\Big) = 0. \tag{1.1.20}$$

To finish the proof, observe that the null set in (1.1.20) contains the set of all $x \in X$ for which $f_{n_k}(x)$ does not converge to $f(x)$.

□

In many situations we are given a sequence of functions and we would like to extract a convergent subsequence. One way to achieve this is via the next theorem, which is a useful variant of Theorem 1.1.11. We first give a relevant definition.

Definition 1.1.12. We say that a sequence of measurable functions $\{f_n\}$ on the measure space (X,μ) is *Cauchy in measure* if for every $\varepsilon > 0$, there exists an $n_0 \in \mathbf{Z}^+$ such that for $n, m > n_0$ we have

$$\mu(\{x \in X : \ |f_m(x) - f_n(x)| > \varepsilon\}) < \varepsilon.$$

Theorem 1.1.13. *Let (X,μ) be a measure space and let $\{f_n\}$ be a complex-valued sequence on X that is Cauchy in measure. Then some subsequence of f_n converges μ-a.e.*

Proof. The proof is very similar to that of Theorem 1.1.11. For all $k=1,2,\dots$ choose n_k inductively such that

$$\mu(\{x\in X:\ |f_{n_k}(x)-f_{n_{k+1}}(x)|>2^{-k}\})<2^{-k} \tag{1.1.21}$$

and such that $n_1<n_2<\dots<n_k<n_{k+1}<\dots$. Define

$$A_k=\{x\in X:\ |f_{n_k}(x)-f_{n_{k+1}}(x)|>2^{-k}\}.$$

As shown in the proof of Theorem 1.1.11, (1.1.21) implies that

$$\mu\Big(\bigcap_{m=1}^{\infty}\bigcup_{k=m}^{\infty}A_k\Big)=0. \tag{1.1.22}$$

For $x\notin\bigcup_{k=m}^{\infty}A_k$ and $i\ge j\ge j_0\ge m$ (and j_0 large enough) we have

$$|f_{n_i}(x)-f_{n_j}(x)|\le\sum_{l=j}^{i-1}|f_{n_l}(x)-f_{n_{l+1}}(x)|\le\sum_{l=j}^{i-1}2^{-l}\le 2^{1-j}\le 2^{1-j_0}.$$

This implies that the sequence $\{f_{n_i}(x)\}_i$ is Cauchy for every x in the set $(\bigcup_{k=m}^{\infty}A_k)^c$ and therefore converges for all such x. We define a function

$$f(x)=\begin{cases}\lim\limits_{j\to\infty}f_{n_j}(x) & \text{when } x\notin\bigcap_{m=1}^{\infty}\bigcup_{k=m}^{\infty}A_k\\ 0 & \text{when } x\in\bigcap_{m=1}^{\infty}\bigcup_{k=m}^{\infty}A_k.\end{cases}$$

Then $f_{n_j}\to f$ almost everywhere. □

1.1.c. A First Glimpse at Interpolation

It is a useful fact that if a function f is in $L^p(X,\mu)$ and in $L^q(X,\mu)$, then it is in $L^r(X,\mu)$ for all $p<r<q$. The usefulness of the spaces $L^{p,\infty}$ can be seen from the following sharpening of this statement:

Proposition 1.1.14. *Let $0<p<q\le\infty$ and let f in $L^{p,\infty}(X,\mu)\cap L^{q,\infty}(X,\mu)$. Then f is in $L^r(X,\mu)$ for all $p<r<q$ and*

$$\big\|f\big\|_{L^r}\le\Big(\frac{r}{r-p}+\frac{r}{q-r}\Big)^{\frac{1}{r}}\big\|f\big\|_{L^{p,\infty}}^{\frac{\frac1r-\frac1q}{\frac1p-\frac1q}}\big\|f\big\|_{L^{q,\infty}}^{\frac{\frac1p-\frac1r}{\frac1p-\frac1q}}, \tag{1.1.23}$$

with the suitable interpretation when $q=\infty$.

Proof. Let us take first $q < \infty$. We know that

$$d_f(\alpha) \le \min\left(\frac{\|f\|_{L^{p,\infty}}^p}{\alpha^p}, \frac{\|f\|_{L^{q,\infty}}^q}{\alpha^q}\right). \tag{1.1.24}$$

Set

$$B = \left(\frac{\|f\|_{L^{q,\infty}}^q}{\|f\|_{L^{p,\infty}}^p}\right)^{\frac{1}{q-p}}. \tag{1.1.25}$$

We now estimate the L^r norm of f. By (1.1.24), (1.1.25), and Proposition 1.1.4 we have

$$\begin{aligned}
\|f\|_{L^r(X,\mu)}^r &= r\int_0^\infty \alpha^{r-1} d_f(\alpha)\,d\alpha \\
&\le r\int_0^\infty \alpha^{r-1}\min\left(\frac{\|f\|_{L^{p,\infty}}^p}{\alpha^p}, \frac{\|f\|_{L^{q,\infty}}^q}{\alpha^q}\right)d\alpha \\
&= r\int_0^B \alpha^{r-1-p}\|f\|_{L^{p,\infty}}^p\,d\alpha + r\int_B^\infty \alpha^{r-1-q}\|f\|_{L^{q,\infty}}^q\,d\alpha \\
&= \frac{r}{r-p}\|f\|_{L^{p,\infty}}^p B^{r-p} + \frac{r}{q-r}\|f\|_{L^{q,\infty}}^q B^{r-q} \\
&= \left(\frac{r}{r-p} + \frac{r}{q-r}\right)\left(\|f\|_{L^{p,\infty}}^p\right)^{\frac{q-r}{q-p}}\left(\|f\|_{L^{q,\infty}}^q\right)^{\frac{r-p}{q-p}}.
\end{aligned} \tag{1.1.26}$$

Observe that the integrals converge since $r - p > 0$ and $r - q < 0$.

The case $q = \infty$ is easier. Since $d_f(\alpha) = 0$ for $\alpha > \|f\|_{L^\infty}$ we only need to use the inequality $d_f(\alpha) \le \alpha^{-p}\|f\|_{L^{p,\infty}}^p$ for $\alpha \le \|f\|_{L^\infty}$ in estimating the first integral in (1.1.26). We obtain

$$\|f\|_{L^r}^r \le \frac{r}{r-p}\|f\|_{L^{p,\infty}}^p\|f\|_{L^\infty}^{r-p},$$

which is nothing other than (1.1.23) when $q = \infty$. This completes the proof. □

Note that (1.1.23) holds with constant 1 if $L^{p,\infty}$ and $L^{q,\infty}$ are replaced by L^p and L^q, respectively. Sometimes it will be convenient to work with functions that are only locally in some L^p space. This leads to the following definition.

Definition 1.1.15. For $0 < p < \infty$, the space $L^p_{\mathrm{loc}}(\mathbf{R}^n, |\cdot|)$ or simply $L^p_{\mathrm{loc}}(\mathbf{R}^n)$ is the set of all Lebesgue-measurable functions f on $\mathbf{R}^n$ such that

$$\int_K |f(x)|^p\,dx < \infty \tag{1.1.27}$$

for any compact subset K of $\mathbf{R}^n$. Functions that satisfy (1.1.27) with $p = 1$ are called *locally integrable* functions on $\mathbf{R}^n$.

The union of all $L^p(\mathbf{R}^n)$ spaces for $1 \le p \le \infty$ is contained in $L^1_{\text{loc}}(\mathbf{R}^n)$. More generally, for $0 < p < q < \infty$ we have the following:

$$L^q(\mathbf{R}^n) \subseteq L^q_{\text{loc}}(\mathbf{R}^n) \subseteq L^p_{\text{loc}}(\mathbf{R}^n).$$

Functions in $L^p(\mathbf{R}^n)$ for $0 < p < 1$ may not be locally integrable. For example, take $f(x) = |x|^{-n-\alpha}\chi_{|x|\le 1}$, which is in $L^p(\mathbf{R}^n)$ when $p < n/(n+\alpha)$, and observe that f is not integrable over any open set in $\mathbf{R}^n$ containing the origin.

Exercises

1.1.1. Suppose f and f_n are measurable functions on (X, μ). Prove that
(a) d_f is right continuous on $[0, \infty)$.
(b) If $|f| \le \liminf_{n\to\infty} |f_n|$ μ-a.e., then $d_f \le \liminf_{n\to\infty} d_{f_n}$.
(c) If $|f_n| \uparrow |f|$, then $d_{f_n} \uparrow d_f$.
[*Hint:* Part (a): Let t_n be a decreasing sequence of positive numbers that tends to zero. Show that $d_f(\alpha_0 + t_n) \uparrow d_f(\alpha_0)$ using a convergence theorem. Part (b): Let $E = \{x \in X : |f(x)| > \alpha\}$ and $E_n = \{x \in X : |f_n(x)| > \alpha\}$. Use that $\mu\big(\bigcap_{n=m}^\infty E_n\big) \le \liminf_{n\to\infty} \mu(E_n)$, and that $E \subseteq \bigcup_{m=1}^\infty \bigcap_{n=m}^\infty E_n$.]

1.1.2. (*Hölder's inequality*) Let $0 < p, p_1, \dots, p_k \le \infty$, where $k \ge 2$, and let f_j be in $L^{p_j} = L^{p_j}(X, \mu)$. Assume that

$$\frac{1}{p} = \frac{1}{p_1} + \cdots + \frac{1}{p_k}.$$

(a) Show that the product $f_1 \dots f_k$ is in L^p and that

$$\big\|f_1 \cdots f_k\big\|_{L^p} \le \big\|f_1\big\|_{L^{p_1}} \cdots \big\|f_k\big\|_{L^{p_k}}.$$

(b) When no p_j is infinite, show that if equality holds in part (a), then $c_1|f_1|^{p_1} = \cdots = c_k|f_k|^{p_k}$ a.e. for some $c_j \ge 0$.
(c) Let $0 < q < 1$. For $r < 0$ and $g > 0$ almost everywhere, let $\big\|g\big\|_{L^r} = \big\|g^{-1}\big\|_{L^{|r|}}^{-1}$. Show that for $f \ge 0$, $g > 0$ a.e. we have

$$\big\|fg\big\|_{L^1} \ge \big\|f\big\|_{L^q}\big\|g\big\|_{L^{q'}}.$$

1.1.3. Let (X, μ) be a probability space [this means $\mu(X) = 1$]. Prove that
(a) If f is in $L^{p_0}(X, \mu)$ for some $p_0 < \infty$, then

$$\lim_{p\to\infty} \big\|f\big\|_{L^p} = \big\|f\big\|_{L^\infty}.$$

(b) (*Jensen's inequality*) Prove that

$$\big\|f\big\|_{L^p} \ge \exp\left(\int_X \log|f(x)|\, d\mu(x)\right)$$

for all $0 < p < \infty$.

(c) If f is in some $L^{p_0}(X,\mu)$ for some $p_0>0$, then

$$\lim_{p\to 0}\|f\|_{L^p}=\exp\left(\int_X \log|f(x)|\,d\mu(x)\right)$$

using the interpretation $e^{-\infty}=0$.
[*Hint:* Part (a): Given $0<\varepsilon<\|f\|_{L^\infty}$, find a measurable set $E\subseteq X$ of positive measure such that $|f(x)|\geq\|f\|_{L^\infty}-\varepsilon$ for all $x\in E$. Then $\|f\|_{L^p}\geq(\|f\|_{L^\infty}-\varepsilon)\mu(E)^{1/p}$ and thus $\liminf_{p\to\infty}\|f\|_{L^p}\geq\|f\|_{L^\infty}-\varepsilon$. Part (b) is a direct consequence of Jensen's inequality $\int_X\log|h|\,d\mu\leq\log\left(\int_X|h|\,d\mu\right)$. Part (c): Fix a sequence $0<p_n<p_0$ such that $p_n\to 0$ and define

$$h_n(x)=(|f_n(x)|^{p_0}-1)/p_0-(|f_n(x)|^{p_n}-1)/p_n.$$

Use that $(t^p-1)/p\downarrow\log t$ as $p\downarrow 0$ for all $t>0$. By the Lebesgue monotone convergence theorem we obtain $\int_X h_n\,d\mu\uparrow\int_X h\,d\mu$, which implies that $\int_X(|f_n|^{p_n}-1)/p_n\,d\mu\to\int_X\log|f|\,d\mu$. (Observe here that the latter could be $-\infty$.) Use Jensen's inequality and the fact that $\log t\leq t-1$ for $t\geq 0$ to obtain $\int_X\log|f|\,d\mu=\frac{1}{p_n}\int_X(|f|^{p_n}-1)\,d\mu$. Now let $n\to\infty$ and use the squeeze property of limits.]

1.1.4. Let a_j be a sequence of positive reals. Show that
(a) $\left(\sum_{j=1}^\infty a_j\right)^\theta\leq\sum_{j=1}^\infty a_j^\theta$, for any $0\leq\theta\leq 1$
(b) $\sum_{j=1}^\infty a_j^\theta\leq\left(\sum_{j=1}^\infty a_j\right)^\theta$, for any $1\leq\theta<\infty$
(c) $\left(\sum_{j=1}^N a_j\right)^\theta\leq N^{\theta-1}\sum_{j=1}^N a_j^\theta$, when $1\leq\theta<\infty$
(d) $\sum_{j=1}^N a_j^\theta\leq N^{1-\theta}\left(\sum_{j=1}^N a_j\right)^\theta$, when $0\leq\theta\leq 1$.

1.1.5. Let $\{f_j\}_{j=1}^\infty$ be a sequence of $L^p(X,\mu)$ functions. Show that
(a) (*Minkowski's inequality*) For $1\leq p\leq\infty$,

$$\Big\|\sum_{j=1}^\infty f_j\Big\|_{L^p}\leq\sum_{j=1}^\infty\|f_j\|_{L^p}.$$

(b) (*Minkowski's inequality*) For $0<p<1$ and $f_j\geq 0$,

$$\sum_{j=1}^\infty\|f_j\|_{L^p}\leq\Big\|\sum_{j=1}^\infty f_j\Big\|_{L^p}.$$

(c) For $0<p<1$,

$$\Big\|\sum_{j=1}^N f_j\Big\|_{L^p}\leq C_N\sum_{j=1}^N\|f_j\|_{L^p},$$

for some constant $C_N>0$ independent of the f_j's.

(d) The smallest possible C_N in part (c) is $N^{\frac{1-p}{p}}$.
[*Hint:* Part (d): Take $\{f_j\}_{j=1}^N$ to be characteristic functions of disjoint sets with the same measure and use Exercise 1.1.4(c).]

1.1.6. (*Minkowski's integral inequality*) Let $1 \le p < \infty$. Let F be a measurable function on the product space $(X,\mu) \times (T,\nu)$. Show that

$$\left[\int_T \left(\int_X |F(x,t)|\, d\mu(x)\right)^p d\nu(t)\right]^{\frac{1}{p}} \le \int_X \left[\int_T |F(x,t)|^p\, d\nu(t)\right]^{\frac{1}{p}} d\mu(x).$$

Moreover, prove that when $0 < p < 1$, then the preceding inequality is reversed.

1.1.7. Let $f_1, \dots, f_N$ be in $L^{p,\infty}(X,\mu)$. Show that
(a) For $1 \le p < \infty$,

$$\Big\|\sum_{j=1}^N f_j\Big\|_{L^{p,\infty}} \le N \sum_{j=1}^N \|f_j\|_{L^{p,\infty}}$$

(b) For $0 < p < 1$,

$$\Big\|\sum_{j=1}^N f_j\Big\|_{L^{p,\infty}} \le N^{\frac{1}{p}} \sum_{j=1}^N \|f_j\|_{L^{p,\infty}}.$$

[*Hint:* Use that $\mu(\{|f_1 + \cdots + f_N| > \alpha\}) \le \sum_{j=1}^N \mu(\{|f_j| > \alpha/N\})$ and Exercise 1.1.4(a) and (c).]

1.1.8. Let $0 < p < \infty$. Prove that $L^p(X,\mu)$ is a complete quasi-normed space. This means that every quasi-norm Cauchy sequence is quasi-norm convergent.
[*Hint:* Let f_n be a Cauchy sequence in L^p. Pass to a subsequence such that $\|f_{n_{i+1}} - f_{n_i}\|_{L^p} \le 2^{-i}$. Then the series $f = f_{n_1} + \sum_{i=1}^\infty (f_{n_{i+1}} - f_{n_i})$ converges in L^p.]

1.1.9. Let (X,μ) be a measure space with $\mu(X) < \infty$. Suppose that a sequence of measurable functions f_n on X converges to f μ-a.e. Prove that f_n converges to f in measure.
[*Hint:* Use that $\varepsilon > 0$, the set $\{x :\in X : f_n(x) \to f(x)\}$ is contained $\bigcup_{m=1}^\infty \bigcap_{n=m}^\infty \{x :\in X : |f_n(x) - f(x)| < \varepsilon\}$.]

1.1.10. Given a measurable function f on (X,μ) and $\gamma > 0$, define $f_\gamma = f\chi_{|f|>\gamma}$ and $f^\gamma = f - f_\gamma = f\chi_{|f|\le\gamma}$.

(a) Prove that

$$d_{f_\gamma}(\alpha) = \begin{cases} d_f(\alpha) & \text{when} \quad \alpha > \gamma, \\ d_f(\gamma) & \text{when} \quad \alpha \le \gamma, \end{cases}$$

$$d_{f^\gamma}(\alpha) = \begin{cases} 0 & \text{when} \quad \alpha \ge \gamma, \\ d_f(\alpha) - d_f(\gamma) & \text{when} \quad \alpha < \gamma. \end{cases}$$

(b) If $f \in L^p(X,\mu)$ then

$$\|f_\gamma\|_{L^p}^p = p\int_\gamma^\infty \alpha^{p-1} d_f(\alpha)\,d\alpha + \gamma^p d_f(\gamma),$$

$$\|f^\gamma\|_{L^p}^p = p\int_0^\gamma \alpha^{p-1} d_f(\alpha)\,d\alpha - \gamma^p d_f(\gamma),$$

$$\int_{\gamma<|f|\le\delta} |f|^p\,d\mu = p\int_\gamma^\delta d_f(\alpha)\alpha^{p-1}\,d\alpha - \delta^p d_f(\delta) + \gamma^p d_f(\gamma).$$

(c) If f is in $L^{p,\infty}(X,\mu)$ prove that f^γ is in $L^q(X,\mu)$ for any $q > p$ and f_γ is in $L^q(X,\mu)$ for any $q < p$. Thus $L^{p,\infty} \subseteq L^{p_0} + L^{p_1}$ when $0 < p_0 < p < p_1 \le \infty$.

1.1.11. Let (X,μ) be a measure space and let E be a subset of X with $\mu(E) < \infty$. Assume that f is in $L^{p,\infty}(X,\mu)$ for some $0 < p < \infty$.
(a) Show that for $0 < q < p$ we have

$$\int_E |f(x)|^q\,d\mu(x) \le \frac{p}{p-q}\mu(E)^{1-\frac{q}{p}}\|f\|_{L^{p,\infty}}^q.$$

(b) Conclude that if $\mu(X) < \infty$ and $0 < q < p$, then

$$L^p(X,\mu) \subseteq L^{p,\infty}(X,\mu) \subseteq L^q(X,\mu).$$

[*Hint:* Part (a): Use $\mu\big(E \cap \{|f| > \alpha\}\big) \le \min\big(\mu(E), \alpha^{-p}\|f\|_{L^{p,\infty}}^p\big)$.]

1.1.12. (*Normability of weak L^p for $p > 1$*) Let (X,μ) be a measure space and let $0 < p < \infty$. Pick $0 < r < p$ and define

$$|||\, f \,|||_{L^{p,\infty}} = \sup_{0<\mu(E)<\infty} \mu(E)^{-\frac{1}{r}+\frac{1}{p}}\left(\int_E |f|^r\,d\mu\right)^{\frac{1}{r}},$$

where the supremum is taken over all measurable subsets E of X of finite measure.
(a) Use the previous Exercise with $q = r$ to conclude that

$$|||\, f \,|||_{L^{p,\infty}} \le \left(\frac{p}{p-r}\right)^{\frac{1}{r}}\|f\|_{L^{p,\infty}}$$

for all f in $L^{p,\infty}(X,\mu)$.
(b) Take $E = \{|f| > \alpha\}$ to deduce that $\|f\|_{L^{p,\infty}} \le |||\, f \,|||_{L^{p,\infty}}$ for all f in $L^{p,\infty}(X,\mu)$.

(c) Show that $L^{p,\infty}(X,\mu)$ is metrizable for all $0 < p < \infty$ and normable when $p > 1$ (by picking $r = 1$).
(d) Use the characterization of the weak L^p quasi-norm obtained in (a) and (b) to prove Fatou's theorem for this space: For all measurable functions g_n on X we have

$$\big\| \liminf_{n\to\infty} |g_n| \big\|_{L^{p,\infty}} \le C_p \liminf_{n\to\infty} \big\| g_n \big\|_{L^{p,\infty}}$$

for some constant C_p that depends only on $0 < p < \infty$.

1.1.13. Consider the $N!$ functions on the line

$$f_\sigma = \sum_{j=1}^{N} \frac{N}{\sigma(j)} \chi_{[\frac{j-1}{N}, \frac{j}{N}]},$$

where σ is a permutation of the set $\{1, 2, \dots, N\}$.
(a) Show that each f_σ satisfies $\big\| f_\sigma \big\|_{L^{1,\infty}} = 1$.
(b) Show that $\big\| \sum_{\sigma \in S_N} f_\sigma \big\|_{L^{1,\infty}} = N!(1 + \frac{1}{2} + \cdots + \frac{1}{N})$.
(c) Conclude that the space $L^{1,\infty}(\mathbf{R})$ is not normable.
(d) Use a similar argument to prove that $L^{1,\infty}(\mathbf{R}^n)$ is not normable by considering the functions $f_\sigma(x_1, \dots, x_n)$

$$\sum_{j_1=1}^{N} \cdots \sum_{j_n=1}^{N} \frac{N^n}{\sigma(\tau(j_1, \dots, j_n))} \chi_{[\frac{j_1-1}{N}, \frac{j_1}{N}]}(x_1) \dots \chi_{[\frac{j_n-1}{N}, \frac{j_n}{N}]}(x_n),$$

where σ is a permutation of the set $\{1, 2, \dots, N^n\}$ and τ is a fixed injective map from the set of all n-tuples of integers with coordinates $1 \le j \le N$ onto the set $\{1, 2, \dots, N^n\}$. We may take

$$\tau(j_1, \dots, j_n) = j_1 + N(j_2 - 1) + N^2(j_3 - 1) + \cdots + N^{n-1}(j_n - 1),$$

for instance.

1.1.14. Let $0 < p < 1$, $0 < s < \infty$ and let (X, μ) be a measure space.
(a) Let f be a measurable function on X. Show that

$$\int_{|f|\le s} |f|\, d\mu \le \frac{s^{1-p}}{1-p} \big\| f \big\|_{L^{p,\infty}}^p.$$

(b) Let f_j, $1 \le j \le m$, be measurable functions on X. Show that

$$\big\| \max_{1\le j\le m} |f_j| \big\|_{L^{p,\infty}}^p \le \sum_{j=1}^{m} \big\| f_j \big\|_{L^{p,\infty}}^p.$$

(c) Conclude that

$$\big\| f_1 + \cdots + f_m \big\|_{L^{p,\infty}}^p \le \frac{2-p}{1-p} \sum_{j=1}^{m} \big\| f_j \big\|_{L^{p,\infty}}^p.$$

The latter estimate is referred to as the p-normability of weak L^p for $p < 1$. [*Hint:* Part (c): First obtain the estimate

$$d_{f_1+\cdots+f_m}(\alpha) \le \mu(\{|f_1+\cdots+f_m|>\alpha, \max|f_j|\le\alpha\}) + d_{\max|f_j|}(\alpha)$$

for all $\alpha > 0$.]

1.1.15. (*Hölder's inequality for weak spaces*) Let f_j be in $L^{p_j,\infty}$ of a measure space X where $0 < p_j < \infty$ and $1 \le j \le k$. Let

$$\frac{1}{p} = \frac{1}{p_1} + \cdots + \frac{1}{p_k}.$$

Prove that

$$\big\|f_1 \cdots f_k\big\|_{L^{p,\infty}} \le p^{-\frac{1}{p}} \prod_{j=1}^{k} p_j^{\frac{1}{p_j}} \prod_{j=1}^{k} \big\|f_j\big\|_{L^{p_j,\infty}}.$$

[*Hint:* Take $\|f_j\|_{L^{p_j,\infty}} = 1$ for all j. Control $d_{f_1\cdots f_k}(\alpha)$ by

$$\begin{aligned}&\mu(\{|f_1|>\alpha/s_1\}) + \cdots + \mu(\{|f_{k-1}|>s_{k-2}/s_{k-1}\}) + \mu(\{|f_k|>s_{k-1}\})\\ &\le (s_1/\alpha)^{p_1} + (s_2/s_1)^{p_2} + \cdots + (s_{k-1}/s_{k-2})^{p_{k-1}} + (1/s_{k-1})^{p_k}.\end{aligned}$$

Set $x_1 = s_1/\alpha$, $x_2 = s_2/s_1, \dots, x_k = 1/s_{k-1}$. Minimize $x_1^{p_1}+\cdots+x_k^{p_k}$ subject to the constraint $x_1 \cdots x_k = 1/\alpha$.]

1.1.16. Let $0 < p_0 < p < p_1 \le \infty$ and let $\frac{1}{p} = \frac{1-\theta}{p_0} + \frac{\theta}{p_1}$ for some $\theta \in [0,1]$. Prove the following:

$$\begin{aligned}\big\|f\big\|_{L^p} &\le \big\|f\big\|_{L^{p_0}}^{1-\theta}\big\|f\big\|_{L^{p_1}}^{\theta},\\ \big\|f\big\|_{L^{p,\infty}} &\le \big\|f\big\|_{L^{p_0,\infty}}^{1-\theta}\big\|f\big\|_{L^{p_1,\infty}}^{\theta}.\end{aligned}$$

1.1.17. (*Loomis and Whitney* [**345**]) Follow the steps below to prove the *isoperimetric inequality*. For $n \ge 2$ and $1 \le j \le n$ define the projection maps π_j from $\mathbf{R}^n$ into $\mathbf{R}^{n-1}$ by setting for $x = (x_1, \dots, x_n)$

$$\pi_j(x) = (x_1, \dots, x_{j-1}, x_{j+1}, \dots, x_n)$$

with the obvious interpretations when $j = 1$ or $j = n$.
(a) For maps $f_j : \mathbf{R}^{n-1} \to \mathbf{C}$ prove that

$$\Lambda(f_1, \dots, f_n) = \int_{\mathbf{R}^n} \prod_{j=1}^{n} |f_j \circ \pi_j|\, dx \le \prod_{j=1}^{n} \big\|f_j\big\|_{L^{n-1}(\mathbf{R}^{n-1})}.$$

(b) Let Ω be a compact set with a rectifiable boundary in $\mathbf{R}^n$ where $n \ge 2$. Show that there is a constant c_n independent of Ω such that

$$|\Omega| \le c_n |\partial\Omega|^{\frac{n}{n-1}},$$

where the expression $|\partial\Omega|$ denotes the $n-1$ dimensional surface measure of the boundary of Ω.

[*Hint:* Part (a): Use induction starting with $n = 2$. Then write

$$\begin{aligned}\Lambda(f_1,\dots,f_n) &\le \int_{\mathbf{R}^{n-1}} P(x_1,\dots,x_{n-1})|f_n(\pi_n(x))|\,dx_1\dots dx_{n-1}\\ &\le \big\|P\big\|_{L^{\frac{n-1}{n-2}}(\mathbf{R}^{n-1})}\big\|f_n\circ\pi_n\big\|_{L^{n-1}(\mathbf{R}^{n-1})},\end{aligned}$$

where $P(x_1,\dots,x_{n-1}) = \int_{\mathbf{R}}|f_1(\pi_1(x))\dots f_{n-1}(\pi_{n-1}(x))|\,dx_n$, and apply the induction hypothesis to the $n-1$ functions

$$\left[\int_{\mathbf{R}} f_j(\pi_j(x))^{n-1}\,dx_n\right]^{\frac{1}{n-2}},$$

for $j = 1,\dots,n-1$, to obtain the required conclusion. Part (b): Specialize part (a) to the case $f_j = \chi_{\pi_j(\Omega)}$ to obtain

$$|\Omega| \le |\pi_1(\Omega)|^{\frac{1}{n-1}}\cdots|\pi_n(\Omega)|^{\frac{1}{n-1}}$$

and then use that $|\pi_j(\Omega)| \le \frac{1}{2}|\partial\Omega|$.]

1.2. Convolution and Approximate Identities

The notion of convolution can be defined on measure spaces endowed with a group structure. It turns out that the most natural environment to define convolution is the context of topological groups. Although the focus of this book is harmonic analysis on Euclidean spaces, we will develop the notion of convolution on general groups. This way we will be able to study this concept on $\mathbf{R}^n$, $\mathbf{Z}^n$, and $\mathbf{T}^n$, in a unified way. Moreover, since the basic properties of convolutions and approximate identities do not require commutativity of the group operation, we will be assuming that the underlying groups are not necessarily abelian. Thus, the results in this section can be also applied to nonabelian structures such as the Heisenberg group.

1.2.a. Examples of Topological Groups

A topological group G is a Hausdorff topological space that is also a group with law

$$(x,y)\to xy \tag{1.2.1}$$

such that the maps $(x,y)\to xy$ and $x\to x^{-1}$ are continuous.

Example 1.2.1. The standard examples are provided by the spaces $\mathbf{R}^n$ and $\mathbf{Z}^n$ with the usual topology and the usual addition of n-tuples. Another example is the space $\mathbf{T}^n$ defined as follows:

$$\mathbf{T}^n = \underbrace{[0,1] \times \cdots \times [0,1]}_{n \text{ times}}$$

with the usual topology and group law addition of n-tuples mod 1, that is,

$$(x_1, \dots, x_n) + (y_1, \dots, y_n) = ((x_1 + y_1) \bmod 1, \dots, (x_n + y_n) \bmod 1).$$

Let G be a locally compact group. It is known that G possesses a positive measure λ on the Borel sets that is nonzero on all nonempty open sets and is left invariant, meaning that

$$\lambda(tA) = \lambda(A), \tag{1.2.2}$$

for all measurable sets A and all $t \in G$. Such a measure λ is called a (left) *Haar measure* on G. For a constructive proof of the existence of Haar measure we refer the reader to Lang [**326**, §16.3]. Furthermore Haar measure is unique up to positive multiplicative constants. If G is abelian then any left Haar measure on G is a constant multiple of any given *right Haar measure* on G, the latter meaning right invariant [i.e., $\lambda(At) = \lambda(A)$, for all measurable $A \subseteq G$ and $t \in G$].

Example 1.2.2. Let $G = \mathbf{R}^* = \mathbf{R}\setminus\{0\}$ with group law the usual multiplication. It is easy to verify that the measure $\lambda = dx/|x|$ is invariant under multiplicative translations, that is,

$$\int_{-\infty}^{\infty} f(tx)\,\frac{dx}{|x|} = \int_{-\infty}^{\infty} f(x)\,\frac{dx}{|x|},$$

for all f in $L^1(G, \mu)$ and all $t \in \mathbf{R}^*$. Therefore, $dx/|x|$ is a Haar measure. [Taking $f = \chi_A$ gives $\lambda(tA) = \lambda(A)$.]

Example 1.2.3. Similarly, on the multiplicative group $G = \mathbf{R}^+$, a Haar measure is dx/x.

Example 1.2.4. Counting measure is a Haar measure on the group $\mathbf{Z}^n$ with group operation the usual addition.

Example 1.2.5. The *Heisenberg group* $\mathbf{H}^n$ is the set $\mathbf{C}^n \times \mathbf{R}$ with the group operation

$$(z_1, \dots, z_n, t)(w_1, \dots, w_n, s) = (z_1 + w_1, \dots, z_n + w_n, t + s + 2\,\mathrm{Im} \sum_{j=1}^{n} z_j \overline{w_j}).$$

It can be easily seen that the identity element of the group is $e = 0 \in \mathbf{C}^n \times \mathbf{R}$ and $(z_1, \dots, z_n, t)^{-1} = (-z_1, \dots, -z_n, -t)$. Topologically the Heisenberg group is identified with $\mathbf{C}^n \times \mathbf{R}$ and both left and right Haar measure on $\mathbf{H}^n$ is Lebesgue measure. The norm

$$|(z_1, \dots, z_n, t)| = \left[\Big(\sum_{j=1}^{n} |z_j|^2 \Big)^2 + t^2 \right]^{\frac{1}{4}}$$

introduces balls $B_r(x) = \{y \in \mathbf{H}^n : |y^{-1}x| < r\}$ on the Heisenberg group which are quite different from Euclidean balls. For x close to the origin the balls $B_r(x)$ are not far from being Euclidean but for x far away from $e = 0$ they look like slanted truncated cylinders. The Heisenberg group can be naturally identified as the boundary of the unit ball in $\mathbf{C}^n$ and plays an important role in quantum mechanics.

1.2.b. Convolution

Throughout the rest of this section, fix a locally compact group G and a left invariant Haar measure λ on G. The spaces $L^p(G, \lambda)$ and $L^{p,\infty}(G, \lambda)$ will be simply denoted by $L^p(G)$ and $L^{p,\infty}(G)$.

Left invariance of λ is equivalent to the fact that for all $t \in G$ and all $f \in L^1(G)$

$$\int_G f(tx)\, d\lambda(x) = \int_G f(x)\, d\lambda(x). \tag{1.2.3}$$

Equation (1.2.3) is a restatement of (1.2.2) if f is a characteristic function. For a general $f \in L^1(G)$ it follows by linearity and approximation.

We are now ready to define the operation of convolution.

Definition 1.2.6. Let f, g be in $L^1(G)$. Define the *convolution* $f * g$ by

$$(f * g)(x) = \int_G f(y) g(y^{-1}x)\, d\lambda(y)\,. \tag{1.2.4}$$

For instance, if $G = \mathbf{R}^n$ with the usual additive structure, then $y^{-1} = -y$ and the integral in (1.2.4) is written as

$$(f * g)(x) = \int_{\mathbf{R}^n} f(y) g(x - y)\, dy\,.$$

Remark 1.2.7. The right-hand side of (1.2.4) is defined a.e. since the following double integral converges absolutely:

$$
\begin{aligned}
&\int_G \int_G |f(y)||g(y^{-1}x)|\, d\lambda(y)\, d\lambda(x) \\
&= \int_G \int_G |f(y)||g(y^{-1}x)|\, d\lambda(x)\, d\lambda(y) \\
&= \int_G |f(y)| \int_G |g(y^{-1}x)|\, d\lambda(x)\, d\lambda(y) \qquad (1.2.5)\\
&= \int_G |f(y)| \int_G |g(x)|\, d\lambda(x)\, d\lambda(y) \qquad \text{by (1.2.2)} \\
&= \|f\|_{L^1(G)} \|g\|_{L^1(G)} < +\infty.
\end{aligned}
$$

By a simple change of variables $z = x^{-1}y$ it can be seen that (1.2.4) is in fact equal to

$$
(f * g)(x) = \int_G f(xz) g(z^{-1})\, d\lambda(z). \tag{1.2.6}
$$

The change of variables $z = x^{-1}y$ satisfies $d\lambda(y) = d\lambda(z)$ because of left invariance.

Example 1.2.8. On $\mathbf{R}$ let $f(x) = 1$ when $-1 \le x \le 1$ and zero otherwise. We see that $(f * f)(x)$ is equal to the length of the intersection of the intervals $[-1, 1]$ and $[x - 1, x + 1]$. It follows that $(f * f)(x) = 2 - |x|$ for $|x| \le 2$ and zero otherwise. Observe that $f * f$ is a smoother function than f. Similarly, we obtain that $f * f * f$ is a smoother function than $f * f$.

We can carry the same calculation when g is the characteristic function of the unit disc $B(0, 1)$ in $\mathbf{R}^2$. ($B(x, r)$ denotes the ball with center x and radius r.) A simple calculation gives

$$
\begin{aligned}
(g * g)(x) &= \big|B(0,1) \cap B(x,1)\big| = \int_{-\sqrt{1-\frac{1}{4}|x|^2}}^{+\sqrt{1-\frac{1}{4}|x|^2}} \Big(2\sqrt{1-t^2} - |x|\Big)\, dt \\
&= 2 \arcsin\left(\sqrt{1 - \tfrac{1}{4}|x|^2}\right) - |x|\sqrt{1 - \tfrac{1}{4}|x|^2}
\end{aligned}
$$

when $x = (x_1, x_2)$ in $\mathbf{R}^2$ satisfies $|x| \le 2$, while $(g * g)(x) = 0$ if $|x| \ge 2$.

A calculation similar to (1.2.5) proves that

$$
\|f * g\|_{L^1(G)} \le \|f\|_{L^1(G)} \|g\|_{L^1(G)}, \tag{1.2.7}
$$

that is, the convolution of two integrable functions is also an integrable function with L^1 norm less than or equal to the product of the L^1 norms.

Proposition 1.2.9. *For all f, g, h in $L^1(G)$, the following properties are valid:*

1. *$f * (g * h) = (f * g) * h$ (associativity)*
2. *$f * g = g * f$, when G is abelian (commutativity)*

3. *$f * (g + h) = f * g + f * h$ and $(f + g) * h = f * h + g * h$ (distributivity)*

Proof. The proofs are omitted. □

Proposition 1.2.9 implies that $L^1(G)$ is a (not necessarily commutative) Banach algebra under the convolution product.

1.2.c. Basic Convolution Inequalities

The most fundamental inequality involving convolutions is the following.

Theorem 1.2.10. *(**Minkowski's inequality**) Let $1 \le p \le \infty$. For f in $L^p(G)$ and g in $L^1(G)$ we have*

$$\|g * f\|_{L^p(G)} \le \|g\|_{L^1(G)} \|f\|_{L^p(G)}. \tag{1.2.8}$$

Proof. Estimate (1.2.8) follows directly from Exercise 1.1.6. Here we give a direct proof. We may assume that $1 < p < \infty$, as the cases $p = 1$ and $p = \infty$ are simple. Clearly,

$$|(g * f)(x)| \le \int_G |f(y^{-1}x)|\,|g(y)|\,d\lambda(y). \tag{1.2.9}$$

Apply Hölder's inequality in (1.2.9) with respect to the measure $|g(y)|\,d\lambda(y)$ to the functions $y \to f(y^{-1}x)$ and 1 with exponents p and $p' = p/(p-1)$, respectively. We obtain

$$|(g * f)(x)| \le \left(\int_G |f(y^{-1}x)|^p |g(y)|\,d\lambda(y)\right)^{\frac{1}{p}} \left(\int_G |g(y)|\,d\lambda(y)\right)^{\frac{1}{p'}}. \tag{1.2.10}$$

Taking L^p norms of both sides of (1.2.10) we deduce

$$\begin{aligned}
\|g * f\|_{L^p} &\le \left(\|g\|_{L^1}^{p-1} \int_G \int_G |f(y^{-1}x)|^p |g(y)|\,d\lambda(y)\,d\lambda(x)\right)^{\frac{1}{p}} \\
&= \left(\|g\|_{L^1}^{p-1} \int_G \int_G |f(y^{-1}x)|^p\,d\lambda(x) |g(y)|\,d\lambda(y)\right)^{\frac{1}{p}} \\
&= \left(\|g\|_{L^1}^{p-1} \int_G \int_G |f(x)|^p\,d\lambda(x) |g(y)|\,d\lambda(y)\right)^{\frac{1}{p}} \qquad \text{by (1.2.3)} \\
&= \left(\|f\|_{L^p}^p \|g\|_{L^1} \|g\|_{L^1}^{p-1}\right)^{\frac{1}{p}} = \|f\|_{L^p} \|g\|_{L^1},
\end{aligned}$$

where the second equality follows by Fubini's theorem. The proof is complete. □

Remark 1.2.11. Theorem 1.2.10 may fail on nonabelian groups if $g * f$ is replaced by $f * g$ in (1.2.8). Note, however, that if

$$\big\|g\big\|_{L^1} = \big\|\widetilde{g}\big\|_{L^1}, \tag{1.2.11}$$

where $\widetilde{g}(x) = g(x^{-1})$, then (1.2.8) holds when $\big\|g*f\big\|_{L^p(G)}$ is replaced by $\big\|f*g\big\|_{L^p(G)}$. To see this, observe that if (1.2.11) holds, then we can use (1.2.6) to conclude that if f in $L^p(G)$ and g in $L^1(G)$, then

$$\big\|f * g\big\|_{L^p(G)} \le \big\|g\big\|_{L^1(G)}\big\|f\big\|_{L^p(G)}. \tag{1.2.12}$$

If the left Haar measure satisfies

$$\lambda(A) = \lambda(A^{-1}) \tag{1.2.13}$$

for all measurable $A \subseteq G$, then (1.2.11) holds and thus (1.2.12) is satisfied for all g in $L^1(G)$. This is, for instance, the case on the Heisenberg group $\mathbf{H}^n$.

Minkowski's inequality (1.2.12) is only a special case of Young's inequality in which the function g can be in any space $L^r(G)$ for $1 \le r \le \infty$.

Theorem 1.2.12. *(**Young's inequality**) Let $1 \le p, q, r \le \infty$ satisfy*

$$\frac{1}{q} + 1 = \frac{1}{p} + \frac{1}{r}. \tag{1.2.14}$$

Then for all f in $L^p(G)$ and all g in $L^r(G)$ satisfying $\big\|g\big\|_{L^r(G)} = \big\|\widetilde{g}\big\|_{L^r(G)}$ we have

$$\big\|f * g\big\|_{L^q(G)} \le \big\|g\big\|_{L^r(G)}\big\|f\big\|_{L^p(G)}. \tag{1.2.15}$$

Proof. Young's inequality is proved in a way similar to Minkowski's inequality. We do a suitable splitting of the product $|f(xy)||g(y^{-1})|$ and apply Hölder's inequality. Observe that when $r < \infty$, the hypotheses on the indices imply that

$$\frac{1}{r'} + \frac{1}{q} + \frac{1}{p'} = 1, \qquad \frac{p}{q} + \frac{p}{r'} = 1, \qquad \frac{r}{q} + \frac{r}{p'} = 1.$$

Using Hölder's inequality with exponents r', q, and p', we obtain

$$\begin{aligned}
|(f * g)(x)| &\le \int_G |f(y)|\,|g(y^{-1}x)|\,d\lambda(y) \\
&\le \int_G |f(y)|^{\frac{p}{r'}} \big(|f(y)|^{\frac{p}{q}}|g(y^{-1}x)|^{\frac{r}{q}}\big)|g(y^{-1}x)|^{\frac{r}{p'}}\,d\lambda(y) \\
&\le \big\|f\big\|_{L^p}^{\frac{p}{r'}} \Big(\int_G |f(y)|^p|g(y^{-1}x)|^r\,d\lambda(y)\Big)^{\frac{1}{q}} \Big(\int_G |g(y^{-1}x)|^r\,d\lambda(y)\Big)^{\frac{1}{p'}} \\
&= \big\|f\big\|_{L^p}^{\frac{p}{r'}} \Big(\int_G |f(y)|^p|g(y^{-1}x)|^r\,d\lambda(y)\Big)^{\frac{1}{q}} \Big(\int_G |\widetilde{g}(x^{-1}y)|^r\,d\lambda(y)\Big)^{\frac{1}{p'}} \\
&= \Big(\int_G |f(y)|^p|g(y^{-1}x)|^r\,d\lambda(y)\Big)^{\frac{1}{q}} \big\|f\big\|_{L^p}^{\frac{p}{r'}} \big\|\widetilde{g}\big\|_{L^r}^{\frac{r}{p'}},
\end{aligned}$$

where we used left invariance. Now take L^q norms (in x) and apply Fubini's theorem to deduce that

$$\begin{aligned}\big\|f*g\big\|_{L^q} &\le \big\|f\big\|_{L^p}^{\frac{p}{r'}}\big\|\widetilde{g}\big\|_{L^r}^{\frac{r}{p'}}\left(\int_G\int_G |f(y)|^p|g(y^{-1}x)|^r\,d\lambda(x)\,d\lambda(y)\right)^{\frac{1}{q}}\\ &= \big\|f\big\|_{L^p}^{\frac{p}{r'}}\big\|\widetilde{g}\big\|_{L^r}^{\frac{r}{p'}}\big\|f\big\|_{L^p}^{\frac{p}{q}}\big\|g\big\|_{L^r}^{\frac{r}{q}}\\ &= \big\|g\big\|_{L^r}\big\|f\big\|_{L^p},\end{aligned}$$

using the hypothesis on g. Finally, observe that if $r=\infty$, the assumptions on p and q imply that $p=1$ and $q=\infty$, in which case the required inequality trivially holds.

□

We now give a version of Theorem 1.2.12 for weak L^p spaces. Theorem 1.2.13 will be improved in Section 1.4.

Theorem 1.2.13. *(**Young's inequality for weak type spaces**) Let G be a locally compact group with left Haar measure λ that satisfies (1.2.13). Let $1\le p<\infty$ and $1<q,r<\infty$ satisfy*

$$\frac{1}{q}+1=\frac{1}{p}+\frac{1}{r}. \tag{1.2.16}$$

Then there exists a constant $C_{p,q,r}>0$ such that for all f in $L^p(G)$ and g in $L^{r,\infty}(G)$ we have

$$\big\|f*g\big\|_{L^{q,\infty}(G)}\le C_{p,q,r}\big\|g\big\|_{L^{r,\infty}(G)}\big\|f\big\|_{L^p(G)}. \tag{1.2.17}$$

Proof. The proof will be based on a suitable splitting of the function g. Let M be a positive real number to be chosen later. Define $g_1=g\chi_{|g|\le M}$ and $g_2=g\chi_{|g|>M}$. In view of Exercise 1.1.10(a) we have

$$d_{g_1}(\alpha)=\begin{cases}0 & \text{if } \alpha\ge M,\\ d_g(\alpha)-d_g(M) & \text{if } \alpha<M.\end{cases} \tag{1.2.18}$$

$$d_{g_2}(\alpha)=\begin{cases}d_g(\alpha) & \text{if } \alpha> M,\\ d_g(M) & \text{if } \alpha\le M.\end{cases} \tag{1.2.19}$$

Proposition 1.1.3 gives

$$d_{f*g}(\alpha)\le d_{f*g_1}(\alpha/2)+d_{f*g_2}(\alpha/2), \tag{1.2.20}$$

and thus it suffices to estimate the distribution functions of $f * g_1$ and $f * g_2$. Since g_1 is the "small" part of g, it is in L^s for any $s > r$. In fact, we have

$$\begin{aligned}\int_G |g_1(x)|^s \, d\lambda(x) &= s\int_0^\infty \alpha^{s-1} d_{g_1}(\alpha)\, d\alpha \\ &= s\int_0^M \alpha^{s-1}(d_g(\alpha) - d_g(M))\, d\alpha \\ &\le s\int_0^M \alpha^{s-1-r}\|g\|_{L^{r,\infty}}^r \, d\alpha - s\int_0^M \alpha^{s-1} d_g(M)\, d\alpha \\ &= \frac{s}{s-r} M^{s-r}\|g\|_{L^{r,\infty}}^r - M^s d_g(M),\end{aligned} \tag{1.2.21}$$

when $s < \infty$.

Similarly, since g_2 is the "large" part of g, it is in L^t for any $t < r$, and

$$\begin{aligned}\int_G |g_2(x)|^t \, d\lambda(x) &= t\int_0^\infty \alpha^{t-1} d_{g_2}(\alpha)\, d\alpha \\ &= t\int_0^M \alpha^{t-1} d_g(M)\, d\alpha + t\int_M^\infty \alpha^{t-1} d_g(\alpha)\, d\alpha \\ &\le M^t d_g(M) + t\int_M^\infty \alpha^{t-1-r}\|g\|_{L^{r,\infty}}^r \, d\alpha \\ &\le M^{t-r}\|g\|_{L^{r,\infty}}^r + \frac{t}{r-t} M^{t-r}\|g\|_{L^{r,\infty}}^r \\ &= \frac{r}{r-t} M^{t-r}\|g\|_{L^{r,\infty}}^r .\end{aligned} \tag{1.2.22}$$

Since $1/r = 1/p' + 1/q$, it follows that $1 < r < p'$. Select $t = 1$ and $s = p'$. Hölder's inequality and (1.2.21) give

$$|(f * g_1)(x)| \le \|f\|_{L^p}\|g_1\|_{L^{p'}} \le \|f\|_{L^p}\left(\frac{p'}{p'-r} M^{p'-r}\|g\|_{L^{r,\infty}}^r\right)^{\frac{1}{p'}}, \tag{1.2.23}$$

when $p' < \infty$, while

$$|(f * g_1)(x)| \le \|f\|_{L^p} M \tag{1.2.24}$$

if $p' = \infty$. Choose an M so that the right-hand side of (1.2.23) [(1.2.24) if $p' = \infty$] is equal to $\alpha/2$. For this choice of M we have that

$$d_{f*g_1}(\alpha/2) = 0.$$

Next by Theorem 1.2.10 and (1.2.22) with $t = 1$ we obtain

$$\|f * g_2\|_{L^p} \le \|f\|_{L^p}\|g_2\|_{L^1} \le \|f\|_{L^p}\frac{r}{r-1} M^{1-r}\|g\|_{L^{r,\infty}}^r . \tag{1.2.25}$$

For the value of M chosen,[1] using (1.2.25) and Chebychev's inequality, we obtain

$$\begin{aligned} d_{f*g}(\alpha) &\le d_{f*g_2}(\alpha/2) \\ &\le (2\big\|f * g_2\big\|_{L^p}\alpha^{-1})^p \\ &\le (2r\big\|f\big\|_{L^p}M^{1-r}\big\|g\big\|_{L^{r,\infty}}^r(r-1)^{-1}\alpha^{-1})^p \\ &= C_{p,q,r}^q\alpha^{-q}\big\|f\big\|_{L^p}^q\big\|g\big\|_{L^{r,\infty}}^q, \end{aligned} \tag{1.2.26}$$

which is the required inequality. This proof gives that the constant $C_{p,q,r}$ blows up like $(r-1)^{-p/q}$ as $r \to 1$.

□

Example 1.2.14. Theorem 1.2.13 may fail at some endpoints:

(1) $r = 1$ and $1 \le p = q \le \infty$. On $\mathbf{R}$ take $g(x) = 1/|x|$ and $f = \chi_{[0,1]}$. Clearly, g is in $L^{1,\infty}$ and f in L^p for all $1 \le p \le \infty$, but the convolution of f and g is identically equal to infinity on the interval $[0,1]$. Therefore, (1.2.17) fails in this case.

(2) $q = \infty$ and $1 < r = p' < \infty$. On $\mathbf{R}$ let $f(x) = (|x|^{1/p}\log|x|)^{-1}$ for $|x| \ge 2$ and zero otherwise, and also let $g(x) = |x|^{-1/r}$. We see that $(f * g)(x) = \infty$ for $|x| \le 1$. Thus (1.2.17) fails in this case also.

(3) $r = q = \infty$ and $p = 1$. Then inequality (1.2.17) trivially holds.

1.2.d. Approximate Identities

We now introduce the notion of approximate identities. The Banach algebra $L^1(G)$ may not have a unit element, that is an element f_0 such that

$$f_0 * f = f = f * f_0 \tag{1.2.27}$$

for all $f \in L^1(G)$. In particular, this is the case when $G = \mathbf{R}$; in fact, the only f_0 that satisfies (1.2.27) for all $f \in L^1(\mathbf{R})$ is not a function but the Dirac delta distribution, which will be introduced in Chapter 2. It is reasonable therefore to introduce the notion of approximate unit or identity, a family of functions k_ε with the property $k_\varepsilon * f \to f$ in L^1 as $\varepsilon \to 0$.

Definition 1.2.15. An *approximate identity* (as $\varepsilon \to 0$) is a family of $L^1(G)$ functions k_ε with the following three properties:

(i) There exists a constant $c > 0$ such that $\big\|k_\varepsilon\big\|_{L^1(G)} \le c$ for all $\varepsilon > 0$.

(ii) $\int_G k_\varepsilon(x)\,d\lambda(x) = 1$ for all $\varepsilon > 0$.

(iii) For any neighborhood V of the identity element e of the group G we have $\int_{V^c} |k_\varepsilon(x)|\,d\lambda(x) \to 0$ as $\varepsilon \to 0$.

---

[1] $M = (\alpha^{p'}2^{-p'}rq^{-1}\|f\|_{L^p}^{-p'}\|g\|_{L^{r,\infty}}^{-r})^{1/(p'-r)}$ if $p' < \infty$ and $M = \alpha/(2\|f\|_{L^1})$ if $p' = \infty$.

The construction of approximate identities on general locally compact groups G is beyond the scope of this book and is omitted. We refer the reader to Hewitt and Ross [**247**] for details. In this book we are only interested in groups with Euclidean structure where approximate identities exist in abundance. See the following examples.

Sometimes we will be thinking of approximate identities as sequences $\{k_n\}_n$. In this case property (iii) holds as $n \to \infty$. It is best to visualize approximate identities as sequences of positive functions k_n that spike near 0 in such a way so that the signed area under the graph of each of function remains constant (equal to one) but the support shrinks to zero. See Figure 1.2.

Example 1.2.16. On $\mathbf{R}^n$ let $k(x)$ be an integrable function with integral one. Let $k_\varepsilon(x) = \varepsilon^{-n}k(\varepsilon^{-1}x)$. It is straightfoward to see that $k_\varepsilon(x)$ is an approximate identity. Property (iii) follows from the fact that

$$\int_{|x|\ge\delta/\varepsilon} |k(x)|\,dx \to 0$$

as $\varepsilon \to 0$ for δ fixed.

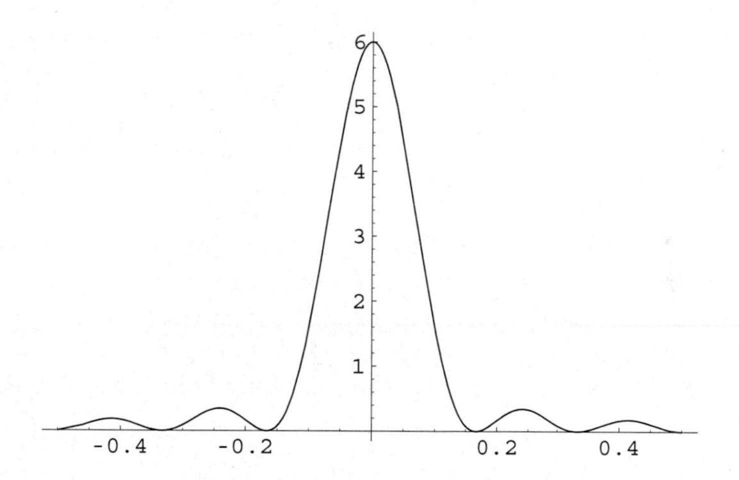

FIGURE 1.2. The Fejér kernel F_5 plotted on the interval $[-\frac{1}{2}, \frac{1}{2}]$.

Example 1.2.17. On $\mathbf{R}$ let $P(x) = (\pi(x^2+1))^{-1}$ and for $\varepsilon > 0$ let $P_\varepsilon(x) = \varepsilon^{-1}P(\varepsilon^{-1}x)$. Since P_ε and P have the same L^1 norm and

$$\int_{-\infty}^{+\infty} \frac{1}{x^2+1}\,dx = \lim_{x\to+\infty} \big[\arctan(x) - \arctan(-x)\big] = (\pi/2) - (-\pi/2) = \pi,$$

property (ii) is satisfied. Property (iii) follows from the fact that

$$\frac{1}{\pi}\int_{|x|\ge\delta} \frac{1}{\varepsilon}\frac{1}{(x/\varepsilon)^2+1}\,dx = 1 - \frac{2}{\pi}\arctan(\delta/\varepsilon) \to 0 \qquad \text{as } \varepsilon \to 0,$$

provided $\delta > 0$. The function P_ε is called the *Poisson kernel.*

Example 1.2.18. On the circle group $\mathbf{T}^1$ let

$$F_N(t) = \sum_{j=-N}^{N} \left(1 - \frac{|j|}{N+1}\right) e^{2\pi i j t} = \frac{1}{N+1}\left(\frac{\sin(\pi(N+1)t)}{\sin(\pi t)}\right)^2. \tag{1.2.28}$$

To check the claimed identity we use that

$$\sin^2(x) = (2 - e^{2ix} - e^{-2ix})/4$$

and we carry out the calculation. F_N is called the *Fejér kernel.* To see that the sequence $\{F_N\}_N$ is an approximate identity, we check conditions (i), (ii), and (iii) in Definition 1.2.15. Property (iii) follows from the expression giving F_N in terms of sines, while property (i) follows from the expression giving F_N in terms of exponentials. Property (ii) is identical to property (i) as F_N is nonnegative.

Next comes our basic theorem about approximate identities.

Theorem 1.2.19. *Let k_ε be an approximate identity on a locally compact group G with left Haar measure λ.*

(1) If $f \in L^p(G)$ for $1 \le p < \infty$, then $\|k_\varepsilon * f - f\|_{L^p(G)} \to 0$ as $\varepsilon \to 0$.

(2) When $p = \infty$, the following is valid: If f is continuous on a compact $K \subseteq G$, then $\|k_\varepsilon * f - f\|_{L^\infty(K)} \to 0$ as $\varepsilon \to 0$.

Proof. We start with the case $1 \le p < \infty$. We recall that continuous functions with compact support are dense in L^p of locally compact spaces. See, for instance, Hewitt and Ross [**247**] (Theorem 12.10). For a continuous function with compact support g we have $|g(h^{-1}x) - g(x)|^p \le (2\|g\|_{L^\infty})^p$ for h in a relatively compact neighborhood of e, and hence by the Lebesgue dominated convergence theorem we obtain

$$\int_G |g(h^{-1}x) - g(x)|^p \, d\lambda(x) \to 0 \tag{1.2.29}$$

as $h \to e$. Now approximate a given f in $L^p(G)$ by a continuous function with compact support g to deduce that

$$\int_G |f(h^{-1}x) - f(x)|^p \, d\lambda(x) \to 0 \qquad \text{as} \qquad h \to e. \tag{1.2.30}$$

Because of (1.2.30), given a $\delta > 0$ there exists a neighborhood V of e such that

$$h \in V \implies \int_G |f(h^{-1}x) - f(x)|^p \, d\lambda(x) < \left(\frac{\delta}{2c}\right)^p, \tag{1.2.31}$$

where c is the constant that appears in Definition 1.2.15 (i). Since k_ε has integral one for all $\varepsilon > 0$, we have

$$(1.2.32)\qquad \begin{aligned} &(k_\varepsilon * f)(x) - f(x) \\ &= (k_\varepsilon * f)(x) - f(x)\int_G k_\varepsilon(y)\,d\lambda(y) \\ &= \int_G (f(y^{-1}x) - f(x))k_\varepsilon(y)\,d\lambda(y) \\ &= \int_V (f(y^{-1}x) - f(x))k_\varepsilon(y)\,d\lambda(y) \\ &\qquad + \int_{V^c} (f(y^{-1}x) - f(x))k_\varepsilon(y)\,d\lambda(y). \end{aligned}$$

Now take L^p norms in x in (1.2.32). In view of (1.2.31),

$$(1.2.33)\qquad \begin{aligned} &\left\| \int_V (f(y^{-1}x) - f(x))k_\varepsilon(y)\,d\lambda(y) \right\|_{L^p(G,d\lambda(x))} \\ &\le \int_V \left\| f(y^{-1}x) - f(x) \right\|_{L^p(G,d\lambda(x))} |k_\varepsilon(y)|\,d\lambda(y) \\ &\le \int_V \frac{\delta}{2c} |k_\varepsilon(y)|\,d\lambda(y) < \frac{\delta}{2}, \end{aligned}$$

while

$$(1.2.34)\qquad \begin{aligned} &\left\| \int_{V^c} (f(y^{-1}x) - f(x))k_\varepsilon(y)\,d\lambda(y) \right\|_{L^p(G,d\lambda(x))} \\ &\le \int_{V^c} 2\|f\|_{L^p(G)} |k_\varepsilon(y)|\,d\lambda(y) < \frac{\delta}{2}, \end{aligned}$$

provided we have that

$$(1.2.35)\qquad \int_{V^c} |k_\varepsilon(x)|\,d\lambda(x) < \frac{\delta}{4\|f\|_{L^p}}.$$

Now choose $\varepsilon_0 > 0$ such that (1.2.35) is valid for $\varepsilon < \varepsilon_0$ by property (iii). Now (1.2.33) and (1.2.34) imply the required conclusion.

The case $p = \infty$ follows similarly. Since f is uniformly continuous on K, given $\delta > 0$ find a neighborhood V of $e \in G$ such that

$$(1.2.36)\qquad h \in V \implies |f(h^{-1}x) - f(x)| < \frac{\delta}{2c} \qquad \text{for all } x \in K,$$

where c is as in Definition 1.2.15(i) and then find an $\varepsilon_0 > 0$ such that for $0 < \varepsilon < \varepsilon_0$ we have

$$(1.2.37)\qquad \int_{V^c} |k_\varepsilon(y)|\,d\lambda(y) < \frac{\delta}{2\|f\|_{L^\infty}}.$$

Using (1.2.36) and (1.2.37), we easily conclude that

$$\begin{aligned}&\sup_{x\in K}|(k_\varepsilon * f)(x) - f(x)|\\ \leq\;& \int_V |k_\varepsilon(y)| \sup_{x\in K}|f(y^{-1}x) - f(x)|\,d\lambda(y) + \int_{V^c} |k_\varepsilon(y)| \sup_{x\in K}|f(y^{-1}x) - f(x)|\,d\lambda(y)\\ \leq\;& \frac{\delta}{2} + \frac{\delta}{2} = \delta\,,\end{aligned}$$

which shows that $k_\varepsilon * f$ converges uniformly to f on K as $\varepsilon \to 0$. □

Remark 1.2.20. Observe that if Haar measure satisfies (1.2.13), then the conclusion of Theorem 1.2.19 also holds for $f * k_\varepsilon$.

A simple variant of the proof of Theorem 1.2.19 yields the following.

Theorem 1.2.21. *Let k_ε be a family of functions on a locally compact group G that satisfies properties (i) and (iii) of Definition 1.2.15 and also*

$$\int_G k_\varepsilon(x)\,d\lambda(x) = a \in \mathbf{C}, \qquad \textit{for all } \varepsilon > 0.$$

Let $f \in L^p(G)$ for some $1 \leq p \leq \infty$.
(a) If $1 \leq p < \infty$, then $\big\|k_\varepsilon * f - af\big\|_{L^p(G)} \to 0$ as $\varepsilon \to 0$.
(b) If $p = \infty$ and f is continuous on a compact $K \subseteq G$, then

$$\big\|k_\varepsilon * f - af\big\|_{L^\infty(K)} \to 0$$

as $\varepsilon \to 0$.

Remark 1.2.22. With the notation of Theorem 1.2.21, if f is continuous and tends to zero at infinity, then $\big\|k_\varepsilon * f - af\big\|_{L^\infty(G)} \to 0$. To see this, simply observe that outside a compact subset of G, both $k_\varepsilon * f$, af have small L^∞ norm, while inside a compact subset of G uniform convergence holds.

Exercises

1.2.1. Let f , g be in $L^1(G)$ be supported in the sets A and B, respectively. Prove that $f * g$ is supported in the algebraic product set AB.

1.2.2. For a function f on a locally compact group G and $t \in G$, let ${}^tf(x) = f(tx)$ and $f^t(x) = f(xt)$. Show that

$${}^tf * g = {}^t(f * g) \qquad \text{and} \qquad f * g^t = (f * g)^t\,.$$

1.2.3. Let $f \in L^p(G)$ and $\tilde{g} \in L^{p'}(G)$, where $1 < p < \infty$. Prove that the mapping $t \to {}^tg$ from G to $L^{p'}(G)$ is right uniformly continuous. Conclude that $f * g$ is right uniformly continuous. Finally, show that for all $\varepsilon > 0$ there exists a compact subset K of G such that $|f * g| < \varepsilon$ outside K.

1.2.4. Let $f \in L^p(G)$ and μ be a finite Borel measure on G with total variation $\|\mu\|$. Define

$$(\mu * f)(x) = \int_G f(y^{-1}x)\, d\mu(y).$$

Show that if μ is an absolutely continuous measure, then the preceding definition extends (1.2.4). Prove that $\|\mu * f\|_{L^p(G)} \le \|\mu\| \|f\|_{L^p(G)}$.

1.2.5. Show that a Haar measure λ for the multiplicative group of all positive real numbers is

$$\lambda(A) = \int_0^\infty \chi_A(t)\, \frac{dt}{t}.$$

1.2.6. Let $G = \mathbf{R}^2$ with group operation $(x, y)(z, w) = (xz, xw + y)$. [Think of G as the group of all 2×2 matrices with bottom row $(0, 1)$.] Show that a left Haar measure on G is

$$\lambda(A) = \int_{-\infty}^{+\infty} \int_{-\infty}^{+\infty} \chi_A(x, y)\, \frac{dx dy}{x^2},$$

while a right Haar measure on G is

$$\rho(A) = \int_{-\infty}^{+\infty} \int_{-\infty}^{+\infty} \chi_A(x, y)\, \frac{dx dy}{|x|}.$$

1.2.7. (*Hardy* [**233**], [**234**]) Use Theorem 1.2.10 to prove that

$$\left(\int_0^\infty \left(\frac{1}{x} \int_0^x |f(t)|\, dt \right)^p dx \right)^{\frac{1}{p}} \le \frac{p}{p-1} \|f\|_{L^p(0,\infty)},$$

$$\left(\int_0^\infty \left(\int_x^\infty |f(t)|\, dt \right)^p dx \right)^{\frac{1}{p}} \le p \left(\int_0^\infty |f(t)|^p t^p\, dt \right)^{\frac{1}{p}}.$$

when $1 < p < \infty$.
[*Hint:* On the multiplicative group $(\mathbf{R}^+, \frac{dt}{t})$ consider the convolution of the function $|f(x)| x^{\frac{1}{p}}$ with the function $x^{-\frac{1}{p'}} \chi_{[1,\infty)}$ and the convolution of the function $|f(x)| x^{1+\frac{1}{p}}$ with $x^{\frac{1}{p}} \chi_{(0,1]}$.]

1.2.8. (*G. H. Hardy*) Let $0 < b < \infty$ and $1 \le p < \infty$. Prove that

$$\left(\int_0^\infty \left(\int_0^x |f(t)|\, dt \right)^p x^{-b-1}\, dx \right)^{\frac{1}{p}} \le \frac{p}{b} \left(\int_0^\infty |f(t)|^p t^{p-b-1}\, dt \right)^{\frac{1}{p}}$$

$$\left(\int_0^\infty \left(\int_x^\infty |f(t)|\, dt \right)^p x^{b-1}\, dx \right)^{\frac{1}{p}} \le \frac{p}{b} \left(\int_0^\infty |f(t)|^p t^{p+b-1}\, dt \right)^{\frac{1}{p}}.$$

[*Hint:* On the multiplicative group $(\mathbf{R}^+, \frac{dt}{t})$ consider the convolution of the function $|f(x)|x^{1-\frac{b}{p}}$ with $x^{-\frac{b}{p}}\chi_{[1,\infty)}$ and of the function $|f(x)|x^{1+\frac{b}{p}}$ with $x^{\frac{b}{p}}\chi_{(0,1]}$.]

1.2.9. On $\mathbf{R}^n$ let $T(f) = f * K$, where K is a positive L^1 function and f is in L^p, $1 \le p \le \infty$. Prove that the operator norm of $T : L^p \to L^p$ is equal to the L^1 norm of K.
[*Hint:* Clearly, $\|T\|_{L^p \to L^p} \le \|K\|_{L^1}$. Conversely, fix $0 < \varepsilon < 1$ and let N be a positive integer. Let $\chi_N = \chi_{B(0,N)}$ and for any $R > 0$ let $K_R = K\chi_{B(0,R)}$, where $B(x,R)$ is the ball of radius R centered at x. Observe that for $|x| \le (1-\varepsilon)N$, we have $B(0, N\varepsilon) \subseteq B(x,N)$; thus $\int_{\mathbf{R}^n} \chi_N(x-y)K_{N\varepsilon}(y)\,dy = \int_{\mathbf{R}^n} K_{N\varepsilon}(y)\,dy = \big\|K_{N\varepsilon}\big\|_{L^1}$. Then

$$\frac{\big\|K * \chi_N\big\|_{L^p}^p}{\big\|\chi_N\big\|_{L^p}^p} \ge \frac{\big\|K_{N\varepsilon} * \chi_N\big\|_{L^p(B(0,(1-\varepsilon)N)}^p}{\big\|\chi_N\big\|_{L^p}^p} \ge \big\|K_{N\varepsilon}\big\|_{L^1}(1-\varepsilon)^n .$$

Let $N \to \infty$ first and then $\varepsilon \to 0$.]

1.2.10. On the multiplicative group $(\mathbf{R}^+, \frac{dt}{t})$ let $T(f) = f * K$, where K is a positive L^1 function and f is in L^p, $1 \le p \le \infty$. Prove that the operator norm of $T : L^p \to L^p$ is equal to the L^1 norm of K. Deduce that the constants $p/(p-1)$ and p/b are sharp in Exercises 1.2.7 and 1.2.8.
[*Hint:* Adapt the idea of the previous Exercise to this setting.]

1.2.11. Let $O(n)$ be the set of all orthogonal transformations on $\mathbf{R}^n$ and let $\mathbf{S}^{n-1}$ be the unit sphere $\{x \in \mathbf{R}^n : |x| = 1\}$. Let $f \in L^q(\mathbf{S}^{n-1})$ for some $1 \le q < \infty$ and $e_1 = (1,0,\dots,0) \in \mathbf{S}^{n-1}$. Prove that for all $\varepsilon > 0$ there exists a $\delta > 0$ such that if $A, B \in O(n)$ and $|A(e_1) - B(e_1)| < \delta$, then the function $x \to f(A(x)) - f(B(x))$ has L^q norm less than ε.
[*Hint:* Suppose first that f is smooth on the sphere $\mathbf{S}^{n-1}$. By uniform continuity, given $\varepsilon > 0$, there exists an $\delta > 0$ such that $|t-s| < \delta$ implies $|f(t) - f(s)| < \varepsilon/\omega_{n-1}^{1/q}$. Note that for x in $\mathbf{S}^{n-1}$ we have $|A(x) - B(x)| = |B^{-1}A(x) - x| = |B^{-1}A(e_1) - e_1| = |A(e_1) - B(e_1)|$. Thus if $|A(e_1) - B(e_1)| < \delta$, then $|f(A(x)) - f(B(x))| < \varepsilon/\omega_{n-1}^{1/q}$ and hence the L^q norm of the function $x \to f(A(x)) - f(B(x))$ is smaller than ε. Now if $f \in L^q(\mathbf{S}^{n-1})$, find a $g \in C^\infty(\mathbf{S}^{n-1})$ such that $\|f - g\|_{L^q(\mathbf{S}^{n-1})} < \varepsilon/3$. Pick a δ that works for g for $\varepsilon/3$ and use the triangle inequality to derive the required conclusion.]

1.2.12. Let $Q_j(t) = c_j(1-t^2)^j$ for $t \in [-1,1]$ and zero elsewhere, where c_j is chosen so that $\int_{-1}^{1} Q_j(t)\,dt = 1$ for all $j = 1, 2, \dots$.
(a) Show that $c_j < \sqrt{j}$.

(b) Use part (a) to show that $\{Q_j\}_j$ is an approximate identity on $\mathbf{R}$ as $j \to \infty$.
(c) Given a continuous function f on $\mathbf{R}$ which vanishes outside the interval $[-1,1]$, show that $f * Q_j$ converges to f uniformly on $[-1,1]$.
(d) (*Weierstrass*) Prove that every continuous function on $[-1,1]$ can be approximated uniformly by polynomials.
[*Hint:* Part (a): Consider the integral $\int_{|t|\le n^{-1/2}} Q_j(t)\,dt$. Part (d): Consider the function $g(t) = f(t) - f(-1) - \frac{t+1}{2}(f(1) - f(-1))$.]

1.2.13. (Christ and Grafakos [**101**]) Let $F \ge 0$, $G \ge 0$ be measurable functions on the sphere $\mathbf{S}^{n-1}$ and let $K \ge 0$ a measurable function on $[-1,1]$. Prove that

$$\int_{\mathbf{S}^{n-1}} \int_{\mathbf{S}^{n-1}} F(\theta)G(\varphi)K(\theta\cdot\varphi)\,d\varphi\,d\theta \le C\|F\|_{L^p(\mathbf{S}^{n-1})}\|G\|_{L^{p'}(\mathbf{S}^{n-1})},$$

where $1 \le p \le \infty$, $\theta\cdot\varphi = \sum_{j=1}^n \theta_j\varphi_j$ and $C = \int_{\mathbf{S}^{n-1}} K(\theta\cdot\varphi)\,d\varphi$, which is independent of θ. Moreover, show that C is the best possible constant in the preceding inequality. Using duality, compute the norm of the linear operator

$$F(\varphi) \to \int_{\mathbf{S}^{n-1}} F(\theta)K(\theta\cdot\varphi)\,d\varphi$$

from $L^p(\mathbf{S}^{n-1})$ into itself.
[*Hint:* Observe that $\int_{\mathbf{S}^{n-1}}\int_{\mathbf{S}^{n-1}} F(\theta)G(\varphi)K(\theta\cdot\varphi)\,d\varphi\,d\theta$ is bounded by the quantity $\{\int_{\mathbf{S}^{n-1}}[\int_{\mathbf{S}^{n-1}} F(\theta)K(\theta\cdot\varphi)\,d\theta]^p\,d\varphi\}^{\frac{1}{p}}\|G\|_{L^{p'}(\mathbf{S}^{n-1})}$. Apply Hölder's inequality to the functions F and 1 with respect to the measure $K(\theta\cdot\varphi)\,d\theta$ to decuce that $\int_{\mathbf{S}^{n-1}} F(\theta)K(\theta\cdot\varphi)\,d\theta$ is controlled by

$$\left(\int_{\mathbf{S}^{n-1}} F(\theta)^p K(\theta\cdot\varphi)\,d\theta\right)^{1/p}\left(\int_{\mathbf{S}^{n-1}} K(\theta\cdot\varphi)\,d\theta\right)^{1/p'}.$$

Use Fubini's theorem to bound the latter by

$$\|F\|_{L^p(\mathbf{S}^{n-1})}\|G\|_{L^{p'}(\mathbf{S}^{n-1})}\int_{\mathbf{S}^{n-1}} K(\theta\cdot\varphi)\,d\varphi.$$

Note that equality is attained if and only if both F and G are constants.]

1.3. Interpolation

The theory of interpolation of operators is vast and extensive. In this section we will only be concerned with a few basic interpolation results that appear in a variety of applications and constitute the foundation of the field. These results are the *Marcinkiewicz interpolation theorem* and the *Riesz-Thorin interpolation theorem.* These theorems are traditionally proved using real and complex variables techniques, respectively. A byproduct of the Riesz-Thorin interpolation theorem, *Stein's theorem on interpolation of analytic families of operators*, has also proved to be an important and useful tool in many applications and is presented at the end of the section.

We begin by setting up the background required to formulate the results of this section. Let (X,μ) and (Y,ν) be two measure spaces. We will be concerned mostly with linear operators T initially defined on the set of simple functions on X, such that for all f simple on X, $T(f)$ is a ν-measurable function on Y. Let $0 < p < \infty$ and $0 < q < \infty$. If there exists a constant $C_{p,q} > 0$ such that for all f simple functions on X we have

$$\big\|T(f)\big\|_{L^q(Y,\nu)} \le C_{p,q}\big\|f\big\|_{L^p(X,\mu)}, \tag{1.3.1}$$

then by density T admits a unique bounded extension from $L^p(X,\mu)$ to $L^q(Y,\nu)$. This extension will also be denoted by T. Operators that map L^p to L^q are called of *strong type* (p,q) and operators that map L^p to $L^{q,\infty}$ are called of *weak type* (p,q).

1.3.a. Real Method: The Marcienkiewicz Interpolation Theorem

Definition 1.3.1. Let T be an operator defined on a linear subspace of the space of all complex-valued measurable functions on a measure space (X,μ) and taking values in the set of all complex-valued measurable functions on a measure space (Y,ν). T is called *linear* if for all f, g and all $\lambda \in \mathbf{C}$, we have

$$T(f+g) = T(f) + T(g) \qquad \text{and} \qquad T(\lambda f) = \lambda T(f). \tag{1.3.2}$$

T is called *sublinear* if for all f, g and all $\lambda \in \mathbf{C}$, we have

$$|T(f+g)| \le |T(f)| + |T(g)| \qquad \text{and} \qquad |T(\lambda f)| = |\lambda||T(f)|. \tag{1.3.3}$$

T is called *quasi-linear* if for all f, g and all $\lambda \in \mathbf{C}$, we have

$$|T(f+g)| \le K(|T(f)| + |T(g)|) \qquad \text{and} \qquad |T(\lambda f)| = |\lambda||T(f)| \tag{1.3.4}$$

for some constant $K > 0$. Sublinearity is a special case of quasi-linearity.

We begin with our first interpolation theorem.

Theorem 1.3.2. *Let (X,μ) and (Y,ν) be two measure spaces and let $0 < p_0 < p_1 \le \infty$. Let T be a sublinear operator defined on the space $L^{p_0}(X) + L^{p_1}(X)$ and taking values in the space of measurable functions on Y. Assume that there exist two positive constants A_0 and A_1 such that*

$$\big\|T(f)\big\|_{L^{p_0,\infty}(Y)} \le A_0\big\|f\big\|_{L^{p_0}(X)} \qquad \textit{for all } f \in L^{p_0}(X), \tag{1.3.5}$$

$$\big\|T(f)\big\|_{L^{p_1,\infty}(Y)} \le A_1\big\|f\big\|_{L^{p_1}(X)} \qquad \textit{for all } f \in L^{p_1}(X). \tag{1.3.6}$$

Then for all $p_0 < p < p_1$ and for all f in $L^p(X)$ we have the estimate

$$\|T(f)\|_{L^p(Y)} \le A\|f\|_{L^p(X)}, \tag{1.3.7}$$

where

$$A = 2\left(\frac{p}{p-p_0} + \frac{p}{p_1-p}\right)^{\frac{1}{p}} A_0^{\frac{\frac{1}{p}-\frac{1}{p_1}}{\frac{1}{p_0}-\frac{1}{p_1}}} A_1^{\frac{\frac{1}{p_0}-\frac{1}{p}}{\frac{1}{p_0}-\frac{1}{p_1}}}. \tag{1.3.8}$$

Proof. Assume that $p_1 < \infty$ first. Fix f a function in $L^p(X)$ and $\alpha > 0$. We split $f = f_0^\alpha + f_1^\alpha$, where f_0^α is in L^{p_0} and f_1^α is in L^{p_1}. The splitting is obtained by cutting $|f|$ at height $\delta\alpha$ for some $\delta > 0$ to be determined later. Set

$$f_0^\alpha(x) = \begin{cases} f(x) & \text{for } |f(x)| > \delta\alpha \\ 0 & \text{for } |f(x)| \le \delta\alpha \end{cases}$$

$$f_1^\alpha(x) = \begin{cases} f(x) & \text{for } |f(x)| \le \delta\alpha \\ 0 & \text{for } |f(x)| > \delta\alpha\,. \end{cases}$$

It can be checked easily that f_0^α (the unbounded part of f) is an L^{p_0} function and that f_1^α (the bounded part of f) is an L^{p_1} function. Indeed, since $p_0 < p$ we have

$$\|f_0^\alpha\|_{L^{p_0}}^{p_0} = \int_{|f|>\delta\alpha} |f|^p |f|^{p_0-p}\, d\mu(x) \le (\delta\alpha)^{p_0-p}\|f\|_{L^p}^p$$

and, similarly, since $p < p_1$

$$\|f_1^\alpha\|_{L^{p_1}}^{p_1} \le (\delta\alpha)^{p_1-p}\|f\|_{L^p}^p.$$

By the sublinearity property (1.3.3) we obtain that

$$|T(f)| \le |T(f_0^\alpha)| + |T(f_1^\alpha)|,$$

which implies

$$\{x : |T(f)(x)| > \alpha\} \subseteq \{x : |T(f_0^\alpha)(x)| > \alpha/2\} \cup \{x : |T(f_1^\alpha)(x)| > \alpha/2\},$$

and therefore

$$d_{T(f)}(\alpha) \le d_{T(f_0^\alpha)}(\alpha/2) + d_{T(f_1^\alpha)}(\alpha/2). \tag{1.3.9}$$

Hypotheses (1.3.5) and (1.3.6) together with (1.3.9) now give

$$d_{T(f)}(\alpha) \le \frac{A_0^{p_0}}{(\alpha/2)^{p_0}} \int_{|f|>\delta\alpha} |f(x)|^{p_0}\, d\mu(x) + \frac{A_1^{p_1}}{(\alpha/2)^{p_1}} \int_{|f|\le\delta\alpha} |f(x)|^{p_1}\, d\mu(x).$$

Using this last estimate and Proposition 1.1.4, we obtain that

$$\begin{aligned}
\big\|T(f)\big\|_{L^p}^p &\le p(2A_0)^{p_0}\int_0^\infty \alpha^{p-1}\alpha^{-p_0}\int_{|f|>\delta\alpha}|f(x)|^{p_0}\,d\mu(x)\,d\alpha\, + \\
&\qquad p(2A_1)^{p_1}\int_0^\infty \alpha^{p-1}\alpha^{-p_1}\int_{|f|\le\delta\alpha}|f(x)|^{p_1}\,d\mu(x)\,d\alpha \\
&\le p(2A_0)^{p_0}\int_X |f(x)|^{p_0}\int_0^{\frac{1}{\delta}|f(x)|}\alpha^{p-1-p_0}\,d\alpha\,d\mu(x)\, + \\
&\qquad p(2A_1)^{p_1}\int_X |f(x)|^{p_1}\int_{\frac{1}{\delta}|f(x)|}^\infty \alpha^{p-1-p_1}\,d\alpha\,d\mu(x) \\
&= \frac{p(2A_0)^{p_0}}{p-p_0}\frac{1}{\delta^{p-p_0}}\int_X |f(x)|^{p_0}|f(x)|^{p-p_0}\,d\mu(x)\, + \\
&\qquad \frac{p(2A_1)^{p_1}}{p_1-p}\frac{1}{\delta^{p-p_1}}\int_X |f(x)|^{p_1}|f(x)|^{p-p_1}\,d\mu(x) \\
&= p\left(\frac{(2A_0)^{p_0}}{p-p_0}\frac{1}{\delta^{p-p_0}}+\frac{(2A_1)^{p_1}}{p_1-p}\delta^{p_1-p}\right)\big\|f\big\|_{L^p}^p,
\end{aligned}$$

and the convergence of the integrals in α is justified from $p_0 < p < p_1$. We pick $\delta > 0$ so that

$$(2A_0)^{p_0}\frac{1}{\delta^{p-p_0}} = (2A_1)^{p_1}\delta^{p_1-p}\,,$$

and observe that the last displayed constant is equal to the pth power of the constant in (1.3.8). We have therefore proved the theorem when $p_1 < \infty$.

We now consider the case $p_1 = \infty$. Write $f = f_0^\alpha + f_1^\alpha$, where

$$\begin{aligned}
f_0^\alpha(x) &= \begin{cases} f(x) & \text{for } \ |f(x)| > \gamma\alpha \\ 0 & \text{for } \ |f(x)| \le \gamma\alpha \end{cases} \\
f_1^\alpha(x) &= \begin{cases} f(x) & \text{for } \ |f(x)| \le \gamma\alpha \\ 0 & \text{for } \ |f(x)| > \gamma\alpha\,. \end{cases}
\end{aligned}$$

We have $\big\|T(f_1^\alpha)\big\|_{L^\infty} \le A_1\big\|f_1^\alpha\big\|_{L^\infty} \le A_1\gamma\alpha = \alpha/2$ provided we choose $\gamma = (2A_1)^{-1}$. It follows that the set $\{x : |T(f_1^\alpha)(x)| > \alpha/2\}$ has measure zero. Therefore,

$$d_{T(f)}(\alpha) \le d_{T(f_0^\alpha)}(\alpha/2).$$

Since T maps L^{p_0} to $L^{p_0,\infty}$, it follows that

$$d_{T(f_0^\alpha)}(\alpha/2) \le \frac{(2A_0)^{p_0}\big\|f_0^\alpha\big\|_{L^{p_0}}^{p_0}}{\alpha^{p_0}} = \frac{(2A_0)^{p_0}}{\alpha^{p_0}}\int_{|f|>\gamma\alpha}|f(x)|^{p_0}\,d\mu(x). \tag{1.3.10}$$

Using (1.3.10) and Proposition 1.1.4, we obtain

$$\begin{aligned}
\|T(f)\|_{L^p}^p &= p\int_0^\infty \alpha^{p-1} d_{T(f)}(\alpha)\,d\alpha \\
&\le p\int_0^\infty \alpha^{p-1} d_{T(f_0^\alpha)}(\alpha/2)\,d\alpha \\
&\le p\int_0^\infty \alpha^{p-1}\frac{(2A_0)^{p_0}}{\alpha^{p_0}}\int_{|f|>\alpha/(2A_1)} |f(x)|^{p_0}\,d\mu(x)\,d\alpha \\
&= p(2A_0)^{p_0}\int_X |f(x)|^{p_0}\int_0^{2A_1|f(x)|} \alpha^{p-p_0-1}\,d\alpha\,d\mu(x) \\
&= \frac{p(2A_1)^{p-p_0}(2A_0)^{p_0}}{p-p_0}\int_X |f(x)|^p\,d\mu(x).
\end{aligned}$$

This proves the theorem with constant

$$A = 2\left(\frac{p}{p-p_0}\right)^{\frac{1}{p}} A_1^{1-\frac{p_0}{p}} A_0^{\frac{p_0}{p}}. \tag{1.3.11}$$

Observe that when $p_1=\infty$, the constant in (1.3.11) coincides with that in (1.3.8).

□

Remark 1.3.3. If T is a linear operator (instead of sublinear), then we can relax the hypotheses of Theorem 1.3.2 by assuming that (1.3.5) and (1.3.6) hold for all f simple functions on X. Then the functions f_0^α and f_1^α constructed in the proof are also simple and we conclude that (1.3.7) holds for all f simple functions on X. By density, T has a unique extension on $L^p(X)$ that also satisfies (1.3.7).

1.3.b. Complex Method: The Riesz-Thorin Interpolation Theorem

The next interpolation theorem assumes stronger endpoint estimates but it gives a more natural bound on the norm of the operator on the intermediate spaces. Unfortunately, it is mostly applicable for linear operators and in some cases for sublinear operators (often via a linearization process). However, it does not apply to quasi-linear operators without some loss in the constant. See the end of this chapter for a short history of this theorem.

Theorem 1.3.4. *Let (X,μ) and (Y,ν) be two measure spaces. Let T be a linear operator defined on the set of all simple functions on X and taking values in the set of measurable functions on Y. Let $1\le p_0,p_1,q_0,q_1\le\infty$ and assume that*

$$\begin{aligned}
\|T(f)\|_{L^{q_0}} &\le M_0\|f\|_{L^{p_0}} \\
\|T(f)\|_{L^{q_1}} &\le M_1\|f\|_{L^{p_1}}
\end{aligned} \tag{1.3.12}$$

for all f simple functions on X. Then for all $0<\theta<1$ we have

$$\|T(f)\|_{L^q} \le M_0^{1-\theta}M_1^{\theta}\|f\|_{L^p} \tag{1.3.13}$$

for all f simple functions on X, where

$$\frac{1}{p}=\frac{1-\theta}{p_0}+\frac{\theta}{p_1} \quad \textit{and} \quad \frac{1}{q}=\frac{1-\theta}{q_0}+\frac{\theta}{q_1}. \tag{1.3.14}$$

By density, T has a unique extension as a bounded operator from $L^p(X,\mu)$ into $L^q(Y,\nu)$ for all p and q as in (1.3.14).

We note that in many applications T may be defined on $L^{p_0}+L^{p_1}$, in which case hypothesis (1.3.12) and conclusion (1.3.13) can be stated in terms of functions in the corresponding Lebesgue spaces.

Proof. Let

$$f=\sum_{k=1}^{m} a_k e^{i\alpha_k}\chi_{A_k}$$

be a simple function on X, where $a_k>0$, α_k real and A_k are pairwise disjoint subsets of X with finite measure.

We need to control

$$\big\|T(f)\big\|_{L^q(Y,\nu)}=\sup\left|\int_Y T(f)(x)g(x)\,d\nu(x)\right|,$$

where the supremum is taken over all simple functions g on Y with $L^{q'}$ norm less than or equal to 1. Write

$$g=\sum_{j=1}^{n} b_j e^{i\beta_j}\chi_{B_j} \tag{1.3.15}$$

where $b_j>0$, β_j real, and B_j are pairwise disjoint subsets of Y with finite measure. Let

$$P(z)=\frac{p}{p_0}(1-z)+\frac{p}{p_1}z \quad \text{and} \quad Q(z)=\frac{q'}{q_0'}(1-z)+\frac{q'}{q_1'}z. \tag{1.3.16}$$

For z in the closed strip $\overline{S}=\{z\in\mathbf{C}:\,0\le \operatorname{Re} z\le 1\}$, define

$$F(z)=\int_Y T(f_z)(x)\,g_z(x)\,d\nu(x),$$

where

$$f_z=\sum_{k=1}^{m} a_k^{P(z)}e^{i\alpha_k}\chi_{A_k}, \quad g_z=\sum_{j=1}^{n} b_j^{Q(z)}e^{i\beta_j}\chi_{B_j}. \tag{1.3.17}$$

By linearity,

$$F(z)=\sum_{k=1}^{m}\sum_{j=1}^{n} a_k^{P(z)}b_j^{Q(z)}e^{i\alpha_k}e^{i\beta_j}\int_Y T(\chi_{A_k})(x)\,\chi_{B_j}(x)\,d\nu(x),$$

and hence F is analytic in z, since $a_k, b_j>0$.

Let us now consider a z with $\operatorname{Re} z = 0$. Note that by the disjointness of the sets A_k we have $\|f_z\|_{L^{p_0}}^{p_0} = \|f\|_{L^p}^p$, since $|a_k^{P(z)}| = a_k^{\frac{p}{p_0}}$. Similarly, by the disjointness of the sets B_j we have that $\|g_z\|_{L^{q_0'}}^{q_0'} = \|g\|_{L^{q'}}^{q'}$, since $|b_j^{Q(z)}| = b_j^{\frac{q'}{q_0'}}$.

By the same token, when $\operatorname{Re} z = 1$, we have $\|f_z\|_{L^{p_1}}^{p_1} = \|f\|_{L^p}^p$ and $\|g_z\|_{L^{q_1'}}^{q_1'} = \|g\|_{L^{q'}}^{q'}$. Hölder's inequality and the hypothesis now give

$$
\begin{aligned}
|F(z)| &\le \|T(f_z)\|_{L^{q_0}} \|g_z\|_{L^{q_0'}} \\
&\le M_0 \|f_z\|_{L^{p_0}} \|g_z\|_{L^{q_0'}} = M_0 \|f\|_{L^p}^{\frac{p}{p_0}} \|g\|_{L^{q'}}^{\frac{q'}{q_0'}},
\end{aligned}
\tag{1.3.18}
$$

when $\operatorname{Re} z = 0$. Similarly, we obtain

$$
|F(z)| \le M_1 \|f\|_{L^p}^{\frac{p}{p_1}} \|g\|_{L^{q'}}^{\frac{q'}{q_1'}},
\tag{1.3.19}
$$

when $\operatorname{Re} z = 1$.

We shall now use the following lemma, known as *Hadamard's three lines lemma*, whose proof we postpone until the end of this section.

Lemma 1.3.5. *Let F be analytic in the open strip $S = \{z \in \mathbf{C} : 0 < \operatorname{Re} z < 1\}$, continuous and bounded on its closure, such that $|F(z)| \le B_0$ when $\operatorname{Re} z = 0$ and $|F(z)| \le B_1$ when $\operatorname{Re} z = 1$, where $0 < B_0, B_1 < \infty$. Then $|F(z)| \le B_0^{1-\theta} B_1^{\theta}$ when $\operatorname{Re} z = \theta$, for any $0 \le \theta \le 1$.*

Assume this lemma for a moment and observe that F is analytic in the open strip S and continuous on its closure. Also F is bounded on the closed unit strip (by some constant that depends on f and g). Therefore, (1.3.18) and (1.3.19) give

$$
\begin{aligned}
|F(z)| &\le \Big(M_0 \|f\|_{L^p}^{\frac{p}{p_0}} \|g\|_{L^{q'}}^{\frac{q'}{q_0'}} \Big)^{1-\theta} \Big(M_1 \|f\|_{L^p}^{\frac{p}{p_1}} \|g\|_{L^{q'}}^{\frac{q'}{q_1'}} \Big)^{\theta} \\
&= M_0^{1-\theta} M_1^{\theta} \|f\|_{L^p} \|g\|_{L^{q'}},
\end{aligned}
$$

when $\operatorname{Re} z = \theta$. Observe that $P(\theta) = Q(\theta) = 1$ and hence

$$
F(\theta) = \int_Y T(f)\, g \, d\nu \,.
$$

Taking the supremum over all g simple functions on Y with $L^{q'}$ norm less than or equal to one, we conclude the proof of the theorem.

□

We now give an application of the theorem just proved.

Example 1.3.6. The Riesz-Thorin interpolation theorem can be used to prove Young's inequality (Theorem 1.2.12). Fix g in L^r and let $T(f) = f * g$. Since $T: L^1 \to L^r$ with norm at most $\|g\|_{L^r}$ and $T: L^{r'} \to L^\infty$ with norm at most $\|g\|_{L^r}$,

Theorem 1.3.4 gives that T maps L^p to L^q with norm at most $\|g\|_{L^r}^{\theta}\|g\|_{L^r}^{1-\theta} = \|g\|_{L^r}$, where

$$\frac{1}{p} = \frac{1-\theta}{1} + \frac{\theta}{r'}, \quad \text{and} \quad \frac{1}{q} = \frac{1-\theta}{r} + \frac{\theta}{\infty}. \tag{1.3.20}$$

Finally, observe that equations (1.3.20) give (1.2.14).

1.3.c. Stein's Interpolation Theorem on Analytic Families of Operators*

Theorem 1.3.4 can now be extended to the case where the interpolated operators are allowed to vary. In particular, if a family of operators depends analytically on a parameter z, then the proof of this theorem can be adapted to work in this setting.

We now describe the setup for this theorem. Let (X,μ) and (Y,ν) be measure spaces. Suppose that for every z in the closed strip $\overline{S} = \{z \in \mathbf{C} : 0 \le \operatorname{Re} z \le 1\}$ there is an associated linear operator T_z defined on the space of simple functions on X and taking values in the space of measurable functions on Y, such that

$$\int_Y |T_z(f)\, g|\, d\nu < \infty \tag{1.3.21}$$

whenever f and g are simple functions on X and Y, respectively. The family $\{T_z\}_z$ is said to be *analytic* if the function

$$z \to \int_Y T_z(f)\, g\, d\nu \tag{1.3.22}$$

is analytic in the open strip $S = \{z \in \mathbf{C} : 0 < \operatorname{Re} z < 1\}$ and continuous on its closure. Finally, the analytic family is of *admissible growth* if there is a constant $a < \pi$ and a constant $C_{f,g}$ such that

$$e^{-a|\operatorname{Im} z|}\left|\int_Y T_z(f)\, g\, d\nu\right| \le C_{f,g} < \infty \tag{1.3.23}$$

for all z satisfying $0 \le \operatorname{Re} z \le 1$. The extension of the Riesz-Thorin interpolation theorem is now stated.

Theorem 1.3.7. *Let T_z be an analytic family of linear operators of admissible growth. Let $1 \le p_0, p_1, q_0, q_1 \le \infty$ and suppose that M_0 and M_1 are real-valued functions such that*

$$\sup_{-\infty<y<+\infty} e^{-b|y|} \log M_j(y) < \infty \tag{1.3.24}$$

for $j = 0, 1$ and some $b < \pi$. Let $0 < \theta < 1$ satisfy

$$\frac{1}{p} = \frac{1-\theta}{p_0} + \frac{\theta}{p_1} \quad \textit{and} \quad \frac{1}{q} = \frac{1-\theta}{q_0} + \frac{\theta}{q_1}. \tag{1.3.25}$$

Suppose that

$$\big\|T_{iy}(f)\big\|_{L^{q_0}} \le M_0(y)\big\|f\big\|_{L^{p_0}} \tag{1.3.26}$$

$$\big\|T_{1+iy}(f)\big\|_{L^{q_1}} \le M_1(y)\big\|f\big\|_{L^{p_1}} \tag{1.3.27}$$

for all simple functions f on X. Then

$$\big\|T_\theta(f)\big\|_{L^q} \le M(\theta)\big\|f\big\|_{L^p} \qquad \textit{when } 0<\theta<1 \tag{1.3.28}$$

for all simple functions f on X, where

$$M(x) = \exp\left\{\frac{\sin(\pi x)}{2}\int_{-\infty}^{\infty}\left[\frac{\log M_0(y)}{\cosh(\pi y)-\cos(\pi x)} + \frac{\log M_1(y)}{\cosh(\pi y)+\cos(\pi x)}\right]dy\right\}.$$

By density, T_θ has a unique extension as a bounded operator from $L^p(X,\mu)$ into $L^q(Y,\nu)$ for all p and q as in (1.3.25).

As expected, the proof of the previous theorem will be based on an extension of Lemma 1.3.5.

Lemma 1.3.8. *Let F be analytic on the open strip $S=\{z\in\mathbf{C} : 0<Re\, z<1\}$ and continuous on its closure, such that*

$$\sup_{z\in\overline{S}} e^{-a|Im\, z|}\log|F(z)| \le A<\infty \tag{1.3.29}$$

for some fixed A and $a<\pi$. Then

$$|F(x)| \le \exp\left\{\frac{\sin(\pi x)}{2}\int_{-\infty}^{\infty}\left[\frac{\log|F(iy)|}{\cosh(\pi y)-\cos(\pi x)} + \frac{\log|F(1+iy)|}{\cosh(\pi y)+\cos(\pi x)}\right]dy\right\}$$

whenever $0<x<1$.

Assuming Lemma 1.3.8, we prove Theorem 1.3.7.

Proof. As in the proof of Theorem 1.3.4, we will be working with simple functions f on X and g on Y. Fix a $0<\theta<1$ and also fix f and g simple functions such that $\big\|f\big\|_{L^p}=1=\big\|g\big\|_{L^{q'}}$. Let

$$f=\sum_{k=1}^{m} a_k e^{i\alpha_k}\chi_{A_k} \quad \text{and} \quad g=\sum_{j=1}^{n} b_j e^{i\beta_j}\chi_{B_j},$$

where $a_k>0$, $b_j>0$, α_k, β_k real, A_k pairwise disjoint subsets of X with finite measure and B_j pairwise disjoint subsets of Y with finite measure. Let f_z and g_z as in the proof of Theorem 1.3.4. Define

$$F(z)=\int_Y T_z(f_z)\, g_z\, d\nu. \tag{1.3.30}$$

It follows from the assumptions about $\{T_z\}_z$ that $F(z)$ is an analytic function that satisfies the hypotheses of Lemma 1.3.8. Moreover,

$$\|f_{iy}\|_{L^{p_0}}^{p_0} = \|f\|_{L^p}^{p} = 1 = \|g\|_{L^{q'}}^{q'} = \|g_{iy}\|_{L^{q_0'}}^{q_0'} \quad \text{when } y \in \mathbf{R}, \tag{1.3.31}$$

$$\|f_{1+iy}\|_{L^{p_1}}^{p_1} = \|f\|_{L^p}^{p} = 1 = \|g\|_{L^{q'}}^{q'} = \|g_{1+iy}\|_{L^{q_1'}}^{q_1'} \quad \text{when } y \in \mathbf{R}. \tag{1.3.32}$$

Hölder's inequality, (1.3.31), and the hypothesis (1.3.26) now give

$$\begin{aligned} |F(iy)| &\le \big\|T_{iy}(f_{iy})\big\|_{L^{q_0}} \big\|g_{iy}\big\|_{L^{q_0'}} \\ &\le M_0(y)\big\|f_{iy}\big\|_{L^{p_0}} \big\|g_{iy}\big\|_{L^{q_0'}} = M_0(y) \end{aligned}$$

for all y real. Similarly, (1.3.32), and (1.3.27) imply

$$\begin{aligned} |F(1+iy)| &\le \big\|T_{1+iy}(f_{1+iy})\big\|_{L^{q_1}} \big\|g_{1+iy}\big\|_{L^{q_1'}} \\ &\le M_1(y)\big\|f_{1+iy}\big\|_{L^{p_1}} \big\|g_{1+iy}\big\|_{L^{q_1'}} = M_1(y). \end{aligned}$$

Therefore, the hypotheses of Lemma 1.3.8 are satisfied. We conclude that

$$\left|\int_Y T_\theta(f)\, g\, d\nu\right| = |F(\theta)| \le M(\theta), \tag{1.3.33}$$

where $M(x)$ is the function given in the hypothesis of the theorem.

Taking the supremum over all g simple functions on Y with $L^{q'}$ norm equal to one, we conclude the proof of the theorem.

□

1.3.d. Proofs of Lemmata 1.3.5 and 1.3.8

Proof of Lemma 1.3.5. Define analytic functions

$$G(z) = F(z)(B_0^{1-z}B_1^{z})^{-1} \qquad \text{and} \qquad G_n(z) = G(z)e^{(z^2-1)/n}\,.$$

Since F is bounded on the closed unit strip and $B_0^{1-z}B_1^{z}$ is bounded from below, we conclude that G is bounded by some constant M on the closed strip. Also, G is bounded by one on its boundary. Since

$$|G_n(x+iy)| \le Me^{-y^2/n}e^{(x^2-1)/n} \le Me^{-y^2/n}\,,$$

we deduce that $G_n(x+iy)$ converges to zero uniformly in $0 \le x \le 1$ as $|y| \to \infty$. Select a y_0 such that for $|y| \ge |y_0|$, $|G_n(x+iy)| \le 1$ uniformly in $0 \le x \le 1$. By the maximum principle we obtain that $|G_n(z)| \le 1$ in the rectangle $[0,1] \times [-|y_0|, |y_0|]$, hence $|G_n(z)| \le 1$ everywhere in the closed strip. Letting $n \to \infty$, we conclude that $|G(z)| \le 1$ in the closed strip.

□

Having disposed of the proof of Lemma 1.3.5, we end this section with the proof of Lemma 1.3.8.

Proof of Lemma 1.3.8. Start with Poisson's formula

$$u(z) = \frac{1}{2\pi} \int_{|\zeta|=R} u(\zeta) \frac{|\zeta|^2 - |z|^2}{|\zeta - z|^2} d\zeta, \tag{1.3.34}$$

which is valid for a harmonic function u defined on the unit disc $D = \{z : |z| < 1\}$ when $|z| < R < 1$. See Rudin [**440**, p. 258].

Consider now a subharmonic function u on D which is continuous on the circle $|\zeta| = R < 1$. The right side of (1.3.34) defines a harmonic function on the set $\{z : |z| < R\}$ which coincides with u on the circle $|\zeta| = R$. The maximum principle for subharmonic functions (Rudin [**440**, p. 362]) implies that for $|z| < R < 1$ we have

$$u(z) \le \frac{1}{2\pi} \int_{|\zeta|=R} u(\zeta) \frac{|\zeta|^2 - |z|^2}{|\zeta - z|^2} d\zeta\,, \tag{1.3.35}$$

for all subharmonic functions u on D that are continuous on the circle $|\zeta| = R$.

It is not difficult to verify that

$$h(\zeta) = \frac{1}{\pi i} \log\left(i \frac{1+\zeta}{1-\zeta} \right)$$

is a conformal map from D onto S. Observe that $i(1+\zeta)/(1-\zeta)$ lies in the upper half-space and the preceding complex logarithm is a well-defined holomorphic function that takes the upper half-space onto the strip $\mathbf{R}\times(0,\pi)$. Since $F\circ h$ is a holomorphic function on D, $\log|F(h(\zeta))|$ is a subharmonic function on D (Rudin [**440**, p. 362].) Apply (1.3.35) to the function $\log|F(h(\zeta))|$ to obtain

$$\log|F(h(z))| \le \frac{1}{2\pi} \int_{|\zeta|=R} \log|F(h(\zeta))| \frac{|\zeta|^2 - |z|^2}{|\zeta - z|^2} d\zeta\,. \tag{1.3.36}$$

Observe that when $|\zeta| = 1$ and $\zeta \ne \pm 1$, $h(\zeta)$ has real part zero or one. It follows from the hypothesis that

$$\log|F(h(\zeta))| \le Ae^{a|\operatorname{Im} h(\zeta)|} = Ae^{a\left|\operatorname{Im} \frac{1}{\pi i} \log\left(i\frac{1+\zeta}{1-\zeta}\right)\right|} \le Ae^{\frac{a}{\pi}\left|\log\left|\frac{1+\zeta}{1-\zeta}\right|\right|}.$$

Therefore, $\log|F(h(\zeta))|$ is bounded by a multiple of $[|1+\zeta|^{-a/\pi} + |1-\zeta|^{-a/\pi}]$ that is integrable over the set $|\zeta| = 1$ since $a < \pi$. Fix now a z with $|z| = \rho < R$ and let $R \to 1$ in (1.3.36). The Lebesgue dominated convergence theorem gives that

$$\log|(F(h(z))| \le \frac{1}{2\pi} \int_{|\zeta|=1} \log|(F(h(\zeta))| \frac{1 - |z|^2}{|\zeta - z|^2} d\zeta.$$

Changing notation, $z = \rho e^{i\theta}$ and $\zeta = e^{i\varphi}$, we write the preceding expression as

$$\log|F(h(\rho e^{i\theta}))| \le \frac{1}{2\pi} \int_{-\pi}^{\pi} \frac{(1-\rho^2)\log|F(h(e^{i\varphi}))|}{1 - 2\rho\cos(\theta - \varphi) + \rho^2} d\varphi\,. \tag{1.3.37}$$

Setting $x = h(\rho e^{i\theta})$, we obtain that

$$\rho e^{i\theta} = h^{-1}(x) = \frac{e^{\pi i x} - i}{e^{\pi i x} + i} = -i\frac{\cos(\pi x)}{1+\sin(\pi x)} = \left\{\frac{\cos(\pi x)}{1+\sin(\pi x)}\right\} e^{-i(\pi/2)},$$

from which it follows that $\rho = (\cos(\pi x))/(1+\sin(\pi x))$ and $\theta = -(\pi/2)$, when $0 < x \le \frac{1}{2}$, while $\rho = -(\cos(\pi x))/(1+\sin(\pi x))$ and $\theta = \pi/2$, when $\frac{1}{2} \le x < 1$. In either case we easily deduce that

$$\frac{1-\rho^2}{1-2\rho\cos(\theta-\varphi)+\rho^2} = \frac{\sin(\pi x)}{1+\cos(\pi x)\sin(\varphi)}.$$

Using this we write (1.3.37) as

$$\log|F(x)| \le \frac{1}{2\pi}\int_{-\pi}^{\pi} \frac{\sin(\pi x)}{1+\cos(\pi x)\sin(\varphi)} \log|F(h(e^{i\varphi}))|\,d\varphi. \tag{1.3.38}$$

We now change variables. On the interval $[-\pi, 0)$ we use the change of variables $iy = h(e^{i\varphi})$ or, equivalently, $e^{i\varphi} = -\tanh(\pi y) - i\,\mathrm{sech}(\pi y)$. Observe that as φ ranges from $-\pi$ to 0, y ranges from $+\infty$ to $-\infty$. Furthermore, $d\varphi = -\pi\,\mathrm{sech}(\pi y)\,dy$. We have

$$\begin{aligned}\frac{1}{2\pi}\int_{-\pi}^{0} &\frac{\sin(\pi x)}{1+\cos(\pi x)\sin(\varphi)} \log|F(h(e^{i\varphi}))|\,d\varphi \\ &= \frac{1}{2}\int_{-\infty}^{\infty} \frac{\sin(\pi x)}{\cosh(\pi y)-\cos(\pi x)} \log|F(iy)|\,dy.\end{aligned} \tag{1.3.39}$$

On the interval $(0, \pi]$ we use the change of variables $1 + iy = h(e^{i\varphi})$ or, equivalently, $e^{i\varphi} = -\tanh(\pi y) + i\,\mathrm{sech}(\pi y)$. Observe that as φ ranges from 0 to π, y ranges from $-\infty$ to $+\infty$. Furthermore, $d\varphi = \pi\,\mathrm{sech}(\pi y)\,dy$. Similarly, we obtain

$$\begin{aligned}\frac{1}{2\pi}\int_{0}^{\pi} &\frac{\sin(\pi x)}{1+\cos(\pi x)\sin(\varphi)} \log|F(h(e^{i\varphi}))|\,d\varphi \\ &= \frac{1}{2}\int_{-\infty}^{\infty} \frac{\sin(\pi x)}{\cosh(\pi y)+\cos(\pi x)} \log|F(1+iy)|\,dy.\end{aligned} \tag{1.3.40}$$

Now add (1.3.39) and (1.3.40) and use (1.3.38) to conclude the proof of the lemma. □

Exercises

1.3.1. Generalize Theorem 1.3.2 to the situation in which T is quasi-linear, that is, it satisfies for some $K > 0$

$$|T(\lambda f)| = |\lambda|\,|T(f)|, \quad \text{and} \quad |T(f+g)| \le K(|T(f)| + |T(g)|),$$

for all $\lambda \in \mathbf{C}$, and all f, g in the domain of T. Prove that in this case, the constant A in (1.3.7) is equal to K times the constant in (1.3.8).

1.3.2. Let $1 < p < r \le \infty$ and suppose that T is a linear operator that maps L^1 into $L^{1,\infty}$ with norm A_0 and L^r into L^r with norm A_1. Prove that T maps L^p into L^p with norm at most

$$8\,(p-1)^{-\frac{1}{p}}\, A_0^{\frac{\frac{1}{p}-\frac{1}{r}}{1-\frac{1}{r}}}\, A_1^{\frac{1-\frac{1}{p}}{1-\frac{1}{r}}}.$$

[*Hint:* First obtain interpolate between $L^{\frac{p+1}{2}}$ and L^r using Theorem 1.3.2 and then interpolate between $L^{\frac{p+1}{2}}$ and L^r using Theorem 1.3.4.]

1.3.3. Let $0 < p_0 < p < p_1 \le \infty$ and let T be an operator as in Theorem 1.3.2, which also satisfies

$$|T(f)| \le T(|f|),$$

for all $f \in L^{p_0} + L^{p_1}$.
(a) If $p_0 = 1$ and $p_1 = \infty$, prove that T maps L^p to L^p with norm at most

$$\frac{p}{p-1} A_0^{\frac{1}{p}} A_1^{1-\frac{1}{p}}.$$

(b) More generally, if $p_0 < p < p_1 = \infty$, prove that the norm of T from L^p to L^p is at most

$$p^{1+\frac{1}{p}} \left[\frac{B(p_0+1, p-p_0)}{p_0^{p_0}(p-p_0)^{p-p_0}} \right]^{\frac{1}{p}} A_0^{\frac{p_0}{p}} A_1^{1-\frac{p_0}{p}},$$

where $B(s,t) = \displaystyle\int_0^1 x^{s-1}(1-x)^{t-1}\,dx$ is the usual Beta function.
(c) When $0 < p_0 < p_1 < \infty$, then the norm of T from L^p into L^p is at most

$$\min_{0<\lambda<1} p^{\frac{1}{p}} \left(\frac{B(p-p_0, p_0+1)}{(\frac{1}{\lambda})^{p_0}} + \frac{\frac{p_1-p+1}{p_1-p}}{(\frac{1}{1-\lambda})^{p_1}} \right)^{\frac{1}{p}} A_0^{\frac{\frac{1}{p}-\frac{1}{p_1}}{\frac{1}{p_0}-\frac{1}{p_1}}} A_1^{\frac{\frac{1}{p_0}-\frac{1}{p}}{\frac{1}{p_0}-\frac{1}{p_1}}}.$$

[*Hint:* Parts (a), (b): The hypothesis $|T(f)| \le T(|f|)$ reduces matters to nonnegative functions. For $f \ge 0$ and for fixed $\alpha > 0$ write $f = f_0 + f_1$, where $f_0 = f - \lambda\alpha/A_1$ for $f \ge \lambda\alpha/A_1$ and zero otherwise for some $0 < \lambda < 1$. Then we have $|\{|T(f)| > \alpha\}| \le |\{|T(f_0)| > (1-\lambda)\alpha\}|$. Use a similar splitting when $p_1 < \infty$.]

1.3.4. Suppose that C, C_0, C_1, A, A_0, A_1 are positive real numbers and that $0 < \gamma_0, \gamma_1 < \pi$. Let T_z be a family of linear operators defined on the strip $S_{a,b} = \{z \in \mathbf{C} : a \le \operatorname{Re} z \le b\}$ that is analytic on the interior of $S_{a,b}$ [in the sense of (1.3.22)], continuous on its closure, and satisfies for all $z \in S_{a,b}$

$$e^{-A|\operatorname{Im} z|/(b-a)} \left| \int_Y T_z(f) g \, d\nu \right| \le C_{f,g},$$

where $A < \pi$. Let $1 \le p_0, q_0, p_1, q_1 \le \infty$. Suppose that T_{a+iy} maps $L^{p_0}(X)$ into $L^{q_0}(Y)$ with bound $M_0(y)$ and T_{b+iy} maps $L^{p_1}(X)$ into $L^{q_1}(Y)$ with bound $M_1(y)$, where

$$|M_j(y)| \le C_j \exp\left(A_j \exp(\gamma_j |y|/(b-a))\right), \quad j = 0, 1.$$

Then for $a < t < b$, T_t maps $L^p(X)$ into $L^q(Y)$, where

$$\frac{1}{p} = \frac{\frac{b-t}{b-a}}{p_0} + \frac{\frac{t-a}{b-a}}{p_1}, \quad \text{and} \quad \frac{1}{q} = \frac{\frac{b-t}{b-a}}{q_0} + \frac{\frac{t-a}{b-a}}{q_1}.$$

1.3.5. (*Stein* [**468**]) On $\mathbf{R}^n$ let $K_\lambda(x_1, \dots, x_n)$ be the function

$$\frac{\pi^{\frac{n-1}{2}} \Gamma(\lambda+1)}{\Gamma(\lambda + \frac{n+1}{2})} \int_{-1}^{+1} e^{2\pi i s (x_1^2 + \cdots + x_n^2)^{1/2}} (1 - s^2)^{\lambda + \frac{n-1}{2}} \, ds,$$

where λ is a complex number and $\Gamma(s) = \int_0^\infty t^{s-1} e^{-t} \, dt$ is the usual gamma function. Let T_λ be the operator given by convolution with K_λ. It can be proved that when $\operatorname{Re}\lambda = 0$, T_λ maps $L^2(\mathbf{R}^n)$ into itself with norm 1. Using this fact and asymptotics for the Bessel functions in Appendix B.6, show that T_λ maps $L^p(\mathbf{R}^n)$ into itself for $\operatorname{Re}\lambda > (n-1)|\frac{1}{2} - \frac{1}{p}|$.
[*Hint:* Use the definition of Bessel functions in Appendix B.1 to rewrite $K_\lambda(x) = c_\lambda |x|^{-\lambda - \frac{n}{2}} J_{\lambda + \frac{n}{2}}(2\pi|x|)$ for some suitable constant. Conclude that K_λ is integrable over $\mathbf{R}^n$ when $\operatorname{Re}\lambda > (n-1)/2$ and then appeal to Theorem 1.3.7.]

1.3.6. Under the same hypotheses as in Theorem 1.3.7, prove the stronger conclusion

$$\left\|T_z(f)\right\|_{L^q} \le B(z) \left\|f\right\|_{L^p}$$

for z in the open strip S, where

$$B(x+iy) = \exp\left\{\frac{\sin(\pi x)}{2} \int_{-\infty}^{\infty} \left[\frac{\log|F(iy)|}{\cosh(\pi(t-y)) - \cos(\pi x)} + \frac{\log|F(1+iy)|}{\cosh(\pi(t-y)) + \cos(\pi x)}\right] dt\right\}.$$

[*Hint:* Show that Lemma 1.3.8 actually yields the stronger conclusion that $|F(z)| \le B(z)$ for z in the strip S.]

1.3.7. (*Yano* [**562**]) Let (X, μ) and (Y, ν) be two measure spaces with $\mu(X) < \infty$ and $\nu(Y) < \infty$. Let T be a sublinear operator that maps $L^p(X)$ into $L^p(Y)$ for every $1 < p \le 2$ with norm $\left\|T\right\|_{L^p \to L^p} \le A(p-1)^{-\alpha}$ for some fixed $A, \alpha > 0$. Prove that for all f measurable on X we have

$$\int_Y |T(f)| \, d\nu \le 6A(1 + \nu(Y))^{\frac{1}{2}} \left[\int_X |f| (\log_2^+ |f|)^\alpha \, d\mu + C_\alpha + \mu(X)^{\frac{1}{2}}\right],$$

where $C_\alpha = \sum_{k=1}^\infty k^\alpha (2/3)^k$. This result provides an example of *extrapolation.*
[*Hint:* Write

$$f = \sum_{k=0}^{\infty} f\chi_{S_k},$$

where $S_k = \{2^k \le |f| < 2^{k+1}\}$ when $k \ge 1$ and $S_0 = \{|f| < 2\}$. Using Hölder's inequality and the hypotheses on T, obtain that

$$\int_Y |T(f\chi_{S_k})|\, d\nu \le 2A\nu(Y)^{\frac{1}{k+1}} 2^k k^\alpha \mu(S_k)^{\frac{k}{k+1}}$$

for $k \ge 1$. Note that for $k \ge 1$ we have $\nu(Y)^{\frac{1}{k+1}} \le \max(1, \nu(Y))^{\frac{1}{2}}$ and consider the cases $\mu(S_k) \ge 3^{-k-1}$ and $\mu(S_k) \le 3^{-k-1}$ when summing in $k \ge 1$. The term with $k = 0$ is easier.]

1.3.8. Prove that

$$\frac{1}{2}\int_{-\infty}^{+\infty} \frac{\sin(\pi x)}{\cosh(\pi y) + \cos(\pi x)}\, dy = x$$

$$\frac{1}{2}\int_{-\infty}^{+\infty} \frac{\sin(\pi x)}{\cosh(\pi y) - \cos(\pi x)}\, dy = 1 - x$$

and conclude that Lemma 1.3.8 is indeed an extension of Lemma 1.3.5.
[*Hint:* In the first integral write $\cosh(\pi y) = \frac{1}{2}(e^{\pi y} + e^{-\pi y})$. Then use the change of variables $z = e^{\pi y}$.]

1.4. Lorentz Spaces*

Suppose that f is a measurable function on a measure space (X, μ). We are interested in constructing another function f^* defined on $[0, \infty)$ that is decreasing and is *equidistributed* with f. By this we mean

$$d_f(\alpha) = d_{f^*}(\alpha) \tag{1.4.1}$$

for all $\alpha \ge 0$. This is achieved via a simple construction discussed in this section.

1.4.a. Decreasing Rearrangements

Definition 1.4.1. Let f be a complex-valued function defined on X. The *decreasing rearrangement* of f is the function f^* defined on $[0, \infty)$ by

$$f^*(t) = \inf\{s > 0 :\ d_f(s) \le t\}. \tag{1.4.2}$$

We adopt the convention $\inf \emptyset = \infty$, thus having $f^*(t) = \infty$ whenever $d_f(\alpha) > t$ for all $\alpha \geq 0$. Observe that f^* is decreasing and supported in $[0, \mu(X)]$ if $\mu(X) < \infty$.

Before we proceed with properties of the function f^*, we will work out three examples.

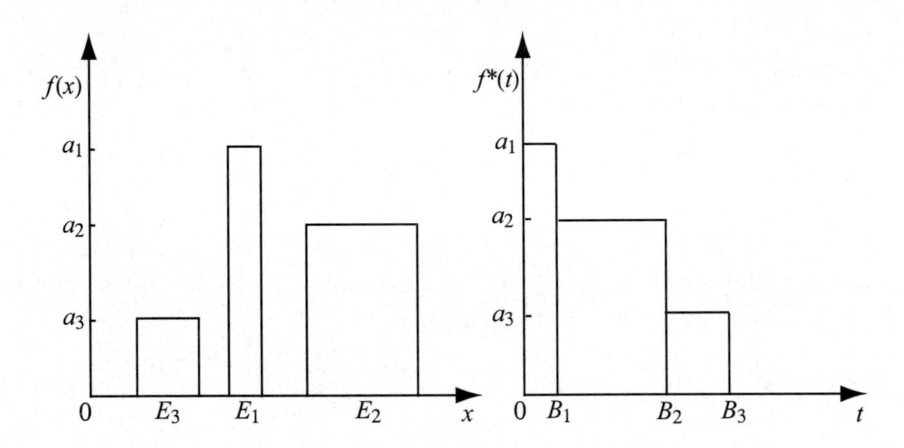

FIGURE 1.3. The graph of a simple function $f(x)$ and its decreasing rearrangement $f^*(t)$.

Example 1.4.2. Consider the simple function of Example 1.1.2

$$f(x) = \sum_{j=1}^{N} a_j \chi_{E_j}(x),$$

where the sets E_j have finite measure and are pairwise disjoint and $a_1 > \cdots > a_N$. We saw in Example 1.1.2 that

$$d_f(\alpha) = \sum_{j=0}^{N} B_j \chi_{[a_{j+1}, a_j)}(\alpha),$$

where

$$B_j = \sum_{i=1}^{j} \mu(E_i)$$

and we set $a_{N+1} = B_0 = 0$. Observe that for $B_0 \leq t < B_1$ the smallest $s > 0$ with $d_f(s) \leq t$ is a_1. Similarly, for $B_1 \leq t < B_2$ the smallest $s > 0$ with $d_f(s) \leq t$ is a_2. Arguing this way, it is not difficult to see that

$$f^*(t) = \sum_{j=1}^{N} a_j \chi_{[B_{j-1}, B_j)}(t).$$

Example 1.4.3. On $(\mathbf{R}^n, dx)$ let

$$f(x) = \frac{1}{1+|x|^p}, \qquad 0 < p < \infty .$$

A computation shows that

$$d_f(\alpha) = \begin{cases} v_n(\frac{1}{\alpha} - 1)^{\frac{n}{p}} & \text{if } \alpha < 1, \\ 0 & \text{if } \alpha \geq 1, \end{cases}$$

and therefore

$$f^*(t) = \frac{1}{(t/v_n)^{p/n} + 1},$$

where v_n is the volume of the unit ball in $\mathbf{R}^n$.

Example 1.4.4. Again on $(\mathbf{R}^n, dx)$ let $g(x) = 1 - e^{-|x|^2}$. We can easily see that $d_g(\alpha) = 0$ if $\alpha \geq 1$ and $d_g(\alpha) = \infty$ if $\alpha < 1$. We conclude that $g^*(t) = 1$ for all $t \geq 0$. This example indicates that a lot of important information about the function is lost at the passage from g to g^*. From the point of the distribution functions and L^p norms, this information is nevertheless irrelevant.

It is clear from the previous examples that f^* is continuous from the right and decreasing. The following are some properties of the function f^*.

Proposition 1.4.5. *For f, g, f_n μ-measurable, $k \in \mathbf{C}$, and $t_1, t_2, t > 0$ we have*

(1) *f^* is right continuous on $[0, \infty)$*

(2) $(kf)^* = |k| f^*$

(3) *$|g| \leq |f|$ μ-a.e. implies that $g^* \leq f^*$*

(4) $(f+g)^*(t_1 + t_2) \leq f^*(t_1) + g^*(t_2)$

(5) $(fg)^*(t_1 t_2) \leq f^*(t_1) + g^*(t_2)$

(6) *$|f| \leq \liminf_{n\to\infty} |f_n|$ μ-a.e. implies $f^* \leq \liminf_{n\to\infty} f_n^*$*

(7) *$|f_n| \uparrow |f|$ μ-a.e. implies $f_n^* \uparrow f^*$*

(8) *$f^*(d_f(\alpha)) \leq \alpha$ whenever $d_f(\alpha) < \infty$*

(9) *$d_f(f^*(t)) = \mu(\{|f| > f^*(t)\}) \leq t \leq \mu(\{|f| \geq f^*(t)\})$, if $f^*(t) < \infty$*

(10) $d_f = d_{f^*}$

(11) *$(|f|^p)^* = (f^*)^p$ when $0 < p < \infty$*

(12) *$\int_X |f|^p \, d\mu = \int_0^\infty f^*(t)^p \, dt$ when $0 < p < \infty$*

(13) $\|f\|_{L^\infty} = f^*(0)$

(14) *$\sup_{t>0} t^s f^*(t) = \sup_{\alpha>0} \alpha \, (d_f(\alpha))^s$ for $0 < s < \infty$.*

Proof. Properties (1), (6), and (7) follow from Exercise 1.1.1. Properties (2), (3), (4), and (5) are left to the reader. To prove (8), just observe that

$$f^*(d_f(\alpha)) = \inf\{s > 0 : \ d_f(s) \le d_f(\alpha)\} \le \alpha.$$

To prove the first inequality in (9), fix $t \ge 0$ and suppose $f^*(t) < \infty$. There exists a sequence α_n decreasing to $f^*(t)$ with $d_f(\alpha_n) \le t$. The right continuity of d_f (Exercise 1.1.1) implies that

$$d_f(f^*(t)) = \lim_{n\to\infty} d_f(\alpha_n) \le t.$$

The second inequality in (9) follows from the definition of f^* which implies that the set $A_n = \{|f| > f^*(t) - 1/n\}$ has measure at least t. The sets A_n form a decreasing sequence as n increases and therefore their intersection would have to have measure at least t. Observe that (10) is immediate for nonnegative simple functions in view of Examples 1.1.2 and 1.4.2. For an arbitrary measurable function f, find a sequence of nonnegative simple functions f_n such that $f_n \uparrow |f|$ and apply (7). Equimeasurability of f and f^* immediately implies (12) in view of Proposition 1.1.4. Also, (11) follows from

$$d_{|f|^p}(\alpha) = d_f(\alpha^{1/p}) = d_{f^*}(\alpha^{1/p}) = d_{(f^*)^p}(\alpha),$$

while (13) is trivial. Finally, (14) is a direct consequence of (8) and (9). □

1.4.b. Lorentz Spaces

Having disposed of the basic properties of the decreasing rearrangement of functions, we proceed with the definition of the Lorentz spaces.

Definition 1.4.6. Given f a measurable function on a measure space (X, μ) and $0 < p, q \le \infty$, define

$$\big\|f\big\|_{L^{p,q}} = \begin{cases} \displaystyle\left(\int_0^\infty \left(t^{\frac{1}{p}} f^*(t)\right)^q \frac{dt}{t}\right)^{\frac{1}{q}} & \text{if } q < \infty, \\ \displaystyle\sup_{t>0} t^{\frac{1}{p}} f^*(t) & \text{if } q = \infty. \end{cases}$$

The set of all f with $\big\|f\big\|_{L^{p,q}} < \infty$ is denoted by $L^{p,q}(X, \mu)$ and is called the *Lorentz space* with indices p and q.

As in L^p and in weak L^p, two functions in $L^{p,q}(X, \mu)$ will be considered equal if they are equal μ-almost everywhere. Observe that the previous definition implies that $L^{\infty,\infty} = L^\infty$, $L^{p,\infty} =$ weak L^p in view of Proposition 1.4.5 (14) and that $L^{p,p} = L^p$.

Remark 1.4.7. Observe that for all $0 < p, r < \infty$ and $0 < q \le \infty$ we have

$$\big\||g|^r\big\|_{L^{p,q}} = \big\|g\big\|_{L^{pr,qr}}^r. \tag{1.4.3}$$

On $\mathbf{R}^n$ let $\delta^\varepsilon(f)(x) = f(\varepsilon x)$, $\varepsilon > 0$ be the dilation operator. It is straightforward that $d_{\delta^\varepsilon(f)}(\alpha) = \varepsilon^{-n} d_f(\alpha)$ and $(\delta^\varepsilon(f))^*(t) = f^*(\varepsilon^n t)$. It follows that Lorentz norms satisfy the following dilation identity:

$$\big\|\delta^\varepsilon(f)\big\|_{L^{p,q}} = \varepsilon^{-n/p}\big\|f\big\|_{L^{p,q}}. \tag{1.4.4}$$

Next, we calculate the Lorentz norms of a simple function.

Example 1.4.8. Using the notation of Example 1.4.2, when $0 < q < \infty$ we have

$$\big\|f\big\|_{L^{p,q}} = \Big(\frac{p}{q}\Big)^{\frac{1}{q}} \Big[a_1^q B_1^{\frac{q}{p}} + a_2^q\Big(B_2^{\frac{q}{p}} - B_1^{\frac{q}{p}}\Big) + \cdots + a_N^q\Big(B_N^{\frac{q}{p}} - B_{N-1}^{\frac{q}{p}}\Big)\Big]^{\frac{1}{q}},$$

$$\big\|f\big\|_{L^{p,\infty}} = \sup_{1\le j\le N} a_j B_j^{\frac{1}{p}}.$$

Remark 1.4.9. It follows from the previous example that the only simple function with finite $L^{\infty,q}$ norm is identically equal to zero. For this reason we have that $L^{\infty,q} = \{0\}$ for any $0 < q < \infty$.

Since $L^{p,p} \subseteq L^{p,\infty}$ we may wonder whether these spaces are nested. The next result shows that for any fixed p, the Lorentz spaces $L^{p,q}$ increase as the exponent q increases.

Proposition 1.4.10. *Suppose $0 < p \le \infty$ and $0 < q < r \le \infty$. Then there exists some constant $c_{p,q,r}$ (that depends on p, q, and r) such that*

$$\big\|f\big\|_{L^{p,r}} \le c_{p,q,r}\big\|f\big\|_{L^{p,q}}. \tag{1.4.5}$$

In other words, $L^{p,q}$ is a subspace of $L^{p,r}$.

Proof. We may assume $p < \infty$ since the case $p = \infty$ is trivial. We have

$$\begin{aligned} t^{1/p}f^*(t) &= \Big\{\frac{q}{p}\int_0^t [s^{1/p}f^*(t)]^q \frac{ds}{s}\Big\}^{1/q} \\ &\le \Big\{\frac{q}{p}\int_0^t [s^{1/p}f^*(s)]^q \frac{ds}{s}\Big\}^{1/q} \qquad \text{since } f^* \text{ is decreasing,} \\ &\le \Big(\frac{q}{p}\Big)^{1/q}\big\|f\big\|_{L^{p,q}}. \end{aligned}$$

Hence, taking the supremum over all $t > 0$, we obtain

$$\big\|f\big\|_{L^{p,\infty}} \le \Big(\frac{q}{p}\Big)^{1/q}\big\|f\big\|_{L^{p,q}}. \tag{1.4.6}$$

This establishes (1.4.5) in the case $r = \infty$. Finally, when $r < \infty$, we have

$$\big\|f\big\|_{L^{p,r}} = \Big\{\int_0^\infty [t^{1/p}f^*(t)]^{r-q+q}\frac{dt}{t}\Big\}^{1/r} \le \big\|f\big\|_{L^{p,\infty}}^{(r-q)/r}\big\|f\big\|_{L^{p,q}}^{q/r}. \tag{1.4.7}$$

(1.4.6) combined with (1.4.7) gives (1.4.5) with $c_{p,q,r} = (q/p)^{(r-q)/rq}$.

□

Unfortunately, the functionals $\|\cdot\|_{L^{p,q}}$ do not satisfy the triangle inequality. For instance, consider the functions $f(t) = t$ and $g(t) = 1-t$ defined on $[0,1]$. Then $f^*(\alpha) = g^*(\alpha) = (1-\alpha)\chi_{[0,1]}(\alpha)$. A simple calculation shows that the triangle inequality for these functions with respect to the norm $\|\cdot\|_{L^{p,q}}$ would be equivalent to

$$\frac{p}{q} \le 2^q \frac{\Gamma(q+1)\Gamma(q/p)}{\Gamma(q+1+q/p)},$$

which fails in general. However, since for all $t > 0$ we have

$$(f+g)^*(t) \le f^*(t/2) + g^*(t/2),$$

the following estimate

$$\|f+g\|_{L^{p,q}} \le c_{p,q}\big(\|f\|_{L^{p,q}} + \|g\|_{L^{p,q}}\big), \tag{1.4.8}$$

where $c_{p,q} = 2^{1/p}\max(1, 2^{(1-q)/q})$, is a consequence of (1.1.4). Also, if $\|f\|_{L^{p,q}} = 0$ then we must have $f = 0$ μ-a.e. Therefore, $L^{p,q}$ is a quasi-normed space for all $0 < p, q \le \infty$. Is this space complete with respect to its quasi-norm? The next theorem answers this question.

Theorem 1.4.11. *Let (X, μ) be a measure space. Then for all $0 < p, q \le \infty$, the spaces $L^{p,q}(X,\mu)$ are complete with respect to their quasi-norm and they are therefore quasi-Banach spaces.*

Proof. We only consider the case $p < \infty$. First we note that convergence in $L^{p,q}$ implies convergence in measure. When $q = \infty$, this is proved in Proposition 1.1.9. When $q < \infty$, in view of Proposition 1.4.5 (14) and (1.4.6), it follows that

$$\sup_{t>0} t^{1/p} f^*(t) = \sup_{\alpha>0} \alpha d_f(\alpha)^{1/p} \le \left(\frac{q}{p}\right)^{1/q} \|f\|_{L^{p,q}}$$

for all $f \in L^{p,q}$, from which the same conclusion follows. Let $\{f_n\}$ be a Cauchy sequence in $L^{p,q}$. Then $\{f_n\}$ is Cauchy in measure and hence it has a subsequence $\{f_{n_k}\}$ that converges almost everywhere to some f by Theorem 1.1.13. Fix k_0 and apply property (6) in Proposition 1.4.5. Since $|f - f_{n_{k_0}}| = \lim_{k\to\infty} |f_{n_k} - f_{n_{k_0}}|$, it follows that

$$(f - f_{n_{k_0}})^*(t) \le \liminf_{k\to\infty} (f_{n_k} - f_{n_{k_0}})^*(t). \tag{1.4.9}$$

Raise (1.4.9) to the power q, multiply by $t^{q/p}$, integrate with respect to $t \in [0,\infty)$, and apply Fatou's lemma to obtain

$$\|f - f_{n_{k_0}}\|_{L^{p,q}}^q \le \liminf_{k\to\infty} \|f_{n_k} - f_{n_{k_0}}\|_{L^{p,q}}^q. \tag{1.4.10}$$

Now let $k_0 \to \infty$ in (1.4.10) and use the fact that $\{f_n\}$ is Cauchy to conclude that f_{n_k} converges to f in $L^{p,q}$. It is a general fact that if a Cauchy sequence has a convergent subsequence in a quasi-normed space, then the sequence is convergent to the same limit. It follows that f_n converges to f in $L^{p,q}$.

$\square$

Remark 1.4.12. It can be shown that the spaces $L^{p,q}$ are normable when p, q are bigger than 1. See Exercise 1.4.3. Therefore, these spaces can be normed to become Banach spaces.

It is natural to ask whether simple functions are dense in $L^{p,q}$. This is in fact the case when $q \neq \infty$.

Theorem 1.4.13. *Simple functions are dense in $L^{p,q}(X,\mu)$ when $0 < q < \infty$.*

Proof. Let $f \in L^{p,q}(X,\mu)$. Assume without loss of generality that $f \geq 0$. Given $n = 1, 2, 3, \dots$, we find a simple function $f_n \geq 0$ such that

$$f_n(x) = 0$$

when $f(x) \leq 1/n$, and

$$f(x) - \frac{1}{n} \leq f_n(x) \leq f(x)$$

when $f(x) > 1/n$, except on a set of measure less than $1/n$. It follows that

$$\mu(\{x \in X : \ |f(x) - f_n(x)| > 1/n\}) < 1/n;$$

hence $(f - f_n)^*(t) \leq 1/n$ for $t \geq 1/n$. Thus

$$(f - f_n)^*(t) \to 0 \quad \text{as } n \to \infty \text{ and } \quad f_n^*(t) \leq f^*(t) \quad \text{for all } t > 0.$$

Since $(f - f_n)^*(t) \leq 2f^*(t/2)$, an application of the Lebesgue dominated convergence theorem gives that $\|f_n - f\|_{L^{p,q}} \to 0$ as $n \to \infty$.

$\square$

Remark 1.4.14. We may wonder whether simple functions are dense in $L^{p,\infty}$. This turns out to be false for all $0 < p \leq \infty$. However, if X is σ-finite, countable linear combinations of characteristic functions of sets with finite measure are dense in $L^{p,\infty}(X,\mu)$. We will call such functions *countably simple*. See Exercise 1.4.4 for details.

1.4.c. Duals of Lorentz Spaces

Given a quasi-Banach space Z with norm $\|\cdot\|_Z$, its dual Z^* is defined as the space of all continuous linear functionals T on Z equipped with the norm

$$\|T\|_{Z^*} = \sup_{\|x\|_Z = 1} |T(x)|.$$

Observe that the dual of quasi-Banach space is always a Banach space.

We are now considering the following question: What are the dual spaces $(L^{p,q})^*$ of $L^{p,q}$? The answer to this question presents some technical difficulties for general measure spaces. In this exposition we will restrict our attention to σ-finite nonatomic measure spaces, where the situation is simpler.

Definition 1.4.15. A subset A of a measure space (X,μ) is called an *atom* if $\mu(A) > 0$ and every subset B of A has measure either equal to zero or equal to $\mu(A)$. A measure space (X,μ) is called *nonatomic* if it contains no atoms. In other words, X is nonatomic if and only if for any $A \subseteq X$ with $\mu(A) > 0$, there exists a proper subset $B \subsetneqq A$ with $\mu(B) > 0$ and $\mu(A \setminus B) > 0$.

For instance, $\mathbf{R}$ with Lebesgue measure is nonatomic, but any measure space with counting measure is atomic. Nonatomic spaces have the property that every measurable subset of them with positive measure contains subsets of any given measure. See Exercise 1.4.5.

Definition 1.4.16. A measure space is called *σ-finite* if there is a sequence of measurable sets K_N with $\mu(K_N) < \infty$ such that

$$\bigcup_{N=1}^{\infty} K_N = X.$$

For instance, $\mathbf{R}^n$ equipped with Lebesgue measure is a σ-finite measure space. So is $\mathbf{Z}^n$ with the usual counting measure.

Theorem 1.4.17. *Suppose that (X,μ) is a nonatomic σ-finite measure space. Then*

(i)	$(L^{p,q})^* = \{0\}$,	*when* $0 < p < 1,\ 0 < q \le \infty$
(ii)	$(L^{p,q})^* = L^\infty$,	*when* $p = 1,\ 0 < q \le 1$
(iii)	$(L^{p,q})^* = \{0\}$,	*when* $p = 1,\ 1 < q < \infty$
(iv)	$(L^{p,q})^* \neq \{0\}$,	*when* $p = 1,\ q = \infty$
(v)	$(L^{p,q})^* = L^{p',\infty}$,	*when* $1 < p < \infty,\ 0 < q \le 1$
(vi)	$(L^{p,q})^* = L^{p',q'}$,	*when* $1 < p < \infty,\ 1 < q < \infty$
(vii)	$(L^{p,q})^* \neq \{0\}$,	*when* $1 < p < \infty,\ q = \infty$
(viii)	$(L^{p,q})^* \neq \{0\}$,	*when* $p = q = \infty$.

Proof. Since X is σ-finite, we have $X = \bigcup_{N=1}^{\infty} K_N$, where K_N is an increasing sequence of sets with $\mu(K_N) < \infty$. Given $T \in (L^{p,q})^*$, where $0 < p < \infty$ and $0 < q \le \infty$, consider the measure $\sigma(E) = T(\chi_E)$. Since σ satisfies $|\sigma(E)| \le (p/q)^{1/q}\|T\|\mu(E)^{1/p}$ when $q < \infty$ and $|\sigma(E)| \le \|T\|\mu(E)^{1/p}$, it follows that σ is absolutely continuous with respect to Lebesgue measure. By the Radon-Nikodym theorem, there exists a complex-valued measurable function g (which satisfies $\int_{K_N} |g|\,d\mu < \infty$ for all N) such that

$$\sigma(E) = T(\chi_E) = \int_X g\,\chi_E\,d\mu\,. \tag{1.4.11}$$

Linearity implies that (1.4.11) holds for any simple function on X. The continuity of T and the density of the simple functions on $L^{p,q}$ (when $q < \infty$) gives

$$T(f) = \int_X g\,f\,d\mu \tag{1.4.12}$$

for every $f \in L^{p,q}$. We now examine each case (i)–(viii) separately.

(i) We first consider the case $0 < p < 1$. Let $f = \sum_n a_n \chi_{E_n}$ be a simple function on X (take f to be countably simple when $q = \infty$). If X is nonatomic, we can split each E_n as a union of N disjoint sets E_{jn} each having measure $N^{-1}\mu(E_n)$. Let $f_j = \sum_n a_n \chi_{E_{jn}}$. We see that $\|f_j\|_{L^{p,q}} = N^{-1/p}\|f\|_{L^{p,q}}$. Now if $T \in (L^{p,q})^*$, it follows that

$$|T(f)| \le \sum_{j=1}^{N} |T(f_j)| \le \|T\| \sum_{j=1}^{N} \|f_j\|_{L^{p,q}} \le \|T\| N^{1-1/p} \|f\|_{L^{p,q}}.$$

Let $N \to \infty$ and use that $p < 1$ to obtain that $T = 0$.

(ii) We now consider the case $p = 1$ and $0 < q \le 1$. Clearly, every $g \in L^\infty$ gives a bounded linear functional on $L^{1,q}$ since

$$\left|\int_X fg\,d\mu\right| \le \|g\|_{L^\infty}\|f\|_{L^1} \le C_q\|g\|_{L^\infty}\|f\|_{L^{1,q}}.$$

Conversely, suppose that $T \in (L^{1,q})^*$ where $q \le 1$. The function g given in (1.4.11) satisfies

$$\left|\int_E g\,d\mu\right| \le \|T\|\,\mu(E)$$

for all $E \subseteq K_N$, and hence $|g| \le \|T\|$ μ-a.e. on every K_N. See Rudin [**440**, p. 31] (Theorem 1.40) for a proof of this fact. It follows that $\|g\|_{L^\infty} \le \|T\|$ and hence $(L^{1,q})^* = L^\infty$.

(iii) Let us now take $p = 1$, $1 < q < \infty$ and suppose that $T \in (L^{1,q})^*$. Then

$$\left|\int_X fg\,d\mu\right| \le \|T\|\;\|f\|_{L^{1,q}}, \tag{1.4.13}$$

where g is the function in (1.4.12). We shall show that $g = 0$ a.e. Suppose that $|g| \ge \delta$ on some set E_0 with $\mu(E_0) > 0$. Let $f = \overline{g}|g|^{-1}\chi_{E_0}h$, where $h \ge 0$. Then (1.4.13) implies that

$$\|h\|_{L^1(E_0)} \le \|T\|\delta^{-1}\|h\|_{L^{1,q}(E_0)}$$

for all $h \ge 0$. Since X is nonatomic, this can't happen unless $T = 0$. See Exercise 1.4.8.

(iv) In the case $p = 1$, $q = \infty$ something interesting happens. Since every continuous linear functional on $L^{1,\infty}$ extends to a continuous linear functional on $L^{1,q}$ for $1 < q < \infty$, it must necessarily vanish on all simple functions by part (iii). However, $(L^{1,\infty})^*$ contains nontrivial linear functionals. For details we refer the reader to articles of Cwikel and Fefferman [**139**], [**140**].

(v) We now take up the case $p > 1$ and $0 < q \le 1$. Using Exercise 1.4.1 and Proposition 1.4.10, we see that if $g \in L^{p',\infty}$, then

$$\begin{aligned}\left|\int_X fg\,d\mu\right| &\le \int_0^\infty t^{\frac{1}{p}} f^*(t) t^{\frac{1}{p'}} g^*(t)\,\frac{dt}{t}\\ &\le \big\|f\big\|_{L^{p,1}} \big\|g\big\|_{L^{p',\infty}}\\ &\le C_{p,q}\big\|f\big\|_{L^{p,q}} \big\|g\big\|_{L^{p',\infty}}.\end{aligned}$$

Conversely, suppose that $T \in (L^{p,q})^*$ when $1 < p < \infty$ and $0 < q \le 1$. Let g satisfy (1.4.12). Taking $f = \overline{g}|g|^{-1}\chi_{|g|>\alpha}$ and using that

$$\left|\int_X fg\,d\mu\right| \le \|T\|\big\|f\big\|_{L^{p,q}}$$

we obtain that

$$\alpha\mu(\{|g| > \alpha\}) \le p^{-1}\big\|T\big\|\mu(\{|g| > \alpha\})^{\frac{1}{p}}.$$

It follows that $\big\|g\big\|_{L^{p',\infty}} \approx \big\|T\big\|$.

(vi) Using Exercise 1.4.2 and Hölder's inequality, we see that every $g \in L^{p',q'}$ gives a bounded linear functional on $L^{p,q}$. Conversely, let $T \in (L^{p,q})^*$. By (1.4.12), T is given by integration against a locally integrable function g. It remains to prove that $g \in L^{p',q'}$. For all all f in $L^{p,q}(X)$ we have

$$\int_0^\infty f^*(t) g^*(t)\,dt = \sup_{h:\ d_h = d_f} \left|\int_X hg\,d\mu\right| \le \big\|T\big\|\ \big\|f\big\|_{L^{p,q}}, \tag{1.4.14}$$

where the equality is a consequence of the fact that X is nonatomic (see Exercise 1.4.5). Pick a function f on X such that

$$f^*(t) = \int_{t/2}^\infty s^{\frac{q'}{p'}-1} g^*(s)^{q'-1}\,\frac{ds}{s}. \tag{1.4.15}$$

This can be achieved again by Exercise 1.4.5. The fact that the integral in (1.4.15) converges is a consequence of the observation that the function f^* defined in (1.4.15) lies in the space $L^q(0,\infty)$ with respect to the measure $t^{q/p-1}dt$. This follows from the inequality

$$\begin{aligned}\|f\|_{L^{p,q}} &= \left(\int_0^\infty t^{\frac{q}{p}}\left[\int_{t/2}^\infty s^{\frac{q'}{p'}} g^*(s)^{q'-1}\frac{ds}{s}\right]^q \frac{dt}{t}\right)^{\frac{1}{q}}\\ &\leq C_1(p,q)\left(\int_0^\infty (t^{\frac{1}{p'}}g^*(t))^{q'}\frac{dt}{t}\right)^{\frac{1}{q}}\\ &= C_1(p,q)\|g\|_{L^{p',q'}}^{q'/q} < \infty,\end{aligned}$$

which is a consequence of Hardy's inequality (Exercise 1.2.8). Using (1.4.14), we conclude that

$$\int_0^\infty f^*(t)g^*(t)\,dt \leq \|T\|\,\|f\|_{L^{p,q}} \leq C_1(p,q)\|T\|\,\|g\|_{L^{p',q'}}^{q'/q}. \tag{1.4.16}$$

On the other hand we have

$$\begin{aligned}\int_0^\infty f^*(t)g^*(t)\,dt &\geq \int_0^\infty \int_{t/2}^t s^{\frac{q'}{p'}-1}g^*(s)^{q'-1}\frac{ds}{s}\,g^*(t)\,dt\\ &\geq \int_0^\infty g^*(t)^{q'}\int_{t/2}^t s^{\frac{q'}{p'}-1}\frac{ds}{s}\,dt\\ &= C_2(p,q)\|g\|_{L^{p',q'}}^{q'}.\end{aligned} \tag{1.4.17}$$

Combining (1.4.16) and (1.4.17), we obtain $\|g\|_{L^{p',q'}} \leq C(p,q)\|T\|$. This estimate is valid only when we have an a priori knowledge that $\|g\|_{L^{p',q'}} < \infty$. Suitably modifying the preceding proof we obtain that $\|g\|_{L^{p',q'}(K_N)} \leq C(p,q)\|T\|$ for all $N = 1, 2, \ldots$. Letting $N \to \infty$, we obtain the required conclusion.

(vii) For a complete characterization of this space, we refer to the article of Cwikel [**138**].

(viii) The dual of L^∞ can be identified with the set of all bounded finitely additive set functions. See Dunford and Schwartz [**166**].

□

Remark 1.4.18. Some parts of Theorem 1.4.17 are false if X is atomic. For instance, the dual of $\ell^p(\mathbf{Z})$ contains l^∞ when $0 < p < 1$ and thus it is not $\{0\}$.

1.4.d. The Off-Diagonal Marcinkiewicz Interpolation Theorem

We now present the main result of this section, the off-diagonal extension of Marcinkiewicz's interpolation theorem (Theorem 1.3.2). Recall that an operator T is called quasi-linear if it satisfies

$$|T(\lambda f)| = |\lambda|\,|T(f)|, \qquad \text{and} \qquad |T(f+g)| \leq K(|T(f)| + |T(g)|),$$

for some $K > 0$, $\lambda \in \mathbf{C}$, and all f, g functions in the domain of T. To avoid trivialities, we will assume that $K \geq 1$.

Theorem 1.4.19. *Let $0 < r \le \infty$, $0 < p_0 \neq p_1 \le \infty$, and $0 < q_0 \neq q_1 \le \infty$ and let (X,μ) and (Y,ν) be two measure spaces. Let T be either a quasi-linear operator defined on $L^{p_0}(X) + L^{p_1}(X)$ and taking values in the set of measurable functions on Y or a linear operator defined on the set of simple functions on X and taking values as before. Assume that for some $M_0, M_1 < \infty$ the following (restricted) weak type estimates hold:*

$$\big\|T(\chi_A)\big\|_{L^{q_0,\infty}} \le M_0\,\mu(A)^{1/p_0} \tag{1.4.18}$$

$$\big\|T(\chi_A)\big\|_{L^{q_1,\infty}} \le M_1\,\mu(A)^{1/p_1} \tag{1.4.19}$$

for all A measurable subsets of X of finite measure. Fix $0 < \theta < 1$ and let

$$\frac{1}{p} = \frac{1-\theta}{p_0} + \frac{\theta}{p_1} \qquad \text{and} \qquad \frac{1}{q} = \frac{1-\theta}{q_0} + \frac{\theta}{q_1}. \tag{1.4.20}$$

Then there exists a constant M, which depends on K, p_0, p_1, q_0, q_1, M_0, M_1, r, and θ such that for all f in the domain of T and in $L^{p,r}(X)$ we have

$$\big\|T(f)\big\|_{L^{q,r}} \le M\big\|f\big\|_{L^{p,r}}. \tag{1.4.21}$$

We note that $L^{p,\infty} \subseteq L^{p_0} + L^{p_1}$ [Exercise 1.1.10(c)], and thus T is well defined on $L^{p,r}$ for all $r \le \infty$. If $r < \infty$ and T is linear and defined on the set of simple functions on X, then T has a unique extension that satisfies (1.4.21) for all f in $L^{p,r}(X)$, since simple functions are dense in this space.

Before we give the proof of Theorem 1.4.19, we state and prove a lemma that is interesting on its own.

Lemma 1.4.20. *Let $0 < p < \infty$ and $0 < q \le \infty$. Let T be either a quasi-linear operator defined on $L^p(X,\mu)$ and taking values in the set of measurable functions of a measure space (Y,ν) or a linear operator initially defined on the space of simple functions on X and taking values as before. Suppose that there exists a constant $L > 0$ such that for all $A \subseteq X$ of finite measure we have*

$$\big\|T(\chi_A)\big\|_{L^{q,\infty}} \le L\mu(A)^{1/p}. \tag{1.4.22}$$

Fix $\alpha_0 < q$ with $0 < \alpha_0 \le \frac{\log 2}{\log 2K}$. Then for all $0 < \alpha \le \alpha_0$ there exists a constant $C(p,q,K,\alpha) > 0$ (depending only on the parameters indicated) such that for all functions f in $L^{p,\alpha}(X)$ that lie in the domain of T, we have the estimate

$$\big\|T(f)\big\|_{L^{q,\infty}} \le C(p,q,K,\alpha)L\,\big\|f\big\|_{L^{p,\alpha}}. \tag{1.4.23}$$

Lemma 1.4.20 is basically saying that if a quasi-linear operator satisfies a $L^{p,1} \to L^{q,\infty}$ estimate uniformly on all characteristic functions, then it must map a Lorentz space $L^{p,\alpha}$ to $L^{q,\infty}$ for some $\alpha < 1$.

Proof. It suffices to prove Lemma 1.4.20 for $f \ge 0$ since we can write a general $f = (f_1 - f_2) + i(f_3 - f_4)$, where $f_j \ge 0$ and use quasi-linearity.

It follows from the Aoki-Rolewicz theorem (Exercise 1.4.6) that for all $f_1, \dots, f_m$ we have the pointwise inequality

$$
\begin{aligned}
|T(f_1 + \cdots + f_m)| &\le 4\Big(\sum_{j=1}^{m} |T(f_j)|^{\alpha_1}\Big)^{\frac{1}{\alpha_1}} \\
&\le 4\Big(\sum_{j=1}^{m} |T(f_j)|^{\alpha}\Big)^{\frac{1}{\alpha}},
\end{aligned}
\tag{1.4.24}
$$

where α_1 satisfies the equation $(2K)^{\alpha_1} = 2$ and $0 < \alpha \le \alpha_1$. The second inequality in (1.4.24) is a simple consequence of the fact that $\alpha \le \alpha_1$. Fix $\alpha_0 > 0$ with

$$
\alpha_0 \le \alpha_1 = \frac{\log 2}{\log 2K} \qquad \text{and} \qquad \alpha_0 < q\,.
$$

This will ensure that the quasi-normed space $L^{q/\alpha,\infty}$ is normable when $\alpha \le \alpha_0$. In fact, Exercise 1.1.12 gives that the space $L^{s,\infty}$ is normable as long as $s > 1$ and for some equivalent norm $|||\, f\, |||_{L^{s,\infty}}$ we have

$$
\big\|f\big\|_{L^{s,\infty}} \le |||\, f\, |||_{L^{s,\infty}} \le \frac{s}{s-1}\big\|f\big\|_{L^{s,\infty}}.
$$

Next we claim that for any $f \ge 0$ we have

$$
\big\|T(f\chi_A)\big\|_{L^{q,\infty}} \le C(q,\alpha) L \mu(A)^{1/p} \big\|f\big\|_{L^\infty}.
\tag{1.4.25}
$$

To prove (1.4.25) first observe that multiplying with a suitable constant we may assume that $f \le 1$. Write

$$
f(x) = \sum_{j=1}^{\infty} d_j(x) 2^{-j}
$$

in binary expansion, where $d_j(x) = 0$ or 1. Let

$$
B_j = \{x \in A :\ d_j(x) = 1\}.
$$

Then $B_j \subseteq A$ and the function $f\chi_A$ can be written as the sum

$$
\sum_{j=1}^{\infty} 2^{-j} \chi_{B_j}\,.
$$

We use (1.4.24) once and (1.4.3) twice in the following argument. We have

$$\begin{aligned}
\big\|T(f\chi_A)\big\|_{L^{q,\infty}} &\le 4\Big\|\Big(\sum_{j=1}^{\infty} 2^{-j\alpha}|T(\chi_{B_j})|^{\alpha}\Big)^{\frac{1}{\alpha}}\Big\|_{L^{q,\infty}} \\
&= 4\Big\|\sum_{j=1}^{\infty} 2^{-j\alpha}|T(\chi_{B_j})|^{\alpha}\Big\|_{L^{q/\alpha,\infty}}^{\frac{1}{\alpha}} \\
&\le 4\,\Big|\!\Big|\!\Big|\sum_{j=1}^{\infty} 2^{-j\alpha}|T(\chi_{B_j})|^{\alpha}\Big|\!\Big|\!\Big|_{L^{q/\alpha,\infty}}^{\frac{1}{\alpha}} \\
&\le 4\Big(\sum_{j=1}^{\infty} 2^{-j\alpha}\,\Big|\!\Big|\!\Big|\,|T(\chi_{B_j})|^{\alpha}\,\Big|\!\Big|\!\Big|_{L^{q/\alpha,\infty}}\Big)^{\frac{1}{\alpha}} \\
&\le 4\Big(\frac{q}{q-\alpha}\Big)^{\frac{1}{\alpha}}\Big(\sum_{j=1}^{\infty} 2^{-j\alpha}\Big\||T(\chi_{B_j})|^{\alpha}\Big\|_{L^{q/\alpha,\infty}}\Big)^{\frac{1}{\alpha}} \\
&= 4\Big(\frac{q}{q-\alpha}\Big)^{\frac{1}{\alpha}}\Big(\sum_{j=1}^{\infty} 2^{-j\alpha}\big\|T(\chi_{B_j})\big\|_{L^{q,\infty}}^{\alpha}\Big)^{\frac{1}{\alpha}} \\
&\le 4\Big(\frac{q}{q-\alpha}\Big)^{\frac{1}{\alpha}}L\Big(\sum_{j=1}^{\infty} 2^{-j\alpha}\mu(B_j)^{\alpha/p}\Big)^{\frac{1}{\alpha}} \\
&\le 2\Big(\frac{q}{q-\alpha}\Big)^{\frac{1}{\alpha}}(1-2^{-\alpha})^{-\frac{1}{\alpha}}L\,\mu(A)^{1/p},
\end{aligned}$$

since $B_j \subseteq A$. This establishes (1.4.25) with

$$C(q,\alpha) = 2\Big(\frac{q}{q-\alpha}\Big)^{\frac{1}{\alpha}}(1-2^{-\alpha})^{-\frac{1}{\alpha}}.$$

Now write the function f as

$$f = \sum_{n=-\infty}^{\infty} f\chi_{A_n},$$

where A_n are measurable sets defined by

$$A_n = \{x \in X :\ f^*(2^{n+1}) < |f(x)| \le f^*(2^n)\}. \tag{1.4.26}$$

Observe that

$$\mu(A_n) = \big|\{t \in \mathbf{R} :\ f^*(2^{n+1}) < f^*(t) \le f^*(2^n)\}\big| = \big|[2^n, 2^{n+1}]\big| = 2^n,$$

since f and f^* are equidistributed. Next we have

$$
\begin{aligned}
\big\|T(f)\big\|_{L^{q,\infty}} &\le 4\Big\|\Big(\sum_{n=-\infty}^{\infty}|T(f\chi_{A_n})|^\alpha\Big)^{\frac{1}{\alpha}}\Big\|_{L^{q,\infty}} \\
&= 4\Big\|\sum_{n=-\infty}^{\infty}|T(f\chi_{A_n})|^\alpha\Big\|_{L^{q/\alpha,\infty}}^{\frac{1}{\alpha}} \\
&\le 4\,\Big|\Big|\Big|\sum_{n=-\infty}^{\infty}|T(f\chi_{A_n})|^\alpha\,\Big|\Big|\Big|_{L^{q/\alpha,\infty}}^{\frac{1}{\alpha}} \\
&\le 4\Big(\sum_{n=-\infty}^{\infty}\Big|\Big|\Big|\,|T(f\chi_{A_n})|^\alpha\,\Big|\Big|\Big|_{L^{q/\alpha,\infty}}\Big)^{\frac{1}{\alpha}} \\
&\le 4\Big(\frac{q}{q-\alpha}\Big)^{\frac{1}{\alpha}}\Big(\sum_{n=-\infty}^{\infty}\Big\||T(f\chi_{A_n})|^\alpha\Big\|_{L^{q/\alpha,\infty}}\Big)^{\frac{1}{\alpha}} \\
&\le 4\Big(\frac{q}{q-\alpha}\Big)^{\frac{1}{\alpha}}\Big(\sum_{n=-\infty}^{\infty}\Big\|T(f\chi_{A_n})\Big\|_{L^{q,\infty}}^\alpha\Big)^{\frac{1}{\alpha}} \\
&\le 8\Big(\frac{q}{q-\alpha}\Big)^{\frac{2}{\alpha}}(1-2^{-\alpha})^{-\frac{1}{\alpha}}L\Big(\sum_{n=-\infty}^{\infty}f^*(2^n)^\alpha 2^{n\alpha/p}\Big)^{\frac{1}{\alpha}} \\
&\le 8\Big(\frac{q}{q-\alpha}\Big)^{\frac{2}{\alpha}}(1-2^{-\alpha})^{-\frac{1}{\alpha}}(\log 2)^{\frac{1}{\alpha}}L\,\big\|f\big\|_{L^{p,\alpha}}.
\end{aligned}
$$

Taking in mind the splitting $f = f_1 - f_2 + if_3 - if_4$, where $f_j \ge 0$, we conclude the proof of the lemma with constant

$$
C(p,q,K,\alpha) = C_p\,K^2\Big(\frac{q}{q-\alpha}\Big)^{2/\alpha}(\log 2)^{\frac{1}{\alpha}}(1-2^{-\alpha})^{-\frac{1}{\alpha}}. \tag{1.4.27}
$$

Recall that we have been assuming that $\alpha < \min\big(\frac{\log 2}{\log 2K}, q\big)$ throughout.

□

We now continue with the proof of Theorem 1.4.19.

Proof. In the following proof we take $p_0 < p_1$. If $p_0 > p_1$, simply reverse the roles of p_0 and p_1. We are assuming first that $p_1 < \infty$. The case $p_1 = \infty$ is discussed at the end. Lemma 1.4.20 implies that

$$
\begin{aligned}
\big\|T(f)\big\|_{L^{q_0,\infty}} &\le A_0\big\|f\big\|_{L^{p_0,m}} \\
\big\|T(f)\big\|_{L^{q_1,\infty}} &\le A_1\big\|f\big\|_{L^{p_1,m}},
\end{aligned} \tag{1.4.28}
$$

where we set $m = \frac{1}{2}\min\left(q_0, q_1, \frac{\log 2}{\log 2K}\right)$,

$$\begin{aligned} A_0 &= C(p_0, q_0, K, m)M_0 \\ A_1 &= C(p_1, q_1, K, m)M_1, \end{aligned}$$

and $C(p, q, K, \alpha)$ is as in (1.4.27).

Fix a function f in $L^{p,r}$. Split $f = f^t + f_t$ as follows:

$$f^t(x) = \begin{cases} f(x) & \text{if } |f(x)| > f^*(t^\gamma) \\ 0 & \text{if } |f(x)| \le f^*(t^\gamma), \end{cases}$$

$$f_t(x) = \begin{cases} 0 & \text{if } |f(x)| > f^*(t^\gamma) \\ f(x) & \text{if } |f(x)| \le f^*(t^\gamma), \end{cases}$$

where γ is the following nonzero real number:

$$\gamma = \frac{\frac{1}{q_0} - \frac{1}{q}}{\frac{1}{p_0} - \frac{1}{p}} = \frac{\frac{1}{q} - \frac{1}{q_1}}{\frac{1}{p} - \frac{1}{p_1}}.$$

Next, observe that the following inequalities are valid:

$$\begin{aligned} (f^t)^*(s) &\le \begin{cases} f^*(s) & \text{if } 0 < s < t^\gamma \\ 0 & \text{if } s \ge t^\gamma \end{cases} \qquad \text{and} \\ (f_t)^*(s) &\le \begin{cases} f^*(t^\gamma) & \text{if } 0 < s < t^\gamma \\ f^*(s) & \text{if } s \ge t^\gamma. \end{cases} \end{aligned} \tag{1.4.29}$$

It follows from these inequalities that $f^t \in L^{p_0,m}$ and $f_t \in L^{p_1,m}$ for all $t > 0$.

The sublinearity of the operator T and (1.4.8) imply

$$\begin{aligned} \|T(f)\|_{L^{q,r}} &= \left\|t^{\frac{1}{q}} T(f)^*(t)\right\|_{L^r(\frac{dt}{t})} \\ &\le K\left\|t^{\frac{1}{q}} (|T(f_t)| + |T(f^t)|)^*(t)\right\|_{L^r(\frac{dt}{t})} \\ &\le K\left\|t^{\frac{1}{q}} T(f_t)^*(\tfrac{t}{2}) + t^{\frac{1}{q}} T(f^t)^*(\tfrac{t}{2})\right\|_{L^r(\frac{dt}{t})} \\ &\le K a_r \left(\left\|t^{\frac{1}{q}} T(f_t)^*(\tfrac{t}{2})\right\|_{L^r(\frac{dt}{t})} + \left\|t^{\frac{1}{q}} T(f^t)^*(\tfrac{t}{2})\right\|_{L^r(\frac{dt}{t})}\right) \\ &\le K a_r \left(\left\|t^{\frac{1}{q} - \frac{1}{q_0}} t^{\frac{1}{q_0}} T(f_t)^*(\tfrac{t}{2})\right\|_{L^r(\frac{dt}{t})} + \left\|t^{\frac{1}{q} - \frac{1}{q_1}} t^{\frac{1}{q_1}} T(f^t)^*(\tfrac{t}{2})\right\|_{L^r(\frac{dt}{t})}\right), \end{aligned} \tag{1.4.30}$$

where

$$a_r = \begin{cases} 1 & \text{when } r \ge 1, \\ 2^{(1-r)/r} & \text{when } r \le 1. \end{cases}$$

It follows from (1.4.28) that

$$t^{\frac{1}{q_0}} T(f^t)^*(\tfrac{t}{2}) \le 2^{\frac{1}{q_0}} \sup_{s>0} s^{\frac{1}{q_0}} T(f^t)^*(s) \le 2^{\frac{1}{q_0}} A_0 \big\| f^t \big\|_{L^{p_0,m}}, \tag{1.4.31}$$

$$t^{\frac{1}{q_1}} T(f_t)^*(\tfrac{t}{2}) \le 2^{\frac{1}{q_1}} \sup_{s>0} s^{\frac{1}{q_1}} T(f_t)^*(s) \le 2^{\frac{1}{q_1}} A_1 \big\| f_t \big\|_{L^{p_1,m}} \tag{1.4.32}$$

for all $t > 0$. Now use (1.4.31) and (1.4.32) to estimate (1.4.30) by

$$K a_r 2^{\frac{1}{q_0}} A_0 \Big\| t^{\frac{1}{q}-\frac{1}{q_0}} \big\| f^t \big\|_{L^{p_0,m}} \Big\|_{L^r(\frac{dt}{t})} + K a_r 2^{\frac{1}{q_1}} A_1 \Big\| t^{\frac{1}{q}-\frac{1}{q_1}} \big\| f_t \big\|_{L^{p_1,m}} \Big\|_{L^r(\frac{dt}{t})},$$

which is the same as

$$\begin{aligned} &K a_r 2^{\frac{1}{q_0}} A_0 \Big\| t^{-\gamma(\frac{1}{p_0}-\frac{1}{p})} \big\| f^t \big\|_{L^{p_0,m}} \Big\|_{L^r(\frac{dt}{t})} \\ &\quad + K a_r 2^{\frac{1}{q_1}} A_1 \Big\| t^{\gamma(\frac{1}{p}-\frac{1}{p_1})} \big\| f_t \big\|_{L^{p_1,m}} \Big\|_{L^r(\frac{dt}{t})}. \end{aligned} \tag{1.4.33}$$

We now calculate. Change variables $u = t^\gamma$ in the first term of (1.4.33) to obtain

$$\begin{aligned} &K a_r 2^{\frac{1}{q_0}} A_0 \Big\| t^{-\gamma(\frac{1}{p_0}-\frac{1}{p})} \big\| f^t \big\|_{L^{p_0,m}} \Big\|_{L^r(\frac{dt}{t})} \\ &\le K a_r \frac{2^{\frac{1}{q_0}} A_0}{|\gamma|^{1/r}} \Bigg\| u^{-(\frac{1}{p_0}-\frac{1}{p})} \left(\int_0^u f^*(s)^m s^{\frac{m}{p_0}} \frac{ds}{s} \right)^{\frac{1}{m}} \Bigg\|_{L^r(\frac{du}{u})} \\ &\le K a_r \frac{2^{\frac{1}{q_0}} A_0}{m|\gamma|^{1/r}} \frac{r}{r(\frac{1}{p_0}-\frac{1}{p})} \left(\int_0^\infty (s^{\frac{1}{p_0}} f^*(s))^r s^{-r(\frac{1}{p_0}-\frac{1}{p})} \frac{ds}{s} \right)^{\frac{1}{r}} \\ &= K a_r \frac{2^{\frac{1}{q_0}} A_0}{m|\gamma|^{1/r}} \frac{1}{\frac{1}{p_0}-\frac{1}{p}} \big\| f \big\|_{L^{p,r}} \end{aligned}$$

where the last inequality is a consequence of Hardy's inequality (Exercise 1.2.8) with $p = r/m \ge 1$ and $b = (1/p_0 - 1/p)r$.

Similarly, change variables $u = t^\gamma$ in the second term of (1.4.33) to obtain

$$
\begin{aligned}
&Ka_r 2^{\frac{1}{q_1}} A_1 \Big\| t^{\gamma(\frac{1}{p}-\frac{1}{p_1})} \big\| f_t \big\|_{L^{p_1,m}} \Big\|_{L^r(\frac{dt}{t})} \\
\leq\ & \frac{Ka_r 2^{\frac{1}{q_1}} A_1}{|\gamma|^{1/r}} \bigg\| u^{\frac{1}{p}-\frac{1}{p_1}} \bigg[\int_0^u f^*(u)^m s^{\frac{m}{p_1}} \frac{ds}{s} + \int_u^\infty f^*(s)^m s^{\frac{m}{p_1}} \frac{ds}{s} \bigg]^{\frac{1}{m}} \bigg\|_{L^r(\frac{du}{u})} \\
\leq\ & \frac{Ka_r^2 2^{\frac{1-m}{m}} 2^{\frac{1}{q_1}} A_1}{|\gamma|^{1/r}} \bigg\{ \frac{p_1}{m} \Big\| u^{\frac{1}{p}-\frac{1}{p_1}} f^*(u) u^{\frac{1}{p_1}} \Big\|_{L^r(\frac{du}{u})} + \\
& \qquad \bigg\| u^{\frac{1}{p}-\frac{1}{p_1}} \bigg(\int_u^\infty f^*(s)^m s^{\frac{m}{p_1}} \frac{ds}{s} \bigg)^{\frac{1}{m}} \bigg\|_{L^r(\frac{du}{u})} \bigg\} \\
\leq\ & \frac{Ka_r^2 2^{\frac{1-m}{m}} 2^{\frac{1}{q_1}} A_1}{|\gamma|^{1/r}} \bigg\{ \frac{p_1}{m} \big\| f \big\|_{L^{p,r}} + \frac{r}{mr(\frac{1}{p}-\frac{1}{p_1})} \Big\| u^{r(\frac{1}{p}-\frac{1}{p_1})} f^*(u)^r u^{\frac{r}{p_1}} \Big\|_{L^r(\frac{du}{u})} \bigg\} \\
=\ & \frac{Ka_r^2 2^{\frac{1-m}{m}} 2^{\frac{1}{q_1}} A_1}{|\gamma|^{1/r}} \bigg\{ \frac{p_1}{m} + \frac{1}{m(\frac{1}{p}-\frac{1}{p_1})} \bigg\} \big\| f \big\|_{L^{p,r}}
\end{aligned}
$$

where the last inequality above is Hardy's inequality (see Exercise 1.2.8) with $p = r/m \geq 1$ and $b = (1/p - 1/p_1)r$.

We have now shown that

$$\big\| T(f) \big\|_{L^{q,r}} \leq M \big\| f \big\|_{L^{p,r}}$$

with constant

$$M = Ka_r \frac{2^{\frac{1}{q_0}} + 2^{\frac{1}{q_1}}}{m|\gamma|^{1/r}} \left(\frac{A_0}{\frac{1}{p_0} - \frac{1}{p}} + a_r 2^{\frac{1-m}{m}} A_1 \Big(p_1 + \frac{1}{\frac{1}{p} - \frac{1}{p_1}} \Big) \right). \tag{1.4.34}$$

We have been tacitly assuming that $r < \infty$. The remaining case is a simple consequence of the result just proved by letting $r \to \infty$, in which case $a_r \to 1$ and $|\gamma|^{1/r} \to 1$.

We now turn to the case $p_1 = \infty$. Hypotheses (1.4.18) and (1.4.19) together with Exercise 1.1.16 imply that

$$\big\| T(\chi_A) \big\|_{L^{q,\infty}} \leq M_0^{1-\theta} M_1^\theta \mu(A)^{1/p}$$

for all $0 \leq \theta \leq 1$. We select $\lambda \in (0,1)$ so that the indices $p = p_\lambda$ and $q = q_\lambda$ defined by (1.4.20) when $\theta = \lambda$ satisfy $p_0 < p < p_\lambda < \infty$ and q_λ is strictly between q_0 and q_1. Then apply the case $p_1 < \infty$ just proved with p_0, q_0 as before and $p_1 = p_\lambda$ and $q_1 = q_\lambda$. The result follows with M as in (1.4.34) except that p_1 is replaced by p_λ and q_1 by q_λ.

□

Corollary 1.4.21. *Let T be as in the statement of Theorem 1.4.19 and let $0 < p_0 \neq p_1 \leq \infty$, and $0 < q_0 \neq q_1 \leq \infty$. If T maps L^{p_0} to $L^{q_0,\infty}$ and L^{p_1} into*

$L^{q_1,\infty}$, and for some $0 < \theta < 1$ we have

$$\frac{1}{p} = \frac{1-\theta}{p_0} + \frac{\theta}{p_1}, \qquad \frac{1}{q} = \frac{1-\theta}{q_0} + \frac{\theta}{q_1}, \qquad \text{and} \qquad p \le q,$$

then T satisfies the strong type estimate $\|T(f)\|_{L^q} \le C\|f\|_{L^p}$ for all f functions in the domain of T. Moreover if T is linear, then it has a bounded extension from $L^p(X,\mu)$ into $L^q(Y,\nu)$.

Proof. Take $r = q$ in the previous theorem. □

Definition 1.4.22. Let $0 < p, q \le \infty$. We call an operator T of *restricted weak type* (p,q) if it satisfies

$$\|T(\chi_A)\|_{L^{q,\infty}} \le C\mu(A)^{1/p}$$

for all A subsets of a measure space (X,μ) with finite measure. Using this terminology, Corollary 1.4.21 is saying that if a quasi-linear operator T is of restricted weak types (p_0,q_0) and (p_1,q_1) for some $p_0 \ne p_1$ and $q_0 \ne q_1$, then it is bounded from L^p to L^q when $p \le q$.

We now give examples to indicate why the assumptions $p_0 \ne p_1$ and $q_0 \ne q_1$ cannot be dropped in Theorem 1.4.19.

Example 1.4.23. Let $X = Y = \mathbf{R}$ and

$$T(f)(x) = |x|^{-1/2} \int_0^1 f(t)\,dt.$$

Then $\alpha|\{x:\ |T(\chi_A)(x)| > \alpha\}|^{1/2} = 2^{1/2}|A \cap [0,1]|$ and thus T is of restricted weak types $(1,2)$ and $(3,2)$. But observe that T does not map $L^2 - L^{2,2}$ to $L^{q,2}$. Thus Theorem 1.4.19 fails if the assumption $q_0 \ne q_1$ is dropped. The dual operator

$$S(f)(x) = \chi_{[0,1]}(x) \int_{-\infty}^{+\infty} f(t)|t|^{-1/2}\,dt$$

satisfies $\alpha|\{x:\ |S(\chi_A)(x)| > \alpha\}|^{1/q} \le c|A|^{1/2}$ when $q = 1$ or 3 and thus it furnishes an example of an operator of restricted weak types $(2,1)$ and $(2,3)$ that is not L^2 bounded. Thus Theorem 1.4.19 fails if the assumption $p_0 \ne p_1$ is dropped.

As an application of Theorem 1.4.19, we give the following strengthening of Theorem 1.2.13.

Theorem 1.4.24. *(**Young's inequality for weak type spaces**) Let G be a locally compact group with left Haar measure λ that satisfies (1.2.13) for all A measurable subsets of G. Let $1 < p, q, r < \infty$ satisfy*

$$\frac{1}{q} + 1 = \frac{1}{p} + \frac{1}{r}. \tag{1.4.35}$$

Then there exists a constant $B_{pqr} > 0$ *such that for all* f *in* $L^p(G)$ *and* g *in* $L^{r,\infty}(G)$ *we have*

$$\|f * g\|_{L^q(G)} \le B_{pqr} \|g\|_{L^{r,\infty}(G)} \|f\|_{L^p(G)}. \tag{1.4.36}$$

Proof. Let's fix $1 < p, q < \infty$. Since p and q range in an open interval, we can find $p_0 < p < p_1$, $q_0 < q < q_1$, and $0 < \theta < 1$ such that (1.4.20) and (1.4.35) hold. Let $T(f) = f * g$, defined for all f functions on G. By Theorem 1.2.13, T extends to a bounded operator from L^{p_0} to $L^{p_1,\infty}$ and from L^{q_0} to $L^{q_1,\infty}$. It follows from the Marcinkiewicz interpolation theorem that T extends to a bounded operator from $L^p(G)$ into $L^q(G)$. □

Exercises

1.4.1. (a) Let g be a nonnegative simple function on (X, μ) and let A be a measurable subset of X. Prove that

$$\int_A g \, d\mu \le \int_0^{\mu(A)} g^*(t) \, dt.$$

(b) (*G. H. Hardy and J. E. Littlewood*) For f and g measurable on (X, μ), prove that

$$\int_X |f(x)g(x)| \, d\mu(x) \le \int_0^\infty f^*(t) g^*(t) \, dt.$$

Compare this result to the classical Hardy-Littlewood result asserting that if $a_j, b_j > 0$, the sum $\sum_j a_j b_j$ is greatest when both a_j and b_j are rearranged in decreasing order (for this see Hardy, Littlewood, and Pólya [**242**, p. 261]).

1.4.2. Prove that if $f \in L^{q_0,\infty} \cap L^{q_1,\infty}$ for some $0 < q_0 < q_1 \le \infty$, then $f \in L^{q,s}$ for all $0 < s \le \infty$ and $q_0 < q < q_1$.

1.4.3. (*Hunt* [**261**]) Given $0 < p, q < \infty$, fix an $r = r(p, q) > 0$ such that $r \le 1$, $r \le q$ and $r < p$. For $t \le \mu(X)$ define

$$f^{**}(t) = \sup_{\mu(E) \ge t} \left(\frac{1}{\mu(E)} \int_E |f|^r \, d\mu \right)^{1/r},$$

while for $t > \mu(X)$ (if $\mu(X) < \infty$) let

$$f^{**}(t) = \left(\frac{1}{t} \int_X |f|^r \, d\mu \right)^{1/r}.$$

Also define

$$|\!|\!| \, f \, |\!|\!|_{L^{p,q}} = \left(\int_0^\infty \left(t^{\frac{1}{p}} f^{**}(t)\right)^q \frac{dt}{t} \right)^{\frac{1}{q}}.$$

(The function f^{**} and the functional $f \to \left|\!\left|\!\left| f \right|\!\right|\!\right|_{L^{p,q}}$ depend on r.)
(a) Prove that the inequality

$$(((f+g)^{**})(t))^r \le (f^{**}(t))^r + (g^{**}(t))^r$$

is valid for all $t \ge 0$. Since $r \le q$, conclude that the functional

$$f \to \left|\!\left|\!\left| f \right|\!\right|\!\right|_{L^{p,q}}^r$$

is subadditive and hence it is a norm when $r = 1$ (this is possible only if $p > 1$).
(b) Show that for all f we have

$$\|f\|_{L^{p,q}} \le \left|\!\left|\!\left| f \right|\!\right|\!\right|_{L^{p,q}} \le \left(\frac{p}{p-r}\right)^{1/r} \|f\|_{L^{p,q}}.$$

(c) In conjunction with Exercise 1.1.12, conclude that $L^{p,q}$ is metrizable for all $0 < p, q \le \infty$ and also normable when $1 < p < \infty$ and $1 \le q \le \infty$.

1.4.4. (a) Show that on σ-finite measure space (X, μ) the set of countable linear combinations of simple functions is dense in $L^{p,\infty}(X)$.
(b) Prove that simple functions are not dense in $L^{p,\infty}(\mathbf{R})$ for any $0 < p \le \infty$.
[*Hint:* Part (b): Show that the function $h(x) = x^{-1/p}\chi_{x>0}$ cannot be approximated by a sequence of simple functions $L^{p,\infty}$. To see this, partition the interval $(0,\infty)$ into small subintervals of length $\varepsilon > 0$ and let f_ε be the step function $\sum_{-[1/\varepsilon]}^{[1/\varepsilon]} f(k\varepsilon)\chi_{[k\varepsilon,(k+1)\varepsilon]}(x)$. Show that for some $c > 0$ we have $\|f_\varepsilon - f\|_{L^{p,\infty}} \ge c$.]

1.4.5. Let (X, μ) be a nonatomic measure space. Prove the following facts:
(a) If $A_0 \subseteq A_1 \subseteq X$, $0 < \mu(A_1) < \infty$, and $\mu(A_0) \le t \le \mu(A_1)$, then there exists an $E_t \subseteq A$ with $\mu(E_t) = t$.
(b) Given $\varphi(t)$ continuous and decreasing on $[0,\infty)$, there exists a measurable function f on X with $f^*(t) = \varphi(t)$ for all $t > 0$.
(c) Given $A \subseteq X$ with $0 < \mu(A) < \infty$ and g an integrable function on X, there exists a set $\widetilde{A}$ with $\mu(\widetilde{A}) = \mu(A)$ such that

$$\int_{\widetilde{A}} g\, d\mu = \int_0^{\mu(A)} g^*(s)\, ds.$$

(d) Given f and g measurable functions on X, we have

$$\sup_{h:\ d_h = d_f} \left| \int_X h\, g\, d\mu \right| = \int_0^\infty f^*(s) g^*(s)\, ds,$$

where the supremum is taken over all h equidistributed with f.
[*Hint:* Part (a): Reduce matters to the situation in which $A_0 = \emptyset$. Consider first the case when for all $A \subseteq X$ there exists a subset B of X satisfying $\frac{1}{10}\mu(A) \le \mu(B) \le \frac{9}{10}\mu(A)$. Then we can find subsets of A_1 of measure

in any arbitrarily small interval and by continuity the required conclusion follows. Next consider the case in which there is a subset A_1 of X such that every $B \subseteq A_1$ satisfies $\mu(B) < \frac{1}{10}\,\mu(A_1)$ or $\mu(B) > \frac{9}{10}\,\mu(A_1)$. Without loss of generality, normalize μ so that $\mu(A_1) = 1$. Let $\mu_1 = \sup\{\mu(C) : C \subseteq A_1,\ \mu(C) < \frac{1}{10}\}$ and pick $B_1 \subseteq A_1$ such that $\frac{1}{2}\,\mu_1 \le \mu(B_1) \le \mu_1$. Set $A_2 = A_1 \setminus B_1$ and define $\mu_2 = \sup\{\mu(C) : C \subseteq A_2,\ \mu(C) < \frac{1}{10}\}$. Continue in this way and define sets $A_1 \supseteq A_2 \supseteq A_3 \supseteq \dots$ and numbers $\frac{1}{10} \ge \mu_1 \ge \mu_2 \ge \mu_3 \ge \dots$. If $C \subseteq A_{n+1}$ with $\mu(C \cup A_{n+1}) < \frac{1}{10}$, then $C \cup B_n \subseteq A_n$ with $\mu(C \cup B_n) < \frac{1}{5} < \frac{9}{10}$, and hence by assumption we must have $\mu(C \cup B_n) < \frac{1}{10}$. Conclude that $\mu_{n+1} \le \frac{1}{2}\mu_n$ and that $\mu(A_n) \ge \frac{4}{5}$ for all $n = 1, 2, \dots$. Then the set $\bigcap_{n=1}^{\infty} A_n$ must be an atom. Part (b): First show that when d is a simple right continuous decreasing function on $[0, \infty)$ there exists a measurable f on X such that $f^* = d$. For general continuous functions, use approximation. Part (c): Let $t = \mu(A)$ and define $A_1 = \{x : |g(x)| > g^*(t)\}$ and $A_2 = \{x : |g(x)| \ge g^*(t)\}$. Then $A_1 \subseteq A_2$ and $\mu(A_1) \le t \le \mu(A_2)$ by Proposition 1.4.5 (9). Pick $\widetilde{A}$ such that $A_1 \subseteq \widetilde{A} \subseteq A_2$ and $\mu(\widetilde{A}) = t = \mu(A)$ by part (a). Then $\int_{\widetilde{A}} g\, d\mu = \int_X g\chi_{\widetilde{A}}\, d\mu = \int_0^\infty (g\chi_{\widetilde{A}})^*\, ds = \int_0^{\mu(\widetilde{A})} g^*(s)\, ds$. Part (d): Let $f = \sum_{j=1}^N a_j \chi_{A_j}$ where $a_1 > a_2 > \dots > a_N > 0$ and the A_j are pairwise disjoint. Write f as $\sum_{j=1}^N b_j \chi_{B_j}$, where $b_j = (a_j - a_{j+1})$ and $B_j = A_1 \cup \dots \cup A_j$. Pick $\widetilde{B_j}$ as in part (c). Then $\widetilde{B_1} \subseteq \dots \subseteq \widetilde{B_N}$ and the function $f_1 = \sum_{j=1}^N b_j \chi_{\widetilde{B_j}}$ has the same distribution function as f. It follows from part (c) that $\int_X f_1 g\, d\mu = \int_0^\infty f^*(s) g^*(s)\, ds$. The case of a general f follows from the case when f is simple using approximation and Exercise 1.4.1.]

1.4.6. (*Aoki* [**8**]/ *Rolewicz* [**431**]) Let $\|\cdot\|$ be a nonnegative functional on a vector space X that satisfies

$$\|x + y\| \le K\big(\|x\| + \|y\|\big)$$

for all x and y in X. (To avoid trivialities, assume that $K \ge 1$.) Then for α defined by the equation

$$(2K)^\alpha = 2, \qquad (\alpha \le 1),$$

we have

$$\|x_1 + \dots + x_n\|^\alpha \le 4\,\big(\|x_1\|^\alpha + \dots + \|x_n\|^\alpha\big)$$

for all $n = 1, 2, \dots$ and all $x_1, x_2, \dots, x_n$ in X.
[*Hint:* Quasi-linearity implies $\|x_1 + \dots + x_n\| \le \max_{1 \le j \le n} [(2K)^j \|x_j\|]$ for all $x_1, \dots, x_n$ in X (use that $K \ge 1$.) Define $H : X \to \mathbf{R}$ by setting $H(0) = 0$ and $H(x) = 2^{j/\alpha}$ if $2^{j-1} < \|x\|^\alpha \le 2^j$. Then $\|x\| \le H(x) \le 2^{1/\alpha}\|x\|$ for all $x \in X$. Prove by induction that $\|x_1 + \dots + x_n\|^\alpha \le 2(H(x_1)^\alpha + \dots + H(x_n)^\alpha)$.

Suppose that this statement is true when $n = m$. To show its validity for $n = m+1$ without loss of generality assume that $\|x_1\| \geq \|x_2\| \geq \cdots \geq \|x_{m+1}\|$. Then $H(x_1) \geq H(x_2) \geq \cdots \geq H(x_{m+1})$. If all the $H(x_j)$'s are distinct, since $H(x_j)^\alpha$ are distinct powers of 2 and therefore they must satisfy $H(x_j)^\alpha \leq 2^{-j+1} H(x_1)^\alpha$. Then

$$\begin{aligned}\|x_1 + \cdots + x_{m+1}\|^\alpha &\leq \big[\max_{1\leq j\leq m+1} (2K)^j \|x_j\|\big]^\alpha \\ &\leq \big[\max_{1\leq j\leq m+1} (2K)^j H(x_j)\big]^\alpha \\ &\leq \big[\max_{1\leq j\leq m+1} (2K)^j 2^{1/\alpha} 2^{-j/\alpha} H(x_1)\big]^\alpha \\ &= 2H(x_1)^\alpha \\ &\leq 2(H(x_1)^\alpha + \cdots + H(x_{m+1})^\alpha),\end{aligned}$$

We now consider the case where $H(x_j) = H(x_{j+1})$ for some $1 \leq j \leq m$. Then for some integer r we must have $2^{r-1} < \|x_{j+1}\|^\alpha \leq \|x_j\|^\alpha \leq 2^r$ and $H(x_j) = 2^{r/\alpha}$. Next note that

$$\|x_j + x_{j+1}\|^\alpha \leq K^\alpha(\|x_j\| + \|x_{j+1}\|)^\alpha \leq K^\alpha (2\, 2^r)^\alpha \leq 2^{r+1}.$$

This implies

$$H(x_j + x_{j+1})^\alpha \leq 2^{r+1} = 2^r + 2^r = H(x_j)^\alpha + H(x_{j+1})^\alpha.$$

Now apply the inductive hypothesis to $x_1, \dots, x_{j-1}, x_j + x_{j+1}, x_{j+1}, \dots, x_m$ and use the previous inequality to obtain the required conclusion.]

1.4.7. (*Stein and Weiss* [**485**]) Let (X, μ) and (Y, ν) be measure spaces. Let Z be a Banach space of complex-valued measurable functions on Y. Assume that Z is closed under absolute values and satisfies $\big\|f\big\|_Z = \big\||f|\big\|_Z$. Suppose that T is a linear operator defined on the space of measurable functions on (X, μ) and taking values in Z. Suppose that for some constant $A > 0$ we have the restricted weak type estimate

$$\big\|T(\chi_E)\big\|_Z \leq A\mu(E)^{1/p}$$

for all A measurable subsets of X. Then there is a constant $C(p) > 0$ such that

$$\big\|T(f)\big\|_Z \leq C(p) A \big\|f\big\|_{L^{p,1}}$$

for all f in the domain of T.
[*Hint:* Let $f = \sum_{j=1}^N a_j \chi_{E_j} \geq 0$, where $a_1 > a_2 > \cdots > a_N > 0$, $\mu(E_j) < \infty$ pairwise disjoint. Let $F_j = E_1 \cup \cdots \cup E_j$, $B_0 = 0$, and $B_j = \mu(F_j)$ for $j \geq 1$.

Write $f = \sum_{j=1}^N (a_j - a_{j+1})\chi_{F_j}$, where $a_{N+1} = 0$. Then

$$\begin{aligned}\big\|T(f)\big\|_Z &= \big\||T(f)|\big\|_Z \\ &\le \sum_{j=1}^N (a_j - a_{j+1})\big\|T(\chi_{F_j})\big\|_Z \\ &\le A\sum_{j=1}^N (a_j - a_{j+1})(\mu(F_j))^{1/p} \\ &= A\sum_{j=0}^{N-1} a_{j+1}(B_{j+1}^{1/p} - B_j^{1/p}) \\ &= C(p)A\big\|f\big\|_{L^{p,1}}.\end{aligned}$$

where the penultimate equality follows summing by parts; see Appendix F.]

1.4.8. Let $0 < p < \infty$ and $0 < q_1 < q_2 \le \infty$. Let $\alpha, \beta, q > 0$.
(a) Show that the function $f(t) = t^{-\alpha}(\log t^{-1})^{-\beta}\chi_{(0,1)}(t)$ lies in $L^{p,q}(\mathbf{R})$ if and only if either $p > 1/\alpha$ or both $p = 1/\alpha$ and $q > 1/\beta$.
(b) Show that the function $t^{-\frac{1}{p}}(\log t^{-1})^{-\frac{1}{q_1}}\chi_{(0,1)}(t)$ lies in $L^{p,q_2}(\mathbf{R})$ but not in $L^{p,q_1}(\mathbf{R})$.
(c) On $\mathbf{R}^n$ construct examples to show that $L^{p,q_1} \subsetneq L^{p,q_2}$.
(d) On a general nonatomic measure space (X,μ) prove the following: There *does not* exist a constant $C(p,q_1,q_2) > 0$ such that for all f in $L^{p,q_2}(X)$ the following is valid

$$\big\|f\big\|_{L^{p,q_1}} \le C(p,q_1,q_2)\big\|f\big\|_{L^{p,q_2}}.$$

1.4.9. (*Stein and Weiss* [**484**]) Let $L^p(\omega)$ denote the weighted L^p space with measure $\omega(x)dx$. Let T be a sublinear operator that maps

$$\begin{aligned} T:\ & L^{p_0}(\omega_0) \to L^{q_0,\infty}(w) \\ T:\ & L^{p_1}(\omega_1) \to L^{q_1,\infty}(w)\end{aligned}$$

for some $p_0 \neq p_1$, where $0 < p_0, p_1, q_0, q_1 \le \infty$ and $\omega_0, \omega_1, \omega$ are positive functions. Suppose that

$$\frac{1}{p_\theta} = \frac{1-\theta}{p_0} + \frac{\theta}{p_1} \qquad \frac{1}{q_\theta} = \frac{1-\theta}{q_0} + \frac{\theta}{q_1}.$$

Then T maps

$$L^{p_\theta}\big(\omega_0^{\frac{1-\theta}{p_0}p_\theta}\omega_1^{\frac{\theta}{p_1}p_\theta}\big) \to L^{q_\theta,\infty}(\omega)$$

[*Hint:* Define

$$L(f) = (\omega_1/\omega_0)^{\frac{1}{p_1-p_0}}$$

and observe that for each $\theta \in [0,1]$, L maps

$$L^{p_\theta}\big(\omega_0^{\frac{1-\theta}{p_0}p_\theta}\omega_1^{\frac{\theta}{p_1}p_\theta}\big) \to L^{p_\theta}\big((\omega_0^{p_1}\omega_1^{-p_0})^{\frac{1}{p_1-p_0}}\big).$$

isometrically. Then apply the classical Marcinkiewicz interpolation theorem to the sublinear operator $T\circ L^{-1}$ and the required conclusion easily follows.]

1.4.10. (*Kalton* [**286**]/*Stein and Weiss* [**488**]) Let λ_n be a sequence of positive numbers with $\sum_n \lambda_n \le 1$ and $\sum_n \lambda_n \log(\frac{1}{\lambda_n}) = K < \infty$.
(a) Let f_n be a sequence of complex-valued functions in $L^{1,\infty}(X)$ such that $\|f_n\|_{L^{1,\infty}} \le 1$ uniformly in n. Prove that $\sum_n \lambda_n f_n$ lies in $L^{1,\infty}(X)$ with norm at most $2(K+2)$. (This property is referred to as the *logconvexity* of $L^{1,\infty}$.)
(b) Let T_n be a sequence of sublinear operators that map $L^1(X)$ to $L^{1,\infty}(Y)$ with norms $\|T_n\|_{L^1\to L^{1,\infty}} \le B$ uniformly in n. Use part (a) to prove that $\sum_n \lambda_n T_n$ maps $L^1(X)$ to $L^{1,\infty}(Y)$ with norm at most $2B(K+2)$.
(c) Given $\delta > 0$ pick $0 < \varepsilon < \delta$ and use the simple estimate

$$\mu(\{\sum_{n=1}^{\infty} 2^{-\delta n} f_n > \alpha\}) \le \sum_{n=1}^{\infty} \mu(\{2^{-\delta n} f_n > (2^{\varepsilon}-1)2^{-\varepsilon n}\alpha\})$$

to obtain a simple proof of the statements in part (a) and (b) when $\lambda_n = 2^{-\delta n}$, $n = 1, 2, \dots$ and zero otherwise.
[*Hint:* Part (a): For fixed $\alpha > 0$, write $f_n = u_n + v_n + w_n$, where $u_n = f_n\chi_{|f_n|\le\frac{\alpha}{2}}$, $v_n = f_n\chi_{|f_n|>\frac{\alpha}{2\lambda_n}}$, and $w_n = f_n\chi_{\frac{\alpha}{2}<|f_n|\le\frac{\alpha}{2\lambda_n}}$. Let $u = \sum_n \lambda_n u_n$, $v = \sum_n \lambda_n v_n$, and $w = \sum_n \lambda_n w_n$. Clearly $|u| \le \alpha/2$. Also $\{v \ne 0\} \subseteq \bigcup_n \{|f_n| > \frac{\alpha}{2\lambda_n}\}$, hence $\mu(\{v \ne 0\}) \le \frac{2}{\alpha}$. Finally

$$\begin{aligned}\int_X |w|\,d\mu &\le \sum_n \lambda_n \int_X |f_n|\chi_{\frac{\alpha}{2}<|f_n|\le\frac{\alpha}{2\lambda_n}}\,d\mu \\ &\le \sum_n \lambda_n \left[\int_{\alpha/2}^{\alpha/(2\lambda_n)} d_{f_n}(\beta)\,d\beta + \int_0^{\alpha/2} d_{f_n}(\alpha/2)\,d\beta\right] \le K+1.\end{aligned}$$

Using $\mu(\{|u+v+w| > \alpha\}) \le \mu(\{|u| > \alpha/2\}) + \mu(\{|v| \ne 0\}) + \mu(\{|w| > \alpha/2\})$, deduce the conclusion.]

1.4.11. Construct a sequence of functions f_k in $L^{1,\infty}(\mathbf{R}^n)$ and a function $f \in L^{1,\infty}$ such that $\|f_k - f\|_{L^\infty} \to 0$ but $\|f_k\|_{L^{1,\infty}} \to \infty$ as $k \to \infty$.

1.4.12. (a) Suppose that X is a quasi-Banach space and let X^* be its dual (which is always a Banach space). Prove that for all $T \in X^*$ we have

$$\|T\|_{X^*} = \sup_{\substack{x\in X \\ \|x\|_X\le 1}} |T(x)|.$$

(b) Now suppose that X is a Banach space. Use the Hahn-Banach theorem to prove that for every $x \in X$ we have

$$\|x\|_X = \sup_{\substack{T \in X^* \\ \|T\|_{X^*} \le 1}} |T(x)|.$$

Observe that this result may fail for quasi-Banach spaces. For example, if $X = L^{1,\infty}$, every linear functional on X^* vanishes on the set of simple functions.
(c) Take $X = L^{p,1}$ and $X^* = L^{p',\infty}$. Then for $1 < p < \infty$ both of these spaces are normable. Conclude that

$$\|f\|_{L^{p,1}} \approx \sup_{\|g\|_{L^{p',\infty}} \le 1} \left| \int_X f g \, d\mu \right|,$$

$$\|f\|_{L^{p,\infty}} \approx \sup_{\|g\|_{L^{p',1}} \le 1} \left| \int_X f g \, d\mu \right|.$$

1.4.13. Let $0 < p, q < \infty$. Prove that any function in $L^{p,q}(X,\mu)$ can be written as

$$f = \sum_{n=-\infty}^{+\infty} c_n f_n,$$

where f_n is a function bounded by $2^{-n/p}$, supported on a set of measure 2^n, and the sequence $\{c_k\}_k$ lies in ℓ^q and satisfies

$$2^{-\frac{1}{p}} (\log 2)^{\frac{1}{q}} \big\|\{c_k\}_k\big\|_{\ell^q} \le \|f\|_{L^{p,q}} \le \big\|\{c_k\}_k\big\|_{\ell^q} 2^{\frac{1}{p}} (\log 2)^{\frac{1}{q}}.$$

[*Hint:* Let $c_n = 2^{n/p} f^*(2^n)$ and $f_n = c_n^{-1} f \chi_{A_n}$ where A_n is as in (1.4.26).]

1.4.14. (*T. Tao*) Let $0 < p < \infty$, $0 < \gamma < 1$, $A > 0$, and let f be a measurable function on a measure space (X, μ).
(a) Suppose that $\|f\|_{L^{p,\infty}} \le A$. Then for every measurable set E of finite measure there exists a measurable subset E' of E with $\mu(E') \ge \gamma \mu(E)$ such that

$$\left| \int_{E'} f \, d\mu \right| \le C_\gamma A \, \mu(E)^{1-\frac{1}{p}}$$

where $C_\gamma = (1-\gamma)^{-1/p}$.
(b) Conversely, if the last condition holds for some $C_\gamma, A < \infty$ and all measurable subsets E of finite measure, then $\|f\|_{L^{p,\infty}} \le c_\gamma A$, where $c_\gamma = C_\gamma 4^{1/p} \gamma^{-1} \sqrt{2}$.

(c) Conclude that

$$\|f\|_{L^{p,\infty}} \approx \sup_{\substack{E\subseteq X \\ 0<\mu(E)<\infty}} \inf_{\substack{E'\subseteq E \\ \mu(E')\geq\frac{1}{2}\mu(E)}} \mu(E)^{-1+\frac{1}{p}} \left| \int_{E'} f\,d\mu \right| .$$

[*Hint:* Part (a): Take $E' = E \setminus \{|f| > A(1-\gamma)^{-\frac{1}{p}}\mu(E)^{-\frac{1}{p}}\}$. Part (b): Given $\alpha > 0$, note that the set $\{|f| > \alpha\}$ is contained in

$$\{\operatorname{Re} f > \tfrac{\alpha}{\sqrt{2}}\} \cup \{\operatorname{Im} f > \tfrac{\alpha}{\sqrt{2}}\} \cup \{\operatorname{Re} f < -\tfrac{\alpha}{\sqrt{2}}\} \cup \{\operatorname{Im} f < -\tfrac{\alpha}{\sqrt{2}}\}.$$

For E any of the preceding four sets, let E' be a subset of it with measure at least $\gamma\,\mu(E)$ such as in the hypothesis. Then $\left|\int_{E'} f\,d\mu\right| \geq \frac{\alpha}{\sqrt{2}}\gamma\,\mu(E)$, which gives $\|f\|_{L^{p,\infty}} \leq 4^{1/p}\gamma^{-1}C_\gamma\sqrt{2}A$.]

1.4.15. (*Wolff* [**558**]) Given a linear operator T defined on the set of measurable functions on a measure space (X,μ) and taking values in the set of measurable functions on a measure space (Y,ν), define its "transpose" T^t via the identity

$$\int_Y T(f)\,g\,d\nu = \int_X T^t(g)\,f\,d\mu$$

for all measurable functions f on X and g on Y, whenever the integrals converge. Let T be such a linear operator given in the form

$$T(f)(y) = \int_X K(y,x)f(x)\,d\mu(x),$$

where K is measurable and bounded by some constant $M > 0$. Suppose that T maps $L^1(X)$ into $L^{1,\infty}(Y)$ with norm $\|T\|$ and T^t maps $L^1(Y)$ into $L^{1,\infty}(X)$ with norm $\|T^t\|$. Show that for all $1 < p < \infty$ there exists a constant C_p that depends only on p and is independent of M such that T maps $L^p(X)$ into $L^p(Y)$ with norm

$$\|T\|_{L^p(X)\to L^p(Y)} \leq C_p\|T\|^{\frac{1}{p}}\|T^t\|^{1-\frac{1}{p}}.$$

[*Hint:* For $R, M > 0$, let $\mathcal{B}_R$ be the set of all (A,B), where A is a measurable subset of X with $\mu(A) \leq R$ and B is a measurable subset of Y with $\nu(B) \leq R$ and let $\mathcal{B}_{R,M}$ be the set of all (A,B) in $\mathcal{B}_R$ such that $|K(x,y)| \leq M$ for all $x \in A$ and $y \in B$. Also let $M_p = M_p(R,M) < \infty$ be the smallest constant such that for all $(A,B) \in \mathcal{B}_{R,M}$ we have $\left|\int_B T(\chi_A)\,d\nu\right| \leq M_p\mu(A)^{\frac{1}{p}}\nu(B)^{\frac{1}{p'}}$. Let $\delta > 0$ and $(A,B) \in \mathcal{B}_{R,M}$. If $\mu(A) \leq \delta\nu(B)$, use the previous Exercise to find a B' with $\nu(B') \geq \frac{1}{2}\nu(B)$ such that $\left|\int_{B'} T(\chi_A)\,d\nu\right| \leq c\|T\|\mu(A) \leq c\,\delta^{\frac{1}{p'}}\|T\|\mu(A)^{\frac{1}{p}}\nu(B)^{\frac{1}{p'}}$. Then $\nu(B \setminus B') \leq \frac{1}{2}\nu(B)$ and we have

$$\left|\int_{B\setminus B'} T(\chi_A)\,d\nu\right| \leq M_p 2^{-\frac{1}{p'}}\mu(A)^{\frac{1}{p}}\nu(B)^{\frac{1}{p'}}.$$

Summing, we obtain $M_p \le M_p 2^{-\frac{1}{p'}} + c\,\delta^{\frac{1}{p'}} \big\|T\big\|$. If $\nu(B) \le \delta^{-1}\mu(A)$, write $\big|\int_B T(\chi_A)\,d\nu\big| = \big|\int_A T^t(\chi_B)\,d\mu\big|$ and use the previous Exercise to find A' with $\mu(A') \ge \frac{1}{2}\mu(A)$ and $\big|\int_{A'} T^t(\chi_B)\,d\mu\big| \le c\big\|T^t\big\|\nu(B)$. Argue similarly to obtain $M_p \le M_p 2^{-\frac{1}{p}} + c\,\delta^{-\frac{1}{p}}\big\|T^t\big\|$. Pick a suitable δ to optimize both expressions. Obtain that M_p is independent of R and M. Considering $B_+ = B \cap \{T(\chi_A) > 0\}$ and $B_- = B \cap \{T(\chi_A) < 0\}$, obtain that $\int_B \big|T(\chi_A)\big|\,d\nu \le 2M_p \mu(A)^{\frac{1}{p}} \nu(B)^{\frac{1}{p'}}$ whenever $(A,B) \in \mathcal{B}_{R,M}$. Use Fatou's lemma to remove the restriction that $(A,B) \in \mathcal{B}_{R,M}$. Finally, use the characterization of $\big\|\cdot\big\|_{L^{p,\infty}}$ obtained in Exercise 1.1.12 with $r = 1$ to conclude $\big\|T(\chi_A)\big\|_{L^{p,\infty}} \le C_p \big\|T\big\|^{\frac{1}{p}} \big\|T^t\big\|^{1-\frac{1}{p}} \mu(A)^{\frac{1}{p}}$.]

APPENDIX: SOME MULTILINEAR INTERPOLATION

Multilinear maps are defined on products on linear spaces and take values in another linear space. We will be concerned with the situation in which these linear spaces are indeed function spaces. Let $(X_1,\mu_1),\dots,(X_m,\mu_m)$ be measure spaces, let $\mathcal{D}_j$ be spaces of measurable functions on X_j, and let T be a map defined on $\mathcal{D}_1 \times \cdots \times \mathcal{D}_m$ and taking values in the set of measurable functions on another measure space (Z,σ). Then T is called multilinear if for all f_j, g_j in $\mathcal{D}_j$ and all scalars λ we have

$$|T(f_1,\dots,\lambda f_j,\dots,f_m)| = |\lambda|\,|T(f_1,\dots,f_j,\dots,f_m)|$$
$$T(f_1,\dots,f_j{+}g_j,\dots,f_m) = T(f_1,\dots,f_j,\dots,f_m) + T(f_1,\dots,g_j,\dots,f_m).$$

If $\mathcal{D}_j$ are dense subspaces of $L^{p_j}(X_j,\mu_j)$ and T is a multlinear map defined on $\prod_{j=1}^m \mathcal{D}_j$ and satisfies

$$\big\|T(f_1,\dots,f_m)\big\|_{L^p(Z)} \le C\big\|f_1\big\|_{L^{p_1}(X_1)} \cdots \big\|f_m\big\|_{L^{p_m}(X_m)},$$

for all $f_j \in \mathcal{D}_j$, then T has a bounded extension from $L^{p_1} \times \cdots \times L^{p_m} \to Z$. The norm of a multilinear map $T:\ L^{p_1} \times \cdots \times L^{p_m} \to Z$ is the smallest constant C such that the preceding inequality holds and will be denoted by

$$\big\|T\big\|_{L^{p_1}\times\cdots\times L^{p_m}\to Z}.$$

Suppose that T is defined on $\prod_{j=1}^m \mathcal{D}_j$, where each $\mathcal{D}_j$ contains the simple functions. We say that T is quasi-multilinear if there is a $K > 0$ such that for all $1 \le j \le m$, all f_j, g_j in $\mathcal{D}_j$, and all $\lambda \in \mathbf{C}$ we have

$$|T(f_1,\dots,\lambda f_j,\dots,f_m)| = |\lambda|\,|T(f_1,\dots,f_j,\dots,f_m)|$$
$$|T(f_1,\dots,f_j{+}g_j,\dots,f_m)| \le K\big(|T(f_1,\dots,f_j,\dots,f_m)| + |T(f_1,\dots,g_j,\dots,f_m)|\big).$$

In the special case in which $K = 1$, T is called submultilinear.

Exercises

1.4.16. Let T be a multilinear map defined on the set of simple functions of the product of m measure spaces $(X_1,\mu_1)\times\cdots\times(X_m,\mu_m)$ and taking values in the set of measurable functions on another measure space (Z,σ). Let $1\le p_{jk}\le\infty$ for $1\le k\le m$ and $j\in\{0,1\}$ and also let $1\le p_j\le\infty$ for $j\in\{0,1\}$. Suppose that T satisfies

$$\big\|T(f_1,\dots,f_m)\big\|_{L^{p_j}}\le M_j\big\|f_1\big\|_{L^{p_{j1}}}\cdots\big\|f_m\big\|_{L^{p_{jm}}},\qquad j=0,1$$

for all simple functions f_k on X_k. Let $(1/q,1/q_1,\dots,1/q_m)$ lie on the open line segment joining $(1/p_0,1/p_{01},\dots,1/p_{0m})$ and $(1/p_1,1/p_{11},\dots,1/p_{1m})$ in $\mathbf{R}^{m+1}$. Then for some $0<\theta<1$ we have

$$\frac{1}{q}=\frac{1-\theta}{p_0}+\frac{\theta}{p_1},\qquad \frac{1}{q_k}=\frac{1-\theta}{p_{0k}}+\frac{\theta}{p_{1k}},\qquad 1\le k\le m.$$

Prove that T has a bounded extension from $L^{q_1}\times\cdots\times L^{q_m}$ into L^q that satisfies

$$\big\|T(f_1,\dots,f_m)\big\|_{L^q}\le M_0^{1-\theta}M_1^{\theta}\big\|f_1\big\|_{L^{q_1}}\cdots\big\|f_m\big\|_{L^{q_m}}$$

for all $f_k\in L^{q_k}(X_k)$.
[*Hint:* Adapt the proof of Theorem 1.3.4.]

1.4.17. Let $(X_1,\mu_1),\dots,(X_m,\mu_m)$ be measure spaces, let $\mathcal{D}_j$ be spaces of measurable functions on X_j that contain the simple functions, and let T be a quasi-multilinear map defined on $\mathcal{D}_1\times\cdots\times\mathcal{D}_m$ that takes values in the set of measurable functions on another measure space (Z,σ). Let $0<p_{jk}\le\infty$ for $1\le j\le m+1$ and $1\le k\le m$, and also let $0<p_j\le\infty$ for $1\le j\le m+1$. Suppose that for all $1\le j\le m+1$, T satisfies

$$\big\|T(\chi_{E_1},\dots,\chi_{E_m})\big\|_{L^{p_j,\infty}}\le M\mu_1(E_1)^{\frac{1}{p_{j1}}}\dots\mu_m(E_m)^{\frac{1}{p_{jm}}}$$

for all sets E_k of finite μ_k measure. Assume that the system

$$\begin{pmatrix}1/p_{11} & 1/p_{12} & \dots & 1/p_{1m} & 1\\ 1/p_{21} & 1/p_{22} & \dots & 1/p_{2m} & 1\\ \vdots & \vdots & \vdots & \vdots & \vdots\\ 1/p_{m1} & 1/p_{m2} & \dots & 1/p_{mm} & 1\\ 1/p_{(m+1)1} & 1/p_{(m+1)2} & \dots & 1/p_{(m+1)m} & 1\end{pmatrix}\begin{pmatrix}\sigma_1\\ \sigma_2\\ \vdots\\ \sigma_m\\ \tau\end{pmatrix}=\begin{pmatrix}1/p_1\\ 1/p_2\\ \vdots\\ 1/p_m\\ 1/p_{m+1}\end{pmatrix}$$

has a *unique* solution $(\sigma_1,\dots,\sigma_m,\tau)\in\mathbf{R}^{m+1}$ with *not all* $\sigma_j=0$. (This assumption implies that the determinant of the displayed square matrix is nonzero.) Suppose $(1/q,1/q_1,\dots,1/q_m)$ lies in the open convex hull of the $m+1$ points $(1/p_j,1/p_{j1},\dots,1/p_{jm})$ in $\mathbf{R}^{m+1}$, $1\le j\le m+1$. Let $0<t_k,t\le\infty$ satisfy

$$\sum_{\sigma_k\neq 0}\frac{1}{t_k}\ge\frac{1}{t}.$$

Prove that there exists a constant C that depends only on the p_{jk}'s, q_k's, p_j's, and on K (but not on M) such that for all f_j in $\mathcal{D}_j$ we have

$$\big\|T(f_1,\dots,f_m)\big\|_{L^{q,t}} \le C\,M\big\|f_1\big\|_{L^{q_1,t_1}}\cdots\big\|f_m\big\|_{L^{q_m,t_m}}.$$

[*Hint:* Prove this result only when $m=2$. In this case one can mimic the proof of Theorem 1.4.19.]

1.4.18. (*O' Neil* [**393**]) Show that

$$\big\|f*g\big\|_{L^{r,s}} \le C_{p,q,s_1,s_2}\big\|f\big\|_{L^{p,s_1}}\big\|g\big\|_{L^{q,s_2}},$$

whenever $1<p,q,r<\infty$, $0<s_1,s_2\le\infty$, $\frac{1}{p}+\frac{1}{q}=\frac{1}{r}+1$, and $\frac{1}{s_1}+\frac{1}{s_2}=\frac{1}{s}$. Also deduce Hölder's inequality for Lorentz spaces

$$\big\|fg\big\|_{L^{r,s}} \le C_{p,q,s_1,s_2}\big\|f\big\|_{L^{p,s_1}}\big\|g\big\|_{L^{q,s_2}},$$

where now $0<p,q,r\le\infty$, $0<s_1,s_2\le\infty$, $\frac{1}{p}+\frac{1}{q}=\frac{1}{r}$, and $\frac{1}{s_1}+\frac{1}{s_2}=\frac{1}{s}$. [*Hint:* Use the previous Exercise.]

1.4.19. (*Grafakos and Tao* [**223**]) Suppose that T is a multilinear operator of the form

$$T(f_1,\dots,f_m)(y)=$$
$$\int_{X_1}\cdots\int_{X_m} K(x_1,\dots,x_m,y)f_1(x_1)\dots f_m(x_m)\,d\mu_1(x_1)\dots d\mu_m(x_m),$$

where the kernel K is bounded by some constant M. The jth transpose of T is the m-linear operator whose kernel is obtained from K by interchanging the variables x_j and y. Suppose that T and all of its transposes map $L^1(X_1)\times\cdots\times L^1(X_m)$ into $L^{1/m,\infty}(Y)$. Conclude that T maps $L^{p_1}(X_1)\times\cdots\times L^{p_m}(X_m)$ into $L^p(Y)$ for all $1<p_j\le\infty$ where $p<\infty$ and $1/p=1/p_1+\cdots+1/p_m$ with a bound independent of the kernel K. [*Hint:* Take $p>1$ and use the same idea as in Exercise 1.4.15. The full range of p's will follow by interpolation.]

HISTORICAL NOTES

The modern theory of measure and integration was founded with the publication of Lebesgue's dissertation [**329**]; see also [**330**]. The theory of the Lebesgue integral reshaped the course of integration. The spaces $L^p([a,b])$, $1<p<\infty$ were first investigated by Riesz [**423**] who obtained many important properties of them. A rigorous treatise of harmonic analysis on general groups can be found in the book of Hewitt and Ross [**247**]. The best constant C_{pqr} in Young's inequality $\|f*g\|_{L^r(\mathbf{R}^n)}\le C_{pqr}\|f\|_{L^p(\mathbf{R}^n)}\|g\|_{L^q(\mathbf{R}^n)}$, $\frac{1}{p}+\frac{1}{q}=\frac{1}{r}+1$, $1<p,q,r<\infty$, was shown by Beckner [**29**] to be $C_{pqr}=(B_pB_qB_{r'})^n$, where $B_p^2=p^{1/p}(p')^{-1/p'}$.

Theorem 1.3.2 first appeared without proof in Marcinkiewicz's brief note [**365**]. After Marcinkiewicz's death in World War II, this theorem seemed to have escaped attention until Zygmund resurfaced it in [**570**]. This reference presents the more difficult off-diagonal version of the theorem, derived by Zygmund. Stein and Weiss [**485**] strengthened Zygmund's theorem by assuming that the initial estimates are of restricted weak type whenever $1 \le p_0, p_1, q_0, q_1 \le \infty$. The extension of this result to the case where $0 < p_0, p_1, q_0, q_1 < 1$ as presented in Theorem 1.4.17 is due to the author; the critical Lemma 1.4.20 was suggested by Kalton. Equivalence of restricted weak type $(1,1)$ and weak type $(1,1)$ properties for certain maximal multipliers was obtained by Moon [**377**]. The following converse of Theorem 1.2.13 is due to Stepanov [**490**]: If a convolution operator maps $L^p(\mathbf{R}^n)$ into $L^{q,\infty}(\mathbf{R}^n)$ for some $1 \le p < q < \infty$ and $1 < r < \infty$ that satisfy $\frac{1}{p} + \frac{1}{r} = \frac{1}{q} + 1$, then its kernel must be in $L^{r,\infty}$.

The extrapolation result of Exercise 1.3.7 is due to Yano [**562**]; see also Zygmund [**572**, pp. 119–120]. We refer to Carro [**88**] for a generalization. See also the related work of Soria [**466**] and Tao [**509**].

The original version of Theorem 1.3.4 was proved by Riesz [**426**] in the context of bilinear forms. This version is called the Riesz convexity theorem as it says that the logarithm of the function $M(\alpha,\beta) = \inf_{x,y} \big|\sum_{j=1}^{n}\sum_{k=1}^{m} a_{jk}x_jy_k\big| \, \|x\|_{\ell^{1/\alpha}}^{-1}\|y\|_{\ell^{1/\beta}}^{-1}$ (where the infimum is taken over all sequences $\{x_j\}_{j=1}^n$ in $\ell^{1/\alpha}$ and $\{y_k\}_{k=1}^m$ in $\ell^{1/\beta}$) is a convex function of (α,β) in the triangle $0 \le \alpha,\beta \le 1$, $\alpha+\beta \ge 1$. Riesz's student Thorin [**520**] extended this triangle to the unit square $0 \le \alpha,\beta \le 1$ and generalized this theorem by replacing the maximum of a bilinear form with the maximum of the modulus of an entire function in many variables. After the end of World War II, Thorin published his thesis [**521**], building the subject and giving a variety of applications. The original proof of Thorin was rather long, but a few years later Tamarkin and Zygmund [**503**] gave a very elegant short proof using the maximum modulus principle in a more efficient way. This theorem is referred to as the Riesz-Thorin interpolation theorem today.

Calderón [**61**] elaborated the complex variables proof of the Riesz-Thorin theorem into a general method of interpolation between Banach spaces. The complex interpolation method can be also be defined for pairs of quasi-Banach spaces although it can reduce to triviality; however, the Riesz-Thorin theorem is true for pairs of L^p spaces (with the "correct" geometric mean constant) for all $0 < p \le \infty$ and also for Lorentz spaces. In this setting duality cannot be used, but a well-developed theory of analytic functions with values in quasi-Banach spaces is crucial. We refer to the articles of Kalton [**287**] and [**288**] for details. Complex interpolation for sublinear maps is also possible; see the article of Calderón and Zygmund [**71**]. Interpolation for analytic families of operators (Theorem 1.3.7) is due to Stein [**468**]. The critical Lemma 1.3.8 used in the proof was previously obtained by Hirschman [**248**].

The fact that nonatomic measure spaces contain subsets of all possible measures is classical. An extension of this result for countably additive vector measures with values in finite-dimensional Banach spaces was obtained by Lyapunov [**350**]; for a proof of this fact, see Diestel and Uhl [**159**, p. 264].

The Aoki-Rolewicz theorem (Exercise 1.4.6) was proved independently by Aoki [**8**] and Rolewicz [**431**]. For a proof of this fact and a variety of its uses in the context of quasi-Banach spaces we refer to the book of Kalton, Peck, and Roberts [**289**].

Decreasing rearrangements of functions were introduced by Hardy and Littlewood [**237**]; the authors attribute their motivation to understanding cricket averages. The $L^{p,q}$ spaces were introduced by Lorentz in [**346**] and in [**347**]. A general treatment of Lorentz spaces is given in the article of Hunt [**261**]. The normability of the spaces $L^{p,q}$ (which holds exactly when $1 < p \le \infty$ and $1 \le q \le \infty$) can be traced back to general principles obtained by Kolmogorov [**310**]. The introduction of the function f^{**} which was used in Exercise 1.4.3 to explicitly define a norm on the normable spaces $L^{p,q}$ is due to Calderón [**61**]. These spaces appear as intermediate spaces in the general interpolation theory of Calderón [**61**] and in that of Lions and Peetre [**338**]. The latter was pointed out by Peetre [**401**]. For a systematic study of the duals of Lorentz spaces we refer to Cwikel [**138**] and Cwikel and Fefferman [**139**], [**140**]. An extension of the Marcinkiewicz interpolation theorem to Lorentz spaces was obtained by Hunt [**260**]. Standard references on interpolation include the books of Bennett and Sharpley [**35**], Berg and Löfström [**36**], Sadosky [**442**], and Chapter 5 in Stein and Weiss [**487**].

There is no extensive literature on multilinear interpolation. The multilinear complex interpolation method (c.f. Exercise 1.4.16) is rather straightforward and is easily obtained by adapting the proof of Theorem 1.3.4 in the setting of many functions; see Zygmund [**572**, p. 106] and Berg and Löfström [**36**]. The multilinear real interpolation method is more involved. References on the subject include (in chronological order) the articles of Strichartz [**491**], Sharpley [**456**] and [**457**], Zafran [**565**], Christ [**92**], Janson [**275**], and Grafakos and Kalton [**213**]. The latter contains, in particular, the proof of Exercise 1.4.17.

CHAPTER 2

Maximal Functions, Fourier Transform, and Distributions

We have already seen that the convolution of a function with a fixed density is a smoothing operation that produces a certain average of the function. Averaging is an important operation in analysis and naturally arises in many situations. The study of averages of functions is better understood and simplified by the introduction of the maximal function. This is defined as the largest average of a function over all balls containing a fixed point. Maximal functions play a key role in differentiation theory, where they are used in obtaining almost everywhere convergence for certain integral averages. The importance of maximal functions lies in the fact that they control crucial quantitative information concerning the given functions, despite their larger size.

Another important operation we study in this chapter is the Fourier transform. This is as fundamental to Fourier analysis as marrow is to the human bone. It is the father of all oscillatory integrals and a powerful transformation that carries a function from its spatial domain to its frequency domain. By doing this, it inverts the function's localization properties. Then magically, if applied one more time, it gives back the function composed with a reflection. More important, it transforms our point of view in harmonic analysis. It changes convolution to multiplication, translation to modulation, and expanding dilation to shrinking dilation, while its decay at infinity encodes information about the local smoothness of the function. The study of the Fourier transform also motivates the launch of a thorough study of general oscillatory integrals. We take a quick look at this topic with emphasis on one-dimensional results.

Distributions changed our view of analysis as they furnished a mathematical framework for many operations that did not exactly qualify to be called functions. These operations found their mathematical place in the world of functionals applied to smooth functions (called test functions). These functionals also introduced the correct interpretation for many physical objects, such as the Dirac delta function. Distributions quickly became an indispensable tool in analysis and brought a broader perspective.

2.1. Maximal Functions

Given a Lebesgue measurable subset A of $\mathbf{R}^n$ we denote by $|A|$ its Lebesgue measure. For $x \in \mathbf{R}^n$ and $r > 0$, we denote by $B(x,r)$ the open ball of radius r centered at x. We will also use the notation $aB(x,\delta) = B(x, a\delta)$, for $a > 0$ for the ball with the same center and radius $a\delta$. Given $\delta > 0$ and $f \in L^1_{\text{loc}}(\mathbf{R}^n)$, a locally integrable function on $\mathbf{R}^n$, let

$$\operatorname*{Avg}_{B(x,\delta)} |f| = \frac{1}{|B(x,\delta)|} \int_{B(x,\delta)} |f(y)|\, dy$$

denote the average of $|f|$ over the ball of radius δ centered at x.

2.1.a. The Hardy-Littlewood Maximal Operator

Definition 2.1.1. The function

$$\mathcal{M}(f)(x) = \sup_{\delta>0} \operatorname*{Avg}_{B(x,\delta)} |f| = \sup_{\delta>0} \frac{1}{v_n \delta^n} \int_{|y|<\delta} |f(x-y)|\, dy$$

is called the *centered Hardy-Littlewood maximal function* of f.

The maximal function is a positive operator since $\mathcal{M}(f) = \mathcal{M}(|f|) \geq 0$. Information concerning cancellation of the function f is lost by passing to $\mathcal{M}(f)$. We will see later that $\mathcal{M}(f)$ pointwise controls f [i.e. $\mathcal{M}(f) \geq |f|$ almost everywhere]. Also note that $\mathcal{M}$ preserves L^∞, that is,

$$\big\|\mathcal{M}(f)\big\|_{L^\infty} \leq \big\|f\big\|_{L^\infty}.$$

Let us compute the Hardy-Littlewood maximal function of a specific function.

Example 2.1.2. On $\mathbf{R}$, let f be the characteristic function of the interval $[a,b]$. For $x \in (a,b)$, clearly $\mathcal{M}(f) = 1$. For $x \geq b$, a simple calculation shows that the largest average of f over all intervals $(x-\delta, x+\delta)$ is obtained when $\delta = x - a$. Similarly, when $x \leq a$, the largest average is obtained when $\delta = b - x$. Therefore,

$$\mathcal{M}(f)(x) = \begin{cases} (b-a)/2|x-b| & \text{when } x \leq a \\ 1 & \text{when } x \in (a,b) \\ (b-a)/2|x-a| & \text{when } x \geq b. \end{cases}$$

Observe that $\mathcal{M}(f)$ has a jump at $x = a$ and $x = b$ equal to one-half that of f.

$\mathcal{M}$ is a sublinear operator and never vanishes. In fact, we have that if $\mathcal{M}(f)(x_0) = 0$ for some $x_0 \in \mathbf{R}^n$, then $f = 0$ a.e. Moreover, if f is compactly supported, say in $|x| \leq R$, then

$$\mathcal{M}(f)(x) \geq \frac{\big\|f\big\|_{L^1}}{v_n} \frac{1}{(|x|+R)^n}, \tag{2.1.1}$$

for $|x| \geq R$, where v_n is the volume of the unit ball in $\mathbf{R}^n$. Equation (2.1.1) implies that $\mathcal{M}(f)$ *is never* in $L^1(\mathbf{R}^n)$ if $f \neq 0$, a strong property that reveals a typical behavior of the maximal function. In fact, if $g \in L^1_{\text{loc}}$ and $\mathcal{M}(g)$ is in $L^1(\mathbf{R}^n)$ then $g = 0$ a.e. To see this use (2.1.1) with $g_R(x) = g(x)\chi_{|x|\leq R}$ to conclude that $g_R(x) = 0$ for almost all x in the ball of radius $R > 0$. Thus $g = 0$ a.e. in $\mathbf{R}^n$. However, as we will show shortly, $\mathcal{M}(f)$ is in $L^{1,\infty}$ when f is in L^1.

A related analogue of $\mathcal{M}(f)$ is its uncentered version $M(f)$ defined as the supremum of all averages of f over all open balls containing a given point.

Definition 2.1.3. The *uncentered Hardy-Littlewood maximal function* of f,

$$M(f)(x) = \sup_{\substack{\delta>0 \\ |y-x|<\delta}} \operatorname*{Avg}_{B(y,\delta)} |f|,$$

is defined as the supremum of the averages of $|f|$ over all open balls $B(y, \delta)$ that contain the point x.

Clearly $\mathcal{M}(f) \leq M(f)$; in other words, M is a larger operator than $\mathcal{M}$. However, the boundedness properties of M are similar to that of $\mathcal{M}$.

Example 2.1.4. On $\mathbf{R}$, let f be the characteristic function of the interval $I = [a, b]$. For $x \in (a, b)$, clearly $M(f) = 1$. For $x > b$, a calculation shows that the largest average of f over all intervals $(x - \delta, x + \delta)$ is obtained when $\delta = x - a$. Similarly, when $x < a$, the largest average is obtained when $\delta = b - x$. We conclude that

$$M(f)(x) = \begin{cases} (b-a)/|x-b| & \text{when } x \leq a \\ 1 & \text{when } x \in (a,b) \\ (b-a)/|x-a| & \text{when } x \geq b. \end{cases}$$

Observe that M does not have a jump at $x = a$ and $x = b$ and that it is comparable to the function $(1 + \frac{\text{dist}(x,I)}{|I|})^{-1}$.

We are now ready to prove some basic properties of the maximal functions. We will need the following simple covering lemma.

Lemma 2.1.5. *Let $\{B_1, B_2, \ldots, B_k\}$ be a finite collection of open balls in $\mathbf{R}^n$. Then there exists a finite subcollection $\{B_{j_1}, \ldots, B_{j_l}\}$ of pairwise disjoint balls such that*

$$\Big|\bigcup_{i=1}^{l} B_{j_i}\Big| \geq 3^{-n}\Big|\bigcup_{i=1}^{k} B_i\Big|. \tag{2.1.2}$$

Proof. Let us reindex the balls so that

$$|B_1| \geq |B_2| \geq \cdots \geq |B_k|.$$

Let $j_1 = 1$. Having chosen $j_1, j_2, \ldots, j_i$, let j_{i+1} be the least index $s > j_i$ such that $\bigcup_{m=1}^{i} B_{j_m}$ is disjoint from B_s. Since we have a finite number of balls, this process will terminate, say after l steps. We have now selected pairwise disjoint balls $B_{j_1}, \dots, B_{j_l}$. If some B_m was not selected, that is, $m \notin \{j_1, \dots, j_l\}$, then B_m must intersect a selected ball B_{j_r} for some $j_r < m$. Then B_m has smaller size than B_{j_r} and we must have $B_m \subseteq 3B_{j_r}$. This shows that the union of the nonselected balls is contained in the union of the triples of the selected balls. Therefore, the union of all balls is contained in the union of the triples of the selected balls. Thus

$$\left|\bigcup_{i=1}^{k} B_i\right| \le \left|\bigcup_{r=1}^{l} 3B_{j_r}\right| \le \sum_{r=1}^{l} |3B_{j_r}| = 3^n \sum_{r=1}^{l} |B_{j_r}| = 3^n \left|\bigcup_{r=1}^{l} B_{j_r}\right|,$$

and the required conclusion follows.

□

We are now ready to prove the main theorem concerning the maximal functions $\mathcal{M}$ and M.

Theorem 2.1.6. *The uncentered Hardy-Littlewood maximal operator M maps $L^1(\mathbf{R}^n)$ into $L^{1,\infty}(\mathbf{R}^n)$ with constant at most 3^n and also $L^p(\mathbf{R}^n)$ into $L^p(\mathbf{R}^n)$ for $1 < p < \infty$ with constant at most $3^{n/p}p(p-1)^{-1}$. The same is true for the centered operator $\mathcal{M}$.*

We note that operators that map L^1 into $L^{1,\infty}$ are called of *weak type* $(1,1)$.

Proof. Since $M(f) \ge \mathcal{M}(f)$, we have

$$\{x \in \mathbf{R}^n : |\mathcal{M}(f)(x)| > \alpha\} \subseteq \{x \in \mathbf{R}^n : |M(f)(x)| > \alpha\},$$

and therefore it suffices to show that

$$|\{x \in \mathbf{R}^n : |M(f)(x)| > \alpha\}| \le 3^n \frac{\|f\|_{L^1}}{\alpha}. \tag{2.1.3}$$

Since $M(f)$ is a supremum of continuous functions, it follows that $M(f)$ is lower semicontinuous and therefore the set

$$E_\alpha = \{x \in \mathbf{R}^n : |M(f)(x)| > \alpha\}$$

is open. Let K be a compact subset of E_α. For each $x \in K$ there exists an open ball B_x containing the point x such that

$$\int_{B_x} |f(y)|\, dy > \alpha |B_x|. \tag{2.1.4}$$

By compactness there exists a finite subcover $\{B_{x_1}, \dots, B_{x_k}\}$ of K. Using Lemma 2.1.5 we find a subcollection of pairwise disjoint balls $B_{x_{j_1}}, \dots, B_{x_{j_l}}$ such that (2.1.2) holds. Then using (2.1.4), (2.1.2), and the disjointness of the $B_{x_{j_i}}$'s, we obtain

$$|K| \le \sum_{i=1}^{k} |B_{x_i}| \le 3^n \sum_{i=1}^{l} |B_{x_{j_i}}| \le \frac{3^n}{\alpha} \sum_{i=1}^{l} \int_{B_{x_{j_i}}} |f(y)|\, dy \le \frac{3^n}{\alpha} \int_{\mathbf{R}^n} |f(y)|\, dy.$$

Taking the supremum over all compact $K \subseteq E_\alpha$, we deduce (2.1.3). We have now proved that M maps $L^1 \to L^{1,\infty}$ with constant 3^n. It is a trivial fact that M maps $L^\infty \to L^\infty$ with constant 1. Recall that M is well defined on L^1_{loc}, which contains $L^1 + L^\infty$. The Marcinkiewicz interpolation theorem 1.3.2 implies that M maps $L^p(\mathbf{R}^n)$ into $L^p(\mathbf{R}^n)$ for all $1 < p < \infty$. Using Exercise 1.3.3, we obtain the following estimate for the operator norm of M on $L^p(\mathbf{R}^n)$:

$$\big\|M\big\|_{L^p \to L^p} \le \frac{p\,3^{\frac{n}{p}}}{p-1}\,. \tag{2.1.5}$$

Observe that a direct application of Theorem 1.3.2 would give the slightly worse bound of $2\big(\frac{p}{p-1}\big)^{\frac{1}{p}} 3^{\frac{n}{p}}$.

□

Remark 2.1.7. The previous proof gives a bound on the operator norm of M on $L^p(\mathbf{R}^n)$ that grows exponentially with the dimension. We may wonder if this bound can be improved to a better bound that does not grow exponentially as $n \to \infty$. This is not possible; see Exercise 2.1.8.

Example 2.1.8. Let $R > 0$. Then there are dimensional constants c_n and c'_n so that

$$\frac{c'_n\, R^n}{(|x| + R)^n} \le M(\chi_{B(0,R)})(x) \le \frac{c_n\, R^n}{(|x| + R)^n}\,. \tag{2.1.6}$$

As these functions are not integrable over $\mathbf{R}^n$, it follows that M does not map $L^1(\mathbf{R}^n)$ into $L^1(\mathbf{R}^n)$.

Next we estimate $M(M(\chi_{B(0,R)}))(x)$. First we write

$$\frac{R^n}{(|x| + R)^n} \le \chi_{B(0,R)} + \sum_{k=0}^{\infty} \frac{R^n}{(R + 2^k R)^n} \chi_{B(0,2^{k+1}R)\setminus B(0,2^k R)}\,.$$

Using the upper estimate in (2.1.6) and the sublinearity of M, we obtain

$$\begin{aligned}
&M\bigg(\frac{R^n}{(|\cdot| + R)^n}\bigg)(x) \\
&\le M(\chi_{B(0,R)})(x) + \sum_{k=0}^{\infty} \frac{1}{(1 + 2^k)^n} M(\chi_{B(0,2^{k+1}R)})(x) \\
&\le \frac{c_n\, R^n}{(|x| + R)^n} + \sum_{k=0}^{\infty} \frac{1}{(1 + 2^k)^n} \frac{c_n\,(2^{k+1}R)^n}{(|x| + 2^{k+1}R)^n} \\
&\le \frac{C_n \log(1 + |x|/R)}{(1 + |x|/R)^n}\,,
\end{aligned}$$

where the last estimate follows by summing separately over k satisfying $2^{k+1} \le |x|/R$ and $2^{k+1} \ge |x|/R$. Note that the presence of the logarithm does not affect the L^p boundedness of this function when $p > 1$.

2.1.b. Control of Other Maximal Operators

We will now study some properties of the Hardy-Littlewood maximal function. We begin with a notational definition that we plan to use throughout this book.

Definition 2.1.9. Given a function g on $\mathbf{R}^n$ and $\varepsilon > 0$, we denote by g_ε the following function:

$$g_\varepsilon(x) = \varepsilon^{-n} g(\varepsilon^{-1} x). \tag{2.1.7}$$

As observed in Example 1.2.16, if g is an integrable function with integral equal to 1, then the family defined by (2.1.7) is an approximate identity. Therefore, convolution with g_ε is an averaging operation. The Hardy-Littlewood maximal function $\mathcal{M}(f)$ is obtained as the supremum of the averages of a function f with respect to the dilates of the kernel $k = v_n^{-1} \chi_{B(0,1)}$ in $\mathbf{R}^n$; here v_n is the volume of the unit ball $B(0,1)$. Indeed, we have

$$\mathcal{M}(f)(x) = \sup_{\varepsilon>0} \frac{1}{v_n \varepsilon^n} \int_{\mathbf{R}^n} |f(x-y)| \, \chi_{B(0,1)}\Big(\frac{y}{\varepsilon}\Big) \, dy = \sup_{\varepsilon>0} |f| * k_\varepsilon \, . \tag{2.1.8}$$

Note that the function $k = v_n^{-1} \chi_{B(0,1)}$ has integral equal to 1 and the operation given by convolution with k_ε is indeed an avegaring operation.

It turns out that the Hardy-Littlewood maximal function controls the averages of a function with respect to any radially decreasing L^1 function. Recall that a function f on $\mathbf{R}^n$ is called *radial* if $f(x) = f(y)$ whenever $|x| = |y|$. Note that a radial function f on $\mathbf{R}^n$ has the form $f(x) = \varphi(|x|)$ for some function φ on $\mathbf{R}^+$. We have the following result.

Theorem 2.1.10. *Let $k \geq 0$ be a function on $[0, \infty)$ which is continuous except at a finite number of points. Suppose that $K(x) = k(|x|)$ is an integrable function on $\mathbf{R}^n$, which satisfies*

$$K(x) \geq K(y), \quad \textit{whenever } |x| \leq |y| \, ; \tag{2.1.9}$$

(i.e. k is decreasing.) Then the following estimate is true:

$$\sup_{\varepsilon>0} (|f| * K_\varepsilon)(x) \leq \big\|K\big\|_{L^1} \mathcal{M}(f)(x) \tag{2.1.10}$$

for all locally integrable functions f on $\mathbf{R}^n$.

Proof. We prove (2.1.10) when K is radial, satisfies (2.1.9), and is compactly supported and continuous. When this case is established, select a sequence K_j of radial, compactly supported, continuous functions that increase to K as $j \to \infty$. This is possible since the function k is continuous except at a finite number of points. If (2.1.10) holds for each K_j, passing to the limit implies that (2.1.10) also holds for K. Next, we observe that it suffices to prove (2.1.10) for $x = 0$. When this case is established, replacing $f(t)$ by $f(t + x)$ implies that (2.1.10) holds for all x.

Let us now fix a radial, compactly supported (say in $|x| \le R$), and continuous function K satisfying (2.1.9). Also fix an $f \in L^1_{\mathrm{loc}}$ and take $x = 0$. Let e_1 be the vector $(1, 0, 0, \dots, 0)$ on the unit sphere $\mathbf{S}^{n-1}$. Polar coordinates give

$$\int_{\mathbf{R}^n} |f(y)| K_\varepsilon(-y)\, dy = \int_0^\infty \int_{\mathbf{S}^{n-1}} |f(r\theta)| K_\varepsilon(re_1) r^{n-1}\, d\theta\, dr. \tag{2.1.11}$$

Set

$$F(r) = \int_{\mathbf{S}^{n-1}} |f(r\theta)|\, d\theta \qquad \text{and} \qquad G(r) = \int_0^r F(s) s^{n-1}\, ds,$$

where $d\theta$ denotes surface measure on $\mathbf{S}^{n-1}$. Using these functions, (2.1.11), and integration by parts, we obtain

$$\begin{aligned} &\int_{\mathbf{R}^n} |f(y)| K_\varepsilon(y)\, dy \\ &= \int_0^\infty F(r) r^{n-1} K_\varepsilon(re_1)\, dr \\ &= G(R) K_\varepsilon(Re_1) - G(0) K_\varepsilon(0) - \int_0^\infty G(r)\, dK_\varepsilon(re_1) \\ &= \int_0^\infty G(r)\, d(-K_\varepsilon(re_1)), \end{aligned} \tag{2.1.12}$$

where two of the integrals are of Lebesgue-Stieltjes type and we used our assumptions that $G(0) = 0$, $K_\varepsilon(0) < \infty$, $G(R) < \infty$, and $K_\varepsilon(Re_1) = 0$. Let v_n be the volume of the unit ball in $\mathbf{R}^n$. Since

$$G(r) = \int_0^r F(s) s^{n-1}\, ds = \int_{|y| \le r} |f(y)|\, dy \le \mathcal{M}(f)(0) v_n r^n,$$

it follows that the last integral in (2.1.12) is dominated by

$$\begin{aligned} &\mathcal{M}(f)(0) v_n \int_0^\infty r^n d(-K_\varepsilon(re_1)) \\ &= \mathcal{M}(f)(0) \int_0^\infty n v_n r^{n-1} K_\varepsilon(re_1)\, dr = \mathcal{M}(f)(0) \big\|K\big\|_{L^1}. \end{aligned}$$

Here we used integration by parts and the fact that the surface measure of the unit sphere $\mathbf{S}^{n-1}$ is equal to nv_n. See Appendix A.3. The theorem is now proved.

□

Remark 2.1.11. Theorem 2.1.10 can be generalized as follows. If K is an L^1 function on $\mathbf{R}^n$ whose absolute value is bounded above by some continuous integrable radial function K_0 that satisfies (2.1.9), then (2.1.10) holds where $\big\|K\big\|_{L^1}$ is replaced by $\big\|K_0\big\|_{L^1}$. Such a K_0 is called a *radial decreasing majorant* of K. This observation can be stated as follows:

Corollary 2.1.12. *If a function φ has an integrable radially decreasing majorant Φ, then the estimate*

$$\sup_{t>0} |(f * \varphi_t)(x)| \le \|\Phi\|_{L^1} \mathcal{M}(f)(x)$$

is valid for all f locally integrable functions on $\mathbf{R}^n$.

Example 2.1.13. Let

$$P(x) = \frac{c_n}{(1+|x|^2)^{\frac{n+1}{2}}},$$

where c_n is a constant so that

$$\int_{\mathbf{R}^n} P(x)\,dx = 1.$$

The function P is called the *Poisson kernel.* We define L^1 dilates P_t of the Poisson kernel by setting

$$P_t(x) = t^{-n} P(t^{-1}x)$$

for $t > 0$. It is straightforward to verify that when $n \ge 2$

$$\frac{d^2}{dt^2} P_t + \sum_{j=1}^{n} \partial_j^2 P_t = 0,$$

that is, $P_t(x_1, \dots, x_n)$ is a *harmonic function* of the variables $(x_1, \dots, x_n, t)$. Therefore, for $f \in L^p(\mathbf{R}^n)$, $1 \le p < \infty$, the function

$$u(x,t) = (f * P_t)(x)$$

is harmonic in $\mathbf{R}^{n+1}_+$ and converges to $f(x)$ in $L^p(dx)$ as $t \to 0$ since $\{P_t\}_{t>0}$ is an approximate identity. If we knew that $f * P_t$ converged to f a.e. as $t \to 0$, then we could say that $u(x,t)$ solves the *Dirichlet problem*

$$\begin{aligned} \sum_{j=1}^{n+1} \partial_j^2 u &= 0 \qquad \text{on } \mathbf{R}^{n+1}_+, \\ u(x,0) &= f(x) \qquad \text{a.e. on } \mathbf{R}^n. \end{aligned} \tag{2.1.13}$$

Solving the Dirichlet problem (2.1.13) motivates the study of the almost everywhere convergence of the expressions $f * P_t$. This will be discussed in the next subsection.

Let us now compute the value of the constant c_n. Denote by ω_{n-1} the surface area of $\mathbf{S}^{n-1}$. Using polar coordinates, we obtain

$$\begin{aligned}
\frac{1}{c_n} &= \int_{\mathbf{R}^n} \frac{dx}{(1+|x|^2)^{\frac{n+1}{2}}} \\
&= \omega_{n-1} \int_0^\infty \frac{r^{n-1}}{(1+r^2)^{\frac{n+1}{2}}} dr \\
&= \omega_{n-1} \int_0^{\pi/2} (\sin\varphi)^{n-1}\, d\varphi \qquad\qquad (r = \tan\varphi) \\
&= \frac{2\pi^{\frac{n}{2}}}{n\Gamma(\frac{n}{2})} \frac{1}{2} \frac{\Gamma(\frac{n}{2})\Gamma(\frac{1}{2})}{\Gamma(\frac{n+1}{2})} \\
&= \frac{\pi^{\frac{n+1}{2}}}{\Gamma(\frac{n+1}{2})},
\end{aligned}$$

where we used the formula for ω_{n-1} in Appendix A3 and the identity in Appendix A.4. We conclude that

$$c_n = \frac{\Gamma(\frac{n+1}{2})}{\pi^{\frac{n+1}{2}}}$$

and that the Poisson kernel on $\mathbf{R}^n$ is given by

$$P(x) = \frac{\Gamma(\frac{n+1}{2})}{\pi^{\frac{n+1}{2}}} \frac{1}{(1+|x|^2)^{\frac{n+1}{2}}}. \tag{2.1.14}$$

Theorem 2.1.10 implies that the solution of the Dirichlet problem (2.1.13) is pointwise bounded by the Hardy-Littlewood maximal function of f.

2.1.c. Applications to Differentiation Theory

We continue this section by obtaining some applications of the boundedness of the Hardy-Littlewood maximal function in differentiation theory.

We will now show that the weak type $(1,1)$ property of the Hardy-Littlewood maximal function implies almost everywhere convergence for a variety of families of functions. We deduce this from the more general fact that a certain weak type property for the supremum of a family of linear operators implies almost everywhere convergence.

Here is our setup. Let (X,μ), (Y,ν) be measure spaces and let $0 < p \le \infty$, $0 < q < \infty$. Suppose that D is a dense subspace of $L^p(X,\mu)$. This means that for all $f \in L^p$ and all $\delta > 0$ there exists a $g \in D$ such that $\big\|f - g\big\|_{L^p} < \delta$. Suppose that for every $\varepsilon > 0$, T_ε is a linear operator defined on $L^p(X,\mu)$ with values in the set of measurable functions on Y. Define a sublinear operator

$$T_*(f)(x) = \sup_{\varepsilon>0} |T_\varepsilon(f)(x)|. \tag{2.1.15}$$

We have the following.

Theorem 2.1.14. *Let $0 < p \le \infty$, $0 < q < \infty$, and T_ε and T_* as previously. Suppose that for some $B > 0$ and all $f \in L^p(X)$ we have*

$$\big\|T_*(f)\big\|_{L^{q,\infty}} \le B\big\|f\big\|_{L^p} \tag{2.1.16}$$

and that for all $f \in D$

$$\lim_{\varepsilon\to 0} T_\varepsilon(f) = T(f) \tag{2.1.17}$$

exists and it is finite ν-a.e. (and defines a linear operator on D). Then for all $f \in L^p(X,\mu)$ the limit (2.1.17) exists and is finite ν-a.e. and defines a linear operator T on $L^p(X)$ (uniquely extending T defined on D) that satisfies

$$\big\|T(f)\big\|_{L^{q,\infty}} \le B\big\|f\big\|_{L^p}\,. \tag{2.1.18}$$

Proof. Given f in L^p, we define the *oscillation* of f

$$O_f(y) = \limsup_{\varepsilon\to 0}\limsup_{\theta\to 0} |T_\varepsilon(f)(y) - T_\theta(f)(y)|.$$

We would like to show that for all $f \in L^p$ and $\delta > 0$,

$$\nu(\{y \in Y:\ O_f(y) > \delta\}) = 0. \tag{2.1.19}$$

Once (2.1.19) is established, given $f \in L^p(X)$, we obtain that $O_f(y) = 0$ for ν-almost all y, which implies that $T_\varepsilon(f)(y)$ is Cauchy for ν-almost all y and it therefore converges ν-a.e. to some $T(f)(y)$ as $\varepsilon \to 0$. The operator T defined this way on $L^p(X)$ is linear and extends T defined on D.

To approximate O_f we use density. Given $\eta > 0$, find a $g \in D$ such that $\big\|f-g\big\|_{L^p} < \eta$. Since $T_\varepsilon(g) \to T(g)$ ν-a.e, it follows that $O_g = 0$ ν-a.e. Using this fact and the linearity of the T_ε's, we conclude that

$$O_f(y) \le O_g(y) + O_{f-g}(y) = O_{f-g}(y) \qquad \nu\text{-a.e.}$$

Now for any $\delta > 0$ we have

$$\begin{aligned}\nu(\{y \in Y:\ O_f(y) > \delta\}) &\le \nu(\{y \in Y:\ O_{f-g}(y) > \delta\})\\ &\le \nu(\{y \in Y:\ 2T_*(f-g)(y) > \delta\})\\ &\le \big(2B\big\|f-g\big\|_{L^p}/\delta\big)^q\\ &\le (2B\eta/\delta)^q\,.\end{aligned}$$

Letting $\eta \to 0$, we deduce (2.1.19). We conclude that $T_\varepsilon(f)$ is a Cauchy sequence and hence it converges ν-a.e. to some $T(f)$. Since $|T(f)| \le |T_*(f)|$, the conclusion (2.1.18) of the theorem follows easily.

□

We now derive some applications. First we return to the issue of almost everywhere convergence of the expressions $f * P_y$, where P is the Poisson kernel.

Example 2.1.15. Fix $1 \le p < \infty$ and $f \in L^p(\mathbf{R}^n)$. Let

$$P(x) = \frac{\Gamma(\frac{n+1}{2})}{\pi^{\frac{n+1}{2}}} \frac{1}{(1+|x|^2)^{\frac{n+1}{2}}}$$

be the Poisson kernel on $\mathbf{R}^n$ and let $P_\varepsilon(x) = \varepsilon^{-n}P(\varepsilon^{-1}x)$. We will deduce from the previous theorem that the family $f * P_\varepsilon$ converges to f a.e. Let D be the set of all continuous functions with compact support on $\mathbf{R}^n$. Since the family $(P_\varepsilon)_{\varepsilon>0}$ is an approximate identity, Theorem 1.2.19 (2) implies that for f in D we have that $f * P_\varepsilon \to f$ uniformly on compact subsets of $\mathbf{R}^n$ and hence a.e. In view of Theorem 2.1.10, the supremum of the family of linear operators $T_\varepsilon(f) = f * P_\varepsilon$ is controlled by the Hardy-Littlewood maximal function and thus it maps L^p to $L^{p,\infty}$ for $1 \le p < \infty$. Theorem 2.1.14 now gives that $f * P_\varepsilon$ converges to f a.e. for all $f \in L^p$.

Here is another application of Theorem 2.1.14. We refer to Exercise 2.1.10 for others.

Corollary 2.1.16. ***(Lebesgue's differentiation theorem)*** *For any locally integrable function f on $\mathbf{R}^n$ we have*

$$\lim_{r\to 0} \frac{1}{|B(x,r)|} \int_{B(x,r)} f(y)\,dy = f(x) \tag{2.1.20}$$

for almost all x in $\mathbf{R}^n$. Consequently we have $|f| \le \mathcal{M}(f)$ a.e.

Proof. Since $\mathbf{R}^n$ is the union of the balls $B(0,N)$ for $N = 1,2,3\dots$, it suffices to prove the required conclusion for almost all x inside the ball $B(0,N)$. Then we may take f supported in a larger ball, thus working with f integrable over the whole space. Let T_ε be the operator given with convolution with k_ε, where $k = v_n^{-1}\chi_{B(0,1)}$. We know that the corresponding maximal operator T_* is controlled by the the centered Hardy-Littlewood maximal function $\mathcal{M}$, which maps L^1 to $L^{1,\infty}$. It is straightforward to verify that (2.1.20) holds for all continuous functions f with compact support. Since the set of these functions is dense in L^1, and T_* maps L^1 to $L^{1,\infty}$, Theorem 2.1.14 implies that (2.1.20) holds for a general f in L^1. □

The following corollaries were inspired by Example 2.1.15.

Corollary 2.1.17. ***(Differentiation theorem for approximate identities)*** *Let K be an L^1 function on $\mathbf{R}^n$ with integral 1 that has a continuous integrable radially decreasing majorant. Then $f*K_\varepsilon \to f$ a.e. as $\varepsilon \to 0$ for all $f \in L^p(\mathbf{R}^n)$, $1 \le p < \infty$.*

Proof. It follows from Example 1.2.16 that K_ε is an approximate identity. Theorem 1.2.19 now implies that $f * K_\varepsilon \to f$ uniformly on compact sets when f is continuous. Let D be the space of all continuous functions with compact support. Then $f * K_\varepsilon \to f$ a.e. for $f \in D$. It follows from Corollary 2.1.12 that

$T_*(f) = \sup_{\varepsilon>0} |f * K_\varepsilon|$ maps L^p to $L^{p,\infty}$ for $1 \le p < \infty$. Using Theorem 2.1.14, we conclude the proof of the corollary.

□

Remark 2.1.18. Fix $f \in L^p(\mathbf{R}^n)$ for some $1 \le p < \infty$. Theorem 1.2.19 implies that $f * K_\varepsilon$ converges to f in L^p and hence some subsequence $f * K_{\varepsilon_n}$ of $f * K_\varepsilon$ converges to f a.e. as $n \to \infty$, $(\varepsilon_n \to 0)$. Compare this result with Corollary 2.1.17, which gives a.e. convergence for the whole family $f * K_\varepsilon$ as $\varepsilon \to 0$.

Corollary 2.1.19. *(**Differentiation theorem for multiples of approximate identities**) Let K be a function on $\mathbf{R}^n$ that has an integrable radially decreasing majorant. Let $a = \int_{\mathbf{R}^n} K(x)\,dx$. Then for all $f \in L^p(\mathbf{R}^n)$ and $1 \le p < \infty$, $(f * K_\varepsilon)(x) \to af(x)$ for almost all $x \in \mathbf{R}^n$ as $\varepsilon \to 0$.*

Proof. Use Theorem 1.2.21 instead of Theorem 1.2.19 in the proof of Corollary 2.1.17.

□

The following application of the Lebesgue differentiation theorem uses a simple *stopping time argument* argument. This is the sort of argument in which a selection procedure stops when it is exhausted at a certain scale and is then repeated at the next scale. A certain refinement of the following proposition will be of fundamental importance in the study of singular integrals given in Chapter 4.

Proposition 2.1.20. *Given a nonnegative integrable function f on $\mathbf{R}^n$ and $\alpha > 0$, there exist disjoint open cubes Q_j such that for almost all $x \in \left(\bigcup_j Q_j\right)^c$ we have $f(x) \le \alpha$ and*

$$\alpha < \frac{1}{|Q_j|} \int_{Q_j} f(t)\,dt \le 2^n \alpha. \tag{2.1.21}$$

Proof. The proof provides an excellent paradigm of a stopping time argument. Start by decomposing $\mathbf{R}^n$ as a union of cubes of equal size, whose interiors are disjoint, and whose diameter is so large that $|Q|^{-1}\int_Q f(x)\,dx \le \alpha$ for every Q in this mesh. This is possible since f is integrable and $|Q|^{-1}\int_Q f(x)\,dx \to 0$ as $|Q| \to \infty$. Call the union of these cubes $\mathcal{E}_0$.

Divide each cube in the mesh into 2^n congruent cubes by bisecting each of the sides. Call the new collection of cubes $\mathcal{E}_1$. Select a cube Q in $\mathcal{E}_1$ if

$$\frac{1}{|Q|} \int_Q f(x)\,dx > \alpha \tag{2.1.22}$$

and call the set of all selected cubes $\mathcal{S}_1$. Now subdivide each cube in $\mathcal{E}_1 \setminus \mathcal{S}_1$ into a 2^n congruent cubes by bisecting each of the sides as before. Call this new collection of cubes $\mathcal{E}_2$. Repeat the same procedure and select a family of cubes $\mathcal{S}_2$ that satisfy (2.1.22). Continue this way ad infinitum and call the cubes in $\bigcup_{m=1}^\infty \mathcal{S}_m$ "selected."

If Q was selected, then there exists Q_1 in $\mathcal{E}_{m-1}$ containing Q that was not selected at the $(m-1)^{\text{th}}$ step for some $m \geq 1$. Therefore,

$$\alpha < \frac{1}{|Q|}\int_Q f(x)\,dx \leq 2^n \frac{1}{|Q_1|}\int_{Q_1} f(x)\,dx \leq 2^n\alpha.$$

Now call F the closure of complement of the union of all selected cubes. If $x \in F$, then there exists a sequence of cubes containing x whose diameter shrinks down to zero such that the average of f over these cubes is less than or equal to α. By Corollary 2.1.16, it follows that $f(x) \leq \alpha$ almost everywhere in F. This proves the proposition.

□

In the proof of Proposition 2.1.20 it was not crucial to assume that f was defined on all $\mathbf{R}^n$ but on a cube. We now give a local version of this result.

Corollary 2.1.21. *Let $f \geq 0$ be an integrable function over a cube Q in $\mathbf{R}^n$ and let $\alpha \geq \frac{1}{|Q|}\int_Q f\,dx$. Then there exist disjoint open subcubes Q_j of Q such that for almost all $x \in Q \setminus \bigcup_j Q_j$ we have $f \leq \alpha$ and (2.1.21) holds for all j.*

Proof. This easily follows by a simple modification of Proposition 2.1.20 in which $\mathbf{R}^n$ is replaced by the fixed cube Q.

□

See Exercise 2.1.4 for an application of Proposition 2.1.20 involving maximal functions.

Exercises

2.1.1. Let μ be a positive measure on $\mathbf{R}^n$ that satisfies the following *doubling condition*: There exists a constant $D(\mu) > 0$ such that for all $x \in \mathbf{R}^n$ and $r > 0$ we have

$$\mu(3B(x,r)) \leq D(\mu)\,\mu(B(x,r)).$$

Prove that the maximal functions $\mathcal{M}_\mu$ and M_μ defined by

$$\mathcal{M}_\mu(f)(x) = \sup_{r>0} \frac{1}{\mu(B(x,r))}\int_{B(x,r)} f(y)\,d\mu(y)$$

$$M_\mu(f)(x) = \sup_{\substack{r>0\\|z-x|\leq r}} \frac{1}{\mu(B(z,r))}\int_{B(z,r)} f(y)\,d\mu(y)$$

map $L^1(\mathbf{R}^n,\mu)$ into $L^{1,\infty}(\mathbf{R}^n,\mu)$ with constant at most $D(\mu)$ and $L^p(\mathbf{R}^n,\mu)$ into itself with constant at most $2(\frac{p}{p-1})^{\frac{1}{p}}D(\mu)^{\frac{1}{p}}$. Prove that the same conclusions hold if the balls are replaced by cubes in the definitions of the doubling measure μ and of the maximal operators $\mathcal{M}_\mu$ and M_μ. Obtain as a consequence a differentiation theorem analogous to Corollary 2.1.16 for

doubling measures.
[*Hint:* Use Theorem 1.3.2.]

2.1.2. On $\mathbf{R}$ consider the maximal function M_μ of Exercise 2.1.1.
(a) (*W. H. Young*) Prove the following covering lemma. Given a finite set $\mathcal{F}$ of open intervals in $\mathbf{R}$, prove that there exist two subfamilies each consisting of pairwise disjoint intervals such that the union of the intervals in the original family is equal to the union of the intervals of both subfamilies. Use this result to show that the maximal function M_μ of Exercise 2.1.1 maps $L^1(\mu) \to L^{1,\infty}(\mu)$ with constant at most 2.
(b) (*Grafakos and Kinnunen* [**216**]) Prove that for any positive measure μ on $\mathbf{R}^n$, $\alpha > 0$, and $f \in L^1_{\text{loc}}(\mathbf{R}^n, \mu)$ we have

$$\frac{1}{\alpha}\int_A |f|\,d\mu - \mu(A) \le \frac{1}{\alpha}\int_{\{|f|>\alpha\}} |f|\,d\mu - \mu(\{|f|>\alpha\}).$$

Use this result and part (a) to prove that for all $\alpha > 0$ and all locally integrable f we have

$$\mu(\{|f|>\alpha\}) + \mu(\{M_\mu(f)>\alpha\}) \le \frac{1}{\alpha}\int_{\{|f|>\alpha\}} |f|\,d\mu + \frac{1}{\alpha}\int_{\{M_\mu(f)>\alpha\}} |f|\,d\mu.$$

(c) Conclude that M_μ maps $L^p(\mu)$ to $L^p(\mu)$, $1 < p < \infty$ with bound at most the unique positive solution A_p of the equation

$$(p-1)x^p - px^{p-1} - 1 = 0.$$

(d) (*Grafakos and Montgomery-Smith* [**220**]) Prove that

$$\big\|M\big\|_{L^p \to L^p} = A_p\,,$$

where A_p is the unique positive solution of the equation in part (c).
[*Hint:* Part (a): Select a subset $\mathcal{G}$ of $\mathcal{F}$ with minimal cardinality such that $\bigcup_{J\in\mathcal{G}} J = \bigcup_{I\in\mathcal{F}} I$. Part (d): One direction follows from part (c). Conversely, note that $M(|x|^{-1/p}) = A_p|x|^{-1/p}$. As this function is not in L^p, consider the family $|x|^{-1/p}\min(|x|^{-\varepsilon}, |x|^{\varepsilon})$ as $\varepsilon \downarrow 0$. Finally, prove that equality is obtained in the second inequality in (b) when $\alpha = 1$ and $f(x) = |x|^{-1/p}$.]

2.1.3. Define the centered Hardy-Littlewood maximal function $\mathcal{M}_c$ using cubes instead of balls in $\mathbf{R}^n$. Also define the uncentered Hardy-Littlewood maximal function M_c using cubes in $\mathbf{R}^n$. Prove that

$$(v_n\sqrt{n}/2)^n \le \frac{M(f)}{M_c(f)} \le (v_n/2)^n\,, \quad (v_n\sqrt{n}/2)^n \le \frac{\mathcal{M}(f)}{\mathcal{M}_c(f)} \le (v_n/2)^n,$$

where v_n is the volume of the unit ball in $\mathbf{R}^n$. Conclude that $\mathcal{M}_c$ and M_c are of weak type (1,1) and they map $L^p(\mathbf{R}^n)$ into itself for $1 < p \le \infty$.

2.1.4. (a) Prove the following estimate:

$$|\{x \in \mathbf{R}^n : \; M(f)(x) > 2\alpha\}| \le \frac{3^n}{\alpha} \int_{\{|f|>\alpha\}} |f(y)|\, dy$$

and conclude that M maps L^p into $L^{p,\infty}$ with norm at most $2 \cdot 3^{n/p}$ for $1 \le p < \infty$.

(b) (*Stein* [**474**]) Apply Proposition 2.1.20 to $|f|$ and $\alpha > 0$ to prove that

$$|\{x \in \mathbf{R}^n : \; M_c(f)(x) > \alpha\}| \ge \frac{2^{-n}}{\alpha} \int_{\{|f|>\alpha\}} |f(y)|\, dy.$$

Then use the previous Exercise to conclude that

$$|\{x \in \mathbf{R}^n : \; M(f)(x) > (\tfrac{\sqrt{n}\, v_n}{2})^n \alpha\}| \ge \frac{2^{-n}}{\alpha} \int_{\{|f|>\alpha\}} |f(y)|\, dy.$$

(c) (*Stein* [**474**]) Suppose that f is integrable and supported in a ball B. Show that $M(f)$ is integrable over B if and only if

$$\int_B |f(x)| \log(1 + |f(x)|)\, dx < \infty.$$

[*Hint:* Part (a): Write $f = f\chi_{|f|>\alpha} + f\chi_{|f|\le\alpha}$.]

2.1.5. (*A. Kolmogorov*) Let S be a sublinear operator that maps $L^1(\mathbf{R}^n)$ into $L^{1,\infty}(\mathbf{R}^n)$ with norm B. Suppose that $f \in L^1(\mathbf{R}^n)$. Prove that for any set A of finite Lebesgue measure and for all $0 < q < 1$ we have

$$\int_A |S(f)(x)|^q\, dx \le (1-q)^{-1} B^q |A|^{1-q} \|f\|_{L^1}^q,$$

and in particular for the Hardy-Littlewood maximal operator

$$\int_A M(f)(x)^q\, dx \le (1-q)^{-1} 3^{nq} |A|^{1-q} \|f\|_{L^1}^q\,.$$

[*Hint:* Use the identity

$$\int_A |S(f)(x)|^q\, dx = \int_0^\infty q\alpha^{q-1} |\{x \in A : \; S(f)(x) > \alpha\}|\, d\alpha$$

and estimate the last measure by $\min(|A|, \frac{B}{\alpha}\|f\|_{L^1})$.]

2.1.6. Let $M_s(f)(x)$ be the supremum of the averages of $|f|$ over all rectangles with sides parallel to the axes containing x. M_s is called the *strong maximal function*.

(a) Prove that M_s maps $L^p(\mathbf{R}^n)$ into itself.

(b) Show that the operator norm of M_s is A_p^n, where A_p is as in Exercise 2.1.2(c).

(c) Prove that M_s is not of weak type (1,1).

2.1.7. Prove that if

$$|\varphi(x_1,\dots,x_n)| \le A(1+|x_1|)^{-1-\varepsilon}\dots(1+|x_n|)^{-1-\varepsilon}$$

for some $A,\varepsilon>0$, and $\varphi_{t_1,\dots,t_n}(x) = t_1^{-1}\dots t_n^{-1}\varphi(t_1^{-1}x_1,\dots t_n^{-1}x_n)$, then the maximal operator

$$f \to \sup_{t_1,\dots,t_n>0} |f*\varphi_{t_1,\dots,t_n}|$$

is pointwise controlled by the strong maximal function of the previous Exercise.

2.1.8. Prove that for any fixed $1<p<\infty$, the operator norm of M on $L^p(\mathbf{R}^n)$ tends to infinity as $n\to\infty$.
[*Hint:* Let f_0 be the characteristic function of the unit ball in $\mathbf{R}^n$. Consider $\operatorname*{Avg}_{B_x} f_0$, where $B_x = B\big(\frac{1}{2}(|x|-|x|^{-1})\frac{x}{|x|},\frac{1}{2}(|x|+|x|^{-1})\big)$ for $|x|>1$.]

2.1.9. (a) In $\mathbf{R}^2$ let $M_0(f)(x)$ be the maximal function obtained by taking the supremum of the averages of $|f|$ over all rectangles (of arbitrary orientation) containing x. Prove that M_0 is not bounded on $L^p(\mathbf{R}^n)$ for $p<2$ and conclude that M_0 is not of weak type $(1,1)$.
(b) Let $M_{00}(f)(x)$ be the maximal function obtained by taking the supremum of the averages of $|f|$ over all rectangles in $\mathbf{R}^2$ of arbitrary orientation but fixed eccentricity containing x. (The eccentricity of a rectangle is the ratio of its longest size over its shortest size.) Using a covering lemma, show that M_{00} is of weak type $(1,1)$ with a bound proportional to the square of the eccentricity.
(c) On $\mathbf{R}^n$ define a maximal function by taking the supremum of the averages of $|f|$ over all products of intervals $I_1\times\dots\times I_n$ containing a point x with $|I_2| = a_2|I_1|,\dots,|I_n| = a_n|I_1|$ and $a_2,\dots,a_n>0$ fixed. Show that this maximal function is of weak type $(1,1)$ with bound independent of the numbers $a_2,\dots,a_n$.
[*Hint:* Part (b): Let b be the eccentricity. If two rectangles with the same eccentricity intersect, then the smaller one is contained in the bigger one scaled $4b$ times. Then use an argument similar to that in Lemma 2.1.5.]

2.1.10. (a) Let p,q,X,Y be as in Theorem 2.1.14. Assume that T_ε is a family of quasi-linear operators defined on $L^p(X)$ [i.e. $|T_\varepsilon(f+g)| \le K(|T_\varepsilon(f)| + |T_\varepsilon(g)|)$ for all $f,g\in L^p(X)$] such that $\lim_{\varepsilon\to0} T_\varepsilon(f) = 0$ for all f in some dense subspace D of $L^p(X)$. Use the argument of Theorem 2.1.14 to prove that $\lim_{\varepsilon\to0} T_\varepsilon(f) = 0$ for all f in $L^p(X)$.
(b) Use the result in part (a) to prove the following improvement of the Lebesgue differentiation theorem: Let $f\in L^p_{\mathrm{loc}}(\mathbf{R}^n)$ for some $1\le p<\infty$.

Then for almost all $x \in \mathbf{R}^n$ we have

$$\lim_{\substack{|B|\to 0\\ B\ni x}} \frac{1}{|B|}\int_B |f(y)-f(x)|^p\,dy = 0\,,$$

where the limit is taken over all open balls B containing x.
[*Hint:* Define

$$T_\varepsilon(f)(x) = \sup_{B(z,\varepsilon)\ni x}\left(\frac{1}{|B(z,\varepsilon)|}\int_{B(z,\varepsilon)}|f(y)-f(x)|^p\,dy\right)^{1/p}$$

and observe that $T_*(f) \le |f| + M(|f|^p)^{\frac{1}{p}}$.]

2.1.11. On $\mathbf{R}$ define the right and left maximal functions M_R and M_L as follows:

$$\begin{aligned} M_L(f)(x) &= \sup_{r>0}\frac{1}{r}\int_{x-r}^{x}|f(t)|\,dt,\\ M_R(f)(x) &= \sup_{r>0}\frac{1}{r}\int_{x}^{x+r}|f(t)|\,dt. \end{aligned}$$

(a) (*Riesz's sunrise lemma* [**424**]) Show that

$$\begin{aligned} |\{x\in\mathbf{R}:\ M_L(f)(x)>\alpha\}| &= \frac{1}{\alpha}\int_{\{M_L(f)>\alpha\}}|f(t)|\,dt\,,\\ |\{x\in\mathbf{R}:\ M_R(f)(x)>\alpha\}| &= \frac{1}{\alpha}\int_{\{M_R(f)>\alpha\}}|f(t)|\,dt\,. \end{aligned}$$

(b) Conclude that M_L and M_R map L^p to L^p with norm at most $p/(p-1)$ for $1<p<\infty$.
(c) Construct examples to show that the operator norms of M_L and M_R on L^p are exactly $p/(p-1)$ for $1<p<\infty$.
(d) (*K. L. Phillips*) Prove that $M = \max(M_R, M_L)$.
(e) (*J. Duoandikoetxea*) Let $N = \min(M_R, M_L)$. Since

$$M(f)^p + N(f)^p = M_L(f)^p + M_R(f)^p$$

integrate over the line and use the following consequence of part (a):

$$\int_{\mathbf{R}} M_L(f)^p + M_R(f)^p\,dx = \frac{p}{p-1}\int_{\mathbf{R}}|f|\left(M(f)^{p-1}+N(f)^{p-1}\right)dx$$

to prove that

$$(p-1)\|M(f)\|_{L^p}^p - p\|f\|_{L^p}\|M(f)\|_{L^p}^{p-1} - \|f\|_{L^p}^p \le 0\,.$$

This provides an alternative proof of the result in Exercise 2.1.2(c).

2.1.12. A cube $Q = [a_1 2^k, (a_1+1)2^k) \times \cdots \times [a_n 2^k, (a_n+1)2^k)$ on $\mathbf{R}^n$ is called *dyadic* if $k, a_1, \cdots, a_n \in \mathbf{Z}$. Observe that two dyadic cubes are either disjoint or one contains the other. Define the *dyadic maximal function*

$$M_d(f)(x) = \sup_Q \frac{1}{|Q|} \int_Q f(x)\,dx,$$

where the supremum is taken over all dyadic cubes containing x.
(a) Prove that M_d maps L^1 to $L^{1,\infty}$ with constant at most one, that is, show that for all $\alpha > 0$ and $f \in L^1(\mathbf{R}^n)$ we have

$$|\{x \in \mathbf{R}^n : \; M_d(f)(x) > \alpha\}| \le \alpha^{-1} \int_{\{M_d(f)>\alpha\}} f(t)\,dt.$$

(b) Conclude from this that M_d maps $L^p(\mathbf{R}^n)$ into itself with constant at most $p/(p-1)$.

2.1.13. Use the following consequence of Exercise 1.1.12.

$$\big\|f\big\|_{L^{p,\infty}} \approx \sup_{0<|E|<\infty} |E|^{-1+\frac{1}{p}} \int_E |f(y)|\,dy$$

to prove that the Hardy-Littlewood maximal operator M maps the space $L^{p,\infty}(\mathbf{R}^n)$ into itself for $1 < p < \infty$.
[*Hint:* Consider the set $E = \{M(f) > \lambda\}$ and use the result of Exercise 2.1.4(a).]

2.1.14. Let $K(x) = (1 + |x|)^{-n-\delta}$ be defined on $\mathbf{R}^n$. Prove that there exists a constant $C_{n,\delta}$ such that for all $\varepsilon_0 > 0$ we have the estimate

$$\sup_{\varepsilon>\varepsilon_0} (|f| * K_\varepsilon)(x) \le C_{n,\delta} \sup_{\varepsilon>\varepsilon_0} \frac{1}{\varepsilon^n} \int_{|y-x|\le\varepsilon} |f(y)|\,dy,$$

for all f locally integrable on $\mathbf{R}^n$.
[*Hint:* Apply only a minor modification to the proof of Theorem 2.1.10.]

2.2. The Schwartz Class and the Fourier Transform

In this section we will introduce the single most important tool in harmonic analysis, the Fourier transform. It is often the case that the Fourier transform is introduced as an operation on L^1 functions. In this exposition we will first define the Fourier transform on a smaller class, the space of Schwartz functions, which turns out to be a very natural environment. Once the basic properties of the Fourier transform are derived, we will extend its definition to other spaces of functions.

We begin with some preliminaries. Given $x = (x_1, \dots, x_n) \in \mathbf{R}^n$, we set $|x| = (x_1^2 + \cdots + x_n^2)^{1/2}$. The partial derivative of a function f on $\mathbf{R}^n$ with respect to the jth variable x_j will be denoted by $\partial_j f$, while the mth derivative with respect to the jth variable will be denoted by $\partial_j^m f$. A *multiindex* α is an ordered n-tuple of nonnegative integers. For a multiindex $\alpha = (\alpha_1, \dots, \alpha_n)$, $\partial^\alpha f$ denotes the derivative $\partial_1^{\alpha_1} \dots \partial_n^{\alpha_n} f$. If $\alpha = (\alpha_1, \dots, \alpha_n)$ is a multiindex, $|\alpha| = \alpha_1 + \cdots + \alpha_n$ denotes its size and $\alpha! = \alpha_1! \dots \alpha_n!$ denotes the product of the factorials of its entries. The number $|\alpha|$ indicates the *total order of differentiation* of $\partial^\alpha f$. The space of functions in $\mathbf{R}^n$ all of whose derivatives of order at most $N \in \mathbf{Z}^+$ are continuous will be denoted by $C^N(\mathbf{R}^n)$ and the space of all *infinitely differentiable functions* on $\mathbf{R}^n$ by $C^\infty(\mathbf{R}^n)$. The space of C^∞ functions with compact support on $\mathbf{R}^n$ will be denoted by $C_0^\infty(\mathbf{R}^n)$. This space is nonempty; see Exercise 2.2.1(a).

For $x \in \mathbf{R}^n$ and $\alpha = (\alpha_1, \dots, \alpha_n)$ a multiindex, we set $x^\alpha = x_1^{\alpha_1} \dots x_n^{\alpha_n}$. It is a simple fact to verify that

$$|x^\alpha| \le c_{n,\alpha} |x|^{|\alpha|}, \tag{2.2.1}$$

for some constant that depends on the dimension n and on α. In fact, $c_{n,\alpha}$ is the maximum of the smooth function $x \to x^\alpha$ on the sphere $\mathbf{S}^{n-1} = \{x \in \mathbf{R}^n : \ |x| = 1\}$. The converse inequality in (2.2.1) fails. However, the following substitute of the converse of (2.2.1) will be of great use:

$$|x|^k \le C_{n,k} \sum_{|\beta|=k} |x^\beta|. \tag{2.2.2}$$

To prove (2.2.2), take $C_{n,k}^{-1}$ to be the minimum of the function

$$x \to \sum_{|\beta|=k} |x^\beta|$$

on $\mathbf{S}^{n-1}$.

We end the preliminaries by noting the validity of the one-dimensional Leibniz rule

$$\frac{d^m}{dt^m}(fg) = \sum_{k=0}^{m} \binom{m}{k} \frac{d^k f}{dt^k} \frac{d^{m-k} g}{dt^{m-k}}, \tag{2.2.3}$$

for all C^m functions f, g on $\mathbf{R}$, and its multidimensional analogue

$$\partial^\alpha (fg) = \sum_{\beta} \binom{\alpha_1}{\beta_1} \cdots \binom{\alpha_n}{\beta_n} (\partial^\beta f)(\partial^{\alpha-\beta} g), \tag{2.2.4}$$

for f, g in $C^{|\alpha|}(\mathbf{R}^n)$ for some multiindex α where the sum in (2.2.4) is taken over all multiindices β with $0 \le \beta_j \le \alpha_j$ for all $1 \le j \le n$. We observe that identity (2.2.4) is easily deduced by repeated application of (2.2.3) which in turn can be easily obtained by induction.

2.2.a. The Class of Schwartz Functions

We now introduce the class of *Schwartz functions* on $\mathbf{R}^n$. Roughly speaking, a function is Schwartz if it is smooth and all of its derivatives decay faster than the reciprocal of any polynomial at infinity. More precisely, we give the following definition.

Definition 2.2.1. A C^∞ complex-valued function f on $\mathbf{R}^n$ is called a Schwartz function if for all multiindices α and β there exist positive constants $C_{\alpha,\beta}$ such that

$$\rho_{\alpha,\beta}(f) = \sup_{x\in\mathbf{R}^n} |x^\alpha \partial^\beta f(x)| = C_{\alpha,\beta} < \infty. \tag{2.2.5}$$

The quantities $\rho_{\alpha,\beta}(f)$ are called the *Schwartz seminorms* of f. The set of all Schwartz functions on $\mathbf{R}^n$ will be denoted by $\mathcal{S}(\mathbf{R}^n)$.

Example 2.2.2. The function $e^{-|x|^2}$ is in $\mathcal{S}(\mathbf{R}^n)$ but $e^{-|x|}$ is not since it fails to be differentiable at the origin. The C^∞ function $g(x) = (1+|x|^4)^{-a}$, $a > 0$ is not in $\mathcal{S}$ since it only decays like the reciprocal of a fixed polynomial at infinity. The set of all smooth functions with compact support, $C_0^\infty(\mathbf{R}^n)$, is contained in $\mathcal{S}(\mathbf{R}^n)$.

Remark 2.2.3. Observe that if f_1 is in $\mathcal{S}(\mathbf{R}^n)$ and f_2 is in $\mathcal{S}(\mathbf{R}^m)$, then $f_1(x_1,\dots,x_n)f_2(x_{n+1},\dots,x_{n+m})$ is in $\mathcal{S}(\mathbf{R}^{n+m})$. If f is in $\mathcal{S}(\mathbf{R}^n)$ and $P(x)$ is a polynomial of n variables, then $P(x)f(x)$ is also in $\mathcal{S}(\mathbf{R}^n)$. If α is a multiindex and f is in $\mathcal{S}(\mathbf{R}^n)$, then $\partial^\alpha f$ is in $\mathcal{S}(\mathbf{R}^n)$. Also note that

$$f \in \mathcal{S}(\mathbf{R}^n) \iff \sup_{x\in\mathbf{R}^n} |\partial^\alpha(x^\beta f(x))| < \infty \qquad \text{for all multiindices } \alpha,\ \beta.$$

Remark 2.2.4. The following alternative characterization of Schwartz functions will be very useful. A C^∞ function f is in $\mathcal{S}(\mathbf{R}^n)$ if and only if for all positive integers N and all multiindices α there exists a positive constant $C_{\alpha,N}$ such that

$$|(\partial^\alpha f)(x)| \le C_{\alpha,N}(1+|x|)^{-N}. \tag{2.2.6}$$

The simple proofs are left to the reader. We now discuss convergence in $\mathcal{S}(\mathbf{R}^n)$.

Definition 2.2.5. Let f_k, f be in $\mathcal{S}(\mathbf{R}^n)$ for $k = 1, 2, \dots$. We say that the sequence f_k converges to f in $\mathcal{S}(\mathbf{R}^n)$ if for all multiindices α and β we have

$$\rho_{\alpha,\beta}(f_k - f) = \sup_{x\in\mathbf{R}^n} |x^\alpha(\partial^\beta(f_k - f))(x)| \to 0 \qquad \text{as} \qquad k \to \infty.$$

This notion of convergence is compatible with a topology on $\mathcal{S}(\mathbf{R}^n)$ under which the operations $(f,g) \to f+g$, $(a,f) \to af$, and $f \to \partial^\alpha f$ are continuous for all complex scalars a and multiindices α, $(f,g \in \mathcal{S}(\mathbf{R}^n))$. A basis for open sets containing 0 in this topology is

$$\{f \in \mathcal{S} :\ \rho_{\alpha,\beta}(f) < r\},$$

for all α, β multiindices and all $r \in \mathbf{Q}^+$. Observe the following: If $\rho_{\alpha,\beta}(f) = 0$, then $f = 0$. This means that $\mathcal{S}(\mathbf{R}^n)$ is a locally convex topological vector space equipped with the family of seminorms $\rho_{\alpha,\beta}$ that separate points. We refer to Reed and Simon [**420**] for the pertinent definitions. Since the origin in $\mathcal{S}(\mathbf{R}^n)$ has a countable base, this space is metrizable. In fact, the following is a metric on $\mathcal{S}(\mathbf{R}^n)$:

$$d(f,g) = \sum_{j=1}^{\infty} 2^{-j} \frac{\rho_j(f-g)}{1+\rho_j(f-g)},$$

where ρ_j is an enumeration of all the seminorms $\rho_{\alpha,\beta}$, α and β multiindices. We can easily show that $\mathcal{S}$ is complete with respect to the metric defined d. In fact, a Cauchy sequence $\{h_j\}_j$ in $\mathcal{S}$ would have to be Cauchy in L^∞ and therefore it would converge uniformly to some function h. The same is true for the sequences $\{\partial^\beta h_j\}_j$ and $\{x^\alpha h_j(x)\}_j$ and the limits of these sequences can be shown to be the functions $\partial^\beta h$ and $x^\alpha h(x)$, respectively. It follows that the sequence $\{h_j\}$ converges to h in $\mathcal{S}$. Therefore, $\mathcal{S}(\mathbf{R}^n)$ is a *Fréchet space (complete metrizable locally convex space).*

We observe that convergence in $\mathcal{S}$ is stronger than convergence in all L^p. We have the following.

Proposition 2.2.6. *Let f, f_k, $k = 1,2,3\dots$ be in $\mathcal{S}(\mathbf{R}^n)$. If $f_k \to f$ in $\mathcal{S}$ then $f_k \to f$ in L^p for all $0 < p \le \infty$. Moreover, there exists a $C_{p,n} > 0$ such that*

$$\big\|\partial^\beta f\big\|_{L^p} \le C_{p,n} \sum_{|\alpha|=[(n+1)/p]+1} \rho_{\alpha,\beta}(f) \tag{2.2.7}$$

for all f for which the right-hand side is finite.

Proof. Set $g = \partial^\beta f$ and observe that

$$\begin{aligned}
\big\|g\big\|_{L^p} &\le \left(\int_{|x|\le 1} \big\|g\big\|_{L^\infty}^p \, dx + \int_{|x|\ge 1} |x|^{n+1}|g(x)|^p|x|^{-(n+1)}\, dx\right)^{1/p} \\
&\le \left(v_n\big\|g\big\|_{L^\infty}^p + \sup_{|x|\ge 1} |x|^{n+1}|g(x)|^p \int_{|x|\ge 1} |x|^{-(n+1)}\, dx\right)^{1/p} \\
&\le C_{p,n}\Big(\big\|g\big\|_{L^\infty} + \sup_{x\in\mathbf{R}^n} \big(|x|^{[(n+1)/p]+1}|g(x)|\big)\Big)\,.
\end{aligned}$$

Now use (2.2.2) to obtain

$$|x|^m|g(x)| \le C_{n,\alpha} \sum_{|\beta|=m} |x^\beta g(x)|\,.$$

Thus the L^p norm of the Schwartz function g is controlled by a constant multiple of a sum of some $\rho_{\alpha,0}$ seminorms of the function. Conclusion (2.2.7) now follows immediately. This shows that convergence in $\mathcal{S}$ implies convergence in L^p. □

We now show that the Schwartz class is closed under certain operations.

Proposition 2.2.7. *Let f, g be in $\mathcal{S}(\mathbf{R}^n)$. Then fg and $f*g$ are in $\mathcal{S}(\mathbf{R}^n)$. Moreover*

$$\partial^\alpha(f*g) = (\partial^\alpha f)*g = f*(\partial^\alpha g) \tag{2.2.8}$$

for all multiindices α.

Proof. Fix f and g in $\mathcal{S}(\mathbf{R}^n)$. Let e_j be the unit vector $(0,\dots,1,\dots,0)$ with 1 in the jth entry and zeros in all the other entries. Since

$$\frac{f(y+he_j)-f(y)}{h} - (\partial_j f)(y) \to 0 \tag{2.2.9}$$

as $h \to 0$, and since the expression in (2.2.9) is pointwise bounded by some constant depending on f, the integral of the expression in (2.2.9) with respect to the measure $g(x-y)\,dy$ converges to zero as $h \to 0$ by the Lebesgue dominated convergence theorem. This proves (2.2.8) when $\alpha = (0,\dots,1,\dots,0)$. The general case follows by repetition of the previous argument and by induction.

We now show that the convolution of two functions in $\mathcal{S}$ is also in $\mathcal{S}$. For each $N > n$ we have

$$|(f*g)(x)| \le C_N \int_{\mathbf{R}^n} (1+|x-y|)^{-N}(1+|y|)^{-N}dy\,. \tag{2.2.10}$$

The piece of the integral in (2.2.10) over the set $\{y:\ \frac{1}{2}|x| \le |y-x|\}$ is bounded by

$$\int_{|y-x|\ge\frac{1}{2}|x|} (1+\tfrac{1}{2}|x|)^{-N}(1+|y|)^{-N}dy \le B_N(1+|x|)^{-N}\,,$$

where B_N is a constant depending on N and on the dimension. When $\frac{1}{2}|x| \ge |y-x|$ it follows that $|y| \ge \frac{1}{2}|x|$ and the piece of the integral in (2.2.10) over the set $\{y:\ |y-x| \le \frac{1}{2}|x|\}$ is bounded by

$$\int_{|y-x|\le\frac{1}{2}|x|} (1+|x-y|)^{-N}(1+\tfrac{1}{2}|x|)^{-N}dy \le B_N(1+|x|)^{-N}\,.$$

This shows that $f*g$ decays like $(1+|x|)^{-N}$ at infinity, but since $N > n$ is arbitrary it follows that $f*g$ decays faster than the reciprocal of any polynomial.

Since $\partial^\alpha(f*g) = (\partial^\alpha f)*g$, replacing f by $\partial^\alpha f$ in the previous argument, we also conclude that all the derivatives of $f*g$ decay faster than the reciprocal of any polynomial at infinity. Using (2.2.6), we conclude that $f*g$ is in $\mathcal{S}$. Finally, the fact that fg is in $\mathcal{S}$ follows directly from Leibniz's rule (2.2.4) and (2.2.4).

□

2.2.b. The Fourier Transform of a Schwartz Function

The Fourier transform is often introduced as an operation on L^1. In that setting problems of convergence arise when certain manipulations of functions are performed. Also, Fourier inversion requires the additional assumption that the Fourier transform is in L^1. Here we shall initially introduce the Fourier transform on the space of Schwartz functions. The rapid decay of Schwartz functions at infinity will allow us to develop its fundamental properties without encountering any convergence problems. As the Fourier transform preserves the Schwartz class and Fourier inversion naturally holds in it, we may argue that this class is a natural environment for it.

For $x = (x_1, \dots, x_n)$, $y = (y_1, \dots, y_n)$ in $\mathbf{R}^n$ we will use the notation

$$x \cdot y = \sum_{j=1}^{n} x_j y_j \, .$$

Definition 2.2.8. Given f in $\mathcal{S}(\mathbf{R}^n)$ we define

$$\widehat{f}(\xi) = \int_{\mathbf{R}^n} f(x) e^{-2\pi i x \cdot \xi} \, dx.$$

We call $\widehat{f}$ the Fourier transform of f.

Example 2.2.9. Let $f(x) = e^{-\pi|x|^2}$ defined on $\mathbf{R}$. Then $\widehat{f}(\xi) = f(\xi)$. First observe that the function

$$s \to \int_{-\infty}^{+\infty} e^{-\pi(x+is)^2} dx, \qquad s \in \mathbf{R}$$

is constant. Indeed, its derivative is

$$\int_{-\infty}^{+\infty} -2\pi i(x+is) e^{-\pi(x+is)^2} dx = \int_{-\infty}^{+\infty} i \frac{d}{dx} (e^{-\pi(x+is)^2}) \, dx = 0.$$

The computation of the Fourier transform of $f(x) = e^{-\pi|x|^2}$ relies on simple completion of squares. We have

$$\begin{aligned}
&\int_{\mathbf{R}^n} e^{-\pi|x|^2} e^{-2\pi i \sum_{j=1}^n x_j \xi_j} \, dx \\
&= \int_{\mathbf{R}^n} e^{-\pi \sum_{j=1}^n (x_j + i\xi_j)^2} e^{\pi \sum_{j=1}^n (i\xi_j)^2} \, dx \\
&= \left(\int_{-\infty}^{+\infty} e^{-\pi x^2} dx \right)^n e^{-\pi|\xi|^2} \\
&= e^{-\pi|\xi|^2},
\end{aligned}$$

where we used that

$$\int_{-\infty}^{+\infty} e^{-x^2} dx = \sqrt{\pi}\,, \tag{2.2.11}$$

a fact that can be found in Appendix A.1.

Remark 2.2.10. It follows from the definition of the Fourier transform that if $f \in \mathcal{S}(\mathbf{R}^n)$ and $g \in \mathcal{S}(\mathbf{R}^m)$, then

$$[f(x_1, \dots, x_n) g(x_{n+1}, \dots, x_{n+m})]\widehat{\ } = \widehat{f}(\xi_1, \dots, \xi_n)\widehat{g}(\xi_{n+1}, \dots, \xi_{n+m}),$$

where the first $\widehat{\ }$ denotes the Fourier transform on $\mathbf{R}^{n+m}$. In other words, the Fourier transform preserves any separation of variables.

Combining this observation with the result in Example 2.2.9, we conclude that the function $f(x) = e^{-\pi|x|^2}$ defined on $\mathbf{R}^n$ is equal to its Fourier transform.

We now continue with some properties of the Fourier transform. Before we do this we shall introduce some notation. For a measurable function f on $\mathbf{R}^n$, $x \in \mathbf{R}^n$, and $a > 0$ we define the *translation*, *dilation*, and *reflection* of f by

$$\begin{aligned} \tau^y(f)(x) &= f(x-y) \\ \delta^a(f)(x) &= f(ax) \\ \widetilde{f}(x) &= f(-x). \end{aligned} \tag{2.2.12}$$

Also recall the notation $f_a = a^{-n}\delta^{1/a}(f)$ introduced in Definition 2.1.9.

Proposition 2.2.11. *Given f, g in $\mathcal{S}(\mathbf{R}^n)$, $y \in \mathbf{R}^n$, $b \in \mathbf{C}$, α multiindex, and $a > 0$, we have*

(1) $\|\widehat{f}\|_{L^\infty} \le \|f\|_{L^1}$

(2) $\widehat{f+g} = \widehat{f} + \widehat{g}$

(3) $\widehat{bf} = b\widehat{f}$

(4) $\widehat{\widetilde{f}} = \widetilde{\widehat{f}}$

(5) $\widehat{\overline{f}} = \overline{\widetilde{\widehat{f}}}$

(6) $\widehat{\tau^y(f)}(\xi) = e^{-2\pi i y\cdot\xi}\widehat{f}(\xi)$

(7) $(e^{2\pi i x\cdot y}f(x))\widehat{\ }(\xi) = \tau^y(\widehat{f})(\xi)$

(8) $\delta^a(f)\widehat{\ } = a^{-n}\delta^{a^{-1}}(\widehat{f}) = (\widehat{f})_a$

(9) $(\partial^\alpha f)\widehat{\ }(\xi) = (2\pi i\xi)^\alpha \widehat{f}(\xi)$

(10) $(\partial^\alpha \widehat{f})(\xi) = ((-2\pi i x)^\alpha f(x))\widehat{\ }(\xi)$

(11) $\widehat{f} \in \mathcal{S}$

(12) $\widehat{f * g} = \widehat{f}\widehat{g}$

(13) $\widehat{f\circ A}(\xi) = \widehat{f}(A\xi)$, *where A is an orthogonal matrix and ξ is a column vector.*

Proof. Property (1) follows from Definition 2.2.8 and implies that the Fourier transform is always bounded. Properties (2)–(5) are trivial. Properties (6)–(8)

require a suitable change of variables but they are left to the reader. Property (9) is proved by integration by parts (which is justified by the fast convergence of the integrals):

$$\begin{aligned}(\partial^\alpha f)\hat{}(\xi) &= \int_{\mathbf{R}^n} (\partial^\alpha f)(x) e^{-2\pi i x\cdot\xi}\, dx \\ &= (-1)^{|\alpha|}\int_{\mathbf{R}^n} f(x)(-2\pi i\xi)^\alpha e^{-2\pi i x\cdot\xi}\, dx \\ &= (2\pi i\xi)^\alpha \widehat{f}(\xi)\,.\end{aligned}$$

To prove (10), let $\alpha = (0,\dots,1,\dots,0)$, where all entries are zero except for the jth entry which is 1. Since

$$\frac{e^{-2\pi i x\cdot(\xi+he_j)} - e^{-2\pi i x\cdot\xi}}{h} - (-2\pi i x_j)e^{-2\pi i x\cdot\xi} \to 0 \tag{2.2.13}$$

as $h \to 0$ and the preceding function is bounded by $C|x|$ for all h and ξ, the Lebesgue dominated convergence theorem implies that the integral of the function in (2.2.13) with respect to the measure $f(x)dx$ converges to zero. Thus we proved (10) for $\alpha = (0,\dots,1,\dots,0)$. For other α's use induction. To prove (11) we use (9), (10), and (1) in the following way

$$\begin{aligned}\big\|x^\alpha(\partial^\beta \widehat{f})(x)\big\|_{L^\infty} &= \frac{(2\pi)^{|\beta|}}{(2\pi)^{|\alpha|}}\big\|(\partial^\alpha(x^\beta f(x)))\hat{}\big\|_{L^\infty} \\ &\le \frac{(2\pi)^{|\beta|}}{(2\pi)^{|\alpha|}}\big\|\partial^\alpha(x^\beta f(x))\big\|_{L^1} < \infty\,.\end{aligned}$$

Identity (12) follows from the following calculation:

$$\begin{aligned}&\widehat{f*g}(\xi) \\ &= \int_{\mathbf{R}^n}\int_{\mathbf{R}^n} f(x-y)g(y)e^{-2\pi i x\cdot\xi}\, dy\, dx \\ &= \int_{\mathbf{R}^n}\int_{\mathbf{R}^n} f(x-y)g(y)e^{-2\pi i(x-y)\cdot\xi}e^{-2\pi i y\cdot\xi}\, dy\, dx \\ &= \int_{\mathbf{R}^n} g(y)\int_{\mathbf{R}^n} f(x-y)e^{-2\pi i(x-y)\cdot\xi} dx\; e^{-2\pi i y\cdot\xi}\, dy \\ &= \widehat{f}(\xi)\widehat{g}(\xi),\end{aligned}$$

where the application of Fubini's theorem is justified by the fast convergence of the integrals. Finally, we prove (13). We have

$$\begin{aligned}\widehat{f\circ A}(\xi) &= \int_{\mathbf{R}^n} f(Ax)e^{-2\pi i x\cdot\xi}\,dx \\ &= \int_{\mathbf{R}^n} f(y)e^{-2\pi i A^{-1}y\cdot\xi}\,dy \\ &= \int_{\mathbf{R}^n} f(y)e^{-2\pi i A^t y\cdot\xi}\,dy \\ &= \int_{\mathbf{R}^n} f(y)e^{-2\pi i y\cdot A\xi}\,dy \\ &= \widehat{f}(A\xi),\end{aligned}$$

where we used the change of variables $y = Ax$ and the fact that $|\det A| = 1$.

□

Corollary 2.2.12. *The Fourier transform of a radial function is radial. Products and convolutions of radial functions are radial.*

Proof. Let ξ_1, ξ_2 in $\mathbf{R}^n$ with $|\xi_1| = |\xi_2|$. Then for some orthogonal matrix A we have $A\xi_1 = \xi_2$. Since f is radial, we have $f = f\circ A$. Then

$$\widehat{f}(\xi_2) = \widehat{f}(A\xi_1) = \widehat{f\circ A}(\xi_1) = \widehat{f}(\xi_1),$$

where we used (13) in Proposition 2.2.11 to justify the second equality. Convolutions of radial functions are easily seen to be radial using the Fourier transform.

□

2.2.c. The Inverse Fourier Transform and Fourier Inversion

We now define the inverse Fourier transform.

Definition 2.2.13. Given a Schwartz function f, we define

$$f^\vee(x) = \widehat{f}(-x),$$

for all $x \in \mathbf{R}^n$. The operation

$$f \to f^\vee$$

is called the *inverse Fourier transform.*

It is straightforward that the inverse Fourier transform shares the same properties as the Fourier transform. As an exercise, the reader may want to list (and prove) properties analogous to those in Proposition 2.2.11 for the inverse Fourier transform.

We now investigate the relation between the Fourier transform and the inverse Fourier transform. In the next theorem, we will prove that one is the inverse operation of the other. This property is referred to as *Fourier inversion.*

Theorem 2.2.14. *Given f, g, and h in $\mathcal{S}(\mathbf{R}^n)$, we have*

(1) $\displaystyle\int_{\mathbf{R}^n} f(x)\widehat{g}(x)\,dx = \int_{\mathbf{R}^n} \widehat{f}(x)g(x)\,dx$

(2) ***(Fourier Inversion)*** $(f^\frown)^\vee = f = (f^\vee)^\frown$

(3) $\displaystyle\int_{\mathbf{R}^n} f(x)\overline{h}(x)\,dx = \int_{\mathbf{R}^n} \widehat{f}(\xi)\overline{\widehat{h}(\xi)}\,d\xi$

(4) ***(Plancherel's identity)*** $\|f\|_{L^2} = \|f^\frown\|_{L^2} = \|f^\vee\|_{L^2}$

(5) $\displaystyle\int_{\mathbf{R}^n} f(x)g(x)\,dx = \int_{\mathbf{R}^n} \widehat{f}(x)g^\vee(x)\,dx\,.$

Proof. (1) follows immediately from the definition of the Fourier transform and Fubini's theorem. To prove (2) we use (1) with

$$g(x) = e^{2\pi i x\cdot t}e^{-\pi|\varepsilon x|^2}\,.$$

By Proposition 2.2.11 (7) and (8) and Example 2.2.9, we have that

$$\widehat{g}(x) = \frac{1}{\varepsilon^n}e^{-\pi|(x-t)/\varepsilon|^2},$$

which is an approximate identity. Now (1) gives

$$\int_{\mathbf{R}^n} f(x)\varepsilon^{-n}e^{-\pi\varepsilon^{-2}|x-t|^2}\,dx = \int_{\mathbf{R}^n} \widehat{f}(x)e^{2\pi i x\cdot t}e^{-\pi|\varepsilon x|^2}\,dx. \tag{2.2.14}$$

Now let $\varepsilon \to 0$ in (2.2.14). The left-hand side of (2.2.14) converges to $f(t)$ uniformly on compact sets by Theorem 1.2.19. The right-hand side of (2.2.14) converges to $(\widehat{f}\,)^\vee(t)$ as $\varepsilon \to 0$ by the Lebesgue dominated convergence theorem. We conclude that $(\widehat{f}\,)^\vee = f$ on $\mathbf{R}^n$. Replacing f by $\widetilde{f}$ and using the result just proved we conclude that $(f^\vee)^\frown = f$.

To prove (3), use (1) with $g = \overline{\widehat{h}}$ and the fact that $\widehat{g} = \overline{h}$, which is a consequence of Proposition 2.2.11 (5) and Fourier inversion. Plancherel's identity is a trivial consequence of (3). [Sometimes the polarized identity (3) is also referred to as Plancherel's identity.] Finally, (5) easily follows from (1).

□

Next we have the following simple corollary of Theorem 2.2.14.

Corollary 2.2.15. *The Fourier transform is a homeomorphism from $\mathcal{S}(\mathbf{R}^n)$ onto itself.*

Proof. The continuity of the Fourier transform (and its inverse) follows from Exercise 2.2.2 while Fourier inversion yields that this map is bijective.

□

2.2.d. The Fourier Transform on $L^1 + L^2$

We have defined the Fourier transform on $\mathcal{S}(\mathbf{R}^n)$. We now extend this definition to the space $L^1(\mathbf{R}^n) + L^2(\mathbf{R}^n)$.

First note that Definition 2.2.8 makes sense for all functions $f \in L^1(\mathbf{R}^n)$ and the Fourier transform is therefore well defined there. Moreover, this operator satisfies properties (1)–(8) in Proposition 2.2.11.

Plancherel's identity (Theorem 2.2.14 (4)) and the density of $\mathcal{S}$ in L^2 allows us to define the Fourier transform $\widehat{g}$ for all g in $L^2(\mathbf{R}^n)$ so that

$$\big\|g\big\|_{L^2} = \big\|\widehat{g}\big\|_{L^2}.$$

This is achieved by taking a sequence of Schwartz functions converging to g in L^2; see Exercise 2.2.8. Furthermore, as a consequence of Exercise 2.2.5, it follows that for $f \in L^2(\mathbf{R}^n) \cap L^1(\mathbf{R}^n)$ we have

$$\widehat{f}(\xi) = \int_{\mathbf{R}^n} f(x)e^{-2\pi i x\cdot\xi}\,dx\,. \tag{2.2.15}$$

We conclude that the definitions of the Fourier transform given on L^1 and L^2 are consistent on the intersection of $L^1 \cap L^2$.

Now for $f \in L^p(\mathbf{R}^n)$, $1 < p < 2$ define $\widehat{f} = \widehat{f_1} + \widehat{f_2}$, where $f_1 \in L^1$, $f_2 \in L^2$ and $f = f_1 + f_2$; we may take, for instance, $f_1 = f\chi_{|f|>1}$ and $f_2 = f\chi_{|f|\le 1}$. This definition is independent of the choice of f_1 and f_2 as if $f_1 + f_2 = h_1 + h_2$ for $f_1, h_1 \in L^1$ and $f_2, h_2 \in L^2$, we have $f_1 - h_1 = h_2 - f_2 \in L^1 \cap L^2$ and using (2.2.15), we obtain $\widehat{f_1} - \widehat{h_1} = \widehat{h_2} - \widehat{f_2}$, which yields $\widehat{f_1 + f_2} = \widehat{h_1 + h_2}$.

We have the following result concerning the action of the Fourier transform on L^p.

Proposition 2.2.16. *(Hausdorff-Young inequality) For every function f in $L^p(\mathbf{R}^n)$ we have the estimate*

$$\big\|\widehat{f}\,\big\|_{L^{p'}} \le \big\|f\big\|_{L^p}$$

whenever $1 \le p \le 2$.

Proof. This follows easily from Theorem 1.3.4. Interpolate between the estimates $\big\|\widehat{f}\,\big\|_{L^\infty} \le \big\|f\big\|_{L^1}$ [Proposition 2.2.11 (1)] and $\big\|\widehat{f}\,\big\|_{L^2} \le \big\|f\big\|_{L^2}$ to obtain $\big\|\widehat{f}\,\big\|_{L^{p'}} \le \big\|f\big\|_{L^p}$. We conclude that the Fourier transform is a bounded operator from $L^p(\mathbf{R}^n)$ into $L^{p'}(\mathbf{R}^n)$ with norm at most 1 when $1 \le p \le 2$. □

Next, we are concerned with the behavior of the Fourier transform at infinity.

Proposition 2.2.17. *(Riemann-Lebesgue lemma) For a function f in $L^1(\mathbf{R}^n)$ we have that*

$$|\widehat{f}(\xi)| \to 0 \qquad \text{as} \qquad |\xi| \to \infty.$$

Proof. Consider the function $\chi_{[a,b]}$ on $\mathbf{R}$. A simple computation gives

$$\widehat{\chi_{[a,b]}}(\xi) = \frac{e^{-2\pi i\xi a} - e^{-2\pi i\xi b}}{2\pi i\xi},$$

which tends to zero as $|\xi| \to \infty$. Likewise if $g = \prod_{j=1}^{n} \chi_{[a_j,b_j]}$ on $\mathbf{R}^n$, then

$$\widehat{g}(\xi) = \prod_{j=1}^{n} \frac{e^{-2\pi i\xi a_j} - e^{-2\pi i\xi b_j}}{2\pi i\xi},$$

which also tends to zero as $|\xi| \to \infty$ in $\mathbf{R}^n$.

To prove the assertion, approximate in the L^1 norm a general integrable function f on $\mathbf{R}^n$ by a finite sum h of "step functions" like g and use

$$|\widehat{f}(\xi)| \le |\widehat{f}(\xi) - \widehat{h}(\xi)| + |\widehat{h}(\xi)| \le \|f - h\|_{L^1} + |\widehat{h}(\xi)|.$$

□

We end this section with an example that illustrates some of the practical uses of the Fourier transform.

Example 2.2.18. We are asked to find a function $f(x_1, x_2, x_3)$ on $\mathbf{R}^3$ that satisfies the partial differential equation

$$f(x) - \partial_1^2\partial_2^2\partial_3^4 f(x) + 4i\partial_1^2 f(x) + \partial_2^7 f(x) = e^{-\pi|x|^2}.$$

Taking the Fourier transform on both sides of this identity and using Proposition 2.2.11 (2), (9) and the result of Example 2.2.9, we obtain

$$\widehat{f}(\xi)\Big[1 - (2\pi i\xi_1)^2(2\pi i\xi_2)^2(2\pi i\xi_3)^4 + 4i(2\pi i\xi_1)^2 + (2\pi i\xi_2)^7\Big] = e^{-\pi|\xi|^2}.$$

Let $p(\xi) = p(\xi_1, \xi_2, \xi_3)$ be the polynomial inside the square brackets. We observe that $p(\xi)$ has no real zeros and we can therefore write

$$\widehat{f}(\xi) = e^{-\pi|\xi|^2} p(\xi)^{-1} \implies f(x) = \left(e^{-\pi|\xi|^2} p(\xi)^{-1}\right)^{\vee}(x).$$

In general, let $P(x) = \sum_{|\alpha|\le N} C_\alpha x^\alpha$ be a polynomial in $\mathbf{R}^n$ with constant complex coefficients C_α indexed by multiindices α. If $P(x)$ has no real zeros, and $u \in \mathcal{S}(\mathbf{R}^n)$, then the partial differential equation

$$P(\partial)f = \sum_{|\alpha|\le N} C_\alpha \partial^\alpha f = u$$

is solved as before to give

$$f = \left(\widehat{u}(\xi)P(\xi)^{-1}\right)^{\vee}.$$

Note that since $P(\xi)$ has no real zeros the function $\widehat{u}(\xi)P(\xi)^{-1}$ is smooth and therefore a Schwartz function. Then f is also a Schwartz function by Proposition 2.2.11 (11).

Exercises

2.2.1. (a) Construct a Schwartz function with compact support.
(b) Construct a $C_0^\infty(\mathbf{R}^n)$ function equal to 1 on the annulus $1 \le |x| \le 2$ and vanishing off the annulus $1/2 \le |x| \le 4$.
(c) Construct a nonnegative nonzero Schwartz function f whose Fourier transform is nonnegative and compactly supported.
[*Hint:* Part (a): Try the construction in dimension one first using the C^∞ function $\eta(x) = e^{-1/x}$ for $x > 0$ and $\eta(x) = 0$ for $x < 0$. Part (c): Take $f = |\varphi * \widetilde{\varphi}|^2$, where $\widehat{\varphi}$ is odd, real valued, and compactly supported.]

2.2.2. If $f_k, f \in \mathcal{S}(\mathbf{R}^n)$ and $f_k \to f$ in $\mathcal{S}(\mathbf{R}^n)$, then $\widehat{f_k} \to \widehat{f}$ and $f_k^\vee \to f^\vee$ in $\mathcal{S}(\mathbf{R}^n)$.

2.2.3. Find the *spectrum* (i.e., the set of all *eigenvalues* of the Fourier transform); that is, all complex numbers λ for which there exist nonzero functions f such that

$$\widehat{f} = \lambda f\,.$$

[*Hint:* Apply the Fourier transform three times to the preceding identity. Consider the functions $xe^{-\pi x^2}$, $(a+bx^2)e^{-\pi x^2}$, and $(cx+dx^3)e^{-\pi x^2}$ for suitable a, b, c, d to show that all fourth roots of unity are indeed eigenvalues of the Fourier transform.]

2.2.4. Use the idea of the proof of Proposition 2.2.7 to show that if the functions f, g defined on $\mathbf{R}^n$ satisfy $|f(x)| \le A(1+|x|)^{-M}$ and $|g(x)| \le B(1+|x|)^{-N}$ for some $M, N > n$, then

$$|(f * g)(x)| \le ABC(1+|x|)^{-L},$$

where $L = \min(N, M)$ and $C = C(N, M) > 0$.

2.2.5. (a) Given f in $L^2(\mathbf{R}^n)$, prove that the sequence of functions

$$\xi \to \int_{|x| \le N} f(x) e^{-2\pi i x \cdot \xi}\, dx$$

converges to $\widehat{f}$ in L^2 as $N \to \infty$.
(b) Conclude that the definition of $\widehat{f}$ given on L^1 coincides with that given on L^2 whenever f lies in $L^1(\mathbf{R}^n) \cap L^2(\mathbf{R}^n)$.

2.2.6. (a) Prove that if $f \in L^1$, then $\widehat{f}$ is uniformly continuous on $\mathbf{R}^n$.
(b) Prove that for $f \in L^1$ and $g \in \mathcal{S}$ we have

$$\int_{\mathbf{R}^n} f(x)\widehat{g}(x)\, dx = \int_{\mathbf{R}^n} \widehat{f}(x) g(x)\, dx.$$

(c) Take $\widehat{g}(x) = \varepsilon^{-n} e^{-\pi\varepsilon^{-2}|x-t|^2}$ in (b) and let $\varepsilon \to 0$ to prove Fourier inversion for L^1: If f and $\widehat{f}$ are both in L^1, then $(\widehat{f}\,)^\vee = f$ a.e.

2.2.7. (a) Prove that if f is continuous at 0, then

$$\lim_{\varepsilon\to 0} \int_{\mathbf{R}^n} \widehat{f}(x) e^{-\pi|\varepsilon x|^2}\, dx = f(0).$$

(b) Prove that if $f \in L^1(\mathbf{R}^n)$, $\widehat{f} \geq 0$, and f is continuous at zero, then $\widehat{f}$ is in L^1 and therefore Fourier inversion holds.
[*Hint:* Part (a): Take $g(x) = e^{-\pi|\varepsilon x|^2}$ in Exercise 2.2.6(b).]

2.2.8. (a) Show that $C_0^\infty(\mathbf{R}^n)$ is dense on $L^p(\mathbf{R}^n)$ for $1 \leq p < \infty$. Conclude that $\mathcal{S}(\mathbf{R}^n)$ is also dense on L^p spaces.
(b) Prove the same density results on L^p for $0 < p < 1$.
[*Hint:* When $1 \leq p < \infty$ you may convolve with an approximate identity. For $0 < p < 1$ you may approximate a compactly supported step function with a smooth function.]

2.2.9. (a) Prove that there exists a constant $B > 0$ such that for all $t > 0$ we have

$$\left| \int_0^t \frac{\sin(\xi)}{\xi}\, d\xi \right| \leq B.$$

(b) If f is an odd L^1 function on the line, conclude that for all $t > 0$

$$\left| \int_0^t \frac{\widehat{f}(\xi)}{\xi}\, d\xi \right| \leq B \big\| f \big\|_{L^1}.$$

(c) Let $g(\xi)$ be a continuous odd function that is equal to $1/\log(\xi)$ for $\xi \geq 2$. Show that there does not exist an L^1 function whose Fourier transform is g.

2.2.10. Let f be in $L^1(\mathbf{R})$. Prove that

$$\int_{-\infty}^{+\infty} f(x)\, dx = \int_{-\infty}^{+\infty} f(x - 1/x)\, dx.$$

2.2.11. (a) Use Exercise 2.2.10 with $f(x) = e^{-tx^2}$ to obtain the *subordination* identity

$$e^{-2t} = \frac{1}{\sqrt{\pi}} \int_0^\infty e^{-y-t^2/y} \frac{dy}{\sqrt{y}}, \qquad \text{where } t > 0.$$

(b) Set $t = \pi|x|$ and integrate with respect to $e^{-2\pi i \xi\cdot x} dx$ to prove that

$$(e^{-2\pi|x|})\widehat{\ }(\xi) = \frac{\Gamma(\frac{n+1}{2})}{\pi^{\frac{n+1}{2}}} \frac{1}{(1+|\xi|^2)^{\frac{n+1}{2}}}.$$

This calculation gives the Fourier transform of the Poisson kernel.

2.2.12. Let $1 \le p \le \infty$ and let p' be its dual index.

(a) Prove that Schwartz functions f on the line satisfy the estimate

$$\|f\|_{L^\infty}^2 \le 2\|f\|_{L^p}\|f'\|_{L^{p'}}\,.$$

(b) Prove that all Schwartz functions f on $\mathbf{R}^n$ satisfy the estimate

$$\|f\|_{L^\infty}^2 \le 2 \sum_{\alpha+\beta=(1,\dots,1)} \|\partial^\alpha f\|_{L^p}\|\partial^\beta f\|_{L^{p'}}\,,$$

where the sum is taken over all multiindices α and β whose sum is $(1,1,\dots,1)$.
[*Hint:* Part (a): Write $f(x)^2 = \int_{-\infty}^x \frac{d}{dt} f(t)^2\,dt$.]

2.2.13. The *uncertainty principle* says that the position and the momentum of a particle cannot be simultaneously localized. Prove the following inequality, which presents a quantitative version of this principle:

$$\|f\|_{L^2(\mathbf{R}^n)}^2 \le \frac{4\pi}{n} \inf_{y\in\mathbf{R}^n}\left[\int_{\mathbf{R}^n} |x-y|^2|f(x)|^2\,dx\right]^{\frac12} \inf_{z\in\mathbf{R}^n}\left[\int_{\mathbf{R}^n} |\xi-z|^2|\widehat{f}(\xi)|^2\,d\xi\right]^{\frac12},$$

where f is a Schwartz function on $\mathbf{R}^n$ (or an L^2 function with sufficient decay at infinity).
[*Hint:* Start with

$$\|f\|_{L^2}^2 = \frac{1}{n}\int_{\mathbf{R}^n} f(x)e^{-2\pi i z\cdot x}\,\overline{f(x)e^{-2\pi i z\cdot x}} \sum_{j=1}^n \partial_j(x_j-y_j)\,dx\,,$$

integrate by parts, and apply the Cauchy-Schwartz inequality.]

2.2.14. Let $-\infty < \alpha < \frac{n}{2} < \beta < +\infty$. Prove the validity of the following inequality:

$$\|g\|_{L^1(\mathbf{R}^n)} \le C\,\big\||x|^\alpha g(x)\big\|_{L^2(\mathbf{R}^n)}^{\frac{\beta-n/2}{\beta-\alpha}} \big\||x|^\beta g(x)\big\|_{L^2(\mathbf{R}^n)}^{\frac{n/2-\alpha}{\beta-\alpha}}$$

for some constant $C = C(n,\alpha,\beta)$ independent of g.
[*Hint:* First prove $\|g\|_{L^1} \le C\big\||x|^\alpha g(x)\big\|_{L^2} + \big\||x|^\beta g(x)\big\|_{L^2}$ and then replace $g(x)$ by $g(\lambda x)$ for a suitable λ.]

2.3. The Class of Tempered Distributions

The fundamental idea of the theory of distributions is that it is generally easier to work with linear functionals acting on spaces of "nice" functions instead of working with "bad" functions directly. The set of "nice" functions we consider is closed under the basic operations in analysis, and these operations are defined to distributions by duality. This wonderful interpretation has proved to be an indispensable tool that has clarified many situations in analysis.

2.3.a. Spaces of Test Functions

We recall the space $C_0^\infty(\mathbf{R}^n)$ of all smooth functions with compact support, and $C^\infty(\mathbf{R}^n)$ of all smooth functions on $\mathbf{R}^n$. We are mainly interested in the following three spaces of "nice" functions on $\mathbf{R}^n$ that are nested as follows:

$$C_0^\infty(\mathbf{R}^n) \subseteq \mathcal{S}(\mathbf{R}^n) \subseteq C^\infty(\mathbf{R}^n).$$

Here $\mathcal{S}(\mathbf{R}^n)$ is the space of Schwartz functions introduced in Section 2.2.

Definition 2.3.1. We now give the definition of convergence of sequences in these spaces. We say

$$f_k \to f \text{ in } C^\infty \iff f_k, f \in C^\infty \text{ and } \lim_{k\to\infty} \sup_{|x|\le N} |\partial^\alpha (f_k - f)(x)| = 0$$
$$\forall\ \alpha \text{ multiindices and all } N = 1, 2, \dots .$$

$$f_k \to f \text{ in } \mathcal{S} \iff f_k, f \in \mathcal{S} \text{ and } \lim_{k\to\infty} \sup_{x\in\mathbf{R}^n} |x^\alpha \partial^\beta (f_k - f)(x)| = 0$$
$$\forall\ \alpha, \beta \text{ multiindices.}$$

$$f_k \to f \text{ in } C_0^\infty \iff f_k, f \in C_0^\infty,\ \text{support}(f_k) \subseteq B \text{ for all } k,\ B \text{ compact,}$$
$$\text{and } \lim_{k\to\infty} \big\|\partial^\alpha (f_k - f)\big\|_{L^\infty} = 0\ \forall\ \alpha \text{ multiindices.}$$

It follows that convergence in $C_0^\infty(\mathbf{R}^n)$ implies convergence in $\mathcal{S}(\mathbf{R}^n)$, which in turn implies convergence in $C^\infty(\mathbf{R}^n)$.

Example 2.3.2. Let φ be a nonzero C_0^∞ function on $\mathbf{R}$. We will call such functions bumps. Define the sequence of bumps $\varphi_k(x) = \varphi(x-k)/k$. Then $\varphi_k(x)$ does not converge to zero in $C_0^\infty(\mathbf{R})$, even though φ_k (and all of its derivatives) converge to zero uniformly. Furthermore, we may see that φ_k does not converge to any function in $\mathcal{S}(\mathbf{R})$. Clearly $\varphi_k \to 0$ in $C^\infty(\mathbf{R})$.

The space $C^\infty(\mathbf{R}^n)$ is equipped with the family of seminorms

$$\tilde{\rho}_{\alpha,N}(f) = \sup_{|x|\le N} |(\partial^\alpha f)(x)|, \qquad \alpha \text{ multiindex, } N = 1, 2, \dots . \tag{2.3.1}$$

As for the space $\mathcal{S}(\mathbf{R}^n)$, it can be shown that $C^\infty(\mathbf{R}^n)$ is complete with respect to this countable family of seminorms i.e., it is a Fréchet space. However, it can be seen that $C_0^\infty(\mathbf{R}^n)$ is not complete with respect to the topology generated by this family of seminorms.

The topology on C_0^∞ given in Definition 2.3.1 is the *inductive limit topology,* and under this topology it can be seen that C_0^∞ is complete. Indeed, $C_0^\infty(\mathbf{R}^n)$ is a countable union of spaces $\bigcup_{k=1}^\infty C_0^\infty(B(0,k))$ and each of these spaces is complete with the respect to the topology generated by the family of seminorms $\widetilde{\rho}_{\alpha,N}$; hence so is $C_0^\infty(\mathbf{R}^n)$. Nevertheless, $C_0^\infty(\mathbf{R}^n)$ is not metrizable. We refer to Reed and Simon [**420**] for details on the topologies of these spaces.

2.3.b. Spaces of Functionals on Test Functions

The dual spaces (i.e., the spaces of continuous linear functionals on the sets of test functions) we introduced will be denoted by

$$\begin{aligned} (C_0^\infty(\mathbf{R}^n))' &= \mathcal{D}'(\mathbf{R}^n) \\ (\mathcal{S}(\mathbf{R}^n))' &= \mathcal{S}'(\mathbf{R}^n) \\ (C^\infty(\mathbf{R}^n))' &= \mathcal{E}'(\mathbf{R}^n)\,. \end{aligned}$$

By definition of the topologies on the dual spaces we have

$$\begin{aligned} T_k \to T \quad \text{in } \mathcal{D}' &\iff T_k, T \in \mathcal{D}' \text{ and } T_k(f) \to T(f) \text{ for all } f \in C_0^\infty. \\ T_k \to T \quad \text{in } \mathcal{S}' &\iff T_k, T \in \mathcal{S}' \text{ and } T_k(f) \to T(f) \text{ for all } f \in \mathcal{S}. \\ T_k \to T \quad \text{in } \mathcal{E}' &\iff T_k, T \in \mathcal{E}' \text{ and } T_k(f) \to T(f) \text{ for all } f \in C^\infty. \end{aligned}$$

The dual spaces are nested as follows:

$$\mathcal{E}'(\mathbf{R}^n) \subseteq \mathcal{S}'(\mathbf{R}^n) \subseteq \mathcal{D}'(\mathbf{R}^n).$$

Definition 2.3.3. Elements of the space $\mathcal{D}'(\mathbf{R}^n)$ are called *distributions.* Elements of $\mathcal{S}'(\mathbf{R}^n)$ are called *tempered distributions.* Elements of the space $\mathcal{E}'(\mathbf{R}^n)$ are called *distributions with compact support.*

Before we work out some examples, we give alternative characterizations of distributions, which are very useful from the practical point of view. The action of a distribution u on a test function f will be denoted by either one of the following two ways:

$$\langle u, f\rangle = u(f)\,.$$

Proposition 2.3.4. *(a) A linear functional u on $C_0^\infty(\mathbf{R}^n)$ is a distribution if and only if for every compact $K \subseteq \mathbf{R}^n$, there exist $C > 0$ and an integer m such that*

$$|\langle u, f\rangle| \le C \sum_{|\alpha|\le m} \big\|\partial^\alpha f\big\|_{L^\infty}, \quad \textit{for all } f \in C^\infty \textit{ with support in } K. \tag{2.3.2}$$

(b) A linear functional u on $\mathcal{S}(\mathbf{R}^n)$ is a tempered distribution if and only if there exist $C > 0$ and k, m integers such that

$$|\langle u, f\rangle| \le C \sum_{\substack{|\alpha|\le m \\ |\beta|\le k}} \rho_{\alpha,\beta}(f), \qquad \textit{for all } f \in \mathcal{S}(\mathbf{R}^n). \tag{2.3.3}$$

(c) A linear functional u on $C^\infty(\mathbf{R}^n)$ is a distribution with compact support if and only if there exist $C > 0$ and N, m integers such that

$$|\langle u, f\rangle| \le C \sum_{|\alpha|\le m} \widetilde{\rho}_{\alpha,N}(f), \qquad \text{for all } f \in C^\infty(\mathbf{R}^n), \tag{2.3.4}$$

where $\rho_{\alpha,\beta}$ and $\widetilde{\rho}_{\alpha,N}$ are defined in (2.2.5) and (2.3.1).

Proof. We only prove (2.3.3) since the proofs of (2.3.2) and (2.3.4) are similar. It is clear that (2.3.3) implies continuity of u. Conversely, it was pointed out in Section 2.2 that the family of sets $\{f \in \mathcal{S}(\mathbf{R}^n) : \rho_{\alpha,\beta}(f) < \delta\}$, where α, β are multiindices and $\delta > 0$, forms a basis for the topology of $\mathcal{S}$. Thus if u is continuous functional on $\mathcal{S}$ there exist integers k, m and a $\delta > 0$ such that

$$|\alpha| \le m, \ |\beta| \le k, \quad \text{and } \rho_{\alpha,\beta}(f) < \delta \implies |\langle u, f\rangle| \le 1. \tag{2.3.5}$$

We see that (2.3.3) follows from (2.3.5) with $C = 1/\delta$. □

Examples 2.3.5. We now discuss some important examples.

1. The *Dirac mass* at the origin δ_0. This is defined by
$$\langle \delta_0, f\rangle = f(0).$$
We claim that δ_0 is in $\mathcal{E}'$. To see this we observe that if $f_k \to f$ in C^∞ then $\langle \delta_0, f_k\rangle \to \langle \delta_0, f\rangle$. The Dirac mass at a point $a \in \mathbf{R}^n$ is defined similarly by the equation
$$\langle \delta_a, f\rangle = f(a).$$
2. Some functions g can be thought as distributions via the identification $g \to L_g$, where L_g is the functional
$$L_g(f) = \int_{\mathbf{R}^n} f(x)g(x)\,dx.$$
Here are some examples: The function 1 is in $\mathcal{S}'$ but not in $\mathcal{E}'$. Compactly supported integrable functions are in $\mathcal{E}'$. The function $e^{|x|^2}$ is in $\mathcal{D}'$ but not in $\mathcal{S}'$.
3. Functions in L^1_{loc} are distributions. To see this, first observe that if $g \in L^1_{\text{loc}}$, then the integral
$$L_g(f) = \int_{\mathbf{R}^n} f(x)g(x)\,dx$$
is well defined for all $f \in \mathcal{D}$ and then note that $f_k \to f$ in $\mathcal{D}$ implies that $L_g(f_k) \to L_g(f)$.
4. Functions in L^p, $1 \le p \le \infty$ are tempered distributions, but they are not in $\mathcal{E}'$ unless they have compact support.

5. Any finite Borel measure μ is a tempered distribution via the identification
$$L_\mu(f) = \int_{\mathbf{R}^n} f(x)\, d\mu(x).$$
To see this observe that $f_k \to f$ in $\mathcal{S}$ implies that $L_\mu(f_k) \to L_\mu(f)$. Finite Borel measures may not be distributions with compact support. Lebesgue measure is also a tempered distribution.
6. Every function g that satisfies $|g(x)| \le C(1+|x|)^k$, for some real number k, is a tempered distribution. To see this, observe that
$$L_g(f) \le \sup_{x\in\mathbf{R}^n} (1+|x|)^m |f(x)| \int_{\mathbf{R}^n} (1+|x|)^{k-m} dx,$$
where $m > n+k$ and the expression $\sup_{x\in\mathbf{R}^n} |(1+|x|)^m f(x)|$ is bounded by a sum of $\rho_{\alpha,\beta}$ seminorms in the Schwartz space.
7. The function $\log|x|$ is a tempered distribution. The integral of this function against Schwartz functions is well defined. More generally, any function that is integrable in a neighborhood of the origin and satisfies $|g(x)| \le C(1+|x|)^k$ for $|x| \ge M$ is a tempered distribution.
8. Here is an example of a compactly supported distribution on $\mathbf{R}$ which is neither a function nor a measure:
$$\langle u, f\rangle = \lim_{\varepsilon\to 0} \int_{\varepsilon\le|x|\le 1} f(x)\frac{dx}{x} = \lim_{\varepsilon\to 0} \int_{\varepsilon\le|x|\le 1} (f(x) - f(0))\frac{dx}{x}.$$
We have that $|\langle u, f\rangle| \le 2\|f'\|_{L^\infty}$ and that if $f_n \to f$ in C^∞, then $\langle u, f_n\rangle \to \langle u, f\rangle$.

2.3.c. The Space of Tempered Distributions

Having set down the basic definitions of distributions, we are now going to concentrate our study on the space of tempered distributions. These distributions are the most useful in harmonic analysis. The main reason for this is that the subject is concerned with boundedness of translation-invariant operators and every such bounded operator from $L^p(\mathbf{R}^n)$ to $L^q(\mathbf{R}^n)$ is given by convolution with a tempered distribution. This fact will be shown in Section 2.5.

Suppose that f and g are Schwartz functions and α a multiindex. Integrating by parts $|\alpha|$ times, we obtain

$$\int_{\mathbf{R}^n} (\partial^\alpha f)(x) g(x)\, dx = (-1)^{|\alpha|} \int_{\mathbf{R}^n} f(x)(\partial^\alpha g)(x)\, dx. \tag{2.3.6}$$

If we wanted to define the derivative of a tempered distribution u, we would have to give a definition that extends the definition of the derivative of the function and that satisfies (2.3.6) for g in $\mathcal{S}'$ and $f \in \mathcal{S}$ if the integrals in (2.3.6) are interpreted as actions of distributions on functions. We simply use equation (2.3.6) to define the derivative of a distribution. We have

Definition 2.3.6. Let $u \in \mathcal{S}'$ and α a multiindex. Define

$$\langle \partial^\alpha u, f \rangle = (-1)^{|\alpha|} \langle u, \partial^\alpha f \rangle. \tag{2.3.7}$$

If u is a function, the derivatives of u in the sense of distributions will be called *distributional derivatives.*

In view of Theorem 2.2.14, it is natural to give the following:

Definition 2.3.7. Let $u \in \mathcal{S}'$. We define the Fourier transform $\widehat{u}$ and the inverse Fourier transform $u^\vee$ of a tempered distribution u by

$$\langle \widehat{u}, f \rangle = \langle u, \widehat{f}\, \rangle \qquad \text{and} \qquad \langle u^\vee, f \rangle = \langle u, f^\vee \rangle, \tag{2.3.8}$$

for all f in $\mathcal{S}$.

Example 2.3.8. We observe that $\widehat{\delta_0} = 1$. More generally, for any multiindex α we have $(\partial^\alpha \delta_0)\widehat{\ }= (2\pi i x)^\alpha$. To see this, observe that for all $f \in \mathcal{S}$ we have

$$\begin{aligned} \langle (\partial^\alpha \delta_0)\widehat{\ }, f \rangle &= \langle \partial^\alpha \delta_0 \,, \widehat{f}\, \rangle = (-1)^{|\alpha|} \langle \delta_0 \,, \partial^\alpha \widehat{f}\, \rangle \\ &= (-1)^{|\alpha|} \langle \delta_0 \,, ((-2\pi i x)^\alpha f(x))\widehat{\ } \rangle \\ &= (-1)^{|\alpha|} ((-2\pi i x)^\alpha f(x))\widehat{\ }(0) \\ &= (-1)^{|\alpha|} \int_{\mathbf{R}^n} (-2\pi i x)^\alpha f(x)\, dx \\ &= \int_{\mathbf{R}^n} (2\pi i x)^\alpha f(x)\, dx. \end{aligned}$$

This calculation indicates that $(\partial^\alpha \delta_0)\widehat{\ }$can be identified with the function $(2\pi i x)^\alpha$.

Example 2.3.9. Recall that $\delta_z(f) = f(z)$. For $z \in \mathbf{R}^n$ we have

$$\langle \widehat{\delta_z}, h \rangle = \langle \delta_z, \widehat{h}\, \rangle = \widehat{h}(z) = \int_{\mathbf{R}^n} h(x) e^{-2\pi i x \cdot z}\, dx, \qquad h \in \mathcal{S}(\mathbf{R}^n),$$

that is $\widehat{\delta_z}$ can be identified with the function $x \to e^{-2\pi i x \cdot z}$. In particular, $\widehat{\delta_0} = 1$.

Example 2.3.10. The function $e^{|x|^2}$ is not in $\mathcal{S}'(\mathbf{R}^n)$ and therefore its Fourier transform is not defined as a distribution. However, the Fourier transform of any locally integrable function with polynomial growth at infinity is defined as a tempered distribution.

Now observe that the following are true whenever f, g are in $\mathcal{S}$.

$$\begin{aligned}\int_{\mathbf{R}^n} g(x)f(x-t)\,dx &= \int_{\mathbf{R}^n} g(x+t)f(x)\,dx\\ \int_{\mathbf{R}^n} g(ax)f(x)\,dx &= \int_{\mathbf{R}^n} g(x)a^{-n}f(a^{-1}x)\,dx,\\ \int_{\mathbf{R}^n} \widetilde{g}(x)f(x)\,dx &= \int_{\mathbf{R}^n} g(x)\widetilde{f}(x)\,dx,\end{aligned} \tag{2.3.9}$$

for all $t \in \mathbf{R}^n$ and $a > 0$. Recall now the definitions of τ^t, δ^a and $\widetilde{\ }$ given in (2.2.12). Motivated by (2.3.9), we give the following:

Definition 2.3.11. The *translation* $\tau^t(u)$, the *dilation* $\delta^a(u)$, and the *reflection* $\widetilde{u}$ of a tempered distribution u are defined as follows:

$$\langle \tau^t(u), f\rangle = \langle u, \tau^{-t}(f)\rangle \tag{2.3.10}$$

$$\langle \delta^a(u), f\rangle = \langle u, a^{-n}\delta^{1/a}(f)\rangle \tag{2.3.11}$$

$$\langle \widetilde{u}, f\rangle = \langle u, \widetilde{f}\,\rangle, \tag{2.3.12}$$

for all $t \in \mathbf{R}^n$ and $a > 0$. Let A be an invertible matrix. The composition of a distribution u with an invertible matrix A is the distribution

$$\langle u^A, \varphi\rangle = (\det \mathrm{A})^{-1}\langle u, \varphi^{A^{-1}}\rangle, \tag{2.3.13}$$

where $\varphi^{A^{-1}}(x) = \varphi(A^{-1}x)$.

The reader is asked to show that the operations of translation, dilation, reflection, and differentiation are continuous on tempered distributions.

Example 2.3.12. The Dirac mass at the origin δ_0 is equal to its reflection, while $\delta^a(\delta_0) = a^{-n}\delta_0$. Also, $\tau^x(\delta_0) = \delta_x$ for any $x \in \mathbf{R}^n$.

Now observe that for f, g, and h in $\mathcal{S}$ we have

$$\int_{\mathbf{R}^n} (h*g)(x)f(x)\,dx = \int_{\mathbf{R}^n} g(x)(\widetilde{h}*f)(x)\,dx. \tag{2.3.14}$$

Motivated by (2.3.14), we define the convolution of a function with a tempered distribution as follows:

Definition 2.3.13. Let $u \in \mathcal{S}'$ and $h \in \mathcal{S}$. Define the convolution $h*u$ by

$$\langle h*u, f\rangle = \langle u, \widetilde{h}*f\rangle, \qquad f \in \mathcal{S}. \tag{2.3.15}$$

Example 2.3.14. Let $u = \delta_{x_0}$ and $f \in \mathcal{S}$. Then $f*\delta_{x_0}$ is the function $x \to f(x-x_0)$, for, if $h \in \mathcal{S}$, we have

$$\langle f*\delta_{x_0}, h\rangle = \langle \delta_{x_0}, \widetilde{f}*h\rangle = (\widetilde{f}*h)(x_0) = \int_{\mathbf{R}^n} f(x-x_0)h(x)\,dx\,;$$

thus $f*\delta_{x_0}$ can be identified with the function $x \to f(x-x_0)$. It follows that convolution with δ_0 is the identity operator.

We now define the product of a function and a distribution.

Definition 2.3.15. Let $u \in \mathcal{S}'$ and let h be a C^∞ function that has at most polynomial growth at infinity and the same is true for all of its derivatives. [This means that it satisfies $|(\partial^\alpha h)(x)| \le C(1+|x|)^{k_\alpha}$ for all α and some $k_\alpha > 0$.] Then define the product hu of h and u by

$$\big\langle hu, f\big\rangle = \big\langle u, hf\big\rangle, \qquad f \in \mathcal{S}. \tag{2.3.16}$$

Note that hf is in $\mathcal{S}$ and thus (2.3.16) is well defined. The product of an arbitrary C^∞ function with a tempered distribution is not defined.

We observe that if a function g is supported in a set K, then for all $f \in C_0^\infty(K^c)$ we have

$$\int_{\mathbf{R}^n} f(x)g(x)\,dx = 0\,. \tag{2.3.17}$$

Moreover the support of g is the intersection of all closed sets K with the property (2.3.17) for all f in $C_0^\infty(K^c)$. Motivated by the preceding observation we give the following:

Definition 2.3.16. Let u be in $\mathcal{D}'(\mathbf{R}^n)$. The *support* of u ($\operatorname{supp} u$) is the intersection of all closed sets K with the property

$$\operatorname{supp}\varphi \subseteq K^c \implies \big\langle u, \varphi\big\rangle = 0\,. \tag{2.3.18}$$

Distributions with compact support are exactly those whose support (as defined in the previous definition) is a compact set. To prove this assertion, we start with a distribution u with compact support as defined in Definition 2.3.3. Then there exist $C, N, m > 0$ such that (2.3.4) holds. For a smooth function f whose support is contained in $B(0,N)^c$, the expression on the right in (2.3.4) vanishes and we must therefore have $\big\langle u, f\big\rangle = 0$. This shows that the support of u is bounded and as it is already closed (as an intersection of closed sets) it must be compact. Conversely, if the support of u as defined in Definition 2.3.16 is a compact set, then there exists an $N > 0$ such that $\operatorname{supp} u$ is contained in $B(0,N)$. We take a smooth function η that is equal to 1 on $B(0,N)$ and vanishes off $B(0,N+1)$. Then the support of $f(1-\eta)$ does not meet the support of u and we must have

$$\big\langle u, f\big\rangle = \big\langle u, f\eta\big\rangle + \big\langle u, f(1-\eta)\big\rangle = \big\langle u, f\eta\big\rangle\,.$$

Taking m to be the integer that corresponds to the compact set $K = \overline{B(0,N+1)}$ in (2.3.2), and using that the L^∞ norm of $f\eta$ is controlled by a finite sum of seminorms of the form $\widetilde{\rho}_{\alpha,N+1}$ with $|\alpha| \le m$, we obtain the validity of (2.3.4).

Example 2.3.17. The support of Dirac mass at x_0 is the set $\{x_0\}$.

Along the same lines, we give the following definition:

Definition 2.3.18. We say that a distribution u in $\mathcal{D}'(\mathbf{R}^n)$ coincides with the function h on an open set Ω if

$$\langle u, f\rangle = \int_{\mathbf{R}^n} f(x)h(x)\,dx \qquad \text{for all } f \text{ in } C_0^\infty(\Omega). \tag{2.3.19}$$

When (2.3.19) occurs we will also occasionally say that u agrees with h away from Ω^c.

This definition implies that the support of the distribution $u - h$ is contained in Ω^c.

Example 2.3.19. The distribution $|x|^2 + \delta_{x_1} + \delta_{x_2}$, where x_1, x_2 are in $\mathbf{R}^n$, coincides with the function $|x|^2$ on any open set not containing the points x_1 and x_2. Also, the distribution in Example 2.3.5 8 coincides with the function $x^{-1}\chi_{|x|\le 1}$ away from the origin.

Having ended the streak of definitions regarding operations with distributions, we now discuss properties of convolutions and Fourier transforms.

Theorem 2.3.20. *If $u \in \mathcal{S}'$ and $\varphi \in \mathcal{S}$, then $\varphi * u$ is a C^∞ function. Moreover, for all α multiindices there exist constants $C_\alpha, k_\alpha > 0$ such that*

$$|\partial^\alpha(\varphi * u)(x)| \le C_\alpha(1 + |x|)^{k_\alpha}.$$

Proof. Let ψ be in $\mathcal{S}(\mathbf{R}^n)$. We have

$$\begin{aligned}(\varphi * u)(\psi) = u(\widetilde{\varphi} * \psi) &= u\left(\int_{\mathbf{R}^n} \widetilde{\varphi}(\,\cdot\, - y)\psi(y)\,dy\right)\\ &= u\left(\int_{\mathbf{R}^n} \tau^y(\widetilde{\varphi})(\,\cdot\,)\psi(y)\,dy\right)\\ &= \int_{\mathbf{R}^n} u\left(\tau^y(\widetilde{\varphi})\right)\psi(y)\,dy,\end{aligned} \tag{2.3.20}$$

where the last step is justified by the continuity of u and by the fact that the Riemann sums of the integral in (2.3.20) converge in L^∞ and therefore in the topology of $\mathcal{S}$. This calculation shows that $\varphi * u$ is the function $x \to u\big(\tau^x(\widetilde{\varphi})\big)$.

We now show that $(\varphi * u)(x)$ is a C^∞ function. Let $e_j = (0, \dots, 1, \dots, 0)$ with 1 in the jth entry and zero elsewhere. Then

$$\frac{\tau^{he_j}(\varphi * u)(x) - (\varphi * u)(x)}{h} = u\left(\frac{\tau^{he_j}(\tau^x(\widetilde{\varphi})) - \tau^x(\widetilde{\varphi})}{h}\right) \to u(\tau^x(\partial_j\widetilde{\varphi}))$$

by the continuity of u and the fact that $\big(\tau^{he_j+x}(\widetilde{\varphi}) - \tau^x(\widetilde{\varphi})\big)/h$ tends to $\tau^x(\partial_j\widetilde{\varphi})$ in $\mathcal{S}$ as $h \to 0$. See Exercise 2.3.5(a). The same calculation for higher order derivatives shows that $\varphi * u \in C^\infty$. It follows from (2.3.3) that for some C, m, and k we have

$$\begin{aligned} |\partial^\alpha(\varphi * u)(x)| &\le C \sum_{\substack{|\gamma|\le m \\ |\beta|\le k}} \sup_{y\in\mathbf{R}^n} |y^\gamma \tau^x (\partial^{\alpha+\beta}\widetilde{\varphi})(y)| \\ &= C \sum_{\substack{|\gamma|\le m \\ |\beta|\le k}} \sup_{y\in\mathbf{R}^n} |(x+y)^\gamma (\partial^{\alpha+\beta}\widetilde{\varphi})(y)| \qquad (2.3.21)\\ &\le 2^m C \sum_{|\beta|\le k} \sup_{y\in\mathbf{R}^n} (|x|^m + |y|^m)|(\partial^{\alpha+\beta}\widetilde{\varphi})(y)| \end{aligned}$$

and this clearly implies that $\partial^\alpha(\varphi * u)$ grows at most polynomially at infinity.

□

Next we have the following important result regarding distributions with compact support:

Theorem 2.3.21. *If u is in $\mathcal{E}'(\mathbf{R}^n)$, then $\widehat{u}$ is a real analytic function on $\mathbf{R}^n$. Moreover, $\widehat{u}$ has a holomorphic extension on $\mathbf{C}^n$. In particular, $\widehat{u}$ is a C^∞ function. Moreover, $\widehat{u}$ and all of its derivatives have polynomial growth at infinity.*

Proof. First observe that since u is a distribution with compact support, the action $u(e^{-2\pi i x\cdot(\cdot)})$ of the distribution u on the C^∞ function $\xi \to e^{-2\pi i x\cdot\xi}$ is a well defined function of x. Let f be in $\mathcal{S}$. Then

$$\langle \widehat{u}, f\rangle = \langle u, \widehat{f}\rangle = u\left(\int_{\mathbf{R}^n} f(x)e^{-2\pi i x\cdot\xi}\,dx\right) = \int_{\mathbf{R}^n} f(x)u(e^{-2\pi i x\cdot(\cdot)})\,dx\,,$$

provided we can justify the passage of u from outside the integral to inside the integral. The idea here is as in the proof of the previous theorem. By the linearity of u, it follows that u can be interchanged with the Riemann sums of the preceding integral. Since these Riemann sums converge to the integral in the topology of $\mathcal{E}$, the conclusion follows by the continuity of u.

Now it is straightforward to verify that the function

$$F(z_1,\dots,z_n) = u\big(e^{-2\pi i \sum_{j=1}^n (\cdot) z_j}\big)$$

defined on $\mathbf{C}^n$ is holomorphic. In fact, the continuity of u shows that F is differentiable and its derivative with respect to z_j is the action of the distribution u to the C^∞ function

$$x \to (-2\pi i x_j)e^{-2\pi i \sum_{j=1}^n x_j z_j}\,.$$

Therefore, the restriction of $\widehat{u}$ on $\mathbf{R}^n$ is real analytic. The fact that $\widehat{u}$ and all of its derivatives have polynomial growth at infinity is a consequence of (2.3.4).

□

Next we give a proposition that extends the properties of the Fourier transform to tempered distributions.

Proposition 2.3.22. *Given u, v in $\mathcal{S}'(\mathbf{R}^n)$, $f \in \mathcal{S}$, $y \in \mathbf{R}^n$, b complex scalar, α multiindex, and $a > 0$, we have*

(1) $\widehat{u+v} = \widehat{u} + \widehat{v}$

(2) $\widehat{bu} = b\widehat{u}$

(3) *If* $u_j \to u$ *in* $\mathcal{S}'$, *then* $\widehat{u}_j \to \widehat{u}$ *in* $\mathcal{S}'$

(4) $(\widetilde{u})\widehat{\ } = (\widehat{u})\widetilde{\ }$

(5) $(\tau^y(u))\widehat{\ } = e^{-2\pi i y\cdot\xi}\widehat{u}$

(6) $(e^{2\pi i x\cdot y}u)\widehat{\ } = \tau^y(\widehat{u})$

(7) $\delta^a(u)\widehat{\ } = (\widehat{u})_a = a^{-n}(\delta^{a^{-1}}(\widehat{u}))$

(8) $(\partial^\alpha u)\widehat{\ } = (2\pi i\xi)^\alpha\widehat{u}$

(9) $\partial^\alpha\widehat{u} = ((-2\pi i x)^\alpha u)\widehat{\ }$

(10) $(\widehat{u})^\vee = u$

(11) $\widehat{f * u} = \widehat{f}\,\widehat{u}$

(12) $\widehat{fu} = \widehat{f} * \widehat{u}$

(13) ***(Leibniz's rule)*** $\partial_j^m(fu) = \sum_{k=0}^m \binom{m}{k}(\partial_j^k f)(\partial_j^{m-k}u)$, $m \in \mathbf{Z}^+$

(14) ***(Leibniz's rule)*** $\partial^\alpha(fu) = \sum_{\gamma_1=0}^{\alpha_1}\cdots\sum_{\gamma_n=0}^{\alpha_n}\binom{\alpha_1}{\gamma_1}\cdots\binom{\alpha_n}{\gamma_n}(\partial^\gamma f)(\partial^{\alpha-\gamma}u)$

(15) *If* u_k, $u \in L^p(\mathbf{R}^n)$ *and* $u_k \to u$ *in* L^p $(1 \le p \le \infty)$, *then* $u_k \to u$ *in* $\mathcal{S}'(\mathbf{R}^n)$. *Therefore, convergence in* $\mathcal{S}$ *implies convergence in* L^p, *which in turn implies convergence in* $\mathcal{S}'(\mathbf{R}^n)$.

Proof. All the statements can be proved easily using duality and the corresponding statements for Schwartz functions. □

We continue with an application of Theorem 2.3.21.

Proposition 2.3.23. *$C_0^\infty(\mathbf{R}^n)$ is dense in $\mathcal{S}'(\mathbf{R}^n)$. In particular, given $u \in \mathcal{S}'(\mathbf{R}^n)$, there exists a sequence of C_0^∞ functions f_k such that $f_k \to u$ in the sense of tempered distributions.*

Proof. Fix a function in $C_0^\infty(\mathbf{R}^n)$ with $\varphi(x) = 1$ in a neighborhood of the origin. Let $\varphi_k(x) = \delta^{1/k}(\varphi)(x) = \varphi(x/k)$. It follows from Exercise 2.3.5(b) that for $u \in \mathcal{S}'(\mathbf{R}^n)$, $\varphi_k u \to u$ in $\mathcal{S}'$. By Proposition 2.3.22 (3), we have that the map $u \to (\varphi_k\widehat{u})^\vee$ is continuous on $\mathcal{S}'(\mathbf{R}^n)$. Now Theorem 2.3.21 gives that $(\varphi_k\widehat{u})^\vee$ is a C^∞ function and therefore $\varphi_j(\varphi_k\widehat{u})^\vee$ is in $C_0^\infty(\mathbf{R}^n)$. As observed, $\varphi_j(\varphi_k\widehat{u})^\vee \to (\varphi_k\widehat{u})^\vee$ in $\mathcal{S}'$ when k is fixed and $j \to \infty$. Exercise 2.3.5(c) gives that the diagonal sequence $\varphi_k(\varphi_k f)\widehat{\ }$ converges to $f\widehat{\ }$ in $\mathcal{S}$ as $k \to \infty$ for all $f \in \mathcal{S}$. Using duality and Exercise

2.2.2, we conclude that the sequence of C_0^∞ functions $\varphi_k(\varphi_k\widehat{u})^\vee$ converges to u in $\mathcal{S}'$ as $k \to \infty$.

□

2.3.d. The Space of Tempered Distributions Modulo Polynomials

Definition 2.3.24. We define $\mathcal{P}$ to be set of all polynomials of n real variables

$$\sum_{|\beta|\le m} c_\beta x^\beta = \sum_{|\beta_1|+\cdots+|\beta_n|\le m} c_{(\beta_1,\dots,\beta_n)} x_1^{\beta_1}\dots x_n^{\beta_n}$$

with complex coefficients c_β and m arbitrary integer. We then define an equivalence relation $\sim$ on $\mathcal{S}'(\mathbf{R}^n)$ by setting

$$u \sim v \iff u - v \in \mathcal{P}\,.$$

The space of all resulting equivalence classes will be denoted by $\mathcal{S}'/\mathcal{P}$.

To avoid cumbersome notation, two elements u, v of the same equivalence class in $\mathcal{S}'/\mathcal{P}$ will be identified and in this case we will be writing $u = v$ in $\mathcal{S}'/\mathcal{P}$. Note that for $u, v \in \mathcal{S}'/\mathcal{P}$ we have

$$u = v \quad \text{in } \mathcal{S}'/\mathcal{P} \iff \begin{cases} \langle\widehat{u},\phi\rangle = \langle\widehat{v},\phi\rangle & \text{for all } \phi\in\mathcal{S}(\mathbf{R}^n) \\ \text{with supp }\phi\cap\big(\mathbf{R}^n\setminus\{0\}\big)=\emptyset\,. \end{cases} \tag{2.3.22}$$

Proposition 2.3.25. *Let $\mathcal{S}_\infty(\mathbf{R}^n)$ the space of all Schwartz functions φ that satisfy*

$$\int_{\mathbf{R}^n} x^\gamma \varphi(x)\,dx = 0$$

for all multiindices γ. Then $\mathcal{S}_\infty(\mathbf{R}^n)$ is a subspace of $\mathcal{S}(\mathbf{R}^n)$ that inherits the same topology as $\mathcal{S}(\mathbf{R}^n)$ and whose dual is $\mathcal{S}'(\mathbf{R}^n)/\mathcal{P}$, that is,

$$\big(\mathcal{S}_\infty(\mathbf{R}^n)\big)' = \mathcal{S}'(\mathbf{R}^n)/\mathcal{P}\,.$$

Proof. Consider the map J that takes an element u of $\mathcal{S}'(\mathbf{R}^n)$ to the equivalence class in $\mathcal{S}'(\mathbf{R}^n)/\mathcal{P}$ that contains it. The kernel of this map is $\mathcal{P}$ and the claimed identification follows. □

It follows that a sequence $u_j \to u$ in $\mathcal{S}'(\mathbf{R}^n)/\mathcal{P}$ if and only if u_j, u are elements of this space and we have

$$\langle u_j,\varphi\rangle \to \langle u,\varphi\rangle$$

for all φ in $\mathcal{S}_\infty(\mathbf{R}^n)$.

Exercises

2.3.1. Show that a positive measure μ that satisfies

$$\int_{\mathbf{R}^n} \frac{d\mu(x)}{(1+|x|)^k} < +\infty,$$

for some $k > 0$, can be identified with a tempered distribution. Note that if we think Lebesgue measure as a tempered distribution, then it coincides with the function 1 (also thought as a tempered distribution).

2.3.2. Let $\varphi, f \in \mathcal{S}(\mathbf{R}^n)$, and for $\varepsilon > 0$ let $\varphi_\varepsilon(x) = \varepsilon^{-n}\varphi(\varepsilon^{-1}x)$. Prove that $\varphi_\varepsilon * f \to b\,f$ in $\mathcal{S}$, where b is the integral of φ.

2.3.3. Prove that for all $a > 0$, $u \in \mathcal{S}'(\mathbf{R}^n)$, and $f \in \mathcal{S}(\mathbf{R}^n)$ we have

$$\delta^a(f) * \delta^a(u) = a^{-n}\delta^a(f * u)$$

2.3.4. (a) Prove that the derivative of $\chi_{[a,b]}$ is $\delta_a - \delta_b$.
(b) Compute the $\partial_j \chi_{B(0,1)}$ on $\mathbf{R}^2$.
(c) Compute the Fourier transforms of the locally integrable functions $\sin x$ and $\cos x$.
(d) Prove that the derivative of the distribution $\log|x| \in \mathcal{S}'(\mathbf{R})$ is the distribution

$$u(\varphi) = \lim_{\varepsilon \to 0} \int_{\varepsilon \le |x|} \varphi(x)\frac{dx}{x}.$$

2.3.5. Let $f \in \mathcal{S}(\mathbf{R}^n)$ and let $\varphi \in C_0^\infty$ be identically equal to 1 in a neighborhood of origin. Define $\varphi_k(x) = \varphi(x/k)$ as in the proof of Proposition 2.3.23.
(a) Prove that $\dfrac{\tau^{he_j}(f) - f}{h} \to \partial_j f$ in $\mathcal{S}$ as $h \to 0$.
(b) Prove that $\varphi_k f \to f$ in $\mathcal{S}$ as $k \to \infty$.
(c) Prove that the sequence $\varphi_k(\varphi_k f)\widehat{\ }$ converges to $\widehat{f}$ in $\mathcal{S}$ as $k \to \infty$.

2.3.6. Use Theorem 2.3.21 to show that there does not exist a nonzero C_0^∞ function whose Fourier transform is also a C_0^∞ function.

2.3.7. Let $f \in L^p(\mathbf{R}^n)$ for some $1 \le p \le \infty$. Show that the sequence of functions

$$g_N(\xi) = \int_{B(0,N)} f(x)e^{-2\pi i x\cdot\xi}\,dx$$

converges to $\widehat{f}$ in $\mathcal{S}'$.

2.3.8. Let $(c_k)_{k\in\mathbf{Z}^n}$ be a sequence that satisfies $|c_k| \le A(1+|k|)^M$ for all k and some fixed M and $A > 0$. Let δ_k denote Dirac mass at the integer k. Show that the sequence of distributions

$$\sum_{|k|\le N} c_k \delta_k$$

converges to some tempered distribution u in $\mathcal{S}'(\mathbf{R}^n)$ as $N \to \infty$. Also show that $\widehat{u}$ is the $\mathcal{S}'$ limit of the sequence of functions

$$h_N(\xi) = \sum_{|k|\le N} c_k e^{-2\pi i \xi\cdot k}.$$

2.3.9. A distribution in $\mathcal{S}'(\mathbf{R}^n)$ is called *homogeneous of degree* $\gamma \in \mathbf{C}$ if

$$\langle u, \delta^\lambda(f)\rangle = \lambda^{-n-\gamma}\langle u, f\rangle, \qquad \text{for all } \lambda > 0.$$

(a) Prove that this definition agrees with the usual definition for functions.
(b) Show that δ_0 is homogeneous of degree $-n$.
(c) Prove that if u is homogeneous of degree γ, then $\partial^\alpha u$ is homogeneous of degree $\gamma - |\alpha|$.
(d) Show that u is homogeneous of degree γ if and only if $\widehat{u}$ is homogeneous of degree $-n-\gamma$.

2.3.10. Show that the functions e^{inx} and e^{-inx} converge to zero in $\mathcal{S}'$ and $\mathcal{D}'$ as $n \to \infty$. Conclude that multiplication of distributions is not a continuous operation even when it is defined. What is the limit of $\sqrt{n}(1+n|x|^2)^{-1}$ in $\mathcal{D}'(\mathbf{R})$ as $n \to \infty$?

2.3.11. (*S. Bernstein*) Let f be a bounded function on $\mathbf{R}^n$ with $\widehat{f}$ supported in the ball $B(0,R)$. Prove that for all multiindices α there exist constants $C_{\alpha,n}$ (depending only on α and on the dimension n) such that

$$\big\|\partial^\alpha f\big\|_{L^\infty} \le C_{\alpha,n} R^{|\alpha|} \big\|f\big\|_{L^\infty}.$$

$\big[$*Hint:* Write $f = f * h_{1/R}$, where h is a Schwartz function h in $\mathbf{R}^n$ whose Fourier transform is equal to one on the ball $B(0,1)$ and vanishes outside the ball $B(0,2)$.$\big]$

2.3.12. Let Φ be a Schwartz function whose Fourier transform has compact support and is equal to 1 in a neighborhood of zero. Also let Θ be a distribution whose Fourier transform is a function that is equal to 1 in a neighborhood of infinity and vanishes in a neighborhood of zero. Prove the following.
(a) For all u in $\mathcal{S}'(\mathbf{R}^n)$ we have

$$\Big(\widehat{\Phi}(\xi/2^N)\widehat{u}\Big)^\vee \to 0 \quad \text{in } \mathcal{S}'(\mathbf{R}^n) \text{ as } N \to \infty.$$

(b) For all u in $\mathcal{S}'(\mathbf{R}^n)/\mathcal{P}$ we have

$$\Big(\widehat{\Theta}(\xi/2^N)\widehat{u}\Big)^\vee \to u \quad \text{in } \mathcal{S}'(\mathbf{R}^n)/\mathcal{P} \text{ as } N \to -\infty.$$

[*Hint:* Prove first the corresponding assertions for functions φ in $\mathcal{S}$ or $\mathcal{S}_\infty$ with convergence in the topology of these spaces.]

2.3.13. Prove that there exists a function in L^p for $2 < p < \infty$ whose distributional Fourier transform is not a locally integrable function.
[*Hint:* Assume the converse and use the closed graph theorem to obtain the inequality $\|\widehat{f}\|_{L^1(B(0,1)} \le C\|f\|_{L^p}$.]

2.4. More about Distributions and the Fourier Transform*

In this section we will discuss some further properties of distributions and Fourier transforms and bring up certain connections that arise between harmonic analysis and partial differential equations.

2.4.a. Distributions Supported at a Point

We begin with the following characterization of distributions supported at a single point.

Proposition 2.4.1. *If $u \in \mathcal{S}'(\mathbf{R}^n)$ is supported in the singleton $\{x_0\}$, then there exists an integer k and complex numbers a_α such that*

$$u = \sum_{|\alpha| \le k} a_\alpha \partial^\alpha \delta_{x_0}.$$

Proof. Without loss of generality we may assume that $x_0 = 0$. By (2.3.3) we have that for some C, m, and k

$$|\langle u, f\rangle| \le C \sum_{\substack{|\alpha|\le m \\ |\beta|\le k}} \sup_{x\in\mathbf{R}^n} |x^\alpha(\partial^\beta f)(x)| \qquad \text{for all } f \in \mathcal{S}(\mathbf{R}^n).$$

We will now prove that if $\varphi \in \mathcal{S}$ satisfies

$$(\partial^\alpha \varphi)(0) = 0 \qquad \text{for all } |\alpha| \le k, \tag{2.4.1}$$

then $\langle u, \varphi\rangle = 0$. To see this, fix a φ satisfying (2.4.1) and let $\zeta(x)$ be a smooth function on $\mathbf{R}^n$ that is equal to 1 when $|x| \ge 2$ and equal to zero for $|x| \le 1$. Let $\zeta^\varepsilon(x) = \zeta(x/\varepsilon)$. Then, using (2.4.1) and the continuity of the derivatives of φ at the origin, it is not hard to show that $\rho_{\alpha,\beta}(\zeta^\varepsilon\varphi - \varphi) \to 0$ as $\varepsilon \to 0$ for all $|\alpha| \le m$ and $|\beta| \le k$. Then

$$|\langle u, \varphi\rangle| \le |\langle u, \zeta^\varepsilon\varphi\rangle| + |\langle u, \zeta^\varepsilon\varphi - \varphi\rangle| \le 0 + C \sum_{\substack{|\alpha|\le m \\ |\beta|\le k}} \rho_{\alpha,\beta}(\zeta^\varepsilon\varphi - \varphi) \to 0,$$

as $\varepsilon \to 0$. This proves our assertion.

Now let $f \in \mathcal{S}(\mathbf{R}^n)$. Let η be a C_0^∞ function on $\mathbf{R}^n$ that is equal to 1 in a neighborhood of the origin. Write

$$f(x) = \eta(x)\Big(\sum_{|\alpha|\le k} \frac{(\partial^\alpha f)(0)}{\alpha!}x^\alpha + h(x)\Big) + (1-\eta(x))f(x), \tag{2.4.2}$$

where $h(x) = O(x^{k+1})$ as $|x| \to 0$. Then ηh satisfies (2.4.1) and hence $\langle u, \eta h\rangle = 0$ by the claim. Also, $\langle u, ((1-\eta)f)\rangle = 0$ by our hypothesis. Applying u to both sides of (2.4.2), we obtain

$$\langle u, f\rangle = \sum_{|\alpha|\le k} \frac{(\partial^\alpha f)(0)}{\alpha!} u(x^\alpha \eta(x)) = \sum_{|\alpha|\le k} a_\alpha (\partial^\alpha \delta_0)(f)\,,$$

with $a_\alpha = (-1)^{|\alpha|} u(x^\alpha \eta(x))/\alpha!$. This proves the proposition.

□

An immediate consequence is the following result.

Corollary 2.4.2. *Let $u \in \mathcal{S}'(\mathbf{R}^n)$. If $\widehat{u}$ is supported in the singleton $\{\xi_0\}$, then u is a finite linear combination of functions $(-2\pi i\xi)^\alpha e^{2\pi i \xi\cdot\xi_0}$, where α is a multiindex. In particular, if $\widehat{u}$ is supported at the origin, then u is a polynomial.*

Proof. Proposition 2.4.1 gives that $\widehat{u}$ is a linear combination of derivatives of Dirac masses at ξ_0. Then property (8) in Proposition 2.3.22 gives the required conclusion.

□

2.4.b. The Laplacian

The *Laplacian* Δ is a partial differential operator acting on tempered distributions on $\mathbf{R}^n$ as follows:

$$\Delta(u) = \sum_{j=1}^n \partial_j^2 u\,.$$

Solutions of Laplace's equation $\Delta(u) = 0$ are called *harmonic* distributions. We have the following:

Corollary 2.4.3. *Let $u \in \mathcal{S}'(\mathbf{R}^n)$ satisfy $\Delta(u) = 0$. Then u is a polynomial.*

Proof. Taking Fourier transforms, we obtain that $\widehat{\Delta(u)} = 0$. Therefore,

$$-4\pi^2|\xi|^2\widehat{u} = 0 \qquad \text{in } \mathcal{S}'.$$

This implies that $\widehat{u}$ is supported at the origin and by Corollary 2.4.2; it follows that u must be polynomial.

□

Liouville's classical theorem, that every bounded harmonic function must be constant, is a consequence of Corollary 2.4.3. See Exercise 2.4.2.

Next we would like to compute the fundamental solutions of Laplace's equation in $\mathbf{R}^n$. A distribution is called a *fundamental solution* of a partial differential operator L if we have $L(u) = \delta_0$. The following result gives the fundamental solution of the Laplacian.

Proposition 2.4.4. *For $n \geq 3$ we have*

$$\Delta(|x|^{2-n}) = -(n-2)\frac{2\pi^{n/2}}{\Gamma(n/2)}\delta_0\,, \tag{2.4.3}$$

while for $n = 2$,

$$\Delta(\log|x|) = 2\pi\delta_0\,. \tag{2.4.4}$$

Proof. We will use Green's identity

$$\int_\Omega v\Delta(u) - u\Delta(v)\,dx = \int_{\partial\Omega}\left(v\frac{\partial u}{\partial \nu} - u\frac{\partial v}{\partial \nu}\right)ds\,,$$

where Ω is an open set in $\mathbf{R}^n$ with smooth boundary and $\dfrac{\partial v}{\partial \nu}$ denotes the derivative of v with respect to the outer unit normal vector. Take $\Omega = \mathbf{R}^n - B(0,\varepsilon)$, $v = |x|^{2-n}$, and $u = f$ a $C_0^\infty(\mathbf{R}^n)$ function in the previous identity. The normal derivative of $f(r\theta)$ is the derivative with respect to the radial variable r. Observe that $\Delta(|x|^{2-n}) = 0$ for $x \neq 0$. We obtain

$$\int_{|x|>\varepsilon}\Delta(f)(x)|x|^{2-n}\,dx = -\int_{|r\theta|=\varepsilon}\left(\varepsilon^{2-n}\frac{\partial f}{\partial r} - f(r\theta)\frac{\partial r^{2-n}}{\partial r}\right)d\theta. \tag{2.4.5}$$

Now observe two things: First, that for some $C = C(f)$ we have

$$\left|\int_{|r\theta|=\varepsilon}\frac{\partial f}{\partial r}\,d\theta\right| \leq C\varepsilon^{n-1}\,;$$

second, that

$$\int_{|r\theta|=\varepsilon} f(r\theta)\varepsilon^{1-n}\,d\theta \to \omega_{n-1}f(0)$$

as $\varepsilon \to 0$. Letting $\varepsilon \to 0$ in (2.4.5), we obtain that

$$\lim_{\varepsilon\to 0}\int_{|x|>\varepsilon}\Delta(f)(x)|x|^{2-n}\,dx = -(n-2)\omega_{n-1}f(0),$$

which implies (2.4.3) in view of the formula for ω_{n-1} given in Appendix A.3.

The proof of (2.4.4) is identical. The only difference is that the quantity $\dfrac{\partial r^{2-n}}{\partial r}$ in (2.4.5) is replaced by the slightly different quantity $\dfrac{\partial \log r}{\partial r}$.

□

2.4.c. Homogeneous Distributions

The fundamental solutions of the Laplacian are locally integrable functions on $\mathbf{R}^n$ and also homogeneous of degree $2-n$ when $n \geq 3$. Since homogeneous distributions arise naturally in applications, it is desirable to pursue their study in general. Here we do not plan to undertake such a study, but we do plan to discuss a few examples of homogeneous distributions.

Definition 2.4.5. For $z \in \mathbf{C}$ we define a distribution u_z as follows:

$$\langle u_z, f\rangle = \int_{\mathbf{R}^n} \frac{\pi^{\frac{z+n}{2}}}{\Gamma(\frac{z+n}{2})} |x|^z f(x)\,dx\,. \tag{2.4.6}$$

Clearly the u_z's coincide with the locally integrable functions

$$\pi^{\frac{z+n}{2}} \Gamma\big(\tfrac{z+n}{2}\big)^{-1} |x|^z$$

when $\operatorname{Re} z > -n$ and the definition makes sense only for that range of z's. It follows from its definition that u_z is a homogeneous distribution of degree z.

We would like to extend the definition of u_z for $z \in \mathbf{C}$. Let $\operatorname{Re} z > -n$ first. Fix N to be a positive integer. Given $f \in \mathcal{S}(\mathbf{R}^n)$, write the integral in (2.4.6) as follows:

$$\begin{aligned}
&\int_{|x|<1} \frac{\pi^{\frac{z+n}{2}}}{\Gamma(\frac{z+n}{2})} \left\{ f(x) - \sum_{|\alpha|\leq N} \frac{(\partial^\alpha f)(0)}{\alpha!} x^\alpha \right\} |x|^z\,dx \\
&\quad + \int_{|x|>1} \frac{\pi^{\frac{z+n}{2}}}{\Gamma(\frac{z+n}{2})} f(x)|x|^z\,dx \\
&\quad + \int_{|x|<1} \frac{\pi^{\frac{z+n}{2}}}{\Gamma(\frac{z+n}{2})} \sum_{|\alpha|\leq N} \frac{(\partial^\alpha f)(0)}{\alpha!} x^\alpha |x|^z\,dx \\
&= \int_{|x|<1} \frac{\pi^{\frac{z+n}{2}}}{\Gamma(\frac{z+n}{2})} \left\{ f(x) - \sum_{|\alpha|\leq N} \frac{(\partial^\alpha f)(0)}{\alpha!} x^\alpha \right\} |x|^z\,dx \\
&\quad + \int_{|x|>1} \frac{\pi^{\frac{z+n}{2}}}{\Gamma(\frac{z+n}{2})} f(x)|x|^z\,dx \\
&\quad + \sum_{|\alpha|\leq N} \frac{(\partial^\alpha f)(0)}{\alpha!} \frac{\pi^{\frac{z+n}{2}}}{\Gamma(\frac{z+n}{2})} \int_{r=0}^1 \int_{\mathbf{S}^{n-1}} (r\theta)^\alpha\, r^{z+n-1}\,dr\,d\theta\,,
\end{aligned}$$

where we switched to polar coordinates in the penultimate integral. Next set

$$b(n,\alpha,z) = \frac{\pi^{\frac{z+n}{2}}}{\Gamma(\frac{z+n}{2})} \frac{1}{\alpha!} \left(\int_{\mathbf{S}^{n-1}} \theta^\alpha\,d\theta \right) \int_{r=0}^1 r^{|\alpha|+n+z-1}\,dr = \frac{\pi^{\frac{z+n}{2}}}{\Gamma(\frac{z+n}{2})} \frac{\frac{1}{\alpha!}\int_{\mathbf{S}^{n-1}} \theta^\alpha\,d\theta}{|\alpha|+z+n}\,.$$

Now recall that the function $\Gamma(\frac{z+n}{2})$ has simple poles at $z = -n$, $z = -(n-1)$, $z = -(n-2)$, and so on (see Appendix A.5). These poles cancel exactly the poles created by $(|\alpha| + z + n)^{-1}$ when $z = -n - |\alpha|$ for $|\alpha| \le N$. We therefore have

$$\begin{aligned} \langle u_z, f \rangle &= \int_{|x|\ge 1} \frac{\pi^{\frac{z+n}{2}}}{\Gamma(\frac{z+n}{2})} f(x)|x|^z\,dx + \sum_{|\alpha|\le N} b(n,\alpha,z)\,\langle \partial^\alpha \delta_0, f\rangle \\ &+ \int_{|x|<1} \frac{\pi^{\frac{z+n}{2}}}{\Gamma(\frac{z+n}{2})} \left\{ f(x) - \sum_{|\alpha|\le N} \frac{(\partial^\alpha f)(0)}{\alpha!} x^\alpha \right\} |x|^z\,dx\,. \end{aligned} \tag{2.4.7}$$

Both integrals converge absolutely when $\operatorname{Re} z > -N - n - 1$, since the expression inside the curly brackets above is bounded by a constant multiple of $|x|^{N+1}$, and the resulting function of z in (2.4.7) is a well-defined analytic function in the range $\operatorname{Re} z > -N - n - 1$.

Since N was arbitrary, $\langle u_z, f\rangle$ has an analytic extension on all of $\mathbf{C}$. Therefore, u_z is a distribution-valued entire function of z.

Next we would like to calculate the Fourier transform of u_z. We know by Exercise 2.3.9 that $\widehat{u_z}$ is a homogeneous distribution of degree $-n - z$. The choice of constant in the definition of u_z was made to justify the following theorem:

Theorem 2.4.6. *For all $z \in \mathbf{C}$ we have $\widehat{u_z} = u_{-n-z}$.*

Proof. The idea of the proof is straightforward. First we show that for a certain range of z's we have

$$\int_{\mathbf{R}^n} |\xi|^z \widehat{\varphi}(\xi)\,d\xi = C(n,z) \int_{\mathbf{R}^n} |x|^{-n-z} \varphi(x)\,dx, \tag{2.4.8}$$

for some fixed constant $C(n,z)$ and all $\varphi \in \mathcal{S}(\mathbf{R}^n)$. Next we pick a specific φ to evaluate the constant $C(n,z)$. Then we use analytic continuation to extend the validity of (2.4.8) for all z's. Use polar coordinates by setting $\xi = \rho\varphi$ and $x = r\theta$ in (2.4.8). We have

$$\begin{aligned} &\int_{\mathbf{R}^n} |\xi|^z \widehat{\varphi}(\xi)\,d\xi \\ &= \int_0^\infty \rho^{z+n-1} \int_0^\infty \int_{\mathbf{S}^{n-1}} \varphi(r\theta) \left(\int_{\mathbf{S}^{n-1}} e^{-2\pi i r\rho(\theta\cdot\varphi)} d\varphi \right) d\theta\, r^{n-1}\,dr\,d\rho \\ &= \int_0^\infty \left(\int_0^\infty \sigma(r\rho)\rho^{z+n-1}\,d\rho \right) \left(\int_{\mathbf{S}^{n-1}} \varphi(r\theta) d\theta \right) r^{n-1}\,dr \\ &= C(n,z) \int_0^\infty r^{-z-n} \left(\int_{\mathbf{S}^{n-1}} \varphi(r\theta)\,d\theta \right) r^{n-1}\,dr \\ &= C(n,z) \int_{\mathbf{R}^n} |x|^{-n-z} \varphi(x)\,dx\,, \end{aligned}$$

where we set

$$\sigma(t) = \int_{\mathbf{S}^{n-1}} e^{-2\pi i t(\theta\cdot\varphi)}\, d\varphi = \int_{\mathbf{S}^{n-1}} e^{-2\pi i t(\varphi_1)}\, d\varphi\,, \tag{2.4.9}$$

$$C(n,z) = \int_0^\infty \sigma(t) t^{z+n-1}\, dt\,, \tag{2.4.10}$$

and the second equality in (2.4.9) is a consequence of rotational invariance. It remains to prove that the integral in (2.4.10) converges for some range of z's.

If $n=1$, then $\sigma(t) = \int_{\mathbf{S}^0} e^{-2\pi i t\varphi} d\varphi = e^{-2\pi i t} + e^{2\pi i t} = 2\cos(2\pi t)$ and the integral in (2.4.10) converges conditionally for $-1 < \operatorname{Re} z < 0$.

Let us therefore assume that $n \ge 2$. Since $|\sigma(t)| \le \omega_{n-1}$, the integral converges near zero when $-n < \operatorname{Re} z$. Let us study the behavior of $\sigma(t)$ for t large. Using the formula in Appendix D.2 and the definition of Bessel functions in Appendix B.1, we write

$$\sigma(t) = \int_{-1}^{1} e^{2\pi i t s} \omega_{n-2}\big(\sqrt{1-s^2}\big)^{n-2} \frac{ds}{\sqrt{1-s^2}} = c_n J_{n-2}(2\pi t),$$

for some constant c_n. Using the asymptotics for Bessel functions (Appendix B.6), we obtain that $|\sigma(t)| \le c t^{-1/2}$ when $n-2 > -1/2$. In either case the integral in (2.4.10) converges absolutely near infinity when $\operatorname{Re} z + n - 1 - 1/2 < -1$ (i.e., when $\operatorname{Re} z < -n + 1/2$).

We have now proved that when $-n < \operatorname{Re} z < -n + 1/2$ we have

$$\widehat{u_z} = C(n,z) u_{-n-z}$$

for some constant $C(n,z)$ that we wish to compute. Insert the function $\varphi(x) = e^{-\pi|x|^2}$ in (2.4.8). Example 2.2.9 gives that this function is equal to its Fourier transform. Use polar coordinates to write

$$\omega_{n-1} \int_0^\infty r^{z+n-1} e^{-\pi r^2} dr = C(n,z)\omega_{n-1} \int_0^\infty r^{-z-n+n-1} e^{-\pi r^2} dr\,.$$

Change variables $s = \pi r^2$ and use the definition of the gamma function to obtain that

$$C(n,z) = \frac{\Gamma(\frac{z+n}{2})}{\Gamma(-\frac{z}{2})} \frac{\pi^{-\frac{z+n}{2}}}{\pi^{\frac{z}{2}}}\,.$$

It follows that $\widehat{u_z} = u_{-n-z}$ for the range of z's considered.

At this point observe that for every $f \in \mathcal{S}(\mathbf{R}^n)$, the function $z \to \langle \widehat{u_z} - u_{-z-n}, f\rangle$ is entire and vanishes for $-n < \operatorname{Re} z < -n + 1/2$. Therefore, it must vanish everywhere and the theorem is proved.

□

Homogeneous distributions were introduced in Exercise 2.3.9. We already saw that the Dirac mass on $\mathbf{R}^n$ is a homogeneous distribution of degree $-n$. There is another important example of a homogeneous distributions of degree $-n$ which we now discuss.

Let Ω be an integrable function on the sphere $\mathbf{S}^{n-1}$ with integral zero. Define a tempered distribution W_Ω on $\mathbf{R}^n$ by setting

$$\langle W_\Omega, f\rangle = \lim_{\varepsilon\to 0}\int_{|x|\ge\varepsilon}\frac{\Omega(x/|x|)}{|x|^n}f(x)\,dx. \tag{2.4.11}$$

Let us see that W_Ω is a well-defined tempered distribution on $\mathbf{R}^n$. Indeed, since $\Omega(x/|x|)/|x|^n$ has integral zero over all annuli centered at the origin, we obtain

$$\begin{aligned}
|\langle W_\Omega,\varphi\rangle| &= \left|\lim_{\varepsilon\to 0}\int_{\varepsilon\le|x|\le 1}\frac{\Omega(x/|x|)}{|x|^n}(\varphi(x)-\varphi(0))\,dx+\int_{|x|\ge 1}\frac{\Omega(x/|x|)}{|x|^n}\varphi(x)\,dx\right|\\
&\le \|\nabla\varphi\|_{L^\infty}\int_{|x|\le 1}\frac{\Omega(x/|x|)}{|x|^{n-1}}\,dx+\Big(\sup_{x\in\mathbf{R}^n}|x|\,|\varphi(x)|\Big)\int_{|x|\ge 1}\frac{\Omega(x/|x|)}{|x|^{n+1}}\,dx\\
&\le C_1\|\nabla\varphi\|_{L^\infty}\|\Omega\|_{L^1(\mathbf{S}^{n-1})}+C_2\sum_{|\alpha|\le 1}\|\varphi(x)x^\alpha\|_{L^\infty}\|\Omega\|_{L^1(\mathbf{S}^{n-1})},
\end{aligned}$$

for suitable constants C_1 and C_2 in view of (2.2.2).

Next leave to the reader to check that $W_\Omega\in\mathcal{S}'(\mathbf{R}^n)$ is a homogeneous distribution of degree $-n$ just like the Dirac mass at the origin. It is an interesting fact that all homogeneous distributions on $\mathbf{R}^n$ of degree $-n$ that coincide with a smooth function away from origin arise in this way. We have the following result.

Proposition 2.4.7. *Suppose that m is a C^∞ function on $\mathbf{R}^n\setminus\{0\}$ that is homogeneous of degree zero. Then there exist a scalar b and a C^∞ function Ω on $\mathbf{S}^{n-1}$ with integral zero such that*

$$m^\vee = b\,\delta_0 + W_\Omega\,, \tag{2.4.12}$$

where W_Ω denotes the distribution defined in (2.4.11).

To prove this result we will need the following Proposition whose proof we postpone until the end of this section.

Proposition 2.4.8. *Suppose that u is a C^∞ function on $\mathbf{R}^n\setminus\{0\}$ that is homogeneous of degree z, $(z\in\mathbf{C})$. Then $\widehat{u}$ is a C^∞ function on $\mathbf{R}^n\setminus\{0\}$.*

We now prove Proposition 2.4.7 using Proposition 2.4.8.

Proof. Let a be the average of the smooth function m over $\mathbf{S}^{n-1}$. The function $m-a$ is homogeneous of degree zero and thus locally integrable on $\mathbf{R}^n$; hence it can be thought as a distribution that we will call $\widehat{u}$ (the Fourier transform of a tempered distribution u). Since $\widehat{u}$ is a C^∞ function on $\mathbf{R}^n\setminus\{0\}$, Proposition 2.4.8 implies that u is also a C^∞ function on $\mathbf{R}^n\setminus\{0\}$. Let Ω be the restriction of u on $\mathbf{S}^{n-1}$. Then Ω is a well-defined C^∞ function on $\mathbf{S}^{n-1}$. Since u is a homogeneous function of degree $-n$ that coincides with the smooth function Ω on $\mathbf{S}^{n-1}$, it follows that $u(x) = \Omega(x/|x|)/|x|^n$ for x in $\mathbf{R}^n \setminus \{0\}$.

We will show that Ω has mean value zero over $\mathbf{S}^{n-1}$. Pick a nonnegative, radial, smooth, and nonzero function ψ on $\mathbf{R}^n$ supported in the annulus $1 < |x| < 2$. Then, switching to polar coordinates, we obtain

$$\begin{aligned} \langle u, \psi \rangle &= \int_{\mathbf{R}^n} \frac{\Omega(x/|x|)}{|x|^n} \psi(x)\, dx = c_\psi \int_{\mathbf{S}^{n-1}} \Omega(\theta)\, d\theta, \\ \langle u, \psi \rangle = \langle \widehat{u}, \widehat{\psi} \rangle &= \int_{\mathbf{R}^n} (m(\xi) - a)\widehat{\psi}(\xi)\, d\xi = c'_\psi \int_{\mathbf{S}^{n-1}} (m(\theta) - a)\, d\theta = 0, \end{aligned}$$

and thus Ω has mean value zero over $\mathbf{S}^{n-1}$ (since $c_\psi \neq 0$).

We can now legitimately define the distribution W_Ω, which coincides with the function $\Omega(x/|x|)/|x|^n$ on $\mathbf{R}^n \setminus \{0\}$. But u also coincides with this function on $\mathbf{R}^n \setminus \{0\}$. It follows that $u - W_\Omega$ is supported at the origin. Proposition 2.4.1 now gives that $u - W_\Omega$ is a sum of derivatives of Dirac masses. Since both distributions are homogeneous of degree $-n$, it follows that

$$u - W_\Omega = c\delta_0.$$

But $u = (m-a)^\vee = m^\vee - a\delta_0$, and thus $m^\vee = (c+a)\delta_0 + W_\Omega$. This proves the proposition.

□

We now turn to the proof of Proposition 2.4.8.

Proof. Let $u \in \mathcal{S}'$ be homogeneous of degree z and C^∞ on $\mathbf{R}^n \setminus \{0\}$. We need to show that $\widehat{u}$ is C^∞ away from the origin. We will prove that $\widehat{u}$ is C^M for all M. Fix $M \in \mathbf{Z}^+$ and let α be any multiindex such that

$$|\alpha| > n + M + \operatorname{Re} z. \tag{2.4.13}$$

Pick a C^∞ function φ on $\mathbf{R}^n$ that is equal to 1 when $|x| \geq 2$ and equal to zero for $|x| \leq 1$. Write $u_0 = (1-\varphi)u$ and $u_\infty = \varphi u$. Then

$$\partial^\alpha u = \partial^\alpha u_0 + \partial^\alpha u_\infty \qquad \text{and thus} \qquad \widehat{\partial^\alpha u} = \widehat{\partial^\alpha u_0} + \widehat{\partial^\alpha u_\infty},$$

where the operations are performed in the sense of distributions. Since u_0 is compactly supported, Theorem 2.3.21 implies that $\widehat{\partial^\alpha u_0}$ is C^∞. Now Leibniz's rule gives that

$$\partial^\alpha u_\infty = v + \varphi \partial^\alpha u,$$

where v is a smooth function supported in the annulus $1 \leq |x| \leq 2$. Then $\widehat{v}$ is C^∞ and we only need to show that $\widehat{\varphi \partial^\alpha u}$ is C^M. The function $\varphi \partial^\alpha u$ is actually C^∞, and by the homogeneity of z (Exercise 2.3.9) we obtain that $(\partial^\alpha u)(x) = |x|^{-|\alpha|+z}(\partial^\alpha u)(x/|x|)$. Since φ is supported away from zero, it follows that

$$|\varphi(x)(\partial^\alpha u)(x)| \leq \frac{C_\alpha}{(1+|x|)^{|\alpha| - \operatorname{Re} z}} \tag{2.4.14}$$

for some $C_\alpha > 0$. It is now straightforward to see that if a function satisfies (2.4.14), then its Fourier transform is C^M whenever (2.4.13) is satisfied. See Exercise 2.4.1.

We conclude that $\widehat{\partial^\alpha u}$ is a C^M function if (2.4.13) holds. Since $\widehat{\partial^\alpha u}(\xi) = (2\pi i\xi)^\alpha \widehat{u}(\xi)$, we can derive smoothness information about $\widehat{u}$ away from the origin. Let $\xi \neq 0$. Pick a neighborhood V of ξ that does not meet the jth coordinate axis for some $1 \le j \le n$. Then $\eta_j \neq 0$ when $\eta \in V$. Let α be the multiindex $(0, \dots, M, \dots, 0)$ with M in the jth coordinate and zeros elsewhere. Then $(2\pi i\eta_j)^M \widehat{u}(\eta)$ is a C^M function on V and thus so is $\widehat{u}(\eta)$ since we can divide by η_j^M. We conclude that $\widehat{u}(\xi)$ is C^M on $\mathbf{R}^n \setminus \{0\}$. Since M is arbitrary, the conclusion follows.

□

We end this section with an example that illustrates a precise use of some of the key ideas discussed in this section.

Example 2.4.9. Let η be a smooth function on $\mathbf{R}^n$ that is equal to 1 on the set $|x| \ge 1/2$ and vanishes on the set $|x| \le 1/4$. Let $0 < \mathrm{Re}(\alpha) < n$. Let

$$g(\xi) = \big(\eta(x)|x|^{-\alpha}\big)^\wedge(\xi)\,.$$

We claim that g decays faster than the reciprocal of any polynomial at infinity and that

$$g(\xi) - \frac{\pi^{\alpha-\frac{n}{2}}\Gamma(\frac{n-\alpha}{2})}{\Gamma(\frac{\alpha}{2})}|\xi|^{\alpha-n}$$

is a C^∞ function on $\mathbf{R}^n$. Therefore, g is integrable on $\mathbf{R}^n$. This example indicates the interplay between the smoothness of a function and the decay of its Fourier transform. The smoothness of the function $\eta(x)|x|^{-\alpha}$ near zero is reflected by the decay of g near infinity. Moreover, the function $\eta(x)|x|^{-\alpha}$ is not affected by the bump η near infinity, and this results in a behavior of $g(\xi)$ near zero similar to that of $(|x|^{-\alpha})^\wedge(\xi)$.

To see these assertions, first observe that $\partial^\gamma(\eta(x)|x|^{-\alpha})$ is integrable and thus $(-2\pi i\xi)^\gamma g(\xi)$ is bounded if γ is large enough. This gives the decay of g near infinity. We now use Theorem 2.4.6 to obtain

$$g(\xi) = \frac{\pi^{\alpha-\frac{n}{2}}\Gamma(\frac{n-\alpha}{2})}{\Gamma(\frac{\alpha}{2})}|\xi|^{\alpha-n} + \widehat{\varphi}(\xi)\,,$$

where $\widehat{\varphi}(\xi) = \big((\eta(x)-1)|x|^{-\alpha}\big)^\wedge(\xi)$, which is C^∞ as the Fourier transform of a compactly supported distribution.

Exercises

2.4.1. Suppose that a function f satisfies the estimate

$$|f(x)| \le \frac{C_\alpha}{(1+|x|)^N}\,,$$

for some $N > n$. Then $\widehat{f}$ is C^M when $1 \le M \le [N-n]$.

2.4.2. Use Corollary 2.4.3 to prove Liouville's theorem that every bounded harmonic function on $\mathbf{R}^n$ must be a constant. Derive as a consequence the *fundamental theorem of algebra* stating that every polynomial on $\mathbf{C}$ must have a complex root.

2.4.3. Prove that e^x is not in $\mathcal{S}'(\mathbf{R})$ but that $e^x e^{ie^x}$ is in $\mathcal{S}'(\mathbf{R})$.

2.4.4. This exercise provides an interesting example of a Schwartz function that coincides with its Fourier transform. Let $f(x) = \operatorname{sech}(\pi x)$, $x \in \mathbf{R}$. Prove that $\widehat{f}(\xi) = f(\xi)$.
[*Hint:* Integrate the function e^{iaz} over the rectangular contour with corners $(-R, 0)$, $(R, 0)$, $(R, i\pi)$, and $(-R, i\pi)$.]

2.4.5. (*Ismagilov* [**272**]) Construct an uncountable family of linearly independent Schwartz functions f_a such that $|f_a| = |f_b|$ and $|\widehat{f_a}| = |\widehat{f_b}|$ for all f_a and f_b in the family.
[*Hint:* Let w be a smooth nonzero function whose Fourier transform is supported in the interval $[-1/2, 1/2]$ and let φ be a real-valued smooth nonconstant periodic function with period 1. Then take $f_a(x) = w(x)e^{i\varphi(x-a)}$ for $a \in \mathbf{R}$.]

2.4.6. Let P_y be the Poison kernel defined in (2.1.14). Prove that for $f \in L^p(\mathbf{R}^n)$, $1 \le p < \infty$, the function

$$(x, y) \to (P_y * f)(x)$$

is a harmonic function on $\mathbf{R}^{n+1}_+$. Using the Fourier transform and Exercise 2.2.11 to prove that $(P_{y_1} * P_{y_2})(x) = P_{y_1+y_2}(x)$ for all $x \in \mathbf{R}^n$.

2.4.7. (a) For a fixed $x_0 \in \mathbf{R}^n$, show that the function

$$v(x; x_0) = \frac{1 - |x|^2}{|x - x_0|^n}$$

is harmonic on $\mathbf{R}^n \setminus \{x_0\}$.
(b) For fixed $x_0 \in \mathbf{S}^{n-1}$, prove that the family of functions $\theta \to v(\theta; rx_0)$, $0 < r < 1$ defined on the sphere satisfies

$$\lim_{r \uparrow 1} \int_{\substack{\theta \in \mathbf{S}^{n-1} \\ |\theta - x_0| > \delta}} v(\theta; rx_0)\, d\theta = 0$$

uniformly in x_0. The function $v(\theta; rx_0)$ is called the *Poisson kernel for the sphere*.
(c) Let f be a continuous function on $\mathbf{S}^{n-1}$. Prove that the function

$$u(x) = \frac{1}{\omega_{n-1}}(1 - |x|^2) \int_{\mathbf{S}^{n-1}} \frac{f(\theta)}{|x - \theta|^n}\, d\theta$$

solves the Dirichlet problem $\Delta(u) = 0$ on $|x| < 1$ and $u = f$ on $|x| = 1$.

2.4.8. Fix a real number λ, $0 < \lambda < n$.
(a) Prove that

$$\int_{\mathbf{S}^n} |\xi - \eta|^{-\lambda}\, d\xi = 2^{n-\lambda} \frac{\pi^{\frac{n}{2}} \Gamma(\frac{n-\lambda}{2})}{\Gamma(n - \frac{\lambda}{2})}.$$

(b) Prove that

$$\int_{\mathbf{R}^n} |x - y|^{-\lambda} (1 + |x|^2)^{\frac{\lambda}{2} - n}\, dx = 2^{n-\lambda} \frac{\pi^{\frac{n}{2}} \Gamma(\frac{n-\lambda}{2})}{\Gamma(n - \frac{\lambda}{2})} (1 + |y|^2)^{-\frac{\lambda}{2}}.$$

[*Hint:* Use the stereographic projection in Appendix D.5.]

2.4.9. Prove the following *beta integral identity:*

$$\int_{\mathbf{R}^n} \frac{dt}{|x - t|^{\alpha_1} |y - t|^{\alpha_2}} = \pi^{\frac{n}{2}} \frac{\Gamma(\frac{n-\alpha_1}{2}) \Gamma(\frac{n-\alpha_2}{2}) \Gamma(\frac{\alpha_1+\alpha_2-n}{2})}{\Gamma(\frac{\alpha_1}{2}) \Gamma(\frac{\alpha_2}{2}) \Gamma(n - \frac{\alpha_1+\alpha_2}{2})} |x - y|^{n-\alpha_1-\alpha_2},$$

where $0 < \alpha_1, \alpha_2 < n$, $\alpha_1 + \alpha_2 > n$.

2.4.10. (a) Prove that if a function f on $\mathbf{R}^n$ ($n \ge 2$) is constant on the spheres $r\mathbf{S}^{n-1}$ for all $r > 0$, then so is its Fourier transform.
(b) If a function on $\mathbf{R}^n$ ($n \ge 2$) is constant on all $(n-2)$–dimensional spheres perpendicular to $e_1 = (1, 0, \dots, 0)$, prove that its Fourier transform possesses the same property.

2.4.11. (*Grafakos and Morpurgo* [**219**]) Suppose that $0 < d_1, d_2, d_3 < n$ satisfy $d_1 + d_2 + d_3 = 2n$. Prove that for any distinct $x, y, z \in \mathbf{R}^n$ we have the identity

$$\int_{\mathbf{R}^n} |x - t|^{-d_2} |y - t|^{-d_3} |z - t|^{-d_1}\, dt$$

$$= \pi^{\frac{n}{2}} \prod_{j=1}^{3} \frac{\Gamma(n - \frac{d_j}{2})}{\Gamma(\frac{d_j}{2})} |x - y|^{d_1-n} |y - z|^{d_2-n} |z - x|^{d_3-n}.$$

[*Hint:* Reduce matters to the case where $z = 0$ and $y = e_1$. Then take the Fourier transform in x and use Exercise 2.4.10.]

2.4.12. (a) Integrate the function e^{iz^2} over the contour consisting of the three pieces $P_1 = \{(x, 0) : 0 \le x \le R\}$, $P_2 = \{(R\cos\theta, R\sin\theta) : 0 \le \theta \le \frac{\pi}{4}\}$, and $P_3 = \{(t, t) : t \text{ between } \frac{R\sqrt{2}}{2} \text{ and } 0\}$ to obtain the *Fresnel integral identity*:

$$\lim_{R\to\infty} \int_{-R}^{+R} e^{ix^2}\, dx = \frac{\sqrt{2\pi}}{2}(1 + i).$$

(b) Use the result in part (a) to show that the Fourier transform of the function $e^{i\pi|x|^2}$ in $\mathbf{R}^n$ is equal to $e^{i\frac{\pi n}{4}}e^{-i\pi|\xi|^2}$.
[*Hint:* Part (a): On P_2 we have $e^{-R^2\sin(2\theta)} \le e^{-\frac{4}{\pi}R^2\theta}$ and the integral over P_2 tends to 0. Part (b): Try first $n = 1$.]

2.5. Convolution Operators on L^p Spaces and Multipliers

In this section we will study the class of operators that commute with translations. We prove in this section that bounded operators that commute with translations must be of convolution type. As we will see in the next chapters, convolution operators arise in many situations, and we would like to know under what circumstances they are bounded.

2.5.a. Operators that Commute with Translations

Definition 2.5.1. A vector space X of measurable functions on $\mathbf{R}^n$ is called *closed under translations* if for $f \in X$ we have $\tau^z(f) \in X$ for all $z \in \mathbf{R}^n$. Let X and Y be vector spaces of measurable functions on $\mathbf{R}^n$ that are closed under translations. Let also T be an operator from X to Y. We say that T *commutes with translations* or is *translation invariant* if

$$T(\tau^y(f)) = \tau^y(T(f))$$

for all $f \in X$ and all $y \in \mathbf{R}^n$.

It is automatic to see that convolution operators commute with translations. One of the main goals of this section is to prove the converse; every bounded linear operator that commutes with translations is of convolution type. We have the following:

Theorem 2.5.2. *Suppose $1 \le p, q \le \infty$. Suppose T is a bounded linear operator from $L^p(\mathbf{R}^n)$ into $L^q(\mathbf{R}^n)$ that commutes with translations. Then there exists a unique tempered distribution v such that*

$$T(f) = f * v \qquad \text{for all } f \in \mathcal{S}.$$

The theorem will be a consequence of the following two results:

Lemma 2.5.3. *Under the hypotheses of Theorem 2.5.2 and for $f \in \mathcal{S}(\mathbf{R}^n)$, the distributional derivatives of $T(f)$ are L^q functions that satisfy*

$$\partial^\alpha T(f) = T(\partial^\alpha f), \qquad \text{for all } \alpha \text{ multiindices.} \tag{2.5.1}$$

Lemma 2.5.4. *Let $1 \le q \le \infty$ and let $h \in L^q(\mathbf{R}^n)$. If all distributional derivatives $\partial^\alpha h$ are also in L^q, then h is almost everywhere equal to a continuous function H satisfying*

$$|H(0)| \le C_{n,q} \sum_{|\alpha| \le n+1} \big\|\partial^\alpha h\big\|_{L^q}. \tag{2.5.2}$$

Proof. Assuming lemmata 2.5.3 and 2.5.4, we prove Theorem 2.5.2. Define a linear functional u on $\mathcal{S}$ by setting

$$\langle u, f\rangle = T(f)(0).$$

By (2.5.1), (2.5.2), (2.2.7), and the boundedness of T, we have

$$\begin{aligned}
\big|\langle u, f\rangle\big| \le\ & C_{n,q} \sum_{|\alpha| \le n+1} \big\|\partial^\alpha T(f)\big\|_{L^q} \\
\le\ & C_{n,q} \sum_{|\alpha| \le n+1} \big\|T(\partial^\alpha f)\big\|_{L^q} \\
\le\ & C_{n,q}\big\|T\big\|_{L^p \to L^q} \sum_{|\alpha| \le n+1} \big\|\partial^\alpha f\big\|_{L^p} \\
\le\ & C_{n,q}\big\|T\big\|_{L^p \to L^q} \sum_{|\alpha|,|\beta| \le N} \rho_{\alpha,\beta}(f)
\end{aligned}$$

which implies that u is in $\mathcal{S}'$. We now set $v = \widetilde{u}$ and we claim that $T(f) = f * v$ for $f \in \mathcal{S}$. To see this by Theorem 2.3.20 and by the translation invariance of T, we have

$$\begin{aligned}
(f * \widetilde{u}\,)(x) &= \langle \widetilde{u}, \tau^{x}(\widetilde{f}\,)\,\rangle = \langle u, \tau^{-x}(f)\rangle \\
&= T(\tau^{-x}(f))(0) = \tau^{-x}(T(f))(0) \\
&= T(f)(x)
\end{aligned}$$

whenever $f \in \mathcal{S}(\mathbf{R}^n)$. This proves the theorem. □

We now return to Lemmata 2.5.3 and 2.5.4. We begin with Lemma 2.5.3.

Proof. Let $\alpha = (0, \dots, 1, \dots, 0)$, where 1 is in the jth entry. Let $f, g \in \mathcal{S}$. As

$$\frac{\tau^{-he_j}(g) - g}{h} - \partial_j g \to 0 \quad \text{in } \mathcal{S} \text{ as } h \to 0, \tag{2.5.3}$$

it follows that (2.5.3) converges to zero in L^p and thus

$$T\bigg(\frac{\tau^{-he_j}(g) - g}{h} - \partial_j g\bigg) \to 0 \quad \text{in } L^q \text{ as } h \to 0. \tag{2.5.4}$$

Therefore, (2.5.4) converges to zero when integrated against a $\varphi \in \mathcal{S} \subseteq L^{q'}$. Taking $\varphi = \partial_j g$, we obtain

$$\begin{aligned}
\langle \partial_j T(f), g\rangle &= -\int_{\mathbf{R}^n} T(f)\, \partial_j g\, dx \\
&= -\lim_{h\to 0}\int_{\mathbf{R}^n} T(f)\Big(\frac{\tau^{-he_j}-I}{h}(g)\Big)\, dx \\
&= \lim_{h\to 0}\int_{\mathbf{R}^n} \Big(\Big(\frac{I-\tau^{he_j}}{h}\Big)\circ T(f)\Big)\, g\, dx \\
&= \lim_{h\to 0}\int_{\mathbf{R}^n} T\Big(\frac{I-\tau^{he_j}}{h}(f)\Big)\, g\, dx \\
&= \int_{\mathbf{R}^n} T(\partial_j f)\, g\, dx\,,
\end{aligned}$$

where we used the fact that T commutes with translations and (2.5.4) . This shows that $\partial_j T(f) = T(\partial_j f)$. The general case follows by induction on $|\alpha|$.

□

We now prove Lemma 2.5.4.

Proof. Let $R \geq 1$. Fix a C_0^∞ function φ_R that is equal to 1 in the ball $|x| \leq R$ and equal to zero when $|x| \geq 2R$. Since h is in $L^q(\mathbf{R}^n)$, it follows that $\varphi_R h$ is in $L^1(\mathbf{R}^n)$. We will show that $\widehat{\varphi_R h}$ is also in L^1. We begin with the inequality

$$1 \leq C_n(1+|x|)^{-(n+1)} \sum_{|\alpha|\leq n+1} |(-2\pi i x)^\alpha|\,, \tag{2.5.5}$$

which is trivial for $|x| \leq 2$ and follows from (2.2.2) when $|x| \geq 2$. Now multiply (2.5.5) by $|\widehat{\varphi_R h}(x)|$ to obtain

$$\begin{aligned}
|\widehat{\varphi_R h}(x)| &\leq C_n(1+|x|)^{-(n+1)} \sum_{|\alpha|\leq n+1} |(-2\pi i x)^\alpha \widehat{\varphi_R h}(x)| \\
&\leq C_n(1+|x|)^{-(n+1)} \sum_{|\alpha|\leq n+1} \big\|(\partial^\alpha(\varphi_R h))\widehat{\ }\big\|_{L^\infty} \\
&\leq C_n(1+|x|)^{-(n+1)} \sum_{|\alpha|\leq n+1} \big\|\partial^\alpha(\varphi_R h)\big\|_{L^1} \\
&\leq C_n(2^n R^n v_n)^{1/q'}(1+|x|)^{-(n+1)} \sum_{|\alpha|\leq n+1} \big\|\partial^\alpha(\varphi_R h)\big\|_{L^q} \\
&\leq C_{n,R}(1+|x|)^{-(n+1)} \sum_{|\alpha|\leq n+1} \big\|\partial^\alpha h\big\|_{L^q}\,,
\end{aligned}$$

where we used Leibniz's rule and the fact that all derivatives of φ_R are bounded by constants (depending on R).

Integrate the previously displayed inequality with respect to x to obtain

$$\big\|\widehat{\varphi_R h}\big\|_{L^1} \le C_{R,n} \sum_{|\alpha|\le n+1} \big\|\partial^\alpha h\big\|_{L^q} < \infty. \tag{2.5.6}$$

Therefore, Fourier inversion holds for $\varphi_R h$ (see Exercise 2.2.6). This implies that $\varphi_R h$ is equal a.e. to a continuous function, namely the inverse Fourier transform of its Fourier transform. Since $\varphi_R = 1$ on the ball $B(0,R)$, we conclude that h is a.e. equal to a continuous function in this ball. Since $R > 0$ was arbitrary, it follows that h is a.e. equal to a continuous function on $\mathbf{R}^n$, which we denote by H. Finally, (2.5.2) is a direct consequence of (2.5.6) with $R = 1$, since $|H(0)| \le \big\|\widehat{\varphi_1 h}\big\|_{L^1}$.

□

2.5.b. The Transpose and the Adjoint of a Linear Operator

We briefly discuss the notions of the transpose and the adjoint of a linear operator. We first recall the complex and real inner products. For f, g measurable functions on $\mathbf{R}^n$, we define the *complex inner product*

$$\langle f \,|\, g \rangle = \int_{\mathbf{R}^n} f(x)\overline{g(x)}\,dx$$

whenever the integral converges absolutely. We will reserve the notation

$$\langle f, g \rangle = \int_{\mathbf{R}^n} f(x) g(x)\,dx$$

for the *real inner product* on $L^2(\mathbf{R}^n)$ and also for the action of a distribution f on a test function g. (This notation also makes sense when a distribution f coincides with a function.)

For a bounded linear operator T from $L^p(X,\mu) \to L^q(Y,\nu)$ we will denote by T^* its *adjoint operator* defined by

$$\langle T(f) \,|\, g \rangle = \int_Y T(f)\,\overline{g}\,d\nu = \int_X f\,\overline{T^*(g)}\,d\mu = \langle f \,|\, T^*(g) \rangle \tag{2.5.7}$$

for f in $L^p(X,\mu)$ and g in $L^{q'}(Y,\nu)$ (or in a dense subspace of it). We also define the *transpose* of T as the unique operator T^t that satisfies

$$\langle T(f), g \rangle = \int_{\mathbf{R}^n} T(f)\,g\,dx = \int_{\mathbf{R}^n} f\,T^t(g)\,dx = \langle f, T^t(g) \rangle$$

for all $f \in L^p(X,\mu)$ and all $g \in L^{q'}(Y,\nu)$.

If T is an integral operator of the form

$$T(f)(x) = \int_X K(x,y) f(y)\,d\mu(y),$$

then T^* and T^t are also integral operators with kernels $K^*(x,y) = \overline{K(y,x)}$ and $K^t(x,y) = K(y,x)$, respectively. If T has the form $T(f) = (\widehat{f}m)^\vee$, that is, it is given by multiplication on the Fourier transform by a (complex-valued) function $m(\xi)$, then T^* is given by multiplication on the Fourier transform by the function $\overline{m(\xi)}$. Indeed for f, g in $\mathcal{S}(\mathbf{R}^n)$ we have

$$\begin{aligned}\int_{\mathbf{R}^n} f\, \overline{T^*(g)}\, dx &= \int_{\mathbf{R}^n} T(f)\, \overline{g}\, dx \\ &= \int_{\mathbf{R}^n} \widehat{T(f)}\, \widehat{\overline{g}}\, d\xi \\ &= \int_{\mathbf{R}^n} \widehat{f}\, \overline{\overline{m}\,\widehat{g}}\, d\xi \\ &= \int_{\mathbf{R}^n} f\, \overline{(\overline{m}\,\widehat{g})^\vee}\, dx\,.\end{aligned}$$

A similar argument [using Theorem 2.2.14 (5)] gives that if T is given by multiplication on the Fourier transform by the function $m(\xi)$, then T^t is given by multiplication on the Fourier transform by the function $m(-\xi)$. As the complex-valued functions $\overline{m(\xi)}$ and $m(-\xi)$ may be different, the operators T^* and T^t may be different in general. Also, if $m(\xi)$ is real valued, then T is *self-adjoint* (i.e., $T = T^*$) while if $m(\xi)$ is even, T is *self-transpose* (i.e., $T = T^t$).

2.5.c. The Spaces $\mathcal{M}^{p,q}(\mathbf{R}^n)$

Definition 2.5.5. Given $1 \le p, q \le \infty$, we shall denote by $\mathcal{M}^{p,q}(\mathbf{R}^n)$ the set of all bounded linear operators from $L^p(\mathbf{R}^n)$ to $L^q(\mathbf{R}^n)$ that commute with translations.

By Theorem 2.5.2 we have that every T in $\mathcal{M}^{p,q}$ is given by convolution with a tempered distribution. We introduce a norm $\|\cdot\|$ on $\mathcal{M}^{p,q}$ by setting

$$\big\|T\big\|_{\mathcal{M}^{p,q}} = \big\|T\big\|_{L^p \to L^q},$$

that is, the norm of T in $\mathcal{M}^{p,q}$ is the operator norm of T as an operator from L^p to L^q. It is a known fact that under this norm $\mathcal{M}^{p,q}$ is a complete normed space (i.e., a Banach space).

Next we show that when $p > q$ the set $\mathcal{M}^{p,q}$ consists of only one element, namely the zero operator $T = 0$. This means that the only interesting classes of operators arise when $p \le q$.

Theorem 2.5.6. $\mathcal{M}^{p,q} = \{0\}$ *whenever* $1 \le q < p \le \infty$.

Proof. Let f be a nonzero C_0^∞ function and let $h \in \mathbf{R}^n$. We have

$$\big\|\tau^h(T(f)) + T(f)\big\|_{L^q} = \big\|T(\tau^h(f) + f)\big\|_{L^q} \le \big\|T\big\|_{L^p \to L^q} \big\|\tau^h(f) + f\big\|_{L^p}.$$

Now let $|h| \to \infty$ and use Exercise 2.5.1. We conclude that

$$2^{\frac{1}{q}} \big\|T(f)\big\|_{L^q} \le \big\|T\big\|_{L^p \to L^q} 2^{\frac{1}{p}} \big\|f\big\|_{L^p},$$

which is impossible if $q < p$ unless T is the zero operator.

□

Next we have a theorem concerning the duals of the spaces $\mathcal{M}^{p,q}(\mathbf{R}^n)$.

Theorem 2.5.7. *Let $1 \le p \le q \le \infty$ and $T \in \mathcal{M}^{p,q}(\mathbf{R}^n)$. Then T can be defined on $L^{q'}(\mathbf{R}^n)$ [coinciding with its previous definition on the subspace $L^p(\mathbf{R}^n) \cap L^{q'}(\mathbf{R}^n)$ of $L^p(\mathbf{R}^n)$] so that it maps $L^{q'}(\mathbf{R}^n)$ to $L^{p'}(\mathbf{R}^n)$ with norm*

$$\big\|T\big\|_{L^{q'}\to L^{p'}} = \big\|T\big\|_{L^p\to L^q}. \tag{2.5.8}$$

(Recall $\infty' = 1$.) In other words, we have the following isometric identification of spaces:

$$\mathcal{M}^{q',p'}(\mathbf{R}^n) = \mathcal{M}^{p,q}(\mathbf{R}^n).$$

Proof. We first observe that if $T : L^p \to L^q$ is given by convolution with $u \in \mathcal{S}'$, then $T^* : L^{q'} \to L^{p'}$ is given by convolution with $\overline{\widetilde{u}} \in \mathcal{S}'$. Indeed, for f in $L^p(\mathbf{R}^n)$ and g in $L^{q'}(\mathbf{R}^n)$ we have

$$\begin{aligned}
\int_{\mathbf{R}^n} f\,\overline{T^*(g)}\,dx &= \int_{\mathbf{R}^n} T(f)\,\overline{g}\,dx \\
&= \int_{\mathbf{R}^n} (f*u)\,\overline{g}\,dx \\
&= \int_{\mathbf{R}^n} f\,(\overline{g}*\widetilde{u})\,dx \\
&= \int_{\mathbf{R}^n} f\,\overline{g*\overline{\widetilde{u}}}\,dx\,.
\end{aligned}$$

Therefore T^* is given by convolution with $\overline{\widetilde{u}}$. Moreover, T^* is well defined on $L^{q'}$. Using the simple identity

$$\overline{f*\overline{\widetilde{u}}} = (\widetilde{\overline{f}}*u)^{\sim} \qquad f \in L^{q'}, \tag{2.5.9}$$

it follows that T is also well defined on $L^{q'}$. It remains to show that T (convolution with u) and T^* (convolution with $\overline{\widetilde{u}}$) map $L^{q'}$ to $L^{p'}$ with the same norm. But this is easily follows from (2.5.9), which implies that

$$\frac{\big\|\widetilde{\overline{f}}*u\big\|_{L^{p'}}}{\big\|f\big\|_{L^{q'}}} = \frac{\big\|f*\overline{\widetilde{u}}\,\big\|_{L^{p'}}}{\big\|f\big\|_{L^{q'}}},$$

for all $f \in L^{q'}$, $f \neq 0$. We conclude that $\big\|T^*\big\|_{L^{q'}\to L^{p'}} = \big\|T\big\|_{L^{q'}\to L^{p'}}$ and therefore $\big\|T\big\|_{L^p\to L^q} = \big\|T\big\|_{L^{q'}\to L^{p'}}$.

□

We next focus attention on the spaces $\mathcal{M}^{p,q}(\mathbf{R}^n)$ whenever $p = q$. These spaces are of particular interest as they include the singular integral operators, which we will study in Chapter 4.

2.5.d. Characterizations of $\mathcal{M}^{1,1}(\mathbf{R}^n)$, $\mathcal{M}^{2,2}(\mathbf{R}^n)$, and $\mathcal{M}^{\infty,\infty}(\mathbf{R}^n)$

It would be desirable to have a characterization of the spaces $\mathcal{M}^{p,p}$ in terms of properties of the convolving distribution. Unfortunately, this is unknown at present (it is not clear if it is possible) except for the cases $p = 1$, $p = 2$, and $p = \infty$ as we will now see.

Theorem 2.5.8. *An operator T is in $\mathcal{M}^{1,1}(\mathbf{R}^n)$ if and only if it is given by convolution with a finite Borel (complex-valued) measure. In this case, the norm of the operator is equal to the total variation of the measure.*

Proof. If T is given with convolution with a finite Borel measure μ, then clearly T maps L^1 into itself and $\|T\|_{L^1\to L^1} \le \|\mu\|_{\mathcal{M}}$, where $\|\mu\|_{\mathcal{M}}$ is the total variation of μ.

Conversely, let T be an operator bounded from L^1 to L^1. By Theorem 2.5.2, T is given by convolution with a tempered distribution u. Let

$$f_\varepsilon(x) = \varepsilon^{-n} e^{-\pi|x/\varepsilon|^2} .$$

Since the functions f_ε are uniformly bounded in L^1, it follows from the boundedness of T that $f_\varepsilon * u$ are also uniformly bounded in L^1. Since L^1 is naturally embedded in the space of finite Borel measures, which is the dual the space C_{00} of continuous functions that tend to zero at infinity, we obtain that the family $f_\varepsilon * u$ lies in a fixed multiple of the unit ball of C_{00}^*. By the Banach-Alaoglu theorem, this is a weak* compact set. Therefore, some subsequence of $f_\varepsilon * u$ converges in the weak* topology to a measure μ. That is, for some $\varepsilon_k \to 0$ and all $g \in C_{00}(\mathbf{R}^n)$ we have

$$\lim_{k\to\infty} \int_{\mathbf{R}^n} g(x)(f_{\varepsilon_k} * u)(x)\,dx = \int_{\mathbf{R}^n} g(x)\,d\mu(x) . \tag{2.5.10}$$

We claim that $u = \mu$. To see this fix $g \in \mathcal{S}$. Equation (2.5.10) implies that

$$\langle u, \widetilde{f_{\varepsilon_k}} * g\rangle = \langle u, f_{\varepsilon_k} * g\rangle \to \langle \mu, g\rangle$$

as $k \to \infty$. Exercise 2.3.2 gives that $g * f_{\varepsilon_k}$ converges to g in $\mathcal{S}$. Therefore,

$$\langle u, f_{\varepsilon_k} * g\rangle \to \langle u, g\rangle .$$

It follows from (2.5.10) that $\langle u, g\rangle = \langle \mu, g\rangle$ and since g was arbitrary, $u = \mu$.

Next, (2.5.10) implies that for all $g \in C_{00}$ we have

$$\left| \int_{\mathbf{R}^n} g(x)\,d\mu(x)\right| \le \|g\|_{L^\infty} \sup_k \|f_{\varepsilon_k} * u\|_{L^1} \le \|g\|_{L^\infty} \|T\|_{L^1\to L^1} . \tag{2.5.11}$$

The Riesz representation theorem gives that the norm of the functional

$$g \to \int_{\mathbf{R}^n} g(x)\,d\mu(x)$$

on C_{00} is exactly $\|\mu\|_{\mathcal{M}}$. It follows from (2.5.11) that $\|T\|_{L^1\to L^1} \ge \|\mu\|_{\mathcal{M}}$. Since the opposite inequality is obvious, we conclude that

$$\|T\|_{L^1\to L^1} = \|\mu\|_{\mathcal{M}}\,.$$

□

The following corollary is a direct consequence of Theorems 2.5.7 and 2.5.8.

Corollary 2.5.9. *An operator T is in $\mathcal{M}^{\infty,\infty}(\mathbf{R}^n)$ if and only if it is given by convolution with a finite Borel measure. If this is the case, the norm of $T : L^\infty \to L^\infty$ is equal to the total variation of the measure.*

We now study the case $p = 2$. We have the following theorem:

Theorem 2.5.10. *An operator T is in $\mathcal{M}^{2,2}(\mathbf{R}^n)$ if and only if it is given by convolution with some $u \in \mathcal{S}'$ whose Fourier transform $\widehat{u}$ is an L^∞ function. In this case the norm of $T : L^2 \to L^2$ is equal to $\|\widehat{u}\|_{L^\infty}$.*

Proof. If $\widehat{u} \in L^\infty$, Plancherel's theorem gives

$$\int_{\mathbf{R}^n} |f * u|^2\,dx = \int_{\mathbf{R}^n} |\widehat{f}(\xi)\widehat{u}(\xi)|^2\,d\xi \le \|\widehat{u}\|_{L^\infty}^2 \|\widehat{f}\|_{L^2}^2\,;$$

therefore, $\|T\|_{L^2\to L^2} \le \|\widehat{u}\|_{L^\infty}$ and hence T is in $\mathcal{M}^{2,2}(\mathbf{R}^n)$.

Now suppose that $T \in \mathcal{M}^{2,2}(\mathbf{R}^n)$ is given by convolution with a tempered distribution u. We will show that $\widehat{u}$ is a bounded function. For $R > 0$ let φ_R be a C_0^∞ function supported inside the ball $B(0, 2R)$ and equal to one on the ball $B(0, R)$. The product of the function φ_R with the distribution $\widehat{u}$ is $\varphi_R\widehat{u} = ((\varphi_R)^\vee * u)\widehat{\ } = T(\varphi_R^\vee)\widehat{\ }$, which is an L^2 function. Since the L^2 function $\varphi_R\widehat{u}$ coincides with the distribution $\widehat{u}$ on the set $B(0, R)$, it follows that $\widehat{u}$ is in $L^2(B(0, R))$ for all $R > 0$ and therefore it is in L^2_{loc}. If $f \in C_0^\infty(\mathbf{R}^n)$, the function $f\widehat{u}$ is in L^2 and therefore Plancherel's theorem and the boundedness of T give

$$\begin{aligned}\int_{\mathbf{R}^n} |f(x)\widehat{u}(x)|^2\,dx &= \int_{\mathbf{R}^n} |T(f^\vee)(x)|^2\,dx\\ &\le \|T\|_{L^2\to L^2}^2 \int_{\mathbf{R}^n} |f(x)|^2\,dx\,,\end{aligned}$$

for all $f \in C_0^\infty(\mathbf{R}^n)$. It follows that

$$\int_{\mathbf{R}^n} (\|T\|_{L^2\to L^2}^2 - |\widehat{u}(x)|^2)\,|f(x)|^2\,dx \ge 0,$$

for all $f \in C_0^\infty(\mathbf{R}^n)$. This implies that $\|T\|_{L^2\to L^2}^2 - |\widehat{u}(x)|^2 \ge 0$ for almost all x. Hence $\widehat{u}$ is in L^∞ and $\|\widehat{u}\|_{L^\infty} \le \|T\|_{L^2\to L^2}$. Combining this with the estimate $\|T\|_{L^2\to L^2} \le \|\widehat{u}\|_{L^\infty}$, which holds if $\widehat{u} \in L^\infty$, we obtain that $\|T\|_{L^2\to L^2} = \|\widehat{u}\|_{L^\infty}$.

□

2.5.e. The Space of Fourier Multipliers $\mathcal{M}_p(\mathbf{R}^n)$

We have now characterized all convolution operators that map L^2 to L^2. Suppose now that T is in $\mathcal{M}^{p,p}$, where $1 < p < 2$. As discussed in Theorem 2.5.7, T also maps $L^{p'}$ to $L^{p'}$. Since $p < 2 < p'$, by Theorem 1.3.4, it follows that T also maps L^2 to L^2. Thus T is given by convolution with a tempered distribution whose Fourier transform is a bounded function.

Definition 2.5.11. Given $1 \le p \le \infty$, we will denote by $\mathcal{M}_p(\mathbf{R}^n)$ the space of all bounded functions m on $\mathbf{R}^n$ such that the operator

$$T_m(f) = (\widehat{f}\, m)^\vee, \qquad f \in \mathcal{S},$$

is bounded on $L^p(\mathbf{R}^n)$ [or is initially defined in a dense subspace of $L^p(\mathbf{R}^n)$ and has a bounded extension on the whole space]. The norm of m in $\mathcal{M}_p(\mathbf{R}^n)$ is defined by

$$\big\|m\big\|_{\mathcal{M}_p} = \big\|T_m\big\|_{L^p \to L^p}. \tag{2.5.12}$$

Definition 2.5.11 implies that $m \in \mathcal{M}_p$ if and only if $T_m \in \mathcal{M}^{p,p}$. Elements of the space $\mathcal{M}_p$ are called *L^p multipliers* or *L^p Fourier multipliers*. It follows from Theorem 2.5.10 that $\mathcal{M}_2$, the set of all L^2 multipliers, is L^∞. Theorem 2.5.8 implies that $\mathcal{M}_1(\mathbf{R}^n)$ is the set of the Fourier transforms of finite Borel measures that is usually denoted by $\mathcal{M}(\mathbf{R}^n)$. Theorem 2.5.7 states that a bounded function m is an L^p multiplier if and only if it is an $L^{p'}$ multiplier and in this case $\big\|m\big\|_{\mathcal{M}_p} = \big\|m\big\|_{\mathcal{M}_{p'}}$. It is a consequence of Theorem 1.3.4 that the normed spaces $\mathcal{M}_p$ are nested, that is, for $1 \le p \le q \le 2$ we have

$$\mathcal{M}_1 \subseteq \mathcal{M}_p \subseteq \mathcal{M}_q \subseteq \mathcal{M}_2 = L^\infty.$$

Moreover, if $m \in \mathcal{M}_p$ and $1 \le p \le 2 \le p'$, Theorem 1.3.4 gives

$$\big\|T_m\big\|_{L^2 \to L^2} \le \big\|T_m\big\|_{L^p \to L^p}^{\frac{1}{2}} \big\|T_m\big\|_{L^{p'} \to L^{p'}}^{\frac{1}{2}} = \big\|T_m\big\|_{L^p \to L^p}, \tag{2.5.13}$$

since $1/2 = (1/2)/p + (1/2)/p'$. Theorem 1.3.4 also gives that

$$\big\|m\big\|_{\mathcal{M}_p} \le \big\|m\big\|_{\mathcal{M}_q}$$

whenever $1 \le q \le p \le 2$. Thus the $\mathcal{M}_p$'s form an increasing family of spaces as p increases from 1 to 2.

Example 2.5.12. The function $m(\xi) = e^{2\pi i \xi \cdot b}$ is an L^p multiplier for all $b \in \mathbf{R}^n$ as the corresponding operator $T_m(f)(x) = f(x + b)$ is bounded on $L^p(\mathbf{R}^n)$. Clearly $\big\|m\big\|_{\mathcal{M}_p} = 1$.

Proposition 2.5.13. *For $1 \le p \le 2$, the normed space $\big(\mathcal{M}_p, \big\|\cdot\big\|_{\mathcal{M}_p}\big)$ is a Banach space. Furthermore, $\mathcal{M}_p$ is closed under pointwise multiplication and is a Banach algebra.*

Proof. It is straightforward that if m_1, m_2 are in $\mathcal{M}_p$ and $b \in \mathbf{C}$ then $m_1 + m_2$ and bm_1 are also in $\mathcal{M}_p$. Observe that m_1m_2 is the multiplier that corresponds to the operator $T_{m_1}T_{m_2} = T_{m_1m_2}$ and thus

$$\big\|m_1m_2\big\|_{\mathcal{M}_p} = \big\|T_{m_1}T_{m_2}\big\|_{L^p \to L^p} \le \big\|m_1\big\|_{\mathcal{M}_p}\big\|m_2\big\|_{\mathcal{M}_p}.$$

This proves that $\mathcal{M}_p$ is an algebra. To show that $\mathcal{M}_p$ is a complete space, take a Cauchy sequence m_j in $\mathcal{M}_p$. It follows from (2.5.13) that m_j is Cauchy in L^∞ and hence it converges to some bounded function m in the L^∞ norm. We have to show that $m \in \mathcal{M}_p$. Fix $f \in \mathcal{S}$. We have

$$T_{m_j}(f)(x) = \int_{\mathbf{R}^n} \widehat{f}(\xi)m_j(\xi)e^{2\pi i x\cdot\xi}\,d\xi \to \int_{\mathbf{R}^n} \widehat{f}(\xi)m(\xi)e^{2\pi i x\cdot\xi}\,d\xi = T_m(f)(x)$$

a.e. by the Lebesgue dominated convergence theorem. Since $\{m_j\}_j$ is a Cauchy sequence in $\mathcal{M}_p$, it is bounded in $\mathcal{M}_p$, thus $\sup_j \big\|m_j\big\|_{\mathcal{M}_p} \le C$. Fatou's lemma now implies that

$$\begin{aligned}
\int_{\mathbf{R}^n} |T_m(f)|^p\,dx &= \int_{\mathbf{R}^n} \liminf_{j\to\infty} |T_{m_j}(f)|^p\,dx \\
&\le \liminf_{j\to\infty} \int_{\mathbf{R}^n} |T_{m_j}(f)|^p\,dx \\
&\le \liminf_{j\to\infty} \big\|m_j\big\|_{\mathcal{M}_p}^p \big\|f\big\|_{L^p}^p \\
&\le C^p \big\|f\big\|_{L^p}^p,
\end{aligned}$$

which implies that $m \in \mathcal{M}_p$. Incidentally, this argument shows that if $\mu_j \in \mathcal{M}_p$ and $\mu_j \to \mu$ a.e., then μ is in $\mathcal{M}_p$ and satisfies

$$\big\|\mu\big\|_{\mathcal{M}_p} \le \liminf_{j\to\infty} \big\|\mu_j\big\|_{\mathcal{M}_p}.$$

Apply this inequality to $\mu_j = m_k - m_j$ and $\mu = m_k - m$ for some fixed k. Then let $k \to \infty$ and use the fact that m_j is a Cauchy sequence in $\mathcal{M}_p$ to obtain that $m_k \to m$ in $\mathcal{M}_p$. This proves that $\mathcal{M}_p$ is a Banach space.

□

The following proposition summarizes some simple properties of multipliers.

Proposition 2.5.14. *For all $m \in \mathcal{M}_p$, $x \in \mathbf{R}^n$ and $h > 0$ we have*

$$\big\|\tau^x(m)\big\|_{\mathcal{M}_p} = \big\|m\big\|_{\mathcal{M}_p} \tag{2.5.14}$$

$$\big\|\delta^h(m)\big\|_{\mathcal{M}_p} = \big\|m\big\|_{\mathcal{M}_p} \tag{2.5.15}$$

$$\big\|\widetilde{m}\big\|_{\mathcal{M}_p} = \big\|m\big\|_{\mathcal{M}_p} \tag{2.5.16}$$

$$\big\|e^{2\pi i(\,\cdot\,)\cdot x}m\big\|_{\mathcal{M}_p} = \big\|m\big\|_{\mathcal{M}_p} \tag{2.5.17}$$

$$\big\|m\circ A\big\|_{\mathcal{M}_p} = \big\|m\big\|_{\mathcal{M}_p}, \qquad A \text{ is an orthogonal matrix.} \tag{2.5.18}$$

Proof. See Exercise 2.5.2. □

Example 2.5.15. We now indicate why $\|\chi_{[a,b]}\|_{\mathcal{M}_p} = \|\chi_{[0,1]}\|_{\mathcal{M}_p}$ whenever $-\infty < a < b < \infty$. Indeed, using property (2.5.14) we obtain that $\|\chi_{[a,b]}\|_{\mathcal{M}_p} = \|\chi_{[0,b-a]}\|_{\mathcal{M}_p}$ and the latter is equal to $\|\chi_{[0,1]}\|_{\mathcal{M}_p}$ in view of property (2.5.15). We will show in Chapter 4 that $\|\chi_{[0,1]}\|_{\mathcal{M}_p} < \infty$ for all $1 < p < \infty$.

We continue with the following interesting result.

Theorem 2.5.16. *Suppose that $m(\xi,\eta) \in \mathcal{M}_p(\mathbf{R}^{n+m})$. Then for almost every $\xi \in \mathbf{R}^n$ the function $\eta \to m(\xi,\eta)$ is in $\mathcal{M}_p(\mathbf{R}^m)$, with*

$$\|m(\xi,\cdot)\|_{\mathcal{M}_p(\mathbf{R}^m)} \le \|m\|_{\mathcal{M}_p(\mathbf{R}^{n+m})}.$$

Proof. If m is only a measurable function, its restriction to lower-dimensional planes is not defined. To avoid technical difficulties of this sort, we first assume that m is continuous at every point. Fix f_1, g_1 in $\mathcal{S}(\mathbf{R}^n)$ and f_2, g_2 in $\mathcal{S}(\mathbf{R}^m)$. Let

$$M(\xi) = \int_{\mathbf{R}^m} m(\xi,\eta)\widehat{f_2}(\eta)\widehat{g_2}(\eta)\,d\eta, \qquad \xi \in \mathbf{R}^n,$$

and observe that

$$\begin{aligned}\left|\int_{\mathbf{R}^n} (M(\cdot)\widehat{f_1})^\vee g_1\,dx\right| &= \left|\int_{\mathbf{R}^n} M(\xi)\widehat{f_1}(\xi)\widehat{g_1}(\xi)\,d\xi\right| \\ &= \left|\iint_{\mathbf{R}^{n+m}} m(\xi,\eta)\widehat{f_1 f_2}(\xi,\eta)\widehat{g_1 g_2}(\xi,\eta)\,d\xi\,d\eta\right| \\ &= \left|\iint_{\mathbf{R}^{n+m}} (m\widehat{f_1 f_2})^\vee g_1 g_2\,d\xi\,d\eta\right| \\ &\le \|m\|_{\mathcal{M}_p(\mathbf{R}^{n+m})}\|f_1\|_{L^p}\|f_2\|_{L^p}\|g_1\|_{L^{p'}}\|g_2\|_{L^{p'}}.\end{aligned}$$

Since by duality we have

$$\|(M(\cdot)\widehat{f_1})^\vee\|_{L^p} = \sup_{g_1:\,\|g_1\|_{L^{p'}}\le 1} \left|\int_{\mathbf{R}^n} (M(\cdot)\widehat{f_1})^\vee g_1\,dx\right|,$$

it follows that $M(\xi)$ is in $\mathcal{M}_p(\mathbf{R}^n)$ with norm

$$\|M\|_{\mathcal{M}_p(\mathbf{R}^n)} \le \|m\|_{\mathcal{M}_p(\mathbf{R}^{n+m})}\|f_2\|_{L^p}\|g_2\|_{L^{p'}}.$$

Since $\|M\|_{L^\infty} \le \|M\|_{\mathcal{M}_p}$ and m is continuous, we obtain that for all $\xi \in \mathbf{R}^n$

$$\left|\int_{\mathbf{R}^m} (m(\xi,\cdot)\widehat{f_2})^\vee g_2\,dy\right| = |M(\xi)| \le \|m\|_{\mathcal{M}_p(\mathbf{R}^{n+m})}\|f_2\|_{L^p}\|g_2\|_{L^{p'}}, \tag{2.5.19}$$

which of course implies the required conclusion for m continuous. The passage to a general m is achieved via a regularization argument. Define the family of functions $m_\varepsilon(\xi,\eta) = (2\varepsilon)^{-n-m}(m * \chi_{|\xi|\le\varepsilon,|\eta|\le\varepsilon})$. By Exercise 2.5.3 we have that

$\|m_\varepsilon\|_{\mathcal{M}_p(\mathbf{R}^{n+m})} \le \|m\|_{\mathcal{M}_p(\mathbf{R}^{n+m})}$ and clearly the m_ε are continuous functions. From this observation and (2.5.19), it follows that

$$\left|\int_{\mathbf{R}^m} m_\varepsilon(\xi,\eta)\widehat{f_2}(\eta)\widehat{g_2}(\eta)\,d\xi\,d\eta\right| \le \|m\|_{\mathcal{M}_p(\mathbf{R}^{n+m})}\|f_2\|_{L^p}\|g_2\|_{L^{p'}}.$$

Now let $\varepsilon \to 0$ and use the Lebesgue dominated convergence theorem. The conclusion follows.

□

Example 2.5.17. (The cone multiplier) On $\mathbf{R}^{n+1}$ define the function

$$m_\lambda(\xi_1,\dots,\xi_{n+1}) = \left(1 - \frac{\xi_1^2+\cdots+\xi_n^2}{\xi_{n+1}^2}\right)_+^\lambda, \qquad \lambda > 0,$$

where the plus sign indicates that $m_\lambda = 0$ if the expression inside the parentheses is negative. The multiplier m_λ is called the cone multiplier with parameter λ. If $m_\lambda \in \mathcal{M}_p(\mathbf{R}^{n+1})$, then the function $b_\lambda(\xi) = (1-|\xi|^2)_+^\lambda$ defined on $\mathbf{R}^n$ is in $\mathcal{M}_p(\mathbf{R}^n)$. Indeed, by Theorem 2.5.16 we have that for some $\xi_{n+1} = h$, $b_\lambda(\xi_1/h,\dots,\xi_n/h)$ is in $\mathcal{M}_p(\mathbf{R}^n)$ and hence so is b_λ by property (2.5.15).

Exercises

2.5.1. Prove that if $f \in L^q(\mathbf{R}^n)$ and $1 \le q < \infty$, then

$$\|\tau^h(f)+f\|_{L^q} \to 2^{1/q}\|f\|_{L^q} \qquad \text{as } |h| \to \infty.$$

2.5.2. Prove Proposition 2.5.14. Also prove that if $\delta_j^{h_j}$ is a dilation operator in the jth variable [for instance $\delta_1^{h_1}(f)(x) = f(h_1x_1, x_2,\dots,x_n)$], then

$$\|\delta_1^{h_1}\dots\delta_n^{h_n}(m)\|_{\mathcal{M}_p} = \|m\|_{\mathcal{M}_p}.$$

2.5.3. Let $m \in \mathcal{M}_p(\mathbf{R}^n)$. (a) If ψ is a function on $\mathbf{R}^n$ whose inverse Fourier transform is an integrable function, then prove that

$$\|\psi m\|_{\mathcal{M}_p} \le \|\psi^\vee\|_{L^1}\|m\|_{\mathcal{M}_p}.$$

(b) If ψ is in $L^1(\mathbf{R}^n)$, then prove that

$$\|\psi * m\|_{\mathcal{M}_p} \le \|\psi\|_{L^1}\|m\|_{\mathcal{M}_p}.$$

2.5.4. Fix a multiindex γ.
(a) Prove that the map $T(f) = f * \partial^\gamma\delta_0$ maps $\mathcal{S}$ continuously into $\mathcal{S}$.
(b) Prove that when $1/q - 1/p \ne |\gamma|/n$, T does not extend to an element of any of the spaces $\mathcal{M}_{p,q}$.

2.5.5. Let $K_\gamma(x) = |x|^{-n+\gamma}$, where $0 < \gamma < n$. Use Theorem 1.4.24 to show that the operator

$$T_\gamma(f) = f * K_\gamma, \qquad f \in \mathcal{S}$$

extends to a bounded operator in $\mathcal{M}^{p,q}$, where $1/p - 1/q = \gamma/n$, $1 < p < q < \infty$. This provides an example of a nontrivial operator in $\mathcal{M}^{p,q}$ when $p < q$.

2.5.6. (a) Use the ideas of the proof of Proposition 2.5.13 to show that if $m_j \in \mathcal{M}_p$, $1 \le p \le \infty$, $\|m_j\|_{\mathcal{M}_p} \le C$ for all $j = 1, 2, \dots$, and $m_j \to m$ a.e., then $m \in \mathcal{M}_p$ and

$$\|m\|_{\mathcal{M}_p} \le \liminf_{j\to\infty} \|m_j\|_{\mathcal{M}_p} \le C .$$

(b) Suppose that for some $1 \le p \le \infty$, $m_t \in \mathcal{M}_p$ for all $0 < t < \infty$. Prove that

$$\int_0^\infty \|m_t\|_{\mathcal{M}_p} \frac{dt}{t} < \infty \implies m(\xi) = \int_0^\infty m_t(\xi) \frac{dt}{t} \in \mathcal{M}_p .$$

(c) Use part (a) to prove that if $m \in \mathcal{M}_p$, then $m_0(x) = \lim_{R\to\infty} m(x/R)$ is also in $\mathcal{M}_p$ and satisfies $\|m_0\|_{\mathcal{M}_p} \le \|m\|_{\mathcal{M}_p}$.
(d) If $m \in \mathcal{M}_p$ has left and right limits at the origin, then prove that

$$\|m\|_{\mathcal{M}_p} \ge \max(|m(0+)|, |m(0-)|).$$

2.5.7. Suppose that $m \in \mathcal{M}_p(\mathbf{R}^n)$ has no zeros. Prove that the operator $T(f) = (\widehat{f} m^{-1})^\vee$ satisfies $\|T(f)\|_{L^p} \ge c_p \|f\|_{L^p}$, where $c_p = \|m\|_{\mathcal{M}_p}^{-1}$.

2.5.8. (a) Prove that if $m \in L^\infty(\mathbf{R}^n)$ satisfies $m^\vee \ge 0$, then

$$\|m\|_{\mathcal{M}_p} = \|m^\vee\|_{L^1} .$$

(b) (*E. Laeng and L. Colzani*) Let $m_1(\xi) = -1$ for $\xi > 0$ and $m_1(\xi) = 1$ for $\xi < 0$. Let $m_2(\xi) = \min(\xi - 1, 0)$ for $\xi > 0$ and $m_2(\xi) = \max(\xi + 1, 0)$ for $\xi < 0$. Prove that

$$\|m_1\|_{\mathcal{M}_p} = \|m_2\|_{\mathcal{M}_p}$$

for all $1 < p < \infty$.
[*Hint:* Part (a): Use Exercise 1.2.9. Part (b): Use part (a) to show that $\|m_2 m_1^{-1}\|_{\mathcal{M}_p} = 1$. Deduce that $\|m_2\|_{\mathcal{M}_p} \le \|m_1\|_{\mathcal{M}_p}$. For the converse use Exercise 2.5.6(c).]

2.5.9. (*de Leeuw* [**158**]) Prove that the following are equivalent:
(a) The operator $f \to \sum_{m\in\mathbf{Z}^n} a_m f(x - m)$ maps $L^p(\mathbf{R}^n)$ to $L^p(\mathbf{R}^n)$ and its norm is equal to $A > 0$.

(b) The function $\sum_{m\in\mathbf{Z}^n} a_m e^{-2\pi i m\cdot x}$ is in $\mathcal{M}_p$ with norm exactly $A>0$.
(c) The operator given by convolution with the sequence $\{a_m\}$ maps $\ell^p(\mathbf{Z}^n)$ into itself with norm exactly $A>0$.

2.5.10. (*Jodeit* [**277**]) Let $m(x)$ in $\mathcal{M}_p(\mathbf{R}^n)$ be supported in $[0,1]^n$. Then the periodic extension of m in $\mathbf{R}^n$,

$$M(x)=\sum_{k\in\mathbf{Z}^n} m(x-k)\,,$$

is also in $\mathcal{M}_p(\mathbf{R}^n)$.

2.5.11. Suppose that u is a C^∞ function on $\mathbf{R}^n\setminus\{0\}$ that is homogeneous of degree $-n+i\tau$, $\tau\in\mathbf{R}$. Prove that the operator given by convolution with u maps $L^2(\mathbf{R}^n)$ to $L^2(\mathbf{R}^n)$.

2.5.12. (*Hahn* [**231**]) Suppose that $m_1\in\mathcal{M}_r$ and $m_2\in\mathcal{M}_s$ for some $2\le r\le\infty$ and $1\le s\le 2$. Prove that $m_1*m_2\in\mathcal{M}_p$, where $1\le p\le 2$ and $2/p=1/s+3/r$. Conclude that if $s=r'$, then $\mathcal{M}_r*\mathcal{M}_{r'}\subseteq\mathcal{M}_p$ when $\left|\frac{1}{2}-\frac{1}{p}\right|=\frac{1}{r}$.
$\big[$*Hint:* Consider the trilinear operator $(m_1,m_2,f)\to\big((m_1*m_2)\widehat{f}\,\big)^\vee$ and apply trilinear interpolation.$\big]$

2.6. Oscillatory Integrals

Oscillatory integrals have played an important role in harmonic analysis from its outset. The Fourier transform, being the father of all oscillatory integrals, provides a prototype of an integral in which the phase (the exponent $2\pi i x\cdot\xi$) is a linear function of the integrating variable. More complicated phases appear naturally in the subject; the reader may check the definition of the Bessel function in Appendix B.1, in which the phase is a sinusoidal function.

In this section we take a quick look at oscillatory integrals. We will mostly concentrate on one-dimensional results, which already require some significant analysis. We only touch on some higher-dimensional results. Our analysis here is far from adequate.

Definition 2.6.1. An *oscillatory integral* is an expression of the form

$$I(\lambda)=\int_{\mathbf{R}^n} e^{i\lambda\varphi(x)}\psi(x)\,dx\,, \tag{2.6.1}$$

where λ is a positive real number, φ is a real-valued function on $\mathbf{R}^n$ called the *phase*, and ψ is a complex-valued and smooth integrable function on $\mathbf{R}^n$, which is often taken to have compact support.

2.6.a. A Simple Case

We begin our study with a the simplest possible one-dimensional case. Suppose that φ and ψ are smooth functions on the real line such that

$$\varphi'(x) \neq 0 \qquad \text{for all } x \in \operatorname{supp} \psi .$$

Then φ' has no zeros and it must be either strictly positive or strictly negative everywhere on the support of ψ. It follows that φ is monotonic and one could change variables

$$u = \varphi(x)$$

in (2.6.1). Then $dx = (\varphi'(x))^{-1} du = (\varphi^{-1})'(u)\, du$, where φ^{-1} is the inverse function of φ. We can therefore transform the integral in (2.6.1) into

$$\int_{\mathbf{R}} e^{i\lambda u} \psi(\varphi^{-1}(u))(\varphi^{-1})'(u)\, du \tag{2.6.2}$$

and we note that the function $\theta(u) = \psi(\varphi^{-1}(u))(\varphi^{-1})'(u)$ is smooth and has compact support on $\mathbf{R}$. We can therefore interpret the integral in (2.6.1) as $\widehat{\theta}(-\lambda/2\pi)$, where $\widehat{\theta}$ is the Fourier transform of θ. Since θ is a smooth function with compact support, it follows that the integral in (2.6.2) has rapid decay as $\lambda \to \infty$.

A quick way to see that the expression $\widehat{\theta}(-\lambda/2\pi)$ has decay of the order λ^{-N} for all $N > 0$ as λ tends to ∞ is the following. Write

$$e^{i\lambda u} = \frac{1}{(i\lambda)^N} \frac{d^N}{du^N}(e^{i\lambda u})$$

and integrate by parts N times to express the integral in (2.6.2) as

$$\frac{(-1)^N}{(i\lambda)^N} \int_{\mathbf{R}} e^{i\lambda u} \frac{d^N \theta}{du^N}\, du$$

from which the assertion follows. It follows that

$$|I(\lambda) = |\widehat{\theta}(-\lambda/2\pi)| \leq C_N \lambda^{-N}, \tag{2.6.3}$$

where $C_N = \left\|\theta^{(N)}\right\|_{L^1}$ which depends on derivatives of φ and ψ.

We now turn to a higher-dimensional analogue of this situation.

Definition 2.6.2. We say that a point x_0 is a *critical point* of a phase function φ if

$$\nabla \varphi(x_0) = \big(\partial_1 \varphi(x_0), \ldots, \partial_n \varphi(x_0)\big) = 0 .$$

Example 2.6.3. Let $\xi \in \mathbf{R}^n \setminus \{0\}$. Then the phase functions $\varphi_1(x) = x \cdot \xi$, $\varphi_2(x) = e^{x \cdot \xi}$ have no critical points while the phase function $\varphi_3(x) = |x|^2 - x \cdot \xi$ has one critical point at $x_0 = \frac{1}{2}\xi$.

We have the following result concerning oscillatory integrals whose phase functions have no critical points.

Proposition 2.6.4. *Suppose that ψ is a compactly supported smooth function on $\mathbf{R}^n$ and that φ is real-valued function on $\mathbf{R}^n$ ($n \ge 1$) that has no critical points on the support of ψ. Then the oscillatory integral*

$$I(\lambda) = \int_{\mathbf{R}^n} e^{i\lambda\varphi(x)}\psi(x)\,dx \tag{2.6.4}$$

obeys a bound of the form $|I(\lambda)| \le C_N \lambda^{-N}$ as $\lambda \to \infty$ for all $N > 0$, where C_N depends on N and on φ and ψ.

Proof. As the case $n = 1$ has already been discussed, we concentrate on dimensions $n \ge 2$. For each y in the support of ψ there is a unit vector θ_y such that

$$\theta_y \cdot \nabla\varphi(y) = |\nabla\varphi(y)|\,.$$

Then there is a small neighborhood $B(y, r_y)$ of y such that for all $x \in B(y, r_y)$ we have

$$\theta_y \cdot \nabla\varphi(x) \ge \frac{1}{2}|\nabla\varphi(y)|\,.$$

Cover the support of ψ by a finite number of balls $B(y_j, r_{y_j})$, $j = 1, \dots, m$ and picking $c = \min_j \frac{1}{2}|\nabla\varphi(y_j)|$, we have

$$\theta_{y_j} \cdot \nabla\varphi(x) \ge c > 0 \tag{2.6.5}$$

for all $x \in B(y_j, r_{y_j})$ and $j = 1, \dots, m$.

Next we find a smooth partition of unity of $\mathbf{R}^n$ such that each member of the partition, ζ_k, is supported in some ball $B(y_j, r_{y_j})$. We can therefore write

$$I(\lambda) = \sum_k \int_{\mathbf{R}^n} e^{i\lambda\varphi(x)}\psi(x)\zeta_k(x)\,dx\,, \tag{2.6.6}$$

where the sum only contains a finite number of indices as only a finite number of the ζ_k's meet the support of ψ. It suffices to show that every term in the sum in (2.6.6) has rapid decay in λ as $\lambda \to \infty$.

To this end, we fix a k and we pick a j such that the support of $\psi\zeta_k$ is contained in some ball $B(y_j, r_{y_j})$. We change variables in the integral

$$I_k(\lambda) = \int_{\mathbf{R}^n} e^{i\lambda\varphi(x)}\psi(x)\zeta_k(x)\,dx$$

by switching the coordinates $x = (x_1, \dots, x_n)$ with respect to the system $y_j + (e_1, \dots, e_n)$ to coordinates $t = (t_1, \dots, t_n)$ with respect to an orthonormal system $y_j + (\theta_{y_j}, \theta_{y_j}^2, \dots, \theta_{y_j}^n)$. The map $x \to t$ is a rotation that fixes y_j and preserves the ball $B(y_j, r_{y_j})$. Defining $\varphi(x) = \varphi^o(t)$, $\psi(x) = \psi^o(t)$, $\zeta_k(x) = \zeta_k^o(t)$, under this new coordinate system we can write

$$I_k(\lambda) = \int_K \left\{ \int_{\mathbf{R}} e^{i\lambda\varphi^o(t_1,\dots,t_n)}\psi^o(t_1, \dots, t_n)\zeta_k^o(t_1, \dots, t_n)\,dt_1 \right\} dt_2 \dots dt_n\,,$$

where K is a compact subset of $\mathbf{R}^{n-1}$. Note that condition (2.6.5) can be rewritten as

$$\frac{\partial \varphi^o(t)}{\partial t_1} \geq c > 0$$

for all $t \in B(y_j, r_{y_j})$ and therefore the inner integral inside the curly brackets is at most $C_N \lambda^{-N}$ in view of estimate (2.6.3). Integrating over K results the same conclusion for $I(\lambda)$ defined in (2.6.4).

□

2.6.b. Sublevel Set Estimates and Van der Corput Lemma

We discuss a sharp decay estimate for one-dimensional oscillatory integrals. This estimate will be obtained as a consequence of delicate size estimates for the Lebesgue measures of the sublevel sets $\{|u| \leq \alpha\}$ for a function u. In what follows $u^{(k)}$ will denote the kth derivative of the function $u(t)$ defined on $\mathbf{R}$. Moreover, C^k denotes the space of all functions whose kth derivative is continuous.

Proposition 2.6.5. *(a) Let u be a real-valued C^k function, $k \in \mathbf{Z}^+$, which satisfies $u^{(k)}(t) \geq 1$ for all $t \in \mathbf{R}$. Then the following estimate is valid for all $\alpha > 0$*

$$\big|\{t \in \mathbf{R} : \ |u(t)| \leq \alpha\}\big| \leq (2e)((k+1)!)^{\frac{1}{k}} \alpha^{\frac{1}{k}} . \tag{2.6.7}$$

(b) There is a constant $C < \infty$ such that for all $k \geq 2$, for any $-\infty < a < b < \infty$, for every real-valued C^k function u that satisfies $u^{(k)}(t) \geq 1$ for $t \in (a,b)$, and every $\lambda > 0$, the following is valid:

$$\left| \int_a^b e^{i\lambda u(t)} dt \right| \leq C\, k\, |\lambda|^{-\frac{1}{k}} . \tag{2.6.8}$$

(c) If $k = 1$, $u'(t)$ is monotonic on (a,b), and $u'(t) \geq 1$ for all $t \in (a,b)$, then

$$\left| \int_a^b e^{i\lambda u(t)} dt \right| \leq 3\, |\lambda|^{-1} . \tag{2.6.9}$$

The proof will be based on the following lemmata which we state and prove first.

Lemma 2.6.6. *Let $k \geq 1$ and suppose that $a_0, \dots, a_k$ are distinct real numbers. Let $a = \min(a_j)$ and $b = \max(a_j)$ and let f be a real-valued C^{k-1} function on $[a,b]$ that is C^k on (a,b). Then there exists a point y in (a,b) such that*

$$\sum_{m=0}^{k} c_m f(a_m) = f^{(k)}(y) ,$$

where $c_m = (-1)^k\, k! \prod_{\substack{\ell=0 \\ \ell \neq m}}^{k} (a_\ell - a_m)^{-1}$.

Proof. Suppose we could find a polynomial $p_k(x) = \sum_{j=0}^{k} b_j x^j$ such that the function

$$\varphi(x) = f(x) - p_k(x) \tag{2.6.10}$$

satisfies $\varphi(a_m) = 0$ for all $0 \le m \le k$. Since the a_j are distinct, then we could apply Rolle's theorem k times to find a point y in (a, b) such that $f^{(k)}(y) = k!\, b_k$.

The existence of a polynomial p_k so that (2.6.10) is satisfied is equivalent to the existence of a solution to the matrix equation

$$\begin{pmatrix} a_0^k & a_0^{k-1} & \dots & a_0 & 1 \\ a_1^k & a_1^{k-1} & \dots & a_1 & 1 \\ \vdots & \vdots & \vdots & \vdots & \vdots \\ a_{k-1}^k & a_{k-1}^{k-1} & \dots & a_{k-1} & 1 \\ a_k^k & a_k^{k-1} & \dots & a_k & 1 \end{pmatrix} \begin{pmatrix} b_k \\ b_{k-1} \\ \vdots \\ b_1 \\ b_0 \end{pmatrix} = \begin{pmatrix} f(a_0) \\ f(a_1) \\ \vdots \\ f(a_{k-1}) \\ f(a_k) \end{pmatrix}.$$

The determinant of the square matrix on the left is called the *Vandermonde determinant* and is equal to

$$\prod_{\ell=0}^{k-1} \prod_{j=\ell+1}^{k} (a_\ell - a_j) \neq 0\,.$$

Since the a_j are distinct, it follows that the system has a unique solution. Using Cramer's rule, we can solve this system to obtain

$$\begin{aligned} b_k &= \sum_{m=0}^{k} (-1)^m f(a_m) \frac{\prod\limits_{\substack{\ell=0 \\ \ell\neq m}}^{k-1} \prod\limits_{\substack{j=\ell+1 \\ j\neq m}}^{k} (a_\ell - a_j)}{\prod\limits_{\ell=0}^{k-1} \prod\limits_{j=\ell+1}^{k} (a_\ell - a_j)} \\ &= \sum_{m=0}^{k} (-1)^m f(a_m) \prod_{\substack{\ell=0 \\ \ell\neq m}}^{k} (a_\ell - a_m)^{-1} (-1)^{k-m}\,. \end{aligned}$$

The required conclusion now follows with c_m as claimed. □

Lemma 2.6.7. *Let E be a measurable subset of $\mathbf{R}$ with finite nonzero Lebesgue measure and let $k \in \mathbf{Z}^+$. Then there exist $a_0, \dots, a_k$ in E such that for all $\ell = 0, 1, \dots, k$ we have*

$$\prod_{\substack{j=0 \\ j\neq \ell}}^{k} |a_j - a_\ell| \geq (|E|/2e)^k\,. \tag{2.6.11}$$

Proof. Given a measurable set E with finite measure, pick a compact subset E' of E such that $|E \setminus E'| < \delta$, for some $\delta > 0$. For $x \in \mathbf{R}$ define $T(x) = |(-\infty, x) \cap E'|$. Then T is a surjective map from $E' \to [0, |E'|]$ and has the distance decreasing property

$$|T(x) - T(y)| \le |x - y|$$

for all $x, y \in E'$. Let a_j be points in E' such that $T(a_j) = \frac{j}{k}|E'|$ for $j = 0, 1, \dots, k$. For k an even integer, we have

$$\prod_{\substack{j=0\\ j\neq \ell}}^{k} |a_j - a_\ell| \ge \prod_{\substack{j=0\\ j\neq \ell}}^{k} \left| \frac{j}{k}|E'| - \frac{\ell}{k}|E'| \right| \ge \prod_{\substack{j=0\\ j\neq \frac{k}{2}}}^{k} \left| \frac{j}{k} - \frac{1}{2} \right| |E'|^k = \prod_{r=0}^{\frac{k}{2}-1} \left(\frac{r - \frac{k}{2}}{k} \right)^2 |E'|^k$$

and it is easily shown that $\big((k/2)!\big)^2 k^{-k} \ge (2e)^{-k}$.

For k an odd integer we have

$$\prod_{\substack{j=0\\ j\neq \ell}}^{k} |a_j - a_\ell| \ge \prod_{\substack{j=0\\ j\neq \ell}}^{k} \left| \frac{j}{k}|E'| - \frac{\ell}{k}|E'| \right| \ge \prod_{\substack{j=0\\ j\neq \frac{k+1}{2}}}^{k} \left| \frac{j}{k} - \frac{k+1}{2k} \right| |E'|^k$$

while the last product is at least

$$\left\{ \frac{1}{k} \cdot \frac{2}{k} \cdots \frac{\frac{k-1}{2}}{k} \right\}^2 \frac{k+1}{2k} \ge (2e)^{-k}.$$

We have therefore proved (2.6.11) with E' replacing E. As $|E \setminus E'| < \delta$ and $\delta > 0$ is arbitrarily small, the required conclusion follows. □

We now turn to the proof of Proposition 2.6.5

Proof. Part (a): Let $E = \{t \in \mathbf{R} : |u(t)| \le \alpha\}$. If $|E|$ is nonzero, then by Lemma 2.6.7 there exist $a_0, a_1, \dots, a_k$ in E such that for all ℓ we have

$$|E|^k \le (2e)^k \prod_{\substack{j=0\\ j\neq \ell}}^{k} |a_j - a_\ell|. \tag{2.6.12}$$

Lemma 2.6.6 implies that there exists a y in the open convex hull of the a_j such that

$$u^{(k)}(y) = (-1)^k \, k! \sum_{m=0}^{k} u(a_m) \prod_{\substack{\ell=0\\ \ell\neq m}}^{k} (a_\ell - a_m)^{-1}. \tag{2.6.13}$$

Using (2.6.12), we obtain that the expression on the right in (2.6.13) is in absolute value at most

$$(k+1)! \max_{0\le j\le k} |u(a_j)| \, (2e)^k \, |E|^{-k} \le (k+1)! \, \alpha \, (2e)^k \, |E|^{-k}$$

since $a_j \in E$. The bound $u^{(k)}(t) \geq 1$ now implies

$$|E|^k \leq (k+1)!\,(2e)^k\,\alpha$$

as claimed. This proves (2.6.7).

Part (b): We now take $k \geq 2$ and we split the interval (a,b) in (2.6.8) into the sets $R_1 = t \in (a,b) : \ |u'(t)| \leq \beta$ and $R_2 = \{t \in (a,b) : \ |u'(t)| > \beta\}$ for some parameter β to be chosen momentarily. The function $v = u'$ satsfies $v^{(k-1)} \geq 1$ on R_2 and $k-1 \geq 1$. It follows from (2.6.7) that

$$\left|\int_{R_1} e^{i\lambda u(t)}\,dt\right| \leq |R_1| \leq 2e \sup_{k\geq 1}(k!)^{\frac{1}{k-1}}\,\beta^{\frac{1}{k-1}} \leq C\,k\,\beta^{\frac{1}{k-1}}\,.$$

To obtain the corresponding estimate over R_2, we note that if $u^{(k)} \geq 1$, then the set $\{|u'| > \beta\}$ is the union of at most a constant multiple of k intervals on each of which u' is monotone. Let (c,d) be such an interval on which u' is monotone. Then u' has a fixed sign on (c,d) and we have

$$\begin{aligned}
\left|\int_c^d e^{i\lambda u(t)}\,dt\right| &= \left|\int_c^d \big(e^{i\lambda u(t)}\big)'\frac{1}{\lambda u'(t)}\,dt\right| \\
&\leq \left|\int_c^d e^{i\lambda u(t)}\Big(\frac{1}{\lambda u'(t)}\Big)'\,dt\right| + \frac{1}{|\lambda|}\left|\frac{e^{i\lambda u(d)}}{u'(d)} - \frac{e^{i\lambda u(c)}}{u'(c)}\right| \\
&\leq \frac{1}{|\lambda|}\int_c^d \left|\Big(\frac{1}{u'(t)}\Big)'\right|\,dt + \frac{2}{|\lambda|\beta} \\
&= \frac{1}{|\lambda|}\left|\int_c^d \Big(\frac{1}{u'(t)}\Big)'\,dt\right| + \frac{2}{|\lambda|\beta} \\
&\leq \frac{1}{|\lambda|}\left|\frac{1}{u'(d)} - \frac{1}{u'(c)}\right| + \frac{2}{|\lambda|\beta} \leq \frac{3}{|\lambda|\beta}\,,
\end{aligned}$$

where we use the monotonicity of $1/u'(t)$ in moving the absolute value from inside the integral to outside. We obtain a similar bound:

$$\left|\int_{R_2} e^{i\lambda u(t)}\,dt\right| \leq \frac{C\,k}{|\lambda|\beta}\,.$$

Optimizing by choosing $\beta = |\lambda|^{-(k-1)/k}$, we deduce the claimed estimate (2.6.8).

Part (c): Repeat the argument in part (b) in the case in which $(c,d) = (a,b)$ and $\beta = 1$.

$\square$

Corollary 2.6.8. *Let (a,b), $u(t)$, $\lambda > 0$, and k be as in Proposition 2.6.5. Then for any function ψ on (a,b) with an integrable derivative and $k \geq 2$, we have*

$$\left|\int_a^b e^{i\lambda u(t)}\psi(t)\,dt\right| \leq C\,k\,\lambda^{-1/k}\left[|\psi(b)| + \int_a^b |\psi'(s)|\,ds\right],$$

where C is the same constant as in (2.6.8). We also have

$$\left| \int_a^b e^{i\lambda u(t)} \psi(t)\, dt \right| \le 3\lambda^{-1} \left[|\psi(b)| + \int_a^b |\psi'(s)|\, ds \right],$$

when $k = 1$ and u' is monotonic on (a, b).

Proof. Set

$$F(x) = \int_a^x e^{i\lambda u(t)}\, dt$$

and use integration by parts to write

$$\int_a^b e^{i\lambda u(t)} \psi(t)\, dt = F(b)\psi(b) - \int_a^b F(t)\psi'(t)\, dt\,.$$

The conclusion easily follows. □

Example 2.6.9. The *Bessel function* of order m can be defined as

$$J_m(z) = \frac{1}{2\pi} \int_0^{2\pi} e^{iz\sin\theta} e^{-im\theta}\, d\theta\,.$$

Here z is a complex number and m is a real number with $\operatorname{Re} m > -\frac{1}{2}$; we refer to Appendix B for an introduction to Bessel functions and their basic properties.

We will use Corollary 2.6.8 to calculate the decay of the Bessel function $J_m(r)$, where r is a real number that tends to ∞. Set

$$\varphi(\theta) = \sin(\theta)$$

and note that $\varphi'(\theta)$ vanishes only at $\theta = \pi/2$ and $3\pi/2$ inside the interval $[0, 2\pi]$ and that $\varphi''(\pi/2) = -1$ while $\varphi''(3\pi/2) = 1$. We now write $1 = \psi_1 + \psi_2 + \psi_3$, where ψ_1 is smooth and compactly supported in a small neighborhood of $\pi/2$ and ψ_2 is smooth and compactly supported in a small neighborhood of $3\pi/2$. Corollary 2.6.8 yields

$$\left| \int_0^{2\pi} e^{ir\sin(\theta)} \big(\psi_j(t) e^{-im\theta}\big)\, d\theta \right| \le 2\, C\, m\, r^{-1/2}$$

for $j = 1, 2$ while the corresponding integral with ψ_3 has arbitrary decay in r in view of (2.6.3) (or Proposition 2.6.4 when $n = 1$).

Exercises

2.6.1. Suppose that u is a C^k function on the line that satisfies $|u^{(k)}(t)| \ge c_0 > 0$ for some $k \ge 2$ and all $t \in (a, b)$. Prove that for $\lambda > 0$ we have

$$\left| \int_a^b e^{i\lambda u(t)}\, dt \right| \le C\, k\, (\lambda c_0)^{-1/k}$$

and that the same conclusion is valid when $k = 1$, provided u' is monotonic. (C here is a universal constant.)

2.6.2. Show that if u' is not monotonic in part (c) of Proposition 2.6.5, then the conclusion may fail.
[*Hint:* Let $\varphi(t)$ be a smooth function on $\mathbf{R}$ that is equal to $10t$ on intervals $[2\pi k+\varepsilon, 2\pi(k+\frac{1}{2})-\varepsilon]$ and equal to t on intervals $[2\pi(k+\frac{1}{2})+\varepsilon, 2\pi(k+1)-\varepsilon]$. Show that the imaginary part of the oscillatory integral in question may tend to infinity over the union of several such intervals.]

2.6.3. Prove that the dependence on k of the constant in part (b) of Proposition 2.6.5 is indeed linear.
[*Hint:* Take $u(t) = t^k/k!$ over the interval $(0, k!)$.]

2.6.4. Follow the steps to give an alternative proof of part (b) of Proposition 2.6.5 with a bigger constant. Assume that the statement is known for some $k \geq 2$ and some constant $C(k)$ (instead of $C\,k$) for all intervals $[a,b]$ and all C^k functions satisfying $u^{(k)} \geq 1$ on $[a,b]$. Let c be the unique point at which the function $u^{(k)}$ attains its minimum in $[a,b]$.
(a) If $u^{(k)}(c) = 0$, then for all $\delta > 0$ we have $u^{(k)}(t) \geq \delta$ outside the interval $(c-\delta, c+\delta)$ and derive the bound

$$\left|\int_a^b e^{i\lambda u(t)}\,dt\right| \leq 2C(k)(\lambda\delta)^{-1/k} + 2\delta\,.$$

(b) If $u^{(k)}(c) \neq 0$, then we must have $c \in \{a,b\}$. Obtain the bound

$$\left|\int_a^b e^{i\lambda u(t)}\,dt\right| \leq C(k)(\lambda\delta)^{-1/k} + \delta\,.$$

(c) Choose a suitable δ to optimize and conclude the induction for $k+1$ with $C(k+1) = 2C(k) + 2 = 5\cdot 2^k - 2$. [Note $C(1) = 3$.]

2.6.5. Show that there is a constant C such that for any $0 < \varepsilon < N < \infty$, for all $\xi,\xi_2 \in \mathbf{R}$, and for all integers $k \geq 2$, we have

$$\left|\int_{\varepsilon\leq|t|\leq N} e^{i(\xi_1 t+\xi_2 t^k)}\,\frac{dt}{t}\right| \leq C\,.$$

[*Hint:* Change variables $s = \xi_1 t$ and consider the cases $|\xi_2| \leq |\xi_1|^k$ and $|\xi_2| > |\xi_1|^k$. In the first case use the inequality $|e^{ix} - 1| \leq |x|$ and in the second case apply Proposition 2.6.5.]

2.6.6. (a) Prove that for all $a \geq 1$ and $\lambda > 0$ the following is valid:

$$\left|\int_{|t|\leq a\lambda} e^{i\lambda\log t}\,dt\right| \leq 6a\,.$$

(b) Prove that there is a constant $c > 0$ so that for all $b > \lambda > 10$ we have

$$\left| \int_0^b e^{i\lambda t \log t} \, dt \right| \leq \frac{c}{\lambda \log \lambda}.$$

[*Hint:* Part (b): Consider the intervals $(0, \delta)$ and $[\delta, b)$ for some δ. On one of these intervals apply Proposition 2.6.5 with $k = 1$ and on the other with $k = 2$. Then optimize over δ.]

2.6.7. Show that there is a constant $C < \infty$ such that for all nonintegers $\gamma > 1$ and all $\lambda, b > 1$ we have

$$\left| \int_0^b e^{i\lambda t^\gamma} \, dt \right| \leq \frac{C}{\lambda^\gamma}.$$

[*Hint:* On the interval $(0, \delta)$ apply Proposition 2.6.5 with $k = [\gamma] + 1$ and on the interval (δ, b) with $k = [\gamma]$. Then optimize by choosing $\delta = \lambda^{-1/\gamma}$.]

HISTORICAL NOTES

The one-dimensional maximal function originated in the work of Hardy and Littlewood [**237**]. Its n-dimensional analogue was introduced by Wiener [**552**] who used Lemma 2.1.5, a variant of the Vitali covering lemma, to derive its L^p boundedness. We refer to Besicovitch [**38**] and de Guzmán [**157**] for extensions and other variants of such covering lemmas. The actual covering lemma proved by Vitali [**542**] says that if a family of closed cubes in $\mathbf{R}^n$ has the property that for every point $x \in A \subseteq \mathbf{R}^n$ there exists a sequence of cubes in the family that tends to x, then it is always possible to extract a sequence of pairwise disjoint cubes E_j from the family such that $|A \setminus \bigcup_j E_j| = 0$. We refer to Saks [**443**] for details and extensions of this theorem.

The class $L \log L$ was introduced by Zygmund to give a sufficient condition on the local integrability of the Hardy-Littlewood maximal operator. The necessity of this condition was observed by Stein [**474**]. Stein [**478**] also showed that the $L^p(\mathbf{R}^n)$ norm of the centered Hardy-Littlewood maximal operator $\mathcal{M}$ is bounded above by some dimension-free constant (which only depends on p for $1 < p < \infty$). See also Stein and Strömberg [**483**]. The situation for the uncentered maximal operator M is different as given any $1 < p < \infty$ there exists $C_p > 1$ such that $\|M\|_{L^p(\mathbf{R}^n) \to L^p(\mathbf{R}^n)} \geq C_p^n$; (see Exercise 2.1.8 for a value of such a constant C_p and also the article of Grafakos and Montgomery-Smith [**220**] for a larger value). The centered maximal function $\mathcal{M}_\mu$ with respect to a general positive measure μ on $\mathbf{R}^n$ is L^p bounded without the additional hypothesis that the measure is doubling, see Fefferman [**193**]. The proof of this result requires the following covering lemma obtained by Besicovitch [**38**]: For all dimensions n there exists number $N(n)$ such that for any family $\mathcal{F}$ of at least $N(n)$ closed balls in $\mathbf{R}^n$ whose radii form a bounded set, there exist $N(n)$ subfamilies of $\mathcal{F}$ consisting of pairwise disjoint balls, such that the center of any ball in the original family $\mathcal{F}$ belongs to some ball in one of these subfamilies. For a proof of this lemma we refer to the book of Ziemer [**568**]. A similar version of this lemma was obtained by Morse [**378**]. The precise value of the operator norm of the uncentered Hardy-Littlewood maximal function on $L^p(\mathbf{R})$ was shown by Grafakos and Montgomery-Smith [**220**] to be the unique positive solution of the equation $(p-1)x^p - px^{p-1} - 1 = 0$. This constant raised to the power n is the operator norm of the strong maximal function M_s on $L^p(\mathbf{R}^n)$ for $1 < p \le \infty$. The best weak type $(1,1)$ constant for the centered Hardy-Littlewood maximal operator was shown by Melas [**372**] to be the largest root of the quadratic equation $12x^2 - 22x + 5 = 0$. The strong maximal operator is not weak type $(1,1)$ but it satisfies the substitute inequality $d_{M_s(f)}(\alpha) \le C \int_{\mathbf{R}^n} \frac{|f(x)|}{\alpha}(1+\log^+ \frac{|f(x)|}{\alpha})^{n-1}\, dx$. This result is due to Jessen, Marcinkiewicz, and Zygmund [**276**], but a geometric proof of it was obtained by Córdoba and Fefferman [**130**].

The basic facts about the Fourier transform go back to Fourier [**195**]. The definition of distributions used here is due to Schwartz [**447**]. For a concise introduction to the theory of distributions we refer the reader to Hörmander [**256**] and Yosida [**564**]. Homogeneous distributions were considered by Riesz [**428**] in the study of the Cauchy problem in partial differential equations, although some earlier accounts can be found in the work of Hadamard. They were later systematically studied by Gelfand and Šilov [**208**], [**209**]. References on the uncertainty principle include the articles of Fefferman [**188**] and Folland and Sitaram [**268**]. The best constant B_p in the Hausdorff-Young inequality $\big\|\widehat{f}\,\big\|_{L^{p'}(\mathbf{R}^n)} \le B_p \big\|f\big\|_{L^p(\mathbf{R}^n)}$ when $1 \le p \le 2$ was shown by Beckner [**29**] to be $B_p = (p^{1/p}(p')^{-1/p'})^{n/2}$. The best constant in the case when p' is an even integer was previously obtained by Babenko [**21**].

A nice treatise of the spaces $\mathcal{M}^{p,q}$ can be found in Hörmander [**253**]. This reference also contains Theorem 2.5.6 which is due to him. Theorem 2.5.16 is due to de Leeuw [**158**], but the proof presented here is taken from Jodeit [**278**]. de Leeuw's result in Exercise 2.5.9 is saying that periodic elements of $\mathcal{M}_p(\mathbf{R}^n)$ can be isometrically identified with elements of $\mathcal{M}(\mathbf{T}^n)$, the latter being the space of all multipliers on $\ell^p(\mathbf{Z}^n)$.

Parts (b) and (c) of Proposition 2.6.5 are due to van der Corput [**535**]. The refinenment in part (a) was subsequently obtained by Arhipov, Karachuba, and Čubarikov [**9**]. The treatment of these results in the text is based on the article of Carbery, Christ, and Wright [**79**], which also investigates higher-dimensional analogues of the theory. Precise asymptotics can be obtained of variety of oscillatory integrals via the method of stationary phase; see Hörmander [**256**]. References on oscillatory integrals include the books of Titchmarsh [**522**], Erdélyi [**174**], Zygmund [**571**], [**572**], Stein [**482**], and Sogge [**464**]. The latter provides a treatment of Fourier integral operators.

CHAPTER 3

Fourier Analysis on the Torus

Principles of Fourier series go back to ancient times. The attempts of the Pythagorean school to explain musical harmony in terms of whole numbers embrace early elements of trigonometric nature. The theory of epicycles in the *Almagest* of Ptolemeus, based on work related to the circles of Appolonius, contains ideas of astronomical periodicities that we would interpret today as harmonic analysis. Early studies of acoustical and optical phenomena, as well as periodic astronomical and geophysical occurrences, provided a stimulus of the physical sciences to the rigorous study of expansions of periodic functions. This study is carefully pursued in this chapter.

The modern theory of Fourier series begins with attempts to solve boundary value problems using trigonometric functions. The work of d'Alembert, Bernoulli, Euler, and Clairaut on the vibrating string led to the belief that it may be possible to represent arbitrary periodic functions as sums of sines and cosines. Fourier announced belief in this possibility in his solution of the problem of heat distribution in spatial bodies (in particular, for the cube $\mathbf{T}^3$) by expanding an arbitrary function of three variables as a triple sine series. Fourier's approach, although heuristic, was appealing and eventually attracted attention. It was carefully studied and further developed by many scientists, but most notably by Laplace and Dirichlet, who were the first to investigate the validity of the representation of a function in terms of its Fourier series. This is the main topic of study in this chapter.

3.1. Fourier Coefficients

We discuss some basic facts of Fourier analysis on the torus $\mathbf{T}^n$. Throughout this chapter, n will be a fixed integer with $n \geq 1$.

3.1.a. The n-torus $\mathbf{T}^n$

The n-torus $\mathbf{T}^n$ is the cube $[0,1]^n$, in which the points $(x_1, \dots, 0, \dots, x_n)$ and $(x_1, \dots, 1, \dots, x_n)$ are identified whenever 0 and 1 appear in the same coordinate.

A more precise definition can be given as follows: For x, y in $\mathbf{R}^n$, we say

$$x \equiv y \tag{3.1.1}$$

if and only if $x - y \in \mathbf{Z}^n$, the additive subgroup of all points in $\mathbf{R}^n$ with integer coordinates. If (3.1.1) holds, then we write $x = y \pmod{\mathbf{1}}$. It is a simple fact that $\equiv$ is an equivalence relation that partitions $\mathbf{R}^n$ into equivalence classes. The n-torus $\mathbf{T}^n$ is then defined as the set $\mathbf{R}^n/\mathbf{Z}^n$ of all such equivalence classes. When $n = 1$, this set can be geometrically viewed as a circle by bending the line segment $[0, 1]$ so that its endpoints are brought together. When $n = 2$, the identification brings together the left and right sides of the unit square $[0, 1]^2$ and then the top and bottom sides as well. The resulting figure is a two-dimensional manifold embedded in $\mathbf{R}^3$ that looks like a donut. See Figure 3.1.

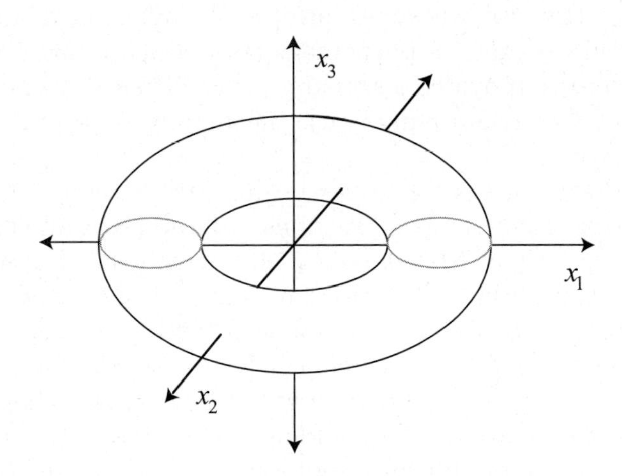

FIGURE 3.1. The graph of the two-dimensional torus $\mathbf{T}^2$.

The n-torus is an additive group, and zero is the identity element of the group, which of course coincides with every $e_j = (0, \dots, 0, 1, 0, \dots, 0)$. To avoid multiple appearances of the identity element in the group, we will often think of the n-torus as the set $[-1/2, 1/2]^n$. Since the group $\mathbf{T}^n$ is additive, the inverse of an element $x \in \mathbf{T}^n$ will be denoted by $-x$. For example, $-(1/3, 1/4) \equiv (2/3, 3/4)$ on $\mathbf{T}^2$, or, equivalently, $-(1/3, 1/4) = (2/3, 3/4) \pmod{\mathbf{1}}$.

The n-torus $\mathbf{T}^n$ can also be thought of as the following subset of $\mathbf{C}^n$,

$$\{(e^{2\pi i x_1}, \dots, e^{2\pi i x_n}) \in \mathbf{C}^n \;:\; (x_1, \dots, x_n) \in [0, 1]^n\} \tag{3.1.2}$$

in a way analogous to which the unit interval $[0, 1]$ can be thought of as the unit circle in $\mathbf{C}$ once 1 and 0 are identified.

Functions on $\mathbf{T}^n$ are functions f on $\mathbf{R}^n$ that satisfy $f(x + m) = f(x)$ for all $x \in \mathbf{R}^n$ and $m \in \mathbf{Z}^n$. Such functions are called 1-*periodic* in every coordinate. Haar measure on the n-torus is the restriction of n-dimensional Lebesgue measure to the set $\mathbf{T}^n = [0,1]^n$. This measure will still be denoted by dx while the measure of a set $A \subseteq \mathbf{T}^n$ will be denoted by $|A|$. Translation invariance of the Lebesgue measure and the periodicity of functions on $\mathbf{T}^n$ imply that, for all f on $\mathbf{T}^n$, we have

$$\int_{\mathbf{T}^n} f(x)\,dx = \int_{[-1/2,1/2]^n} f(x)\,dx = \int_{[a_1,1+a_1]\times\cdots\times[a_n,1+a_n]} f(x)\,dx \tag{3.1.3}$$

for any $a_1,\dots,a_n$ real numbers. The L^p spaces on $\mathbf{T}^n$ are nested and L^1 contains all L^p spaces for $p \ge 1$.

Elements of $\mathbf{Z}^n$ will be denoted by $m = (m_1,\dots,m_n)$. For $m \in \mathbf{Z}^n$, we define the *total size* of m will be the number $|m| = (m_1^2 + \cdots + m_n^2)^{1/2}$. Recall that for $x = (x_1,\dots,x_n)$ and $y = (y_1,\dots,y_n)$ in $\mathbf{R}^n$,

$$x \cdot y = x_1 y_1 + \cdots + x_n y_n$$

denotes the usual dot product. Finally, for $x \in \mathbf{T}^n$, $|x|$ will denote the usual Euclidean norm of x. If we identify $\mathbf{T}^n$ with $[-1/2,1/2]^n$, then $|x|$ can be interpreted as the distance of the element x from the origin, and then we have that $0 \le |x| \le \sqrt{n}/2$ for all $x \in \mathbf{T}^n$.

3.1.b. Fourier Coefficients

Definition 3.1.1. For a complex-valued function f in $L^1(\mathbf{T}^n)$ and m in $\mathbf{Z}^n$, we define

$$\widehat{f}(m) = \int_{\mathbf{T}^n} f(x) e^{-2\pi i m\cdot x}\,dx\,. \tag{3.1.4}$$

We call $\widehat{f}(m)$ the mth *Fourier coefficient* of f. We note that $\widehat{f}(m)$ is not defined for noninteger ξ as the function $x \to e^{-2\pi i \xi\cdot x}$ is not periodic and therefore not well defined on $\mathbf{T}^n$.

The *Fourier series* of f at $x \in \mathbf{T}^n$ is the series

$$\sum_{m\in\mathbf{Z}^n} \widehat{f}(m) e^{2\pi i m\cdot x}. \tag{3.1.5}$$

It is not clear at present in which sense and for which $x \in \mathbf{T}^n$ (3.1.5) converges. The study of convergence of Fourier series is the main topic of study in this chapter.

We quickly recall the notation we introduced in Chapter 2. We denote by $\overline{f}$ the complex conjugate of the function f, by $\widetilde{f}$ the function $\widetilde{f}(x) = f(-x)$, and by $\tau^y(f)$ the function $\tau^y(f)(x) = f(x-y)$ for all $y \in \mathbf{T}^n$. We mention some elementary properties of Fourier coefficients.

Proposition 3.1.2. *Let f, g be in $L^1(\mathbf{T}^n)$. Then for all $m, k \in \mathbf{Z}^n$, $\alpha \in \mathbf{C}$, $y \in \mathbf{T}^n$, and all multiindices α we have*

(1) $\widehat{f+g}(m) = \widehat{f}(m) + \widehat{g}(m)$

(2) $\widehat{\alpha f}(m) = \alpha \widehat{f}(m)$

(3) $\widehat{\overline{f}}\,(m) = \overline{\widehat{f}(-m)}$

(4) $\widehat{\widetilde{f}}\,(m) = \widehat{f}(-m)$

(5) $\widehat{\tau^y(f)}(m) = \widehat{f}(m)e^{-2\pi i m\cdot y}$

(6) $(e^{2\pi i k(\cdot)} f)\widehat{\ }(m) = \widehat{f}(m-k)$

(7) $\widehat{f}(0) = \displaystyle\int_{\mathbf{T}^n} f(x)\,dx$

(8) $\displaystyle\sup_{m\in\mathbf{Z}^n} |\widehat{f}(m)| \le \|f\|_{L^1(\mathbf{T}^n)}$

(9) $\widehat{f * g}(m) = \widehat{f}(m)\widehat{g}(m)$

(10) $\widehat{\partial^\alpha f}(m) = (2\pi i m)^\alpha \widehat{f}(m)$, *whenever* $f \in C^\alpha$.

Proof. The proof of Proposition 3.1.2 is obvious and is left to the reader. We only sketch the proof of (9). We have

$$\widehat{f * g}(m) = \int_{\mathbf{T}^n}\int_{\mathbf{T}^n} f(x-y)g(y)e^{-2\pi i m\cdot(x-y)}e^{-2\pi i m\cdot y}\,dy\,dx = \widehat{f}(m)\widehat{g}(m),$$

where the interchange of integrals is justified by the absolute convergence of the integrals and Fubini's theorem. □

Remark 3.1.3. The Fourier coefficients have the following property. For a function f_1 on $\mathbf{T}^{n_1}$ and a function f_2 on $\mathbf{T}^{n_2}$, $f_1 f_2$ is a periodic function on $\mathbf{T}^{n_1+n_2}$ whose Fourier coefficients are

$$\widehat{f_1 f_2}(m_1, m_2) = \widehat{f_1}(m_1)\widehat{f_2}(m_2), \tag{3.1.6}$$

for all $m_1 \in \mathbf{Z}^{n_1}$ and $m_2 \in \mathbf{Z}^{n_2}$.

Definition 3.1.4. A *trigonometric polynomial* on $\mathbf{T}^n$ is a function of the form

$$P(x) = \sum_{m\in\mathbf{Z}^n} a_m e^{2\pi i m\cdot x}, \tag{3.1.7}$$

where $\{a_m\}_{m\in\mathbf{Z}^n}$ is a finitely supported sequence in $\mathbf{Z}^n$. The *degree* of P is the largest number $|q_1| + \cdots + |q_n|$ such that a_q is nonzero, where $q = (q_1, \dots, q_n)$.

Example 3.1.5. A *trigonometric monomial* is a function of the form $P(x) = a\,e^{2\pi i(q_1x_1+\cdots+q_nx_n)}$ for some $q = (q_1, \dots, q_n) \in \mathbf{Z}^n$ and $a \in \mathbf{C}$. Observe that $\widehat{P}(q) = a$ and $\widehat{P}(m) = 0$ for $m \ne q$.

Let $P(x) = \sum_{|m|\le N} a_m e^{2\pi i m\cdot x}$ be a trigonometric polynomial and let f in $L^1(\mathbf{T}^n)$. Exercise 3.1.1 gives that $(f*P)(x) = \sum_{|m|\le N} a_m \widehat{f}(m) e^{2\pi i m\cdot x}$. This implies that the partial sums $\sum_{|m|\le N} \widehat{f}(m) e^{2\pi i m\cdot x}$ of the Fourier series of f given in (3.1.5) can be obtained by convolving f with the functions

$$D_N(x) = \sum_{|m|\le N} e^{2\pi i m\cdot x}. \tag{3.1.8}$$

These expressions are named after Dirichlet as the following definition indicates.

3.1.c. The Dirichlet and Fejér Kernels

Definition 3.1.6. Let $0 \le R < \infty$. The square *Dirichlet kernel* on $\mathbf{T}^n$ is the function

$$D(n,R)(x) = \sum_{\substack{m\in\mathbf{Z}^n \\ |m_j|\le R}} e^{2\pi i m\cdot x}. \tag{3.1.9}$$

The circular (or spherical) Dirichlet kernel on $\mathbf{T}^n$ is the function

$$\widetilde{D}(n,R)(x) = \sum_{\substack{m\in\mathbf{Z}^n \\ |m|<R}} e^{2\pi i m\cdot x}. \tag{3.1.10}$$

In dimension 1, the family of functions $D(1,R) = \widetilde{D}(1,R)$ (for $R \ge 0$) will be referred to as the Dirichlet kernel and will be denoted by D_R as in (3.1.8). The function D_5 is plotted in Figure 3.2.

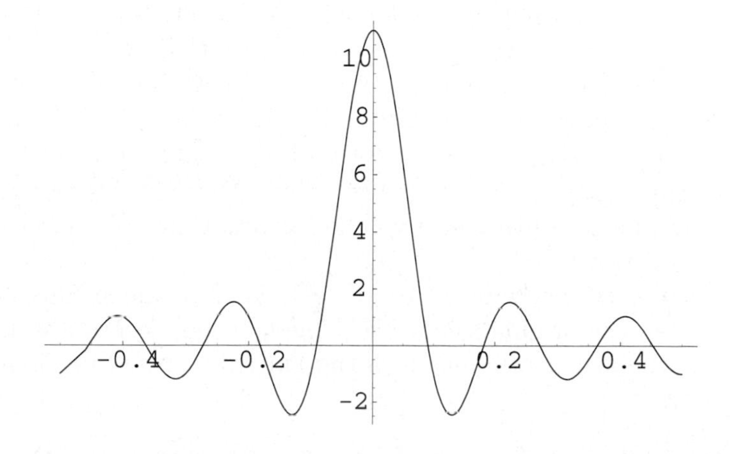

FIGURE 3.2. The graph of the Dirichlet kernel D_5 plotted on the interval $[-1/2, 1/2]$.

Both the square and circular (or spherical) Dirichlet kernels are trigonometric polynomials. The square Dirichlet kernel on $\mathbf{T}^n$ is equal to a product of one-dimensional Dirichlet kernels, that is,

$$D(n,R)(x_1,\dots,x_n) = D_R(x_1)\dots D_R(x_n). \tag{3.1.11}$$

We have the following two equivalent ways to write the Dirichlet kernel D_N:

$$D_N(x) = \sum_{|m|\le N} e^{2\pi i m\cdot x} = \frac{\sin((2N+1)\pi x)}{\sin(\pi x)}, \qquad x\in[0,1]. \tag{3.1.12}$$

To verify the validity of (3.1.12), sum the geometric series on the left in (3.1.12) to obtain

$$e^{-2\pi i N x}\frac{e^{2\pi i(2N+1)x}-1}{e^{2\pi i x}-1} = \frac{e^{2\pi i(N+1)x}-e^{-2\pi i N x}}{e^{\pi i x}(e^{\pi i x}-e^{-\pi i x})} = \frac{\sin((2N+1)\pi x)}{\sin(\pi x)}.$$

It follows that for $R\in\mathbf{R}^+\cup\{0\}$ we have

$$D_R(x) = \frac{\sin(\pi x(2[R]+1))}{\sin(\pi x)}. \tag{3.1.13}$$

It is reasonable to ask whether the family $\{D_R\}_{R>0}$ forms an approximate identity as $R\to\infty$. Using (3.1.12) we see that each D_R is integrable over $[-1/2,1/2]$ with integral equal to 1. But we can easily obtain from (3.1.12) that for all $\delta>0$ there is a constant $c_\delta>0$ such that

$$\int_{1/2\ge|x|\ge\delta} |D_R(x)|\,dx \ge c_\delta$$

for all $R>0$. Therefore the family $\{D_R\}_{R>0}$ does not satisfy property (iii) in Definition 1.2.15. More important, it follows from Exercise 3.1.8 that $\|D_R\|_{L^1}\approx \log R$ as $R\to\infty$ and therefore property (i) in Definition 1.2.15 also fails for D_R. We conclude that the family $\{D_R\}_{R>0}$ is not an approximate identity on $\mathbf{T}^1$, a fact that significantly complicates the study of Fourier series. It follows immediately that the family $\{D(n,R)\}_{R>0}$ does not form an approximate identity on $\mathbf{T}^n$. The same is true for the family of circular (or spherical) Dirichlet kernels $\{\widetilde{D}(n,R)\}_{R>0}$, although this is harder to prove, but it will be a consequence of the results in Section 3.4.

A typical situation encountered in analysis is that the means of a sequence behaves better than the original sequence. This fact led Cesàro and independently Fejér to consider the arithmetic means of the Dirichlet kernel in dimension 1, that is, the expressions

$$F_N(x) = \frac{1}{N+1}\big[D_0(x)+D_1(x)+D_2(x)+\dots+D_N(x)\big]. \tag{3.1.14}$$

It can be checked (see Exercise 3.1.3) that (3.1.14) is in fact equal to the Fejér kernel given in Example 1.2.18, that is,

$$F_N(x) = \sum_{j=-N}^{N} \left(1 - \frac{|j|}{N+1}\right) e^{2\pi i j x} = \frac{1}{N+1} \left(\frac{\sin(\pi(N+1)x)}{\sin(\pi x)} \right)^2, \tag{3.1.15}$$

whenever N is a nonnegative integer. Identity (3.1.15) implies that the mth Fourier coefficient of F_N is $\left(1 - \frac{|m|}{N+1}\right)$ if $|m| \le N$ and zero otherwise.

Definition 3.1.7. Let N be a nonnegative integer. The *Fejér kernel* $F(n, N)$ on $\mathbf{T}^n$ is defined as the average of the product of the Dirichlet kernels in each variable, precisely

$$\begin{aligned} F(n,N)(x_1, \dots, x_n) &= \frac{1}{(N+1)^n} \sum_{k_1=0}^{N} \cdots \sum_{k_n=0}^{N} D_{k_1}(x_1) \dots D_{k_n}(x_n) \\ &= \prod_{j=1}^{n} \left(\frac{1}{N+1} \sum_{k=0}^{N} D_k(x_j) \right) = \prod_{j=1}^{n} F_N(x_j) . \end{aligned}$$

So $F(n, N)$ is equal to the product of the Fejér kernels in each variable. Note that $F(n, N)$ is a trigonometric polynomial of degree nN.

Remark 3.1.8. Using the first expression for F_N in (3.1.15), we can write

$$F(n,N)(x) = \sum_{\substack{m \in \mathbf{Z}^n \\ |m_j| \le N}} \left(1 - \frac{|m_1|}{N+1}\right) \cdots \left(1 - \frac{|m_n|}{N+1}\right) e^{2\pi i m \cdot x} \tag{3.1.16}$$

for $N \ge 0$ an integer. Observe that $F(n, 0)(x) = 1$ for all $x \in \mathbf{T}^n$.

Remark 3.1.9. To verify that the Fejér kernel $F(n, N)$ is an approximate identity on $\mathbf{T}^n$, we use the second expression for $F(1, N)$ in (3.1.15) to obtain

$$F(n,N)(x_1, \dots, x_n) = \frac{1}{(N+1)^n} \prod_{j=1}^{n} \left(\frac{\sin(\pi(N+1)x_j)}{\sin(\pi x_j)} \right)^2 . \tag{3.1.17}$$

Properties (i) and (iii) of approximate identities (see Definition 1.2.15) can be proved using the identity (3.1.17), while property (ii) follows from identity (3.1.16). See Exercise 3.1.3 for details.

Having introduced the Fejér kernel, let us see how we can use it to obtain some interesting results.

3.1.d. Reproduction of Functions from Their Fourier Coefficients

Proposition 3.1.10. *The set of trigonometric polynomials is dense in* $L^p(\mathbf{T}^n)$ *for* $1 \le p < \infty$.

Proof. Given f in $L^p(\mathbf{T}^n)$ for $1 \le p < \infty$, consider $f * F(n,N)$. Because of Exercise 3.1.1, $f * F(n,N)$ is also a trigonometric polynomial. By Theorem 1.2.19 (1), $f * F(n,N)$ converges to f in L^p as $N \to \infty$.

□

Corollary 3.1.11. *(**Weierstrass approximation theorem for trigonometric polynomials**) Every continuous function on the torus is a uniform limit of trigonometric polynomials.*

Proof. Since f is continuous on $\mathbf{T}^n$ and $\mathbf{T}^n$ is a compact set, Theorem 1.2.19 (2) gives that $f * F(n,N)$ converges uniformly to f as $N \to \infty$. Since $f * F(n,N)$ is a trigonometric polynomial, we conclude that every continuous function on $\mathbf{T}^n$ can be uniformly approximated by trigonometric polynomials.

□

We now define partial sums of Fourier series.

Definition 3.1.12. For $R \ge 0$ the expressions

$$(f * D(n,R))(x) = \sum_{\substack{m \in \mathbf{Z}^n \\ |m_j| \le R}} \widehat{f}(m) e^{2\pi i m \cdot x}$$

are called the *square partial sums of the Fourier series* of f and the expressions

$$(f * \widetilde{D}(n,R))(x) = \sum_{\substack{m \in \mathbf{Z}^n \\ |m| \le R}} \widehat{f}(m) e^{2\pi i m \cdot x}$$

are called the *circular (or spherical) partial sums* of the Fourier series of f. Similarly for $N \in \mathbf{Z}^+ \cup \{0\}$ the expressions

$$(f * F(n,N))(x) = \sum_{\substack{m \in \mathbf{Z}^n \\ |m_j| \le N}} \left(1 - \frac{|m_1|}{N+1}\right) \cdots \left(1 - \frac{|m_n|}{N+1}\right) \widehat{f}(m) e^{2\pi i m \cdot x}$$

are called the *square Cesàro means* (or *square Fejér means*) of f. Finally, for $R \ge 0$ the expressions

$$(f * \widetilde{F}(n,R))(x) = \sum_{\substack{m \in \mathbf{Z}^n \\ |m| \le R}} \left(1 - \frac{|m|}{R}\right) \widehat{f}(m) e^{2\pi i m \cdot x}$$

are called the *circular Cesàro means* (or *circular Fejér means*) of f.

Observe that $f * \widetilde{F}(n,R)$ is equal to the average of the expressions $f * \widetilde{D}(n,R)$ from 0 to R in the following sense:

$$(f * \widetilde{F}(n,R))(x) = \frac{1}{R} \int_0^R (\widetilde{D}(n,r) * f)(x) \, dr.$$

This is analogous to the fact that the Fejér kernel F_N is the average of the Dirichlet kernels $D_0, D_1, \dots, D_N$. Also observe that $f * F(n, R)$ can also be defined for $R \geq 0$, but it would be constant on intervals of the form $[a, a+1)$, where $a \in \mathbf{Z}^+$.

A fundamental problem in harmonic analysis is in what sense the partial sums of the Fourier series converge back to the function as $N \to \infty$. This problem is of central importance in harmonic analysis and will be partly investigated in this chapter.

We now ask the question of whether the Fourier coefficients uniquely determine the function. The answer is affirmative and simple.

Proposition 3.1.13. *If $f, g \in L^1(\mathbf{T}^n)$ satisfy $\widehat{f}(m) = \widehat{g}(m)$ for all m in $\mathbf{Z}^n$, then $f = g$ a.e.*

Proof. By linearity of the problem, it suffices to assume that $g = 0$. If $\widehat{f}(m) = 0$ for all $m \in \mathbf{Z}^n$, Exercise 3.1.1 implies that $F(n, N) * f = 0$ for all $N \in \mathbf{Z}^+$. The sequence $\{F(n, N)\}_{N \in \mathbf{Z}^+}$ is an approximate identity as $N \to \infty$. Therefore, $\big\|f - F(n, N) * f\big\|_{L^1} \to 0$ as $N \to \infty$, hence $\big\|f\big\|_{L^1} = 0$, from which we conclude that $f = 0$ a.e.

□

A useful consequence of the result just proved is the following.

Proposition 3.1.14. ***(Fourier inversion)*** *Suppose that $f \in L^1(\mathbf{T}^n)$ and that*

$$\sum_{m \in \mathbf{Z}^n} |\widehat{f}(m)| < \infty.$$

Then

$$f(x) = \sum_{m \in \mathbf{Z}^n} \widehat{f}(m) e^{2\pi i m \cdot x} \qquad a.e., \tag{3.1.18}$$

and therefore f is almost everywhere equal to a continuous function.

Proof. It is straightforward to check that both functions in (3.1.18) are well defined and have the same Fourier coefficients. Therefore, they must be almost everywhere equal by Proposition 3.1.13. Moreover, the function on the right in (3.1.18) is everywhere continuous.

□

We continue with a short discussion of Fourier series of square summable functions.

Let H be a separable Hilbert space with complex inner product $\langle \cdot \,|\, \cdot \rangle$. Recall that a subset E of H is called *orthonormal* if $\langle f \,|\, g \rangle = 0$ for all f, g in E with $f \neq g$, while $\langle f \,|\, f \rangle = 1$ for all f in E. A *complete orthonormal system* is a subset of H having the additional property that the only vector orthogonal to all of its elements is the zero vector. We refer the reader to Rudin [**440**] for the relevant definitions and theorems and in particular for the proof of the following proposition:

Proposition 3.1.15. *Let H be a separable Hilbert space and let $\{\varphi_n\}_{n\in\mathbf{Z}}$ be an orthonormal system in H. Then the following are equivalent:*
(1) *$\{\varphi_n\}_{n\in\mathbf{Z}}$ is a complete orthonormal system.*
(2) *For every $f \in H$ we have*

$$\|f\|_H^2 = \sum_{k\in\mathbf{Z}} |\langle f \,|\, \varphi_k\rangle|^2.$$

(3) *For every $f \in H$ we have*

$$f = \lim_{N\to\infty} \sum_{|k|\le N} \langle f \,|\, \varphi_k\rangle \varphi_k,$$

where the series converges in H.

Now consider the Hilbert space space $L^2(\mathbf{T}^n)$ with inner product

$$\langle f \,|\, g\rangle = \int_{\mathbf{T}^n} f(t)\overline{g(t)}\, dt.$$

Let φ_m be the sequence of functions $\xi \to e^{2\pi i m\cdot\xi}$ indexed by $m \in \mathbf{Z}^n$. The orthonormality of the sequence $\{\varphi_m\}$ is a consequence of the following simple but powerful identity:

$$\int_{[0,1]^n} e^{2\pi i m\cdot x}\overline{e^{2\pi i k\cdot x}}\, dx = \begin{cases} 1 & \text{when } m = k, \\ 0 & \text{when } m \neq k. \end{cases}$$

The completeness of the sequence $\{\varphi_m\}$ is also evident. Since $\langle f \,|\, \varphi_m\rangle = \widehat{f}(m)$ for all $f \in L^2(\mathbf{T}^n)$, it follows from Proposition 3.1.13 that if $\langle f \,|\, \varphi_m\rangle = 0$ for all $m \in \mathbf{Z}^n$, then $f = 0$ a.e.

The next result is a consequence of Proposition 3.1.15.

Proposition 3.1.16. *The following are valid for $f, g \in L^2(\mathbf{T}^n)$:*
(1) ***(Plancherel's identity)***

$$\|f\|_{L^2}^2 = \sum_{m\in\mathbf{Z}^n} |\widehat{f}(m)|^2 .$$

(2) *The function f is a.e. equal to the $L^2(\mathbf{T}^n)$ limit of the sequence*

$$f = \lim_{M\to\infty} \sum_{|m|\le M} \widehat{f}(m) e^{2\pi i m\cdot\xi} .$$

(3) ***(Parseval's identity)***

$$\int_{\mathbf{T}^n} f(t)\overline{g(t)}\, dt = \sum_{m\in\mathbf{Z}^n} \widehat{f}(m)\overline{\widehat{g}(m)}.$$

(4) *The map $f \to \{\widehat{f}(m)\}_{m\in\mathbf{Z}^n}$ is an isometry from $L^2(\mathbf{T}^n)$ onto ℓ^2.*

(5) *For all* $k \in \mathbf{Z}^n$ *we have*

$$\widehat{fg}(k) = \sum_{m\in\mathbf{Z}^n} \widehat{f}(m)\widehat{g}(k-m) = \sum_{m\in\mathbf{Z}^n} \widehat{f}(k-m)\widehat{g}(m).$$

Proof. (1) and (2) follow from the corresponding statements in Proposition 3.1.15. Parseval's identity (3) follows from polarization. First replace f by $f+g$ in (1) and expand the squares. We obtain that the real parts of the expressions in (3) are equal. Next replace f by $f+ig$ in (1) and expand the squares. We obtain that the imaginary parts of the expressions in (3) are equal. Thus (3) holds. Next we prove (4). We already know that the map $f \to \{\widehat{f}(m)\}_{m\in\mathbf{Z}^n}$ is an injective isometry. It remains to show that it is onto. Given a square summable sequence $\{a_m\}_{m\in\mathbf{Z}^n}$ of complex numbers, define

$$f_N(t) = \sum_{|m|\le N} a_m e^{2\pi i m\cdot\xi}.$$

Observe that f_N is a Cauchy sequence in $L^2(\mathbf{T}^n)$ and it therefore converges to some $f \in L^2(\mathbf{T}^n)$. Then we have $\widehat{f}(m) = a_m$ for all $m \in \mathbf{Z}^n$. Finally, (5) is a consequence of (3) and Proposition 3.1.2 (6) and (3).

□

3.1.e. The Poisson Summation Formula

We end this section with a useful result that connects Fourier analysis on the torus with Fourier analysis on $\mathbf{R}^n$. Suppose that f is an integrable function on $\mathbf{R}^n$ and let $\widehat{f}$ be its Fourier transform. Restrict $\widehat{f}$ on $\mathbf{Z}^n$ and form the "Fourier series" (assuming that it converges)

$$\sum_{m\in\mathbf{Z}^n} \widehat{f}(m)e^{2\pi i m\cdot x}.$$

What does this series represent? Since the preceding function is 1-periodic in every variable, it follows that it cannot be equal to f in general. However, it should not come as a surprise that in many cases it is equal to the periodization of f on $\mathbf{R}^n$. More precisely, we have the following.

Theorem 3.1.17. ***(Poisson summation formula)*** *Suppose that* $f, \widehat{f} \in L^1(\mathbf{R}^n)$ *satisfy*

$$|f(x)| + |\widehat{f}(x)| \le C(1+|x|)^{-n-\delta}$$

for some $C, \delta > 0$. *Then* f *and* $\widehat{f}$ *are both continuous and for all* $x \in \mathbf{R}^n$ *we have*

$$\sum_{m\in\mathbf{Z}^n} \widehat{f}(m)e^{2\pi i m\cdot x} = \sum_{m\in\mathbf{Z}^n} f(x+m), \tag{3.1.19}$$

and in particular $\sum_{m\in\mathbf{Z}^n} \widehat{f}(m) = \sum_{m\in\mathbf{Z}^n} f(m)$.

Proof. Since $\widehat{f}$ is integrable on $\mathbf{R}^n$, inversion holds and f can be identified with a continuous function. Define a 1-periodic function on $\mathbf{T}^n$ by setting

$$F(x) = \sum_{m \in \mathbf{Z}^n} f(x+m).$$

It is straightforward to verify that $F \in L^1(\mathbf{T}^n)$. The calculation

$$\widehat{F}(m) = \int_{\mathbf{T}^n} F(x) e^{-2\pi i m \cdot x}\, dx = \sum_{k \in \mathbf{Z}^n} \int_{[-\frac{1}{2},\frac{1}{2}]^n - k} f(x) e^{-2\pi i m \cdot x}\, dx = \widehat{f}(m)$$

gives that the sequence of the Fourier coefficients of F coincides with the restriction of the Fourier transform of f on $\mathbf{Z}^n$. Since we have that

$$\sum_{m \in \mathbf{Z}^n} |\widehat{F}(m)| = \sum_{m \in \mathbf{Z}^n} |\widehat{f}(m)| \le C \sum_{m \in \mathbf{Z}^n} \frac{1}{(1+|m|)^{n+\delta}} < \infty,$$

Proposition 3.1.14 implies conclusion (3.1.19).

□

Example 3.1.18. We have seen earlier (see Exercise 2.2.11) that the following identity gives the Fourier transform of the Poisson kernel in $\mathbf{R}^n$:

$$(e^{-2\pi|x|})\widehat{\ }(\xi) = \frac{\Gamma(\frac{n+1}{2})}{\pi^{\frac{n+1}{2}}} \frac{1}{(1+|\xi|^2)^{\frac{n+1}{2}}}.$$

The Poisson summation formula yields the identity

$$\frac{\Gamma(\frac{n+1}{2})}{\pi^{\frac{n+1}{2}}} \sum_{k \in \mathbf{Z}^n} \frac{\varepsilon}{(\varepsilon^2 + |k+x|^2)^{\frac{n+1}{2}}} = \sum_{k \in \mathbf{Z}^n} e^{-2\pi\varepsilon|k|} e^{-2\pi i k \cdot x}. \tag{3.1.20}$$

It follows that

$$\sum_{k \in \mathbf{Z}^n \setminus \{0\}} \frac{1}{(\varepsilon^2 + |k|^2)^{\frac{n+1}{2}}} = \frac{1}{\varepsilon} \left(\frac{\pi^{\frac{n+1}{2}}}{\Gamma(\frac{n+1}{2})} \sum_{k \in \mathbf{Z}^n} e^{-2\pi\varepsilon|k|} - \frac{1}{\varepsilon^n} \right)$$

from which we obtain the identity

$$\sum_{k \in \mathbf{Z}^n \setminus \{0\}} \frac{1}{|k|^{n+1}} = \lim_{\varepsilon \to 0} \frac{1}{\varepsilon} \left(\frac{\pi^{\frac{n+1}{2}}}{\Gamma(\frac{n+1}{2})} \sum_{k \in \mathbf{Z}^n} e^{-2\pi\varepsilon|k|} - \frac{1}{\varepsilon^n} \right). \tag{3.1.21}$$

The limit in (3.1.21) can be calculated easily in dimension 1 since the sum inside the parentheses in (3.1.21) is a geometric series. Carrying out the calculation, we obtain

$$\sum_{k \neq 0} \frac{1}{k^2} = \frac{\pi^2}{3}.$$

Example 3.1.19. Let η and g be as in Example 2.4.9. Let $0 < \text{Re}\,\alpha < n$ and $x \in [-\frac{1}{2}, \frac{1}{2})^n$. The Poisson summation formula gives

$$\sum_{m\in\mathbf{Z}^n\setminus\{0\}} \frac{e^{2\pi i m\cdot x}}{|m|^\alpha} = \sum_{m\in\mathbf{Z}^n} \frac{\eta(|m|)e^{2\pi i m\cdot x}}{|m|^\alpha} = g(x) + \sum_{m\in\mathbf{Z}^n\setminus\{0\}} g(x+m)\,.$$

But it was shown in Example 2.4.9 that $g(\xi)$ decays faster than any polynomial at infinity and is equal to $\pi^{\alpha-\frac{n}{2}}\Gamma(\frac{n-\alpha}{2})\Gamma(\frac{\alpha}{2})^{-1}|\xi|^{\alpha-n} + h(\xi)$, where h is a smooth function on $\mathbf{R}^n$. Then, for $x \in [-\frac{1}{2}, \frac{1}{2})^n$, the function

$$\sum_{m\in\mathbf{Z}^n\setminus\{0\}} g(x+m)$$

is also smooth and we conclude that

$$\sum_{m\in\mathbf{Z}^n\setminus\{0\}} \frac{e^{2\pi i m\cdot x}}{|m|^\alpha} = \frac{\pi^{\alpha-\frac{n}{2}}\Gamma(\frac{n-\alpha}{2})}{\Gamma(\frac{\alpha}{2})}|x|^{\alpha-n} + h_1(x)\,,$$

where $h_1(x)$ is a C^∞ function on $\mathbf{T}^n$.

For other applications of the Poisson summation formula related to lattice points, see Exercises 3.1.12 and 3.1.13.

Exercises

3.1.1. Let P be a trigonometric polynomial on $\mathbf{T}^n$.
(a) Prove that $P(x) = \sum \widehat{P}(m)e^{2\pi i m\cdot x}$.
(b) Let f be in $L^1(\mathbf{T}^n)$. Prove that $(f * P)(x) = \sum \widehat{P}(m)\widehat{f}(m)e^{2\pi i m\cdot x}$.

3.1.2. On $\mathbf{T}^1$ let P be a trigonometric polynomial of degree $N > 0$. Show that P has at most $2N$ zeros. Construct a trigonometric polynomial with exactly $2N$ zeros.

3.1.3. Prove the identities (3.1.15) (3.1.16), and (3.1.17) about the Fejér kernel $F(n, N)$ on $\mathbf{T}^n$. Deduce from them that the family $\{F_N\}_N$ is an approximate identity as $N \to \infty$.
[*Hint:* Express the functions $\sin^2(\pi x)$ and $\sin^2(\pi(N+1)x)$ in terms of exponentials.]

3.1.4. (*de la Vallée Poussin kernel*). On $\mathbf{T}^1$ define

$$V_N(x) = 2F_{2N+1}(x) - F_N(x)\,.$$

(a) Show that the sequence V_N is an approximate identity.
(b) Prove that $\widehat{V_N}(m) = 1$ when $|m| \le N+1$, and $\widehat{V_N}(m) = 0$ when $|m| \ge 2N+2$.

3.1.5. (*Hausdorff-Young inequality*) Prove that when $f \in L^p$, $1 \le p \le 2$, the sequence of Fourier coefficients of f is in $l^{p'}$ and

$$\Big(\sum_{m\in\mathbf{Z}^n} |\widehat{f}(m)|^{p'}\Big)^{1/p'} \le \|f\|_{L^p}.$$

Also observe that 1 is the best constant in the preceding inequality.

3.1.6. Use without proof that there exists a constant $C>0$ such that for all $t \in \mathbf{R}$ we have

$$\left|\sum_{k=2}^{N} e^{ik\log k} e^{ikt}\right| \le C\sqrt{N}, \qquad N = 2,3,4,\dots,$$

to construct an example of a continuous function g on $\mathbf{T}^1$ with

$$\sum_{m\in\mathbf{Z}} |\widehat{g}(m)|^q = \infty$$

for all $q<2$. Thus the Hausdorff-Young inequality of the previous Exercise fails for $p>2$.

[*Hint:* Consider $g(x) = \sum_{k=2}^{\infty} \frac{e^{ik\log k}}{k^{1/2}(\log k)^2} e^{2\pi i k x}$. For a proof of the previous estimate, see Zygmund [**571**, Theorem (4.7) p. 199].]

3.1.7. The Poisson kernel on $\mathbf{T}^n$ is the function

$$P_{r_1,\dots,r_n}(x) = \sum_{m\in\mathbf{Z}^n} r_1^{|m_1|}\dots r_n^{|m_n|} e^{2\pi i m\cdot x}$$

and is defined for $r_1,\dots,r_n < 1$. Prove that $P_{r_1,\dots,r_n}$ can be written as

$$\begin{aligned} P_{r_1,\dots,r_n}(x_1,\dots,x_n) &= \prod_{j=1}^{n} \operatorname{Re}\left(\frac{1+r_j e^{2\pi i x_j}}{1-r_j e^{2\pi i x_j}}\right) \\ &= \prod_{j=1}^{n} \frac{1-r_j^2}{1-2r\cos(2\pi x_j)+r_j^2}, \end{aligned}$$

and conclude that $P_{r,\dots,r}(x)$ is an approximate identity as $r \uparrow 1$.

3.1.8. Let $D_N = D(1,N)$ be the Dirichlet kernel on $\mathbf{T}^1$. Prove that

$$\frac{4}{\pi^2}\sum_{k=1}^{N}\frac{1}{k} \le \|D_N\|_{L^1} \le 2 + \frac{\pi}{4} + \frac{4}{\pi^2}\sum_{k=1}^{N}\frac{1}{k}.$$

Conclude that the numbers $\|D_N\|_{L^1}$ grow logarithmically as $N\to\infty$ and therefore the family $\{D_N\}_N$ is not an approximate identity on $\mathbf{T}^1$. The

numbers $\|D_N\|_{L^1}$, $N = 1, 2, \dots$ are called the *Lebesgue constants.*
[*Hint:* Use that $\left|\frac{1}{\sin(\pi x)} - \frac{1}{\pi x}\right| \le \frac{\pi}{4}$ when $|x| \le \frac{1}{2}$.]

3.1.9. Let D_N be the Dirichlet kernel on $\mathbf{T}^1$. Prove that for all $1 < p < \infty$ there exist two constants $C_p, c_p > 0$ such that

$$c_p\,(2N+1)^{1/p'} \le \|D_N\|_{L^p} \le C_p\,(2N+1)^{1/p'}.$$

[*Hint:* Consider the two closest zeros of D_N near the origin and split the integral into the intervals thus obtained.]

3.1.10. (*S. Bernstein*) Let $P(x)$ be a trigonometric polynomial of degree N on $\mathbf{T}^1$. Prove that $\|P'\|_{L^\infty} \le 4\pi N \|P\|_{L^\infty}$.
[*Hint:* Prove first that $P'(x)/2\pi i N$ is equal to

$$\big((e^{-2\pi i N(\cdot)}P) * F_{N-1}\big)(x)\,e^{2\pi i N x} - \big((e^{2\pi i N(\cdot)}P) * F_{N-1}\big)(x)\,e^{-2\pi i N x}$$

and then take L^∞ norms.]

3.1.11. (*Fejér and F. Riesz*) Let $P(\xi) = \sum_{k=-N}^{N} a_k e^{2\pi i k\xi}$ be a trigonometric polynomial on $\mathbf{T}^1$ of degree N such that $P(\xi) > 0$ for all ξ. Prove that there exists a trigonometric polynomial $Q(\xi)$ of the form $\sum_{k=0}^{N} b_k e^{2\pi i k\xi}$ such that $P(\xi) = |Q(\xi)|^2$.
[*Hint:* Note that N zeros of the polynomial $R(z) = \sum_{k=-N}^{N} a_k z^{k+N}$ lie inside the unit circle and the other N lie outside.]

3.1.12. (*Landau* [**325**]) Points in $\mathbf{Z}^n$ are called *lattice points*. Follow the following steps to obtain the number of lattice points $N(R)$ inside a closed ball of radius R in $\mathbf{R}^n$. Let B be the closed unit ball in $\mathbf{R}^n$, χ_B its characteristic function, and v_n its volume.
(a) Using the results in Appendices B.5 and B.6, observe that there is a constant C_n such that for all $\xi \in \mathbf{R}^n$ we have

$$|\widehat{\chi_B}(\xi)| \le C_n(1+|\xi|)^{-\frac{n+1}{2}}.$$

(b) For $0 < \varepsilon < \frac{1}{10}$ let $\Phi^\varepsilon = \chi_{(1-\frac{\varepsilon}{2})B} * \zeta_\varepsilon$, where $\zeta_\varepsilon(x)\frac{1}{\varepsilon^n}\zeta(\frac{x}{\varepsilon})$ and ζ is a smooth function that is supported in $|x| \le \frac{1}{2}$ and has integral equal to 1. Also let $\Psi^\varepsilon = \chi_{(1+\frac{\varepsilon}{2})B} * \zeta_\varepsilon$. Prove that

$$\Phi^\varepsilon(x) = 1 \text{ when } |x| \le 1-\varepsilon \quad\text{and}\quad \Phi^\varepsilon(x) = 0 \text{ when } |x| \ge 1$$
$$\Psi^\varepsilon(x) = 1 \text{ when } |x| \le 1 \quad\text{and}\quad \Psi^\varepsilon(x) = 0 \text{ when } |x| \ge 1+\varepsilon$$

and also that

$$\big|\widehat{\Phi^\varepsilon}(\xi)\big| + \big|\widehat{\Psi^\varepsilon}(\xi)\big| \le C_{n,N}(1+|\xi|)^{-\frac{n+1}{2}}(1+\varepsilon|\xi|)^{-N}$$

for every $\xi \in \mathbf{R}^n$ and N large positive number.

(c) Use the result in (b) and the Poisson summation formula to obtain

$$\sum_{m\in\mathbf{Z}^n} \chi_B(\tfrac{m}{R}) \ge \sum_{m\in\mathbf{Z}^n} \Phi^\varepsilon(\tfrac{m}{R}) = R^n\widehat{\Phi^\varepsilon}(0) + \sum_{m\in\mathbf{Z}^n\setminus\{0\}} R^n\widehat{\Phi^\varepsilon}(Rm)$$

$$\ge v_n(1-\varepsilon)^n - C_{n,N} \sum_{m\in\mathbf{Z}^n\setminus\{0\}} R^n(1+R|m|)^{-\frac{n+1}{2}}(1+\varepsilon R|m|)^{-N}.$$

Now use $(1-\varepsilon)^n \ge 1-n\varepsilon$ and pick ε so that $\varepsilon R^n = \varepsilon^{-\frac{n-1}{2}}$ to deduce the estimate $N(R) \ge v_n R^n + O(R^{n\frac{n-1}{n+1}})$ as $R\to\infty$. Argue similarly with Ψ^ε to obtain the identity

$$N(R) = v_n R^n + O(R^{n\frac{n-1}{n+1}}),$$

as $R\to\infty$.

3.1.13. (*Minkowski*) Let S be an open convex symmetric set in $\mathbf{R}^n$ and assume that the Fourier transform of its characteristic function satisfies the decay estimate

$$|\widehat{\chi_S}(\xi)| \le C(1+|\xi|)^{-\frac{n+1}{2}}.$$

(This is the case if the boundary of S has nonzero Gaussian curvature). Assume that $|S| > 2^n$. Prove that S contains at least one lattice point other than the origin.
[*Hint:* Assume the contrary, set $f = \chi_{\frac{1}{2}S} * \chi_{\frac{1}{2}S}$, and apply the Poisson summation formula to f to prove that $f(0) \ge \widehat{f}(0)$.]

3.2. Decay of Fourier Coefficients

In this section we investigate the interplay between the smoothness of a function and the decay of its Fourier coefficients.

3.2.a. Decay of Fourier Coefficients of Arbitrary Integrable Functions

We begin with the classical result asserting that the Fourier coefficients of any integrable function tend to zero at infinity. The reader should compare the following proposition with Proposition 2.2.17.

Proposition 3.2.1. ***(Riemann-Lebesgue lemma)*** *Let f be in $L^1(\mathbf{T}^n)$. Then $|\widehat{f}(m)| \to 0$ as $|m|\to\infty$.*

Proof. Given $f \in L^1(\mathbf{T}^n)$ and $\varepsilon > 0$, let P be a trigonometric polynomial such that $\|f - P\|_{L^1} < \varepsilon$. If $|m| >$ degree (P), then $\widehat{P}(m) = 0$ and thus

$$|\widehat{f}(m)| = |\widehat{f}(m) - \widehat{P}(m)| \le \|f - P\|_{L^1} < \varepsilon.$$

This proves that $|\widehat{f}(m)| \to 0$ as $|m| \to \infty$. □

Several questions are naturally raised. How fast may the Fourier coefficients of an L^1 function tend to zero? Does additional smoothness of the function imply faster decay of the Fourier coefficients? Can such a decay be quantitatively expressed in terms of the smoothness of the function?

We answer the first question. Fourier coefficients of an L^1 function can tend to zero arbitrarily slowly, that is, slower than any given rate of decay.

Theorem 3.2.2. *Let $(d_m)_{m \in \mathbf{Z}^n}$ be a sequence of positive real numbers with $d_m \to 0$ as $|m| \to \infty$. Then there exists a $g \in L^1(\mathbf{T}^n)$ such that $|\widehat{g}(m)| \ge d_m$ for all $m \in \mathbf{Z}^n$. In other words, given any rate of decay, there exists an integrable function on the torus whose Fourier coefficients have slower rate of decay.*

We will first prove this theorem when $n = 1$ and then extend it to higher dimensions. We will first need the following two lemmata.

Lemma 3.2.3. *Given a sequence of positive real numbers $\{a_m\}_{m=0}^{\infty}$ that tends to zero as $m \to \infty$, there exists a sequence $\{c_m\}_{m=0}^{\infty}$ that satisfies*

$$c_m \ge a_m, \qquad c_m \downarrow 0, \qquad \text{and} \quad c_{m+2} + c_m \ge 2c_{m+1}$$

for all $m = 0, 1, \dots$. We will call such sequences convex.

Lemma 3.2.4. *Given a convex decreasing sequence of positive real numbers $\{c_m\}_{m=0}^{\infty}$ with $\lim_{m\to\infty} c_m = 0$ and $s \ge 0$ a fixed integer, we have that*

$$\sum_{r=0}^{\infty} (r+1)(c_{r+s} + c_{r+s+2} - 2c_{r+s+1}) = c_s. \tag{3.2.1}$$

We first prove Lemma 3.2.3.

Proof. Let $k_0 = 0$ and suppose that $a_m \le M$ for all $m \ge 0$. Find $k_1 > k_0$ such that for $m \ge k_1$ we have $a_m \le M/2$. Now find $k_2 > k_1 + \frac{k_1 - k_0}{2}$ such that for $m \ge k_2$ we have $a_m \le M/4$. Next find $k_3 > k_2 + \frac{k_2 - k_1}{2}$ such that for $m \ge k_3$ we have $a_m \le M/8$. Continue inductively this way and construct a subsequence $k_0 < k_1 < k_2 < \dots$ of the integers such that for $m \ge k_j$ we have $a_m \le 2^{-j}M$ and $k_{j+1} > k_j + \frac{k_j - k_{j-1}}{2}$ for $j \ge 1$. Join the points $(k_0, 2M)$, (k_1, M), $(k_2, M/2)$, $(k_3, M/4), \dots$ by straight lines and note that by the choice of the subsequence $\{k_j\}_{j=0}^{\infty}$ the resulting piecewise linear function h is convex on $[0, \infty)$. Define $c_m = h(m)$ and observe that the sequence $\{c_m\}_{m=0}^{\infty}$ satisfies the required properties. See also Exercise 3.2.1 for alternative proof. □

We now prove Lemma 3.2.4.

Proof. We have that

$$\sum_{r=0}^{N}(r+1)(c_{r+s}+c_{r+s+2}-2c_{r+s+1}) \tag{3.2.2}$$
$$= c_s-(N+1)(c_{s+N+1}-c_{s+N+2})-c_{s+N+1}\,.$$

To show that the last expression tends to c_s as $N\to\infty$, we take $M=[\frac{N}{2}]$ and we use convexity $\big(c_{s+M+j}-c_{s+M+j+1}\ge c_{s+M+j+1}-c_{s+M+j+2}\big)$ to obtain

$$\begin{aligned} c_{s+M+1}-c_{s+N+2} &= c_{s+M+1}-c_{s+M+2}\\ &\quad + c_{s+M+2}-c_{s+M+3}\\ &\quad + \dots\\ &\quad + c_{s+N+1}-c_{s+N+2}\\ &\ge (M+1)(c_{s+N+1}-c_{s+N+2})\\ &\ge \tfrac{N+1}{2}(c_{s+N+1}-c_{s+N+2})\ge 0\,. \end{aligned}$$

The preceding calculation implies that $(N+1)(c_{s+N+1}-c_{s+N+2})$ tends to zero as $N\to\infty$ and thus the expression in (3.2.2) converges to c_s as $N\to\infty$.

□

We now continue with the proof of Theorem 3.2.2 when $n=1$.

Proof. We are given a sequence of positive numbers $\{a_m\}_{m\in\mathbf{Z}}$ that converges to zero as $|m|\to\infty$ and we would like to find an integrable function on $\mathbf{T}^1$ with $|\widehat{f}(m)|\ge a_m$ for all $m\in\mathbf{Z}$. Apply Lemma 3.2.3 to the sequence $\{a_m+a_{-m}\}_{m\ge0}$ to find a convex sequence $\{c_m\}_{m\ge0}$ that dominates $\{a_m+a_{-m}\}_{m\ge0}$ and decreases to zero as $m\to\infty$. Extend c_m for $m<0$ by setting $c_m=c_{|m|}$. Now define

$$f(x)=\sum_{j=0}^{\infty}(j+1)(c_j+c_{j+2}-2c_{j+1})F_j(x), \tag{3.2.3}$$

where F_j is the (one-dimensional) Fejér kernel. The convexity of the sequence c_m and the positivity of the Fejér kernel imply that $f\ge0$. Lemma 3.2.4 with $s=0$ gives that

$$\sum_{j=0}^{\infty}(j+1)(c_j+c_{j+2}-2c_{j+1})\big\|F_j\big\|_{L^1}=c_0<\infty, \tag{3.2.4}$$

since $\big\|F_j\big\|_{L^1}=1$ for all j. Therefore (3.2.3) defines an integrable function f on $\mathbf{T}^1$. We now compute the Fourier coefficients of f. Since the series in (3.2.3) converges in L^1, for $m \in \mathbf{Z}$ we have

$$\begin{aligned}
\widehat{f}(m) &= \sum_{j=0}^{\infty}(j+1)(c_j + c_{j+2} - 2c_{j+1})\widehat{F_j}(m) \\
&= \sum_{j=|m|}^{\infty}(j+1)(c_j + c_{j+2} - 2c_{j+1})\left(1 - \frac{|m|}{j+1}\right) \qquad (3.2.5) \\
&= \sum_{r=0}^{\infty}(r+1)(c_{r+|m|} + c_{r+|m|+2} - 2c_{r+|m|+1}) = c_{|m|} = c_m\,,
\end{aligned}$$

where we used Lemma 3.2.4 with $s = |m|$.

Let us now extend this result on $\mathbf{T}^n$. Let $(d_m)_{m\in\mathbf{Z}^n}$ be a positive sequence with $d_m \to 0$ as $|m| \to \infty$. By Exercise 3.2.2, there exists a positive sequence $(a_j)_{j\in\mathbf{Z}}$ with $a_{m_1}\dots a_{m_n} \geq d_{(m_1,\dots,m_n)}$ and $a_j \to 0$ as $|j| \to \infty$. Let

$$g(x_1,\dots,x_n) = f(x_1)\cdots f(x_n),$$

where f is the function previously constructed when $n = 1$. It can be seen easily using (3.1.6) that $\widehat{g}(m) \geq d_m$.

$\square$

3.2.b. Decay of Fourier Coefficients of Smooth Functions

We next study the decay of the Fourier coefficients of functions that possess a certain amount of smoothness. We will see that the decay of the Fourier coefficients reflects the smoothness of the function in a rather precise quantitative way. Conversely, if the Fourier coefficients of an integrable function have polynomial decay faster than the dimension, then a certain amount of smoothness can be inferred about the function.

Definition 3.2.5. For $0 \leq \gamma < 1$ define

$$\big\|f\big\|_{\dot{\Lambda}_\gamma} = \sup_{x,h\in\mathbf{T}^n} \frac{|f(x+h) - f(x)|}{|h|^\gamma}$$

and

$$\dot{\Lambda}_\gamma(\mathbf{T}^n) = \{f : \mathbf{T}^n \to \mathbf{C} \text{ with } \big\|f\big\|_{\dot{\Lambda}_\gamma} < \infty\}.$$

We will call $\dot{\Lambda}_\gamma(\mathbf{T}^n)$ the *homogeneous Lipschitz space* of order γ on the torus. Functions f on $\mathbf{T}^n$ with $\big\|f\big\|_{\dot{\Lambda}_\gamma} < \infty$ will be called *homogeneous Lipschitz* functions of order γ.

Some remarks are in order.

Remark 3.2.6. $\dot{\Lambda}_\gamma(\mathbf{T}^n)$ is called the *homogeneous Lipschitz space* of order γ on $\mathbf{T}^n$ in contrast with the space $\Lambda_\gamma(\mathbf{T}^n)$, which is called the Lipschitz space of order γ. The latter space is defined as

$$\Lambda_\gamma(\mathbf{T}^n) = \{f : \mathbf{T}^n \to \mathbf{C} \text{ with } \|f\|_{\Lambda_\gamma} < \infty\},$$

where

$$\|f\|_{\Lambda_\gamma} = \|f\|_{L^\infty} + \|f\|_{\dot{\Lambda}_\gamma}.$$

Remark 3.2.7. The positive functional $\|\cdot\|_{\dot{\Lambda}_\gamma}$ satisfies the triangle inequality but it does not satisfy the property $\|f\|_{\dot{\Lambda}_\gamma} = 0 \implies f = 0$ a.e. required to be a norm. It is therefore a seminorm on $\dot{\Lambda}_\gamma(\mathbf{T}^n)$. However, if we identify functions whose difference is a constant, we form the space of all equivalence classes $\dot{\Lambda}_\gamma(\mathbf{T}^n)/\{\text{constants}\}$ (defined for $0 \le \gamma < 1$) on which the functional $f \to \|f\|_{\dot{\Lambda}_\gamma}$ is a norm.

Remark 3.2.8. Observe that homogeneous Lipschitz functions of order $0 < \gamma < 1$ are continuous and thus bounded. Therefore, $\dot{\Lambda}_\gamma(\mathbf{T}^n) \subseteq L^\infty(\mathbf{T}^n)$ set theoretically. However, the norm inequality $\|f\|_{L^\infty} \le C\|f\|_{\dot{\Lambda}_\gamma}$ fails for any constant C independent of all functions f. Take, for example, $f = N + \sin(2\pi x)$ on $\mathbf{T}^1$ with $N \to \infty$ to obtain a counterexample. Nevertheless, under the identification of functions whose difference is a constant, we have that

$$\|f\|_{L^\infty} \approx \|f\|_{\dot{\Lambda}_0/\{\text{constants}\}}$$

and

$$\|f\|_{L^\infty} \le C_\gamma \|f\|_{\dot{\Lambda}_\gamma/\{\text{constants}\}}$$

for all $0 < \gamma < 1$. In other words, the space $\dot{\Lambda}_\gamma/\{\text{constants}\}$ embeds in $L^\infty(\mathbf{T}^n)$.

The following theorem clearly indicates how the smoothness of a function is reflected by the decay of its Fourier coefficients.

Theorem 3.2.9. *Let $s \in \mathbf{Z}$ with $s \ge 0$.*
(a) Suppose that $\partial^\alpha f$ exist and are integrable for all $|\alpha| \le s$. Then

$$|\widehat{f}(m)| \le \left(\frac{\sqrt{n}}{2\pi}\right)^s \frac{\sup_{|\alpha|=s} |\widehat{\partial^\alpha f}(m)|}{|m|^s}, \qquad m \neq 0, \tag{3.2.6}$$

and thus $|\widehat{f}(m)|(1 + |m|^s) \to 0$ as $|m| \to \infty$.
(b) Suppose that $\partial^\alpha f$ exist for all $|\alpha| \le s$ and whenever $|\alpha| = s$, $\partial^\alpha f$ are in $\dot{\Lambda}_\gamma(\mathbf{T}^n)$ for some $0 \le \gamma < 1$. Then

$$|\widehat{f}(m)| \le \frac{(\sqrt{n})^{s+\gamma}}{(2\pi)^s 2^{\gamma+1}} \frac{\sup_{|\alpha|=s} \|\partial^\alpha f\|_{\dot{\Lambda}_\gamma}}{|m|^{s+\gamma}}, \qquad m \neq 0. \tag{3.2.7}$$

Proof. Fix $m \in \mathbf{Z}^n \setminus \{0\}$ and pick a j such that $|m_j| = \sup_{1\le k\le n}|m_k|$. Then clearly $m_j \neq 0$. Integrating by parts s times with respect to the variable x_j, we obtain

$$\widehat{f}(m) = \int_{\mathbf{T}^n} f(x)e^{-2\pi i x\cdot m}\,dx = \int_{\mathbf{T}^n} (\partial_j^s f)(x)\frac{e^{-2\pi i x\cdot m}}{(-2\pi i m_j)^s}\,dx, \tag{3.2.8}$$

where the boundary terms all vanish because of the periodicity of the integrand. Taking absolute values and using that $|m| \le \sqrt{n}\,|m_j|$, we obtain assertion (3.2.6).

We now turn to the second part of theorem. Let $e_j = (0,\dots,1,\dots,0)$ be the element of the torus $\mathbf{T}^n$ whose jth coordinate is one and all the others are zero. A simple change of variables together with the fact that $e^{\pi i} = -1$ gives that

$$\int_{\mathbf{T}^n} (\partial_j^s f)(x)e^{-2\pi i x\cdot m}\,dx = -\int_{\mathbf{T}^n} (\partial_j^s f)(x - \tfrac{e_j}{2m_j})e^{-2\pi i x\cdot m}\,dx\,,$$

which implies that

$$\int_{\mathbf{T}^n} (\partial_j^s f)(x)e^{-2\pi i x\cdot m}\,dx = \frac{1}{2}\int_{\mathbf{T}^n} \big[(\partial_j^s f)(x) - (\partial_j^s f)(x - \tfrac{e_j}{2m_j})\big]e^{-2\pi i x\cdot m}\,dx\,.$$

Now use the estimate

$$|(\partial_j^s f)(x) - (\partial_j^s f)(x - \tfrac{e_j}{2m_j})| \le \frac{\|\partial_j^s f\|_{\dot{\Lambda}_\gamma}}{(2|m_j|)^\gamma}$$

and identity (3.2.8) to conclude the proof of (3.2.7). □

The following is an immediate consequence.

Corollary 3.2.10. *Let $s \in \mathbf{Z}$ with $s \ge 0$.*
(a) Suppose that $\partial^\alpha f$ exist and are integrable for all $|\alpha| \le s$. Then for some constant $c_{n,s}$ we have

$$|\widehat{f}(m)| \le c_{n,s}\frac{\max\big(\|f\|_{L^1}, \sup_{|\alpha|=s}|\widehat{\partial^\alpha f}(m)|\big)}{(1+|m|)^s}\,. \tag{3.2.9}$$

(b) Suppose that $\partial^\alpha f$ exist for all $|\alpha| \le s$ and whenever $|\alpha| = s$, $\partial^\alpha f$ are in $\dot{\Lambda}_\gamma(\mathbf{T}^n)$ for some $0 \le \gamma < 1$. Then for some constant $c'_{n,s}$ we have

$$|\widehat{f}(m)| \le c'_{n,s}\frac{\max\big(\|f\|_{L^1}, \sup_{|\alpha|=s}\|\partial^\alpha f\|_{\dot{\Lambda}_\gamma}\big)}{(1+|m|)^{s+\gamma}}\,. \tag{3.2.10}$$

Remark 3.2.11. Observe that the conclusions of Theorem 3.2.9 and Corollary 3.2.10 are also valid when $\gamma = 1$. In this case the spaces $\dot{\Lambda}_\gamma$ should be replaced by the space Lip 1 equipped with the seminorm

$$\|f\|_{\mathrm{Lip}\,1} = \sup_{x,h\in\mathbf{T}^n}\frac{|f(x+h)-f(x)|}{|h|}\,.$$

There is a slight lack of uniformity in the notation here, as in the theory of Lipschitz spaces the notation $\dot{\Lambda}_1$ is usually reserved for the space with seminorm

$$\|f\|_{\dot{\Lambda}_1} = \sup_{x,h\in\mathbf{T}^n} \frac{|f(x+h)+f(x-h)-2f(x)|}{|h|}.$$

The following proposition provides a partial converse to Theorem 3.2.9.

Proposition 3.2.12. *Let $s>0$ be a positive real number and let $[s]$ be the integer part of s. Suppose that f is an integrable function on the torus with*

$$|\widehat{f}(m)| \le C(1+|m|)^{-s-n}$$

for all $m\in\mathbf{Z}^n$. Then
(a) If $s\notin\mathbf{Z}$, then f has partial derivatives of all orders $|\alpha|\le[s]$. Moreover, $\partial^\alpha f\in\dot{\Lambda}_\gamma$ for all multiindices $|\alpha|=[s]$ and for all $0\le\gamma<s-[s]$.
(b) If $s\in\mathbf{Z}$, then f has partial derivatives of all orders $|\alpha|\le s-1$ and $\partial^\alpha f\in\dot{\Lambda}_\gamma$ for all multiindices $|\alpha|=s-1$ and all $0\le\gamma<1$.

Proof. Since f has an absolutely convergent Fourier series, Proposition 3.1.14 gives that

$$f(x) = \sum_{m\in\mathbf{Z}^n} \widehat{f}(m) e^{2\pi i x\cdot m}, \tag{3.2.11}$$

for almost all $x\in\mathbf{T}^n$.

If $s\notin\mathbf{Z}$, the series in (3.2.11) can be differentiated α times where $|\alpha|=[s]$ since

$$\sum_{m\in\mathbf{Z}^n} \widehat{f}(m)\partial^\alpha e^{2\pi i x\cdot m} = \sum_{m\in\mathbf{Z}^n} \widehat{f}(m)(2\pi i m)^\alpha e^{2\pi i x\cdot m}$$

and the last series converges absolutely because of the decay assumptions on the Fourier coefficients of f; (m^α is shorthand for $m_1^{\alpha_1}\dots m_n^{\alpha_n}$). Moreover, we have

$$(\partial^\alpha f)(x) = \sum_{m\in\mathbf{Z}^n} \widehat{f}(m)(2\pi i m)^\alpha e^{2\pi i x\cdot m}$$

for all multiindices $(\alpha_1,\dots,\alpha_n)$ with $|\alpha|=[s]$. Now suppose that $0\le\gamma<s-[s]$. Then

$$\begin{aligned}
&|(\partial^\alpha f)(x+h)-(\partial^\alpha f)(x)| \\
&= \Big|\sum_{m\in\mathbf{Z}^n} \widehat{f}(m)(2\pi i m)^\alpha e^{2\pi i x\cdot m}\big(e^{2\pi i m\cdot h}-1\big)\Big| \\
&\le 2^{1-\gamma}(2\pi)^s \sum_{m\in\mathbf{Z}^n} |m|^{[s]}\frac{|h|^\gamma|m|^\gamma}{(1+|m|)^{n+s}} = C'_{\gamma,n,s}|h|^\gamma,
\end{aligned}$$

where we used the fact that $n+s-[s]-\gamma>n$ and that

$$|e^{2\pi i m\cdot h}-1| \le \min(2,|m|\,|h|) \le 2^{1-\gamma}|m|^\gamma|h|^\gamma.$$

The case when $s\in\mathbf{Z}$ is proved similarly.

□

We have seen that if a function on $\mathbf{T}^1$ has an integrable derivative, then its Fourier coefficients tend to zero when divided by $|m|^{-1}$. In this case we say that the Fourier coefficients of f are $o(|m|^{-1})$ as $|m| \to \infty$. Let us denote by L_1^1 the class of all functions on $\mathbf{T}^1$ whose derivative is also in L^1. Next we introduce a slightly larger class of functions on $\mathbf{T}^1$ whose Fourier coefficients decay like $|m|^{-1}$ as $|m| \to \infty$.

Definition 3.2.13. A measurable function f on $\mathbf{T}^1$ is called of *bounded variation* if it is defined everywhere and

$$\mathrm{Var}(f) = \sup\Big\{ \sum_{j=1}^{M} |f(x_j) - f(x_{j-1})| : \, 0 = x_0 < x_1 < \ldots < x_M = 1 \Big\} < \infty,$$

where the supremum is taken over all partitions of the interval $[0,1]$. The expression $\mathrm{Var}(f)$ is called the *total variation* of f. The class of functions of bounded variation will be denoted by BV.

The following result concerns functions of bounded variation.

Proposition 3.2.14. *If f is in $BV(\mathbf{T}^1)$, then*

$$|\widehat{f}(m)| \le \frac{\mathrm{Var}(f)}{2\pi|m|}$$

whenever $m \neq 0$.

Proof. If f is a function of bounded variation, then the Lebesgue-Stieltjes integral with respect to f is well defined. An integration by parts gives

$$\widehat{f}(m) = \int_{\mathbf{T}^1} f(x) e^{-2\pi i m x}\, dx = \int_{\mathbf{T}^1} \frac{e^{-2\pi i m x}}{-2\pi i m}\, df\,,$$

where the boundary terms vanish because of periodicity. The conclusion of the theorem follows from the fact that the norm of the measure df is the total variation of f.

□

The inclusions are valid ,

$$L_1^1(\mathbf{T}^1) \subseteq BV(\mathbf{T}^1) \subseteq L^\infty(\mathbf{T}^1),$$

and we have proved the following rate of decay for the Fourier coefficients of functions in the previous spaces, respectively as $|m| \to \infty$

$$o(|m|^{-1}), \quad O(|m|^{-1}), \quad o(1).$$

3.2.c. Functions with Absolutely Summable Fourier Coefficients

Decay for the Fourier coefficients can also be indirectly deduced from a certain knowledge about the summability of these coefficients. The simplest such kind of summability is in the sense of ℓ^1. It is therefore natural to consider the class of functions on the torus whose Fourier coefficients form an absolutely summable series.

Definition 3.2.15. An integrable function f on the torus is said to have an *absolutely convergent* Fourier series if

$$\sum_{m\in\mathbf{Z}^n} |\widehat{f}(m)| < +\infty.$$

We will denote by $A(\mathbf{T}^n)$ the space of all integrable functions on the torus $\mathbf{T}^n$ whose Fourier series are absolutely convergent. We then introduce a norm on $A(\mathbf{T}^n)$ by setting

$$\big\|f\big\|_{A(\mathbf{T}^n)} = \sum_{m\in\mathbf{Z}^n} |\widehat{f}(m)|\,.$$

It is straightforward that every function in $A(\mathbf{T}^n)$ must be bounded. The following theorem gives us a sufficient condition for a function to be in $A(\mathbf{T}^n)$.

Theorem 3.2.16. *Let s be a nonnegative integer and let $0 \le \alpha < 1$. Assume that f is a function defined on $\mathbf{T}^n$ all of whose partial derivatives of order s lie in the space $\dot{\Lambda}_\alpha$. Suppose that $s+\alpha > n/2$. Then $f \in A(\mathbf{T}^n)$ and*

$$\big\|f\big\|_{A(\mathbf{T}^n)} \le C \sup_{|\beta|=s} \big\|\partial^\beta f\big\|_{\dot{\Lambda}_\alpha},$$

where C depends on n, α, and s.

Proof. For $1 \le j \le n$, let e_j be the element of $\mathbf{R}^n$ with zero entries except for the jth coordinate, which is 1. Let l be a positive integer and let $h_j = 2^{-l-2}e_j$.

Then for a multiindex $m = (m_1, \dots, m_n)$ satisfying $2^l \le |m| \le 2^{l+1}$ and for $j \in \{1, \dots, n\}$ selected so that $|m_j| = \sup_k |m_k|$ we have

$$\frac{|m_j|}{2^l} \ge \frac{|m|}{2^l\sqrt{n}} \ge \frac{1}{\sqrt{n}}.$$

We use the elementary fact that $|t| \le \pi \implies |e^{it}-1| \ge 2|t|/\pi$ to obtain

$$|e^{2\pi i m\cdot h_j} - 1| = |e^{2\pi i m_j 2^{-l-2}} - 1| \ge \frac{2}{\pi}\frac{|2\pi m_j|}{2^{l+2}} = \frac{|m_j|}{2^l} \ge \frac{1}{\sqrt{n}}$$

whenever $\dfrac{|2\pi m_j|}{2^{l+2}} \le \pi$, which is always true since $\dfrac{|2\pi m_j|}{2^{l+2}} \le \dfrac{2\pi 2^{l+1}}{2^{l+2}} \le \pi$.

We now have

$$
\begin{aligned}
&\Big(\sum_{2^l\le|m|<2^{l+1}} |\widehat{f}(m)|\Big)^2\\
&\le \Big(\sum_{2^l\le|m|<2^{l+1}} 1^2\Big)\Big(\sum_{2^l\le|m|<2^{l+1}} |\widehat{f}(m)|^2\Big)\\
&\le C_n 2^{ln}\sum_{j=1}^{n}\sum_{\substack{2^l\le|m|<2^{l+1}\\ |m_j|=\sup_k|m_k|}} |\widehat{f}(m)|^2\\
&\le C_n' 2^{ln}\sum_{j=1}^{n}\sum_{\substack{2^l\le|m|<2^{l+1}\\ |m_j|=\sup_k|m_k|}} |e^{2\pi i m\cdot h_j}-1|^2|\widehat{f}(m)|^2\frac{|2\pi m_j|^{2s}}{|2\pi m_j|^{2s}}\\
&\le C_{n,s}2^{l(n-2s)}\sum_{j=1}^{n}\sum_{m\in\mathbf{Z}^n} |e^{2\pi i m\cdot h_j}-1|^2|\widehat{\partial_j^s f}(m)|^2\\
&= C_{n,s}2^{l(n-2s)}\sum_{j=1}^{n}\big\|\partial_j^s f-\partial_j^s f(\cdot+h_j)\big\|_{L^2}^2\\
&\le C_{n,s}' 2^{l(n-2s)}(2^{-(l+3)})^{2\alpha}\sup_{|\beta|=s}\big\|\partial^\beta f\big\|_{\dot\Lambda_\alpha}^2 .
\end{aligned}
$$

Taking square roots, summing over all positive integers l, and using that $s+\alpha > n/2$, we obtain the desired conclusion.

□

Exercises

3.2.1. Given a sequence $\{a_n\}_{n=0}^\infty$ of positive numbers such that $a_n \to 0$ as $n \to \infty$, find a nonnegative integrable function h on $[0,1]$ such that

$$\int_0^1 h(t)t^m\,dt \ge a_m.$$

Use this result to deduce a different proof of Lemma 3.2.3.
$\Big[$*Hint:* Try $h = e\sum_{k=0}^{\infty}(\sup_{j\ge k} a_j - \sup_{j\ge k+1} a_j)(k+2)\chi_{[\frac{k+1}{k+2},1]}$.$\Big]$

3.2.2. Prove that given a positive sequence $\{d_m\}_{m\in\mathbf{Z}^n}$ with $d_m \to 0$ as $|m| \to \infty$, there exists a positive sequence $\{a_j\}_{j\in\mathbf{Z}}$ with $a_{m_1}\dots a_{m_n} \ge d_{(m_1,\dots,m_n)}$ and $a_j \to 0$ as $|j| \to \infty$.

3.2.3. (a) Use the idea of the proof of Lemma 3.2.4 to prove that if a twice continuously differentiable function $f \ge 0$ is defined on $(0,\infty)$ and satisfies $f'(x) \le 0$ and $f''(x) \ge 0$ for all $x > 0$, then $\lim_{x\to\infty} xf'(x) = 0$.
(b) Suppose that a twice continuously differentiable function g is defined on $(0,\infty)$ and satisfies $g \ge 0$, $g' \le 0$, and $\int_1^\infty g(x)\,dx < +\infty$. Prove that

$$\lim_{x\to\infty} xg(x) = 0.$$

3.2.4. Prove that for $0 \le \gamma < \delta < 1$ we have $\|f\|_{\dot{\Lambda}_\gamma} \le C_{n,\gamma,\delta}\|f\|_{\dot{\Lambda}_\delta}$ for all functions f and thus $\dot{\Lambda}_\delta$ is a subspace of $\dot{\Lambda}_\gamma$.

3.2.5. Prove the inclusions $L_1^1(\mathbf{T}^1) \subseteq BV(\mathbf{T}^1) \subseteq L^\infty(\mathbf{T}^1)$ as follows.
(a) If $f \in L_1^1(\mathbf{T}^1)$, then $\mathrm{Var}(f) \le \|f'\|_{L^1}$.
(b) If $f \in BV(\mathbf{T}^1)$, then $\|f\|_{L^\infty} \le \mathrm{Var}(f) + |f(0)|$.

3.2.6. Suppose that f is a differentiable function on $\mathbf{T}^1$ whose derivative $f' \in L^2(\mathbf{T}^1)$. Prove that $f \in A(\mathbf{T}^1)$ and that

$$\|f\|_{A(\mathbf{T}^1)} \le \|f\|_{L^1} + \frac{1}{2\pi}\Big(\sum_{j\neq 0} j^{-2}\Big)^{1/2}\|f'\|_{L^2}.$$

3.2.7. (a) Prove that the product of two functions in $A(\mathbf{T}^n)$ is also in $A(\mathbf{T}^n)$ and that

$$\|fg\|_{A(\mathbf{T}^n)} \le \|f\|_{A(\mathbf{T}^n)}\|g\|_{A(\mathbf{T}^n)}.$$

(b) Prove that the convolution of two square integrable functions on $\mathbf{T}^n$ always gives a function in $A(\mathbf{T}^n)$.

3.2.8. Fix $0 < \alpha < 1$ and define f on $\mathbf{T}^1$ by setting

$$f(x) = \sum_{k=0}^\infty 2^{-\alpha k} e^{2\pi i 2^k x}.$$

Prove that $f \in \dot{\Lambda}_\alpha$. Conclude that the decay $|\widehat{f}(m)| \le C|m|^{-\alpha}$ is best possible for $f \in \dot{\Lambda}_\alpha$.

3.2.9. Use without proof that there exists a constant $C > 0$ such that

$$\sup_{t\in\mathbf{R}}\left|\sum_{k=2}^N e^{ik\log k} e^{ikt}\right| \le C\sqrt{N}, \qquad N = 2,3,4,\dots$$

to prove that the function

$$g(x) = \sum_{k=2}^\infty \frac{e^{2\pi i k\log k}}{k} e^{2\pi i k x}$$

is in $\dot{\Lambda}_{1/2}(\mathbf{T}^1)$ but not in $A(\mathbf{T}^1)$. Conclude that the restriction $s > 1/2$ in Theorem 3.2.16 is sharp.

3.2.10. Use the open mapping theorem to prove that there exist sequences $\{a_m\}_{m\in\mathbf{Z}^n}$ that tend to zero as $|m| \to \infty$ for which there do not exist functions f in $L^1(\mathbf{T}^n)$ with $\widehat{f}(m) = a_m$ for all m.

3.3. Pointwise Convergence of Fourier Series

In this section we will be concerned with the pointwise convergence of the square partial sums and the Fejér means of a function defined on the torus.

3.3.a. Pointwise Convergence of the Fejér Means

We saw in Section 3.1 that the Fejér kernel is an approximate identity. This implies that the Fejér (or Cesàro) means of an L^p function on $\mathbf{T}^n$ converge to the function in L^p for any $1 \le p < \infty$. Moreover, if f is continuous at x_0, then the means $(F(n,N) * f)(x_0)$ converge to $f(x_0)$ as $N \to \infty$ in view of Theorem 1.2.19 (2). Although this is a satisfactory result, it is restrictive, since it only applies to continuous functions. It is natural to ask what happens for more general functions.

Using properties of the Fejér kernel, we obtain the following one-dimensional result regarding the convergence of the Fejér means:

Theorem 3.3.1. *(Fejér) If a function f in $L^1(\mathbf{T}^1)$ has left and right limits at a point x_0, denoted by $f(x_0-)$ and $f(x_0+)$, respectively, then*

$$(F_N * f)(x_0) \to \frac{1}{2}\big(f(x_0+) + f(x_0-)\big), \qquad \text{as} \qquad N \to \infty. \tag{3.3.1}$$

In particular this is the case for functions of bounded variation.

Proof. Let us identify $\mathbf{T}^1$ with $[-1/2, 1/2]$. Given $\varepsilon > 0$, find $\delta > 0$ such that $0 < t \le \delta$ implies

$$\max\big(|f(x_0+t) - f(x_0+)|\,,\, |f(x_0-t) - f(x_0-)|\big) < \varepsilon\,.$$

Then we have

$$0 < |t| \le \delta \implies \left| \frac{f(x_0+t) + f(x_0-t)}{2} - \frac{f(x_0+) + f(x_0-)}{2} \right| < \varepsilon. \tag{3.3.2}$$

Set $V = [-\delta, \delta]$. Using the second expression for F_N in (3.1.15), we can find an $N_0 > 0$ such that for $N \ge N_0$ we have

$$\sup_{t\in\mathbf{T}^1\setminus V} F_N(t) < \varepsilon. \tag{3.3.3}$$

We now have

$$\begin{aligned}(F_N * f)(x_0) - f(x_0+) &= \int_{\mathbf{T}^1} F_N(-t)\big(f(x_0+t) - f(x_0+)\big)\,dt\,,\\ (F_N * f)(x_0) - f(x_0-) &= \int_{\mathbf{T}^1} F_N(t)\big(f(x_0-t) - f(x_0-)\big)\,dt\,.\end{aligned}$$

Averaging these two identities and using that F_N is an even function, we obtain

$$\begin{aligned}&(F_N * f)(x_0) - \frac{f(x_0+)+f(x_0-)}{2}\\ &= \int_{\mathbf{T}^1} F_N(t)\left(\frac{f(x_0+t)+f(x_0-t)}{2} - \frac{f(x_0+)+f(x_0-)}{2}\right)dt\,.\end{aligned} \tag{3.3.4}$$

We split the integral in (3.3.4) in two pieces, the integral over V and the integral over $\mathbf{T}^1 \setminus V$. By (3.3.2), the integral over V is controlled by $\varepsilon \int_{\mathbf{T}^1} F_N(t)\,dt = \varepsilon$. Also (3.3.3) gives that for $N \ge N_0$

$$\begin{aligned}&\left|\int_{\mathbf{T}^1\setminus V} F_N(t)\left(\frac{f(x_0-t)+f(x_0+t)}{2} - \frac{f(x_0-)+f(x_0+)}{2}\right)dt\right|\\ &\le \frac{\varepsilon}{2}\left(\big\|f - f(x_0-)\big\|_{L^1} + \big\|f - f(x_0+)\big\|_{L^1}\right) = \varepsilon\, c(f,x_0)\,,\end{aligned}$$

where $c(f,x_0)$ is a constant depending on f and x_0. We have now proved that given $\varepsilon > 0$ there exists an N_0 such that for $N \ge N_0$ the second expression in (3.3.4) is bounded by $\varepsilon\,(c(f,x_0)+1)$. This proves the required conclusion.

Functions of bounded variation can be written as differences of increasing functions, and since increasing functions have left and right limits everywhere, (3.3.1) holds for these functions.

□

We continue with an elementary but very useful proposition. We refer to Exercise 3.3.2 for some of its applications.

Proposition 3.3.2. *(a) Let f be in $L^1(\mathbf{T}^n)$. If x_0 is a point of continuity of f and the square partial sums of the Fourier series of f converge at x_0, then they must converge to $f(x_0)$.*
(b) In dimension 1, if $f(x)$ has left and right limits as $x \to x_0$ and the partial sums of the Fourier series of f converge, then they must converge to $\frac{1}{2}\big(f(x_0+)+f(x_0-)\big)$.

Proof. (a) We observed before that if $f \in L^1(\mathbf{T}^n)$ is continuous at x_0, then $(F(n,N) * f)(x_0) \to f(x_0)$ as $N \to \infty$. If $(D(n,N) * f)(x_0) \to A(x_0)$ as $N \to \infty$, then the arithmetic means of this sequence must converge to the same number as the sequence. Therefore,

$$(F(n,N) * f)(x_0) \to A(x_0)$$

as $N \to \infty$ and thus $A(x_0) = f(x_0)$. Part (b) is proved using the same argument and the result of Theorem 3.3.1.

□

3.3.b. Almost Everywhere Convergence of the Fejér Means

We have seen that the Fejér means of a relatively nice function (such as of bounded variation) converge everywhere. What can we say about the Fejér means of a general integrable function? Since the Fejér kernel is a well-behaved approximate identity, the following result should not come as a surprise.

Theorem 3.3.3. *(a) For $f \in L^1(\mathbf{T}^n)$, let*

$$\mathcal{H}(f) = \sup_{N \in \mathbf{Z}^+} |f * F(n, N)|.$$

Then $\mathcal{H}$ maps $L^1(\mathbf{T}^n)$ to $L^{1,\infty}(\mathbf{T}^n)$ and $L^p(\mathbf{T}^n)$ into itself for $1 < p \le \infty$.
(b) For any function $f \in L^1(\mathbf{T}^n)$, we have

$$(F(n, N) * f)(x) \to f(x)$$

as $N \to \infty$ for almost all $x \in \mathbf{T}^n$.

Proof. It is an elementary fact that $|t| \le \frac{\pi}{2} \implies |\sin t| \ge \frac{2}{\pi}|t|$; see Appendix E. Using this fact and the expression (3.1.15) we obtain for all t in $[-\frac{1}{2}, \frac{1}{2}]$

$$\begin{aligned}
|F_N(t)| &= \frac{1}{N+1}\left|\frac{\sin(\pi(N+1)t)}{\sin(\pi t)}\right|^2 \\
&\le \frac{N+1}{4}\left|\frac{\sin(\pi(N+1)t)}{(N+1)t}\right|^2 \\
&\le \frac{N+1}{4}\min\left(\pi^2, \frac{1}{(N+1)^2t^2}\right) \\
&\le \frac{\pi^2}{2}\frac{N+1}{1+(N+1)^2|t|^2}.
\end{aligned}$$

For $t \in \mathbf{R}$ let us set $\varphi(t) = (1+|t|^2)^{-1}$ and $\varphi_\varepsilon(t) = \frac{1}{\varepsilon}\varphi(\frac{t}{\varepsilon})$ for $\varepsilon > 0$. For $x = (x_1, \dots, x_n) \in \mathbf{R}^n$ and $\varepsilon > 0$ we also set

$$\Phi(x) = \varphi(x_1)\dots\varphi(x_n)$$

and $\Phi_\varepsilon(x) = \varepsilon^{-n}\Phi(\varepsilon^{-1}x)$. Then for $|t| \le \frac{1}{2}$ we have $|F_N(t)| \le \frac{\pi^2}{2}\varphi_\varepsilon(t)$ with $\varepsilon = (N+1)^{-1}$ and for $y \in [-\frac{1}{2}, \frac{1}{2}]^n$ we have

$$|F(n, N)(y)| \le (\tfrac{\pi^2}{2})^n\, \Phi_\varepsilon(y), \qquad \text{with } \varepsilon = (N+1)^{-1}.$$

Now let f be an integrable function on $\mathbf{T}^n$ and let f_0 denote its periodic extension on $\mathbf{R}^n$. For $x \in [-\frac{1}{2}, \frac{1}{2}]^n$ we have

$$\begin{aligned} \mathcal{H}(f)(x) &\le \sup_{N>0} \left| \int_{\mathbf{T}^n} F(n,N)(y) f(x-y)\, dy \right| \\ &\le (\tfrac{\pi^2}{2})^n \sup_{\varepsilon>0} \int_{[-\frac{1}{2},\frac{1}{2}]^n} |\Phi_\varepsilon(y)|\, |f_0(x-y)|\, dy \\ &\le 5^n \sup_{\varepsilon>0} \int_{\mathbf{R}^n} |\Phi_\varepsilon(y)|\, |(f_0\chi_Q)(x-y)|\, dy \\ &= 5^n \mathcal{G}(f_0\chi_Q)(x), \end{aligned} \tag{3.3.5}$$

where Q is the cube $[-1,1]^n$ and $\mathcal{G}$ is the operator

$$\mathcal{G}(h) = \sup_{\varepsilon>0} |h| * \Phi_\varepsilon\,.$$

If we can show that $\mathcal{G}$ maps $L^1(\mathbf{R}^n)$ to $L^{1,\infty}(\mathbf{R}^n)$, the corresponding conclusion for $\mathcal{H}$ on $\mathbf{T}^n$ will follow from the fact $\mathcal{H}(f) \le 5^n \mathcal{G}(f_0\chi_Q)$ proved in (3.3.5) and the sequence of inequalities

$$\begin{aligned} \big\|\mathcal{H}(f)\big\|_{L^{1,\infty}(\mathbf{T}^n)} &\le 5^n \big\|\mathcal{G}(f_0\chi_Q)\big\|_{L^{1,\infty}(\mathbf{R}^n)} \\ &\le 5^n C \big\|f_0\chi_Q\big\|_{L^1(\mathbf{R}^n)} \\ &= C' \big\|f\big\|_{L^1(\mathbf{T}^n)}. \end{aligned}$$

Moreover, the L^p conclusion about $\mathcal{H}$ follows from the weak type $(1,1)$ result and the trivial L^∞ inequality, in view of the Marcinkiewicz interpolation Theorem 1.3.2. The required weak type $(1,1)$ estimate for $\mathcal{G}$ on $\mathbf{R}^n$ will be a consequence of Lemma 3.3.4. This will complete the proof of the statement in part (a) of the theorem. To prove the statement in part (b) observe that for $f \in C^\infty(\mathbf{T}^n)$, which is a dense subspace L^1, we have $F(n,N) * f \to f$ uniformly on $\mathbf{T}^n$ as $N \to \infty$, since the sequence $\{F_N\}_N$ is an approximate identity. Since by part (a) $\mathcal{H}$ maps $L^1(\mathbf{T}^n)$ into $L^{1,\infty}(\mathbf{T}^n)$, Theorem 2.1.14 applies and gives that for all $f \in L^1(\mathbf{T}^n)$, $F(n,N)*f \to f$ a.e.

□

We now prove the weak type $(1,1)$ boundedness of $\mathcal{G}$ used earlier.

Lemma 3.3.4. *Let $\Phi(x_1,\dots,x_n) = (1+|x_1|^2)^{-1}\cdots(1+|x_n|^2)^{-1}$ and $\Phi_\varepsilon(x) = \varepsilon^{-n}\Phi(\varepsilon^{-1}x)$ for any $\varepsilon > 0$. Then the maximal operator*

$$\mathcal{G}(f) = \sup_{\varepsilon>0} |f| * \Phi_\varepsilon$$

maps $L^1(\mathbf{R}^n)$ into $L^{1,\infty}(\mathbf{R}^n)$.

Proof. Let $I_0 = [-1,1]$ and $I_k = \{t \in \mathbf{R} : 2^{k-1} \le |t| \le 2^k\}$ for $k = 1,2,\dots$. Also, let $\widetilde{I_k}$ to be the convex hull of I_k, that is, the interval $[-2^k, 2^k]$. For $a_2,\dots,a_n$ fixed positive numbers, let $M_{a_2,\dots,a_n}$ be the maximal operator obtained by averaging a function on $\mathbf{R}^n$ over all products of closed intervals $J_1 \times \cdots \times J_n$ containing a given point with

$$|J_1| = 2^{a_2}|J_2| = \cdots = 2^{a_n}|J_n|.$$

In view of Exercise 2.1.9(c), we have that $M_{a_2,\dots,a_n}$ maps L^1 into $L^{1,\infty}$ with some constant independent of the a_j's. (This is due to the nice doubling property of this family of rectangles.) For a fixed $\varepsilon > 0$ we need to estimate the expression

$$(\Phi_\varepsilon * |f|)(0) = \int_{\mathbf{R}^n} \frac{|f(-\varepsilon y)|\, dy}{(1+y_1^2)\cdots(1+y_n^2)}.$$

Split $\mathbf{R}^n$ into $n!$ of regions of the form $|y_{j_1}| \ge \cdots \ge |y_{j_n}|$, where $\{j_1, \dots, j_n\}$ is a permutation of the set $\{1, \dots, n\}$. By the symmetry of the problem, let us look at the region $\mathcal{R}$ where $|y_1| \ge \cdots \ge |y_n|$. Then for some constant $C > 0$ we have

$$\int_{\mathcal{R}} \frac{|f(-\varepsilon y)|\, dy}{(1+y_1^2)\cdots(1+y_n^2)} \le C \sum_{k_1=0}^{\infty} \sum_{k_2=0}^{k_1} \cdots \sum_{k_n=0}^{k_{n-1}} 2^{-(2k_1+\cdots+2k_n)} \int_{I_{k_1}} \cdots \int_{I_{k_n}} |f(-\varepsilon y)|\, dy$$

and the last expression can be trivially controlled by the corresponding expression, where the I_k's are replaced by the $\widetilde{I}_k$'s. This, in turn, is controlled by

$$C' \sum_{k_1=0}^{\infty} \sum_{k_2=0}^{k_1} \cdots \sum_{k_n=0}^{k_{n-1}} 2^{-(k_1+\cdots+k_n)} M_{k_1-k_2,\dots,k_1-k_n}(f)(0). \tag{3.3.6}$$

Now set $s_2 = k_1 - k_2, \dots, s_n = k_1 - k_n$, observe that $s_j \ge 0$, use that

$$2^{-(k_1+\cdots+k_n)} \le 2^{-k_1/2} 2^{-s_2/2n} \cdots 2^{-s_n/2n},$$

and change the indices of summation to estimate the expression in (3.3.6) by

$$C'' \sum_{k_1=0}^{\infty} \sum_{s_2=0}^{\infty} \cdots \sum_{s_n=0}^{\infty} 2^{-k_1/2} 2^{-s_2/2n} \cdots 2^{-s_n/2n} M_{s_2,\dots,s_n}(f)(0).$$

Argue similarly for the remaining regions $|y_{j_1}| \ge \cdots \ge |y_{j_n}|$ and translate matters to an arbitrary point x to obtain the estimate

$$|(\Phi_\varepsilon * f)(x)| \le C'' n! \sum_{s_2=0}^{\infty} \cdots \sum_{s_n=0}^{\infty} 2^{-s_2/2n} \cdots 2^{-s_n/2n} M_{s_2,\dots,s_n}(f)(x).$$

Now take the supremum over all $\varepsilon > 0$ and use the fact that the maximal functions $M_{s_2,\dots,s_n}$ map L^1 into $L^{1,\infty}$ uniformly in $s_2, \dots, s_n$ as well as the result of Exercise 1.4.10 to obtain the desired conclusion for $\mathcal{G}$.

□

3.3.c. Pointwise Divergence of the Dirichlet Means

We now pass to the more difficult question of convergence of the square partial sums of a Fourier series. It is natural to start our investigation with the class of continuous functions. Do the partial sums of the Fourier series of continuous functions converge pointwise? The following simple proposition gives us a certain warning about the behavior of partial sums.

Proposition 3.3.5. *(duBois Reymond) There exists a continuous function f on $\mathbf{T}^n$ and an $x_0 \in \mathbf{T}^n$ such that the sequence of square partial sums*

$$(D(n,N) * f)(x_0) = \sum_{\substack{m \in \mathbf{Z}^n \\ |m_j| \le N}} \widehat{f}(m) e^{2\pi i x_0 \cdot m}$$

satisfy

$$\limsup_{N \to \infty} |(D(n,N) * f)(x_0)| = \infty$$

and hence they diverge as $N \to \infty$.

Proof. It suffices to prove the proposition when $n = 1$. The one-dimensional example can be transferred easily to n dimensions by extending the function to be constant in the remaining $n-1$ variables.

We give a functional analytic proof. For a constructive proof, see Exercise 3.3.6. Let $C(\mathbf{T}^1)$ be the set of all continuous functions on the circle. Consider the continuous linear functionals

$$f \to T_N(f) = (D_N * f)(0)$$

on $C(\mathbf{T}^1)$ for $N = 1, 2, \dots$. We will show that the norms of the T_N's on $C(\mathbf{T}^1)$ converge to infinity as $N \to \infty$. To see this, given any integer $N \ge 100$, let $\varphi_N(x)$ be a continuous even function on $[-\frac{1}{2}, \frac{1}{2}]$ that is bounded by 1 and is equal to the sign of $D_N(x)$ except at small intervals of length $(2N+1)^{-2}$ around the $2N+1$ zeros of D_N. Call the union of all these intervals B_N and set $A_N = [-\frac{1}{2}, \frac{1}{2}] \setminus B_N$. Then

$$\int_{B_N} |D_N(x)|\, dx + \left| \int_{B_N} \varphi_N(x) D_N(x)\, dx \right| \le 2\, |B_N| (2N+1) = 2.$$

Using this estimate we obtain

$$\begin{aligned}
\|T_N\|_{C(\mathbf{T}^1) \to \mathbf{C}} &\ge |T_N(\varphi_N)(0)| = \left| \int_{\mathbf{T}^1} D_N(-x) \varphi_N(x)\, dx \right| \\
&\ge \int_{A_N} |D_N(x)|\, dx - \left| \int_{B_N} D_N(x) \varphi_N(x)\, dx \right| \\
&= \int_{\mathbf{T}^1} |D_N(x)|\, dx - \left| \int_{B_N} D_N(x) \varphi_N(x)\, dx \right| - \int_{B_N} |D_N(x)|\, dx \\
&\ge \frac{4}{\pi^2} \sum_{k=1}^{N} \frac{1}{k} - 2.
\end{aligned}$$

It follows that the norms of the linear functionals T_N are not uniformly bounded. The uniform boundedness principle now implies the existence of an $f \in C(\mathbf{T}^1)$ and of a sequence $N_j \to \infty$ such that $|T_{N_j}(f)| \to \infty$ as $j \to \infty$. The Fourier series of this f diverges at $x = 0$.

□

3.3.d. Pointwise Convergence of the Dirichlet Means

We have seen that continuous functions may have divergent Fourier series. How about Lipschitz continuous functions? As it turns out, there is a more general condition due to Dini that implies convergence for the Fourier series of functions that satisfy a certain integrability condition.

Theorem 3.3.6. *(**Dini**) Let f be an integrable function on $\mathbf{T}^n$ and let $a = (a_1, \dots, a_n) \in \mathbf{T}^n$. If*

$$\int_{\mathbf{T}^n} \frac{|f(x) - f(a)|}{|x_1 - a_1| \dots |x_n - a_n|} \, dx < \infty, \tag{3.3.7}$$

then we have $(D(n, N) * f)(a) \to f(a)$.

Proof. Replacing f by $f(\cdot - a)$, we may assume that $a = 0$. Replacing f by $f - f(0)$, we may also assume that $f(0) = 0$. Using identities (3.1.12) and (3.1.11), we can write

$$\begin{aligned} (D(n, N) * f)(0) &= \int_{\mathbf{T}^n} f(-x) \prod_{j=1}^{n} \frac{\sin((2N+1)\pi x_j)}{\sin(\pi x_j)} \, dx \\ &= \int_{\mathbf{T}^n} f(-x) \prod_{j=1}^{n} \left(\frac{\sin(2N\pi x_j)\cos(\pi x_j)}{\sin(\pi x_j)} + \cos(2N\pi x_j) \right) dx. \end{aligned} \tag{3.3.8}$$

Expanding out the product, we obtain a sum of terms each of which contains a factor of $\cos(2N\pi x_j)$ or $\sin(2N\pi x_j)$ and a term of the form

$$f(-x) \prod_{j \in I} \frac{\cos(\pi x_j)}{\sin(\pi x_j)}, \tag{3.3.9}$$

where I is a subset of $\{1, 2, \dots, n\}$. The function in (3.3.9) is integrable on $[-\frac{1}{2}, \frac{1}{2}]^n$ except possibly in a neighborhood of the origin. But condition (3.3.7) with $a = 0$ guarantees that any function of the form (3.3.9) is also integrable in a neighborhood of the origin. It is now a consequence of the Riemann-Lebesgue Lemma 3.2.1 that the expression in (3.3.8) tends to zero as $N \to \infty$.

□

The following are consequences of Dini's test.

Corollary 3.3.7. *(**Riemann's principle of localization**) Let $f \in L^1(\mathbf{T}^n)$ and assume that f vanishes on an open ball B contained in $\mathbf{T}^n$. Then $D(n, N) * f$ converges to zero on the ball B.*

Proof. Simply observe that (3.3.7) holds in this case. □

Corollary 3.3.8. *Let $a \in \mathbf{T}^n$ and suppose that $f \in L^1(\mathbf{T}^n)$ satisfies*

$$|f(x) - f(a)| \le C|x_1 - a_1|^{\delta_1} \dots |x_n - a_n|^{\delta_n}$$

for some $C, \delta_j > 0$. (When $n = 1$, this is saying that f is Lipschitz continuous.) Then the square partial sums $(D(n,N) * f)(a)$ converge to $f(a)$.

Proof. Note that condition (3.3.7) holds. □

Corollary 3.3.9. *(**Dirichlet**) If f is defined on $\mathbf{T}^1$ and is a differentiable function at a point a in $\mathbf{T}^1$, then $(D_N * f)(a) \to f(a)$.*

Proof. Again condition (3.3.7) holds. □

Exercises

3.3.1. Identify $\mathbf{T}^1$ with $[-1/2, 1/2)$ and fix $0 < b < 1/2$. Prove the following:

(a) The mth Fourier coefficient of the function x is $i\dfrac{(-1)^m}{2\pi m}$ when $m \neq 0$ and zero when $m = 0$.

(b) The mth Fourier coefficient of the function $\chi_{[-b,b]}$ is $\dfrac{\sin(2\pi b m)}{m\pi}$.

(c) The mth Fourier coefficient of the function $\left(1 - \dfrac{|x|}{b}\right)_+$ is $\dfrac{\sin^2(\pi b m)}{b m^2 \pi^2}$.

(d) The mth Fourier coefficient of the function $|x|$ is $-\dfrac{1}{2m^2\pi^2} + \dfrac{(-1)^m}{2m^2\pi^2}$ when $m \neq 0$ and $\dfrac{1}{4}$ when $m = 0$.

(e) The mth Fourier coefficient of the function x^2 is $\dfrac{(-1)^m}{2m^2\pi^2}$ when $m \neq 0$ and $\dfrac{1}{12}$ when $m = 0$.

(f) The mth Fourier coefficient of the function $\cosh(2\pi x)$ is $\dfrac{(-1)^m}{1+m^2} \sinh \pi$.

(g) The mth Fourier coefficient of the function $\sinh(2\pi x)$ is $\dfrac{im(-1)^m}{1+m^2} \sinh \pi$.

3.3.2. Use the previous Exercise and Proposition 3.3.2 to prove that

$$\sum_{k\in\mathbf{Z}} \frac{1}{(2k+1)^2} = \frac{\pi^2}{4}, \qquad \sum_{k\in\mathbf{Z}\setminus\{0\}} \frac{1}{k^2} = \frac{\pi^2}{3},$$

$$\sum_{k\in\mathbf{Z}\setminus\{0\}} \frac{(-1)^{k+1}}{k^2} = \frac{\pi^2}{6}, \qquad \sum_{k\in\mathbf{Z}} \frac{(-1)^k}{k^2+1} = \frac{2}{e^{\pi}-e^{-\pi}}.$$

3.3.3. Let $M > N$ be given positive integers.
(a) For $f \in L^1(\mathbf{T}^1)$, prove the following identity:

$$\begin{aligned}(D_N * f)(x) &= \frac{M+1}{M-N}(F_M * f)(x) - \frac{N+1}{M-N}(F_N * f)(x) \\ &\quad - \frac{M+1}{M-N} \sum_{N<|j|\le M} \left(1 - \frac{|j|}{M+1}\right) \widehat{f}(j) e^{2\pi i j x}.\end{aligned}$$

(b) (*G. H. Hardy*) Suppose that a function f on $\mathbf{T}^1$ satisfies the following condition: Given any $\varepsilon > 0$, there exists an $a > 0$ and $k_0 > 0$ such that for $k \ge k_0$

$$\sum_{k<|m|\le[ak]} |\widehat{f}(m)| < \varepsilon.$$

Use part (a) to prove that if $(F_N * f)(x)$ converges to $A(x)$ as $N \to \infty$, then $(D_N * f)(x)$ also converges to $A(x)$ as $N \to \infty$.

3.3.4. Use the previous Exercise to prove that f is a function of bounded variation on $\mathbf{T}^1$, then

$$(D_N * f)(x) \to \frac{1}{2}(f(x+0) + f(x-0))$$

for every $t \in \mathbf{T}^1$. Apply this result to the function $\chi_{[-b,b]}$ of Exercise 3.3.1(b) to obtain that

$$\lim_{N\to\infty} \sum_{m=-N}^{N} \frac{\sin(2\pi b m)}{m\pi} e^{2\pi i b m} = \frac{1}{2} - 2b.$$

3.3.5. (a) Prove that the Riemann-Lebesgue lemma holds uniformly on compact subsets of $L^1(\mathbf{T}^n)$. This means that given any compact subset of $L^1(\mathbf{T}^n)$ and $\varepsilon > 0$ there exists an $N_0 > 0$ such that for $|m| \ge N_0$ we have $|\widehat{f}(m)| \le \varepsilon$ for all $f \in K$.
(b) Use part (a) to prove the following sharpening of the localization theorem. If f vanishes on an open ball B in $\mathbf{T}^n$, then $D(n, N) * f$ converges to zero uniformly on compact subsets of B.

3.3.6. Follow the steps given to obtain a constructive proof of a continuous function whose Fourier series diverges at a point. On $\mathbf{T}^1$ let

$$g(x) = -2\pi i(x - \frac{1}{2}).$$

(a) Prove that $\widehat{g}(m) = \frac{1}{m}$ when $m \neq 0$ and zero otherwise.
(b) Prove that for all nonnegative integers M and N we have

$$\big((e^{2\pi i N(\cdot)}(g * D_N)) * D_M\big)(x) = e^{2\pi i N x} \sum_{1 \le |r| \le N} \frac{1}{r} e^{2\pi i r x}$$

when $M \geq 2N$ and equal to

$$e^{2\pi i N x} \sum_{\substack{-N \le r \le M-N \\ r \neq 0}} \frac{1}{r} e^{2\pi i r x}$$

when $M < 2N$. Conclude that there exists a constant $C > 0$ such that for all M, N, and $x \neq 0$ we have

$$\big|\big(e^{2\pi i N(\cdot)}(g * D_N) * D_M\big)(x)\big| \le \frac{C}{|x|}.$$

(c) Show that there exists a constant $C_1 > 0$ such that

$$\sup_{N>0} \sup_{x \in \mathbf{T}^1} \big|(g * D_N)(x)\big| = \sup_{N>0} \sup_{x \in \mathbf{T}^1} \left| \sum_{1 \le |r| \le N} \frac{1}{r} e^{2\pi i r x} \right| \le C_1 < \infty.$$

(d) Let $\lambda_k = 1 + e^{e^k}$. Define

$$f(x) = \sum_{k=1}^{\infty} \frac{1}{k^2} e^{2\pi i \lambda_k x} (g * D_{\lambda_k})(x)$$

and prove that f is continuous on $\mathbf{T}^1$ and that its Fourier series converges at every $x \neq 0$, but $\limsup_{M \to \infty} |(f * D_M)(0)| = \infty$.
$\big[$*Hint:* Take $M = e^{e^m}$ with $m \to \infty$.$\big]$

3.4. Divergence of Fourier Series and Bochner-Riesz Summability*

We saw in the previous section that the Fourier series of a continuous function may diverge at a point. As expected, the situation can only get worse as the functions get worse. In this section we will present an example, due to A. N. Kolmogorov, of an integrable function on $\mathbf{T}^1$ whose Fourier series diverges almost everywhere. Using this example, we may construct integrable functions on $\mathbf{T}^n$ whose square Dirichlet means diverge a.e.; see Exercise 3.4.1.

3.4.a. Motivation for Bochner-Riesz Summability

We now consider an analogous question for the circular Dirichlet means of integrable functions on $\mathbf{T}^n$. In dimension 1 we saw that the Fejér means of integrable functions are better behaved than their Dirichlet means. We investigate if there is a similar phenomenon in higher dimensions. Recall that the circular (or spherical) partial sums of the Fourier series of f are given by

$$(f * \widetilde{D}(n,R))(x) = \sum_{\substack{m\in\mathbf{Z}^n\\|m|\le R}} \widehat{f}(m)e^{2\pi i m\cdot x},$$

where $R \geq 0$. Taking the averages of these expressions, we obtain

$$\frac{1}{R}\int_0^R (f * \widetilde{D}(n,r))(x)\,dr = \sum_{\substack{m\in\mathbf{Z}^n\\|m|\le R}} \big(1-\tfrac{|m|}{R}\big)\widehat{f}(m)e^{2\pi i m\cdot x},$$

and we call these expressions the *circular Cesàro means* (or *circular Fejér means*) of f. It will turn out that the circular Cesàro means of integrable functions on $\mathbf{T}^2$ always converge in L^1, but in dimension 3, this may fail. Theorem 3.4.6 gives an example of an integrable function f on $\mathbf{T}^3$ whose circular Cesàro means diverge a.e. However, as we will see, this is not the case if the circular Cesàro means of a function f in $L^1(\mathbf{T}^3)$ are replaced by the only slightly different-looking means

$$\sum_{\substack{m\in\mathbf{Z}^n\\|m|\le R}} \big(1-\tfrac{|m|}{R}\big)^{1+\varepsilon}\widehat{f}(m)e^{2\pi i m\cdot x},$$

for some $\varepsilon > 0$. The previous discussion suggests that the preceding expressions behave better as ε increases, but for a fixed ε they get worse as the dimension increases. To study this situation more carefully, we define the family of operators where the exponent $1+\varepsilon$ is replaced by a general nonnegative index $\alpha \geq 0$.

Definition 3.4.1. Let $\alpha \geq 0$. The *Bochner-Riesz means* of order α of an integrable function f on $\mathbf{T}^n$ are defined as follows:

$$B_R^\alpha(f)(x) = \sum_{\substack{m\in\mathbf{Z}^n\\|m|\le R}} \big(1-\tfrac{|m|^2}{R^2}\big)^{\alpha}\widehat{f}(m)e^{2\pi i m\cdot x}. \tag{3.4.1}$$

This family of operators forms a natural "spherical" analogue of the Cesàro-Fejér sums. It turns out that there is no different behavior of the means if the expression $\big(1-\frac{|m|^2}{R^2}\big)^\alpha$ in (3.4.1) is replaced by the expression $\big(1-\frac{|m|}{R}\big)^\alpha$. See Exercise 3.6.1 on the equivalence of means generated by these two expressions. The advantage of the quadratic expression in (3.4.1) is that it has an easily computable kernel. Moreover, the appearance of the quadratic term in the definition of the Bochner-Riesz means is responsible for the following reproducing formula:

$$(3.4.2)\quad B_R^\alpha(f) = \frac{2\Gamma(\alpha+1)}{\Gamma(\alpha-\beta)\Gamma(\beta+1)}\,\frac{1}{R}\int_0^R\left(1-\frac{r^2}{R^2}\right)^{\alpha-\beta-1}\left(\frac{r^2}{R^2}\right)^{\beta+\frac{1}{2}} B_r^\beta(f)\,dr\,,$$

which precisely quantifies the way in which B_R^α is smoother than B_R^β when $\alpha > \beta$. Identity (3.4.2) also says that when $\alpha > \beta$, the operator $B_R^\alpha(f)$ is an average of the operators $B_r^\beta(f)$, $0 < r < R$ with respect to a certain density.

Note that the Bochner-Riesz means of order zero coincide with the circular (or spherical) Dirichlet means and, as we have seen, these converge in $L^2(\mathbf{T}^n)$. We now indicate why the Bochner-Riesz means $B_R^\alpha(f)$ converge to f in $L^1(\mathbf{T}^n)$ as $R \to \infty$ when $\alpha > (n-1)/2$. Consider the function

$$m_\alpha(\xi) = (1-|\xi|^2)_+^\alpha$$

defined for ξ in $\mathbf{R}^n$. Using the result proved in Appendix B.5, we have that

$$(3.4.3)\qquad (m_\alpha)^\vee(x) = K^\alpha(x) = \frac{\Gamma(\alpha+1)}{\pi^\alpha}\,\frac{J_{\frac{n}{2}+\alpha}(2\pi|x|)}{|x|^{\frac{n}{2}+\alpha}}\,,$$

where J_λ is the Bessel function of order λ. Using the asymptotics in Appendix B.6, we obtain that if $\alpha > (n-1)/2$, then the function K^α satisfies an estimate of the form

$$(3.4.4)\qquad |K^\alpha(x)| \le C_{n,\alpha}(1+|x|)^{-n-(\alpha-\frac{n-1}{2})}$$

and hence it is in $L^1(\mathbf{R}^n)$. Using the Poisson summation formula, we write

$$\begin{aligned} B_R^\alpha(f)(x) &= \sum_{l\in\mathbf{Z}^n} m_\alpha(\tfrac{l}{R})\widehat{f}(l)e^{2\pi i l\cdot x} \\ &= \sum_{l\in\mathbf{Z}^n} (f*(K^\alpha)_{1/R})(x+l) \\ &= (f*(L^\alpha)_{1/R})(x)\,, \end{aligned}$$

where $L^\alpha(x) = \sum_{k\in\mathbf{Z}^n} K^\alpha(x+k)$ and $g_{1/R}(x) = R^n g(Rx)$. Using (3.4.4), we show easily that the function L^α is an integrable 1-periodic function on $\mathbf{T}^n$. Moreover,

$$\int_{\mathbf{T}^n} L^\alpha(t)\,dt = \int_{\mathbf{R}^n} K^\alpha(x)\,dx = \widehat{m_\alpha}(0) = 1\,.$$

Therefore, the family $\{(L^\alpha)_\varepsilon\}_{\varepsilon>0}$ is an approximate identity on $\mathbf{T}^n$ as $\varepsilon \to 0$. Using Theorem 1.2.19, we obtain that for $\alpha > (n-1)/2$ we have

(a) For $f \in L^p(\mathbf{T}^n)$, $1 \le p < \infty$, $B_R^\alpha(f)$ converge to f in L^p as $R \to \infty$.
(b) For f continuous on $\mathbf{T}^n$, $B_R^\alpha(f)$ converge to f uniformly as $R \to \infty$.

We may wonder if there are analogous results for $\alpha \le (n-1)/2$. Theorem 3.4.6 warns us that the Bochner-Riesz means may diverge when $\alpha = (n-1)/2$. For this reason the number $\alpha = (n-1)/2$ is referred to as the critical index. The optimal range for α's for which the Bochner-Riesz means of order α converge in $L^p(\mathbf{T}^n)$ when $1 < p < \infty$ will be investigated in Chapter 10.

3.4.b. Divergence of Fourier Series of Integrable Functions

It is natural to start our investigation with the case $n = 1$. We begin with the following important result:

Theorem 3.4.2. *There exists an integrable function on the circle $\mathbf{T}^1$ whose Fourier series diverges almost everywhere.*

Proof. The proof of this theorem is a bit involved, and we will need a sequence of lemmata, which we prove first.

Lemma 3.4.3. *(**Kronecker**) Suppose that $n \in \mathbf{Z}^+$ and*

$$\{x_1, x_2, \dots, x_n, 1\}$$

is a linearly independent set over the rationals. Then for any $\varepsilon > 0$ and any $z_1, z_2, \dots, z_n$ complex numbers with $|z_j| = 1$, there exists an integer $m \in \mathbf{Z}$ such that

$$|e^{2\pi i m x_j} - z_j| < \varepsilon \qquad \textit{for all} \quad 1 \le j \le n.$$

Proof. Identifying $\mathbf{T}^n$ with the set $\{(e^{2\pi i t_1}, \dots, e^{2\pi i t_n}) : 0 \le t_j \le 1\}$, the required conclusion will be a consequence of the fact that for a fixed $x = (x_1, \dots, x_n)$ the set $\{mx : m \in \mathbf{Z}\}$ is dense in $\mathbf{T}^n$. If this were not the case, then there would exist an open set U in $\mathbf{T}^n$ that contains no elements of the set $\{mx : m \in \mathbf{Z}\}$. Pick a smooth, nonzero, and nonnegative function f on $\mathbf{T}^n$ supported in U. Then $f(mx) = 0$ for all $m \in \mathbf{Z}$, but

$$\widehat{f}(0) = \int_{\mathbf{T}^n} f(x)\, dx > 0\,.$$

Then we have

$$\begin{aligned}
0 = \frac{1}{N}\sum_{m=0}^{N-1} f(mx) &= \frac{1}{N}\sum_{m=0}^{N-1}\Big(\sum_{l\in\mathbf{Z}^n} \widehat{f}(l) e^{2\pi i l\cdot mx}\Big)\\
&= \sum_{l\in\mathbf{Z}^n} \widehat{f}(l)\Big(\frac{1}{N}\sum_{m=0}^{N-1} e^{2\pi i m(l\cdot x)}\Big)\\
&= \sum_{l\in\mathbf{Z}^n\setminus\{0\}} \widehat{f}(l)\Big(\frac{1}{N}\frac{e^{2\pi i N(l\cdot x)}-1}{e^{2\pi i(l\cdot x)}-1}\Big) + \widehat{f}(0)\,.
\end{aligned}$$

In the last identity we used the fact that $e^{2\pi i(l\cdot x)} \neq 1$ since by assumption the set $\{x_1, x_2, \dots, x_n, 1\}$ is linearly independent over the rationals. But the expression inside the parentheses above is bounded by 1 and tends to 0 as $N \to \infty$. Since $|\widehat{f}(l)| \le C(f,n)(1+|l|)^{-100n}$, taking limits as $N \to \infty$ and using the Lebesgue dominated convergence theorem, we obtain that

$$0 = \lim_{N\to\infty} \sum_{l\in\mathbf{Z}^n\setminus\{0\}} \widehat{f}(l)\Big(\frac{1}{N}\frac{e^{2\pi i N(l\cdot x)}-1}{e^{2\pi i(l\cdot x)}-1}\Big) + \widehat{f}(0) = \widehat{f}(0)\,,$$

which contradicts our assumption on f.

□

Lemma 3.4.4. *Let N be a large positive integer. Then there exists a positive measure μ_N on $\mathbf{T}^1$ with $\mu_N(\mathbf{T}^1) = 1$ such that*

$$\sup_{L\geq 1} \big|(\mu_N * D_L)(x)\big| = \sup_{L\geq 1} \bigg| \sum_{k=-L}^{L} \widehat{\mu_N}(k) e^{2\pi i k x}\bigg| \geq c \log N \tag{3.4.5}$$

for almost all $x \in \mathbf{T}^1$ (c is a fixed constant).

Proof. We choose points $0 \leq x_1 < x_2 < \cdots < x_N \leq 1$ such that if $x_{N+1} = x_1 + 1$, then

$$\frac{1}{2N} \leq |x_{j+1} - x_j| \leq \frac{2}{N}, \qquad 1 \leq j \leq N\,, \tag{3.4.6}$$

and the set $\{x_1, \dots, x_N, 1\}$ is linearly independent over the rationals. Let

$$E_N = \Big\{ x \in [0,1] : \ \{x - x_1, \dots, x - x_N, 1\} \quad \text{is linearly independent over } \mathbf{Q} \Big\}$$

and observe that almost all[1] x in $\mathbf{T}^1$ belongs to E_N. Define the probability measure

$$\mu_N = \frac{1}{N} \sum_{j=1}^{N} \delta_{x_j}\,.$$

For this measure we have

$$\begin{aligned}
\bigg| \sum_{k=-L}^{L} \widehat{\mu_N}(k) e^{2\pi i k x} \bigg| &= \bigg| \sum_{k=-L}^{L} \Big(\frac{1}{N} \sum_{j=1}^{N} e^{-2\pi i k x_j} \Big) e^{2\pi i k x} \bigg| \\
&= \bigg| \frac{1}{N} \sum_{j=1}^{N} D_L(x - x_j) \bigg| \\
&= \bigg| \frac{1}{N} \sum_{j=1}^{N} \frac{\sin(2\pi (L+\frac{1}{2})(x - x_j))}{\sin(\pi(x - x_j))} \bigg| \\
&= \bigg| \frac{1}{N} \sum_{j=1}^{N} \frac{\operatorname{Im}\big[e^{2\pi i (L+\frac{1}{2})(x - x_j)}\big] \operatorname{sgn}\big(\sin(\pi(x - x_j))\big)}{|\sin(\pi(x - x_j))|} \bigg|,
\end{aligned} \tag{3.4.7}$$

where the signum function is defined as $\operatorname{sgn} a = 1$ for $a > 0$, -1 for $a < 0$ and zero if $a = 0$. By Lemma 3.4.3, for all $x \in E_N$ there exists an $L \in \mathbf{Z}^+$ such that

$$\big| e^{2\pi i L (x - x_j)} - i e^{-2\pi i \frac{1}{2}(x - x_j)} \operatorname{sgn}\big(\sin(\pi(x - x_j))\big) \big| < \frac{1}{2}\,,$$

---

[1] Every x in $[0,1] \setminus \mathbf{Q}[x_1, \dots, x_N]$ belongs to E_N. Here $\mathbf{Q}[x_1, \dots, x_N]$ denotes the field extension of $\mathbf{Q}$ obtained by attaching to it the linearly independent elements $\{x_1, \dots, x_N\}$.

which can be equivalently written as

$$\big|e^{2\pi i(L+\frac{1}{2})(x-x_j)}\mathrm{sgn}\big(\sin(\pi(x-x_j))\big)-i\big|<\frac{1}{2}\,. \tag{3.4.8}$$

It follows from (3.4.8) that

$$\mathrm{Im}\,\big[e^{2\pi i(L+\frac{1}{2})(x-x_j)}\big]\mathrm{sgn}\big(\sin(\pi(x-x_j))\big)>\frac{1}{2}\,.$$

Combining this with the result of the calculation in (3.4.7), we obtain that

$$\bigg|\sum_{k=-L}^{L}\widehat{\mu_N}(k)e^{2\pi ikx}\bigg|>\frac{1}{2N}\sum_{j=1}^{N}\frac{1}{|\sin(\pi(x-x_j))|}\geq\frac{1}{2\pi N}\sum_{j=1}^{N}\frac{1}{|x-x_j|}\,.$$

But for every $x\in[0,1]$ there exists a j_0 such that $x\in[x_{j_0},x_{j_0+1})$. It follows from (3.4.6) that $|x-x_j|\leq C(|j-j_0|+1)N^{-1}$ and thus

$$\sum_{j=1}^{N}\frac{1}{|x-x_j|}\geq c'N\log N\,.$$

Thus for every $x\in E_N$ there exists an $L\in\mathbf{Z}^+$ such that

$$\bigg|\sum_{k=-L}^{L}\widehat{\mu_N}(k)e^{2\pi ikx}\bigg|>c\log N\,,$$

which proves the required conclusion. □

Lemma 3.4.5. *For each $0<M<\infty$ there exists a trigonometric polynomial g_M and a measurable subset A_M of $\mathbf{T}^1$ with measure $|A_M|>1-2^{-M}$ such that $\big\|g_M\big\|_{L^1}=1$, and such that*

$$\inf_{x\in A_M}\sup_{L\geq1}\big|(D_L*g_M)(x)\big|=\inf_{x\in A_M}\sup_{L\geq1}\bigg|\sum_{k=-L}^{L}\widehat{g_M}(k)e^{2\pi ikx}\bigg|>2^M\,. \tag{3.4.9}$$

Proof. Given an M with $0<M<\infty$, we pick an integer $N(M)$ such that $c\log N(M)>2^{M+2}$, where c is as in (3.4.5), and we also pick the measure $\mu_{N(M)}$, which satisfies (3.4.5). By Fatou's lemma we have

$$\begin{aligned}1\;&=\big|\{x\in\mathbf{T}^1:\ \lim_{L\to\infty}\sup_{1\leq j\leq L}|(D_j*\mu_{N(M)})(x)|\geq2^{M+1}\}\big|\\&\leq\liminf_{L\to\infty}\big|\{x\in\mathbf{T}^1:\ \sup_{1\leq j\leq L}|(D_j*\mu_{N(M)})(x)|\geq2^{M+1}\}\big|\end{aligned}$$

and thus we can find a positive integer $L(M)$ such that the set

$$A_M=\{x\in\mathbf{T}^1:\ \sup_{1\leq j\leq L(M)}|(D_j*\mu_{N(M)})(x)|\geq2^{M+1}\}$$

has measure greater than $1 - 2^{-M}$. We pick a positive integer $K(M)$ such that

$$\sup_{1 \le j \le L(M)} \big\| F_{K(M)} * D_j - D_j \big\|_{L^\infty} \le 1,$$

where F_K is the Fejér kernel. This is possible since the Fejér kernel is an approximate identity and $\{D_j : 1 \le j \le L\}$ is a finite family of continuous functions. Then we define $g_M = \mu_{N(M)} * F_{K(M)}$. Since $\mu_{N(M)}$ is a probability measure, we obtain

$$|(D_j * g_M)(x) - (D_j * \mu_{N(M)})(x)| \le \big\| D_j * F_{K(M)} - D_j \big\|_{L^\infty} \le 1$$

for all $x \in [0,1]$ and $1 \le j \le L$. It follows that for $x \in A_M$ and $1 \le j \le L$ we have

$$|(D_j * g_M)(x)| \ge |(D_j * \mu_{N(M)})(x)| - 1 \ge 2^{M+2} - 1 \ge 2^{M+1}.$$

Therefore, (3.4.9) is satisfied for this g_M and A_M. Since μ_N is a probability measure and $F_{K(M)}$ is nonnegative and has L^1 norm 1, we have that

$$\big\| g_M \big\|_{L^1} = \big\| \mu_{N(M)} * F_{K(M)} \big\|_{L^1} = \big\| \mu_{N(M)} \big\|_{\mathcal{M}} \big\| F_{K(M)} \big\|_{L^1} = 1.$$

□

We now have the tools we need to construct an example of a function whose Fourier series diverges almost everywhere. The example is given as a series of functions each of which has a behavior that worsens as its index becomes bigger. The function we wish to construct will be a sum of the form

$$g = \sum_{j=1}^{\infty} \varepsilon_j g_{M_j}, \tag{3.4.10}$$

for a choice of sequences $\varepsilon_j \to 0$ and $M_j \to \infty$, where g_M are as in Lemma 3.4.5.

Let us be specific. We set $\varepsilon_0 = M_0 = d_0 = 1$. Assume that we have defined ε_j, M_j, and d_j for all $0 \le j < N$. We first set

$$\varepsilon_N = 2^{-N}(3d_{N-1})^{-1}. \tag{3.4.11}$$

Then we pick M_N such that

$$\varepsilon_N 2^{M_N} \ge 2^N + d_{N-1} + 1. \tag{3.4.12}$$

Finally, we set

$$d_N = \max_{1 \le s \le N} \text{degree}\,(g_{M_s}), \tag{3.4.13}$$

where g_M is the trigonometric polynomial of Lemma 3.4.5. This defines ε_N, M_N, and d_N for a given N provided these numbers are known for all $j < N$. By induction we define ε_N, M_N, and d_N for all natural numbers N.

We observe that the selections of ε_j and M_j force the inequalities $\varepsilon_j \le 2^{-j}$ and $d_j \le d_{j+1}$ for all $j \ge 0$. Since each g_{M_j} has L^1 norm 1 and $\varepsilon_j \le 2^{-j}$, the function g in (3.4.10) is integrable and has L^1 norm at most 1.

For a given $j \ge 0$ and $x \in A_{M_j}$, by Lemma 3.4.5 there exists an $L \ge 1$ such that $|(D_L * g_{M_j})(x)| > 2^{M_j}$. Set $k = k(x) = \min(L, d_j)$. Then we have

$$|(D_k * g)(x)| \ge \varepsilon_j |(D_k * g_{M_j})(x)| - \sum_{1<s<j} \varepsilon_s |(D_k * g_{M_s})(x)| - \sum_{s>j} \varepsilon_s |(D_k * g_{M_s})(x)|.$$

We make the following observations:

(i) $|(D_k * g_{M_j})(x)| = |(D_L * g_{M_j})(x)| > 2^{M_j}$.

(ii) $|(D_k * g_{M_s})(x)| = |(D_{\min(d_s,k)} * g_{M_s})(x)| \le \big\|D_{\min(d_s,L)}\big\|_{L^\infty} \le 3d_s$, whenever $s < j$.

(iii) $|(D_k * g_{M_s})(x)| = |(D_{\min(d_s,k)} * g_{M_s})(x)| \le \big\|D_{\min(d_j,L)}\big\|_{L^\infty} \le 3d_j$, whenever $s > j$.

In these estimates we have used that $k = \min(L, d_j)$, $\big\|D_m\big\|_{L^\infty} \le 2m+1 \le 3m$, and that

$$D_r * g_{M_s} = D_{\min(r,d_s)} * g_{M_s},$$

which follows easily by examining the corresponding Fourier coefficients.

Using the estimates in (i), (ii), and (iii), for this x in A_{M_j} and $k = k(x)$ we obtain

$$|(D_k * g)(x)| \ge \varepsilon_j 2^{M_j} - 3 \sum_{1<s<j} \varepsilon_s d_s - 3 \sum_{s>j} \varepsilon_s d_j . \tag{3.4.14}$$

Our selection of ε_j and M_j now ensures that (3.4.14) is a large number. In fact, we have

$$3 \sum_{s>j} \varepsilon_s d_j \le \sum_{s>j} 2^{-s} d_j (d_{s-1})^{-1} \le \sum_{s>j} 2^{-s} \le 1$$

and

$$3 \sum_{1<s<j} \varepsilon_s d_s \le 3d_{j-1} \sum_{1<s<j} \varepsilon_s \le d_{j-1} \sum_{1<s<j} 2^{-s} (d_{s-1})^{-1} \le d_{j-1} .$$

Therefore, the expression in (3.4.14) is at least $\varepsilon_j 2^{M_j} - d_{j-1} - 1 \ge 2^j$. It follows that for every $j \ge 0$ and every $x \in A_{M_j}$ there exists a $k = k(x)$ such that

$$|(D_k * g)(x)| \ge 2^j . \tag{3.4.15}$$

We conclude that for every $j \ge 0$ and $x \in A_{M_j}$ we have

$$\sup_{k\ge1} |(D_k * g)(x)| \ge 2^j .$$

Thus, for all x in the set $A = \bigcap_{j=0}^{\infty} \bigcup_{r=j}^{\infty} A_{M_r}$ we have

$$\sup_{k\ge1} |(D_k * g)(x)| = \infty . \tag{3.4.16}$$

But A is a countable intersection of subsets of $\mathbf{T}^1$ of full measure. Therefore, A has measure 1 and the required conclusion follows.

□

3.4.c. Divergence of Bochner-Riesz Means of Integrable Functions

We now turn to the corresponding n-dimensional problem for spherical summability of Fourier series. Things here are quite similar at the critical index $\alpha = \frac{n-1}{2}$.

Theorem 3.4.6. *Let $n > 1$. There exists an integrable function f on $\mathbf{T}^n$ such that*

$$\limsup_{R\to\infty} |B_R^{\frac{n-1}{2}}(f)(x)| = \limsup_{R\to\infty} \Bigg| \sum_{\substack{m\in\mathbf{Z}^n \\ |m|\le R}} \big(1 - \tfrac{|m|^2}{R^2}\big)^{\frac{n-1}{2}} \widehat{f}(m) e^{2\pi i m\cdot x} \Bigg| = \infty$$

for almost all $x \in \mathbf{T}^n$. Furthermore, such a function can be constructed so that it is supported in an arbitrarily small given neighborhood of the origin.

Proof. We start by defining the set

$$S = \big\{x \in \mathbf{R}^n : \ \{|x - m| : \ m \in \mathbf{Z}^n\} \text{ is linearly independent over } \mathbf{Q}\big\}.$$

We will show that S has full measure in $\mathbf{R}^n$. Indeed, if $x \in \mathbf{R}^n \setminus S$, then there exist $k \in \mathbf{Z}^+$, $m_1, \dots m_k \in \mathbf{Z}^n$ and $a_{m_1}, \dots, a_{m_k}$ nonzero rational numbers such that

$$\sum_{j=1}^{k} a_{m_j} |x - m_j| = 0. \tag{3.4.17}$$

Since the function

$$t \to \sum_{j=1}^{k} a_{m_j} |t - m_j|$$

is nonzero and real analytic on $\mathbf{R}^n \setminus \mathbf{Z}^n$, it must vanish only on a set of Lebesgue measure zero. Therefore, there exists a set $A_{m_1,\dots,m_k,a_{m_1},\dots,a_{m_k}}$ of Lebesgue measure zero so that (3.4.17) holds exactly when x is in this set. Then

$$\mathbf{R}^n \setminus S \subseteq \bigcup_{k=1}^{\infty} \bigcup_{m_1,\dots,m_k\in\mathbf{Z}^n} \bigcup_{a_{m_1},\dots,a_{m_k}\in\mathbf{Q}} A_{m_1,\dots,m_k,a_{m_1},\dots,a_{m_k}}$$

from which it follows that $\mathbf{R}^n \setminus S$ has Lebesgue measure zero.

Let us set

$$K_R^\alpha(x) = \sum_{|m|\le R} \big(1 - \tfrac{|m|^2}{R^2}\big)^\alpha e^{2\pi i m\cdot x}.$$

We will need the following lemma regarding K_R^α:

Lemma 3.4.7. *For each $x \in S \cap \mathbf{T}^n$, $n \geq 2$ we have*

$$\limsup_{R\to\infty} |K_R^{\frac{n-1}{2}}(x)| = \infty .$$

It is noteworthy to compare the result of this lemma with the analogous one-dimensional statement,

$$\limsup_{R\to\infty} |D_R(x)| = \infty ,$$

for the Dirichlet kernel, which holds exactly when $x = 0$. Thus the uniform ill behavior of the kernel $K_R^{\frac{n-1}{2}}$ reflects in some sense its lack of localization.

Proof. Using (3.4.3) and the Poisson summation formula (Theorem 3.1.17), we obtain the identity

$$K_R^\alpha(x) = \frac{\Gamma(\alpha+1)}{\pi^\alpha} R^{\frac{n}{2}-\alpha} \sum_{m\in\mathbf{Z}^n} \frac{J_{\frac{n}{2}+\alpha}(2\pi R|x-m|)}{|x-m|^{\frac{n}{2}+\alpha}}, \tag{3.4.18}$$

which is valid for all $x \in \mathbf{T}^n \setminus \mathbf{Z}^n$. Because of the asymptotics in Appendix B.6 the sum (3.4.18) converges for $\alpha > \frac{n-1}{2}$. The same asymptotics imply that for $x \notin \mathbf{Z}^n$ and $R \geq 1$ we have

$$\begin{aligned} J_{\frac{n}{2}+\alpha}(2\pi R|x-m|) &= \frac{e^{2\pi i R|x-m|} e^{-i\frac{\pi}{2}(\frac{n}{2}+\alpha)-i\frac{\pi}{4}} + e^{-2\pi i R|x-m|} e^{i\frac{\pi}{2}(\frac{n}{2}+\alpha)+i\frac{\pi}{4}}}{\pi\sqrt{R|x-m|}} \\ &\quad + O\big((R|x-m|)^{-\frac{3}{2}}\big) \end{aligned}$$

for all $\alpha > 0$. It is not possible to let $\alpha \to \frac{n-1}{2}$ in (3.4.18) since the series on the right of that identity will diverge for this value of α. It is a remarkable fact, however, that if we average over R first, we obtain an oscillatory factor that will allow us to let $\alpha = \frac{n-1}{2}$ in the previous identity. Now for $x \notin \mathbf{Z}^n$ and $T > 1$ we obtain

$$\begin{aligned} &\frac{1}{T}\int_1^T K_R^\alpha(x) e^{2\pi i \lambda R}\, dR \\ = &\frac{\Gamma(\alpha+1)}{\pi^\alpha} \sum_{m\in\mathbf{Z}^n} \frac{e^{-i\frac{\pi}{2}(\frac{n}{2}+\alpha)-i\frac{\pi}{4}}}{|x-m|^{\frac{n+1}{2}+\alpha}} \frac{1}{T}\int_1^T e^{2\pi i R(\lambda+|x-m|)} R^{\frac{n-1}{2}-\alpha}\, dR \\ + &\frac{\Gamma(\alpha+1)}{\pi^\alpha} \sum_{m\in\mathbf{Z}^n} \frac{e^{i\frac{\pi}{2}(\frac{n}{2}+\alpha)+i\frac{\pi}{4}}}{|x-m|^{\frac{n+1}{2}+\alpha}} \frac{1}{T}\int_1^T e^{2\pi i R(\lambda-|x-m|)} R^{\frac{n-1}{2}-\alpha}\, dR \\ + &\frac{\Gamma(\alpha+1)}{\pi^\alpha} \sum_{m\in\mathbf{Z}^n} O\bigg(\frac{1}{|x-m|^{\frac{n+3}{2}+\alpha}}\bigg) \frac{1}{T}\int_1^T R^{\frac{n-3}{2}-\alpha}\, dR . \end{aligned}$$

We now let $\alpha \to \frac{n-1}{2}$ in the preceding expression. Then we have

$$\begin{aligned}(3.4.19)\qquad & \frac{1}{T}\int_1^T K_R^{\frac{n-1}{2}}(x)e^{2\pi i\lambda R}\,dR \\ &= \frac{\Gamma(\frac{n+1}{2})}{\pi^{\frac{n-1}{2}}}\sum_{m\in\mathbf{Z}^n}\frac{e^{-i\frac{\pi}{2}(\frac{2n-1}{2})-i\frac{\pi}{4}}}{|x-m|^n}\frac{1}{T}\int_1^T e^{2\pi iR(\lambda+|x-m|)}\,dR \\ &+ \frac{\Gamma(\frac{n+1}{2})}{\pi^{\frac{n-1}{2}}}\sum_{m\in\mathbf{Z}^n}\frac{e^{i\frac{\pi}{2}(\frac{2n-1}{2})+i\frac{\pi}{4}}}{|x-m|^n}\frac{1}{T}\int_1^T e^{2\pi iR(\lambda-|x-m|)}\,dR \\ &+ \frac{\Gamma(\frac{n+1}{2})}{\pi^{\frac{n-1}{2}}}\sum_{m\in\mathbf{Z}^n}O\Big(\frac{1}{|x-m|^{n+1}}\Big)\frac{1}{T}\int_1^T\frac{dR}{R},\end{aligned}$$

and the wonderful fact is that the first two sums converge because of the appearance of the oscillatory factors. See Exercise 3.4.8(c). It follows from the previous identity that if $\lambda > 0$ and $\lambda \neq \pm|x-m_0|$ for any $m_0 \in \mathbf{Z}^n$, then the expression in (3.4.19) converges to zero as $T \to \infty$ while it converges to

$$\frac{\Gamma(\frac{n+1}{2})}{\pi^{\frac{n-1}{2}}}\frac{e^{\pm i(\frac{\pi}{2}(\frac{2n-1}{2})+i\frac{\pi}{4})}}{|x-m_0|^n} = \frac{\Gamma(\frac{n+1}{2})}{\pi^{\frac{n-1}{2}}}\frac{e^{\pm i\frac{\pi n}{2}}}{|x-m_0|^n}$$

if $\lambda = \pm|x-m_0|$ for some $m_0 \in \mathbf{Z}^n$. We now fix $x_0 \in S\cap\mathbf{T}^n$ and we set

$$\Lambda_{x_0} = \{|x_0-m| : \ m\in\mathbf{Z}^n\} = \{\lambda_1,\lambda_2,\lambda_3,\dots\},$$

where $0<\lambda_1<\lambda_2<\lambda_3<\dots$. Observe that

$$(3.4.20)\qquad \sum_{j=1}^{\infty}\frac{1}{\lambda_j^n} = \infty.$$

We have shown that

$$(3.4.21)\qquad \lim_{T\to\infty}\frac{1}{T}\int_1^T K_t^{\frac{n-1}{2}}(x_0)e^{2\pi i\lambda t}\,dt = \begin{cases}\frac{\Gamma(\frac{n+1}{2})}{\pi^{\frac{n-1}{2}}}\frac{e^{i\frac{\pi n}{2}}}{\lambda_j^n} & \text{if } \lambda=\lambda_j,\\ 0 & \text{if } \lambda\neq\pm\lambda_j,\\ \frac{\Gamma(\frac{n+1}{2})}{\pi^{\frac{n-1}{2}}}\frac{e^{-i\frac{\pi n}{2}}}{\lambda_j^n} & \text{if } \lambda=-\lambda_j.\end{cases}$$

Since $x_0 \in S\cap\mathbf{T}^n$, the λ_j are linearly independent and thus no expression of the form $\pm\lambda_{j_1}\pm\cdots\pm\lambda_{j_s}$ is equal to any other λ_j. It follows from (3.4.21) that

$$\begin{aligned}\lim_{T\to\infty}\frac{1}{T}\int_1^T K_t^{\frac{n-1}{2}}(x_0)\prod_{j=1}^{N}\Big[1+\frac{e^{-i\frac{\pi n}{2}}e^{2\pi i\lambda_j t}+e^{i\frac{\pi n}{2}}e^{-2\pi i\lambda_j t}}{2}\Big]dt \\ = \frac{\Gamma(\frac{n+1}{2})}{\pi^{\frac{n-1}{2}}}\sum_{j=1}^{N}\frac{1}{\lambda_j^n}.\end{aligned}$$

Suppose that for $x_0 \in S \cap \mathbf{T}^n$ we had

$$\sup_{R\ge 1} |K_R^{\frac{n-1}{2}}(x_0)| \le A_{x_0} < \infty .$$

Then it would follow from the previous identity that

$$\begin{aligned}\frac{\Gamma(\frac{n+1}{2})}{\pi^{\frac{n-1}{2}}}\sum_{j=1}^{N}\frac{1}{\lambda_j^n} &\le A_{x_0}\lim_{T\to\infty}\frac{1}{T}\int_1^T\prod_{j=1}^{N}\left[1+\frac{e^{-i\frac{\pi n}{2}}e^{2\pi i\lambda_j t}+e^{i\frac{\pi n}{2}}e^{-2\pi i\lambda_j t}}{2}\right]dt\\ &= A_{x_0},\end{aligned}$$

which contradicts (3.4.20). We conclude that for every $x_0 \in S \cap \mathbf{T}^n$ we have $\sup_{R\ge 1} |K_R^{\frac{n-1}{2}}(x_0)| = \infty$, which concludes the proof of Lemma 3.4.7.

□

We now proceed with the proof of Theorem 3.4.6. This part of the proof is similar to the proof of Theorem 3.4.2. Lemma 3.4.7 says that the means $B_R^{\frac{n-1}{2}}(\delta_0)(x)$, where δ_0 is the Dirac mass at 0, do not converge for almost all $x \in \mathbf{T}^n$. Our goal will be to replace this Dirac mass by a series of integrable functions on $\mathbf{T}^n$ that have a peak at the origin.

Let us fix a nonnegative C^∞ radial function $\widehat{\Phi}$ on $\mathbf{R}^n$ that is supported in the unit ball $|\xi| \le 1$ and has integral equal to 1. We now set

$$\varphi_\varepsilon(x) = \sum_{m\in\mathbf{Z}^n} \tfrac{1}{\varepsilon^n}\widehat{\Phi}(\tfrac{x+m}{\varepsilon}) = \sum_{m\in\mathbf{Z}^n} \Phi(\varepsilon m)e^{2\pi i m\cdot x},$$

where the identity is valid because of the Poisson summation formula. It follows that the mth Fourier coefficient of φ_ε is $\Phi(\varepsilon m)$. Therefore, we have the estimate

$$\sup_{x\in\mathbf{T}^n}\sup_{R>0}|B_R^{\frac{n-1}{2}}(\varphi_\varepsilon)(x)| \le \sum_{m\in\mathbf{Z}^n}|\Phi(\varepsilon m)| \le \sum_{m\in\mathbf{Z}^n}\frac{C_n'}{(1+\varepsilon|m|)^{n+1}} \le \frac{C_n}{\varepsilon^n}. \tag{3.4.22}$$

For any $j \ge 1$, we shall construct measurable subsets E_j of $\mathbf{T}^n$ that satisfy $|E_j| \ge 1-\frac{1}{j}$, a sequence of positive numbers $0 < R_1 < R_2 < \dots$, and two sequences of positive numbers $\varepsilon_j \le \delta_j$ such that

$$\sup_{R\le R_j}\Big|B_R^{\frac{n-1}{2}}\big(\sum_{s=1}^{\infty}2^{-s}(\varphi_{\varepsilon_s}-\varphi_{\delta_s})\big)(x)\Big| \ge j \qquad \text{for } x \in E_j. \tag{3.4.23}$$

We pick $E_1 = \emptyset$, $R_1 = 1$ and $\varepsilon_1 = \delta_1 = 1$. Let $k > 1$ and suppose that we have selected E_j, R_j, δ_j, and ε_j for all $1 \le j \le k-1$ such that (3.4.23) is satisfied. We will construct E_k, R_k, δ_k, and ε_k so that (3.4.23) is satisfied with $j = k$. We begin by choosing δ_k. Let B be a constant such that

$$|\Phi(x) - \Phi(y)| \le B|x-y|$$

for all $x, y \in \mathbf{R}^n$. Pick δ_k small enough so that

$$B\delta_k \sum_{|m|\le R_{k-1}} |m| \le 1\,. \tag{3.4.24}$$

Then we let

$$A_k = C_n 2^{-k}\delta_k^{-n} + C_n \sum_{j=1}^{k-1} 2^{-j}(\varepsilon_j^{-n} + \delta_j^{-n})\,,$$

where C_n is the constant in (3.4.22), and observe that in view of (3.4.22) we have

$$\sup_{x\in\mathbf{T}^n}\sup_{R>0} \big|B_R^{\frac{n-1}{2}}\big(- 2^{-k}\varphi_{\delta_k} + \sum_{j=1}^{k-1} 2^{-j}(\varphi_{\varepsilon_j} - \varphi_{\delta_j})\big)(x)\big| \le A_k\,. \tag{3.4.25}$$

Let δ_0 be the Dirac mass at the origin in $\mathbf{T}^n$. Since by Fatou's lemma and Lemma 3.4.7 we have

$$\liminf_{N\to\infty} \Big|\Big\{x\in\mathbf{T}^n :\ \sup_{0<R\le N} \big|B_R^{\frac{n-1}{2}}(\delta_0)(x)\big| > A_k + k + 2\Big\}\Big| = 1\,,$$

there exists an $R_k > R_{k-1}$ such that the set

$$E_k = \Big\{x\in\mathbf{T}^n :\ \sup_{0<R\le R_k} \big|B_R^{\frac{n-1}{2}}(\delta_0)(x)\big| > A_k + k + 2\Big\}$$

has measure at least $1-\frac{1}{k}$. We now choose $\varepsilon_k \le \delta_k$ such that

$$\sup_{x\in\mathbf{T}^n} \big|B_R^{\frac{n-1}{2}}(\delta_0)(x) - B_R^{\frac{n-1}{2}}(\varphi_{\varepsilon_k})(x)\big| \le \sum_{|m|\le R_k} \Big(1-\frac{|m|^2}{R_k^2}\Big)^{\frac{n-1}{2}} |1-\widehat{\varphi_{\varepsilon_k}}(m)| \le 1\,.$$

This is possible as the preceding expression in the middle tends to zero as $\varepsilon_k \to 0$. Then for $x\in E_k$ we have

$$\inf_{x\in E_k}\sup_{R\le R_k} 2^{-k}|B_R^{\frac{n-1}{2}}(\varphi_{\varepsilon_k})(x)| \ge A_k + k + 1\,. \tag{3.4.26}$$

Observe that the construction of δ_k gives the estimate

$$\begin{aligned} &\sup_{x\in\mathbf{T}^n}\sup_{R\le R_{k-1}} |B_R^{\frac{n-1}{2}}(\varphi_{\varepsilon_k} - \varphi_{\delta_k})(x)| \\ \le &\sum_{|m|\le R_{k-1}} |\Phi(\varepsilon_k m) - \Phi(\delta_k m)| \\ \le &\ B(\delta_k - \varepsilon_k) \sum_{|m|\le R_{k-1}} |m| \le B\delta_k \sum_{|m|\le R_{k-1}} |m| \le 1. \end{aligned} \tag{3.4.27}$$

using (3.4.24). The inductive selection of the parameters can be described schematically as follows:

$$\delta_1, R_1, E_1, \varepsilon_1 \implies \delta_2 \implies A_2 \implies R_2, E_2 \implies \varepsilon_2 \implies \delta_3 \implies \text{etc.}$$

Let us now prove (3.4.23) for $j = k$. Write

$$\begin{aligned} B_R^{\frac{n-1}{2}}\Big(\sum_{s=1}^{\infty} 2^{-s}(\varphi_{\varepsilon_s}-\varphi_{\delta_s})\Big)(x) &= B_R^{\frac{n-1}{2}}\Big(-2^{-k}\varphi_{\delta_k}+\sum_{s=1}^{k-1} 2^{-s}(\varphi_{\varepsilon_s}-\varphi_{\delta_s})\Big)(x) \\ &\quad + B_R^{\frac{n-1}{2}}\big(2^{-k}\varphi_{\varepsilon_k}\big)(x) \\ &\quad + B_R^{\frac{n-1}{2}}\Big(\sum_{s=k+1}^{\infty} 2^{-s}(\varphi_{\varepsilon_s}-\varphi_{\delta_s})\Big)(x)\,. \end{aligned}$$

In view of (3.4.25), and (3.4.26), and (3.4.27) for all $x \in E_k$, we obtain

$$\sup_{R \le R_{k-1}} \Big| B_R^{\frac{n-1}{2}}\Big(\sum_{s=1}^{\infty} 2^{-s}(\varphi_{\varepsilon_s}-\varphi_{\delta_s})\Big)(x)\Big| \ge k\,,$$

which clearly implies (3.4.23) (with $j = k$) since $R_k > R_{k-1}$. Setting

$$f = \sum_{s=1}^{\infty} 2^{-s}(\varphi_{\varepsilon_s}-\varphi_{\delta_s}) \in L^1(\mathbf{T}^n)$$

we have now proved that $\sup_{R>0} \big|B_R^{\frac{n-1}{2}}(f)(x)\big| = \infty$ for all x in

$$\bigcap_{k=1}^{\infty}\bigcup_{r=k}^{\infty} E_r\,.$$

Since the later set has full measure in $\mathbf{T}^n$, the required conclusion follows.

By taking ε_1 arbitrarily small (instead of picking $\varepsilon_1 = 1$), we force f to be supported in an arbitrarily small neighborhood of the origin.

□

The previous argument shows that the Bochner-Riesz means B_R^{α} are badly behaved on $L^1(\mathbf{T}^n)$ when $\alpha = \frac{n-1}{2}$. It follows that the "rougher" spherical Dirichlet means $\widetilde{D}(n,N) * f$ (which correspond to $\alpha = 0$) are also ill behaved on $L^1(\mathbf{T}^n)$. See Exercise 3.4.5. In Chapter 10 we will show the stronger negative result that the spherical Dirichlet means of L^p functions may also diverge in L^p when $p \neq 2$.

Exercises

3.4.1. Prove that if $f \in L^1(\mathbf{T}^1)$ satisfies $\limsup_{N\to\infty} |(D_N * f)(x)| = \infty$ for almost all $x \in \mathbf{T}^1$, then for the function

$$F(x_1, \dots, x_n) = f(x_1)\dots f(x_n)$$

on $\mathbf{T}^n$ we have $\limsup_{N\to\infty} |(D(n,N) * F)(x)| = \infty$ for almost all $x \in \mathbf{T}^n$.

3.4.2. (*H. Weyl*) A sequence $\{a_k\}_{k=0}^{\infty}$ with values in $\mathbf{T}^n$ is called *equidistributed* if for every square Q in $\mathbf{T}^n$ we have

$$\lim_{N\to\infty}\frac{\#\{k:\ 0\le k\le N-1,\quad a_k\in Q\}}{N}=|Q|\,.$$

Show that the following are equivalent:
(a) The sequence $\{a_k\}_{k=0}^{\infty}$ is equidistributed
(b) For every smooth function f on $\mathbf{T}^n$ we have that

$$\lim_{N\to\infty}\frac{1}{N}\sum_{k=0}^{N-1}f(a_k)=\int_{\mathbf{T}^n}f(x)\,dx\,.$$

(c) For every $m\in\mathbf{Z}^n\setminus\{0\}$ we have

$$\lim_{N\to\infty}\frac{1}{N}\sum_{k=0}^{N-1}e^{2\pi i m\cdot a_k}=0\,.$$

[*Hint:* Prove that $(a)\implies(b)\implies(c)\implies(b)\implies(a)$. In proving $(a)\iff(b)$, approximate f by step functions. In proving $(c)\implies(b)$, use Fourier inversion.]

3.4.3. Suppose that $x=(x_1,\dots,x_n)\in\mathbf{T}^n$ and $m\cdot x$ is irrational for all $m\in\mathbf{Z}^n\setminus\{0\}$. Use the previous Exercise to show that the sequence $\{([kx_1],\dots,[kx_n])\}_{k=0}^{\infty}$ is equidistributed. (In dimension 1 the hypothesis is satisfied if x is irrational.)

3.4.4. Use the beta function identity (Appendix A.2)

$$t^{\alpha}=\frac{1}{B(\alpha-\beta,\beta+1)}\int_0^t(t-s)^{\alpha-\beta-1}s^{\beta}\,ds$$

to show that the function $K_R^{\alpha}(x)=\sum_{|m|\le R}(1-\frac{|m|^2}{R^2})^{\alpha}e^{2\pi i m\cdot x}$ satisfies (3.4.2). [*Hint:* Take $t=1-\frac{|m|^2}{R^2}$ and change variables $s=\frac{r^2-|m|^2}{R^2}$ in the previous beta function identity.]

3.4.5. Use the previous Exercise to obtain that if for some $x_0\in\mathbf{T}^n$ we have

$$\limsup_{R\to\infty}|K_R^{\alpha}(x_0)|<\infty\,,$$

then for all $\beta>\alpha$ we have

$$\sup_{R>0}|K_R^{\beta}(x_0)|<\infty\,.$$

Conclude that the circular (spherical) Dirichlet means of the function f constructed in the proof of Theorem 3.4.6 diverge a.e. The same conclusion is true for the Bochner-Riesz means of f of every order $\alpha\le\frac{n-1}{2}$.

3.4.6. For $t \in [0,\infty)$ let

$$N(t) = \#\{m \in \mathbf{Z}^n : \ |m| \le t\}.$$

Let $0 = r_0 < r_1 < r_2 < \dots$ be the sequence all of numbers r for which there exist $m \in \mathbf{Z}^n$ such that $|m| = r$.

(a) Observe that N is right continuous and constant on intervals of the form $[r_j, r_{j+1})$.

(b) Show that the distributional derivative of N is the measure

$$\mu(t) = \#\{m \in \mathbf{Z}^n : \ |m| = t\},$$

defined via the identity $\langle \mu, \varphi \rangle = \sum_{j=0}^{\infty} \#\{m \in \mathbf{Z}^n : \ |m| = r_j\}\, \varphi(r_j)$.

3.4.7. Let $f \in C^1([0,\infty))$ and $0 \le a < b < \infty$ not equal to any r_j as defined in the previous Exercise. Use that Exercise to derive the useful identity

$$\sum_{\substack{m \in \mathbf{Z}^n \\ a \le |m| \le b}} f(|m|) = f(b)N(b) - f(a)N(a) - \int_a^b f'(x)N(x)\,dx.$$

3.4.8. (a) Let $0 < \lambda < \infty$ and fix a transcendental number γ in $(0,1)$. Prove that for $k \in \mathbf{Z}^+$ we have

$$\sum_{\substack{m \in \mathbf{Z}^n \\ k+\gamma \le |m| \le k+1+\gamma}} \frac{e^{i|m|}}{|m|^\lambda} = \frac{-i\omega_{n-1}e^{i(k+1+\gamma)}}{(k+1+\gamma)^{\lambda-(n-1)}} - \frac{-i\omega_{n-1}e^{i(k+\gamma)}}{(k+\gamma)^{\lambda-(n-1)}}$$

$$+ O(k^{-\lambda+(n-1)-\frac{n-1}{n+1}}),$$

as $k \to \infty$, where ω_{n-1} is the volume of $\mathbf{S}^{n-1}$.

(b) Use part (a) to show that if $\lambda \le n-1$, the limit

$$\lim_{R\to\infty} \sum_{\substack{m \in \mathbf{Z}^n \setminus \{0\} \\ |m| \le R}} \frac{e^{i|m|}}{|m|^\lambda}$$

does not exist.

(c) Show that if $\lambda > n - \frac{n-1}{n+1}$, the following series converges:

$$\sum_{m \in \mathbf{Z}^n \setminus \{0\}} \frac{e^{i|m|}}{|m|^\lambda},$$

where the infinite sum is interpreted as the limit in part (b).

[*Hint:* Part (a): You will need Exercises 3.4.7 and 3.1.12. Part (b): Suppose the limit exists and let $\beta_k = \frac{-i\omega_{n-1}e^{i(k+1+\gamma)}}{(k+1+\gamma)^{\lambda-(n-1)}} - \frac{-i\omega_{n-1}e^{i(k+\gamma)}}{(k+\gamma)^{\lambda-(n-1)}}$. If the series $\sum_{m\in\mathbf{Z}^n\setminus\{0\}} \frac{e^{i|m|}}{|m|^\lambda}$ converged, then $\beta_k \to 0$ as $k \to \infty$. But then $k^{\lambda-(n-1)}\beta_k$ would also tend to zero, which gives a contradiction.]

3.4.9. (*Pinsky, Stanton, and Trapa* [**415**]) Prove that the spherical partial sums of the Fourier series of the characteristic function of the ball $B(0, \frac{1}{2\pi})$ in $\mathbf{T}^n$ diverges at $x = 0$ when $n \geq 3$.
[*Hint:* Use the idea of the previous Exercise with $\lambda = \frac{n+1}{2}$.]

3.5. The Conjugate Function and Convergence in Norm

In this section we will address the following fundamental question: Do Fourier series converge in norm? We will begin with some abstract necessary and sufficient conditions that guarantee such a convergence. In one dimension, we will be able to reduce matters to the study of the so-called conjugate function on the circle, a sister operator of the Hilbert transform, which will be the center of study of the next chapter. In higher dimensions the situation is more complicated, but we will be able to give a positive answer in the case of square summability.

3.5.a. Equivalent Formulations of Convergence in Norm

The question we pose is for which $1 \leq p \leq \infty$ we have

$$\big\|D(n,N) * f - f\big\|_{L^p(\mathbf{T}^n)} \to 0 \qquad \text{as } N \to \infty, \tag{3.5.1}$$

and similarly for the circular Dirichlet kernel $\widetilde{D}(n,N)$. We will be able to tackle this question by looking at an equivalent formulation of it.

Theorem 3.5.1. *Fix $1 \leq p \leq \infty$ and $\{a_m\}$ in $\ell^\infty(\mathbf{Z}^n)$. For each $R \geq 0$, let $\{a_m(R)\}_{m\in\mathbf{Z}^n}$ be a finitely supported sequence (whose support depends on R) that satisfies $\lim\limits_{R\to\infty} a_m(R) = a_m$. For $f \in L^p(\mathbf{T}^n)$ define*

$$S_R(f)(x) = \sum_{m\in\mathbf{Z}^n} a_m(R)\widehat{f}(m)e^{2\pi i m\cdot x}\,.$$

Then for all $f \in L^p(\mathbf{T}^n)$ the sequence $S_R(f)$ converges in L^p as $R \to \infty$ if and only if there exists a constant $K < \infty$ such that

$$\sup_{R\geq 0} \big\|S_R\big\|_{L^p\to L^p} \leq K. \tag{3.5.2}$$

For $f \in C^\infty(\mathbf{T}^n)$ define $A(f)$ as an absolutely convergent series by

$$A(f)(x) = \sum_{m\in\mathbf{Z}^n} a_m\widehat{f}(m)e^{2\pi i m\cdot x}.$$

Furthermore, if (3.5.2) holds, then for the same constant K we have

$$\sup_{0\neq f\in C^\infty} \frac{\big\|A(f)\big\|_{L^p}}{\big\|f\big\|_{L^p}} \leq K\,, \tag{3.5.3}$$

and then A extends to a bounded operator from $L^p(\mathbf{T}^n)$ into itself. Moreover, for every $f \in L^p(\mathbf{T}^n)$ we have that $S_R(f) \to A(f)$ in L^p as $R \to \infty$.

Proof. If $S_R(f)$ converges in L^p, then $\|S_R(f)\|_{L^p} \le C_f$ for some constant C_f that depends on f. The uniform boundedness theorem now gives that the operator norms of S_R from L^p to L^p are bounded uniformly in R. This proves (3.5.2).

Conversely, assume (3.5.2). For $f \in C^\infty(\mathbf{T}^n)$ Fatou's lemma gives

$$\|A(f)\|_{L^p} = \|\lim_{R\to\infty} S_R(f)\|_{L^p} \le \liminf_{R\to\infty} \|S_R(f)\|_{L^p} \le K\|f\|_{L^p};$$

hence (3.5.3) holds. Thus A extends to a bounded operator on $L^p(\mathbf{T}^n)$ by density. We will show that for all $f \in L^p(\mathbf{T}^n)$ we have $S_R(f) \to A(f)$ in L^p as $R \to \infty$. Fix $f \in L^p(\mathbf{T}^n)$ and let $\varepsilon > 0$ be given. Pick a trigonometric polynomial P satisfying $\|f - P\|_{L^p} \le \varepsilon$. Let d be the degree of P. Then

$$\begin{aligned}\|S_R(P) - A(P)\|_{L^p} &\le \|S_R(P) - A(P)\|_{L^\infty} \\ &\le \sum_{|m_1|+\cdots+|m_n|\le d} |a_m(R) - a_m|\,|\widehat{P}(m)| \le \varepsilon\end{aligned}$$

provided $R > R_0$, since $a_m(R) \to a_m$ for every m with $|m_1| + \cdots + |m_n| \le d$. Then

$$\begin{aligned}&\|S_R(f)-A(f)\|_{L^p} \\ &\le \|S_R(f)-S_R(P)\|_{L^p} + \|S_R(P)-A(P)\|_{L^p} + \|A(P)-A(f)\|_{L^p} \\ &\le K\varepsilon + \varepsilon + K\varepsilon = (2K+1)\varepsilon\,.\end{aligned}$$

This proves that $S_R(f)$ converges to $A(f)$ in L^p as $R \to \infty$.

□

The most interesting situation arises, of course, when $a_m = 1$ for all $m \in \mathbf{Z}^n$. In this case we expect the operators $S_R(f)$ to converge back to f as $R \to \infty$. We should keep in mind the following three examples: first, the sequence $a_m(R) = 1$ when $\max_{1\le j\le n} |m_j| \le R$ and zero otherwise, in which case the S_R of Theorem 3.5.1 is

$$S_R(f) = f * D(n,R)\,; \tag{3.5.4}$$

second, the sequence $a_m(R) = 1$ when $|m| \le R$ and zero otherwise, in which case the S_R of Theorem 3.5.1 is

$$\widetilde{S}_R(f) = f * \widetilde{D}(n,R)\,; \tag{3.5.5}$$

finally, the sequence $a_m(R) = (1 - \frac{|m|^2}{R^2})_+^\alpha$, in which case $S_R = B_R^\alpha$.

Corollary 3.5.2. *Let $1 \le p \le \infty$ and $\alpha \ge 0$. Let S_R and $\widetilde{S}_R$ be as in (3.5.4) and (3.5.5), respectively, and let B_R^α be the Bochner-Riesz means as defined in*

(3.4.1). Then

$$\forall\, f \in L^p(\mathbf{T}^n),\ D(n,R) * f \to f \quad \text{in } L^p \iff \sup_{R\geq 0} \left\| S_R \right\|_{L^p \to L^p} < \infty.$$

$$\forall\, f \in L^p(\mathbf{T}^n),\ \widetilde{D}(n,R) * f \to f \quad \text{in } L^p \iff \sup_{R\geq 0} \left\| \widetilde{S}_R \right\|_{L^p \to L^p} < \infty.$$

$$\forall\, f \in L^p(\mathbf{T}^n),\ B_R^\alpha(f) \to f \quad \text{in } L^p \iff \sup_{R\geq 0} \left\| B_R^\alpha \right\|_{L^p \to L^p} < \infty.$$

Example 3.5.3. We investigate the one-dimensional case in some detail. We take $n = 1$ and we define $a_m(N) = 1$ for all $-N \leq m \leq N$ and zero otherwise. Then $S_N(f) = \widetilde{S}_N(f) = D_N * f$, where D_N is the Dirichlet kernel. The expressions $\left\| S_N \right\|_{L^p \to L^p}$ can always be estimated from above by the L^1 norm of D_N. But the latter blows up as $N \to \infty$ and it is therefore not appropriate in using Corollary 3.5.2. We will later find an alternative way to prove that the expressions $\left\| S_N \right\|_{L^p \to L^p}$ are uniformly bounded in N when $1 < p < \infty$.

This reasoning allows us to deduce that for some function g, $S_N(g)$ may not converge in L^1, which is also a consequence of the proof of Theorem 3.4.2; see (3.4.16). Note that since the Fejér kernel F_M has L^1 norm 1, we have

$$\left\| S_N \right\|_{L^1 \to L^1} \geq \lim_{M\to\infty} \left\| D_N * F_M \right\|_{L^1} = \left\| D_N \right\|_{L^1}.$$

This implies that the expressions $\left\| S_N \right\|_{L^1 \to L^1}$ are not uniformly bounded in N, and therefore Corollary 3.5.2 gives that for some $f \in L^1(\mathbf{T}^1)$, $S_N(f)$ does not converge to f (nor to any other integrable function) in L^1.

Although convergence of the partial sums of Fourier series fails in L^1, it is a consequence of Plancherel's theorem that it holds in L^2. More precisely, if $f \in L^2(\mathbf{T}^n)$, then

$$\left\| \widetilde{D}_N * f - f \right\|_{L^2}^2 = \sum_{|m|>N} |\widehat{f}(m)|^2 \to 0$$

as $N \to \infty$ and the same result is true for D_N. The following question is therefore naturally raised. Does L^p convergence hold for $p \neq 2$? This question was answered in the affirmative by M. Riesz in dimension 1. In higher dimensions a certain interesting dichotomy appears. Although it is a consequence of the one-dimensional result that the square partial sums $D(n,N) * f$ converge to f in $L^p(\mathbf{T}^n)$, this is not the case for the circular partial sums as there exists $f \in L^p(\mathbf{T}^n)$ such that $\widetilde{D}(n,N) * f$ do not converge in L^p if $1 < p \neq 2 < \infty$. We will study this issue in Chapter 10.

Let us begin our discussion with the one-dimensional situation.

Definition 3.5.4. For $f \in C^\infty(\mathbf{T}^1)$ define the *conjugate function* $\widetilde{f}$ by

$$\widetilde{f}(x) = -i \sum_{m \in \mathbf{Z}^1} \operatorname{sgn}(m) \widehat{f}(m) e^{2\pi i m x},$$

where $\operatorname{sgn}(m) = 1$ for $m > 0$, -1 for $m < 0$, and 0 for $m = 0$. Also define the *Riesz projections* P_+ and P_- by

$$P_+(f)(x) = \sum_{m=1}^{\infty} \widehat{f}(m)e^{2\pi imx}, \tag{3.5.6}$$

$$P_-(f)(x) = \sum_{m=-\infty}^{-1} \widehat{f}(m)e^{2\pi imx}. \tag{3.5.7}$$

Observe that $f = P_+(f) + P_-(f) + \widehat{f}(0)$ while $\widetilde{f} = -iP_+(f) + iP_-(f)$ when f is in $C^\infty(\mathbf{T}^1)$. The following is a consequence of Theorem 3.5.1.

Proposition 3.5.5. *Let $1 \le p \le \infty$. Then the expressions $S_N(f) = D_N * f$ converge to f in $L^p(\mathbf{T}^1)$ as $N \to \infty$ if and only if there exists a constant $C_p > 0$ such that for all smooth f we have $\|\widetilde{f}\|_{L^p(\mathbf{T}^1)} \le C_p \|f\|_{L^p(\mathbf{T}^1)}$.*

Proof. Observe that

$$P_+(f) = \frac{1}{2}(f + i\widetilde{f}) - \frac{1}{2}\widehat{f}(0)$$

and therefore the L^p boundedness of the operator $f \to \widetilde{f}$ is equivalent to that of the operator $f \to P_+(f)$.

Next, note the validity of the identity

$$e^{-2\pi iNx} \sum_{m=0}^{2N} \big(f(\cdot)e^{2\pi iN(\cdot)}\big)\widehat{\ }(m)e^{2\pi imx} = \sum_{m=-N}^{N} \widehat{f}(m)e^{2\pi imx}.$$

Since multiplication by exponentials does not affect L^p norms, this identity implies that the norm of the operator $S_N(f) = D_N * f$ from $L^p \to L^p$ is equal to that of the operator

$$S_N'(g)(x) = \sum_{m=0}^{2N} \widehat{g}(m)e^{2\pi imx}$$

from $L^p \to L^p$. Therefore,

$$\sup_{N\ge 0} \|S_N\|_{L^p \to L^p} < \infty \iff \sup_{N \ge 0} \|S_N'\|_{L^p \to L^p} < \infty. \tag{3.5.8}$$

Suppose now that for all $f \in L^p(\mathbf{T}^1)$, $S_N(f) \to f$ in L^p as $N \to \infty$. Then $\sup_{N\ge 0}\|S_N\|_{L^p\to L^p} < \infty$ by Corollary 3.5.2; hence $\sup_{N\ge 0}\|S_N'\|_{L^p\to L^p} < \infty$ by (3.5.8). Theorem 3.5.1 applied to the sequence $a_m(R) = 1$ for $0 \le m \le R$ and $a_m(R) = 0$ otherwise gives that the operator $A(f) = P_+(f) + \widehat{f}(0)$ is bounded on $L^p(\mathbf{T}^1)$. Hence so is P_+.

Conversely, suppose that P_+ extends to a bounded operator from $L^p(\mathbf{T}^1)$ into itself. For all f smooth we can write

$$\begin{aligned} S_N'(f)(x) &= \sum_{m=0}^{\infty} \widehat{f}(m)e^{2\pi imx} - \sum_{m=2N+1}^{\infty} \widehat{f}(m)e^{2\pi imx} \\ &= \sum_{m=0}^{\infty} \widehat{f}(m)e^{2\pi imx} - e^{2\pi i(2N+1)x}\sum_{m=0}^{\infty} \widehat{f}(m+2N+1)e^{2\pi imx} \\ &= P_+(f)(x) - e^{2\pi i(2N+1)x}P_+(e^{-2\pi i(2N+1)(\cdot)}f) - \widehat{f}(0)(1-e^{2\pi i(2N+1)x}). \end{aligned}$$

The previous identity implies that

$$\sup_{N\ge 0} \big\|S_N'(f)\big\|_{L^p} \le \big(2\big\|P_+\big\|_{L^p\to L^p} + 2\big)\big\|f\big\|_{L^p} \tag{3.5.9}$$

for all f smooth, and by density for all $f \in L^p(\mathbf{T}^1)$. [Note that S_N' is well defined on $L^p(\mathbf{T}^1)$.] In view of (3.5.8), estimate (3.5.9) also holds for S_N. Theorem 3.5.1 applied again gives that $S_N(f) \to f$ in L^p for all $f \in L^p(\mathbf{T}^1)$.

□

3.5.b. The L^p Boundedness of the Conjugate Function

We know now that convergence of Fourier series in L^p is equivalent to the L^p boundedness of the conjugate function or any of one of the two Riesz projections. It is natural to ask whether these operators are L^p bounded.

Theorem 3.5.6. *Given $1 < p < \infty$, there is a constant $A_p > 0$ such that for all $f \in C^\infty(\mathbf{T}^1)$ we have*

$$\big\|\widetilde{f}\big\|_{L^p} \le A_p\big\|f\big\|_{L^p}. \tag{3.5.10}$$

Consequently, the Fourier series of L^p functions on the circle converge back to the functions in L^p when $1 < p < \infty$.

Proof. We will present a relatively short proof of this theorem due to S. Bochner. Let $f(t)$ be a real valued trigonometric polynomial on $\mathbf{T}^1$ and assume that $\widehat{f}(0) = 0$. Then $\widetilde{f}$ is also real-valued [see Exercise 3.5.4(b)] and the polynomial $f + i\widetilde{f}$ contains only positive frequencies. Thus for $k \in \mathbf{Z}^+$ we have

$$\int_{\mathbf{T}^1} (f(t) + i\widetilde{f}(t))^{2k}\,dt = 0.$$

Expanding the $2k$ power and taking real parts, we obtain

$$\sum_{j=0}^{k} (-1)^{k-j}\binom{2k}{2j}\int_{\mathbf{T}^1} \widetilde{f}(t)^{2k-2j}f(t)^{2j}\,dt = 0,$$

where we used that f is real valued. Therefore,

$$\begin{aligned}\big\|\widetilde{f}\,\big\|_{L^{2k}}^{2k} &\le \sum_{j=1}^{k}\binom{2k}{2j}\int_{\mathbf{T}^1}\widetilde{f}(t)^{2k-2j}f(t)^{2j}\,dt\\ &\le \sum_{j=1}^{k}\binom{2k}{2j}\big\|\widetilde{f}\,\big\|_{L^{2k}}^{2k-2j}\big\|f\big\|_{L^{2k}}^{2j},\end{aligned}$$

by applying Hölder's inequality with exponents $2k/(2k-2j)$ and $2k/(2j)$ to the jth term of the sum. Dividing the last inequality by $\big\|f\big\|_{L^{2k}}^{2k}$, we obtain

$$R^{2k} \le \sum_{j=1}^{k}\binom{2k}{2j}R^{2k-2j}, \tag{3.5.11}$$

where $R = \big\|\widetilde{f}\,\big\|_{L^{2k}}/\big\|f\big\|_{L^{2k}}$. It is an elementary fact that if $R > 0$ satisfies (3.5.11), then there exists a positive constant C_{2k} such that $R \le C_{2k}$. We conclude that

$$\big\|\widetilde{f}\,\big\|_{L^p} \le C_p\big\|f\big\|_{L^p} \qquad \text{when } p = 2k. \tag{3.5.12}$$

We can now remove the assumption that $\widehat{f}(0) = 0$. Apply (3.5.12) to $f - \widehat{f}(0)$, observe that the conjugate function of a constant is zero, and use the triangle inequality and the fact that $|\widehat{f}(0)| \le \big\|f\big\|_{L^1} \le \big\|f\big\|_{L^p}$ to obtain $\big\|\widetilde{f}\,\big\|_{L^p} \le 2C_p\big\|f\big\|_{L^p}$ when $p = 2k$ and f is a real-valued trigonometric polynomial. Since a general trigonometric polynomial can be written as $P + iQ$, where P and Q are real-valued trigonometric polynomials, we obtain the inequality $\big\|\widetilde{f}\,\big\|_{L^p} \le 4C_p\big\|f\big\|_{L^p}$ for all trigonometric polynomials f when $p = 2k$. Since trigonometric polynomials are dense in L^p, it follows that (3.5.10) holds for all smooth functions when $p = 2k$. It also follows that the conjugate function has a bounded extension on $L^p(\mathbf{T}^1)$ when $p = 2k$ and, in particular, this extension is well defined for simple functions.

Every real number $p \ge 2$ lies in an interval of the form $[2k, 2k+2]$, for some $k \in \mathbf{Z}^+$. Theorem 1.3.4 gives that

$$\big\|\widetilde{f}\,\big\|_{L^p} \le A_p\big\|f\big\|_{L^p} \tag{3.5.13}$$

for some $A_p > 0$ and all $2 \le p < \infty$ when f is a simple function. By density the same result is valid for all L^p functions when $p \ge 2$. Finally, we observe that the adjoint operator of $f \to \widetilde{f}$ is $f \to -\widetilde{f}$. By duality, estimate (3.5.13) is also valid for $1 < p \le 2$ with constant $A_{p'} = A_p$.

□

Next we prove that convergence in norm extends to higher dimensions for the square partial sums of a Fourier series.

Theorem 3.5.7. *Let $1 < p < \infty$ and $f \in L^p(\mathbf{T}^n)$. Then $D(n,N) * f$ converges to f in L^p as $N \to \infty$.*

Proof. Let us prove this theorem in dimension $n = 2$. The same proof can be adjusted to work in every dimension. In view of Corollary 3.5.2, it suffices to prove that for all f smooth on $\mathbf{T}^2$ we have

$$\sup_{N\geq 0}\int_0^1\int_0^1\bigg|\sum_{|m_1|\leq N}\sum_{|m_2|\leq N} e^{2\pi i(m_1x_1+m_2x_2)}\widehat{f}(m_1,m_2)\bigg|^p dx_1\,dx_2 \leq K^{2p}\big\|f\big\|_{L^p(\mathbf{T}^2)}^p .$$

For fixed $f \in C^\infty(\mathbf{T}^2)$, $N \geq 0$, and $x_2 \in [0,1]$, define a trigonometric polynomial g_{N,x_2} on $\mathbf{T}^1$ by setting

$$\sum_{|m_2|\leq N} e^{2\pi i m_2 x_2}\widehat{f}(m_1,m_2) = \widehat{g_{N,x_2}}(m_1)$$

for all $m_1 \in \mathbf{Z}$. Then we have

$$(D_N * g_{N,x_2})(x_1) = \sum_{|m_1|\leq N} e^{2\pi i m_1 x_1}\widehat{g_{N,x_2}}(m_1)$$

and also

$$g_{N,x_2}(x_1) = \sum_{|m_2|\leq N} e^{2\pi i m_2 x_2}\bigg[\sum_{m_1\in\mathbf{Z}} e^{2\pi i m_1 x_1}\widehat{f}(m_1,m_2)\bigg] = (D_N * f_{x_1})(x_2),$$

where f_{x_1} is the function defined by $f_{x_1}(y) = f(x_1,y)$. We have

$$\begin{aligned}
&\int_0^1\int_0^1\bigg|\sum_{|m_1|\leq N}\sum_{|m_2|\leq N} e^{2\pi i(m_1x_1+m_2x_2)}\widehat{f}(m_1,m_2)\bigg|^p dx_1\,dx_2\\
&= \int_0^1\int_0^1\Big|(D_N * g_{N,x_2})(x_1)\Big|^p dx_1\,dx_2\\
&\leq K^p\int_0^1\int_0^1 |g_{N,x_2}(x_1)|^p\,dx_1\,dx_2\\
&= K^p\int_0^1\int_0^1\Big|(D_N * f_{x_1})(x_2)\Big|^p dx_2\,dx_1\\
&\leq K^{2p}\int_0^1\int_0^1 |f_{x_1}(x_2)|^p\,dx_2\,dx_1 = K^{2p}\big\|f\big\|_{L^p(\mathbf{T}^2)}^p.
\end{aligned}$$

We used twice the fact that the one-dimensional partial sums are uniformly bounded in L^p when $1 < p < \infty$, a consequence of Corollary 3.5.2, Proposition 3.5.5, and Theorem 3.5.6.

□

Exercises

3.5.1. If $f \in C^\infty(\mathbf{T}^n)$, then show that $D(n,N) * f$ and $\widetilde{D}(n,N) * f$ converge to f uniformly and in L^p for $1 \leq p \leq \infty$.

3.5.2. Prove that the norms of the Riesz projections on $L^2(\mathbf{T}^1)$ are at most 1, while the operation of conjugation $f \to \widetilde{f}$ is an isometry on $L^2(\mathbf{T}^1)$.

3.5.3. Let $-\infty \le a_j < b_j \le +\infty$ for $1 \le j \le n$. Consider the rectangular projection operator defined on $C^\infty(\mathbf{T}^n)$ by

$$P(f)(x) = \sum_{a_j \le m_j \le b_j} \widehat{f}(m) e^{2\pi i(m_1 x_1 + \cdots + m_n x_n)} .$$

Prove that when $1 < p < \infty$, P extends to a bounded operator from $L^p(\mathbf{T}^n)$ into itself with bounds independent of the a_j, b_j.
[*Hint:* Express P in terms of the Riesz projection P_+.]

3.5.4. Let $P_r(t)$ be the Poisson kernel on $\mathbf{T}^1$ as defined in Exercise 3.1.7. For $0 < r < 1$, define the *conjugate Poisson kernel* $Q_r(t)$ on the circle by

$$Q_r(t) = -i \sum_{m=-\infty}^{+\infty} \operatorname{sgn}(m)\, r^{|m|} e^{2\pi i m t} .$$

(a) For $0 < r < 1$, prove the identity

$$Q_r(t) = \frac{2r \sin(2\pi t)}{1 - 2r\cos(2\pi t) + r^2} .$$

(b) Prove that $\widetilde{f}(t) = \lim_{r \to 1} (Q_r * f)(t)$ whenever f is smooth. Conclude that if f is real valued, then so is $\widetilde{f}$.
(c) Let $f \in L^1(\mathbf{T}^1)$. Prove that the functions $z \to (P_r * f)(t)$ and $z \to (Q_r * f)(t)$ are harmonic functions of $z = re^{2\pi i t}$ in the region $|z| < 1$.
(d) Let $f \in L^1(\mathbf{T}^1)$. Prove that the function

$$z \to (P_r * f)(t) + i(Q_r * f)(t)$$

is analytic in $z = re^{2\pi i t}$ and thus $(P_r * f)(t)$ and $(Q_r * f)(t)$ are conjugate harmonic functions.

3.5.5. Let $f \in C^\infty(\mathbf{T}^1)$. Prove that the conjugate function $\widetilde{f}$ can be written as

$$\begin{aligned} \widetilde{f}(x) &= \lim_{\varepsilon \to 0} \int_{\varepsilon \le |t| \le 1/2} f(x-t)\cot(\pi t)\, dt \\ &= \int_{|t| \le 1/2} \big(f(x-t) - f(x)\big)\cot(\pi t)\, dt. \end{aligned}$$

[*Hint:* Use part (b) of Exercise 3.5.4 and the fact that Q_r has integral zero over the circle to write $(f * Q_r)(x) = \big((f - f(x)) * Q_r\big)(x)$, allowing use of the Lebesgue dominated convergence theorem.]

3.5.6. Suppose that f is a real-valued function on $\mathbf{T}^1$ and $0 \le \lambda < \pi/2$.

(a) Prove that

$$\int_{\mathbf{T}^1} e^{\lambda \widetilde{f}(t)}\, dt \le \frac{1}{\cos(\lambda)}.$$

(b) Prove that for $0 \le \lambda < \pi/2$ we have

$$\int_{\mathbf{T}^1} e^{\lambda |\widetilde{f}(t)|}\, dt \le \frac{2}{\cos(\lambda)}.$$

[*Hint:* Part (a): Consider the analytic function $F(z)$ on the disc $|z| < 1$ defined by $F(z) = -i(P_r * f)(\theta) + (Q_r * f)(\theta)$, where $z = re^{2\pi i\theta}$. Then $\operatorname{Re} e^{\lambda F(z)}$ is harmonic and its average over the circle $|z| = r$ is equal to its value at the origin, which is $\cos(\lambda f(0)) \le 1$. Let $r \uparrow 1$ and use that for $z = e^{2\pi it}$ on the circle we have $\operatorname{Re} e^{\lambda F(z)} \ge e^{\lambda \widetilde{f}(t)} \cos(\lambda)$.]

3.5.7. Use that $\cot(\pi t) = \frac{1}{\pi t} + b(t)$, where $b(t)$ is a bounded function, to prove that if $f \in \dot{\Lambda}_\alpha(\mathbf{T}^1)$ for some $0 < \alpha < 1$, then $\widetilde{f} \in \dot{\Lambda}_\alpha(\mathbf{T}^1)$.
[*Hint:* Write the difference $\widetilde{f}(x+h) - \widetilde{f}(x)$ as

$$\int_{|t|\le 1/2} \big(f(x+h-t) - f(x+h) - f(x-t) + f(x)\big)\frac{dt}{\pi t} + O(|h|^\alpha)$$

and break up the integral over the sets $0 \le |t| \le 5|h|$ and $5|h| \le |t| \le 1/2$.]

3.5.8. (a) Show that for M, N positive integers we have

$$(F_M * D_N)(x) = \begin{cases} F_M(x) & \text{for } M \le N \\ F_N(x) + \frac{M-N}{(M+1)(N+1)} \sum\limits_{|k|\le N} |k|\, e^{2\pi ikx} & \text{for } M > N. \end{cases}$$

(b) Prove that for some constant $c > 0$ we have

$$\int_{\mathbf{T}^1} \bigg| \sum_{|k|\le N} |k|\, e^{2\pi ikx} \bigg|\, dx \ge c\, N \log N$$

as $N \to \infty$.
[*Hint:* Part (b): Show that for $x \in [-\frac{1}{2}, \frac{1}{2}]$ we have

$$\sum_{|k|\le N} |k|\, e^{2\pi ikx} = \frac{N+1}{2} \frac{\sin((2N+1)\pi x)}{\sin(\pi x)} + \frac{\cos(2\pi(N+1)x) - 1}{4\sin^2(\pi x)}$$

and use the result of Exercise 3.1.8 and the fact that the L^1 norm of the second fraction is at most a constant multiple of N.]

3.5.9. Show that the Fourier series of the integrable functions

$$f_1(x) = \sum_{j=0}^{\infty} 2^{-j} F_{2^{2^j}}(x), \qquad f_2(x) = \sum_{j=1}^{\infty} \frac{1}{j^2} F_{2^{2^j}}(x) \qquad x \in \mathbf{T}^1$$

do not converge in $L^1(\mathbf{T}^1)$.

$\big[$*Hint:* Let $M_j = 2^{2^{2^j}}$ or $M_j = 2^{2^j}$ depending on the situation. For fixed N let j_N be the least integer j such that $M_j > N$. Then for $j \ge j_N+1$ we have $M_j \ge M_{j_N}^2 > N^2 \ge 2N+1$, hence $\frac{M_j - N}{M_j+1} \ge \frac{1}{2}$. Split the summation indices in the sets $j \ge j_N+1$, $j = j_N$, and $j < j_N$. Conclude that both $\big\|f_1 * D_N\big\|_{L^1}$ and $\big\|f_2 * D_N\big\|_{L^1}$ tend to infinity as $N \to \infty$ using Exercise 3.5.8.$\big]$

3.5.10. (*Stein* [**468**]) Note that if $\alpha \ge 0$, then B_R^α are bounded on $L^2(\mathbf{T}^n)$ uniformly in $R > 0$. Show that if $\alpha > \frac{n-1}{2}$, then B_R^α are bounded on $L^1(\mathbf{T}^n)$ uniformly in $R > 0$. Use complex interpolation to prove that for $\alpha > \frac{n-1}{2}\big|\frac{1}{p} - \frac{1}{2}\big|$, then B_R^α are bounded on $L^p(\mathbf{T}^n)$ uniformly in $R > 0$. Compare this problem with Exercise 1.3.5.

3.6. Multipliers, Transference, and Almost Everywhere Convergence*

In Chapter 2 we saw that bounded operators from $L^p(\mathbf{R}^n)$ into $L^q(\mathbf{R}^n)$ that commute with translations are given by convolution with tempered distributions on $\mathbf{R}^n$. In particular, if $p = q$, these tempered distributions have bounded Fourier transforms. We called these Fourier transforms multipliers, and we denoted the space of all L^p multipliers on $\mathbf{R}^n$ by $\mathcal{M}_p(\mathbf{R}^n)$. In this section we will do something analogous on the torus.

3.6.a. Multipliers on the Torus

In analogy with the nonperiodic case, we could identify convolution operators on $\mathbf{T}^n$ with appropriate distributions on the torus. See Exercise 3.6.2 for an introduction to distributions on the torus. However, as we will see, it is simpler to avoid this point of view and consider the multipliers directly. This is because of the following theorem.

Theorem 3.6.1. *Suppose that T is a linear operator that commutes with translations and maps $L^p(\mathbf{T}^n)$ into $L^q(\mathbf{T}^n)$ for some $1 \le p, q \le \infty$. Then there exists a bounded sequence $\{a_m\}_{m\in\mathbf{Z}^n}$ such that*

$$T(f)(x) = \sum_{m\in\mathbf{Z}^n} a_m \widehat{f}(m) e^{2\pi i m\cdot x} \tag{3.6.1}$$

for all $f \in C^\infty(\mathbf{T}^n)$. Moreover, we have $\big\|\{a_m\}\big\|_{\ell^\infty} \le \big\|T\big\|_{L^p \to L^q}$.

Proof. Consider the functions $e_m(x) = e^{2\pi i m\cdot x}$ defined on $\mathbf{T}^n$ for m in $\mathbf{Z}^n$. Since T is translation invariant for all $h \in \mathbf{T}^n$, we have

$$T(e_m)(x-h) = T(\tau^h(e_m)) = e^{-2\pi i m\cdot h}T(e_m)(x) \qquad \text{a.e. in } x.$$

By Fubini's theorem we have $T(e_m)(x_0-h) = e^{-2\pi i m\cdot h}T(e_m)(x_0)$ for some $x_0 \in \mathbf{T}^n$ and almost all $h \in \mathbf{T}^n$. Replacing $x_0 - h$ by x, we obtain

$$T(e_m)(x) = e^{2\pi i m\cdot x}\big(e^{-2\pi i m\cdot x_0}T(e_m)(x_0)\big) = a_m e_m(x) \tag{3.6.2}$$

for almost all $x \in \mathbf{T}^n$, where we set $a_m = e^{-2\pi i m\cdot x_0}T(e_m)(x_0)$, for $m \in \mathbf{Z}^n$. Taking L^q norms in (3.6.2), we obtain $|a_m| = \big\|T(e_m)\big\|_{L^q} \le \big\|T\big\|_{L^p\to L^q}$ and thus a_m is bounded. Moreover, since $T(e_m) = a_m e_m$ for all m in $\mathbf{Z}^n$, it follows that (3.6.1) holds for all trigonometric polynomials. By density this extends to all $f \in C^\infty(\mathbf{T}^n)$ and the theorem is proved.

□

Definition 3.6.2. Motivated by Theorem 3.6.1, we will call a bounded sequence $\{a_m\}_{m\in\mathbf{Z}^n}$ an *(L^p, L^q) multiplier*, if the corresponding operator given by (3.6.1) maps $L^p(\mathbf{T}^n)$ into $L^q(\mathbf{T}^n)$. If $p = q$, (L^p, L^p) multipliers will be simply called L^p multipliers. The space of all L^p multipliers on $\mathbf{T}^n$ will be denoted by $\mathcal{M}_p(\mathbf{Z}^n)$. Here we follow the standard notation $\mathcal{M}_p(\widehat{G})$ for multipliers on $L^p(G)$, where G is a locally compact group and $\widehat{G}$ is its dual group. The norm of element $\{a_m\}$ in $\mathcal{M}_p(\mathbf{Z}^n)$ is the norm of the operator T given by (3.6.1) from $L^p(\mathbf{T}^n)$ into itself. This norm will be denoted by $\big\|\{a_m\}\big\|_{\mathcal{M}_p}$.

It is not difficult to see that the same basic properties of $\mathcal{M}_p(\mathbf{R}^n)$ are also true for $\mathcal{M}_p(\mathbf{Z}^n)$. In particular, $\mathcal{M}_p(\mathbf{Z}^n)$ is a closed subspace of ℓ^∞ and thus a Banach space itself. Moreover, sums, scalar multiples, and products of elements of $\mathcal{M}_p(\mathbf{Z}^n)$ are also in $\mathcal{M}_p(\mathbf{Z}^n)$, which makes this space a Banach algebra. Also, as we saw in the nonperiodic case, we have $\mathcal{M}_p(\mathbf{Z}^n) = \mathcal{M}_{p'}(\mathbf{Z}^n)$ when $1 \le p \le \infty$. We now examine some special cases. We begin with the case $p = q = 2$. As expected, it turns out that $\mathcal{M}_2(\mathbf{Z}^n) = \ell^\infty(\mathbf{Z}^n)$.

Theorem 3.6.3. *A linear operator T that commutes with translations maps $L^2(\mathbf{T}^n)$ into itself if and only if there exists a sequence $\{a_m\}_{m\in\mathbf{Z}^n}$ in ℓ^∞ such that*

$$T(f)(x) = \sum_{m\in\mathbf{Z}^n} a_m \widehat{f}(m) e^{2\pi i m\cdot x} \tag{3.6.3}$$

for all $f \in C^\infty(\mathbf{T}^n)$. Moreover, in this case we have $\big\|T\big\|_{L^2\to L^2} = \big\|\{a_m\}\big\|_{\ell^\infty}$.

Proof. The existence of such a sequence is guaranteed by Theorem 3.6.1, which also gives $\big\|\{a_m\}\big\|_{\ell^\infty} \le \big\|T\big\|_{L^2\to L^2}$. Conversely, any operator given by the form (3.6.3) satisfies

$$\big\|T(f)\big\|_{L^2}^2 = \sum_{m\in\mathbf{Z}^n} |a_m\widehat{f}(m)|^2 \le \big\|\{a_m\}\big\|_{\ell^\infty}^2 \sum_{m\in\mathbf{Z}^n} |\widehat{f}(m)|^2,$$

and thus $\|T\|_{L^2\to L^2} \le \|\{a_m\}\|_{\ell^\infty}$.

□

We continue with the case $p=q=1$. Recall the definition of a finite Borel measure on $\mathbf{T}^n$. Given such a measure μ, its Fourier coefficients are defined by

$$\widehat{\mu}(m) = \int_{\mathbf{T}^n} e^{-2\pi i x\cdot m}\,d\mu(x), \qquad m \in \mathbf{Z}^n.$$

Clearly all the Fourier coefficients of the measure μ are bounded by the total variation $\|\mu\|$ of μ. See Exercise 3.6.3 for basic properties of Fourier transforms of distributions on the torus.

Theorem 3.6.4. *A linear operator T that commutes with translations maps $L^1(\mathbf{T}^n)$ into itself if and only if there exists a finite Borel measure μ on the torus such that*

$$T(f)(x) = \sum_{m\in\mathbf{Z}^n} \widehat{\mu}(m)\widehat{f}(m)e^{2\pi i m\cdot x} \tag{3.6.4}$$

for all $f \in C^\infty(\mathbf{T}^n)$. Moreover, in this case we have $\|T\|_{L^1\to L^1} = \|\mu\|$. In other words, $\mathcal{M}_1(\mathbf{Z}^n)$ is the set of all sequences given by Fourier coefficients of finite Borel measures on $\mathbf{T}^n$.

Proof. Fix $f \in L^1(\mathbf{T}^n)$. If (3.6.4) is valid, then $\widehat{T(f)}(m) = \widehat{f}(m)\widehat{\mu}(m)$ for all $m \in \mathbf{Z}^n$. But Exercise 3.6.3 gives that $\widehat{f*\mu}(m) = \widehat{f}(m)\widehat{\mu}(m)$ for all $m \in \mathbf{Z}^n$; therefore, the integrable functions $f*\mu$ and $T(f)$ have the same Fourier coefficients and they must be equal. Thus $T(f) = f*\mu$, which implies that T is bounded on L^1 and $\|T(f)\|_{L^1} \le \|\mu\|\|f\|_{L^1}$.

To prove the converse direction, we suppose that T commutes with translations and maps $L^1(\mathbf{T}^n)$ into itself. We recall the following identity obtained in (3.1.20):

$$P_\varepsilon(x) = \sum_{m\in\mathbf{Z}^n} e^{-2\pi|m|\varepsilon}e^{2\pi i m\cdot x} = \frac{\Gamma(\frac{n+1}{2})}{\pi^{\frac{n+1}{2}}} \sum_{m\in\mathbf{Z}^n} \frac{\varepsilon^{-n}}{(1+|\frac{x+m}{\varepsilon}|^2)^{\frac{n+1}{2}}} \ge 0 \tag{3.6.5}$$

for all $x \in \mathbf{T}^n$. We can integrate the series term by term in (3.6.5) to conclude that $\|P_\varepsilon\|_{L^1(\mathbf{T}^n)} = 1$. It follows that $\|T(P_\varepsilon)\|_{L^1(\mathbf{T}^n)} \le \|T\|_{L^1\to L^1}$ for all $\varepsilon > 0$. The Banach-Alaoglu theorem gives that there exists a sequence $\varepsilon_j \downarrow 0$ and a finite Borel measure μ on $\mathbf{T}^n$ such that $T(P_{\varepsilon_j})$ tends to μ weakly as $j \to \infty$. This means that for all g continuous functions on $\mathbf{T}^n$ we have

$$\lim_{j\to\infty} \int_{\mathbf{T}^n} g(x)T(P_{\varepsilon_j})(x)\,dx = \int_{\mathbf{T}^n} g(x)\,d\mu(x). \tag{3.6.6}$$

It follows from (3.6.6) that for all g continuous on $\mathbf{T}^n$ we have

$$\left|\int_{\mathbf{T}^n} g(x)\,d\mu(x)\right| \le \sup_j \|T(P_{\varepsilon_j})\|_{L^1}\|g\|_{L^\infty} \le \|T\|_{L^1\to L^1}\|g\|_{L^\infty}.$$

Since by the Riesz representation theorem we have that the norm of the linear functional $g \to \int_{\mathbf{T}^n} g(x)\, d\mu(x)$ on $C(\mathbf{T}^n)$ is $\|\mu\|$, it follows that

$$\|\mu\| \le \|T\|_{L^1 \to L^1}. \tag{3.6.7}$$

It remains to prove that T has the form given in (3.6.4). By Theorem 3.6.1 we have that there exists a bounded sequence $\{a_m\}$ on $\mathbf{Z}^n$ such that (3.6.1) is satisfied. Taking $g(x) = e^{-2\pi i k \cdot x}$ in (3.6.6) and using the representation for T in (3.6.1), we obtain

$$\widehat{\mu}(k) = \int_{\mathbf{T}^n} e^{-2\pi i k\cdot x}\, d\mu(x) = \lim_{j\to\infty} \int_{\mathbf{T}^n} e^{-2\pi i k\cdot x} \sum_{m\in\mathbf{Z}^n} a_m e^{-2\pi\varepsilon_j |m|} e^{2\pi i m\cdot x}\, dx = a_k\,.$$

This proves assertion (3.6.4). It follows from (3.6.4) that $T(f) = f * \mu$ and thus $\|T\|_{L^1\to L^1} \le \|\mu\|$. This fact combined with (3.6.7) gives $\|T\|_{L^1\to L^1} = \|\mu\|$.

□

The case $p = \infty$ follows immediately from the case $p = 1$ by duality. The details of the proof of the next corollary are left to the reader.

Corollary 3.6.5. *A linear operator T that commutes with translations maps $L^\infty(\mathbf{T}^n)$ into itself if and only if there exists a finite Borel measure μ on the torus such that*

$$T(f)(x) = \sum_{m\in\mathbf{Z}^n} \widehat{\mu}(m)\widehat{f}(m) e^{2\pi i m\cdot x}$$

for all $f \in C^\infty(\mathbf{T}^n)$. Moreover, in this case we have $\|T\|_{L^\infty\to L^\infty} = \|\mu\|$.

3.6.b. Transference of Multipliers

It is clear by now that multipliers on $L^1(\mathbf{T}^n)$ and $L^1(\mathbf{R}^n)$ are very similar, and the same is true for $L^2(\mathbf{T}^n)$ and $L^2(\mathbf{R}^n)$. These similarities became obvious when we characterized L^1 and L^2 multipliers on both $\mathbf{R}^n$ and $\mathbf{T}^n$. So far, there is no known nontrivial characterization of $\mathcal{M}_p(\mathbf{R}^n)$, but we might ask whether this space is related to $\mathcal{M}_p(\mathbf{Z}^n)$. We will see that there are several connections of this type and that there are general ways to produce multipliers on the torus from multipliers on $\mathbf{R}^n$ and vice versa. General methods of this sort are called transference of multipliers.

We begin with a useful definition.

Definition 3.6.6. A bounded function b on $\mathbf{R}^n$ is called *regulated* iff

$$\lim_{\varepsilon\to 0} \frac{1}{\varepsilon^n} \int_{|t|\le\varepsilon} \big(b(m-t) - b(m)\big)\, dt = 0\,, \text{for all } m \in \mathbf{Z}^n. \tag{3.6.8}$$

Condition (3.6.8) is saying that every point in $\mathbf{Z}^n$ is a Lebesgue point of b. This condition in particular holds if the function b is continuous at every point $m \in \mathbf{Z}^n$. Another instance when condition (3.6.8) holds is when $b(m-t) = -b(m+t)$ in a small neighborhood of m. When $m = 0$, this happens if b is odd near the origin. The second conclusion of the theorem is a consequence of the first conclusion ($R = 1$), since the functions $b(\xi/R)$ and $b(\xi)$ have the same norm in $\mathcal{M}_p(\mathbf{R}^n)$.

Our first transference result is the following.

Theorem 3.6.7. *Suppose that b is a regulated function that lies in $\mathcal{M}_p(\mathbf{R}^n)$ for some $1 \le p \le \infty$. Then the sequence $\{b(m)\}_{m\in\mathbf{Z}^n}$ is in $\mathcal{M}_p(\mathbf{Z}^n)$ and moreover $\big\|\{b(m)\}\big\|_{\mathcal{M}_p(\mathbf{Z}^n)} \le \big\|b\big\|_{\mathcal{M}_p(\mathbf{R}^n)}$. Also, for all $R > 0$, the sequences $\{b(m/R)\}_{m\in\mathbf{Z}^n}$ are in $\mathcal{M}_p(\mathbf{Z}^n)$ and we have*

$$\sup_{R>0} \big\|\{b(m/R)\}\big\|_{\mathcal{M}_p(\mathbf{Z}^n)} \le \big\|b\big\|_{\mathcal{M}_p(\mathbf{R}^n)}\,.$$

Before we begin the proof, we will need the following lemma.

Lemma 3.6.8. *Let T be the operator on $\mathbf{R}^n$ whose multiplier is $b(\xi)$, and let S be the operator on $\mathbf{T}^n$ whose multiplier is the sequence $\{b(m)\}_{m\in\mathbf{Z}^n}$. Assume that $b(\xi)$ satisfies the regularity condition (3.6.8) at every point $m \in \mathbf{Z}^n$. Suppose that P and Q are trigonometric polynomials on $\mathbf{T}^n$ and let $L_\varepsilon(x) = e^{-\pi\varepsilon|x|^2}$ for $x \in \mathbf{R}^n$ and $\varepsilon > 0$. Then the following identity is valid whenever $\alpha, \beta > 0$ and $\alpha + \beta = 1$:*

$$\lim_{\varepsilon\to 0} \varepsilon^{\frac{n}{2}} \int_{\mathbf{R}^n} T(PL_{\varepsilon\alpha})(x)\overline{Q(x)G_{\varepsilon\beta}(x)}\,dx = \int_{\mathbf{T}^n} S(P)(x)\overline{Q(x)}\,dx. \tag{3.6.9}$$

Proof. It suffices to prove the required assertion for $P(x) = e^{2\pi i m\cdot x}$ and $Q(x) = e^{2\pi i k\cdot x}$, $k, m \in \mathbf{Z}^n$, since the general case will easily follow from this case by linearity. In view of Parseval's identity [Proposition 3.1.16 (3)], we have

$$\begin{aligned} \int_{\mathbf{T}^n} S(P)(x)\overline{Q(x)}\,dx &= \sum_{r\in\mathbf{Z}^n} b(r)\widehat{P}(r)\overline{\widehat{Q}(r)} \\ &= \begin{cases} b(m) & \text{when } k = m, \\ 0 & \text{when } k \neq m. \end{cases} \end{aligned} \tag{3.6.10}$$

On the other hand, using the identity in Theorem 2.2.14 (3), we obtain

$$\begin{aligned} &\varepsilon^{\frac{n}{2}} \int_{\mathbf{R}^n} T(PL_{\varepsilon\alpha})(x)\overline{Q(x)L_{\varepsilon\beta}(x)}\,dx \\ &= \varepsilon^{\frac{n}{2}} \int_{\mathbf{R}^n} b(\xi)\widehat{PL_{\varepsilon\alpha}}(\xi)\overline{\widehat{QL_{\varepsilon\beta}}(\xi)}\,d\xi \\ &= \varepsilon^{\frac{n}{2}} \int_{\mathbf{R}^n} b(\xi)(\varepsilon\alpha)^{-\frac{n}{2}} e^{-\pi\frac{|\xi-m|^2}{\varepsilon\alpha}} (\varepsilon\beta)^{-\frac{n}{2}} e^{-\pi\frac{|\xi-k|^2}{\varepsilon\beta}}\,d\xi \\ &= (\varepsilon\alpha\beta)^{-\frac{n}{2}} \int_{\mathbf{R}^n} b(\xi) e^{-\pi\frac{|\xi-m|^2}{\varepsilon\alpha}} e^{-\pi\frac{|\xi-k|^2}{\varepsilon\beta}}\,d\xi\,. \end{aligned} \tag{3.6.11}$$

Now if $m = k$, since $\alpha + \beta = 1$, the expression in (3.6.11) is equal to

$$(\varepsilon\alpha\beta)^{-\frac{n}{2}} \int_{\mathbf{R}^n} b(\xi) e^{-\pi \frac{|\xi - m|^2}{\varepsilon\alpha\beta}} \, d\xi \,, \tag{3.6.12}$$

which tends to $b(m)$ if b is continuous at m, since the family $\varepsilon^{-\frac{n}{2}} e^{-\pi \frac{|\xi|^2}{\varepsilon}}$ is an approximate identity as $\varepsilon \to 0$. If now b is not continuous at m but satisfies the condition (3.6.8), then still the expression in (3.6.12) tends to $b(m)$ as $\varepsilon \to 0$. See Exercise 3.6.6 for details.

We now consider the case $m \neq k$ in (3.6.11). If $|m - k| \geq 1$, then every ξ in $\mathbf{R}^n$ must satisfy either $|\xi - m| \geq 1/2$ or $|\xi - k| \geq 1/2$. Therefore, the expression in (3.6.11) is controlled by

$$(\varepsilon\alpha\beta)^{-\frac{n}{2}} \bigg(\int_{|\xi - m| \geq \frac{1}{2}} b(\xi) e^{-\frac{\pi}{4\varepsilon\alpha}} e^{-\pi \frac{|\xi - k|^2}{\varepsilon\beta}} \, d\xi + \int_{|\xi - k| \geq \frac{1}{2}} b(\xi) e^{-\frac{\pi}{4\varepsilon\beta}} e^{-\pi \frac{|\xi - m|^2}{\varepsilon\alpha}} \, d\xi \bigg),$$

which is in turn controlled by

$$\big\| b \big\|_{L^\infty} \big(\alpha^{-\frac{n}{2}} e^{-\frac{\pi}{4\varepsilon\alpha}} + \beta^{-\frac{n}{2}} e^{-\frac{\pi}{4\varepsilon\beta}} \big) \,,$$

which tends to zero as $\varepsilon \to 0$. This proves that the expression in (3.6.10) is equal to the limit of the expression in (3.6.11) as $\varepsilon \to 0$.

□

Having proved Lemma 3.6.8, we turn to the proof of Theorem 3.6.7.

Proof. The case $p = 1$ can be proved easily using Theorems 2.5.8, 3.6.4, and Exercise 3.6.4 and is left to the reader. The case $p = \infty$ can be reduced to the case $p = 1$ by duality. Let us therefore consider the case $1 < p < \infty$. We are assuming that T maps $L^p(\mathbf{R}^n)$ into itself and we need to show that S maps $L^p(\mathbf{T}^n)$ into itself. We will prove this using duality. For P and Q trigonometric polynomials we have

$$\begin{aligned}
&\bigg| \int_{\mathbf{T}^n} S(P)(x) \overline{Q(x)} \, dx \bigg| \\
&= \bigg| \lim_{\varepsilon \to 0} \varepsilon^{\frac{n}{2}} \int_{\mathbf{R}^n} T(PL_{\varepsilon/p})(x) \overline{Q(x) L_{\varepsilon/p'}(x)} \, dx \bigg| \\
&\leq \big\| T \big\|_{L^p \to L^p} \limsup_{\varepsilon \to 0} \varepsilon^{\frac{n}{2}} \big\| PL_{\varepsilon/p} \big\|_{L^p(\mathbf{R}^n)} \big\| QL_{\varepsilon/p'} \big\|_{L^{p'}(\mathbf{R}^n)} \\
&= \big\| T \big\|_{L^p \to L^p} \limsup_{\varepsilon \to 0} \bigg(\varepsilon^{\frac{n}{2}} \int_{\mathbf{R}^n} |P(x)|^p e^{-\varepsilon\pi|x|^2} \, dx \bigg)^{\frac{1}{p}} \bigg(\varepsilon^{\frac{n}{2}} \int_{\mathbf{R}^n} |Q(x)|^{p'} e^{-\varepsilon\pi|x|^2} \, dx \bigg)^{\frac{1}{p'}} \\
&= \big\| T \big\|_{L^p \to L^p} \bigg(\int_{\mathbf{T}^n} |P(x)|^p \, dx \bigg)^{\frac{1}{p}} \bigg(\int_{\mathbf{T}^n} |Q(x)|^{p'} \, dx \bigg)^{\frac{1}{p'}},
\end{aligned}$$

provided for all continuous functions g on $\mathbf{T}^n$ we have that

$$\lim_{\varepsilon \to 0} \varepsilon^{\frac{n}{2}} \int_{\mathbf{R}^n} g(x) e^{-\varepsilon\pi|x|^2} \, dx = \int_{\mathbf{T}^n} g(x) \, dx. \tag{3.6.13}$$

Assuming (3.6.13) for the moment, we take the supremum over all Q trigonometric polynomials on $\mathbf{T}^n$ with $L^{p'}$ norm at most 1 to obtain that S maps $L^p(\mathbf{T}^n)$ into itself with norm at most $\|T\|_{L^p\to L^p}$, proving Theorem 3.6.7.

We now prove (3.6.13). Use the Poisson summation formula to write the left-hand side of (3.6.13) as

$$\begin{aligned}
\varepsilon^{\frac{n}{2}} \sum_{k\in\mathbf{Z}^n} \int_{\mathbf{T}^n} g(x-k)e^{-\varepsilon\pi|x|^2}\,dx &= \int_{\mathbf{T}^n} g(x)\varepsilon^{\frac{n}{2}} \sum_{k\in\mathbf{Z}^n} e^{-\varepsilon\pi|x-k|^2}\,dx \\
= \int_{\mathbf{T}^n} g(x) \sum_{k\in\mathbf{Z}^n} e^{-\pi|k|^2/\varepsilon} e^{2\pi i x\cdot k}\,dx &= \int_{\mathbf{T}^n} g(x)\,dx + A_\varepsilon,
\end{aligned}$$

where $|A_\varepsilon| \le \|g\|_{L^\infty} \sum_{|k|\ge 1} e^{-\pi|k|^2/\varepsilon} \to 0$ as $\varepsilon\to 0$. This completes the proof. □

We will now obtain a converse of Theorem 3.6.7. If $b(\xi)$ is a bounded function on $\mathbf{R}^n$ and the sequence $\{b(m)\}_{m\in\mathbf{Z}^n}$ is in $\mathcal{M}_p(\mathbf{Z}^n)$, then we cannot necessarily obtain that b is in $\mathcal{M}_p(\mathbf{R}^n)$ since such a conclusion would have to depend on all the values of b and not on the values of b on a set of measure zero. However, a converse can be formulated if we assume that for all $R>0$, the sequences $\{b(Rm)\}_{m\in\mathbf{Z}^n}$ are in $\mathcal{M}_p(\mathbf{Z}^n)$ uniformly in R. Then we obtain that $b(R\xi)$ is in $\mathcal{M}_p(\mathbf{R}^n)$ uniformly in $R>0$, which is equivalent to saying that $b\in\mathcal{M}_p(\mathbf{R}^n)$ since dilations of multipliers on $\mathbf{R}^n$ do not affect their norms, see Proposition 2.5.14. These remarks can be formulated precisely into the following theorem.

Theorem 3.6.9. *Suppose that $b(\xi)$ is a bounded function* defined for all $\xi\in\mathbf{R}^n$ *that is regulated in sense of (3.6.8). Suppose that the sequences $\{b(\frac{m}{R})\}_{m\in\mathbf{Z}^n}$ are in $\mathcal{M}_p(\mathbf{Z}^n)$ uniformly in $R>0$ for some $1\le p\le\infty$. Then b is in $\mathcal{M}_p(\mathbf{R}^n)$ and we have*

$$\|b\|_{\mathcal{M}_p(\mathbf{R}^n)} \le \sup_{R>0} \|\{b(\tfrac{m}{R})\}_{m\in\mathbf{Z}^n}\|_{\mathcal{M}_p(\mathbf{Z}^n)}. \tag{3.6.14}$$

Proof. Let us first assume that b is continuous. We will be able to dispose of this assumption later. Suppose that f and g are smooth functions with compact support on $\mathbf{R}^n$. Then for $R\ge R_0$ we have that the functions $F_R(x)=f(Rx)$ and $G_R(x)=g(Rx)$ are supported in $[-1/2,1/2]^n$ and they can be viewed as functions on $\mathbf{T}^n$ once they are periodized. Observe that the mth Fourier coefficient of F_R is $R^{-n}\widehat{f}(m/R)$ and of G_R is $R^{-n}\widehat{g}(m/R)$. Set $C_b=\sup_{R>0}\|\{b(m/R)\}_{m\in\mathbf{Z}^n}\|_{\mathcal{M}_p(\mathbf{Z}^n)}$.

Now for $R \geq R_0$ we have

$$\left| \sum_{m \in \mathbf{Z}^n} b(m/R) \widehat{f}(m/R) \overline{\widehat{g}(m/R)} \,\text{Volume}\left([\tfrac{m}{R}, \tfrac{m+1}{R}]^n \right) \right| \tag{3.6.15}$$

$$= \left| R^n \sum_{m \in \mathbf{Z}^n} b(m/R) \widehat{F_R}(m) \overline{\widehat{G_R}(m)} \right|$$

$$= \left| R^n \int_{\mathbf{T}^n} \left(\sum_{m \in \mathbf{Z}^n} b(m/R) \widehat{F_R}(m) e^{2\pi i m \cdot x} \right) \overline{G_R(x)} \, dx \right|$$

$$\leq R^n \big\| \{ b(m/R) \}_m \big\|_{\mathcal{M}_p(\mathbf{Z}^n)} \big\| F_R \big\|_{L^p(\mathbf{T}^n)} \big\| G_R \big\|_{L^{p'}(\mathbf{T}^n)}$$

$$\leq C_b \, R^n \big\| F_R \big\|_{L^p(\mathbf{R}^n)} \big\| G_R \big\|_{L^{p'}(\mathbf{R}^n)}$$

$$= C_b \, \big\| f \big\|_{L^p(\mathbf{R}^n)} \big\| g \big\|_{L^{p'}(\mathbf{R}^n)} . \tag{3.6.16}$$

Observe that the expressions in (3.6.15) tend to

$$\left| \int_{\mathbf{R}^n} b(\xi) \widehat{f}(\xi) \overline{\widehat{g}(\xi)} \, d\xi \right|$$

as $R \to \infty$ by the definition of the Riemann integral since b, $\widehat{f}$, and $\widehat{g}$ are all continuous functions. We conclude that

$$\int_{\mathbf{R}^n} b(\xi) \widehat{f}(\xi) \overline{\widehat{g}(\xi)} \, d\xi = \int_{\mathbf{R}^n} (b \widehat{f})^{\vee}(x) \overline{g(x)} \, dx$$

is controlled in absolute value by the expression in (3.6.16). This implies the conclusion of the theorem if b is continuous.

We now dispose of the assumption that b is continuous. Consider the approximate identity $K_\varepsilon(x) = \varepsilon^{-n} e^{-\pi |x/\varepsilon|^2}$ on $\mathbf{R}^n$ as $\varepsilon \to 0$. Then the functions $K_\varepsilon * b$ are continuous, bounded by $\big\| b \big\|_{L^\infty}$, and $(K_\varepsilon * b)(\xi) \to b(\xi)$ for all $\xi \in \mathbf{R}^n$ as $\varepsilon \to 0$. The last fact follows from Exercise 3.6.6 and by assumption (3.6.8). If we had that

$$\sup_{\varepsilon, R > 0} \big\| \{ (K_\varepsilon * b)(\tfrac{m}{R}) \}_{m \in \mathbf{Z}^n} \big\|_{\mathcal{M}_p(\mathbf{Z}^n)} \leq C_b , \tag{3.6.17}$$

then inequality (3.6.14), applied to the continuous function $K_\varepsilon * b$, would give that $\big\| K_\varepsilon * b \big\|_{\mathcal{M}_p(\mathbf{R}^n)} \leq C_b$. We would then have a sequence of multipliers with uniformly bounded norms in $\mathcal{M}_p(\mathbf{R}^n)$ that converges pointwise to b. Applying Exercise 2.5.6(a) we would obtain that b is in $\mathcal{M}_p(\mathbf{R}^n)$ and satisfies (3.6.14). It suffices therefore to prove the crucial inequality (3.6.17). The Poisson summation formula gives that

$$U_\varepsilon(y) = \frac{1}{\varepsilon^n} \sum_{k \in \mathbf{Z}^n} e^{-\pi \frac{|y-k|^2}{\varepsilon^2}} = \sum_{k \in \mathbf{Z}^n} e^{-\pi \varepsilon^2 |k|^2} e^{2\pi i k \cdot y} ,$$

which implies that U_ε is a periodic function with $L^1(\mathbf{T}^n)$ norm equal to 1. Since $\widehat{K_\varepsilon * b}(\frac{m}{R}) = e^{-\pi\varepsilon^2|m|^2/R^2} b(\frac{m}{R})$, to prove (3.6.17), it suffices to show that

$$\Big\| \sum_{m \in \mathbf{Z}^n} \widehat{f}(m) e^{-\pi\varepsilon^2|m|^2/R^2} b(m/R) e^{2\pi i m\cdot x} \Big\|_{L^p(\mathbf{T}^n)} \le C_b \big\|f\big\|_{L^p(\mathbf{T}^n)} . \tag{3.6.18}$$

If $f \in L^p(\mathbf{T}^n)$, then the Fourier coefficients of the periodic function $f * U_{\varepsilon/R}$ are $\widehat{f}(m)e^{-\pi\varepsilon^2|m|^2/R^2}$. Thus (3.6.18) immediately follows from

$$\Big\| \sum_{m \in \mathbf{Z}^n} \widehat{f * U_{\frac{\varepsilon}{R}}}(m) b(m/R) e^{2\pi i m\cdot x} \Big\|_{L^p(\mathbf{T}^n)} \le C_b \big\|f * U_{\frac{\varepsilon}{R}}\big\|_{L^p(\mathbf{T}^n)} \le C_b \big\|f\big\|_{L^p(\mathbf{T}^n)} ,$$

and this proves (3.6.17) and the theorem.

□

3.6.c. Applications of Transference

Having established our two main transference theorems, we turn to an application.

Corollary 3.6.10. *Let $1 \le p \le \infty$ and $f \in L^p(\mathbf{T}^n)$ and $\alpha \ge 0$. Then*
(a) $\big\|D(n,R) * f - f\big\|_{L^p(\mathbf{T}^n)} \to 0$ as $R \to \infty$ if and only if $\chi_{[-1,1]^n} \in \mathcal{M}_p(\mathbf{R}^n)$
(b) $\big\|\widetilde{D}(n,R) * f - f\big\|_{L^p(\mathbf{T}^n)} \to 0$ as $R \to \infty$ if and only if $\chi_{B(0,1)} \subset \mathcal{M}_p(\mathbf{R}^n)$
(c) $\big\|B_R^\alpha(f) - f\big\|_{L^p(\mathbf{T}^n)} \to 0$ as $R \to \infty$ if and only if $(1-|\xi|^2)_+^\alpha \in \mathcal{M}_p(\mathbf{R}^n)$

Proof. First observe that in view of Corollary 3.5.2, the first statements in (a), (b), and (c) are equivalent to the statements

$$\sup_{R>0} \big\|D(n,R) * f\big\|_{L^p(\mathbf{T}^n)} \le C_p \big\|f\big\|_{L^p(\mathbf{T}^n)},$$

$$\sup_{R>0} \big\|\widetilde{D}(n,R) * f\big\|_{L^p(\mathbf{T}^n)} \le C_p \big\|f\big\|_{L^p(\mathbf{T}^n)},$$

$$\sup_{R>0} \big\|B_R^\alpha(f)\big\|_{L^p(\mathbf{T}^n)} \le C_p \big\|f\big\|_{L^p(\mathbf{T}^n)},$$

for some constant $0 < C_p < \infty$ and all f in $L^p(\mathbf{T}^n)$. Now define

$$\widetilde{\chi}_{B(0,1)}(x) = \begin{cases} 1 & \text{when } |x| < 1, \\ 1/2 & \text{when } |x| = 1, \\ 0 & \text{when } |x| > 1, \end{cases}$$

and

$$\widetilde{\chi}_{[-1,1]^n}(x_1,\dots,x_n) = \begin{cases} 1 & \text{when all } |x_j| < 1, \\ 1/2 & \text{when some but not all } |x_j| = 1, \\ 1/2^n & \text{when all } |x_j| = 1, \\ 0 & \text{when some } |x_j| > 1. \end{cases}$$

It is not difficult to see that the functions $\widetilde{\chi}_{B(0,1)}$ and $\widetilde{\chi}_{[-1,1]^n}$ are regulated; see Exercise 3.6.7. The function $(1-|\xi|^2)_+^\alpha$ is continuous and therefore regulated. Theorems 3.6.7 and 3.6.9 imply that the uniform (in $R > 0$) boundedness of the operators $D(n,R)$, $\widetilde{D}(n,R)$, and B_R^α on $L^p(\mathbf{T}^n)$ is equivalent to the statements that the functions $\widetilde{\chi}_{B(0,1)}$, $\widetilde{\chi}_{[-1,1]^n}$, and $(1-|\xi|^2)_+^\alpha$ are in $\mathcal{M}_p(\mathbf{T}^n)$, respectively. Since $\widetilde{\chi}_{[-1,1]^n} = \chi_{[-1,1]^n}$ a.e. and $\widetilde{\chi}_{B(0,1)} = \chi_{B(0,1)}$ a.e., the required conclusion follows. □

3.6.d. Transference of Maximal Multipliers

We will now prove a theorem concerning maximal multipliers analogous to Theorems 3.6.7 and 3.6.9. This will enable us to reduce problems related to almost everywhere convergence of Fourier series on the torus to problems of boundedness of maximal operators on $\mathbf{R}^n$.

Let b be a bounded regulated function function that is defined on all of $\mathbf{R}^n$ and that lies in $\mathcal{M}_p(\mathbf{R}^n)$ for some $1 < p < \infty$. For $R > 0$, we introduce multiplier operators

$$S_{b,R}(F)(x) = \sum_{m\in\mathbf{Z}^n} \widehat{F}(m)b(m/R)e^{2\pi i m\cdot x} \tag{3.6.19}$$

$$T_{b,R}(f)(x) = \int_{\mathbf{R}^n} \widehat{f}(\xi)b(\xi/R)e^{2\pi i\xi\cdot x}\,d\xi \tag{3.6.20}$$

initially defined for smooth functions with compact support f on $\mathbf{R}^n$ and smooth functions F on $\mathbf{T}^n$.

In view of Theorems 3.6.7 and 3.6.9, $S_{b,R}$ admits a bounded extension on $L^p(\mathbf{T}^n)$ and $T_{b,R}$ admits a bounded extension on $L^p(\mathbf{R}^n)$. These extensions will be denoted in the same way. We introduce maximal operators

$$M_b(F)(x) = \sup_{R>0} \big|S_{b,R}(F)(x)\big| \tag{3.6.21}$$

$$N_b(f)(x) = \sup_{R>0} \big|T_{b,R}(f)(x)\big| \tag{3.6.22}$$

and we have the following result concerning them:

Theorem 3.6.11. *Suppose that $b(\xi)$ is a bounded function* defined for all $\xi \in \mathbf{R}^n$ *that is regulated in the sense of (3.6.8). Let $1 < p < \infty$, and suppose that b*

lies in $\mathcal{M}_p(\mathbf{R}^n)$. *Let* M_b *and* N_b *be as in (3.6.21) and (3.6.22). Then the following assertions are equivalent for some finite constant* C_p:

$$\big\|M_b(F)\big\|_{L^p(\mathbf{T}^n)} \le C_p\big\|F\big\|_{L^p(\mathbf{T}^n)}, \qquad F \in L^p(\mathbf{T}^n), \tag{3.6.23}$$

$$\big\|N_b(f)\big\|_{L^p(\mathbf{R}^n)} \le C_p\big\|f\big\|_{L^p(\mathbf{R}^n)}, \qquad f \in L^p(\mathbf{R}^n). \tag{3.6.24}$$

Proof. Using Exercise 3.6.9, it will suffice to prove the required equivalences for the maximal operators

$$\begin{aligned} M_b^{\mathcal{F}}(F)(x) &= \sup_{t_1<\cdots<t_k} \big|S_{b,t_j}(F)(x)\big| \\ N_b^{\mathcal{F}}(f)(x) &= \sup_{t_1<\cdots<t_k} \big|T_{b,t_j}(f)(x)\big|, \end{aligned}$$

uniformly in the choice of the finite subset

$$\mathcal{F} = \{t_1, \dots, t_k\}$$

of $\mathbf{R}^+$. Then $M_b^{\mathcal{F}}$ may be viewed as an operator defined on $L^p(\mathbf{T}^n)$ and taking values in $L^p(\mathbf{T}^n, l^\infty(\mathcal{F}))$ and $N_b^{\mathcal{F}}$ defined on $L^p(\mathbf{R}^n)$ with values in $L^p(\mathbf{R}^n, l^\infty(\mathcal{F}))$. Using this reduction and duality, inequalities (3.6.23) and (3.6.24) are equivalent to the pair of inequalities

$$\Big\|\sum_{j=1}^k \sum_{m\in\mathbf{Z}^n} \widehat{F_j}(m) b(m/t_j) e^{2\pi i m\cdot x}\Big\|_{L^{p'}(\mathbf{T}^n)} \le C_p \Big\|\sum_{j=1}^k |F_j|\Big\|_{L^{p'}(\mathbf{T}^n)} \tag{3.6.25}$$

$$\Big\|\sum_{j=1}^k \int_{\mathbf{R}^n} \widehat{f_j}(\xi) b(\xi/t_j) e^{2\pi i \xi\cdot x}\Big\|_{L^{p'}(\mathbf{R}^n)} \le C_p \Big\|\sum_{j=1}^k |f_j|\Big\|_{L^{p'}(\mathbf{R}^n)}, \tag{3.6.26}$$

where f_j are functions on $\mathbf{R}^n$ and F_j are functions on $\mathbf{T}^n$. In proving the equivalence of (3.6.25) and (3.6.26), by density, we may work with smooth functions with compact support f_j and trigonometric polynomials F_j. Suppose that (3.6.25) holds and let $f_1, \dots, f_k, g$ be smooth functions with compact support on $\mathbf{R}^n$. Then for $R \ge R_0$ we have that the functions $F_{j,R}(x) = f_j(Rx)$ and $G_R(x) = g(Rx)$ are supported in $[-1/2, 1/2]^n$ and they can be viewed as functions on $\mathbf{T}^n$ once they are periodized. As before, the mth Fourier coefficient of $F_{j,R}$ is $R^{-n}\widehat{f_j}(m/R)$ and of G_R is $R^{-n}\widehat{g}(m/R)$.

As in the proof of Theorem 3.6.9, for $R \geq R_0$ we have

$$
\begin{aligned}
(3.6.27)\qquad & \left|\sum_{j=1}^{k}\sum_{m\in\mathbf{Z}^n} b(m/Rt_j)\widehat{f_j}(m/R)\overline{\widehat{g}(m/R)}\,\text{Volume}\left([\tfrac{m}{R},\tfrac{m+1}{R}]^n\right)\right| \\
&= \left|R^n\int_{\mathbf{T}^n}\left(\sum_{j=1}^{k}\sum_{m\in\mathbf{Z}^n} b(m/R)\widehat{F_{j,R}}(m)e^{2\pi i m\cdot x}\right)\overline{G_R(x)}\,dx\right| \\
&\leq C_p R^n \Big\|\sum_{j=1}^{k}|F_{j,R}|\Big\|_{L^{p'}(\mathbf{T}^n)}\big\|G_R\big\|_{L^p(\mathbf{T}^n)} \\
&\leq C_p\, R^n \Big\|\sum_{j=1}^{k}|F_{j,R}|\Big\|_{L^{p'}(\mathbf{R}^n)}\big\|G_R\big\|_{L^p(\mathbf{R}^n)} \\
&= C_p \Big\|\sum_{j=1}^{k}|f_j|\Big\|_{L^{p'}(\mathbf{R}^n)}\big\|g\big\|_{L^p(\mathbf{R}^n)}.
\end{aligned}
$$

Set $\delta^{t_j^{-1}}(b)(\xi) = b(\xi/t_j)$. Letting $R\to\infty$ in (3.6.27), we obtain

$$
\left|\int_{\mathbf{R}^n}\sum_{j=1}^{k}(\delta^{t_j^{-1}}(b)\,\widehat{f_j})^{\vee}(x)\overline{g(x)}\,dx\right| \leq C_p\Big\|\sum_{j=1}^{k}|f_j|\Big\|_{L^{p'}(\mathbf{R}^n)}\big\|g\big\|_{L^p(\mathbf{R}^n)}
$$

and taking the supremum over all smooth functions with compact support g whose L^p norm is at most 1, we obtain (3.6.26).

We now turn to the converse. Assume that (3.6.26) holds. Let $P_1,\dots,P_k$ and Q be trigonometric polynomials on $\mathbf{T}^n$. Then with $L_\varepsilon(x) = e^{-\pi\varepsilon|x|^2}$ we have

$$
\begin{aligned}
& \left|\int_{\mathbf{T}^n}\left(\sum_{j=1}^{k}\sum_{m\in\mathbf{Z}^n}\widehat{P_j}(m)b(m/Rt_j)e^{2\pi i m\cdot x}\right)\overline{Q(x)}\,dx\right| \\
&= \left|\lim_{\varepsilon\to 0}\varepsilon^{\frac{n}{2}}\int_{\mathbf{R}^n}\left(\sum_{j=1}^{k}\int_{\mathbf{R}^n}P_j(\xi)L_{\varepsilon/p'}(\xi)b(\xi/Rt_j)e^{2\pi i\xi\cdot x}d\xi\right)\overline{Q(x)L_{\varepsilon/p}(x)}\,dx\right| \\
&\leq C_p\limsup_{\varepsilon\to 0}\varepsilon^{\frac{n}{2}}\Big\|\sum_{j=1}^{k}|P_jL_{\varepsilon/p'}|\Big\|_{L^{p'}(\mathbf{R}^n)}\big\|QL_{\varepsilon/p'}\big\|_{L^p(\mathbf{R}^n)} \\
&= C_p\limsup_{\varepsilon\to 0}\ \varepsilon^{\frac{n}{2p'}}\Big\|\sum_{j=1}^{k}|P_j|L_{\varepsilon/p'}\Big\|_{L^{p'}(\mathbf{R}^n)}\left(\varepsilon^{\frac{n}{2}}\int_{\mathbf{R}^n}|Q(x)|^pe^{-\varepsilon\pi|x|^2}dx\right)^{\frac{1}{p}} \\
&= C_p\Big\|\sum_{j=1}^{k}|P_j|\Big\|_{L^{p'}(\mathbf{R}^n)}\big\|Q\big\|_{L^p(\mathbf{T}^n)},
\end{aligned}
$$

where we used (3.6.13) in the last equality. Taking the supremum over all trigonometric polynomials Q with L^p norm 1, we obtain (3.6.25), and this completes the proof of the theorem.

□

Remark 3.6.12. Under the hypotheses of Theorem 3.6.11, the following two inequalities are also equivalent:

$$\big\|M_b(F)\big\|_{L^{p,\infty}(\mathbf{T}^n)} \le C_p\big\|F\big\|_{L^p(\mathbf{T}^n)}, \qquad F \in L^p(\mathbf{T}^n), \tag{3.6.28}$$

$$\big\|N_b(f)\big\|_{L^{p,\infty}(\mathbf{R}^n)} \le C_p\big\|f\big\|_{L^p(\mathbf{R}^n)}, \qquad f \in L^p(\mathbf{R}^n). \tag{3.6.29}$$

Indeed, Exercise 1.4.12 gives that the pair of inequalities (3.6.28) and (3.6.29) is equivalent to the pair of inequalities

$$\Big\|\sum_{j=1}^{k}\sum_{m\in\mathbf{Z}^n}\widehat{F_j}(m)b(m/t_j)e^{2\pi i m\cdot x}\Big\|_{L^{p'}(\mathbf{T}^n)} \le C_p\Big\|\sum_{j=1}^{k}|F_j|\Big\|_{L^{p',1}(\mathbf{T}^n)} \tag{3.6.30}$$

$$\Big\|\sum_{j=1}^{k}\int_{\mathbf{R}^n}\widehat{f_j}(\xi)b(\xi/t_j)e^{2\pi i \xi\cdot x}\Big\|_{L^{p'}(\mathbf{R}^n)} \le C_p\Big\|\sum_{j=1}^{k}|f_j|\Big\|_{L^{p',1}(\mathbf{R}^n)}, \tag{3.6.31}$$

where $L^{p',1}$ is the Lorentz space.

Now (3.6.31) follows from (3.6.30) in exactly the same way that (3.6.26) follows from (3.6.25). Conversely, assuming (3.6.31), in order to prove (3.6.30) it will suffice to know that

$$\lim_{\varepsilon\to 0}\varepsilon^{\frac{n}{2p'}}\Big\|\sum_{j=1}^{k}|P_j|L_{\varepsilon/p'}\Big\|_{L^{p',1}(\mathbf{R}^n)} = \Big\|\sum_{j=1}^{k}|P_j|\Big\|_{L^{p',1}(\mathbf{T}^n)}. \tag{3.6.32}$$

For this we refer to Exercise 3.6.8.

3.6.e. Transference and Almost Everywhere Convergence

Next we consider the issue of almost everywhere convergence of Fourier series of functions on $\mathbf{T}^1$. The following results are valid.

Theorem 3.6.13. *There exists a constant $C > 0$ such that for any function F in $L^2(\mathbf{T}^1)$ we have*

$$\big\|\sup_{N>0}|F * D_N|\,\big\|_{L^{2,\infty}} \le C\big\|F\big\|_{L^2}.$$

Consequently, for any $f \in L^2(\mathbf{T}^1)$, we have

$$\lim_{N\to\infty}\sum_{|m|\le N}\widehat{F}(m)e^{2\pi i m x} = F(x) \qquad \textit{for almost every } x \in [0,1].$$

This theorem can be extended to L^p functions on $\mathbf{T}^1$ for $1 < p < \infty$.

Theorem 3.6.14. *For every $1 < p < \infty$ there exists a finite constant C_p such that for all $F \in L^p(\mathbf{T}^1)$ we have*

$$\Big\| \sup_{N>0} |F * D_N| \Big\|_{L^p} \le C_p \big\|F\big\|_{L^p}.$$

Consequently, for any $f \in L^p(\mathbf{T}^1)$, we have

$$\lim_{N\to\infty} \sum_{|m|\le N} \widehat{F}(m) e^{2\pi i m x} = F(x) \qquad \text{for almost every } x \in [0,1].$$

The proofs of Theorems 3.6.13 and 3.6.14 are lengthy and involved. They will follow from results in Chapter 10 (Theorems 10.5.1 and 10.6.1, respectively). We now discuss the relationship between the aforementioned four theorems.

Consider the function defined on $\mathbf{R}$

$$b(x)(x) = \begin{cases} 1 & \text{when } |x| < 1, \\ 1/2 & \text{when } |x| = 1, \\ 0 & \text{when } |x| > 1. \end{cases} \tag{3.6.33}$$

Then b is easily seen to be regulated. Let $\{D_R\}_{R>0}$ be the family of Dirichlet kernels as defined in (3.1.13). As $D_R = D_{R+\varepsilon}$ whenever $0 < \varepsilon < 1$, for all $F \in L^p(\mathbf{T}^1)$, $1 < p < \infty$ we have

$$\sup_{R>0} \big\|F * D_R\big\|_{L^p(\mathbf{T}^1)} = \sup_{N\in\mathbf{Z}^+} \big\|F * D_N\big\|_{L^p(\mathbf{T}^1)} \le C_p \big\|F\big\|_{L^p(\mathbf{T}^1)}, \tag{3.6.34}$$

where the last estimate follows from Theorem 3.5.1, Proposition 3.5.5, and Theorem 3.5.6. (The constant C_p naturally only depends on p.)

Let $S_{b,R}$ is as in (3.6.19). For an integrable function F on $\mathbf{T}^1$ we have

$$S_{b,R}(F) = F * D_R + Q_R(F),$$

where

$$Q_R(F)(x) = \begin{cases} \frac{1}{2}\widehat{F}(R)e^{2\pi i x R} + \frac{1}{2}\widehat{F}(-R)e^{-2\pi i x R} & \text{when } R \in \mathbf{Z}^+ \\ 0 & \text{when } R \in \mathbf{R}^+ \setminus \mathbf{Z}^+. \end{cases}$$

As Q_R is bounded on $L^p(\mathbf{T}^1)$ with norm 1, using (3.6.34) we conclude that

$$\sup_{R>0} \big\|S_{b,R}(F)\big\|_{L^p(\mathbf{T}^1)} \le (C_p + 1)\big\|F\big\|_{L^p(\mathbf{T}^1)}$$

for all F in $L^p(\mathbf{T}^1)$. Appealing to Theorem 3.6.9, we deduce that the function b defined in (3.6.33) lies in $\mathcal{M}_p(\mathbf{R})$ (i.e., it is an L^p Fourier multiplier).

Next we discuss the boundedness of the corresponding maximal multipliers. If M_b is as in (3.6.21), then

$$M_b(F)(x) = \sup_{R>0} |(F * D_R)(x) + Q_R(F)(x)|,$$

whenever F is a function on $\mathbf{T}^1$ and

$$N_b(f)(x) = \sup_{R>0} \left| \int_{\mathbf{R}} \widehat{f}(\xi) b(\xi/R) e^{2\pi i x\xi} \, d\xi \right| = \sup_{R>0} \left| \int_{-R}^{+R} \widehat{f}(\xi) e^{2\pi i x\xi} \, d\xi \right|$$

for f in $L^p(\mathbf{R})$, $1 < p < \infty$. Both integrals may not be absolutely convergent for all $f \in L^p(\mathbf{R})$ but they should be interpreted as the quantity $T_{b,R}(f)(x)$, which of course coincides with them for nice f. [$T_{b,R}$ is defined in (3.6.20).]

As the sublinear operator $F \to \sup_{R>0} |Q_R(F)(x)|$ is clearly bounded on $L^p(\mathbf{T}^1)$, it follows from Theorem 3.6.11 that the boundedness of the maximal operator M_b on $L^p(\mathbf{T}^1)$ is equivalent to that of the maximal operator N_b on $L^p(\mathbf{R})$. [N_b is defined in (3.6.22) and is associated with the function b in (3.6.33).]

The maximal operator N_b is called the *Carleson operator* and will be denoted by $\mathcal{C}$. Then

$$\mathcal{C}(f)(x) = \sup_{R>0} \left| \int_{-R}^{+R} \widehat{f}(\xi) e^{2\pi i x\xi} \, d\xi \right| .$$

The boundedness of this operator on $L^p(\mathbf{R})$ will be obtained in Sections 10.5 and 10.6.

The extension of Theorem 3.6.14 to higher dimensions is a rather straightforward consequence of the one-dimensional result.

Theorem 3.6.15. *For every $1 < p < \infty$, there exists a finite constant $C_{p,n}$ such that for all $f \in L^p(\mathbf{T}^n)$ we have*

$$\left\| \sup_{N>0} |D(n,N) * f| \right\|_{L^p(\mathbf{T}^n)} \le C_{p,n} \|f\|_{L^p(\mathbf{T}^n)} \tag{3.6.35}$$

and consequently

$$\lim_{N\to\infty} \sum_{\substack{m\in\mathbf{Z}^n \\ |m_j|\le N}} \widehat{f}(m) e^{2\pi i m\cdot x} = f(x)$$

for almost every $x \in \mathbf{T}^n$.

Proof. We prove Theorem 3.6.15 when $n = 2$. Fix a p with $1 < p < \infty$. Since the Riesz projection P_+ is bounded on $L^p(\mathbf{T}^1)$, transference gives that the function $\chi_{(0,\infty)}$ is in $\mathcal{M}_p(\mathbf{R})$. It follows that the characteristic function of any half-space of the form $x_j > 0$ in $\mathbf{R}^2$ is in $\mathcal{M}_p(\mathbf{R}^2)$. (These functions have to be suitably defined on the line $x_j = 0$ to be regulated.) Since rotations of multipliers do not affect their norms, it follows that the characteristic function of the half-space $x_2 > x_1$ is in $\mathcal{M}_p(\mathbf{R}^2)$. Products of two multipliers is a multiplier; thus the characteristic function of the truncated cone $|x_1| \le |x_2| \le L$ is also in $\mathcal{M}_p(\mathbf{R}^2)$. Transference gives that the sequence $\{a_{m_1,m_2}\}_{m_1,m_2}$ defined by $a_{m_1,m_2} = 1$ when $|m_1| \le |m_2| \le L$ and zero otherwise is in $\mathcal{M}_p(\mathbf{Z}^2)$ with norm independent of $L > 0$. This means that for some constant B_p we have the following inequality for all f in $L^p(\mathbf{T}^2)$:

$$\int_{\mathbf{T}^2}\bigg|\sum_{\substack{m_2\in\mathbf{Z}\\|m_2|\le L}}\sum_{\substack{m_1\in\mathbf{Z}\\|m_1|\le|m_2|}}\widehat{f}(m_1,m_2)e^{2\pi i(m_1x_1+m_2x_2)}\bigg|^p dx_2\,dx_1 \le B_p^p\|f\|^p_{L^p(\mathbf{T}^n)}, \tag{3.6.36}$$

where the constant B_p is independent of $L>0$. Now let $1<p<\infty$ and suppose that $f\in L^p(\mathbf{T}^2)$. For fixed $x_1\in\mathbf{T}^1$ define a function f_{x_1} on $\mathbf{T}^1$ as follows:

$$f_{x_1}(x_2) = \sum_{m_2\in\mathbf{Z}\,\|m_2|\le L}\;\sum_{\substack{m_1\in\mathbf{Z}\\|m_1|\le|m_2|}}\Big[\widehat{f}(m_1,m_2)e^{2\pi i m_1x_1}\Big]e^{2\pi i m_2x_2}.$$

Then $f_{x_1}\in L^p(\mathbf{T}^1)$ and its Fourier coefficients are zero for $|m_2|>L$ and equal to

$$\widehat{f_{x_1}}(m_2) = \sum_{|m_1|\le|m_2|}\widehat{f}(m_1,m_2)e^{2\pi i m_1x_1}$$

for $|m_2|\le L$. We now have

$$\begin{aligned}
&\int_{\mathbf{T}^1}\int_{\mathbf{T}^1}\sup_{0<N\le L}\bigg|\sum_{|m_1|\le N}\sum_{|m_2|\le N}\widehat{f}(m_1,m_2)e^{2\pi i m_1x_1}e^{2\pi i m_2x_2}\bigg|^p dx_2\,dx_1\\
&\le 2\int_{\mathbf{T}^1}\int_{\mathbf{T}^1}\sup_{0<N\le L}\bigg|\sum_{|m_2|\le N}\Big[\sum_{|m_1|\le|m_2|}\widehat{f}(m_1,m_2)e^{2\pi i m_1x_1}\Big]e^{2\pi i m_2x_2}\bigg|^p dx_2\,dx_1\\
&= 2\int_{\mathbf{T}^1}\int_{\mathbf{T}^1}\sup_{0<N\le L}\big|(D_N*f_{x_1})(x_2)\big|^p dx_2\,dx_1\\
&\le 2C_p^p\int_{\mathbf{T}^1}\int_{\mathbf{T}^1}|f_{x_1}(x_2)|^p\,dx_2\,dx_1 \le 2C_p^pB_p^p\|f\|^p_{L^p(\mathbf{T}^2)},
\end{aligned}$$

where we used Theorem 3.6.14 in the penultimate inequality and estimate (3.6.36) in the last inequality. Since the last estimate we obtained is independent of $L>0$, letting $L\to\infty$ and applying Fatou's lemma, we obtain the conclusion (3.6.35) for $n=2$. The case of a general dimension $n\ge3$ presents no additional difficulties. □

Exercises

3.6.1. Let $\alpha\ge0$. Prove that the function $(1-|\xi|^2)_+^\alpha$ is in $\mathcal{M}_p(\mathbf{R}^n)$ if and only if the function $(1-|\xi|)_+^\alpha$ is in $\mathcal{M}_p(\mathbf{R}^n)$.
[*Hint:* Use that smooth functions with compact support lie in $\mathcal{M}_p$.]

3.6.2. The purpose of this exercise is to introduce distributions on the torus. The set of test functions on the torus is $C^\infty(\mathbf{T}^n)$ equipped with the following topology. Given f_j, f be in $C^\infty(\mathbf{T}^n)$, we say that $f_j\to f$ in $C^\infty(\mathbf{T}^n)$ if and only if

$$\|\partial^\alpha f_j - \partial^\alpha f\|_{L^\infty(\mathbf{T}^n)} \to 0 \quad \text{as } j \to \infty,\ \forall\ \alpha.$$

Under this notion of convergence, $C^\infty(\mathbf{T}^n)$ is a topological vector space with topology induced by the family of seminorms $\rho_\alpha(\varphi) = \sup_{x\in\mathbf{T}^n} |(\partial^\alpha f)(x)|$, where α ranges over all multiindices. The dual space of $C^\infty(\mathbf{T}^n)$ under this topology is the set of all distributions on $\mathbf{T}^n$ and is denoted by $\mathcal{D}'(\mathbf{T}^n)$. The definition implies that for u_j and u in $\mathcal{D}'(\mathbf{T}^n)$ we have $u_j \to u$ in $\mathcal{D}'(\mathbf{T}^n)$ if and only if

$$\langle u_j, f\rangle \to \langle u, f\rangle \quad \text{as } j \to \infty \text{ for all } f \in C^\infty(\mathbf{T}^n).$$

The following operations can be defined on elements of $\mathcal{D}'(\mathbf{T}^n)$: Differentiation (as in Definition 2.3.6), translation and reflection (as in Definition 2.3.11), convolution with a C^∞ function (as in Definition 2.3.13), multiplication with a C^∞ function (as in Definition 2.3.15), the support of a distribution (as in Definition 2.3.16). Use the same ideas as in $\mathbf{R}^n$ to prove the following:
(a) Prove that if $u \in \mathcal{D}'(\mathbf{T}^n)$ and $f \in C^\infty(\mathbf{T}^n)$, then $f*u$ is the C^∞ function $x \to \langle u, \tau^x(\widetilde{f})\rangle$.
(b) Unlike in $\mathbf{R}^n$, the convolution of two distributions on the torus can be defined. For $u, v \in \mathcal{D}'(\mathbf{T}^n)$ and $f \in C^\infty(\mathbf{T}^n)$ define

$$\langle u * v, f\rangle = \langle u, f * \widetilde{v}\rangle .$$

Check that convolution of distributions on $\mathcal{D}'(\mathbf{T}^n)$ is associative, commutative, and distributive.
(c) Prove the analogue of Proposition 2.3.23. $C^\infty(\mathbf{T}^n)$ is dense in $\mathcal{D}'(\mathbf{T}^n)$.

3.6.3. For $u \in \mathcal{D}'(\mathbf{T}^n)$ and $m \in \mathbf{Z}^n$ define the Fourier coefficient $\widehat{u}(m)$ by

$$\widehat{u}(m) = u(e^{-2\pi i m\cdot(\,\cdot\,)}) = \langle u, e^{-2\pi i m\cdot(\,\cdot\,)}\rangle .$$

Prove properties (1), (2), (4), (5), (6), (8), (9), (11), and (12) of Proposition 2.3.22 regarding the Fourier coefficients of distributions on the circle. Moreover, prove that for any u, v in $\mathcal{D}'(\mathbf{T}^n)$ we have $(u*v)\widehat{\ }(m) = \widehat{u}(m)\,\widehat{v}(m)$. In particular, this is valid for finite Borel measures.

3.6.4. Let μ be a finite Borel measure on $\mathbf{R}^n$ and let ν be the periodization of μ, that is ν is a measure on $\mathbf{T}^n$ defined by

$$\nu(A) = \sum_{m\in\mathbf{Z}^n} \mu(A+m)$$

for all measurable subsets A of $\mathbf{T}^n$. Prove that the restriction of the Fourier transform of μ on $\mathbf{Z}^n$ coincides with the sequence of the Fourier coefficients of the measure ν.

3.6.5. Let T be an operator that commutes with translations and maps $L^p(\mathbf{T}^n)$ into $L^q(\mathbf{T}^n)$ for some $1 \le p, q \le \infty$. Prove that there exists a distribution u on $\mathbf{T}^n$ such that $T(f) = f * u$.

3.6.6. (*G. Weiss*) Suppose that the function b on $\mathbf{R}^n$ is regulated at the point x_0 in the sense

$$\lim_{\varepsilon\to 0} \frac{1}{\varepsilon^n} \int_{|t|\le\varepsilon} \big(b(x_0 - t) - b(x_0)\big)\, dt = 0.$$

Let $K_\varepsilon(x) = \varepsilon^{-n} e^{-\pi|x/\varepsilon|^2}$ for $\varepsilon > 0$. Prove that $(b * K_\varepsilon)(x_0) \to b(x_0)$ as $\varepsilon \to 0$.
[*Hint:* Prove that for all $\delta > 0$ we have

$$\begin{aligned}\big|(b * K_\varepsilon)(x_0) - b(x_0)\big| \;&\le 2\left\|b\right\|_{L^\infty} \int_{|y|\ge\delta/\varepsilon} e^{-\pi|y|^2}\, dy \\ &\quad + |F_{x_0}(\delta)|\, \frac{\delta^n}{\varepsilon^n}\, e^{-\pi\delta^2/\varepsilon^2} \\ &\quad + 2\pi \sup_{0<r\le\delta} |F_{x_0}(r)| \int_0^{\frac{\delta}{\varepsilon}} r^{n+1} e^{-\pi r^2}\, dr,\end{aligned}$$

where $F_{x_0}(\delta) = \frac{1}{\delta^n} \int_{|t|\le\delta} \big(b(x_0 - t) - b(x_0)\big)\, dt.$]

3.6.7. Let v_n be the volume of the unit ball in $\mathbf{R}^n$ and $e_1 = (1, 0, \dots, 0)$. Prove that

$$\lim_{\varepsilon\to 0} \frac{1}{v_n \varepsilon^n} \int_{|x-e_1|\le\varepsilon} \chi_{|x|\le 1}\, dx \to \frac{1}{2}\,.$$

Conclude that the function

$$\widetilde{\chi}_{B(0,1)}(x) = \begin{cases} 1 & \text{when } |x| < 1, \\ 1/2 & \text{when } |x| = 1, \\ 0 & \text{when } |x| > 1 \end{cases}$$

is regulated.

3.6.8. Let $L_\varepsilon(x) = e^{-\pi\varepsilon|x|^2}$ for $\varepsilon > 0$. Given a continuous function g on $\mathbf{T}^n$, prove that

$$\lim_{\varepsilon\to 0} \varepsilon^{\frac{n}{2q}} \left\|g L_{\varepsilon/q}\right\|_{L^{q,1}(\mathbf{R}^n)} = \left\|g\right\|_{L^{q,1}(\mathbf{T}^n)}$$

for all $1 < q < \infty$.

3.6.9. Suppose that $\{f_t\}_{t\in\mathbf{R}}$ is a family of measurable functions on a measure space X that satisfies

$$\big\|\sup_{t\in F} |f_t|\big\|_{L^p} \le b < \infty$$

for every finite subset F of $\mathbf{R}$. Prove that for any t there is a measurable function $\widetilde{f_t}$ on X that is a.e. equal to f_t such that

$$\big\| \sup_{t\in\mathbf{R}} |\widetilde{f_t}| \big\|_{L^p} \le b.$$

[*Hint:* Let $a = \sup_F \big\| \sup_{t\in F} |f_t| \big\|_{L^p} \le b$, where the supremum is taken over all finite subsets F of $\mathbf{R}$. Pick a sequence of sets F_n so that $\big\| \sup_{t\in F_n} |f_t| \big\|_{L^p}$ converges to a as $n \to \infty$. Let $g = \sup_n \sup_{t\in F_n} |f_t|$ and note that $\big\| g \big\|_p = a$. The for any $s \in \mathbf{R}$ we have

$$\big\| \sup(|f_s|, \sup_{1\le k\le n} \sup_{t\in F_k} |f_t|) \big\|_{L^p} \le a.$$

This implies $\big\| \max(|f_s|, g) \big\|_{L^p} \le a = \big\| g \big\|_{L^p}$ so that $|f_s| \le g$ a.e. for all $s \in \mathbf{R}$. This means g is an a.e. upper bound for all f_t.]

3.6.10. (*E. Prestini*) Let $p \ge 2$ and $k > 0$. Show that for $f \in L^p(\mathbf{T}^2)$ we have that

$$\sum_{\substack{|m_1|\le N \\ |m_2|\le N^k}} \widehat{f}(m_1, m_2) e^{2\pi i (m_1 x_1 + m_2 x_2)} \to f(x_1, x_2)$$

for almost all (x_1, x_2) in $\mathbf{T}^2$.
[*Hint:* It suffices to consider the case $p = 2$. Use the splitting $\widehat{f}(m_1, m_2) = \widehat{f}(m_1, m_2)\chi_{|m_2|\le|m_1|^k} + \widehat{f}(m_1, m_2)\chi_{|m_2|>|m_1|^k}$ and apply the idea of the proof of Theorem 3.6.15.]

3.7. Lacunary Series*

In this section we take a quick look at lacunary series. These series provide examples of functions that possess some remarkable properties.

3.7.a. Definition and Basic Properties of Lacunary Series

We begin by defining lacunary sequences.

Definition 3.7.1. A sequence of positive integers $\{\lambda_k\}_{k=1}^{\infty}$ is called *lacunary* if there exists a constant $A > 1$ such that $\lambda_{k+1} \ge A\lambda_k$ for all $k \in \mathbf{Z}^+$.

Examples of lacunary sequences are provided by exponential sequences, such as $\lambda_k = 2^k, 3^k, 4^k$. Observe that polynomial sequences such as $\lambda_k = 1 + k^2$ are not lacunary. Also observe that lacunary sequences tend to infinity as $k \to \infty$.
We begin with the following result.

Proposition 3.7.2. *Let λ_k be a lacunary sequence and let f be an integrable function on the circle that is differentiable at a point and has Fourier coefficients*

$$\widehat{f}(m) = \begin{cases} a_m & \text{when } m = \lambda_k, \\ 0 & \text{when } m \neq \lambda_k. \end{cases} \tag{3.7.1}$$

Then we have

$$\lim_{k\to+\infty} \widehat{f}(\lambda_k)\lambda_k = 0.$$

Proof. Applying translation, we may assume that the point at which f is differentiable is the origin. Replacing f by

$$g(t) = f(t) - f(0)\cos(t) - f'(0)\sin(t) = f(t) - f(0)\frac{e^{it}+e^{-it}}{2} - f'(0)\frac{e^{it}-e^{-it}}{2i}$$

we may assume that $f(0) = f'(0) = 0$. [We have $\widehat{g}(m) = \widehat{f}(m)$ for $|m| \geq 2$ and thus the final conclusion for f is equivalent to that for g.]

Using the lacunarity condition and (3.7.1), we obtain that

$$1 \leq |m - \lambda_k| < \min(A-1, 1-A^{-1})\lambda_k \implies \widehat{f}(m) = 0\,. \tag{3.7.2}$$

Given $\varepsilon > 0$, pick a positive integer k_0 such that if $[\min(A-1,1-A^{-1})\lambda_{k_0}] = 2N_0$, then $N_0^{-2} < \varepsilon$, and

$$\sup_{|x|<N_0^{-\frac{1}{4}}} \left| \frac{f(x)}{x} \right| < \varepsilon\,. \tag{3.7.3}$$

The expression in (3.7.3) can be made arbitrarily small since f is differentiable at the origin. Now take an integer k with $k \geq k_0$ and set $2N = [\min(A-1,1-A^{-1})\lambda_k]$, which is of course at least $2N_0$. Using (3.7.2), we obtain that for any trigonometric polynomial K_N of degree $2N$ with $\widehat{K_N}(0) = 1$ we have

$$\widehat{f}(\lambda_k) = \int_{|x|\leq\frac{1}{2}} f(x)K_N(x)e^{-2\pi i\lambda_k x}\,dx\,. \tag{3.7.4}$$

We take $K_N = (F_N/\|F_N\|_{L^2})^2$, where F_N is the Fejér kernel. Using (3.1.15), we obtain first the identity

$$\|F_N\|_{L^2}^2 = \sum_{j=-N}^{N} \left(1 - \frac{|j|}{N+1}\right)^2 = 1 + \frac{1}{3}\frac{N(2N+1)}{N+1} > \frac{N}{3} \tag{3.7.5}$$

and also the estimate

$$F_N(x)^2 \leq \left(\frac{1}{N+1}\frac{1}{4x^2}\right)^2, \tag{3.7.6}$$

which is valid for $|x| \leq 1/2$. In view of (3.7.5) and (3.7.6), we have the estimate

$$K_N(x) \leq \frac{3}{16}\frac{1}{N^3}\frac{1}{x^4}\,. \tag{3.7.7}$$

We now use (3.7.4) to obtain

$$\lambda_k \widehat{f}(\lambda_k) = \lambda_k \int_{|x|\le \frac{1}{2}} f(x) K_N(x) e^{-2\pi i \lambda_k x}\, dx = I_k^1 + I_k^2 + I_k^3,$$

where

$$\begin{aligned} I_k^1 &= \lambda_k \int_{|x|\le N^{-1}} f(x) K_N(x) e^{-2\pi i \lambda_k x}\, dx\,, \\ I_k^2 &= \lambda_k \int_{N^{-1}<|x|\le N^{-\frac{1}{4}}} f(x) K_N(x) e^{-2\pi i \lambda_k x}\, dx\,, \\ I_k^3 &= \lambda_k \int_{N^{-\frac{1}{4}}<|x|\le \frac{1}{2}} f(x) K_N(x) e^{-2\pi i \lambda_k x}\, dx\,. \end{aligned}$$

Since $\|K_N\|_{L^1} = 1$, it follows that

$$|I_k^1| \le \frac{\lambda_k}{N} \sup_{|x|<N^{-1}} \left|\frac{f(x)}{x}\right| \le \frac{(2N+1)\,\varepsilon}{\min(A-1, 1-A^{-1})N}\,,$$

which can be made arbitrarily small if ε is small. Also, using (3.7.7), we obtain

$$|I_k^2| \le \frac{3\lambda_k}{16N^3} \sup_{|x|<N^{-\frac{1}{4}}} \left|\frac{f(x)}{x}\right| \int_{N^{-1}<|x|\le N^{-\frac{1}{4}}} \frac{dx}{x^3} \le \frac{3\lambda_k}{16N} \sup_{|x|<N^{-\frac{1}{4}}} \left|\frac{f(x)}{x}\right|,$$

which, as observed, is bounded by a constant multiple of ε. Finally, using again (3.7.7), we obtain

$$|I_k^3| \le \frac{3}{16N^3} \frac{1}{N^{-\frac{1}{4}}} \int_{N^{-\frac{1}{4}}<|x|\le \frac{1}{2}} |f(x)|\, dx \le \frac{3}{16N^2} \|f\|_{L^1} < \frac{3\varepsilon}{16} \|f\|_{L^1}\,.$$

It follows that for all $k \ge k_0$ we have

$$|\lambda_k \widehat{f}(\lambda_k)| \le |I_k^1| + |I_k^2| + |I_k^3| \le C(f)\,\varepsilon$$

for some fixed constant $C(f)$. This proves the required conclusion. □

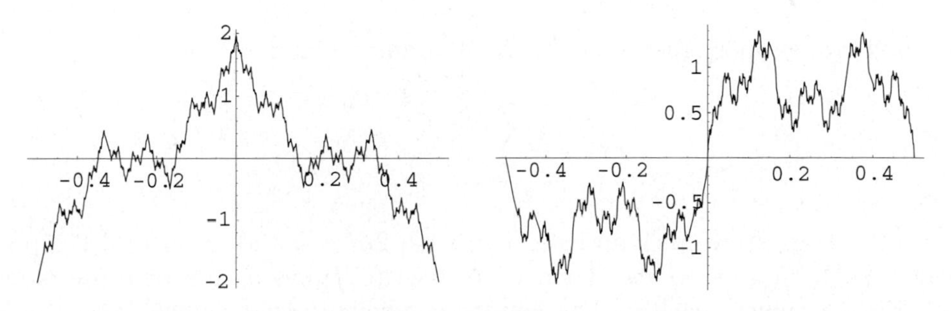

FIGURE 3.3. The graph of the real and imaginary parts of the function $f(t) = \sum_{k=0}^{\infty} 2^{-k} e^{2\pi i 3^k t}$.

Corollary 3.7.3. ***(Weierstrass)*** *There exists a continuous function on the circle which is nowhere differentiable.*

Proof. Consider the 1-periodic function

$$f(t)=\sum_{k=0}^{\infty}2^{-k}e^{2\pi i3^kt}\,.$$

Then f is continuous since the series defining it converges absolutely and uniformly. If f were differentiable at a point, then by Proposition 3.7.2 we would have that $2^{-k}\widehat{f}(k)$ tends to zero as $k\to\infty$. But since $\widehat{f}(k)=3^k$ for $k\geq 0$, this is impossible. Therefore, f is nowhere differentiable. The real and imaginary parts of this function are displayed in Figure 3.3.

□

3.7.b. Equivalence of L^p Norms of Lacunary Series

We now turn to one of the most important properties of lacunary series, equivalence of their norms. It is a remarkable result that lacunary Fourier series have comparable L^p norms for $1\leq p<\infty$. More precisely, we have the following theorem:

Theorem 3.7.4. *Let $0<\lambda_1<\lambda_2<\lambda_3<\dots$ be a lacunary sequence with constant $A>1$. Set $\Lambda=\{\lambda_k:\ k\in\mathbf{Z}^+\}$. Then for all $1<p<\infty$ there exists a constant $C_p(A)$ such that for all $f\in L^1(\mathbf{T}^1)$ with $\widehat{f}(k)=0$ when $k\in\mathbf{Z}\setminus\Lambda$ we have*

$$\big\|f\big\|_{L^p(\mathbf{T}^1)}\leq C_p(A)\big\|f\big\|_{L^1(\mathbf{T}^1)}\,. \tag{3.7.8}$$

Note that the converse inequality to (3.7.8) is trivial. Therefore, L^p norms of lacunary Fourier series are all equivalent for $1\leq p<\infty$.

Proof. We suppose first that $f\in L^2(\mathbf{T}^1)$ and we define

$$f_N(x)=\sum_{j=1}^{N}\widehat{f}(\lambda_j)e^{2\pi i\lambda_jx}\,. \tag{3.7.9}$$

Given a $2\leq p<\infty$, we pick an integer m with $2m>p$ and we also pick a positive integer r such that $A^r>m$. Then we can write f_N as a sum of r functions φ_s, $s=1,2,\dots,r$, where each φ_s has Fourier coefficients that vanish except possibly on the lacunary set

$$\{\lambda_{kr+s}:\ k\in\mathbf{Z}^+\cup\{0\}\}=\{\mu_1,\mu_2,\mu_3,\dots\}\,.$$

It is a simple fact that the sequence $\{\mu_k\}_k$ is lacunary with constant A^r. Then we have

$$\int_0^1 |\varphi_s(x)|^{2m}\,dx = \sum_{\substack{1\le j_1,\dots,j_m,k_1,\dots,k_m\le N\\ \mu_{j_1}+\cdots+\mu_{j_m}=\mu_{k_1}+\cdots+\mu_{k_m}}} \widehat{\varphi_s}(\mu_{j_1})\cdots\widehat{\varphi_s}(\mu_{j_m})\overline{\widehat{\varphi_s}(\mu_{k_1})}\cdots\overline{\widehat{\varphi_s}(\mu_{k_m})}\,.$$

We claim that if $\mu_{j_1}+\cdots+\mu_{j_m} = \mu_{k_1}+\cdots+\mu_{k_m}$, then $\max(\mu_{j_1},\dots,\mu_{j_m}) = \max(\mu_{k_1},\dots,\mu_{k_m})$. Indeed, if $\max(\mu_{j_1},\dots,\mu_{j_m}) > \max(\mu_{k_1},\dots,\mu_{k_m})$, then

$$\max(\mu_{j_1},\dots,\mu_{j_m}) \le \mu_{k_1}+\cdots+\mu_{k_m} \le m\max(\mu_{k_1},\dots,\mu_{k_m})\,.$$

But since

$$A^r\max(\mu_{k_1},\dots,\mu_{k_m}) < \max(\mu_{j_1},\dots,\mu_{j_m})\,,$$

it would follow that $A^r \le m$, which contradicts our choice of r. Likewise, we eliminate the case where $\max(\mu_{j_1},\dots,\mu_{j_m}) < \max(\mu_{k_1},\dots,\mu_{k_m})$. We conclude that these numbers are equal. We can now continue the same reasoning by induction to conclude that if $\mu_{j_1}+\cdots+\mu_{j_m} = \mu_{k_1}+\cdots+\mu_{k_m}$, then

$$\{\mu_{k_1},\dots,\mu_{k_m}\} = \{\mu_{j_1},\dots,\mu_{j_m}\}\,.$$

Using this fact in the evaluation of the previous multiple sum, we obtain

$$\int_0^1 |\varphi_s(x)|^{2m}\,dx = \sum_{j_1=1}^{N}\cdots\sum_{j_m=1}^{N} |\widehat{\varphi_s}(\mu_{j_1})|^2\dots|\widehat{\varphi_s}(\mu_{j_m})|^2 = \left(\big\|\varphi_s\big\|_{L^2}^2\right)^m,$$

which implies that $\big\|\varphi_s\big\|_{L^{2m}} = \big\|\varphi_s\big\|_{L^2}$ for all $s\in\{1,2,\dots,r\}$. Thus we have

$$\big\|f_N\big\|_{L^p} \le \big\|f_N\big\|_{L^{2m}} \le \sqrt{r}\,\Big(\sum_{s=1}^{r}\big\|\varphi_s\big\|_{L^{2m}}^2\Big)^{\frac12} = \sqrt{r}\,\Big(\sum_{s=1}^{r}\big\|\varphi_s\big\|_{L^2}^2\Big)^{\frac12} = \sqrt{r}\,\big\|f_N\big\|_{L^2}\,,$$

since the functions φ_s are orthogonal on L^2. Since r can be chosen to be $[\log_A m]+1$ and m can be taken to be $[\frac{p}{2}]+1$, we have now established the inequality

$$\big\|f_N\big\|_{L^p(\mathbf{T}^1)} \le C_p(A)\big\|f_N\big\|_{L^2(\mathbf{T}^1)}\,,\qquad p\ge 2 \tag{3.7.10}$$

with $C_p(A) = \sqrt{1+\big[\log_A[\frac{p}{2}]+1\big]}$ for all f_N that have the form (3.7.9). To extend (3.7.10) to all $f\in L^2(\mathbf{T}^1)$, we observe that $f_N\to f$ in L^2 and some subsequence of them f_{N_j} tends to f a.e. Then Fatou's lemma gives

$$\begin{aligned}\int_0^1 |f(x)|^p\,dx &= \int_0^1 \liminf_{j\to\infty} |f_{N_j}(x)|^p\,dx \\ &\le \liminf_{j\to\infty} \int_0^1 |f_{N_j}(x)|^p\,dx \\ &\le C_p(A)^p \liminf_{j\to\infty} \big\|f_{N_j}\big\|_{L^2}^p \\ &= C_p(A)^p \big\|f\big\|_{L^2}^p\,,\end{aligned}$$

which proves (3.7.10) for all $f \in L^2$. To obtain (3.7.8) for $1 < p < 2$, we use interpolation, as in

$$\big\|f\big\|_{L^2} \le \big\|f\big\|_{L^4}^{\frac{2}{3}}\big\|f\big\|_{L^1}^{\frac{1}{3}} \le \big([\log_A 3]+1\big)^{\frac{1}{2}\cdot\frac{2}{3}}\big\|f\big\|_{L^2}^{\frac{2}{3}}\big\|f\big\|_{L^1}^{\frac{1}{3}}\,,$$

which implies that for $1 < p < 2$ we have

$$\big\|f\big\|_{L^p(\mathbf{T}^1)} \le \big\|f\big\|_{L^2(\mathbf{T}^1)} \le \big([\log_A 3]+1\big)\big\|f\big\|_{L^1(\mathbf{T}^1)}\,.$$

Combining this with (3.7.10), which now holds for all $f \in L^2$, yields (3.7.8).

□

Theorem 3.7.4 describes the equivalence of the L^p norms of lacunary Fourier series for $p < \infty$. The question that remains is whether there is a similar characterization of the L^∞ norms of lacunary Fourier series. Such a characterization is given in Theorem 3.7.6. Before we state and prove this theorem, we will need a classical tool, referred to as a Riesz product.

Definition 3.7.5. A *Riesz product* is a function of the form

$$P_N(x) = \prod_{j=1}^{N} \big(1 + a_j \cos(2\pi\lambda_j x + 2\pi\gamma_j)\big)\,, \tag{3.7.11}$$

where N is a positive integer, $\lambda_1 < \lambda_2 < \cdots < \lambda_N$ is a lacunary sequence of integers, a_j are complex numbers, and $\gamma_j \in [0,1]$.

We make a few observations about Riesz products. A simple calculation gives that if $P_{N,j}(x) = 1 + a_j \cos(2\pi\lambda_j x + 2\pi\gamma_j)$, then

$$\widehat{P_{N,j}}(m) = \begin{cases} 1 & \text{when } m = 0 \\ \frac{1}{2}a_j e^{2\pi i\gamma_j} & \text{when } m = \lambda_j, \\ \frac{1}{2}a_j e^{-2\pi i\gamma_j} & \text{when } m = -\lambda_j, \\ 0 & \text{when } m \notin \{0, \lambda_j, -\lambda_j\}. \end{cases} \tag{3.7.12}$$

Let us assume that the lacunarity constant A associated with the sequence $\lambda_1 < \lambda_2 < \cdots < \lambda_N$ is at least 3. Then each integer m has at most one representation as a sum

$$m = \varepsilon_1\lambda_1 + \cdots + \varepsilon_N\lambda_N\,,$$

where $\varepsilon_j \in \{-1, 1, 0\}$. See Exercise 3.7.1. We will now calculate the Fourier coefficients of the Riesz product defined in (3.7.11). For a fixed integer b, let us denote by δ_b the sequence of integers that is equal to 1 at b and zero otherwise. Then, using (3.7.12), we obtain that

$$\widehat{P_{N,j}} = \delta_0 + \tfrac{1}{2}a_j e^{2\pi i \gamma_j}\delta_{\lambda_j} + \tfrac{1}{2}a_j e^{-2\pi i \gamma_j}\delta_{-\lambda_j},$$

and thus $\widehat{P_N}$ is the N-fold convolution of these functions. Using that $\delta_a * \delta_b = \delta_{a+b}$, we obtain

$$\widehat{P_N}(m) = \begin{cases} 1 & \text{when } m = 0, \\ \prod_{j=1}^{N} \frac{1}{2}a_j e^{2\pi i \varepsilon_j \gamma_j} & \text{when } m = \sum_{j=1}^{N} \varepsilon_j \lambda_j \text{ and } \sum_{j=1}^{N} |\varepsilon_j| > 0, \\ 0 & \text{otherwise.} \end{cases}$$

It follows that $\widehat{P_N}(\lambda_j) = \frac{1}{2}a_j e^{2\pi i \gamma_j}$ for $1 \le j \le N$ and that $\widehat{P_N}(\lambda_j) = 0$ for $j \ge N+1$ since each λ_j can be written uniquely as a sum of λ_k's as $0 \cdot \lambda_1 + \cdots + 0 \cdot \lambda_{j-1} + 1 \cdot \lambda_j$. See Exercise 3.7.1.

We recall the space $A(\mathbf{T}^1)$ of all functions with absolutely summable Fourier coefficients with norm the ℓ^1 norm of the coefficients.

Theorem 3.7.6. *Let $0 < \lambda_1 < \lambda_2 < \lambda_3 < \dots$ be a lacunary sequence of integers with constant $A > 1$. Set $\Lambda = \{\lambda_k : k \in \mathbf{Z}^+\}$. Then there exists a constant $C(A)$ such that for all $f \in L^\infty(\mathbf{T}^1)$ with $\widehat{f}(k) = 0$ when $k \in \mathbf{Z} \setminus \Lambda$ we have*

$$\big\|f\big\|_{A(\mathbf{T}^1)} = \sum_{k \in \Lambda} |\widehat{f}(k)| \le C(A)\big\|f\big\|_{L^\infty(\mathbf{T}^1)}. \tag{3.7.13}$$

Proof. Let us assume first that $A \ge 3$. Also fix $f \in L^\infty(\mathbf{T}^1)$. We consider the Riesz product

$$P_N(x) = \prod_{j=1}^{N} \big(1 + \cos(2\pi\lambda_j x + 2\pi\gamma_j)\big),$$

where γ_j is defined via the identity $|\widehat{f}(\lambda_j)| = e^{2\pi i \gamma_j}\overline{\widehat{f}(\lambda_j)}$. Then $P_N \ge 0$ and since $\widehat{P_N}(0) = 1$, it follows that $\big\|P_N\big\|_{L^1} = 1$. By Parserval's identity we obtain

$$\Big|\sum_{m \in \mathbf{Z}} \widehat{P_N}(m)\overline{\widehat{f}(m)}\Big| = \Big|\int_0^1 P_N(x)\overline{f(x)}\,dx\Big| \le \big\|f\big\|_{L^\infty}, \tag{3.7.14}$$

and the sum in (3.7.14) is finite since the Fourier coefficients of $\widehat{P_N}$ form a finitely supported sequence. But $\widehat{f}(m) = 0$ for $m \notin \Lambda$, while $\widehat{P_N}(\lambda_j) = \frac{1}{2}e^{2\pi i \gamma_j}$ for $1 \le j \le N$. Moreover, $\widehat{P_N}(\lambda_j) = 0$ for $j \ge N+1$, as observed earlier. Thus (3.7.14) reduces to

$$\frac{1}{2}\sum_{j=1}^{N} |\widehat{f}(\lambda_j)| = \Big|\sum_{j=1}^{N} \frac{1}{2}e^{2\pi i \gamma_j}\overline{\widehat{f}(\lambda_j)}\Big| \le \big\|f\big\|_{L^\infty}.$$

Letting $N \to \infty$, we obtain that $\sum_{j=1}^{\infty} |\widehat{f}(\lambda_j)| \le 2\|f\|_{L^\infty}$, which proves (3.7.13) when $A \ge 3$.

To prove the theorem for $1 < A < 3$, we pick a positive integer r with $A^r \ge 3$ (take $r = [\log_A 3] + 1$). We now consider the sequences

$$\{\lambda_{kr+s}\}_k \,, \qquad k \in \mathbf{Z}^+ \cup \{0\},$$

and we observe that each such sequence is lacunary with constant A^r. The preceding construction gives

$$\sum_{j=1}^{\infty} |\widehat{f}(\lambda_{jr+s})| \le 2\|f\|_{L^\infty} \,.$$

Summing over s in the set $\{1, 2, \dots, r\}$, we obtain the required conclusion with $C(A) = 2r = 2[\log_A 3] + 2$.

□

It follows from Theorem 3.7.6 that if $\Lambda = \{\lambda_k : \, k \in \mathbf{Z}^+\}$ is a lacunary set and f is a bounded function on the circle that satisfies $\widehat{f}(k) = 0$ when $k \in \mathbf{Z} \setminus \Lambda$, then we have

$$f(x) = \sum_{k \in \Lambda} \widehat{f}(k) e^{2\pi i k x} \,.$$

This is a consequence of the inversion result in Proposition 3.1.14.

Given a subset Λ of the integers, we denote by C_Λ the space of all continuous functions on $\mathbf{T}^1$ such that

$$m \in \mathbf{Z} \setminus \Lambda \implies \widehat{f}(m) = 0 \,. \tag{3.7.15}$$

It is straightforward that C_Λ is a closed subspace of all bounded functions on the circle $\mathbf{T}^1$ with the standard L^∞ norm.

Definition 3.7.7. A set of integers Λ is called a *Sidon set* if every function in C_Λ has an absolutely convergent Fourier series.

Example 3.7.8. Every lacunary set is a Sidon set. Indeed, if f satisfies (3.7.15), then Theorem 3.7.6 gives that

$$\sum_{m \in \Lambda} |\widehat{f}(m)| \le C(A)\|f\|_{L^\infty} \,;$$

hence f has an absolutely convergent Fourier series.

Example 3.7.9. There exist subsets of $\mathbf{R}$ that are not Sidon. For example $\mathbf{Z} \setminus \{0\}$, is not a Sidon set. See Exercise 3.7.2.

Exercises

3.7.1. Suppose that $0 < \lambda_1 < \lambda_2 < \cdots < \lambda_N$ is a lacunary sequence of integers with lacunarity constant $A \geq 3$. Prove that for every integer m there exist at most one N-tuple $(\varepsilon_1, \dots, \varepsilon_N)$ with each $\varepsilon_j \in \{-1, 1, 0\}$ such that

$$m = \varepsilon_1 \lambda_1 + \cdots + \varepsilon_N \lambda_N .$$

[*Hint:* Suppose there exist two such N-tuples. Pick the largest k such that the coefficients of λ_k are different.]

3.7.2. Consider the 1-periodic continuous function $h(t) = \cos(2\pi t)$. Then we have $\widehat{h}(0) = 0$, but show that $\sum_k |\widehat{h}(k)| = \infty$. Thus $\mathbf{Z} \setminus \{0\}$ is not a Sidon set.

3.7.3. Suppose that $0 < \lambda_1 < \lambda_2 < \dots$ is a lacunary sequence and let f be a bounded function on the circle that satisfies $\widehat{f}(m) = 0$ whenever $m \in \mathbf{Z} \setminus \{\lambda_1, \lambda_2, \dots\}$. Suppose also that

$$\sup_{t \neq 0} \frac{|f(t) - f(0)|}{|t|^\alpha} = B < \infty$$

for some $0 < \alpha < 1$.
(a) Prove that there is a constant C such that $|\widehat{f}(\lambda_k)| \leq C B \lambda_k^{-\alpha}$ for all $k \geq 1$.
(b) Prove that $f \in \dot{\Lambda}_\alpha(\mathbf{T}^1)$.
[*Hint:* Let $2N = [\min(A-1, 1-A^{-1})\lambda_k]$ and let K_N be as in the proof of Proposition 3.7.2. Write

$$\begin{aligned} \widehat{f}(\lambda_k) &= \int_{|x| \leq N^{-1}} (f(x) - f(0)) e^{-2\pi i \lambda_k x} K_N(x)\, dx \\ &\quad + \int_{N^{-1} \leq |x| \leq \frac{1}{2}} (f(x) - f(0)) e^{-2\pi i \lambda_k x} K_N(x)\, dx \,. \end{aligned}$$

Use that $\big\|K_N\big\|_{L^1} = 1$ and also the estimate (3.7.6). Part (b): Use the estimate in part (a).]

3.7.4. Let f be an integrable function on the circle whose Fourier coefficients vanish outside a lacunary set $\Lambda = \{\lambda_1, \lambda_2, \lambda_3, \dots\}$. Suppose that f vanishes identically in a small neighborhood of the origin. Show that f is in $C^\infty(\mathbf{T}^1)$.
[*Hint:* Let $2N = [\min(A-1, 1-A^{-1})\lambda_k]$ and let K_N be as in the proof of Proposition 3.7.2. Write

$$\widehat{f}(\lambda_k) = \int_{|x| \leq \frac{1}{2}} f(x) e^{-2\pi i \lambda_k x} K_N(x)\, dx$$

and use estimate (3.7.6) to obtain that f is in C^2. Continue by induction.]

3.7.5. Let $0 < a, b < \infty$. Consider the 1-periodic function

$$f(x) = \sum_{k=0}^{\infty} a^{-k} e^{2\pi i b^k x}.$$

Prove that the following statements are equivalent:
(a) f is differentiable at a point.
(b) $a < b$.
(c) f is differentiable everywhere.

3.7.6. Let Λ be a subset of the integers such that for any sequence of complex numbers $\{d_\lambda\}_{\lambda \in \Lambda}$ with $|d_\lambda| = 1$ there is a finite Borel measure μ on $\mathbf{T}^1$ such that

$$|\widehat{\mu}(\lambda) - d_\lambda| < \frac{1}{2}$$

for all $\lambda \in \Lambda$. Show that Λ is a Sidon set.

3.7.7. Let $\Lambda \subseteq \mathbf{Z}^+$. Suppose that there is a constant $A < \infty$ such that for any $n \in \Lambda \cup \{0\}$ the number of elements in the set

$$\Big\{(\varepsilon_1, \dots, \varepsilon_m) \in \{-1, 1\}^m : \; n = \sum_{j=1}^{m} \varepsilon_j n_j, \; n_1 < \cdots < n_m, \; n_j \in \Lambda\Big\}$$

is at most A^m. Show that Λ is a Sidon set.
[*Hint:* Construct a suitable measure μ and use the previous Exercise.]

3.7.8. Show that the set

$$\big\{3^{2^{m+2}} + 3^{2^m + k}, \; 0 \le k \le 2^{m-1}, \; m \ge 1\big\}$$

is a Sidon set.
[*Hint:* Use Exercise 3.7.7.]

HISTORICAL NOTES

Trigonometric series in one dimension were first considered in the study of the vibrating string problem and are implicitly contained in the work of d'Alembert, D. Bernoulli, Clairaut, and Euler. The analogous problem for vibrating higher-dimensional bodies naturally suggested the use of multiple trigonometric series. However, it was the work of Fourier on steady-state heat conduction that incited the subsequent systematic development of such series. Fourier announced his results in 1811, although his classical book "Théorie de la chaleur" was published in 1822. This book contains several examples of heuristic use of trigonometric expansions and motivated other mathematicians to carefully study such expansions.

The fact that the Fourier series of a continuous function can diverge was first observed by DuBois Reymond in 1876. The Riemann-Lebesgue lemma was first proved by Riemann in his memoir on trigonometric series (appeared between 1850 and 1860). It carries Lebesgue's name today, because Lebesgue later extended it to his notion of integral. The rebuilding of the theory of Fourier series based on Lebesgue's integral was mainly achieved by de la Vallée-Poussin and Fatou.

Theorem 3.2.16 was obtained by S. Bernstein in dimension $n = 1$. Higher-dimensional analogues of the Hardy-Littlewood series of Exercise 3.2.9 were studied by Wainger [**544**]. These series can be used to produce examples indicating that the restriction $s > \alpha + n/2$ in Bernstein's theorem is sharp even in higher dimensions. Part (b) of Theorem 3.3.3 is due to Lebesgue when $n = 1$ and Marcinkiewicz and Zygmund [**368**] when $n = 2$. Marcinkiewicz and Zygmund's proof also extends to higher dimensions. The proof given here is based on Lemma 3.3.4 proved by Stein [**479**] in a different context. The proof of Lemma 3.3.4 presented here was suggested by T. Tao.

The development of the complex methods in the study of Fourier series was pioneered by the Russian school, especially Luzin and his students, Kolmogorov, Menshov, and Privalov. The existence of an integrable function on $\mathbf{T}^1$ whose Fourier series diverges almost everywhere (Theorem 3.4.2) is due to Kolmogorov [**306**]. An example of an integrable function whose Fourier series diverges everywhere was also produced by Kolmogorov [**309**] three years later. Localization of the Bochner-Riesz means at the critical exponent $\alpha = \frac{n-1}{2}$ fails for L^1 functions on $\mathbf{T}^n$ (see Bochner [**43**]) but holds for functions f such that $|f| \log^+ |f|$ is integrable over $\mathbf{T}^n$ (see Stein [**470**]). The latter article also contains the L^p boundedness of the maximal Bochner-Riesz operator $\sup_{R>0} |B_R^\alpha(f)|$ for $1 < p < \infty$ when $\alpha > |\frac{1}{p} - \frac{1}{2}|$. Theorem 3.4.6 is also due to Stein [**472**]. The technique that involves the points for which the set $\{|x - m| : m \in \mathbf{Z}^n\}$ is linearly independent over the rationals was introduced by Bochner [**43**].

The boundedness of the conjugate function on the circle (Theorem 3.5.6) and hence the L^p convergence of one dimensional Fourier series was announced by Riesz in [**425**], but its proof appeared a little later in [**426**]. Luzin's conjecture [**349**] on almost everywhere convergence of the Fourier series of continuous functions was announced in 1913 and settled by Carleson [**84**] in 1965 for the more general class of square summable functions (Theorem 3.6.13). Carleson's theorem was later extended by Hunt [**262**] for the class of L^p functions for all $1 < p < \infty$ (Theorem 3.6.14). Sjölin [**459**] sharpened this result by showing that the Fourier series of functions f with $|f|(\log^+ |f|)(\log^+ \log^+ |f|)$ integrable over $\mathbf{T}^1$ converge almost everywhere. Antonov [**6**] improved Sjölin's result by extending it to functions f with $|f|(\log^+ |f|)(\log^+ \log^+ \log^+ |f|)$ integrable over $\mathbf{T}^1$. The reader may also consult the related results of Soria [**466**] and Arias de Reyna [**10**]. The book [**11**] of Arias de Reyna contains a historically motivated comprehensive study of topics related to the Carleson-Hunt theorem. Counterexamples due to Konyagin [**311**] show that Fourier series of functions f with $|f|(\log^+ |f|)^{\frac{1}{2}}(\log^+ \log^+ |f|)^{-\frac{1}{2}-\varepsilon}$ integrable over $\mathbf{T}^1$ may diverge when $\varepsilon > 0$. Examples of continuous functions whose Fourier series diverge exactly on given sets of measure zero are given in Katznelson [**298**] and Kahane and Katznelson [**284**].

The extension of the Carleson-Hunt theorem to higher dimensions for square summability of Fourier series (Theorem 3.6.15) is a rather straightforward consequence of the one-dimensional result and was independently obtained by Fefferman [**183**], Sjölin [**459**], and Tevzadze [**517**]. An example showing that the circular partial sums of a Fourier series may not converge in $L^p(\mathbf{T}^n)$ for $n \geq 2$ and $p \neq 2$ was obtained by Fefferman [**185**]. This example also shows that there exist L^p functions on $\mathbf{T}^n$ for $n \geq 2$ whose circular partial sums do not converge almost everywhere when $1 \le p < 2$. Indeed, if the opposite happened, then the maximal operator $f \to \sup_{N\ge 0} |\widetilde{D}(n,N) * f|$ would have to be finite a.e. for all $f \in L^p(\mathbf{T}^n)$ and by Stein's theorem [**472**] it would have to be of weak type (p,p) for some $1 < p < 2$. But this would contradict Fefferman's counterexample on L^{p_1} for some $p < p_1 < 2$. On the other hand, almost everywhere is valid for the square partial sums of functions f with $|f|(\log^+ |f|)^n(\log^+ \log^+ \log^+ |f|)$ integrable over $\mathbf{T}^n$, as shown by Antonov [**7**]; see also Sjölin and Soria [**462**].

Transference of regulated multipliers originated in the article of de Leeuw [**158**]. The methods of transference in Section 3.6 were beautifully placed into the framework of a general theory by Coifman and Weiss [**121**]. Transference of maximal multipliers (Theorem 3.6.11) was first obtained by Kenig and Tomas [**303**] and later elaborated by Asmar, Berkson, and Gillespie [**15**], [**16**].

The main references for trigonometric series are the books of Bary [**28**] and Zygmund [**571**], [**572**]. Other references for one-dimensional Fourier series include the books of Edwards [**173**], Dym and McKean [**172**], Katznelson [**299**], Körner [**314**], and the first eight chapters in Torchinsky [**525**]. The reader is also referred to the book of Krantz [**317**] for a historical introduction to the subject of Fourier series.

A classical treatment of multiple Fourier series can be found in the last chapter of Bochner's book [**44**] and in parts of his other book [**45**]. Other references include the last chapter in Zygmund [**572**], the books of Yanushauskas [**563**] (in Russian) and Zhizhiashvili [**567**], the last chapter in Stein and Weiss [**487**], and the article of Alimov, Ashurov, and Pulatov in [**4**]. A brief survey article on the subject was written by Ash [**14**]. More extensive expositions were written by Shapiro [**455**], Igari [**270**], and Zhizhiashvili [**566**]. Finally, a short note on the history of Fourier series was written by Zygmund [**574**].

CHAPTER 4

Singular Integrals of Convolution Type

In this chapter we take up the one of the fundamental topics covered in this book, that of singular integrals. This topic is motivated by its intimate connection with some of the most important problems in Fourier analysis, such as convergence of Fourier series. As we have seen, the L^p boundedness of the conjugate function on the circle is equivalent to the L^p convergence of Fourier series of L^p functions. And as the Hilbert transform on the line is just a version of the conjugate function, it plays the same role to convergence of Fourier integrals on the line as the conjugate function does on the circle.

The Hilbert transform is the prototype of all singular integrals, and a careful study of it provided the insight and inspiration for subsequent development of the subject. Historically, the theory of the Hilbert transform depended on complex analysis techniques. With the development of the Calderón-Zygmund school, real variable methods slowly replaced complex analysis, and this led to the introduction of singular integrals in other areas of mathematics. Singular integrals are nowadays intimately connected with partial differential equations, operator theory, several complex variables, and other fields. There are two kinds of singular integral operators, those of convolution type and those of nonconvolution type. In this chapter we will systematically study singular integrals of convolution type.

4.1. The Hilbert Transform and the Riesz Transforms

We begin the presentation of the subject with a careful study of the Hilbert transform. This study provides a great model for the development of the general theory of singular integrals, which will be discussed in the remaining sections and in Chapter 8.

4.1.a. Definition and Basic Properties of the Hilbert Transform

There are several equivalent ways to define the Hilbert transform, and we must choose one to begin. In this exposition we first define the Hilbert transform as a convolution operator with a certain principal value distribution, but we later discuss other equivalent definitions.

We begin by defining a distribution W_0 in $\mathcal{S}'(\mathbf{R})$ as follows:

$$\langle W_0, \varphi \rangle = \frac{1}{\pi} \lim_{\varepsilon \to 0} \int_{\varepsilon \le |x| \le 1} \frac{\varphi(x)}{x}\, dx + \frac{1}{\pi} \int_{|x| \ge 1} \frac{\varphi(x)}{x}\, dx\,, \tag{4.1.1}$$

for φ in $\mathcal{S}(\mathbf{R})$. The function $1/x$ integrated over $[-1-\varepsilon] \cup [\varepsilon, 1]$ has mean value zero, and we may replace $\varphi(x)$ by $\varphi(x) - \varphi(0)$ in the first integral in (4.1.1). Since $(\varphi(x) - \varphi(0))x^{-1}$ is controlled by $\left\|\varphi'\right\|_{L^\infty}$, it follows that the limit in (4.1.1) exists. To see that W_0 is indeed in $\mathcal{S}'(\mathbf{R})$, we go an extra step in the previous reasoning and obtain the estimate

$$\left|\langle W_0, \varphi \rangle\right| \le \frac{2}{\pi}\left\|\varphi'\right\|_{L^\infty} + \frac{2}{\pi} \sup_{x \in \mathbf{R}} |x\varphi(x)|\,. \tag{4.1.2}$$

This guarantees that $W_0 \in \mathcal{S}'(\mathbf{R})$.

Definition 4.1.1. The *truncated Hilbert transform* of $f \in \mathcal{S}(\mathbf{R})$ (at height ε) is defined by

$$H^{(\varepsilon)}(f)(x) = \frac{1}{\pi} \int_{|y| \ge \varepsilon} \frac{f(x-y)}{y}\, dy = \frac{1}{\pi} \int_{|x-y| \ge \varepsilon} \frac{f(y)}{x-y}\, dy. \tag{4.1.3}$$

The *Hilbert transform* of $f \in \mathcal{S}(\mathbf{R})$ is defined by

$$H(f)(x) = (W_0 * f)(x) = \lim_{\varepsilon \to 0} H^{(\varepsilon)}(f)(x)\,. \tag{4.1.4}$$

The integral

$$\int_{-\infty}^{+\infty} \frac{f(x-y)}{y}\, dy$$

does not converge absolutely but is defined as a limit of the absolutely convergent integrals

$$\int_{|y| \ge \varepsilon} \frac{f(x-y)}{y}\, dy\,,$$

as $\varepsilon \to 0$. Such limits are called *principal value integrals* and will be denoted by the letters p.v. Using this notation, the Hilbert transform is

$$H(f)(x) = \frac{1}{\pi}\,\text{p.v.} \int_{-\infty}^{+\infty} \frac{f(x-y)}{y}\, dy = \frac{1}{\pi}\,\text{p.v.} \int_{-\infty}^{+\infty} \frac{f(y)}{x-y}\, dy\,. \tag{4.1.5}$$

Remark 4.1.2. Note that for given $x \in \mathbf{R}$, $H(f)(x)$ is defined for all integrable functions f on $\mathbf{R}$ that satisfy a Hölder condition near the point x, that is,

$$|f(x) - f(y)| \le C_x |x-y|^{\varepsilon_x}$$

for some $C_x > 0$ and $\varepsilon_x > 0$ whenever $|y - x| < \delta_x$. Indeed, suppose that this is the case. Then

$$\begin{aligned} H^{(\varepsilon)}(f)(x) &= \frac{1}{\pi}\int_{\varepsilon<|x-y|<\delta_x} \frac{f(y)}{x-y}\,dy + \frac{1}{\pi}\int_{|x-y|\geq\delta_x} \frac{f(y)}{x-y}\,dy \\ &= \frac{1}{\pi}\int_{\varepsilon<|x-y|<\delta_x} \frac{f(y)-f(x)}{x-y}\,dy + \frac{1}{\pi}\int_{|x-y|\geq\delta_x} \frac{f(y)}{x-y}\,dy. \end{aligned}$$

Both integrals converge absolutely, and hence the limit of $H^{(\varepsilon)}(f)(x)$ exists as $\varepsilon \to 0$. Therefore, the Hilbert transform of a piecewise smooth integrable function is well defined at all points of Hölder-Lipschitz continuity of the function. On the other hand, observe that $H^{(\varepsilon)}(f)$ is well defined for all $f \in L^p$, $1 \leq p < \infty$, as it follows from Hölder's inequality, since $1/x$ is in $L^{p'}$ on the set $|x| \geq \varepsilon$.

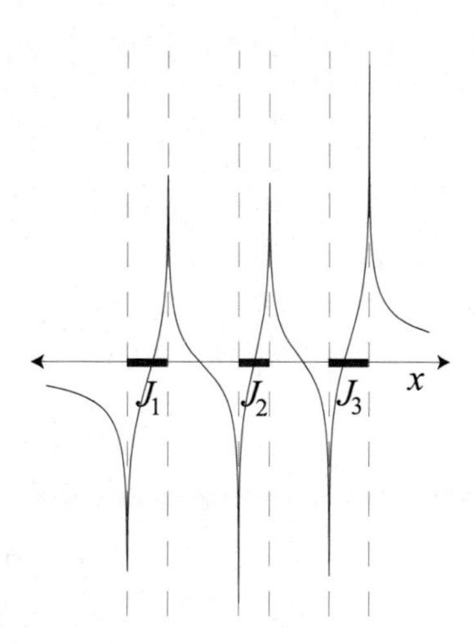

FIGURE 4.1. The graph of the function $H(\chi_E)$ when E is a union of three disjoint intervals $J_1 \cup J_2 \cup J_3$.

Example 4.1.3. Consider the characteristic function $\chi_{[a,b]}$ of an interval $[a, b]$. It is a simple calculation to show that

$$H(\chi_{[a,b]})(x) = \frac{1}{\pi}\log\frac{|x-a|}{|x-b|}. \tag{4.1.6}$$

Let us verify this identity. Pick $\varepsilon < \min(|x - a|, |x - b|)$. To show (4.1.6) consider the three cases $0 < x - b$, $x - a < 0$, and $x - b < 0 < x - a$. In the first two cases, (4.1.6) follows immediately. In the third case we have

$$H(\chi_{[a,b]})(x) = \frac{1}{\pi}\lim_{\varepsilon\to 0}\left(\log\frac{|x-a|}{\varepsilon} + \log\frac{\varepsilon}{|x-b|}\right), \tag{4.1.7}$$

which gives (4.1.6). It is crucial to observe how the cancellation of the odd kernel $1/x$ is manifested in (4.1.7). Note that $H(\chi_{[a,b]})(x)$ blows up logarithmically for x near the points a and b and decays like $|x|^{-1}$ as $x \to \infty$. See Figure 4.1.

Example 4.1.4. Let $\log^+ x = \log x$ when $x \geq 1$ and zero otherwise. Observe that the calculation in the previous example actually gives

$$H^{(\varepsilon)}(\chi_{[a,b]})(x) = \begin{cases} \dfrac{1}{\pi} \log^+ \dfrac{|x-a|}{\max(\varepsilon, |x-b|)} & \text{when } x > b, \\[2ex] -\dfrac{1}{\pi} \log^+ \dfrac{|x-b|}{\max(\varepsilon, |x-a|)} & \text{when } x < a, \\[2ex] \dfrac{1}{\pi} \log^+ \dfrac{|x-a|}{\varepsilon} - \dfrac{1}{\pi} \log^+ \dfrac{|x-b|}{\varepsilon} & \text{when } a < x < b. \end{cases}$$

We shall now give an alternative characterization of the Hilbert transform using the Fourier transform. To achieve this we need to compute the Fourier transform of the distribution W_0 defined in (4.1.1). Fix a Schwartz function φ on $\mathbf{R}$. Then

$$\begin{aligned} \langle \widehat{W_0}, \varphi \rangle \; = \; \langle W_0, \widehat{\varphi} \rangle &= \frac{1}{\pi} \lim_{\varepsilon \to 0} \int_{|\xi| \geq \varepsilon} \widehat{\varphi}(\xi) \, \frac{d\xi}{\xi} \\ &= \frac{1}{\pi} \lim_{\varepsilon \to 0} \int_{\frac{1}{\varepsilon} \geq |\xi| \geq \varepsilon} \int_{\mathbf{R}} \varphi(x) e^{-2\pi i x \xi} \, dx \, \frac{d\xi}{\xi} \\ &= \lim_{\varepsilon \to 0} \int_{\mathbf{R}} \varphi(x) \left[\frac{1}{\pi} \int_{\frac{1}{\varepsilon} \geq |\xi| \geq \varepsilon} e^{-2\pi i x \xi} \, \frac{d\xi}{\xi} \right] dx \\ &= \lim_{\varepsilon \to 0} \int_{\mathbf{R}} \varphi(x) \left[\frac{-i}{\pi} \int_{\frac{1}{\varepsilon} \geq |\xi| \geq \varepsilon} \sin(2\pi x \xi) \, \frac{d\xi}{\xi} \right] dx. \end{aligned} \tag{4.1.8}$$

Now use the results (a) and (b) of Exercise 4.1.1 to deduce that the terms inside the square brackets in (4.1.8) are uniformly bounded by 8 and they converge to

$$\lim_{\varepsilon \to 0} \int_{\frac{1}{\varepsilon} \geq |\xi| \geq \varepsilon} \sin(2\pi x \xi) \frac{d\xi}{\xi} = \pi \operatorname{sgn} x = \begin{cases} \pi & \text{when } x > 0, \\ 0 & \text{when } x = 0, \\ -\pi & \text{when } x < 0. \end{cases} \tag{4.1.9}$$

The Lebesgue dominated convergence theorem and these facts allow us to pass the limit inside the integral in (4.1.8) to obtain that

$$\langle \widehat{W_0}, \varphi \rangle = \int_{\mathbf{R}} \varphi(x) (-i \operatorname{sgn}(x)) \, dx \,, \tag{4.1.10}$$

which implies that

$$\widehat{W_0}(\xi) = -i \operatorname{sgn} \xi \,. \tag{4.1.11}$$

In particular, identity (4.1.11) is saying that $\widehat{W_0}$ is a (bounded) function.

We now use identity (4.1.11) to write

$$H(f)(x) = \big(\widehat{f}(\xi)(-i\,\mathrm{sgn}\,\xi)\big)^{\vee}(x). \tag{4.1.12}$$

This formula can be used to give an alternative definition of the Hilbert transform. An immediate consequence of (4.1.12) is that

$$\big\|H(f)\big\|_{L^2} = \big\|f\big\|_{L^2}, \tag{4.1.13}$$

that is, H is an isometry on $L^2(\mathbf{R})$. Moreover, H satisfies

$$H^2 = HH = -I, \tag{4.1.14}$$

where I is the identity operator. Equation (4.1.14) is a simple consequence of the fact that $(-i\,\mathrm{sgn}\,\xi)^2 = -1$. The adjoint operator H^* of H is uniquely defined via the identity

$$\big\langle f \,|\, H(g)\big\rangle = \int_{\mathbf{R}^n} f\,\overline{H(g)}\,dx = \int_{\mathbf{R}^n} H^*(f)\,\overline{g}\,dx = \big\langle H^*(f)\,|\,g\big\rangle\,,$$

and we can easily obtain that H^* has multiplier $\overline{-i\,\mathrm{sgn}\,\xi} = i\,\mathrm{sgn}\,\xi$. We conclude that $H^* = -H$. Likewise, we obtain $H^t = -H$.

4.1.b. Connections with Analytic Functions

We will now investigate connections of the Hilbert transform with the Poisson kernel. Recall the definition of the Poisson kernel P_y given in Example 1.2.17. Then for $f \in L^p(\mathbf{R})$, $1 \le p \le \infty$ we have

$$(P_y * f)(x) = \frac{y}{\pi}\int_{-\infty}^{+\infty} \frac{f(t)}{(x-t)^2 + y^2}\,dt, \tag{4.1.15}$$

and the integral in (4.1.15) converges absolutely by Hölder's inequality since the function $t \to ((x-t)^2 + y^2)^{-1}$ is in $L^{p'}(\mathbf{R})$ whenever $y > 0$.

Let $\mathrm{Re}\,z$ and $\mathrm{Im}\,z$ denote the real and imaginary parts of a complex number z. Observe that

$$(P_y * f)(x) = \mathrm{Re}\left(\frac{i}{\pi}\int_{-\infty}^{+\infty} \frac{f(t)}{x - t + iy}\,dt\right) = \mathrm{Re}\left(\frac{i}{\pi}\int_{-\infty}^{+\infty} \frac{f(t)}{z-t}\,dt\right),$$

where $z = x + iy$. The function

$$F_f(z) = \frac{i}{\pi}\int_{-\infty}^{+\infty} \frac{f(t)}{z-t}\,dt$$

defined on

$$\mathbf{R}^2_+ = \{z = x + iy :\ y > 0\}$$

is analytic since its $\partial/\partial\bar{z}$ derivative is zero. The real part of $F_f(x+iy)$ is $(P_y * f)(x)$. The imaginary part of $F_f(x+iy)$ is

$$\mathrm{Im}\left(\frac{i}{\pi}\int_{-\infty}^{+\infty}\frac{f(t)}{x-t+iy}\,dt\right) = \frac{1}{\pi}\int_{-\infty}^{+\infty}\frac{f(t)(x-t)}{(x-t)^2+y^2}\,dt = (f * Q_y)(x),$$

where Q_y is called the *conjugate Poisson kernel* and is given by

$$Q_y(x) = \frac{1}{\pi}\frac{x}{x^2+y^2}. \tag{4.1.16}$$

The functions $u_f(x+iy) = (f * P_y)(x)$ and $v_f(x+iy) = (f * Q_y)(x)$ are *conjugate harmonic functions* since $u_f + iv_f$ is analytic. Since the family P_y, $y > 0$ is an approximate identity, it follows from Theorem 1.2.19 that, $P_y * f \to f$ in $L^p(\mathbf{R})$ as $y \to 0$. The following question therefore arises: What is the limit of $f * Q_y$ as $y \to 0$? As we show next this limit has to be $H(f)$.

Theorem 4.1.5. *Let $1 \le p < \infty$. For any $f \in L^p(\mathbf{R})$ we have*

$$f * Q_\varepsilon - H^{(\varepsilon)}(f) \to 0 \tag{4.1.17}$$

in L^p and almost everywhere as $\varepsilon \to 0$.

Proof. We see that

$$(Q_\varepsilon * f)(x) - \frac{1}{\pi}\int_{|t|\ge\varepsilon}\frac{f(x-t)}{t}\,dt = \frac{1}{\pi}(f * \psi_\varepsilon)(x),$$

where $\psi_\varepsilon(x) = \varepsilon^{-1}\psi(\varepsilon^{-1}x)$ and

$$\psi(t) = \begin{cases} \frac{t}{t^2+1} - \frac{1}{t} & \text{when } |t| \ge 1 \\ \frac{t}{t^2+1} & \text{when } |t| < 1. \end{cases} \tag{4.1.18}$$

Note that ψ has integral zero. Furthermore, the integrable function

$$\Psi(t) = \begin{cases} \frac{1}{t^2+1} & \text{when } |t| \ge 1 \\ 1 & \text{when } |t| < 1 \end{cases} \tag{4.1.19}$$

is a radially decreasing majorant of ψ. It follows from Theorem 1.2.21 and Corollary 2.1.19 (with $a = 0$) that $f * \psi_\varepsilon \to 0$ in L^p and almost everywhere as $\varepsilon \to 0$. □

Remark 4.1.6. For $f \in \mathcal{S}(\mathbf{R})$ we know that $\lim_{\varepsilon\to 0} H^{(\varepsilon)}(f) = H(f)$, and we can therefore conclude from (4.1.17) that

$$Q_\varepsilon * f \to H(f) \qquad \text{a.e.}$$

as $\varepsilon \to 0$. This convergence is also true for a general $f \in L^p$, but this will be shown in Theorem 4.1.12. In the same theorem we will show that $H^{(\varepsilon)}(f)$ converges to $H(f)$ in L^p when $1 < p < \infty$.

4.1.c. L^p Boundedness of the Hilbert Transform

The reader is encouraged to solve Exercise 4.1.4. As a consequence of the result in that Exercise and of the fact that

$$x \le \tfrac{1}{2}(e^x - e^{-x}),$$

we obtain that

$$|\{x:\ |H(\chi_E)(x)| > \alpha\}| \le \frac{2}{\pi}\frac{|E|}{\alpha}, \qquad \alpha > 0, \tag{4.1.20}$$

for all E subsets of the real line of finite measure. Theorem 1.4.19 with $p_0 = q_0 = 1$ and $p_1 = q_1 = 2$ now implies that H is bounded on L^p for $1 < p < 2$. Duality gives that $H^* = -H$ is bounded on L^p for $2 < p < \infty$ and hence so is H.

We give another proof of the boundedness of the Hilbert transform H on $L^p(\mathbf{R})$ that has the advantage that it gives the best possible constant in the resulting norm inequality when p is a power of 2.

Theorem 4.1.7. *For all $1 < p < \infty$, there exists a positive constant C_p such that*

$$\big\|H(f)\big\|_{L^p} \le C_p\big\|f\big\|_{L^p}$$

for all f in $\mathcal{S}(\mathbf{R})$. Moreover, the C_p satisfies $C_p \le 2p$ for $2 \le p < \infty$ and $C_p \le 2p/(p-1)$ for $1 < p \le 2$. Therefore, the Hilbert transform H admits an extension to a bounded operator on $L^p(\mathbf{R})$ when $1 < p < \infty$.

Proof. The proof we give is based on the interesting identity valid for

$$H(f)^2 = f^2 + 2H(fH(f)), \tag{4.1.21}$$

whenever f is a real-valued Schwartz function. Before we prove (4.1.21), we discuss its origin. As the function $f + iH(f)$ has a holomorphic extension on $\mathbf{R}^2_+$, so does $(f+iH(f))^2 = f^2 - H(f)^2 + i\,2fH(f)$. Then $f^2 - H(f)^2$ has a harmonic extension u on the upper half-space whose conjugate harmonic function v must have boundary values $H(f^2 - H(f)^2)$. Thus $H(f^2 - H(f)^2) = 2fH(f)$, which implies (4.1.21) as $H^2 = -I$.

To give an alternative proof of (4.1.21) we take Fourier transforms. Let $m(\xi) = -i\,\mathrm{sgn}\,\xi$ be the symbol of the Hilbert transform. We have

$$\begin{aligned}
&\widehat{f^2}(\xi) + 2[H(fH(f))]^\wedge(\xi)\\
&= (\widehat{f} * \widehat{f})(\xi) + 2m(\xi)(\widehat{f} * \widehat{H(f)})(\xi)
\end{aligned}$$

$$= \int_{\mathbf{R}} \widehat{f}(\eta)\widehat{f}(\xi-\eta)\,d\eta + 2m(\xi)\int_{\mathbf{R}} \widehat{f}(\eta)\widehat{f}(\xi-\eta)m(\eta)\,d\eta \tag{4.1.22}$$

$$= \int_{\mathbf{R}} \widehat{f}(\eta)\widehat{f}(\xi-\eta)\,d\eta + 2m(\xi)\int_{\mathbf{R}} \widehat{f}(\eta)\widehat{f}(\xi-\eta)m(\xi-\eta)\,d\eta. \tag{4.1.23}$$

Averaging (4.1.22) and (4.1.23) we obtain

$$\widehat{f^2}(\xi) + 2[H(fH(f))]\widehat{\ }(\xi) = \int_{\mathbf{R}} \widehat{f}(\eta)\widehat{f}(\xi-\eta)\big[1 + m(\xi)\big(m(\eta) + m(\xi-\eta)\big)\big]\,d\eta\,.$$

But the last displayed expression is equal to

$$\int_{\mathbf{R}} \widehat{f}(\eta)\widehat{f}(\xi-\eta)m(\eta)m(\xi-\eta)\,d\eta = (\widehat{H(f)} * \widehat{H(f)})(\xi)$$

in view of the identity

$$m(\eta)m(\xi-\eta) = 1 + m(\xi)m(\eta) + m(\xi)m(\xi-\eta),$$

which is valid for the function $m(\xi) = -i\operatorname{sgn}\xi$.

Having established (4.1.21), we can easily obtain L^p bounds for H when $p = 2^k$ is a power of 2. We already know that H is bounded on L^p with norm one when $p = 2^k$ and $k = 1$. Suppose that H is bounded on L^p with bound c_p for $p = 2^k$ for some $k \in \mathbf{Z}^+$. Then

$$\begin{aligned}\big\|H(f)\big\|_{L^{2p}} = \big\|H(f)^2\big\|_{L^p}^{\frac{1}{2}} &\le \big(\big\|f^2\big\|_{L^p} + \big\|2H(fH(f))\big\|_{L^p}\big)^{\frac{1}{2}}\\ &\le \big(\big\|f\big\|_{L^{2p}}^2 + 2c_p\big\|fH(f)\big\|_{L^p}\big)^{\frac{1}{2}}\\ &\le \big(\big\|f\big\|_{L^{2p}}^2 + 2c_p\big\|f\big\|_{L^{2p}}\big\|H(f)\big\|_{L^{2p}}\big)^{\frac{1}{2}}.\end{aligned}$$

We obtain that

$$\left(\frac{\big\|H(f)\big\|_{L^{2p}}}{\big\|f\big\|_{L^{2p}}}\right)^2 - 2c_p\frac{\big\|H(f)\big\|_{L^{2p}}}{\big\|f\big\|_{L^{2p}}} - 1 \le 0.$$

If follows that

$$\frac{\big\|H(f)\big\|_{L^{2p}}}{\big\|f\big\|_{L^{2p}}} \le c_p + \sqrt{c_p^2 + 1}\,,$$

and from this we conclude that H is bounded on L^{2p} with bound

$$c_{2p} \le c_p + \sqrt{c_p^2 + 1}\,. \tag{4.1.24}$$

This completes the induction. We have proved that H maps L^p to L^p when $p = 2^k$, $k = 1, 2, \dots$. Interpolation now gives that H maps L^p to L^p for all $p \ge 2$. Since $H^* = -H$, duality gives that H is also bounded on L^p for $1 < p \le 2$. The reader is welcome to compare this proof with the one given in Theorem 3.5.6.

The previous proof of the boundedness of the Hilbert transform provides us with some useful information about the norm of this operator on $L^p(\mathbf{R})$. Let us begin with the identity

$$\cot\frac{x}{2} = \cot x + \sqrt{1 + \cot^2 x},$$

valid for $0 < x < \frac{\pi}{2}$. If $c_p \le \cot\frac{\pi}{2p}$, then (4.1.24) gives that

$$c_{2p} \le c_p + \sqrt{c_p^2 + 1} \le \cot\frac{\pi}{2p} + \sqrt{1 + \cot^2\frac{\pi}{2p}} = \cot\frac{\pi}{2\cdot 2p}\,,$$

and since $1 = \cot \frac{\pi}{4} = \cot \frac{\pi}{2\cdot 2}$, we obtain by induction that the numbers $\cot \frac{\pi}{2p}$ are indeed bounds for the norm of H on L^p when $p = 2^k$, $k = 1, 2, \dots$. Duality now gives that the numbers $\tan \frac{\pi}{2p}$ are bounds for norm of H on L^p when $p = \frac{2^k}{2^k - 1}$, $k = 1, 2, \dots$. These bounds allow us to derive good estimates for the norm $\big\|H\big\|_{L^p \to L^p}$ as $p \to 1$ and $p \to \infty$. Indeed, since $\cot \frac{\pi}{2p} \le p$ when $p \ge 2$, the Riesz-Thorin interpolation theorem gives that $\big\|H\big\|_{L^p \to L^p} \le 2p$ for $2 \le p < \infty$ and by duality $\big\|H\big\|_{L^p \to L^p} \le \frac{2p}{p-1}$ for $1 < p \le 2$.

□

Remark 4.1.8. The numbers $\cot \frac{\pi}{2p}$ for $2 \le p < \infty$ and $\tan \frac{\pi}{2p}$ for $1 < p \le 2$ are indeed equal to the norms of the Hilbert transform H on $L^p(\mathbf{R})$. This requires a more delicate argument, for which we refer to Exercise 4.1.13.

Remark 4.1.9. We may wonder what happens when $p = 1$ or $p = \infty$. The Hilbert transform of $\chi_{[a,b]}$ computed in Example 4.1.3 is easily seen to be unbounded and not integrable since it behaves like $1/|x|$ as $x \to \infty$. This behavior near infinity suggests that the Hilbert transform may map L^1 to $L^{1,\infty}$. This is indeed the case, but this will not be shown until Section 4.3.

We now introduce the maximal Hilbert transform.

Definition 4.1.10. The *maximal Hilbert transform* is the operator

$$H^{(*)}(f)(x) = \sup_{\varepsilon > 0} \left| H^{(\varepsilon)}(f)(x) \right| \tag{4.1.25}$$

defined for all f in L^p, $1 \le p < \infty$. For such f, $H^{(\varepsilon)}(f)$ is well defined by Hölder's inequality and so is $H^{(*)}(f)$. Observe that for some values of x, $H^{(*)}(f)(x)$ may be infinite.

Example 4.1.11. Using the result of Example 4.1.4, we see that

$$H^{(*)}(\chi_{[a,b]})(x) = \frac{1}{\pi} \left| \log \frac{|x-a|}{|x-b|} \right| . \tag{4.1.26}$$

We see that, in general, $H^{(*)}(f)(x) \neq |H(f)(x)|$ by taking f to be the characteristic function of the a union of two disjoint closed intervals.

The definition of H gives that $H^{(\varepsilon)}(f)$ converges pointwise to $H(f)$ if f is a smooth function with compact support. If we knew an estimate of the type $\big\|H^{(*)}(f)\big\|_{L^p} \le C_p \big\|f\big\|_{L^p}$, it would follow from Theorem 2.1.14 that $H^{(\varepsilon)}(f)$ converges to $H(f)$ a.e. as $\varepsilon \to 0$ for any $f \in L^p$. This realization of $H(f)$ as an a.e. limit of a sequence of L^p functions provides an expression $H(f)$ for general f in L^p. Observe that Theorem 4.1.7 only implies that H has a (unique) extension bounded on L^p, but it does not give a way to describe $H(f)$ when f is a general L^p function.

The next theorem is a simple consequence of these ideas.

Theorem 4.1.12. *There exists a constant C such that for all $1 < p < \infty$ we have*

$$\big\|H^{(*)}(f)\big\|_{L^p} \le C\max\big(p,(p-1)^{-1}\big)\big\|f\big\|_{L^p}. \tag{4.1.27}$$

Moreover, for all f in $L^p(\mathbf{R})$, $H^{(\varepsilon)}(f)$ converges to $H(f)$ a.e. and in L^p.

Proof. The following proof yields the slightly weaker bound $C\max\big(p,(p-1)^{-2}\big)$. For another proof of this theorem with the asserted bound in (4.1.27) we refer the reader to Theorem 8.2.2.

Recall the kernels P_ε and Q_ε defined in (4.1.15) and (4.1.16). Fix $1 < p < \infty$ and suppose momentarily that

$$f * Q_\varepsilon = H(f) * P_\varepsilon\,, \qquad \varepsilon > 0 \tag{4.1.28}$$

holds whenever f is an L^p function. Then we have

$$H^{(\varepsilon)}(f) = H^{(\varepsilon)}(f) - f * Q_\varepsilon + H(f) * P_\varepsilon\,. \tag{4.1.29}$$

Using the identity

$$H^{(\varepsilon)}(f)(x) - (f * Q_\varepsilon)(x) = -\int_{\mathbf{R}} f(x-t)\psi_\varepsilon(t)\,dt\,, \tag{4.1.30}$$

where ψ is as in (4.1.18) and Corollary 2.1.12, we obtain the estimate

$$\sup_{\varepsilon>0}\big|H^{(\varepsilon)}(f)(x) - (f * Q_\varepsilon)(x)\big| \le \big\|\Psi\big\|_{L^1} M(f)(x) \tag{4.1.31}$$

Ψ is as in (4.1.19) and M is the Hardy-Littlewood maximal function. Using (4.1.29) and (4.1.31), we obtain for $f \in L^p(\mathbf{R}^n)$

$$|H^{(*)}(f)(x)| \le \big\|\Psi\big\|_{L^1} M(f)(x) + M(H(f))(x)\,. \tag{4.1.32}$$

It follows immediately from (4.1.32) that $H^{(*)}$ is L^p bounded with norm at most $C\max\big(p,(p-1)^{-2}\big)$.

It suffices therefore to establish (4.1.28). In the proof of (4.1.28), we might as well assume that f is a Schwartz function. Taking Fourier transforms, we see that (4.1.28) will be a consequence of the identity

$$\big((-i\operatorname{sgn}\xi)e^{-2\pi|\xi|}\big)^\vee(x) = \frac{1}{\pi}\frac{x}{x^2+1}. \tag{4.1.33}$$

Writing the inverse Fourier transform as an integral from $-\infty$ to $+\infty$ and then changing this to an integral from 0 to ∞, we obtain that (4.1.33) is equivalent to the identity

$$(-i)(2i)\int_0^\infty e^{-2\pi\xi}\sin(2\pi x\xi)\,d\xi = \frac{1}{\pi}\frac{x}{x^2+1},$$

which can be checked easily using integration by parts twice.

The statement in the theorem about the almost everywhere convergence of $H^{(\varepsilon)}(f)$ to $H(f)$ is a consequence of (4.1.27), of the fact that the alleged convergence holds for Schwartz functions, and of Theorem 2.1.14. Finally, the L^p convergence follows from the almost everywhere convergence and the Lebesgue dominated theorem in view of the validity of (4.1.32).

□

4.1.d. The Riesz Transforms

We will now study an n-dimensional analogue of the Hilbert transform. It turns out that there exist n operators in $\mathbf{R}^n$, called the Riesz transforms, with properties analogous to those of the Hilbert transform on $\mathbf{R}$.

To define the Riesz transforms, we first introduce tempered distributions W_j on $\mathbf{R}^n$, for $1 \le j \le n$, as follows. For $\varphi \in \mathcal{S}(\mathbf{R}^n)$, let

$$\langle W_j, \varphi \rangle = \frac{\Gamma(\frac{n+1}{2})}{\pi^{\frac{n+1}{2}}} \lim_{\varepsilon \to 0} \int_{|y| \ge \varepsilon} \frac{y_j}{|y|^{n+1}} \varphi(y)\, dy.$$

The reader should check that indeed $W_j \in \mathcal{S}'(\mathbf{R}^n)$. Observe that the normalization of W_j is similar to that of the Poisson kernel.

Definition 4.1.13. For $1 \le j \le n$, the jth *Riesz transform* of f is given by convolution with the distribution W_j, that is,

$$\begin{aligned} R_j(f)(x) &= (f * W_j)(x) \\ &= \frac{\Gamma(\frac{n+1}{2})}{\pi^{\frac{n+1}{2}}} \text{ p.v.} \int_{\mathbf{R}^n} \frac{x_j - y_j}{|x-y|^{n+1}} f(y)\, dy\,, \end{aligned} \tag{4.1.34}$$

for all $f \in \mathcal{S}(\mathbf{R}^n)$. Definition 4.1.13 makes sense for any integrable function f that has the property that for all x there exist $C_x > 0$, $\varepsilon_x > 0$, and $\delta_x > 0$ such that for y satisfying $|y - x| < \delta_x$ we have $|f(x) - f(y)| \le C_x |x - y|^{\varepsilon_x}$. The principal value integral in (4.1.34) is as in Definition 4.1.1.

We now give a characterization of R_j using the Fourier transform. For this we will need to compute the Fourier transform of W_j.

Proposition 4.1.14. *The Riesz transforms R_j are given on the Fourier transform side by multiplication with the functions $-i\xi_j/|\xi|$. That is, for any f in $\mathcal{S}(\mathbf{R}^n)$ we have*

$$R_j(f)(x) = \Big(-\frac{i\xi_j}{|\xi|}\widehat{f}(\xi)\Big)^{\vee}(x). \tag{4.1.35}$$

Proof. The proof is essentially a reprise of the corresponding proof for the Hilbert transform, but it involves a few technical difficulties. Fix a Schwartz function φ on $\mathbf{R}^n$. Then for $1 \le j \le n$ we have

$$\begin{aligned}
\langle \widehat{W_j}, \varphi \rangle &= \langle W_j, \widehat{\varphi} \rangle = \frac{\Gamma(\frac{n+1}{2})}{\pi^{\frac{n+1}{2}}} \lim_{\varepsilon \to 0} \int_{|\xi| \ge \varepsilon} \widehat{\varphi}(\xi) \frac{\xi_j}{|\xi|^{n+1}}\, d\xi \\
&= \frac{\Gamma(\frac{n+1}{2})}{\pi^{\frac{n+1}{2}}} \lim_{\varepsilon \to 0} \int_{\frac{1}{\varepsilon} \ge |\xi| \ge \varepsilon} \int_{\mathbf{R}^n} \varphi(x) e^{-2\pi i x \cdot \xi}\, dx\, \frac{\xi_j}{|\xi|^{n+1}}\, d\xi \\
&= \lim_{\varepsilon \to 0} \int_{\mathbf{R}^n} \varphi(x) \left[\frac{\Gamma(\frac{n+1}{2})}{\pi^{\frac{n+1}{2}}} \int_{\frac{1}{\varepsilon} \ge |\xi| \ge \varepsilon} e^{-2\pi i x \cdot \xi} \frac{\xi_j}{|\xi|^{n+1}}\, d\xi \right] dx \\
&= \lim_{\varepsilon \to 0} \int_{\mathbf{R}^n} \varphi(x) \left[\frac{\Gamma(\frac{n+1}{2})}{\pi^{\frac{n+1}{2}}} \int_{\mathbf{S}^{n-1}} \int_{\varepsilon \le r \le \frac{1}{\varepsilon}} e^{-2\pi i r x \cdot \theta} \frac{r}{r^{n+1}} r^{n-1} dr\, \theta_j d\theta \right] dx \\
&= \int_{\mathbf{R}^n} \varphi(x) \left[-i \frac{\Gamma(\frac{n+1}{2})}{\pi^{\frac{n+1}{2}}} \int_{\mathbf{S}^{n-1}} \int_0^\infty \sin(2\pi r x \cdot \theta) \frac{dr}{r}\, \theta_j\, d\theta \right] dx \\
&= \int_{\mathbf{R}^n} \varphi(x) \left[-i \frac{\pi}{2} \frac{\Gamma(\frac{n+1}{2})}{\pi^{\frac{n+1}{2}}} \int_{\mathbf{S}^{n-1}} \operatorname{sgn}(x \cdot \theta) \theta_j\, d\theta \right] dx \\
&= \int_{\mathbf{R}^n} -i \varphi(x) \frac{x_j}{|x|}\, dx\,,
\end{aligned}$$

where in the penultimate equality we used the identity $\int_0^\infty \frac{\sin t}{t} dt = \frac{\pi}{2}$ for which we refer to Exercise 4.1.1, while in the last equality we used the identity

$$-i \frac{\pi}{2} \frac{\Gamma(\frac{n+1}{2})}{\pi^{\frac{n+1}{2}}} \int_{\mathbf{S}^{n-1}} \operatorname{sgn}(x \cdot \theta) \theta_j\, d\theta = -i \frac{x_j}{|x|}\,, \tag{4.1.36}$$

which we need to establish. The passage of the limit inside the integral in the previous calculation is a consequence of the Lebesgue dominated convergence theorem, which is justified from the fact that

$$\left| \int_\varepsilon^{1/\varepsilon} \frac{\sin(2\pi r \theta)}{r}\, dr \right| \le 4 \tag{4.1.37}$$

for all $\varepsilon > 0$. For a proof of (4.1.37) we again refer to Exercise 4.1.1.

□

It remains to establish (4.1.36). At this point we will need a lemma. Let us recall that $O(n)$ is the set of all orthogonal $n \times n$ matrices with real entries. An invertible matrix A is called orthogonal if its transpose A^t is equal to its inverse A^{-1}, that is, $AA^t = A^t A = I$.

Lemma 4.1.15. *Let $m(x) = (m_1(x), \dots, m_n(x))^t$ be an $n \times 1$ column vector of real-valued measurable functions defined on $\mathbf{R}^n$. Suppose that $m(x)$ is homogeneous* *of degree zero function [i.e., $m(\delta x) = m(x)$ for all $\delta > 0$] and that m commutes with all orthogonal matrices, that is,*

$$(m \circ A)(x) = (A \circ m)(x)$$

for all $x \in \mathbf{R}^n$ and $A \in O(n)$. Then there exists a constant c such that

$$m(x) = c\,\frac{x}{|x|}\,. \tag{4.1.38}$$

Assuming Lemma 4.1.15, we prove (4.1.36). Let

$$m_j(x) = \int_{\mathbf{S}^{n-1}} \operatorname{sgn}(x \cdot \theta)\theta_j \, d\theta$$

and m be the corresponding $n \times 1$ column vector. It is straightforward that m is homogeneous of degree zero and that it commutes with rotations. The last assertion follows from a change of variables, which gives

$$m_j(Ax) = \int_{\mathbf{S}^{n-1}} \operatorname{sgn}(Ax \cdot \theta)\theta_j \, d\theta = \int_{\mathbf{S}^{n-1}} \operatorname{sgn}(x \cdot \theta)(A\theta)_j \, d\theta = A(m(x))_j,$$

where the last term denotes the jth entry of the vector $A(m(x))$. Therefore, Lemma 4.1.15 implies that $m_j(x) = c\,x_j\,|x|^{-1}$ for some c. It now suffices to evaluate c. Let e_j be the element of $\mathbf{S}^{n-1}$ with one on the jth entry and zero elsewhere. Then $c = m_j(e_j)$ and it suffices to compute $m_j(e_j)$. We have

$$\begin{aligned}
c &= m_j(e_j) = \int_{\mathbf{S}^{n-1}} |\theta_j| \, d\theta \\
&= \int_{-1}^{1} |s| \int_{\sqrt{1-s^2}\,\mathbf{S}^{n-2}} d\varphi \, \frac{ds}{(1-s^2)^{\frac{1}{2}}} \qquad \text{by Appendix D.2} \\
&= \omega_{n-2} \int_{-1}^{1} |s|(1-s^2)^{\frac{n-3}{2}} \, ds \\
&= \omega_{n-2} \int_{0}^{1} u^{\frac{n-3}{2}} \, du = \frac{2\omega_{n-2}}{n-1} = \frac{2\pi^{\frac{n-1}{2}}}{\Gamma(\frac{n-1}{2})(\frac{n-1}{2})} \qquad \text{by Appendix A.3} \\
&= \frac{2\pi^{\frac{n-1}{2}}}{\Gamma(\frac{n+1}{2})}\,.
\end{aligned}$$

This implies (4.1.36), and the proof of the Proposition 4.1.14 is complete.

We now prove Lemma 4.1.15.

Proof. By homogeneity it suffices to take $x \in \mathbf{S}^{n-1}$ in (4.1.38). Let e_j be the usual unit vectors. Pick an $A \in O(n)$ that fixes e_1 and takes the vector $m(e_1)$ to a vector on the x_1-axis. That is, pick A so that $A(e_1) = e_1$ and $A(m(e_1)) = (c, 0, \dots, 0)$. Since $A(m(x)) = m(A(x))$, it follows that $A(m(e_1)) = m(A(e_1)) =$

$m(e_1)$ and hence $c = m_1(e_1)$ and $m_j(e_1) = 0$ for $j \geq 2$. Let now $B = (b_{jk})$ be any other orthogonal matrix in $O(n)$. Then

$$m_j(B(e_1)) = \sum_{k=1}^{n} b_{jk} m_k(e_1) = c\, b_{j1}.$$

But if $B(e_1) = x$, then $b_{j1} = x_j$ so $m_j(x) = cx_j$, which proves the lemma since $|x| = 1$.

□

Proposition 4.1.16. *The Riesz transforms satisfy*

$$-I = \sum_{j=1}^{n} R_j^2, \tag{4.1.39}$$

where I is the identity operator.

Proof. Use the Fourier transform and the identity $\sum_{j=1}^n (-i\xi_j/|\xi|)^2 = -1$ to obtain that $\sum_{j=1}^n R_j^2(f) = -f$ for any f in the Schwartz class.

□

We end this section by discussing a use of the Riesz transforms to partial differential equations.

Example 4.1.17. Suppose that f is a given Schwartz function on $\mathbf{R}^n$ and that u is a distribution that solves *Laplace's equation*

$$\Delta(u) = f\,.$$

Then we can express all second-order derivatives of u in terms of the Riesz transforms of f. First we note that

$$(-4\pi^2|\xi|^2)\,\widehat{u}(\xi) = \widehat{f}(\xi)\,.$$

It follows that for all $1 \leq j,k \leq n$ we have

$$\begin{aligned}
\partial_j\partial_k u &= \big[(2\pi i\xi_j)(2\pi i\xi_k)\widehat{u}(\xi)\big]^{\vee} \\
&= \Big[(2\pi i\xi_j)(2\pi i\xi_k)\frac{\widehat{f}(\xi)}{-4\pi^2|\xi|^2}\Big]^{\vee} \\
&= -\,R_jR_k(f)
\end{aligned}$$

and, in particular, we conclude that $\partial_j\partial_k u$ are functions.

Thus the Riesz transforms provide an explicit way to recover second-order derivatives in terms of the Laplacian. Such representations are useful in precisely controlling quantitative expressions (such as norms) of second-order derivatives in terms of the corresponding expressions for the Laplacian. For instance, this is the case with the L^p norm; the L^p boundedness of the Riesz transforms will be one of the main results of the next section. We refer to Exercises 4.2.10 and 4.2.11 for similar applications.

Exercises

4.1.1. (a) Show that for all $0 < a < b < \infty$ we have

$$\left| \int_a^b \frac{\sin x}{x} \, dx \right| \le 4 .$$

(b) For $a > 0$ define

$$I(a) = \int_0^\infty \frac{\sin x}{x} e^{-ax} \, dx$$

and show that $I(a)$ is continuous at zero. Differentiate in a and look at the behavior of $I(a)$ as $a \to \infty$ to obtain the identity

$$I(a) = \frac{\pi}{2} - \arctan(a) .$$

From this deduce that $I(0) = \frac{\pi}{2}$ and hence the following identity used in (4.1.9):

$$\int_{-\infty}^{+\infty} \frac{\sin(bx)}{x} dx = \pi \operatorname{sgn}(b)$$

(c) Argue as in part (b) to prove for $a \ge 0$ the identity

$$\int_0^\infty \frac{1 - \cos x}{x^2} e^{-ax} \, dx = \frac{\pi}{2} - \arctan(a) + a \log \frac{a}{\sqrt{1 + a^2}} .$$

[*Hint:* Part (a): Consider the cases $b \le 1$, $a \le 1 \le b$, $1 \le a$. When $a \ge 1$, integrate by parts.]

4.1.2. (a) Let φ be a compactly supported C^{m+1} function on $\mathbf{R}$ for some m in $\mathbf{Z}^+ \cup \{0\}$. Prove thatif $\varphi^{(m)}$ is the mth derivative of φ, then

$$|H(\varphi^{(m)})(x)| \le C_{m,\varphi} (1 + |x|)^{-m-1}$$

for some $C_{m,\varphi} > 0$.
(b) Let φ be a compactly supported C^{m+1} function on $\mathbf{R}^n$ for some $m \in \mathbf{Z}^+$. Show that

$$|R_j(\partial^\alpha \varphi)(x)| \le C_{n,m,\varphi} (1 + |x|)^{-n-m}$$

for some $C_{n,m,\varphi} > 0$ and all multi-indices α with $|\alpha| = m$.
(c) Let I be an interval on the line and assume that a function h is equal to 1 on the left half of I, equal to -1 on the right half of I, and vanishes outside I. Prove that for $x \notin \frac{3}{2} I$ we have

$$|H(h)(x)| \le 4 |I|^2 |x - \text{center}(I)|^{-2} .$$

[*Hint:* Use that when $|t| \le \frac{1}{2}$ we have $\log(1+t) = t + R_1(t)$, where $|R_1(t)| \le 2|t|^2$.]

4.1.3. (a) Using equation (4.1.12), extend the definition of the Hilbert transform to any function f on $\mathbf{R}$ with at most polynomial growth at infinity. Then $H(f)$ is a well-defined tempered distribution for any such f. Prove that

$$\begin{aligned} H(e^{ix}) &= -ie^{ix}\,, \\ H(\cos x) &= \sin x\,, \\ H(\sin x) &= -\cos x\,, \\ H(\sin(\pi x)/\pi x) &= (\cos(\pi x)-1)/\pi x\,. \end{aligned}$$

(b) Show that the operators given by convolution with the smooth function $\sin(t)/t$ and the distribution p.v. $\cos(t)/t$ are bounded on $L^p(\mathbf{R})$ whenever $1<p<\infty$.

4.1.4. (*Stein and Weiss* [**485**]) Show that the distribution function of the Hilbert transform of a characteristic function of a measurable subset E of $\mathbf{R}$ of finite measure is

$$d_{H(\chi_E)}(\alpha) = \frac{4|E|}{e^{\pi\alpha}-e^{-\pi\alpha}}\,, \qquad \alpha>0\,.$$

[*Hint:* First take $E=\bigcup_{j=1}^N (a_j,b_j)$, where $b_j<a_{j+1}$. Show that the equation $H(\chi_E)(x)=\pi\alpha$ has exactly one root ρ_j in each open interval (a_j,b_j) for $1\le j\le N$ and exactly one root r_j in each interval (b_j,a_{j+1}) for $1\le j\le N$, $(a_{N+1}=\infty)$. Then $|\{x\in\mathbf{R}:\ H(\chi_E)(x)>\pi\alpha\}|=\sum_{j=1}^N r_j-\sum_{j=1}^N \rho_j$, and this can be expressed in terms of $\sum_{j=1}^N a_j$ and $\sum_{j=1}^N b_j$. Argue similarly for the set $\{x\in\mathbf{R}:\ H(\chi_E)(x)<-\pi\alpha\}$. For a general measurable set E, find sets E_n such that each E_n is a finite union of intervals and that $\chi_{E_n}\to\chi_E$ in L^2. Then $H(\chi_{E_n})\to H(\chi_E)$ in measure; thus $H(\chi_{E_{n_k}})\to H(\chi_E)$ a.e. for some subsequence n_k. The Lebesgue dominated convergence theorem gives $d_{H(\chi_{E_{n_k}})}\to d_{H(\chi_E)}$. See Figure 4.1.]

4.1.5. Let $1\le p<\infty$. Suppose that there exists a constant $C>0$ such that for all $f\in\mathcal{S}(\mathbf{R})$ with L^p norm one we have

$$|\{x:\ |H(f)(x)|>1\}|\le C.$$

Using only this inequality, prove that H maps $L^p(\mathbf{R})$ to $L^{p,\infty}(\mathbf{R})$.
[*Hint:* Try functions of the form $\lambda^{-1/p}f(\lambda^{-1}x)$.]

4.1.6. Let φ be in $\mathcal{S}(\mathbf{R})$. Prove that

$$\begin{aligned} \lim_{N\to\infty}\ \text{p.v.}\int_{\mathbf{R}}\frac{e^{2\pi iNx}}{x}\varphi(x)\,dx &= \varphi(0)\pi i, \\ \lim_{N\to-\infty}\ \text{p.v.}\int_{\mathbf{R}}\frac{e^{2\pi iNx}}{x}\varphi(x)\,dx &= -\varphi(0)\pi i. \end{aligned}$$

4.1.7. Use Exercise 4.1.6 to find an explicit formula for the operator T_α given by convolution with the distribution whose Fourier transform is the function

$$u_\alpha(\xi) = e^{-\pi i \alpha \operatorname{sgn} \xi}.$$

Conclude that the T_α's are isometries on $L^2(\mathbf{R})$ that satisfy

$$(T_\alpha)^{-1} = T_{2\pi - \pi\alpha}.$$

4.1.8. Let $Q_y^{(j)}$ be the jth *conjugate Poisson kernel* of P_y defined by

$$Q_y^{(j)}(x) = \frac{\pi^{\frac{n+1}{2}}}{\Gamma(\frac{n+1}{2})} \frac{x_j}{(|x|^2 + y^2)^{\frac{n+1}{2}}}.$$

Prove that $(Q_y^{(j)})^\wedge(\xi) = -i\xi_j e^{-2\pi|\xi|}/|\xi|$. Conclude that $R_j(P_y) = Q_y^{(j)}$ and that for f in $L^2(\mathbf{R}^n)$ we have $R_j(f) * P_y = f * Q_y^{(j)}$. These results are analogous to the statements $\widehat{Q_y}(\xi) = -i \operatorname{sgn}(\xi)\widehat{P_y}(\xi)$, $H(P_y) = Q_y$, and $H(f) * P_y = f * Q_y$.

4.1.9. Let $f_0, f_1, \dots, f_n$ all belong to $L^2(\mathbf{R}^n)$ and let $u_j = P_y * f_j$ be their corresponding Poisson integrals for $0 \le j \le n$. Show that a necessary and sufficient condition for

$$f_j = R_j(f_0), \qquad j = 1, \dots, n$$

is that the following system of generalized Cauchy-Riemann equations holds:

$$\sum_{j=0}^{n} \frac{\partial u_j}{\partial x_j} = 0,$$

$$\frac{\partial u_j}{\partial x_k} = \frac{\partial u_k}{\partial x_j}, \qquad j \neq k, \qquad x_0 = y.$$

4.1.10. Prove the distributional identity

$$\partial_j |x|^{-n+1} = (1-n)\text{p.v.}\,\frac{x_j}{|x|^{n+1}}.$$

Then take Fourier transforms of both sides and use Theorem 2.4.6 to obtain another proof of Proposition 4.1.14.

4.1.11. (a) Prove that if T is a bounded operator on $L^2(\mathbf{R})$ that commutes with translations and dilations and anticommutes with the reflection $f(x) \to \widetilde{f}(x) = f(-x)$, then T is a constant multiple of the Hilbert transform.
(b) Prove that if T is a bounded operator on $L^2(\mathbf{R})$ that commutes with translations and dilations and vanishes when applied to functions whose Fourier transform is supported in $[0, \infty)$, then T is a constant multiple of the operator $f \to (\widehat{f}\chi_{(-\infty,0]})^\vee$.

4.1.12. Use Lemma 4.1.15 to prove that if T_j is bounded on $L^2(\mathbf{R}^n)$ for $1 \le j \le n$ and commutes with translations and dilations, and for each $A \in O(n)$ we have $A(T(f)) = T(A(f))$, then there exists a constant c such that $T_j = cR_j$. Here $T = (T_1, \dots, T_n)^t$ and $A(f)(x) = f(A(x))$.

4.1.13. (*Pichorides* [**413**]) Fix $1 < p \le 2$.
(a) Show that the function $(x, y) \to \mathrm{Re}\,(|x| + iy)^p$ is subharmonic on $\mathbf{R}^2$.
(b) Prove that for f in $C_0^\infty(\mathbf{R})$ we have

$$\int_{\mathbf{R}} \mathrm{Re}\,(|f(x)| + iH(f)(x))^p\, dx \ge 0.$$

(c) Prove that for all a and b reals we have

$$|b|^p \le \big(\tan \frac{\pi}{2p}\big)^p |a|^p - D_p \mathrm{Re}\,(|a| + ib)^p$$

for some $D_p > 0$. Then use part (b) to conclude that

$$\big\|H\big\|_{L^p \to L^p} \le \tan \frac{\pi}{2p}.$$

(d) To deduce that this constant is sharp, take $\pi/2p' < \gamma < \pi/2p$ and let $f_\gamma(x) = (x+1)^{-1}|x+1|^{2\gamma/\pi}|x-1|^{-2\gamma/\pi}\cos\gamma$. Then

$$H(f_\gamma)(x) = \begin{cases} \frac{1}{x+1}\frac{|x+1|}{|x-1|}^{2\gamma/\pi} \sin\gamma & \text{when } |x| > 1, \\ -\frac{1}{x+1}\frac{|x+1|}{|x-1|}^{2\gamma/\pi} \sin\gamma & \text{when } |x| < 1. \end{cases}$$

[*Hint:* Part (b): Let C_R be the circle of radius R centered at $(0, R)$ in $\mathbf{R}^2$. Use that the integral of the subharmonic function

$$(x, y) \to \mathrm{Re}\,(|(P_y * f)(x)| + i(Q_y * f)(x))^p$$

over C_R is at least $2\pi R\,\mathrm{Re}\,(|(P_R * f)(0)| + i(Q_R * f)(0))^p$ and let $R \to \infty$. Part (d): This is best seen by considering the restriction of the analytic function

$$F(z) = (z+1)^{-1}\left(\frac{iz+i}{z-1}\right)^{2\gamma/\pi}$$

on $\mathbf{R} \times \{0\}$.]

4.2. Homogeneous Singular Integrals and the Method of Rotations

So far we have introduced the Hilbert and the Riesz transforms and we have derived the L^p boundedness of the first. The boundedness properties of the Riesz transforms will be a consequence of the results discussed in this section.

4.2.a. Homogeneous Singular Integrals

We introduce singular integral operators on $\mathbf{R}^n$ that appropriately generalize the Riesz transforms on $\mathbf{R}^n$. Here is the setup. We fix Ω to be an integrable function of the unit sphere $\mathbf{S}^{n-1}$ with mean value zero. Observe that the kernel

$$K_\Omega(x) = \frac{\Omega(x/|x|)}{|x|^n}, \qquad x \neq 0, \tag{4.2.1}$$

is homogeneous of degree $-n$ just like the functions $x_j/|x|^{n+1}$. Since K_Ω is not in $L^1(\mathbf{R}^n)$, convolution with K_Ω cannot be defined as an operation on Schwartz functions on $\mathbf{R}^n$. For this reason we introduce a distribution W_Ω in $\mathcal{S}'(\mathbf{R}^n)$ by setting

$$\langle W_\Omega, \varphi\rangle = \lim_{\varepsilon\to 0}\int_{|x|\geq\varepsilon} K_\Omega(x)\varphi(x)\,dx = \lim_{\varepsilon\to 0}\int_{\varepsilon\leq|x|\leq\varepsilon^{-1}} K_\Omega(x)\varphi(x)\,dx \tag{4.2.2}$$

for $\varphi \in \mathcal{S}(\mathbf{R}^n)$. Using the fact that Ω has mean value zero, we can easily see that W_Ω is a well-defined tempered distribution on $\mathbf{R}^n$. Indeed, since K_Ω has integral zero over all annuli centered at the origin, we have

$$\begin{aligned}
&\big|\langle W_\Omega, \varphi\rangle\big| \\
&= \left|\lim_{\varepsilon\to 0}\int_{\varepsilon\leq|x|\leq 1}\frac{\Omega(x/|x|)}{|x|^n}(\varphi(x)-\varphi(0))\,dx + \int_{|x|\geq 1}\frac{\Omega(x/|x|)}{|x|^n}\varphi(x)\,dx\right| \\
&\leq \big\|\nabla\varphi\big\|_{L^\infty}\int_{|x|\leq 1}\frac{\Omega(x/|x|)}{|x|^{n-1}}\,dx + \sup_{y\in\mathbf{R}^n}|y|\,|\varphi(y)|\int_{|x|\geq 1}\frac{\Omega(x/|x|)}{|x|^{n+1}}\,dx \\
&\leq C_1\big\|\nabla\varphi\big\|_{L^\infty}\big\|\Omega\big\|_{L^1} + C_2\sum_{|\alpha|\leq 1}\big\|\varphi(x)x^\alpha\big\|_{L^\infty}\big\|\Omega\big\|_{L^1},
\end{aligned}$$

for suitable C_1 and C_2, where we used (2.2.2) in the last estimate. Note that the distribution W_Ω coincides with the function K_Ω on $\mathbf{R}^n \setminus \{0\}$.

Definition 4.2.1. Let Ω be integrable on the sphere $\mathbf{S}^{n-1}$ with mean value zero. For $0 < \varepsilon < N$ and $f \in \bigcup_{1\leq p<\infty} L^p(\mathbf{R}^n)$ we define the *truncated singular integral*

$$T_\Omega^{(\varepsilon,N)}(f)(x) = \int_{\varepsilon\leq|y|\leq N} f(x-y)\,\frac{\Omega(y/|y|)}{|y|^n}\,dy\,. \tag{4.2.3}$$

Note that for $f \in L^p(\mathbf{R}^n)$ we have

$$\big\|T_\Omega^{(\varepsilon,N)}(f)\big\|_{L^p} \leq \big\|\Omega\big\|_{L^1}\log(N/\varepsilon)\,\big\|f\big\|_{L^p(\mathbf{R}^n)},$$

which implies that the (4.2.3) is finite a.e. and therefore well defined. We denote by T_Ω the singular integral operator whose kernel is the distribution W_Ω, that is,

$$T_\Omega(f)(x) = (f * W_\Omega)(x) = \lim_{\substack{\varepsilon\to 0\\ N\to\infty}} T_\Omega^{(\varepsilon,N)}(f)(x),$$

defined for $f \in \mathcal{S}(\mathbf{R}^n)$. The associated *maximal singular integral* is defined by

$$T_\Omega^{(*)}(f) = \sup_{0<N<\infty}\sup_{0<\varepsilon<N} \big|T_\Omega^{(\varepsilon,N)}(f)\big|. \tag{4.2.4}$$

We note that if Ω is bounded, there is no need to use the upper truncations in the definition of $T_\Omega^{(\varepsilon,N)}$ given in (4.2.3). In this case the maximal singular integrals could be defined as

$$T_\Omega^{(*)}(f)(x) = \sup_{\varepsilon>0}\left|\int_{|y|\ge\varepsilon} f(x-y)\frac{\Omega(y/|y|)}{|y|^n}\,dy\right|. \tag{4.2.5}$$

This is the case with the Hilbert transform and the Riesz transforms. Indeed, suppose Ω is bounded. It is clear that

$$\left|\int_{\varepsilon\le|y|\le N} f(x-y)\frac{\Omega(y/|y|)}{|y|^n}\,dy\right| \le \sup_{0<N<\infty}\big|T_\Omega^{(\varepsilon,N)}(f)(x)\big|. \tag{4.2.6}$$

Then for $f \in L^p(\mathbf{R}^n)$, $1 \le p < \infty$, we let $N \to \infty$ on the left in (4.2.6) and we note that the limit exists in view of the assumptions on f and Ω. Then we take the supremum over $\varepsilon > 0$ to deduce that the expression in (4.2.5) is controlled by the expression for $T_\Omega^{(*)}$ given in (4.2.4). In this sense the definition in (4.2.4) is more general.

The Hilbert transform and the Riesz transforms are examples of these general operators T_Ω. For instance, $\Omega(\theta) = \frac{\theta}{\pi|\theta|} = \frac{1}{\pi}\operatorname{sgn}\theta$ defined on the unit sphere $\mathbf{S}^0 = \{-1,1\} \subseteq \mathbf{R}$ gives rise to the Hilbert transform, while $\Omega(\theta) = \frac{\Gamma(\frac{n+1}{2})}{\pi^{\frac{n+1}{2}}}\frac{\theta_j}{|\theta|}$ defined on $\mathbf{S}^{n-1} \subseteq \mathbf{R}^n$ generates the jth Riesz transform.

A certain class of multipliers can be realized as singular integral operators of the kind discussed. Recall from Proposition 2.4.7 that if m is homogeneous of degree 0 and infinitely differentiable on the sphere, then $m^\vee$ is given by

$$m^\vee = c\,\delta_0 + W_\Omega,$$

for some c complex constant and some smooth Ω on $\mathbf{S}^{n-1}$ with mean value zero. Therefore, all convolution operators whose multipliers are homogeneous of degree zero smooth functions on $\mathbf{S}^{n-1}$ can be realized as a constant multiple of the identity plus an operator of the form T_Ω.

Example 4.2.2. Let $P(\xi) = \sum_{|\alpha|=k} b_\alpha \xi^\alpha$ be a homogeneous polynomial of degree k in $\mathbf{R}^n$ that vanishes only at the origin. Let α be a multi-index of order k.

Then the function

$$m(\xi) = \frac{\xi^\alpha}{P(\xi)} \tag{4.2.7}$$

is infinitely differentiable on the sphere and homogeneous of degree zero. The operator given by multiplication on the Fourier transform by $m(\xi)$ is a constant multiple of the identity plus an operator given by convolution with a distribution of the form W_Ω for some Ω in $C^\infty(\mathbf{S}^{n-1})$ with mean value zero. In this section we will establish the L^p boundedness of such operators when Ω has appropriate smoothness on the sphere. This, in particular, will imply that $m(\xi)$ defined by (4.2.7) lies in the space $\mathcal{M}_p(\mathbf{R}^n)$ defined in Section 2.5.

4.2.b. L^2 Boundedness of Homogeneous Singular Integrals

Next we would like to compute the Fourier transform of W_Ω. This will give information as to whether the operator given by convolution with K_Ω is L^2 bounded. We have the following result.

Proposition 4.2.3. *Let $n \ge 2$ and $\Omega \in L^1(\mathbf{S}^{n-1})$ have mean value zero. Then the Fourier transform of W_Ω is a (finite a.e.) function given by the formula*

$$\widehat{W_\Omega}(\xi) = \int_{\mathbf{S}^{n-1}} \Omega(\theta)\left(\log\frac{1}{|\xi\cdot\theta|} - \frac{i\pi}{2}\operatorname{sgn}\,(\xi\cdot\theta)\right)d\theta. \tag{4.2.8}$$

Remark 4.2.4. We need to show that the function of ξ on the right in (4.2.8) is well defined and finite for almost all ξ in $\mathbf{R}^n$. Write $\xi = |\xi|\xi'$ where $\xi' \in \mathbf{S}^{n-1}$ and decompose $\log\frac{1}{|\xi\cdot\theta|}$ as $\log\frac{1}{|\xi|} + \log\frac{1}{|\xi'\cdot\theta|}$. Since Ω has mean value zero, the term $\log\frac{1}{|\xi|}$ multiplied by $\Omega(\theta)$ vanishes when integrated over the sphere.

We need to show that

$$\int_{\mathbf{S}^{n-1}} |\Omega(\theta)| \log\frac{1}{|\xi'\cdot\theta|}\,d\theta < \infty \tag{4.2.9}$$

for almost all $\xi' \in \mathbf{S}^{n-1}$. Integrate (4.2.9) over $\xi' \in \mathbf{S}^{n-1}$ and apply Fubini's theorem to obtain

$$\begin{aligned}
&\int_{\mathbf{S}^{n-1}} |\Omega(\theta)| \int_{\mathbf{S}^{n-1}} \log\frac{1}{|\xi'\cdot\theta|}\,d\xi'\,d\theta \\
&= \int_{\mathbf{S}^{n-1}} |\Omega(\theta)| \int_{\mathbf{S}^{n-1}} \log\frac{1}{|\xi_1|}\,d\xi\,d\theta \\
&= \omega_{n-2}\int_{\mathbf{S}^{n-1}} |\Omega(\theta)| \int_{-1}^{+1}\left(\log\frac{1}{|s|}\right)(1-s^2)^{\frac{n-3}{2}}\,ds\,d\theta \\
&= C_n\|\Omega\|_{L^1(\mathbf{S}^{n-1})} < \infty,
\end{aligned}$$

since we are assuming that $n \ge 2$. (The second to last identity follows from the result in Appendix D.2.) We conclude that (4.2.9) holds for almost all $\xi' \in \mathbf{S}^{n-1}$.

Since the function of ξ on the right in (4.2.8) is homogeneous of degree zero, it follows that it is a locally integrable function on $\mathbf{R}^n$.

Before we return to the proof of Proposition 4.2.3, we will need the following lemma:

Lemma 4.2.5. *Let a be a nonzero real number. Then for $0 < \varepsilon < N < \infty$ we have*

$$\lim_{\substack{\varepsilon\to 0\\ N\to\infty}} \int_\varepsilon^N \frac{\cos(ra)-\cos(r)}{r}\,dr = \log\frac{1}{|a|}\,, \tag{4.2.10}$$

$$\left|\int_\varepsilon^N \frac{\cos(ra)-\cos(r)}{r}\,dr\right| \le 2\left|\log\frac{1}{|a|}\right| \qquad \textit{for all } N > \varepsilon > 0\ , \tag{4.2.11}$$

$$\lim_{\substack{\varepsilon\to 0\\ N\to\infty}} \int_\varepsilon^N \frac{e^{-ira}-\cos(r)}{r}\,dr = \log\frac{1}{|a|} - i\frac{\pi}{2}\operatorname{sgn} a\,, \tag{4.2.12}$$

$$\left|\int_\varepsilon^N \frac{e^{-ira}-\cos(r)}{r}\,dr\right| \le 2\left|\log\frac{1}{|a|}\right| + 4 \qquad \textit{for all } N > \varepsilon > 0\ . \tag{4.2.13}$$

Proof. We first prove (4.2.10) and (4.2.11). By the fundamental theorem of calculus we can write

$$\begin{aligned}
\int_\varepsilon^N \frac{\cos(ra)-\cos(r)}{r}\,dr &= \int_\varepsilon^N \frac{\cos(r|a|)-\cos(r)}{r}\,dr \\
&= -\int_\varepsilon^N \int_1^{|a|} \sin(tr)\,dt\,dr \\
&= -\int_1^{|a|} \int_\varepsilon^N \sin(tr)\,dr\,dt \\
&= -\int_1^{|a|} \frac{\cos(\varepsilon t)}{t}\,dt + \int_N^{N|a|} \frac{\cos(t)}{t}\,dt\,,
\end{aligned}$$

and from this expression, we clearly obtain (4.2.11). But the first integral converges to $-\log|a|$ as $\varepsilon \to 0$ while the second integral converges to zero as $N \to \infty$ by an integration by parts. This proves (4.2.10).

To prove (4.2.12) and (4.2.13) we will need to know that the expressions

$$\left|\int_\varepsilon^N \frac{\sin(ra)}{r}\,dr\right| = \left|\int_{\varepsilon|a|}^{N|a|} \frac{\sin(r)}{r}\,dr\right| \tag{4.2.14}$$

tend to $\frac{\pi}{2}$ as $\varepsilon \to 0$ and $N \to \infty$ and are bounded by 4. Both statements follow from Exercise 4.4.1.

□

Let us now prove Proposition 4.2.3.

Proof. Let us set $\xi = |\xi|\xi'$. We have the following:

$$
\begin{aligned}
&\langle \widehat{W_\Omega}, \varphi \rangle = \langle W_\Omega, \widehat{\varphi} \rangle \\
&= \lim_{\varepsilon \to 0} \int_{|x| \geq \varepsilon} \frac{\Omega(x/|x|)}{|x|^n} \widehat{\varphi}(x)\, dx \\
&= \lim_{\substack{\varepsilon \to 0 \\ N \to \infty}} \int_{\mathbf{R}^n} \varphi(\xi) \int_{\varepsilon \leq |x| \leq N} \frac{\Omega(x/|x|)}{|x|^n} e^{-2\pi i x \cdot \xi}\, dx\, d\xi \\
&= \lim_{\substack{\varepsilon \to 0 \\ N \to \infty}} \int_{\mathbf{R}^n} \varphi(\xi) \int_{\mathbf{S}^{n-1}} \Omega(\theta) \int_{\varepsilon \leq r \leq N} e^{-2\pi i r \theta \cdot \xi} \frac{dr}{r}\, d\theta\, d\xi \\
&= \lim_{\substack{\varepsilon \to 0 \\ N \to \infty}} \int_{\mathbf{R}^n} \varphi(\xi) \int_{\mathbf{S}^{n-1}} \Omega(\theta) \int_{\varepsilon \leq r \leq N} (e^{-2\pi r |\xi| i \theta \cdot \xi'} - \cos(2\pi r |\xi|)) \frac{dr}{r}\, d\theta\, d\xi \\
&= \lim_{\substack{\varepsilon \to 0 \\ N \to \infty}} \int_{\mathbf{R}^n} \varphi(\xi) \int_{\mathbf{S}^{n-1}} \Omega(\theta) \int_{\frac{\varepsilon}{2\pi|\xi|} \leq r \leq \frac{N}{2\pi|\xi|}} \frac{e^{-ir\theta \cdot \xi'} - \cos(r)}{r}\, dr\, d\theta\, d\xi \\
&= \int_{\mathbf{R}^n} \varphi(\xi) \int_{\mathbf{S}^{n-1}} \Omega(\theta) \Big(\log \frac{1}{|\xi \cdot \theta|} - \frac{i\pi}{2} \operatorname{sgn}(\xi \cdot \theta) \Big)\, d\theta\, d\xi,
\end{aligned}
$$

by the Lebesgue dominated convergence theorem, Lemma 4.2.5, and Remark 4.2.4. We were able to subtract $\cos(2\pi r|\xi|)$ from the r integral in the previous calculation since Ω has mean value zero over the sphere.

□

The following is an immediate consequence of Proposition 4.2.3:

Corollary 4.2.6. *Let $\Omega \in L^1(\mathbf{S}^{n-1})$ have mean value zero. Then the associated operator T_Ω maps $L^2(\mathbf{R}^n)$ into itself if and only if*

$$
\begin{aligned}
&\operatorname*{essup}_{\xi \in \mathbf{R}^n} \left| \int_{\mathbf{S}^{n-1}} \Omega(\theta) \log \frac{1}{|\xi \cdot \theta|} d\theta \right| \\
&= \operatorname*{essup}_{\xi' \in \mathbf{S}^{n-1}} \left| \int_{\mathbf{S}^{n-1}} \Omega(\theta) \log \frac{1}{|\xi' \cdot \theta|} d\theta \right| < \infty,
\end{aligned}
\tag{4.2.15}
$$

where essup *in (4.2.15) denotes the essential supremum of the respective quantities.*

By picking an Ω such that the integral in (4.2.15) diverges we see that not all $\Omega \in L^1(\mathbf{S}^{n-1})$ with mean value zero give rise to bounded operators on $L^2(\mathbf{R}^n)$. Observe, however, that for Ω odd [i.e., $\Omega(-\theta) = -\Omega(\theta)$ for all $\theta \in \mathbf{S}^{n-1}$], (4.2.15) trivially holds since $\log \frac{1}{|\xi \cdot \theta|}$ is even and its product against an odd function must have integral zero over $\mathbf{S}^{n-1}$. We conclude that singular integrals with odd kernels are always L^2 bounded.

4.2.c. The Method of Rotations

Having settled the issue of L^2 boundedness for singular integrals of the form T_Ω with Ω odd, we turn our attention to their L^p boundedness. A simple procedure called the method of rotations plays a crucial role in the study of operators T_Ω when Ω is an odd function.

Theorem 4.2.7. *If Ω is odd and integrable over $\mathbf{S}^{n-1}$, then T_Ω and $T_\Omega^{(*)}$ are L^p bounded for all $1 < p < \infty$. More precisely, T_Ω initially defined on Schwartz functions has a bounded extension on $L^p(\mathbf{R}^n)$ (which is also denoted by T_Ω).*

Proof. We introduce the directional Hilbert transforms. Fix a unit vector θ in $\mathbf{R}^n$. For a Schwartz function f on $\mathbf{R}^n$ let

$$\mathcal{H}_\theta(f)(x) = \frac{1}{\pi}\,\text{p.v.}\int_{-\infty}^{+\infty} f(x - t\theta)\frac{dt}{t}.$$

We call $\mathcal{H}_\theta(f)$ the *directional Hilbert transform* of f in the direction θ. Let e_j be the usual unit vectors in $\mathbf{S}^{n-1}$. Then $\mathcal{H}_{e_1}$ is simply obtained by applying the Hilbert transform in the first variable followed by the identity operator in the remaining variables. Clearly, $\mathcal{H}_{e_1}$ is bounded on $L^p(\mathbf{R}^n)$ with norm equal to that of the Hilbert transform on $L^p(\mathbf{R})$. Next observe that the following identity is valid for all matrices $A \in O(n)$:

$$\mathcal{H}_{A(e_1)}(f)(x) = \mathcal{H}_{e_1}(f \circ A)(A^{-1}x)\,. \tag{4.2.16}$$

This implies that the L^p boundedness of $\mathcal{H}_\theta$ can be reduced to that of $\mathcal{H}_{e_1}$. We conclude that $\mathcal{H}_\theta$ is L^p bounded for $1 < p < \infty$ with norm bounded by the norm of the Hilbert transform on $L^p(\mathbf{R})$ for every $\theta \in \mathbf{S}^{n-1}$.

Likewise, we can define the *directional maximal Hilbert transforms.* For a function f in $\bigcup_{1\le p<\infty} L^p(\mathbf{R}^n)$ and $0 < \varepsilon < N < \infty$ we let

$$\mathcal{H}_\theta^{(\varepsilon,N)}(f)(x) = \frac{1}{\pi}\int_{\varepsilon\le|t|\le N} f(x - t\theta)\frac{dt}{t},$$

$$\mathcal{H}_\theta^{(*)}(f)(x) = \sup_{0<\varepsilon<N<\infty}\left|\mathcal{H}_\theta^{(\varepsilon,N)}(f)(x)\right|.$$

We observe that for any fixed $0 < \varepsilon < N < \infty$ and $f \in L^p$, $\mathcal{H}_\theta^{(\varepsilon,N)}(f)$ is well defined almost everywhere. Indeed, by Minkowski's integral inequality we obtain

$$\big\|\mathcal{H}_\theta^{(\varepsilon,N)}(f)\big\|_{L^p(\mathbf{R}^n)} \le \frac{1}{\pi}\big\|f\big\|_{L^p(\mathbf{R}^n)}\log\frac{N}{\varepsilon} < \infty\,,$$

which implies that $\mathcal{H}_\theta^{(\varepsilon,N)}(f)(x)$ is finite for almost all $x \in \mathbf{R}^n$. Thus $\mathcal{H}_\theta^{(*)}(f)$ is well defined a.e. for f in $\bigcup_{1\le p<\infty} L^p(\mathbf{R}^n)$.

Identity (4.2.16) is also valid for $\mathcal{H}_\theta^{(\varepsilon,N)}$ and $\mathcal{H}_\theta^{(*)}$. It follows that $\mathcal{H}_\theta^{(*)}$ is L^p bounded for $1 < p < \infty$ with norm at most that of the operator $H^{(*)}$ on $L^p(\mathbf{R})$.

Next we realize a general singular integral T_Ω with Ω odd as an average of the directional Hilbert transforms $\mathcal{H}_\theta$. We start with f in $\bigcup_{1\le p<\infty} L^p(\mathbf{R}^n)$ and the following identities:

$$\begin{aligned}\int_{\varepsilon\le|y|\le N}\frac{\Omega(y/|y|)}{|y|^n}f(x-y)\,dy &= +\int_{\mathbf{S}^{n-1}}\Omega(\theta)\int_{r=\varepsilon}^{N}f(x-r\theta)\,\frac{dr}{r}\,d\theta\\ &= -\int_{\mathbf{S}^{n-1}}\Omega(\theta)\int_{r=\varepsilon}^{N}f(x+r\theta)\,\frac{dr}{r}\,d\theta,\end{aligned}$$

where the first follows by switching to polar coordinates and the second one is a consequence of the first one and of the fact that Ω is odd via the change variables $\theta \to -\theta$. Averaging the two identities, we obtain

$$\begin{aligned}&\int_{\varepsilon\le|y|\le N}\frac{\Omega(y/|y|)}{|y|^n}f(x-y)\,dy\\ &=\frac{1}{2}\int_{\mathbf{S}^{n-1}}\Omega(\theta)\int_{r=\varepsilon}^{N}\frac{f(x-r\theta)-f(x+r\theta)}{r}\,dr\,d\theta\\ &=\frac{\pi}{2}\int_{\mathbf{S}^{n-1}}\Omega(\theta)\,\mathcal{H}_\theta^{(\varepsilon,N)}(f)(x)\,d\theta.\end{aligned}\tag{4.2.17}$$

It follows from the identity in (4.2.17) that

$$\int_{\varepsilon\le|y|\le N}\frac{\Omega(y/|y|)}{|y|^n}f(x-y)\,dy=\frac{\pi}{2}\int_{\mathbf{S}^{n-1}}\Omega(\theta)\,\mathcal{H}_\theta^{(\varepsilon,N)}(f)(x)\,d\theta\,,\tag{4.2.18}$$

from which we conclude that

$$T_\Omega^{(*)}(f)(x)\le\frac{\pi}{2}\int_{\mathbf{S}^{n-1}}|\Omega(\theta)|\,\mathcal{H}_\theta^{(*)}(f)(x)\,d\theta\,.\tag{4.2.19}$$

Using the Lebesgue dominated convergence theorem, we see that for f in $\mathcal{S}(\mathbf{R}^n)$, we can pass the limits as $\varepsilon\to 0$ and $N\to\infty$ inside the integral in (4.2.18), concluding that

$$T_\Omega(f)(x)=\frac{\pi}{2}\int_{\mathbf{S}^{n-1}}\Omega(\theta)\,\mathcal{H}_\theta(f)(x)\,d\theta\,,\qquad f\in\mathcal{S}(\mathbf{R}^n)\,.\tag{4.2.20}$$

The L^p boundedness of T_Ω and $T_\Omega^{(*)}$ for Ω odd are then trivial consequences of (4.2.20) and (4.2.19) via Minkowski's integral inequality.

□

Corollary 4.2.8. *The Riesz transforms and the maximal Riesz transforms are bounded on $L^p(\mathbf{R}^n)$ for $1 < p < \infty$.*

Proof. The Riesz transforms have odd kernels.

□

Remark 4.2.9. It follows from the proof of Theorem 4.2.7 and from Theorem 4.1.12 that whenever Ω is an odd function on $\mathbf{S}^{n-1}$, we have

$$\big\|T_\Omega\big\|_{L^p\to L^p} \le \big\|\Omega\big\|_{L^1}\begin{cases} a\,p & \text{when } p\ge 2,\\ a\,(p-1)^{-1} & \text{when } 1<p\le 2,\end{cases}$$

$$\big\|T_\Omega^{(*)}\big\|_{L^p\to L^p} \le \big\|\Omega\big\|_{L^1}\begin{cases} a\,p & \text{when } p\ge 2,\\ a\,(p-1)^{-1} & \text{when } 1<p\le 2,\end{cases}$$

for some $a>0$ independent of p and of the dimension.

4.2.d. Singular Integrals with Even Kernels*

Since a general integrable function Ω on $\mathbf{S}^{n-1}$ with mean value zero can be written as a sum of an odd and of an even function, it suffices to study singular integral operators T_Ω with even kernels. For the rest of this section, fix an integrable even function Ω on $\mathbf{S}^{n-1}$ with mean value zero. The following idea will be crucial in treating such singular integrals. Proposition 4.1.16 implies that

$$T_\Omega = -\sum_{j=1}^{n} R_j R_j T_\Omega\,. \tag{4.2.21}$$

If R_jT_Ω were another singular integral operator of the form T_{Ω_j} for some odd Ω_j, then the boundedness of T_Ω would follow from that of T_{Ω_j} via the identity (4.2.21) and Theorem 4.2.7. It turns out that R_jT_Ω does have an odd kernel, but it may not necessarily be integrable on $\mathbf{S}^{n-1}$ unless Ω itself possesses an additional amount of integrability. The amount of extra integrability needed is logarithmic, more precisely of this sort:

$$c_\Omega = \int_{\mathbf{S}^{n-1}} |\Omega(\theta)|\log^+|\Omega(\theta)|\,d\theta < \infty. \tag{4.2.22}$$

Observe that

$$\big\|\Omega\big\|_{L^1} \le \frac{1}{\log 2}\big(c_\Omega + 2\omega_{n-1}\big) \le C_n(c_\Omega+1)\,,$$

which says that the norm $\big\|\Omega\big\|_{L^1}$ is always controlled by a dimensional multiple of the constant $c_\Omega+1$. The following theorem is the main result of this section.

Theorem 4.2.10. *Let Ω be an even integrable function on $\mathbf{S}^{n-1}$ with mean value zero that satisfies (4.2.22). Then the corresponding singular integral T_Ω is bounded on $L^p(\mathbf{R}^n)$ for $1<p<\infty$ with norm at most a dimensional multiple of* $\max\big((p-1)^{-2},p^2\big)(c_\Omega+1)$.

We note that if the operator T_Ω in Theorem 4.2.10 is of weak type $(1,1)$, then the estimate on the L^p operator norm of T_Ω can be improved to $\big\|T_\Omega\big\|_{L^p\to L^p}\le C_n(p-1)^{-1}$ as $p\to 1$. This is indeed the case; see the historical comments at the end of this chapter.

Proof. Let W_Ω be the distributional kernel of T_Ω. Using Proposition 4.2.3 and the fact that Ω is an even function, we obtain the formula

$$\widehat{W_\Omega}(\xi) = \int_{\mathbf{S}^{n-1}} \Omega(\theta) \log \frac{1}{|\xi\cdot\theta|}\, d\theta\,, \tag{4.2.23}$$

which implies that $\widehat{W_\Omega}$ is itself an even function. Now, using Exercise 4.2.3 and condition (4.2.22), we conclude that $\widehat{W_\Omega}$ is a bounded function. Therefore, T_Ω is L^2 bounded. To obtain the L^p boundedness of T_Ω, we will use the idea mentioned earlier involving the Riesz transforms. In view of (4.1.39), we have that

$$T = -\sum_{i=1}^{n} R_j T_j, \tag{4.2.24}$$

where $T_j = R_j T_\Omega$. Equality (4.2.24) makes sense as an operator identity on $L^2(\mathbf{R}^n)$, since T_Ω and each R_j are well defined and bounded on $L^2(\mathbf{R}^n)$.

The kernel of the operator T_j is the inverse Fourier transform of the distribution $-i\frac{\xi_j}{|\xi|}\widehat{W_\Omega}(\xi)$, which we denote by K_j. At this point we only know that K_j is a tempered distribution whose Fourier transform is the function $-i\frac{\xi_j}{|\xi|}\widehat{W_\Omega}(\xi)$. Our first goal will be to show that K_j coincides with an integrable function on an annulus. To prove this assertion we write

$$W_\Omega = W_\Omega^0 + W_\Omega^1 + W_\Omega^\infty\,,$$

where W_Ω^0 is a distribution and $W_\Omega^1, W_\Omega^\infty$ are functions defined by

$$\begin{aligned}
\langle W_\Omega^0, \varphi\rangle &= \lim_{\varepsilon\to\infty} \int_{\varepsilon<|x|\le\frac12} \frac{\Omega(x/|x|)}{|x|^n}\varphi(x)\,dx\,,\\
W_\Omega^1(x) &= \frac{\Omega(x/|x|)}{|x|^n}\chi_{\frac12\le|x|\le 2}\,,\\
W_\Omega^\infty(x) &= \frac{\Omega(x/|x|)}{|x|^n}\chi_{2<|x|}\,.
\end{aligned}$$

We now fix a $j\in\{1,2,\dots,n\}$ and we write

$$K_j = K_j^0 + K_j^1 + K_j^\infty\,,$$

where

$$\begin{aligned}
K_j^0 &= \big(-i\tfrac{\xi_j}{|\xi|}\widehat{W_\Omega^0}(\xi)\big)^\vee\,,\\
K_j^1 &= \big(-i\tfrac{\xi_j}{|\xi|}\widehat{W_\Omega^1}(\xi)\big)^\vee\,,\\
K_j^\infty &= \big(-i\tfrac{\xi_j}{|\xi|}\widehat{W_\Omega^\infty}(\xi)\big)^\vee\,.
\end{aligned}$$

Define the annulus

$$A = \{x\in\mathbf{R}^n :\ 2/3<|x|<3/2\}.$$

For $x \in A$, the convolution of W_Ω^0 with the kernel of the Riesz transform R_j can be written as the convergent integral inside the absolute value:

$$
\begin{aligned}
\text{(4.2.25)}\qquad &\left|\frac{\Gamma(\frac{n+1}{2})}{\pi^{\frac{n+1}{2}}}\lim_{\varepsilon\to 0}\int_{\varepsilon<|y|<\frac{1}{2}}\frac{x_j-y_j}{|x-y|^{n+1}}\frac{\Omega(y/|y|)}{|y|^n}\,dy\right| \\
&= \frac{\Gamma(\frac{n+1}{2})}{\pi^{\frac{n+1}{2}}}\left|\int_{|y|<\frac{1}{2}}\left(\frac{x_j-y_j}{|x-y|^{n+1}}-\frac{x_j}{|x|^{n+1}}\right)\frac{\Omega(y/|y|)}{|y|^n}\,dy\right| \\
&\le \int_{|y|\le\frac{1}{2}} C_n|y|\,\frac{|\Omega(y/|y|)|}{|y|^n}\,dy = C_n'\big\|\Omega\big\|_{L^1},
\end{aligned}
$$

where we used the fact that $\Omega(y/|y|)|y|^{-n}$ has integral zero over annuli of the form $\varepsilon<|y|<\frac{1}{2}$, the mean value theorem applied to the function $x_j|x|^{-(n+1)}$, and the fact that $|x-y|\ge 1/6$ for x in the annulus A. We conclude that on A, K_j^0 coincides with the bounded function inside the absolute value in (4.2.25).

Likewise, for $x\in A$ we have

$$
\begin{aligned}
\text{(4.2.26)}\qquad &\frac{\Gamma(\frac{n+1}{2})}{\pi^{\frac{n+1}{2}}}\left|\int_{|y|>2}\frac{x_j-y_j}{|x-y|^{n+1}}\frac{\Omega(y/|y|)}{|y|^n}\,dy\right| \\
&\le \frac{\Gamma(\frac{n+1}{2})}{\pi^{\frac{n+1}{2}}}\int_{|y|>2}\frac{1}{|x-y|^n}\frac{|\Omega(y/|y|)|}{|y|^n}\,dy \\
&\le \frac{\Gamma(\frac{n+1}{2})}{\pi^{\frac{n+1}{2}}}\int_{|y|>2}\frac{4^n}{|y|^{2n}}|\Omega(y)|\,dy = C\big\|\Omega\big\|_{L^1},
\end{aligned}
$$

from which it follows that on A, K_j^∞ coincides with the bounded function inside the absolute value in (4.2.26).

Now observe that condition (4.2.22) gives that the function W_Ω^1 satisfies

$$
\begin{aligned}
&\int_{|x|\le 2}|W_\Omega^1(x)|\,\log^+|W_\Omega^1(x)|\,dx \\
&\le \omega_{n-1}\int_{1/2}^{2}\int_{\mathbf{S}^{n-1}}\frac{|\Omega(\theta)|}{r^n}\log^+[2^n|\Omega(\theta)|]\,d\theta r^{n-1}\frac{dr}{r} \\
&\le \omega_{n-1}(\log 4)\,\big[n(\log 2)\big\|\Omega\big\|_{L^1}+c_\Omega\big] < \infty.
\end{aligned}
$$

Since the Riesz transform R_j maps L^p to L^p with norm at most $4(p-1)^{-1}$ for $1<p<2$, it follows from Exercise 1.3.7 that $K_j^1 = R_j(W_\Omega^1)$ is integrable over the ball $|x|\le 3/2$ and, moreover, it satisfies

$$
\int_A |K_j^1(x)|\,dx \le C_n\bigg[\int_{|x|\le 2}|W_\Omega^1(x)|\,\log^+|W_\Omega^1(x)|\,dx+1\bigg] \le C_n'(c_\Omega+1)\,.
$$

We have proved that K_j is a distribution that coincides with an integrable function on the annulus A. Furthermore, since $\widehat{K_j}$ is homogeneous of degree zero, we have that K_j is a homogeneous distribution of degree $-n$ (Exercise 2.3.9). This means that for all test functions φ and all $\lambda > 0$ we have

$$\left\langle K_j, \delta^\lambda(\varphi)\right\rangle = \left\langle K_j, \varphi\right\rangle .$$

But then for φ supported in the annulus $3/4 < |x| < 4/3$ and for λ in $(8/9, 9/8)$ we have that $\delta^\lambda(\varphi)$ is supported in A and thus

$$\int K_j(x)\varphi(\lambda x)\, dx = \left\langle K_j, \delta^\lambda(\varphi)\right\rangle = \left\langle K_j, \varphi\right\rangle = \int \lambda^{-n} K_j(\lambda^{-1}x)\varphi(x)\, dx .$$

From this we conclude that $K_j(x) = \lambda^{-n}K_j(\lambda^{-1}x)$ for $3/4 < |x| < 4/3$ and $8/9 < \lambda < 9/8$. Thus for $8/9 < |x| < 9/8$ we have

$$K_j(x) = |x|^{-n}K_j(x/|x|) = |x|^{-n}\Omega_j(x/|x|)\,, \tag{4.2.27}$$

where we defined Ω_j to be the restriction of K_j over $\mathbf{S}^{n-1}$. The integrability of K_j over the annulus $8/9 < |x| < 9/8$ implies the integrability (and hence finiteness a.e.) of Ω_j over $\mathbf{S}^{n-1}$ via (4.2.27).

Pick a nonnegative, radial, smooth, and nonzero function ψ on $\mathbf{R}^n$ supported in $8/9 < |x| < 9/8$. Then by switching to polar coordinates we obtain

$$\begin{aligned}
\left\langle K_j, \psi\right\rangle &= \int_{\mathbf{R}^n} \frac{\Omega_j(x/|x|)}{|x|^n}\psi(x)\, dx = c_\psi \int_{\mathbf{S}^{n-1}} \Omega_j(\theta)\, d\theta, \\
\left\langle K_j, \psi\right\rangle &= \left\langle \widehat{K_j}, \widehat{\psi}\,\right\rangle = \int_{\mathbf{R}^n} \frac{-i\xi_j}{|\xi|}\widehat{W_\Omega}(\xi)\widehat{\psi}(\xi)\, d\xi \\
&= c'_\psi \int_{\mathbf{S}^{n-1}} \frac{-i\theta_j}{|\theta|}\widehat{W_\Omega}(\theta)\, d\theta = 0,
\end{aligned}$$

since by (4.2.23), $\frac{-i\xi_j}{|\xi|}\widehat{W_\Omega}(\xi)$ is an odd function. We conclude that Ω_j has mean value zero over $\mathbf{S}^{n-1}$. We can now define the distribution W_{Ω_j}, and we claim that

$$K_j = W_{\Omega_j}\,. \tag{4.2.28}$$

To establish the claim, we use (4.2.27) to obtain that the homogeneous distributions K_j and W_{Ω_j} agree on the open set $8/9 < |x| < 9/8$ and, thus, they must agree everywhere on $\mathbf{R}^n \setminus \{0\}$ (check that $\left\langle K_j, \varphi\right\rangle = \left\langle W_{\Omega_j}, \varphi\right\rangle$ for all $\varphi \in C_0^\infty(\mathbf{R}^n \setminus \{0\})$ by dilating and translating their support). Therefore, $K_j - W_{\Omega_j}$ is supported at the origin and since it is homogeneous of degree $-n$, it must be equal to $b\delta_0$, a constant multiple of the Dirac mass. But $\widehat{K_j}$ is an odd function and hence K_j is also odd. It follows that W_{Ω_j} is an odd function on $\mathbf{R}^n \setminus \{0\}$, which implies that Ω_j is an odd function. Defining odd distributions in the natural way, we obtain that $K_j - W_{\Omega_j}$ is an odd distribution and thus the previous multiple of the Dirac mass must be an odd distribution. But if $b\delta_0$ is odd, then $b = 0$. We conclude that for each j there exists an odd integrable function Ω_j on $\mathbf{S}^{n-1}$ with $\left\|\Omega_j\right\|_{L^1}$ controlled by a multiple of $c_\Omega + 1$ such that (4.2.28) holds.

Then we use (4.2.24) and (4.2.28) to write

$$T_\Omega = \sum_{j=1}^{n} R_j T_{\Omega_j} \, ,$$

and we can appeal to the boundedness of each T_{Ω_j} (Theorem 4.2.7) and to that of the Riesz transforms to obtain the required L^p boundedness for T_Ω.

□

Before we proceed with the corresponding result for maximal singular integrals, we note that Theorem 4.2.10 holds for all $\Omega \in L^1(\mathbf{S}^{n-1})$ that satisfy (4.2.22), not necessarily even Ω. Simply write $\Omega = \Omega_e + \Omega_o$, where Ω_e is even and Ω_o is odd, and check that condition (4.2.22) holds for Ω_e.

4.2.e. Maximal Singular Integrals with Even Kernels*

We have the corresponding theorem for maximal singular integrals.

Theorem 4.2.11. *Let Ω be an even integrable function on $\mathbf{S}^{n-1}$ with mean value zero that satisfies (4.2.22). Then the corresponding maximal singular integral $T_\Omega^{(*)}$ is bounded on $L^p(\mathbf{R}^n)$ for $1 < p < \infty$ with norm at most a dimensional multiple of* $\max(p^2, (p-1)^{-2})(c_\Omega + 1)$.

Proof. For $f \in L^1_{\text{loc}}(\mathbf{R}^n)$, define the maximal function of f in the direction θ by setting

$$M_\theta(f)(x) = \sup_{a>0} \frac{1}{2a} \int_{|r|\le a} |f(x - r\theta)| \, dr. \tag{4.2.29}$$

By Exercise 4.2.6(a) we have that M_θ is bounded on $L^p(\mathbf{R}^n)$ with norm at most $3\,p\,(p-1)^{-1}$.

Fix Φ a smooth radial function such that $\Phi(x) = 0$ for $|x| < 1/4$, $\Phi(x) = 1$ for $|x| > 3/4$, and $0 \le \Phi(x) \le 1$ for all x in $\mathbf{R}^n$. For $f \in L^p(\mathbf{R}^n)$ and $0 < \varepsilon < N \le \infty$ we introduce the smoothly truncated singular integral

$$\widetilde{T}_\Omega^{(\varepsilon,N)}(f)(x) = \int_{\mathbf{R}^n} \frac{\Omega\big(\frac{x-y}{|x-y|}\big)}{|x-y|^n} \Big(\Phi\big(\tfrac{x-y}{\varepsilon}\big) - \Phi\big(\tfrac{x-y}{N}\big) \Big) f(y)\, dy$$

and the corresponding maximal singular integral operator

$$\widetilde{T}_\Omega^{(*)}(f) = \sup_{0<N<\infty} \sup_{0<\varepsilon<N} |\widetilde{T}_\Omega^{(\varepsilon,N)}(f)| \, .$$

For f in $L^p(\mathbf{R}^n)$ (for some $1 < p < \infty$), we have

$$\begin{aligned}
&\sup_{0<\varepsilon<N<\infty} |\widetilde{T}_\Omega^{(\varepsilon,N)}(f)(x) - T_\Omega^{(\varepsilon,N)}(f)(x)| \\
&= \sup_{0<\varepsilon<N<\infty} \bigg| \int_{\frac{\varepsilon}{4}\le|y|\le\varepsilon} \frac{\Omega(\frac{y}{|y|})}{|y|^n} \Phi\left(\tfrac{y}{\varepsilon}\right) f(x-y)\,dy \\
&\qquad\qquad - \int_{\frac{N}{4}\le|y|\le N} \frac{\Omega(\frac{y}{|y|})}{|y|^n} \Phi\left(\tfrac{y}{N}\right) f(x-y)\,dy \bigg| \\
&\le \sup_{0<\varepsilon<N<\infty} \left[\int_{\frac{\varepsilon}{4}\le|y|\le\varepsilon} \frac{|\Omega(\frac{y}{|y|})|}{|y|^n} |f(x-y)|\,dy + \int_{\frac{N}{4}\le|y|\le N} \frac{|\Omega(\frac{y}{|y|})|}{|y|^n} |f(x-y)|\,dy \right] \\
&\le \sup_{0<\varepsilon<N<\infty} \int_{\mathbf{S}^{n-1}} |\Omega(\theta)| \left[\frac{4}{\varepsilon} \int_{\frac{\varepsilon}{4}}^{\varepsilon} |f(x-r\theta)|\,dr + \frac{4}{N} \int_{\frac{N}{4}}^{N} |f(x-r\theta)|\,dr \right] d\theta \\
&\le 16 \int_{\mathbf{S}^{n-1}} |\Omega(\theta)|\, M_\theta(f)(x)\,d\theta\,.
\end{aligned}$$

Using the result of Exercise 4.2.6(a) we conclude that

$$\big\|\widetilde{T}_\Omega^{(*)}(f) - T_\Omega^{(*)}(f)\big\|_{L^p} \le 96\, \big\|\Omega\big\|_{L^1} \max(1, (p-1)^{-1})\, \big\|f\big\|_{L^p}\,.$$

This implies that it suffices to obtain the required L^p bound for the smoothly truncated maximal singular integral operator $\widetilde{T}_\Omega^{(*)}$.

Next we observe that for f in the Schwartz class the operator $\widetilde{T}_\Omega^{(\varepsilon,\infty)}$ is well defined and satisfies an estimate

$$\big\|\widetilde{T}_\Omega^{(\varepsilon,\infty)}(f)\big\|_{L^p} \le C_{p,n}\varepsilon^{-\frac{n}{p}} \big\|f\big\|_{L^p}$$

since the corresponding operator $T_\Omega^{(\varepsilon,\infty)}$ does so. It follows that for each $\varepsilon > 0$, $\widetilde{T}_\Omega^{(\varepsilon,\infty)}$ admits an extension that is L^p bounded. We denote this operator still by $\widetilde{T}_\Omega^{(\varepsilon,\infty)}$. It follows by density that the identity

$$\widetilde{T}_\Omega^{(\varepsilon,N)}(f) = \widetilde{T}_\Omega^{(\varepsilon,\infty)}(f) - \widetilde{T}_\Omega^{(N,\infty)}(f)$$

holds for all $f \in L^p(\mathbf{R}^n)$. To estimate $\widetilde{T}_\Omega^{(*)}$, it suffices therefore to estimate the operator

$$f \to \sup_{\varepsilon>0} |\widetilde{T}_\Omega^{(\varepsilon,\infty)}(f)|$$

on $L^p(\mathbf{R}^n)$.

Let K_j, Ω_j, and T_j be as in the previous theorem, and let F_j be the Riesz transform of the function $\Omega(x/|x|)\Phi(x)|x|^{-n}$. A calculation with $f \in L^p(\mathbf{R}^n)$ with compact support gives the identity

$$\begin{aligned}\widetilde{T}_\Omega^{(\varepsilon,\infty)}(f)(x) &= \int_{\mathbf{R}^n} \frac{1}{\varepsilon^n}\frac{\Omega(\frac{y}{\varepsilon}/|\frac{y}{\varepsilon}|)}{|\frac{y}{\varepsilon}|^n}\Phi(\tfrac{y}{\varepsilon})f(x-y)\,dy \\ &= -\Big[\sum_{j=1}^n \tfrac{1}{\varepsilon^n}F_j\left(\tfrac{\cdot}{\varepsilon}\right) * R_j(f)\Big](x),\end{aligned} \tag{4.2.30}$$

where in the last step we used Proposition 4.1.16. Therefore, for such f we can write

$$\begin{aligned}-\widetilde{T}_\Omega^{(\varepsilon,\infty)}(f)(x) &= \sum_{j=1}^n \tfrac{1}{\varepsilon^n}\int_{\mathbf{R}^n} F_j\left(\tfrac{x-y}{\varepsilon}\right) R_j(f)(y)\,dy \\ &= A_1^{(\varepsilon)}(f)(x) + A_2^{(\varepsilon)}(f)(x) + A_3^{(\varepsilon)}(f)(x),\end{aligned}$$

where

$$\begin{aligned}A_1^{(\varepsilon)}(f)(x) &= \sum_{j=1}^n \frac{1}{\varepsilon^n}\int_{|x-y|\le\varepsilon} F_j\left(\tfrac{x-y}{\varepsilon}\right) R_j(f)(y)\,dy, \\ A_2^{(\varepsilon)}(f)(x) &= \sum_{j=1}^n \frac{1}{\varepsilon^n}\int_{|x-y|>\varepsilon} \left\{F_j\left(\tfrac{x-y}{\varepsilon}\right) - K_j\left(\tfrac{x-y}{\varepsilon}\right)\right\} R_j(f)(y)\,dy, \\ A_3^{(\varepsilon)}(f)(x) &= \sum_{j=1}^n \frac{1}{\varepsilon^n}\int_{|x-y|>\varepsilon} K_j\left(\tfrac{x-y}{\varepsilon}\right) R_j(f)(y)\,dy.\end{aligned}$$

It follows from the definitions of F_j and K_j that

$$\begin{aligned}&F_j(z) - K_j(z) \\ &= \frac{\Gamma(\frac{n+1}{2})}{\pi^{\frac{n+1}{2}}}\lim_{\varepsilon\to 0}\int_{\varepsilon\le|y|}\frac{\Omega(y/|y|)}{|y|^n}(\Phi(y)-1)\frac{z_j-y_j}{|z-y|^{n+1}}\,dy \\ &= \frac{\Gamma(\frac{n+1}{2})}{\pi^{\frac{n+1}{2}}}\int_{|y|\le\frac34}\frac{\Omega(y/|y|)}{|y|^n}(\Phi(y)-1)\left\{\frac{z_j-y_j}{|z-y|^{n+1}} - \frac{z_j}{|z|^{n+1}}\right\}dy\end{aligned}$$

whenever $|z|\ge 1$. But using the mean value theorem, the last expression is easily seen to be bounded by

$$C_n\int_{|y|\le\frac34}\frac{\Omega(y/|y|)}{|y|^n}\frac{|y|}{|z|^{n+1}}\,dy = C_n'\big\|\Omega\big\|_{L^1}|z|^{-(n+1)},$$

whenever $|z|\ge 1$. Using this estimate, we obtain that the jth term in $A_2^{(\varepsilon)}(f)(x)$ is controlled by a dimensional multiple of

$$\frac{\big\|\Omega\big\|_{L^1}}{\varepsilon^n}\int_{|x-y|>\varepsilon}\frac{|R_j(f)(y)|\,dy}{(|x-y|/\varepsilon)^{n+1}} \le \frac{2\big\|\Omega\big\|_{L^1}}{2^{-n}\varepsilon^n}\int_{\mathbf{R}^n}\frac{|R_j(f)(y)|\,dy}{\left(1+\frac{|x-y|}{\varepsilon}\right)^{n+1}}.$$

It follows that

$$\sup_{\varepsilon>0} |A_2^{(\varepsilon)}(f)| \le C_n \|\Omega\|_{L^1} M(R_j(f))$$

in view of Theorem 2.1.10. (M here is the Hardy-Littlewood maximal operator.) Since by Theorem 2.1.6 M maps $L^p(\mathbf{R}^n)$ into itself with norm at most a dimensional multiple of $\max(1,(p-1)^{-1})$, and since by Remark 4.2.9 the norm $\|R_j\|_{L^p\to L^p}$ is controlled by a dimensional multiple of $\max(p,(p-1)^{-1})$, it follows that

$$\|\sup_{\varepsilon>0} |A_2^{(\varepsilon)}(f)|\|_{L^p} \le C_n \|\Omega\|_{L^1} \max(p,(p-1)^{-1}) \|f\|_{L^p}. \tag{4.2.31}$$

Next, recall that in the proof of Theorem 4.2.10 we showed that

$$K_j(x) = \frac{\Omega_j(x/|x|)}{|x|^n},$$

where Ω_j are integrable functions on $\mathbf{S}^{n-1}$ that satisfy

$$\|\Omega_j\|_{L^1} \le C_n(c_\Omega + 1). \tag{4.2.32}$$

Thus $|A_3^{(\varepsilon)}(f)|$ is bounded by a sum of operators of the form $T_{\Omega_j}^{(\varepsilon,\infty)}(R_j(f))$ (the latter is well defined on L^p), and by Remark 4.2.9 this term has L^p norm at most a dimensional multiple of $\|\Omega_j\|_{L^1} \max(p,(p-1)^{-1}) \|R_j(f)\|_{L^p}$. It follows that

$$\|\sup_{\varepsilon>0} |A_3^{(\varepsilon)}(f)|\|_{L^p} \le C_n \max(p^2,(p-1)^{-2})(c_\Omega + 1)\, \|f\|_{L^p}. \tag{4.2.33}$$

Finally, we turn our attention to term $A_1^{(\varepsilon)}(f)$. To prove the required estimate, we will need to show that there exist nonnegative homogeneous of degree zero functions G_j on $\mathbf{R}^n$ that satisfy

$$|F_j(x)| \le G_j(x) \qquad \text{when } |x| \le 2 \tag{4.2.34}$$

and

$$\int_{\mathbf{S}^{n-1}} |G_j(\theta)|\, d\theta \le C_n(c_\Omega + 1). \tag{4.2.35}$$

To prove (4.2.34), first note that if $|x| \le 1/8$, then

$$\begin{aligned} |F_j(x)| &= \frac{\Gamma(\frac{n+1}{2})}{\pi^{\frac{n+1}{2}}} \left| \int_{\mathbf{R}^n} \frac{\Omega(y/|y|)}{|y|^n} \Phi(y) \frac{x_j - y_j}{|x-y|^{n+1}}\, dy \right| \\ &\le C_n \int_{|y|\ge\frac14} \frac{|\Omega(y/|y|)|}{|y|^{2n}}\, dy \le C_n' \|\Omega\|_{L^1}. \end{aligned}$$

We now fix an x satisfying $1/8 \le |x| \le 1$ and we write

$$\begin{aligned}
&|F_j(x)| \le \Phi(x)|K_j(x)| + |F_j(x) - \Phi(x)K_j(x)| \\
&\le |K_j(x)| + \frac{\Gamma(\frac{n+1}{2})}{\pi^{\frac{n+1}{2}}}\left| \lim_{\varepsilon\to 0}\int_{|y|>\varepsilon} \frac{x_j - y_j}{|x-y|^{n+1}}\big(\Phi(y) - \Phi(x)\big)\frac{\Omega(y/|y|)}{|y|^n}dy\right| \\
&= |K_j(x)| + \frac{\Gamma(\frac{n+1}{2})}{\pi^{\frac{n+1}{2}}}\big(P_1(x) + P_2(x) + P_3(x)\big),
\end{aligned}$$

where

$$\begin{aligned}
P_1(x) &= \left|\int_{|y|\le\frac{1}{16}}\left(\frac{x_j - y_j}{|x-y|^{n+1}} - \frac{x_j}{|x|^{n+1}}\right)\big(\Phi(y) - \Phi(x)\big)\frac{\Omega(y/|y|)}{|y|^n}\,dy\right|, \\
P_2(x) &= \left|\int_{\frac{1}{16}\le|y|\le 2}\frac{x_j - y_j}{|x-y|^{n+1}}\big(\Phi(y) - \Phi(x)\big)\frac{\Omega(y/|y|)}{|y|^n}\,dy\right|, \\
P_3(x) &= \left|\int_{|y|\ge 2}\frac{x_j - y_j}{|x-y|^{n+1}}\big(\Phi(y) - \Phi(x)\big)\frac{\Omega(y/|y|)}{|y|^n}\,dy\right|.
\end{aligned}$$

But since $1/8 \le |x| \le 1$, we see that

$$P_1(x) \le C_n \int_{|y|\le\frac{1}{16}} \frac{|y|}{|x|^{n+1}}\frac{|\Omega(y/|y|)|}{|y|^n}\,dy \le C_n'\big\|\Omega\big\|_{L^1}.$$

and that

$$P_3(x) \le C_n \int_{|y|\ge 2} \frac{|\Omega(y/|y|)|}{|y|^{2n}}\,dy \le C_n'\big\|\Omega\big\|_{L^1}.$$

For $P_2(x)$ we use the estimate $|\Phi(y) - \Phi(x)| \le C|x-y|$ to obtain

$$\begin{aligned}
P_2(x) &\le \int_{\frac{1}{16}\le|y|\le 2}\frac{C}{|x-y|^{n-1}}\frac{|\Omega(y/|y|)|}{|y|^n}\,dy \\
&\le 4C\int_{\frac{1}{16}\le|y|\le 2}\frac{|\Omega(y/|y|)|}{|x-y|^{n-1}|y|^{n-\frac{1}{2}}}\,dy \\
&\le 4C\int_{\mathbf{R}^n}\frac{|\Omega(y/|y|)|}{|x-y|^{n-1}|y|^{n-\frac{1}{2}}}\,dy.
\end{aligned}$$

Recall that $K_j(x) = \Omega_j(x/|x|)|x|^{-n}$. We now set

$$G_j(x) = C_n\left(\big\|\Omega\big\|_{L^1} + \Omega_j\Big(\frac{x}{|x|}\Big) + |x|^{n-\frac{3}{2}}\int_{\mathbf{R}^n}\frac{|\Omega(y/|y|)|\,dy}{|x-y|^{n-1}|y|^{n-\frac{1}{2}}}\right) \tag{4.2.36}$$

and we observe that G_j is a homogeneous of degree zero function, it satisfies (4.2.34), and it is integrable over the annulus $\frac{8}{9} \le |x| \le \frac{9}{8}$. Indeed,

$$\int_{\frac{8}{9}\le|x|\le\frac{9}{8}}\int_{\mathbf{R}^n}\frac{|\Omega(y/|y|)|\,dy\,dx}{|x-y|^{n-1}|y|^{n-\frac{1}{2}}} = \int_{\mathbf{R}^n}\frac{|\Omega(y/|y|)|}{|y|^{n-\frac{1}{2}}}\left\{\int_{\frac{8}{9}\le|x|\le\frac{9}{8}}\frac{dx}{|x-y|^{n-1}}\right\}dy,$$

and since the integral inside the curly brackets is at most $C\min(1,|y|^{-(n-1)})$, it follows from (4.2.36) and (4.2.32) that

$$\int_{\frac{8}{9}\le|x|\le\frac{9}{8}} |G_j(x)|\,dx \le C_n\big(\|\Omega\|_{L^1}+\|\Omega_j\|_{L^1}+\|\Omega\|_{L^1}\big) \le C_n(c_\Omega+1).$$

Since G_j is homogeneous of degree zero, (4.2.35) follows.

To complete the proof, we return to the sublinear map $f \to \sup_{\varepsilon>0}|A_1^{(\varepsilon)}(f)|$. Using the definition of $A_1^{(\varepsilon)}(f)$, we obtain

$$\begin{aligned}
&\sup_{\varepsilon>0}|A_1^{(\varepsilon)}(f)(x)| \\
&\le \sup_{\varepsilon>0}\sum_{j=1}^n \frac{1}{\varepsilon^n}\int_{|z|\le\varepsilon}|F_j(z)|\,|R_j(f)(x-z)|\,dz \\
&\le \sup_{\varepsilon>0}\sum_{j=1}^n \frac{1}{\varepsilon^n}\int_{r=0}^{\varepsilon}\int_{\mathbf{S}^{n-1}}|F_j(r\theta)|\,|R_j(f)(x-r\theta)|\,r^{n-1}\,d\theta\,dr \\
&\le \sum_{j=1}^n\int_{\mathbf{S}^{n-1}}|G_j(\theta)|\bigg\{\sup_{\varepsilon>0}\frac{1}{\varepsilon^n}\int_{r=0}^{\varepsilon}|R_j(f)(x-r\theta)|\,r^{n-1}\,dr\bigg\}\,d\theta \\
&\le 2\sum_{j=1}^n\int_{\mathbf{S}^{n-1}}|G_j(\theta)|M_\theta(R_j(f))(x)\,d\theta\,.
\end{aligned}$$

Using (4.2.35) together with the L^p boundedness of the Riesz transforms and of M_θ we obtain

$$\Big\|\sup_{\varepsilon>0}|A_1^{(\varepsilon)}(f)|\Big\|_{L^p} \le C_n\max(p,(p-1)^{-2})(c_\Omega+1)\big\|f\big\|_{L^p}\,. \tag{4.2.37}$$

Combining (4.2.37), (4.2.31), and (4.2.33), we obtain the required conclusion.

□

The following corollary is a consequence of Theorem 4.2.11.

Corollary 4.2.12. *Let Ω be as in Theorem 4.2.11. Then for $1<p<\infty$ and $f\in L^p(\mathbf{R}^n)$ the functions $T_\Omega^{(\varepsilon,N)}(f)$ converge to $T_\Omega(f)$ almost everywhere and in L^p as $\varepsilon\to0$ and $N\to\infty$.*

Proof. The a.e. convergence follows from Theorem 2.1.14. The L^p convergence is a consequence of the Lebesgue dominated convergence theorem since for a fixed $f\in L^p(\mathbf{R}^n)$ we have that $|T_\Omega^{(\varepsilon)}(f)|\le T_\Omega^{(*)}(f)$ for all $\varepsilon>0$ and $T_\Omega^{(*)}(f)$ is in $L^p(\mathbf{R}^n)$.

□

Exercises

4.2.1. Show that the directional Hilbert transform $\mathcal{H}_\theta$ is given by convolution with the distribution w_θ in $\mathcal{S}'(\mathbf{R}^n)$ defined by

$$\langle w_\theta, \varphi\rangle = \frac{1}{\pi}\,\text{p.v.}\int_{-\infty}^{+\infty} \frac{\varphi(t\theta)}{t}\,dt.$$

Compute the Fourier transform of w_θ. Identify the operator $f \to w_\theta * f$ and conclude that T_Ω may be L^p bounded even when Ω is not an L^1 function.

4.2.2. Extend the definitions of W_Ω and T_Ω to $\Omega = d\mu$ a finite signed Borel measure on $\mathbf{S}^{n-1}$ with mean value zero. Compute the Fourier transform of such W_Ω and find a necessary and sufficient condition on measures $\Omega = d\mu$ so that T_Ω be L^2 bounded.

4.2.3. Use the inequality $AB \le A\log A + e^B$ for $A \ge 1$ and $B > 0$ to prove that if Ω satisfies (4.2.22) then it must satisfy (4.2.15). Conclude that if $|\Omega|\log^+|\Omega|$ is in $L^1(\mathbf{S}^{n-1})$, then T_Ω is L^2 bounded.
[*Hint:* Use that $\int_{\mathbf{S}^{n-1}} |\xi\cdot\theta|^{-\alpha}\,d\theta$ converges when $\alpha < 1$. See Appendix D.3.]

4.2.4. Let Ω be a nonzero integrable function on $\mathbf{S}^{n-1}$ with mean value zero. Let $f \ge 0$ be nonzero and integrable over $\mathbf{R}^n$. Prove that $T_\Omega(f)$ in not in $L^1(\mathbf{R}^n)$.
[*Hint:* Show that $\widehat{T_\Omega(f)}$ cannot be continuous at zero.]

4.2.5. Use the idea of the boundedness of $\mathcal{H}_\theta$ to show that M_θ maps $L^p(\mathbf{R}^n)$ into itself with the same norm as the norm of the centered Hardy-Littlewood maximal operator on $L^p(\mathbf{R})$.

4.2.6. (a) Let $\theta \in \mathbf{S}^{n-1}$. Take inspiration from identity (4.2.16) to show that the maximal operators

$$\sup_{a>0}\frac{1}{a}\int_0^a |f(x-r\theta)|\,dr\,, \qquad \sup_{a>0}\frac{1}{2a}\int_{-a}^{+a} |f(x-r\theta)|\,dr$$

are $L^p(\mathbf{R}^n)$ bounded for $1 < p < \infty$ with norm at most $3\,p\,(p-1)^{-1}$.
(b) For $\Omega \in L^1(\mathbf{S}^{n-1})$ and f locally integrable on $\mathbf{R}^n$, define

$$M_\Omega(f)(x) = \sup_{R>0}\frac{1}{v_nR^n}\int_{|y|\le R} |\Omega(y/|y|)|\,|f(x-y)|\,dy\,.$$

Apply the method of rotations to prove that M_Ω maps $L^p(\mathbf{R}^n)$ into itself for $1 < p < \infty$.

4.2.7. Let $\Omega(x,\theta)$ be a function on $\mathbf{R}^n \times \mathbf{S}^{n-1}$ satisfying
(a) $\Omega(x,-\theta) = -\Omega(x,\theta)$ for all x and θ.

(b) $\displaystyle\int_{\mathbf{S}^{n-1}} \Omega(x,\theta)\,d\theta = 0$ for all $x \in \mathbf{R}^n$.

(c) $\sup_x |\Omega(x,\theta)|$ is in $L^1(\mathbf{S}^{n-1})$. Use the method of rotations to prove that

$$T_\Omega(f)(x) = \text{p.v.} \int_{\mathbf{R}^n} \frac{\Omega(x,y/|y|)}{|y|^n} f(x-y)\,dy$$

is bounded on $L^p(\mathbf{R}^n)$ for $1 < p < \infty$.

4.2.8. Let $\Omega \in L^q(\mathbf{S}^{n-1})$. Prove that the function $\widehat{W_\Omega}$ defined by (4.2.8) is uniformly continuous on $\mathbf{S}^{n-1}$. You may want to use Exercise 1.2.11.

4.2.9. Let $\Omega \in L^1(\mathbf{S}^{n-1})$ have mean value zero. Prove that if T_Ω maps $L^p(\mathbf{R}^n)$ into $L^q(\mathbf{R}^n)$, then $p = q$.
[*Hint:* Use dilations.]

4.2.10. Prove that for all $1 < p < \infty$ there exists a constant $A_p > 0$ such that for all complex-valued $C^2(\mathbf{R}^2)$ function f with compact support we have the bound

$$\big\|\partial_{x_1} f\big\|_{L^p} + \big\|\partial_{x_2} f\big\|_{L^p} \le A_p \big\|\partial_{x_1} f + i\partial_{x_2} f\big\|_{L^p}.$$

4.2.11. (a) Let $\Delta = \sum_{j=1}^n \partial_{x_j}^2$ be the usual Laplacian on $\mathbf{R}^n$. Prove that for all $1 < p < \infty$ there exists a constant $A_p > 0$ such that for all C^2 functions f with compact support we have the bound

$$\big\|\partial_{x_j}\partial_{x_k} f\big\|_{L^p} \le A_p \big\|\Delta(f)\big\|_{L^p}.$$

(b) Let $\Delta^m = \overbrace{\Delta\circ\cdots\circ\Delta}^{m \text{ times}}$. Show that for any $1 < p < \infty$ there exists a $C_p > 0$ such that for all f of class C^{2m} with compact support and all differential monomials ∂_x^α of order $|\alpha| = 2m$ we have

$$\big\|\partial_x^\alpha f\big\|_{L^p} \le C_p \big\|\Delta^m(f)\big\|_{L^p}.$$

4.2.12. Use the same idea as in Lemma 4.2.5 to show that if f is a continuous function on $[0,\infty)$ that is differentiable in its interior and satisfies

$$\lim_{N\to\infty} \int_N^{Na} f(u)\,du = 0$$

for all $a > 0$, then

$$\lim_{\substack{\varepsilon\to 0\\ N\to\infty}} \int_\varepsilon^N \frac{f(at)-f(t)}{t}\,dt = f(0)\log\frac{1}{a}\,.$$

4.2.13. Let Ω_o be an odd integrable function on $\mathbf{S}^{n-1}$ and Ω_e an even function on $\mathbf{S}^{n-1}$ that satisfies (4.2.22). Let f be a function supported in a ball B in

$\mathbf{R}^n$. Prove that

(a) If $|f|\log^+|f|$ is integrable over a ball B, then $T_{\Omega_o}(f)$ and $T^{(*)}_{\Omega_o}(f)$ are in integrable over B.

(b) If $|f|(\log^+|f|)^2$ is integrable over a ball B then $T_{\Omega_e}(f)$ and $T^{(*)}_{\Omega_e}(f)$ are in integrable over B.

[*Hint:* Use Exercise 1.3.7.]

4.2.14. (*Sjögren and Soria* [**458**]) Let Ω be integrable on $\mathbf{S}^{n-1}$ with mean value zero. Use Jensen's inequality to show that for some $C > 0$ and all $f \in L^2(\mathbf{R}^n)$ radial functions we have

$$\big\|T_\Omega(f)\big\|_{L^2} \le C\big\|f\big\|_{L^2}.$$

This inequality subsumes that T_Ω is well defined for f radial.

4.3. The Calderón-Zygmund Decomposition and Singular Integrals

The behavior of singular integral operators on $L^1(\mathbf{R}^n)$ is a more subtle issue than that on L^p for $1 < p < \infty$. It turns out that singular integrals are not bounded from L^1 to L^1. See Example 4.1.3 and also Exercise 4.2.4. In this section we will see that singular integrals map L^1 into the larger space $L^{1,\infty}$. This result strengthens their L^p boundedness.

4.3.a. The Calderón-Zygmund Decomposition

To prove that singular integrals are of weak type $(1,1)$ we will need to introduce the Calderón-Zygmund decomposition. This is a powerful stopping time construction that has many other interesting applications. We have already seen an example of a stopping time argument in Section 2.1.

Theorem 4.3.1. *Let $f \in L^1(\mathbf{R}^n)$ and $\alpha > 0$. Then there exist functions g and b on $\mathbf{R}^n$ such that*

(1) $f = g + b$.

(2) $\big\|g\big\|_{L^1} \le \big\|f\big\|_{L^1}$ *and* $\big\|g\big\|_{L^\infty} \le 2^n\alpha$.

(3) $b = \sum_j b_j$, *where each b_j is supported in a dyadic cube Q_j. Furthermore, the cubes Q_k and Q_j have disjoint interiors when $j \ne k$.*

(4) $\displaystyle\int_{Q_j} b_j(x)\,dx = 0.$

(5) $\|b_j\|_{L^1} \leq 2^{n+1}\alpha|Q_j|$.

(6) $\sum_j |Q_j| \leq \alpha^{-1}\|f\|_{L^1}$.

Remark 4.3.2. This is called the *Calderón-Zygmund decomposition* of f at height α. The function g is called the *good function* of the decomposition since it is both integrable and bounded; hence the letter g. The function b is called the *bad function* since it contains the singular part of f (hence the letter b), but it is carefully chosen to have mean value zero. It follows from (5) and (6) that the bad function b is integrable and

$$\|b\|_{L^1} \leq \sum_j \|b_j\|_{L^1} \leq 2^{n+1}\alpha \sum_j |Q_j| \leq 2^{n+1}\|f\|_{L^1}.$$

By (2) the good function is integrable and bounded; hence it is in all the L^p spaces for $1 \leq p \leq \infty$. More specifically, we have the following estimate:

$$\|g\|_{L^p} \leq \|g\|_{L^1}^{\frac{1}{p}}\|g\|_{L^\infty}^{1-\frac{1}{p}} \leq \|f\|_{L^1}^{\frac{1}{p}}(2^n\alpha)^{1-\frac{1}{p}} = 2^{\frac{n}{p'}}\alpha^{\frac{1}{p'}}\|f\|_{L^1}^{\frac{1}{p}}. \tag{4.3.1}$$

Proof. Recall that a dyadic cube in $\mathbf{R}^n$ is a cube of the form

$$[2^k m_1, 2^k(m_1+1)) \times \cdots \times [2^k m_n, 2^k(m_n+1)),$$

where $k, m_1, \dots, m_n$ are integers. Decompose $\mathbf{R}^n$ into a mesh of equal size disjoint dyadic cubes so that

$$|Q| \geq \frac{1}{\alpha}\|f\|_{L^1}$$

for every cube Q in the mesh. Subdivide each cube in the mesh into 2^n congruent cubes by bisecting each of its sides. We now have a new mesh of dyadic cubes. Select a cube in the new mesh if

$$\frac{1}{|Q|}\int_Q |f(x)|\,dx > \alpha. \tag{4.3.2}$$

Let S be the set of all selected cubes. Now subdivide each nonselected cube into 2^n congruent subcubes by bisecting each side as before. Then select one of these new cubes if (4.3.2) holds. Put all selected cubes of this generation into the set S. Repeat this procedure indefinitely.

The set of all selected cubes S is exactly the set of the cubes Q_j proclaimed in the proposition. Let us observe that these cubes are disjoint, for otherwise some Q_k would be a proper subset of some Q_j, which is impossible since the selected cube Q_j was never subdivided. Now define

$$b_j = \left(f - \frac{1}{|Q_j|}\int_{Q_j} f\,dx\right)\chi_{Q_j},$$

$b = \sum_j b_j$ and $g = f - b$.

For a selected cube Q_j there exists a unique nonselected cube Q' with twice its side length that contains Q_j. Let us call this cube the parent of Q_j. Since its parent Q' was not selected, we have $|Q'|^{-1}\int_{Q'}|f|\,dx \le \alpha$. Then

$$\frac{1}{|Q_j|}\int_{Q_j}|f(x)|\,dx \le \frac{1}{|Q_j|}\int_{Q'}|f(x)|\,dx = \frac{2^n}{|Q'|}\int_{Q'}|f(x)|\,dx \le 2^n\alpha.$$

Consequently,

$$\int_{Q_j}|b_j|\,dx \le \int_{Q_j}|f|\,dx + |Q_j|\left|\frac{1}{|Q_j|}\int_{Q_j}f\,dx\right| \le 2\int_{Q_j}|f|\,dx \le 2^{n+1}\alpha|Q_j|,$$

which proves (5). To prove (6), simply observe that

$$\sum_j |Q_j| \le \frac{1}{\alpha}\sum_j \int_{Q_j}|f|\,dx = \frac{1}{\alpha}\int_{\cup_j Q_j}|f|\,dx \le \frac{1}{\alpha}\|f\|_{L^1}.$$

Next we need to obtain the estimates on g. Write $\mathbf{R}^n = \cup_j Q_j \cup F$, where F is a closed set. Since $b = 0$ on F and $f - b_j = |Q_j|^{-1}\int_{Q_j} f\,dx$, we have

$$g = \begin{cases} f & \text{on } F, \\ \frac{1}{|Q_j|}\int_{Q_j} f\,dx & \text{on } Q_j. \end{cases} \tag{4.3.3}$$

On the cube Q_j, g is equal to the constant $|Q_j|^{-1}\int_{Q_j} f\,dx$, and this is bounded by $2^n\alpha$. It suffices to show that g is bounded on the set F. Given $x \in F$, we have that x does not belong to any selected cube. Therefore, there exists a sequence of cubes $Q^{(k)}$ whose closures contain x and whose side lengths tend to zero as $k \to \infty$. Since the cubes $Q^{(k)}$ were never selected, we have

$$\left|\frac{1}{|Q^{(k)}|}\int_{Q^{(k)}} f\,dx\right| \le \frac{1}{|Q^{(k)}|}\int_{Q^{(k)}}|f|\,dx \le \alpha.$$

Using a version of Corollary 2.1.16 where the balls are replaced with cubes, we conclude that

$$|f(x)| = \left|\lim_{k\to\infty}\frac{1}{|Q^{(k)}|}\int_{Q^{(k)}} f\,dx\right| \le \alpha$$

whenever $x \in F$. But since $g = f$ a.e. on F, if follows that g is bounded by α on F. Finally, it follows from (4.3.3) that $\|g\|_{L^1} \le \|f\|_{L^1}$. This finishes the proof of the theorem.

□

We will now apply the Calderón-Zygmund decomposition to obtain weak type $(1,1)$ bounds for a wide class of singular integral operators that includes the operators T_Ω we studied in the previous section.

4.3.b. General Singular Integrals

Let K be a measurable function defined on $\mathbf{R}^n \setminus \{0\}$ that we will assume satisfies the mild "size" condition

$$\sup_{R>0} \int_{R\le|x|\le 2R} |K(x)|\, dx = A_1 < \infty\,. \tag{4.3.4}$$

This condition is less restrictive than the standard size estimate

$$\sup_{x\in\mathbf{R}^n} |x|^n |K(x)| < \infty\,, \tag{4.3.5}$$

but it is general enough to capture size properties of kernels $K(x) = \Omega(x/|x|)/|x|^n$, where $\Omega \in L^1(\mathbf{S}^{n-1})$. We also note that condition (4.3.4) is equivalent to

$$\sup_{R>0} \frac{1}{R} \int_{|x|\le R} |K(x)|\,|x|\, dx < \infty\,. \tag{4.3.6}$$

See Exercise 4.3.1.

The size condition (4.3.4) is sufficient to make K a tempered distribution away from the origin. Indeed, for $\varphi \in \mathcal{S}(\mathbf{R}^n)$ we have

$$\begin{aligned}\int_{|x|\ge 1} |K(x)\varphi(x)|\, dx \;&\le \sum_{m=0}^{\infty} \int_{2^{m+1}\ge|x|\ge 2^m} \frac{|K(x)|(1+|x|)^N|\varphi(x)|}{(1+2^m)^N}\, dx \\ &\le \sum_{m=0}^{\infty} \frac{A_1}{(1+2^m)^N} \sup_{x\in\mathbf{R}^n} (1+|x|)^N |\varphi(x)|\,,\end{aligned}$$

and the latter is controlled by a finite sum of Schwartz seminorms of φ.

We are interested in tempered distributions W on $\mathbf{R}^n$ that extend the function K defined on $\mathbf{R}^n \setminus \{0\}$ and that have the form

$$W(\varphi) = \lim_{j\to\infty} \int_{|x|\ge\delta_j} K(x)\varphi(x)\, dx, \qquad \varphi \in \mathcal{S}(\mathbf{R}^n), \tag{4.3.7}$$

for some sequence $\delta_j \downarrow 0$ as $j \to \infty$. It is not hard to see that there exists a tempered distribution W satisfying (4.3.7) for all $\varphi \in \mathcal{S}(\mathbf{R}^n)$ if and only if

$$\lim_{j\to\infty} \int_{1\ge|x|\ge\delta_j} K(x)\, dx = L \tag{4.3.8}$$

exists. See Exercise 4.3.2. If such a distribution W exists it may not be unique since it depends on the choice of the sequence δ_j. Two different sequences tending to zero may give two different tempered distributions W of the form (4.3.7), both coinciding with the function K on $\mathbf{R}^n \setminus \{0\}$. See Example 4.4.2 and Remark 4.4.3. Furthermore not all functions K on $\mathbf{R}^n \setminus \{0\}$ give rise to distributions W defined by (4.3.7); take, for example, $K(x) = |x|^{-n}$.

If condition (4.3.8) is satisfied, we can define

$$W(\varphi) = \lim_{j\to\infty} \int_{j\geq|x|\geq\delta_j} K(x)\varphi(x)\,dx \tag{4.3.9}$$

and the limit exists as $j \to \infty$ for all $\varphi \in \mathcal{S}(\mathbf{R}^n)$ and is equal to

$$W(\varphi) = \int_{|x|\leq 1} K(x)(\varphi(x) - \varphi(0))\,dx + \varphi(0)L + \int_{|x|\geq 1} K(x)\varphi(x)\,dx.$$

Moreover, the previous calculations show that W is an element of $\mathcal{S}'(\mathbf{R}^n)$.

Next we assume that the given function K on $\mathbf{R}^n \setminus \{0\}$ satisfies a certain smoothness condition. There are three kinds of smoothness conditions that we shall encounter in the sequel: first, the *gradient condition*

$$|\nabla K(x)| \leq A_2|x|^{-n-1}, \qquad x \neq 0, \tag{4.3.10}$$

next, the more general *Lipschitz condition*,

$$|K(x-y) - K(x)| \leq A_2\frac{|y|^\delta}{|x|^{n+\delta}}, \qquad \text{whenever } |x| \geq 2|y|; \tag{4.3.11}$$

and finally the more general smoothness condition

$$\sup_{y\neq 0} \int_{|x|\geq 2|y|} |K(x-y) - K(x)|\,dx = A_2, \tag{4.3.12}$$

for some $A_2 < \infty$. The reader should easily verify that (4.3.12) is a weaker condition than (4.3.11), which in turn is weaker than (4.3.10). Condition (4.3.12) is often referred to as *Hörmander's condition.*

4.3.c. L^r Boundedness Implies Weak Type $(1,1)$ Boundedness

This next theorem provides a classical application of the Calderón-Zygmund decomposition.

Theorem 4.3.3. *Assume that K is defined on $\mathbf{R}^n \setminus \{0\}$ and satisfies (4.3.12) for some $A_2 > 0$. Let $W \in \mathcal{S}'(\mathbf{R}^n)$ be as in (4.3.7) coinciding with K on $\mathbf{R}^n \setminus \{0\}$. Suppose that the operator T given by convolution with W maps $L^r(\mathbf{R}^n)$ into itself with norm B for some $1 < r \leq \infty$. Then T has an extension that maps $L^1(\mathbf{R}^n)$ into $L^{1,\infty}(\mathbf{R}^n)$ with norm*

$$\big\|T\big\|_{L^1\to L^{1,\infty}} \leq C_n(A_2 + B), \tag{4.3.13}$$

and T also extends to a bounded operator from $L^p(\mathbf{R}^n)$ into itself for $1 < p < \infty$ with norm

$$\big\|T\big\|_{L^p\to L^p} \leq C_n' \max\left(p, (p-1)^{-1}\right)(A_2 + B), \tag{4.3.14}$$

where C_n, C_n' are constants that depend on the dimension but not on r or p.

Proof. We first explain the idea of the proof. We write $f = g + b$; hence

$$T(f) = T(g) + T(b).$$

The function $T(g)$ is in L^r and thus it satisfies a weak type L^r estimate. The bad part of f is a sum of functions with mean value zero. Cancellation is used to subtract a suitable term from every piece of the bad function that allows us to use Hörmander's condition (4.3.12). Let us proceed with the details. We will work out the case $r < \infty$, and we refer the reader to Exercise 4.3.7 for the case $r = \infty$.

Fix $\alpha > 0$ and let f be in $L^1(\mathbf{R}^n)$. We will assume that f is in the Schwartz class since $T(f)$ may not be a priori defined for $f \in L^1(\mathbf{R}^n)$. Once (4.3.13) is obtained for f in $\mathcal{S}(\mathbf{R}^n)$, a density argument will give that T admits an extension on L^1 that also satisfies (4.3.13). Apply the Calderón-Zygmund decomposition to f at height $\gamma\alpha$, where γ is a positive constant to be chosen later. That is, write the function f as the sum

$$f = g + b,$$

where conditions (1)–(6) of Theorem 4.3.1 are satisfied with the constant α replaced by $\gamma\alpha$. We denote by $\ell(Q)$ the side length of a cube Q. Let Q_j^* be the unique cube with sides parallel to the axes having the same center as Q_j and having side length

$$\ell(Q_j^*) = 2\sqrt{n}\,\ell(Q_j)\,.$$

We have

$$\begin{aligned}
&\big|\{x \in \mathbf{R}^n : \ |T(f)(x)| > \alpha\}\big| \\
\le\ & \Big|\Big\{x \in \mathbf{R}^n : \ |T(g)(x)| > \frac{\alpha}{2}\Big\}\Big| + \Big|\Big\{x \in \mathbf{R}^n : \ |T(b)(x)| > \frac{\alpha}{2}\Big\}\Big| \\
\le\ & \frac{2^r}{\alpha^r}\big\|T(g)\big\|_{L^r}^r + \Big|\bigcup_j Q_j^*\Big| + \Big|\Big\{x \notin \bigcup_j Q_j^* : \ |T(b)(x)| > \frac{\alpha}{2}\Big\}\Big| \\
\le\ & \frac{2^r}{\alpha^r}B^r\big\|g\big\|_{L^r}^r + \sum_j |Q_j^*| + \frac{2}{\alpha}\int_{(\bigcup_j Q_j^*)^c} |T(b)(x)|\,dx \\
\le\ & \frac{2^r}{\alpha^r}2^{\frac{nr}{r'}}B^r(\gamma\alpha)^{\frac{r}{r'}}\big\|f\big\|_{L^1} + (2\sqrt{n})^n\frac{\big\|f\big\|_{L^1}}{\gamma\alpha} + \frac{2}{\alpha}\sum_j\int_{(Q_j^*)^c}|T(b_j)(x)|\,dx \\
\le\ & \left(\frac{(2^{n+1}B\gamma)^r}{2^n\gamma} + \frac{(2\sqrt{n})^n}{\gamma}\right)\frac{\big\|f\big\|_{L^1}}{\alpha} + \frac{2}{\alpha}\sum_j\int_{(Q_j^*)^c}|T(b_j)(x)|\,dx\,.
\end{aligned}$$

It suffices to show that the last sum is bounded by some constant multiple of the L^1 norm of f. It is here where we will use the fact that b_j has mean value zero and

Hörmander's condition (4.3.12). Let y_j be the center of the cube Q_j. We have

$$\begin{aligned}
\sum_j \int_{(Q_j^*)^c} |T(b_j)(x)|\,dx &= \sum_j \int_{(Q_j^*)^c} \left| \int b_j(y) K(x-y)\,dy \right| dx \\
&= \sum_j \int_{(Q_j^*)^c} \left| \int_{Q_j} b_j(y) \big(K(x-y) - K(x-y_j) \big)\,dy \right| dx \\
&\le \sum_j \int_{Q_j} |b_j(y)| \int_{(Q_j^*)^c} |K(x-y) - K(x-y_j)|\,dx\,dy \\
&= \sum_j \int_{Q_j} |b_j(y)| \int_{-y_j+(Q_j^*)^c} |K(x-(y-y_j)) - K(x)|\,dx\,dy \\
&\le \sum_j \int_{Q_j} |b_j(y)| \int_{|x|\ge 2|y-y_j|} |K(x-(y-y_j)) - K(x)|\,dx\,dy \\
&\le A_2 \sum_j \int_{Q_j} |b_j(y)|\,dy = A_2 \sum_j \big\|b_j\big\|_{L^1} \le A_2 2^{n+1} \big\|f\big\|_{L^1},
\end{aligned}$$

where we used the fact that if $x \in -y_j + (Q_j^*)^c$ then $|x| \ge \frac{1}{2}\ell(Q_j^*) = \sqrt{n}\,\ell(Q_j)$. But since $y - y_j \in -y_j + Q_j$ we have $|y - y_j| \le \frac{\sqrt{n}}{2}\,\ell(Q_j)$, which implies that $|x| \ge 2|y - y_j|$. Here we used the fact that the diameter of a cube is equal to $\sqrt{n}$ times its side length. See Figure 4.2.

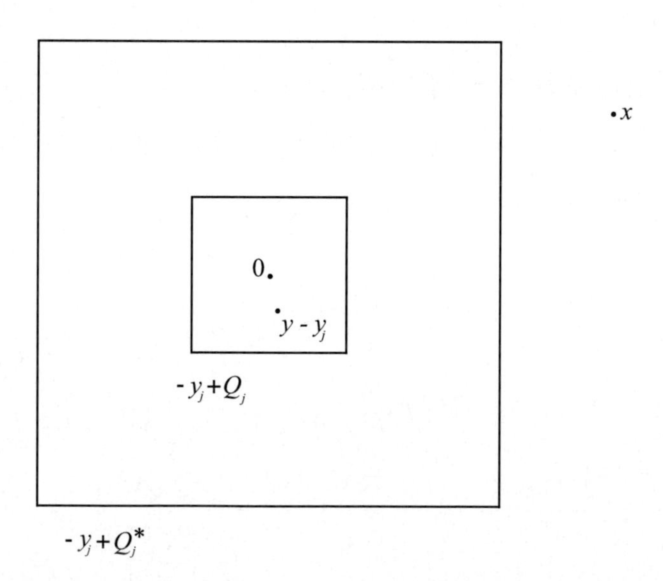

FIGURE 4.2. The cubes $-y_j + Q_j$ and $-y_j + Q_j^*$. The point $y - y_j$ lies in the smaller cube, but x is not in the larger cube.

The weak type $(1,1)$ estimate (4.3.13) for T now follows by choosing $\gamma = 2^{-(n+1)}B^{-1}$ and $C_n = 2 + 2^{n+1}(2\sqrt{n})^n + 2^{n+2}$.

In view of Exercise 1.3.2, we have that T maps L^p into L^p with bound at most $C_n(A_2 + B)(p-1)^{-1}$ whenever $1 < p < r$. This proves (4.3.14) for $1 < p < r$. To obtain a similar conclusion for $r < p < \infty$ we use duality. We first note that the adjoint operator T^* of T defined by

$$\langle T(f)\,|\,g\rangle = \langle f\,|\,T^*(g)\rangle$$

has a distributional kernel that coincides with the function $K^*(x) = \overline{K(-x)}$ on $\mathbf{R}^n \setminus \{0\}$. Next we notice that since K satisfies (4.3.12), then so does K^* and with the same bound. Therefore, T^*, which maps $L^{r'}$ into $L^{r'}$, has a kernel that satisfies Hörmander's condition. It must therefore map L^1 into $L^{1,\infty}$ and $L^{p'}$ into $L^{p'}$ for $1 < p' < r'$ by the argument just shown with norm at most $C_n(A_2 + B)(p'-1)^{-1}$. It follows that T maps L^p into L^p with norm at most $C_n(A_2 + B)p$ for $r < p < \infty$. This proves (4.3.14) for $r < p < \infty$.

□

Suppose now that K satisfies the size condition (4.3.4) and let $0 < \varepsilon < N < \infty$. Define the truncated kernel $K^{(\varepsilon,N)}$ by setting

$$K^{(\varepsilon,N)}(x) = K(x)\chi_{\varepsilon \le |x| \le N}(x)\,. \tag{4.3.15}$$

We now define the truncated singular integrals $T^{(\varepsilon,N)}$ by

$$T^{(\varepsilon,N)}(f) = f * K^{(\varepsilon,N)}\,,$$

and we observe that these operators are well defined at least for $f \in \mathcal{S}(\mathbf{R}^n)$. We also observe that solely under condition (4.3.4), the operators $T^{(\varepsilon,N)}(f)$ are not well defined as convergent integrals when f is in L^p for $p < \infty$ (despite the double truncation).

Assuming, however, *that $T^{(\varepsilon,N)}$ have bounded extensions on $L^2(\mathbf{R}^n)$*, a fact that will be obtained in the next section using some additional assumptions on K, we can define the *maximal singular integral operator* $T^{(*)}$ associated with K by

$$T^{(*)}(f) = \sup_{0<\varepsilon<N<\infty} |T^{(\varepsilon,N)}(f)|\,. \tag{4.3.16}$$

Therefore, for all $x \in \mathbf{R}^n$, $T^{(*)}(f)(x)$ is also well defined, but potentially infinite, for all functions f in $L^2(\mathbf{R}^n)$.

Exercise 4.3.3 says that if a function K satisfies Hörmander's condition (4.3.12), then so do all of its truncations $K^{(\varepsilon,N)}$. Using this result and Theorem 4.3.3, we obtain the following:

Corollary 4.3.4. *Assume that K satisfies (4.3.12) for some $A_2 > 0$. Let W in $\mathcal{S}'(\mathbf{R}^n)$ be as in (4.3.7), extending K. Suppose that the operators $T^{(\varepsilon,N)}$ given by convolution with $K^{(\varepsilon,N)}$ admit extensions that are uniformly bounded on $L^2(\mathbf{R}^n)$ with norm B. Then for every $0 < \varepsilon < N < \infty$, $T^{(\varepsilon,N)}$ maps $L^1(\mathbf{R}^n)$ into $L^{1,\infty}(\mathbf{R}^n)$*

with norm at most $C_n(A_2 + B)$ and also $L^p(\mathbf{R}^n)$ into itself with norm at most $C_n \max\left(p, (p-1)^{-1}\right)(A_2+B)$, whenever $1 < p < \infty$, where C_n is some dimensional constant.

4.3.d. L^r Boundedness for Maximal Singular Integrals Implies Weak Type $(1,1)$ Boundedness

We now state and prove a result analogous to that in Theorem 4.3.3 for maximal singular integrals.

Theorem 4.3.5. *Let $K(x)$ be function on $\mathbf{R}^n \setminus \{0\}$ satisfying (4.3.4) with constant $A_1 < \infty$ and Hörmander's condition (4.3.12) with constant $A_2 < \infty$. Suppose that the operator $T^{(*)}$ as defined in (4.3.16) maps $L^2(\mathbf{R}^n)$ into itself with norm B. Then $T^{(*)}$ maps $L^1(\mathbf{R}^n)$ into $L^{1,\infty}(\mathbf{R}^n)$ with norm*

$$\left\|T^{(*)}\right\|_{L^1 \to L^{1,\infty}} \le C_n(A_1 + A_2 + B),$$

where C_n is some dimensional constant.

We note that if (4.3.12) in the previous theorem is replaced by the slightly stronger condition (4.3.11), then the following proof may be simplified significantly; see Exercise 4.4.6(d).

Proof. The proof of this theorem is only a little more involved than the proof of Theorem 4.3.3. We fix an $L^1(\mathbf{R}^n)$ function f. We apply the Calderón-Zygmund decomposition of f at height $\gamma\alpha$ for some $\gamma, \alpha > 0$. We then write $f = g + b$, where $b = \sum_j b_j$ and each b_j is supported in some cube Q_j. We define Q_j^* as the cube with the same center as Q_j and with sides parallel to the sides of Q_j having length $\ell(Q_j^*) = 5\sqrt{n}\,\ell(Q_j)$. This is only a minor change compared with the definition of Q_j in Theorem 4.3.3. The main change of the proof is in the treatment of the term

$$\left|\left\{x \in \Big(\bigcup_j Q_j^*\Big)^c : \; |T^{(*)}(b)(x)| > \frac{\alpha}{2}\right\}\right|. \tag{4.3.17}$$

We will show that for all $\gamma \le (2^{n+5}A_1)^{-1}$ we have

$$\left|\left\{x \in \Big(\bigcup_j Q_j^*\Big)^c : \; |T^{(*)}(b)(x)| > \frac{\alpha}{2}\right\}\right| \le 2^{n+8} A_2 \,\frac{\left\|f\right\|_{L^1}}{\alpha}. \tag{4.3.18}$$

Let us conclude the proof of the theorem assuming for the moment the validity of (4.3.18). As in the proof of Theorem 4.3.3, we can show that

$$\left|\left\{x \in \mathbf{R}^n : \; |T^{(*)}(g)(x)| > \frac{\alpha}{2}\right\}\right| + \left|\bigcup_j Q_j^*\right| \le \left(2^{n+2}B^2\gamma + \frac{(5\sqrt{n})^n}{\gamma}\right)\frac{\left\|f\right\|_{L^1}}{\alpha}.$$

Combining this estimate with (4.3.18) and choosing

$$\gamma = (2^{n+5}(A_1 + A_2 + B))^{-1},$$

we obtain the required estimate

$$\big|\{x \in \mathbf{R}^n : \; |T^{(*)}(f)(x)| > \alpha\}\big| \le C_n(A_1 + A_2 + B)\frac{\|f\|_{L^1}}{\alpha}$$

with $C_n = 2^{-3} + (5\sqrt{n})^n 2^{n+5} + 2^{n+8}$.

It remains to prove (4.3.18). This estimate will be a consequence of the fact that for $x \in \big(\bigcup_j Q_j^*\big)^c$ we have the key inequality

$$T^{(*)}(b)(x) \le 4E_1(x) + 2^{n+2}\alpha\gamma E_2(x) + 2^{n+3}\alpha\gamma A_1\,, \tag{4.3.19}$$

where

$$\begin{aligned} E_1(x) &= \sum_j \int_{Q_j} |K(x-y) - K(x-y_j)|\,|b_j(y)|\,dy\,, \\ E_2(x) &= \sum_j \int_{Q_j} |K(x-y) - K(x-y_j)|\,dy\,, \end{aligned}$$

and y_j is the center of Q_j.

If we had (4.3.19), then we could easily derive (4.3.18). Indeed, fix a γ satisfying $\gamma \le (2^{n+5}A_1)^{-1}$. Then we have $2^{n+3}\alpha\gamma A_1 < \frac{\alpha}{3}$ and, using (4.3.19), we obtain

$$\begin{aligned} &\Big|\Big\{x \in \Big(\bigcup_j Q_j^*\Big)^c : \; |T^{(*)}(b)(x)| > \frac{\alpha}{2}\Big\}\Big| \\ &\le \Big|\Big\{x \in \Big(\bigcup_j Q_j^*\Big)^c : \; 4E_1(x) > \frac{\alpha}{12}\Big\}\Big| \\ &\qquad\qquad + \Big|\Big\{x \in \Big(\bigcup_j Q_j^*\Big)^c : \; 2^{n+2}\alpha\gamma E_2(x) > \frac{\alpha}{12}\Big\}\Big| \\ &\le \frac{48}{\alpha}\int_{(\bigcup_j Q_j^*)^c} E_1(x)\,dx + 2^{n+6}\gamma \int_{(\bigcup_j Q_j^*)^c} E_2(x)\,dx, \end{aligned} \tag{4.3.20}$$

since $\frac{\alpha}{2} = \frac{\alpha}{3} + \frac{\alpha}{12} + \frac{\alpha}{12}$. We have

$$\begin{aligned} &\int_{(\bigcup_j Q_j^*)^c} E_1(x)\,dx \\ &\le \sum_j \int_{Q_j} |b_j(y)| \int_{(Q_j^*)^c} |(K(x-y) - K(x-y_j)|\,dx\,dy \\ &\le \sum_j \int_{Q_j} |b_j(y)| \int_{|x-y_j| \ge 2|y-y_j|} |K(x-y) - K(x-y_j)|\,dx\,dy \\ &\le A_2 \sum_j \int_{Q_j} |b_j(y)|\,dy = A_2 \sum_j \|b_j\|_{L^1} \le A_2 2^{n+1}\|f\|_{L^1}\,, \end{aligned} \tag{4.3.21}$$

where we used the fact that if $x \in (Q_j^*)^c$, then $|x - y_j| \ge \frac{1}{2}\ell(Q_j^*) = \frac{5}{2}\sqrt{n}\,\ell(Q_j)$. But since $|y - y_j| \le \frac{\sqrt{n}}{2}\,\ell(Q_j)$ this implies that $|x - y_j| \ge 2|y - y_j|$. Here we used the fact that the diameter of a cube is equal to $\sqrt{n}$ times its side length. Likewise, we obtain that

$$\int_{(\bigcup_j Q_j^*)^c} E_2(x)\,dx \le A_2 \sum_j |Q_j| \le A_2 \frac{\|f\|_{L^1}}{\alpha\gamma}. \tag{4.3.22}$$

Combining (4.3.21) and (4.3.22) with (4.3.20) yields (4.3.18).

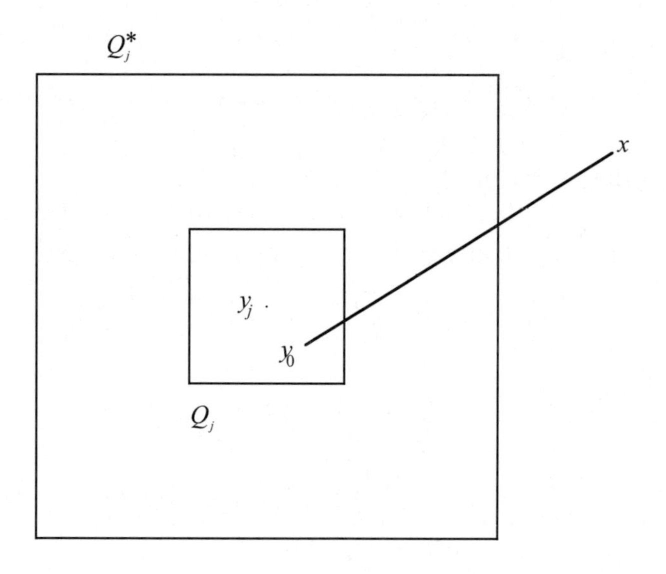

FIGURE 4.3. The cubes Q_j and Q_j^*.

Therefore, the main task of the proof is to show (4.3.19). Recall that $b = \sum_j b_j$ and to estimate $T^{(*)}(b)$ it suffices to estimate each $|T^{(\varepsilon,N)}(b_j)|$ uniformly in ε and N. To achieve this we will use that

$$|T^{(\varepsilon,N)}(b_j)| \le |T^{(\varepsilon,\infty)}(b_j)| + |T^{(N,\infty)}(b_j)|. \tag{4.3.23}$$

We work with $T^{(\varepsilon,\infty)}$ and we note that $T^{(N,\infty)}$ can be treated similarly. For fixed $x \notin \bigcup_j Q_j^*$ and $\varepsilon > 0$ we define

$$\begin{aligned} J_1(x,\varepsilon) &= \{j : \ \forall y \in Q_j \ \text{ we have } |x - y| < \varepsilon\}, \\ J_2(x,\varepsilon) &= \{j : \ \forall y \in Q_j \ \text{ we have } |x - y| > \varepsilon\}, \\ J_3(x,\varepsilon) &= \{j : \ \exists y \in Q_j \ \text{ we have } |x - y| = \varepsilon\}. \end{aligned}$$

Note that $T^{\varepsilon,\infty}(b_j)(x) = 0$ whenever $x \notin \bigcup_j Q_j^*$ and $j \in J_1(x,\varepsilon)$. Also note that $K^{\varepsilon,\infty}(x-y) = K(x-y)$ whenever $x \notin \bigcup_j Q_j^*$, $j \in J_2(x,\varepsilon)$ and $y \in Q_j$. Therefore,

$$\sup_{\varepsilon>0} |T^{\varepsilon,\infty}(b)(x)| \le \sup_{\varepsilon>0} \Big| \sum_{j\in J_2(x,\varepsilon)} T(b_j)(x)\Big| + \sup_{\varepsilon>0} \Big| \sum_{j\in J_3(x,\varepsilon)} T(b_j\chi_{|x-\cdot|\ge\varepsilon})(x)\Big|$$

but since

$$\sup_{\varepsilon>0} \Big| \sum_{j\in J_2(x,\varepsilon)} T(b_j)(x)\Big| \le \sum_j |T(b_j)(x)| = E_1(x)\,, \tag{4.3.24}$$

it suffices to estimate the term

$$\sup_{\varepsilon>0} \Big| \sum_{j\in J_3(x,\varepsilon)} T(b_j\chi_{|x-\cdot|\ge\varepsilon})(x)\Big|\,.$$

We now make some geometric observations; see Figure 4.3. Fix $\varepsilon > 0$, an x in $(\bigcup_j Q_j^*)^c$, and also a cube Q_j with $j \in J_3(x,\varepsilon)$. Then we have

$$\varepsilon \ge \frac{1}{2}\big(\ell(Q_j^*) - \ell(Q_j)\big) = \frac{1}{2}(5\sqrt{n}-1)\ell(Q_j) \ge 2\sqrt{n}\,\ell(Q_j)\,. \tag{4.3.25}$$

Since $j \in J_3(x,\varepsilon)$, there exists a $y_0 \in Q_j$ with $|x-y_0| = \varepsilon$. Using (4.3.25), we obtain that for any $y \in Q_j$ we have

$$\begin{aligned}
&\frac{\varepsilon}{2} \le \varepsilon - \sqrt{n}\,\ell(Q_j) \le |x-y_0| - |y-y_0| \le |x-y|\,,\\
&|x-y| \le |x-y_0| + |y-y_0| \le \varepsilon + \sqrt{n}\,\ell(Q_j) \le \frac{3\varepsilon}{2}\,.
\end{aligned}$$

Therefore, we have proved that

$$\bigcup_{j\in J_3(x,\varepsilon)} Q_j \subseteq B(x,\tfrac{3\varepsilon}{2}) \setminus B(x,\tfrac{\varepsilon}{2})\,.$$

Letting

$$c_j(\varepsilon) = |Q_j|^{-1} \int_{Q_j} b_j(y)\chi_{|x-y|\ge\varepsilon}(y)\,dy\,,$$

we note that in view of property (5) of the Calderón-Zygmund decomposition (Theorem 4.3.1), the estimate $|c_j(\varepsilon)| \le 2^{n+1}\alpha\gamma$ holds. Then

$$
\begin{aligned}
&\sup_{\varepsilon>0}\bigg|\sum_{j\in J_3(x,\varepsilon)}\int_{Q_j} K(x-y)b_j(y)\chi_{|x-y|\ge\varepsilon}(y)\,dy\bigg| \\
&\le \sup_{\varepsilon>0}\bigg|\sum_{j\in J_3(x,\varepsilon)}\int_{Q_j} K(x-y)\big(b_j(y)\chi_{|x-y|\ge\varepsilon}(y)-c_j(\varepsilon)\big)\,dy\bigg| \\
&\qquad\qquad + \sup_{\varepsilon>0}\bigg|\sum_{j\in J_3(x,\varepsilon)} c_j(\varepsilon)\int_{Q_j} K(x-y)\,dy\bigg| \\
&\le \sup_{\varepsilon>0}\bigg|\sum_{j\in J_3(x,\varepsilon)}\int_{Q_j} \big(K(x-y)-K(x-y_j)\big)\big(b_j(y)\chi_{|x-y|\ge\varepsilon}(y)-c_j(\varepsilon)\big)\,dy\bigg| \\
&\qquad\qquad + 2^{n+1}\alpha\gamma\sup_{\varepsilon>0}\int_{B(x,\frac{3\varepsilon}{2})\setminus B(x,\frac{\varepsilon}{2})}|K(x-y)|\,dy \\
&\le \sum_j \int_{Q_j}\big|K(x-y)-K(x-y_j)\big|\big(|b_j(y)|+2^{n+1}\alpha\gamma\big)\,dy \\
&\qquad\qquad + 2^{n+1}\alpha\gamma\sup_{\varepsilon>0}\int_{\frac{\varepsilon}{2}\le|x-y|\le\frac{3\varepsilon}{2}}|K(x-y)|\,dy \\
&\le E_1(x)+2^{n+1}\alpha\gamma E_2(x)+2^{n+1}\alpha\gamma(2A_1)\,.
\end{aligned}
$$

The last estimate, together with (4.3.24), with (4.3.23), and with the analogous estimate for $\sup_{N>0}|T^{N,\infty}(b)(x)|$ (which is obtained entirely similarly), yields (4.3.19). □

The value of the previous theorem lies in the fact that if for some sequences $\varepsilon_j \downarrow 0$, $N_j \uparrow \infty$ the pointwise limit $T^{(\varepsilon_j,N_j)}(f)$ exists a.e. for all f in a dense subclass of L^1, then $T^{(\varepsilon_j,N_j)}(f)$ exists a.e. for all f in $L^1(\mathbf{R}^n)$.

If the singular integrals have kernels of the form $\Omega(x/|x|)|x|^{-n}$ with Ω in L^∞, such as the Hilbert transform and the Riesz transforms, then the upper truncations are not need for K in (4.3.15). In this case

$$
T_\Omega^{(\varepsilon)}(f)(x) = \int_{|y|\ge\varepsilon} f(x-y)\frac{\Omega(y/|y|)}{|y|^n}\,dy
$$

is well defined for $f\in\bigcup_{1\le p<\infty}L^p(\mathbf{R}^n)$ by Hölder's inequality and is equal to

$$
\lim_{N\to\infty}\int_{\varepsilon\le|y|\le N} f(x-y)\frac{\Omega(y/|y|)}{|y|^n}\,dy\,.
$$

Corollary 4.3.6. *The maximal Hilbert transform $H^{(*)}$ and the maximal Riesz transforms $R_j^{(*)}$ are of weak type $(1,1)$. Therefore, the limits $\lim_{\varepsilon\to 0} H^{(\varepsilon)}(f)$ and $\lim_{\varepsilon\to 0} R_j^{(\varepsilon)}(g)$ exist a.e. for all $f \in L^1(\mathbf{R})$ and $g \in L^1(\mathbf{R}^n)$, as $\varepsilon \to 0$.*

Proof. Since the kernels $1/x$ on $\mathbf{R}$ and $x_j/|x|^n$ on $\mathbf{R}^n$ satisfy (4.3.10), the first statement in the corollary is an immediate consequence of Theorem 4.3.5. The second statement follows from Theorem 2.1.14 and Corollary 4.2.8 since these limits exist for Schwartz functions.

□

Corollary 4.3.7. *Under the hypotheses of Theorem 4.3.5, $T^{(*)}$ maps $L^p(\mathbf{R}^n)$ into itself for $1 < p < 2$ with norm*

$$\big\|T^{(*)}\big\|_{L^p\to L^p} \le \frac{C_n(A_1 + A_2 + B)}{p-1},$$

where C_n is some dimensional constant.

Exercises

4.3.1. Let A_1 be defined in (4.3.4). Prove that

$$\frac{1}{2}A_1 \le \sup_{R>0} \frac{1}{R}\int_{|x|\le R} |K(x)|\,|x|\,dx \le 2A_1\,;$$

thus the expressions in (4.3.6) and (4.3.4) are equivalent.

4.3.2. Suppose that K is a locally integrable function on $\mathbf{R}^n \setminus \{0\}$ that satisfies (4.3.4). Suppose that $\delta_j \downarrow 0$. Prove that the principal value operation

$$W(\varphi) = \lim_{j\to\infty} \int_{\delta_j \le |x| \le 1} K(x)\varphi(x)\,dx$$

defines a distribution in $\mathcal{S}'(\mathbf{R}^n)$ if and only if the limit exists:

$$\lim_{j\to\infty} \int_{\delta_j \le |x| \le 1} K(x)\,dx\,.$$

4.3.3. Suppose that a function K on $\mathbf{R}^n \setminus \{0\}$ satisfies Hörmander's condition with constant A_2.
(a) Show that the functions $K(x)\chi_{|x|\ge\varepsilon}$ also satisfy Hörmander's condition uniformly in $\varepsilon > 0$ with constant $C_n A_2$ for some dimensional constant C_n.
(b) Obtain a similar conclusion for the upper truncations $K(x)\chi_{|x|\le N}$.
(c) Deduce the same conclusion for the double truncations $K^{(\varepsilon,N)}(x) = K(x)\chi_{\varepsilon\le|x|\le N}$.

4.3.4. Modify the proof of Theorem 4.3.5 to prove that if $T^{(*)}$ maps $L^r \to L^r$ for some $1 < r < \infty$, and K satisfies Hörmander's condition (4.3.12), then $T^{(*)}$ maps L^1 to $L^{1,\infty}$.

4.3.5. Assume that T is a linear operator acting on measurable functions on $\mathbf{R}^n$ such that whenever a function f is supported in a cube Q, then $T(f)$ is supported in a fixed multiple of Q.
(a) Suppose that T maps L^p into itself for some $1 < p < \infty$ with norm B. Prove that T extends to a bounded operator from L^1 into $L^{1,\infty}$ with norm a constant multiple of B.
(b) Suppose that T maps L^p into L^q for some $1 < q < p < \infty$ with norm B. Prove that T extends to a bounded operator from L^1 into $L^{s,\infty}$ with norm a multiple of B, where

$$\frac{1}{p'} + \frac{1}{q} = \frac{1}{s}.$$

4.3.6. (a) Prove that the good function g in the Calderón-Zygmund decomposition of $f = g + b$ at height α lies in the Lorentz space $L^{q,1}$ for $1 \le q < \infty$. Moreover for some dimensional constant C_n we have $\big\|g\big\|_{L^{q,1}} \le C_n \alpha^{1/q'}$.
(b) Using this result, prove the following generalization of Theorem 4.3.3: If T maps $L^{q,1} \to L^{q,\infty}$ with norm B for some $1 < q < \infty$, then T is of weak type $(1,1)$ with norm at most a multiple of $A_2 + B$.
(c) When $1 < q < \infty$, use the results of Exercise 1.1.12 and Exercise 1.4.7 to prove that if

$$|\{x : \; |T(\chi_E)(x)| > \alpha\}| \le B\,\frac{|E|}{\alpha^q}$$

for all subsets E of $\mathbf{R}^n$ with finite measure, then T is of weak type $(1,1)$ with norm at most a multiple of $A_2 + B$.

4.3.7. Let K satisfy (4.3.12) for some $A_2 > 0$, let $W \in \mathcal{S}'(\mathbf{R}^n)$ be an extension of K on $\mathbf{R}^n$ as in (4.3.7), and let T be the operator given by convolution with W. Obtain the case $r = \infty$ in Theorem 4.3.3. Precisely, prove that if T maps $L^\infty(\mathbf{R}^n)$ into itself, then T has an extension on $L^1 + L^\infty$ that satisfies

$$\big\|T\big\|_{L^1 \to L^{1,\infty}} \le C_n' A_2,$$

and

$$\big\|T\big\|_{L^p \to L^p} \le C_n\,\frac{p}{p-1}\,(A_2 + B),$$

where C_n, C_n' are constants that depend only on the dimension.
[*Hint:* Apply the Calderón-Zygmund decomposition $f = g + b$ at height $\alpha\gamma$, where $\gamma = 2^{-n-1}\big\|T\big\|_{L^\infty \to L^\infty}$. Since $|g| \le 2^n \alpha\gamma$ observe that

$$|\{x : \; |T(f)(x)| > \alpha\}| \le |\{x : \; |T(b)(x)| > \alpha/2\}|\,.]$$

4.3.8. (*Calderón-Zygmund decomposition on* L^q) Fix a function $f \in L^q(\mathbf{R}^n)$ for some $1 \le q < \infty$ and let $\alpha > 0$. Then there exist functions g and b on $\mathbf{R}^n$ such that

(1) $f = g + b$.

(2) $\|g\|_{L^q} \le \|f\|_{L^q}$ and $\|g\|_{L^\infty} \le 2^{\frac{n}{q}}\alpha$.

(3) $b = \sum_j b_j$ where each b_j is supported in a cube Q_j. Furthermore, the cubes Q_k and Q_j have disjoint interiors when $j \neq k$.

(4) $\|b_j\|_{L^q}^q \le 2^{n+q}\alpha^q |Q_j|$.

(5) $\int_{Q_j} b_j(x)\,dx = 0$.

(6) $\sum_j |Q_j| \le \alpha^{-q}\|f\|_{L^q}^q$.

(7) $\|b\|_{L^q} \le 2^{\frac{n+q}{q}}\|f\|_{L^q}$ and $\|b\|_{L^1} \le 2^{\frac{n+q}{q}}\alpha^{1-q}\|f\|_{L^q}^q$.

[*Hint:* Imitate the basic idea of the proof of Theorem 4.3.1, but select a cube Q if $\left(\frac{1}{|Q|}\int_Q |f(x)|^q\,dx\right)^{1/q} > \alpha$. Define g and b as in the proof of Theorem 4.3.1.]

4.3.9. Let $f \in L^1(\mathbf{R}^n)$. Then for any $\alpha > 0$, prove that there exist disjoint cubes Q_j in $\mathbf{R}^n$ such that the set $E_\alpha = \{x \in \mathbf{R}^n : M_c(f)(x) > \alpha\}$ is contained in $\bigcup_j 3Q_j$ and $\frac{\alpha}{4^n} < \frac{1}{|Q_j|}\int_{Q_j} |f(t)|\,dt \le \frac{\alpha}{2^n}$.

[*Hint:* For given $\alpha > 0$, select all maximal dyadic cubes $Q_j(\alpha)$ so that the average of f over them is bigger than α. Given $x \in E_\alpha$, pick a cube R that contains x so that the average of $|f|$ over R is bigger than α and find a dyadic cube Q such that $2^{-n}|Q| < |R| \le |Q|$ and that $\int_{R\cap Q} |f|\,dx > 2^{-n}\alpha|R|$. Conclude that Q is contained in some $Q_k(4^{-n}\alpha)$ and thus R is contained in $3Q_k(4^{-n}\alpha)$. The collection of all $Q_j = Q_j(4^{-n}\alpha)$ is the required one.]

4.4. Sufficient Conditions for L^p Boundedness

So far we have used the Calderón-Zygmund decomposition to prove weak type $(1,1)$ boundedness for singular integral and maximal singular integral operators, assuming that these operators are already L^2 bounded. It is therefore natural to ask for sufficient conditions that imply L^2 boundedness for such operators. Precisely, what are sufficient conditions on functions K on $\mathbf{R}^n \setminus \{0\}$ so that the corresponding singular and maximal singular integral operators associated with K are L^2 bounded?

We saw in Section 4.2 that if K has the special form $K(x) = \Omega(x/|x|)/|x|^n$ for some $\Omega \in L^1(\mathbf{S}^{n-1})$ with mean value zero, then condition (4.2.15) is necessary and sufficient for the L^2 boundedness of T, while the L^2 boundedness of $T^{(*)}$ requires the stronger smoothness condition (4.2.22).

For the general K considered in this section (for which the corresponding operator does not necessarily commute with dilations), we will only give some sufficient conditions for L^2 boundedness of T and $T^{(*)}$.

Throughout this section K will be locally integrable function on $\mathbf{R}^n \setminus \{0\}$ that satisfies the "size" condition

$$\sup_{R>0} \int_{R\le|x|\le 2R} |K(x)|\, dx = A_1 < \infty\,, \tag{4.4.1}$$

the "smoothness" condition

$$\sup_{y\neq 0} \int_{|x|\ge 2|y|} |K(x-y) - K(x)|\, dx = A_2 < \infty\,, \tag{4.4.2}$$

and the "cancellation" condition

$$\sup_{0<R_1<R_2<\infty} \left| \int_{R_1<|x|<R_2} K(x)\, dx \right| = A_3 < \infty\,, \tag{4.4.3}$$

for some $A_1, A_2, A_3 > 0$. As mentioned earlier, condition (4.4.2) is often referred to as Hörmander's condition. In this section we will show that these three conditions give rise to convolution operators that are bounded on L^p.

4.4.a. Sufficient Conditions for L^p Boundedness of Singular Integrals

We first note that under conditions (4.4.1), (4.4.2), and (4.4.3), there exists a tempered distribution W that coincides with K on $\mathbf{R}^n \setminus \{0\}$. Indeed, condition (4.4.3) implies that there exists a sequence $\delta_j \downarrow 0$ such that

$$\lim_{j\to\infty} \int_{\delta_j<|x|\le 1} K(x)\, dx = L$$

exists. Using (4.3.8), we conclude that there exists such a tempered distribution W. Note that we must have $|L| \le A_3$.

We observe that the difference of two distributions W and W' that coincide with K on $\mathbf{R}^n \setminus \{0\}$ must be supported at the origin.

Theorem 4.4.1. *Assume that K satisfies (4.4.1), (4.4.2), and (4.4.3), and let W be a tempered distribution that coincides with K on $\mathbf{R}^n \setminus \{0\}$. Then we have*

$$\sup_{0<\varepsilon<N<\infty}\ \sup_{\xi\neq 0} |(K\chi_{\varepsilon<|\cdot|<N})\widehat{\ }(\xi)| \le 15(A_1+A_2+A_3). \tag{4.4.4}$$

Thus the operator given by convolution with W maps $L^2(\mathbf{R}^n)$ into itself with norm at most $15(A_1+A_2+A_3)$. Consequently, it also maps $L^1(\mathbf{R}^n)$ into $L^{1,\infty}(\mathbf{R}^n)$ with *bound at most a dimensional multiple of* $A_1 + A_2 + A_3$ *and* $L^p(\mathbf{R}^n)$ *into itself with bound at most* $C_n \max(p, (p-1)^{-1})(A_1 + A_2 + A_3)$*, for some dimensional constant* C_n*, whenever* $1 < p < \infty$.

Proof. Let us set $K^{(\varepsilon,N)}(x) = K(x)\chi_{\varepsilon<|x|<N}$. If we prove (4.4.4), then for all f in $\mathcal{S}(\mathbf{R}^n)$ we would have the estimate

$$\big\|f * K^{(\delta_j,j)}\big\|_{L^2} \le 15(A_1 + A_2 + A_3)\big\|f\big\|_{L^2}$$

uniformly in j. Using this, (4.3.9), and Fatou's lemma, we can obtain that

$$\big\|f * W\big\|_{L^2} \le 15(A_1 + A_2 + A_3)\big\|f\big\|_{L^2}$$

thus proving the second conclusion of the theorem.

Let us now fix a ξ with $\varepsilon < |\xi|^{-1} < N$ and prove (4.4.4). Write $\widehat{K^{(\varepsilon,N)}}(\xi) = I_1(\xi) + I_2(\xi)$, where

$$\begin{aligned} I_1(\xi) &= \int_{\varepsilon<|x|<|\xi|^{-1}} K(x)e^{-2\pi i x\cdot\xi}\,dx \\ I_2(\xi) &= \int_{|\xi|^{-1}<|x|<N} K(x)e^{-2\pi i x\cdot\xi}\,dx. \end{aligned}$$

We now have

$$I_1(\xi) = \int_{\varepsilon<|x|<|\xi|^{-1}} K(x)\,dx + \int_{\varepsilon<|x|<|\xi|^{-1}} K(x)\,(e^{-2\pi i x\cdot\xi} - 1)\,dx. \tag{4.4.5}$$

It follows that

$$|I_1(\xi)| \le A_3 + 2\pi|\xi| \int_{|x|<|\xi|^{-1}} |x|\,|K(x)|\,dx \le A_3 + 2\pi(2A_1)$$

uniformly in ε. Let us now examine $I_2(\xi)$. Let $z = \dfrac{\xi}{2|\xi|^2}$ so that $e^{2\pi i z\cdot\xi} = -1$ and $2|z| = |\xi|^{-1}$. By changing variables $x = x' - z$, rewrite I_2 as

$$I_2(\xi) = -\int_{|\xi|^{-1}<|x'-z|<N} K(x'-z)\,e^{-2\pi i x'\cdot\xi}\,dx'\,;$$

hence averaging gives

$$I_2(\xi) = \frac{1}{2}\int_{|\xi|^{-1}<|x|<N} K(x)e^{-2\pi i x\cdot\xi}dx - \frac{1}{2}\int_{|\xi|^{-1}<|x-z|<N} K(x-z)\,e^{-2\pi i x\cdot\xi}\,dx\,.$$

Now use that

$$\int_A F\,dx - \int_B G\,dx = \int_B (F-G)\,dx + \int_{A\setminus B} F\,dx - \int_{B\setminus A} F\,dx \tag{4.4.6}$$

to write $I_2(\xi) = J_1(\xi) + J_2(\xi) + J_3(\xi) + J_4(\xi) + J_5(\xi)$, where

$$J_1(\xi) = +\frac{1}{2} \int_{|\xi|^{-1}<|x-z|<N} (K(x) - K(x-z))e^{-2\pi i x\cdot\xi}\,dx, \tag{4.4.7}$$

$$J_2(\xi) = +\frac{1}{2} \int_{\substack{|\xi|^{-1}<|x|<N \\ |x-z|\le|\xi|^{-1}}} K(x)\,e^{-2\pi i x\cdot\xi}\,dx, \tag{4.4.8}$$

$$J_3(\xi) = +\frac{1}{2} \int_{\substack{|\xi|^{-1}<|x|<N \\ |x-z|\ge N}} K(x)\,e^{-2\pi i x\cdot\xi}\,dx, \tag{4.4.9}$$

$$J_4(\xi) = -\frac{1}{2} \int_{\substack{|\xi|^{-1}<|x-z|<N \\ |x|\le|\xi|^{-1}}} K(x)\,e^{-2\pi i x\cdot\xi}\,dx, \tag{4.4.10}$$

$$J_5(\xi) = -\frac{1}{2} \int_{\substack{|\xi|^{-1}<|x-z|<N \\ |x|\ge N}} K(x)\,e^{-2\pi i x\cdot\xi}\,dx. \tag{4.4.11}$$

Since $2|z| = |\xi|^{-1}$, $J_1(\xi)$ is bounded in absolute value by $\frac{1}{2}A_2$, in view of (4.4.3).

Next observe that $|\xi|^{-1} \le |x| \le \frac{3}{2}|\xi|^{-1}$ in (4.4.8), while $\frac{1}{2}|\xi|^{-1} \le |x| \le |\xi|^{-1}$ in (4.4.10); hence both of these terms are bounded by $\frac{1}{2}A_1$. Finally, we have $\frac{1}{2}N < |x| < N$ in (4.4.9) (since $|x| > N - \frac{1}{2}|\xi|^{-1}$), and similarly we have $N < |x| < \frac{3}{2}N$ in (4.4.11). Thus both J_3 and J_5 are bounded above by $\frac{1}{2}A_1$.

We are left to consider the case where either $\varepsilon < N \le |\xi|^{-1}$ or $|\xi|^{-1} \le \varepsilon < N$. In the first case we estimate

$$\int_{\varepsilon<|x|<N} K(x)e^{-2\pi i x\cdot\xi}\,dx$$

by adapting the previous argument for the term I_1, while in the second case we run the argument used for the term I_2 to complete the proof.

□

4.4.b. An Example

We now give an example of a distribution that satisfies conditions (4.4.1), (4.4.2), and (4.4.3).

Example 4.4.2. Let τ be a nonzero real number and let $K(x) = \dfrac{1}{|x|^{n+i\tau}}$ defined for $x \in \mathbf{R}^n \setminus \{0\}$. For a sequence $\delta_k \downarrow 0$ and φ a Schwartz function on $\mathbf{R}^n$, define

$$\langle W, \varphi \rangle = \lim_{k\to\infty} \int_{\delta_k \le |x|} \varphi(x) \frac{dx}{|x|^{n+i\tau}}, \tag{4.4.12}$$

whenever the limit exists. We claim that for some choices of sequences δ_k, W is well-defined a tempered distribution on $\mathbf{R}^n$. Take, for example, $\delta_k = e^{-2\pi k/\tau}$. For this sequence δ_k, observe that

$$\int_{\delta_k \le |x|} \frac{1}{|x|^{n+i\tau}}\, dx = \omega_{n-1} \frac{1-(e^{-2\pi k/\tau})^{-i\tau}}{-i\tau} = 0\,,$$

and thus

$$\langle W, \varphi \rangle = \int_{|x|\le 1} (\varphi(x) - \varphi(0)) \frac{dx}{|x|^{n+i\tau}} + \int_{|x|\ge 1} \varphi(x) \frac{dx}{|x|^{n+i\tau}}\,, \tag{4.4.13}$$

which implies that $W \in \mathcal{S}'(\mathbf{R}^n)$ since

$$|\langle W, \varphi \rangle| \le C\|\nabla\varphi\|_{L^\infty} + C\|\varphi\|_{L^\infty}.$$

If φ is supported in $\mathbf{R}^n \setminus \{0\}$, then

$$\langle W, \varphi \rangle = \int K(x)\varphi(x)\, dx\,.$$

Therefore W coincides with the function K away from the origin. Moreover, (4.4.1) and (4.4.2) are clearly satisfied for K, while (4.4.3) is also satisfied since

$$\left| \int_{R_1<|x|<R_2} \frac{dx}{|x|^{n+i\tau}}\, dx \right| = \omega_{n-1} \left| \frac{R_1^{-i\tau} - R_2^{-i\tau}}{-i\tau} \right| \le \frac{2\omega_{n-1}}{|\tau|}\,.$$

Remark 4.4.3. It is important to emphasize that the limit in (4.4.12) *may not* exist for all sequences $\delta_k \to 0$. For example, the limit in (4.4.12) does not exist if $\delta_k = e^{-\pi k/\tau}$. Moreover, for a different choice of a sequence δ_k for which the limit in (4.4.12) exists (for example, $\delta_k = e^{-\pi(2k+1)/\tau}$), we obtain a different distribution W_1 that coincides with the function $K(x) = |x|^{-n-i\tau}$.

We discuss a point of caution. We can directly check that the distributions W defined by (4.4.12) are not homogeneous distributions of degree $-n-i\tau$. In fact, the only homogeneous distribution of degree $-n-i\tau$ that coincides with the function $|x|^{-n-i\tau}$ away from zero is a multiple of the distribution $u_{-n-i\tau}$, where u_z is defined in (2.4.6). Let us investigate the relationship between $u_{-n-i\tau}$ and W defined in (4.4.13). Recall that (2.4.7) gives

$$\begin{aligned}\langle u_{-n-i\tau}, \varphi \rangle = & \int_{|x|\ge 1} \varphi(x) \frac{\pi^{-i\frac{\tau}{2}}}{\Gamma(-i\frac{\tau}{2})} |x|^{-n-i\tau}\, dx \\ & + \int_{|x|\le 1} (\varphi(x) - \varphi(0)) \frac{\pi^{-i\frac{\tau}{2}}}{\Gamma(-i\frac{\tau}{2})} |x|^{-n-i\tau}\, dx + \frac{\omega_{n-1}\pi^{-i\frac{\tau}{2}}}{-i\tau\Gamma(-i\frac{\tau}{2})}\varphi(0)\,.\end{aligned}$$

Using (4.4.13), we conclude that $u_{-n-i\tau}-c_1W=c_2\delta_0$ for suitable nonzero constants c_1 and c_2. Since the Dirac mass at the origin is not a homogeneous distribution of degree $-n-i\tau$, it follows that W is neither.

Since $\widehat{u_{-n-i\tau}}=u_{i\tau}=c_3|\xi|^{i\tau}$, the equation $u_{-n-i\tau}-c_1u=c_2\delta_0$ can be used to obtain a formula for the Fourier transform of W and thus obtain a different proof that convolution with W is a bounded operator on $L^2(\mathbf{R}^n)$.

4.4.c. Necessity of Condition (4.4.3)

Although conditions (4.4.1), (4.4.2), and (4.4.3) are sufficient for L^2 boundedness, they are not necessary. However, (4.4.3) is also necessary. We have the following:

Proposition 4.4.4. *Suppose that K is a function on $\mathbf{R}^n\setminus\{0\}$ that satisfies (4.4.1). Let W be a tempered distribution on $\mathbf{R}^n$ extending K given by (4.3.7). If the operator $T(f)=f*W$ maps $L^2(\mathbf{R}^n)$ into itself (equivalently if $\widehat{W}$ is an L^∞ function), then the function K must satisfy (4.4.3).*

Proof. Pick a radial C^∞ function φ supported in the ball $|x|\le 2$ with $0\le\varphi\le 1$, and $\varphi(x)=1$ when $|x|\le 1$. For $R>0$ let $\varphi^R(x)=\varphi(x/R)$. Fourier inversion for distributions gives the second equality:

$$(W*\varphi^R)(0)=\langle W,\varphi^R\rangle=\langle \widehat{W},\widehat{\varphi^R}\rangle=\int_{\mathbf{R}^n}\widehat{W}(\xi)R^n\widehat{\varphi}(R\xi)\,d\xi$$

and the preceding identity implies that

$$|(W*\varphi^R)(0)|\le \big\|\widehat{W}\big\|_{L^\infty}\big\|\widehat{\varphi}\big\|_{L^1}=\big\|T\big\|_{L^2\to L^2}\big\|\widehat{\varphi}\big\|_{L^1}$$

uniformly in $R>0$. Fix $0<R_1<R_2<\infty$. If $R_2\le 2R_1$, we have

$$\left|\int_{R_1<|x|<R_2}K(x)\,dx\right|\le\int_{R_1<|x|<2R_1}|K(x)|\,dx\le A_1\,,$$

which implies the required conclusion. We may therefore assume that $2R_1<R_2$. Since the part of the integral in (4.4.3) over the set $R_1<|x|<2R_1$ is controlled by A_1, it suffices to control the integral of $K(x)$ over the set $2R_1<|x|<R_2$. Since the function $\varphi^{R_2}-\varphi^{R_1}$ is supported away from the origin, the action of the distribution W on it can be written as integration against the function K. We have

$$\int_{\mathbf{R}^n}K(x)\ \ (\varphi^{R_2}(x)-\varphi^{R_1}(x))\,dx=\int_{2R_1<|x|<R_2}K(x)\,dx$$

$$+\int_{R_1<|x|<2R_1}K(x)(1-\varphi^{R_1}(x))\,dx+\int_{R_2<|x|<2R_2}K(x)\varphi^{R_2}(x)\,dx.$$

The sum of the last two integrals is bounded by $3A_1$ (since $0 \le \varphi \le 1$) while the first integral is equal to $(W * \varphi^{R_2})(0) - (W * \varphi^{R_1})(0)$ and is therefore bounded by $2\|T\|_{L^2\to L^2}\|\widehat{\varphi}\|_{L^1}$. We conclude that the function K must satisfy (4.4.3) with constant

$$A_3 \le 2A_1 + 2\|\widehat{\varphi}\|_{L^1}\|T\|_{L^2\to L^2} \le c\big(A_1 + \|T\|_{L^2\to L^2}\big)\,.$$

□

4.4.d. Sufficient Conditions for L^p Boundedness of Maximal Singular Integrals

We now discuss the analogous result to Theorem 4.4.1 for the maximal singular integral operator $T^{(*)}$. One of the problems here is how to define $T^{(*)}$ for general $f \in L^p(\mathbf{R}^n)$. We may not define the maximal singular integral operator corresponding to T on L^p under the general conditions (4.4.1) and (4.4.2) on K. Indeed, the problem is that the integral

$$\int_{|x-y|\ge\varepsilon} K(x-y)f(y)\,dy$$

does not converge absolutely, even for f in $L^\infty(\mathbf{R}^n)$. Moreover, even the doubly truncated integral

$$\int_{\varepsilon\le|x-y|\le N} K(x-y)f(y)\,dy$$

does not converge absolutely for $f \in L^p(\mathbf{R}^n)$ if $p < \infty$. To avoid these difficulties, we define things as follows:

Given K satisfying (4.4.1), (4.4.2), and (4.4.3) and $0 < \varepsilon < N \le \infty$, we set

$$K^{(\varepsilon,N)}(x) = K(x)\chi_{\varepsilon\le|x|\le N}(x)$$

and for f in $\mathcal{S}(\mathbf{R}^n)$ we define

$$T^{(\varepsilon,N)}(f) = f * K^{(\varepsilon,N)}\,.$$

In view of Theorem 4.4.1, the operators $T^{(\varepsilon,N)}$ admit a bounded extension on $L^2(\mathbf{R}^n)$ [with norm at most $15(A_1 + A_2 + A_3)$] that is independent of ε and N. Using Exercise 4.3.3, we obtain that the functions $K^{(\varepsilon,N)}$ also satisfy Hörmander's condition with constant only a dimensional multiple of A_2 (in particular, also independent of ε and N). Using Theorem 4.3.3, we conclude that $T^{(\varepsilon,N)}$ have extensions that map L^p to L^p and are of weak type $(1,1)$ with norms independent of ε and N. We denote these extensions also by $T^{(\varepsilon,N)}$. Therefore, for $f \in \bigcup_{1\le p<\infty} L^p(\mathbf{R}^n)$ the functions $T^{(\varepsilon,N)}(f)$ are well defined. We now set

$$T^{(*)}(f) = \sup_{0<\varepsilon<N<\infty} |T^{(\varepsilon,N)}(f)| \tag{4.4.14}$$

for $f \in \bigcup_{1\le p<\infty} L^p(\mathbf{R}^n)$. Our goal is to obtain the boundedness of $T^{(*)}$. Our main result claims the L^p boundedness of the maximal singular integrals as defined in (4.4.14) in this general setting.

Theorem 4.4.5. *Suppose that K satisfies (4.4.1), (4.4.2), and (4.4.3) and let $T^{(*)}$ be as in (4.4.14). Then $T^{(*)}$ is bounded on $L^p(\mathbf{R}^n)$, $1 < p < \infty$, with norm*

$$\|T^{(*)}\|_{L^p\to L^p} \le C_n \max(p,(p-1)^{-1})(A_1 + A_2 + A_3),$$

where C_n is a dimensional constant.

Proof. We first define an operator T associated with K that satisfies (4.4.1), (4.4.2), and (4.4.3). Because of condition (4.4.3), there exists a sequence $\delta_j \downarrow 0$ such that

$$\lim_{j\to\infty} \int_{\delta_j<|x|\le 1} K(x)\,dx$$

exists. Therefore, for $\varphi \in \mathcal{S}(\mathbf{R}^n)$ we can define a tempered distribution

$$\langle W, \varphi\rangle = \lim_{j\to\infty} \int_{\delta_j\le|x|\le j} K(x)\varphi(x)\,dx$$

and an operator T given by $T(f) = f * W$ for $f \in \mathcal{S}(\mathbf{R}^n)$. By Theorems 4.4.1 and 4.3.3, T admits an extension that is L^p bounded for all $1 < p < \infty$ with bound

$$\|T\|_{L^p\to L^p} \le c_n \max(p,(p-1)^{-1})(A_1 + A_2 + A_3) \tag{4.4.15}$$

and is of weak type $(1,1)$. This extension is still denoted by T.

Fix $1 < p < \infty$ and a function $f \in L^p(\mathbf{R}^n) \cap L^\infty(\mathbf{R}^n)$, which we take initially to have compact support. We have

$$\begin{aligned}
&T^{(\varepsilon,N)}(f)(x)\\
&= \int_{\varepsilon\le|x-y|<N} K(x-y)f(y)\,dy = T^{(\varepsilon,\infty)}(f)(x) - T^{N,\infty}(f)(x)\\
&= \int_{\varepsilon\le|x-y|} K(x-y)f(y)\,dy - \int_{N\le|x-y|} K(x-y)f(y)\,dy\\
&= \int_{\varepsilon\le|x-y|} (K(x-y)-K(z_1-y))f(y)\,dy + \int_{\varepsilon\le|x-y|} K(z_1-y)f(y)\,dy\\
&\qquad - \int_{N\le|x-y|} (K(x-y)-K(z_2-y))f(y)\,dy - \int_{N\le|x-y|} K(z_2-y)f(y)\,dy\\
&= \int_{\varepsilon\le|x-y|} (K(x-y)-K(z_1-y))f(y)\,dy + T(f)(z_1) - T(f\chi_{|x-\cdot|<\varepsilon})(z_1)\\
&\qquad - \int_{N\le|x-y|} (K(x-y)-K(z_2-y))f(y)\,dy - T(f)(z_2) + T(f\chi_{|x-\cdot|<N})(z_2),
\end{aligned}$$

where z_1 and z_2 are arbitrary points in $\mathbf{R}^n$ that satisfy $|z_1 - x| \le \frac{\varepsilon}{2}$ and $|z_2 - x| \le \frac{N}{2}$. We used that f has compact support in order to be able to write $T^{(\varepsilon,\infty)}(f)$ and $T^{(N,\infty)}(f)$ as convergent integrals.

At this point we take absolute values, average over $|z_1 - x| \le \frac{\varepsilon}{2}$ and $|z_2 - x| \le \frac{N}{2}$, and we apply Hölder's inequality in two terms. We obtain the estimate

$$
\begin{aligned}
&|T^{(\varepsilon,N)}(f)(x)| \\
&\le \frac{1}{v_n}\Big(\frac{2}{\varepsilon}\Big)^n \int_{|z_1-x|\le\frac{\varepsilon}{2}} \int_{|x-y|\ge\varepsilon} |K(x-y) - K(z_1-y)|\,|f(y)|\,dy\,dz_1 \\
&\quad + \frac{1}{v_n}\Big(\frac{2}{\varepsilon}\Big)^n \int_{|z_1-x|\le\frac{\varepsilon}{2}} |T(f)(z_1)|\,dz_1 \\
&\quad + \bigg(\frac{1}{v_n}\Big(\frac{2}{\varepsilon}\Big)^n \int_{|z_1-x|\le\frac{\varepsilon}{2}} |T(f\chi_{|x-\cdot|<\varepsilon})(z_1)|^p\,dz_1\bigg)^{\frac{1}{p}} \\
&\quad + \frac{1}{v_n}\Big(\frac{2}{N}\Big)^n \int_{|z_2-x|\le\frac{N}{2}} \int_{|x-y|\ge N} |K(x-y) - K(z_2-y)|\,|f(y)|\,dy\,dz_2 \\
&\quad + \frac{1}{v_n}\Big(\frac{2}{N}\Big)^n \int_{|z_2-x|\le\frac{N}{2}} |T(f)(z_2)|\,dz_2 \\
&\quad + \bigg(\frac{1}{v_n}\Big(\frac{2}{N}\Big)^n \int_{|z_2-x|\le\frac{N}{2}} |T(f\chi_{|x-\cdot|<N})(z_2)|^p\,dz_2\bigg)^{\frac{1}{p}},
\end{aligned}
$$

where v_n is the volume of the unit ball in $\mathbf{R}^n$. Applying condition (4.4.2) and estimate (4.4.15), we obtain for f in $L^p(\mathbf{R}^n) \cap L^\infty(\mathbf{R}^n)$ with compact support

$$
\begin{aligned}
&|T^{(\varepsilon,N)}(f)(x)| \\
&\le \frac{1}{v_n}\Big(\frac{2}{\varepsilon}\Big)^n \int_{|z_1-x|\le\frac{\varepsilon}{2}} |T(f)(z_1)|\,dz_1 + \frac{1}{v_n}\Big(\frac{2}{N}\Big)^n \int_{|z_2-x|\le\frac{N}{2}} |T(f)(z_2)|\,dz_2 \\
&\quad + c_n\Big(\sum_{j=1}^3 A_j\Big)\max(p,(p-1)^{-1})\bigg(\frac{1}{v_n}\Big(\frac{2}{\varepsilon}\Big)^n \int_{|z_1-x|\le\varepsilon} |f(z_1)|^p\,dz_1\bigg)^{\frac{1}{p}} \\
&\quad + c_n\Big(\sum_{j=1}^3 A_j\Big)\max(p,(p-1)^{-1})\bigg(\frac{1}{v_n}\Big(\frac{2}{N}\Big)^n \int_{|z_2-x|\le N} |f(z_2)|^p\,dz_2\bigg)^{\frac{1}{p}} \\
&\quad + 2A_2\big\|f\big\|_{L^\infty}.
\end{aligned}
$$

We now use density to remove the compact support condition on f and obtain the last displayed estimate for all functions f in $L^p(\mathbf{R}^n) \cap L^\infty(\mathbf{R}^n)$. Taking the supremum over all $0 < \varepsilon < N$ and over all $N > 0$, we deduce that for all f in $L^p(\mathbf{R}^n) \cap L^\infty(\mathbf{R}^n)$ we have the estimate

$$
T^{(*)}(f)(x) \le 2A_2\big\|f\big\|_{L^\infty} + S_p(f)(x), \tag{4.4.16}
$$

where S_p is the sublinear operator defined by

$$S_p(f)(x) = 2M(T(f))(x) + 2^{n+1}c_n\Big(\sum_{j=1}^{3} A_j\Big)\max(p,(p-1)^{-1})(M(|f|^p)(x))^{\frac{1}{p}},$$

and M is the Hardy-Littlewood maximal operator.

Recalling that M maps L^1 into $L^{1,\infty}$ with bound at most 3^n and also L^p into $L^{p,\infty}$ with bound at most $2\cdot 3^{n/p}$ for $1<p<\infty$ (Exercise 2.1.4), we conclude that S_p maps $L^p(\mathbf{R}^n)$ into $L^{p,\infty}(\mathbf{R}^n)$ with norm at most

$$\big\|S_p\big\|_{L^p\to L^{p,\infty}} \le \widetilde{c}_n(A_1+A_2+A_3)\max(p,(p-1)^{-1}), \tag{4.4.17}$$

where $\widetilde{c}_n$ is another dimensional constant.

Now write $f = f_\alpha + f^\alpha$, where

$$f_\alpha = f\chi_{|f|\le \alpha/(16A_2)} \qquad \text{and} \qquad f^\alpha = f\chi_{|f|>\alpha/(16A_2)}.$$

The function f_α is in $L^\infty \cap L^p$ and f^α is in $L^1\cap L^p$. Moreover, we see that

$$\big\|f^\alpha\big\|_{L^1} \le (16A_2/\alpha)^{p-1}\big\|f\big\|_{L^p}^p. \tag{4.4.18}$$

Apply the Calderón-Zygmund decomposition (Theorem 4.3.1) to the function f^α at height $\alpha\gamma$ to write $f^\alpha = g^\alpha + b^\alpha$, where g^α is the good function and b^α is the bad function of this decomposition. Using (4.3.1), we obtain

$$\big\|g^\alpha\big\|_{L^p} \le 2^{n/p'}(\alpha\gamma)^{1/p'}\big\|f^\alpha\big\|_{L^1}^{1/p} \le 2^{(n+4)/p'}(A_2\gamma)^{1/p'}\big\|f\big\|_{L^p}. \tag{4.4.19}$$

We now use (4.4.16) to get

$$|\{x\in\mathbf{R}^n:\ T^{(*)}(f)(x)>\alpha\}| \le b_1+b_2+b_3, \tag{4.4.20}$$

where

$$\begin{aligned} b_1 &= |\{x\in\mathbf{R}^n:\ 2A_2\big\|f_\alpha\big\|_{L^\infty} + S_p(f_\alpha)(x) > \alpha/4\}|,\\ b_2 &= |\{x\in\mathbf{R}^n:\ 2A_2\big\|g^\alpha\big\|_{L^\infty} + S_p(g^\alpha)(x) > \alpha/4\}|,\\ b_3 &= |\{x\in\mathbf{R}^n:\ T^{(*)}(b^\alpha)(x) > \alpha/2\}|. \end{aligned}$$

Observe that $2A_2\big\|f_\alpha\big\|_{L^\infty} \le \alpha/8$. Selecting $\gamma = 2^{-n-5}(A_1+A_2)^{-1}$ and using property (2) in Theorem 4.3.1, we obtain

$$2A_2\big\|g^\alpha\big\|_{L^\infty} \le A_2 2^{n+1}\alpha\gamma \le \alpha 2^{-4} < \frac{\alpha}{8}$$

and therefore

$$\begin{aligned} b_1 &\le |\{x\in\mathbf{R}^n:\ S(f_\alpha)(x) > \alpha/8\}|,\\ b_2 &\le |\{x\in\mathbf{R}^n:\ S(g^\alpha)(x) > \alpha/8\}|. \end{aligned} \tag{4.4.21}$$

Since $\gamma \le (2^{n+5}A_1)^{-1}$, it follows from (4.3.18) that

$$b_3 \le \Big|\bigcup_j Q_j^*\Big| + 2^{n+8}A_2\frac{\big\|f^\alpha\big\|_{L^1}}{\alpha} \le \left(\frac{(5\sqrt{n})^n}{\gamma} + 2^{n+8}A_2\right)\frac{\big\|f^\alpha\big\|_{L^1}}{\alpha}$$

and, using (4.4.18), we obtain

$$b_3 \le C_n (A_1 + A_2)^p \alpha^{-p} \big\|f\big\|_{L^p}^p .$$

Using Chebychev's inequality in (4.4.21) and (4.4.17), we finally obtain that

$$b_1 + b_2 \le (8/\alpha)^p\, (\widetilde{c}_n)^p (A_1 + A_2 + A_3)^p \max(p, (p-1)^{-1})^p \big(\big\|f\big\|_{L^p}^p + \big\|g^\alpha\big\|_{L^p}^p\big).$$

Combining the estimates for b_1, b_2, and b_3 and using (4.4.19), we deduce

$$\big\|T^{(*)}(f)\big\|_{L^{p,\infty}} \le C_n (A_1 + A_2 + A_3) \max(p, (p-1)^{-1}) \big\|f\big\|_{L^p(\mathbf{R}^n)} . \tag{4.4.22}$$

Finally, we need to obtain a similar estimate to (4.4.22), in which the weak L^p norm on the left is replaced by the L^p norm. This will be a consequence of the Marcinkiewicz interpolation Theorem 1.3.2 between the estimates $L^{\frac{p+1}{2}} \to L^{\frac{p+1}{2},\infty}$ and $L^{2p} \to L^{2p,\infty}$ for $2 < p < \infty$ and between the estimates $L^{2p} \to L^{2p,\infty}$ and $L^1 \to L^{1,\infty}$ for $1 < p < 2$. The latter estimate follows from Theorem 4.3.5. See also Corollary 4.3.7.

□

Exercises

4.4.1. Suppose that T is a convolution operator that is L^2 bounded. Suppose that $f \in L^1(\mathbf{R}^n) \cap L^2(\mathbf{R}^n)$ has vanishing integral and that $T(f)$ is integrable. Prove that $T(f)$ has also vanishing integral.

4.4.2. Let K satisfy (4.4.1), (4.4.2), and (4.4.3) and let $W \in \mathcal{S}'$ be an extension of K on $\mathbf{R}^n$. Let f be a Schwartz function on $\mathbf{R}^n$ with mean value zero. Prove that the function $f * W$ is in $L^1(\mathbf{R}^n)$.

4.4.3. Suppose K is a function on $\mathbf{R}^n \setminus \{0\}$ that satisfies (4.4.1), (4.4.2), and (4.4.3). Let $K^{(\varepsilon,N)}(x) = K(x)\chi_{\varepsilon<|x|<N}$ for $0 < \varepsilon < N < \infty$ and let $T^{(\varepsilon,N)}$ be the operator given by convolution with $K^{(\varepsilon,N)}$. Use Theorem 4.4.5 to prove that $T^{(\varepsilon,N)}(f)$ converges to $T(f)$ in $L^p(\mathbf{R}^n)$ and almost everywhere whenever $1 < p < \infty$ and $f \in L^p(\mathbf{R}^n)$ as $\varepsilon \to 0$ and $N \to \infty$.

4.4.4. (a) Prove that for all $x, y \in \mathbf{R}^n$ that satisfy $|x| \ge 2|y|$ we have

$$\left| \frac{x-y}{|x-y|} - \frac{x}{|x|} \right| \le 2\, \frac{|y|}{|x|} .$$

(b) Let Ω be an integrable function with mean value zero on the sphere $\mathbf{S}^{n-1}$. Suppose that Ω satisfies a *Lipschitz (Hölder) condition* of order $0 < \alpha < 1$ on $\mathbf{S}^{n-1}$. This means that

$$|\Omega(\theta_1) - \Omega(\theta_2)| \le B_0 |\theta_1 - \theta_2|^\alpha$$

for all $\theta_1, \theta_2 \in \mathbf{S}^{n-1}$. Prove that the function $K(x) = \Omega(x/|x|)/|x|^n$ satisfies Hörmander's condition with constant at most a multiple of $B_0 + \big\|\Omega\big\|_{L^\infty}$.

4.4.5. Let Ω be an L^1 function on $\mathbf{S}^{n-1}$ with mean value zero.
(a) Let $\omega_\infty(t) = \sup\{|\Omega(\theta_1) - \Omega(\theta_2)| : \theta_1, \theta_2 \in \mathbf{S}^{n-1}, |\theta_1 - \theta_2| \le t\}$ and suppose that the following *Dini condition* holds:

$$\int_0^1 \omega_\infty(t) \frac{dt}{t} < \infty .$$

Prove that the function $K(x) = \Omega(x/|x|)|x|^{-n}$ satisfies Hörmander's condition.
(b) (*A. Calderón and A. Zygmund*) For $A \in O(n)$, let

$$\|A\| = \sup\{|\theta - A(\theta)| : \theta \in \mathbf{S}^{n-1}\} .$$

Suppose that Ω satisfies the more general Dini-type condition

$$\int_0^1 \omega_1(t) \frac{dt}{t} < \infty ,$$

where

$$\omega_1(t) = \sup_{\substack{A \in O(n) \\ \|A\| \le t}} \int_{\mathbf{S}^{n-1}} |\Omega(A(\theta)) - \Omega(\theta)| \, d\theta .$$

Prove the same conclusion as in part (a).
[*Hint:* Part (b): Use the result in part (a) of the previous Exercise and switch to polar coordinates.]

4.4.6. (*Cotlar* [**133**]) Suppose that K is defined on $\mathbf{R}^n \setminus \{0\}$ and satisfies $|K(x)| \le A_1|x|^{-n}$, $|K(x-y) - K(x)| \le A_2|y|^\delta |x|^{-n-\delta}$ whenever $|x| \ge 2|y| > 0$, and $\sup_{r<R<\infty} \big| \int_{r \le |x| \le R} K(x)\, dx \big| \le A_3$. Let φ be a radially decreasing smooth function supported in the unit ball of $\mathbf{R}^n$ with integral 1.
(a) Set $K_{\varepsilon^{-1}}(x) = \varepsilon^n K(\varepsilon x)$ and prove that $K_{\varepsilon^{-1}}$ satisfies the same estimates as K uniformly in $\varepsilon > 0$.
(b) Recall that $K^{(1)}(x) = K(x)\chi_{|x|\ge 1}$. Prove that

$$f * K^{(\varepsilon)} = f * K * \varphi_\varepsilon + f * \big((K_{\varepsilon^{-1}})^{(1)} - K_{\varepsilon^{-1}} * \varphi\big)_\varepsilon ,$$

where $g_\varepsilon(x) = \varepsilon^{-n} g(\varepsilon^{-1}x)$ for any function g.
(c) Use Corollary 2.1.12 to show that

$$\sup_{\varepsilon>0} \big| f * \big((K_{\varepsilon^{-1}})^{(1)} - K_{\varepsilon^{-1}} * \varphi\big)_\varepsilon \big| \le c\,(A_1 + A_2 + A_3) M(f)$$

by proving that

$$\big| \big((K_{\varepsilon^{-1}})^{(1)} - K_{\varepsilon^{-1}} * \varphi\big)(x) \big| \le C(A_1 + A_2 + A_3)(1 + |x|)^{-n-\delta}$$

uniformly in $\varepsilon > 0$.
(d) Prove *Cotlar's inequality*:

$$T^{(*)}(f)(x) \le M(f * K) + c\,(A_1 + A_2 + A_3) M(f)$$

and conclude the L^p boundedness of $T^{(*)}$ from that of T.
[*Hint:* Part (c): Consider the cases $|x| \geq 2$ and $|x| \leq 2$. When $|x| \leq 2$, estimate $(K_{\varepsilon^{-1}})^{(1)}(x)$ by A_1 and $|(K_{\varepsilon^{-1}} * \varphi)(x)|$ by a multiple of $A_1 + A_3$.]

4.5. Vector-Valued Inequalities*

Certain nonlinear expressions that appear in Fourier analysis, such as maximal functions and square functions, can be viewed as linear quantities taking values in some Banach space. This point of view provides the motivation for a systematic study of Banach-valued operators. Let us illustrate this line of thinking via an example. Let T be a linear operator acting on L^p of some measure space (X,μ) and taking values in the set of measurable functions of another measure space (Y,ν). The seemingly nonlinear inequality

$$\Big\|\Big(\sum_j |T(f_j)|^2\Big)^{\frac{1}{2}}\Big\|_{L^p} \leq C_p \Big\|\Big(\sum_j |f_j|^2\Big)^{\frac{1}{2}}\Big\|_{L^p} \tag{4.5.1}$$

can be transformed to a linear one with only a slight change of view. Let us denote by $L^p(X,\ell^2)$ the Banach space of all sequences $\{f_j\}_j$ of measurable functions on X that satisfy

$$\big\|\{f_j\}_j\big\|_{L^p(X,\ell^2)} = \Big(\int_X \Big(\sum_j |f_j|^2\Big)^{\frac{p}{2}} d\mu\Big)^{\frac{1}{p}} < \infty\,. \tag{4.5.2}$$

Define a linear operator acting on such sequences by setting

$$\vec{T}(\{f_j\}_j) = \{T(f_j)\}_j. \tag{4.5.3}$$

Then 4.5.1 is equivalent to the inequality

$$\big\|\vec{T}(\{f_j\}_j)\big\|_{L^p(Y,\ell^2)} \leq C_p \big\|\{f_j\}_j\big\|_{L^p(X,\ell^2)} \tag{4.5.4}$$

in which $\vec{T}$ is thought as a linear operator acting on the L^p space of ℓ^2-valued functions on X. This is the basic idea of vector-valued inequalities. A nonlinear inequality such as (4.5.1) can be viewed as a linear norm estimate for an operator acting and taking values in suitable Banach spaces.

4.5.a. ℓ^2-Valued Extensions of Linear Operators

The following result is classical and fundamental in the subject of vector-valued inequalities.

Theorem 4.5.1. *Let $0 < p,q < \infty$ and let (X,μ) and (Y,ν) be two measure spaces. The following are valid:*

(a) Suppose that T is a bounded linear operator from $L^p(X)$ to $L^q(Y)$ with norm A. Then T has an ℓ^2-valued extension, that is, for all complex-valued functions f_j in $L^p(X)$ we have

$$\Big\| \Big(\sum_j |T(f_j)|^2 \Big)^{\frac{1}{2}} \Big\|_{L^q} \le C_{p,q} A \Big\| \Big(\sum_j |f_j|^2 \Big)^{\frac{1}{2}} \Big\|_{L^p} \tag{4.5.5}$$

for some constant $C_{p,q}$ that depends only on p and q. Moreover, the constant $C_{p,q}$ satisfies $C_{p,q} = 1$ if $p \le q$.
(b) Suppose that T is a bounded linear operator from $L^p(X)$ to $L^{q,\infty}(Y)$ with norm A. Then T has an ℓ^2-valued extension, that is,

$$\Big\| \Big(\sum_j |T(f_j)|^2 \Big)^{\frac{1}{2}} \Big\|_{L^{q,\infty}} \le D_{p,q} A \Big\| \Big(\sum_j |f_j|^2 \Big)^{\frac{1}{2}} \Big\|_{L^p} \tag{4.5.6}$$

for some constant $D_{p,q}$ that depends only on p and q.

To prove this theorem, we will need the following identities.

Lemma 4.5.2. *For any $0 < r < \infty$, define constants*

$$A_r = \left(\frac{\Gamma(\frac{r+1}{2})}{\pi^{\frac{r+1}{2}}} \right)^{\frac{1}{r}} \quad \text{and} \quad B_r = \left(\frac{\Gamma(\frac{r}{2}+1)}{\pi^{\frac{r}{2}}} \right)^{\frac{1}{r}}. \tag{4.5.7}$$

Then for any $\lambda_1, \lambda_2, \dots, \lambda_n \in \mathbf{R}$ we have

$$\left(\int_{\mathbf{R}^n} |\lambda_1 x_1 + \cdots + \lambda_n x_n|^r e^{-\pi|x|^2} \, dx \right)^{\frac{1}{r}} = A_r \, (\lambda_1^2 + \cdots + \lambda_n^2)^{\frac{1}{2}} \tag{4.5.8}$$

and for all $w_1, w_2, \dots, w_n \in \mathbf{C}$ we have

$$\left(\int_{\mathbf{C}^n} |w_1 z_1 + \cdots + w_n z_n|^r e^{-\pi|z|^2} \, dz \right)^{\frac{1}{r}} = B_r (|w_1|^2 + \cdots + |w_n|^2)^{\frac{1}{2}} . \tag{4.5.9}$$

Proof. Dividing both sides of (4.5.8) by $(\lambda_1^2 + \cdots + \lambda_n^2)^{\frac{1}{2}}$, we reduce things to the situation where $\lambda_1^2 + \cdots + \lambda_n^2 = 1$. Let $e_1 = (1, 0, \dots, 0)$ be the usual unit vector on $\mathbf{S}^{n-1}$ and find an orthogonal $n \times n$ matrix $A \in O(n)$ (orthogonal means real matrix satisfying $A^t = A^{-1}$) such that $A^{-1} e_1 = (\lambda_1, \dots, \lambda_n)$. Then the first coordinate of Ax is

$$(Ax)_1 = Ax \cdot e_1 = x \cdot A^t e_1 = x \cdot A^{-1} e_1 = \lambda_1 x_1 + \cdots + \lambda_n x_n .$$

Now change variables $y = Ax$ in the integral in (4.5.8) and use the fact that $|Ax| = |x|$ to obtain

$$\begin{aligned}
&\left(\int_{\mathbf{R}^n} |\lambda_1 x_1 + \cdots + \lambda_n x_n|^r e^{-\pi|x|^2}\, dx\right)^{\frac{1}{r}} \\
&= \left(\int_{\mathbf{R}^n} |y_1|^r e^{-\pi|y|^2}\, dy\right)^{\frac{1}{r}} = \left(2\int_0^\infty t^r e^{-\pi t^2}\, dt\right)^{\frac{1}{r}} \\
&= \left(\int_0^\infty s^{\frac{r-1}{2}} e^{-\pi s}\, ds\right)^{\frac{1}{r}} = \left(\frac{\Gamma(\frac{r+1}{2})}{\pi^{\frac{r+1}{2}}}\right)^{\frac{1}{r}} \\
&= A_r\,,
\end{aligned}$$

which proves (4.5.8).

The proof of (4.5.9) is almost identical. We normalize by assuming that

$$|w_1|^2 + \cdots + |w_n|^2 = 1\,,$$

and we let ϵ_1 be the element of $\mathbf{C}^n$ having 1 in the first entry and zero elsewhere. We find a hermitian $n \times n$ matrix $\mathcal{A}$ (hermitian means $\mathcal{A}^{-1} = \mathcal{A}^*$, where $\mathcal{A}^*$ is the adjoint matrix of $\mathcal{A}$, i.e., the unique matrix that satisfies $\mathcal{A}z \cdot \overline{w} = z \cdot \overline{\mathcal{A}^* w}$ for all z, w in $\mathbf{C}^n$) such that $\mathcal{A}^{-1}\epsilon_1 = (w_1, \dots, w_n)$. Then $(\mathcal{A}z)_1 = w_1 z_1 + \cdots + w_n z_n$ and also $|\mathcal{A}z| = |z|$; therefore, changing variables $\zeta = \mathcal{A}z$ in the integral in (4.5.9), we can rewite that integral as

$$\begin{aligned}
\left(\int_{\mathbf{C}^n} |\zeta_1|^r e^{-\pi|\zeta|^2}\, d\zeta\right)^{\frac{1}{r}} &= \left(\int_{\mathbf{C}} |\zeta_1|^r e^{-\pi|\zeta_1|^2}\, d\zeta_1\right)^{\frac{1}{r}} \\
&= \left(2\pi \int_0^\infty t^r e^{-\pi t^2}\, t\, dt\right)^{\frac{1}{r}} \\
&= \left(\pi \int_0^\infty s^{\frac{r}{2}} e^{-\pi s}\, ds\right)^{\frac{1}{r}} \\
&= B_r\,.
\end{aligned}$$

□

Let us now continue with the proof of Theorem 4.5.1.

Proof. If T maps real-valued functions to real-valued functions, then we may use conclusion (4.5.8) of Lemma 4.5.2. In general, T maps complex-valued functions to complex-valued functions, and we use conclusion (4.5.9).

Let us begin with part (a). Let B_r be the constant appearing in (4.5.7). Let us first consider the case $q \le p$. Assume without loss of generality that a sequence $\{f_j\}_j$ is indexed by the set of positive integers; also fix a positive integer n. Use successively identity (4.5.9), the boundedness of T, the Cauchy-Schwarz inequality, and identity (4.5.9) again to decuce

$$\begin{aligned}
\Big\|\Big(\sum_{j=1}^n |T(f_j)|^2\Big)^{\frac{1}{2}}\Big\|_{L^q(Y)}^q &= (B_q)^{-q}\int_Y\int_{\mathbf{C}^n} |z_1T(f_1)+\cdots+z_nT(f_n)|^q e^{-\pi|z|^2}\,dz\,d\nu \\
&= (B_q)^{-q}\int_{\mathbf{C}^n}\int_Y |T(z_1f_1+\cdots+z_nf_n)|^q\,d\nu\,e^{-\pi|z|^2}\,dz \\
&\le (B_q)^{-q}A^q\int_{\mathbf{C}^n}\Big(\int_X |z_1f_1+\cdots+z_nf_n|^p\,d\mu\Big)^{\frac{q}{p}} e^{-\pi|z|^2}\,dz \\
&\le (B_q)^{-q}A^q\Big(\int_{\mathbf{C}^n}\int_X |z_1f_1+\cdots+z_nf_n|^p\,d\mu\,e^{-\pi|z|^2}\,dz\Big)^{\frac{q}{p}} \\
&= (B_q)^{-q}A^q\Big(B_p^p\int_X\Big(\sum_{j=1}^n|f_j|^2\Big)^{\frac{p}{2}}d\mu\Big)^{\frac{q}{p}} \\
&= (B_pB_q^{-1})^qA^q\Big\|\Big(\sum_{j=1}^n|f_j|^2\Big)^{\frac{1}{2}}\Big\|_{L^p(X)}^q .
\end{aligned}$$

Now, letting $n\to\infty$ in the previous inequality, we obtain the required conclusion with $C_{p,q}=B_pB_q^{-1}$. The crucial point in the preceding argument lies in the fact that the constants B_p and B_q are independent of the number n.

We now turn to the case $q>p$. Using similar reasoning, we obtain

$$\begin{aligned}
\Big\|\Big(\sum_{j=1}^n |T(f_j)|^2\Big)^{\frac{1}{2}}\Big\|_{L^q(Y)}^q &= (B_q)^{-q}\int_Y\int_{\mathbf{C}^n} |z_1T(f_1)+\cdots+z_nT(f_n)|^q e^{-\pi|z|^2}\,dz\,d\nu \\
&= (B_q)^{-q}\int_{\mathbf{C}^n}\int_Y |T(z_1f_1+\cdots+z_nf_n)|^q\,d\nu\,e^{-\pi|z|^2}\,dz \\
&\le (AB_q^{-1})^q\int_{\mathbf{R}^n}\Big(\int_X |z_1f_1+\cdots+z_nf_n|^p\,d\mu\Big)^{\frac{q}{p}} e^{-\pi|z|^2}\,dz \\
&= (AB_q^{-1})^q\Big\|\int_X |z_1f_1+\cdots+z_nf_n|^p\,d\mu\Big\|_{L^{\frac{q}{p}}(\mathbf{C}^n,e^{-\pi|z|^2}dz)}^{q/p} \\
&\le (AB_q^{-1})^q\Big\{\int_X \Big\||z_1f_1+\cdots+z_nf_n|^p\Big\|_{L^{\frac{q}{p}}(\mathbf{C}^n,e^{-\pi|z|^2}dz)}\,d\mu\Big\}^{\frac{q}{p}} \\
&= (AB_q^{-1})^q\Big\{\int_X\Big(\int_{\mathbf{C}^n}|z_1f_1+\cdots+z_nf_n|^q e^{-\pi|z|^2}dz\Big)^{\frac{p}{q}}d\mu\Big\}^{\frac{q}{p}} \\
&= (AB_q^{-1})^q\Big\{\int_X (B_q)^p\Big(\sum_{j=1}^n|f_j|^2\Big)^{\frac{p}{2}}d\mu\Big\}^{\frac{q}{p}} \\
&= A^q\Big\|\Big(\sum_{j=1}^n|f_j|^2\Big)^{\frac{1}{2}}\Big\|_{L^p(X)}^q .
\end{aligned}$$

Note that we made use of Minkowski's integral inequality (Exercise 1.1.6) in the last inequality.

We now turn our attention to part (b) of the theorem. Inequality (4.5.6) will be a consequence of (4.5.5) and of the following result of Exercise 1.4.3 (see also Exercise 1.1.12):

$$\big\|g\big\|_{L^{q,\infty}} \le \sup_{0<\nu(E)<\infty} \nu(E)^{\frac{1}{q}-\frac{1}{r}}\Big(\int_E |g|^r\,d\nu\Big)^{\frac{1}{r}} \le \Big(\frac{q}{q-r}\Big)^{\frac{1}{r}}\big\|g\big\|_{L^{q,\infty}}, \tag{4.5.10}$$

where $0 < r < q$ and the supremum is taken over all E subsets of Y of finite measure. Using (4.5.10), we obtain

$$\begin{aligned}
&\Big\|\Big(\sum_j |T(f_j)|^2\Big)^{\frac{1}{2}}\Big\|_{L^{q,\infty}(Y)} \\
&\le \sup_{0<\nu(E)<\infty} \nu(E)^{\frac{1}{q}-\frac{1}{r}}\Big(\int_E \Big(\sum_j |T(f_j)|^2\Big)^{\frac{r}{2}}\,d\nu\Big)^{\frac{1}{r}} \\
&= \sup_{0<\nu(E)<\infty} \nu(E)^{\frac{1}{q}-\frac{1}{r}}\Big(\int_Y \Big(\sum_j |\chi_E T(f_j)|^2\Big)^{\frac{r}{2}}\,d\nu\Big)^{\frac{1}{r}} \\
&\le \sup_{0<\nu(E)<\infty} \nu(E)^{\frac{1}{q}-\frac{1}{r}}\big\|T_E\big\|_{L^p\to L^r}\Big(\int_X \Big(\sum_j |f_j|^2\Big)^{\frac{p}{2}}\,d\mu\Big)^{\frac{1}{p}}, \qquad (4.5.11)
\end{aligned}$$

where T_E is defined by $T_E(f) = \chi_E T(f)$. But since for any function f in $L^p(X)$ we have

$$\nu(E)^{\frac{1}{q}-\frac{1}{r}}\big\|T_E(f)\big\|_{L^r} \le \Big(\frac{q}{q-r}\Big)^{\frac{1}{r}}\big\|T(f)\big\|_{L^{q,\infty}} \le \Big(\frac{q}{q-r}\Big)^{\frac{1}{r}}A\big\|f\big\|_{L^p},$$

it follows that for any measurable set E of finite measure the estimate

$$\nu(E)^{\frac{1}{q}-\frac{1}{r}}\big\|T_E\big\|_{L^p\to L^r} \le \Big(\frac{q}{q-r}\Big)^{\frac{1}{r}}A \tag{4.5.12}$$

is valid. Inserting (4.5.12) in (4.5.11), we obtain the required conclusion.

□

4.5.b. Applications and ℓ^r-Valued Extensions of Linear Operators

Here is an application of Theorem 4.5.1:

Example 4.5.3. On the real line consider the intervals $I_j = [b_j, \infty)$ for $j \in \mathbf{Z}$. Let T_j be the operator given by multiplication on the Fourier transform by the characteristic function of I_j. Then we have the following two inequalities:

$$\Big\|\Big(\sum_{j\in\mathbf{Z}}|T_j(f_j)|^2\Big)^{\frac{1}{2}}\Big\|_{L^p} \le C_p\Big\|\Big(\sum_{j\in\mathbf{Z}}|f_j|^2\Big)^{\frac{1}{2}}\Big\|_{L^p} \tag{4.5.13}$$

$$\Big\|\Big(\sum_{j\in\mathbf{Z}}|T_j(f_j)|^2\Big)^{\frac{1}{2}}\Big\|_{L^{1,\infty}} \le C\Big\|\Big(\sum_{j\in\mathbf{Z}}|f_j|^2\Big)^{\frac{1}{2}}\Big\|_{L^1} \tag{4.5.14}$$

for $1 < p < \infty$. To prove these, first observe that the operator $T = \frac{1}{2}(I + iH)$ is given on the Fourier transform by multiplication by the characteristic function of the half-axis $[0, \infty)$. Moreover, each T_j is given by

$$T_j(f)(x) = e^{2\pi i b_j x}T(e^{-2\pi i b_j(\cdot)}f)(x)$$

and thus with $g_j(x) = e^{-2\pi i b_j(x)}f(x)$, (4.5.13) and (4.5.14) can be written respectively as

$$\Big\|\Big(\sum_{j\in\mathbf{Z}}|T(g_j)|^2\Big)^{\frac{1}{2}}\Big\|_{L^p} \le C_p\Big\|\Big(\sum_{j\in\mathbf{Z}}|g_j|^2\Big)^{\frac{1}{2}}\Big\|_{L^p},$$

$$\Big\|\Big(\sum_{j\in\mathbf{Z}}|T(g_j)|^2\Big)^{\frac{1}{2}}\Big\|_{L^{1,\infty}} \le C\Big\|\Big(\sum_{j\in\mathbf{Z}}|g_j|^2\Big)^{\frac{1}{2}}\Big\|_{L^1}.$$

Theorem 4.5.1 gives that both of the previous estimates are valid by in view of the boundedness of $T = \frac{1}{2}(I + iH)$ from $L^p \to L^p$ and from $L^1 \to L^{1,\infty}$. For a slight generalization and an extension to higher dimensions, see Exercise 4.6.1.

We have now seen that bounded operators from L^p to L^q (or to $L^{q,\infty}$) always admit ℓ^2-valued extensions. It is natural to ask whether they also admit ℓ^r-valued extensions for some $r \neq 2$. For some values of r we may answer this question. Here is a straightforward corollary of Theorem 4.5.1.

Corollary 4.5.4. *Suppose that T is a linear bounded operator from $L^p(X)$ to $L^p(Y)$ with norm A for some $1 \le p < \infty$. Let r be a number between p and 2. Then we have*

$$\Big\|\Big(\sum_{j}|T(f_j)|^r\Big)^{\frac{1}{r}}\Big\|_{L^p} \le A\Big\|\Big(\sum_{j}|f_j|^r\Big)^{\frac{1}{r}}\Big\|_{L^p}. \tag{4.5.15}$$

Proof. The endpoint case $r = 2$ is a consequence of Theorem 4.5.1, while the endpoint case $r = p$ is trivial. Interpolation (see Exercise 4.5.2) gives the required conclusion for r between p and 2.

□

We note that Exercise 4.5.2 and Corollary 4.5.4 are also valid for indices less than 1.

Example 4.5.5. The result of Corollary 4.5.4 may fail if r does not lie in the interval with endpoints p and 2. Let us take, for example, $1 < p < 2$ and consider an $r < p$. Take $X = Y = \mathbf{R}$ and define a linear operator T by setting

$$T(f)(x) = \widehat{f}(x)\chi_{[0,1]}.$$

Then T is L^p bounded since $\|T(f)\|_{L^p} \le \|T(f)\|_{L^{p'}} \le \|f\|_{L^p}$. Now take $f_j = \chi_{[j,j+1]}$ for $j = 1, \dots, N$. A simple calculation gives

$$\Big(\sum_{j=1}^{N} |T(f_j)|^r\Big)^{\frac{1}{r}} = N^{\frac{1}{r}} \left| \frac{e^{-2\pi i\xi} - 1}{-2\pi i\xi}\,\chi_{[0,1]} \right|$$

while

$$\Big(\sum_{j=1}^{N} |f_j|^r\Big)^{\frac{1}{r}} = \chi_{[0,N]}\,.$$

It follows that $N^{1/r} \le CN^{1/p}$ for all $N > 1$, and hence (4.5.15) cannot hold if $p > r$.

We have now seen that ℓ^r-valued extensions for $r \neq 2$ may fail in general. But do they fail for some specific operators of interest in Fourier analysis? For instance, is the following inequality

$$\Big\|\Big(\sum_{j\in\mathbf{Z}} |H(f_j)|^r\Big)^{\frac{1}{r}}\Big\|_{L^p} \le C_{p,r}\Big\|\Big(\sum_{j\in\mathbf{Z}} |f_j|^r\Big)^{\frac{1}{r}}\Big\|_{L^p} \tag{4.5.16}$$

true for the Hilbert transform H whenever $1 < p, r < \infty$? The answer to this question is affirmative. Inequality (4.5.16) is indeed valid and was first proved using complex function theory. In the next section we plan to study inequalities such as (4.5.16) for general singular integrals using the Calderón-Zygmund theory of the previous section applied to the context of Banach space-valued functions.

4.5.c. General Banach-Valued Extensions

We now set up the background required to state the main results of this section. Although the Banach spaces of most interest to us will be ℓ^r for $1 \le r \le \infty$, we will introduce the basic notions we need in general.

Let $\mathcal{B}$ be a Banach space over the field of complex numbers with norm $\|\ \|_{\mathcal{B}}$ and let $\mathcal{B}^*$ be its dual (with norm $\|\ \|_{\mathcal{B}^*}$). A function F defined on a σ-finite measure space (X, μ) and taking values in $\mathcal{B}$ is called $\mathcal{B}$-measurable if there exists a measurable subset X_0 of X so that $\mu(X \setminus X_0) = 0$ and $F[X \setminus X_0]$ is contained in some separable subspace $\mathcal{B}_0$ of $\mathcal{B}$, and for every $u^* \in \mathcal{B}^*$ the complex-valued map

$$x \to \langle u^*, F(x)\rangle$$

is measurable. A consequence of this definition is that the positive function $x \to \big\|F(x)\big\|_{\mathcal{B}}$ on X is measurable since it can be written as the supremum of the measurable functions $\big|\langle u^*, F(x)\rangle\big|$ over all u^* in the unit ball of $\mathcal{B}^*$. For $0 < p \le \infty$, denote by $L^p(X, \mathcal{B})$ the space of all $\mathcal{B}$-measurable functions F on X satisfying

$$\left(\int_X \|F(x)\|_{\mathcal{B}}^p \, d\mu(x)\right)^{\frac{1}{p}} < \infty, \tag{4.5.17}$$

with the obvious modification when $p = \infty$. Similarly define $L^{p,\infty}(X, \mathcal{B})$ as the space of all $\mathcal{B}$-measurable functions F on X satisfying

$$\Big\| \|F(\cdot)\|_{\mathcal{B}} \Big\|_{L^{p,\infty}(X)} < \infty. \tag{4.5.18}$$

$L^p(X, \mathcal{B})$ [respectively, $L^{p,\infty}(X, \mathcal{B})$] is called the L^p (respectively, $L^{p,\infty}$) space of functions on X with values in $\mathcal{B}$. Similarly, we can define other Lorentz spaces of $\mathcal{B}$-valued functions. The quantity in (4.5.17) [respectively, in (4.5.18)] is the *norm* of F in $L^p(X, \mathcal{B})$ [respectively, in $L^{p,\infty}(X, \mathcal{B})$].

Let us denote by $L^p(X)$ the space $L^p(X, \mathbf{C})$. Let $L^p(X)\otimes\mathcal{B}$ be the set of all finite linear combinations of elements of $\mathcal{B}$ with coefficients in $L^p(X)$, that is, elements of the form

$$F = f_1 u_1 + \cdots + f_m u_m, \tag{4.5.19}$$

where $f_j \in L^p(X)$, $u_j \in \mathcal{B}$, and $m \in \mathbf{Z}^+$.

We will show that every function F in $L^p(X, \mathcal{B})$ is a limit (in norm) of a sequence of elements in $L^p(X)\otimes\mathcal{B}$. More precisely, we will show that given any F in $L^p(X, \mathcal{B})$ and $\varepsilon > 0$, then there exist measurable sets $\{A_j\}_{j=1}^N$ of X with $\mu(A_j) < \infty$ and $u_k \in \mathcal{B}$ such that

$$\Big\|\sum_{j=1}^N \chi_{A_j} u_j - F\Big\|_{L^p(X,\mathcal{B})} < \varepsilon.$$

To see this assertion, given $\varepsilon > 0$, first use that X is σ-finite to find a subset X_0 of X with $0 < \mu(X_0) < \infty$ so that

$$\int_{X \setminus X_0} \|F\|_{\mathcal{B}}^p \, d\mu < \frac{\varepsilon^p}{3}.$$

Let F_0 be the restriction of F to X_0. As F_0 takes values in a separable subspace $\mathcal{B}_0$ of $\mathcal{B}$, there is a countable dense subset $\{u_k\}_{k=1}^\infty$ of the unit ball U_0 of $\mathcal{B}_0$. Find u_k^* in $\mathcal{B}_0^*$ so that $\langle u_\ell^*, u_k\rangle = \delta_{\ell k}$, where the latter is equal to 1 when $\ell = k$ and zero otherwise. Then

$$U_0 = \bigcap_{k=1}^\infty \{u \in \mathcal{B} : \ |\langle u_k^*, u\rangle| \le 1\},$$

which implies that the set $F_0^{-1}[U_0]$ is measurable in X. Let $\delta = \varepsilon(3\mu(X_0))^{-\frac{1}{p}}$. Using an inductive argument, we can find elements $\{v_k\}_{k=1}^\infty$ in $\mathcal{B}_0$ such that the sets

$v_k + \delta\, U_0$ are pairwise disjoint and their union is $\mathcal{B}_0$. Then for $k = 1, 2, 3, \dots$ we define sets

$$A_k = F_0^{-1}\big[v_k + \delta\, U_0\big]\,.$$

We note that the A_k are disjoint measurable subsets of X_0 that satisfy

$$X_0 = \bigcup_{k=1}^{\infty} A_k\,.$$

Next we find an integer N such that

$$\int_{\bigcup_{k=N+1}^{\infty} A_k} \|F\|_{L^p(X)}^p \, d\mu < \frac{\varepsilon^p}{3}\,.$$

Then we can see easily that

$$\left(\int_X \Big\|F - \sum_{k=1}^{N} \chi_{A_k} u_k\Big\|_{\mathcal{B}}^p \, d\mu\right)^{\frac{1}{p}} < \varepsilon\,.$$

If F is an element of $L^1 \otimes \mathcal{B}$ given as in (4.5.19), we can define its integral (which will be an element of $\mathcal{B}$) by setting

$$\int_X F(x)\, d\mu(x) = \sum_{j=1}^{m} \left(\int_{\mathbf{R}^n} f_j(x)\, d\mu(x)\right) u_j.$$

Observe that for every $F \in L^1 \otimes \mathcal{B}$ we have

$$\begin{aligned}
\Big\|\int_X F(x)\, d\mu(x)\Big\|_{\mathcal{B}} &= \sup_{\|u^*\|_{\mathcal{B}^*}\le 1} \left|\left\langle u^*, \sum_{j=1}^{m} \left(\int_X f_j\, d\mu\right) u_j\right\rangle\right| \\
&= \sup_{\|u^*\|_{\mathcal{B}^*}\le 1} \left|\int_X \langle u^*, \sum_{j=1}^{m} f_j u_j\rangle\, d\mu\right| \\
&\le \int_X \sup_{\|u^*\|_{\mathcal{B}^*}\le 1} \big|\langle u^*, \sum_{j=1}^{m} f_j u_j\rangle\big|\, d\mu = \|F\|_{L^1(X,\mathcal{B})}.
\end{aligned}$$

Thus the linear operator

$$F \to I_F = \int_X F(x)\, d\mu(x)$$

is bounded from $L^1(X) \otimes \mathcal{B}$ into $\mathcal{B}$. As every element of $\mathcal{B}$ is a (norm) limit of a sequence of elements in $L^1(X)\otimes\mathcal{B}$, by continuity, the operator $F \to I_F$ has a unique extension on $L^1(X, \mathcal{B})$ that we call the *Bochner integral* of F and denote by

$$\int_{\mathbf{R}^n} F(x)\, d\mu(x)\,.$$

It is not difficult to show that the Bochner integral of F is the only element of $\mathcal{B}$ that satisfies

$$\left\langle u^*, \int_{\mathbf{R}^n} F(x)\, d\mu(x) \right\rangle = \int_X \langle u^*, F(x) \rangle \, d\mu(x)$$

for all $u^* \in \mathcal{B}^*$.

Definition 4.5.6. Let T be a linear operator that maps $L^p(\mathbf{R}^n)$ into $L^q(\mathbf{R}^n)$ [respectively, $L^{q,\infty}(\mathbf{R}^n)$] for some $0 < p, q \le \infty$. We define another operator $\vec{T}$ acting on $L^p \otimes \mathcal{B}$ by setting

$$\vec{T}\Big(\sum_{j=1}^m f_j u_j\Big) = \sum_{j=1}^m T(f_j)\, u_j .$$

If $\vec{T}$ happens to be bounded from $L^p(\mathbf{R}^n, \mathcal{B})$ to $L^q(\mathbf{R}^n, \mathcal{B})$ [respectively, $L^{q,\infty}(\mathbf{R}^n)$], then we say that T has a bounded $\mathcal{B}$-*valued extension.* In this case we call the operator $\vec{T}$ the $\mathcal{B}$-valued extension of T.

Example 4.5.7. Let $\mathcal{B} = \ell^r$ for some $1 \le r < \infty$. Then a measurable function $F : X \to \mathcal{B}$ is just a sequence $\{f_j\}_j$ of measurable functions $f_j : X \to \mathbf{C}$. The space $L^p(X, \ell^r)$ consists of all measurable complex-valued sequences $\{f_j\}_j$ on X that satisfy

$$\big\|\{f_j\}_j\big\|_{L^p(X,\ell^r)} = \Big\|\Big(\sum_j |f_j|^r\Big)^{\frac{1}{r}}\Big\|_{L^p(X)} < \infty .$$

The space $L^p(X) \otimes \ell^r$ is the set of all finite sums

$$\sum_{j=1}^m (a_{j1}, a_{j2}, a_{j3}, \dots)\, g_j$$

where $g_j \in L^p(X)$ and $(a_{j1}, a_{j2}, a_{j3}, \dots) \in \ell^r$ which is nothing else than the set of all sequences $(f_1, f_2, \dots)$ of $L^p(X)$ functions f_j.

If T is a linear operator bounded from $L^p(X)$ to $L^q(Y)$, then $\vec{T}$ is defined by

$$\vec{T}(\{f_j\}_j) = \{T(f_j)\}_j .$$

According to Definition 4.5.6, T has an ℓ^r-extension if and only if the inequality

$$\Big\|\Big(\sum_j |T(f_j)|^r\Big)^{\frac{1}{r}}\Big\|_{L^q} \le C \Big\|\Big(\sum_j |f_j|^r\Big)^{\frac{1}{r}}\Big\|_{L^p}$$

is valid.

An operator T acting on measurable functions is called *positive* if it satisfies $f \geq 0 \implies T(f) \geq 0$. It is straightforward to verify that positive operators satisfy

$$
\begin{aligned}
&f \leq g \implies T(f) \leq T(g)\\
&|T(f)| \leq T(|f|) \qquad\qquad (4.5.20)\\
&\sup_j |T(f_j)| \leq T(\sup_j |f_j|)
\end{aligned}
$$

for all f, g, f_j measurable functions. We have the following result regarding vector-valued extensions of positive operators:

Proposition 4.5.8. *Let $0 < p, q \leq \infty$ and (X, μ), (Y, ν) be two measure spaces. Let T be a positive linear operator mapping $L^p(X)$ into $L^q(Y)$ [respectively, into $L^{q,\infty}(Y)$] with norm A. Let $\mathcal{B}$ be a Banach space. Then T has a $\mathcal{B}$-valued extension $\vec{T}$ that maps $L^p(X, \mathcal{B})$ into $L^q(Y, \mathcal{B})$ [respectively, into $L^{q,\infty}(Y, \mathcal{B})$] with the same norm.*

Proof. Let us first understand this theorem when $\mathcal{B} = \ell^r$ for $1 \leq r \leq \infty$. The two endpoint cases $r = 1$ and $r = \infty$ can be checked easily using the properties in (4.5.20). For instance, for $r = 1$ we have

$$
\Big\| \sum_j |T(f_j)| \Big\|_{L^q} \leq \Big\| \sum_j T(|f_j|) \Big\|_{L^q} = \Big\| T\Big(\sum_j |f_j| \Big) \Big\|_{L^q} \leq A \Big\| \sum_j |f_j| \Big\|_{L^p},
$$

while for $r = \infty$ we have

$$
\Big\| \sup_j |T(f_j)| \Big\|_{L^q} \leq \Big\| T(\sup_j |f_j|) \Big\|_{L^q} \leq A \,\| \sup_j |f_j| \|_{L^p}.
$$

The required inequality for $1 < r < \infty$

$$
\Big\| \Big(\sum_j |T(f_j)|^r \Big)^{\frac{1}{r}} \Big\|_{L^q} \leq A \Big\| \Big(\sum_j |f_j|^r \Big)^{\frac{1}{r}} \Big\|_{L^p}
$$

follows from the Riesz-Thorin interpolation theorem (see Exercise 4.5.2).

The result for a general Banach space $\mathcal{B}$ can be proved using the following inequality:

$$
\|\vec{T}(F)(x)\|_{\mathcal{B}} \leq T\Big(\|F\|_{\mathcal{B}} \Big)(x), \qquad x \in \mathbf{R}^n \tag{4.5.21}
$$

by simply taking L^q norms. To prove (4.5.21), let us take $F = \sum_{j=1}^{n} f_j u_j$. Then

$$
\begin{aligned}
\big\|\vec{T}(F)(x)\big\|_{\mathcal{B}} &= \Big\|\sum_{j=1}^{n} T(f_j)(x)u_j\Big\|_{\mathcal{B}} \\
&= \sup_{\|u^*\|_{\mathcal{B}^*}\le 1} \Big|\Big\langle u^*, \sum_{j=1}^{n} T(f_j)(x)u_j\Big\rangle\Big| \\
&= \sup_{\|u^*\|_{\mathcal{B}^*}\le 1} \Big|T\Big(\sum_{j=1}^{n} f_j\langle u^*, u_j\rangle\Big)(x)\Big| \\
&\le T\Big(\sup_{\|u^*\|_{\mathcal{B}^*}\le 1} \Big|\Big\langle u^*, \sum_{j=1}^{n} f_j u_j\Big\rangle\Big|\Big)(x) \\
&\le T\Big(\Big\|\sum_{j=1}^{n} f_j u_j\Big\|_{\mathcal{B}}\Big)(x) \\
&= T\Big(\big\|F\big\|_{\mathcal{B}}\Big)(x),
\end{aligned}
$$

where in the first inequality we used that T is a positive operator.

□

We end this section with a simple extension of Theorem 4.5.1.

Proposition 4.5.9. *Let $\mathcal{H}$ be a Hilbert space and let $0 < p < \infty$. Then every bounded linear operator T from $L^p(\mathbf{R}^d)$ into $L^p(\mathbf{R}^d)$ has an $\mathcal{H}$-valued extension. In particular for all measurable families of functions $\{f_t\}_{t\in\mathbf{R}^d}$ and for all positive measures μ on $\mathbf{R}^d$ the following estimate is valid:*

$$
\Big\|\Big(\int_{\mathbf{R}^d} |T(f_t)|^2\, d\mu(t)\Big)^{\frac{1}{2}}\Big\|_{L^p} \le \big\|T\big\|_{L^p\to L^p}\Big\|\Big(\int_{\mathbf{R}^d} |f_t|^2\, d\mu(t)\Big)^{\frac{1}{2}}\Big\|_{L^p}.
$$

Proof. If the Hilbert space $\mathcal{H}$ is finite dimensional, then it is isometrically isomorphic to $\ell^2(\{1,2,\dots,N\})$ for some positive integer N. If $\mathcal{H}$ is infinite dimensional and separable, then it is isometrically isomorphic to $\ell^2(\mathbf{Z})$. By Theorem 4.5.1, the linear operator T has an ℓ^2-valued extension and in view of the isometry with $\mathcal{H}$, it must also have an $\mathcal{H}$-valued extension. If the Hilbert space $\mathcal{H}$ is not separable, we obtain a vector-valued extension of T for all separable subspaces of $\mathcal{H}$ with norm independent of the subspace.

□

Exercises

4.5.1. Let $\mathcal{B}$ be a Banach space and $1 \le p \le \infty$. Let $F \in L^p(\mathbf{R}^n, \mathcal{B})$.
(a) For $G \in L^{p'}(\mathbf{R}^n, \mathcal{B}^*)$ show that the function $x \to \langle G(x), F(x) \rangle$ is integrable over $\mathbf{R}^n$ and, moreover,

$$\big\|G\big\|_{L^{p'}(\mathbf{R}^n,\mathcal{B}^*)} = \sup_{\|F\|_{L^p(\mathbf{R}^n,\mathcal{B})}\le 1} \left| \int_{\mathbf{R}^n} \langle G(x), F(x) \rangle \, dx \right| .$$

(b) Conclude that

$$L^{p'}(\mathbf{R}^n, \mathcal{B}^*) \subseteq (L^p(\mathbf{R}^n, \mathcal{B}))^* ,$$

that is, $L^{p'}(\mathbf{R}^n, \mathcal{B}^*)$ is isometrically contained in the dual of $L^p(\mathbf{R}^n, \mathcal{B})$. Furthermore, if $\mathcal{B}$ is reflexive and $1 \le p < \infty$, we have that

$$L^{p'}(\mathbf{R}^n, \mathcal{B}^*) = L^p(\mathbf{R}^n, \mathcal{B})^* .$$

(c) If $1 < p < \infty$, prove that for any $\varepsilon > 0$ there exists a functional G in $L^{p'}(\mathbf{R}^n, \mathcal{B}^*)$ with norm 1 such that

$$\big\|F\big\|_{L^p(\mathbf{R}^n,\mathcal{B})} \le \left| \int_{\mathbf{R}^n} \langle G(x), F(x) \rangle \, dx \right| + \varepsilon .$$

4.5.2. Prove the following version of the Riesz-Thorin interpolation theorem. Let $1 \le p_0, q_0, , p_1, q_1, r_0, s_0, r_1, s_1 \le \infty$ and $0 < \theta < 1$ satisfy

$$\frac{1-\theta}{p_0} + \frac{\theta}{p_1} = \frac{1}{p} \qquad \frac{1-\theta}{q_0} + \frac{\theta}{q_1} = \frac{1}{q}$$
$$\frac{1-\theta}{r_0} + \frac{\theta}{r_1} = \frac{1}{r} \qquad \frac{1-\theta}{s_0} + \frac{\theta}{s_1} = \frac{1}{s} .$$

Suppose that $\vec{T}$ is a linear operator that maps $L^{p_0}(\mathbf{R}^n, \ell^{r_0})$ into $L^{q_0}(\mathbf{R}^n, \ell^{s_0})$ with norm A_0 and $L^{p_1}(\mathbf{R}^n, \ell^{r_1})$ into $L^{q_1}(\mathbf{R}^n, \ell^{s_2})$ with norm A_1. Prove that $\vec{T}$ maps $L^p(\mathbf{R}^n, \ell^r)$ into $L^q(\mathbf{R}^n, \ell^s)$ with norm at most $A_0^{1-\theta} A_1^{\theta}$.

4.5.3. (a) Prove the following version of the Marcinkiewicz interpolation theorem. Let $0 < p_0 < p < p_1 \le \infty$ and $0 < \theta < 1$ satisfy

$$\frac{1-\theta}{p_0} + \frac{\theta}{p_1} = \frac{1}{p}$$

Suppose that $\vec{T}$ is a sublinear operator, that is, it satisfies

$$\big\|\vec{T}(F+G)\big\|_{\mathcal{B}_2} \le \big\|\vec{T}(F)\big\|_{\mathcal{B}_2} + \big\|\vec{T}(G)\big\|_{\mathcal{B}_2} ,$$

for all F and G. Assume that $\vec{T}$ maps $L^{p_0}(\mathbf{R}^n, \mathcal{B}_1)$ into $L^{p_0,\infty}(\mathbf{R}^n, \mathcal{B}_2)$ with norm A_0 and $L^{p_1}(\mathbf{R}^n, \mathcal{B}_1)$ into $L^{p_1,\infty}(\mathbf{R}^n, \mathcal{B}_2)$ with norm A_1. Show that $\vec{T}$ maps $L^p(\mathbf{R}^n, \mathcal{B}_1)$ into $L^p(\mathbf{R}^n, \mathcal{B}_2)$ with norm at most $2\big(\frac{p}{p-p_0} + \frac{p}{p_1-p}\big)^{\frac{1}{p}} A_0^{1-\theta} A_1^{\theta}$.

(b) Let $p_0 = 1$. If $\vec{T}$ is linear and maps $L^1(\mathbf{R}^n, \mathcal{B}_1)$ into $L^{1,\infty}(\mathbf{R}^n, \mathcal{B}_2)$ with norm A_0 and $L^{p_1}(\mathbf{R}^n, \mathcal{B}_1)$ into $L^{p_1}(\mathbf{R}^n, \mathcal{B}_2)$ with norm A_1, show that the constant in part (a) can be improved to $8(p-1)^{-1/p} A_0^{1-\theta} A_1^{\theta}$; see also Exercise 1.3.2.

4.5.4. Suppose that all $x \in \mathbf{R}^n$, $K(x)$ is a bounded linear operator from $\mathcal{B}_1$ to $\mathcal{B}_2$ and let $\vec{T}(F)(x) = \int_{\mathbf{R}^n} K(x-y)F(y)\,dy$ be the vector-valued operator given by convolution with K.

(a) Suppose that K satisfies

$$\int_{\mathbf{R}^n} \big\|K(x)\big\|_{\mathcal{B}_1 \to \mathcal{B}_2}\,dx = C < \infty\,.$$

Prove that the operator $\vec{T}F$ maps $L^p(\mathbf{R}^n, \mathcal{B}_1)$ into $L^p(\mathbf{R}^n, \mathcal{B}_2)$ with norm at most C for $1 \le p \le \infty$.

(b) (*Young's inequality*) Suppose that K satisfies

$$\bigg(\int_{\mathbf{R}^n} \big\|K(x)\big\|^s_{\mathcal{B}_1 \to \mathcal{B}_2}\,dx\bigg)^{1/s} = C < \infty\,.$$

Prove that $\vec{T}(F)$ maps $L^p(\mathbf{R}^n, \mathcal{B}_1)$ into $L^q(\mathbf{R}^n, \mathcal{B}_2)$ with norm at most C whenever $1 \le p, q, s \le \infty$ and $1/q + 1 = 1/s + 1/p$.

(c) (*Young's inequality for weak type spaces*) Suppose that K satisfies

$$\Big\|\big\|K(\,\cdot\,)\big\|_{\mathcal{B}_1 \to \mathcal{B}_2}\Big\|_{L^{s,\infty}} < \infty\,.$$

Prove that $\vec{T}(F)$ maps $L^p(\mathbf{R}^n, \mathcal{B}_1)$ into $L^q(\mathbf{R}^n, \mathcal{B}_2)$ whenever $1 \le p < \infty$, $1 < p, s < \infty$ and $1/q + 1 = 1/s + 1/p$.

4.5.5. Prove the following (slight) generalization of the previous Exercise when $p = 1$. Suppose that K satisfies

$$\int_{\mathbf{R}^n} \big\|K(x)u\big\|_{\mathcal{B}_2}\,dx \le C\big\|u\big\|_{\mathcal{B}_1}$$

for all $u \in \mathcal{B}_1$. Then $\vec{T}$ maps $L^1(\mathbf{R}^n, \mathcal{B}_1)$ into $L^1(\mathbf{R}^n, \mathcal{B}_2)$ with norm at most C. Show, however, that the preceding condition is not strong enough to imply L^p boundedness for $\vec{T}$ for $1 < p < \infty$.

4.5.6. Use the inequality for the Rademacher functions in Appendix C.2 instead of Lemma 4.5.2 to prove part (a) of Theorem 4.5.1 in the special case $p = q$.

4.5.7. Prove the following extension of Theorem 4.4.1. If T is a bounded linear operator from L^p to the Lorentz $L^{q,s}$, then it has an ℓ^2-valued extension. Here $0 < p, q, s \le \infty$.

4.5.8. Let $T_j(f)(x) = f(x-j)$ and $f_j(x) = \chi_{[-j,1-j]}$ for $j = 1, 2, \dots, N$. Use these functions and operators to show that the inequality

$$\Big\| \Big(\sum_j |T_j(f_j)|^2 \Big)^{\frac{1}{2}} \Big\|_{L^p} \le C_p \Big\| \Big(\sum_j |f_j|^2 \Big)^{\frac{1}{2}} \Big\|_{L^p}$$

may be false in general although the linear operators T_j are uniformly bounded from $L^p(\mathbf{R})$ to $L^p(\mathbf{R})$.

4.5.9. Suppose that T is a linear operator that takes real-valued functions to real-valued functions. Prove that

$$\sup_{\substack{f \text{ real-valued} \\ f \neq 0}} \frac{\big\|T(f)\big\|_{L^p}}{\big\|f\big\|_{L^p}} = \sup_{\substack{f \text{ complex-valued} \\ f \neq 0}} \frac{\big\|T(f)\big\|_{L^p}}{\big\|f\big\|_{L^p}} .$$

[*Hint:* Use Theorem 4.5.1 (a) with $p = q$.]

4.6. Vector-Valued Singular Integrals

We now discuss some results about vector-valued singular integrals. By this we mean singular integral operators taking values in Banach spaces. At this point we will restrict our attention to the situation where $X = Y = \mathbf{R}^n$.

4.6.a. Banach-Valued Singular Integral Operators

We consider a kernel $\vec{K}$ defined on $\mathbf{R}^n \setminus \{0\}$ that takes values in the space $L(\mathcal{B}_1, \mathcal{B}_2)$ of all bounded linear operators from $\mathcal{B}_1$ to $\mathcal{B}_2$. In other words, for all $x \in \mathbf{R}^n \setminus \{0\}$, $\vec{K}(x)$ is a bounded linear operator from $\mathcal{B}_1$ to $\mathcal{B}_2$ whose norm we will denote by $\big\|\vec{K}(x)\big\|_{\mathcal{B}_1 \to \mathcal{B}_2}$. We assume that $\vec{K}(x)$ is $L(\mathcal{B}_1, \mathcal{B}_2)$-measurable and locally integrable away from the origin so that the integral

$$\vec{T}(F)(x) = \int_{\mathbf{R}^n} \vec{K}(x-y) F(y)\, dy \tag{4.6.1}$$

is well defined as an element of $\mathcal{B}_2$ for all $F \in L^\infty(\mathbf{R}^n, \mathcal{B}_1)$ with compact support when x lies outside the support of F.

We will assume that the kernel $\vec{K}$ satisfies Hörmander's condition:

$$\int_{|x| \ge 2|y|} \big\|\vec{K}(x-y) - \vec{K}(x)\big\|_{\mathcal{B}_1 \to \mathcal{B}_2} dx \le A < \infty, \qquad y \in \mathbf{R}^n \setminus \{0\}, \tag{4.6.2}$$

which is a certain form of regularity for $\vec{K}$ familiar to us from the scalar case.

The following vector-valued extension of Theorem 4.3.3 is the main result of this section.

Theorem 4.6.1. *Let $\mathcal{B}_1$ and $\mathcal{B}_2$ be Banach spaces. Suppose that $\vec{T}$ given by (4.6.1) is a bounded linear operator from $L^r(\mathbf{R}^n,\mathcal{B}_1)$ into $L^r(\mathbf{R}^n,\mathcal{B}_2)$ with norm $B = B(r)$ for some $1 < r \le \infty$. Assume that $\vec{K}$ satisfies Hörmander's condition (4.6.2) for some $A > 0$. Then $\vec{T}$ has well-defined extensions on $L^p(\mathbf{R}^n,\mathcal{B}_1)$ for all $1 \le p < \infty$. Moreover, there exist dimensional constants C_n and C_n' such that*

$$\big\|\vec{T}(F)\big\|_{L^{1,\infty}(\mathbf{R}^n,\mathcal{B}_2)} \le C_n'(A+B)\big\|F\big\|_{L^1(\mathbf{R}^n,\mathcal{B}_1)} \tag{4.6.3}$$

for all F in $L^1(\mathbf{R}^n,\mathcal{B}_1)$ and

$$\big\|\vec{T}(F)\big\|_{L^p(\mathbf{R}^n,\mathcal{B}_2)} \le C_n \max\big(p,(p-1)^{-1}\big)(A+B)\big\|F\big\|_{L^p(\mathbf{R}^n,\mathcal{B}_1)} \tag{4.6.4}$$

whenever $1 < p < \infty$ and F is in $L^p(\mathbf{R}^n,\mathcal{B}_1)$.

Proof. We prove the weak type estimate (4.6.3) by applying the Calderón-Zygmund decomposition just as in the scalar case to the function $x \to \big\|F(x)\big\|_{\mathcal{B}_1}$ defined on $\mathbf{R}^n$. The proof of Theorem 4.3.3 is directly applicable here, and an identical repetition of the arguments given in the scalar case with suitable norms replacing absolute values yields (4.6.3).

We interpolate between the estimates $\vec{T}: L^1(\mathbf{R}^n,\mathcal{B}_1) \to L^{1,\infty}(\mathbf{R}^n,\mathcal{B}_2)$ and $\vec{T}: L^r(\mathbf{R}^n,\mathcal{B}_1) \to L^r(\mathbf{R}^n,\mathcal{B}_2)$. Using Exercise 4.5.3, we obtain for $1 < p < r$

$$\big\|\vec{T}(F)\big\|_{L^p(\mathbf{R}^n,\mathcal{B}_2)} \le C_n \max(1,(p-1)^{-1})(A+B)\big\|F\big\|_{L^p(\mathbf{R}^n,\mathcal{B}_1)}, \tag{4.6.5}$$

where C_n is independent of r, p, $\mathcal{B}_1$, and $\mathcal{B}_2$ (and only depends on n).

Next we obtain (4.6.4) for p near infinity. Since $\vec{K}(x) \in L(\mathcal{B}_1,\mathcal{B}_2)$, it follows that its adjoint $\vec{K}^*(x) \in L(\mathcal{B}_2^*,\mathcal{B}_1^*)$ has the same norm. Therefore, Hörmander's condition (4.6.2) can be written as

$$\sup_{y\in\mathbf{R}^n\setminus\{0\}} \int_{|x|\ge 2|y|} \big\|\vec{K}^*(x-y)-\vec{K}^*(x)\big\|_{\mathcal{B}_2^*\to\mathcal{B}_1^*}\,dx = A < \infty\,. \tag{4.6.6}$$

The operator corresponding to the kernel $\vec{K}^*$ is

$$\vec{T}^*(F)(x) = \int_{\mathbf{R}^n} \vec{K}^*(x-y)F(y)\,dy$$

initially defined for $F \in L^\infty(\mathbf{R}^n,\mathcal{B}_2^*)$ with compact support and for x outside the support of F. The assumption on $\vec{T}$ gives that $\vec{T}^*$ is bounded from $L^{r'}(\mathbf{R}^n,\mathcal{B}_2^*)$ into $L^{r'}(\mathbf{R}^n,\mathcal{B}_1^*)$. Then

$$\big\|\vec{T}^*(F)\big\|_{L^{1,\infty}(\mathbf{R}^n,\mathcal{B}_1^*)} \le C_n(A+B)\big\|F\big\|_{L^1(\mathbf{R}^n,\mathcal{B}_2^*)},$$

and therefore by interpolation we obtain for $1 < p' < r'$

$$\big\|\vec{T}^*(F)\big\|_{L^{p'}(\mathbf{R}^n,\mathcal{B}_1^*)} \le C_n \max(1,p-1)(A+B)\big\|F\big\|_{L^{p'}(\mathbf{R}^n,\mathcal{B}_2^*)}, \tag{4.6.7}$$

as $(p'-1)^{-1} = p-1$.

We now fix $r < p < \infty$. Let F lie in some dense subspace of $L^p(\mathbf{R}^n, \mathcal{B}_1)$ so that $\|\vec{T}(F)\|_{L^p(\mathbf{R}^n,\mathcal{B}_2)} < \infty$. For a given $\varepsilon > 0$, using Exercise 4.5.1(c), we can find a G in $L^{p'}(\mathbf{R}^n, \mathcal{B}_2^*)$ of norm 1 such that

$$\begin{aligned}
\|\vec{T}(F)\|_{L^p(\mathbf{R}^n,\mathcal{B}_2)} &\leq \left| \int_{\mathbf{R}^n} \langle G(x), \vec{T}(F)(x) \rangle \, dx \right| + \varepsilon \\
&= \left| \int_{\mathbf{R}^n} \langle \vec{T}^*(G)(x), F(x) \rangle \, dx \right| + \varepsilon \\
&\leq \|\vec{T}^*(G)\|_{L^{p'}(\mathbf{R}^n,\mathcal{B}_1^*)} \|F\|_{L^p(\mathbf{R}^n,\mathcal{B}_1)} + \varepsilon \\
&\leq C_n \max(1,p)(A+B) \|G\|_{L^{p'}(\mathbf{R}^n,\mathcal{B}_1^*)} \|F\|_{L^p(\mathbf{R}^n,\mathcal{B}_1)} + \varepsilon \\
&= C_n \max(1,p)(A+B) \|F\|_{L^p(\mathbf{R}^n,\mathcal{B}_1)} + \varepsilon \,,
\end{aligned}$$

where we used (4.6.7). Since ε was aribtrary and F lies in some dense subspace of $L^p(\mathbf{R}^n, \mathcal{B}_1)$, the required conclusion follows. This combined with (4.6.5) implies the required conclusion whenever $1 < r < \infty$. Observe that the case $r = \infty$ is easier and only requires Exercise 4.3.7 adapted to the Banach-valued setting.

□

4.6.b. Applications

We proceed with some applications. An important consequence of Theorem 4.6.1 is the following:

Corollary 4.6.2. *Let W_j be a sequence of tempered distributions on $\mathbf{R}^n$ whose Fourier transforms are uniformly bounded functions (i.e., $|\widehat{W_j}| \leq B$ for some B). Suppose that each W_j coincides with some locally integrable function K_j on $\mathbf{R}^n \setminus \{0\}$ that satisfies*

$$\int_{|x|\geq 2|y|} \sup_j |K_j(x-y) - K_j(x)| \, dx \leq A, \qquad y \in \mathbf{R}^n \setminus \{0\}. \tag{4.6.8}$$

Then there are constants $C_n, C_n' > 0$ so that for all $1 < p, r < \infty$ we have

$$\Big\| \Big(\sum_j |W_j * f_j|^r \Big)^{\frac{1}{r}} \Big\|_{L^{1,\infty}} \leq C_n' \max(r, (r-1)^{-1})(A+B) \Big\| \Big(\sum_j |f_j|^r \Big)^{\frac{1}{r}} \Big\|_{L^1},$$

$$\Big\| \Big(\sum_j |W_j * f_j|^r \Big)^{\frac{1}{r}} \Big\|_{L^p} \leq C_n \, c(p,r)(A+B) \Big\| \Big(\sum_j |f_j|^r \Big)^{\frac{1}{r}} \Big\|_{L^p},$$

where $c(p,r) = \max(p, (p-1)^{-1}) \max(r, (r-1)^{-1})$.

Proof. Let T_j be the operator given by convolution with the distribution W_j. It follows from Theorem 4.3.3 that the T_j's are weak type $(1,1)$ and also bounded on L^r with bounds at most a dimensional multiple of $\max(r,(r-1)^{-1})(A+B)$, uniformly in j. Naturally, set $\mathcal{B}_1 = \mathcal{B}_2 = \ell^r$ and define

$$\vec{T}(\{f_j\}_j) = \{W_j * f_j\}_j$$

for $\{f_j\}_j \in L^r(\mathbf{R}^n, \ell^r)$. Summing gives that $\vec{T}$ maps $L^r(\mathbf{R}^n, \ell^r)$ into itself with norm at most a dimensional multiple of $\max(r,(r-1)^{-1})(A+B)$.

The operator $\vec{T}$ has the form

$$\vec{T}(F)(x) = \int_{\mathbf{R}^n} \vec{K}(x-y)F(y)\,dy$$

for $F \in L^r(\mathbf{R}^n, \ell^r)$ with compact support and $x \notin \text{support}(F)$, where $\vec{K}(x)$ in $L(\ell^r, \ell^r)$ is the following operator:

$$\vec{K}(x)(\{t_j\}_j) = \{K_j(x)t_j\}_j, \qquad \{t_j\}_j \in \ell^r.$$

Clearly,

$$\big\|\vec{K}(x-y) - \vec{K}(y)\big\|_{\ell^r \to \ell^r} \le \sup_j |K_j(x-y) - K_j(x)|\,,$$

and therefore Hörmander's condition holds for $\vec{K}$ as a consequence of (4.6.8). The desired conclusion follows from Theorem 4.6.1.

□

If all the W_j's are equal, we obtain the following corollary, which contains in particular the result (4.5.16) mentioned earlier.

Corollary 4.6.3. *Let W be an element of $\mathcal{S}'(\mathbf{R}^n)$ whose Fourier transform is a function bounded in absolute value by some $B > 0$. Suppose that W coincides with some locally integrable function K on $\mathbf{R}^n \setminus \{0\}$ that satisfies Hörmander's condition:*

$$\int_{|x|\ge 2|y|} |K(x-y) - K(x)|\,dx \le A, \qquad y \in \mathbf{R}^n \setminus \{0\}. \tag{4.6.9}$$

Let T be the operator given by convolution with W. Then there exist constants $C_n, C_n' > 0$ such that for all $1 < p, r < \infty$ we have that

$$\Big\|\Big(\sum_j |T(f_j)|^r\Big)^{\frac{1}{r}}\Big\|_{L^{1,\infty}} \le C_n' \max(r,(r-1)^{-1})(A+B)\Big\|\Big(\sum_j |f_j|^r\Big)^{\frac{1}{r}}\Big\|_{L^1},$$

$$\Big\|\Big(\sum_j |T(f_j)|^r\Big)^{\frac{1}{r}}\Big\|_{L^p} \le C_n c(p,r) \max(r,(r-1)^{-1})(A+B)\Big\|\Big(\sum_j |f_j|^r\Big)^{\frac{1}{r}}\Big\|_{L^p},$$

where $c(p,r) = \max(p,(p-1)^{-1})\max(r,(r-1)^{-1})$. In particular, these inequalities are valid for the Hilbert transform and the Riesz transforms.

Interestingly enough, we can use the very statement of Theorem 4.6.1 to obtain its corresponding vector-valued version.

Proposition 4.6.4. *Let let $1 < p, r < \infty$ and let $\mathcal{B}_1$ and $\mathcal{B}_2$ be two Banach spaces. Suppose that $\vec{T}$ given by (4.6.1) is a bounded linear operator from $L^r(\mathbf{R}^n, \mathcal{B}_1)$ into $L^r(\mathbf{R}^n, \mathcal{B}_2)$ with norm $B = B(r)$. Also assume that $\vec{K}$ satisfies Hörmander's condition (4.6.2) for some $A > 0$. Then there exist positive constants C_n, C_n' such that for all $\mathcal{B}_1$-valued functions F_j we have*

$$\Big\| \Big(\sum_j \|\vec{T}(F_j)\|_{\mathcal{B}_2}^r \Big)^{\frac{1}{r}} \Big\|_{L^{1,\infty}(\mathbf{R}^n, \mathcal{B}_2)} \le C_n'(A+B) \Big\| \Big(\sum_j \|F_j\|_{\mathcal{B}_1}^r \Big)^{\frac{1}{r}} \Big\|_{L^1(\mathbf{R}^n, \mathcal{B}_1)},$$

$$\Big\| \Big(\sum_j \|\vec{T}(F_j)\|_{\mathcal{B}_2}^r \Big)^{\frac{1}{r}} \Big\|_{L^p(\mathbf{R}^n, \mathcal{B}_2)} \le C_n(A+B)c(p) \Big\| \Big(\sum_j \|F_j\|_{\mathcal{B}_1}^r \Big)^{\frac{1}{r}} \Big\|_{L^p(\mathbf{R}^n, \mathcal{B}_1)},$$

where $c(p) = \max(p, (p-1)^{-1})$.

Proof. Let us denote by $\ell^r(\mathcal{B}_1)$ the Banach space of all $\mathcal{B}_1$-valued sequences $\{t_j\}_j$ that satisfy

$$\big\|\{t_j\}_j\big\|_{\ell^r(\mathcal{B}_1)} = \big(\sum_j \|t_j\|_{\mathcal{B}_1}^r \big)^{\frac{1}{r}} < \infty.$$

Now consider the operator $\vec{S}$ defined by

$$\vec{S}(\{F_j\}_j) = \{\vec{T}(F_j)\}_j\,.$$

It is obvious that $\vec{S}$ maps $L^r(\mathbf{R}^n, \ell^r(\mathcal{B}_1))$ into $L^r(\mathbf{R}^n, \ell^r(\mathcal{B}_2))$ with norm at most B. Moreover, $\vec{S}$ has kernel $\widetilde{K}(x) \in L(\ell^r(\mathcal{B}_1), \ell^r(\mathcal{B}_2))$ given by

$$\widetilde{K}(x)(\{t_j\}_j) = \{\vec{K}(x)t_j\}_j,$$

where $\vec{K}$ is the kernel of $\vec{T}$. It is not hard to see that the operator norms of $\vec{K}$ and $\widetilde{K}$ coincide and therefore

$$\big\|\vec{K}(x-y) - \vec{K}(x)\big\|_{\mathcal{B}_1 \to \mathcal{B}_2} = \big\|\widetilde{K}(x-y) - \widetilde{K}(x)\big\|_{\ell^r(\mathcal{B}_1) \to \ell^r(\mathcal{B}_2)}.$$

We conclude that $\widetilde{K}$ satisfies the hypotheses of Theorem 4.6.1. The conclusions of Theorem 4.6.1 for $\vec{S}$ are the desired inequalities for $\vec{T}$.

□

4.6.c. Vector-Valued Estimates for Maximal Functions

Next, we discuss applications of vector-valued inequalities to some nonlinear operators. We fix an integrable function Φ on $\mathbf{R}^n$ and for $t > 0$ define $\Phi_t(x) = t^{-n}\Phi(t^{-1}x)$. We suppose that Φ satisfies the following *regularity* condition:

$$\int_{|x| \ge 2|y|} \sup_{t>0} |\Phi_t(x-y) - \Phi_t(x)|\,dx = A_\Phi < \infty, \qquad y \in \mathbf{R}^n \setminus \{0\}. \tag{4.6.10}$$

We consider the maximal operator

$$M_\Phi(f)(x) = \sup_{t>0} |(f * \Phi_t)(x)|$$

defined for f in $L^1 + L^\infty$. We are interested in obtaining L^p estimates for M_Φ. It is reasonable to start with $p = \infty$, which yields the easiest of all the L^p estimates for M_Φ, the trivial estimate

$$\big\|M_\Phi(f)\big\|_{L^\infty} \le \big\|\Phi\big\|_{L^1}\big\|f\big\|_{L^\infty}. \tag{4.6.11}$$

We think of M_Φ as a linear operator taking values in a Banach space. Indeed, it is natural to set

$$\mathcal{B}_1 = \mathbf{C} \qquad \text{and} \qquad \mathcal{B}_2 = L^\infty(\mathbf{R}^+)$$

and view M_Φ as the linear operator $f \to \{f * \Phi_\delta\}_{\delta>0}$ that maps $\mathcal{B}_1$-valued functions to $\mathcal{B}_2$-valued functions.

To do this precisely, we define a $\mathcal{B}_2$-valued kernel

$$\vec{K}_\Phi(x) = \{\Phi_\delta(x)\}_{\delta\in\mathbf{R}^+}$$

and a $\mathcal{B}_2$ -valued linear operator

$$\vec{M}_\Phi(f) = f * \vec{K}_\Phi = \{f * \Phi_\delta\}_{\delta\in\mathbf{R}^+}$$

acting on complex-valued functions on $\mathbf{R}^n$. We know that $\vec{M}_\Phi$ maps

$$L^\infty(\mathbf{R}^n, \mathcal{B}_1) \to L^\infty(\mathbf{R}^n, \mathcal{B}_2)$$

with norm at most $\big\|\Phi\big\|_{L^1}$. Now condition (4.6.10) implies Hörmander's condition (4.6.2) for the kernel $\vec{K}_\Phi$. Applying Theorem 4.6.1, we obtain for $1 < p < \infty$

$$\big\|M_\Phi(f)\big\|_{L^p} \le C_n \max(p, (p-1)^{-1})(A_\Phi + \big\|\Phi\big\|_{L^1})\big\|f\big\|_{L^p}, \tag{4.6.12}$$

which can be immediately improved to

$$\big\|M_\Phi(f)\big\|_{L^r(\mathbf{R}^n)} \le C_n \max(1, (r-1)^{-1})(A_\Phi + \big\|\Phi\big\|_{L^1})\big\|f\big\|_{L^r(\mathbf{R}^n)} \tag{4.6.13}$$

via interpolation with estimate (4.6.11) for all $1 < r < \infty$.

Next we use estimate (4.6.13) to obtain vector-valued estimates for the sublinear operator M_Φ.

Corollary 4.6.5. *Let Φ be an integrable function on $\mathbf{R}^n$ that satisfies (4.6.10). Then there exist dimensional constants C_n and C_n' such that for all $1 < p, r < \infty$ the following vector-valued inequalities are valid:*

$$\Big\|\Big(\sum_j |M_\Phi(f_j)|^r\Big)^{\frac{1}{r}}\Big\|_{L^{1,\infty}} \le C_n' c(r)(A_\Phi + \big\|\Phi\big\|_{L^1})\Big\|\Big(\sum_j |f_j|^r\Big)^{\frac{1}{r}}\Big\|_{L^1}, \tag{4.6.14}$$

where $c(r) = 1 + (r-1)^{-1}$ and

$$\Big\|\Big(\sum_j |M_\Phi(f_j)|^r\Big)^{\frac{1}{r}}\Big\|_{L^p} \le C_n c(p,r)(A_\Phi + \big\|\Phi\big\|_{L^1})\Big\|\Big(\sum_j |f_j|^r\Big)^{\frac{1}{r}}\Big\|_{L^p}, \tag{4.6.15}$$

where $c(p,r) = \big(1+(r-1)^{-1}\big)\big(p+(p-1)^{-1}\big)$.

Proof. We set $\mathcal{B}_1 = \mathbf{C}$ and $\mathcal{B}_2 = \ell^\infty(\mathbf{R}^+)$. We use estimate (4.6.13) as a starting point in Proposition 4.6.4 that immediately yields the required conclusions (4.6.14) and (4.6.15).

□

Similar estimates hold for the Hardy-Littlewood maximal operator.

Theorem 4.6.6. *For* $1 < p, r < \infty$ *the Hardy-Littlewood maximal function* M *satisfies the vector-valued inequalities*

$$\Big\|\Big(\sum_j |M(f_j)|^r\Big)^{\frac{1}{r}}\Big\|_{L^{1,\infty}} \le C_n'\big(1+(r-1)^{-1}\big)\Big\|\Big(\sum_j |f_j|^r\Big)^{\frac{1}{r}}\Big\|_{L^1}, \tag{4.6.16}$$

$$\Big\|\Big(\sum_j |M(f_j)|^r\Big)^{\frac{1}{r}}\Big\|_{L^p} \le C_n\, c(p,r)\Big\|\Big(\sum_j |f_j|^r\Big)^{\frac{1}{r}}\Big\|_{L^p}, \tag{4.6.17}$$

where $c(p,r) = \big(1+(r-1)^{-1}\big)\big(p+(p-1)^{-1}\big)$.

Proof. Let us fix an even positive symmetrically decreasing Schwartz function Φ on $\mathbf{R}^n$ that satisfies $\Phi(x) \ge 1$ when $|x| \le 1$. Then the Hardy-Littlewood maximal function $M(f)$ is pointwise controlled by a constant multiple of the function $M_\Phi(|f|)$. In view of Corollary 4.6.5, it suffices to check that for such a Φ, (4.6.10) holds. First observe that in view of the decreasing character of Φ, we have

$$\sup_j |f| * \Phi_{2^j} \le M_\Phi(|f|) \le 2^n \sup_j |f| * \Phi_{2^j},$$

and for this reason we choose to work with the easier dyadic maximal operator

$$M_\Phi^d(f) = \sup_j |f * \Phi_{2^j}|.$$

We observe the validity of the simple inequalties

$$M(f) \le M_\Phi(|f|) \le 2^n M_\Phi^d(|f|). \tag{4.6.18}$$

If we can show that

$$\sup_{y\in\mathbf{R}^n\setminus\{0\}} \int_{|x|\ge 2|y|} \sup_{j\in\mathbf{Z}} |\Phi_{2^j}(x-y) - \Phi_{2^j}(x)|\, dx = C_n < \infty, \tag{4.6.19}$$

then (4.6.14) and (4.6.15) will be satisfied with M_Φ^d replacing M_Φ. We therefore turn our attention to (4.6.19). We have

$$
\begin{aligned}
&\int_{|x|\ge 2|y|} \sup_{j\in\mathbf{Z}} |\Phi_{2^j}(x-y)-\Phi_{2^j}(x)|\,dx \\
&\le \sum_{j\in\mathbf{Z}} \int_{|x|\ge 2|y|} |\Phi_{2^j}(x-y)-\Phi_{2^j}(x)|\,dx \\
&\le \sum_{2^j>|y|} \int_{|x|\ge 2|y|} \frac{|y|\,|\nabla\Phi(\frac{x-\theta y}{2^j})|}{2^{(n+1)j}}\,dx + \sum_{2^j\le|y|} \int_{|x|\ge 2|y|} (|\Phi_{2^j}(x-y)|+|\Phi_{2^j}(x)|)\,dx \\
&\le \sum_{2^j>|y|} \int_{|x|\ge 2|y|} \frac{|y|}{2^{(n+1)j}} \frac{C_N\,dx}{(1+|2^{-j}(x-\theta y)|)^N} + 2\sum_{2^j\le|y|} \int_{|x|\ge|y|} |\Phi_{2^j}(x)|\,dx \\
&\le \sum_{2^j>|y|} \int_{|x|\ge 2|y|} \frac{|y|}{2^{(n+1)j}} \frac{C_N}{(1+|2^{-j-1}x|)^N}\,dx + 2\sum_{2^j\le|y|} \int_{|x|\ge 2^{-j}|y|} |\Phi(x)|\,dx \\
&\le \sum_{2^j>|y|} \int_{|x|\ge 2^{-j}|y|} \frac{|y|}{2^j} \frac{C_N}{(1+|x|)^N}\,dx + 2\sum_{2^j\le|y|} C_N(2^{-j}|y|)^{-N} \\
&\le C_N \sum_{2^j>|y|} \frac{|y|}{2^j} + C_N = 2C_N\,,
\end{aligned}
$$

where $C_N>0$ depends on $N>n$, $\theta\in[0,1]$ and $|x-\theta y|\ge |x|/2$ when $|x|\ge 2|y|$.

Now apply (4.6.14) and (4.6.15) to M_Φ^d and use (4.6.18) to obtain the desired vector-valued inequalities.

□

Remark 4.6.7. Observe that (4.6.16) and (4.6.17) also hold for $r=\infty$. These endpoint estimates can be proved directly by observing that

$$\sup_j M(f_j) \le M(\sup_j |f_j|)\,.$$

The same is true for estimates (4.6.14) and (4.6.15).

Exercises

4.6.1. (a) For all $j\in\mathbf{Z}$, let I_j be an interval in $\mathbf{R}$ and let T_j be the operator given on the Fourier transform by multiplication with the characteristic function of I_j. Prove that there exists a constant $C>0$ such that for all $1<p,r<\infty$

and for all integrable functions f_j on $\mathbf{R}$ we have

$$\Big\|\Big(\sum_j |T_j(f_j)|^r\Big)^{\frac{1}{r}}\Big\|_{L^p} \le C\, c(p,r)\Big\|\Big(\sum_j |f_j|^r\Big)^{\frac{1}{r}}\Big\|_{L^p},$$

$$\Big\|\Big(\sum_j |T_j(f_j)|^r\Big)^{\frac{1}{r}}\Big\|_{L^{1,\infty}} \le C \max\big(r,(r-1)^{-1}\big)\Big\|\Big(\sum_j |f_j|^r\Big)^{\frac{1}{r}}\Big\|_{L^1},$$

where $c(p,r) = \max\big(r,(r-1)^{-1}\big)\max\big(p,(p-1)^{-1}\big)$.
(b) Let R_j be arbitrary rectangles on $\mathbf{R}^n$ with sides parallel to the axes and let S_j be the operators given on the Fourier transform by multiplication with the characteristic functions of R_j. Prove that there exists a dimensional constant $C_n > 0$ such that for all indices $1 < p, r < \infty$ and for all functions f_j in $L^p(\mathbf{R}^n)$ we have

$$\Big\|\Big(\sum_j |S_j(f_j)|^r\Big)^{\frac{1}{r}}\Big\|_{L^p} \le C_n c(p,r)^n\Big\|\Big(\sum_j |f_j|^r\Big)^{\frac{1}{r}}\Big\|_{L^p},$$

where $c(p,r)$ is as in part (a).
$\big[$*Hint:* Use Theorem 4.5.1 and the fact that the operator whose multiplier is the characteristic function of the interval (a,b) on the line is equal to $\frac{i}{2}\big(M^aHM^{-a} - M^bHM^{-b}\big)$, where $M^a(f)(x) = f(x)e^{2\pi i a x}$ and H is the Hilbert transform.$\big]$

4.6.2. For every $t \in \mathbf{R}^d$, let $R(t)$ be a rectangle with sides parallel to the axes in $\mathbf{R}^n$ so that the map $t \to R(t)$ is measurable. Then there is a constant $C_n > 0$ such that for all $1 < p < \infty$, for all σ-finite measures μ on $\mathbf{R}^d$, and for all families of measurable functions f_t on $\mathbf{R}^n$ we have

$$\Big\|\Big(\int_{\mathbf{R}^d} |(\widehat{f_t}\chi_{R(t)})^\vee|^2\, d\mu(t)\Big)^{\frac{1}{2}}\Big\|_{L^p} \le C_n\, c(p)^n\Big\|\Big(\int_{\mathbf{R}^d} |f_t|^2\, d\mu(t)\Big)^{\frac{1}{2}}\Big\|_{L^p},$$

where $c(p) = \max(p,(p-1)^{-1})$.
$\big[$*Hint:* Reduce this estimate to Corollary 4.5.9. Observe that when $d = 1$, this provides a continuous version of the result of the previous Exercise.$\big]$

4.6.3. (a) Let Φ is a radially decreasing function on $\mathbf{R}^n$ that satisfies

$$\int_{\mathbf{R}^n} |\Phi(x-y) - \Phi(x)|\, dx \le \eta(y), \qquad \int_{|x|\ge R} |\Phi(x)|\, dx \le \eta(R^{-1})$$

for all $R > 1$, where η is an increasing function with $\eta(0) = 0$ such that

$$\int_0^1 \frac{\eta(t)}{t}\, dt < \infty .$$

Prove that (4.6.19) holds.
[*Hint:* Modify the calculation in the proof of Theorem 4.6.6.]
(b) Use Theorem 4.6.1 with $r = \infty$ to conclude that the maximal function $f \to \sup_{j\in\mathbf{Z}} |f * \Phi_{2^j}|$ maps $L^p(\mathbf{R}^n)$ into itself for $1 < p \le \infty$.

4.6.4. (a) On $\mathbf{R}$, take $f_j = \chi_{[2^{j-1},2^j]}$ to prove that inequality (4.6.17) fails when $p = \infty$ and $1 < r < \infty$.
(b) Again on $\mathbf{R}$, take $N > 2$ and $f_j = \chi_{[\frac{j-1}{N},\frac{j}{N}]}$ for $j = 1, 2, \dots$ to prove that (4.6.17) fails when $1 < p < \infty$ and $r = 1$.

4.6.5. Prove that the vector-valued inequality

$$\Big\|\Big(\sum_j |K * f_j|^q\Big)^{\frac{1}{q}}\Big\|_{L^p} \le C_{p,q}\Big\|\Big(\sum_j |f_j|^q\Big)^{\frac{1}{q}}\Big\|_{L^p}$$

may fail in general when $p > q$ even when $K \ge 0$.
[*Hint:* Take $K = \chi_{[-1,1]}$ and $f_j = \chi_{[\frac{j-1}{N},\frac{j}{N}]}$ for $1 \le j \le N$.]

4.6.6. Let $\{Q_j\}_j$ be a countable collection of cubes in $\mathbf{R}^n$ with disjoint interiors. Let c_j be the center of the cube Q_j and d_j its diameter. For $\varepsilon > 0$, define the *Marcinkiewicz function* associated with the family $\{Q_j\}_j$ as follows:

$$M_\varepsilon(x) = \sum_j \frac{d_j^{n+\varepsilon}}{|x - c_j|^{n+\varepsilon} + d_j^{n+\varepsilon}}.$$

Prove the following estimates.

$$\big\|M_\varepsilon\big\|_{L^p} \le C_{n,\varepsilon,p}\Big(\sum_j |Q_j|\Big)^{\frac{1}{p}} \qquad p > \frac{n}{n+\varepsilon},$$

$$\big\|M_\varepsilon\big\|_{L^{\frac{n}{n+\varepsilon},\infty}} \le C_{n,\varepsilon}\Big(\sum_j |Q_j|\Big)^{\frac{n+\varepsilon}{n}}.$$

Conclude that for some $C_n > 0$ we have

$$\int_{\mathbf{R}^n} M_\varepsilon(x)\,dx \le C_{n,\varepsilon}\sum_j |Q_j|.$$

[*Hint:* Verify that

$$\frac{d_j^{n+\varepsilon}}{|x - c_j|^{n+\varepsilon} + d_j^{n+\varepsilon}} \le CM(\chi_{Q_j})(x)^{\frac{n+\varepsilon}{n}}$$

and use Corollary 4.6.5.]

HISTORICAL NOTES

The L^p boundedness of the conjugate function on the circle was announced in 1924 by Riesz [**425**], but its first proof appeared three years later in [**427**]. In view of the identification of the Hilbert transform with the conjugate function, the L^p boundedness of the Hilbert transform is also attributed to M. Riesz. Riesz's proof was first given for $p = 2k$, $k \in \mathbf{Z}^+$, via an argument similar to that in the proof of Theorem 3.5.6. For $p \neq 2k$ this proof relied on interpolation and was completed with the simultaneous publication of Riesz's article on interpolation of bilinear forms [**426**]. The weak type $(1,1)$ property of the Hilbert transform is due to Kolmogorov [**308**]. Additional proofs of the boundedness of the Hilbert transform have been obtained by Stein [**489**], Loomis [**344**], and Calderón [**59**]. The proof of Theorem 4.1.7, based on identity (4.1.21), is a refinement of a proof given by Cotlar [**133**].

The norm of the conjugate function on $L^p(\mathbf{T}^1)$, and consequently that of the Hilbert transform on $L^p(\mathbf{R})$, was shown by Gohberg and Krupnik [**211**] to be $\cot(\pi/2p)$ when p is a power of 2. Duality gives that this norm is $\tan(\pi/2p)$ for $1 < p \leq 2$ whenever p' is a power of 2. Pichorides [**413**] extended this result to all $1 < p < \infty$ by refining Calderón's proof of Riesz's theorem. This result was also independently obtained by B. Cole (unpublished). The direct and simplified proof for the Hilbert transform given in Exercise 4.1.13 is in Grafakos [**212**]. The norm of the operators $\frac{1}{2}(I \pm iH)$ for real-valued functions was found to be $\frac{1}{2}\big[\min(\cos(\pi/2p), \sin(\pi/2p))\big]^{-1}$ by Verbitsky [**538**] and later independently by Essén [**175**]. The norm of the same operators for complex-valued functions was shown to be equal to $[\sin(\pi/p)]^{-1}$ by Hollenbeck and Verbitsky [**252**]. The best constant in the weak type $(1,1)$ estimate for the Hilbert transform is equal to $(1 + \frac{1}{3^2} + \frac{1}{5^2} + \dots)(1 - \frac{1}{3^2} + \frac{1}{5^2} - \dots)^{-1}$ as shown by Davis [**153**] using Brownian motion; an alternative proof was later obtained by Baernstein [**22**]. Iwaniec and Martin [**273**] showed that the norm of the Riesz transforms on $L^p(\mathbf{R}^n)$ coincide with that of the Hilbert transform on $L^p(\mathbf{R})$.

Operators of the kind T_Ω as well as the stopping time decomposition of Theorem 4.3.1 were introduced by Calderón and Zygmund [**69**]. In the same article Calderón and Zygmund used this decomposition to prove Theorem 4.3.3 for operators of the form T_Ω when Ω satisfies a certain weak smoothness condition. The more general condition (4.3.12) first appeared in Hörmander article [**253**]. Theorems 4.2.10 and 4.2.11 are also due to Calderón and Zygmund [**72**]. The latter article contains the method of rotations. Algebras of operators of the form T_Ω were studied in [**73**]. For more information on algebras of singular integrals see the article of Calderón [**64**]. Theorem 4.4.1 is due to Benedek, Calderón, and Panzone [**32**] while Example 4.4.2 is taken from Muckenhoupt [**380**]. Theorem 4.4.5 is due to Riviere [**429**]. A weaker version of this theorem, applicable for smoother singular integrals such as the maximal Hilbert transform, was obtained by Cotlar [**133**] (Exercise 4.4.6). For a general overview of singular integrals and their applications, consult the expository article of Calderón [**63**].

Part (a) of Theorem 4.5.1 is due to Marcinkiewicz and Zygmund [**367**], although the case $p = q$ was proved earlier by Paley [**399**] with a larger constant. The values of r for which a general linear operator of weak or strong type (p, q) admits bounded ℓ^r extensions are described in Rubio de Francia and Torrea [**438**]. The L^p and weak L^p spaces in Theorem 4.5.1 can be replaced by general Banach lattices, as shown by Krivine [**319**] using Grothendieck's inequality. Hilbert space-valued estimates for singular integrals were obtained by Benedek, Calderón, and Panzone [**32**]. Other operator-valued singular integral operators were studied by Rubio de Francia, Ruiz, and Torrea [**439**]. Banach space-valued singular integrals have been studied in great detail in the book of García-Cuerva and Rubio de Francia [**204**], which provides an excellent presentation of the subject. The ℓ^r-valued estimates (4.5.16) for the Hilbert transform were first obtained by Boas and Bochner [**42**]. The corresponding vector-valued estimates for the Hardy-Littlewood maximal function in Theorem 4.6.6 are due to Fefferman and Stein [**190**]. Conditions of the form (4.6.10) have been applied to several situations and can be traced in Zo [**569**].

The sharpness of the logarithmic condition (4.2.22) was indicated by Weiss and Zygmund [**548**], who constructed an example of an integrable function Ω on $\mathbf{S}^1$ that satisfies $\int_{\mathbf{S}^{n-1}} |\Omega(\theta)| \log^+ |\Omega(\theta)| \big(\log(2+\log(2+|\Omega(\theta)|))\big)^{-\delta} \, d\theta = \infty$ for all $\delta > 0$ and for which there exists a continuous function in $L^p(\mathbf{R}^2)$ for all $1 < p < \infty$ such that $\limsup_{\varepsilon \to 0} |T_\Omega^{(\varepsilon)}(f)(x)| = \infty$ for almost all $x \in \mathbf{R}^2$. The proofs of Theorems 4.2.10 and 4.2.11 can be modified to give that if Ω is in the Hardy space H^1 of $\mathbf{S}^{n-1}$, then T_Ω and $T_\Omega^{(*)}$ map L^p to L^p for $1 < p < \infty$. For T_Ω this fact was proved by Connett [**125**] and independently by Ricci and Weiss [**422**]; for $T_\Omega^{(*)}$ this was proved by Fan and Pan [**180**] and independently by Grafakos and Stefanov [**221**]. The last authors [**222**] also obtained that the logarithmic condition $\operatorname{essup}_{|\xi|=1} \int_{\mathbf{S}^{n-1}} |\Omega(\theta)| (\log \frac{1}{|\xi \cdot \theta|} |)^{1+\alpha} \, d\theta < \infty$, $\alpha > 0$ implies L^p boundedness for T_Ω and $T_\Omega^{(*)}$ for some $p \neq 2$. See also Fan, Guo, and Pan [**179**] as well as Ryabogin and Rubin [**441**] for extensions. The relatively weak condition $|\Omega| \log^+ |\Omega| \in L^1(\mathbf{S}^{n-1})$ also implies weak type $(1,1)$ boundedness for operators T_Ω. This was obtained by Seeger [**452**] and later extended by Tao [**507**] to situations when there is no Fourier transform structure. Earlier partial results are in Christ and Rubio de Francia [**103**] and in the simultaneous work of Hofmann [**249**], both inspired by the work of Christ [**96**]. Soria and Sjogren [**458**] showed that for arbitrary Ω in $L^1(\mathbf{S}^{n-1})$, T_Ω is weak type $(1,1)$ when restricted to radial functions. Examples due to Christ (published in [**221**]) indicate that even for bounded functions Ω on $\mathbf{S}^{n-1}$, T_Ω may not map the endpoint Hardy space $H^1(\mathbf{R}^n)$ into $L^1(\mathbf{R}^n)$. However, Tao and Seeger [**511**] have showed that T_Ω always maps the Hardy space $H^1(\mathbf{R}^n)$ into the Lorentz space $L^{1,2}(\mathbf{R}^n)$ when $|\Omega|(\log^+ |\Omega|)^2$ is integrable over $\mathbf{S}^{n-1}$. This result is sharp in the sense that for such Ω, T_Ω may not map $H^1(\mathbf{R}^n)$ into $L^{1,q}(\mathbf{R}^n)$ when $q < 2$ in general. If T_Ω maps $H^1(\mathbf{R}^n)$ into itself, Daly and Phillips [**145**] (in dimension $n = 2$) and Daly [**144**] (in dimensions $n \geq 3$) have showed that Ω must lie in the Hardy space $H^1(\mathbf{S}^{n-1})$. There are also results concerning the singular maximal operator $M_\Omega(f)(x) = \sup_{r>0} \frac{1}{v_n r^n} \int_{|y| \leq r} |f(x-y)| \, |\Omega(y)| \, dy$ where Ω is an integrable function on $\mathbf{S}^{n-1}$ of not necessarily vanishing integral. Such operators were studied by Fefferman [**192**], Christ [**96**], and Hudson [**259**].

CHAPTER 5

Littlewood-Paley Theory and Multipliers

In this section we will be concerned with orthogonality properties of the Fourier transform. This orthogonality is easily understood on L^2, but at this point it is not clear how it manifests itself on other spaces. Square functions introduce a way to express and quantify orthogonality of the Fourier transform on L^p and other function spaces. The introduction of square functions in this setting was pioneered by Littlewood and Paley and the theory that subsequently developed is named after them. The extent to which Littlewood-Paley theory characterizes function spaces is very remarkable. This topic will be investigated in the next chapter.

Historically, Littlewood-Paley theory first appeared in the context of one-dimensional Fourier series and depended on complex function theory. With the development of real variable methods, the whole theory became independent of complex methods and was extended to $\mathbf{R}^n$. This is the approach that we will follow in this chapter. It turns out that the Littlewood-Paley theory is intimately related to the Calderón-Zygmund theory introduced in the previous chapter. This connection is deep and far reaching, but its central feature is that we can derive the main results of one theory from the other.

The thrust and power of the Littlewood-Paley theory will become apparent in some of the applications we will discuss in this chapter. Such applications include sufficient conditions for bounded functions to be L^p multipliers. Theorems of this sort are called multiplier theorems and will be discussed in this chapter as well. As a consequence of Littlewood-Paley theory we may also prove that the lacunary partial Fourier integrals $\int_{|\xi|\le 2^N} \widehat{f}(\xi)e^{2\pi i x\cdot\xi}\,d\xi$ converge almost everywhere to an L^p function f on $\mathbf{R}^n$.

5.1. Littlewood-Paley Theory

Let us begin by examining more closely what we mean by orthogonality of the Fourier transform. If the functions f_j defined on $\mathbf{R}^n$ have Fourier transforms $\widehat{f_j}$ supported in disjoint sets, then they are *orthogonal* in the sense that

$$\|\sum_j f_j\|_{L^2}^2 = \sum_j \|f_j\|_{L^2}^2. \tag{5.1.1}$$

Unfortunately, when 2 is replaced by some $p \neq 2$ in (5.1.1), the previous quantities may not even be comparable, as we will show in Examples 5.1.8 and 5.1.9. The Littlewood-Paley theorem provides a substitute inequality to (5.1.1) expressing the fact that certain orthogonality considerations also valid in $L^p(\mathbf{R}^n)$.

5.1.a. The Littlewood-Paley Theorem

The orthogonality we are searching for is best seen in the context of one-dimensional Fourier series (which was the setting in which Littlewood and Paley formulated their result). The primary observation is that the exponential $e^{2\pi i 2^k x}$ oscillates half as much as $e^{2\pi i 2^{k+1} x}$ and is therefore nearly constant in each period of the latter. This observation was instrumental in the proof of Theorem 3.7.4, which implied in particular that for all $1 < p < \infty$ we have

$$\Big\| \sum_{j=1}^{N} a_k e^{2\pi i 2^k x} \Big\|_{L^p[0,1]} \approx \Big(\sum_{j=1}^{N} |a_k|^2 \Big)^{\frac{1}{2}} . \tag{5.1.2}$$

In other words, we can calculate the L^p norm of $\sum_{j=1}^{N} a_k e^{2\pi i 2^k x}$ in almost a precise fashion to obtain (modulo multiplicative constants) the same answer as in the L^2 case. Similar calculations are valid for more general blocks of exponentials in the dyadic range $\{2^k + 1, \dots, 2^{k+1} - 1\}$ as the exponentials in each such block behave independently from those in each previous block. In particular, the L^p integrability of a function on $\mathbf{T}^1$ is not affected by the randomization of the sign of its Fourier coefficients in the previous dyadic blocks. This is the intuition behind the Littlewood-Paley theorem.

Motivated by this discussion we introduce the Littlewood-Paley operators in the continuous setting.

Definition 5.1.1. Let Ψ be an integrable function on $\mathbf{R}^n$ and $j \in \mathbf{Z}$. We define the *Littlewood-Paley operator* Δ_j associated with Ψ by

$$\Delta_j(f) = f * \Psi_{2^{-j}},$$

where $\Psi_{2^{-j}}(x) = 2^{jn}\Psi(2^j x)$ for all x in $\mathbf{R}^n$. Thus we have

$$\widehat{\Psi_{2^{-j}}}(\xi) = \widehat{\Psi}(2^{-j}\xi)$$

for all ξ in $\mathbf{R}^n$. We note that whenever Ψ is a Schwartz function and f is a tempered distribution, the quantity $\Delta_j(f)$ is a well-defined function.

These operators depend on the choice of the function Ψ, which in most applications we will choose to be a smooth function with compactly supported Fourier transform. Observe that if $\widehat{\Psi}$ is supported in some annulus $0 < c_1 < |\xi| < c_2 < \infty$, then the Fourier transform of Δ_j is supported in the annulus $c_1 2^j < |\xi| < c_2 2^j$; in other words, it is localized near the frequency $|\xi| \approx 2^j$. Thus the purpose of Δ_j is to isolate the part of frequency of a function concentrated near $|\xi| \approx 2^j$.

The *square function* associated with the Littlewood-Paley operators Δ_j is defined as

$$f \to \Big(\sum_{j\in\mathbf{Z}} |\Delta_j(f)|^2\Big)^{\frac{1}{2}}.$$

It turns out that this quadratic expression captures crucial orthogonality information about the function f. Precisely, we have the following theorem.

Theorem 5.1.2. *(Littlewood-Paley theorem) Suppose that Ψ is an integrable C^1 function on $\mathbf{R}^n$ with mean value zero that satisfies*

$$|\Psi(x)| + |\nabla\Psi(x)| \le B(1+|x|)^{-n-1}. \tag{5.1.3}$$

Then there exists a constant $C = C_n < \infty$ such that for all $1 < p < \infty$ and all f in $L^p(\mathbf{R}^n)$ we have

$$\Big\|\Big(\sum_{j\in\mathbf{Z}} |\Delta_j(f)|^2\Big)^{\frac{1}{2}}\Big\|_{L^p(\mathbf{R}^n)} \le CB\max\big(p,(p-1)^{-1}\big)\big\|f\big\|_{L^p(\mathbf{R}^n)}. \tag{5.1.4}$$

There also exists a $C' = C'_n < \infty$ such that for all f in $L^1(\mathbf{R}^n)$ we have

$$\Big\|\Big(\sum_{j\in\mathbf{Z}} |\Delta_j(f)|^2\Big)^{\frac{1}{2}}\Big\|_{L^{1,\infty}(\mathbf{R}^n)} \le C'_n B\big\|f\big\|_{L^1(\mathbf{R}^n)}. \tag{5.1.5}$$

Conversely, suppose that Ψ is a Schwartz function that satisfies either

$$\sum_{j\in\mathbf{Z}} |\widehat{\Psi}(2^{-j}\xi)|^2 = 1, \qquad \xi\in\mathbf{R}^n\setminus\{0\}, \tag{5.1.6}$$

or

$$\sum_{j\in\mathbf{Z}} \widehat{\Psi}(2^{-j}\xi) = 1, \qquad \xi\in\mathbf{R}^n\setminus\{0\}, \tag{5.1.7}$$

and that f is a tempered distribution so that the function $\big(\sum_{j\in\mathbf{Z}}|\Delta_j(f)|^2\big)^{\frac{1}{2}}$ is in $L^p(\mathbf{R}^n)$ for some $1 < p < \infty$, then there exists a unique polynomial Q such that the distribution $f-Q$ coincides with an L^p function and we have

$$\big\|f-Q\big\|_{L^p(\mathbf{R}^n)} \le C\,B\max\big(p,(p-1)^{-1}\big)\Big\|\Big(\sum_{j\in\mathbf{Z}} |\Delta_j(f)|^2\Big)^{\frac{1}{2}}\Big\|_{L^p(\mathbf{R}^n)} \tag{5.1.8}$$

for some constant $C = C_{n,\Psi}$.

Proof. We first prove (5.1.4) when $p = 2$. Using Plancherel's theorem, we see that (5.1.4) will be a consequence of the inequality

$$\sum_{j} |\widehat{\Psi}(2^{-j}\xi)|^2 \le C_n B \tag{5.1.9}$$

for some $C_n < \infty$. Because of (5.1.3), Fourier inversion holds for Ψ. Furthermore, Ψ has mean value zero and we can write

$$\widehat{\Psi}(\xi) = \int_{\mathbf{R}^n} e^{-2\pi i x\cdot\xi}\Psi(x)\,dx = \int_{\mathbf{R}^n} (e^{-2\pi i x\cdot\xi} - 1)\Psi(x)\,dx\,, \tag{5.1.10}$$

from which we obtain the estimate

$$|\widehat{\Psi}(\xi)| \le \sqrt{4\pi|\xi|}\int_{\mathbf{R}^n} |x|^{\frac{1}{2}}|\Psi(x)|\,dx \le C_n B|\xi|^{\frac{1}{2}}\,. \tag{5.1.11}$$

For $\xi = (\xi_1, \dots, \xi_n) \neq 0$, let j be such that $|\xi_j| \ge |\xi_k|$ for all $k \in \{1, \dots, n\}$. Integrate by parts with respect to ∂_j in (5.1.10) to obtain

$$\widehat{\Psi}(\xi) = -\int_{\mathbf{R}^n} (-2\pi i\xi_j)^{-1} e^{-2\pi i x\cdot\xi}(\partial_j\Psi)(x)\,dx,$$

from which we deduce the estimate

$$|\widehat{\Psi}(\xi)| \le \sqrt{n}\,|\xi|^{-1}\int_{\mathbf{R}^n} |\nabla\Psi(x)|\,dx \le C_n B|\xi|^{-1}. \tag{5.1.12}$$

We now break up the sum into (5.1.9) in the parts where $2^{-j}|\xi| \le 1$ and $2^{-j}|\xi| \ge 1$ and use (5.1.11) and (5.1.12), respectively, to obtain (5.1.9). (See also Exercise 5.1.2.) This proves (5.1.4) when $p = 2$.

We now turn our attention to the case $p \neq 2$ in (5.1.4). We view (5.1.4) and (5.1.5) as vector-valued inequalities in the spirit of section 4.5. Define an operator $\vec{T}$ acting on functions on $\mathbf{R}^n$ as follows:

$$\vec{T}(f)(x) = \{\Delta_j(f)(x)\}_j\,.$$

The inequalities (5.1.4) and (5.1.5) we wish to prove are simply saying that $\vec{T}$ is a bounded operator from $L^p(\mathbf{R}^n, \mathbf{C})$ into $L^p(\mathbf{R}^n, \ell^2)$ and from $L^1(\mathbf{R}^n, \mathbf{C})$ into $L^{1,\infty}(\mathbf{R}^n, \ell^2)$. We just proved that this statement is true when $p = 2$ and therefore the first hypothesis of Theorem 4.6.1 is satisfied. We now observe that the operator $\vec{T}$ can be written in the form

$$\vec{T}(f)(x) = \left\{\int_{\mathbf{R}^n} \Psi_{2^{-j}}(x-y)f(y)\,dy\right\}_j = \int_{\mathbf{R}^n} \vec{K}(x-y)(f(y))\,dy,$$

where for each $x \in \mathbf{R}^n$, $\vec{K}(x)$ is a bounded linear operator from C to ℓ^2 given by

$$\vec{K}(x)(a) = \{\Psi_{2^{-j}}(x)a\}_j. \tag{5.1.13}$$

We clearly have that $\big\|\vec{K}(x)\big\|_{\mathbf{C}\to\ell^2} = \big(\sum_j |\Psi_{2^{-j}}(x)|^2\big)^{\frac{1}{2}}$ and to be able to apply Theorem 4.6.1 we will need to know that

$$\int_{|x|\ge 2|y|} \big\|\vec{K}(x-y) - \vec{K}(x)\big\|_{\mathbf{C}\to\ell^2}\,dx \le C_n B, \qquad y \neq 0. \tag{5.1.14}$$

Since Ψ is a C^1 function, for $|x| \ge 2|y|$ we have

$$\begin{aligned} & |\Psi_{2^{-j}}(x-y) - \Psi_{2^{-j}}(x)| \\ & \le 2^{(n+1)j} |\nabla\Psi(2^j(x-\theta y))|\,|y| \qquad \text{for some } \theta\in[0,1], \\ & \le B2^{(n+1)j}\big(1+2^j|x-\theta y|\big)^{-(n+1)}|y| \\ & \le B2^{(n+1)j}\big(1+2^{j-1}|x|\big)^{-(n+1)}|y| \qquad \text{since } |x-\theta y|\ge \tfrac12|x|. \end{aligned} \tag{5.1.15}$$

This estimate implies that

$$|\Psi_{2^{-j}}(x-y) - \Psi_{2^{-j}}(x)| \le B2^{(n+1)j}|y|\,. \tag{5.1.16}$$

We also have that

$$\begin{aligned} & |\Psi_{2^{-j}}(x-y) - \Psi_{2^{-j}}(x)| \\ & \le 2^{nj}|\Psi(2^j(x-y))| + 2^{jn}|\Psi(2^jx)| \\ & \le B2^{nj}\big(1+2^j|x|\big)^{-(n+1)} + B2^{jn}\big(1+2^{j-1}|x|\big)^{-(n+1)} \\ & \le 2B2^{nj}\big(1+2^{j-1}|x|\big)^{-(n+1)}. \end{aligned} \tag{5.1.17}$$

Taking the geometric mean of (5.1.15) and (5.1.17), we obtain

$$|\Psi_{2^{-j}}(x-y) - \Psi_{2^{-j}}(x)| \le 2B|y|^{\frac12}2^{(n+\frac12)j}\big(1+2^{j-1}|x|\big)^{-(n+1)}\,. \tag{5.1.18}$$

We now use estimate (5.1.16) when $2^j < \frac{2}{|x|}$ and (5.1.18) when $2^j \ge \frac{2}{|x|}$. We obtain

$$\big\|\vec{K}(x-y) - \vec{K}(x)\big\|_{\mathbf{C}\to\ell^2} = \Big(\sum_{j\in\mathbf{Z}} \big|\Psi_{2^{-j}}(x-y) - \Psi_{2^{-j}}(x)\big|^2\Big)^{\frac12}$$

$$\begin{aligned} & \le \sum_{j\in\mathbf{Z}} \big|\Psi_{2^{-j}}(x-y) - \Psi_{2^{-j}}(x)\big| \\ & \le 2B\Big(|y| \sum_{2^j<\frac{2}{|x|}} 2^{(n+1)j} + |y|^{\frac12} \sum_{2^j\ge\frac{2}{|x|}} 2^{(n+\frac12)j}2^{-(n+1)(j-1)}|x|^{-(n+1)}\Big) \\ & \le C_nB\big(|y||x|^{-n-1} + |y|^{\frac12}|x|^{-n-\frac12}\big) \end{aligned}$$

whenever $|x| \ge 2|y|$. Using this bound, we can easily deduce (5.1.14) by integrating over the region $|x| \ge 2|y|$. Finally, using Theorem 4.6.1 we conclude the proof of (5.1.4) and (5.1.5).

The proof in (5.1.8) follows by duality. Let Δ_j^* be the adjoint operator of Δ_j given by $\widehat{\Delta_j^* f} = \widehat{f}\,\overline{\widehat{\Psi_{2^{-j}}}}$. Let f in $\mathcal{S}'(\mathbf{R}^n)$. Then the series $\sum_{j\in\mathbf{Z}} \Delta_j^*\Delta_j(f)$ converges in L^p and hence in $\mathcal{S}'(\mathbf{R}^n)$. (Convergence in L^p follows by testing against $L^{p'}$ functions g, applying the Cauchy-Schwarz inequality in the sum and Hölder's inequality in the integral, using the hypothesis on f; this argument is implicitly used in the subsequent alignment.) Then the Fourier transform of the distribution $f - \sum_{j\in\mathbf{Z}} \Delta_j^*\Delta_j(f)$ is supported at the origin. This implies that there exists a polynomial Q such that $f-Q=\sum_{j\in\mathbf{Z}}\Delta_j^*\Delta_j(f)$. Now let g be a Schwartz function. We have

$$\begin{aligned}
|\langle f-Q\,,\overline{g}\,\rangle| &= |\langle \sum_{j\in\mathbf{Z}}\Delta_j^*\Delta_j(f),\overline{g}\,\rangle| \\
&= \Big|\sum_{j\in\mathbf{Z}}\langle \Delta_j(f),\overline{\Delta_j(g)}\,\rangle\Big| \\
&= \left|\int_{\mathbf{R}^n}\sum_{j\in\mathbf{Z}}\Delta_j(f)\,\overline{\Delta_j(g)}\,dx\right| \\
&\le \int_{\mathbf{R}^n}\Big(\sum_{j\in\mathbf{Z}}|\Delta_j(f)|^2\Big)^{\frac12}\Big(\sum_{j\in\mathbf{Z}}|\Delta_j(g)|^2\Big)^{\frac12}dx \\
&\le \Big\|\Big(\sum_{j\in\mathbf{Z}}|\Delta_j(f)|^2\Big)^{\frac12}\Big\|_{L^p}\Big\|\Big(\sum_{j\in\mathbf{Z}}|\Delta_j(g)|^2\Big)^{\frac12}\Big\|_{L^{p'}} \\
&\le \Big\|\Big(\sum_{j\in\mathbf{Z}}|\Delta_j(f)|^2\Big)^{\frac12}\Big\|_{L^p} C_nB\max\big(p',(p'-1)^{-1}\big)\big\|g\big\|_{L^{p'}},
\end{aligned}$$

where we used Theorem 2.2.14 (3) twice, the Cauchy-Schwarz inequality, Hölder's inequality, and (5.1.4). Taking the supremum over all g in $L^{p'}$ with norm at most one, we obtain that the tempered distribution $f-Q$ is a bounded linear functional on $L^{p'}$. By the Riesz representation theorem, $f-Q$ coincides with an L^p function whose norm satisfies the estimate

$$\big\|f-Q\big\|_{L^p}\le C_nB\max\big(p,(p-1)^{-1}\big)\Big\|\Big(\sum_{j\in\mathbf{Z}}|\Delta_j(f)|^2\Big)^{\frac12}\Big\|_{L^p}.$$

We now show uniqueness. If Q_1 is another polynomial, with $f-Q_1\in L^p$, then $Q-Q_1$ must be an L^p function; but the only polynomial that lies in L^p is the zero polynomial. This completes the proof of the converse of the theorem under hypothesis (5.1.6). To obtain the same conclusion under the hypothesis (5.1.7) we argue in a similar way but we leave the details as an exercise. (The reader may adapt the argument in the proof of Corollary 5.1.7 in this setting.)

□

Remark 5.1.3. Let us make some comments. If $\widehat{\Psi}$ is real valued, then the operators Δ_j are self-adjoint. Indeed,

$$\begin{aligned}
\int_{\mathbf{R}^n}\Delta_j(f)\,\overline{g}\,dx &= \int_{\mathbf{R}^n}\widehat{f}\,\widehat{\Psi_{2^{-j}}}\,\overline{\widehat{g}}\,d\xi \\
&= \int_{\mathbf{R}^n}\widehat{f}\,\overline{\widehat{\Psi_{2^{-j}}}\,\widehat{g}}\,d\xi = \int_{\mathbf{R}^n}f\,\overline{\Delta_j(g)}\,dx.
\end{aligned}$$

Moreover, if Ψ is a radial function, we see that the operators Δ_j are self-transpose, that is, they satisfy

$$\int_{\mathbf{R}^n} \Delta_j(f)\, g\, dx = \int_{\mathbf{R}^n} f\, \Delta_j(g)\, dx.$$

We will assume that Ψ is both radial and has a real-valued Fourier transform. Suppose that also Ψ satisfies (5.1.3) and that it has mean value zero. Then the inequality

$$\Big\|\sum_{j\in\mathbf{Z}} \Delta_j(f_j)\Big\|_{L^p} \le C_n B \max\big(p,(p-1)^{-1}\big)\Big\|\Big(\sum_{j\in\mathbf{Z}} |f_j|^2\Big)^{\frac{1}{2}}\Big\|_{L^p} \tag{5.1.19}$$

is true for all sequences of functions $\{f_j\}_j$. To see this we use duality. Let $\vec{T}(f) = \{\Delta_j(f)\}_j$. Then $\vec{T}^*(\{g_j\}_j) = \sum_j \Delta_j(g_j)$. Inequality (5.1.4) is saying that $\vec{T}$ maps $L^p(\mathbf{R}^n, \mathbf{C})$ into $L^p(\mathbf{R}^n, \ell^2)$ and its dual statement is that $\vec{T}^*$ maps $L^{p'}(\mathbf{R}^n, \ell^2)$ into $L^{p'}(\mathbf{R}^n, \mathbf{C})$. This is exactly the statement in (5.1.19) if p is replaced by p'. Since p can be any number between in $(1,\infty)$, (5.1.19) is proved.

5.1.b. A Vector-Valued Analogue

We now obtain a vector-valued extension of Theorem 5.1.2. We have the following.

Proposition 5.1.4. *Let Ψ be an integrable C^1 function on $\mathbf{R}^n$ with mean value zero that satisfies (5.1.3) and let Δ_j be the Littlewood-Paley operator associated with Ψ. Then there exists a constant $C_n < \infty$ such that for all $1 < p < \infty$ and all sequences of L^p functions f_j we have*

$$\Big\|\Big(\sum_{j\in\mathbf{Z}}\sum_{k\in\mathbf{Z}} |\Delta_k(f_j)|^2\Big)^{\frac{1}{2}}\Big\|_{L^p(\mathbf{R}^n)} \le C_n B \max(p,(p-1)^{-1})\Big\|\Big(\sum_{j\in\mathbf{Z}} |f_j|^2\Big)^{\frac{1}{2}}\Big\|_{L^p(\mathbf{R}^n)}$$

and, in particular,

$$\Big\|\Big(\sum_{j\in\mathbf{Z}} |\Delta_j(f_j)|^2\Big)^{\frac{1}{2}}\Big\|_{L^p(\mathbf{R}^n)} \le C_n B \max(p,(p-1)^{-1})\Big\|\Big(\sum_{j\in\mathbf{Z}} |f_j|^2\Big)^{\frac{1}{2}}\Big\|_{L^p(\mathbf{R}^n)}.$$

Moreover, for some $C_n' > 0$ and all sequences of L^p functions f_j we have

$$\Big\|\Big(\sum_{j\in\mathbf{Z}}\sum_{k\in\mathbf{Z}} |\Delta_k(f_j)|^2\Big)^{\frac{1}{2}}\Big\|_{L^{1,\infty}(\mathbf{R}^n)} \le C_n' B\Big\|\Big(\sum_{j\in\mathbf{Z}} |f_j|^2\Big)^{\frac{1}{2}}\Big\|_{L^1(\mathbf{R}^n)}.$$

Proof. Define $\mathcal{B}_1 = \mathbf{C}$, $\mathcal{B}_2 = \ell^2$ and $\vec{T}$ as follows:

$$\vec{T}(f) = \{\Delta_k(f)\}_k.$$

In the proof of Theorem 5.1.2 we saw that $\vec{T}$ has a kernel that satisfies Hörmander's condition. Furthermore, $\vec{T}$ maps $L^2(\mathbf{R}^n, \mathbf{C})$ into $L^2(\mathbf{R}^n, \ell^2)$. Applying Proposition 4.6.4, we obtain the required result.

□

5.1.c. L^p Estimates for Square Functions Associated with Dyadic Sums

Let us pick a Schwartz function Ψ whose Fourier transform is compactly supported in the annulus $2^{-1} \le |\xi| \le 2^2$ so that (5.1.6) is satisfied. [Clearly (5.1.6) has no chance of being satisfied if $\widehat{\Psi}$ is only supported in the annulus $1 \le |\xi| \le 2$.] The Littlewood-Paley operation $f \to \Delta_j(f)$ represents the smoothly truncated frequency localization of a function f near the dyadic annulus $|\xi| \approx 2^j$. Theorem 5.1.2 is saying that the square function formed by these localizations has L^p norm comparable to that of original function. In other words, this square function can characterize the L^p norm of a function. This is the whole point of Littlewood-Paley theory.

We may ask the question of whether Theorem 5.1.2 still holds if the Littlewood-Paley operators Δ_j are replaced by their nonsmooth versions

$$f \to \big(\chi_{2^j \le |\xi| < 2^{j+1}} \widehat{f}(\xi)\big)^{\vee}(x). \tag{5.1.20}$$

This question has a surprising answer that already signals that there may be some fundamental differences between one-dimensional and higher-dimensional Fourier analysis. The square function formed by the operators in (5.1.20) can be used to characterize $L^p(\mathbf{R})$ in the same way Δ_j did, but not $L^p(\mathbf{R}^n)$ when $n > 1$ and $p \neq 2$. The problem lies in the fact that the characteristic function of the unit disc is not an L^p multiplier on $\mathbf{R}^n$ when $n \ge 2$ unless $p = 2$; this fact will be discussed in detail in Section 10.1. The one-dimensional result we alluded to earlier is the following.

For $j \in \mathbf{Z}$ we introduce the one-dimensional operator

$$\Delta_j^{\flat}(f)(x) = (\widehat{f}\chi_{I_j})^{\vee}(x)\,, \tag{5.1.21}$$

where I_j is the set $[2^j, 2^{j+1}) \cup (-2^{j+1}, -2^j]$. $\Delta_j^{\flat}$ is a version of the operator Δ_j in which the characteristic function of the set $2^j \le |\xi| < 2^{j+1}$ is replacing the function $\widehat{\Psi}(2^{-j}\xi)$.

Theorem 5.1.5. *There exists a constant C_1 such that for all $1 < p < \infty$ and all $f \in L^p(\mathbf{R})$ we have*

$$\frac{\|f\|_{L^p}}{C_1(p+\frac{1}{p-1})^2} \le \Big\| \Big(\sum_{j\in\mathbf{Z}} |\Delta_j^{\flat}(f)|^2 \Big)^{\frac{1}{2}} \Big\|_{L^p} \le C_1(p+\tfrac{1}{p-1})^2 \|f\|_{L^p}. \tag{5.1.22}$$

Proof. Pick a Schwartz function ψ on the line whose Fourier transform is supported in the set $2^{-1} \le |\xi| \le 2^2$ and is equal to 1 on the set $1 \le |\xi| \le 2$. Let Δ_j be the Littlewood-Paley operator associated with ψ. Observe that $\Delta_j \Delta_j^{\flat} = \Delta_j^{\flat} \Delta_j = \Delta_j^{\flat}$

since $\widehat{\psi}$ is equal to one on the support of $\Delta_j^\flat(f)\widehat{\ }$. We now use Exercise 4.6.1(a) to obtain

$$\begin{aligned}\Big\|\Big(\sum_{j\in\mathbf{Z}}|\Delta_j^\flat(f)|^2\Big)^{\frac{1}{2}}\Big\|_{L^p} &= \Big\|\Big(\sum_{j\in\mathbf{Z}}|\Delta_j^\flat\Delta_j(f)|^2\Big)^{\frac{1}{2}}\Big\|_{L^p}\\ &\le C\max(p,(p-1)^{-1})\Big\|\Big(\sum_{j\in\mathbf{Z}}|\Delta_j(f)|^2\Big)^{\frac{1}{2}}\Big\|_{L^p}\\ &\le CB\max(p,(p-1)^{-1})^2\big\|f\big\|_{L^p},\end{aligned}$$

where the last inequality follows from Theorem 5.1.2. The reverse inequality for $1<p<\infty$ follows just like the reverse inequality (5.1.8) of Theorem 5.1.2 by simply replacing the Δ_j's by the $\Delta_j^\flat$'s and setting the polynomial Q equal to zero. (There is no need to use the Riesz representation theorem here, just the fact that the L^p norm of f can be realized as the supremum of expressions $\big|\langle f,g\rangle\big|$ where g has $L^{p'}$ norm at most 1.)

□

There is a higher-dimensional version of Theorem 5.1.5 with dyadic rectangles replacing the dyadic intervals. As it has already been pointed out the higher-dimensional version with dyadic annuli replacing the dyadic intervals is false.

Let us introduce some notation. For $j\in\mathbf{Z}$, we will denote by I_j the dyadic set $[2^j,2^{j+1})\cup(-2^{j+1},-2^j]$ as in the statement of Theorem 5.1.5. For $j_1,\dots j_n\in\mathbf{Z}$ let $R_{j_1,\dots,j_n}$ be the dyadic rectangle $I_{j_1}\times\cdots\times I_{j_n}$ in $\mathbf{R}^n$. In fact $R_{j_1,\dots,j_n}$ is a union of 2^n rectangles but with some abuse of language we still call it a rectangle. We will denote the rectangle $R_{j_1,\dots,j_n}$ by $R_{\boldsymbol{j}}$ where $\boldsymbol{j}=(j_1,\dots,j_n)\in\mathbf{Z}^n$. Observe that for different $\boldsymbol{j},\boldsymbol{j}'\in\mathbf{Z}^n$ the rectangles $R_{\boldsymbol{j}}$ and $R_{\boldsymbol{j}'}$ have disjoint interiors and that the union of all the $R_{\boldsymbol{j}}$'s is equal to $\mathbf{R}^n\setminus\{0\}$. In other words, the family of $R_{\boldsymbol{j}}$'s, where $\boldsymbol{j}\in\mathbf{Z}^n$ forms a *tiling* of $\mathbf{R}^n$, which we will call the *dyadic decomposition* of $\mathbf{R}^n$. We now introduce operators

$$\Delta_{\boldsymbol{j}}^\flat(f)(x)=(\widehat{f}\chi_{R_{\boldsymbol{j}}})^\vee(x)\tag{5.1.23}$$

and we have the following n-dimensional extension of Theorem 5.1.5.

Theorem 5.1.6. *For a Schwartz function ψ on the line with integral zero we define the operator*

$$\Delta_{\boldsymbol{j}}(f)(x)=\big(\widehat{\psi}(2^{-j_1}\xi_1)\dots\widehat{\psi}(2^{-j_n}\xi_n)\widehat{f}(\xi)\big)^\vee(x),\tag{5.1.24}$$

where $\boldsymbol{j}=(j_1,\dots,j_n)\in\mathbf{Z}^n$. Then there is a dimensional constant C_n such that

$$\Big\|\Big(\sum_{\boldsymbol{j}\in\mathbf{Z}^n}|\Delta_{\boldsymbol{j}}(f)|^2\Big)^{\frac{1}{2}}\Big\|_{L^p}\le C_n\big(p+\tfrac{1}{p-1}\big)^n\big\|f\big\|_{L^p}.\tag{5.1.25}$$

Let $\Delta_j^\flat$ be the operators defined in (5.1.23). Then there exists a positive constant C_n such that for all $1 < p < \infty$ and all $f \in L^p(\mathbf{R}^n)$ we have

$$\frac{\|f\|_{L^p}}{C_n(p+\frac{1}{p-1})^{2n}} \le \Big\|\Big(\sum_{j\in\mathbf{Z}^n}|\Delta_j^\flat(f)|^2\Big)^{\frac{1}{2}}\Big\|_{L^p} \le C_n(p+\tfrac{1}{p-1})^{2n}\|f\|_{L^p}. \tag{5.1.26}$$

Proof. We first prove (5.1.25). Note that if $j = (j_1,\dots,j_n) \in \mathbf{Z}^n$, then the operator Δ_j is equal to

$$\Delta_j(f) = \Delta_{j_1}^{(j_1)}\dots\Delta_{j_n}^{(j_n)},$$

where the $\Delta_{j_r}^{(j_r)}$ are one-dimensional operators given on the Fourier transform by multiplication with $\widehat{\psi}(2^{-j_r}\xi_r)$ with the remaining variables are fixed. Inequality in (5.1.25) is a consequence of the one-dimensional case. For instance, we discuss the case $n = 2$. Using Proposition 5.1.4, we obtain

$$\begin{aligned}
&\Big\|\Big(\sum_{j\in\mathbf{Z}^2}|\Delta_j(f)|^2\Big)^{\frac{1}{2}}\Big\|_{L^p(\mathbf{R}^2)}^p \\
&= \int_{\mathbf{R}}\Big[\int_{\mathbf{R}}\Big(\sum_{j_1\in\mathbf{Z}}\sum_{j_2\in\mathbf{Z}}|\Delta_{j_1}^{(1)}\Delta_{j_2}^{(2)}(f)(x_1,x_2)|^2\Big)^{\frac{p}{2}}dx_1\Big]dx_2 \\
&\le C^p\max(p,(p-1)^{-1})^p\int_{\mathbf{R}}\Big[\int_{\mathbf{R}}\Big(\sum_{j_2\in\mathbf{Z}}|\Delta_{j_2}^{(2)}(f)(x_1,x_2)|^2\Big)^{\frac{p}{2}}dx_1\Big]dx_2 \\
&= C^p\max(p,(p-1)^{-1})^p\int_{\mathbf{R}}\Big[\int_{\mathbf{R}}\Big(\sum_{j_2\in\mathbf{Z}}|\Delta_{j_2}^{(2)}(f)(x_1,x_2)|^2\Big)^{\frac{p}{2}}dx_2\Big]dx_1 \\
&\le C^{2p}\max(p,(p-1)^{-1})^{2p}\int_{\mathbf{R}}\Big[\int_{\mathbf{R}}|f(x_1,x_2)|^p\,dx_2\Big]dx_1 \\
&= C^{2p}\max(p,(p-1)^{-1})^{2p}\|f\|_{L^p(\mathbf{R}^2)}^p,
\end{aligned}$$

where we also used Theorem 5.1.5 in the calculation. We can now easily obtain higher-dimensional versions of this estimate by induction.

We now turn to the upper inequality in (5.1.26). We pick a Schwartz function ψ so that its Fourier transform is supported in the union of two intervals $[-2^{j+2},-2^{j-1}]\cup[2^{j-1},-2^{j+2}]$ and is equal to 1 on $[-2^{j+1},-2^{j}]\cup[2^{j},-2^{j+1}]$. Then we clearly have

$$\Delta_j^\flat = \Delta_j^\flat\Delta_j$$

since $\widehat{\psi}(2^{-j_1}\xi_1)\dots\widehat{\psi}(2^{-j_n}\xi_n)$ is equal to 1 on the rectangle R_j. We now use Exercise 4.6.1 (b) and estimate (5.1.25) to obtain

$$\Big\|\Big(\sum_{j\in\mathbf{Z}^n}|\Delta_j^\flat(f)|^2\Big)^{\frac{1}{2}}\Big\|_{L^p}=\Big\|\Big(\sum_{j\in\mathbf{Z}}|\Delta_j^\flat\Delta_j(f)|^2\Big)^{\frac{1}{2}}\Big\|_{L^p}$$
$$\leq C\max(p,(p-1)^{-1})^n\Big\|\Big(\sum_{j\in\mathbf{Z}}|\Delta_j(f)|^2\Big)^{\frac{1}{2}}\Big\|_{L^p}$$
$$\leq CB\max(p,(p-1)^{-1})^{2n}\big\|f\big\|_{L^p}.$$

The reverse inequality for $1<p<\infty$ is proved in the same way the corresponding inequality (5.1.8) was obtained in Theorem 5.1.2 (by replacing the Δ_j's by the $\Delta_j^\flat$'s).

□

Next we observe that if the Schwartz function ψ is suitably chosen, then estimate (5.1.25) also holds if the inequality is reversed. More precisely, suppose $\widehat{\psi}(\xi)$ is an even smooth real-valued function supported in the set $\frac{9}{10}\leq|\xi|\leq\frac{21}{10}$ in $\mathbf{R}$ that satisfies

$$\sum_{j\in\mathbf{Z}}\widehat{\psi}(2^{-j}\xi)=1,\qquad \xi\in\mathbf{R}\setminus\{0\}; \tag{5.1.27}$$

then we have the following.

Corollary 5.1.7. *Suppose that ψ satisfies (5.1.27) and let Δ_j be as in (5.1.24). Let f be a tempered distribution on $\mathbf{R}^n$ such that the function $\big(\sum_{j\in\mathbf{Z}^n}|\Delta_j(f)|^2\big)^{\frac{1}{2}}$ is in $L^p(\mathbf{R}^n)$. Then there is a unique polynomial Q such that $f-Q$ is an L^p function and we have lower estimate*

$$\frac{\big\|f-Q\big\|_{L^p}}{C_n(p+\frac{1}{p-1})^n}\leq\Big\|\Big(\sum_{j\in\mathbf{Z}^n}|\Delta_j(f)|^2\Big)^{\frac{1}{2}}\Big\|_{L^p} \tag{5.1.28}$$

for some constant C_n that depends only on the dimension and ψ.

Proof. If we had $\sum_{j\in\mathbf{Z}}|\widehat{\psi}(2^{-j}\xi)|^2=1$ instead of (5.1.27), then we could apply the method used in the lower estimate of Theorem 5.1.2 to obtain the required conclusion. In this case we provide another argument that is very similar in spirit.

Let f be a tempered distribution on $\mathbf{R}^n$. Then the series $\sum_{j\in\mathbf{Z}^n}\Delta_j(f)$ converges in L^p and hence in $\mathcal{S}'(\mathbf{R}^n)$. (Convergence in L^p follows by testing against $L^{p'}$ functions g, applying the Cauchy-Schwarz inequality in the sum and Hölder's inequality in the integral, using the hypothesis on f.) Then the Fourier transform of the distribution $f-\sum_{j\in\mathbf{Z}^n}\Delta_j(f)$ is supported at the origin. This implies that there exists a polynomial Q such that

$$f-Q=\sum_{j\in\mathbf{Z}^n}\Delta_j(f).$$

Now let g be a Schwartz function. We have

$$
\begin{aligned}
\big|\langle f-Q,\overline{g}\,\rangle\big| &= \Big|\Big\langle \sum_{\boldsymbol{j}\in\mathbf{Z}^n}\Delta_{\boldsymbol{j}}(f),\sum_{\boldsymbol{k}\in\mathbf{Z}^n}\overline{\Delta_{\boldsymbol{k}}(g)}\,\Big\rangle\Big| \\
&\le \sum_{\boldsymbol{j}\in\mathbf{Z}^n}\sum_{\boldsymbol{k}\in\mathbf{Z}^n}\big|\big\langle \Delta_{\boldsymbol{j}}(f),\overline{\Delta_{\boldsymbol{k}}(g)}\,\big\rangle\big| \\
&\le \sum_{\boldsymbol{j}\in\mathbf{Z}^n}\sum_{k_r\in\{j_r-1,j_r,j_r+1\}}\big|\big\langle \Delta_{\boldsymbol{j}}(f),\overline{\Delta_{\boldsymbol{k}}(g)}\,\big\rangle\big| \\
&\le \int_{\mathbf{R}^n}\sum_{\boldsymbol{j}\in\mathbf{Z}^n}\sum_{k_r\in\{j_r-1,j_r,j_r+1\}}|\Delta_{\boldsymbol{j}}(f)|\,|\Delta_{\boldsymbol{k}}(g)|\,dx \\
&\le 3^n\int_{\mathbf{R}^n}\Big(\sum_{\boldsymbol{j}\in\mathbf{Z}^n}|\Delta_{\boldsymbol{j}}(f)|^2\Big)^{\frac12}\Big(\sum_{\boldsymbol{k}\in\mathbf{Z}^n}|\Delta_{\boldsymbol{k}}(g)|^2\Big)^{\frac12}dx \\
&\le 3^n\Big\|\Big(\sum_{\boldsymbol{j}\in\mathbf{Z}^n}|\Delta_{\boldsymbol{j}}(f)|^2\Big)^{\frac12}\Big\|_{L^p}\Big\|\Big(\sum_{\boldsymbol{k}\in\mathbf{Z}^n}|\Delta_{\boldsymbol{k}}(g)|^2\Big)^{\frac12}\Big\|_{L^{p'}} \\
&\le C_n^{-1}\max\big(p',(p'-1)^{-1}\big)^n\|g\|_{L^{p'}}\Big\|\Big(\sum_{\boldsymbol{j}\in\mathbf{Z}^n}|\Delta_{\boldsymbol{j}}(f)|^2\Big)^{\frac12}\Big\|_{L^p},
\end{aligned}
$$

where we used the fact that $\Delta_{\boldsymbol{j}}(f)$ and $\Delta_{\boldsymbol{k}}(g)$ are orthogonal operators unless some coordinate of $\boldsymbol{k}$ is within 1 unit from the corresponding coordinate of $\boldsymbol{j}$, a fact that follows easily from the support properties of $\widehat{\psi}$. We now take the supremum over all g in $L^{p'}$ with norm at most 1. It follows that f is a bounded linear functional on $L^{p'}$. By the Riesz representation theorem we conclude that the tempered distribution f coincides with an L^p function that satisfies the claimed estimate. The uniqueness of the polynomial Q follows as in Theorem 5.1.2 and this completes the proof of the corollary.

□

5.1.d. Lack of Orthogonality on L^p

Before we end this section we discuss two examples indicating why (5.1.1) cannot hold if the exponent 2 is replaced by some other exponent $q \neq 2$. More precisely, we will show that if the functions f_j have Fourier transforms supported in disjoint sets, then the inequality

$$
\Big\|\sum_j f_j\Big\|_{L^p}^p \le C_p\sum_j\|f_j\|_{L^p}^p \tag{5.1.29}
$$

cannot hold if $p > 2$, and similarly the inequality

$$\sum_j \|f_j\|_{L^p}^p \le C_p \Big\| \sum_j f_j \Big\|_{L^p}^p \tag{5.1.30}$$

cannot hold if $p < 2$. In both (5.1.29) and (5.1.30) the constants C_p are supposed to be independent of the functions f_j.

Example 5.1.8. Pick a Schwartz function ζ whose Fourier transform is positive and supported in the interval $|\xi| \le 1/4$. Let N be a large integer and let $f_j(x) = e^{2\pi i j x}\zeta(x)$. Then $\widehat{f_j}(\xi) = \widehat{\zeta}(\xi - j)$ and the $\widehat{f_j}$'s have disjoint Fourier transforms. We obviously have

$$\sum_{j=0}^{N} \|f_j\|_{L^p}^p = (N+1)\|\zeta\|_{L^p}^p.$$

On the other hand, the estimate holds

$$\begin{aligned} \Big\| \sum_{j=0}^{N} f_j \Big\|_{L^p}^p &= \int_{\mathbf{R}} \Big| \tfrac{e^{2\pi i (N+1)x} - 1}{e^{2\pi i x} - 1} \Big|^p |\zeta(x)|^p \, dx \\ &\ge c \int_{|x| < .01(N+1)^{-1}} \tfrac{(N+1)^p |x|^p}{|x|^p} |\zeta(x)|^p \, dx = C_\zeta (N+1)^{p-1} \end{aligned}$$

since ζ does not vanish in a neighborhood of zero. We conclude that (5.1.29) cannot hold for this choice of f_j's for $p > 2$.

Example 5.1.9. We now indicate why (5.1.30) cannot hold for $p < 2$. We pick a smooth function Ψ on the line whose Fourier transform $\widehat{\Psi}$ is supported in $[\frac{7}{8}, \frac{17}{8}]$, is nonnegative, is equal to 1 on $[\frac{9}{8}, \frac{15}{8}]$, and satisfies

$$\sum_{j \in \mathbf{Z}} \widehat{\Psi}(2^{-j}\xi)^2 = 1, \qquad \xi > 0.$$

Extend Ψ to be an even function on the whole line and let Δ_j be the Littlewood-Paley operator associated with Ψ. Also pick a nonzero Schwartz function φ on the real line whose Fourier transform is nonnegative and supported in the set $[\frac{11}{8}, \frac{13}{8}]$. Fix N a large positive integer and let

$$f_j(x) = e^{2\pi i \frac{12}{8} 2^j x} \varphi(x), \tag{5.1.31}$$

for $j = 1, 2, \dots, N$. Then the function $\xi \to \widehat{\varphi}(\xi - \frac{12}{8}2^j)$ is supported in the set $[\frac{11}{8} + \frac{12}{8}2^j, \frac{13}{8} + \frac{12}{8}2^j]$ that is contained in $[\frac{9}{8}2^j + \frac{15}{8}2^j]$ for $j \ge 1$. In other words $\widehat{\Psi}(2^{-j}\xi)$ is equal to 1 on the support of $\xi \to \widehat{\varphi}(\xi - \frac{12}{8}2^j)$. This implies that $\Delta_j(f_j) = f_j$ for $j \ge 1$. This observation together with the result in Remark 5.1.3 give

$$\Big\| \sum_{j=1}^{N} f_j \Big\|_{L^p} = \Big\| \sum_{j=1}^{N} \Delta_j(f_j) \Big\|_{L^p} \le C_p \Big\| \Big(\sum_{j=1}^{N} |f_j|^2 \Big)^{\frac{1}{2}} \Big\|_{L^p} = C_p \|\varphi\|_{L^p} N^{\frac{1}{2}}$$

for $1 < p < \infty$. On the other hand, (5.1.31) easily implies that

$$\sum_{j=1}^{N} \|f_j\|_{L^p} = \|\varphi\|_{L^p} N^{\frac{1}{p}}.$$

By letting $N \to \infty$ we see that (5.1.30) cannot hold for $p < 2$ even when the f_j's have Fourier transforms supported in disjoint sets.

The same example illustrates the necessity of the exponent 2 in both (5.1.4) and (5.1.8) even under the assumption (5.1.6) on Ψ. To see this let Ψ and Δ_j be as in Example 5.1.9. Let us fix $1 < p < \infty$. We first show that the inequality

$$\Big\|\Big(\sum_{j\in\mathbf{Z}} |\Delta_j(f)|^q\Big)^{\frac{1}{q}}\Big\|_{L^p} \le C_p\|f\|_{L^p}. \tag{5.1.32}$$

cannot hold for all functions f when $q < 2$. Take f to be $\sum_{j=1}^{N} f_j$, where f_j are as in Example 5.1.8. Then the left-hand side of (5.1.32) is equal to $\|\varphi\|_{L^p}N^{1/q}$, while the right hand side is bounded by $\|\varphi\|_{L^p}N^{1/2}$. If N is large we see that (5.1.32) is impossible when $q < 2$.

Next we claim that the lower inequality

$$\|g\|_{L^p} \le C\Big\|\Big(\sum_{j\in\mathbf{Z}} |\Delta_j(g)|^q\Big)^{\frac{1}{q}}\Big\|_{L^p} \tag{5.1.33}$$

cannot hold for all functions g when $q > 2$. Let us suppose that inequality (5.1.33) did hold for some $q > 2$. Then duality and the self-adjointness of the Δ_j's would give

$$\begin{aligned}
&\Big\|\Big(\sum_{k\in\mathbf{Z}} |\Delta_k(g)|^{q'}\Big)^{\frac{1}{q'}}\Big\|_{L^{p'}} \\
&= \sup_{\|\|\{h_k\}_k\|_{\ell^q}\|_{L^p}\le 1} \Big|\int_{\mathbf{R}^n} \sum_{k\in\mathbf{Z}} \Delta_k(g)\,\overline{h_k}\,dx\Big| \\
&\le \|g\|_{L^{p'}} \sup_{\|\|\{h_k\}_k\|_{\ell^q}\|_{L^p}\le 1} \Big\|\sum_{k\in\mathbf{Z}} \overline{\Delta_k(h_k)}\Big\|_{L^p} \\
&\le C\|g\|_{L^{p'}} \sup_{\|\|\{h_k\}_k\|_{\ell^q}\|_{L^p}\le 1} \Big\|\Big(\sum_{k\in\mathbf{Z}}\Big|\Delta_k\Big(\sum_{l\in\mathbf{Z}}\Delta_l(h_l)\Big)\Big|^q\Big)^{\frac{1}{q}}\Big\|_{L^q} \qquad \text{by (5.1.33)} \\
&\le C'\|g\|_{L^{p'}} \sup_{\|\|\{h_k\}_k\|_{\ell^q}\|_{L^p}\le 1} \Big\{\sum_{l=-1}^{1}\Big\|\Big(\sum_{k\in\mathbf{Z}}|\Delta_k\Delta_l(h_l)|^q\Big)^{\frac{1}{q}}\Big\|_{L^q}\Big\} \\
&\le C''\|g\|_{L^{p'}} \sup_{\|\|\{h_k\}_k\|_{\ell^q}\|_{L^p}\le 1} \Big\|\Big(\sum_{k\in\mathbf{Z}}|h_k|^q\Big)^{\frac{1}{q}}\Big\|_{L^q} \le C''\|g\|_{L^{p'}},
\end{aligned}$$

where the previous to last inequality follows from Proposition 5.1.4 applied twice, while the one before follows from support considerations. But since $q' < 2$, this exactly proves (5.1.32) previously shown to be false, a contradiction.

Thus (5.1.32) cannot hold for $q < 2$ and (5.1.33) cannot hold for $q > 2$. We conclude that both assertions (5.1.4) and (5.1.8) of Theorem 5.1.2 can only be valid for the exponent $q = 2$. See also Exercise 5.1.6, indicating the crucial use of the fact that ℓ^2 is a Hilbert space in the converse inequality (5.1.8) of Theorem 5.1.2.

Exercises

5.1.1. Construct a Schwartz function Ψ that satisfies (5.1.6) and whose Fourier transform is supported in the annulus $\frac{8}{9} \le |x| \le \frac{9}{4}$.
[*Hint:* Set $\widehat{\Psi}(\xi) = \eta(\xi)\big(\sum_{k\in\mathbf{Z}} \eta(2^{-k}\xi)\big)^{-1}$, where η is a suitable smooth bump.]

5.1.2. Suppose that a function Ψ satisfies $\big|\widehat{\Psi}(\xi)\big| \le B \min\big(|\xi|^{\delta}, |\xi|^{-\varepsilon}\big)$ for some $\delta, \varepsilon > 0$. Show that for some dimensional constant C_n we have

$$\sum_{j\in\mathbf{Z}} \big|\widehat{\Psi}(2^{-j}\xi)\big| \le C_n B\,.$$

5.1.3. Let Ψ be an integrable function on $\mathbf{R}^n$ with mean value zero that satisfies

$$|\Psi(x)| \le B|x|^{-n-\varepsilon}, \qquad \int_{\mathbf{R}^n} |\Psi(x-y) - \Psi(x)|\, dx \le B|y|^{\varepsilon},$$

for some $B, \varepsilon > 0$ and all $y \neq 0$.
(a) Prove that $\big|\widehat{\Psi}(\xi)\big|$ is at most a multiple of $B \min\big(|\xi|^{\frac{\varepsilon}{2}}, |\xi|^{-\varepsilon}\big)$ and, using the previous Exercise, conclude that (5.1.4) holds for $p = 2$.

(b) Prove that if $\vec{K}$ is defined by (5.1.13), then (5.1.14) holds and therefore conclude that (5.1.4) and (5.1.5) are valid.
[*Hint:* Part (a): Write

$$\widehat{\Psi}(\xi) = \int_{\mathbf{R}^n} e^{-2\pi i x\cdot\xi}\,\Psi(x)\, dx = -\int_{\mathbf{R}^n} e^{-2\pi i x\cdot\xi}\,\Psi(x-y)\, dx\,,$$

where $y = \frac{1}{2}\frac{\xi}{|\xi|^2}$ when $|\xi| \ge 1$. For $|\xi| \le 1$ use the mean value property of Ψ. Part (b): Split the sum

$$\sum_{j\in\mathbf{Z}} \int_{|x|\ge 2|y|} \big|\Psi_{2^{-j}}(x-y) - \Psi_{2^{-j}}(x)\big|\, dx$$

into the parts $\sum_{2^j \le |y|^{-1}}$ and $\sum_{2^j > |y|^{-1}}$.]

5.1.4. Under the same hypotheses of Theorem 5.1.2, prove the following continuous versions of its conclusions: Show that there exist constants C_n, C'_n such that for all $1 < p < \infty$ and for all $f \in L^p(\mathbf{R}^n)$ we have

$$\Big\|\Big(\int_0^\infty |f * \Psi_t|^2 \frac{dt}{t}\Big)^{\frac{1}{2}}\Big\|_{L^p(\mathbf{R}^n)} \le C_n B \max(p,(p-1)^{-1})\big\|f\big\|_{L^p(\mathbf{R}^n)}$$

and also for all $f \in L^1(\mathbf{R}^n)$ we have

$$\Big\|\Big(\int_0^\infty |f * \Psi_t|^2 \frac{dt}{t}\Big)^{\frac{1}{2}}\Big\|_{L^{1,\infty}} \le C'_n B\big\|f\big\|_{L^1}.$$

Under the additional hypothesis that $\int_0^\infty |\widehat{\varphi}(t\xi)|^2 \frac{dt}{t} > 0$, prove the validity of the converse inequality

$$\big\|f\big\|_{L^p(\mathbf{R}^n)} \le C_n B \max(p,(p-1)^{-1})\Big\|\Big(\int_0^\infty |f * \Psi_t|^2 \frac{dt}{t}\Big)^{\frac{1}{2}}\Big\|_{L^p(\mathbf{R}^n)}$$

for all $f \in L^p(\mathbf{R}^n)$.

5.1.5. Prove the following generalization of Theorem 5.1.2. Suppose that $\{K_j\}_j$ is a sequence of tempered distributions on $\mathbf{R}^n$ that coincide with locally integrable functions away from the origin that satisfy

$$\sup_{y \in \mathbf{R}^n \setminus \{0\}} \int_{|x| \ge 2|y|} \Big(\sum_j |K_j(x-y) - K_j(x)|^2\Big)^{\frac{1}{2}} dx \le A < \infty.$$

If the Fourier transforms of K_j are functions satisfying

$$\sum_{j \in \mathbf{Z}} |\widehat{K_j}(\xi)|^2 \le B^2,$$

then the operator

$$f \to \Big(\sum_{j \in \mathbf{Z}} |K_j * f|^2\Big)^{\frac{1}{2}}$$

maps $L^p(\mathbf{R}^n)$ into itself and is of weak type $(1,1)$.

5.1.6. Suppose that $\mathcal{H}$ is a Hilbert space with inner product $\langle \cdot,\cdot \rangle_{\mathcal{H}}$ and that $T: L^2(\mathbf{R}^n) \to L^2(\mathbf{R}^n, \mathcal{H})$ is a multiple of an isometry, that is,

$$\big\|T(f)\big\|_{L^2(\mathbf{R}^n,\mathcal{H})} = A\big\|f\big\|_{L^2(\mathbf{R}^n)}$$

for all f. Then the inequality $\|T(f)\|_{L^p(\mathbf{R}^n,\mathcal{H})} \le C_p\|f\|_{L^p(\mathbf{R}^n)}$ for all $f \in L^p(\mathbf{R}^n)$ and some $p \in (1,\infty)$ implies

$$\big\|f\big\|_{L^{p'}(\mathbf{R}^n)} \le C_{p'} A^{-2}\big\|T(f)\big\|_{L^{p'}(\mathbf{R}^n,\mathcal{H})}$$

for all f in $L^{p'}(\mathbf{R}^n)$.
[*Hint:* Use the inner product structure and polarization to obtain

$$A^2\left|\int_{\mathbf{R}^n} f(x)\overline{g(x)}\,dx\right| = \left|\int_{\mathbf{R}^n} \big\langle T(f)(x), T(g)(x)\big\rangle_{\mathcal{H}}\,dx\right|$$

and then argue as in the proof of inequality (5.1.8).]

5.1.7. (a) Let Ψ be a Schwartz function whose Fourier transform is supported in a compact set that does not contain the origin, and let Δ_j be the Littlewood-Paley operator associated with Ψ. Use duality and Theorem 5.1.2 to prove that there exists a constant $C_{n,p} > 0$ such that for all $S \subseteq \mathbf{Z}$ and all f in $L^p(\mathbf{R}^n)$ we have

$$\Big\|\sum_{j\in S}\Delta_j^2(f)\Big\|_{L^p(\mathbf{R}^n)} \le C_{n,p}\Big\|\Big(\sum_{j\in S}|\Delta_j(f)|^2\Big)^{\frac{1}{2}}\Big\|_{L^p(\mathbf{R}^n)}.$$

(b) Let Ψ be as in part (a) but also assume that $\widehat{\Psi}$ is real valued and satisfies

$$\sum_{j\in\mathbf{Z}}\widehat{\Psi}(2^{-j}\xi)^2 = 1.$$

Then for each $f \in \mathcal{S}(\mathbf{R}^n)$ we have $f = \sum_j \Delta_j^2(f)$. Use the Lebesgue dominated convergence theorem and the estimate of part (a) for the set $S = \{j : |j| > N\}$ to conclude that $\sum_{|j|\le N}\Delta_j^2(f) \to f$ in L^p as $N \to \infty$ for all $f \in \mathcal{S}(\mathbf{R}^n)$. Obtain as a corollary that Schwartz functions whose Fourier transforms have compact support that does not contain the origin are dense in $L^p(\mathbf{R}^n)$.

5.1.8. Suppose that $\{m_j\}_{j\in\mathbf{Z}}$ is a sequence of L^p multipliers on the line (for some $1 < p < \infty$) that are supported in the intervals $[2^j, 2^{j+1}]$. Let $T_j(f) = (\widehat{f}m_j)^\vee$ be the corresponding operators. Assume that the vector-valued inequality

$$\Big\|\Big(\sum_j|T_j(f_j)|^2\Big)^{\frac{1}{2}}\Big\|_{L^p} \le A_p\Big\|\Big(\sum_j|f_j|^2\Big)^{\frac{1}{2}}\Big\|_{L^p}$$

is valid. Prove that for some constant $C_p > 0$ we have

$$\Big\|\sum_{j\in\mathbf{Z}}m_j\Big\|_{\mathcal{M}_p} \le C_p\,A_p.$$

[*Hint:* Use that $\big\langle\sum_j T_j(f), g\big\rangle = \sum_j\langle T_j\Delta_j^\flat(f), \Delta_j^\flat(g)\rangle$ and then apply the Cauchy-Schwarz inequality.]

5.1.9. Fix a nonzero Schwartz function h on the line whose Fourier transform is supported in the interval $[-\frac{1}{8},\frac{1}{8}]$. For $\{a_j\}$ a sequence of numbers, set

$$f(x)=\sum_{j=1}^{\infty} a_j e^{2\pi i 2^j x} h(x)\,.$$

Prove that for all $1<p<\infty$ there exists a constant C_p such that

$$\big\|f\big\|_{L^p(\mathbf{R})}\le C_p\Big(\sum_j |a_j|^2\Big)^{\frac{1}{2}}\big\|h\big\|_{L^p}\,.$$

[*Hint:* Write $f=\sum_{j=1}^{\infty}\Delta_j(a_j e^{2\pi i 2^j(\cdot)}h)$, where Δ_j is given by convolution with $\varphi_{2^{-j}}$ for some φ whose Fourier transform is supported in the interval $[-\frac{6}{8},\frac{10}{8}]$ and is equal to 1 on $[-\frac{7}{8},\frac{9}{8}]$. Then use (5.1.19).]

5.1.10. Let Ψ be a Schwartz function whose Fourier transform is supported in the annulus $\frac{1}{2}\le|\xi|\le 2$ and which satisfies (5.1.7). Define a Schwartz function Φ by setting

$$\widehat{\Phi}(\xi)=\begin{cases}\sum_{j\le 0}\widehat{\Psi}(2^{-j}\xi) & \text{when } \xi\neq 0,\\ 1 & \text{when } \xi=0.\end{cases}$$

Let S_0 be the operator given by convolution with Φ.
(a) Prove that for all $f\in\mathcal{S}'(\mathbf{R}^n)$ we have

$$S_0(f)+\sum_{j=1}^{N}\Delta_j(f)\to f$$

in $\mathcal{S}'(\mathbf{R}^n)$.
(a) Prove that for all $f\in\mathcal{S}'(\mathbf{R}^n)/\mathcal{P}$ we have

$$\sum_{j=-N}^{N}\Delta_j(f)\to f$$

in $\mathcal{S}'(\mathbf{R}^n)/\mathcal{P}$.
[*Hint:* Use Exercise 2.3.12.]

5.1.11. Let Δ_j and S_0 be as in the previous Exercise. Then for $1<p<\infty$ we have

$$\big\|f\big\|_{L^p}\approx\big\|S_0(f)\big\|_{L^p}+\Big\|\Big(\sum_{j=1}^{\infty}|\Delta_j(f)|^2\Big)^{\frac{1}{2}}\Big\|_{L^p},$$

where one of the two inequalities should be interpreted as follows: If for a tempered distribution f, the right-hand side is finite, then f can be identified with an L^p function whose norm satisfies an appropriate estimate.
[*Hint:* Use Theorem 5.1.2 (do not reprove it), together with the identity $S_0+\sum_{j=1}^{\infty}\Delta_j=I$, which holds in $\mathcal{S}'(\mathbf{R}^n)$ by the previous Exercise.]

5.2. Two Multiplier Theorems

We now return to the spaces $\mathcal{M}_p$ introduced in Section 2.5. We seek sufficient conditions on L^∞ functions defined on $\mathbf{R}^n$ to be elements of $\mathcal{M}_p$. In this section we will be concerned with two fundamental theorems that provide such sufficient conditions. These results are called the Marcinkiewicz and the Hörmander-Mihlin multiplier theorems. Both multiplier theorems will be a consequence of the Littlewood-Paley theory introduced in the previous section.

Using the dyadic decomposition of $\mathbf{R}^n$, we can write any L^∞ function m as the sum

$$m = \sum_{j\in\mathbf{Z}^n} m\chi_{R_j},$$

where $j = (j_1, \dots, j_n)$ and R_j is the dyadic rectangle $I_{j_1} \times \cdots \times I_{j_n}$ and $I_k = [2^k, 2^{k+1}] \cup [-2^{k+1}, -2^k]$. For $j \in \mathbf{Z}^n$ let us set $m_j = m\chi_{R_j}$. A consequence of the ideas developed so far is the following characterization of $\mathcal{M}_p(\mathbf{R}^n)$ in terms of a vector-valued inequality.

Proposition 5.2.1. *Let $m \in L^\infty(\mathbf{R}^n)$ and let $m_j = m\chi_{R_j}$. Then m lies in $\mathcal{M}_p(\mathbf{R}^n)$, that is, for some c_p we have*

$$\|(\widehat{f}m)^\vee\|_{L^p} \le c_p\|f\|_{L^p}, \qquad f \in L^p(\mathbf{R}^n),$$

if and only if for some $C_p > 0$ we have

$$\Big\|\Big(\sum_{j\in\mathbf{Z}^n} |(\widehat{f_j}m_j)^\vee|^2\Big)^{\frac{1}{2}}\Big\|_{L^p} \le C_p\Big\|\Big(\sum_{j\in\mathbf{Z}^n} |f_j|^2\Big)^{\frac{1}{2}}\Big\|_{L^p} \tag{5.2.1}$$

for all sequences of functions f_j in $L^p(\mathbf{R}^n)$.

Proof. Suppose that $m \in \mathcal{M}_p(\mathbf{R}^n)$. Then Exercise 4.6.1 gives the first inequality:

$$\begin{aligned}\Big\|\Big(\sum_{j\in\mathbf{Z}^n} |(\chi_{R_j} m\widehat{f_j})^\vee|^2\Big)^{\frac{1}{2}}\Big\|_{L^p} &\le C_p\Big\|\Big(\sum_{j\in\mathbf{Z}^n} |(m\widehat{f_j})^\vee|^2\Big)^{\frac{1}{2}}\Big\|_{L^p}\\ &\le C_p\Big\|\Big(\sum_{j\in\mathbf{Z}^n} |f_j|^2\Big)^{\frac{1}{2}}\Big\|_{L^p},\end{aligned}$$

while the second inequality follows from Theorem 4.5.1. (Observe that when $p = q$ in Theorem 4.5.1, then $C_{p,q} = 1$.) Conversely, suppose that (5.2.1) holds for all sequences of functions f_j. Fix a function f and apply (5.2.1) to the sequence $(\widehat{f}\chi_{R_j})^\vee$, where R_j is the dyadic rectangle indexed by $j = (j_1, \dots, j_n) \in \mathbf{Z}^n$. We obtain

$$\Big\|\Big(\sum_{j\in\mathbf{Z}^n} |(\widehat{f}\, m\chi_{R_j})^\vee|^2\Big)^{\frac{1}{2}}\Big\|_{L^p} \le C_p\Big\|\Big(\sum_{j\in\mathbf{Z}^n} |(\widehat{f}\chi_{R_j})^\vee|^2\Big)^{\frac{1}{2}}\Big\|_{L^p}.$$

Using Theorem 5.1.6, we obtain that the previous inequality is equivalent to the inequality

$$\big\|(\widehat{f}\,m)^{\vee}\big\|_{L^p} \le c_p \big\|f\big\|_{L^p},$$

which implies that $m \in \mathcal{M}_p(\mathbf{R}^n)$.

□

5.2.a. The Marcinkiewicz Multiplier Theorem on R

Proposition 5.2.1 suggests that the behavior of m on each dyadic rectangle R_j should play a crucial role in determining whether m is an L^p multiplier. The Marcinkiewicz multiplier theorem provides such sufficient conditions on m restricted to any dyadic rectangle R_j. Before we state this theorem, let us illustrate the main idea behind it via the following example. Suppose that m is a bounded function that vanishes near $-\infty$, which is differentiable at every point, and whose derivative is integrable. Then we can write

$$m(\xi) = \int_{-\infty}^{\xi} m'(t)\,dt = \int_{-\infty}^{+\infty} \chi_{[t,\infty)}(\xi) m'(t)\,dt$$

from which it follows that for a Schwartz function f we have

$$(\widehat{f}m)^{\vee} = \int_{\mathbf{R}} (\widehat{f}\chi_{[t,\infty)})^{\vee} m'(t)\,dt.$$

Since the operators $f \to (\widehat{f}\chi_{[t,\infty)})^{\vee}$ map $L^p(\mathbf{R})$ into itself independently of t, it follows that

$$\big\|(\widehat{f}m)^{\vee}\big\|_{L^p} \le C_p \big\|m'\big\|_{L^1} \big\|f\big\|_{L^p},$$

thus obtaining that m is in $\mathcal{M}_p(\mathbf{R})$. The next multiplier theorem is an improvement of this result and is based on the Littlewood-Paley theorem. We begin with the one-dimensional case, which already captures most of the ideas of the general situation.

Theorem 5.2.2. *(Marcinkiewicz multiplier theorem) Let $m : \mathbf{R} \to \mathbf{R}$ be a bounded function which is C^1 in every dyadic set $(2^j, 2^{j+1}) \cup (-2^{j+1}, -2^j)$ for $j \in \mathbf{Z}$. Assume that the derivative m' of m satisfies*

$$\sup_j \left[\int_{-2^{j+1}}^{-2^j} |m'(\xi)|\,d\xi + \int_{2^j}^{2^{j+1}} |m'(\xi)|\,d\xi \right] \le A < \infty. \tag{5.2.2}$$

Then for all $1 < p < \infty$ we have that $m \in \mathcal{M}_p(\mathbf{R})$ and for some $C > 0$ we have

$$\big\|m\big\|_{\mathcal{M}_p(\mathbf{R})} \le C \max\left(p, (p-1)^{-1}\right)^6 \left(\big\|m\big\|_{L^\infty} + A\right). \tag{5.2.3}$$

Proof. Since the function m has an integrable derivative on $(2^j, 2^{j+1})$, it has bounded variation in this interval and hence it is a difference of two increasing functions. Therefore, m has left and right limits at the points 2^j and 2^{j+1} and by redefining m at these points we may assume that it is right continuous at 2^j.

Recall the notation $I_j = [2^j, 2^{j+1}) \cup (-2^{j+1}, -2^j]$, although we will need to work with the intervals $I_j^+ = [2^j, 2^{j+1})$ whenever $j \in \mathbf{Z}$. Given an interval I in $\mathbf{R}$, we introduce an operator Δ_I defined by $\Delta_I(f) = (\widehat{f}\chi_I)^\vee$. With this notation $\Delta_{I_j^+}(f)$ is "half" of the operator $\Delta_j^\flat$ introduced in the previous section. Given a Schwartz function f, we also set $f_+ = (\widehat{f}\chi_{(0,\infty)})^\vee$ and $f_- = (\widehat{f}\chi_{(-\infty,0)})^\vee$. Given m as in the statement of the theorem, let us write $m(\xi) = m_+(\xi) + m_-(\xi)$, where $m_+(\xi) = m(\xi)\chi_{\xi>0}$ and $m_-(\xi) = m(\xi)\chi_{\xi\le 0}$. We will show that both m_+ and m_- are L^p multipliers. Since m' is integrable over all intervals of the form $[2^j, \xi]$ when $2^j \le \xi < 2^{j+1}$, the fundamental theorem of calculus gives

$$m(\xi) = m(2^j) + \int_{2^j}^{\xi} m'(t)\,dt, \qquad \text{for } 2^j \le \xi < 2^{j+1},$$

from which it follows that for a Schwartz function f on the real line we have

$$m(\xi)\widehat{f}(\xi)\chi_{I_j^+}(\xi) = m(2^j)\widehat{f}(\xi)\chi_{I_j^+}(\xi) + \int_{2^j}^{2^{j+1}} \widehat{f}(\xi)\chi_{[t,\infty)}(\xi)\chi_{I_j^+}(\xi)\, m'(t)\,dt\,.$$

We therefore obtain the identity

$$(\widehat{f}\chi_{I_j}m_+)^\vee = (\widehat{f}m\chi_{I_j^+})^\vee = m(2^j)\Delta_{I_j^+}(f) + \int_{2^j}^{2^{j+1}} \Delta_{[t,\infty)}\Delta_{I_j^+}(f)\, m'(t)\,dt\,,$$

from which it follows that

$$|(\widehat{f}\chi_{I_j}m_+)^\vee| \le \|m\|_{L^\infty}|\Delta_{I_j^+}(f)| + A^{\frac12}\bigg(\int_{2^j}^{2^{j+1}} \big|\Delta_{[t,\infty)}\Delta_{I_j}(f_+)\big|^2\,|m'(t)|\,dt\bigg)^{\frac12},$$

using the hypothesis (5.2.2). Taking $\ell^2(\mathbf{Z})$ norms we obtain

$$\begin{aligned}\Big(\sum_{j\in\mathbf{Z}}|(\widehat{f}\chi_{I_j}m_+)^\vee|^2\Big)^{\frac12} &\le \|m\|_{L^\infty}\Big(\sum_{j\in\mathbf{Z}}|\Delta_{I_j}(f_+)|^2\Big)^{\frac12} \\ &\quad + A^{\frac12}\bigg(\int_0^\infty \big|\Delta_{[t,\infty)}\Delta^\flat_{[\log_2 t]}(f_+)\big|^2\,|m'(t)|\,dt\bigg)^{\frac12}.\end{aligned}$$

Exercise 4.6.2 now gives that

$$\begin{aligned}&A^{\frac12}\bigg\|\bigg(\int_0^\infty \big|\Delta_{[t,\infty)}\Delta^\flat_{[\log_2 t]}(f_+)\big|^2|m'(t)|\,dt\bigg)^{\frac12}\bigg\|_{L^p} \\ &\qquad\le C\max(p,(p-1)^{-1})A^{\frac12}\bigg\|\bigg(\int_0^\infty \big|\Delta^\flat_{[\log_2 t]}(f_+)\big|^2|m'(t)|\,dt\bigg)^{\frac12}\bigg\|_{L^p},\end{aligned}$$

while the last L^p norm is

$$\bigg\|\Big(\sum_{j\in\mathbf{Z}}|\Delta_{I_j}(f_+)|^2\int_{I_j^+}|m'(t)|\,dt\Big)^{\frac12}\bigg\|_{L^p} \le A^{\frac12}\bigg\|\Big(\sum_j|\Delta_{I_j}(f_+)|^2\Big)^{\frac12}\bigg\|_{L^p}.$$

Using Theorem 5.1.5 we obtain that the L^p norm of $(\sum_j |\Delta_{I_j}(f_+)|^2)^{\frac{1}{2}}$ is controlled by $C\max(p,(p-1)^{-1})^2\|f_+\|_{L^p} \le C'\max(p,(p-1)^{-1})^3\|f\|_{L^p}$ by the boundedness of the Hilbert transform on L^p. Putting things together we deduce that

$$\Big\|\Big(\sum_j |(\widehat{f}\chi_{I_j}m_+)^\vee|^2\Big)^{\frac{1}{2}}\Big\|_{L^p} \tag{5.2.4}$$
$$\le C'\max(p,(p-1)^{-1})^4(A+\|m\|_{L^\infty})\|f\|_{L^p},$$

from which we obtain the estimate

$$\|(\widehat{f}m_+)^\vee\|_{L^p} \le C\max(p,(p-1)^{-1})^6(A+\|m\|_{L^\infty})\|f\|_{L^p},$$

using the lower estimate of Theorem 5.1.5. This proves (5.2.3) for m_+. A similar argument also works for m_-, and this concludes the proof by summing the corresponding estimates for m_+ and m_-.

□

We remark that the same proof applies under the more general assumption that m is a function of bounded variation on every interval $[2^j,2^{j+1}]$ and $[-2^{j+1},-2^j]$. In this case the measure $|m'(t)|\,dt$ should be replaced by the total variation $|dm(t)|$ of the Lebesgue-Stieljes measure $dm(t)$.

Example 5.2.3. Any bounded function that is constant on dyadic intervals is an L^p multiplier. Also, the function

$$m(\xi) = |\xi|2^{-[\log_2|\xi|]}$$

is an L^p multiplier on $\mathbf{R}$ for $1<p<\infty$.

5.2.b. The Marcinkiewicz Multiplier Theorem on $\mathbf{R}^n$

We will now extend this theorem on $\mathbf{R}^n$. As usually, we will denote the coordinates of a point $\xi \in \mathbf{R}^n$ by $(\xi_1,\dots,\xi_n)$. We will also use the notation $I_j = (-2^{j+1},-2^j)\cup(2^j,2^{j+1})$ and $R_{\boldsymbol{j}} = I_{j_1}\times\cdots\times I_{j_n}$ whenever $\boldsymbol{j}=(j_1,\dots,j_n)\in\mathbf{Z}^n$.

Theorem 5.2.4. *Let m be a bounded function on $\mathbf{R}^n$ that is C^n in all regions $R_{\boldsymbol{j}}$. Assume that there is a positive constant A such that for all $k\in\{1,\dots,n\}$, all $j_1,\dots,j_k\in\{1,2,\dots,n\}$, all $l_{j_1},\dots,l_{j_k}\in\mathbf{Z}$, and all $\xi_s\in J_{l_s}$ for $s\notin\{j_1,\dots,j_k\}$, $1\le s\le n$, we have*

$$\int_{I_{l_{j_1}}}\cdots\int_{I_{l_{j_k}}} \big|(\partial_{j_1}\dots\partial_{j_k}m)(\xi_1,\dots,\xi_n)\big|\,d\xi_{j_1}\dots d\xi_{j_k} \le A. \tag{5.2.5}$$

Then m is in $\mathcal{M}_p(\mathbf{R}^n)$ whenever $1<p<\infty$ and there is a constant $C_n<\infty$ such that

$$\|m\|_{\mathcal{M}_p(\mathbf{R}^n)} \le C_n(A+\|m\|_{L^\infty})\max\big(p,(p-1)\big)^{6n}. \tag{5.2.6}$$

Proof. We only prove this theorem in dimension $n = 2$, since the general case presents absolutely no differences but only some notational inconvenience. We first write the given function m as

$$m(\xi) = m_{++}(\xi) + m_{-+}(\xi) + m_{+-}(\xi) + m_{--}(\xi),$$

where each of the last four terms is supported in one of the four quadrants. For instance, the function $m_{+-}(\xi_1, \xi_2)$ is supported in the quadrant where $\xi_1 > 0$ and $\xi_2 < 0$. As in the one-dimensional case, we will work with each of these pieces separately. By symmetry we choose to work with m_{++} in the following argument. We also set

$$f_{++} = \big(\widehat{f}(\xi_1, \xi_2)\chi_{(0,\infty)}(\xi_1)\chi_{(0,\infty)}(\xi_2)\big)^{\vee}$$

for a Schwartz function f.

We will redefine m on a set of measure zero to make it continuous and C^{n-1} on any dyadic rectangle of the form $[2^{j_1}, 2^{j_1+1}) \times [2^{j_2}, 2^{j_2+1})$. To achieve this we fix a point (ξ_1^0, ξ_2^0) in $(2^{j_1}, 2^{j_1+1}) \times (2^{j_2}, 2^{j_2+1})$ and we use the identity

$$m(b_1, b_2) = m(\xi_1^0, \xi_2^0) - \sum_{k=1}^{2} \int_{b_k}^{\xi_k^0} (\partial_k m)(t)\, dt, \tag{5.2.7}$$

which is valid for any other point (b_1, b_2) in $(2^{j_1}, 2^{j_1+1}) \times (2^{j_2}, 2^{j_2+1})$ by the fundamental theorem of calculus. By condition (5.2.5), the right-hand side in (5.2.7) has a limit as (b_1, b_2) tends to a point in the set

$$\overline{R_j} \cap (\mathbf{R}^+)^2 \cap \{(t_1, t_2) : \ t_1 = 2^{j_1} \text{ or } t_2 = 2^{j_2}\}. \tag{5.2.8}$$

We take this limit to be the value of m on the left and bottom part of the boundary of $R_j \cap (\mathbf{R}^+)^2$. Similarly, we can define $\partial^\beta m(b_1, b_2)$ for (b_1, b_2) in the set in (5.2.8) for all multiindices β with $|\beta| \le n-1$. Note that by assumption we have $m \in C^n(R_j)$.

Recall that we are working with $n = 2$, although we may write n in general. Using the fundamental theorem of calculus, we obtain the following simple identity valid for $2^{j_1} \le \xi_1 < 2^{j_1+1}$ and $2^{j_2} \le \xi_2 < 2^{j_2+2}$:

$$\begin{aligned} m(\xi_1, \xi_2) &= m(2^{j_1}, 2^{j_2}) \\ &\quad + \int_{2^{j_1}}^{\xi_1} (\partial_1 m)(t_1, 2^{j_2})\, dt_1 \\ &\quad + \int_{2^{j_2}}^{\xi_2} (\partial_2 m)(2^{j_1}, t_2)\, dt_2 \\ &\quad + \int_{2^{j_1}}^{\xi_1} \int_{2^{j_2}}^{\xi_2} (\partial_1 \partial_2 m)(t_1, t_2)\, dt_2\, dt_1 . \end{aligned}$$

We introduce operators $\Delta_I^{(r)}$, $r \in \{1, 2\}$ acting in the rth variable (with the other variable remaining fixed) given by multiplication on the Fourier transform by the characteristic function of the interval I. Likewise, we introduce operators $\Delta_j^{\flat(r)}$, $r \in \{1,2\}$ (also acting in the rth variable) given by multiplication on the Fourier transform by the characteristic function of the set $[-2^{j+1},-2^j] \cup [2^j, 2^{j+1}]$.

Multiplying both sides of the previously displayed identity by the function

$$\widehat{f}\chi_{R_j}\chi_{(0,\infty)^2}$$

and taking inverse Fourier transforms yields

$$\begin{aligned}
(\widehat{f}\chi_{R_j} m_{++})^\vee &= \\
& m(2^{j_1},2^{j_2})\Delta_{j_1}^{\flat(1)}\Delta_{j_2}^{\flat(2)}(f_{++}) \\
&+ \int_{2^{j_1}}^{2^{j_1+1}} \Delta_{j_2}^{\flat(2)}\Delta_{[t_1,\infty)}^{(1)}\Delta_{j_1}^{\flat(1)}(f_{++})\,(\partial_1 m)(t_1,2^{j_2})\,dt_1 \\
&+ \int_{2^{j_2}}^{2^{j_2+1}} \Delta_{j_1}^{\flat(1)}\Delta_{[t_2,\infty)}^{(2)}\Delta_{j_2}^{\flat(2)}(f_{++})\,(\partial_2 m)(t_2,2^{j_1})\,dt_2 \\
&+ \int_{2^{j_1}}^{2^{j_1+1}}\int_{2^{j_2}}^{2^{j_2+1}} \Delta_{[t_1,\infty)}^{(1)}\Delta_{j_1}^{\flat(1)}\Delta_{[t_2,\infty)}^{(2)}\Delta_{j_2}^{\flat(2)}(f_{++})\,(\partial_1\partial_2 m)(t_1,t_2)\,dt_2\,dt_1\,.
\end{aligned}$$

We now take absolute values in the previous identity, we apply the Cauchy-Schwarz inequality in the last three terms, we use hypothesis (5.2.5), we sum over $j \in \mathbf{Z}^2$, then we take square roots and use the subadditivity of square roots, to deduce the pointwise estimate

$$\begin{aligned}
\Big(\sum_{j\in\mathbf{Z}^2} |(\widehat{f}\chi_{R_j} m_{++})^\vee|^2\Big)^{\frac12} &\le \\
& \|m\|_{L^\infty}\Big(\sum_{j\in\mathbf{Z}^2} |\Delta_j^\flat(f_{++})|^2\Big)^{\frac12} \\
&+ A^{\frac12}\Big(\int_0^\infty\int_0^\infty |\Delta_{[t_1,\infty)}^{(1)}\Delta_{[\log_2 t_2]}^{\flat(2)}\Delta_{[\log_2 t_1]}^{\flat(1)}(f_{++})|^2\,|(\partial_1 m)(t_1,2^{[\log_2 t_2]})|\,dt_1 dt_2\Big)^{\frac12} \\
&+ A^{\frac12}\Big(\int_0^\infty\int_0^\infty |\Delta_{[t_2,\infty)}^{(2)}\Delta_{[\log_2 t_1]}^{\flat(1)}\Delta_{[\log_2 t_2]}^{\flat(2)}(f_{++})|^2\,|(\partial_2 m)(2^{[\log_2 t_1]},t_2)|\,dt_1 dt_2\Big)^{\frac12} \\
&+ A^{\frac12}\Big(\int_0^\infty\int_0^\infty |\Delta_{[t_1,\infty)}^{(1)}\Delta_{[t_2,\infty)}^{(2)}\Delta_{[\log_2 t_1]}^{\flat(1)}\Delta_{[\log_2 t_2]}^{\flat(2)}(f_{++})|^2|(\partial_1\partial_2 m)(t_1,t_2)|\,dt_1 dt_2\Big)^{\frac12}.
\end{aligned}$$

We now take L^p norms and we estimate separately the contribution of each of the four terms on the right side. Using Exercise 4.6.2 we obtain

$$\begin{aligned}
&\Big\|\Big(\sum_{j\in\mathbf{Z}^n}|(\widehat{f}\chi_{R_j}m_{++})^\vee|^2\Big)^{\frac12}\Big\|_{L^p} \le \|m\|_{L^\infty}\Big\|\Big(\sum_{j\in\mathbf{Z}^n}|\Delta_j^\flat(f_{++})|^2\Big)^{\frac12}\Big\|_{L^p}\\
&+ C_nA^{\frac12}\max\big(p,(p-1)^{-1}\big)^n\\
&\Big\{\Big\|\Big(\int_0^\infty\int_0^\infty|\Delta^{\flat(2)}_{[\log_2 t_2]}\Delta^{\flat(1)}_{[\log_2 t_1]}(f_{++})|^2\,|(\partial_1 m)(t_1,2^{[\log_2 t_2]})|\,dt_1\,dt_2\Big)^{\frac12}\Big\|_{L^p}\\
&+\Big\|\Big(\int_0^\infty\int_0^\infty|\Delta^{\flat(1)}_{[\log_2 t_1]}\Delta^{\flat(2)}_{[\log_2 t_2]}(f_{++})|^2\,|(\partial_2 m)(2^{[\log_2 t_1]},t_2)|\,dt_1\,dt_2\Big)^{\frac12}\Big\|_{L^p}\\
&+\Big\|\Big(\int_0^\infty\int_0^\infty|\Delta^{\flat(1)}_{[\log_2 t_1]}\Delta^{\flat(2)}_{[\log_2 t_2]}(f_{++})|^2|(\partial_1\partial_2 m)(t_1,t_2)|\,dt_1dt_2\Big)^{\frac12}\Big\|_{L^p}\Big\},
\end{aligned}$$

where $n=2$. But the functions $(t_1,t_2)\to\Delta^{\flat(1)}_{[\log_2 t_1]}\Delta^{\flat(2)}_{[\log_2 t_2]}(f_{++})$ are constant on products of intervals of the form $[2^{j_1},2^{j_1+1})\times[2^{j_2},2^{j_2+1})$, hence using hypothesis (5.2.5) we deduce the estimate (recall $n=2$)

$$\begin{aligned}
&\Big\|\Big(\sum_{j\in\mathbf{Z}^n}|(\widehat{f}\chi_{R_j}m_{++})^\vee|^2\Big)^{\frac12}\Big\|_{L^p}\\
&\le C_n\big(\|m\|_{L^\infty}+A\big)\max\big(p,(p-1)^{-1}\big)^n\Big\|\Big(\sum_{j\in\mathbf{Z}^n}|\Delta_j^\flat(f_{++})|^2\Big)^{\frac12}\Big\|_{L^p}\\
&\le C_n\big(\|m\|_{L^\infty}+A\big)\max\big(p,(p-1)^{-1}\big)^{3n}\|(\widehat{f}\chi_{(0,\infty)^n})^\vee\|_{L^p}\\
&\le C_n\big(\|m\|_{L^\infty}+A\big)\max\big(p,(p-1)^{-1}\big)^{4n}\|f\|_{L^p},
\end{aligned}$$

where the penultimate estimate follows from Theorem 5.1.6 and the last estimate by the boundedness of the Hilbert transform. We now appeal to the lower estimate of Theorem 5.1.6, which yields the required estimate for m_{++}. A similar argument also works for the remaining parts of m and this concludes the proof of estimate (5.2.6).

□

We now give a condition that implies (5.2.5) and is well suited for a variety of applications.

Corollary 5.2.5. *Let m be a bounded function defined away from the coordinate axes on $\mathbf{R}^n$ that is C^n in that region. Assume furthermore that for all $k\in\{1,\dots,n\}$, all $j_1,\dots,j_k\in\{1,2,\dots,n\}$, and all $\xi_r\in\mathbf{R}$ for $r\notin\{j_1,\dots,j_k\}$ we have*

$$|(\partial_{j_1}\dots\partial_{j_k}m)(\xi_1,\dots,\xi_n)|\le A\,|\xi_{j_1}|^{-1}\dots|\xi_{j_k}|^{-1}. \tag{5.2.9}$$

Then m satisfies (5.2.6).

Proof. Simply observe that condition (5.2.9) implies (5.2.5). □

Example 5.2.6. The following are examples of functions that satisfy the hypotheses of Corollary 5.2.5.

$$m_1(\xi) = \frac{\xi_1}{\xi_1 + i(\xi_2^2 + \cdots + \xi_n^2)},$$

$$m_2(\xi) = \frac{|\xi_1|^{\alpha_1} \dots |\xi_n|^{\alpha_n}}{(\xi_1^2 + \xi_2^2 + \cdots + \xi_n^2)^{\alpha/2}},$$

where $\alpha_1 + \alpha_2 + \cdots + \alpha_n = \alpha$, $\alpha_j > 0$,

$$m_3(\xi) = \frac{\xi_2 \xi_3^2}{i\xi_1 + \xi_2^2 + \xi_3^4}.$$

The functions m_1 and m_2 are defined on $\mathbf{R}^n$ and m_3 on $\mathbf{R}^3$.

5.2.c. The Hörmander-Mihlin Multiplier Theorem on $\mathbf{R}^n$

We now discuss our second multiplier theorem.

Theorem 5.2.7. *Let $m(\xi)$ be a complex-valued bounded function on $\mathbf{R}^n \setminus \{0\}$ that satisfies*
(a) either Mihlin's condition

$$|\partial_\xi^\alpha m(\xi)| \le A|\xi|^{-|\alpha|} \tag{5.2.10}$$

for all multiindices $|\alpha| \le [\frac{n}{2}] + 1$,
(b) or Hörmander's condition

$$\sup_{R>0} R^{-n+2|\alpha|} \int_{R<|\xi|<2R} |\partial_\xi^\alpha m(\xi)|^2 \, d\xi \le A^2 < \infty \tag{5.2.11}$$

for all multiindices $|\alpha| \le [n/2] + 1$.
Then for all $1 < p < \infty$, m lies in $\mathcal{M}_p(\mathbf{R}^n)$ and the following estimate is valid:

$$\|m\|_{\mathcal{M}_p} \le C_n \max(p, (p-1)^{-1})(A + \|m\|_{L^\infty}). \tag{5.2.12}$$

Moreover, the operator $f \to (\widehat{f} m)^\vee$ maps $L^1(\mathbf{R}^n)$ into $L^{1,\infty}(\mathbf{R}^n)$ with norm at most a dimensional multiple of $A + \|m\|_{L^\infty}$.

Proof. First we observe that condition (5.2.11) is a generalization of (5.2.10) and therefore it suffices to work with (5.2.11).

Since m is a bounded function, the operator given by convolution with $W = m^\vee$ is bounded on $L^2(\mathbf{R}^n)$. To prove that this operator maps $L^1(\mathbf{R}^n)$ into $L^{1,\infty}(\mathbf{R}^n)$, it suffices to prove that the distribution W coincides with a function K on $\mathbf{R}^n \setminus \{0\}$ that satisfies Hörmander's condition.

Let $\widehat{\zeta}$ be a smooth function supported in the annulus $\frac{1}{2} \le |\xi| \le 2$ such that

$$\sum_{j\in\mathbf{Z}} \widehat{\zeta}(2^{-j}\xi) = 1, \qquad \text{when } \xi \neq 0.$$

Set $m_j(\xi) = m(\xi)\widehat{\zeta}(2^{-j}\xi)$ for $j \in \mathbf{Z}$ and $K_j = m_j^\vee$. We begin by observing that $\sum_{-N}^{N} K_j$ converges to W in $\mathcal{S}'(\mathbf{R}^n)$. Indeed, for all $\varphi \in \mathcal{S}(\mathbf{R}^n)$ we have

$$\Big\langle \sum_{j=-N}^{N} K_j, \varphi \Big\rangle = \Big\langle \sum_{j=-N}^{N} m_j, \widehat{\varphi} \Big\rangle \to \langle m, \widehat{\varphi} \rangle = \langle W, \varphi \rangle .$$

We set $n_0 = [\frac{n}{2}] + 1$. We claim that for any $0 < \varepsilon < \min(n_0 - \frac{n}{2}, 1)$ there is a constant $C_{n,\varepsilon}$ such that

$$\sup_{j\in\mathbf{Z}} \int_{\mathbf{R}^n} |K_j(x)|\,(1 + 2^j|x|)^{\varepsilon}\,dx \le C_{n,\varepsilon} A\,, \tag{5.2.13}$$

$$\sup_{j\in\mathbf{Z}} 2^{-j} \int_{\mathbf{R}^n} |\nabla K_j(x)|\,(1 + 2^j|x|)^{\varepsilon}\,dx \le C_{n,\varepsilon} A\,. \tag{5.2.14}$$

To prove (5.2.13) we multiply and divide the integrand in (5.2.13) by the expression $(1 + 2^j|x|)^{n_0}$. Applying the Cauchy-Schwarz inequality to $|K_j(x)|\,(1 + 2^j|x|)^{n_0}$ and $(1 + 2^j|x|)^{-n_0+\varepsilon}$, we control the integral in (5.2.13) by the product

$$\Big(\int_{\mathbf{R}^n} |K_j(x)|^2 (1 + 2^j|x|)^{2n_0}\,dx\Big)^{\frac{1}{2}} \Big(\int_{\mathbf{R}^n} (1 + 2^j|x|)^{-2n_0+2\varepsilon}\,dx\Big)^{\frac{1}{2}}. \tag{5.2.15}$$

We now note that since $\varepsilon < n_0 - \frac{n}{2}$ the second expression in (5.2.15) is equal to a constant multiple of $2^{-jn/2}$. To estimate the first integral in (5.2.15) we use the simple fact that

$$(1 + 2^j|x|)^{n_0} \le C_n \sum_{|\gamma|\le n_0} |(2^j x)^{\gamma}|\,.$$

We now have that the expression inside the supremum in (5.2.13) is controlled by

$$C_{n,\varepsilon} 2^{-jn/2} \sum_{|\gamma|\le n_0} \Big(\int_{\mathbf{R}^n} |K_j(x)|^2 2^{2j|\gamma|} |x^{\gamma}|^2\,dx\Big)^{\frac{1}{2}} \tag{5.2.16}$$

which is equal to

$$2^{-jn/2} \sum_{|\gamma|\le n_0} C_{\gamma,\varepsilon} 2^{j|\gamma|} \Big(\int_{\mathbf{R}^n} |(\partial^{\gamma} m_j)(\xi)|^2\,d\xi\Big)^{\frac{1}{2}} \tag{5.2.17}$$

using Plancherel's theorem. Now for any $|\gamma| \le n_0$ we use Leibnitz's rule to obtain

$$\begin{aligned}\int_{\mathbf{R}^n} |(\partial^\gamma m_j)(\xi)|^2\,d\xi &\le \sum_{|\delta|\le|\gamma|} C_\delta \int_{\mathbf{R}^n} \Big|2^{-j|\gamma-\delta|}(\partial_\xi^{\gamma-\delta}\widehat{\zeta})(2^{-j}\xi)(\partial_\xi^\delta m)(\xi)\Big|^2\,d\xi \\ &\le \sum_{|\delta|\le|\gamma|} C_\delta 2^{-2j|\gamma|}2^{2j|\delta|} \int_{2^{j-1}\le|\xi|\le 2^{j+1}} \Big|(\partial_\xi^\delta m)(\xi)\Big|^2\,d\xi \\ &\le \sum_{|\delta|\le|\gamma|} C_\delta 2^{-2j|\gamma|}2^{2j|\delta|}2A^2 2^{jn}2^{-2j|\delta|} = C_n A^2 2^{jn} 2^{-2j|\gamma|}\,,\end{aligned}$$

which is all we need to obtain (5.2.13). To obtain (5.2.14) we repeat the same argument for every derivative $\partial_r K_j$. Since the Fourier transform of $(\partial_r K_j)(x)\,x^\gamma$ is equal to a constant multiple of $\partial^\gamma\big(\xi_r m(\xi)\widehat{\zeta}(2^{-j}\xi)\big)$, we observe that the extra factor 2^{-j} in (5.2.14) can be combined with ξ_r to write $2^{-j}\partial^\gamma\big(\xi_r m(\xi)\widehat{\zeta}(2^{-j}\xi)\big)$ as $\partial^\gamma\big(m(\xi)\widehat{\zeta_r}(2^{-j}\xi)\big)$, where $\widehat{\zeta_r}(\xi) = \xi_r\widehat{\zeta}(\xi)$. The previous calculation with $\widehat{\zeta_r}$ replacing $\widehat{\zeta}$ can then be used to complete the proof of (5.2.14).

We now show that for all $x \neq 0$, the series $\sum_{j\in\mathbf{Z}} K_j(x)$ converges to a function that we will denote by $K(x)$. Indeed, it is trivial to see that for all $x \in \mathbf{R}^n$ we have

$$|K_j(x)| \le C_n 2^{jn}\big\|m\big\|_{L^\infty}\,,$$

which shows that the function

$$\sum_{j\le 0}|K_j(x)|$$

is bounded. Moreover, as a consequence of (5.2.13) we have that

$$(1+2^j\delta)^\varepsilon \int_{|x|\ge\delta} |K_j(x)|\,dx \le C_{n,\varepsilon}A\,,$$

for any $\delta > 0$, which implies that the function

$$\sum_{j>0}|K_j(x)|$$

is integrable away from the origin and thus finite almost everywhere there. We conclude that the series $\sum_{j\in\mathbf{Z}} K_j(x)$ represents a well-defined function $K(x)$ away from the origin that coincides with the distribution $W = m^\vee$.

We will now prove that the function $K = \sum_{j\in\mathbf{Z}} K_j$ (defined on $\mathbf{R}^n\setminus\{0\}$) satisfies Hörmander's condition. It suffices to prove that for all $y \neq 0$ we have

$$\sum_{j\in\mathbf{Z}} \int_{|x|\ge 2|y|} |K_j(x-y) - K_j(x)|\,dx \le C_n A\,. \tag{5.2.18}$$

Fix a $y \in \mathbf{R}^n \setminus \{0\}$ and pick a $k \in \mathbf{Z}$ such that $2^{-k} \le |y| \le 2^{-k+1}$. The part of the sum in (5.2.18) where $j > k$ is bounded by

$$
\begin{aligned}
&\sum_{j>k} \int_{|x|\ge 2|y|} |K_j(x-y)| + |K_j(x)|\, dx \\
&\le 2 \sum_{j>k} \int_{|x|\ge |y|} |K_j(x)|\, dx \\
&\le 2 \sum_{j>k} \int_{|x|\ge |y|} |K_j(x)| \frac{(1+2^j|x|)^\varepsilon}{(1+2^j|x|)^\varepsilon}\, dx \\
&\le \sum_{j>k} \frac{2C_{n,\varepsilon}A}{(1+2^j|y|)^\varepsilon} \le \sum_{j>k} \frac{2C_{n,\varepsilon}A}{(1+2^j 2^{-k})^\varepsilon} = C_n A\,,
\end{aligned}
$$

where we used (5.2.13). The part of the sum in (5.2.18) where $j \le k$ is bounded by

$$
\begin{aligned}
&\sum_{j\le k} \int_{|x|\ge 2|y|} |K_j(x-y) - K_j(x)|\, dx \\
&\le \sum_{j\le k} \int_{|x|\ge 2|y|} \int_0^1 |-y\cdot \nabla K_j(x-\theta y)|\, d\theta\, dx \\
&\le \int_0^1 \sum_{j\le k} 2^{-k+1} \int_{|x|\ge 2|y|} |\nabla K_j(x-\theta y)| \frac{(1+2^j|x-\theta y|)^\varepsilon}{(1+2^j|x-\theta y|)^\varepsilon}\, dx\, d\theta \\
&\le \int_0^1 \sum_{j\le k} 2^{-k+1} \frac{C_{n,\varepsilon}A 2^j}{(1+2^j 2^{-k})^\varepsilon}\, d\theta \le C_n A\,,
\end{aligned}
$$

since $|x-\theta y| \ge |x| - |\theta y| \ge |y| \ge 2^{-k}$ when $|x| \ge 2|y|$. We used (5.2.14) and the fact that $0 < \varepsilon < 1$. Hörmander's condition is satisfied for K and we appeal to Theorem 4.3.3 to complete the proof of (5.2.12).

□

Corollary 5.2.8. *Let $\{m_k\}_{k\in\mathbf{Z}}$ be bounded functions on $\mathbf{R}^n$ whose L^∞ norms are uniformly controlled by a constant A. Suppose that*

$$
\sup_{R>0} R^{-n+2|\alpha|} \sum_{k\in\mathbf{Z}} \int_{R<|\xi|<2R} |\partial_\xi^\alpha m_k(\xi)|^2\, d\xi \le A^2 \qquad \text{for all } |\alpha| \le [\tfrac{n}{2}]+1.
$$

Then for some $C_n < \infty$ and for all functions f_k we have

$$
\Big\| \Big(\sum_{k\in\mathbf{Z}} |(\widehat{f_k} m_k)^\vee|^2 \Big)^{\frac12} \Big\|_{L^p} \le C_n (p + (p-1)^{-1}) A \Big\| \Big(\sum_{k\in\mathbf{Z}} |f_k|^2 \Big)^{\frac12} \Big\|_{L^p}. \tag{5.2.19}
$$

Proof. Write each $K^k = m_k^\vee$ as a sum $\sum_j K_j^k$ as in the proof of Theorem 5.2.7. Using our hypothesis, we can prove that

$$\sup_{j\in\mathbf{Z}} \int_{\mathbf{R}^n} \sup_k |K_j^k(x)|\,(1+2^j|x|)^\varepsilon\,dx \le C_{n,\varepsilon}A\,, \tag{5.2.20}$$

$$\sup_{j\in\mathbf{Z}} 2^{-j} \int_{\mathbf{R}^n} \sup_k |\nabla K_j^k(x)|\,(1+2^j|x|)^\varepsilon\,dx \le C_{n,\varepsilon}A\,, \tag{5.2.21}$$

by estimating the supremum in k in (5.2.20) and (5.2.21) by the sum in k. These estimates are proved by following the proof of estimates (5.2.13) and (5.2.14) verbatim while carrying through the extra sum on k. Using (5.2.20) and (5.2.21), we can now easily derive that

$$\int_{|x|\ge 2|y|} \sup_k |K^k(x-y) - K^k(x)|\,dx \le C_n A$$

as we did in the proof of Theorem 5.2.7. This is Hörmander condition we need in this setting, which allows us to use Corollary 4.6.2. The proof of (5.2.19) follows. □

Example 5.2.9. Suppose that τ is a real number. Then the function $|\xi|^{i\tau}$ is in $\mathcal{M}_p$ for all $1 < p < \infty$. Indeed, condition (5.2.10) is satisfied.

We end this section by comparing Theorems 5.2.2 and 5.2.7. It is obvious that in dimension $n = 1$, Theorem 5.2.2 is stronger than Theorem 5.2.7. But in higher dimensions neither theorem includes the other. For, condition (5.2.10) for all $|\alpha| \le n$ is less general than condition (5.2.9). Thus for functions that are C^n away from the origin, it is better to use use Theorem 5.2.2. However, in Theorem 5.2.7 the function m is only assumed to be $C^{[n/2]+1}$, thus requiring almost half the amount of differentiability required in condition (5.2.9).

It should be noted that both theorems have their shortcomings. In particular, they are not L^p sensitive (i.e., delicate enough to detect whether m is bounded on some L^p but not on some other L^p).

Exercises

5.2.1. Let $1 \le k < n$. Use the same idea as in the proof of Proposition 5.2.1 to prove that $m \in \mathcal{M}_p(\mathbf{R}^n)$ if and only if (5.2.1) is satisfied with m_j replaced by

$$m_j(\xi) = m(\xi)\psi(2^{-j_1}\xi_1)\dots\psi(2^{-j_k}\xi_k),$$

where $\psi(\xi)$ is a smooth compactly function equal to 1 on the interval $[2^{-1}, 4]$ that satisfies

$$\sum_{j\in\mathbf{Z}} \psi(2^{-j}\xi) = 1, \qquad \xi \neq 0.$$

5.2.2. (*Calderón reproducing formula*) Let Ψ and Φ be radial Schwartz functions whose Fourier transforms are real valued and compactly supported away from the origin and satisfy

$$\sum_{j\in\mathbf{Z}} \widehat{\Psi}(2^{-j}\xi)\widehat{\Phi}(2^{-j}\xi) = 1$$

for all $\xi \neq 0$. Prove that for every function f in $\mathcal{S}_\infty(\mathbf{R}^n)$ we have

$$\sum_{j\in\mathbf{Z}} f * \Psi_{2^{-j}} * \Phi_{2^{-j}} = f\,,$$

where the series converges in $\mathcal{S}_\infty(\mathbf{R}^n)$. Conclude that the identity

$$\sum_{j\in\mathbf{Z}} \Delta_j^\Psi \Delta_j^\Phi = I$$

holds in the sense of $\mathcal{S}'(\mathbf{R}^n)/\mathcal{P}$. Here Δ_j^Ψ is the operator given by convolution with $\Psi_{2^{-j}}$ and Δ_j^Φ is defined likewise.

5.2.3. Consider the differential operators

$$\begin{aligned} L_1 &= \partial_1 - \partial_2^2 + \partial_3^4 \\ L_2 &= \partial_1 + \partial_2^2 + \partial_3^2\,. \end{aligned}$$

Prove that for every $1 < p < \infty$ there exists a constant $C_p < \infty$ such that for all Schwartz functions f on $\mathbf{R}^3$ we have

$$\begin{aligned} \big\|\partial_2\partial_3^2 f\big\|_{L^p} &\le C_p\big\|L_1(f)\big\|_{L^p}\,, \\ \big\|\partial_1 f\big\|_{L^p} &\le C_p\big\|L_2(f)\big\|_{L^p}\,. \end{aligned}$$

[*Hint:* Use Corollary 5.2.5. What is the relevance of multipliers m_1 and m_3 in Example 5.2.6?]

5.2.4. (a) Suppose that $m(\xi)$ is real valued and satisfies $|\partial^\alpha m(\xi)| \le C_\alpha|\xi|^{-\alpha}$ for all $|\alpha| \le [\frac{n}{2}] + 1$ and all $\xi \in \mathbf{R}^n \setminus \{0\}$. Prove that $e^{im(\xi)}$ is in $\mathcal{M}_p(\mathbf{R}^n)$ for $1 < p < \infty$.
(b) Suppose that $m(\xi)$ is real valued and satisfies (5.2.9). Prove that $c^{im(\xi)}$ is in $\mathcal{M}_p(\mathbf{R}^n)$ for $1 < p < \infty$.

5.2.5. Suppose that $\varphi(\xi)$ is a smooth function on $\mathbf{R}^n$ that vanishes in a neighborhood of the origin and is equal to 1 in a neighborhood of infinity. Prove that the function $e^{i\xi_j|\xi|^{-1}}\varphi(\xi)$ is in $\mathcal{M}_p(\mathbf{R}^n)$ for $1 < p < \infty$ for every ξ_j.

5.2.6. Let $\tau, \tau_1, \dots, \tau_n$ be real numbers. Prove that the following functions are L^p multipliers on $\mathbf{R}^n$:

$$|\xi_1|^{i\tau_1} \dots |\xi_n|^{i\tau_n},$$
$$(|\xi_1|^{\tau_1} + \dots + |\xi_n|^{\tau_n})^{i\tau},$$
$$|\xi_1|^{i\tau_1} \dots |\xi_n|^{i\tau_n}(|\xi_1| + \dots + |\xi_n|)^{i\tau}.$$

5.2.7. Let $\widehat{\zeta}(\xi)$ be a smooth function on the line that is supported in a compact set that does not contain the origin and let a_j be a bounded sequence of complex numbers. Prove that the function

$$m(\xi) = \sum_{j \in \mathbf{Z}} a_j \widehat{\zeta}(2^{-j}\xi)$$

is in $\mathcal{M}_p(\mathbf{R})$ for all $1 < p < \infty$.

5.2.8. Let ζ be as in the previous Exercise and let $\Delta_j^\zeta(f) = (\widehat{f}(\xi)\widehat{\zeta}(2^{-j}\xi))^\vee$. Show that the operator

$$f \to \sup_{N>0} \Big| \sum_{j<N} \Delta_j^\zeta(f) \Big|$$

is bounded on $L^p(\mathbf{R})$ when $1 < p < \infty$.
[*Hint:* Pick a Schwartz function φ with $\widehat{\varphi}$ compactly supported satisfying $\sum_{j \in \mathbf{Z}} \widehat{\varphi}(2^{-j}\xi) = 1$ on $\mathbf{R}^n \setminus \{0\}$. Write

$$f = \sum_{j \in \mathbf{Z}} \Delta_j^\varphi(f),$$

where Δ_j^φ is given by convolution with $\varphi_{2^{-j}}$, and control the maximal operator in question by a multiple of the function

$$M\Big(\sum_{j \in \mathbf{Z}} \Delta_j^\zeta(f)\Big),$$

where M is the Hardy-Littlewood maximal function.]

5.3. Applications of Littlewood-Paley Theory

We will now turn our attention to some important applications of Littlewood-Paley theory. We are interested in obtaining bounds for singular and maximal operators. These bounds are obtained by controlling the corresponding operators by quadratic expressions.

5.3.a. Estimates for Maximal Operators

One way to control the maximal operator $\sup_k |T_k(f)|$ is by introducing a good averaging function φ and using the majorization

$$\begin{aligned}\sup_k |T_k(f)| &\le \sup_k |T_k(f) - f * \varphi_{2^{-k}}| + \sup_k |f * \varphi_{2^{-k}}| \\ &\le \Big(\sum_k |T_k(f) - f * \varphi_{2^{-k}}|^2\Big)^{\frac{1}{2}} + C_\varphi M(f)\end{aligned} \tag{5.3.1}$$

for some constant C_φ depending on φ. We apply this idea to prove the following theorem.

Theorem 5.3.1. *Let m be a bounded function on $\mathbf{R}^n$ that is C^1 in a neighborhood of the origin and satisfies $m(0) = 1$ and $|m(\xi)| \le C|\xi|^{-\varepsilon}$ for some $C, \varepsilon > 0$ and all $\xi \neq 0$. For each $k \in \mathbf{Z}$ define $T_k(f)(x) = (\widehat{f}(\xi)m(2^{-k}\xi))^\vee(x)$. Then there is a constant C_n such that for all f Schwartz functions on $\mathbf{R}^n$ we have*

$$\big\|\sup_{k\in\mathbf{Z}} |T_k(f)|\big\|_{L^2} \le C_n \big\|f\big\|_{L^2}. \tag{5.3.2}$$

Proof. Select a Schwartz function φ such that $\widehat{\varphi}(0) = 1$. Then there are constants C_1 and C_2 such that $|m(\xi) - \widehat{\varphi}(\xi)| \le C_1|\xi|^{-\varepsilon}$ for $|\xi|$ away from zero and $|m(\xi) - \widehat{\varphi}(\xi)| \le C_2|\xi|$ for $|\xi|$ near zero. These two inequalities imply that

$$\sum_k |m(2^{-k}\xi) - \widehat{\varphi}(2^{-k}\xi)|^2 \le C_3 < \infty$$

from which the L^2 boundedness of the operator

$$f \to \Big(\sum_k |T_k(f) - f * \varphi_{2^{-k}}|^2\Big)^{1/2}$$

follows easily. Using estimate (5.3.1) and the well-known L^2 estimate for the Hardy-Littlewood maximal function, we obtain (5.3.2).

□

If $m(\xi)$ is the characteristic function of a square or a ball that contains the origin, then this result can be extended to L^p.

Theorem 5.3.2. *Let $1 < p < \infty$ and $m(\xi)$ be the characteristic function of any bounded open set in $\mathbf{R}^n$ that contains the origin. For each $k \in \mathbf{Z}$ define $T_k(f)(x) = (\widehat{f}(\xi)m(2^{-k}\xi))^\vee(x)$. Then there is a constant $C_{p,n}$ such that for all Schwartz functions f on $\mathbf{R}^n$ we have*

$$\big\|\sup_{k\in\mathbf{Z}} |T_k(f)|\big\|_{L^p(\mathbf{R}^n)} \le C_{p,n} \big\|f\big\|_{L^p(\mathbf{R}^n)}.$$

Proof. Take a Schwartz function ψ such that $\widehat{\psi}(0) = 0$ and $\widehat{\psi}(\xi) = 1$ for ξ inside an open annulus that contains the boundary of the support of m. Then there is a

Schwartz function φ such that $\widehat{\varphi} = (1-\widehat{\psi}\,)m$. This implies that $m - \widehat{\varphi} = m\widehat{\psi}$, from which it follows that for all f in Schwartz class we have

$$T_k(f) = T_k(f) - f * \varphi_{2^{-k}} + f * \varphi_{2^{-k}} = T_k(f * \psi_{2^{-k}}) + f * \varphi_{2^{-k}}\,. \tag{5.3.3}$$

Then the L^p norm of the square function $f \to \big(\sum_k |T_k(f) - f * \varphi_{2^{-k}}|^2\big)^{1/2}$ is controlled by that of the square function $\big(\sum_k |T_k(f * \psi_{2^{-k}})|^2\big)^{1/2}$. We now observe that if $m_k(\xi) = \widehat{\psi}(2^{-k}\xi)$, then the hypothesis of Corollary 5.2.8 holds. Indeed, for all $|\alpha| \le [\frac{n}{2}]+1$ and a fixed $R>0$ we have

$$\begin{aligned}
&R^{-n+2|\alpha|} \sum_{k\in\mathbf{Z}} \int_{R<|\xi|<2R} |\partial_\xi^\alpha m_k(\xi)|^2\, d\xi \\
&\le CR^{-n+2|\alpha|} \sum_{\substack{k\in\mathbf{Z}\\ 2^k\approx R}} \int_{\substack{R<|\xi|<2R\\ |\xi|\approx 2^k}} |2^{-k\alpha}(\partial_\xi^\alpha \widehat{\psi})(2^{-k}\xi)|^2\, d\xi \\
&\le C_{n,\alpha} R^{-n+2|\alpha|} \sum_{\substack{k\in\mathbf{Z}\\ 2^k\approx R}} 2^{kn} 2^{-2k|\alpha|} \le C'_{n,\alpha}\,.
\end{aligned}$$

The conclusion of Corollary 5.2.8 gives that the L^p norm of the square function

$$\Big\|\Big(\sum_k |T_k(f * \psi_{2^{-k}})|^2\Big)^{\frac{1}{2}}\Big\|_{L^p} \le C_{p,n} \Big\|\Big(\sum_k |f * \psi_{2^{-k}}|^2\Big)^{\frac{1}{2}}\Big\|_{L^p}\,.$$

Finally, the upper inequality of the Littlewood-Paley theorem gives that the last square function is controlled by a multiple of $\|f\|_{L^p}$. Using (5.3.3), we conclude that the L^p norm of the maximal operator $\sup_{k\in\mathbf{Z}} |T_k(f)|$ is controlled by a multiple of $\|M(f)\|_{L^p} + \|f\|_{L^p}$, which yields the required conclusion.

□

The following lacunary version of the Carleson-Hunt theorem is yet another indication of the powerful techniques of Littlewood-Paley theory.

Corollary 5.3.3. *Let f be in $\bigcup_{1<p<\infty} L^p(\mathbf{R}^n)$. Then we have*

$$\lim_{k\to\infty} \int_{|\xi|\le 2^k} \widehat{f}(\xi) e^{2\pi i x\cdot\xi}\, d\xi = f(x)$$

and

$$\lim_{k\to\infty} \int_{\sup_j |\xi_j|\le 2^k} \widehat{f}(\xi) e^{2\pi i x\cdot\xi}\, d\xi = f(x)$$

for all almost all $x \in \mathbf{R}^n$.

Proof. The aforementioned convergence easily holds if f is a function of the Schwartz class. To obtain almost everywhere convergence for all f in L^p we appeal to Theorem 2.1.14. The required control of the corresponding maximal operator is a consequence of Theorem 5.3.2 with $m(\xi)$ being either $\chi_{(-1,1)^n}$ or $\chi_{B(0,1)}$, where $B(0,1)$ is the unit ball in $\mathbf{R}^n$.

□

5.3.b. Estimates for Singular Integrals with Rough Kernels

We now turn to another application of the Littlewood-Paley theory involving singular integrals.

Theorem 5.3.4. *Suppose that μ is a finite Borel measure on $\mathbf{R}^n$ with compact support that satisfies $|\widehat{\mu}(\xi)| \le B\min\left(|\xi|^{-b}, |\xi|^{b}\right)$ for some $b > 0$ and all $\xi \neq 0$. Define measures μ_j by setting $\widehat{\mu_j}(\xi) = \mu(2^{-j}\xi)$. Then the operator*

$$T_\mu(f)(x) = \sum_{j\in\mathbf{Z}} (f * \mu_j)(x)$$

is bounded on $L^p(\mathbf{R}^n)$ for all $1 < p < \infty$.

Proof. It is natural to begin with the L^2 boundedness of T_μ. The estimate on $\widehat{\mu}$ implies that

$$\sum_{j\in\mathbf{Z}} |\widehat{\mu}(2^{-j}\xi)| \le \sum_{j\in\mathbf{Z}} B\min\left(|2^{-j}\xi|^{b}, |2^{-j}\xi|^{-b}\right) \le C_b B < \infty \,. \tag{5.3.4}$$

The L^2 boundedness of T_μ is an immediate consequence of (5.3.4).

We now turn to the L^p boundedness of T_μ for $1 < p < \infty$. We fix a radial Schwartz function ψ whose Fourier transform is supported in the annulus $\frac{1}{2} < |\xi| < 2$ and that satisfies

$$\sum_{j\in\mathbf{Z}} \widehat{\psi}(2^{-j}\xi) = 1 \tag{5.3.5}$$

whenever $\xi \neq 0$. We let $\psi_{2^{-k}}(x) = 2^{kn}\psi(2^k x)$, so that $\widehat{\psi_{2^{-k}}}(\xi) = \widehat{\psi}(2^{-k}\xi)$, and we observe that the identity

$$K_j = \sum_{k\in\mathbf{Z}} \mu_j * \psi_{2^{-j-k}}$$

is valid by taking Fourier transforms and using (5.3.5). We now define operators S_k by setting

$$S_k(f) = \sum_{j\in\mathbf{Z}} \mu_j * \psi_{2^{-j-k}} * f = \sum_{j\in\mathbf{Z}} (\mu * \psi_{2^{-k}})_{2^{-j}} * f \,.$$

Then for nice f we have that

$$T_\mu(f) = \sum_{j\in\mathbf{Z}} \mu_j * f = \sum_{j,k\in\mathbf{Z}} \mu_j * \psi_{2^{-j-k}} * f = \sum_{k\in\mathbf{Z}} S_k(f) \,.$$

It suffices therefore to obtain L^p boundedness for the sum of the S_k's. We begin by investigating the L^2 boundedness of each S_k. Since the product

$$\widehat{\psi_{2^{-j-k}}}\widehat{\psi_{2^{-j'-k}}}$$

is nonzero only when $j' \in \{j-1, j, j+1\}$, it follows that

$$\begin{aligned}
\big\|S_k(f)\big\|_{L^2}^2 &\le \sum_{j\in\mathbf{Z}}\sum_{j'\in\mathbf{Z}}\int_{\mathbf{R}^n} |\widehat{\mu_j}(\xi)\widehat{\mu_{j'}}(\xi)\widehat{\psi}(2^{-j-k}\xi)\widehat{\psi}(2^{-j'-k}\xi)|\,|\widehat{f}(\xi)|^2\,d\xi \\
&\le C_1\sum_{j\in\mathbf{Z}}\sum_{j'=j-1}^{j+1}\int_{|\xi|\approx 2^{j+k}} |\widehat{\mu_j}(\xi)\widehat{\mu_{j'}}(\xi)|\,|\widehat{f}(\xi)|^2\,d\xi \\
&\le C_2\sum_{j\in\mathbf{Z}}\int_{|\xi|\approx 2^{j+k}} B^2\min(|2^{-j}\xi|^b, |2^{-j}\xi|^{-b})^2|\widehat{f}(\xi)|^2\,d\xi \\
&\le C_3^2B^2 2^{-2|k|b}\sum_{j\in\mathbf{Z}}\int_{|\xi|\approx 2^{j+k}} |\widehat{f}(\xi)|^2\,d\xi \\
&= C_3^2B^2\,2^{-2|k|b}\big\|f\big\|_{L^2}^2 .
\end{aligned}$$

We have therefore obtained that for all $k\in\mathbf{Z}$ and $f\in\mathcal{S}(\mathbf{R}^n)$ we have

$$\big\|S_k(f)\big\|_{L^2} \le C_3B2^{-b|k|}\big\|f\big\|_{L^2} . \tag{5.3.6}$$

Next we will show that the kernel of each S_k satisfies Hörmander's condition with constant at most a multiple of $(1+|k|)$. Fix $y\neq 0$. Then

$$\begin{aligned}
&\int_{|x|\ge 2|y|}\Big|\sum_{j\in\mathbf{Z}}(\mu*\psi_{2^{-k}})_{2^{-j}}(x-y)-(\mu*\psi_{2^{-k}})_{2^{-j}}(x)\Big|\,dx \\
&\le \sum_{j\in\mathbf{Z}}\int_{|x|\ge 2|y|} 2^{jn}\big|(\mu*\psi_{2^{-k}})(2^jx-2^jy)-(\mu*\psi_{2^{-k}})(2^jx)\big|\,dx \\
&= \sum_{j\in\mathbf{Z}} I_{j,k}(y) ,
\end{aligned}$$

where

$$I_{j,k}(y) = \int_{|x|\ge 2^{j+1}|y|}\big|(\mu*\psi_{2^{-k}})(x-2^jy)-(\mu*\psi_{2^{-k}})(x)\big|\,dx .$$

We observe that $I_{j,k}(y) \le C_4\|\mu\|_{\mathcal{M}}$. To obtain a more delicate estimate for $I_{j,k}(y)$ we argue as follows:

$$\begin{aligned}
I_{j,k}(y) &\le \int\limits_{|x|\ge 2^{j+1}|y|} \int_{\mathbf{R}^n} \big|\psi_{2^{-k}}(x-2^j y - z) - \psi_{2^{-k}}(x-z)\big|\, d\mu(z)\, dx \\
&= \int_{\mathbf{R}^n} 2^{kn} \int\limits_{|x|\ge 2^{j+1}|y|} \big|\psi(2^k x - 2^k z - 2^{j+k}y) - \psi(2^k x - 2^k z)\big|\, dx\, d\mu(z) \\
&\le C_5 \int\limits_{|x|\ge 2^{j+1}|y|} \int_{\mathbf{R}^n} 2^{kn} 2^{j+k}|y| \big|\nabla\psi(2^k x - 2^k z - \theta)\big|\, d\mu(z)\, dx \\
&\le C_6 2^{j+k} \int_{\mathbf{R}^n} \int\limits_{|x|\ge 2^{j+1}|y|} 2^{kn}|y| \big(1+|2^k x - 2^k z - \theta|\big)^{-n-2} dx\, d\mu(z) \\
&= C_6 2^{j+k}|y| \int_{\mathbf{R}^n} \int\limits_{|x|\ge 2^{j+k+1}|y|} \big(1+|x - 2^k z - \theta|\big)^{-n-2} dx\, d\mu(z)\,,
\end{aligned}$$

where $|\theta| \le 2^{j+k}|y|$. From this and from $I_{j,k}(y) \le C_4\|\mu\|_{\mathcal{M}}$ we obtain

$$I_{j,k}(y) \le C_7 \,\|\mu\|_{\mathcal{M}} \min\big(1, 2^{j+k}|y|\big), \tag{5.3.7}$$

which is valid for all j,k and $y \ne 0$. To estimate the last double integral even more delicately, we consider the following two cases: $|x| \ge 2^{k+2}|z|$ and $|x| < 2^{k+2}|z|$. In the first case we have $|x - 2^k z - \theta| \ge \frac{1}{4}|x|$ given the fact that $|x| \ge 2^{j+k+1}|y|$. In the second case we have that $|x| \le 2^{k+2}R$, where $B(0,R)$ contains the support of μ. Applying these observations in the last double integral, we obtain the following estimate:

$$\begin{aligned}
I_{j,k}(y) &\le C_8 2^{j+k}|y| \int_{\mathbf{R}^n} \Bigg[\int\limits_{\substack{|x|\ge 2^{j+k+1}|y| \\ |x|\ge 2^{k+2}|z|}} \frac{dx}{\big(1+\frac{1}{4}|x|\big)^{n+2}} + \int\limits_{\substack{|x|\ge 2^{j+k+1}|y| \\ |x|< 2^{k+2}R}} dx \Bigg] d\mu(z) \\
&\le C_9 2^{j+k}|y| \|\mu\|_{\mathcal{M}} \left[\frac{1}{(2^{j+k}|y|)^2} + 0\right] = C_9 (2^{j+k}|y|)^{-1} \|\mu\|_{\mathcal{M}}\,,
\end{aligned}$$

provided $2^j|y| \ge 2R$. Combining this estimate with (5.3.7), we obtain

$$I_{j,k}(y) \le C_{10}\|\mu\|_{\mathcal{M}} \begin{cases} \min\big(1, 2^{j+k}|y|\big) & \text{for all } j,k \text{ and } y, \\ (2^{j+k}|y|)^{-1} & \text{when } 2^j|y| \ge 2R. \end{cases} \tag{5.3.8}$$

We now estimate $\sum_j I_{j,k}(y)$. Using (5.3.8), for $k \geq -\log_2(2R)$, we obtain

$$\sum_j I_{j,k}(y) \leq \sum_{2^j \leq \frac{1}{2^k|y|}} 2^{j+k}|y| + \sum_{\frac{1}{2^k|y|} \leq 2^j \leq \frac{2R}{|y|}} 1 + \sum_{2^j \geq \frac{2R}{|y|}} (2^{j+k}|y|)^{-1}$$

$$\leq C_{11}\|\mu\|_{\mathcal{M}}(1+|k|), \qquad (C_{11} \text{ grows like } \log R.)$$

Also, for $k < -\log_2(2R)$ the middle sum does not appear and we obtain

$$\sum_j I_{j,k}(y) \leq \sum_{2^j \leq \frac{1}{2^k|y|}} 2^{j+k}|y| + \sum_{2^j \geq \frac{2R}{|y|}} (2^{j+k}|y|)^{-1} \leq C_{12}\|\mu\|_{\mathcal{M}}.$$

This gives

$$\sum_j I_{j,k}(y) \leq C_{13}\|\mu\|_{\mathcal{M}}(1+|k|), \tag{5.3.9}$$

where all constants C_1 through C_{13} depend on the dimension and R. We now use estimate (5.3.6) and (5.3.9) and Theorem 4.3.3 to obtain that each S_k maps $L^1(\mathbf{R}^n)$ into $L^{1,\infty}(\mathbf{R}^n)$ with constant at most

$$C_n(2^{-b|k|} + 1 + |k|)\|\mu\|_{\mathcal{M}} \leq C_n(2+|k|)\|\mu\|_{\mathcal{M}}.$$

It follows from the Marcinkiewicz interpolation theorem 1.3.2 that S_k maps $L^p(\mathbf{R}^n)$ into itself for $1 < p < 2$ with bound at most $C_{p,n}2^{-b|k|\theta_p}(1+|k|)^{1-\theta_p}$, where $\frac{1}{p} = \frac{\theta_p}{2} + 1 - \theta_p$. Summing over all $k \in \mathbf{Z}$, we obtain that T_μ maps $L^p(\mathbf{R}^n)$ into itself for $1 < p < \infty$. The boundedness of T_μ for the remaining p's follows by duality.

□

An immediate consequence of the previous result is the following.

Corollary 5.3.5. *Let μ_j be as in the previous theorem. Then the square function*

$$G(f) = \Big(\sum_j |\mu_j * f|^2\Big)^{\frac{1}{2}} \tag{5.3.10}$$

maps $L^p(\mathbf{R}^n)$ into itself whenever $1 < p < \infty$.

Proof. To obtain the boundedness of the square function in (5.3.10) we use the Rademacher functions $r_j(t)$ in Appendix C.1. For each t we introduce the operators

$$T_\mu^t(f) = \sum_j r_j(t)(f * \mu_j).$$

Next we observe that for each t in $[0,1]$ the operators T_μ^t map $L^p(\mathbf{R}^n)$ into itself with the same constant as the operator T_μ, which is in particular independent of

t. Using that square function in (5.3.10) raised to the power p is controlled by a multiple of

$$\int_0^1 \Big|\sum_j r_j(t)(f * \mu_j)\Big|^p \, dt\,,$$

a fact that can be found in Appendix C.2, we obtain the required conclusion by integrating over $\mathbf{R}^n$.

□

5.3.c. An Almost Orthogonality Principle on L^p

We end this section with another application of the Littlewood-Paley theorem. Suppose that T_j are multiplier operators given by $T_j(f) = (\widehat{f} m_j)^\vee$, for some multipliers m_j. If the functions m_j have disjoint supports and they are bounded uniformly in j, then the operator

$$T = \sum_j T_j$$

is bounded on L^2. The following theorem gives an L^p analogue of this theorem.

Theorem 5.3.6. *Suppose that $1 < p \le 2 \le q < \infty$. Let $T_j(f) = (\widehat{f} m_j)^\vee$ where m_j are Schwartz functions supported in the annuli $2^{j-1} \le |\xi| \le 2^{j+1}$. Then for each $f \in L^p(\mathbf{R}^n)$, the series*

$$T(f) = \sum_j T_j(f)$$

converges in the L^q norm and there exists a constant $C_{p,q,n} < \infty$ such that

$$\big\|T\big\|_{L^p \to L^q} \le C_{p,q,n} \sup_j \big\|T_j\big\|_{L^p \to L^q}. \tag{5.3.11}$$

Proof. Fix a function φ in $C_0^\infty(\mathbf{R}^n \setminus \{0\})$ that is equal to 1 on the annulus $\frac{1}{2} \le |\xi| \le 2$ and vanishes outside the annulus $\frac{1}{4} \le |\xi| \le 4$. We set $\varphi_{2^{-j}}(x) = 2^{jn}\zeta(2^j x)$ so that $\widehat{\varphi_{2^{-j}}}$ is equal to 1 on the support of each m_j. Setting $\Delta_j(f) = f * \varphi_{2^{-j}}$ we have that

$$T_j = \Delta_j T_j \Delta_j$$

for all $j \in \mathbf{Z}$. For a positive integer N we set

$$T^N = \sum_{|j| \le N} \Delta_j T_j \Delta_j$$

and we observe that for any N and every $f \in L^p$, $T^N(f)$ is in L^q. Using (5.1.19) and Proposition 5.1.4, for any fixed $f \in L^p(\mathbf{R}^n)$, we obtain

$$\begin{aligned}
\big\|T^N(f)\big\|_{L^q} &= \Big\| \sum_{|j|\le N} \Delta_j T_j \Delta_j(f)\Big\|_{L^q} \\
&\le C_q \Big\|\Big(\sum_{j\in\mathbf{Z}} |\Delta_j T_j \Delta_j(f)|^2\Big)^{\frac12}\Big\|_{L^q} \\
&\le C_q' \Big\|\Big(\sum_{j\in\mathbf{Z}} |T_j \Delta_j(f)|^2\Big)^{\frac12}\Big\|_{L^q} \\
&\le C_q' \Big\|\sum_{j\in\mathbf{Z}} |T_j \Delta_j(f)|^2\Big\|_{L^{q/2}}^{\frac12} \\
&\le C_q' \Big(\sum_{j\in\mathbf{Z}} \Big\||T_j \Delta_j(f)|^2\Big\|_{L^{q/2}}\Big)^{\frac12} \\
&\le C_q' \Big(\sum_{j\in\mathbf{Z}} \big\|T_j \Delta_j(f)\big\|_{L^q}^2\Big)^{\frac12},
\end{aligned}$$

where in the penultimate inequality we used Minkowski's inequality since $q/2 \ge 1$. Using the boundedness of each T_j from L^p to L^q, we obtain that

$$\begin{aligned}
C_q'\Big(\sum_{j\in\mathbf{Z}} \big\|T_j\Delta_j(f)\big\|_{L^q}^2\Big)^{\frac12} &\le C_q' \sup_{j\in\mathbf{Z}} \big\|T_j\big\|_{L^p\to L^q}\Big(\sum_{j\in\mathbf{Z}} \big\|\Delta_j(f)\big\|_{L^p}^2\Big)^{\frac12} \\
&= C_q' \sup_{j\in\mathbf{Z}} \big\|T_j\big\|_{L^p\to L^q}\Big(\sum_{j\in\mathbf{Z}} \big\||\Delta_j(f)|^2\big\|_{L^{p/2}}\Big)^{\frac12} \\
&\le C_q' \sup_{j\in\mathbf{Z}} \big\|T_j\big\|_{L^p\to L^q}\Big(\Big\|\sum_{j\in\mathbf{Z}} |\Delta_j(f)|^2\Big\|_{L^{p/2}}\Big)^{\frac12} \\
&= C_q' \sup_{j\in\mathbf{Z}} \big\|T_j\big\|_{L^p\to L^q}\Big\|\Big(\sum_{j\in\mathbf{Z}} |\Delta_j(f)|^2\Big)^{\frac12}\Big\|_{L^p} \\
&\le C_q' C_p \sup_j \big\|T_j\big\|_{L^p\to L^q}\big\|f\big\|_{L^p(\mathbf{R}^n)},
\end{aligned}$$

where we used the result of Exercise 1.1.5(b) since $p \le 2$ and Theorem 5.1.2. We conclude that the operators T^N are uniformly bounded from $L^p(\mathbf{R}^n)$ to $L^q(\mathbf{R}^n)$.

If $\widehat{h}$ is compactly supported in a subset of $\mathbf{R}^n \setminus \{0\}$, then the sequence $T^N(h)$ becomes independent of N for N large enough and hence it is Cauchy in L^q. But in view of Exercise 5.1.7, the set of all such h is dense in $L^p(\mathbf{R}^n)$. Combining these two results with the uniform boundedness of the T^N's from L^p to L^q, a simple $\frac{\varepsilon}{3}$ argument gives that for all $f \in L^p$ the sequence $T^N(f)$ is Cauchy in L^q. Therefore, for all $f \in L^p$ the sequence $\{T^N(f)\}_N$ converges in L^q to some $T(f)$ in L^q. Fatou's lemma gives

$$\|T(f)\|_{L^q} \le C'_q C_p \sup_j \|T_j\|_{L^p \to L^q} \|f\|_{L^p},$$

which proves (5.3.11). □

Exercises

5.3.1. (*The g-function*) Let $P_t(x) = \frac{\Gamma(\frac{n+1}{2})}{\pi^{\frac{n+1}{2}}} \frac{t}{(t^2+|x|^2)^{\frac{n+1}{2}}}$ be the Poisson kernel.
(a) Use Exercise 5.1.4 with $\Psi(x) = \frac{\partial}{\partial t} P_t(x)\big|_{t=1}$ to obtain that the operator

$$f \to \left(\int_0^\infty t \left| \frac{\partial}{\partial t}(P_t * f)(x) \right|^2 dt \right)^{1/2}$$

is L^p bounded for $1 < p < \infty$.
(b) Use Exercise 5.1.3 with $\Psi(x) = \partial_k P_1(x)$ to obtain that the operator

$$f \to \left(\int_0^\infty t |\partial_k (P_t * f)(x)|^2 \, dt \right)^{1/2}$$

is L^p bounded for $1 < p < \infty$.
(c) Conclude that the g-function

$$g(f)(x) = \left(\int_0^\infty t |\nabla_{x,t} (P_t * f)(x)|^2 \, dt \right)^{1/2}$$

is L^p bounded for $1 < p < \infty$.

5.3.2. Suppose that μ is a finite Borel measure on $\mathbf{R}^n$ with compact support that satisfies $\widehat{\mu}(0) = 0$ and $|\widehat{\mu}(\xi)| \le C|\xi|^{-a}$ for some $a > 0$ and all $\xi \neq 0$. Define measures μ_j by setting $\widehat{\mu_j}(\xi) = \widehat{\mu}(2^{-j}\xi)$. Show that the operator

$$T_\mu(f)(x) = \sum_{j \in \mathbf{Z}} (f * \mu_j)(x)$$

is bounded on L^p for all $1 < p < \infty$.

5.3.3. (*Calderón* [**76**]/*Coifman and Weiss* [**123**]) (a) Suppose that μ is a finite Borel measure with compact support that satisfies $|\widehat{\mu}(\xi)| \le C|\xi|^{-a}$ for some $a > 0$ and all $\xi \neq 0$. Then the maximal function

$$\mathcal{M}_\mu(f)(x) = \sup_{j \in \mathbf{Z}} \left| \int_{\mathbf{R}^n} f(x - 2^j y) \, d\mu(y) \right|$$

is bounded on L^p for all $1 < p < \infty$.
(b) Let μ be the surface measure on the sphere $\mathbf{S}^{n-1}$ when $n \ge 2$. Conclude that the *dyadic spherical maximal function* $\mathcal{M}_\mu$ is bounded on $L^p(\mathbf{R}^n)$ for all $1 < p < \infty$ whenever $n \ge 2$.

[*Hint:* Pick φ a compactly supported smooth function on $\mathbf{R}^n$ with $\widehat{\varphi}(0) = 1$. Then the measure $\sigma = \mu - \widehat{\mu}(0)\varphi$ satisfies the hypotheses of Corollary 5.3.5. But it is straightforward that

$$\mathcal{M}_\mu(f)(x) \le \Big(\sum_j |(\sigma_j * f)(x)|^2\Big)^{1/2} + |\widehat{\mu}(0)| M(f)(x)$$

from which it follows that $\mathcal{M}_\mu$ is bounded on $L^p(\mathbf{R}^n)$ whenever $1 < p < \infty$. Now let $\mu = d\sigma$ be surface measure on $\mathbf{S}^{n-1}$. It follows from the results in Appendices B.4 and B.6 that $|\widehat{d\sigma}(\xi)| \le C|\xi|^{-\frac{n-1}{2}}$.]

5.3.4. Let Ω be in $L^q(\mathbf{S}^{n-1})$ for some $1 < q < \infty$ and define the absolutely continuous measure

$$d\mu(x) = \frac{\Omega(x/|x|)}{|x|^n}\chi_{1<|x|\le 2}\, dx\,.$$

Show that for all $a < 1/q'$ we have that $|\widehat{\mu}(\xi)| \le C|\xi|^{-a}$. Under the additional hypothesis that Ω has mean value zero, conclude that the singular integral operator

$$T_\Omega(f)(x) = \text{p.v.} \int_{\mathbf{R}^n} \frac{\Omega(y/|y|)}{|y|^n} f(x-y)\, dy = \sum_j f * \mu_j$$

is L^p bounded for all $1 < p < \infty$. This provides an alternative proof of Theorem 4.2.10 under the hypothesis that $\Omega \in L^q(\mathbf{S}^{n-1})$.

5.3.5. For a function F on $\mathbf{R}$ define

$$u(F)(x) = \bigg(\int_0^\infty |F(x+t) + F(x-t) - 2F(x)|^2 \frac{dt}{t^3}\bigg)^{1/2}.$$

Given $f \in L^1_{\text{loc}}(\mathbf{R})$ we denote by F the indefinite integral of f, that is,

$$F(x) = \int_0^x f(t)\, dt\,.$$

Prove that for all $1 < p < \infty$ there exist constants c_p and C_p such that

$$c_p \big\|f\big\|_{L^p} \le \big\|u(F)\big\|_{L^p} \le C_p \big\|f\big\|_{L^p}\,.$$

[*Hint:* Let $\varphi = \chi_{[-1,0]} - \chi_{[0,1]}$. Then

$$(\varphi_t * f)(x) = \frac{1}{t}\big(F(x+t) + F(x-t) - 2F(x)\big)$$

and you may use Exercise 5.1.4.]

5.3.6. Let $m \in \mathcal{M}_p(\mathbf{R}^n)$. Define an operator T_t by setting $\widehat{T_t(f)}(\xi) = \widehat{f}(\xi)m(t\xi)$. Show that the maximal operator

$$\sup_{N>0} \left(\frac{1}{N} \int_0^N \left| T_t(f)(x) \right|^2 dt \right)^{\frac{1}{2}}$$

maps $L^p(\mathbf{R}^n)$ into itself for all $1 < p < \infty$.
[*Hint:* Majorize this maximal operator by a constant multiple of the sum $M(f)(x) + \left(\int_0^\infty |T_t(f)(x) - (f * \varphi_t)(x)|^2 \frac{dt}{t} \right)^{\frac{1}{2}}$ for a suitable function φ.]

5.4. The Haar System, Conditional Expectation, and Martingales*

There is a very strong connection between the Littlewood-Paley operators and certain notions from probability, such as conditional expectation and martingale difference operators. The conditional expectation we will be concerned with is with respect to the increasing σ-algebra of all dyadic cubes on $\mathbf{R}^n$.

5.4.a. Conditional Expectation and Dyadic Martingale Differences

We begin by defining dyadic cubes.

Definition 5.4.1. A *dyadic interval* in $\mathbf{R}$ is an interval of the form

$$[m2^{-k}, (m+1)2^{-k})$$

where m, k are integers. A *dyadic cube* in $\mathbf{R}^n$ is a product of dyadic intervals of the same length. That is, a dyadic cube is a set of the form

$$\prod_{j=1}^{n} [m_j 2^{-k}, (m_j+1)2^{-k})$$

for some integers $m_1, \dots, m_n, k$.

We defined dyadic intervals to be closed on the left and open on the right, so that different dyadic intervals of the same length are always disjoint sets.

Given a cube Q in $\mathbf{R}^n$ we will denote by $|Q|$ its Lebesgue measure and by $\ell(Q)$ its side length. We clearly have $|Q| = \ell(Q)^n$. We introduce some more notation.

Definition 5.4.2. For $k \in \mathbf{Z}$ we denote by $\mathcal{D}_k$ the set of all dyadic cubes in $\mathbf{R}^n$ whose side length is 2^{-k}. We also denote by $\mathcal{D}$ the set of all dyadic cubes in $\mathbf{R}^n$. Then we have

$$\mathcal{D} = \bigcup_{k \in \mathbf{Z}} \mathcal{D}_k$$

and moreover, the σ-algebra $\sigma(\mathcal{D}_k)$ of measurable subsets of $\mathbf{R}^n$ formed by countable unions and complements of elements of $\mathcal{D}_k$ is increasing as k increases.

We observe the fundamental property of dyadic cubes, which clearly justifies their usefulness. Any two dyadic intervals of the same side length are either disjoint or they coincide. Moreover, any two dyadic intervals are either disjoint, or one contains the other. Similarly, any two dyadic cubes are either disjoint, or one contains the other.

Definition 5.4.3. Given a locally integrable function f on $\mathbf{R}^n$, we let

$$\operatorname*{Avg}_{Q} f = \frac{1}{|Q|}\int_Q f(t)\,dt$$

denote the average of f over a cube Q.

The *conditional expectation* of a locally integrable function f on $\mathbf{R}^n$ with respect to the increasing family of σ-algebras $\sigma(\mathcal{D}_k)$ generated by $\mathcal{D}_k$ is defined as

$$E_k(f)(x) = \sum_{Q\in\mathcal{D}_k} (\operatorname*{Avg}_{Q} f)\,\chi_Q(x),$$

for all $k \in \mathbf{Z}$. We also define the *dyadic martingale difference operator* D_k as follows:

$$D_k(f) = E_k(f) - E_{k-1}(f),$$

also for $k \in \mathbf{Z}$.

Next we introduce the family of Haar functions.

Definition 5.4.4. For a dyadic interval $I = [m2^{-k}, (m+1)2^{-k})$ we define $I_L = [m2^{-k}, (m+\frac{1}{2})2^{-k})$ and $I_R = [(m+\frac{1}{2})2^{-k}, (m+1)2^{-k})$ to be the left and right parts of I, respectively. The function

$$h_I(x) = |I|^{-\frac{1}{2}}\chi_{I_L} - |I|^{-\frac{1}{2}}\chi_{I_R}$$

is called the *Haar function associated with the interval I*.

We remark that Haar functions are constructed in such a way so that they have L^2 norm equal to 1. Moreover, the Haar functions have the following fundamental orthogonality property:

$$\int_{\mathbf{R}} h_I(x)h_{I'}(x)\,dx = \begin{cases} 0 & \text{when } I \neq I', \\ 1 & \text{when } I = I'. \end{cases} \tag{5.4.1}$$

To see this, observe that the Haar functions have L^2 norm equal to 1 by construction. Moreover, if $I \neq I'$, then they must have different lengths, say $|I'| < |I|$, and if they are not disjoint, then I' is contained either in the left or in the right half of I on either of which h_I is constant. Thus (5.4.1) follows. At this point we recall our notation

$$\langle f, g\rangle = \int_{\mathbf{R}} f(x)g(x)\,dx$$

valid for square integrable functions. Under this notation, (5.4.1) can be rewritten as $\langle h_I, h_{I'}\rangle = \delta_{I,I'}$, where the latter is 1 when $I = I'$ and zero otherwise.

5.4.b. Relation Between Dyadic Martingale Differences and Haar Functions

We have the following result relating the Haar functions to the dyadic martingale difference operators.

Proposition 5.4.5. *For every locally integrable function f on* **R** *and for all $k \in \mathbf{Z}$ we have the identity*

$$D_k(f) = \sum_{I\in\mathcal{D}_{k-1}} \langle f, h_I\rangle\, h_I \tag{5.4.2}$$

and also

$$\big\|D_k(f)\big\|_{L^2}^2 = \sum_{I\in\mathcal{D}_{k-1}} |\langle f, h_I\rangle|^2. \tag{5.4.3}$$

Proof. We observe that every interval J in $\mathcal{D}_k$ is either an I_L or an I_R for some unique $I \in \mathcal{D}_{k-1}$. Thus we can write

$$\begin{aligned} E_k(f) &= \sum_{J\in\mathcal{D}_k} (\operatorname*{Avg}_J f)\,\chi_J \\ &= \sum_{I\in\mathcal{D}_{k-1}} \bigg[\Big(\frac{2}{|I|}\int_{I_L} f(t)\,dt\Big)\chi_{I_L} + \Big(\frac{2}{|I|}\int_{I_R} f(t)\,dt\Big)\chi_{I_R}\bigg]. \end{aligned} \tag{5.4.4}$$

But we also have

$$\begin{aligned} E_{k-1}(f) &= \sum_{I\in\mathcal{D}_{k-1}} (\operatorname*{Avg}_I f)\,\chi_I \\ &= \sum_{I\in\mathcal{D}_{k-1}} \Big(\frac{1}{|I|}\int_{I_L} f(t)\,dt + \frac{1}{|I|}\int_{I_R} f(t)\,dt\Big)\big(\chi_{I_L} + \chi_{I_R}\big). \end{aligned} \tag{5.4.5}$$

Now taking the difference between (5.4.4) and (5.4.5) we obtain

$$\begin{aligned} D_k(f) = \sum_{I\in\mathcal{D}_{k-1}} \bigg[&\Big(\frac{1}{|I|}\int_{I_L} f(t)\,dt\Big)\chi_{I_L} - \Big(\frac{1}{|I|}\int_{I_R} f(t)\,dt\Big)\chi_{I_L} \\ &+ \Big(\frac{1}{|I|}\int_{I_R} f(t)\,dt\Big)\chi_{I_R} - \Big(\frac{1}{|I|}\int_{I_L} f(t)\,dt\Big)\chi_{I_R}\bigg], \end{aligned}$$

which is easily checked to be equal to

$$\sum_{I\in\mathcal{D}_{k-1}} \Big(\int_I f(t)h_I(t)\,dt\Big) h_I = \sum_{I\in\mathcal{D}_{k-1}} \langle f, h_I\rangle\, h_I.$$

Finally, (5.4.3) is a consequence of (5.4.1). □

Theorem 5.4.6. *Every function $f \in L^2(\mathbf{R}^n)$ can be written as*

$$f = \sum_{k\in\mathbf{Z}} D_k(f)\,, \tag{5.4.6}$$

where the series converges almost everywhere and in L^2. We also have

$$\big\|f\big\|_{L^2}^2 = \sum_{k\in\mathbf{Z}} \big\|D_k(f)\big\|_{L^2}^2. \tag{5.4.7}$$

Moreover, when $n=1$ we have the representation

$$f = \sum_{I\in\mathcal{D}} \langle f, h_I\rangle h_I\,, \tag{5.4.8}$$

where the sum converges a.e. and in L^2 and also

$$\big\|f\big\|_{L^2(\mathbf{R})}^2 = \sum_{I\in\mathcal{D}} |\langle f, h_I\rangle|^2\,. \tag{5.4.9}$$

Proof. In view of the Lebesgue differentiation theorem, the analogue of Corollary 2.1.16 for cubes, given a function $f\in L^2(\mathbf{R}^n)$ there is a set N_f of measure zero on $\mathbf{R}^n$ such that for all $x\in\mathbf{R}^n\setminus N_f$ we have that

$$\operatorname*{Avg}_{Q_j} f \to f(x)$$

whenever Q_j is sequence of decreasing cubes such that $\bigcap_j \overline{Q_j} = \{x\}$. Given x in $\mathbf{R}^n\setminus N_f$ there exists a unique sequence of dyadic cubes $Q_j(x)\in\mathcal{D}_j$ such that $\bigcap_{j=0}^\infty \overline{Q_j(x)} = \{x\}$. Then for all $x\in\mathbf{R}^n\setminus N_f$ we have

$$\lim_{j\to\infty} E_j(f)(x) = \lim_{j\to\infty} \sum_{Q\in\mathcal{D}_j} (\operatorname*{Avg}_{Q} f)\,\chi_Q(x) = \lim_{j\to\infty} \operatorname*{Avg}_{Q_j(x)} f = f(x)\,.$$

From this we conclude that $E_j(f)\to f$ a.e. as $j\to\infty$. We also observe that since $|E_j(f)|\le M_c f$, where M_c denotes the uncentered maximal function with respect to cubes, we have that $|E_j(f)-f|\le 2M_c(f)$, which allows us to obtain from the Lebesgue dominated convergence theorem that $E_j(f)\to f$ in L^2 as $j\to\infty$.

Next we study convergence of $E_j(f)$ as $j\to-\infty$. For a given $x\in\mathbf{R}^n$ and $Q_j(x)$ as before we have that

$$|E_j(f)(x)| = \big|\operatorname*{Avg}_{Q_j(x)} f\big| \le \bigg(\frac{1}{|Q_j(x)|}\int_{Q_j(x)} |f(t)|^2\,dt\bigg)^{\frac12} \le 2^{\frac{jn}{2}} \big\|f\big\|_{L^2}^2\,,$$

which tends to zero as $j\to-\infty$ since the side length of each $Q_j(x)$ is 2^{-j}. Since $|E_j(f)|\le M_c(f)$, the Lebesgue dominated convergence theorem allows us to conclude that $E_j(f)\to 0$ in L^2 as $j\to-\infty$. To obtain the conclusion asserted in (5.4.6) we simply observe that

$$\sum_{k=M}^{N} D_k(f) = E_N(f) - E_{M-1}(f) \to f$$

in L^2 and almost everywhere as $N \to \infty$ and $M \to -\infty$.

To prove (5.4.7) we first observe that we can rewrite $D_k(f)$ as

$$\begin{aligned} D_k(f) &= \sum_{Q\in\mathcal{D}_k} (\operatorname*{Avg}_Q f)\,\chi_Q - \sum_{R\in\mathcal{D}_{k-1}} (\operatorname*{Avg}_R f)\,\chi_R \\ &= \sum_{R\in\mathcal{D}_{k-1}} \Big[\sum_{\substack{Q\in\mathcal{D}_k \\ Q\subseteq R}} (\operatorname*{Avg}_Q f)\,\chi_Q - (\operatorname*{Avg}_R f)\,\chi_R \Big] \\ &= \sum_{R\in\mathcal{D}_{k-1}} \Big[\sum_{\substack{Q\in\mathcal{D}_k \\ Q\subseteq R}} (\operatorname*{Avg}_Q f)\,\chi_Q - \frac{1}{2^n} \sum_{\substack{Q\in\mathcal{D}_k \\ Q\subseteq R}} (\operatorname*{Avg}_Q f)\,\chi_R \Big] \\ &= \sum_{R\in\mathcal{D}_{k-1}} \sum_{\substack{Q\in\mathcal{D}_k \\ Q\subseteq R}} (\operatorname*{Avg}_Q f)\,\big(\chi_Q - 2^{-n}\chi_R\big)\,. \end{aligned} \tag{5.4.10}$$

Using this identity we obtain that for given integers $k' > k$ we have

$$\begin{aligned} &\int_{\mathbf{R}^n} D_k(f)(x)\, D_{k'}(f)(x)\, dx \\ &= \sum_{R\in\mathcal{D}_{k-1}} \sum_{\substack{Q\in\mathcal{D}_k \\ Q\subseteq R}} (\operatorname*{Avg}_Q f) \sum_{R'\in\mathcal{D}_{k'-1}} \sum_{\substack{Q'\in\mathcal{D}_{k'} \\ Q'\subseteq R'}} (\operatorname*{Avg}_{Q'} f) \int \big(\chi_Q - 2^{-n}\chi_R\big)\big(\chi_{Q'} - 2^{-n}\chi_{R'}\big)\, dx\,. \end{aligned}$$

Since $k' > k$ the last integral is nonzero only when $R' \subsetneqq R$. If this is the case, then $R' \subseteq Q_{R'}$ for some dyadic cube $Q_{R'} \in \mathcal{D}_k$ with $Q_{R'} \subseteq R$.

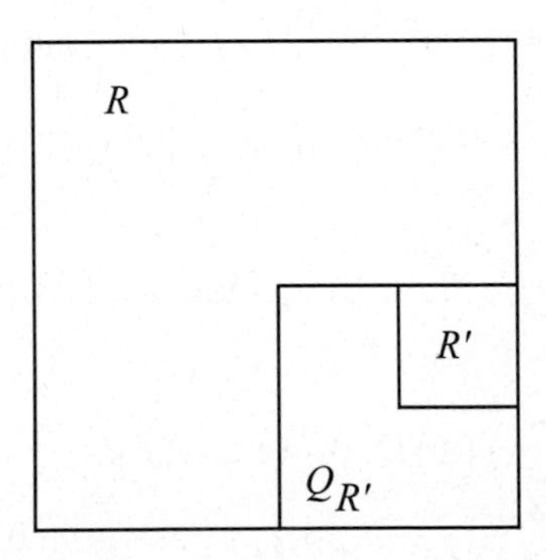

FIGURE 5.1. Picture of the cubes R, R', and $Q_{R'}$.

Then the function $\chi_{Q'} - 2^{-n}\chi_{R'}$ is supported in the cube $Q_{R'}$ and the function $\chi_Q - 2^{-n}\chi_R$ is constant on any dyadic subcube Q of R (of half its side length) and in particular is constant on $Q_{R'}$. Then

$$\sum_{\substack{Q'\in\mathcal{D}_{k'}\\ Q'\subseteq R'}} \big(\operatorname*{Avg}_{Q'} f\big) \int_{Q_{R'}} \chi_{Q'} - 2^{-n}\chi_{R'}\, dx = \sum_{\substack{Q'\in\mathcal{D}_{k'}\\ Q'\subseteq R'}} \big(\operatorname*{Avg}_{Q'} f\big) \big(|Q'| - 2^{-n}|R'|\big) = 0$$

since $|R'| = 2^n|Q'| = 2^{k'n}$. We conclude that $\big\langle D_k(f), D_{k'}(f)\big\rangle = 0$ whenever $k \neq k'$, from which we easily derive (5.4.7).

Now observe that (5.4.8) is a direct consequence of (5.4.2) and (5.4.9) is a direct consequence of (5.4.3).

□

5.4.c. The Dyadic Martingale Square Function

As a consequence of identity (5.4.7) proved in the previous subsection we obtain that

$$\Big\|\Big(\sum_{k\in\mathbf{Z}} |D_k(f)|^2\Big)^{\frac{1}{2}}\Big\|_{L^2(\mathbf{R}^n)} = \big\|f\big\|_{L^2(\mathbf{R}^n)}, \tag{5.4.11}$$

which says that the *dyadic martingale square function*

$$S(f) = \Big(\sum_{k\in\mathbf{Z}} |D_k(f)|^2\Big)^{\frac{1}{2}}$$

is L^2 bounded. It is natural to ask whether there exist L^p analogues of this result, and this is the purpose of the following theorem.

Theorem 5.4.7. *For any $1 < p < \infty$ there exists a constant $c_{p,n}$ such that for every function f in $L^p(\mathbf{R}^n)$ we have*

$$\frac{1}{c_{p',n}}\big\|f\big\|_{L^p(\mathbf{R}^n)} \le \big\|S(f)\big\|_{L^p(\mathbf{R}^n)} \le c_{p,n}\big\|f\big\|_{L^p(\mathbf{R}^n)}. \tag{5.4.12}$$

The lower inequality subsumes the fact that if $\big\|S(f)\big\|_{L^p(\mathbf{R}^n)} < \infty$, then f must be an L^p function.

Proof. We use the Rademacher functions r_j (Appendix C.1) to rewrite the upper estimate in (5.4.12) as

$$\int_0^1 \int_{\mathbf{R}^n} \Big|\sum_{k\in\mathbf{Z}} r_k(\omega)D_k(f)(x)\Big|^p dx\, d\omega \le C_p^p\big\|f\big\|_{L^p}^p, \tag{5.4.13}$$

where in this estimate we used an enumeration of the functions r_k indexed by $k \in \mathbf{Z}$. We will show a stronger estimate than (5.4.13), namely that for all $\omega \in [0, 1]$ we have

$$\int_{\mathbf{R}^n} \big|T_\omega(f)(x)\big|^p dx \le C_p^p\big\|f\big\|_{L^p}^p, \tag{5.4.14}$$

where

$$T_\omega(f)(x) = \sum_{k\in\mathbf{Z}} r_k(\omega) D_k(f)(x)\,.$$

In view of the L^2 estimate (5.4.11), we have that the operator T_ω is L^2 bounded with norm 1. We will show that T_ω is weak type $(1,1)$.

To show that T_ω is weak type $(1,1)$ we fix a function $f \in L^1$ and $\alpha > 0$. We apply the Calderón-Zygmund decomposition (Theorem 4.3.1) to f at height α to write

$$f = g + b, \qquad b = \sum_j (f - \operatorname*{Avg}_{Q_j} f)\chi_{Q_j}\,,$$

where Q_j are dyadic cubes which satisfy $\sum_j |Q_j| \le \frac{1}{\alpha}\big\|f\big\|_{L^1}$ and g has L^2 norm at most $(2^n\alpha\big\|f\big\|_{L^1})^{\frac{1}{2}}$; see (4.3.1). To achieve this decomposition, we apply the proof of Theorem 4.3.1 starting with a dyadic mesh of large cubes so that $|Q| \ge \frac{1}{\alpha}\big\|f\big\|_{L^1}$ for all Q in the mesh. Then we subdivide each Q in the mesh by halving each side and we select those cubes for which the average of $|f$ over them is bigger than α (and thus at most $2^n\alpha$). Since the original mesh consists of dyadic cubes, the stopping time argument of Theorem 4.3.1 ensures that each selected cube is dyadic.

We observe (and this is the key observation) that $T_\omega(b)$ is supported in $\bigcup_j Q_j$. To see this, we use identity (5.4.10) to write $T_\omega(b)$ as

$$\sum_j \Bigg[\sum_k r_k(\omega) \sum_{R\in\mathcal{D}_{k-1}} \sum_{\substack{Q\in\mathcal{D}_k\\ Q\subseteq R}} \operatorname*{Avg}_{Q}[(f - \operatorname*{Avg}_{Q_j} f)\chi_{Q_j}]\,\big(\chi_Q - 2^{-n}\chi_R\big)\Bigg]\,. \tag{5.4.15}$$

We consider the following three cases for the cubes Q that appear in the inner sum in (5.4.15): (i) $Q_j \subseteq Q$, (ii) $Q_j \cap Q = \emptyset$, and (iii) $Q \subsetneqq Q_j$. It is simple to see that in cases (i) and (ii) we have $\operatorname{Avg}_Q[(f - \operatorname{Avg}_{Q_j} f)\chi_{Q_j}] = 0$. Therefore the inner sum in (5.4.15) is taken over all Q that satisfy $Q \subsetneqq Q_j$. But then we must have that the unique dyadic parent R of Q is also contained in Q_j. It follows that the expression inside the square brackets in (5.4.15) is supported in R and therefore in Q_j. We conclude that $T_\omega(b)$ is supported in $\bigcup_j Q_j$. Using Exercise 4.3.5(a) we obtain that T_ω is of weak type $(1,1)$ with norm at most

$$\frac{\alpha\big|\{|T_\omega(g)| > \frac{\alpha}{2}\}\big| + \alpha\big|\bigcup_j Q_j\big|}{\big\|f\big\|_{L^1}} \le \frac{\alpha 4\alpha^{-2}\big\|g\big\|_{L^2}^2 + \big\|f\big\|_{L^1}}{\big\|f\big\|_{L^1}} \le 2^{n+2} + 1\,.$$

We have now established that T_ω is of weak type $(1,1)$. Since T_ω is L^2 bounded with norm 1, it follows by interpolation that T_ω is L^p bounded for all $1 < p < 2$. The L^p boundedness of T_ω for the remaining $p > 2$ follows by duality. (Note that the operators D_k and E_k are self-transpose.) We conclude the validity of (5.4.14), which implies that of (5.4.13). As observed, this is equivalent to the upper estimate in (5.4.12).

Finally, we notice that the lower estimate in (5.4.12) is a consequence of the upper estimate as in the case of the Littlewood-Paley operators Δ_j. Indeed, we need to observe that in view of (5.4.6) we have

$$\begin{aligned}
|\langle f,g\rangle| &= |\langle \sum_k D_k(f), \sum_{k'} D_{k'}(g)\rangle| \\
&= \Big|\sum_k \sum_{k'} \langle D_k(f), D_{k'}(g)\rangle\Big| \\
&= \Big|\sum_k \langle D_k(f), D_k(g)\rangle\Big| && \text{[Exercise 5.4.6(a)]} \\
&\le \int_{\mathbf{R}^n} \sum_k |D_k(f)(x)|\,|D_k(g)(x)|\,dx \\
&\le \int_{\mathbf{R}^n} S(f)(x)\,S(g)(x)\,dx && \text{(Cauchy-Schwarz inequality)} \\
&\le \big\|S(f)\big\|_{L^p} \big\|S(g)\big\|_{L^{p'}} && \text{(Hölder's inequality)} \\
&\le \big\|S(f)\big\|_{L^p} c_{p',n} \big\|g\big\|_{L^{p'}}
\end{aligned}$$

Taking the supremum over all functions g on $\mathbf{R}^n$ with $L^{p'}$ norm at most 1, we obtain that f gives rise to bounded linear functional on $L^{p'}$. It follows by the Riesz representation theorem that f must be an L^p function that satisfies the lower estimate in (5.4.12).

□

5.4.d. Almost Orthogonality between the Littlewood-Paley Operators and the Dyadic Martingale Difference Operators

Next, we discuss connections between the Littlewood-Paley operators Δ_j and the dyadic martingale difference operators D_k. It turns out that these operators are almost orthogonal in the sense that the L^2 operator norm of the composition $D_k\Delta_j$ decays exponentially as the indices j and k get farther away from each other.

For the purposes of the next theorem we define the Littlewood-Paley operators Δ_j as convolution operators with the function $\Psi_{2^{-j}}$, where

$$\widehat{\Psi}(\xi) = \widehat{\Phi}(\xi) - \widehat{\Phi}(2\xi)$$

and Φ is a fixed radial Schwartz function whose Fourier transform $\widehat{\Phi}$ is real valued, supported in the ball $|\xi| < 2$, and is equal to 1 on the ball $|\xi| < 1$. In this case we clearly have the identity

$$\sum_{j\in\mathbf{Z}} \widehat{\Psi}(2^{-j}\xi) = 1, \qquad \xi \neq 0 .$$

Then we have the following theorem.

Theorem 5.4.8. *There exists a constant C such that for every k, j in $\mathbf{Z}$ the following estimate on the operator norm of $D_k\Delta_j : L^2(\mathbf{R}^n) \to L^2(\mathbf{R}^n)$ is valid*

$$\big\|D_k\Delta_j\big\|_{L^2\to L^2} = \big\|\Delta_j D_k\big\|_{L^2\to L^2} \le C2^{-|j-k|}. \tag{5.4.16}$$

Proof. Since Ψ is a radial function it follows that Δ_j is equal to its transpose operator on L^2. Moreover, the operator D_k is also equal to its transpose. Thus $(D_k\Delta_j)^t = \Delta_j D_k$ and it therefore suffices to only prove that

$$\big\|D_k\Delta_j\big\|_{L^2\to L^2} \le C2^{-|j-k|}. \tag{5.4.17}$$

By a simple dilation argument it suffices to prove (5.4.17) when $k = 0$. In this case we have the estimate

$$\begin{aligned}\big\|D_0\Delta_j\big\|_{L^2\to L^2} &= \big\|E_0\Delta_j - E_{-1}\Delta_j\big\|_{L^2\to L^2}\\ &\le \big\|E_0\Delta_j - \Delta_j\big\|_{L^2\to L^2} + \big\|E_{-1}\Delta_j - \Delta_j\big\|_{L^2\to L^2}\end{aligned}$$

and also by the self-transposeness of the D_k's and Δ_j's we have

$$\begin{aligned}\big\|D_0\Delta_j\big\|_{L^2\to L^2} &= \big\|\Delta_j D_0\big\|_{L^2\to L^2} = \big\|\Delta_j E_0 - \Delta_j E_{-1}\big\|_{L^2\to L^2}\\ &\le \big\|\Delta_j E_0 - E_0\big\|_{L^2\to L^2} + \big\|\Delta_j E_{-1} - E_0\big\|_{L^2\to L^2}.\end{aligned}$$

The required estimate (5.4.17) (when $k = 0$) will be a consequence of the pair of inequalities

$$\big\|E_0\Delta_j - \Delta_j\big\|_{L^2\to L^2} + \big\|E_{-1}\Delta_j - \Delta_j\big\|_{L^2\to L^2} \le C2^{j} \quad \text{for } j \le 0, \tag{5.4.18}$$

$$\big\|\Delta_j E_0 - E_0\big\|_{L^2\to L^2} + \big\|\Delta_j E_{-1} - E_0\big\|_{L^2\to L^2} \le C2^{-j} \quad \text{for } j \ge 0. \tag{5.4.19}$$

We start by proving (5.4.18). We only consider the term $E_0\Delta_j - \Delta_j$ since the term $E_{-1}\Delta_j - \Delta_j$ is similar. Let $f \in L^2(\mathbf{R}^n)$. Then

$$\begin{aligned}\big\|E_0\Delta_j(f) - \Delta_j(f)\big\|_{L^2}^2 &= \sum_{Q\in\mathcal{D}_0} \big\|f * \Psi_{2^{-j}} - \operatorname*{Avg}_Q(f * \Psi_{2^{-j}})\big\|_{L^2(Q)}^2\\
&\le \sum_{Q\in\mathcal{D}_0} \int_Q\int_Q |(f * \Psi_{2^{-j}})(x) - (f * \Psi_{2^{-j}})(t)|^2\, dt\, dx\\
&\le \sum_{Q\in\mathcal{D}_0} \int_Q\int_Q \bigg(\int_{3Q} |f(y)||\Psi_{2^{-j}}(x-y)|\, dy\bigg)^2 dt\, dx\\
&\quad + \sum_{Q\in\mathcal{D}_0} \int_Q\int_Q \bigg(\int_{3Q} |f(y)||\Psi_{2^{-j}}(t-y)|\, dy\bigg)^2 dt\, dx\\
&\quad + \sum_{Q\in\mathcal{D}_0} \int_Q\int_Q \bigg(\int_{(3Q)^c} |f(y)|2^{jn+j}|\nabla\Psi(2^j(\xi_{x,t} - y))|\, dy\bigg)^2 dt\, dx,\end{aligned}$$

where $\xi_{x,t}$ lies on the line segment between x and t. It is a simple fact that the sum of the last three expressions is bounded by

$$C2^{2jn}\sum_{Q\in\mathcal{D}_0}\int_{3Q}|f(y)|^2\,dy+C_M2^{2j}\sum_{Q\in\mathcal{D}_0}\int_Q\bigg(\int_{\mathbf{R}^n}\frac{2^{jn}|f(y)|\,dy}{(1+2^j|x-y|)^M}\bigg)^2dx$$

which is clearly controlled by $C2^{2j}\big\|f\big\|_{L^2}^2$. This estimate is useful when $j\le 0$.

We now turn to the proof of (5.4.19). We set $S_j=\sum_{k\le j}\Delta_j$. Since Δ_j is the difference of two S_j's, it will suffice to prove (5.4.19), where Δ_j is replaced by S_j. We only work with the term $S_jE_0-E_0$ since the other term can be treated similarly. We have

$$\begin{aligned}\big\|S_jE_0(f)-E_0(f)\big\|_{L^2}^2&=\Big\|\sum_{Q\in\mathcal{D}_0}(\operatorname*{Avg}_Q f)\,(\Phi_{2^{-j}}*\chi_Q-\chi_Q)\Big\|_{L^2}^2\\&\le 2\Big\|\sum_{Q\in\mathcal{D}_0}(\operatorname*{Avg}_Q f)\,(\Phi_{2^{-j}}*\chi_Q-\chi_Q)\chi_{3Q}\Big\|_{L^2}^2\\&\quad+2\Big\|\sum_{Q\in\mathcal{D}_0}(\operatorname*{Avg}_Q f)\,(\Phi_{2^{-j}}*\chi_Q)\chi_{(3Q)^c}\Big\|_{L^2}^2.\end{aligned}$$

Since the functions appearing inside the sum in the first term have supports with bounded overlap, we obtain

$$\Big\|\sum_{Q\in\mathcal{D}_0}(\operatorname*{Avg}_Q f)(\Phi_{2^{-j}}*\chi_Q-\chi_Q)\chi_{3Q}\Big\|_{L^2}^2\le C\sum_{Q\in\mathcal{D}_0}(\operatorname*{Avg}_Q|f|)^2\big\|\Phi_{2^{-j}}*\chi_Q-\chi_Q\big\|_{L^2}^2,$$

and the crucial observation is that

$$\big\|\Phi_{2^{-j}}*\chi_Q-\chi_Q\big\|_{L^2}\le C2^{-j}|Q|^{1/2},$$

which can be easily checked using the Fourier transform. Using the bounded overlap property of the cubes $3Q$, we obtain

$$\Big\|\sum_{Q\in\mathcal{D}_0}(\operatorname*{Avg}_Q f)\,(\Phi_{2^{-j}}*\chi_Q-\chi_Q)\chi_{3Q}\Big\|_{L^2}^2\le C2^{-2j}\big\|f\big\|_{L^2}^2,$$

and the required conclusion will be proved if we can show that

$$\Big\|\sum_{Q\in\mathcal{D}_0}(\operatorname*{Avg}_Q f)\,(\Phi_{2^{-j}}*\chi_Q)\chi_{(3Q)^c}\Big\|_{L^2}^2\le C2^{-2j}\big\|f\big\|_{L^2}^2.\tag{5.4.20}$$

We prove (5.4.20) by using a purely size estimate. Let c_Q be the center of the dyadic cube Q. For $x\notin 3Q$ we have the estimate

$$|(\Phi_{2^{-j}}*\chi_Q)(x)|\le\frac{C_M2^{jn}}{(1+2^j|x-c_Q|)^M}\le\frac{C_M2^{jn}}{(1+2^j)^{M/2}}\frac{1}{(1+|x-c_Q|)^{M/2}}$$

since both $2^j \geq 1, |x - c_Q| \geq 1$. We now control the left hand side of (5.4.20) by

$$
\begin{aligned}
&2^{j(2n-M)} \sum_{Q\in\mathcal{D}_0} \sum_{Q'\in\mathcal{D}_0} (\operatorname{Avg}_Q |f|)(\operatorname{Avg}_{Q'} |f|) \int_{\mathbf{R}^n} \frac{C_M\, dx}{(1+|x-c_Q|)^{\frac{M}{2}} (1+|x-c_{Q'}|)^{\frac{M}{2}}} \\
&\leq 2^{j(2n-M)} \sum_{Q\in\mathcal{D}_0} \sum_{Q'\in\mathcal{D}_0} \frac{(\operatorname{Avg}_Q |f|)(\operatorname{Avg}_{Q'} |f|)}{(1+|c_Q - c_{Q'}|)^{\frac{M}{4}}} \int_{\mathbf{R}^n} \frac{C_M\, dx}{(1+|x-c_Q|)^{\frac{M}{4}} (1+|x-c_{Q'}|)^{\frac{M}{4}}} \\
&\leq 2^{j(2n-M)} \sum_{Q\in\mathcal{D}_0} \sum_{Q'\in\mathcal{D}_0} \frac{C_M}{(1+|c_Q - c_{Q'}|)^{\frac{M}{4}}} \left(\int_Q |f(y)|^2\, dy + \int_{Q'} |f(y)|^2\, dy \right) \\
&\leq C_M 2^{j(2n-M)} \sum_{Q\in\mathcal{D}_0} \int_Q |f(y)|^2\, dy = C_M 2^{j(2n-M)} \big\| f \big\|_{L^2}^2 .
\end{aligned}
$$

By taking M large enough, we obtain (5.4.20) and thus (5.4.19). □

Exercises

5.4.1. (a) Prove that no dyadic cube in $\mathbf{R}^n$ contains the point 0 in its interior.
(b) Prove that every interval in $\mathbf{R}$ is contained in the union of two dyadic intervals of at most its length.
(c) Prove that every cube in $\mathbf{R}^n$ is contained in the union of 2^n dyadic cubes of at most its side length.

5.4.2. The set $[m2^{-k}, (m+2)2^{-k})$ is a dyadic interval if and only if m is an even integer. More generally, the set $[m2^{-k}, (m+s)2^{-k})$ is a dyadic interval if and only if s is a power of 2 and m is an integer multiple of s.

5.4.3. Let Σ be the set of all $\sigma = (\sigma_1, \dots, \sigma_n)$ that satisfy $\sigma_j \in \{0, \frac{1}{2}, \frac{2}{3}\}$ for all j. Show that every cube Q in $\mathbf{R}^n$ is contained in cube of the form $\sigma + R$, where σ is in Σ and R is dyadic and has side length comparable to that of Q.

5.4.4. Show that the martingale maximal function $f \to \sup_k |E_k(f)|$ is weak type $(1,1)$ with constant at most 1.

5.4.5. (a) Show that $E_N(f) \to f$ a.e. as $N \to \infty$ for all $f \in L^1_{\mathrm{loc}}(\mathbf{R}^n)$.
(b) Prove that $E_N(f) \to f$ in L^p as $N \to \infty$ for all $f \in L^p(\mathbf{R}^n)$ whenever $1 < p \leq \infty$.

5.4.6. (a) Show that for all functions f and g if $k \neq k'$, then we have

$$\big\langle D_k(f), D_{k'}(g) \big\rangle = 0 .$$

(b) Conclude that for all functions f_j we have

$$\Big\|\sum_j D_j(f_j)\Big\|_{L^2} = \Big(\sum \|D_j(f_j)\|_{L^2}^2\Big)^{\frac{1}{2}}.$$

(c) Use Theorem 5.4.8 to show that

$$\Big\|\sum_j D_j\Delta_{j+r}D_j(f)\Big\|_{L^2} \le C2^{-|r|}\|f\|_{L^2}.$$

5.4.7. (*Grafakos and Kalton* [**214**]) Let D_j, Δ_j be as in the previous Exercise.
(a) Use that Exercise to obtain that the operator

$$V_r = \sum_{j\in\mathbf{Z}} D_j\Delta_{j+r}$$

is L^2 bounded with norm at most a multiple of $2^{-|r|}$.
(b) Use duality to show that V_r is L^p bounded for all $1 < p < \infty$ with a constant depending only on p and n.
(c) Combine the results in parts (a) and (b) to obtain that V_r is L^p bounded for all $1 < p < \infty$ with norm at most a multiple of $2^{-(1-|1-\frac{2}{p}|)\,|r|}$.
[*Hint:* Parts (a) and (b): Write $V_r = \sum_{j\in\mathbf{Z}} D_j\Delta_{j+r}\widetilde{\Delta}_{j+r}$, where $\widetilde{\Delta}_j$ is another family of Littlewood-Paley operators that satisfies $\Delta_j\widetilde{\Delta}_j = \Delta_j$.]

5.5. The Spherical Maximal Function*

In this section we discuss yet another consequence of the Littlewood-Paley theory, the boundedness of the spherical maximal operator.

5.5.a. Introduction of the Spherical Maximal Function

We denote throughout this section by $d\sigma$ the normalized Lebesgue measure on the sphere $\mathbf{S}^{n-1}$. For f in $L^p(\mathbf{R}^n)$, $1 \le p \le \infty$, we define the maximal operator

$$\mathcal{M}(f)(x) = \sup_{t>0}\left|\int_{\mathbf{S}^{n-1}} f(x-t\theta)\,d\sigma(\theta)\right| \tag{5.5.1}$$

and we observe that by Minkowski's integral inequality each expression inside the supremum in (5.5.1) is well defined for $f \in L^p$ for almost all $x \in \mathbf{R}^n$. The operator $\mathcal{M}$ is called the *spherical maximal function*. It is unclear at this point, for which functions f we have $\mathcal{M}(f) < \infty$ a.e. and for which values of p (if any) we have the L^p maximal inequality

$$\|\mathcal{M}(f)\|_{L^p(\mathbf{R}^n)} \le C_p\|f\|_{L^p(\mathbf{R}^n)}. \tag{5.5.2}$$

The introduction of the spherical maximal function is motivated by the fact that the spherical average

$$\int_{\mathbf{S}^2} f(x-ty)\,d\sigma(y) \tag{5.5.3}$$

is a solution of the *wave equation*

$$\begin{aligned}\frac{\partial^2 u}{\partial t^2}(x,t) &= \Delta(u)(x,t)\\ u(x,0) &= 0\\ \frac{\partial^2 u}{\partial t^2}(x,0) &= f(x)\end{aligned}$$

in $\mathbf{R}^3$. It is rather remarkable that the Fourier transform can be used to study almost everywhere convergence for several kinds of maximal averaging operators such as the spherical averages (5.5.3). This is achieved via the boundedness of the corresponding maximal operator; the maximal operator controlling spherical averages on $\mathbf{R}^n$ is given in (5.5.1).

Before we begin the analysis of the spherical maximal function, we notice that in view of the discussion in Appendices B.4 and B.6, we have that the Fourier transform of the spherical measure $d\sigma$ satisfies an estimate

$$|\widehat{d\sigma}(\xi)| \le \frac{C_n}{(1+|\xi|)^{\frac{n-1}{2}}}\,. \tag{5.5.4}$$

This fact will be the single most important ingredient in the proof of the next theorem.

Theorem 5.5.1. *Let $n \ge 3$. For each $\frac{n}{n-1} < p \le \infty$, there is a constant C_p such that*

$$\big\|\mathcal{M}(f)\big\|_{L^p(\mathbf{R}^n)} \le C_p\big\|f\big\|_{L^p(\mathbf{R}^n)} \tag{5.5.5}$$

holds for all f in $L^p(\mathbf{R}^n)$. It follows that for all $\frac{n}{n-1} < p \le \infty$ and $f \in L^p(\mathbf{R}^n)$ we have

$$\lim_{t\to 0}\int_{\mathbf{S}^{n-1}} f(x-t\theta)\,d\sigma(\theta) = f(x) \tag{5.5.6}$$

for almost all $x \in \mathbf{R}^n$.

The proof of this theorem will be presented in the rest of this section. The operator we need to study is the maximal multiplier operator

$$\sup_{t>0}\big|\big(\widehat{f}(\xi)\, m(t\xi)\big)^{\vee}\big|,$$

where we set $m(\xi) = \widehat{d\sigma}(\xi)$. We decompose the multiplier $m(\xi)$ into radial pieces as follows: We fix a radial C^∞ function φ_0 in $\mathbf{R}^n$ such that $\varphi_0(\xi) = 1$ when $|\xi| \le 1$

and $\varphi_0(\xi) = 0$ when $|\xi| \geq 2$. For $j \geq 1$ we let

$$\varphi_j(\xi) = \varphi_0(2^{-j}\xi) - \varphi_0(2^{1-j}\xi) \tag{5.5.7}$$

and we observe that $\varphi_j(\xi)$ is localized near $|\xi| \approx 2^j$. Then we have

$$\sum_{j=0}^{\infty} \varphi_j = 1\,,$$

which implies

$$m = \sum_{j=0}^{\infty} m_j\,,$$

where $m_j = \varphi_j m$ for all $j \geq 0$. Also, the following estimate is valid:

$$\mathcal{M}(f) \leq \sum_{j=0}^{\infty} \mathcal{M}_j(f)\,,$$

where

$$\mathcal{M}_j(f)(x) = \sup_{t>0} \big|\big(\widehat{f}(\xi)\, m_j(t\xi)\big)^\vee(x)\big|\,.$$

Since the function m_0 is C^∞ with compact support, we have that $\mathcal{M}_0$ maps L^p into itself for all $1 < p \leq \infty$. (See Exercise 5.5.1.)

We define *g-functions* associated with m_j as follows:

$$G_j(f)(x) = \bigg(\int_0^\infty |A_{j,t}(f)(x)|^2 \,\frac{dt}{t}\bigg)^{\frac{1}{2}},$$

where $A_{j,t}(f)(x) = \big(\widehat{f}(\xi)\, m_j(t\xi)\big)^\vee(x)$.

5.5.b. The First Key Lemma

We have the following lemma:

Lemma 5.5.2. *There is a constant $C = C(n) < \infty$ such that for any $j \geq 1$ we have the estimate*

$$\big\|\mathcal{M}_j(f)\big\|_{L^2} \leq C 2^{(\frac{1}{2}-\frac{n-1}{2})j}\big\|f\big\|_{L^2}$$

for all functions f in $L^2(\mathbf{R}^n)$.

Proof. We define a function

$$\widetilde{m}_j(\xi) = \xi \cdot \nabla m_j(\xi)$$

and we let $\widetilde{A}_{j,t}(f)(x) = \big(\widehat{f}(\xi)\, M_j(t\xi)\big)^\vee(x)$ and

$$\widetilde{G}_j(f)(x) = \bigg(\int_0^\infty |\widetilde{A}_{j,t}(f)(x)|^2 \,\frac{dt}{t}\bigg)^{\frac{1}{2}},$$

be the associated g-function. The identity

$$t\,\frac{dA_{j,t}}{dt}(f) = \widetilde{A}_{j,t}(f)$$

is clearly valid for all j, t, and f Schwartz functions. Since

$$\lim_{s\to\infty} A_{j,s}(f) = \lim_{s\to\infty} A_{j,s}(f) = 0$$

for Schwartz functions f, it follows that

$$\begin{aligned} A_{j,t}(f)^2(x) &= 2\int_0^\infty A_{j,s}(f)(x)\, s\frac{dA_{j,s}}{ds}(f)(x)\,\frac{ds}{s} \\ &= 2\int_0^\infty A_{j,s}(f)(x)\widetilde{A}_{j,s}(f)(x)\,\frac{ds}{s}\,, \end{aligned}$$

from which we obtain the estimate

$$|A_{j,t}(f)(x)|^2 \le 2\int_0^\infty |A_{j,s}(f)(x)|\,|\widetilde{A}_{j,s}(f)(x)|\,\frac{ds}{s} \tag{5.5.8}$$

for all f in the Schwartz class. By density the same estimate holds for all $f \in \bigcup_{1<p<\infty} L^p$.

Taking the supremum over all $t > 0$ on the left-hand side in (5.5.8) and integrating over $\mathbf{R}^n$, we obtain the estimate

$$\begin{aligned} \|\mathcal{M}_j(f)\|_{L^2}^2 &\le 2\int_{\mathbf{R}^n}\left|\int_0^\infty A_{j,s}(f)(x)\widetilde{A}_{j,s}(f)(x)\,\frac{ds}{s}\right|^2 dx \\ &\le 2\,\|G_j(f)\|_{L^2}\|\widetilde{G}_j(f)\|_{L^2}\,, \end{aligned}$$

where the last inequality follows by applying the Cauchy-Schwarz inequality twice. Using the facts that $\|m_j\|_{L^\infty} \le 2^{-j\frac{n-1}{2}}$ and $\|\widetilde{m}_j\|_{L^\infty} \le 2^{j(1-\frac{n-1}{2})}$ and that the functions m_j and $\widetilde{m}_j$ are supported in the annuli $2^{j-1} \le |\xi| \le 2^{j+1}$, we obtain that the g-functions G_j and $\widetilde{G}_j$ are L^2 bounded with norms at most a constant multiples of the quantities $2^{-j\frac{n-1}{2}}$ and $2^{j(1-\frac{n-1}{2})}$, respectively; see Exercise 5.5.2. Note that since $n \ge 3$, both exponents are negative. We conclude that

$$\|\mathcal{M}_j(f)\|_{L^2} \le C2^{j(\frac{1}{2}-\frac{n-1}{2})}\|f\|_{L^2}\,,$$

which is what we needed to prove.

□

5.5.c. The Second Key Lemma

Next we will need the following lemma.

Lemma 5.5.3. *There exists a constant $C = C(n) < \infty$ such that for all $j \geq 1$ and for all f in $L^1(\mathbf{R}^n)$ we have*

$$\big\|\mathcal{M}_j(f)\big\|_{L^{1,\infty}} \leq C\, 2^j \big\|f\big\|_{L^1}.$$

Proof. Let $K^{(j)} = (\varphi_j)^\vee * d\sigma = \Phi_{2^{-j}} * d\sigma$, where Φ is a Schwartz function. Setting

$$(K^{(j)})_t(x) = t^{-n} K^{(j)}(x/t)$$

we have that

$$\mathcal{M}_j(f) = \sup_{t>0} |(K^{(j)})_t * f|. \tag{5.5.9}$$

The proof of the lemma is based on the following estimate:

$$\mathcal{M}_j(f) \leq C\, 2^j\, M(f) \tag{5.5.10}$$

and the weak type $(1,1)$ boundedness of the Hardy-Littlewood maximal operator M (Theorem 2.1.6). To establish (5.5.10), it will suffice to show that for any $N > 0$ there is a constant $C_N < \infty$ such that

$$|K^{(j)}(x)| = |(\Phi_{2^{-j}} * d\sigma)(x)| \leq \frac{C_N\, 2^j}{(1+|x|)^N}. \tag{5.5.11}$$

Then Theorem 2.1.10 will yield (5.5.10) and hence the required conclusion.

Using the fact that Φ is a Schwartz function, we obtain that

$$|(\Phi_{2^{-j}} * d\sigma)(x)| \leq C_N \int_{\mathbf{S}^{n-1}} \frac{2^{nj}\, d\sigma(y)}{(1+2^j|x-y|)^N}.$$

We split the last integral in the regions

$$S_{-1} = \mathbf{S}^{n-1} \cap \{y \in \mathbf{R}^n :\ 2^j|x-y| \leq 1\}$$

and for $r \geq 0$

$$S_r = \mathbf{S}^{n-1} \cap \{y \in \mathbf{R}^n :\ 2^r < 2^j|x-y| \leq 2^{r+1}\}.$$

The key observation is that whenever $B = B(x,R)$ is a ball in $\mathbf{R}^n$, then the spherical measure of the set $\mathbf{S}^{n-1} \cap B(x,R)$ is at most a dimensional multiple of R^{n-1}. Using this observation, we obtain the following estimate for the expression $|(\Phi_{2^{-j}} * d\sigma)(x)|$:

$$\sum_{r=-1}^{j}\int_{S_r}\frac{C_N 2^{nj}\,d\sigma(y)}{(1+2^j|x-y|)^N}+\sum_{r=j+1}^{\infty}\int_{S_r}\frac{C_N 2^{nj}\,d\sigma(y)}{(1+2^j|x-y|)^N}$$

$$\le C_N' 2^{nj}\Bigg[\sum_{r=-1}^{j}\frac{d\sigma(S_r)\chi_{B(0,5)}}{2^{rN}}+\sum_{r=j+1}^{\infty}\frac{d\sigma(S_r)\,\chi_{B(0,2^{r-j+2})}}{2^{rN}}\Bigg]$$

$$\le C_N' 2^{nj}\Bigg[\sum_{r=-1}^{j}\frac{c_n 2^{(r-j+1)(n-1)}\chi_{B(0,5)}}{2^{rN}}+\sum_{r=j+1}^{\infty}\frac{c_n 2^{(r-j+1)(n-1)}\,\chi_{B(0,2^{r-j+2})}}{2^{rN}}\Bigg]$$

$$\le C_{N,n}\Bigg[2^j\chi_{B(0,5)}+2^{(n-N)j}\sum_{s=1}^{\infty}2^{(n-1-N)s}\,\chi_{B(0,4\cdot 2^s)}\Bigg]$$

$$\le C_{N,n}'\Bigg[2^j\chi_{B(0,5)}+\frac{1}{(1+|x|)^{N-n+1}}\Bigg]$$

$$\le \frac{C_{N,n}'' 2^j}{(1+|x|)^{N-n+1}}\,.$$

Taking N large enough, we establish (5.5.11).

□

5.5.d. Completion of the Proof

It remains to combine Lemmata 5.5.2 and 5.5.3 to obtain the proof of the theorem. This is a simple matter. Interpolating between the $L^2 \to L^2$ and $L^1 \to L^{1,\infty}$ estimates obtained in Lemmata 5.5.2 and 5.5.3, we obtain

$$\big\|\mathcal{M}_j(f)\big\|_{L^p(\mathbf{R}^n)} \le C_p 2^{(\frac{n}{p}-(n-1))j}\big\|f\big\|_{L^p(\mathbf{R}^n)}$$

for all $1 < p \le 2$. When $p > \frac{n}{n-1}$ the series $\sum_{j=1}^{\infty} 2^{(\frac{n}{p}-(n-1))j}$ converges and we conclude that $\mathcal{M}$ is L^p bounded for these p's. The boundedness of $\mathcal{M}$ on L^p for $p > 2$ follows by interpolation between L^q for $q < 2$ and the trivial estimate $\mathcal{M} : L^\infty \to L^\infty$.

Exercises

5.5.1. (a) Suppose that the bounded function m whose inverse Fourier transform satisfies $|m^\vee(x)| \le C(1+|x|)^{-n-\delta}$ for some $\delta > 0$. Show that the maximal multiplier

$$\mathcal{M}_m(f)(x) = \sup_{t>0}\big|\big(\widehat{f}(\xi)\,m(t\xi)\big)^\vee(x)\big|$$

is L^p bounded for all $1 < p < \infty$.

(b) Obtain the same conclusion when $\xi^\alpha m(\xi)$ is in $L^1(\mathbf{R}^n)$ for all multiindices α with $|\alpha| \le [\frac{n}{2}] + 1$.

[*Hint:* Control $\mathcal{M}_m$ by the Hardy-Littlewood maximal operator.]

5.5.2. Suppose that the function m is supported in the annulus $R \le |\xi| \le 2R$ and is bounded by A. Show that the g-function

$$G(f)(x) = \left(\int_0^\infty |(m(t\xi)\widehat{f}(\xi))^\vee(x)|^2 \frac{dt}{t} \right)^{\frac{1}{2}}$$

maps $L^2(\mathbf{R}^n)$ into $L^2(\mathbf{R}^n)$ with bound at most $A\sqrt{\log 2}$.

5.5.3. (*Rubio de Francia* [**436**]) Use the idea of Lemma 5.5.2 to show that if $m(\xi)$ satisfies $|m(\xi)| \le (1+|\xi|)^{-a}$ and $|\nabla m(\xi)| \le (1+|\xi|)^{-b}$ and $a+b>1$, then the maximal operator

$$\mathcal{M}_m(f)(x) = \sup_{t>0} \left| \left(\widehat{f}(\xi)\, m(t\xi) \right)^\vee (x) \right|$$

is bounded from $L^2(\mathbf{R}^n)$ into itself.

[*Hint:* Use $\mathcal{M}_m \le \sum_{j=0}^\infty \mathcal{M}_{m,j}$, where each $\mathcal{M}_{m,j}$ corresponds to the multiplier $\varphi_j m$, φ_j as in (5.5.7). Show that

$$\big\|\mathcal{M}_{m,j}(f)\big\|_{L^2} \le C\big(\|\varphi_j m\|_{L^\infty} \|\varphi_j \widetilde{m}\|_{L^\infty}\big)^{\frac{1}{2}} \big\|f\big\|_{L^2} \le C 2^{j\frac{1-(a+b)}{2}} \big\|f\big\|_{L^2},$$

where $\widetilde{m}(\xi) = \xi \cdot \nabla m(\xi)$.]

5.5.4. (*Rubio de Francia* [**436**]) Observe that the proof of Theorem 5.5.1 gives the following more general result: If $m(\xi)$ is the Fourier transform of a compactly supported Borel measure and satisfies $|m(\xi)| \le (1+|\xi|)^{-a}$ for some $a > 0$ and all $\xi \in \mathbf{R}^n$, then the maximal operator of the previous Exercise maps $L^p(\mathbf{R}^n)$ into itself when $p > \frac{2a+1}{2a}$.

5.5.5. Show that Theorem 5.5.1 is false when $n = 1$, that is, show that the maximal operator

$$\mathcal{M}_1(f)(x) = \sup_{t>0} \frac{|f(x+t) + f(x-t)|}{2}$$

is unbounded on $L^p(\mathbf{R})$ for all $p < \infty$.

5.5.6. Show that Theorem 5.5.1 is false when $n \ge 2$ and $p \le \frac{n}{n-1}$.

[*Hint:* Choose a compactly supported and radial function that is equal to $|y|^{1-n}(-\log|y|)^{-1}$ when $|y| \le 1$.]

5.6. Wavelets

We will be concerned with orthonormal bases of $L^2(\mathbf{R})$ generated by translations and dilations of a single function such as the Haar functions we encountered in Section 5.4. The Haar functions are generated by integer translations and dyadic dilations of the single function $\chi_{[0,\frac{1}{2})} - \chi_{[\frac{1}{2},1)}$. This function is not smooth, and the main question addressed in this section is whether there exist smooth analogues of the Haar functions.

Definition 5.6.1. A square integrable function φ on $\mathbf{R}^n$ is called a *wavelet* if the family of functions

$$\varphi_{\nu,k}(x) = 2^{\frac{\nu n}{2}}\varphi(2^\nu x - k)\,,$$

where ν ranges over $\mathbf{Z}$ and k over $\mathbf{Z}^n$, is an orthonormal basis of $L^2(\mathbf{R}^n)$. Note that the Fourier transform of $\varphi_{\nu,k}$ is given by

$$\widehat{\varphi_{\nu,k}}(\xi) = 2^{-\frac{\nu n}{2}}\widehat{\varphi}(2^{-\nu}\xi)e^{-2\pi i 2^{-\nu}\xi\cdot k}\,. \tag{5.6.1}$$

Rephrasing the question posed earlier, the main issue that will be addressed in this section is whether smooth wavelets actually exist. Before we embark on this topic, let us recall that we have already encountered examples of nonsmooth wavelets.

Example 5.6.2. (The Haar wavelet) Recall the family of functions

$$h_I(x) = |I|^{-\frac{1}{2}}(\chi_{I_L} - \chi_{I_R})\,,$$

where I ranges over $\mathcal{D}$ (the set of all dyadic intervals) and I_L is the left part of I and I_R is the right part of I. Note that if $I = [2^{-\nu}k, 2^{-\nu}(k+1))$, then

$$h_I(x) = 2^{\frac{\nu}{2}}\varphi(2^\nu x - k)\,,$$

where

$$\varphi(x) = \chi_{[0,\frac{1}{2})} - \chi_{[\frac{1}{2},1)}\,. \tag{5.6.2}$$

The single function φ in (5.6.2) therefore generates the Haar basis by taking translations and dilations. Moreover, we observed in Section 5.4 that the family $\{h_I\}_I$ is orthonormal. In Theorem 5.4.6 we obtained the representation

$$f = \sum_{I\in\mathcal{D}}\langle f, h_I\rangle h_I \qquad \text{in } L^2\,,$$

which proves the completeness of the system $\{h_I\}_{I\in\mathcal{D}}$ in $L^2(\mathbf{R})$.

5.6.a. Some Preliminary Facts

Before we look at more examples, we make some observations. We begin with the following useful fact.

Proposition 5.6.3. *Let $g \in L^1(\mathbf{R}^n)$. Then*

$$\widehat{g}(m) = 0 \qquad \text{for all } m \in \mathbf{Z}^n \setminus \{0\},$$

if and only if

$$\sum_{k\in\mathbf{Z}^n} g(x+k) = \int_{\mathbf{R}^n} g(t)\,dt$$

for almost all $x \in \mathbf{T}^n$.

Proof. We define the periodized function

$$G(x) = \sum_{k\in\mathbf{Z}^n} g(x+k)\,,$$

which is easily shown to be in $L^1(\mathbf{T}^n)$. Moreover, we have

$$\widehat{G}(m) = \widehat{g}(m)$$

for all $m \in \mathbf{Z}^n$, where $\widehat{G}(m)$ denotes the mth Fourier coefficient of G and $\widehat{g}(m)$ denotes the Fourier transform of g at $\xi = m$. If $\widehat{g}(m) = 0$ for all $m \in \mathbf{Z}^n \setminus \{0\}$, then all the Fourier coefficients of G (except for $m = 0$) vanish, which means that the sequence $\{\widehat{G}\}_{m\in\mathbf{Z}^n}$ lies in $\ell^1(\mathbf{Z}^n)$ and hence Fourier inversion applies. We conclude that for almost all $x \in \mathbf{T}^n$ we have

$$G(x) = \sum_{m\in\mathbf{Z}^n} \widehat{G}(m)e^{2\pi i m\cdot x} = \widehat{G}(0) = \widehat{g}(0) = \int_{\mathbf{R}^n} g(t)\,dt\,.$$

Conversely, if G is a constant, then $\widehat{G}(m) = 0$ for all $m \in \mathbf{Z}^n \setminus \{0\}$ and so the same holds for g. □

A consequence of the previous proposition is the following.

Proposition 5.6.4. *Let $\varphi \in L^1(\mathbf{R}^n)$. Then the sequence*

$$\{\varphi(x-k)\}_{k\in\mathbf{Z}^n} \tag{5.6.3}$$

forms an orthonormal set in $L^2(\mathbf{R}^n)$ if and only if

$$\sum_{k\in\mathbf{Z}^n} |\widehat{\varphi}(\xi+k)|^2 = 1 \tag{5.6.4}$$

for almost all $\xi \in \mathbf{R}^n$.

Proof. Observe that either hypothesis (5.6.3) or (5.6.4) implies that $\big\|\varphi\big\|_{L^2} = 1$. Also the orthonormality condition

$$\int_{\mathbf{R}^n} \varphi(x-j)\varphi(x-k)\,dx = \begin{cases} 1 & \text{when } j = k, \\ 0 & \text{when } j \neq k \end{cases}$$

is equivalent to

$$\int_{\mathbf{R}^n} e^{-2\pi i k\cdot\xi}\widehat{\varphi}(\xi)\overline{e^{-2\pi i j\cdot\xi}\widehat{\varphi}(\xi)}\,dx = (|\widehat{\varphi}|^2)^\frown(k-j) = \begin{cases} 1 & \text{when } j = k, \\ 0 & \text{when } j \neq k \end{cases}$$

in view of Parserval's identity. Proposition 5.6.3 gives that the latter is equivalent to

$$\sum_{k\in\mathbf{Z}^n} |\widehat{\varphi}(\xi+k)|^2 = \int_{\mathbf{R}^n} |\widehat{\varphi}(t)|^2\, dt = 1$$

for almost all $\xi \in \mathbf{R}^n$. □

Corollary 5.6.5. *Let $\varphi \in L^1(\mathbf{R}^n)$ and suppose that the sequence*

$$\{\varphi(x-k)\}_{k\in\mathbf{Z}^n} \tag{5.6.5}$$

forms an orthonormal set in $L^2(\mathbf{R}^n)$. Then the measure of the support of $\widehat{\varphi}$ is at least 1, that is,

$$|\operatorname{supp}\widehat{\varphi}| \geq 1\,. \tag{5.6.6}$$

If $|\operatorname{supp}\widehat{\varphi}| = 1$, then $|\widehat{\varphi}(\xi)| = 1$ for almost all $\xi \in \operatorname{supp}\widehat{\varphi}$.

Proof. It follows from (5.6.4) that $|\widehat{\varphi}| \leq 1$ for almost all $\xi \in \mathbf{R}^n$. Therefore,

$$|\operatorname{supp}\widehat{\varphi}| \geq \int_{\mathbf{R}^n} |\widehat{\varphi}(\xi)|^2\, d\xi = \int_{\mathbf{T}^n} \sum_{k\in\mathbf{Z}^n} |\widehat{\varphi}(\xi+k)|^2\, d\xi = \int_{\mathbf{T}^n} 1\, d\xi = 1\,.$$

It follows from the previous series of inequalities that if equality holds in (5.6.6), then $|\widehat{\varphi}(\xi)| = 1$ for almost all ξ in $\operatorname{supp}\widehat{\varphi}$. □

5.6.b. Construction of a Nonsmooth Wavelet

Having established these preliminary facts, we now start our search for examples of wavelets. It follows from Corollary 5.6.5 that the support of the Fourier transform of a wavelet must have measure at least 1. It is reasonable to ask whether this support can have measure exactly 1. Example 5.6.6 indicates that this can indeed happen. As dictated by the same corollary, the Fourier transform of such a wavelet must satisfy $|\widehat{\varphi}(\xi)| = 1$ for almost all $\xi \in \operatorname{supp}\widehat{\varphi}$, so it is natural to look for a wavelet φ such that $\widehat{\varphi} = \chi_A$ for some set A. We can start by asking whether the function

$$\widehat{\varphi} = \chi_{[-\frac{1}{2},\frac{1}{2}]}$$

on $\mathbf{R}$ is an appropriate Fourier transform of a wavelet, but a moment's thought shows that the functions $\varphi_{\mu,0}$ and $\varphi_{\nu,0}$ cannot be orthogonal to each other when $\mu \neq 0$. The problem here is that the Fourier transforms of the functions $\varphi_{\nu,k}$ cluster near the origin and do not allow for the needed orthogonality. We can fix this problem by considering a function whose Fourier transform vanishes near the origin. Among such functions, a natural candidate is

$$\chi_{[-1,-\frac{1}{2})} + \chi_{[\frac{1}{2},1)}\,, \tag{5.6.7}$$

which we will show shortly that it is indeed the Fourier transform of a wavelet.

Example 5.6.6. Let $A = [-1,-\frac{1}{2}) \cup [\frac{1}{2},1)$ and define a function φ on $\mathbf{R}^n$ by setting

$$\widehat{\varphi} = \chi_{A^n}\,.$$

Then we assert that the family of functions $\{2^{\nu n/2}\varphi(2^\nu x - k)\}_{k\in\mathbf{Z}^n,\nu\in\mathbf{Z}}$, is an orthonormal basis of $L^2(\mathbf{R}^n)$ (i.e., the function φ is a wavelet). This is an example of a wavelet with *minimally supported frequency.*

To see this assertion, first note that $\{\varphi_{0,k}\}_{k\in\mathbf{Z}^n}$ is an orthonormal set since (5.6.4) is easily seen to hold. Dilating by 2^ν, it follows that $\{\varphi_{\nu,k}\}_{k\in\mathbf{Z}^n}$ is also an orthonormal set for every fixed $\nu\in\mathbf{Z}$. Second, observe that if $\mu\neq\nu$, then

$$\operatorname{supp}\widehat{\varphi_{\nu,k}} \cap \operatorname{supp}\widehat{\varphi_{\mu,l}} = \emptyset\,. \tag{5.6.8}$$

This implies that the family $\{2^{\nu n/2}\varphi(2^\nu x - k)\}_{k\in\mathbf{Z}^n,\nu\in\mathbf{Z}}$ is also orthonormal.

Finally, we need to show completeness. Here we use Exercise 5.6.2 to write

$$(\varphi_{2^{-\nu}} * f)\widehat{\ }(\xi) = 2^{-\nu n}\sum_{k\in\mathbf{Z}^n}(\varphi_{2^{-\nu}} * f)(-\tfrac{k}{2^\nu})e^{2\pi i\frac{k}{2^\nu}\xi}\,, \qquad \xi\in A^n\,, \tag{5.6.9}$$

where the series converges in $L^2(A^n)$. Next we observe that the following identity holds for $\widehat{\varphi}$:

$$\sum_{\nu\in\mathbf{Z}}|\widehat{\varphi}(2^\nu\xi)|^2 = 1\,, \qquad \xi\neq 0\,. \tag{5.6.10}$$

This implies that for all f in $\mathcal{S}(\mathbf{R}^n)$ we have

$$f = \sum_{\nu\in\mathbf{Z}}\varphi_{2^{-\nu}} * \varphi_{2^{-\nu}} * f = \Big[\sum_{\nu\in\mathbf{Z}}\widehat{\varphi_{2^{-\nu}}}\,(\varphi_{2^{-\nu}} * f)\widehat{\ }\Big]^\vee\,, \tag{5.6.11}$$

where the series converges in L^2. Inserting in (5.6.11) the value of $(\varphi_{2^{-\nu}} * f)\widehat{\ }$ given in identity (5.6.9), we obtain

$$\begin{aligned}
f(x) &= \sum_{\nu\in\mathbf{Z}}\sum_{k\in\mathbf{Z}^n}(\varphi_{2^{-\nu}} * f)(-\tfrac{k}{2^\nu})\big[\widehat{\varphi_{2^{-\nu}}}(\xi)e^{2\pi i\frac{k}{2^\nu}\xi}\big]^\vee(x) \\
&= 2^{-\nu n}\sum_{\nu\in\mathbf{Z}}\sum_{k\in\mathbf{Z}^n}(\varphi_{2^{-\nu}} * f)(-\tfrac{k}{2^\nu})\varphi_{2^{-\nu}}(x+\tfrac{k}{2^\nu}) \\
&= \sum_{\nu\in\mathbf{Z}}\sum_{k\in\mathbf{Z}^n}\langle f,\varphi_{\nu,k}\rangle\varphi_{\nu,k}(x)\,,
\end{aligned}$$

where the double series converges in $L^2(\mathbf{R}^n)$. This shows that every Schwartz function can be written as an L^2 sum of $\varphi_{\nu,k}$'s and by density the same is true for every square integrable f.

5.6.c. Construction of a Smooth Wavelet*

The wavelet basis of $L^2(\mathbf{R}^n)$ we constructed in Example 5.6.6 is forced to have slow decay at infinity since the Fourier transforms of the elements of the basis are nonsmooth. Smoothing out the function $\widehat{\varphi}$ but still expecting φ to be wavelet is a bit tricky as property (5.6.8) may be violated when $\mu \neq \nu$ and moreover (5.6.4) may be destroyed. These two obstacles are overcome by the careful construction of the next theorem.

Theorem 5.6.7. *There exists a Schwartz function φ on the real line that is a wavelet [i.e., the collection of functions $\{2^{\frac{\nu}{2}}\varphi(2^\nu x - k)\}_{k,\nu \in \mathbf{Z}}$ is an orthonormal basis of $L^2(\mathbf{R})$]. Moreover, the function φ can be constructed so that its Fourier transform satisfies*

$$\operatorname{supp}\widehat{\varphi} \subseteq [-\tfrac{4}{3}, -\tfrac{1}{3}] \cup [\tfrac{1}{3}, \tfrac{4}{3}]. \tag{5.6.12}$$

Note that in view of condition (5.6.12), the function φ must have vanishing moments of all orders.

Proof. We start with an odd smooth real-valued function Θ on the real line such that $\Theta(t) = \frac{\pi}{4}$ for $t \geq \frac{1}{6} - 10^{-10}$ and such that Θ is increasing on the interval $[-\frac{1}{6}, \frac{1}{6}]$. We set

$$\alpha(t) = \sin(\Theta(t) + \tfrac{\pi}{4}), \qquad \beta(t) = \cos(\Theta(t) + \tfrac{\pi}{4})$$

and we observe that

$$\alpha(t)^2 + \beta(t)^2 = 1$$

and that

$$\alpha(-t) = \beta(t)$$

for all real t. Next we introduce the smooth function ω defined via

$$\omega(t) = \begin{cases} \beta(-\frac{t}{2} - \frac{1}{2}) = \alpha(\frac{t}{2} + \frac{1}{2}) & \text{when } t \in [-\frac{4}{3}, -\frac{2}{3}], \\ \alpha(-t - \frac{1}{2}) = \beta(t + \frac{1}{2}) & \text{when } t \in [-\frac{2}{3}, -\frac{1}{3}], \\ \alpha(t - \frac{1}{2}) & \text{when } t \in [\frac{1}{3}, \frac{2}{3}], \\ \beta(\frac{t}{2} - \frac{1}{2}) & \text{when } t \in [\frac{2}{3}, \frac{4}{3}], \end{cases}$$

on the interval $[-\frac{4}{3}, -\frac{1}{3}] \cup [\frac{1}{3}, \frac{4}{3}]$. Note that ω is an even function. Finally we define the function φ by letting

$$\widehat{\varphi}(\xi) = e^{-\pi i \xi}\omega(\xi)$$

and we note that

$$\varphi(x) = \int_{\mathbf{R}} \omega(\xi) e^{2\pi i \xi (x - \frac{1}{2})} d\xi = 2\int_0^\infty \omega(\xi)\cos\big(2\pi(x - \tfrac{1}{2})\xi\big)\, d\xi\,.$$

It follows that the function φ is symmetric about the number $\frac{1}{2}$, that is, we have

$$\varphi(x) = \varphi(1 - x)$$

for all $x \in \mathbf{R}$. Note that φ is a Schwartz function whose Fourier transform is supported in the set $[-\frac{4}{3}, -\frac{1}{3}] \cup [\frac{1}{3}, \frac{4}{3}]$.

Having defined φ we proceed by showing that it is a wavelet. In view of identity (5.6.1) we have that $\widehat{\varphi_{\nu,k}}$ is supported in the set $\frac{1}{3}2^\nu \leq |\xi| \leq \frac{4}{3}2^\nu$ while $\widehat{\varphi_{\mu,k}}$ is supported in the set $\frac{1}{3}2^\mu \leq |\xi| \leq \frac{4}{3}2^\mu$. The intersection of these sets has measure zero when $|\mu - \nu| \geq 2$, which implies that such wavelets are orthogonal to each other. Therefore, it suffices to verify orthogonality between adjacent scales (i.e., when $\nu = \mu$ and $\nu = \mu + 1$).

We begin with the case $\nu = \mu$, which, by a simple dilation, is reduced to the case $\nu = \mu = 0$. Thus to obtain the orthogonality of the functions $\varphi_{0,k}(x) = \varphi(x-k)$ and $\varphi_{0,j}(x) = \varphi(x-j)$, in view of Proposition 5.6.4, it will suffice to show that

$$\sum_{k\in\mathbf{Z}} |\widehat{\varphi}(\xi + k)|^2 = 1\,. \tag{5.6.13}$$

Since the sum in (5.6.13) is 1-periodic, we only check that is equal to 1 for $\xi \in [\frac{1}{3}, \frac{4}{3}]$. First for $\xi \in [\frac{1}{3}, \frac{2}{3}]$, the sum in (5.6.13) is equal to

$$\begin{aligned} |\widehat{\varphi}(\xi)|^2 + |\widehat{\varphi}(\xi - 1)|^2 &= \omega(\xi)^2 + \omega(\xi-1)^2 \\ &= \alpha(\xi - \tfrac{1}{2})^2 + \beta((\xi-1) + \tfrac{1}{2})^2 \\ &= 1 \end{aligned}$$

from the definition of ω. A similar argument also holds for $\xi \in [\frac{2}{3}, \frac{4}{3}]$, and this completes the proof of (5.6.13). As a consequence of this identity we also obtain that the functions $\varphi_{0,k}$ have L^2 norm equal to 1 and therefore so have the functions $\varphi_{\nu,k}$, via a standard dilation.

Next we prove the orthogonality of the functions $\varphi_{\nu,k}$ and $\varphi_{\nu+1,j}$ for general $\nu, k, j \in \mathbf{Z}$. We begin by observing the validity of the following identity:

$$\widehat{\varphi}(\xi)\overline{\widehat{\varphi}(\tfrac{\xi}{2})} = \begin{cases} e^{-\pi i \xi/2}\beta(\frac{\xi}{2} - \frac{1}{2})\alpha(\frac{\xi}{2} - \frac{1}{2}) & \text{when } \frac{2}{3} \leq \xi \leq \frac{4}{3}, \\ e^{-\pi i \xi/2}\alpha(\frac{\xi}{2} + \frac{1}{2})\beta(\frac{\xi}{2} + \frac{1}{2}) & \text{when } -\frac{4}{3} \leq \xi \leq -\frac{2}{3}. \end{cases} \tag{5.6.14}$$

Indeed, from the definition of φ, it follows that

$$\widehat{\varphi}(\xi)\overline{\widehat{\varphi}(\tfrac{\xi}{2})} = e^{-\pi i \xi/2}\omega(\xi)\omega(\tfrac{\xi}{2})\,.$$

This function is supported in

$$\{\xi \in \mathbf{R} : \tfrac{1}{3} \leq |\xi| \leq \tfrac{4}{3}\} \cap \{\xi \in \mathbf{R} : \tfrac{2}{3} \leq |\xi| \leq \tfrac{8}{3}\} = \{\xi \in \mathbf{R} : \tfrac{2}{3} \leq |\xi| \leq \tfrac{4}{3}\}$$

and on this set it is equal to

$$e^{-\pi i \xi/2}\begin{cases} \beta(\frac{\xi}{2} - \frac{1}{2})\alpha(\frac{\xi}{2} - \frac{1}{2}) & \text{when } \frac{2}{3} \leq \xi \leq \frac{4}{3}, \\ \alpha(\frac{\xi}{2} + \frac{1}{2})\beta(\frac{\xi}{2} + \frac{1}{2}) & \text{when } -\frac{4}{3} \leq \xi \leq -\frac{2}{3}, \end{cases}$$

by the definition of ω. This establishes (5.6.14).

We now turn to the orthogonality of the functions $\varphi_{\nu,k}$ and $\varphi_{\nu+1,j}$ for general $\nu, k, j \in \mathbf{Z}$. Using (5.6.1) and (5.6.14) we have

$$\begin{aligned}
\langle \varphi_{\nu,k} \,|\, \varphi_{\nu+1,j} \rangle &= \langle \widehat{\varphi_{\nu,k}} \,|\, \widehat{\varphi_{\nu+1,j}} \rangle \\
&= \int_{\mathbf{R}} 2^{-\frac{\nu}{2}} \widehat{\varphi}(2^{-\nu}\xi) e^{-2\pi i \frac{\xi k}{2^\nu}} \overline{2^{-\frac{\nu+1}{2}} \widehat{\varphi}(2^{-\frac{\nu+1}{2}}\xi) e^{-2\pi i \frac{\xi j}{2^{\nu+1}}}} \, d\xi \\
&= \frac{1}{\sqrt{2}} \int_{\mathbf{R}} \widehat{\varphi}(\xi) \overline{\widehat{\varphi}(\tfrac{\xi}{2})} e^{-2\pi i \xi (k - \frac{j}{2})} \, d\xi \\
&= \frac{1}{\sqrt{2}} \int_{-\frac{4}{3}}^{-\frac{2}{3}} \alpha(\tfrac{\xi}{2} + \tfrac{1}{2}) \beta(\tfrac{\xi}{2} + \tfrac{1}{2}) e^{-2\pi i \xi (k - \frac{j}{2} + \frac{1}{4})} \, d\xi \\
&\quad + \frac{1}{\sqrt{2}} \int_{\frac{2}{3}}^{\frac{4}{3}} \alpha(\tfrac{\xi}{2} - \tfrac{1}{2}) \beta(\tfrac{\xi}{2} - \tfrac{1}{2}) e^{-2\pi i \xi (k - \frac{j}{2} + \frac{1}{4})} \, d\xi \\
&= 0 \,,
\end{aligned}$$

where the last identity follows from the change of variables $\xi = \xi' - 2$ in the second to last integral, which transforms its range of integration to $[\frac{2}{3}, \frac{4}{3}]$ and its integrand to negative of that of the last displayed integral.

Our final task is to show that the orthonormal system $\{\varphi_{\nu,k}\}_{\nu,k \in \mathbf{Z}}$ is complete. We will show this by proving that whenever a square-integrable function f satisfies

$$\langle f \,|\, \varphi_{\nu,k} \rangle = 0 \tag{5.6.15}$$

for all $\nu, k \in \mathbf{Z}$, then f must be zero. Suppose that (5.6.15) holds. Then by Plancherel's identity

$$\int_{\mathbf{R}} \widehat{f}(\xi) \overline{2^{-\frac{\nu}{2}} \widehat{\varphi}(2^{-\nu}\xi) e^{-2\pi i 2^{-\nu} \xi k}} \, d\xi = 0$$

for all ν, k and thus

$$\int_{\mathbf{R}} \widehat{f}(2^\nu \xi) \widehat{\varphi}(\xi) e^{2\pi i \xi k} \, d\xi = \big(\widehat{f}(2^\nu(\cdot))\, \widehat{\varphi}\big)\widehat{\ }(-k) = 0 \tag{5.6.16}$$

for all $\nu, k \in \mathbf{Z}$. It follows from Proposition 5.6.3 and (5.6.16) (with $k = 0$) that

$$\sum_{k \in \mathbf{Z}} \widehat{f}(2^\nu(\xi + k)) \widehat{\varphi}(\xi + k) = \int_{\mathbf{R}} \widehat{f}(2^\nu \xi)\, \widehat{\varphi}(\xi) \, d\xi = \big(\widehat{f}(2^\nu(\cdot))\, \widehat{\varphi}\big)\widehat{\ }(0) = 0$$

for all $\nu \in \mathbf{Z}$.

We will next show that the identity

$$\sum_{k \in \mathbf{Z}} \widehat{f}(2^\nu(\xi + k)) \widehat{\varphi}(\xi + k) = 0 \tag{5.6.17}$$

for all $\nu \in \mathbf{Z}$ implies that $\widehat{f}$ must be identically equal to zero. Suppose that $\frac{1}{3} \le \xi \le \frac{2}{3}$. In this case the support properties of $\widehat{\varphi}$ imply that the only terms in the sum in (5.6.17) that do not vanish are $k=0$ and $k=-1$. Thus for $\frac{1}{3}\le\xi\le\frac{2}{3}$, the identity in (5.6.17) reduces to

$$\begin{aligned}0 &= \widehat{f}(2^\nu(\xi-1))\widehat{\varphi}(\xi-1)+\widehat{f}(2^\nu\xi)\widehat{\varphi}(\xi)\\ &= \widehat{f}(2^\nu(\xi-1))e^{-\pi i(\xi-1)}\beta((\xi-1)+\tfrac{1}{2})+\widehat{f}(2^\nu\xi)e^{-\pi i\xi}\alpha(\xi-\tfrac{1}{2}),\end{aligned}$$

hence

$$-\widehat{f}(2^\nu(\xi-1))\beta(\xi-\tfrac{1}{2})+\widehat{f}(2^\nu\xi)\alpha(\xi-\tfrac{1}{2})=0,\qquad \tfrac{1}{3}\le\xi\le\tfrac{2}{3}. \tag{5.6.18}$$

Next we observe that when $\frac{2}{3}\le\xi\le\frac{4}{3}$ only the terms with $k=0$ and $k=-2$ survive in the identity in (5.6.17). This is because when $k=-1$, $\xi+k=\xi-1\in[-\frac{1}{3},\frac{1}{3}]$ and this interval has null intersection with the support of $\widehat{\varphi}$. Therefore, (5.6.17) reduces to

$$\begin{aligned}0 &= \widehat{f}(2^\nu(\xi-2))\widehat{\varphi}(\xi-2)+\widehat{f}(2^\nu\xi)\widehat{\varphi}(\xi)\\ &= \widehat{f}(2^\nu(\xi-2))e^{-\pi i(\xi-2)}\alpha(\tfrac{\xi-2}{2}+\tfrac{1}{2})+\widehat{f}(2^\nu\xi)e^{-\pi i\xi}\beta(\tfrac{\xi}{2}-\tfrac{1}{2});\end{aligned}$$

hence

$$\widehat{f}(2^\nu(\xi-2))\alpha(\tfrac{\xi}{2}-\tfrac{1}{2})+\widehat{f}(2^\nu\xi)\beta(\tfrac{\xi}{2}-\tfrac{1}{2})=0,\qquad \tfrac{2}{3}\le\xi\le\tfrac{4}{3}. \tag{5.6.19}$$

Replacing first ν by $\nu-1$ and then $\frac{\xi}{2}$ by ξ in (5.6.19) we obtain

$$\widehat{f}(2^\nu(\xi-1))\alpha(\xi-\tfrac{1}{2})+\widehat{f}(2^\nu\xi)\beta(\xi-\tfrac{1}{2})=0,\qquad \tfrac{1}{3}\le\xi\le\tfrac{2}{3}. \tag{5.6.20}$$

Now consider the 2×2 system of equations given by (5.6.18) and (5.6.20) with unknown $\widehat{f}(2^\nu(\xi-1))$ and $\widehat{f}(2^\nu\xi)$. The determinant of the system is

$$\det\begin{pmatrix}-\beta(\xi-1/2) & \alpha(\xi-1/2)\\ \alpha(\xi-1/2) & \beta(\xi-1/2)\end{pmatrix}=-1\neq 0.$$

Therefore, the system has the unique solution

$$\widehat{f}(2^\nu(\xi-1))=\widehat{f}(2^\nu\xi)=0,$$

which is valid for all $\nu\in\mathbf{Z}$ and all $\xi\in[\frac{1}{3},\frac{2}{3}]$. We conclude that $\widehat{f}(\xi)=0$ for all $\xi\in\mathbf{R}^n$ and thus $f=0$. This proves the completeness of the system $\{\varphi_{\nu,k}\}$. We conclude that the function φ is a wavelet.

□

5.6.d. A Sampling Theorem

We end this section by discussing how one can recover a band-limited function by its values at a countable number of points.

Definition 5.6.8. An integrable function on $\mathbf{R}^n$ is called *band limited* if its Fourier transform has compact support.

For every band-limited function there is a $B > 0$ such that its Fourier transform is supported in the cube $[-B, B]^n$. In such a case we say that the function is band limited on the cube $[-B, B]^n$.

It is an interesting observation that such functions are completely determined by their values at the points $x = k/2B$, where $k \in \mathbf{Z}^n$. We have the following result.

Theorem 5.6.9. *Let f be band limited on the cube $[-B, B]^n$. Then f can be sampled by its values at the points $x = k/2B$, where $k \in \mathbf{Z}^n$. In particular, we have*

$$f(x_1, \dots, x_n) = \sum_{k \in \mathbf{Z}^n} f\Big(\frac{k}{2B}\Big) \prod_{j=1}^{n} \frac{\sin(2\pi B x_j - \pi k_j)}{2\pi B x_j - \pi k_j} \tag{5.6.21}$$

for all $x \in \mathbf{R}^n$.

Proof. Since the function $\widehat{f}$ is supported in $[-B, B]^n$, we can use Exercise 5.6.2 to obtain

$$\widehat{f}(\xi) = \frac{1}{(2B)^n} \sum_{k \in \mathbf{Z}^n} \widehat{\widehat{f}}\Big(\frac{k}{2B}\Big) e^{2\pi i \frac{k}{2B} \cdot \xi} = \frac{1}{(2B)^n} \sum_{k \in \mathbf{Z}^n} f\Big(-\frac{k}{2B}\Big) e^{2\pi i \frac{k}{2B} \cdot \xi} .$$

Inserting this identity in the inversion formula

$$f(x) = \int_{[-B,B]^n} \widehat{f}(\xi) e^{2\pi i x \cdot \xi} \, d\xi ,$$

which holds since $\widehat{f}$ is continuous and therefore integrable over $[-B, B]^n$, we obtain

$$\begin{aligned} f(x) &= \int_{[-B,B]^n} \frac{1}{(2B)^n} \sum_{k \in \mathbf{Z}^n} f\Big(-\frac{k}{2B}\Big) e^{2\pi i \frac{k}{2B} \cdot \xi} e^{2\pi i x \cdot \xi} \, d\xi \\ &= \sum_{k \in \mathbf{Z}^n} f\Big(-\frac{k}{2B}\Big) \frac{1}{(2B)^n} \int_{[-B,B]^n} e^{2\pi i (\frac{k}{2B} + x) \cdot \xi} \, d\xi \\ &= \sum_{k \in \mathbf{Z}^n} f\Big(-\frac{k}{2B}\Big) \prod_{j=1}^{n} \frac{\sin(2\pi B x_j + \pi k_j)}{2\pi B x_j + \pi k_j} . \end{aligned}$$

This is exactly (5.6.21) when we change k to $-k$. □

Remark 5.6.10. Identity (5.6.21) holds for any $B' > B$. In particular, we have

$$\sum_{k \in \mathbf{Z}^n} f\Big(\frac{k}{2B}\Big) \prod_{j=1}^{n} \frac{\sin(2\pi B x_j - \pi k_j)}{2\pi B x_j - \pi k_j} = \sum_{k \in \mathbf{Z}^n} f\Big(\frac{k}{2B'}\Big) \prod_{j=1}^{n} \frac{\sin(2\pi B' x_j - \pi k_j)}{2\pi B' x_j - \pi k_j}$$

for all $x \in \mathbf{R}^n$ whenever f is band-limited in $[-B, B]^n$. In particular band-limited functions in $[-B, B]^n$ can be sampled by their values at the points $k/2B'$ for any $B' \geq B$.

However, band-limited functions in $[-B, B]^n$ cannot be sampled by the points $k/2B'$ for any $B' < B$, as the following example indicates.

Example 5.6.11. For $0 < B' < B$, let $f(x) = g(x)\sin(2\pi B'x)$, where $\widehat{g}$ is supported in the interval $[-(B-B'), B-B']$. Then f is band limited in $[-B,B]$, but it cannot be sampled by its values at the points $k/2B'$ since it vanishes at these points and f is not identically zero if g is not the zero function.

Exercises

5.6.1. (a) Let $A = [-1,-\frac{1}{2}) \cup [\frac{1}{2},1)$. Show that the family $\{e^{2\pi i m x}\}_{m\in\mathbf{Z}}$ is an orthonormal basis of $L^2(A)$.
(b) Obtain the same conclusion for the family $\{e^{2\pi i m\cdot x}\}_{m\in\mathbf{Z}^n}$ in $L^2(A^n)$.
[*Hint:* To show completeness, given $f \in L^2(A)$, define h on $[0,1]$ by setting $h(x) = f(x-1)$ for $x \in [0,\frac{1}{2})$ and $h(x) = f(x)$ for $x \in [\frac{1}{2},1)$. Observe that $\widehat{h}(m) = \widehat{f}(m)$ for all $m \in \mathbf{Z}$ and expand h in Fourier series.]

5.6.2. (a) Suppose that g is supported in $[-b,b]^n$ for some $b > 0$ and that the sequence $\{\widehat{g}(k/2b)\}_{k\in\mathbf{Z}^n}$ lies in $\ell^2(\mathbf{Z}^n)$. Show that

$$g(x) = (2b)^{-n} \sum_{k\in\mathbf{Z}^n} \widehat{g}(\tfrac{k}{2b}) e^{2\pi i \frac{k}{2b}\cdot x}$$

when $x \in [-b,b]^n$, where the series converges in $L^2(\mathbf{R}^n)$.
(b) Suppose that g is supported in $[0,b]^n$ for some $b > 0$ and that the sequence $\{\widehat{g}(k/b)\}_{k\in\mathbf{Z}^n}$ lies in $\ell^2(\mathbf{Z}^n)$. Show that

$$g(x) = b^{-n} \sum_{k\in\mathbf{Z}^n} \widehat{g}(\tfrac{k}{b}) e^{2\pi i \frac{k}{b}\cdot x}$$

for $x \in [0,b]^n$, where the series converges in $L^2(\mathbf{R}^n)$.
(c) When $n = 1$, obtain the same as conclusion in part (b) for $x \in [-b,-\frac{b}{2}) \cup [\frac{b}{2},b)$ provided g is supported in this set.
[*Hint:* All the results follow by dilations. Part (c): Use the result in the previous Exercise.]

5.6.3. Show that the sequence of functions

$$H_k(x_1,\dots,x_n) = (2B)^{\frac{n}{2}} \prod_{j=1}^{n} \frac{\sin\left(\pi(2Bx_j - k_j)\right)}{\pi(2Bx_j - k_j)}, \qquad k \in \mathbf{Z}^n$$

is orthonormal in $L^2(\mathbf{R}^n)$.
[*Hint:* Interpret the previous functions as the Fourier transforms of known functions.]

5.6.4. Prove the following spherical multidimensional version of Theorem 5.6.9. Suppose that $\widehat{f}$ is supported in the ball $|\xi| \le R$. Show that

$$f(x) = \sum_{k \in \mathbf{Z}^n} f\big(-\tfrac{k}{2R}\big) \frac{1}{2^n} \frac{J_{\frac{n}{2}}(2\pi |Rx + \frac{k}{2}|)}{|Rx + \frac{k}{2}|^{\frac{n}{2}}},$$

where J_a is the Bessel function of order a.
[*Hint:* Use the result in Appendix B.5.]

5.6.5. (*X. H. Wang*) For ψ a function on $\mathbf{R}$ set $\psi_{j,k}(x) = 2^{j/2}\psi(2^j x - k)$. Show that the equality

$$\|f\|_{L^2}^2 = \sum_{j,k \in \mathbf{Z}} |\langle f \,|\, \psi_{j,k}\rangle|^2$$

holds for all f in $L^2(\mathbf{R})$ exactly when

$$\sum_{j \in \mathbf{Z}} |\widehat{\psi}(2^j \xi)|^2 = 1 \qquad \text{a.e.}$$

and

$$\sum_{j \ge 0} \widehat{\psi}(2^j \xi)\overline{\widehat{\psi}(2^j(\xi + q))} = 0 \qquad \text{a.e.}$$

for all q odd integers.

5.6.6. Let $\{a_k\}_{k \in \mathbf{Z}^n}$ be in ℓ^p for some $1 < p < \infty$. Show that the sum

$$\sum_{k \in \mathbf{Z}^n} a_k \prod_{j=1}^{n} \frac{\sin(2\pi B x_j - \pi k_j)}{2\pi B x_j - \pi k_j}$$

converges in $\mathcal{S}'(\mathbf{R}^n)$ to an L^p function A on $\mathbf{R}^n$ that is band limited in $[-B, B]^n$. Moreover the L^p norm of A is controlled by a constant multiple of the ℓ^p norm of $\{a_k\}_k$.

5.6.7. (a) Suppose that f is a tempered distribution on $\mathbf{R}^n$ whose Fourier transform is supported in the ball $B(0, (1 - \varepsilon)\frac{1}{2})$ for some $\varepsilon > 0$. Show that for all $0 < p \le \infty$ there is a constant $C_{n,p,\varepsilon}$ such that

$$\big\|f\big\|_{L^p(\mathbf{R}^n)} \le C_{n,p,\varepsilon} \big\|\{f(k)\}_k\big\|_{\ell^p(\mathbf{Z}^n)}.$$

In particular, if the values $\{f(k)\}_{k \in \mathbf{Z}^n}$ form an ℓ^p sequence, then f must coincide with an L^p function.
(b) Consider functions of the form $\sin(\pi x)/(\pi x)$ on $\mathbf{R}$ to construct a counterexample to the statement in part (a) when $\varepsilon = 0$.
[*Hint:* Take a Schwartz function Φ whose Fourier transform is supported in $B(0, \frac{1}{2})$ and that is identically equal to 1 on the support of $\widehat{f}$. Then $f = f * \Phi$.

Apply Theorem 5.6.9 to the function $f * \Phi$ and use the rapid decay of Φ to sum the series.]

5.6.8. (a) Let $\psi(x)$ be a nonzero continuous integrable function on $\mathbf{R}$ that satisfies $\int_{\mathbf{R}} \psi(x)\, dx = 0$ and

$$C_\psi = 2\pi \int_{-\infty}^{+\infty} \frac{|\widehat{\psi}(t)|^2}{|t|}\, dt < \infty\,.$$

Define the *wavelet transform* of f in $L^2(\mathbf{R})$ by setting

$$W(f;a,b)(x) = \frac{1}{\sqrt{|a|}} \int_{-\infty}^{+\infty} f(x)\overline{\psi\Big(\frac{x-b}{a}\Big)}\, dx$$

when $a \neq 0$ and $W(f;0,b) = 0$. Show that for any $f \in L^2(\mathbf{R})$ the following inversion formula holds:

$$f(x) = \frac{1}{C_\psi} \int_{-\infty}^{+\infty} \int_{-\infty}^{+\infty} \frac{1}{|a|^{\frac{1}{2}}} \psi\Big(\frac{x-b}{a}\Big) W(f;a,b)\, db\, \frac{da}{a^2}\,.$$

(b) State and prove an analogous wavelet transform inversion property on $\mathbf{R}^n$.

[*Hint:* Apply Theorem 2.2.14 (5) in the b-integral to reduce matters to Fourier inversion.]

5.6.9. (*P. Casazza*) Let $\tau^a(\varphi)(x) = \varphi(x-a)$ be the translation operator. Let $\varphi \in L^2(\mathbf{R})$ and $b \in \mathbf{R} \setminus \{0\}$. For $\gamma \in [0,1]$ define

$$\Phi_b(\gamma) = \sum_{n \in \mathbf{Z}} |\widehat{\varphi}(\tfrac{\gamma+n}{b})|^2\,,$$

$$f = \sum_{n \in \mathbf{Z}} a_n \tau^{nb}(\varphi)\,,$$

and

$$\Psi(f)(\gamma) = \sum_{n \in \mathbf{Z}} a_n e^{2\pi i n \gamma}\,.$$

Show that

$$\|f\|_{L^2(\mathbf{R})}^2 = \frac{1}{b} \int_0^1 |\Psi(f)(\gamma)|^2 \Phi_b(\gamma)\, d\gamma$$

and

$$\sum_{m \in \mathbf{Z}} |\langle f, \tau^{mb}(\varphi)\rangle|^2 = \frac{1}{b^2} \int_0^1 |\Psi(f)(\gamma)|^2 \Phi_b(\gamma)^2\, d\gamma\,.$$

(a) Conclude that the sequence $\{\tau^{nb}(\varphi)\}_{n\in\mathbf{Z}}$ is a *frame*, that is, it satisfies for some $0 < A \le B < \infty$ and all f in $L^2(\mathbf{R})$

$$A\big\|f\big\|_{L^2}^2 \le \sum_{n\in\mathbf{Z}} \big|\langle f, \tau^{nb}(\varphi)\rangle\big|^2 \le B\big\|f\big\|_{L^2}^2,$$

if and only if $\frac{1}{b}A \le \Phi_b \le \frac{1}{b}B$.
(b) Show that when $a/b \in \mathbf{Z}$, then the sequence $\{\tau^{nb}(\varphi)\}_{n\in\mathbf{Z}}$ is a frame if and only if the sequence $\{\tau^{na}(\varphi)\}_{n\in\mathbf{Z}}$ is a frame.
(c) If $0 < |b| < |a|$ and $a/b \notin \mathbf{Z}$, then there exists a function φ such the sequence $\{\tau^{nb}(\varphi)\}_{n\in\mathbf{Z}}$ is a frame but the sequence $\{\tau^{na}(\varphi)\}_{n\in\mathbf{Z}}$ is not a frame.

5.6.10. (*P. Casazza*) On $\mathbf{R}^n$ let e_j be the vector whose coordinates are zero everywhere except for the jth entry that is 1. Set $q_j = e_j - \sum_{k=1}^n e_k$ for $1 \le j \le n$ and also $q_{n+1} = n^{-1/2}\sum_{k=1}^n e_k$. Prove that

$$\sum_{j=1}^{n+1} |q_j \cdot x| = |x|^2$$

for all $x \in \mathbf{R}^n$. This provides an example of a *tight frame* on $\mathbf{R}^n$.

HISTORICAL NOTES

An early account of square functions in the context of Fourier series appears in the work of Kolmogorov [**307**], who proved the almost everywhere convergence of lacunary partial sums of Fourier series of periodic square-integrable functions. This result was systematically studied and extended to L^p functions, $1 < p < \infty$, by Littlewood and Paley [**340**], [**341**], [**342**] using complex analysis techniques. The real variable treatment of the Littlewood and Paley theorem was pioneered by Stein [**471**] and allowed the higher-dimensional extension of the theory. The use of vector-valued inequalities in the proof of Theorem 5.1.2 is contained in Benedek, Calderón, and Panzone [**32**]. A Littlewood-Paley theorem for lacunary sectors in $\mathbf{R}^2$ was obtained by Nagel, Stein, and Wainger [**388**].

An interesting Littlewood-Paley estimate holds for $2 \le p < \infty$: There exists a constant C_p such that for all families of disjoint open intervals I_j in $\mathbf{R}$ the following estimate $\big\|(\sum_j |(\widehat{f}\chi_{I_j})^\vee|^2)^{\frac{1}{2}}\big\|_{L^p} \le C_p\big\|f\big\|_{L^p}$ holds for all functions $f \in L^p(\mathbf{R})$. This was shown by Rubio de Francia [**435**] but the special case in which $I_j = (j, j+1)$ was previously obtained by Carleson [**85**]. An alternative proof of Rubio de Francia's theorem was obtained by Bourgain [**47**]. A higher-dimensional analogue of this estimate for arbitrary disjoint open rectangles in $\mathbf{R}^n$ with sides parallel to the axes was obtained by Journé [**283**]. Easier proofs of the higher-dimensional result were subsequently obtained by Sjölin [**460**], Soria [**465**], and Sato [**445**].

Part (a) of Theorem 5.2.7 is due to Mihlin [**375**] and the generalization in part (b) to Hörmander [**253**]. The multiplier Theorem 5.2.2 can be found in Marcinkiewicz's article [**366**] in the context of one-dimensional Fourier series. The power 6 in estimate (5.2.3) that appears in the statement of Theorem 5.2.2 is not optimal. Tao and Wright [**513**] proved that the best power of $(p-1)^{-1}$ in this theorem is $\frac{3}{2}$ as $p \to 1$. An improvement of the Marcinkiewicz multiplier theorem in one dimension was obtained by Coifman, Rubio de Francia, and Semmes [**120**]. Weighted norm estimates for Hörmander-Mihlin multipliers were obtained by Kurtz and Wheeden [**321**] and for Marcinkiwiecz multipliers by Kurtz [**320**].

The method of proof of Theorem 5.3.4 is adapted from Duoandikoetxea and Rubio de Francia [**168**]. The method in this article is rather general and can be used to obtain L^p boundedness for a variety of rough singular integrals. A version of Theorem 5.3.6 was used by Christ [**93**] to obtain L^p smoothing estimates for Cantor-Lebesgue measures. When $p = q \neq 2$, Theorem 5.3.6 is false in general but it is true for all r satisfying $|\frac{1}{r} - \frac{1}{2}| < |\frac{1}{p} - \frac{1}{2}|$ under the additional assumption that the m_j's are Lipschitz functions uniformly at all scales. This result was independently obtained by Carbery [**78**] and Seeger [**451**].

The probabilistic notions of conditional expectations and martingales have a strong connection with the Littlewood-Paley theory discussed in this chapter. For the purposes of our exposition we only considered the case of the sequence of σ-algebras generated by the dyadic cubes of side length 2^{-k} in $\mathbf{R}^n$. The L^p boundedness of the maximal conditional expectation (Doob [**160**]) is analogous to the L^p boundedness of the dyadic maximal function; likewise with the corresponding weak type $(1,1)$ estimate. The L^p boundedness of the dyadic martingale square function (Burkholder [**54**]) is analogous to Theorem 5.1.2. Moreover, the estimate $\big\| \sup_k |E_k(f)| \big\|_{L^p} \approx \big\| S(f) \big\|_{L^p}$, $0 < p < \infty$, obtained by Burkholder and Gundy [**55**] and also by Davis [**152**], is analogous to the square-function characterization of the H^p norm that will be discussed in the next chapter. For an exposition on the different and unifying aspects of Littlewood-Paley theory we refer to Stein [**475**]. The proof of Theorem 5.4.8, which quantitatively expresses the almost orthogonality of the Littlewood-Paley and the dyadic martingale difference operators, is taken from Grafakos and Kalton [**214**].

The use of quadratic expressions in the study of certain maximal operators has a long history. We refer to the article of Stein [**477**] for a historical survey. Theorem 5.5.1 was first proved by Stein [**476**]. The proof in the text is taken from an article of Rubio de Francia [**436**]. Another proof when $n \geq 3$ is due to Cowling and Mauceri [**134**]. The more difficult case $n = 2$ was settled by Bourgain [**48**] about 10 years later. Alternative proofs when $n = 2$ were given by Mockenhaupt, Seeger, and Sogge [**376**] as well as Schlag [**448**]. Weighted norm inequalities for the spherical maximal operator were obtained by Duoandikoetxea and Vega [**169**]. The discrete spherical maximal function was studied by Magyar, Stein, and Wainger [**351**].

Much of the theory of square functions and the ideas associated with them has analogues in the dyadic setting. A dyadic analogue of the theory discussed here can be obtained. For an introduction to the area of dyadic harmonic analysis, we refer to Pereyra [**409**].

The idea of expressing (or reproducing) a signal as a weighted average of translations and dilations of a single function appeared in early work of Calderón [**61**]. This idea is in some sense a forerunner of wavelets. An early example of a wavelet was constructed by Strömberg [**496**] in his search for unconditional bases for Hardy spaces. Another example of a wavelet basis was obtained by Meyer [**360**]. The construction of an orthonormal wavelet presented in Theorem 5.6.7 is in Lemarié and Meyer [**331**]. A compactly supported wavelet was constructed by Daubechies [**146**]. Mallat [**352**] introduced the notion of multiresolution analysis that led to a systematic production of wavelets. The area of wavelets has taken off significantly since its inception spurred by these early results. A general theory of wavelets and its use in Fourier analysis was carefully developed in the two-volume monograph of Meyer [**361**], [**362**] and its successor Meyer and Coifman [**363**]. For further study and a complete account on the recent developments on the subject we refer the reader to the books of Daubechies [**147**], Chui [**104**], Wickerhauser [**554**], Kaiser [**285**], Benedetto and Frazier [**33**], Hérnandez and Weiss [**245**], Wojtaszczyk [**557**], Mallat [**353**], Meyer [**364**], Frazier [**196**], Gröchenig [**229**], and the references therein.

CHAPTER 6

Smoothness and Function Spaces

In this chapter we study differentiability and smoothness of functions. There are several ways to interpret smoothness and numerous ways to describe it and quantify it. A fundamental fact is that smoothness can be measured and fine-tuned using the Fourier transform, and this point of view will be of great importance to us. In fact, we will base our investigation of the subject on this point. It is not surprising, therefore, that Littlewood-Paley theory will play a crucial and deep role in our study.

Certain spaces of functions are introduced to serve the purpose of measuring smoothness. The main function spaces we will be interested in are Lipschitz, Sobolev, and Hardy spaces, although the latter only measure smoothness within the realm of rough distributions. Hardy spaces also serve as a substitute for L^p when $p < 1$. We also take a quick look at Besov-Lipschitz and Triebel-Lizorkin spaces, which provide an appropriate framework that unifies the scope and breadth of the subject. One of the main achievements of this chapter is the characterization of these spaces using Littlewood-Paley theory. Another major accomplishment of this chapter is the atomic characterization of these function spaces. This is obtained from the Littlewood-Paley characterization of these spaces in a single way for all of them.

Before we start our study of function spaces we will need to understand differentiability and smoothness in terms of the Fourier transform. This can be achieved using the Laplacian and the potential operators and is discussed in the first section.

6.1. Riesz Potentials, Bessel Potentials, and Fractional Integrals

Recall the Laplacian operator

$$\Delta = \partial_1^2 + \cdots + \partial_n^2 ,$$

which may act on functions or tempered distributions. The Fourier transform of a Schwartz function (or even a tempered distribution f) satisfies the following identity:

$$-\widehat{\Delta(f)}(\xi) = 4\pi^2|\xi|^2\widehat{f}(\xi) .$$

Motivated by this identity, we replace the exponent 2 by a complex exponent z and we define $(-\Delta)^{\frac{z}{2}}$ as the operator given by the multiplication with the function $(2\pi|\xi|)^z$ on the Fourier transform. More precisely, for $z \in \mathbf{C}$ and Schwartz functions f we define

$$(-\Delta)^{\frac{z}{2}}(f)(x) = ((2\pi|\xi|)^z \widehat{f}(\xi))^\vee(x)\,. \tag{6.1.1}$$

Roughly speaking, the operator $(-\Delta)^{\frac{z}{2}}$ is acting as a derivative of order z if z is a positive integer. If z is a complex number with real part less than $-n$, then the function $|\xi|^z$ is not locally integrable on $\mathbf{R}^n$ and so (6.1.1) may not be well defined. For this reason, whenever we write (6.1.1), we will assume that either $\operatorname{Re} z > -n$, or $\operatorname{Re} z < -n$ and that $\widehat{f}$ vanishes to sufficiently high order at the origin so that the expression $|\xi|^z \widehat{f}(\xi)$ is locally integrable. Note that the family of operators $(-\Delta)^z$ satisfies the semigroup property

$$(-\Delta)^z(-\Delta)^w = (-\Delta)^{z+w}\,, \qquad \text{for all } z, w \in \mathbf{C},$$

when acting on spaces of suitable functions.

The operator $(-\Delta)^{\frac{z}{2}}$ is given by convolution with the inverse Fourier transform of $(2\pi)^z|\xi|^z$. Theorem 2.4.6 gives that this inverse Fourier transform is equal to

$$(2\pi)^z(|\xi|^z)^\vee(x) = (2\pi)^z \frac{\pi^{-\frac{z}{2}}}{\pi^{\frac{z+n}{2}}} \frac{\Gamma(\frac{n+z}{2})}{\Gamma(\frac{-z}{2})} |x|^{-z-n}\,. \tag{6.1.2}$$

The expression in (6.1.2) is in $L^1_{\mathrm{loc}}(\mathbf{R}^n)$ only when $-\operatorname{Re} z - n > -n$, that is when $\operatorname{Re} z < 0$. In general, (6.1.2) is a distribution. Thus only in the range $-n < \operatorname{Re} z < 0$ both the function $|\xi|^z$ and its inverse Fourier transform are locally integrable functions.

6.1.a. Riesz Potentials

When z is a negative real number, the operation $f \to (-\Delta)^{\frac{z}{2}}(f)$ is not really "differentiating" f, but "integrating" it instead. For this reason, we introduce a slightly different notation in this case by replacing z by $-s$.

Definition 6.1.1. Let s be a complex number with $\operatorname{Re} s > 0$. The *Riesz potential* of order s is the operator

$$I_s = (-\Delta)^{-\frac{s}{2}}.$$

Using identity (6.1.2), we see that I_s is actually given in the form

$$I_s(f)(x) = 2^{-s}\pi^{-\frac{n}{2}} \frac{\Gamma(\frac{n-s}{2})}{\Gamma(\frac{s}{2})} \int_{\mathbf{R}^n} f(x-y)|y|^{-n+s}\,dy$$

and the integral is convergent if f is a function in the Schwartz class.

We begin with a simple, yet interesting, remark concerning the homogeneity of the operator I_s.

Remark 6.1.2. Suppose that we had an estimate

$$\|I_s f\|_{L^q(\mathbf{R}^n)} \le C(p,q,n,s)\|f\|_{L^p(\mathbf{R}^n)} \tag{6.1.3}$$

for some positive indices p,q and all $f \in L^p(\mathbf{R}^n)$. Then p and q must be related by

$$\frac{1}{p}-\frac{1}{q}=\frac{s}{n}. \tag{6.1.4}$$

This follows by applying (6.1.3) to the dilation $\delta^a(f)(x)=f(ax)$ of the function f, $a>0$, in lieu of f, for some fixed f, say $f(x)=e^{-x^2}$. Indeed, replacing f by $\delta^a(f)$ in (6.1.3) and carrying out some algebraic manipulations using the identity $I_s(\delta^a(f))=a^{-s}\delta^a(I_s(f))$, we obtain

$$a^{-\frac{n}{q}-s}\|I_s(f)\|_{L^q(\mathbf{R}^n)} \le C(p,q,n,s)a^{-\frac{n}{p}}\|f\|_{L^p(\mathbf{R}^n)}\,. \tag{6.1.5}$$

Suppose now that $\frac{1}{p}>\frac{1}{q}+\frac{s}{n}$. Then we can write (6.1.5) as

$$\|I_s(f)\|_{L^q(\mathbf{R}^n)} \le C(p,q,n,s)a^{\frac{n}{q}-\frac{n}{p}+s}\|f\|_{L^p(\mathbf{R}^n)} \tag{6.1.6}$$

and let $a\to\infty$ to obtain that $I_s(f)=0$, a contradiction. Similarly, if $\frac{1}{p}<\frac{1}{q}+\frac{s}{n}$, we could write (6.1.5) as

$$a^{-\frac{n}{q}+\frac{n}{p}-s}\|I_s(f)\|_{L^q(\mathbf{R}^n)} \le C(p,q,n,s)\|f\|_{L^p(\mathbf{R}^n)} \tag{6.1.7}$$

and let $a\to 0$ to obtain that $\|f\|_{L^p}=\infty$, again a contradiction. It follows that (6.1.4) must necessarily hold.

We conclude that the homogeneity (or dilation structure) of an operator dictates a relationship on the indices p and q for which it (may) map L^p into L^q.

As we saw in Remark 6.1.2, if the Riesz potentials map L^p into L^q for some p,q, then we must have $q>p$. Such operators that improve the integrability of a function are called *smoothing*. The importance of the Riesz potentials lies in the fact that they are indeed smoothing operators. This is the essence of the *Hardy-Littlewood-Sobolev theorem on fractional integration,* which we now formulate and prove.

Theorem 6.1.3. *Let s be a real number with $0<s<n$ and let $1\le p<q<\infty$ satisfy (6.1.4). Then there exist constants $C(n,s,p)<\infty$ such that for all f in $L^p(\mathbf{R}^n)$ we have*

$$\|I_s(f)\|_{L^q} \le C(n,s,p)\|f\|_{L^p}$$

when $p>1$, and also $\|I_s(f)\|_{L^{q,\infty}} \le C(n,s)\|f\|_{L^1}$ when $p=1$.

We note that the $L^p \to L^{q,\infty}$ estimate in Theorem 6.1.3 is a consequence of Theorem 1.2.13, for, the kernel $|x|^{-n+s}$ of I_s lies in the space $L^{r,\infty}$ when $r=\frac{n}{n-s}$ and (1.2.16) is satisfied for this r. Applying Theorem 1.4.19, we obtain the required conclusion. Nevertheless, for matters of exposition, we choose to give another proof of Theorem 6.1.3.

Proof. We begin by observing that the function $I_s(f)$ is well defined whenever f is bounded and has some decay at infinity. This makes the operator I_s well defined on a dense subclass of all the L^p spaces with $p < \infty$. Second, we may assume that $f \geq 0$ since $|I_s(f)| \leq I_s(|f|)$.

Under these assumptions we write the convolution

$$\int_{\mathbf{R}^n} f(x-y)|y|^{s-n}\,dy = J_1(f)(x) + J_2(f)(x),$$

where, in the spirit of interpolation, J_1 and J_2 are defined by

$$\begin{aligned} J_1(f)(x) &= \int_{|y|<R} f(x-y)|y|^{s-n}\,dy, \\ J_2(f)(x) &= \int_{|y|\geq R} f(x-y)|y|^{s-n}\,dy, \end{aligned}$$

for some R to be determined later. Observe that J_1 is given by convolution with the function $|y|^{-n+s}\chi_{|y|<R}(y)$, which is a radial, integrable, and symmetrically decreasing about the origin. It follows from Theorem 2.1.10 that

$$J_1(f)(x) \leq M(f)(x)\int_{|y|<R} |y|^{-n+s}\,dy = \frac{\omega_{n-1}}{s}R^s M(f)(x), \tag{6.1.8}$$

where M is the Hardy-Littlewood maximal function. Now Hölder's inequality gives that

$$\begin{aligned} |J_2(f)(x)| &\leq \left(\int_{|y|\geq R} (|y|^{-n+s})^{p'}\,dy\right)^{\frac{1}{p'}} \|f\|_{L^p(\mathbf{R}^n)} \\ &= \left(\frac{q\omega_{n-1}}{p'n}\right)^{\frac{1}{p'}} R^{-\frac{n}{q}} \|f\|_{L^p(\mathbf{R}^n)}, \end{aligned} \tag{6.1.9}$$

and note that this estimate is also valid when $p = 1$ (in which case $q = \frac{n}{n-s}$), provided the $L^{p'}$ norm is interpreted as L^∞ norm and the constant $\left(\frac{q\omega_{n-1}}{p'n}\right)^{\frac{1}{p'}}$ as 1. Combining (6.1.8) and (6.1.9), we obtain that

$$I_s(f)(x) \leq C'_{n,s,p}\big(R^s M(f)(x) + R^{-\frac{n}{q}}\|f\|_{L^p}\big) \tag{6.1.10}$$

for all $R > 0$. We now choose an $R > 0$ to minimize the expression in (6.1.10), that is,

$$R = \frac{(M(f)(x))^{-\frac{p}{n}}}{\|f\|_{L^p}^{-\frac{p}{n}}}.$$

This choice of R yields the estimate

$$I_s(f)(x) \leq C_{n,s,p}\, M(f)(x)^{\frac{p}{q}} \|f\|_{L^p}^{1-\frac{p}{q}}. \tag{6.1.11}$$

The required inequality for $p > 1$ follows by raising to the power q, integrating over $\mathbf{R}^n$ and using the boundedness of the Hardy-Littlewood maximal operator M on

$L^p(\mathbf{R}^n)$. The case $p=1$, $q=\frac{n}{n-s}$ also follows from (6.1.11) by using the weak type $(1,1)$ estimate for M. Indeed,

$$\begin{aligned}\left|\{C_{n,s,1}M(f)^{\frac{n-s}{n}}\|f\|_{L^1}^{\frac{s}{n}}>\lambda\}\right| &= \left|\left\{M(f)>\left(\frac{\lambda}{C_{n,s,1}\|f\|_{L^1}^{\frac{s}{n}}}\right)^{\frac{n}{n-s}}\right\}\right|\\ &\le 3^n\left(\frac{C_{n,s,1}\|f\|_{L^1}^{\frac{s}{n}}}{\lambda}\right)^{\frac{n}{n-s}}\|f\|_{L^1}\\ &= C(n,s)\left(\frac{\|f\|_{L^1}}{\lambda}\right)^{\frac{n}{n-s}}.\end{aligned}$$

We now give an alternative proof of the case $p=1$ that corresponds to $q=\frac{n}{n-s}$. Without loss of generality we may assume that $f\ge 0$ has L^1 norm 1. Once this case is proved, the general case will follow by scaling. Observe that

$$\int_{\mathbf{R}^n} f(y)|x-y|^{s-n}\,dy \le \sum_{j\in\mathbf{Z}} 2^{j(s-n)}\int_{|y|\le 2^j} f(x-y)\,dy. \tag{6.1.12}$$

Let $E_\lambda=\{x:\ I_s(f)(x)>\lambda\}$. Then

$$\begin{aligned}|E_\lambda| &\le \frac{1}{\lambda}\int_{E_\lambda} I_s(f)(x)\,dx = \frac{1}{\lambda}\int_{E_\lambda}\int_{\mathbf{R}^n}|y|^{s-n}f(x-y)\,dy\,dx\\ &\le \frac{1}{\lambda}\int_{E_\lambda}\sum_{j\in\mathbf{Z}}2^{j(s-n)}\int_{|y|\le 2^j}f(x-y)\,dy\,dx\\ &= \frac{1}{\lambda}\sum_{j\in\mathbf{Z}}2^{j(s-n)}\int_{E_\lambda}\int_{|y|\le 2^j}f(x-y)\,dy\,dx\\ &\le \frac{1}{\lambda}\sum_{j\in\mathbf{Z}}2^{j(s-n)}\min(|E_\lambda|,v_n2^{jn})\\ &= \frac{1}{\lambda}\sum_{2^j>|E_\lambda|^{\frac{1}{n}}}2^{j(s-n)}|E_\lambda| + \frac{v_n}{\lambda}\sum_{2^j\le|E_\lambda|^{\frac{1}{n}}}2^{j(s-n)}2^{jn}\\ &\le \frac{C}{\lambda}\left(|E_\lambda|^{\frac{s-n}{n}}|E_\lambda|+|E_\lambda|^{\frac{s}{n}}\right) = \frac{2C}{\lambda}|E_\lambda|^{\frac{s}{n}}.\end{aligned} \tag{6.1.13}$$

It follows that $|E_\lambda|^{\frac{n-s}{n}}\le\frac{2C}{\lambda}$, which implies the weak type $(1,\frac{n}{n-s})$ estimate for I_s. □

6.1.b. Bessel Potentials

While the behavior of the kernels $|x|^{-n+s}$ as $|x|\to 0$ is well suited for their smoothing properties, their decay as $|x|\to\infty$ gets worse as s increases. We can slightly adjust the Riesz potentials so that we maintain their essential behavior near zero but achieve exponential decay at infinity. The simplest way to achieve this is by replacing the "nonnegative" operator $-\Delta$ by the "strictly positive" operator $I-\Delta$. Here the terms nonnegative and strictly positive, as one may have surmised, refer to the Fourier transforms of these expressions.

Definition 6.1.4. Let s be a complex number with $0 < \operatorname{Re} s < \infty$. The *Bessel potential* of order s is the operator

$$\mathcal{J}_s = (I-\Delta)^{-\frac{s}{2}}$$

whose action on functions is given by

$$\mathcal{J}_s(f) = \big(\widehat{f}\,\widehat{G_s}\big)^{\vee} = f * G_s\,,$$

where

$$G_s(x) = \big((1+4\pi^2|\xi|^2)^{-\frac{s}{2}}\big)^{\vee}(x)\,.$$

Let us see why this adjustment yields exponential decay for G_s at infinity.

Proposition 6.1.5. *Let s be a positive real number. Then*

$$G_s(x) > 0 \qquad \textit{for all } x \in \mathbf{R}^n. \tag{6.1.14}$$

Moreover, there exist constants $0 < C(s,n), c(s,n) < \infty$ such that

$$G_s(x) \le C(s,n) e^{-\frac{|x|}{2}}, \qquad \textit{when } |x| \ge 2, \tag{6.1.15}$$

and such that

$$\frac{1}{c(s,n)} \le \frac{G_s(x)}{H_s(x)} \le c(s,n)\,,$$

when $|x| \le 2$, where H_s is a function that satisfies

$$H_s(x) = \begin{cases} |x|^{s-n} + 1 + o(|x|^{s-n+2}) & \textit{for } 0 < s < n, \\ \log\frac{2}{|x|} + 1 + o(|x|^2) & \textit{for } s = n, \\ 1 + o(|x|^{s-n}) & \textit{for } s > n, \end{cases}$$

as $|x| \to 0$.

Proof. For $A, s > 0$ we have the gamma function identity

$$A^{-\frac{s}{2}} = \frac{1}{\Gamma(\frac{s}{2})} \int_0^\infty e^{-tA} t^{\frac{s}{2}} \frac{dt}{t}\,,$$

which we use to obtain

$$(1+4\pi^2|\xi|^2)^{-\frac{s}{2}} = \frac{1}{\Gamma(\frac{s}{2})} \int_0^\infty e^{-t} e^{-\pi|2\sqrt{\pi t}\,\xi|^2} t^{\frac{s}{2}} \frac{dt}{t}\,.$$

Note that the previous integral converges at both ends. Now take the inverse Fourier transform in ξ and use the fact that the function $e^{-\pi|\xi|^2}$ is equal to its Fourier transform (Example 2.2.9) to obtain

$$G_s(x) = \frac{(2\sqrt{\pi}\,)^{-n}}{\Gamma(\frac{s}{2})} \int_0^\infty e^{-t} e^{-\frac{|x|^2}{4t}} t^{\frac{s-n}{2}} \frac{dt}{t}\,.$$

This proves (6.1.14). Now suppose $|x| \ge 2$. Then $t+\frac{|x|^2}{4t} \ge t+\frac{1}{t}$ and also $t+\frac{|x|^2}{4t} \ge |x|$. This implies that

$$-t-\frac{|x|^2}{4t} \le -\frac{t}{2}-\frac{1}{2t}-\frac{|x|}{2},$$

from which it follows that when $|x| \ge 2$

$$|G_s(x)| \le \frac{(2\sqrt{\pi}\,)^{-n}}{\Gamma(\frac{s}{2})}\left(\int_0^\infty e^{-\frac{t}{2}}e^{-\frac{1}{2t}}t^{\frac{s-n}{2}}\frac{dt}{t}\right)e^{-\frac{|x|}{2}} = C_{s,n}e^{-\frac{|x|}{2}}.$$

This proves (6.1.15).

Suppose now that $|x| \le 2$. Write $G_s(x) = G_s^1(x)+G_s^2(x)+G_s^3(x)$, where

$$\begin{aligned}
G_s^1(x) &= \frac{(2\sqrt{\pi}\,)^{-n}}{\Gamma(\frac{s}{2})}\int_0^{|x|^2} e^{-t}e^{-\frac{|x|^2}{4t}}t^{\frac{s-n}{2}}\frac{dt}{t},\\
G_s^2(x) &= \frac{(2\sqrt{\pi}\,)^{-n}}{\Gamma(\frac{s}{2})}\int_{|x|^2}^{4} e^{-t}e^{-\frac{|x|^2}{4t}}t^{\frac{s-n}{2}}\frac{dt}{t},\\
G_s^3(x) &= \frac{(2\sqrt{\pi}\,)^{-n}}{\Gamma(\frac{s}{2})}\int_4^{\infty} e^{-t}e^{-\frac{|x|^2}{4t}}t^{\frac{s-n}{2}}\frac{dt}{t}.
\end{aligned}$$

Since $t|x|^2 \le 4$ in G_s^1, we have $e^{-t|x|^2} = 1+o(t|x|^2)$ as $|x| \to 0$; thus after changing variables we can write

$$\begin{aligned}
G_s^1(x) &= |x|^{s-n}\frac{(2\sqrt{\pi}\,)^{-n}}{\Gamma(\frac{s}{2})}\int_0^1 e^{-t|x|^2}e^{-\frac{1}{4t}}t^{\frac{s-n}{2}}\frac{dt}{t}\\
&= |x|^{s-n}\frac{(2\sqrt{\pi}\,)^{-n}}{\Gamma(\frac{s}{2})}\int_0^1 e^{-\frac{1}{4t}}t^{\frac{s-n}{2}}\frac{dt}{t}+\frac{o(|x|^{s-n+2})}{\Gamma(\frac{s}{2})}\int_0^1 e^{-\frac{1}{4t}}t^{\frac{s-n}{2}}\,dt\\
&= c_{s,n}^1|x|^{s-n}+o(|x|^{s-n+2}) \qquad\qquad \text{as } |x| \to 0.
\end{aligned}$$

Since $0 \le \frac{|x|^2}{4t} \le \frac{1}{4}$ and $0 \le t \le 4$ in G_s^2, we have $e^{-\frac{17}{16}} \le e^{-t-\frac{|x|^2}{4t}} \le 1$; thus as $|x| \to 0$ we obtain

$$G_s^2(x) \approx \int_{|x|^2}^4 t^{\frac{s-n}{2}}\frac{dt}{t} = \begin{cases} \frac{|x|^{s-n}}{n-s}-\frac{2^{s-n+1}}{n-s} & \text{for } s<n,\\ 2\log\frac{2}{|x|} & \text{for } s=n,\\ \frac{1}{s-n}2^{s-n+1} & \text{for } s>n. \end{cases}$$

Finally, we have $e^{-\frac{1}{4}} \le e^{-\frac{|x|^2}{4t}} \le 1$ in G_s^3, which yields that $G_s^3(x)$ is bounded above and below by fixed positive constants. Combining the estimates for $G_s^1(x)$, $G_s^2(x)$, and $G_s^3(x)$, we obtain the required conclusion.

□

We end this section with a result analogous to that of Theorem 6.1.3 for the operator $\mathcal{J}_s$.

Corollary 6.1.6. *(a) For all $0 < s < \infty$, the operator $\mathcal{J}_s$ maps $L^r(\mathbf{R}^n)$ into itself with norm 1 for all $1 \le r \le \infty$.*
(b) Let $0 < s < n$ and $1 \le p < q < \infty$ satisfy (6.1.4). Then there exist constants $C_{p,q,n} < \infty$ such that for all f in $L^p(\mathbf{R}^n)$ we have

$$\big\|\mathcal{J}_s(f)\big\|_{L^q} \le C_{n,s,p}\big\|f\big\|_{L^p}$$

when $p > 1$, and also $\big\|\mathcal{J}_s(f)\big\|_{L^{q,\infty}} \le C_{n,s}\big\|f\big\|_{L^1}$ when $p = 1$.

Proof. (a) Since $\widehat{G_s}(0) = 1$ and $G_s > 0$, it follows that G_s has L^1 norm 1. The operator $\mathcal{J}_s$ is given by convolution with the positive function G_s, which has L^1 norm 1; thus it maps $L^r(\mathbf{R}^n)$ into itself with norm 1 for all $1 \le r \le \infty$ (see Exercise 1.2.9).
(b) In the special case $0 < s < n$ we have that the kernel G_s of $\mathcal{J}_s$ satisfies

$$G_s(x) \approx \begin{cases} |x|^{-n+s} & \text{when } |x| \le 2, \\ e^{-\frac{|x|}{2}} & \text{when } |x| \ge 2. \end{cases}$$

Then we can write

$$\begin{aligned} \mathcal{J}_s(f)(x) &\le C_{n,s}\bigg[\int_{|y|\le 2} |f(x-y)|\,|y|^{-n+s}\,dy + \int_{|y|\ge 2} |f(x-y)|\,e^{-\frac{|y|}{2}}\,dy\bigg] \\ &\le C_{n,s}\bigg[I_s(|f|)(x) + \int_{\mathbf{R}^n} |f(x-y)|\,e^{-\frac{|y|}{2}}\,dy\bigg]. \end{aligned}$$

We now use that the function $y \to e^{-\frac{|y|}{2}}$ is in L^r for all $r < \infty$, Young's inequality (Theorem 1.2.12), and Theorem 6.1.3 to complete the proof of the corollary.

□

Exercises

6.1.1. (a) Let $0 < s, t < \infty$ be such that $s + t < n$. Show that

$$I_s I_t = I_{s+t}\,.$$

(b) Prove the operator identities

$$I_s(-\Delta)^z = (-\Delta)^z I_s = I_{s-2z} = (-\Delta)^{z-\frac{s}{2}}$$

whenever $\operatorname{Re} s > 2\operatorname{Re} z$.
(c) Prove that for all $z \in \mathbf{C}$ we have

$$\langle(-\Delta)^z(f)\,|\,(-\Delta)^{-z}(g)\rangle$$

whenever the Fourier transforms of f and g vanish to sufficiently high order at the origin.
(d) Given $\operatorname{Re} s > 0$, find an $\alpha \in \mathbf{C}$ such that

$$\langle I_s(f)\,|\,f\rangle = \big\|(-\Delta)^\alpha(f)\big\|_{L^2}^2$$

when f is as in part (b).

6.1.2. Use Exercise 2.2.14 to prove that for $-\infty < \alpha < n/2 < \beta < \infty$ we have

$$\|f\|_{L^\infty(\mathbf{R}^n)} \le C\|\Delta^{\alpha/2}(f)\|_{L^2(\mathbf{R}^n)}^{\frac{\beta-n/2}{\beta-\alpha}}\|\Delta^{\beta/2}(f)\|_{L^2(\mathbf{R}^n)}^{\frac{n/2-\alpha}{\beta-\alpha}},$$

where C depends only on α, n, β.

6.1.3. Show that when $0 < s < n$ we have

$$\sup_{\|f\|_{L^1(\mathbf{R}^n)}=1} \Big[\|I_s(f)\|_{L^{\frac{n}{n-s}}(\mathbf{R}^n)} + \|\mathcal{J}_s(f)\|_{L^{\frac{n}{n-s}}(\mathbf{R}^n)}\Big] = \infty .$$

Thus I_s and $\mathcal{J}_s$ are not of strong type $(1, \frac{n}{n-s})$.
[*Hint:* Consider an approximate identity.]

6.1.4. Let $0 < s < n$. Consider the function $h(x) = |x|^{-s}(\log \frac{1}{|x|})^{-\frac{s}{n}(1+\delta)}$ for $|x| \le 1$ and zero otherwise. Prove that when $0 < \delta < \frac{n-s}{s}$ we have $h \in L^{\frac{n}{s}}$ but that $\lim_{x\to 0} I_s(h)(x) = \infty$. Conclude that I_s does not map $L^{\frac{n}{s}}(\mathbf{R}^n)$ into $L^\infty(\mathbf{R}^n)$.

6.1.5. For $1 \le p \le \infty$ and $0 < s < \infty$ define the *Bessel potential space* $\mathcal{L}^p_s(\mathbf{R}^n)$ as the space of all functions $f \in L^p(\mathbf{R}^n)$ for which there exists another function f_0 in $L^p(\mathbf{R}^n)$ such that $\mathcal{J}_s(f_0) = f$. Define a norm on these spaces by setting $\|f\|_{\mathcal{L}^p_s} = \|f_0\|_{L^p}$. Prove the following properties of these spaces:
(a) $\|f\|_{\mathcal{L}^p_s} \le \|f\|_{L^p}$, hence $\mathcal{L}^p_s(\mathbf{R}^n)$ is a subspace of $L^p(\mathbf{R}^n)$.
(b) For all $0 < t, s < \infty$ we have $G_s * G_t = G_{s+t}$ and thus

$$\mathcal{L}^p_s(\mathbf{R}^n) * \mathcal{L}^q_t(\mathbf{R}^n) \subseteq \mathcal{L}^r_{s+t}(\mathbf{R}^n),$$

where $1 \le p, q, r \le \infty$ and $\frac{1}{p} + \frac{1}{q} = \frac{1}{r} + 1$.
(c) The sequence of norms $\|f\|_{\mathcal{L}^p_s}$ increases, and therefore the spaces $\mathcal{L}^p_s(\mathbf{R}^n)$ decrease as s increases.
(d) The map $\mathcal{I}_t$ is an isomorphism from the space $\mathcal{L}^p_s(\mathbf{R}^n)$ onto $\mathcal{L}^p_{s+t}(\mathbf{R}^n)$.
[Note: Note that the Bessel potential space $\mathcal{L}^p_s(\mathbf{R}^n)$ coincides with the Sobolev space $L^p_s(\mathbf{R}^n)$, which will be introduced in Section 6.2.]

6.1.6. For $0 \le s < n$ define the *fractional maximal function*

$$M^s(f)(x) = \sup_{t>0} \frac{1}{(v_n t^n)^{\frac{n-s}{n}}} \int_{|y|\le t} |f(x-y)|\, dy$$

where v_n is the volume of the unit ball in $\mathbf{R}^n$.
(a) Show that for some constant C we have

$$M^s(f) \le C I_s(f)$$

for all $f \ge 0$ and conclude that M^s maps L^p into L^q whenever I_s does.

(b) (*Adams* [**1**]) Let $s>0$, $1<p<\frac{n}{s}$, $1\le q\le\infty$ such that $\frac{1}{r}=\frac{1}{p}-\frac{s}{n}+\frac{sp}{nq}$. Show that there is a constant $C>0$ (depending on the previous parameters) such that for all positive functions f we have

$$\big\|I_s(f)\big\|_{L^r}\le C\big\|M^{n/p}(f)\big\|_{L^q}^{\frac{sp}{n}}\big\|f\big\|_{L^p}^{1-\frac{sp}{n}}.$$

[*Hint:* For $f\neq 0$, write $I_s(f)=I_1+I_2$, where

$$I_1=\int_{|x-y|\le\delta}f(y)\,|y|^{s-n}\,dy\qquad I_2=\int_{|x-y|>\delta}f(y)\,|y|^{s-n}\,dy.$$

Show that $I_1\le C\delta^s M^0(f)$ and that $I_2(f)\le C\delta^{s-\frac{n}{p}}M^{n/p}(f)$. Optimize over $\delta>0$ to obtain

$$I_s(f)\le CM^{n/p}(f)^{\frac{sp}{n}}M^0(f)^{1-\frac{sp}{n}}$$

from which the required conclusion follows easily.]

6.1.7. Suppose that a function K satisfies $|K(y)|\le C(1+|y|)^{-s+n}$, where $0<s<n$. Prove that the maximal operator

$$\sup_{t>0}t^{-n+s}\bigg|\int_{\mathbf{R}^n}f(x-y)K(y/t)\,dy\bigg|$$

maps $L^p(\mathbf{R}^n)$ into $L^q(\mathbf{R}^n)$ whenever I_s maps $L^p(\mathbf{R}^n)$ into $L^q(\mathbf{R}^n)$.
[*Hint:* Control this operator by M^s.]

6.1.8. Let $0<s<n$. Use the following steps to obtain a simpler proof of Theorem 6.1.3 based on more delicate interpolation.
(a) Prove that $\big\|I_s(\chi_E)\big\|_{L^\infty}\le|E|^{\frac{s}{n}}$ for any set E of finite measure.
(b) For any two sets E and F of finite measure show that

$$\int_F|I_s(\chi_E)(x)|\,dx\le|E|\,|F|^{\frac{s}{n}}.$$

(c) Use Exercise 1.1.12 to obtain that

$$\big\|I_s(\chi_E)\big\|_{L^{\frac{n}{n-s},\infty}}\le C_{ns}|E|.$$

(d) Use (a), (c) and Theorem 1.4.19 to obtain another proof of Theorem 6.1.3.
[*Hint:* Parts (a) and (b): Use that when $\lambda>0$, the integral $\int_E|y|^{-\lambda}\,dy$ becomes largest when E is a ball centered at the origin equimeasurable to E.]

6.1.9. (*Welland* [**549**]) Let $0<\alpha<n$ and suppose $0<\varepsilon<\min(\alpha,n-\alpha)$. Show that there exists a constant depending only on α,ε and n such that for all compactly supported bounded functions f we have

$$|I_\alpha(f)|\le C\sqrt{M^{\alpha-\varepsilon}(f)M^{\alpha+\varepsilon}(f)}$$

where $M^\beta(f)$ is the fractional maximal function of Exercise 6.1.6.
[*Hint:* Write

$$|I_\alpha(f)| \le \int_{|x-y|<s} \frac{|f(y)|\,dy}{|x-y|^{n-\alpha}} + \int_{|x-y|\ge s} \frac{|f(y)|\,dy}{|x-y|^{n-\alpha}}$$

and split each integral as a sum of integrals over annuli centered at x to obtain the estimate

$$|I_\alpha(f)| \le C\big(s^\varepsilon M^{\alpha-\varepsilon}(f) + s^{-\varepsilon} M^{\alpha+\varepsilon}(f)\big)\,.$$

Then optimize over s.]

6.1.10. Show that the *discrete fractional integral operator*

$$\{a_j\}_{j\in\mathbf{Z}^n} \to \Big\{\sum_{k\in\mathbf{Z}^n} \frac{a_k}{(|j-k|+1)^{n-\alpha}}\Big\}_{j\in\mathbf{Z}^n}$$

maps $\ell^s(\mathbf{Z}^n) \to \ell^t(\mathbf{Z}^n)$ when $0<\alpha<n$, $1<s<t$, and

$$\frac{1}{s} - \frac{1}{t} = \frac{\alpha}{n}\,.$$

6.1.11. Show that the bilinear operator

$$B_\alpha(f,g)(x) = \int_{\mathbf{R}^n}\int_{\mathbf{R}^n} f(y)g(z)(|x-y|+|x-z|)^{-2n+\alpha}\,dy\,dz$$

maps $L^p(\mathbf{R}^n)\times L^q(\mathbf{R}^n)$ into $L^r(\mathbf{R}^n)$ when $1<p,q<\infty$ and

$$\frac{1}{p}+\frac{1}{q} = \frac{\alpha}{n}+\frac{1}{r}\,.$$

[*Hint:* Control it by the product of two fractional integrals.]

6.1.12. (*Grafakos and Kalton* [**213**]/*Kenig and Stein* [**302**]) (a) Prove that the bilinear operator

$$S(f,g)(x) = \int_{|t|\le 1} |f(x+t)g(x-t)|\,dt$$

maps $L^1(\mathbf{R}^n)\times L^1(\mathbf{R}^n)$ into $L^{\frac{1}{2}}(\mathbf{R}^n)$.
(b) For $0<\alpha<n$ prove that the *bilinear fractional integral operator*

$$I_\alpha(f,g)(x) = \int_{\mathbf{R}^n} f(x+t)g(x-t)|t|^{-n+\alpha}\,dt$$

maps $L^1(\mathbf{R}^n)\times L^1(\mathbf{R}^n)$ into $L^{\frac{n}{2n-\alpha},\infty}(\mathbf{R}^n)$.
[*Hint:* Part (a): Write $f=\sum_{k\in\mathbf{Z}^n} f_k$, where each f_k is supported in the cube $k+[0,1]^n$ and similarly for g. Observe that the resulting double sum reduces to a single sum and use that $(\sum_j a_j)^{1/2} \le \sum_j a_j^{1/2}$ for $a_j\ge 0$. Part (b): Use part (a) and adjust the argument in (6.1.13) to a bilinear setting.]

6.2. Sobolev Spaces

In this section we study a quantitative way of measuring smoothness of functions. Sobolev spaces serve exactly this purpose. They measure the smoothness of a given function in terms of the integrability of its derivatives. We begin with the classical definition of Sobolev spaces.

Definition 6.2.1. Let k be a nonnegative integer and let $1 < p < \infty$. The *Sobolev space* $L^p_k(\mathbf{R}^n)$ is defined as the space of functions f in $L^p(\mathbf{R}^n)$ all of whose distributional derivatives $\partial^\alpha f$ are also in $L^p(\mathbf{R}^n)$ for all multiindices α that satisfy $|\alpha| \le k$. This space is normed by the expression

$$\|f\|_{L^p_k} = \sum_{|\alpha|\le k} \|\partial^\alpha f\|_{L^p}, \tag{6.2.1}$$

where $\partial^{(0,\dots,0)} f = f$.

Sobolev spaces measure smoothness of functions. The index k indicates the "degree" of smoothness of a given function in L^p_k. As k increases the functions become smoother. Equivalently, these spaces form a decreasing sequence

$$L^p \supset L^p_1 \supset L^p_2 \supset L^p_3 \supset \dots$$

meaning that each $L^p_{k+1}(\mathbf{R}^n)$ is a subspace of $L^p_k(\mathbf{R}^n)$. This property, which coincides with our intuition of smoothness, is a consequence of the definition of the Sobolev norms.

We next observe that the space $L^p_k(\mathbf{R}^n)$ is complete. Indeed, if f_j is a Cauchy sequence in the norm given by (6.2.1), then $\{\partial^\alpha f_j\}_j$ are Cauchy sequences for all $|\alpha| \le k$. By the completeness of L^p, there exist functions f_α such that $\partial^\alpha f_j \to f_\alpha$ in L^p. This implies that for all φ in the Schwartz class we have

$$(-1)^{|\alpha|} \int_{\mathbf{R}^n} f_j\,(\partial^\alpha \varphi)\,dx = \int_{\mathbf{R}^n} (\partial^\alpha f_j)\,\varphi\,dx \to \int_{\mathbf{R}^n} f_\alpha\,\varphi\,dx.$$

Since the first expression converges to

$$(-1)^{|\alpha|} \int_{\mathbf{R}^n} f_0\,(\partial^\alpha \varphi)\,dx\,,$$

it follows that the distributional derivative $\partial^\alpha f_0$ is f_α. This implies that $f_j \to f_0$ in $L^p_k(\mathbf{R}^n)$ and proves the completeness of this space.

Our goal in this section is to investigate relations between these spaces and the Riesz and Bessel potentials discussed in the previous section and to obtain a Littlewood-Paley characterization of them. Before we embark on this study, we note that we can extend the definition of Sobolev spaces to the case where the index k is not necessarily an integer. In fact, we will extend the definition of the spaces $L^p_k(\mathbf{R}^n)$ to the case where the number k is real.

6.2.a. Definition and Basic Properties of General Sobolev Spaces

Definition 6.2.2. Let s be a real number and let $1 < p < \infty$. The *inhomogeneous Sobolev space* $L^p_s(\mathbf{R}^n)$ is defined as the space of all tempered distributions u in $\mathcal{S}'(\mathbf{R}^n)$ with the property that

$$((1+|\xi|^2)^{\frac{s}{2}}\widehat{u})^\vee \tag{6.2.2}$$

is an element of $L^p(\mathbf{R}^n)$. For such distributions u we define

$$\big\|u\big\|_{L^p_s} = \big\|((1+|\cdot|^2)^{\frac{s}{2}}\,\widehat{u})^\vee\big\|_{L^p(\mathbf{R}^n)}\,.$$

Note that the function $(1+|\xi|^2)^{\frac{s}{2}}$ is C^∞ and has at most polynomial growth at infinity. Since $\widehat{u}\in\mathcal{S}'(\mathbf{R}^n)$, the product in (6.2.2) is well defined.

Several observations are in order. First, we note that when $s=0$, $L^p_s = L^p$. It is natural to ask whether elements of L^p_s are always L^p functions. We show that this is the case when $s \ge 0$ but not when $s<0$. We also show that the space L^p_s coincides with the space L^p_k given in Definition 6.2.1 when $s=k$ and k is an integer.

To prove that elements of L^p_s are indeed L^p functions when $s\ge 0$, we simply note that if $f_s = ((1+|\xi|^2)^{s/2}\widehat{f})^\vee$, then

$$f = \mathcal{J}_s(f_s) = f_s * G_s\,,$$

where $\mathcal{J}_s$ is the Bessel potential and G_s is given in Definition 6.1.4. It follows from Corollary 6.1.6 that $\big\|f\big\|_{L^p} \le \big\|f_s\big\|_{L^p} = \big\|f\big\|_{L^p_s}$.

We now prove that if $s=k$ is a nonnegative integer and $1<p<\infty$, then the norm of the space L^p_k as given in Definition 6.2.1 is comparable to that in Definition 6.2.2. Suppose that $f\in L^p_k$ according to Definition 6.2.2. Then for all $|\alpha|\le k$ we have

$$\partial^\alpha f = c_\alpha(\widehat{f}(\xi)\xi^\alpha)^\vee = c_\alpha\bigg(\widehat{f}(\xi)(1+|\xi|^2)^{\frac{k}{2}}\frac{\xi^\alpha}{(1+|\xi|^2)^{\frac{k}{2}}}\bigg)^\vee . \tag{6.2.3}$$

Theorem 5.2.7 gives that the function

$$\frac{\xi^\alpha}{(1+|\xi|^2)^{k/2}}$$

is an L^p multiplier. Since by assumption $\big(\widehat{f}(\xi)(1+|\xi|^2)^{\frac{k}{2}}\big)^\vee$ is in $L^p(\mathbf{R}^n)$, it follows from (6.2.3) that $\partial^\alpha f$ is in L^p and also that

$$\sum_{|\alpha|\le k}\big\|\partial^\alpha f\big\|_{L^p} \le C_{p,n,k}\big\|((1+|\cdot|^2)^{\frac{k}{2}}\,\widehat{f})^\vee\big\|_{L^p}.$$

Conversely, suppose that $f\in L^p_k$ according to Definition 6.2.1; then

$$(1+\xi_1^2+\cdots+\xi_n^2)^{\frac{k}{2}} = \sum_{|\alpha|\le k}\frac{k!}{\alpha_1!\cdots\alpha_n!}\,\xi^\alpha\,\frac{\xi^\alpha}{(1+|\xi|^2)^{\frac{k}{2}}}\,.$$

As we have already observed, the functions $m_\alpha(\xi) = \xi^\alpha(1+|\xi|^2)^{-\frac{k}{2}}$ are L^p multipliers whenever $|\alpha| \le k$. Since

$$((1+|\xi|^2)^{\frac{k}{2}}\widehat{f}\,)^\vee = \sum_{|\alpha|\le k} c_{\alpha,k}(m_\alpha(\xi)\xi^\alpha\widehat{f}\,)^\vee = \sum_{|\alpha|\le k} c'_{\alpha,k}(m_\alpha(\xi)\widehat{\partial^\alpha f}\,)^\vee,$$

it follows that

$$\big\|(\widehat{f}(\xi)(1+|\xi|^2)^{\frac{k}{2}})^\vee\big\|_{L^p} \le C_{p,n,k}\sum_{|\gamma|\le k}\big\|(\widehat{f}(\xi)\xi^\gamma)^\vee\big\|_{L^p}.$$

Example 6.2.3. Every Schwartz function lies in $L^p_s(\mathbf{R}^n)$ for s real. Sobolev spaces with negative indices s can indeed contain tempered distributions that are not locally integrable functions. For example, Dirac mass at the origin δ_0 is an element of $L^p_{-s}(\mathbf{R}^n)$ for all $s > n/p'$. Indeed, when $0 < s < n$, Proposition 6.1.5 gives that G_s [i.e., the inverse Fourier transform of $(1+|\xi|^2)^{-\frac{s}{2}}$] is integrable to the power p as long as $(s-n)p > -n$ (i.e. $s > n/p'$). When $s \ge n$, G_s is integrable to any positive power.

We now continue with the *Sobolev embedding theorem.*

Theorem 6.2.4. *(a) Let $0 < s < \frac{n}{p}$ and $1 < p < \infty$. Then the Sobolev space $L^p_s(\mathbf{R}^n)$ continuously embeds in $L^q(\mathbf{R}^n)$ when*

$$\frac{1}{p} - \frac{1}{q} = \frac{s}{n}.$$

(b) Let $0 < s = \frac{n}{p}$ and $1 < p < \infty$. Then $L^p_s(\mathbf{R}^n)$ continuously embeds in $L^q(\mathbf{R}^n)$ for any $\frac{n}{s} < q < \infty$.
(c) Let $\frac{n}{p} < s < \infty$ and $1 < p < \infty$. Then every element of $L^p_s(\mathbf{R}^n)$ can be modified on a set of measure zero so that the resulting function is uniformly continuous.

Proof. (a) If $f \in L^p_s$, then $f_s(x) = ((1+|\xi|^2)^{\frac{s}{2}}\widehat{f}\,)^\vee(x)$ is in $L^p(\mathbf{R}^n)$. Thus

$$f(x) = ((1+|\xi|^2)^{-\frac{s}{2}}\widehat{f_s}\,)^\vee(x);$$

hence $f = G_s * f_s$. Since $s < n$, Proposition 6.1.5 gives that

$$|G_s(x)| \le C_{s,n}|x|^{s-n}$$

for all $x \in \mathbf{R}^n$. This implies that $|f| = |G_s * f_s| \le C_{s,n} I_s(|f_s|)$. Theorem 6.1.3 now yields the required conclusion

$$\big\|f\big\|_{L^q} \le C'_{s,n}\big\|I_s(|f_s|)\big\|_{L^p} = C''_{s,n}\big\|f\big\|_{L^p_s}.$$

(b) Given any $\frac{s}{n} < q < \infty$ we can find t such that

$$1 + \frac{1}{q} = \frac{s}{n} + \frac{1}{t} = \frac{1}{p} + \frac{1}{t}.$$

Then $1 < \frac{s}{n}+\frac{1}{t}$, which implies that $(-n+s)t > -n$. Thus the function $|x|^{-n+s}\chi_{|x|\le 2}$ is integrable to the tth power, which implies that G_s is in L^t. Since $f = G_s * f_s$, Young's inequality gives that

$$\big\|f\big\|_{L^q(\mathbf{R}^n)} \le \big\|f_s\big\|_{L^p(\mathbf{R}^n)}\big\|G_s\big\|_{L^t(\mathbf{R}^n)} = C_{n,s}\big\|f\big\|_{L^p_{n/p}}.$$

(c) As before, $f = G_s * f_s$. If $s \ge n$, then Proposition 6.1.5 gives that the function G_s is in $L^{p'}(\mathbf{R}^n)$. Now if $n > s$, then $G_s(x)$ looks like $|x|^{-n+s}$ near zero. This function is integrable to the power p' near the origin if and only if $s > n/p$, which is what we are assuming. Thus f is given as the convolution of an L^p and of an $L^{p'}$ function and it can be identified with a uniformly continuous function (c.f. Exercise 1.2.3).

□

We now introduce the homogeneous Sobolev spaces $\dot{L}^p_s$. The main difference with the inhomogeneous spaces L^p_s is that elements of $\dot{L}^p_s$ may not themselves be elements of L^p. Another difference is that elements of homogeneous Sobolev spaces are not tempered distributions but equivalent classes of tempered distributions.

We would expect the homogeneous Sobolev space $\dot{L}^p_s$ to be the space of all distributions u in $\mathcal{S}'(\mathbf{R}^n)$ for which the expression

$$(|\xi|^s\widehat{u})^\vee \tag{6.2.4}$$

is an L^p function. Since the function $|\xi|^s$ is not (always) smooth at the origin, some care is needed in defining the product in (6.2.4). The idea is that when u lies in $\mathcal{S}'/\mathcal{P}$, then the value of $\widehat{u}$ at the origin is irrelevant as we may add to $\widehat{u}$ a distribution supported at the origin and obtain another element of the equivalence class of u (Proposition 2.4.1). It is because of this irrelevance that we are able to multiply $\widehat{u}$ by a function that may be nonsmooth at the origin (and which has polynomial growth at infinity).

To do this, we fix a smooth function $\eta(\xi)$ on $\mathbf{R}^n$ that is equal to 1 when $|\xi| \ge 2$ and vanishes when $|\xi| \le 1$. Then for $s \in \mathbf{R}$, $u \in \mathcal{S}'(\mathbf{R}^n)/\mathcal{P}$, and $\varphi \in \mathcal{S}(\mathbf{R}^n)$ we define

$$\big\langle |\xi|^s\widehat{u}, \varphi\big\rangle = \lim_{\varepsilon\to 0}\big\langle \widehat{u}, \eta(\tfrac{\xi}{\varepsilon})|\xi|^s\varphi(\xi)\big\rangle$$

provided that the last limit exists. Note that this defines $|\xi|^s\widehat{u}$ as another element of $\mathcal{S}'/\mathcal{P}$ and this definition is independent of the function η as it easily follows from (2.3.22).

Definition 6.2.5. Let s be a real number and let $1 < p < \infty$. The *homogeneous Sobolev space* $\dot{L}^p_s(\mathbf{R}^n)$ is defined as the space of all tempered distributions modulo polynomials u in $\mathcal{S}'(\mathbf{R}^n)/\mathcal{P}$ for which the expression

$$(|\xi|^s\widehat{u})^\vee$$

is a function in $L^p(\mathbf{R}^n)$. For distributions u in $\dot{L}^p_s(\mathbf{R}^n)$ we define

$$\big\|u\big\|_{\dot{L}^p_s} = \big\|(|\cdot|^s\widehat{u})^\vee\big\|_{L^p(\mathbf{R}^n)}\,. \tag{6.2.5}$$

As noted earlier, to avoid working with equivalence classes of functions, we will identify two distributions in $\dot{L}^p_s(\mathbf{R}^n)$ whose difference is a polynomial. In view of this identification, the quantity in (6.2.5) is a norm.

6.2.b. Littlewood-Paley Characterization of Inhomogeneous Sobolev Spaces

We now present the first main result of this section, the characterization the inhomogeneous Sobolev spaces using Littlewood-Paley theory.

For the purposes of the next theorem we will need the following setup. We fix a radial Schwartz function Ψ on $\mathbf{R}^n$ whose Fourier transform is nonnegative, supported in the annulus $\frac{1}{2}+\frac{1}{10}\le|\xi|\le 2-\frac{1}{10}$, and satisfies

$$\sum_{j\in\mathbf{Z}}\widehat{\Psi}(2^{-j}\xi)=1 \tag{6.2.6}$$

for all $\xi\neq 0$. Associated with this bump, we define the Littlewood-Paley operators Δ_j given by multiplication on the Fourier transform with the function $\widehat{\Psi}(2^{-j}\xi)$, that is,

$$\Delta_j(f)=\Delta_j^{\Psi}(f)=\Psi_{2^{-j}}*f\,. \tag{6.2.7}$$

Notice that the support properties of the Δ_j's yield the simple identity

$$\Delta_j=\big(\Delta_{j-1}+\Delta_j+\Delta_{j+1}\big)\Delta_j$$

for all $j\in\mathbf{Z}$. We also define a Schwartz function Φ so that

$$\widehat{\Phi}(\xi)=\begin{cases}\sum_{j\le 0}\widehat{\Psi}(2^{-j}\xi) & \text{when } \xi\neq 0,\\ 1 & \text{when } \xi=0.\end{cases} \tag{6.2.8}$$

Note that $\widehat{\Phi}(\xi)$ is equal to 1 for $|\xi|\le 1$, vanishes when $|\xi|\ge 2$, and is chosen so that

$$\widehat{\Phi}(\xi)+\sum_{j=1}^{\infty}\widehat{\Psi}(2^{-j}\xi)=1 \tag{6.2.9}$$

for all ξ in $\mathbf{R}^n$. We now introduce an operator S_0 by setting

$$S_0(f)=\Phi*f\,. \tag{6.2.10}$$

The identity of functions (6.2.9) yields the operator identity

$$S_0+\sum_{j=1}^{\infty}\Delta_j=I$$

in which the series converges in $\mathcal{S}'(\mathbf{R}^n)$; see Exercise 2.3.12. [Note that $S_0(f)$ and $\Delta_j(f)$ are well-defined functions when f is a tempered distribution.]

Having introduced the relevant background, we are now ready to state and prove the following result.

Theorem 6.2.6. *Let Φ, Ψ satisfy (6.2.6) and (6.2.8) and let Δ_j, S_0 be as in (6.2.7) and (6.2.10). Fix $s \in \mathbf{R}$ and all $1 < p < \infty$. Then there exists a constant C_1 that depends only on n, s, p, Φ, and Ψ such that for all $f \in L^p_s$ we have*

$$
\big\|S_0(f)\big\|_{L^p} + \Big\|\Big(\sum_{j=1}^{\infty}(2^{js}|\Delta_j(f)|)^2\Big)^{\frac{1}{2}}\Big\|_{L^p} \le C_1\big\|f\big\|_{L^p_s}\,. \tag{6.2.11}
$$

Conversely, there exists a constant C_2 that depends on the parameters n, s, p, Φ, and Ψ, such that every tempered distribution f that satisfies

$$
\big\|S_0(f)\big\|_{L^p} + \Big\|\Big(\sum_{j=1}^{\infty}(2^{js}|\Delta_j(f)|)^2\Big)^{\frac{1}{2}}\Big\|_{L^p} < \infty
$$

is an element of the Sobolev space L^p_s with norm

$$
\big\|f\big\|_{L^p_s} \le C_2\Big(\big\|S_0(f)\big\|_{L^p} + \Big\|\Big(\sum_{j=1}^{\infty}(2^{js}|\Delta_j(f)|)^2\Big)^{\frac{1}{2}}\Big\|_{L^p}\Big)\,. \tag{6.2.12}
$$

Proof. We will denote by C a generic constant that depends on the parameters n, s, p, Φ, and Ψ and that may vary in different occurrences. For a given tempered distribution f we define another tempered distribution f_s by setting

$$
f_s = \big((1+|\cdot|^2)^{\frac{s}{2}}\widehat{f}\,\big)^{\vee}
$$

so that we have $\big\|f\big\|_{L^p_s} = \big\|f_s\big\|_{L^p}$ if $f \in L^p_s$.

We first assume that the expression on the right in (6.2.12) is finite and we show that the tempered distribution f lies in the space L^p_s by controlling its Sobolev norm by a multiple of this expression. We begin by writing

$$
\big\|f\big\|_{L^p_s} = \big\|f_s\big\|_{L^p} \le \big\|(\widehat{\Phi}\,\widehat{f_s}\,)^{\vee}\big\|_{L^p} + \big\|((1-\widehat{\Phi})\,\widehat{f_s}\,)^{\vee}\big\|_{L^p}\,,
$$

and we plan to show that both quantities on the right are finite. Pick a smooth function with compact support η_0 that is equal to 1 on the support of $\widehat{\Phi}$. It is a simple fact that for all $s \in \mathbf{R}$ the function $(1+|\xi|^2)^{\frac{s}{2}}\eta_0(\xi)$ is in $\mathcal{M}_p(\mathbf{R}^n)$ (i.e., it is an L^p Fourier multiplier). Since

$$
(\widehat{\Phi}\,\widehat{f_s})^{\vee}(x) = \big\{\big((1+|\xi|^2)^{\frac{s}{2}}\eta_0(\xi)\big)\,\widehat{S_0(f)}(\xi)\big\}^{\vee}(x)\,, \tag{6.2.13}
$$

we have the estimate

$$
\big\|(\widehat{\Phi}\,\widehat{f_s}\,)^{\vee}\big\|_{L^p} \le C\|S_0(f)\|_{L^p}\,. \tag{6.2.14}
$$

We now introduce a smooth function η_∞ that vanishes in a neighborhood of the origin and is equal to 1 on the support of $1-\widehat{\Phi}$. Using Theorem 5.2.7, we can easily see that the function

$$
\frac{(1+|\xi|^2)^{\frac{s}{2}}}{|\xi|^s}\eta_\infty(\xi)
$$

is in $\mathcal{M}_p(\mathbf{R}^n)$ (with constant depending on n, p, η_∞, and s). Since

$$\big((1+|\xi|^2)^{\frac{s}{2}}(1-\widehat{\Phi}(\xi))\,\widehat{f}\,\big)^\vee(x) = \Big(\frac{(1+|\xi|^2)^{\frac{s}{2}}\eta_\infty(\xi)}{|\xi|^s}\,|\xi|^s(1-\widehat{\Phi}(\xi))\,\widehat{f}\,\Big)^\vee(x)$$

we obtain the estimate

$$\big\|\big((1-\widehat{\Phi})\,\widehat{f_s}\,\big)^\vee\big\|_{L^p} \le C\big\|f_\infty\big\|_{L^p}, \tag{6.2.15}$$

where f_∞ is another tempered distribution defined via

$$f_\infty = \big(|\xi|^s(1-\widehat{\Phi}(\xi))\,\widehat{f}\,\big)^\vee.$$

We are going to show that the quantity $\|f_\infty\|_{L^p}$ is finite using Littlewood-Paley theory. To achieve this, we introduce a smooth bump $\widehat{\zeta}$ supported in the annulus $\frac{1}{2} \le |\xi| \le 2$ and equal to 1 on the annulus $\frac{1}{2}+\frac{1}{10} \le |\xi| \le 2-\frac{1}{10}$, which contains the support of $\widehat{\Psi}$. Then we define $\widehat{\theta}(\xi) = |\xi|^s\widehat{\zeta}(\xi)$ and we introduce Littlewood-Paley operators

$$\Delta_j^\theta(g) = g * \theta_{2^{-j}},$$

where $\theta_{2^{-j}}(t) = 2^{jn}\theta(2^j t)$. Recalling that

$$1-\widehat{\Phi}(\xi) = \sum_{k\ge 1}\widehat{\Psi}(2^{-k}\xi),$$

we obtain that

$$\widehat{f_\infty} = \sum_{j=1}^{\infty}|\xi|^s\widehat{\Psi}(2^{-j}\xi)\widehat{\zeta}(2^{-j}\xi)\widehat{f} = \sum_{j=1}^{\infty}2^{js}\widehat{\Psi}(2^{-j}\xi)\widehat{\theta}(2^{-j}\xi)\widehat{f}$$

and hence

$$f_\infty = \sum_{j=1}^{\infty}\Delta_j^\theta(2^{js}\Delta_j(f)).$$

Using estimate (5.1.19), we obtain

$$\big\|f_\infty\big\|_{L^p} \le C\Big\|\Big(\sum_{j=1}^{\infty}|2^{js}\Delta_j(f)|^2\Big)^{\frac{1}{2}}\Big\|_{L^p} < \infty. \tag{6.2.16}$$

Combining (6.2.14), (6.2.15), and (6.2.16), we deduce the estimate in (6.2.12). (Incidentally, this argument shows that f_∞ is a function.)

To obtain the converse inequality (6.2.11) we essentially have to reverse our steps. Here we assume that $f \in L_s^p$ and we will show the validity of (6.2.11). First we have the estimate

$$\big\|S_0(f)\big\|_{L^p} \le C\big\|f_s\big\|_{L^p} = C\big\|f\big\|_{L_s^p} \tag{6.2.17}$$

since we can obtain the Fourier transform of $S_0(f) = \Phi * f$ by multiplying $\widehat{f_s}$ with the L^p Fourier multiplier $(1+|\xi|^2)^{-\frac{s}{2}}\widehat{\Phi}(\xi)$. Second, setting $\widehat{\sigma}(\xi) = |\xi|^{-s}\widehat{\Psi}(\xi)$ and letting Δ_j^σ be the Littlewood-Paley operator associated with the bump $\widehat{\sigma}(2^{-j}\xi)$, we have

$$2^{js}\widehat{\Psi}(2^{-j}\xi)\widehat{f} = \widehat{\sigma}(2^{-j}\xi)|\xi|^s\widehat{f} = \widehat{\sigma}(2^{-j}\xi)|\xi|^s(1-\widehat{\Phi}(\xi))\widehat{f},$$

when $j \geq 2$ [since $\widehat{\Phi}$ vanishes on the support of $\widehat{\sigma}(2^{-j}\xi)$ when $j \geq 2$]. This yields the operator identity

$$2^{js}\Delta_j(f) = \Delta_j^\sigma(f_\infty). \tag{6.2.18}$$

Using identity (6.2.18) and Proposition 5.1.4, we obtain

$$\Big\|\Big(\sum_{j=2}^{\infty}|2^{js}\Delta_j(f)|^2\Big)^{\frac{1}{2}}\Big\|_{L^p} = \Big\|\Big(\sum_{j=2}^{\infty}|\Delta_j^\sigma(f_\infty)|^2\Big)^{\frac{1}{2}}\Big\|_{L^p} \leq C\big\|f_\infty\big\|_{L^p}, \tag{6.2.19}$$

where the last inequality follows by Theorem 5.1.2. Notice that

$$f_\infty = \big(|\xi|^s(1-\widehat{\Phi}(\xi))\,\widehat{f}\,\big)^\vee = \Big(\frac{|\xi|^s(1-\widehat{\Phi}(\xi))}{(1+|\xi|^2)^{\frac{s}{2}}}\,\widehat{f_s}\Big)^\vee$$

and since the function $|\xi|^s(1-\widehat{\Phi}(\xi))(1+|\xi|^2)^{-\frac{s}{2}}$ is in $\mathcal{M}_p(\mathbf{R}^n)$ by Theorem 5.2.7, it follows that

$$\big\|f_\infty\big\|_{L^p} \leq C\big\|f_s\big\|_{L^p} = C\big\|f\big\|_{L^p_s},$$

which combined with (6.2.19) yields

$$\Big\|\Big(\sum_{j=2}^{\infty}|2^{js}\Delta_j(f)|^2\Big)^{\frac{1}{2}}\Big\|_{L^p} \leq C\big\|f\big\|_{L^p_s}. \tag{6.2.20}$$

Finally, we have

$$2^s\Delta_1(f) = 2^s\big(\widehat{\Psi}(\tfrac{1}{2}\xi)(1+|\xi|^2)^{-\frac{s}{2}}(1+|\xi|^2)^{\frac{s}{2}}\,\widehat{f}\,\big)^\vee = 2^s\big(\widehat{\Psi}(\tfrac{1}{2}\xi)(1+|\xi|^2)^{-\frac{s}{2}}\,\widehat{f_s}\,\big)^\vee$$

and since the function $\widehat{\Psi}(\frac{1}{2}\xi)(1+|\xi|^2)^{-\frac{s}{2}}$ is smooth with compact support and thus in $\mathcal{M}_p$, it follows that

$$\big\|2^s\Delta_1(f)\big\|_{L^p} \leq C\big\|f_s\big\|_{L^p} = C\big\|f\big\|_{L^p_s}. \tag{6.2.21}$$

Combining estimates (6.2.17), (6.2.20), and (6.2.21), we conclude the proof of (6.2.11).

□

6.2.c. Littlewood-Paley Characterization of Homogeneous Sobolev Spaces

We now state and prove the homogeneous version of the previous theorem.

Theorem 6.2.7. *Let Ψ satisfy (6.2.6) and let Δ_j be the Littlewood-Paley operator associated with Ψ. Let $s \in \mathbf{R}$ and $1 < p < \infty$. Then there exists a constant C_1 that depends only on n, s, p, and Ψ such that for all $f \in \dot{L}^p_s(\mathbf{R}^n)$ we have*

$$\Big\|\Big(\sum_{j\in\mathbf{Z}}(2^{js}|\Delta_j(f)|)^2\Big)^{\frac{1}{2}}\Big\|_{L^p} \le C_1\big\|f\big\|_{\dot{L}^p_s}. \tag{6.2.22}$$

Conversely, there exists a constant C_2 that depends on the parameters n, s, p, and Ψ, such that every element f of $\mathcal{S}'(\mathbf{R}^n)/\mathcal{P}$ that satisfies

$$\Big\|\Big(\sum_{j\in\mathbf{Z}}(2^{js}|\Delta_j(f)|)^2\Big)^{\frac{1}{2}}\Big\|_{L^p} < \infty$$

lies in the homogeneous Sobolev space $\dot{L}^p_s$ and we have

$$\big\|f\big\|_{\dot{L}^p_s} \le C_2\Big\|\Big(\sum_{j\in\mathbf{Z}}(2^{js}|\Delta_j(f)|)^2\Big)^{\frac{1}{2}}\Big\|_{L^p}. \tag{6.2.23}$$

Proof. The proof of the theorem is similar but a bit simpler than that of Theorem 6.2.6. To obtain (6.2.22) we start with $f \in \dot{L}^p_s$ and we note that

$$2^{js}\Delta_j(f) = 2^{js}\big(|\xi|^s|\xi|^{-s}\widehat{\Psi}(2^{-j}\xi)\,\widehat{f}\,\big)^\vee = \big(\widehat{\sigma}(2^{-j}\xi)\,\widehat{f_s}\,\big)^\vee = \Delta_j^\sigma(f_s)\,,$$

where $\widehat{\sigma}(\xi) = \widehat{\Psi}(\xi)|\xi|^{-s}$ and Δ_j^σ is the Littlewood-Paley operator given on the Fourier transform by multiplication with the function $\widehat{\sigma}(2^{-j}\xi)$. We have

$$\Big\|\Big(\sum_{j\in\mathbf{Z}}|2^{js}\Delta_j(f)|^2\Big)^{\frac{1}{2}}\Big\|_{L^p} = \Big\|\Big(\sum_{j\in\mathbf{Z}}|\Delta_j^\sigma(f_s)|^2\Big)^{\frac{1}{2}}\Big\|_{L^p} \le C\big\|f_s\big\|_{L^p} = C\big\|f\big\|_{\dot{L}^p_s},$$

where the last inequality follows from Theorem 5.1.2. This proves (6.2.22).

Next we show that if the expression on the right in (6.2.23) is finite, then the distribution f in $\mathcal{S}'(\mathbf{R}^n)/\mathcal{P}$ must lie the in the homogeneous Sobolev space $\dot{L}^p_s$ with norm controlled by a multiple of this expression.

Define Littlewood-Paley operators Δ_j^η given by convolution with $\eta_{2^{-j}}$, where $\widehat{\eta}$ is a smooth bump supported in the annulus $\frac{3}{5} \le |\xi| \le \frac{5}{3}$ that satisfies

$$\sum_{k\in\mathbf{Z}}\widehat{\eta}(2^{-k}\xi) = 1, \qquad \xi \ne 0, \tag{6.2.24}$$

or, in operator form,

$$\sum_{k\in\mathbf{Z}}\Delta_j^\eta = I\,,$$

where the convergence is in the sense of $\mathcal{S}'/\mathcal{P}$ in view of Exercise 2.3.12. We note that (6.2.24) is possible since $2 \cdot \frac{3}{5} < \frac{5}{3}$. (Draw a picture.) We introduce another family of Littlewood-Paley operators Δ_j^θ given by convolution with $\theta_{2^{-j}}$, where $\widehat{\theta}(\xi) = \widehat{\eta}(\xi)|\xi|^s$. Given $f \in \mathcal{S}'(\mathbf{R}^n)/\mathcal{P}$, we set $f_s = \big(|\xi|^s\widehat{f}\,\big)^\vee$, which is also an element of $\mathcal{S}'(\mathbf{R}^n)/\mathcal{P}$. Because of the validity of (6.2.24) we can use the reverse estimate (5.1.8) in the Littlewood-Paley Theorem 5.1.2 to obtain for some polynomial Q

$$\|f\|_{\dot{L}^p_s} = \|f_s - Q\|_{L^p} \le C\Big\|\Big(\sum_{j\in\mathbf{Z}} |\Delta_j^\eta(f_s)|^2\Big)^{\frac{1}{2}}\Big\|_{L^p}$$

$$= C\Big\|\Big(\sum_{j\in\mathbf{Z}} |2^{js}\Delta_j^\theta(f)|^2\Big)^{\frac{1}{2}}\Big\|_{L^p}.$$

Recalling the definition of Δ_j (see the discussion before the statement of Theorem 6.2.6), we notice that the function

$$\widehat{\Psi}(\tfrac{1}{2}\xi) + \widehat{\Psi}(\xi) + \widehat{\Psi}(2\xi)$$

is equal to 1 on the support of η. It follows that

$$\Delta_j^\theta = (\Delta_{j-1} + \Delta_j + \Delta_{j+1})\Delta_j^\theta\,.$$

We therefore have the estimate

$$\Big\|\Big(\sum_{j\in\mathbf{Z}} |2^{js}\Delta_j^\theta(f)|^2\Big)^{\frac{1}{2}}\Big\|_{L^p} \le \sum_{r=-1}^{1} \Big\|\Big(\sum_{j\in\mathbf{Z}} |\Delta_j^\theta\Delta_{j+r}(2^{js}f)|^2\Big)^{\frac{1}{2}}\Big\|_{L^p}$$

and applying Proposition 5.1.4 we can control the last expression (and thus $\|f\|_{\dot{L}^p_s}$) by a constant multiple of

$$\Big\|\Big(\sum_{j\in\mathbf{Z}} |\Delta_j(2^{js}f)|^2\Big)^{\frac{1}{2}}\Big\|_{L^p}.$$

This proves that the homogeneous Sobolev norm of f is controlled by a multiple of the expression in (6.2.23). In particular, the distribution f lies in the homogeneous Sobolev space $\dot{L}^p_s$. This ends the proof of the converse direction and completes the proof of the theorem.

□

Exercises

6.2.1. Show that the spaces $\dot{L}^p_s$ and L^p_s are complete and that the latter are decreasing as s increases.

6.2.2. Let $f \in L^p_s$ for some $1 < p < \infty$.
(a) Suppose that φ is in $\mathcal{S}(\mathbf{R}^n)$. Prove that φf is also an element of L^p_s.
(b) Let v be a function whose Fourier transform is a bounded function. Prove that vf is in L^p_s.

6.2.3. Let $s > 0$ and α a fixed multiindex. Find the set of p in $(1,\infty)$ such that the distribution $\partial^\alpha\delta_0$ belongs to L^p_{-s}.

6.2.4. Let I be the identity operator, I_1 the Riesz potential of order 1 and R_j the usual Riesz transform. Prove that

$$I = \sum_{j=1}^{n} I_1 R_j \partial_j .$$

and use this identity to obtain Theorem 6.2.4 when $s = 1$.
[*Hint:* Take the Fourier transform.]

6.2.5. Let f be in L^p_s for some $1 < p < \infty$. Prove that $\partial^\alpha f$ is in $L^p_{s-|\alpha|}$.

6.2.6. Prove that for all C^1 functions f that are supported in a ball B we have

$$|f(x)| \le \frac{1}{\omega_{n-1}} \int_B |\nabla f(y)| |x-y|^{-n+1} \, dy \, ,$$

where $\omega_{n-1} = |\mathbf{S}^{n-1}|$. For such functions obtain the local Sobolev inequality

$$\big\|f\big\|_{L^q(B)} \le C_{q,r,n} \big\|\nabla f\big\|_{L^p(B)}$$

where $1 < p < q < \infty$ and $1/p = 1/q + 1/n$.
[*Hint:* Start from $f(x) = \int_0^\infty \nabla f(x - t\theta) \cdot \theta \, dt$ and integrate over $\theta \in \mathbf{S}^{n-1}$.]

6.2.7. Show that there is a constant C so that for all C^1 functions f that are supported in a ball B we have

$$\frac{1}{|B'|} \int_{B'} |f(x) - f(z)| \, dz \le C \int_B |\nabla f(y)| |x-y|^{-n+1} \, dy \, .$$

for all B' balls contained in B and all $x \in B'$.
[*Hint:* Start with $f(z) - f(x) = \int_0^1 \nabla f(x + t(z-x)) \cdot (z-x) \, dt$.]

6.2.8. Let $1 < p < \infty$ and $s > 0$. Show that

$$f \in L^p_s \iff f \in L^p \quad \text{and} \quad f \in \dot{L}^p_s \, .$$

Conclude that $\dot{L}^p_s \cap L^p = L^p_s$ and obtain an estimate for the corresponding norms.
[*Hint:* If f is in $\dot{L}^p_s \cap L^p$ use Theorem 5.2.7 to obtain that $\big\|f\big\|_{L^p_s}$ is controlled by a multiple of $(\widehat{f}(\xi)(1 + |\xi|^s))^\vee$. Use the same theorem to show that $\big\|f\big\|_{\dot{L}^p_s} \le C \big\|f\big\|_{L^p_s}$.]

6.2.9. (*Gagliardo* [**201**]/*Nirenberg* [**394**]) Prove that all Schwartz functions on $\mathbf{R}^n$ satisfy the estimate

$$\big\|f\big\|_{L^q} \le \prod_{j=1}^{n} \big\|\partial_j f\big\|_{L^1}^{1/n} \, ,$$

where $1/q + 1/n = 1$.
[*Hint:* Use induction beginning with the case $n = 1$. Assuming the inequality is valid for $n-1$, set $I_j(x_1) = \int_{\mathbf{R}^{n-1}} |\partial_j f(x_1, x')|\, dx'$ for $j = 2, \dots, n$, where $x = (x_1, x') \in \mathbf{R}^n$ and $I_1(x') = \int_{\mathbf{R}^1} |\partial_1 f(x_1, x')|\, dx_1$. Apply the induction hypothesis to obtain

$$\big\|f(x_1, \cdot\,)\big\|_{L^{q'}} \le \prod_{j=2}^{n} I_j(x_1)^{1/(n-1)}$$

and use the fact that $|f|^q \le I_1(x')^{1/(n-1)}|f|$ and Hölder's inequality to calculate $\big\|f\big\|_{L^q}$.]

6.2.10. Let $f \in L^2_1(\mathbf{R}^n)$. Prove that there is a constant $c_n > 0$ such that

$$\int_{\mathbf{R}^n}\int_{\mathbf{R}^n} \frac{|f(x+t) + f(x-t) - 2f(x)|^2}{|t|^{n+2}}\, dx\, dt = c_n \int_{\mathbf{R}^n} \sum_{j=1}^{n} |\partial_j f(x)|^2\, dx.$$

6.2.11. (*Christ* [**91**]) Let $0 \le \beta < \infty$ and let

$$C_0 = \int_{\mathbf{R}^n} |\widehat{g}(\xi)|^2 (1 + |\xi|)^n \big(\log(2 + |\xi|)\big)^{-\beta}\, d\xi\,.$$

(a) Prove that there is a constant $C(n, \beta, C_0)$ such that for every $q > 2$ we have

$$\big\|g\big\|_{L^q(\mathbf{R}^n)} \le C(n, \beta, C_0) q^{\frac{\beta+1}{2}}\,.$$

(b) Conclude that for any compact subset K of $\mathbf{R}^n$ we have

$$\int_K e^{|g(x)|^\gamma}\, dx < \infty$$

whenever $\gamma < \frac{2}{\beta+1}$.
[*Hint:* Part (a): For $q > 2$ control $\big\|g\big\|_{L^q(\mathbf{R}^n)}$ by $\big\|\widehat{g}\big\|_{L^{q'}(\mathbf{R}^n)}$ and apply Hölder's inequality with exponents $\frac{2}{q'}$ and $\frac{2(q-1)}{q-2}$. Part (b): Expand the exponential in Taylor series.]

6.2.12. Suppose that $m \in L^2_s(\mathbf{R}^n)$ for some $s > \frac{n}{2}$ and let $\lambda > 0$. Define the operator T_λ by setting $\widehat{T_\lambda(f)}(\xi) = m(\lambda\xi)\widehat{f}(\xi)$. Show that there exists a constant $C = C(n, s)$ such that for all f and u and $\lambda > 0$ we have

$$\int_{\mathbf{R}^n} |T_\lambda(f)(x)|^2\, u(x)\, dx \le C \int_{\mathbf{R}^n} |f(x)|^2\, M(u)(x)\, dx\,.$$

6.3. Lipschitz Spaces

The classical definition says that a function f on $\mathbf{R}^n$ is Lipschitz (or Hölder) continuous of order $\gamma > 0$, if there is constant $C < \infty$ such that for all $x, y \in \mathbf{R}^n$ we have

$$|f(x+y) - f(x)| \le C|y|^\gamma. \tag{6.3.1}$$

It turns out that only constant functions satisfy (6.3.1) when $\gamma > 1$ and the corresponding definition needs to be suitably adjusted in this case. This will be discussed in this section. The key point is that any function f that satisfies (6.3.1) possesses a certain amount of smoothness "measured" by the quantity γ. The Lipschitz norm of a function is introduced to serve this purpose, that is, to precisely quantify and exactly measure this smoothness. In this section we formalize these concepts and we explore connections they have with the orthogonality considerations of the previous chapter. The main achievement of this section will be a characterization of Lipschitz spaces using Littlewood-Paley theory.

6.3.a. Introduction of Lipschitz Spaces

Definition 6.3.1. Let $0 < \gamma < 1$. A function f on $\mathbf{R}^n$ is said to be *Lipschitz of order* γ if it is bounded and satisfies (6.3.1) for some $C < \infty$. In this case we let

$$\big\|f\big\|_{\Lambda_\gamma(\mathbf{R}^n)} = \big\|f\big\|_{L^\infty} + \sup_{x\in\mathbf{R}^n}\sup_{h\in\mathbf{R}^n\setminus\{0\}} \frac{|f(x+h)-f(x)|}{|h|^\gamma}$$

and we set

$$\Lambda_\gamma(\mathbf{R}^n) = \{f : \mathbf{R}^n \to \mathbf{C} \text{ continuous} : \ \big\|f\big\|_{\Lambda_\gamma(\mathbf{R}^n)} < \infty\}.$$

Note that functions in $\Lambda_\gamma(\mathbf{R}^n)$ are automatically continuous when $\gamma < 1$, so we did not need to make this part of the definition. We call $\Lambda_\gamma(\mathbf{R}^n)$ the *inhomogeneous Lipschitz space* of order γ. For reasons of uniformity we also set

$$\Lambda_0(\mathbf{R}^n) = L^\infty(\mathbf{R}^n) \cap C(\mathbf{R}^n),$$

where $C(\mathbf{R}^n)$ is the space of all continuous functions on $\mathbf{R}^n$. See Exercise 6.3.2.

Example 6.3.2. The function $h(x) = \cos(x\cdot a)$ for some fixed $a \in \mathbf{R}^n$ is in Λ_γ for all $\gamma < 1$. Simply notice that $|h(x) - h(y)| \le \min(2, |a|\,|x-y|)$.

We now extend this definition to indices $\gamma \ge 1$.

Definition 6.3.3. For $h \in \mathbf{R}^n$ define the *difference operator* D_h by setting

$$D_h(f)(x) = f(x+h) - f(x)$$

for a continuous function $f: \mathbf{R}^n \to \mathbf{C}$. We may check that

$$\begin{aligned} D_h^2(f)(x) &= D_h(D_h f)(x) = f(x+2h) - 2f(x+h) + f(x), \\ D_h^3(f)(x) &= D_h(D_h^2 f)(x) = f(x+3h) - 3f(x+2h) + 3f(x+h) - f(x) \end{aligned}$$

and, in general, that $D_h^{k+1}(f) = D_h^k(D_h(f))$ is given by

$$D_h^{k+1}(f)(x) = \sum_{s=1}^{k+1} (-1)^{k+s+1} \binom{k+1}{s} f(x+s\,h) \tag{6.3.2}$$

for a nonnegative integer k. See Exercise 6.3.3. For $\gamma > 0$ define

$$\|f\|_{\Lambda_\gamma} = \|f\|_{L^\infty} + \sup_{x \in \mathbf{R}^n} \sup_{h \in \mathbf{R}^n \setminus \{0\}} \frac{|D_h^{[\gamma]+1}(f)(x)|}{|h|^\gamma},$$

where $[\gamma]$ denotes the integer part of γ and set

$$\Lambda_\gamma = \{f : \mathbf{R}^n \to \mathbf{C} \text{ continuous} : \|f\|_{\Lambda_\gamma} < \infty\}.$$

We call $\Lambda_\gamma(\mathbf{R}^n)$ the inhomogeneous *Lipschitz space* of order $\gamma \in \mathbf{R}^+$.

For a tempered distribution u we also define another distribution $D_h^k(u)$ via the identity

$$\langle D_h^k(u), \varphi \rangle = \langle u, D_{-h}^k(\varphi) \rangle$$

for all φ in the Schwartz class.

At this point the reader should check that the quantity $\|\cdot\|_{\Lambda_\gamma}$ is a norm and that Λ_γ is a closed subspace of L^∞, and thus complete.

Finally, we define the homogeneous Lipschitz spaces. We will adhere to the usual convention of using a dot on a space to indicate its homogeneous nature.

Definition 6.3.4. For $\gamma > 0$ we define

$$\|f\|_{\dot{\Lambda}_\gamma} = \sup_{x \in \mathbf{R}^n} \sup_{h \in \mathbf{R}^n \setminus \{0\}} \frac{|D_h^{[\gamma]+1}(f)(x)|}{|h|^\gamma}$$

and we also let $\dot{\Lambda}_\gamma$ be the space of all continuous functions f on $\mathbf{R}^n$ that satisfy $\|f\|_{\dot{\Lambda}_\gamma} < \infty$. We call $\dot{\Lambda}_\gamma$ the *homogeneous Lipschitz space* of order γ. We note that elements of $\dot{\Lambda}_\gamma$ have at most polynomial growth at infinity and thus they are elements of $\mathcal{S}'(\mathbf{R}^n)$.

A few observations are in order here. Constant functions f satisfy $D_h(f)(x) = 0$ for all $h, x \in \mathbf{R}^n$ and therefore the homogeneous quantity $\|\cdot\|_{\dot{\Lambda}_\gamma}$ is insensitive to constants. Similarly the expressions $D_h^{k+1}(f)$ and $\|f\|_{\dot{\Lambda}_\gamma}$ do not recognize polynomials of degree up to k. Moreover, polynomials are the only continuous functions with this property; see Exercise 6.3.1. This means that the quantity $\|f\|_{\dot{\Lambda}_\gamma}$ is not a norm but only a seminorm. To make it a norm, we will need to consider functions modulo polynomials as we did in the case of homogeneous Sobolev spaces. For this reason we think of $\dot{\Lambda}_\gamma$ as a subspace of $\mathcal{S}'(\mathbf{R}^n)/\mathcal{P}$.

We will make use of the following proposition concerning properties of the difference operators D_h^k.

Proposition 6.3.5. *Let f be a C^k function on $\mathbf{R}^n$ for some $k \in \mathbf{Z}^+$. Then for all $h = (h_1, \dots, h_n)$ and $x \in \mathbf{R}^n$ there is a $\theta_1 \in [0,1]$ so that the identity*

$$D_h(f)(x) = \sum_{m=1}^{n} h_m \, (\partial_m f)(x + \theta_1 h) \tag{6.3.3}$$

is valid. More generally, we have

$$D_h^k(f)(x) = \sum_{m_1=1}^{n} \cdots \sum_{m_k=1}^{n} h_{m_1} \dots h_{m_k} \, (\partial_{m_1} \dots \partial_{m_k} f)(x + \theta_k h)\,, \tag{6.3.4}$$

for some $\theta_k \in [0,1]$.

Proof. Identity (6.3.3) is a simple consequence of the mean value theorem applied to the function $t \to f((1-t)x + t(x+h))$ on $[0,1]$. Identity (6.3.4) follows by induction.

□

6.3.b. Littlewood-Paley Characterization of Homogeneous Lipschitz Spaces

We will now characterize the homogeneous Lipschitz spaces using the Littlewood-Paley operators Δ_j. We recall the function Ψ introduced in the previous section that was a radial Schwartz function on $\mathbf{R}^n$ whose Fourier transform was nonnegative, was supported in the annulus $\frac{1}{2} + \frac{1}{10} \le |\xi| \le 2 - \frac{1}{10}$, was at least $\frac{1}{4}$ in the smaller annulus $\frac{3}{5} \le |\xi| \le \frac{5}{3}$ and satisfied

$$\sum_{j \in \mathbf{Z}} \widehat{\Psi}(2^{-j}\xi) = 1 \tag{6.3.5}$$

for all $\xi \neq 0$. Associated with Ψ, we defined the Littlewood-Paley operators $\Delta_j = \Delta_j^\Psi$ given by multiplication on the Fourier transform with the smooth bump $\widehat{\Psi}(2^{-j}\xi)$.

Theorem 6.3.6. *Let Δ_j be as previously and $\gamma > 0$. Then there is a constant $C = C(n, \gamma)$ so that for every f in $\dot{\Lambda}_\gamma$ we have the estimate*

$$\sup_{j \in \mathbf{Z}} 2^{j\gamma} \big\| \Delta_j(f) \big\|_{L^\infty} \le C \big\| f \big\|_{\dot{\Lambda}_\gamma} . \tag{6.3.6}$$

Conversely, every element f of $\mathcal{S}'(\mathbf{R}^n)/\mathcal{P}$ that satisfies

$$\sup_{j \in \mathbf{Z}} 2^{j\gamma} \big\| \Delta_j(f) \big\|_{L^\infty} < \infty \tag{6.3.7}$$

is an element of $\dot{\Lambda}_\gamma$ *with norm*

$$\|f\|_{\dot{\Lambda}_\gamma} \le C' \sup_{j\in\mathbf{Z}} 2^{j\gamma} \|\Delta_j(f)\|_{L^\infty} \tag{6.3.8}$$

for some constant $C' = C'(n,\gamma)$.

Note that condition (6.3.7) remains invariant if a polynomial is added to the function f that is consistent with this property of $\|f\|_{\dot{\Lambda}_\gamma}$.

Proof. We begin with the proof of (6.3.8). Let $k = [\gamma]$ be the integer part of γ. Let us pick a Schwartz function η on $\mathbf{R}^n$ whose Fourier transform is nonnegative, supported in the annulus $\frac{3}{5} \le |\xi| \le \frac{5}{3}$, and satisfies

$$\sum_{j\in\mathbf{Z}} \widehat{\eta}(2^{-j}\xi)^2 = 1 \tag{6.3.9}$$

for all $\xi \neq 0$. Associated with η, we define the Littlewood-Paley operators Δ_j^η given by multiplication on the Fourier transform with the smooth bump $\widehat{\eta}(2^{-j}\xi)$. We also set

$$\widehat{\Theta}(\xi) = \widehat{\Psi}(\tfrac{1}{2}\xi) + \widehat{\Psi}(\xi) + \widehat{\Psi}(2\xi)$$

and we denote by $\Delta_j^\Theta = \Delta_{j-1} + \Delta_j + \Delta_{j+1}$ the Littlewood-Paley operator given by multiplication on the Fourier transform with the smooth bump $\widehat{\Theta}(2^{-j}\xi)$.

The fact that the previous function is equal to 1 on the support of $\widehat{\eta}$ together with the functional identity (6.3.9) yields the operator identity

$$I = \sum_{j\in\mathbf{Z}} (\Delta_j^\eta)^2 = \sum_{j\in\mathbf{Z}} \Delta_j^\Theta \Delta_j^\eta \Delta_j^\eta\,,$$

with convergence in the sense of the space $\mathcal{S}'(\mathbf{R}^n)/\mathcal{P}$. Since convolution is a linear operation we have $D_h^{k+1}(F * G) = F * D_h^{k+1}(G)$, from which we deduce

$$\begin{aligned} D_h^{k+1}(f) = & \sum_{j\in\mathbf{Z}} \Delta_j^\Theta(f) \, * \, D_h^{k+1}(\eta_{2^{-j}}) \, * \, \eta_{2^{-j}} \\ = & \sum_{j\in\mathbf{Z}} D_h^{k+1}(\Delta_j^\Theta(f)) \, * \, (\eta * \eta)_{2^{-j}} \end{aligned} \tag{6.3.10}$$

for all tempered distributions f. The convergence of the series in (6.3.10) is in the sense of $\mathcal{S}'/\mathcal{P}$ in view of Exercise 5.2.2. The convergence of the series in (6.3.10) in the L^∞ norm is a consequence of condition (6.3.7) and is contained in the following argument.

Using (6.3.2), we easily obtain the estimate

$$\|D_h^{k+1}(\Delta_j^\Theta(f)) * (\eta * \eta)_{2^{-j}}\|_{L^\infty} \le 2^{k+1} \|\eta * \eta\|_{L^1} \|\Delta_j^\Theta(f)\|_{L^\infty}\,. \tag{6.3.11}$$

Also using Proposition 6.3.5 we have

$$\begin{aligned} & D_h^{k+1}(\eta_{2^{-j}})(x) \\ &= \sum_{m_1=1}^{n} \cdots \sum_{m_{k+1}=1}^{n} h_{m_1} \dots h_{m_{k+1}} (\partial_{m_1} \dots \partial_{m_{k+1}} \eta_{2^{-j}})(x+\theta h) \\ &= 2^{j(k+1)} \sum_{m_1=1}^{n} \cdots \sum_{m_{k+1}=1}^{n} h_{m_1} \dots h_{m_{k+1}} (\partial_{m_1} \dots \partial_{m_{k+1}} \eta)_{2^{-j}}(x+\theta h) \end{aligned}$$

and from this we deduce the estimate

$$\begin{aligned} & \big\|\Delta_j^{\Theta}(f) * D_h^{k+1}(\eta_{2^{-j}}) * \eta_{2^{-j}}\big\|_{L^\infty} \\ & \le \big\|\Delta_j^{\Theta}(f)\big\|_{L^\infty} \big\|D_h^{k+1}(\eta_{2^{-j}}) * \eta_{2^{-j}}\big\|_{L^1} \\ & \le \big\|\Delta_j^{\Theta}(f)\big\|_{L^\infty} |2^j h|^{k+1} \sum_{|\alpha|\le k+1} \big\|\partial^\alpha \eta\big\|_{L^\infty} \big\|\eta\big\|_{L^1} . \end{aligned} \tag{6.3.12}$$

Combining (6.3.11) and (6.3.12), we obtain

$$\begin{aligned} & \big\|\Delta_j^{\Theta}(f) * D_h^{k+1}(\eta_{2^{-j}}) * \eta_{2^{-j}}\big\|_{L^\infty} \\ & \qquad \le C_{\eta,n,k} \big\|\Delta_j^{\Theta}(f)\big\|_{L^\infty} \min\big(1, |2^j h|^{k+1}\big) . \end{aligned} \tag{6.3.13}$$

Combining (6.3.10) and (6.3.13) we deduce

$$\frac{\big\|D_h^{k+1}(f)\big\|_{L^\infty}}{|h|^\gamma} \le \frac{C'}{|h|^\gamma} \sum_{j\in\mathbf{Z}} 2^{j\gamma} \big\|\Delta_j^{\Theta}(f)\big\|_{L^\infty} \min\big(2^{-j\gamma}, 2^{j(k+1-\gamma)}|h|^{k+1}\big) ,$$

from which it follows that

$$\begin{aligned} \big\|f\big\|_{\dot{\Lambda}_\gamma} & \le \sup_{h\in\mathbf{R}^n\setminus\{0\}} \frac{C'}{|h|^\gamma} \sum_{j\in\mathbf{Z}} 2^{j\gamma} \big\|\Delta_j^{\Theta}(f)\big\|_{L^\infty} \min\big(2^{-j\gamma}, 2^{j(k+1-\gamma)}|h|^{k+1}\big) \\ & \le C' \sup_{j\in\mathbf{Z}} 2^{j\gamma} \big\|\Delta_j^{\Theta}(f)\big\|_{L^\infty} \sup_{h\neq 0} \sum_{j\in\mathbf{Z}} \min\big(|h|^{-\gamma} 2^{-j\gamma}, 2^{j(k+1-\gamma)}|h|^{k+1-\gamma}\big) \\ & \le C' \sup_{j\in\mathbf{Z}} 2^{j\gamma} \big\|\Delta_j^{\Theta}(f)\big\|_{L^\infty} \end{aligned}$$

since the last numerical series converges ($\gamma < k+1 = [\gamma]+1$.) This proves (6.3.8) except that instead of Δ_j we have Δ_j^{Θ} on the right. The passage to Δ_j is a trivial matter since $\Delta_j^{\Theta} = \Delta_{j-1} + \Delta_j + \Delta_{j+1}$.

Having established (6.3.8), we now turn to the proof of (6.3.6). We first consider the case $0 < \gamma < 1$, which is very simple. Since each Δ_j is given by convolution with a function with mean value zero, we may write

$$\begin{aligned}\Delta_j(f)(x) &= \int_{\mathbf{R}^n} f(x-y)\Psi_{2^{-j}}(y)\,dy \\ &= \int_{\mathbf{R}^n} (f(x-y)-f(x))\Psi_{2^{-j}}(y)\,dy \\ &= 2^{-j\gamma}\int_{\mathbf{R}^n} \frac{D_{-y}(f)(x)}{|y|^\gamma}|2^j y|^\gamma 2^{jn}\Psi(2^j y)\,dy\end{aligned}$$

and the previous expression is easily seen to be controlled by a constant multiple of $2^{-j\gamma}\|f\|_{\dot\Lambda_\gamma}$. This proves (6.3.6) when $0<\gamma<1$. In the case when $\gamma \ge 1$ we have to work a bit harder.

As before set $k=[\gamma]$. We observe that we can write $D_h^{k+1}(\Psi_{2^{-j}})$ in the form

$$D_h^{k+1}(\Psi_{2^{-j}}) = \big(\widehat{\Psi}(2^{-j}\xi)\,(e^{2\pi i\xi\cdot h}-1)^{k+1}\big)^\vee .$$

If the support of $\widehat{\Psi}(2^{-j}\xi)$ did not intersect the set of all ξ for which $\xi\cdot h$ is an integer, then we would be able to define the function

$$w = \big(\widehat{\Psi}(2^{-j}\xi)\,(e^{2\pi i\xi\cdot h}-1)^{-(k+1)}\big)^\vee .$$

To achieve this, we write $\widehat{\Psi}$ as a finite sum of functions $\widehat{\Psi^{(r)}}$, where each $\widehat{\Psi^{(r)}}$ is supported in a small open subset U_r of the annulus $\frac{1}{2}<|\xi|<2$ so that

$$\xi\in U_r \implies \tfrac{1}{2}\le|\xi\cdot e_r|\le 2$$

for some fixed unit vector e_r. For a fixed r and $h=\frac{1}{8}2^{-j}e_r$ we have

$$\begin{aligned}\Psi^{(r)}_{2^{-j}} * f &= \big(\widehat{\Psi^{(r)}}(2^{-j}\xi)\,(e^{2\pi i\xi\cdot h}-1)^{-(k+1)}(e^{2\pi i\xi\cdot h}-1)^{k+1}\widehat{f}(\xi)\big)^\vee \\ &= \big(\widehat{\Psi^{(r)}}(2^{-j}\xi)\,(e^{2\pi i 2^{-j}\xi\cdot\frac{1}{8}e_r}-1)^{-(k+1)}\widehat{D_h^{k+1}(f)}(\xi)\big)^\vee\end{aligned} \tag{6.3.14}$$

and note that

$$2^{-j}\xi\in U_r \implies \tfrac{1}{16}\le|2^{-j}\xi\cdot\tfrac{1}{8}e_r|\le\tfrac{1}{2}.$$

Since the function $\widehat{\zeta^{(r)}} = \widehat{\Psi^{(r)}}(\xi)\,(e^{2\pi i\xi\cdot\frac{1}{8}e_r}-1)^{-(k+1)}$ is well defined and smooth with compact support, it follows that

$$\Psi^{(r)}_{2^{-j}} * f = \zeta^{(r)}_{2^{-j}} * D^{k+1}_{2^{-j}\frac{1}{8}e_r}(f),$$

which implies that

$$\|\Psi^{(r)}_{2^{-j}} * f\|_{L^\infty} \le \|\zeta^{(r)}_{2^{-j}}\|_{L^1}\|D^{k+1}_{2^{-j}\frac{1}{8}e_r}(f)\|_{L^\infty} \le \|\zeta^{(r)}\|_{L^1}\|f\|_{\dot\Lambda_\gamma}2^{-j\gamma}.$$

Summing over the finite number of r, we obtain the estimate

$$\|\Delta_j(f)\|_{L^\infty}\le C\|f\|_{\dot\Lambda_\gamma}2^{-j\gamma},$$

which concludes the proof of the theorem. □

6.3.c. Littlewood-Paley Characterization of Inhomogeneous Lipschitz Spaces

We have seen that quantities involving the Littlewood-Paley operators Δ_j characterize homogeneous Lipschitz spaces. We now address the same question for inhomogeneous spaces.

As in the Littlewood-Paley characterization of inhomogeneous Sobolev spaces, we will need to treat the contribution of the frequencies near zero separately. For this reason we recall the Schwartz function Φ introduced in the previous section that satisfied

$$\widehat{\Phi}(\xi) = \begin{cases} \sum_{j\le 0} \widehat{\Psi}(2^{-j}\xi) & \text{when } \xi \neq 0, \\ 1 & \text{when } \xi = 0. \end{cases} \tag{6.3.15}$$

Note that $\widehat{\Phi}(\xi)$ is equal to 1 for $|\xi| \le 1$ and vanishes when $|\xi| \ge 2$. As before we also define an operator $S_0(f) = \Phi * f$. The reader will not be surprised to find out that a theorem similar to 6.2.7 is valid for Lipschitz spaces as well. We have the following.

Theorem 6.3.7. *Let Ψ and Δ_j be as in the Theorem 6.3.6, Φ as in (6.3.15), and $\gamma > 0$. Then there is a constant $C = C(n,\gamma)$ so that for every f in Λ_γ we have the estimate*

$$\big\|S_0(f)\big\|_{L^\infty} + \sup_{j\ge 1} 2^{j\gamma}\big\|\Delta_j(f)\big\|_{L^\infty} \le C\big\|f\big\|_{\Lambda_\gamma}. \tag{6.3.16}$$

Conversely, every tempered distribution f, which satisfies

$$\big\|S_0(f)\big\|_{L^\infty} + \sup_{j\ge 1} 2^{j\gamma}\big\|\Delta_j(f)\big\|_{L^\infty} < \infty \tag{6.3.17}$$

can be identified with an element of Λ_γ. Moreover, there is a constant $C' = C'(n,\gamma)$ so that for all f that satisfy (6.3.17) we have

$$\big\|f\big\|_{\Lambda_\gamma} \le C'\Big(\big\|S_0(f)\big\|_{L^\infty} + \sup_{j\ge 1} 2^{j\gamma}\big\|\Delta_j(f)\big\|_{L^\infty}\Big). \tag{6.3.18}$$

Proof. The proof of (6.3.16) is immediate since we trivially have

$$\big\|S_0(f)\big\|_{L^\infty} = \big\|f * \Phi\big\|_{L^\infty} \le \big\|\Phi\big\|_{L^1}\big\|f\big\|_{L^\infty} \le C\big\|f\big\|_{\Lambda_\gamma}$$

and also

$$\sup_{j\ge 1} 2^{j\gamma}\big\|\Delta_j(f)\big\|_{L^\infty} \le C\big\|f\big\|_{\dot{\Lambda}_\gamma} \le C\big\|f\big\|_{\Lambda_\gamma}$$

by the previous theorem.

Therefore, the main part of the argument is contained in the proof of the converse estimate (6.3.18). Here we introduce Schwartz functions ζ, η so that

$$\widehat{\zeta}(\xi)^2 + \sum_{j=1}^{\infty} \widehat{\eta}(2^{-j}\xi)^2 = 1$$

and such that $\widehat{\eta}$ is supported in the annulus $\frac{3}{5} \le |\xi| \le \frac{5}{3}$ and $\widehat{\zeta}$ is supported in the ball $|\xi| \le \frac{4}{5}$. We associate Littlewood-Paley operators Δ_j^η given by convolution with the functions $\eta_{2^{-j}}$ and we also let $\Delta_j^\Theta = \Delta_{j-1} + \Delta_j + \Delta_{j+1}$. Observe that $\widehat{\Phi}$ is equal to one on the ball $|\xi| \le 1$ and thus on the support of $\widehat{\zeta}$. Moreover, $\Delta_j^\Theta \Delta_j^\eta = \Delta_j^\eta$, hence we have the identity

$$f = \zeta * \zeta * \Phi * f + \sum_{j=1}^{\infty} \eta_{2^{-j}} * \eta_{2^{-j}} * \Delta_j^\Theta(f) \tag{6.3.19}$$

for tempered distributions f where the series converges in $\mathcal{S}'(\mathbf{R}^n)$. With $k = [\gamma]$ we write

$$\frac{D_h^{k+1}(f)}{|h|^\gamma} = \zeta * \frac{D_h^{k+1}(\zeta)}{|h|^\gamma} * \Phi * f + \sum_{j=1}^{\infty} \eta_{2^{-j}} * \frac{D_h^{k+1}(\eta_{2^{-j}})}{|h|^\gamma} * \Delta_j^\Theta(f) \tag{6.3.20}$$

and we use Proposition 6.3.5 to estimate the L^∞ norm of the term $\zeta * \frac{D_h^{k+1}(\zeta)}{|h|^\gamma} * \Phi * f$ in the previous sum as follows:

$$\begin{aligned} \Big\| \zeta * \tfrac{D_h^{k+1}(\zeta)}{|h|^\gamma} * \Phi * f \Big\|_{L^\infty} &\le \Big\| \tfrac{D_h^{k+1}(\zeta)}{|h|^\gamma} \Big\|_{L^\infty} \big\| \zeta * \Phi * f \big\|_{L^1} \\ &\le C \min\big(\tfrac{1}{|h|^\gamma}, \tfrac{|h|^{k+1}}{|h|^\gamma}\big) \big\| \Phi * f \big\|_{L^\infty} \\ &\le C \big\| \Phi * f \big\|_{L^\infty} . \end{aligned} \tag{6.3.21}$$

The corresponding L^∞ estimates for $\Delta_j^\Theta(f) * \eta_{2^{-j}} * D_h^{k+1}(\eta_{2^{-j}})$ have already been obtained in (6.3.13). Indeed, we obtained

$$\big\| D_h^{k+1}(\eta_{2^{-j}}) * \eta_{2^{-j}} * \Delta_j^\Theta(f) \big\|_{L^\infty} \le C_{\eta,n,k} \big\| \Delta_j^\Theta(f) \big\|_{L^\infty} \min\big(1, |2^j h|^{k+1}\big) ,$$

from which it follows that

$$\begin{aligned} &\Big\| \sum_{j=1}^{\infty} \eta_{2^{-j}} * \frac{D_h^{k+1}(\eta_{2^{-j}})}{|h|^\gamma} * \Delta_j^\Theta(f) \Big\|_{L^\infty} \\ &\le C \Big(\sup_{j \ge 1} 2^{j\gamma} \big\| \Delta_j^\Theta(f) \big\|_{L^\infty} \Big) \sum_{j=1}^{\infty} 2^{-j\gamma} |h|^{-\gamma} \min\big(1, |2^j h|^{k+1}\big) \\ &\le C \Big(\sup_{j \ge 1} 2^{j\gamma} \big\| \Delta_j(f) \big\|_{L^\infty} \Big) \sum_{j=1}^{\infty} \min\big(|2^j h|^{-\gamma}, |2^j h|^{k+1-\gamma}\big) \\ &\le C \sup_{j \ge 1} 2^{j\gamma} \big\| \Delta_j(f) \big\|_{L^\infty} , \end{aligned} \tag{6.3.22}$$

since the last series is easily seen to converge uniformly in $h \in \mathbf{R}^n$ since $k + 1 = [\gamma] + 1 > \gamma$. We now combine identity (6.3.20) with estimates (6.3.21) and (6.3.22) to obtain that the expression on the right in (6.3.19) has a bounded L^∞ norm. This implies that f can be identified with a bounded function that satisfies (6.3.18).

□

Next, we obtain consequences of the Littlewood-Paley characterization of Lipschitz spaces. In the following corollary we identify Λ_0 with L^∞.

Corollary 6.3.8. *For $0 \le \gamma \le \delta < \infty$ there is a constant $C_{n,\gamma,\delta} < \infty$ such that for all $f \in \Lambda_\delta(\mathbf{R}^n)$ we have*

$$\big\|f\big\|_{\Lambda_\gamma} \le C_{n,\gamma,\delta}\big\|f\big\|_{\Lambda_\delta}.$$

In other words, the space $\Lambda_\delta(\mathbf{R}^n)$ can be identified with a subspace of $\Lambda_\gamma(\mathbf{R}^n)$.

Proof. If $0 < \gamma \le \delta$ and $j \ge 0$, then we must have $2^{j\gamma} \le 2^{j\delta}$ and thus

$$\sup_{j\ge 1} 2^{j\gamma}\big\|\Delta_j(f)\big\|_{L^\infty} \le \sup_{j\ge 1} 2^{j\delta}\big\|\Delta_j(f)\big\|_{L^\infty}.$$

Adding $\big\|f\big\|_{L^\infty}$ and using Theorem 6.3.7, we obtain the required conclusion. The case $\gamma = 0$ is trivial.

□

Remark 6.3.9. We proved estimates (6.3.18) and (6.3.8) using the Littlewood-Paley operators Δ_j constructed by a fixed choice of the function Ψ; Φ also depended on Ψ. It should be noted that the specific choice of the functions Ψ and Φ was unimportant in those estimates. In particular, if we know (6.3.18) and (6.3.8) for some choice of Littlewood-Paley operators $\widetilde{\Delta}_j$ and some Schwartz function $\widetilde{\Phi}$ whose Fourier transform is supported in a neighborhood of the origin, then (6.3.18) and (6.3.8) would also hold for our fixed choice of Δ_j and Φ. This situation is illustrated in the next corollary.

Corollary 6.3.10. *Let $\gamma > 0$ and let α be a multiindex with $|\alpha| < \gamma$. If $f \in \Lambda_\gamma$, then the distributional derivative $\partial^\alpha f$ (of f) lies in $\Lambda_{\gamma-|\alpha|}$. Likewise, if $f \in \dot{\Lambda}_\gamma$, then $\partial^\alpha f \in \dot{\Lambda}_{\gamma-|\alpha|}$. Precisely, we have the norm estimates*

$$\big\|\partial^\alpha f\big\|_{\Lambda_{\gamma-|\alpha|}} \le C_{n,\alpha}\big\|f\big\|_{\Lambda_\gamma} \tag{6.3.23}$$

$$\big\|\partial^\alpha f\big\|_{\dot{\Lambda}_{\gamma-|\alpha|}} \le C_{n,\alpha}\big\|f\big\|_{\dot{\Lambda}_\gamma}. \tag{6.3.24}$$

In particular, elements of Λ_γ and $\dot{\Lambda}_\gamma$ are in C^α for all $|\alpha| < \gamma$.

Proof. Let α be a multiindex with $|\alpha| < \gamma$. We denote by $\Delta_j^{\partial^\alpha\Psi}$ the Littlewood-Paley operator associated with the bump $(\partial^\alpha\Psi)_{2^{-j}}$. It is straightforward to check that the identity

$$\Delta_j(\partial^\alpha f) = 2^{j|\alpha|}\Delta_j^{\partial^\alpha\Psi}(f)$$

is valid for any tempered distribution f. Using the support properties of Ψ, we obtain

$$2^{j(\gamma-|\alpha|)}\Delta_j(\partial^\alpha f) = 2^{j\gamma}\Delta_j^{\partial^\alpha\Psi}(\Delta_{j-1} + \Delta_j + \Delta_{j+1})(f) \tag{6.3.25}$$

and from this it easily follows that

$$\sup_{j\in\mathbf{Z}} 2^{j(\gamma-|\alpha|)}\big\|\Delta_j(\partial^\alpha f)\big\|_{L^\infty} \le \big\|\partial^\alpha\Psi\big\|_{L^1} \sup_{j\in\mathbf{Z}} 2^{j\gamma}\big\|\Delta_j(f)\big\|_{L^\infty}.$$

Using Theorem 6.3.6, we obtain that if $f \in \dot{\Lambda}_\gamma$, then $\partial^\alpha f \in \dot{\Lambda}_{\gamma-|\alpha|}$ and we also obtain (6.3.24).

To derive the inhomogeneous version, we note that

$$\Phi * (\partial^\alpha f) = (-1)^{|\alpha|}(\partial^\alpha \Phi * f) = (-1)^{|\alpha|}\big(\partial^\alpha \Phi * (\Phi + \Psi_{2^{-1}}) * f\big)$$

since the function $\widehat{\Phi} + \widehat{\Psi_{2^{-1}}}$ is equal to 1 on the support of $\widehat{\partial^\alpha \Phi}$. Taking L^∞ norms, we obtain

$$\begin{aligned}\big\|\Phi * \partial^\alpha f\big\|_{L^\infty} &\le \big\|\partial^\alpha \Phi\big\|_{L^1}\big(\big\|\Phi * f\big\|_{L^\infty} + \big\|\Psi_{2^{-1}} * f\big\|_{L^\infty}\big) \\ &\le \big\|\partial^\alpha \Phi\big\|_{L^1}\big(\big\|S_0(f)\big\|_{L^\infty} + \sup_{j\ge 1}\big\|\Psi_{2^{-j}} * f\big\|_{L^\infty}\big),\end{aligned}$$

which, combined with (6.3.25), yields the estimate $\big\|\partial^\alpha f\big\|_{\Lambda_{\gamma-|\alpha|}} \le C\big\|f\big\|_{\Lambda_\gamma}$.

□

Exercises

6.3.1. Fix $k \in \mathbf{Z}^+$. Show that

$$D_h^k(f)(x) = 0$$

for all x, h in $\mathbf{R}^n$ if and only if f is a polynomial of degree at most $k-1$. [*Hint:* One direction may be proved by direct verification. For the converse direction, show that $\widehat{f}$ is supported at the origin and use Proposition 2.4.1.]

6.3.2. (a) Extend Definition 6.3.1 to the case $\gamma = 0$ and show that for all continuous functions f we have

$$\big\|f\big\|_{L^\infty} \le \big\|f\big\|_{\Lambda_0} \le 3\big\|f\big\|_{L^\infty};$$

hence the space $\Lambda_0(\mathbf{R}^n)$ can be identified with $L^\infty(\mathbf{R}^n) \cap C(\mathbf{R}^n)$.
(b) Given a measurable function f on $\mathbf{R}^n$ we define

$$\big\|f\big\|_{\dot{L}^\infty} = \inf\big\{\big\|f + c\big\|_{L^\infty} :\ c \in \mathbf{C}\big\}$$

and we let $\dot{L}^\infty(\mathbf{R}^n)$ be the space of all f with $\big\|f\big\|_{\dot{L}^\infty} < \infty$. Identify all functions in $\dot{L}^\infty(\mathbf{R}^n)$ whose difference is a constant. Show that for all continuous functions f we have

$$\big\|f\big\|_{\dot{L}^\infty} \le \sup_{x,h\in\mathbf{R}^n} |f(x+h) - f(x)| \le 2\big\|f\big\|_{\dot{L}^\infty};$$

thus $\dot{\Lambda}_0(\mathbf{R}^n)$ can be identified with $\dot{L}^\infty(\mathbf{R}^n) \cap C(\mathbf{R}^n)$.

6.3.3. (a) For a continuous function f prove the identity

$$D_h^{k+1}(f)(x) = \sum_{s=1}^{k+1} (-1)^{k+s+1}\binom{k+1}{s} f(x+sh)$$

for all $x, h \in \mathbf{R}^n$ and $k \in \mathbf{Z}^+ \cup \{0\}$.
(b) Prove that $D_h^k D_h^l = D_h^{k+l}$ for all $k, l \in \mathbf{Z}^+ \cup \{0\}$.

6.3.4. Let

$$f(x) = \sum_{k=1}^{\infty} 2^{-k} e^{2\pi i 2^k x}, \qquad x \in \mathbf{R}.$$

(a) Prove that $f \in \Lambda_\gamma(\mathbf{R})$ for all $0 < \gamma < 1$.
(b) Prove that there is an $A < \infty$ so that

$$\sup_{x, t \neq 0} |f(x+t) + f(x-t) - 2f(x)|\, |t|^{-1} \le A;$$

thus $f \in \Lambda_1(\mathbf{R})$.
(c) Show, however, that

$$\sup_{x, t \neq 0} |f(x+t) - f(x)|\, |t|^{-1} = \infty;$$

thus f fails to be differentiable.
[*Hint:* Part (c): Use that $f(x)$ is 1-periodic and thus

$$\int_0^1 |f(x+t) - f(x)|^2\, dx = \sum_{k=1}^{\infty} 2^{-2k} |e^{2\pi i 2^k t} - 1|^2.$$

Observe that when $2^k|t| \le \frac{1}{2}$ we have $|e^{2\pi i 2^k t} - 1| \ge 2^{k+2}|t|$.]

6.3.5. For $0 < a, b < \infty$ and $x \in \mathbf{R}$ let

$$h_{ab}(x) = \sum_{k=1}^{\infty} 2^{-ak} e^{2\pi i 2^{bk} x}.$$

Show that h_{ab} lies in $\Lambda_{\frac{a}{b}}(\mathbf{R}) \cap \dot{\Lambda}_{\frac{a}{b}}(\mathbf{R})$.
[*Hint:* Use the estimate $|D_h^L(e^{2\pi i 2^{bk} x})| \le C \min\big(1, (2^{bk}|h|)^L\big)$ with $L = [a/b] + 1$ and split the sum into two parts.]

6.3.6. Let $\gamma > 0$ and let $k = [\gamma]$.
(a) Use Exercise 6.3.3(b) to prove that if $|D_h^k(f)(x)| \le C|h|^\gamma$ for all $x, h \in \mathbf{R}^n$, then $|D_h^{k+l}(f)(x)| \le C 2^l |h|^\gamma$ for all $l \ge 1$.
(b) Conversely, assuming that for some $l \ge 1$ we have

$$\sup_{x, h \in \mathbf{R}^n} \frac{|D_h^{k+l}(f)(x)|}{|h|^\gamma} < \infty,$$

show that $f \in \Lambda_\gamma$.
[*Hint:* Part (b): Use (6.3.14) but replace $k+1$ by $k+l$.]

6.3.7. Let Ψ and Δ_j be as in Theorem 6.3.7. Define a continuous operator Q_t by setting

$$Q_t(f) = f * \Psi_t, \qquad \Psi_t(x) = t^{-n}\Psi(t^{-1}x).$$

Show that all tempered distributions f satisfy

$$\sup_{t>0} t^{-\gamma}\big\|Q_t(f)\big\|_{L^\infty} \approx \sup_{j\in\mathbf{Z}} 2^{j\gamma}\big\|\Delta_j(f)\big\|_{L^\infty}$$

with the interpretation that if either term is finite, then it controls the other term by a constant multiple of itself.
[*Hint:* Observe that $Q_t = Q_t(\Delta_{j-2} + \Delta_{j-1} + \Delta_j + \Delta_{j+1} + \Delta_{j+2})$ when $2^j \le t \le 2^{j+1}$.]

6.3.8. (a) Let $0 \le \gamma < 1$ and suppose that $\partial_j f \in \dot\Lambda_\gamma$ for all $1 \le j \le n$. Show that for some constant C we have

$$\big\|f\big\|_{\dot\Lambda_{\gamma+1}} \le C\sum_{j=1}^{n}\big\|\partial_j f\big\|_{\dot\Lambda_\gamma}$$

and conclude that $f \in \dot\Lambda_{\gamma+1}$.
(b) Let $\gamma \ge 0$. If we have $\partial^\alpha f \in \dot\Lambda_\gamma$ for all multiindices α with $|\alpha| = r$, then there is an estimate

$$\big\|f\big\|_{\dot\Lambda_{\gamma+r}} \le C_\gamma \sum_{|\alpha|=r}\big\|\partial^\alpha f\big\|_{\dot\Lambda_\gamma}$$

and thus $f \in \dot\Lambda_{\gamma+r}$.
(c) Use Corollary 6.3.10 to obtain that the estimates in both (a) and (b) can be reversed.
[*Hint:* Part (a): Write

$$D_h^2(f)(x) = \int_0^1 \sum_{j=1}^{n} \big[\partial_j f(x+th+2h) - \partial_j f(x+th+h)\big]\, h_j\, dt\,.$$

Part (b): Use induction.]

6.4. Hardy Spaces*

Having been able to characterize L^p spaces, Sobolev spaces, and Lipschitz spaces using Littlewood-Paley theory, it should not come as a surprise that the theory can be used to characterize other spaces as well. This is the case with the Hardy spaces $H^p(\mathbf{R}^n)$ that form a family of spaces with some remarkable properties in which the integrability index p can go all the way down to zero.

There exists a plethora of equivalent characterizations for Hardy spaces of which only a few representative are discussed in this section. If a reader is interested in going through the material quickly, he orshe may want to define the Hardy space H^p as the space of all tempered distributions f modulo polynomials for which the quantity is finite

$$\|f\|_{H^p} = \Big\| \Big(\sum_{j\in\mathbf{Z}} |\Delta_j(f)|^2 \Big)^{\frac{1}{2}} \Big\|_{L^p} < \infty \tag{6.4.1}$$

whenever $0 < p \le 1$. An atomic decomposition for Hardy spaces can be obtained from this definition (see Section 6.6), and once this is at hand, the analysis of these spaces is simplified significantly. For historical reasons, however, we chose to define Hardy spaces using a more classical approach and, as a result, we have to go through a considerable amount of work to obtain the characterization alluded to in (6.4.1).

6.4.a. Definition of Hardy Spaces

To give the definition of Hardy spaces on $\mathbf{R}^n$, we will need some background material. We say that a tempered distribution v is *bounded* if $\varphi * v \in L^\infty(\mathbf{R}^n)$ whenever φ is in $\mathcal{S}(\mathbf{R}^n)$. We observe that if v is a bounded tempered distribution and $h \in L^1(\mathbf{R}^n)$, then the convolution $h * v$ can be defined as a distribution via the convergent integral

$$\langle h * v, \varphi \rangle = \langle \widetilde{\varphi} * v, \widetilde{h} \rangle = \int_{\mathbf{R}^n} (\widetilde{\varphi} * v)(x) \widetilde{h}(x)\, dx,$$

where φ is a Schwartz function and, as usual, we set $\widetilde{\varphi}(x) = \varphi(-x)$.

Let us recall the Poisson kernel P introduced in (2.1.14):

$$P(x) = \frac{\Gamma(\frac{n+1}{2})}{\pi^{\frac{n+1}{2}}} \frac{1}{(1+|x|^2)^{\frac{n+1}{2}}}\,. \tag{6.4.2}$$

For $t > 0$ we let $P_t(x) = t^{-n} P(t^{-1}x)$. If v is a bounded tempered distribution, then $P_t * v$ is a well-defined distribution since P_t is in L^1. We claim that $P_t * v$ can be identified with a well-defined bounded function. To see this write $1 = \widehat{\varphi}(\xi) + \widehat{\psi}(\xi)$, where $\widehat{\varphi}$ has compact support and $\widehat{\psi}$ vanishes in a neighborhood of the origin. Then

$$\widehat{P_t}(\xi) = e^{-2\pi t|\xi|} = e^{-2\pi t|\xi|}\widehat{\varphi}(t\xi) + e^{-2\pi t|\xi|}\widehat{\psi}(t\xi)$$

is a sum of a compactly supported and a Schwartz function. Then

$$P_t * v = P_t * (\varphi_t * v) + \psi_t * v,$$

but $\varphi_t * v$ and $\psi_t * v$ are bounded functions since φ_t and ψ_t are in the Schwartz class. The last identity proves that $P_t * v$ is a bounded function.

Before we define Hardy spaces we introduce some notation.

Definition 6.4.1. Let $a, b > 0$. Let Φ be a Schwartz function and let f be a tempered distribution on $\mathbf{R}^n$. We define the *smooth maximal function of f with respect to Φ* as

$$M(f;\Phi)(x) = \sup_{t>0} |(\Phi_t * f)(x)|\,.$$

We define the *nontangential maximal function (with aperture a) of f with respect to Φ* as

$$M_a^*(f;\Phi)(x) = \sup_{t>0} \sup_{\substack{y\in\mathbf{R}^n \\ |y-x|\le at}} |(\Phi_t * f)(y)|\,.$$

We also define the *auxiliary maximal function*

$$M_b^{**}(f;\Phi)(x) = \sup_{t>0} \sup_{y\in\mathbf{R}^n} \frac{|(\Phi_t * f)(x-y)|}{(1+t^{-1}|y|)^b}$$

and we observe that

$$M(f;\Phi) \le M_a^*(f;\Phi) \le (1+a)^b M_b^{**}(f;\Phi) \tag{6.4.3}$$

for all $a, b > 0$. We note that if Φ is merely integrable, for example, if Φ is the Poisson kernel, the maximal functions $M(f;\Phi)$, $M_a^*(f;\Phi)$, and $M_b^{**}(f;\Phi)$ are only well defined for bounded tempered distributions f on $\mathbf{R}^n$.

For a fixed positive integer m and a Schwartz function φ we define the normalization quantity

$$\mathfrak{N}_m(\varphi) = \int_{\mathbf{R}^n} (1+|x|)^m \sum_{|\alpha|\le m+1} |\partial^\alpha \varphi(x)|\, dx\,. \tag{6.4.4}$$

We now define

$$\mathcal{F}_N = \Big\{\varphi \in \mathcal{S}(\mathbf{R}^n):\ \mathfrak{N}_N(\varphi) \le 1\Big\} \tag{6.4.5}$$

and we also define the *grand maximal function of f (with respect to N)* as

$$\mathcal{M}_N(f)(x) = \sup_{\varphi\in\mathcal{F}_N} M_1^*(f;\varphi)(x)\,.$$

Having introduced a variety of smooth maximal operators that will be useful in the development of the theory, we proceed with the definition of Hardy spaces.

Definition 6.4.2. Let f be a bounded tempered distribution on $\mathbf{R}^n$ and let $0 < p < \infty$. We say that f lies in the *Hardy space* $H^p(\mathbf{R}^n)$ if and only if the *Poisson maximal function*

$$M(f;P)(x) = \sup_{t>0} |(P_t * f)(x)| \tag{6.4.6}$$

is in $L^p(\mathbf{R}^n)$. If this is the case, we set

$$\big\|f\big\|_{H^p} = \big\|M(f;P)\big\|_{L^p}.$$

At this point we don't know whether these spaces coincide with any other known spaces for some values of p's. In the next theorem we show that this is the case when $1 < p < \infty$.

Theorem 6.4.3. *(a) Let $1 < p < \infty$. Then every bounded tempered distribution f in H^p is an element of L^p. Moreover, there is a constant $C_{n,p}$ such that for all such f we have*

$$\big\|f\big\|_{L^p} \le \big\|f\big\|_{H^p} \le C_{n,p}\big\|f\big\|_{L^p}$$

and therefore $H^p(\mathbf{R}^n)$ coincides with $L^p(\mathbf{R}^n)$.
(b) When $p = 1$, every element of H^1 is an integrable function. In other words, $H^1(\mathbf{R}^n) \subseteq L^1(\mathbf{R}^n)$ and for all $f \in H^1$ we have

$$\big\|f\big\|_{L^1} \le \big\|f\big\|_{H^1}\,. \tag{6.4.7}$$

Proof. (a) Let $f \in H^p(\mathbf{R}^n)$. The set $\{P_t * f : t > 0\}$ lies in a multiple of the unit ball of L^p. By the Banach-Alaoglu-Bourbaki theorem there exists a sequence $t_j \to 0$ such that $P_{t_j} * f$ converges to some L^p function f_0 in the weak* topology of L^p. On the other hand, we see that $P_t * \varphi \to \varphi$ in $\mathcal{S}(\mathbf{R}^n)$ as $t \to 0$ for all φ in $\mathcal{S}(\mathbf{R}^n)$. Thus

$$P_t * f \to f \qquad \text{in } \mathcal{S}'(\mathbf{R}^n) \tag{6.4.8}$$

and it follows that the distribution f coincides with the L^p function f_0. Since the family $\{P_t\}_{t>0}$ is an approximate identity, Theorem 1.2.19 gives that

$$\big\|P_t * f - f\big\|_{L^p} \to 0 \qquad \text{as } t \to 0,$$

from which it follows that

$$\big\|f\big\|_{L^p} \le \big\|\sup_{t>0} |P_t * f|\big\|_{L^p} = \big\|f\big\|_{H^p}\,. \tag{6.4.9}$$

The converse inequality is a consequence of the fact that

$$\sup_{t>0} |P_t * f| \le M(f)\,,$$

where M is the Hardy-Littlewood maximal operator. (See Corollary 2.1.12.)

(b) The case $p = 1$ only requires a small modification of the case $p > 1$. Embedding L^1 into the space of finite Borel measures $\mathcal{M}$ whose unit ball is weak* compact, we can extract a sequence $t_j \to 0$ such that $P_{t_j} * f$ converges to some measure μ in the topology of measures. In view of (6.4.8), it follows that the distribution f can be identified with the measure μ.

It remains to show that μ is absolutely continuous with respect to Lebesgue measure, which would imply that it coincides with some L^1 function. Let $|\mu|$ be the total variation of μ. We will show that μ is absolutely continuous by showing that for all subsets E of $\mathbf{R}^n$ we have $|E| = 0 \implies |\mu|(E) = 0$. Given an $\varepsilon > 0$, there exists a $\delta > 0$ such that for any measurable subset F of $\mathbf{R}^n$ we have

$$|F| < \delta \implies \int_F \sup_{t>0} |P_t * f|\, dx < \varepsilon\,.$$

Given E with $|E| = 0$, we can find an open set U such that $E \subseteq U$ and $|U| < \delta$. Then for any g continuous function supported in U we have

$$\begin{aligned} \left| \int_{\mathbf{R}^n} g \, d\mu \right| &= \lim_{j\to\infty} \left| \int_{\mathbf{R}^n} g(x) \, (P_{t_j} * f)(x) \, dx \right| \\ &\le \big\|g\big\|_{L^\infty} \int_U \sup_{t>0} |P_t * f| \, dx < \varepsilon \big\|g\big\|_{L^\infty} . \end{aligned}$$

But we have

$$|\mu(U)| = \sup \left\{ \left| \int_{\mathbf{R}^n} g \, d\mu \right| : \; g \text{ continuous function supported in } U \right\},$$

which implies that $|\mu(U)| < \varepsilon$. Since ε was arbitrary, it follows that $|\mu|(E) = 0$; hence μ is absolutely continuous with respect to Lebesgue measure. Finally, (6.4.7) is a consequence of (6.4.9), which is also valid for $p = 1$.

□

We may wonder whether H^1 coincides with L^1. We will show in Theorem 6.7.4 that elements of H^1 have integral zero; thus H^1 is a proper subspace of L^1.

We now proceed to obtain some characterizations of these spaces.

6.4.b. Quasi-norm Equivalence of Several Maximal Functions

It is a fact that all the maximal functions have comparable L^p quasi-norms for all $0 < p < \infty$. This is the essence of the following theorem.

Theorem 6.4.4. *Let $0 < p < \infty$. Then the following statements are valid:*
(a) There exists a Schwartz function Φ with $\int_{\mathbf{R}^n} \Phi(x) \, dx \neq 0$ and a constant C_1 (that does not depend on any parameters) such that

$$\big\|M(f;\Phi)\big\|_{L^p} \le C_1 \big\|f\big\|_{H^p} \tag{6.4.10}$$

for all bounded $f \in \mathcal{S}'(\mathbf{R}^n)$.
(b) For every $a > 0$ and Φ in $\mathcal{S}(\mathbf{R}^n)$ there exists a constant $C_2(n,p,a,\Phi)$ such that

$$\big\|M_a^*(f;\Phi)\big\|_{L^p} \le C_2(n,p,a,\Phi) \big\|M(f;\Phi)\big\|_{L^p} \tag{6.4.11}$$

for all $f \in \mathcal{S}'(\mathbf{R}^n)$.
(c) For every $a > 0$, $b > n/p$ and Φ in $\mathcal{S}(\mathbf{R}^n)$ there exists a constant $C_3(n,p,a,b,\Phi)$ such that

$$\big\|M_b^{**}(f;\Phi)\big\|_{L^p} \le C_3(n,p,a,b,\Phi) \big\|M_a^*(f;\Phi)\big\|_{L^p} \tag{6.4.12}$$

for all $f \in \mathcal{S}'(\mathbf{R}^n)$.
(d) For every $b > 0$ and Φ in $\mathcal{S}(\mathbf{R}^n)$ with $\int_{\mathbf{R}^n} \Phi(x) \, dx \neq 0$ there exists a constant $C_4(b,\Phi)$ such that if $N = [b] + 1$ we have

$$\big\|\mathcal{M}_N(f)\big\|_{L^p} \le C_4(b,\Phi) \big\|M_b^{**}(f;\Phi)\big\|_{L^p} \tag{6.4.13}$$

for all $f \in \mathcal{S}'(\mathbf{R}^n)$.
(e) For every positive integer N there exists a constant $C_5(n,N)$ such that every distribution f with $\big\|\mathcal{M}_N(f)\big\|_{L^p} < \infty$ is bounded and satisfies

$$\big\|f\big\|_{H^p} \le C_5(n,N)\big\|\mathcal{M}_N(f)\big\|_{L^p}, \tag{6.4.14}$$

that is, it lies in the Hardy space H^p.

It follows that all the quasi-norms in statements $(a)-(e)$ are all equivalent. This furnishes a variety of characterizations of Hardy spaces. The proof of this theorem is based on the following lemma, whose proof is postponed until the end of this section.

Lemma 6.4.5. *Let $m \in \mathbf{Z}^+$ and let Φ in $\mathcal{S}(\mathbf{R}^n)$ satisfy $\int_{\mathbf{R}^n}\Phi(x)\,dx = 1$. Then there exists a constant C_0 (that depends only on Φ and m) such that for any Ψ in $\mathcal{S}(\mathbf{R}^n)$, there exist Schwartz functions $\Theta^{(s)}$, $0<s<1$, with the properties*

$$\Psi(x) = \int_0^1 (\Theta^{(s)} * \Phi_s)(x)\,ds \tag{6.4.15}$$

and

$$\int_{\mathbf{R}^n} (1+|x|)^m |\Theta^{(s)}(x)|\,dx \le C_0\, s^m\, \mathfrak{N}_m(\Psi). \tag{6.4.16}$$

Assuming Lemma 6.4.5, we prove Theorem 6.4.4.

Proof. Proof of (a) We pick a continuous and integrable function $\psi(s)$ on the interval $[1,\infty)$ that decays faster than the reciprocal of any polynomial (i.e., $|\psi(s)| \le C_N s^{-N}$ for all $N>0$) such that

$$\int_1^\infty s^k\,\psi(s)\,ds = \begin{cases} 1 & \text{if } k=0 \\ 0 & \text{if } k=1,2,3,\dots. \end{cases} \tag{6.4.17}$$

Such a function exists; in fact, we may take

$$\psi(s) = \frac{e}{\pi}\frac{1}{s}\,\mathrm{Im}\left(e^{(\frac{\sqrt{2}}{2}-i\frac{\sqrt{2}}{2})(s-1)^{\frac14}}\right). \tag{6.4.18}$$

See Exercise 6.4.4. We now define a function

$$\Phi(x) = \int_1^\infty \psi(s)P_s(x)\,ds\,, \tag{6.4.19}$$

where P_s is the Poisson kernel. The Fourier transform Φ is

$$\widehat{\Phi}(\xi) = \int_1^\infty \psi(s)\widehat{P_s}(\xi)\,ds = \int_1^\infty \psi(s)e^{-2\pi s|\xi|}\,ds$$

(c.f. Exercise 2.2.11), which is easily seen to be rapidly decreasing as $|\xi| \to \infty$. The same is true for all the derivatives of $\widehat{\Phi}$. The function $\widehat{\Phi}$ is clearly smooth on

$\mathbf{R}^n \setminus \{0\}$. Moreover,

$$\partial_j \widehat{\Phi}(\xi) = \sum_{k=0}^{L-1} (-2\pi)^{k+1} \frac{|\xi|^k}{k!} \frac{\xi_j}{|\xi|} \int_1^\infty s^{k+1} \psi(s)\, ds + O(|\xi|^L) = O(|\xi|^L)$$

as $|\xi| \to 0$, which implies that the distributional derivative $\partial_j \widehat{\Phi}$ is continuous at the origin. Since

$$\partial_\xi^\alpha (e^{-2\pi s|\xi|}) = s^{|\alpha|} p_\alpha(\xi) |\xi|^{-m_\alpha} e^{-2\pi s|\xi|}$$

for some $m_\alpha \in \mathbf{Z}^+$ and some polynomial p_α, choosing L sufficiently large gives that every derivative of $\widehat{\Phi}$ is also continuous at the origin. We conclude that the function $\widehat{\Phi}$ is in the Schwartz class, and thus so is Φ. It also follows from (6.4.17) and (6.4.19) that

$$\int_{\mathbf{R}^n} \Phi(x)\, dx = 1 \neq 0\,.$$

Finally we have the estimate

$$\begin{aligned} M(f;\Phi)(x) &= \sup_{t>0} |(\Phi_t * f)(x)| \\ &= \sup_{t>0} \left| \int_1^\infty \psi(s)(f * P_{ts})(x)\, ds \right| \\ &\le \int_1^\infty |\psi(s)|\, ds\, M(f;P)(x)\,, \end{aligned}$$

and the required conclusion follows with $C_1 = \int_1^\infty |\psi(s)|\, ds$. Note that we actually obtained the stronger pointwise estimate $M(f;\Phi) \le C_1 M(f;P)$ rather than (6.4.10).

Proof of (b) The control of the nontagential maximal function $M_a^*(\cdot\,;\Phi)$ in terms of the vertical maximal function $M(\,\cdot\,;\Phi)$ is the hardest and most technical part of the proof. For matters of exposition, we only present the proof in the case when $a = 1$ and we note that the case of general $a > 0$ presents only notational differences. We begin by noting that (6.4.11) will be a consequence of the estimate

$$\left(\int_{\mathbf{R}^n} M_1^*(f;\Phi)^{\varepsilon,N}(x)^p dx \right)^{\frac{1}{p}} \le c_2(n,p,\Phi) \big\| M(f;\Phi) \big\|_{L^p}\,, \tag{6.4.20}$$

where N is a fixed large enough integer (depending on f and Φ), $0 < \varepsilon < 1$, and

$$M_1^*(f;\Phi)^{\varepsilon,N}(x) = \sup_{0<t<\frac{1}{\varepsilon}} \sup_{|y-x|\le t} |(\Phi_t * f)(y)| \Big(\frac{t}{t+\varepsilon} \Big)^N \frac{1}{(1+\varepsilon|y|)^N}\,.$$

Indeed, once (6.4.20) is established, Fatou's lemma yields (6.4.11) by letting $\varepsilon \to 0$.

Let us a fix an element f in $\mathcal{S}'(\mathbf{R}^n)$ such that $M(f;\Phi) \in L^p$. We first show that $M_1^*(f;\Phi)^{\varepsilon,N}$ lies in $L^p(\mathbf{R}^n) \cap L^\infty(\mathbf{R}^n)$. Indeed, using (2.3.21) (with $\alpha = 0$), we obtain the following estimate for some constants m and l (depending on f):

$$\begin{aligned} |(\Phi_t * f)(y)| &\le C \sum_{|\beta|\le l} \sup_{z\in\mathbf{R}^n} (|y|^m + |z|^m)|(\partial^\beta \widetilde{\Phi}_t)(z)| \\ &\le C(1+|y|^m) \sum_{|\beta|\le l} \sup_{z\in\mathbf{R}^n} (1+|z|^m)|(\partial^\beta \Phi_t)(-z)| \\ &\le C\frac{(1+|y|^m)}{t^n \min(1,t^l)} \sum_{|\beta|\le l} \sup_{z\in\mathbf{R}^n} (1+|z|^m)|(\partial^\beta \Phi)(-z/t)| \\ &\le C\frac{(1+|y|^m)}{\min(t^n,t^{n+l})}(1+t^m) \sum_{|\beta|\le l} \sup_{z\in\mathbf{R}^n} (1+|z/t|^m)|(\partial^\beta \Phi)(-z/t)| \\ &\le C(f,\Phi)(1+\varepsilon|y|^m)(1+\varepsilon^{-m})(1+t^m)(t^{-n}+t^{-n-l}). \end{aligned}$$

Multiplying by $(\frac{t}{t+\varepsilon})^N(1+\varepsilon|y|)^{-N}$ and taking $0<t<\frac{1}{\varepsilon}$ and $|y-x|\le t$, we obtain $1+\varepsilon|x|\le 2(1+\varepsilon|y|)$ and also

$$\big|(\Phi_t * f)(y)\big|\Big(\frac{t}{t+\varepsilon}\Big)^N \frac{1}{(1+\varepsilon|y|)^N} \le C(f,\Phi)\frac{(1+\varepsilon^{-m})^2\varepsilon^{-N}(\varepsilon^{n-N}+\varepsilon^{n+l-N})}{(1+\varepsilon|y|)^{N-m}},$$

which imply that

$$M_1^*(f;\Phi)^{\varepsilon,N}(x) \le C(f,\Phi)\frac{C_\varepsilon}{(1+\varepsilon|x|)^{N-m}}.$$

Taking $N>(m+n)/p$, we obtain that $M_1^*(f;\Phi)^{\varepsilon,N}$ lies in $L^p(\mathbf{R}^n)\cap L^\infty(\mathbf{R}^n)$.

We now introduce another parameter L and functions

$$V(f;\Phi)^{\varepsilon,N,L}(x) = \sup_{0<t<\frac{1}{\varepsilon}} \sup_{y\in\mathbf{R}^n} \big|(\Phi_t * f)(y)\big| \frac{\big(\frac{t}{t+\varepsilon}\big)^N}{(1+\varepsilon|y|)^N}\Big(\frac{t}{t+|x-y|}\Big)^L$$

and

$$U(f;\Phi)^{\varepsilon,N}(x) = \sup_{0<t<\frac{1}{\varepsilon}} \sup_{|y-x|\le t} t\big|\nabla(\Phi_t * f)(y)\big|\Big(\frac{t}{t+\varepsilon}\Big)^N \frac{1}{(1+\varepsilon|y|)^N}.$$

To prove of estimate (6.4.20), we will need to make use of the norm estimate

$$\big\|V(f;\Phi)^{\varepsilon,N,L}\big\|_{L^p} \le C_{n,p}\big\|M_1^*(f;\Phi)^{\varepsilon,N}\big\|_{L^p} \tag{6.4.21}$$

and the pointwise estimate

$$U(f;\Phi)^{\varepsilon,N} \le C_\Phi\, V(f;\Phi)^{\varepsilon,N,L}. \tag{6.4.22}$$

To prove (6.4.21) we observe that when $z\in B(y,t)\subseteq B(x,|x-y|+t)$ we have

$$\big|(\Phi_t * f)(y)\big|\Big(\frac{t}{t+\varepsilon}\Big)^N \frac{1}{(1+\varepsilon|y|)^N} \le M_1^*(f;\Phi)^{\varepsilon,N}(z),$$

from which it follows for any $0 < q < \infty$ and $y \in \mathbf{R}^n$

$$
\begin{aligned}
&\big|(\Phi_t * f)(y)\big|\Big(\frac{t}{t+\varepsilon}\Big)^N \frac{1}{(1+\varepsilon|y|)^N} \\
&\le \bigg(\frac{1}{|B(y,t)|}\int_{B(y,t)} M_1^*(f;\Phi)^{\varepsilon,N}(z)^q\,dz\bigg)^{\frac{1}{q}} \\
&\le \Big(\frac{|x-y|+t}{t}\Big)^{\frac{n}{q}}\bigg(\frac{1}{|B(x,|x-y|+t)|}\int_{B(x,|x-y|+t)} M_1^*(f;\Phi)^{\varepsilon,N}(z)^q\,dz\bigg)^{\frac{1}{q}} \\
&\le \Big(\frac{|x-y|+t}{t}\Big)^{L} M\Big(\big[M_1^*(f;\Phi)^{\varepsilon,N}\big]^q\Big)^{\frac{1}{q}}(x)\,,
\end{aligned}
$$

where we took $L > n/p$. We now take $0 < q < p$ and we use the boundedness of the Hardy-Littlewood maximal operator M on $L^{p/q}$ to obtain (6.4.21).

In proving (6.4.22) we may assume that Φ has integral 1; otherwise we can multiply Φ with a suitable constant to arrange for this to happen. We note that

$$
t\big|\nabla(\Phi_t * f)\big| = \big|(\nabla\Phi)_t * f\big| = \Big(\sum_{j=1}^n |(\partial_j\Phi)_t * f|^2\Big)^{\frac{1}{2}}
$$

and it suffices to work with each partial derivative $\partial_j\Phi$ of Φ. Using Lemma 6.4.5 we can write

$$
\partial_j\Phi = \int_0^1 \Theta^{(s)} * \Phi_s\,ds
$$

for suitable Schwartz functions $\Theta^{(s)}$. Then we have

$$
\begin{aligned}
&\big|((\partial_j\Phi)_t * f)(y)\big|\Big(\frac{t}{t+\varepsilon}\Big)^N \frac{1}{(1+\varepsilon|y|)^N} \\
&= \Big(\frac{t}{t+\varepsilon}\Big)^N \frac{1}{(1+\varepsilon|y|)^N}\bigg|\int_0^1 ((\Theta^{(s)})_t * \Phi_{st} * f)(y)\,ds\bigg| \qquad (6.4.23)\\
&\le \Big(\frac{t}{t+\varepsilon}\Big)^N \int_0^1\int_{\mathbf{R}^n} t^{-n}\big|\Theta^{(s)}(t^{-1}z)\big|\,\frac{\big|(\Phi_{st} * f)(y-z)\big|}{(1+\varepsilon|y|)^N}\,dz\,ds\,.
\end{aligned}
$$

Inserting the factor 1 written as

$$
\Big(\frac{ts}{ts+|x-(y-z)|}\Big)^L\Big(\frac{ts}{ts+\varepsilon}\Big)^N\Big(\frac{ts+|x-(y-z)|}{ts}\Big)^L\Big(\frac{ts+\varepsilon}{ts}\Big)^N
$$

in the preceding z-integral and using that

$$
\frac{1}{(1+\varepsilon|y|)^N} \le \frac{(1+\varepsilon|z|)^N}{(1+\varepsilon|y-z|)^N}\,,
$$

we deduce the estimate

$$\begin{aligned}
&\Big(\frac{t}{t+\varepsilon}\Big)^N \int_0^1\int_{\mathbf{R}^n} t^{-n}\big|\Theta^{(s)}(t^{-1}z)\big|\,\frac{\big|(\Phi_{st}*f)(y-z)\big|}{(1+\varepsilon|y|)^N}\,dz\,ds\\
&\le V(f;\Phi)^{\varepsilon,N,L}(x)\int_0^1\int_{\mathbf{R}^n}(1+\varepsilon|z|)^N\Big(\frac{ts+|x-(y-z)|}{ts}\Big)^L t^{-n}\big|\Theta^{(s)}(t^{-1}z)\big|\,dz\,\frac{ds}{s^N}\\
&\le CV(f;\Phi)^{\varepsilon,N,L}(x)\int_0^1\int_{\mathbf{R}^n} s^{-L-N}(1+|z|)^{N+L}\big|\Theta^{(s)}(z)\big|\,dz\,ds\\
&\le CC_0V(f;\Phi)^{\varepsilon,N,L}(x)\,\mathfrak{N}_{N+L}(\partial_j\Phi)\\
&= C_\Phi V(f;\Phi)^{\varepsilon,N,L}(x)\,,
\end{aligned}$$

in view of conclusion (6.4.16) of Lemma 6.4.5. Combining this estimate with (6.4.23), we deduce (6.4.22). Having established both (6.4.21) and (6.4.22), we conclude that

$$\big\|U(f;\Phi)^{\varepsilon,N}\big\|_{L^p}\le C_{n,p,\Phi}\big\|M_1^*(f;\Phi)^{\varepsilon,N}\big\|_{L^p}\,. \tag{6.4.24}$$

We now set

$$E_\varepsilon=\big\{x\in\mathbf{R}^n:\ U(f;\Phi)^{\varepsilon,N}(x)\le KM_1^*(f;\Phi)^{\varepsilon,N}(x)\big\}$$

for some constant K to be determined shortly. Then we have

$$\begin{aligned}
\int_{E_\varepsilon^c}\big[M_1^*(f;\Phi)^{\varepsilon,N}(x)\big]^p\,dx &\le\frac{1}{K^p}\int_{E_\varepsilon^c}\big[U(f;\Phi)^{\varepsilon,N}(x)\big]^p\,dx\\
&\le\frac{1}{K^p}\int_{\mathbf{R}^n}\big[U(f;\Phi)^{\varepsilon,N}(x)\big]^p\,dx\\
&\le\frac{C_{n,p,\Phi}^p}{K^p}\int_{\mathbf{R}^n}\big[M_1^*(f;\Phi)^{\varepsilon,N}(x)\big]^p\,dx\\
&\le\frac12\int_{\mathbf{R}^n}\big[M_1^*(f;\Phi)^{\varepsilon,N}(x)\big]^p\,dx
\end{aligned}\tag{6.4.25}$$

provided we choose K so that $K^p=2C_{n,p,\Phi}^p$. It remains to estimate the contribution of the integral of $\big[M_1^*(f;\Phi)^{\varepsilon,N}(x)\big]^p$ over the set E_ε. We claim that the following pointwise estimate is valid:

$$M_1^*(f;\Phi)^{\varepsilon,N}(x)\le C_{n,r}M\big(M_1^*(f;\Phi)^r\big)^{\frac1r}(x)\,,\qquad x\in E_\varepsilon \tag{6.4.26}$$

for any $r>0$. To prove (6.4.26) we fix $x\in E_\varepsilon$. By the definition of $M_1^*(f;\Phi)^{\varepsilon,N}(x)$ there exists a $(y_0,t)\in\mathbf{R}_+^{n+1}$ such that $|x-y_0|\le\frac1t<\frac1\varepsilon$ and

$$\big|(\Phi_t*f)(y_0)\big|\Big(\frac{t}{t+\varepsilon}\Big)^N\frac{1}{(1+\varepsilon|y|)^N}\ge\frac12 M_1^*(f;\Phi)^{\varepsilon,N}(x)\,. \tag{6.4.27}$$

It also follows from the definitions of E_ε and $U(f;\Phi)^{\varepsilon,N}$ that

$$t\big|\nabla(\Phi_t * f)(z)\big|\Big(\frac{t}{t+\varepsilon}\Big)^N \frac{1}{(1+\varepsilon|y|)^N} \le K M_1^*(f;\Phi)^{\varepsilon,N}(x) \tag{6.4.28}$$

for all z satisfying $|z-x| \le t < \frac{1}{\varepsilon}$ whenever $x \in E_\varepsilon$. It follows from (6.4.27) and (6.4.28) that

$$t\big|\nabla(\Phi_t * f)(z)\big| \le 2K\big|(\Phi_t * f)(y_0)\big| \tag{6.4.29}$$

for all z satisfying $|z-x| \le t < \frac{1}{\varepsilon}$. Let us fix such a z. Applying the mean value theorem and using (6.4.29), we obtain

$$\begin{aligned}
\big|(\Phi_t * f)(z) - (\Phi_t * f)(y_0)\big| &= \big|\nabla(\Phi_t * f)(\xi)\big|\,|z-y_0| \\
&\le \frac{2K}{t}\big|(\Phi_t * f)(y_0)\big|\,|z-y_0| \\
&\le \frac{1}{2}\big|(\Phi_t * f)(y_0)\big|
\end{aligned}$$

for some ξ between y_0 and z, provided z is also chosen to satisfy $|z-y_0| \le (4K)^{-1}t$ (in addition to $|z-x| \le t$). Therefore, for z satisfying $|z-y_0| \le (4K)^{-1}t$ and $|z-x| \le t$ we have

$$\big|(\Phi_t * f)(z)\big| \ge \frac{1}{2}\big|(\Phi_t * f)(y_0)\big| \ge \frac{1}{4} M_1^*(f;\Phi)^{\varepsilon,N}(x)\,,$$

where the second inequality is a consequence of (6.4.27).

Using this estimate, we obtain

$$\begin{aligned}
M\big(M(f;\Phi)^r\big)(x) &\ge \frac{1}{|B(x,t)|}\int_{B(x,t)} \big[M(f;\Phi)(w)\big]^r\,dw \\
&\ge \frac{1}{|B(x,t)|}\int_{B(x,t)\cap B(y_0,\frac{t}{4K})} \big[M(f;\Phi)(w)\big]^r\,dw \\
&\ge \frac{1}{|B(x,t)|}\int_{B(x,t)\cap B(y_0,\frac{t}{4K})} \frac{1}{4^r}\big[M_1^*(f;\Phi)^{\varepsilon,N}(x)\big]^r\,dw \\
&\ge \frac{|B(x,t)\cap B(y_0,\frac{t}{4K})|}{|B(x,t)|}\,\frac{1}{4^r}\big[M_1^*(f;\Phi)^{\varepsilon,N}(x)\big]^r \\
&\ge C_{n,K}4^{-r}\big[M_1^*(f;\Phi)^{\varepsilon,N}(x)\big]^r,
\end{aligned}$$

where we used the simple geometric fact that if $|x-y_0| \le t$ and $\delta > 0$, then

$$\frac{|B(x,t)\cap B(y_0,\delta t)|}{|B(x,t)|} \ge c_{n,\delta} > 0\,,$$

the minimum of this constant is obtained when $|x-y_0| = t$. See Figure 6.1.

This proves (6.4.26), which in turn implies (by taking $r < p$) that

$$\int_{E_\varepsilon} \big[M_1^*(f;\Phi)^{\varepsilon,N}(x)\big]^p\,dx \le C_{n,p,\Phi}\int_{\mathbf{R}^n} M_1^*(f;\Phi)(x)^p\,dx\,. \tag{6.4.30}$$

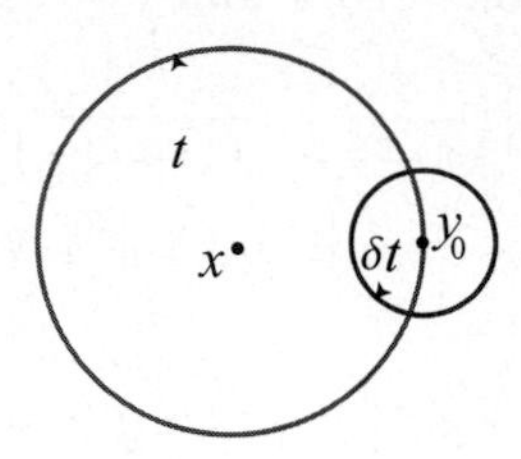

FIGURE 6.1. The ball $B(y_0, \delta t)$ captures at least a fixed proportion of the ball $B(x, t)$.

Combining this estimate with (6.4.25), we obtain

$$\int_{\mathbf{R}^n} \big[M_1^*(f;\Phi)^{\varepsilon,N}\big]^p\,dx \le C_{n,p,\Phi} \int_{\mathbf{R}^n} M_1^*(f;\Phi)^p\,dx + \frac{1}{2}\int_{\mathbf{R}^n} \big[M_1^*(f;\Phi)^{\varepsilon,N}\big]^p\,dx\,.$$

and using the fact (obtained earlier) that the function $M_1^*(f;\Phi)^{\varepsilon,N}$ lies in L^p we obtain the required conclusion (6.4.11). It is because of this step that we needed to work with the truncated maximal function $M_1^*(f;\Phi)^{\varepsilon,N}$ instead of the function $M_1^*(f;\Phi)$.

Proof of (c) Let us denote by $B(x,R)$ the closed ball centered at x with radius R. It follows from the definition of $M_a^*(f;\Phi)$ that

$$|(\Phi_t * f)(y)| \le M_a^*(f;\Phi)(z)\,, \qquad \text{if } \ z \in B(y,at)\,.$$

But the ball $B(y,at)$ is contained in the ball $B(x,|x-y|+at)$; hence it follows that

$$\begin{aligned}
|(\Phi_t * f)(y)|^{\frac{n}{b}} &\le \frac{1}{|B(y,at)|}\int_{B(y,at)} M_a^*(f;\Phi)(z)^{\frac{n}{b}}\,dz \\
&\le \frac{1}{|B(y,at)|}\int_{B(x,|x-y|+at)} M_a^*(f;\Phi)(z)^{\frac{n}{b}}\,dz \\
&\le \left(\frac{|x-y|+at}{at}\right)^n M\big(M_a^*(f;\Phi)^{\frac{n}{b}}\big)(x) \\
&\le \max(1,a^{-n})\left(\frac{|x-y|}{t}+1\right)^n M\big(M_a^*(f;\Phi)^{\frac{n}{b}}\big)(x)
\end{aligned}$$

from which we conclude that for all $x \in \mathbf{R}^n$ we have

$$M_b^{**}(f;\Phi)(x) \le \max(1,a^{-n})\Big\{M\big(M_a^*(f;\Phi)^{\frac{n}{b}}\big)(x)\Big\}^{\frac{b}{n}}.$$

Raising to the power p, using the fact that $p > n/b$ and the boundedness of the Hardy-Littlewood maximal operator M on $L^{pb/n}$, we obtain the required conclusion (6.4.12).

Proof of (d) In proving (d) we may replace b by $b_0 = [b]+1$ since this would yield a stronger estimate than the one we wish to obtain. Let Φ be a Schwartz function with nonvanishing integral. By multiplying Φ by a constant we can assume that Φ has integral equal to 1. Applying Lemma 6.4.5 with $m = b_0$, we can write any function φ in $\mathcal{F}_N$ as

$$\varphi(y) = \int_0^1 (\Theta^{(s)} * \Phi_s)(y)\, ds$$

for some choice of Schwartz functions $\Theta^{(s)}$. Then we have

$$\varphi_t(y) = \int_0^1 ((\Theta^{(s)})_t * \Phi_{ts})(y)\, ds$$

for all $t > 0$. Fix $x \in \mathbf{R}^n$. Then for y in $B(x,t)$ we have

$$\begin{aligned}
|(\varphi_t * f)(y)| &\le \int_0^1 \int_{\mathbf{R}^n} |(\Theta^{(s)})_t(z)|\, |(\Phi_{ts} * f)(y-z)|\, dz\, ds \\
&\le \int_0^1 \int_{\mathbf{R}^n} |(\Theta^{(s)})_t(z)|\, M_{b_0}^{**}(f;\Phi)(x) \left(\frac{|x-(y-z)|}{st} + 1\right)^{b_0} dz\, ds \\
&\le \int_0^1 s^{-b_0} \int_{\mathbf{R}^n} |(\Theta^{(s)})_t(z)|\, M_{b_0}^{**}(f;\Phi)(x) \left(\frac{|x-y|}{t} + \frac{|z|}{t} + 1\right)^{b_0} dz\, ds \\
&\le 2^{b_0} M_{b_0}^{**}(f;\Phi)(x) \int_0^1 s^{-b_0} \int_{\mathbf{R}^n} |\Theta^{(s)}(w)|\, \big(|w|+1\big)^{b_0}\, dw\, ds \\
&\le 2^{b_0} M_{b_0}^{**}(f;\Phi)(x) \int_0^1 s^{-b_0} C_0\, s^{b_0}\, \mathfrak{N}_{b_0}(\varphi)\, ds\,,
\end{aligned}$$

where we applied conclusion (6.4.16) of Lemma 6.4.5. Setting $N = b_0 = [b]+1$, we obtain for y in $B(x,t)$ and $\varphi \in \mathcal{F}_N$

$$|(\varphi_t * f)(y)| \le C_0\, M_{b_0}^{**}(f;\Phi)(x)\,,$$

where C_0 is independent of φ; in fact, C_0 only depends on Φ and b. Taking the supremum over all y in $B(x,t)$, over all $t>0$, and over all φ in $\mathcal{F}_N$, we obtain the pointwise estimate

$$\mathcal{M}_N(f)(x) \le C_0\, M_{b_0}^{**}(f;\Phi)(x) \qquad x \in \mathbf{R}^n\,,$$

which is clearly stronger than (6.4.13). Here we take $C_4 = C_0$, which depends only on b and Φ.

Proof of (e) We fix an $f \in \mathcal{S}'(\mathbf{R}^n)$ that satisfies $\big\|\mathcal{M}_N(f)\big\|_{L^p} < \infty$ for some fixed positive integer N. To show that f is a bounded distribution, we fix a Schwartz function φ and we observe that for some positive constant $c = c_\varphi$, we have that $c\,\varphi$

is an element of $\mathcal{F}_N$ and thus $M_1^*(f;c\varphi) \le \mathcal{M}_N(f)$. Then

$$\begin{aligned} c\,|(\varphi * f)(x)|^p &\le \inf_{|y-x|\le 1} \sup_{|z-y|\le 1} |(c\varphi * f)(z)|^p \\ &\le \inf_{|y-x|\le 1} M_1^*(f;c\varphi)(y)^p \\ &\le \frac{1}{v_n}\int_{|y-x|\le 1} M_1^*(f;c\varphi)(y)^p\,dy \\ &\le \frac{1}{v_n}\int_{\mathbf{R}^n} M_1^*(f;c\varphi)(y)^p\,dy \\ &\le \frac{1}{v_n}\int_{\mathbf{R}^n} \mathcal{M}_N(f)(y)^p\,dy < \infty\,; \end{aligned}$$

which implies that $\varphi * f$ is a bounded function. We conclude that f is a bounded distribution. We now proceed to showing that f is an element of H^p. We fix a smooth function with compact support θ such that

$$\theta(x) = \begin{cases} 1 & \text{if} \quad |x| < 1, \\ 0 & \text{if} \quad |x| > 2. \end{cases}$$

We observe that the identity

$$\begin{aligned} P(x) &= P(x)\theta(x) + \sum_{k=1}^{\infty} \big(\theta(2^{-k}x)P(x) - \theta(2^{-(k-1)}x)P(x)\big) \\ &= P(x)\theta(x) + \frac{\Gamma(\frac{n+1}{2})}{\pi^{\frac{n+1}{2}}} \sum_{k=1}^{\infty} 2^{-k} \bigg(\frac{\theta(\,\cdot\,) - \theta(2(\,\cdot\,))}{(2^{-2k} + |\cdot|^2)^{\frac{n+1}{2}}} \bigg)_{2^k}(x) \end{aligned}$$

is valid for all $x \in \mathbf{R}^n$. Setting

$$\Phi^{(k)}(x) = \big(\theta(x) - \theta(2x)\big)\frac{1}{(2^{-2k} + |x|^2)^{\frac{n+1}{2}}}\,,$$

we note that for some fixed constant $c_0 = c_0(n,N)$, the functions $c_0\,\theta\,P$ and $c_0\Phi^{(k)}$ lie in $\mathcal{F}_N$ uniformly in $k = 1,2,3,\dots$. Combining this observation with the identity for $P(x)$ obtained just earlier, we conclude that

$$\begin{aligned} \sup_{t>0} |P_t * f| &\le \sup_{t>0} |(\theta P)_t * f| + \frac{1}{c_0}\frac{\Gamma(\frac{n+1}{2})}{\pi^{\frac{n+1}{2}}} \sup_{t>0} \sum_{k=1}^{\infty} 2^{-k} \big|(c_0\Phi^{(k)})_{2^k t} * f\big| \\ &\le C_5(n,N)\mathcal{M}_N(f)\,, \end{aligned}$$

which proves the required conclusion (6.4.14).

We observe that the last estimate also yields the stronger estimate

$$M_1^*(f;P)(x) = \sup_{t>0} \sup_{\substack{y\in\mathbf{R}^n \\ |y-x|\le at}} |(P_t * f)(y)| \le C_5(n,N)\mathcal{M}_N(f)(x)\,. \tag{6.4.31}$$

It follows that the quasi-norm $\|M_1^*(f;P)\|_{L^p(\mathbf{R}^n)}$ is also equivalent to $\|f\|_{H^p}$. This will be useful to us later. □

Remark 6.4.6. To simplify the understanding of the equivalences just proved, a first-time reader may wish to define the H^p quasi-norm of a distribution f as

$$\|f\|_{H^p} = \|M_1^*(f;P)\|_{L^p}$$

and then only study the implications (a) $\implies$ (c), (c) $\implies$ (d), (d) $\implies$ (e), and (e) $\implies$ (a) in the proof of Theorem 6.4.4. In this way we avoid passing through the statement in part (b). See Exercise 6.4.5 for details. For many applications, the identification of $\|f\|_{H^p}$ with $\|M_1^*(f;\Phi)\|_{L^p}$ for some Schwartz function Φ (with nonvanishing integral) suffices.

We also remark that the proof of Theorem 6.4.4 yields

$$\|f\|_{H^p(\mathbf{R}^n)} \approx \|\mathcal{M}_N(f)\|_{H^p(\mathbf{R}^n)};$$

where $N = [\frac{n}{p}] + 1$.

6.4.c. Consequences of the Characterizations of Hardy Spaces

In this subsection we look at a few consequences of Theorem 6.4.4. In many applications we will need to be working with dense subspaces of H^p. It turns out that both $H^p \cap L^2$ and $H^p \cap L^1$ are dense in H^p.

Proposition 6.4.7. *Let $0 < p \le 1$. Then for any r with $p \le r \le \infty$, $L^r \cap H^p$ is dense in H^p. In particular, $H^p \cap L^2$ and $H^p \cap L^1$ are dense in H^p.*

Proof. Let f be a distribution in $H^p(\mathbf{R}^n)$. Recall the Poisson kernel $P(x)$ and set $N = [\frac{n}{p}] + 1$. For any fixed $x \in \mathbf{R}^n$ and $t > 0$ we have

$$|(P_t * f)(x)| \le M_1^*(f;P)(y) \le C\mathcal{M}_N(f)(y) \tag{6.4.32}$$

for any $|y - x| \le t$. Indeed, the first estimate in (6.4.32) follows from the definition of $M_1^*(f;P)$ and the second estimate by (6.4.31). Raising (6.4.32) to the power p and averaging over the ball $B(x,t)$, we obtain

$$|(P_t * f)(x)|^p \le \frac{C^p}{v_n t^n} \int_{B(x,t)} \mathcal{M}_N(f)(y)^p \, dy \le \frac{C_1^p}{t^n} \|f\|_{H^p}^p .$$

It follows that the function $P_t * f$ is in $L^\infty(\mathbf{R}^n)$ with norm at most a constant multiple of $t^{-n/p}\|f\|_{H^p}$. Moreover, this function is also in $L^p(\mathbf{R}^n)$ as it is controlled by $M(f;P)$. Therefore, the functions $P_t * f$ lie in $L^r(\mathbf{R}^n)$ for all $r \le p \le \infty$. It remains to show that $P_t * f$ also lie in H^p and that $P_t * f \to f$ in H^p as $t \to 0$.

To see that $P_t * f$ lies in H^p, we use the semigroup formula $P_t * P_s = P_{t+s}$ for the Poisson kernel, which is a consequence of the fact that $\widehat{P_t}(\xi) = e^{-2\pi t|\xi|}$ by applying the Fourier transform. Therefore, for any $t > 0$ we have

$$\sup_{s>0} |P_s * P_t * f| = \sup_{s>0} |P_{s+t} * f| \le \sup_{s>0} |P_s * f| \, ,$$

which implies that

$$\big\| P_t * f \big\|_{H^p} \le \big\| f \big\|_{H^p}$$

for all $t > 0$. We now need to show that $P_t * f \to f$ in H^p as $t \to 0$. This will be a consequence of the Lebesgue dominated convergence theorem once we know that

$$\sup_{s>0} |(P_s * P_t * f - P_s * f)(x)| \to 0 \qquad \text{as} \quad t \to 0 \tag{6.4.33}$$

pointwise for all $x \in \mathbf{R}^n$ and also

$$\sup_{s>0} |P_s * P_t * f - P_s * f| \le 2 \sup_{s>0} |P_s * f| \in L^p(\mathbf{R}^n) \, . \tag{6.4.34}$$

Statement (6.4.34) is a trivial consequence of the Poisson semigroup formula. As far as (6.4.33) is concerned, we note that for all $x \in \mathbf{R}^n$ the function

$$s \to |(P_s * P_t * f)(x) - (P_s * f)(x)| = |(P_{s+t} * f)(x) - (P_s * f)(x)|$$

is bounded by a constant multiple of $s^{-n/p}$ and therefore tends to zero as $s \to \infty$. Given any $\varepsilon > 0$, there exists an $M > 0$ such that for all $t > 0$ we have

$$\sup_{s>M} |(P_s * P_t * f - P_s * f)(x)| < \frac{\varepsilon}{2} \, . \tag{6.4.35}$$

Moreover, the function $t \to \sup_{0 \le s \le M} |(P_s * P_t * f - P_s * f)(x)|$ is continuous in t. Therefore, there exists a $t_0 > 0$ such that for $t < t_0$ we have

$$\sup_{0 \le s \le M} |(P_s * P_t * f - P_s * f)(x)| < \frac{\varepsilon}{2} \, . \tag{6.4.36}$$

Combining (6.4.35) and (6.4.36) proves (6.4.33).

□

Next we observe the following consequence of Theorem 6.4.4.

Corollary 6.4.8. *For any two Schwartz functions Φ and Θ with nonvanishing integral we have*

$$\big\| \sup_{t>0} |\Theta_t * f| \big\|_{L^p} \approx \big\| \sup_{t>0} |\Phi_t * f| \big\|_{L^p} \approx \big\| f \big\|_{H^p}$$

for all $f \in \mathcal{S}'(\mathbf{R}^n)$, with constants depending only on n, p, Φ, and Θ.

Proof. This follows easily from Theorem 6.4.4.

□

Next we define a *norm*, of a Schwartz function φ which will be relevant in the theory of Hardy spaces:

$$\mathfrak{N}_N(\varphi; x_0, R) = \int_{\mathbf{R}^n} \left(1 + |\tfrac{x - x_0}{R}|\right)^N \sum_{|\alpha| \le N+1} R^{|\alpha|} |\partial^\alpha \varphi(x)| \, dx \, .$$

Note that $\mathfrak{N}_N(\varphi;0,1)=\mathfrak{N}_N(\varphi)$.

Corollary 6.4.9. *For any $1<p<\infty$, any $f\in H^p(\mathbf{R}^n)$, and any $\varphi\in\mathcal{S}(\mathbf{R}^n)$ we have*

$$|\langle f,\varphi\rangle|\le \mathfrak{N}_N(\varphi)\inf_{|z|\le 1}\mathcal{M}_N(f)(z)\,, \tag{6.4.37}$$

where $N=[\frac{n}{p}]+1$. More generally, for any $x_0\in\mathbf{R}^n$ and $R>0$ we have

$$|\langle f,\varphi\rangle|\le \mathfrak{N}_N(\varphi;x_0,R)\inf_{|z-x_0|\le R}\mathcal{M}_N(f)(z)\,. \tag{6.4.38}$$

Proof. Set $\psi(x)=\varphi(-Rx+x_0)$. It follows directly from Definition 6.4.1 that for any fixed z with $|z-x_0|\le R$ we have

$$\begin{aligned}|\langle f,\varphi\rangle| &= R^n|(f*\psi_R)(x_0)|\\ &\le \sup_{y:\ |y-z|\le R} R^n|(f*\psi_R)(y)|\\ &\le R^n\bigg[\int_{\mathbf{R}^n}(1+|w|)^N\sum_{|\alpha|\le N+1}|\partial^\alpha\psi(w)|\,dw\bigg]\mathcal{M}_N(f)(z)\,,\end{aligned}$$

from which the second assertion in the corollary follows easily by the change of variables $x=-Rw+x_0$. Taking the infimum over all z with $|z-x_0|\le R$ yields the required conclusion.

□

Proposition 6.4.10. *Let $0<p\le 1$. Then the following statements are valid:*
(a) Convergence in H^p is a stronger notion than convergence in $\mathcal{S}'$.
(b) H^p is a complete quasi-normed metrizable space.

Proof. Part (a) says that if a sequence $f_j\to f$ in $H^p(\mathbf{R}^n)$, then $f_j\to f$ in $\mathcal{S}'(\mathbf{R}^n)$. But this easily follows from the estimate

$$|\langle f,\varphi\rangle|\le C_\varphi\inf_{|z|\le 1}\mathcal{M}_N(f)(z)\le \frac{C_\varphi}{v_n}\int_{\mathbf{R}^n}\mathcal{M}_N(f)^p\,dz\le C_\varphi C_{n,p}\big\|f\big\|_{H^p}^p\,,$$

which is a direct consequence of (6.4.37) for all φ in $\mathcal{S}(\mathbf{R}^n)$. As before, here $N=[\frac{n}{p}]+1$.

To obtain the statement in (b), we first observe that the quantity $f\to\big\|f\big\|_{H^p}^p$ is a metric on H^p that generates the same topology as the quasi-norm $f\to\big\|f\big\|_{H^p}$. To show that H^p is a complete space, it will suffice to show that for any sequence of functions f_j that satisfies

$$\sum_j\int_{\mathbf{R}^n}\mathcal{M}_N(f_j)^p\,dx<\infty\,,$$

the series $\sum_j f_j$ converges in $H^p(\mathbf{R}^n)$. The partial sums of this series are Cauchy in $H^p(\mathbf{R}^n)$ and therefore are Cauchy in $\mathcal{S}'(\mathbf{R}^n)$ by part (a). It follows that the sequence

$\sum_{-k}^{k} f_j$ converges to some tempered distribution f in $\mathcal{S}'(\mathbf{R}^n)$. Sublinearity gives

$$\int_{\mathbf{R}^n} \mathcal{M}_N(f)^p\,dx = \int_{\mathbf{R}^n} \mathcal{M}_N\Big(\sum_j f_j\Big)^p\,dx \le \sum_j \int_{\mathbf{R}^n} \mathcal{M}_N(f_j)^p\,dx < \infty\,,$$

which implies that $f \in H^p$. Finally,

$$\int_{\mathbf{R}^n} \mathcal{M}_N\Big(f - \sum_{j=-k}^{k} f_j\Big)^p dx \le \sum_{|j|\ge k+1} \int_{\mathbf{R}^n} \mathcal{M}_N(f_j)^p\,dx \to 0$$

as $k \to \infty$; thus the series converges in H^p.

□

6.4.d. Vector-Valued H^p and its Characterizations

We now obtain a vector-valued analogue of Theorem 6.4.4 that will be crucial in our goal to characterize Hardy spaces using Littlewood-Paley theory. To state this analogue we need to extend the definitions of the maximal operators to sequences of distributions. Let $a, b > 0$ and Φ be a Schwartz function on $\mathbf{R}^n$. In accordance with Definition 6.4.1, we give the following sequence of definitions.

Definition 6.4.11. For a sequence $\vec{f} = \{f_j\}_{j\in\mathbf{Z}}$ of tempered distributions on $\mathbf{R}^n$ we define the *smooth maximal function of $\vec{f}$ with respect to Φ* as

$$\boldsymbol{M}(\vec{f}\,;\Phi)(x) = \sup_{t>0} \big\|\{(\Phi_t * f_j)(x)\}_j\big\|_{\ell^2}\,.$$

We define the *nontangential maximal function (with aperture a) of f with respect to Φ* as

$$\boldsymbol{M}_a^*(\vec{f}\,;\Phi)(x) = \sup_{t>0} \sup_{\substack{y\in\mathbf{R}^n\\ |y-x|\le at}} \big\|\{(\Phi_t * f_j)(y)\}_j\big\|_{\ell^2}\,.$$

We also define the *auxiliary maximal function*

$$\boldsymbol{M}_b^{**}(\vec{f}\,;\Phi)(x) = \sup_{t>0} \sup_{y\in\mathbf{R}^n} \frac{\big\|\{(\Phi_t * f_j)(x-y)\}_j\big\|_{\ell^2}}{(1+t^{-1}|y|)^b}\,.$$

We note that if Φ is merely integrable, for example, if Φ is the Poisson kernel, the maximal functions $\boldsymbol{M}(\vec{f}\,;\Phi)$, $\boldsymbol{M}_a^*(\vec{f}\,;\Phi)$, and $\boldsymbol{M}_b^{**}(\vec{f}\,;\Phi)$ are only well defined for sequences $\vec{f} = \{f_j\}_j$ whose terms are bounded tempered distributions on $\mathbf{R}^n$.

For a fixed positive integer N we define the *grand maximal function of $\vec{f}$ (with respect to N)* as

$$\boldsymbol{\mathcal{M}}_N(\vec{f}\,) = \sup_{\varphi\in\mathcal{F}_N} \boldsymbol{M}_1^*(\vec{f}\,;\varphi)\,, \tag{6.4.39}$$

where

$$\mathcal{F}_N = \Big\{\varphi \in \mathcal{S}(\mathbf{R}^n) :\ \mathfrak{N}_N(\varphi) \le 1\Big\}$$

is as defined in (6.4.5).

We note that as in the scalar case we have the sequence of simple inequalities

$$M(\vec{f};\Phi) \le M_a^*(\vec{f};\Phi) \le (1+a)^b M_b^{**}(\vec{f};\Phi)\,. \tag{6.4.40}$$

We now define the vector-valued Hardy space $H^p(\mathbf{R}^n,\ell^2)$.

Definition 6.4.12. Let $\vec{f} = \{f_j\}_j$ be a sequence of bounded tempered distributions on $\mathbf{R}^n$ and let $0<p<\infty$. We say that $\vec{f}$ lies in the vector-valued Hardy space $H^p(\mathbf{R}^n,\ell^2)$ if and only if the *Poisson maximal function*

$$M(\vec{f};P)(x) = \sup_{t>0} \big\|\{(P_t * f_j)(x)\}_j\big\|_{\ell^2}$$

lies in $L^p(\mathbf{R}^n)$. If this is the case, we set

$$\big\|\vec{f}\big\|_{H^p(\mathbf{R}^n,\ell^2)} = \big\|M(\vec{f};P)\big\|_{L^p(\mathbf{R}^n)} = \bigg\|\sup_{\varepsilon>0}\Big(\sum_j |f_j * P_\varepsilon|^2\Big)^{\frac12}\bigg\|_{L^p(\mathbf{R}^n)}.$$

The next theorem provides a vector-valued analogue of Theorem 6.4.4.

Theorem 6.4.13. *Let $0<p<\infty$. Then the following statements are valid:*
(a) There exists a Schwartz function Φ with $\int_{\mathbf{R}^n}\Phi(x)\,dx \neq 0$ and a constant C_1 (that does not depend on any parameters) such that

$$\big\|M(\vec{f};\Phi)\big\|_{L^p(\mathbf{R}^n,\ell^2)} \le C_1\big\|\vec{f}\big\|_{H^p(\mathbf{R}^n,\ell^2)} \tag{6.4.41}$$

for every sequence $\vec{f}=\{f_j\}_j$ of tempered distributions.
(b) For every $a>0$ and Φ in $\mathcal{S}(\mathbf{R}^n)$ there exists a constant $C_2(n,p,a,\Phi)$ such that

$$\big\|M_a^*(\vec{f};\Phi)\big\|_{L^p(\mathbf{R}^n,\ell^2)} \le C_2(n,p,a,\Phi)\big\|M(\vec{f};\Phi)\big\|_{L^p(\mathbf{R}^n,\ell^2)} \tag{6.4.42}$$

for every sequence $\vec{f}=\{f_j\}_j$ of tempered distributions.
(c) For every $a>0$, $b>n/p$ and Φ in $\mathcal{S}(\mathbf{R}^n)$ there exists a constant $C_3(n,p,a,b,\Phi)$ such that

$$\big\|M_b^{**}(\vec{f};\Phi)\big\|_{L^p(\mathbf{R}^n,\ell^2)} \le C_3(n,p,a,b,\Phi)\big\|M_a^*(\vec{f};\Phi)\big\|_{L^p(\mathbf{R}^n,\ell^2)} \tag{6.4.43}$$

for every sequence $\vec{f}=\{f_j\}_j$ of tempered distributions.
(d) For every $b>0$ and Φ in $\mathcal{S}(\mathbf{R}^n)$ with $\int_{\mathbf{R}^n}\Phi(x)\,dx\neq 0$ there exists a constant $C_4(b,\Phi)$ such that if $N=[\frac{n}{p}]+1$ we have

$$\big\|\mathcal{M}_N(\vec{f})\big\|_{L^p(\mathbf{R}^n,\ell^2)} \le C_4(b,\Phi)\big\|M_b^{**}(\vec{f};\Phi)\big\|_{L^p(\mathbf{R}^n,\ell^2)} \tag{6.4.44}$$

for every sequence $\vec{f}=\{f_j\}_j$ of tempered distributions.
(e) For every positive integer N there exists a constant $C_5(n,N)$ such that every sequence $\vec{f}=\{f_j\}_j$ of tempered distributions that satisfies $\big\|\mathcal{M}_N(\vec{f})\big\|_{L^p(\mathbf{R}^n,\ell^2)}<\infty$ consists of bounded distributions and satisfies

$$\big\|\vec{f}\big\|_{H^p(\mathbf{R}^n,\ell^2)} \le C_5(n,N)\big\|\mathcal{M}_N(\vec{f})\big\|_{L^p(\mathbf{R}^n,\ell^2)}\,, \tag{6.4.45}$$

that is, it lies in the Hardy space $H^p(\mathbf{R}^n, \ell^2)$.

Moreover, all the constants C_1, C_2, C_3, C_4, C_5 *are the same as the corresponding constants in Theorem 6.4.4.*

Proof. The proof of this theorem is obtained via a step-by-step repetition of the proof of Theorem 6.4.4 in which the scalar absolute values are replaced by ℓ^2 norms. This small change in one's point of view yields a self-improvement of Theorem 6.4.4 without any loss in the constants. Moreover, it provides an example of the power of Hilbert space techniques. The details of this verification are left to the interested reader.

□

We end this subsection by observing the validity of the following vector-valued analogue of (6.4.38):

$$\Big(\sum_j |\langle f_j, \varphi\rangle|^2\Big)^{\frac{1}{2}} \le \mathfrak{N}_N(\varphi; x_0, R) \inf_{|z-x_0|\le R} \boldsymbol{\mathcal{M}}_N(\vec{f})(z)\,. \tag{6.4.46}$$

The proof of (6.4.46) is identical to the corresponding estimate for scalar-valued functions. Set $\psi(x) = \varphi(-Rx + x_0)$. It follows directly from Definition 6.4.11 that for any fixed z with $|z - x_0| \le R$ we have

$$\begin{aligned}\Big(\sum_j |\langle f_j, \varphi\rangle|^2\Big)^{\frac{1}{2}} &= R^n \big\|\{(f_j * \psi_R)(x_0)\}_j\big\|_{\ell^2} \\ &\le \sup_{y:\,|y-z|\le R} R^n \big\|\{(f_j * \psi_R)(y)\}_j\big\|_{\ell^2} \\ &\le R^n\, \mathfrak{N}_N(\psi)\, \boldsymbol{\mathcal{M}}_N(\vec{f})(z)\,,\end{aligned}$$

which, combined with the observation

$$R^n\, \mathfrak{N}_N(\psi) = \mathfrak{N}_N(\varphi; x_0, R)\,,$$

yields the required conclusion by taking the infimum over all z with $|z - x_0| \le R$.

6.4.e. Singular Integrals on Hardy Spaces

To obtain the Littlewood-Paley characterization of Hardy spaces, we will need a multiplier theorem for vector-valued Hardy spaces.

Suppose that $K_j(x)$ is a family of functions defined on $\mathbf{R}^n \setminus \{0\}$ that satisfies the following: There exist constants $A, B < \infty$ and an integer N such that for all multiindices α with $|\alpha| \le N$ we have

$$\Big|\sum_{j\in\mathbf{Z}} \partial^\alpha K_j(x)\Big| \le A\,|x|^{-n-|\alpha|} < \infty \tag{6.4.47}$$

and such that

$$\sup_{\xi\in\mathbf{R}^n}\Big|\sum_{j\in\mathbf{Z}}\widehat{K_j}(\xi)\Big|\le B<\infty\,. \tag{6.4.48}$$

Theorem 6.4.14. *Suppose that a sequence of kernels $\{K_j\}_j$ satisfies (6.4.47) and (6.4.48) with $N=[\frac{n}{p}]+1$, for some $0<p\le 1$. Then there exists a constant $C_{n,p}$ that depends only on the dimension n, on p such that for all sequences of tempered distributions $\{f_j\}_j$ we have the estimate*

$$\Big\|\sum_j K_j*f_j\Big\|_{H^p(\mathbf{R}^n)}\le C_{n,p}(A+B)\big\|\{f_j\}_j\big\|_{H^p(\mathbf{R}^n,\ell^2)}\,.$$

Proof. We fix a smooth positive function Φ supported in the unit ball $B(0,1)$ with $\int_{\mathbf{R}^n}\Phi(x)\,dx=1$ and we consider the sequence of smooth maximal functions

$$M\big(\sum_j K_j*f_j;\Phi\big)=\sup_{\varepsilon>0}\big|\Phi_\varepsilon*\sum_j K_j*f_j\big|\,,$$

which we need to show that lies in $L^p(\mathbf{R}^n,\ell^2)$. We will be working with a fixed sequence of integrable functions $\vec{f}=\{f_j\}_j$ since such functions are dense in $L^p(\mathbf{R}^n,\ell^2)$ in view of Proposition 6.4.7.

We now fix a $\lambda>0$ and we set $N=[\frac{n}{p}]+1$. We also fix $\gamma>0$ to be chosen later and we define the set

$$\Omega_\lambda=\{x\in\mathbf{R}^n:\ \boldsymbol{\mathcal{M}}_N(\vec{f})(x)>\gamma\lambda\}\,.$$

The set Ω_λ is open, and by the Proposition in Appendix J we can write it is a union of cubes Q_k such that

(a) $\bigcup_k Q_k=\Omega_\lambda$ and the Q_k's have disjoint interiors.

(b) $\sqrt{n}\,\ell(Q_k)\le \operatorname{dist}(Q_k,(\Omega_\lambda)^c)\le 4\sqrt{n}\,\ell(Q_k)$.

We denote by $c(Q_k)$ the center of the cube Q_k. For each k we set

$$d_k=\operatorname{dist}(Q_k,(\Omega_\lambda)^c)+2\sqrt{n}\,\ell(Q_k)\approx\ell(Q_k)$$

so that

$$B(c(Q_k),d_k)\cap(\Omega_\lambda)^c\neq\emptyset\,.$$

We now introduce a partition of unity $\{\varphi_k\}_k$ adapted to the sequence of cubes $\{Q_k\}_k$ such that

(c) $\chi_{\Omega_\lambda}=\sum_k\varphi_k$ and each φ_k satisfies $0\le\varphi_k\le 1$.

(d) Each φ_k is supported in $\frac{6}{5}\,Q_k$ and satisfies $\int_{\mathbf{R}^n}\varphi_k\,dx\approx d_k^n$

(e) $\big\|\partial^\alpha\varphi_k\big\|_{L^\infty}\le C_\alpha d_k^{-|\alpha|}$ for all multiindices α and some constants C_α.

We decompose each f_j as

$$f_j=g_j+\sum_k b_{j,k}\,,$$

where g_j is the *good function* of the decomposition defined as follows:

$$g_j = f_j \chi_{\mathbf{R}^n \setminus \Omega_\lambda} + \sum_k \frac{\int_{\mathbf{R}^n} f_j \varphi_k \, dx}{\int_{\mathbf{R}^n} \varphi_k \, dx} \varphi_k$$

and $b_j = \sum_k b_{j,k}$ is the *bad function* of the decomposition given by

$$b_{j,k} = \left(f_j - \frac{\int_{\mathbf{R}^n} f_j \varphi_k \, dx}{\int_{\mathbf{R}^n} \varphi_k \, dx} \right) \varphi_k \, .$$

We note that each $b_{j,k}$ has integral zero. We define

$$\vec{g} = \{g_j\}_j \qquad \text{and} \qquad \vec{b} = \{b_j\}_j \, .$$

At this point we appeal to (6.4.46) and to properties (d) and (e) to obtain

$$\text{(6.4.49)} \qquad \Big(\sum_j \Big| \frac{\int_{\mathbf{R}^n} f_j \varphi_k \, dx}{\int_{\mathbf{R}^n} \varphi_k \, dx} \Big|^2 \Big)^{\frac{1}{2}} \le \frac{\mathfrak{N}_N(\varphi_k; c(Q_k), d_k)}{\int_{\mathbf{R}^n} \varphi_k \, dx} \inf_{|z - c(Q_k)| \le d_k} \boldsymbol{\mathcal{M}}_N(\vec{f})(z) \, .$$

But since

$$\frac{\mathfrak{N}_N(\varphi_k; c(Q_k), d_k)}{\int_{\mathbf{R}^n} \varphi_k \, dx} \le \left[\int_{Q_k} \Big(1 + \frac{|x - c(Q_k)|}{d_k}\Big)^N \sum_{|\alpha| \le N+1} \frac{d_k^{|\alpha|} C_\alpha d_k^{-|\alpha|}}{\int_{\mathbf{R}^n} \varphi_k \, dx} \, dx \right] \le C_{N,n} \, ,$$

if follows that (6.4.49) is at most a constant multiple of λ since the ball $B(c(Q_k), d_k)$ meets the complement of Ω_λ. We conclude that

$$\text{(6.4.50)} \qquad \|\vec{g}\|_{L^\infty(\Omega_\lambda, \ell^2)} \le C_{N,n} \, \gamma \, \lambda \, .$$

We now turn to estimating $M(\sum_j K_j * b_{j,k}; \Phi)$. For fixed k and $\varepsilon > 0$ we have

$$\begin{aligned}
&\Big(\Phi_\varepsilon * \sum_j K_j * b_{j,k}\Big)(x) \\
&= \int_{\mathbf{R}^n} \Phi_\varepsilon * \sum_j K_j(x-y) \left[f_j(y) \varphi_k(y) - \frac{\int_{\mathbf{R}^n} f_j \varphi_k \, dx}{\int_{\mathbf{R}^n} \varphi_k \, dx} \varphi_k(y) \right] dy \\
&= \int_{\mathbf{R}^n} \sum_j \Big\{ (\Phi_\varepsilon * K_j)(x-z) - \int_{\mathbf{R}^n} (\Phi_\varepsilon * K_j)(x-y) \frac{\varphi_k(y)}{\int_{\mathbf{R}^n} \varphi_k \, dx} \, dy \Big\} \varphi_k(z) f_j(z) \, dz \\
&= \int_{\mathbf{R}^n} \sum_j R_{j,k}(x,z) \varphi_k(z) f_j(z) \, dz
\end{aligned}$$

where we set $R_{j,k}(x,z)$ for the expression inside the curly brackets. Using (6.4.38). we obtain

$$\begin{aligned}&\Big|\int_{\mathbf{R}^n}\sum_j R_{j,k}(x,z)\varphi_k(z)f_j(z)\,dz\Big|\\ &\le \sum_j \mathfrak{N}_N(R_{j,k}(x,\,\cdot\,)\varphi_k; c(Q_k),d_k)\inf_{|z-c(Q_k)|\le d_k}\mathcal{M}_N(f_j)(z)\\ &\le \sum_j \mathfrak{N}_N(R_{j,k}(x,\,\cdot\,)\varphi_k; c(Q_k),d_k)\inf_{|z-c(Q_k)|\le d_k}\boldsymbol{\mathcal{M}}_N(\vec{f}\,)(z)\,.\end{aligned}\tag{6.4.51}$$

Since $\varphi_k(z)$ is supported in $\frac{6}{5}Q_k$, the term $(1+\frac{|z-c(Q_k)|}{d_k})^N$ contributes only a constant factor in the integral giving $\mathfrak{N}_N(R_{j,k}(x,\,\cdot\,)\varphi_k; c(Q_k),d_k)$, and we obtain

$$\begin{aligned}&\mathfrak{N}_N(R_{j,k}(x,\,\cdot\,)\varphi_k; c(Q_k),d_k)\\ &\le C_{N,n}\int_{\frac{6}{5}Q_k}\sum_{|\alpha|\le N+1} d_k^{|\alpha|+n}\Big|\frac{\partial^\alpha}{\partial z^\alpha}\big(R_{j,k}(x,z)\varphi_k(z)\big)\Big|\,dz\,.\end{aligned}\tag{6.4.52}$$

For notational convenience we set

$$K_j^\varepsilon=\Phi_\varepsilon * K_j\,.$$

We observe that the family $\{K_j^\varepsilon\}_j$ satisfies (6.4.47) and (6.4.48) with constants A' and B' that are only multiples of A and B, respectively, uniformly in ε. We now obtain a pointwise estimate for $\mathfrak{N}_N(R_{j,k}(x,\,\cdot\,)\varphi_k; c(Q_k),d_k)$ when $x\in\mathbf{R}^n\setminus\Omega_\lambda$. We have

$$R_{j,k}(x,z)\varphi_k(z)=\int_{\mathbf{R}^n}\varphi_k(z)\Big\{K_j^\varepsilon(x-z)-K_j^\varepsilon(x-y)\Big\}\frac{\varphi_k(y)\,dy}{\int_{\mathbf{R}^n}\varphi_k\,dx}$$

from which it follows that

$$\Big|\frac{\partial^\alpha}{\partial z^\alpha}R_{j,k}(x,z)\varphi_k(z)\Big|\le\int_{\mathbf{R}^n}\Big|\frac{\partial^\alpha}{\partial z^\alpha}\Big\{\varphi_k(z)\Big[K_j^\varepsilon(x-z)-K_j^\varepsilon(x-y)\Big]\Big\}\Big|\frac{\varphi_k(y)\,dy}{\int_{\mathbf{R}^n}\varphi_k\,dx}\,.$$

Using hypothesis (6.4.47), we can now easily obtain the estimate

$$\sum_j\Big|\frac{\partial^\alpha}{\partial z^\alpha}\Big\{\varphi_k(z)\Big\{K_{i,j}^\varepsilon(x-z)-K_{i,j}^\varepsilon(x-y)\Big\}\Big\}\Big|\le C_{N,n}A\frac{d_k d_k^{-|\alpha|}}{|x-c(Q_k)|^{n+1}}$$

for all $|\alpha|\le N$ and for $x\in\mathbf{R}^n\setminus\Omega_\lambda$ since for such x we have $|x-c(Q_k)|\ge c_n\,d_k$. It follows that

$$d_k^{|\alpha|+n}\sum_j\Big|\frac{\partial^\alpha}{\partial z^\alpha}\{R_{j,k}(x,z)\varphi_k(z)\}\Big|\le C_{N,n}A\,d_k^n\Big(\frac{d_k}{|x-c(Q_k)|^{n+1}}\Big)\,.$$

Inserting this estimate in the summation of (6.4.52) over all j yields

$$\sum_j\mathfrak{N}_N(R_{j,k}(x,\,\cdot\,)\varphi_k; c(Q_k),d_k)\le C_{N,n}A\Big(\frac{d_k^{n+1}}{|x-c(Q_k)|^{n+1}}\Big)\,.\tag{6.4.53}$$

Combining (6.4.53) with (6.4.51) gives for $x \in \mathbf{R}^n \setminus \Omega_\lambda$

$$\sum_j \left| \int_{\mathbf{R}^n} R_{i,j,k}(x,z)\varphi_k(z) f_j(z)\, dz \right| \le \frac{C_{N,n} A\, d_k^{n+1}}{|x-c(Q_k)|^{n+1}} \inf_{|z-c(Q_k)|\le d_k} \mathcal{M}_N(\vec{f})(z)\,,$$

which provides the estimate

$$\sup_{\varepsilon>0} \big| \sum_j (K_j^\varepsilon * b_{j,k})(x) \big| \le \frac{C_{N,n} A\, d_k^{n+1}}{|x-c(Q_k)|^{n+1}}\, \gamma\, \lambda$$

for all $x \in \mathbf{R}^n \setminus \Omega_\lambda$ since the ball $B(c(Q_k), d_k)$ is intersects $(\Omega_\lambda)^c$. Summing over k results

$$M\Big(\sum_j K_j * b_j; \Phi\Big)(x) \le \sum_k \frac{C_{N,n} A\,\gamma\,\lambda\, d_k^{n+1}}{|x-c(Q_k)|^{n+1}} \le \sum_k \frac{C_{N,n} A\,\gamma\,\lambda\, d_k^{n+1}}{(d_k + |x-c(Q_k)|)^{n+1}}$$

for all $x \in (\Omega_\lambda)^c$. The last sum is known as the *Marcinkiewicz function*. It is a simple fact that

$$\int_{\mathbf{R}^n} \sum_k \frac{d_k^{n+1}}{(d_k + |x-c(Q_k)|)^{n+1}}\, dx \le C_n \sum_k |Q_k| = C_n\, |\Omega_\lambda|;$$

see Exercise 4.6.6. We have therefore shown that

$$\int_{\mathbf{R}^n} M(\vec{K} * \vec{b}; \Phi)(x)\, dx \le C_{N,n} A \gamma\, \lambda\, |\Omega_\lambda|\,, \tag{6.4.54}$$

where we used the notation $\vec{K} * \vec{b} = \sum_j K_j * b_j$.

We now combine the information we have acquired so far. First we have

$$\big|\{M(\vec{K} * \vec{f}; \Phi) > \lambda\}\big| \le \big|\{M(\vec{K} * \vec{g}; \Phi) > \tfrac{\lambda}{2}\}\big| + \big|\{M(\vec{K} * \vec{b}; \Phi) > \tfrac{\lambda}{2}\}\big|\,.$$

For the good function $\vec{g}$ we have the estimate

$$\begin{aligned}
\big|\{M(\vec{K} * \vec{g}; \Phi) > \tfrac{\lambda}{2}\}\big| &\le \frac{4}{\lambda^2} \int_{\mathbf{R}^n} M(\vec{K} * \vec{g}; \Phi)(x)^2\, dx \\
&\le \frac{4}{\lambda^2} \sum_j \int_{\mathbf{R}^n} M(K_j * g_j)(x)^2\, dx \\
&\le \frac{C_n B^2}{\lambda^2} \int_{\mathbf{R}^n} \sum_j |g_j(x)|^2\, dx \\
&\le \frac{C_n B^2}{\lambda^2} \int_{\Omega_\lambda} \sum_j |g_j(x)|^2\, dx + \frac{C_n B^2}{\lambda^2} \int_{(\Omega_\lambda)^c} \sum_j |f_j(x)|^2\, dx \\
&\le B^2 C_{N,n} \gamma^2\, |\Omega_\lambda| + \frac{C_n B^2}{\lambda^2} \int_{(\Omega_\lambda)^c} \mathcal{M}_N(\vec{f})(x)^2\, dx\,,
\end{aligned}$$

where we used Corollary 2.1.12, the L^2 boundedness of the Hardy-Littlewood maximal operator, hypothesis (6.4.48), the fact that $f_j = g_j$ on $(\Omega_\lambda)^c$, estimate (6.4.50), and the fact that $\|\vec{f}\|_{\ell^2} \le \mathcal{M}_N(\vec{f})$ in the sequence of estimates.

On the other hand, estimate (6.4.54) and Chebychev's inequality gives

$$\big|\{M(\vec{K} * \vec{b}\,;\Phi) > \tfrac{\lambda}{2}\}\big| \le C_{N,n} A\,\gamma\,|\Omega_\lambda|\,,$$

which, combined with the previously obtained estimate for $\vec{g}$, gives

$$\big|\{M(\vec{K} * \vec{f}\,;\Phi) > \lambda\}\big| \le C_{N,n}(A\,\gamma + B^2\,\gamma^2)\,|\Omega_\lambda| + \frac{C_n B^2}{\lambda^2}\int_{(\Omega_\lambda)^c} \mathcal{M}_N(\vec{f})(x)^2\,dx\,.$$

Multiplying this estimate by $p\lambda^{p-1}$, recalling that $\Omega_\lambda = \{\mathcal{M}_N(\vec{f}) > \gamma\,\lambda\}$, and integrating in λ from 0 to ∞, we can easily obtain

$$\big\|M(\vec{K} * \vec{f}\,;\Phi)\big\|^p_{L^p(\mathbf{R}^n,\ell^2)} \le C_{N,n}(A\gamma^{1-p} + B^2\gamma^{2-p})\big\|\mathcal{M}_N(\vec{f})\big\|^p_{L^p(\mathbf{R}^n,\ell^2)}\,. \tag{6.4.55}$$

Choosing $\gamma = (A+B)^{-1}$ and recalling that $N = [\frac{n}{p}]+1$ gives the required conclusion for some constant $C_{n,p}$ that depends only on n and p.

Finally, use density to extend this estimate to all $\vec{f}$ in $H^p(\mathbf{R}^n,\ell^2)$.

□

6.4.f. The Littlewood-Paley Characterization of Hardy Spaces

We now come to an important characterization of Hardy spaces in terms of Littlewood-Paley square functions. The vector-valued Hardy spaces and the action of singular integrals on them, studied in the previous subsection, will be crucial in obtaining this characterization.

We first discuss the setup. We fix a radial Schwartz function Ψ on $\mathbf{R}^n$ whose Fourier transform is nonnegative, supported in the annulus $\frac{1}{2} + \frac{1}{10} \le |\xi| \le 2 - \frac{1}{10}$, and satisfies

$$\sum_{j\in\mathbf{Z}} \widehat{\Psi}(2^{-j}\xi) = 1 \tag{6.4.56}$$

for all $\xi \neq 0$. Associated with this bump, we define the Littlewood-Paley operators Δ_j given by multiplication on the Fourier transform with the function $\widehat{\Psi}(2^{-j}\xi)$, that is,

$$\Delta_j(f) = \Delta_j^{\Psi}(f) = \Psi_{2^{-j}} * f\,. \tag{6.4.57}$$

We have the following.

Theorem 6.4.15. *Let Ψ be as previously and let Δ_j be the Littlewood-Paley operators associated with Ψ. Let $0 < p \le 1$ Then there exists a constant $C = C_{n,p,\Psi}$*

such that for all $f \in H^p(\mathbf{R}^n)$ we have

$$\Big\|\Big(\sum_{j\in\mathbf{Z}}|\Delta_j(f)|^2\Big)^{\frac{1}{2}}\Big\|_{L^p} \le C\|f\|_{H^p}. \tag{6.4.58}$$

Conversely, suppose that a tempered distribution f satisfies

$$\Big\|\Big(\sum_{j\in\mathbf{Z}}|\Delta_j(f)|^2\Big)^{\frac{1}{2}}\Big\|_{L^p} < \infty. \tag{6.4.59}$$

Then there exists a unique polynomial $Q(x)$ such that $f-Q$ lies in the Hardy space H^p and satisfies the estimate

$$\frac{1}{C}\|f-Q\|_{H^p} \le \Big\|\Big(\sum_{j\in\mathbf{Z}}|\Delta_j(f)|^2\Big)^{\frac{1}{2}}\Big\|_{L^p}. \tag{6.4.60}$$

Proof. We fix $\Phi \in \mathcal{S}(\mathbf{R}^n)$ with integral equal to 1 and we take $f \in H^p \cap L^1$ and $M \in \mathbf{Z}^+$. Let r_j be the Rademacher functions introduced in Appendix C.1. We begin with the estimate

$$\Big|\Phi_\varepsilon * \sum_{j=-M}^{M} r_j(\omega)\Delta_j(f)\Big| \le \sup_{\varepsilon>0}\Big|\Phi_\varepsilon * \sum_{j=-M}^{M} r_j(\omega)\Delta_j(f)\Big|,$$

which holds since $\{\Phi_\varepsilon\}_{\varepsilon>0}$ is an approximate identity. We raise this inequality to the power p, we integrate over $x \in \mathbf{R}^n$ and $\omega \in [0,1]$, and we use the maximal function characterization of H^p [Theorem 6.4.4 (a)] to obtain

$$\int_0^1\int_{\mathbf{R}^n}\Big|\sum_{j=-M}^{M} r_j(\omega)\Delta_j(f)(x)\Big|^p dx\,d\omega \le C_{p,n}^p\Big\|\sum_{j=-M}^{M} r_j(\omega)\Delta_j(f)\Big\|_{H^p}^p.$$

The lower inequality for the Rademacher functions in Appendix C.2 gives

$$\int_{\mathbf{R}^n}\Big(\sum_{j=-M}^{M}|\Delta_j(f)(x)|^2\Big)^{\frac{p}{2}} dx \le C_p^pC_{p,n}^p\Big\|\sum_{j=-M}^{M} r_j(\omega)\Delta_j(f)\Big\|_{H^p}^p,$$

where the second estimate is a consequence of Theorem 6.4.14 (we only need the scalar version here) since the kernel

$$\sum_{k=-M}^{M} r_k(\omega)\Psi_{2^{-k}}(x)$$

satisfies (6.4.47) and (6.4.48) with constants A and B depending only on n and Ψ (and, in particular, independent of M). We have now proved that

$$\Big\|\Big(\sum_{j=-M}^{M}|\Delta_j(f)|^2\Big)^{\frac{1}{2}}\Big\|_{L^p} \le C_{n,p,\Psi}\|f\|_{H^p},$$

from which (6.4.58) follows directly by letting $M \to \infty$. We have now established (6.4.58) for $f \in H^p \cap L^1$. Using density, we can extend this estimate to all $f \in H^p$.

To obtain the converse estimate, for $r \in \{0,1,2\}$ we consider the sets

$$3\mathbf{Z} + r = \{3k + r : \ k \in \mathbf{Z}\}$$

and we observe that for $j, k \in 3\mathbf{Z}+r$ the Fourier transforms of $\Delta_j(f)$ and $\Delta_k(f)$ are disjoint if $j \neq k$. We fix a Schwartz function η whose Fourier transform is compactly supported away from the origin so that for all $j, k \in 3\mathbf{Z}$ we have

$$\Delta_j^\eta \Delta_k = \begin{cases} \Delta_j & \text{when } j = k \\ 0 & \text{when } j \neq k, \end{cases} \tag{6.4.61}$$

where Δ_j^η is the Littlewood-Paley operator associated with η [i.e., $\Delta_j^\eta(f) = f * \eta_{2^{-j}}$]. It follows from Theorem 6.4.14 that the map

$$\{f_j\}_{j\in\mathbf{Z}} \to \sum_{j\in 3\mathbf{Z}} \Delta_j^\eta(f_j)$$

maps $H^p(\mathbf{R}^n, \ell^2)$ into $H^p(\mathbf{R}^n)$. Indeed, we can see easily that

$$\Big|\sum_{j\in 3\mathbf{Z}} \widehat{\eta}(2^{-j}\xi)\Big| \leq B$$

and

$$\sum_{j\in 3\mathbf{Z}} \big|\partial^\alpha\big(2^{jn}\eta(2^j x)\big)\big| \leq A_\alpha |x|^{-n-|\alpha|}$$

for all multiindices α and for constants depending only on B and A_α. Applying this estimate with $f_j = \Delta_j(f)$ and using (6.4.61) yields the estimate

$$\Big\|\sum_{j\in 3\mathbf{Z}} \Delta_j(f)\Big\|_{H^p} \leq C_{n,p,\Psi} \Big\|\Big(\sum_{j\in 3\mathbf{Z}} |\Delta_j(f)|^2\Big)^{\frac{1}{2}}\Big\|_{L^p}$$

for all distributions f that satisfy (6.4.59). Applying the same idea with $3\mathbf{Z}+1$ and $3\mathbf{Z}+2$ replacing $3\mathbf{Z}$ and summing the corresponding estimates gives

$$\Big\|\sum_{j\in \mathbf{Z}} \Delta_j(f)\Big\|_{H^p} \leq 3^{\frac{1}{p}} C_{n,p,\Psi} \Big\|\Big(\sum_{j\in \mathbf{Z}} |\Delta_j(f)|^2\Big)^{\frac{1}{2}}\Big\|_{L^p}.$$

But note that $f - \sum_j \Delta_j(f)$ is equal to a polynomial $Q(x)$ since its Fourier transform is supported at the origin. It follows that $f - Q$ lies in H^p and satisfies (6.4.60).

□

We will show in the next section that the characterization of H^p is independent of the choice of the function Ψ.

6.4.g. The Proof of Lemma 6.4.5

It remains to prove Lemma 6.4.5.

Proof. We start with a smooth function ζ supported in $[0,1]$ that satisfies

$$\begin{aligned} 0 \le \zeta(s) &\le \frac{s^m}{m!} && \text{for all } 0 \le s \le 1 \\ \zeta(s) &= \frac{s^m}{m!} && \text{for all } 0 \le s \le \frac{1}{2}, \\ \zeta(s) &= 0 && \text{for all } \frac{8}{9} \le s \le 1. \end{aligned}$$

We claim that the identity

$$\Psi = \int_0^1 \left[(-1)^{m+1}\zeta(s)\, \frac{d^{m+1}(\overbrace{\Phi_s * \cdots * \Phi_s}^{m+2 \text{ terms}} * \Psi)}{ds^{m+1}} - \frac{d^{m+1}\zeta(s)}{ds^{m+1}} (\overbrace{\Phi_s * \cdots * \Phi_s}^{m+2 \text{ terms}} * \Psi) \right] ds$$

is valid. Indeed, to prove it we integrate by parts $m+1$ times. Since ζ has vanishing derivatives up to order $\le m-1$ at both endpoints, the only nonzero boundary term results from the $(m+1)$st integration by parts and is equal to

$$\lim_{s\to 0} \left[\frac{d^m\zeta(s)}{ds^m} (\Phi_s * \cdots * \Phi_s * \Psi) \right] = \lim_{s\to 0} (\Phi * \cdots * \Phi)_s * \Psi = \Psi\,.$$

The last equality is due to the fact that the family $\{(\Phi * \cdots * \Phi)_s\}_{s>0}$ is an approximate identity as $s \to 0$ since Φ has integral 1 by assumption. This proves the identity in question. Motivated by this identity, we define

$$\Theta^{(s)} = (-1)^{m+1}\zeta(s)\, \frac{d^{m+1}(\Omega_s * \Psi)}{ds^{m+1}} - \frac{d^{m+1}\zeta(s)}{ds^{m+1}} (\Omega_s * \Psi)\,, \tag{6.4.62}$$

where $\Omega = \Phi * \cdots * \Phi$ [$(m+1)$ times] and we observe that (6.4.15) holds for this choice of $\Theta^{(s)}$. We now prove estimate (6.4.16). For the second term on the right in (6.4.62), we note that the $(m+1)$st derivative of $\zeta(s)$ is zero for s outside the interval $[\frac{1}{2}, \frac{8}{9}]$. We use this observation in the following calculation to write

$$\begin{aligned} &\int_{\mathbf{R}^n} (1+|x|)^m \Big| \frac{d^{m+1}\zeta(s)}{ds^{m+1}} \Big| \, |\Omega_s * \Psi(x)| \, dx \\ &\le C\, \chi_{[\frac{1}{2},\frac{8}{9}]}(s) \int_{\mathbf{R}^n} (1+|x|)^m \left[\int_{\mathbf{R}^n} \tfrac{1}{s^n} |\Omega(\tfrac{x-y}{s})| \, |\Psi(y)| \, dy \right] dx \\ &\le \chi_{[\frac{1}{2},\frac{8}{9}]}(s) \int_{\mathbf{R}^n} (1+|x|)^m \int_{\mathbf{R}^n} \tfrac{1}{s^n} |\Omega(\tfrac{x-y}{s})| \, |\Psi(y)| \, dy \, dx \\ &\le C\, \chi_{[\frac{1}{2},\frac{8}{9}]}(s) \int_{\mathbf{R}^n} \int_{\mathbf{R}^n} (1+|x+sy|)^m |\Omega(x)| \, |\Psi(y)| \, dy \, dx \\ &\le C\, \chi_{[\frac{1}{2},\frac{8}{9}]}(s) \int_{\mathbf{R}^n} \int_{\mathbf{R}^n} (1+|x|)^m |\Omega(x)| \, (1+|y|)^m |\Psi(y)| \, dy \, dx \\ &\le C' \, s^m \, \mathfrak{N}_m(\Psi)\,. \end{aligned}$$

To obtain a similar estimate for the first term on the right in (6.4.62), we argue as follows:

$$
\begin{aligned}
&\int_{\mathbf{R}^n} (1+|x|)^m |\zeta(s)| \left| \frac{d^{m+1}(\Omega_s * \Psi)}{ds^{m+1}}(x) \right| dx \\
&= \int_{\mathbf{R}^n} (1+|x|)^m \, |\zeta(s)| \left| \frac{d^{m+1}}{ds^{m+1}} \int_{\mathbf{R}^n} \Omega(\tfrac{x-y}{s}) \Psi(y)\, dy \right| dx \\
&= \int_{\mathbf{R}^n} (1+|x|)^m \, |\zeta(s)| \left| \int_{\mathbf{R}^n} \Omega(y) \frac{d^{m+1}\Psi(x-sy)}{ds^{m+1}}\, dy \right| dx \\
&\le \int_{\mathbf{R}^n} (1+|x|)^m \, |\zeta(s)| \int_{\mathbf{R}^n} |\Omega(y)| \Big[\sum_{|\alpha| \le m+1} |\partial^\alpha \Psi(x-sy)| \, |y|^{m+1} \Big] \, dy\, dx \\
&\le C|\zeta(s)| \int_{\mathbf{R}^n} \int_{\mathbf{R}^n} (1+|x+sy|)^m |\Omega(y)| \sum_{|\alpha| \le m+1} |\partial^\alpha \Psi(x)| \, |y|^{m+1}\, dy\, dx \\
&\le C s^m \int_{\mathbf{R}^n} (1+s|y|)^m \, |\Omega(y)| \, |y|^{m+1}\, dy \int_{\mathbf{R}^n} (1+|x|)^m \sum_{|\alpha| \le m+1} |\partial^\alpha \Psi(x)|\, dx \\
&\le C'' s^m \, \mathfrak{N}_m(\Psi)
\end{aligned}
$$

since $s \le 1$. We now set $C_0 = C' + C''$ and we note that C_0 only depends on m and Φ. These calculations prove estimate (6.4.16).

□

Exercises

6.4.1. Prove that if v is a bounded tempered distribution and h_1, h_2 are in $\mathcal{S}(\mathbf{R}^n)$, then

$$(h_1 * h_2) * v = h_1 * (h_2 * v).$$

6.4.2. (a) Show that H^1 norm remains invariant under the L^1 dilation $f \to f_t = \frac{1}{t^n} f(\frac{\cdot}{t})$.
(b) Show that H^p norm remains invariant under the L^p dilation $f \to t^{n-n/p} f_t$ interpreted in the sense of distributions.

6.4.3. (a) Show that the function $f(x) = \chi_{(-1,0)} - \chi_{(0,1)}$ lies in the Hardy space $H^1(\mathbf{R})$.
(b) Determine all $0 < p \le 1$ for which f lies in $H^p(\mathbf{R})$.
[*Hint:* Show that $M(f;\Phi)$ lies in L^1 for a smooth function Φ supported in the $(-1,1)$ with integral 1.]

6.4.4. Show that the function $\psi(s)$ defined in (6.4.18) is continuous and integrable over $[1,\infty)$, decays faster than the reciprocal of any polynomial, and satisfies

(6.4.17), that is,

$$\int_1^\infty s^k\,\psi(s)\,ds = \begin{cases} 1 & \text{if } k=0 \\ 0 & \text{if } k=1,2,3,\dots. \end{cases}$$

[*Hint:* Apply Cauchy's theorem over a suitable contour.]

6.4.5. Let $0<a<\infty$ be fixed. Show that a bounded tempered distribution f lies in H^p if and only if the nontangential Poisson maximal function

$$M_a^*(f;P)(x) = \sup_{t>0} \sup_{\substack{y\in\mathbf{R}^n \\ |y-x|\le at}} |(P_t*f)(y)|$$

lies in L^p and in this case we have $\big\|f\big\|_{H^p} \approx \big\|M_a^*(f;P)\big\|_{L^p}$.
[*Hint:* Observe that $M(f;P)$ can be replaced with $M_a^*(f;P)$ in the proof of parts (a) and (e) of Theorem 6.4.4).]

6.4.6. Show that $H^p(\mathbf{R}^n,\ell^2) = L^p(\mathbf{R}^n,\ell^2)$ whenever $1<p<\infty$ and that $H^1(\mathbf{R}^n,\ell^2)$ is contained in $L^1(\mathbf{R}^n,\ell^2)$.

6.4.7. Show that the space of all Schwartz functions whose Fourier transform is supported away from a neighborhood of the origin is dense in H^p.
[*Hint:* Use the square function characterization of H^p.]

6.4.8. (a) Suppose that $f\in H^p(\mathbf{R}^n)$ for some $0<p\le 1$ and Φ in $\mathcal{S}(\mathbf{R}^n)$. Then show that for all $t>0$ the function Φ_t*f belongs to $L^r(\mathbf{R}^n)$ for all $p\le r\le\infty$. Find an estimate for the L^r norm of Φ_t*f in terms of $\big\|f\big\|_{H^p}$ and $t>0$.
(b) Let $0<p\le 1$. Show that there exists a constant $C_{n,p}$ such that for all f be in $H^p(\mathbf{R}^n)$ we have

$$|\widehat{f}(\xi)| \le C_{n,p}|\xi|^{\frac{n}{p}-n}\big\|f\big\|_{H^p}$$

[*Hint:* Obtain that $\big\|\Phi_t*f\big\|_{L^1} \le Ct^{-n/p+n}\big\|f\big\|_{H^p}$ Using an idea from the proof of Proposition 6.4.7.]

6.4.9. For a sequence of tempered distributions $\vec{f} = \{f_j\}_j$, define the following variation of the grand maximal function

$$\widetilde{\boldsymbol{\mathcal{M}}}_N(\vec{f})(x) = \sup_{\{\varphi_j\}_j\in\boldsymbol{\mathcal{F}}_N} \sup_{\varepsilon>0} \sup_{\substack{y\in\mathbf{R}^n \\ |y-x|<\varepsilon}} \Big(\sum_j |((\varphi_j)_\varepsilon * f_j)(y)|^2\Big)^{\frac12},$$

where $N\ge[\frac{n}{p}]+1$ and

$$\boldsymbol{\mathcal{F}}_N = \Big\{\{\varphi_j\}_j\in\mathcal{S}(\mathbf{R}^n):\ \sum_j \mathfrak{N}_N(\varphi_j)\le 1\Big\}.$$

Show that for all sequences of tempered distributions $\vec{f} = \{f_j\}_j$ we have

$$\big\|\widetilde{\mathcal{M}}_N(\vec{f})\big\|_{L^p(\mathbf{R}^n,\ell^2)} \approx \big\|\mathcal{M}_N(\vec{f})\big\|_{L^p(\mathbf{R}^n,\ell^2)}$$

with constants depending only on n and p.
[*Hint:* Fix Φ in $\mathcal{S}(\mathbf{R}^n)$ with integral 1. Using Lemma 6.4.5, write

$$(\varphi_j)_t(y) = \int_0^1 ((\Theta_j^{(s)})_t * \Phi_{ts})(y)\,ds$$

and apply a vector-valued extension of the proof of part (d) of Theorem 6.4.4 to obtain the pointwise estimate

$$\widetilde{\mathcal{M}}_N(\vec{f}) \le C_{n,p} M_m^{**}(\vec{f};\Phi)\,,$$

where $m > n/p$.]

6.5. Besov-Lipschitz and Triebel-Lizorkin Spaces*

The main achievement of the previous sections was the characterization of Sobolev, Lipschitz, and Hardy spaces using the Littlewood-Paley operators Δ_j. These remarkable characterizations provide a solid motivation for the introduction of a multiple scale of function spaces defined in terms of expressions involving the operators Δ_j. This scale furnishes a general framework within which we can launch a study of function spaces from a unified perspective.

We have encountered two kinds of expressions involving the operators Δ_j in the characterizations of the function spaces we obtained in the previous sections. Some spaces were characterized by some L^p norm of the Littlewood-Paley square function

$$\Big(\sum_j |2^{j\alpha}\Delta_j(f)|^2\Big)^{\frac{1}{2}}$$

and other spaces were characterized by the ℓ^q norm of the sequence of norms $\big\|2^{j\alpha}\Delta_j(f)\big\|_{L^p}$. Examples of spaces in the first category are the homogeneous Sobolev spaces, Hardy spaces, and naturally L^p spaces. We have only studied one example of spaces in the second category, Lipschitz spaces, in which case $q = \infty$. These examples motivate the two basic categories of spaces we plan to investigate, called the Triebel-Lizorkin and Besov-Lipschitz spaces, respectively.

6.5.a. Introduction of Function Spaces

Before we give the pertinent definitions, we recall the setup we developed in Section 6.2 and used in Section 6.3. Throughout this section we fix a radial Schwartz function Ψ on $\mathbf{R}^n$ whose Fourier transform is nonnegative, is supported in the annulus $\frac{1}{2} \le |\xi| \le 2$, and satisfies

$$\sum_{j\in\mathbf{Z}} \widehat{\Psi}(2^{-j}\xi) = 1 \qquad \xi \neq 0. \tag{6.5.1}$$

Associated with this bump, we define the Littlewood-Paley operators $\Delta_j = \Delta_j^\Psi$ given by multiplication on the Fourier transform with the function $\widehat{\Psi}(2^{-j}\xi)$. We also define a Schwartz function Φ so that

$$\widehat{\Phi}(\xi) = \begin{cases} \sum_{j\le 0} \widehat{\Psi}(2^{-j}\xi) & \text{when } \xi \neq 0, \\ 1 & \text{when } \xi = 0. \end{cases} \tag{6.5.2}$$

Note that $\widehat{\Phi}(\xi)$ is equal to 1 for $|\xi| \le 1$ and vanishes when $|\xi| \ge 2$. It follows from these definitions that

$$S_0 + \sum_{j=0}^{\infty} \Delta_j = I, \tag{6.5.3}$$

where $S_0 = S_0^\Psi$ is the operator given by convolution with the bump Φ and the convergence of the series in (6.5.3) is in $\mathcal{S}'(\mathbf{R}^n)$. Moreover, we also have the identity

$$\sum_{j\in\mathbf{Z}} \Delta_j = I, \tag{6.5.4}$$

where the convergence of the series in (6.5.4) is in the sense of $\mathcal{S}'(\mathbf{R}^n)/\mathcal{P}$.

Definition 6.5.1. Let $\alpha \in \mathbf{R}$ and $0 < p, q \le \infty$. For $f \in \mathcal{S}'(\mathbf{R}^n)$ we set

$$\|f\|_{B^p_{\alpha,q}} = \|S_0(f)\|_{L^p} + \Big(\sum_{j=1}^{\infty} (2^{j\alpha}\|\Delta_j(f)\|_{L^p})^q\Big)^{\frac{1}{q}}$$

with the obvious modification when $p, q = \infty$. When $p, q < \infty$ we also set

$$\|f\|_{F^p_{\alpha,q}} = \|S_0(f)\|_{L^p} + \Big\|\Big(\sum_{j=1}^{\infty} (2^{j\alpha}|\Delta_j(f)|)^q\Big)^{\frac{1}{q}}\Big\|_{L^p}.$$

The space of all tempered distributions f for which the quantity $\|f\|_{B^p_{\alpha,q}}$ is finite is called the (inhomogeneous) *Besov-Lipschitz* space with indices α, p, q and is denoted by $B^p_{\alpha,q}$. The space of all tempered distributions f for which the quantity $\|f\|_{F^p_{\alpha,q}}$ is finite is called the (inhomogeneous) *Triebel-Lizorkin* space with indices α, p, q and is denoted by $F^p_{\alpha,q}$.

We now define the corresponding homogeneous versions of these spaces. For an element f of $\mathcal{S}'(\mathbf{R}^n)/\mathcal{P}$ we let

$$\|f\|_{\dot{B}^p_{\alpha,q}} = \Big(\sum_{j\in\mathbf{Z}} (2^{j\alpha}\|\Delta_j(f)\|_{L^p})^q\Big)^{\frac{1}{q}}$$

and

$$\big\|f\big\|_{\dot F^{p}_{\alpha,q}} = \Big\|\Big(\sum_{j\in\mathbf{Z}} (2^{j\alpha}|\Delta_j(f)|)^q\Big)^{\frac{1}{q}}\Big\|_{L^p}\,.$$

The space of all f in $\mathcal{S}'(\mathbf{R}^n)/\mathcal{P}$ for which the quantity $\big\|f\big\|_{\dot B^{p}_{\alpha,q}}$ is finite is called the (homogeneous) *Besov-Lipschitz* space with indices α, p, q and is denoted by $\dot B^{p}_{\alpha,q}$. The space of f in $\mathcal{S}'(\mathbf{R}^n)/\mathcal{P}$ for which the quantity $\big\|f\big\|_{\dot F^{p}_{\alpha,q}}$ is finite is called the (homogeneous) *Triebel-Lizorkin* space with indices α, p, q and is denoted by $\dot F^{p}_{\alpha,q}$.

We now make several observations related to these definitions. First we note that the expressions $\|\cdot\|_{\dot F^{p}_{\alpha,q}}$, $\|\cdot\|_{F^{p}_{\alpha,q}}$, $\|\cdot\|_{\dot B^{p}_{\alpha,q}}$, and $\|\cdot\|_{B^{p}_{\alpha,q}}$ are built in terms of L^p quasi-norms of ℓ^q quasi-norms of $2^{j\alpha}\Delta_j$ or ℓ^q quasi-norms of L^p quasi-norms of the same expressions. As a result, we can see that these quantities satisfy the triangle inequality with a constant (which may be taken to be 1 when $1 \le p, q < \infty$). To determine whether these quantities are indeed quasi-norms, we need to check whether the following property holds:

$$\big\|f\big\|_X = 0 \implies f = 0\,, \tag{6.5.5}$$

where X is one of the $\dot F^{p}_{\alpha,q}$, $F^{p}_{\alpha,q}$, $\dot B^{p}_{\alpha,q}$, and $B^{p}_{\alpha,q}$. Since these are spaces of distributions, the identity $f = 0$ in (6.5.5) should be interpreted in the sense of distributions. If $\big\|f\big\|_X = 0$ for some inhomogeneous space X, then $S_0(f) = 0$ and $\Delta_j(f) = 0$ for all $j \ge 0$. Using (6.5.3), we conclude that $f = 0$; thus the quantities $\|\cdot\|_{F^{p}_{\alpha,q}}$ and $\|\cdot\|_{B^{p}_{\alpha,q}}$ are indeed quasi-norms. Let us investigate what happens when $\big\|f\big\|_X = 0$ for some homogeneous space X. In this case we must have $\Delta_j(f) = 0$, and using (6.5.4) we conclude that $\widehat{f}$ must be supported at the origin. Proposition 2.4.1 yields that f must be a polynomial and thus f must be zero (since distributions whose difference is a polynomial are identified in homogeneous spaces).

Remark 6.5.2. We interpret the previous definition in certain cases. According to what we have seen so far, we have

$$\begin{array}{rclr}
\dot F^{p}_{0,2} &\approx& F^{p}_{0,2} \approx L^p & 1<p<\infty\\
\dot F^{p}_{0,2} &\approx& H^p & 0<p\le 1\\
F^{p}_{s,2} &\approx& L^p_s & 1<p<\infty\\
\dot F^{p}_{s,2} &\approx& \dot L^p_s & 1<p<\infty\\
B^{\infty}_{\gamma,\infty} &\approx& \Lambda_\gamma & \gamma>0\\
\dot B^{\infty}_{\gamma,\infty} &\approx& \dot\Lambda_\gamma & \gamma>0\,,
\end{array}$$

where $\approx$ indicates that the corresponding norms are equivalent.

To avoid unnecessary complications, we have purposely avoided to define the Triebel-Lizorkin spaces when $p = \infty$. When $p = \infty$, $\dot F^{\infty}_{0,q}$ can be defined as the space of all $f \in \mathcal{S}'/\mathcal{P}$, tempered distributions modulo polynomials, that satisfy

$$\|f\|_{\dot{F}^{\infty}_{\alpha,q}} = \sup_{Q \text{ dyadic cube}} \int_Q \frac{1}{|Q|}\Big(\sum_{j=-\log_2 \ell(Q)}^{\infty} (2^{j\alpha}|\Delta_j(f)|)^q\Big)^{\frac{1}{q}} < \infty .$$

In the particular case $q = 2$ and $\alpha = 0$, the space obtained this way is called BMO. This space, which will be introduced and studied in the next chapter, serves as a substitute for L^∞ and plays a fundamental role in analysis. It should be now clear that several important spaces in analysis can be thought as members of the scale of Triebel-Lizorkin spaces.

We also remark that contrary to some authors notation, we denote Besov-Lipschitz and Triebel-Lizorkin spaces by $B^p_{\alpha,q}$ and $F^p_{\alpha,q}$ to keep the placement of the upper and lower indices in accordance with those of the previously defined Lebesgue, Sobolev, Lipschitz, and Hardy spaces. We would like to acknowledge, however, that most authors who work on the subject prefer the notation $B^{\alpha,q}_p$ and $F^{\alpha,q}_p$ instead of ours $B^p_{\alpha,q}$ and $F^p_{\alpha,q}$.

6.5.b. Equivalence of Definitions

It is not clear from the definitions whether the finiteness of the quasi-norms defining the spaces $B^p_{\alpha,q}$, $F^p_{\alpha,q}$,$\dot{B}^p_{\alpha,q}$, and $\dot{F}^p_{\alpha,q}$ depends on the choice of the function Ψ (recall Φ is determined by Ψ). We will show that if Ω is another function that satisfies (6.5.1) and Θ is defined in terms of Ω in the same way that Φ is defined in terms of Ψ, [i.e., via (6.5.2)], then the norms defined in 6.5.1 with respect to the pairs (Φ, Ψ) and (Θ, Ω) are comparable. To prove this we will need the following lemma.

Lemma 6.5.3. *Let $0 < c_0 < \infty$ and $0 < r < \infty$. Then there exist constants C_1 and C_2 (that depend only on n, c_0, and r) such that for all $t > 0$ and for all C^1 functions u on $\mathbf{R}^n$ whose Fourier transform is supported in the ball $|\xi| \le c_0 t$ and that satisfy $|u(z)| \le B(1+|z|)^{\frac{n}{r}}$ for some $B > 0$ we have the estimate*

$$\sup_{z\in\mathbf{R}^n} \frac{1}{t}\frac{|\nabla u(x-z)|}{(1+t|z|)^{\frac{n}{r}}} \le C_1 \sup_{z\in\mathbf{R}^n} \frac{|u(x-z)|}{(1+t|z|)^{\frac{n}{r}}} \le C_2 M(|u|^r)(x)^{\frac{1}{r}} , \tag{6.5.6}$$

where M denotes the Hardy-Littlewood maximal operator. (The constants C_1 and C_2 are independent of B.)

Proof. Select a Schwartz function ψ whose Fourier transform is supported in the ball $|\xi| \le 2c_0$ and is equal to 1 on the smaller ball $|\xi| \le c_0$. Then $\widehat{\psi}(\frac{\xi}{t})$ is equal to 1 on the support of $\widehat{u}$ and we can write

$$u(x-z) = \int_{\mathbf{R}^n} t^n \psi(t(x-z-y))u(y)\,dy .$$

Taking partial derivatives and using that ψ is a Schwartz function we obtain

$$|\nabla u(x-z)| \le C_N \int_{\mathbf{R}^n} t^{n+1}(1+t|x-z-y|)^{-N}|u(y)|\,dy\,,$$

where N is arbitrarily large. Using that for all $x,y,z \in \mathbf{R}^n$ we have

$$1 \le (1+t|x-z-y|)^{\frac{n}{r}} \frac{(1+t|z|)^{\frac{n}{r}}}{(1+t|x-y|)^{\frac{n}{r}}}\,,$$

we obtain

$$\frac{1}{t}\frac{|\nabla u(x-z)|}{(1+t|z|)^{\frac{n}{r}}} \le C_N \int_{\mathbf{R}^n} t^n (1+t|x-z-y|)^{\frac{n}{r}-N} \frac{|u(y)|}{(1+t|x-y|)^{\frac{n}{r}}}\,dy\,,$$

from which the first estimate in (6.5.6) follows easily.

Let $|y| \le \delta$ for some $\delta > 0$ to be chosen later. We now use the mean value theorem to write

$$u(x-z) = (\nabla u)(x-z-\xi_y)\cdot y + u(x-z-y)$$

for some ξ_y satisfying $|\xi_y| \le |y| \le \delta$. This implies that

$$|u(x-z)| \le \sup_{|w|\le|z|+\delta} |(\nabla u)(x-w)|\,\delta + |u(x-z-y)|.$$

Raising to the power r, averaging over the ball $|y| \le \delta$, and then raising to the power $\frac{1}{r}$ yields

$$|u(x-z)| \le c_r \left[\sup_{|w|\le|z|+\delta} |(\nabla u)(x-w)|\,\delta + \left(\frac{1}{v_n\delta^n}\int_{|y|\le\delta} |u(x-z-y)|^r\,dy\right)^{\frac{1}{r}}\right]$$

with $c_r = \max(2^{1/r}, 2^r)$. Here v_n is the volume of the unit ball in $\mathbf{R}^n$. Then

$$\frac{|u(x-z)|}{(1+t|z|)^{\frac{n}{r}}} \le c_r \left[\sup_{|w|\le|z|+\delta} \frac{|(\nabla u)(x-w)|}{(1+t|z|)^{\frac{n}{r}}}\,\delta\, \frac{\left(\frac{1}{v_n\delta^n}\int_{|y|\le\delta+|z|} |u(x-y)|^r\,dy\right)^{\frac{1}{r}}}{(1+t|z|)^{\frac{n}{r}}}\right].$$

We now set $\delta = \varepsilon/t$ for some $\varepsilon \le 1$. Then we have

$$|w| \le |z| + \frac{\varepsilon}{t} \implies \frac{1}{1+t|z|} \le \frac{2}{1+t|w|}$$

and we can use this to obtain the estimate

$$\frac{|u(x-z)|}{(1+t|z|)^{\frac{n}{r}}} \le c_{r,n} \left[\sup_{w\in\mathbf{R}^n} \frac{1}{t}\frac{|(\nabla u)(x-w)|}{(1+t|w|)^{\frac{n}{r}}}\,\varepsilon\, \frac{\left(\frac{t^n}{v_n\varepsilon^n}\int_{|y|\le\frac{1}{t}+|z|} |u(x-y)|^r\,dy\right)^{\frac{1}{r}}}{(1+t|z|)^{\frac{n}{r}}}\right]$$

with $c_{r,n} = c_r 2^{n/r} = \max(2^{1/r}, 2^r)2^{n/r}$. It follows that

$$\sup_{z\in\mathbf{R}^n} \frac{|u(x-z)|}{(1+t|z|)^{\frac{n}{r}}} \le c_{r,n}\left[\sup_{w\in\mathbf{R}^n} \frac{1}{t}\frac{|(\nabla u)(x-w)|}{(1+t|w|)^{\frac{n}{r}}}\,\varepsilon + \varepsilon^{-\frac{n}{r}} M(|u|^r)(x)^{\frac{1}{r}}\right].$$

Taking $\varepsilon = \frac{1}{2}(c_{r,n} C_1)^{-1}$, where C_1 is the constant in (6.5.6), we obtain the second estimate in (6.5.6) with $C_2 = 2\varepsilon^{-n/r}$. At this step we used the hypothesis that

$$\sup_{z\in\mathbf{R}^n} \frac{|u(x-z)|}{(1+t|z|)^{\frac{n}{r}}} \le \sup_{z\in\mathbf{R}^n} \frac{B(1+|x|+|z|)^{\frac{n}{r}}}{(1+t|z|)^{\frac{n}{r}}} < \infty .$$

□

Remark 6.5.4. The reader is reminded that $\widehat{u}$ in Lemma 6.5.3 may not be a function; for example, this is the case when u is a polynomial (say of degree $[n/r]$). If $\widehat{u}$ were an integrable function, then u would be a bounded function, and condition $|u(x)| \le B(1+|x|)^{\frac{n}{r}}$ would not be needed.

We now return to a point alluded to earlier, that changing the bump Ψ by another bump Ω that satisfies similar properties yields equivalent norms for the function spaces given in Definition 6.5.1. Suppose that Ω is another bump that whose Fourier transform is supported in the annulus $\frac{1}{2} \le |\xi| \le 2$ and that satisfies (6.5.1). The support properties of Ψ and Ω imply the identity

$$\Delta_j^{\Omega} = \Delta_j^{\Omega}(\Delta_{j-1}^{\Psi} + \Delta_j^{\Psi} + \Delta_{j+1}^{\Psi}) . \tag{6.5.7}$$

Let $0 < p < \infty$. Then for some $r < p$ and $N > \frac{n}{r} + n$ we have

$$\begin{aligned} \big|\Delta_j^{\Omega}\Delta_j^{\Psi}(f)(x)\big| &\le C_{N,\Omega} \int_{\mathbf{R}^n} \frac{\big|\Delta_j^{\Psi}(f)(x-z)\big|}{(1+2^j|z|)^{\frac{n}{r}}} \frac{2^{jn}dz}{(1+2^j|z|)^{N-\frac{n}{r}}} \\ &\le C_{N,\Omega} \sup_{z\in\mathbf{R}^n} \frac{\big|\Delta_j^{\Psi}(f)(x-z)\big|}{(1+2^j|z|)^{\frac{n}{r}}} \int_{\mathbf{R}^n} \frac{2^{jn}dz}{(1+2^j|z|)^{N-\frac{n}{r}}} \\ &\le C_{N,r,\Omega}(M(|\Delta_j^{\Psi}(f)|^r)(x)^{\frac{1}{r}} \end{aligned} \tag{6.5.8}$$

in view of Lemma 6.5.3. The latter is easily shown bounded on L^p for all $0 < p < \infty$. The same estimate is also valid for $\Delta_j^{\Omega}\Delta_{j\pm1}^{\Psi}$. Using estimate (6.5.8), the analogous estimate for $\Delta_j^{\Omega}\Delta_{j\pm1}^{\Psi}$, identity (6.5.7), and the boundedness of the Hardy-Littlewood maximal operator on $L^{p/r}$, we obtain that the homogeneous Besov-Lipschitz norm defined using the bump Ψ is controlled by a constant multiple of that defined using Ψ. A similar argument applies for the inhomogeneous Besov-Lipschitz norms. The equivalence constants depend on Ψ, Ω, n, p, q, and α.

The corresponding equivalence of norms for Triebel-Lizorkin spaces is more difficult to obtain, and it will be a consequence of the characterization of these spaces proved later.

Definition 6.5.5. For $b > 0$ and $j \in \mathbf{R}$ we introduce the notation

$$M_{b,j}^{**}(f;\Psi)(x) = \sup_{y\in\mathbf{R}^n} \frac{|(\Psi_{2^{-j}} * f)(x-y)|}{(1+2^j|y|)^b}$$

so that we have

$$M_b^{**}(f;\Psi) = \sup_{t>0} M_{b,t}^{**}(f;\Psi) ,$$

in accordance with the notation in the previous section. The function $M_b^{**}(f;\Psi)$ is called the *Peetre maximal function of f (with respect to Ψ).*

We clearly have

$$|\Delta_j^\Psi(f)| \le M_{b,j}^{**}(f;\Psi)\,,$$

but the next result shows that a certain converse is also valid.

Theorem 6.5.6. *Let $b > n(\min(p,q))^{-1}$ and $0 < p,q < \infty$. Let Ψ and Ω be Schwartz functions whose Fourier transforms are supported in the annulus $\frac{1}{2} \le |\xi| \le 2$ and satisfy (6.5.1). Then we have*

$$\Big\|\Big(\sum_{j\in\mathbf{Z}} |2^{j\alpha} M_{b,j}^{**}(f;\Omega)|^q\Big)^{\frac{1}{q}}\Big\|_{L^p} \le C\Big\|\Big(\sum_{j\in\mathbf{Z}} |2^{j\alpha}\Delta_j^\Psi(f)|^q\Big)^{\frac{1}{q}}\Big\|_{L^p} \tag{6.5.9}$$

for all $f \in \mathcal{S}'(\mathbf{R}^n)$, where $C = C_{\alpha,p,q,n,b,\Psi,\Omega}$.

Proof. We start with a Schwartz function Θ whose Fourier transform is nonnegative, supported in the annulus $\frac{3}{5} \le |\xi| \le \frac{5}{3}$, and satisfies

$$\sum_{j\in\mathbf{Z}} \widehat{\Theta}(2^{-j}\xi)^2 = 1, \qquad \xi \in \mathbf{R}^n \setminus \{0\}\,. \tag{6.5.10}$$

Using (6.5.10), we have

$$\Omega_{2^{-k}} * f = \sum_{j\in\mathbf{Z}} (\Omega_{2^{-k}} * \Theta_{2^{-j}}) * (\Theta_{2^{-j}} * f)\,.$$

It follows that

$$\begin{aligned}
&2^{k\alpha}\frac{|(\Omega_{2^{-k}} * f)(x-z)|}{(1+2^k|z|)^b}\\
&\le \sum_{j\in\mathbf{Z}} 2^{k\alpha}\int_{\mathbf{R}^n} |(\Omega_{2^{-k}} * \Theta_{2^{-j}})(y)|\,\frac{|(\Theta_{2^{-j}} * f)(x-z-y)|}{(1+2^k|z|)^b}\,dy\\
&= \sum_{j\in\mathbf{Z}} 2^{k\alpha}\int_{\mathbf{R}^n} 2^{kn}|(\Omega * \Theta_{2^{-(j-k)}})(2^k y)|\,\frac{(1+2^j|y+z|)^b}{(1+2^k|z|)^b}\,\frac{|(\Theta_{2^{-j}} * f)(x-z-y)|}{(1+2^j|y+z|)^b}\,dy\\
&\le \sum_{j\in\mathbf{Z}} 2^{k\alpha}\int_{\mathbf{R}^n} |(\Omega * \Theta_{2^{-(j-k)}})(y)|\,\frac{(1+2^j|2^{-k}y+z|)^b}{(1+2^k|z|)^b}\,\frac{|(\Theta_{2^{-j}} * f)(x-z-y)|}{(1+2^j|y+z|)^b}\,dy\\
&\le \sum_{j\in\mathbf{Z}} 2^{(k-j)\alpha}\int_{\mathbf{R}^n} |(\Omega * \Theta_{2^{-(j-k)}})(y)|\,\frac{(1+2^{j-k}|y|+2^j|z|)^b}{(1+2^k|z|)^b}\,dy\,2^{j\alpha}M_{b,j}^{**}(f;\Theta)(x)\\
&\le \sum_{j\in\mathbf{Z}} 2^{(k-j)\alpha}\int_{\mathbf{R}^n} |(\Omega * \Theta_{2^{-(j-k)}})(y)|\,(1+2^{j-k})^b(1+2^{j-k}|y|)^b dy 2^{j\alpha}M_{b,j}^{**}(f;\Theta)(x)\,.
\end{aligned}$$

We conclude that

$$2^{k\alpha}M_{b,k}^{**}(f;\Omega)(x) \le \sum_{j\in\mathbf{Z}} V_{k-j}\, 2^{j\alpha}M_{b,j}^{**}(f;\Theta)(x)\,, \tag{6.5.11}$$

where

$$V_j = 2^{-j\alpha}(1+2^j)^b \int_{\mathbf{R}^n} |(\Omega * \Theta_{2^{-j}})(y)|\,(1+2^j|y|)^b\,dy\,.$$

We now use the facts that both Ω and Θ have vanishing moments of all orders and the result in Appendix K.2 to obtain

$$|(\Omega * \Theta_{2^{-j}})(y)| \le C_{L,N,n,\Theta,\Omega}\frac{2^{-|j|L}}{(1+2^{\min(0,j)}|y|)^N}$$

for all $L, N > 0$. We deduce the estimate

$$|V_j| \le C_{L,M,n,\Theta,\Omega}2^{-|j|M}$$

for all M sufficiently large, which, in turn, yields the estimate

$$\sum_{j\in\mathbf{Z}} |V_j|^{\min(1,q)} < \infty\,.$$

We deduce from (6.5.11) that for all $x \in \mathbf{R}^n$ we have

$$\big\|\{2^{k\alpha}M_{b,k}^{**}(f;\Omega)(x)\}_k\big\|_{\ell^q} \le C_{\alpha,p,q,n,\Psi,\Omega}\big\|\{2^{k\alpha}M_{b,k}^{**}(f;\Theta)(x)\}_k\big\|_{\ell^q}\,.$$

We now appeal to Lemma 6.5.3, which gives

$$2^{k\alpha}M_{b,k}^{**}(f;\Theta) \le C2^{k\alpha}M(|\Delta_k^{\Theta}(f)|^r)^{\frac{1}{r}} = CM(|2^{k\alpha}\Delta_k^{\Theta}(f)|^r)^{\frac{1}{r}}$$

with $b = n/r$. We choose $r < \min(p,q)$. We now use the $L^{p/r}(\mathbf{R}^n, \ell^{q/r})$ into $L^{p/r}(\mathbf{R}^n, \ell^{q/r})$ boundedness of the Hardy-Littlewood maximal operator, Theorem 4.6.6, to complete the proof of (6.5.9) with the exception that the function Ψ on the right-hand side of (6.5.9) is replaced by Θ. The passage to Ψ is a simple matter (at least when $p \ge 1$) since

$$\Delta_j^{\Psi} = \Delta_j^{\Psi}(\Delta_{j-1}^{\Theta} + \Delta_j^{\Theta} + \Delta_{j+1}^{\Theta})\,.$$

For general $0 < p < \infty$ the conclusion follows with the use of (6.5.8).

□

We obtain as a corollary that a different choice of bumps gives equivalent Triebel-Lizorkin norms.

Corollary 6.5.7. *Let Ψ, Ω be Schwartz functions whose Fourier transforms are supported in the annulus $\frac{1}{2} \le |\xi| \le 2$ and that satisfy (6.5.1). Let Φ be as in (6.5.2) and let*

$$\widehat{\Theta}(\xi) = \begin{cases} \sum_{j\le 0}\widehat{\Omega}(2^{-j}\xi) & \text{when } \xi \neq 0, \\ 1 & \text{when } \xi = 0. \end{cases}$$

Then the Triebel-Lizorkin quasi-norms defined with respect to the pairs (Ψ,Φ) *and* (Ω,Θ) *are equivalent.*

Proof. We note that the quantity on the left in (6.5.9) is bigger than or equal to

$$\Big\|\Big(\sum_{j\in\mathbf{Z}}|2^{j\alpha}\Delta_j^{\Omega}(f)|^q\Big)^{\frac{1}{q}}\Big\|_{L^p}$$

for all $f\in\mathcal{S}'(\mathbf{R}^n)$. This shows that the homogeneous Triebel-Lizorkin norm defined using Ω is bounded by a constant multiple of that defined using Ψ. This proves the equivalence of norms in the homogeneous case.

In the case of the inhomogeneous spaces, we let S_0^{Ψ} and S_0^{Ω} be the operators given by convolution with the bumps Φ and Θ, respectively (recall these are defined in terms of Ψ and Ω). Then for $f\in\mathcal{S}'(\mathbf{R}^n)$ we have

$$\Theta*f=\Theta*(\Phi*f)+\Theta*(\Psi_{2^{-1}}*f)\,, \tag{6.5.12}$$

since the Fourier transform of the function $\Phi+\Psi_{2^{-1}}$ is equal to 1 on the support of $\widehat{\Theta}$. Applying Lemma 6.5.3 (with $t=1$), we obtain that

$$|\Theta*(\Phi*f)|\le C_rM(|\Phi*f|^r)^{\frac{1}{r}}$$

and also

$$|\Theta*(\Psi_{2^{-1}}*f)|\le C_rM(|\Psi_{2^{-1}}*f|^r)^{\frac{1}{r}}$$

for any $0<r<\infty$. Picking $r<p$, we obtain that

$$\big\|\Theta*(\Phi*f)\big\|_{L^p}\le C\big\|S_0^{\Psi}(f)\big\|_{L^p}$$

and also

$$\big\|\Theta*(\Psi_{2^{-1}}*f)\big\|_{L^p}\le C\big\|\Delta_1^{\Psi}(f)\big\|_{L^p}\,.$$

Combining the last two estimates with (6.5.12), we obtain that $\big\|S_0^{\Omega}(f)\big\|_{L^p}$ is controlled by a multiple of the Triebel-Lizorkin norm of f defined using Ψ. This gives the equivalence of norms in the inhomogeneous case.

□

Several other properties of these spaces are valid. We refer to the Exercises for a short description of some of these properties.

Exercises

6.5.1. Let $0<q_0<q_1\le\infty$ and $\alpha\in\mathbf{R}$.

(a) Prove that

$$B^p_{\alpha,q_0}\subseteq B^p_{\alpha,q_1}\,,\qquad\text{for } 0<p\le\infty$$

and that

$$F^p_{\alpha+\varepsilon,q_0}\subseteq B^p_{\alpha,q_1}\,,\qquad\text{for } 0<p<\infty.$$

(b) Let $\varepsilon > 0$. Obtain the inclusions

$$B^p_{\alpha,q_0} \subseteq B^p_{\alpha,q_1}, \qquad \text{for } 0 < p \le \infty$$

and

$$F^p_{\alpha+\varepsilon,q_0} \subseteq B^p_{\alpha,q_1}, \qquad \text{for } 0 < p < \infty.$$

6.5.2. Let $0 < q \le \infty$, $0 < p < \infty$, and $\alpha \in \mathbf{R}$. Show that

$$B^p_{\alpha,\min(p,q)} \subseteq F^p_{\alpha,q} \subseteq B^p_{\alpha,\max(p,q)}.$$

[*Hint:* Consider the cases $p \ge q$ and $p < q$ and use the triangle inequality in the spaces $L^{p/q}$ and $\ell^{q/p}$, respectively.]

6.5.3. (a) Let $0 < p, q \le \infty$ and $\alpha \in \mathbf{R}$. Show that $\mathcal{S}(\mathbf{R}^n)$ is continuously embedded in $B^p_{\alpha,q}(\mathbf{R}^n)$ and that the latter is continuously embedded in $\mathcal{S}'(\mathbf{R}^n)$.
(b) Obtain the same conclusion for $F^p_{\alpha,q}(\mathbf{R}^n)$ when $p, q < \infty$.

6.5.4. $0 < p, q < \infty$ and $\alpha \in \mathbf{R}$. Show that the Schwartz functions are dense in all the spaces $B^p_{\alpha,q}(\mathbf{R}^n)$ and $F^p_{\alpha,q}(\mathbf{R}^n)$.
[*Hint:* Every Cauchy sequence $\{f_k\}_k$ in $B^p_{\alpha,q}$ is also Cauchy in $\mathcal{S}'(\mathbf{R}^n)$ and hence converges to some f in $\mathcal{S}'(\mathbf{R}^n)$. Then $\Delta_j(f_k) \to \Delta_j(f)$ in $\mathcal{S}'(\mathbf{R}^n)$. But $\Delta_j(f_k)$ is also Cauchy in L^p and therefore converges to $\Delta_j(f)$ in L^p. Argue similarly for $F^p_{\alpha,q}(\mathbf{R}^n)$.]

6.5.5. Let $\alpha \in \mathbf{R}$, let $0 < p, q < \infty$ and let $N = [\frac{n}{2} + \frac{n}{\min(p,q)}] + 1$. Assume that m is a C^N function on $\mathbf{R}^n \setminus \{0\}$ that satisfies

$$|\partial^\gamma m(\xi)| \le C_\gamma |\xi|^{-|\gamma|}$$

for all $|\gamma| \le N$. Show that there exists a constant C such that for all $f \in \mathcal{S}'(\mathbf{R}^n)$ we have

$$\big\|(m\widehat{f})^\vee\big\|_{\dot{B}^p_{\alpha,q}} \le C\big\|f\big\|_{\dot{B}^p_{\alpha,q}}.$$

[*Hint:* Pick $r < \min(p,q)$ so that $N > \frac{n}{2} + \frac{n}{r}$. Write $m = \sum_j m_j$, where $\widehat{m_j}(\xi) = \widehat{\Theta}(2^{-j}\xi)m(\xi)$ and $\widehat{\Theta}(2^{-j}\xi)$ is supported in an annulus $2^j \le |\xi| \le 2^{j+1}$. Obtain the estimate

$$\begin{aligned} &\sup_{z\in\mathbf{R}^n} \frac{\big|\big(m_j\widehat{\Delta_j(f)}\big)^\vee(x-z)\big|}{(1+2^j|z|)^{\frac{n}{r}}} \\ &\le C \sup_{z\in\mathbf{R}^n} \frac{|\Delta_j(f)(x-z)|}{(1+2^j|z|)^{\frac{n}{r}}} \int_{\mathbf{R}^n} |m_j^\vee(y)|(1+2^j|y|)^{\frac{n}{r}}\,dy \end{aligned}$$

and control the previous integral by

$$C\bigg(\int_{\mathbf{R}^n} |m_j(2^j(\,\cdot\,))^\vee(y)|^2(1+|y|)^{2N}\,dy\bigg)^{\frac{1}{2}}.$$

Then use the hypothesis on m and apply Lemma 6.5.3.]

6.5.6. (*Peetre* [**406**]) Let m be as in the previous Exercise. Show that there exists a constant C such that for all $f \in \mathcal{S}'(\mathbf{R}^n)$ we have

$$\|(m\,\widehat{f})^\vee\|_{\dot F^p_{\alpha,q}} \le C\|f\|_{\dot F^p_{\alpha,q}}\,.$$

[*Hint:* Use the hint in the previous Exercise and Theorem 4.6.6.]

6.5.7. (a) Suppose that $B^{p_0}_{\alpha_0,q_0} = B^{p_1}_{\alpha_1,q_1}$ with equivalent norms. Prove that $\alpha_0 = \alpha_1$ and $p_0 = p_1$. Prove the same result for the F scale of spaces.
(b) Suppose that $B^{p_0}_{\alpha_0,q_0} = B^{p_1}_{\alpha_1,q_1}$ with equivalent norms. Prove that $q_0 = q_1$. Argue similarly with the F scale of spaces.
[*Hint:* Part (a): Test the corresponding norms on the function $\Psi(2^j x)$, where Ψ is chosen so that its Fourier transform is supported in $\frac{1}{2} \le |\xi| \le 2$. Part (b): Try a function f of the form $\widehat{f}(\xi) = \sum_{j=1}^N a_j \varphi(\xi_1 - 2^j, \xi_2, \dots, \xi_n)$, where φ is a C_0^∞ function whose Fourier transform is supported in a small neighborhood of the origin.]

6.6. Atomic Decomposition*

In this section we focus our attention on the homogeneous Triebel-Lizorkin spaces $\dot F^p_{\alpha,q}$ as the analysis for the other spaces is similar or simpler. These include the Hardy spaces discussed in Section 6.4. We note that many of the results discussed in this section can be carried through for the inhomogeneous spaces as well as the Besov-Lipschitz spaces. We refer the interested reader to the relevant literature on the subject at the end of this chapter.

6.6.a. The Space of Sequences $\dot f^p_{\alpha,q}$

To provide more intuition in our understanding of the homogeneous Triebel-Lizorkin spaces we introduce a related space consisting of sequences of scalars. This space is denoted by $\dot f^p_{\alpha,q}$ and is related to $\dot F^p_{\alpha,q}$ in a way similar to that in which $\ell^2(\mathbf{Z})$ is related to $L^2([0,1])$.

Definition 6.6.1. Let $0 < q \le \infty$. Let $\mathcal{D}$ be the set of all dyadic cubes in $\mathbf{R}^n$. We consider the set of all sequences $\{s_Q\}_{Q\in\mathcal{D}}$ such that the function

$$g^{\alpha,q}(\{s_Q\}_Q) = \Big(\sum_{Q\in\mathcal{D}} (|Q|^{-\frac{\alpha}{n}-\frac{1}{2}}\,|s_Q|\,\chi_Q)^q\Big)^{\frac{1}{q}} \tag{6.6.1}$$

is in $L^p(\mathbf{R}^n)$. For such sequences $s = \{s_Q\}_Q$ we set

$$\big\|s\big\|_{\dot f^p_{\alpha,q}} = \big\|g^{\alpha,q}(s)\big\|_{L^p(\mathbf{R}^n)}\,.$$

6.6.b. The Smooth Atomic Decomposition of $\dot F^p_{\alpha,q}$

We now discuss the smooth atomic decomposition of these spaces. We begin with the definition of smooth atoms on $\mathbf{R}^n$.

Definition 6.6.2. Let Q be a dyadic cube. A C^∞ function a_Q on $\mathbf{R}^n$ is called a *smooth L-atom for Q* if it satisfies

(a) a_Q is supported in $3Q$ (the cube concentric with Q having three times its side length).

(b) $\displaystyle\int_{\mathbf{R}^n} x^\gamma a_Q(x)\,dx = 0$ for all multiindices $|\gamma| \le L$.

(c) $|\partial^\gamma a_Q| \le |Q|^{-\frac{|\gamma|}{n}-\frac12}$ for all multiindices γ.

We now state and prove a theorem saying that elements of $\dot F^p_{\alpha,q}$ can be decomposed as sums of smooth atoms.

Theorem 6.6.3. *Let $0 < p,q < \infty$, $\alpha \in \mathbf{R}$ and L be a nonnegative integer satisfying $L \ge [n\max(1,\frac1p,\frac1q) - n - \alpha]$. Then there is a constant $C_{n,p,q,\alpha}$ such that for every sequence of smooth L-atoms $\{a_Q\}_{Q\in\mathcal{D}}$ and every sequence of complex scalars $\{s_Q\}_{Q\in\mathcal{D}}$ we have*

$$\Big\|\sum_{Q\in\mathcal{D}} s_Q a_Q\Big\|_{\dot F^p_{\alpha,q}} \le C_{n,p,q,\alpha}\big\|\{s_Q\}_Q\big\|_{\dot f^p_{\alpha,q}}\,. \tag{6.6.2}$$

Conversely, there is a constant $C'_{n,p,q,\alpha}$ such that given any distribution f in $\dot F^p_{\alpha,q}$ and any $L \ge 0$, there exists a sequence of smooth L-atoms $\{a_Q\}_{Q\in\mathcal{D}}$ and a sequence of complex scalars $\{s_Q\}_{Q\in\mathcal{D}}$ such that

$$f = \sum_{Q\in\mathcal{D}} s_Q a_Q\,,$$

where the sum converges in the sense of distributions modulo polynomials and, moreover,

$$\big\|f\big\|_{\dot F^p_{\alpha,q}} \le C'_{n,p,q,\alpha}\big\|\{s_Q\}_Q\big\|_{\dot f^p_{\alpha,q}}\,. \tag{6.6.3}$$

Proof. We let Δ_j^Ψ be the Littlewood-Paley operator associated with a Schwartz function Ψ whose Fourier is compactly supported away from the origin in $\mathbf{R}^n$. Let a_Q be a smooth L-atom supported in a cube $3Q$ with center C_Q and let the side length $\ell(Q) = 2^{-\mu}$. It follows from Definition 6.6.2 that for all multiindices γ and all $N>0$, a_Q trivially satisfies

$$|\partial_y^\gamma a_Q(y)| \le C_{N,n} 2^{-\frac{\mu n}{2}} \frac{2^{\mu|\gamma|+\mu n}}{(1+2^\mu|y-c_Q|)^N}\,.$$

Moreover, the function $y \to \Psi_{2^{-j}}(y-x)$ satisfies

$$|\partial_y^\delta \Psi_{2^{-j}}(y-x)| \le C_{N,n,\delta} \frac{2^{j|\delta|+jn}}{(1+2^j|y-x|)^N}$$

for all multiindices δ. Using these estimates and the facts that a_Q has vanishing moments of order at most $L=(L+1)-1$ and Ψ has vanishing moments of all orders, we apply twice the result in Appendix K.2 to deduce the following estimate for all $N>0$:

$$\big|\Delta_j^\Psi(a_Q)(x)\big| \le C_{N,n,L'}\, 2^{-\frac{\mu n}{2}} \frac{2^{\min(j,\mu)n-|\mu-j|L'}}{(1+2^{\min(j,\mu)}|x-c_Q|)^N}\,, \tag{6.6.4}$$

where

$$L' = \begin{cases} L+1 & \text{when } j<\mu,\\ \text{as large as necessary} & \text{when } \mu \le j.\end{cases}$$

Now fix $0<b<\min(1,p,q)$ so that

$$L+1 > \tfrac{n}{b} - n - \alpha\,. \tag{6.6.5}$$

This can be achieved by taking b close enough to $\min(1,p,q)$ since our assumption $L \ge [n\max(1,\frac{1}{p},\frac{1}{q}) - n - \alpha]$ implies $L+1 > n\max(1,\frac{1}{p},\frac{1}{q}) - n - \alpha$.

Using Exercise 6.6.6, we obtain

$$\sum_{\substack{Q\in\mathcal{D}\\ \ell(Q)=2^{-\mu}}} \frac{|s_Q|}{(1+2^{\min(j,\mu)}|x-c_Q|)^N} \le c\, 2^{\max(\mu-j,0)\frac{n}{b}} \Big\{M\Big(\sum_{\substack{Q\in\mathcal{D}\\ \ell(Q)=2^{-\mu}}} |s_Q|^b\chi_Q\Big)(x)\Big\}^{\frac{1}{b}}$$

whenever $N>n/b$, where M is the Hardy-Littlewood maximal operator. It follows from the previous estimate and (6.6.4) that

$$2^{j\alpha}\sum_{\mu\in\mathbf{Z}}\sum_{\substack{Q\in\mathcal{D}\\ \ell(Q)=2^{-\mu}}} |s_Q|\,\big|\Delta_j^\Psi(a_Q)(x)\big| \le C\sum_{\mu\in\mathbf{Z}} 2^{\min(j,\mu)n}2^{-|j-\mu|L'}$$

$$2^{-\mu n}2^{(j-\mu)\alpha}2^{\max(\mu-j,0)\frac{n}{b}}\Big\{M\Big(\sum_{\substack{Q\in\mathcal{D}\\ \ell(Q)=2^{-\mu}}} \big(|s_Q|\,|Q|^{-\frac{1}{2}-\frac{\alpha}{n}}\big)^b\chi_Q\Big)(x)\Big\}^{\frac{1}{b}}.$$

Raise the preceding inequality to the power q and sum over $j \in \mathbf{Z}$; then raise to the power $1/q$ and take $\big\|\cdot\big\|_{L^p}$ norms in x. We obtain

$$\|f\|_{\dot F^p_{\alpha,q}} \le \bigg\|\bigg\{\sum_{j\in\mathbf{Z}}\bigg[\sum_{\mu\in\mathbf{Z}} d(j-\mu)\Big\{M\Big(\sum_{\substack{Q\in\mathcal{D}\\ \ell(Q)=2^{-\mu}}} \big(|s_Q|\,|Q|^{-\frac{1}{2}-\frac{\alpha}{n}}\big)^b\chi_Q\Big)\Big\}^{\frac{1}{b}}\bigg]^q\bigg\}^{\frac{1}{q}}\bigg\|_{L^p},$$

where $f = \sum_{Q\in\mathcal{D}} s_Q a_Q$ and

$$d(j-\mu) = C\,2^{\min(j-\mu,0)(n-\frac{n}{b})+(j-\mu)\alpha-|j-\mu|L'}.$$

We can now estimate the expression inside the last L^p norm by

$$\Big\{\sum_{j\in\mathbf{Z}} d(j-\mu)^{\min(1,q)}\Big\}^{\frac{1}{q}}\bigg\{\sum_{\mu\in\mathbf{Z}}\Big\{M\Big(\sum_{\substack{Q\in\mathcal{D}\\ \ell(Q)=2^{-\mu}}} \big(|s_Q|\,|Q|^{-\frac{1}{2}-\frac{\alpha}{n}}\big)^b\chi_Q\Big)\Big\}^{\frac{q}{b}}\bigg\}^{\frac{1}{q}}$$

and we note that the first term is a constant in view of (6.6.5). We conclude that

$$\begin{aligned}
\Big\|\sum_{Q\in\mathcal{D}} s_Q a_Q\Big\|_{\dot F^p_{\alpha,q}} &\le C\bigg\|\bigg\{\sum_{\mu\in\mathbf{Z}}\Big\{M\Big(\sum_{\substack{Q\in\mathcal{D}\\ \ell(Q)=2^{-\mu}}} \big(|s_Q|\,|Q|^{-\frac{1}{2}-\frac{\alpha}{n}}\big)^b\chi_Q\Big)\Big\}^{\frac{q}{b}}\bigg\}^{\frac{1}{q}}\bigg\|_{L^p}\\
&= C\bigg\|\bigg\{\sum_{\mu\in\mathbf{Z}}\Big\{M\Big(\sum_{\substack{Q\in\mathcal{D}\\ \ell(Q)=2^{-\mu}}} \big(|s_Q|\,|Q|^{-\frac{1}{2}-\frac{\alpha}{n}}\big)^b\chi_Q\Big)\Big\}^{\frac{q}{b}}\bigg\}^{\frac{b}{q}}\bigg\|_{L^{\frac{p}{b}}}^{\frac{1}{b}}\\
&\le C'\bigg\|\bigg\{\sum_{\mu\in\mathbf{Z}}\Big\{\sum_{\substack{Q\in\mathcal{D}\\ \ell(Q)=2^{-\mu}}} \big(|s_Q|\,|Q|^{-\frac{1}{2}-\frac{\alpha}{n}}\big)^b\chi_Q\Big\}^{\frac{q}{b}}\bigg\}^{\frac{b}{q}}\bigg\|_{L^{\frac{p}{b}}}^{\frac{1}{b}}\\
&= C'\bigg\|\bigg\{\sum_{\mu\in\mathbf{Z}}\sum_{\substack{Q\in\mathcal{D}\\ \ell(Q)=2^{-\mu}}} \big(|s_Q|\,|Q|^{-\frac{1}{2}-\frac{\alpha}{n}}\big)^q\chi_Q\bigg\}^{\frac{1}{q}}\bigg\|_{L^p}\\
&= C'\big\|\{s_Q\}_Q\big\|_{\dot f^p_{\alpha,q}},
\end{aligned}$$

where in the last inequality we used Theorem 4.6.6, which is valid under the assumption $1 < \frac{p}{b}, \frac{q}{b} < \infty$. This proves (6.6.2).

We now turn to the second statement of the theorem. It is not difficult to see that there exist Schwartz functions Ψ (unrelated to the previous one) and Θ such that $\widehat{\Psi}$ is supported in the annulus $\frac{1}{2} \le |\xi| \le 2$, $\widehat{\Psi}$, is at least $c > 0$ in the smaller annulus $\frac{3}{5} \le |\xi| \le \frac{5}{3}$, and Θ is supported in the ball $|x| \le 1$ and satisfies $\int_{\mathbf{R}^n} x^\gamma \Theta(x)\,dx = 0$ for all $|\gamma| \le L$, such that the identity

$$\sum_{j\in\mathbf{Z}} \widehat{\Psi}(2^{-j}\xi)\widehat{\Theta}(2^{-j}\xi) = 1 \tag{6.6.6}$$

holds for all $\xi \in \mathbf{R}^n \setminus \{0\}$. (See Exercise 6.6.1.)

Using identity (6.6.6), we can write

$$f = \sum_{j\in\mathbf{Z}} \Psi_{2^{-j}} * \Theta_{2^{-j}} * f\,.$$

Setting $\mathcal{D}_j = \{Q \in \mathcal{D}:\ \ell(Q) = 2^{-j}\}$, we now have

$$f = \sum_{j\in\mathbf{Z}}\sum_{Q\in\mathcal{D}_j} \int_Q \Theta_{2^{-j}}(x-y)(\Psi_{2^{-j}} * f)(y)\,dy = \sum_{j\in\mathbf{Z}}\sum_{Q\in\mathcal{D}_j} s_Q a_Q\,,$$

where we also set

$$s_Q = |Q|^{\frac{1}{2}} \sup_{y\in Q} |(\Psi_{2^{-j}} * f)(y)| \sup_{|\gamma|\le L} \big\|\partial^\gamma \Theta\big\|_{L^1}$$

for Q in $\mathcal{D}_j$ and

$$a_Q(x) = \frac{1}{s_Q}\int_Q \Theta_{2^{-j}}(x-y)(\Psi_{2^{-j}} * f)(y)\,dy\,.$$

It is straightforward to verify that a_Q is supported in $3Q$ and that it has vanishing moments up to order L. Moreover, we have

$$|\partial^\gamma a_Q| \le \frac{1}{s_Q}\big\|\partial^\gamma\Theta\big\|_{L^\infty} 2^{j(n+|\gamma|)} \sup_Q |\Psi_{2^{-j}} * f| \le |Q|^{-\frac{1}{2}-\frac{|\gamma|}{n}}\,,$$

which makes the function a_Q a smooth L-atom. Now note

$$\begin{aligned}
&\sum_{\ell(Q)=2^{-j}} \big(|Q|^{-\frac{\alpha}{n}-\frac{1}{2}} s_Q \chi_Q(x)\big)^q \\
&= C \sum_{\ell(Q)=2^{-j}} \big(2^{j\alpha} \sup_{y\in Q} |(\Psi_{2^{-j}} * f)(y)| \chi_Q(x)\big)^q \\
&\le C \sup_{|z|\le\sqrt{n}2^{-j}} \big(2^{j\alpha}(1+2^j|z|)^{-b}|(\Psi_{2^{-j}} * f)(x-z)|\big)^q (1+2^j|z|)^{bq} \\
&\le C\big(2^{j\alpha} M^{**}_{b,j}(f,\Psi)(x)\big)^q\,,
\end{aligned}$$

where we used the fact that in the first inequality there is only one nonzero term in the sum because of the appearance of the characteristic function. Summing over all $j \in \mathbf{Z}^n$, raising to the power $\frac{1}{q}$, and taking L^p norms yields the estimate

$$\big\|\{s_Q\}_Q\big\|_{\dot f^p_{\alpha,q}} \le C\Big\|\Big(\sum_{j\in\mathbf{Z}} |2^{j\alpha} M^{**}_{b,j}(f;\Psi)|^q\Big)^{\frac{1}{q}}\Big\|_{L^p} \le C\big\|f\big\|_{\dot F^p_{\alpha,q}}\,,$$

where the last inequality follows from Theorem 6.5.6. This proves (6.6.3).

□

6.6.c. The Nonsmooth Atomic Decomposition of $\dot{F}^p_{\alpha,q}$

We now discuss the main theorem of this section, the nonsmooth atomic decomposition of the homogeneous Triebel-Lizorkin spaces $\dot{F}^p_{\alpha,q}$, which in particular includes that of the Hardy spaces H^p. We begin this task with a definition.

Definition 6.6.4. Let $0 < p \le 1$ and $1 \le q \le \infty$. A sequence of complex numbers $r = \{r_Q\}_{Q\in\mathcal{D}}$ is called an ∞-atom for $\dot{f}^p_{\alpha,q}$ if there exists a dyadic cube Q_0 such that

(a) $r_Q = 0$ if $Q \not\subseteq Q_0$.

(b) $\big\|g^{\alpha,q}(r)\big\|_{L^\infty} \le |Q_0|^{-\frac{1}{p}}$.

We observe that every ∞-atom $r = \{r_Q\}$ for $\dot{f}^p_{\alpha,q}$ satisfies $\big\|r\big\|_{\dot{f}^p_{\alpha,q}} \le 1$. Indeed,

$$\big\|r\big\|^p_{\dot{f}^p_{\alpha,q}} = \int_{Q_0} |g^{\alpha,q}(r)|^p\,dx \le |Q_0|^{-1}|Q_0| = 1\,.$$

The following theorem concerns the atomic decomposition of the spaces $\dot{f}^p_{\alpha,q}$.

Theorem 6.6.5. *Suppose $\alpha \in \mathbf{R}$, $0 < q < \infty$, $0 < p < \infty$ and $s = \{s_Q\}_Q$ is in $\dot{f}^p_{\alpha,q}$. Then there exist $C_{n,p,q} > 0$, a sequence of scalars λ_j, and a sequence of ∞-atoms $r_j = \{r_{j,Q}\}_Q$ for $\dot{f}^p_{\alpha,q}$ such that*

$$s = \{s_Q\}_Q = \sum_{j=1}^{\infty} \lambda_j \{r_{j,Q}\}_Q = \sum_{j=1}^{\infty} \lambda_j r_j$$

and such that

$$\Big(\sum_{j=1}^{\infty} |\lambda_j|^p\Big)^{\frac{1}{p}} \le C_{n,p,q}\big\|s\big\|_{\dot{f}^p_{\alpha,q}}\,. \tag{6.6.7}$$

Proof. We fix α, p, q and a sequence $s = \{s_Q\}_Q$ as in the statement of the theorem. For a dyadic cube R in $\mathcal{D}$ we define the function

$$g^{\alpha,q}_R(s)(x) = \Big(\sum_{\substack{Q\in\mathcal{D}\\ R\subseteq Q}} \big(|Q|^{\frac{\alpha}{n}-\frac{1}{2}}|s_Q|\,\chi_Q(x)\big)^q\Big)^{\frac{1}{q}}$$

and we observe that this function is constant on R. We also note that for dyadic cubes R_1 and R_2 with $R_1 \subseteq R_2$ we have

$$g^{\alpha,q}_{R_2}(s) \le g^{\alpha,q}_{R_1}(s)\,.$$

Finally, we observe that

$$\lim_{\substack{\ell(R)\to\infty\\ x\in R}} g^{\alpha,q}_R(s)(x) = 0$$

$$\lim_{\substack{\ell(R)\to 0\\ x\in R}} g^{\alpha,q}_R(s)(x) = g^{\alpha,q}(s)(x)\,,$$

where $g^{\alpha,q}(s)$ is the function defined in (6.6.1).

For $k \in \mathbf{Z}$ we set

$$\mathcal{A}_k = \big\{R \in \mathcal{D}: \ g_R^{\alpha,q}(s)(x) > 2^k \quad \text{for all } x \in R\big\}.$$

We note that $\mathcal{A}_{k+1} \subseteq \mathcal{A}_k$ for all k in $\mathbf{Z}$ and that

$$\{x \in \mathbf{R}^n: \ g^{\alpha,q}(s)(x) > 2^k\} = \bigcup_{R \in \mathcal{A}_k} R. \tag{6.6.8}$$

Moreover, we have for all $k \in \mathbf{Z}$

$$\Big(\sum_{Q \in \mathcal{D} \setminus \mathcal{A}_k} \big(|Q|^{-\frac{\alpha}{n}-\frac{1}{2}} |s_Q| \, \chi_Q(x)\big)^q\Big)^{\frac{1}{q}} \le 2^k, \qquad \text{for all } x \in \mathbf{R}^n. \tag{6.6.9}$$

To prove (6.6.9) we assume that $g^{\alpha,q}(s)(x) > 2^k$; otherwise, the conclusion is trivial. Then there exists a maximal dyadic cube $R_{\max}$ in $\mathcal{A}_k$ such that $x \in R_{\max}$. Letting R_0 be the unique dyadic cube that contains $R_{\max}$ and has twice its side length, we have that the left-hand side of (6.6.9) is equal to $g_{R_0}^{\alpha,q}(s)(x)$, which is at most 2^k since R_0 is not contained in $\mathcal{A}_k$.

Since $g^{\alpha,q}(s) \in L^p(\mathbf{R}^n)$, by our assumption, and $g^{\alpha,q}(s) > 2^k$ for all $x \in Q$ if $Q \in \mathcal{A}_k$, the cubes in $\mathcal{A}_k$ must have size bounded above by some constant. We set

$$\mathcal{B}_k = \big\{Q \in \mathcal{D}: \quad Q \text{ is a maximal dyadic cube in } \mathcal{A}_k \setminus \mathcal{A}_{k+1}\big\}.$$

For J in $\mathcal{B}_k$ we define a sequence $t(k,J) = \{t(k,J)_Q\}_{Q \in \mathcal{D}}$ by setting

$$t(k,J)_Q = \begin{cases} s_Q & \text{if } Q \subseteq J \text{ and } Q \in \mathcal{A}_k \setminus \mathcal{A}_{k+1}, \\ 0 & \text{otherwise.} \end{cases}$$

We can see that if

$$Q \notin \bigcup_{k \in \mathbf{Z}} \mathcal{A}_k, \qquad \text{then} \qquad s_Q = 0,$$

and the identity

$$s = \sum_{k \in \mathbf{Z}} \sum_{J \in \mathcal{B}_k} t(k,J) \tag{6.6.10}$$

is valid. For all $x \in \mathbf{R}^n$ we have

$$\begin{aligned} \big|g^{\alpha,q}(t(k,J))(x)\big| &= \Big(\sum_{\substack{Q \subseteq J \\ Q \in \mathcal{A}_k \setminus \mathcal{A}_{k+1}}} \big(|Q|^{-\frac{\alpha}{n}-\frac{1}{2}} |s_Q| \, \chi_Q(x)\big)^q\Big)^{\frac{1}{q}} \\ &\le \Big(\sum_{\substack{Q \subseteq J \\ Q \in \mathcal{D} \setminus \mathcal{A}_{k+1}}} \big(|Q|^{-\frac{\alpha}{n}-\frac{1}{2}} |s_Q| \, \chi_Q(x)\big)^q\Big)^{\frac{1}{q}} \\ &\le 2^{k+1}, \end{aligned} \tag{6.6.11}$$

where we used (6.6.9) in the last estimate. We now define our atoms $r(k,J) = \{r(k,J)_Q\}_{Q\in\mathcal{D}}$ by setting

$$r(k,J)_Q = 2^{-k-1}|J|^{-\frac{1}{p}}t(k,J)_Q \tag{6.6.12}$$

and we also define scalars

$$\lambda_{k,J} = 2^{k+1}|J|^{\frac{1}{p}}.$$

To see that each $r(k,J)$ is an ∞-atom for $\dot{f}^p_{\alpha,q}$, we observe that $r(k,J)_Q = 0$ if $Q \nsubseteq J$ and that

$$\left|g^{\alpha,q}(t(k,J))(x)\right| \le |J|^{-\frac{1}{p}}, \qquad \text{for all } x \in \mathbf{R}^n,$$

in view of (6.6.11). Also using (6.6.10) and (6.6.12), we obtain that

$$s = \sum_{k\in\mathbf{Z}}\sum_{J\in\mathcal{B}_k} \lambda_{k,J}\, r(k,J),$$

which says that s can be written as a linear combination of atoms. Finally, we estimate the sum of the pth power of the coefficients $\lambda_{k,J}$. We have

$$\begin{aligned}
\sum_{k\in\mathbf{Z}}\sum_{J\in\mathcal{B}_k} |\lambda_{k,J}|^p &= \sum_{k\in\mathbf{Z}} 2^{(k+1)p}\sum_{J\in\mathcal{B}_k}|J| \\
&\le 2^p\sum_{k\in\mathbf{Z}} 2^{kp}\Big|\bigcup_{Q\in\mathcal{A}_k} Q\Big| \\
&= 2^p\sum_{k\in\mathbf{Z}} 2^{k(p-1)}2^k|\{x\in\mathbf{R}^n:\ g^{\alpha,q}(s)(x) > 2^k\}| \\
&\le 2^p\sum_{k\in\mathbf{Z}}\int_{2^k}^{2^{k+1}} 2^{k(p-1)}|\{x\in\mathbf{R}^n:\ g^{\alpha,q}(s)(x) > \tfrac{\lambda}{2}\}|\,d\lambda \\
&\le 2^p\sum_{k\in\mathbf{Z}}\int_{2^k}^{2^{k+1}} \lambda^{p-1}|\{x\in\mathbf{R}^n:\ g^{\alpha,q}(s)(x) > \tfrac{\lambda}{2}\}|\,d\lambda \\
&= \frac{2^{2p}}{p}\big\|g^{\alpha,q}(s)\big\|^p_{L^p} = \frac{2^{2p}}{p}\big\|s\big\|^p_{\dot{f}^p_{\alpha,q}}.
\end{aligned}$$

Taking the pth root yields (6.6.7). The proof of the theorem is now complete. □

We now deduce a corollary regarding a new characterization of the space $\dot{f}^p_{\alpha,q}$.

Corollary 6.6.6. *Suppose $\alpha\in\mathbf{R}$, $0<p\le 1$ and $p\le q\le\infty$. Then we have*

$$\big\|s\big\|_{\dot{f}^p_{\alpha,q}} \approx \inf\left\{\Big(\sum_{j=1}^\infty|\lambda_j|^p\Big)^{\frac{1}{p}}:\ s=\sum_{j=1}^\infty\lambda_j r_j,\quad r_j \textit{ is an } \infty\textit{-atom for } \dot{f}^p_{\alpha,q}\right\}.$$

Proof. One direction in the previous estimate is a direct consequence of (6.6.7). The other direction uses the observation made after Definition 6.6.4 that every ∞-atom r for $\dot{f}^p_{\alpha,q}$ satisfies $\|r\|_{\dot{f}^p_{\alpha,q}} \leq 1$ and that for $p \leq 1$ and $p \leq q$ the quantity $s \to \|s\|^p_{\dot{f}^p_{\alpha,q}}$ is subadditive, see Exercise 6.6.2. Then each $s = \sum_{j=1}^\infty \lambda_j r_j$ (with r_j ∞-atoms for $\dot{f}^p_{\alpha,q}$ and $\sum_{j=1}^\infty |\lambda_j|^p < \infty$) must be an element of $\dot{f}^p_{\alpha,q}$ since

$$\Big\|\sum_{j=1}^\infty \lambda_j r_j\Big\|^p_{\dot{f}^p_{\alpha,q}} \leq \sum_{j=1}^\infty |\lambda_j|^p \,\|r_j\|^p_{\dot{f}^p_{\alpha,q}} \leq \sum_{j=1}^\infty |\lambda_j|^p < \infty .$$

□

The theorem we just proved allows us to obtain an atomic decomposition for the space $\dot{F}^p_{\alpha,q}$ as well. Indeed, we have the following result:

Corollary 6.6.7. *Let $\alpha \in \mathbf{R}$, $0 < p \leq 1$, $L \geq [\frac{n}{p} - n - \alpha]$ and let q satisfy $p \leq q < \infty$. Then we have the following representation:*

$$\|f\|_{\dot{F}^p_{\alpha,q}} \approx \inf\Big\{\Big(\sum_{j=1}^\infty |\lambda_j|^p\Big)^{\frac{1}{p}} :\ f = \sum_{j=1}^\infty \lambda_j A_j,\quad A_j = \sum_{Q\in\mathcal{D}} r_Q a_Q\,,\quad a_Q \text{ are}$$
$$\text{smooth } L\text{-atoms for } \dot{F}^p_{\alpha,q} \text{ and } \{r_Q\}_Q \text{ is an } \infty\text{-atom for } \dot{f}^p_{\alpha,q}\Big\}.$$

Proof. Let $f = \sum_{j=1}^\infty \lambda_j A_j$ as described previously. Using Exercise 6.6.2, we have

$$\|f\|^p_{\dot{F}^p_{\alpha,q}} \leq \sum_{j=1}^\infty |\lambda_j|^p \|A_j\|^p_{\dot{F}^p_{\alpha,q}} \leq c_{n,p} \sum_{j=1}^\infty |\lambda_j|^p \|r\|^p_{\dot{f}^p_{\alpha,q}}\,,$$

where in the last estimate we used Theorem 6.6.3. Using the fact that every ∞-atom $r = \{r_Q\}$ for $\dot{f}^p_{\alpha,q}$ satisfies $\|r\|_{\dot{f}^p_{\alpha,q}} \leq 1$, we conclude that every element f in $\mathcal{S}'(\mathbf{R}^n)$ that has the form $\sum_{j=1}^\infty \lambda_j A_j$ lies in the homogeneous Triebel-Lizorkin space $\dot{F}^p_{\alpha,q}$ [and has norm controlled by a constant multiple of $\sum_{j=1}^\infty |\lambda_j|^p)^{\frac{1}{p}}$].

Conversely, Theorem 6.6.3 gives that every element of $\dot{F}^p_{\alpha,q}$ has a smooth atomic decomposition. Then we can write

$$f = \sum_{Q\in\mathcal{D}} s_Q a_Q\,,$$

where each a_Q is a smooth L-atom for the cube Q. Using Theorem 6.6.5 we can now write $s = \{s_Q\}_Q$ as a sum of ∞-atoms for $\dot{f}^p_{\alpha,q}$, that is,

$$s = \sum_{j=1}^\infty \lambda_j r_j\,,$$

where

$$\Big(\sum_{j=1}^{\infty}|\lambda_j|^p\Big)^{\frac{1}{p}} \le c\big\|s\big\|_{\dot{f}^p_{\alpha,q}} \le c\big\|f\big\|_{\dot{F}^p_{\alpha,q}},$$

where the last step uses Theorem 6.6.3 again. It is simple to see that

$$f = \sum_{Q\in\mathcal{D}}\sum_{j=1}^{\infty}\lambda_j r_{j,Q} a_Q = \sum_{j=1}^{\infty}\lambda_j\Big(\sum_{Q\in\mathcal{D}} r_{j,Q} a_Q\Big),$$

and we set the expression inside the parentheses equal to A_j.

□

6.6.d. Atomic Decomposition of Hardy Spaces

We now pass to one of the main theorems of this chapter, the atomic decomposition of $H^p(\mathbf{R}^n)$ for $0 < p \le 1$. We begin by defining atoms for H^p.

Definition 6.6.8. Let $1 < q \le \infty$. A function A is called *an L^q-atom for $H^p(\mathbf{R}^n)$* if there exists a cube Q such that

(a) A is supported in Q.

(b) $\big\|A\big\|_{L^q} \le |Q|^{\frac{1}{q}-\frac{1}{p}}$.

(c) $\displaystyle\int x^\gamma A(x)\,dx = 0$ for all multiindices γ with $|\gamma| \le [\frac{n}{p} - n]$.

Notice that any L^r-atom for H^p is also an L^q-atom for H^p whenever $0 < p \le 1 < q < r \le \infty$. It is also simple to verify that an L^q-atom A for H^p is in fact in H^p. We prove this result in the next theorem for $p = 2$, and we refer the reader to Exercise 6.6.4 for the case of a general q.

Theorem 6.6.9. *Let $0 < p \le 1$. There is a constant $C = C_{n,p}$ such that every L^2-atom A for $H^p(\mathbf{R}^n)$ satisfies*

$$\big\|A\big\|_{H^p} \le C_{n,p}\,.$$

Proof. We can prove this theorem in two ways. We can either show that the smooth maximal function $M(A;\Phi)$ is in L^p or we can show that the square function $\big(\sum_j|\Delta_j(A)|^2\big)^{\frac{1}{2}}$ is in L^p. The operators Δ_j here are as in Theorem 5.1.2. Both proofs are similar in spirit; we only present the second and we refer the reader to Exercise 6.6.3 for the first.

Let $A(x)$ be an atom that we assume is supported in a cube Q centered at the origin [otherwise apply the argument to the atom $A(x - c_Q)$, where c_Q is the center of Q]. We control the L^p (quasi)norm of $\big(\sum_j|\Delta_j(A)|^2\big)^{\frac{1}{2}}$ by estimating it over the cube Q^* and over $(Q^*)^c$, where $Q^* = 2\sqrt{n}\,Q$. We have

$$\Big(\int_{Q^*}\big(\sum_j |\Delta_j(A)|^2\big)^{\frac{p}{2}}\,dx\Big)^{\frac{1}{p}} \le \Big(\int_{Q^*}\sum_j |\Delta_j(A)|^2\,dx\Big)^{\frac{1}{2}}|Q^*|^{\frac{1}{p(2/p)'}}.$$

Using that the square function $f \to \big(\sum_j |\Delta_j(f)|^2\big)^{\frac{1}{2}}$ is L^2 bounded, we obtain

$$\begin{aligned}\Big(\int_{Q^*}\big(\sum_j |\Delta_j(A)|^2\big)^{\frac{p}{2}}\,dx\Big)^{\frac{1}{p}} &\le C_n\big\|A\big\|_{L^2}|Q^*|^{\frac{1}{p(2/p)'}}\\ &\le C_n(2\sqrt{n}\,)^{\frac{n}{p}-\frac{n}{2}}|Q|^{\frac{1}{2}-\frac{1}{p}}|Q|^{\frac{1}{p}-\frac{1}{2}}\\ &= C_n'\,.\end{aligned} \tag{6.6.13}$$

To estimate the contribution of square function outside Q^*, we use the cancellation of the atoms. Let $k = [\frac{n}{p} - n] + 1$. We have

$$\begin{aligned}\Delta_j(A)(x) &= \int_Q A(y)\Psi_{2^{-j}}(x-y)\,dy\\ &= 2^{jn}\int_Q A(y)\Big[\Psi(2^jx - 2^jy) - \sum_{|\beta|\le k-1}(\partial^\beta\Psi)(2^jx)\frac{(2^jy)^\beta}{\beta!}\Big]\,dy\\ &= 2^{jn}\int_Q A(y)\Big[\sum_{|\beta|=k}(\partial^\beta\Psi)(2^jx - 2^j\theta y)\frac{(2^jy)^\beta}{\beta!}\Big]\,dy\,,\end{aligned}$$

where $0 \le \theta \le 1$. Taking absolute values, using the fact that $\partial^\beta\Psi$ are Schwartz functions, and that $|x - \theta y| \ge |x| - |y| \ge \frac{1}{2}|x|$ whenever $y \in Q$ and $x \notin Q^*$, we obtain the estimate

$$\begin{aligned}|\Delta_j(A)(x)| &\le 2^{jn}\int_Q |A(y)|\sum_{|\beta|=k}\frac{C_N}{(1+2^j\frac{1}{2}|x|)^N}\frac{|2^jy|^k}{\beta!}\,dy\\ &\le \frac{C_{N,p,n}2^{j(k+n)}}{(1+2^j|x|)^N}\Big(\int_Q |A(y)|^2\,dy\Big)^{\frac{1}{2}}\Big(\int_Q |y|^{2k}\,dy\Big)^{\frac{1}{2}}\\ &\le \frac{C'_{N,p,n}2^{j(k+n)}}{(1+2^j|x|)^N}|Q|^{\frac{1}{2}-\frac{1}{p}}|Q|^{\frac{k}{n}+\frac{1}{2}}\\ &= \frac{C_{N,p,n}2^{j(k+n)}}{(1+2^j|x|)^N}|Q|^{1+\frac{k}{n}-\frac{1}{p}}\end{aligned}$$

for $x \in (Q^*)^c$. For such x we now have

$$\Big(\sum_{j\in\mathbf{Z}}|\Delta_j(A)(x)|^2\Big)^{\frac{1}{2}} \le C_{N,p,n}|Q|^{1+\frac{k}{n}-\frac{1}{p}}\Big(\sum_{j\in\mathbf{Z}}\frac{2^{2j(k+n)}}{(1+2^j|x|)^{2N}}\Big)^{\frac{1}{2}}. \tag{6.6.14}$$

It is a simple fact that the series in (6.6.14) converges. Indeed, considering the cases $2^j \le \frac{1}{|x|}$ and $2^j > \frac{1}{|x|}$ we see that both terms in the second series in (6.6.14) contribute at most a fixed multiple of $|x|^{-2k-2n}$. It remains to estimate the L^p quasi-norm of the square root of the second series in (6.6.14) raised over $(Q^*)^c$. This is bounded by a constant multiple of

$$\bigg(\int_{(Q^*)^c} \frac{1}{|x|^{p(k+n)}}\,dx\bigg)^{\frac{1}{p}} \le C_{n,p}\bigg(\int_{c|Q|^{\frac{1}{n}}} r^{-p(k+n)+n-1}\,dr\bigg)^{\frac{1}{p}},$$

for some constant c, and the latter is easily seen to be bounded above by a constant multiple of $|Q|^{-1-\frac{k}{n}+\frac{1}{p}}$. Here we use the fact that $p(k+n) > n$ or, equivalently, $k > \frac{n}{p} - n$, which is certainly true since k was chosen to be $[\frac{n}{p} - n]+1$. Combining this estimate with the one obtained in (6.6.13), we conclude the proof of the theorem. □

We now know that L^q-atoms for H^p are indeed elements of H^p. The main result of this section is to obtain the converse (i.e., every element of H^p can be decomposed as a sum of L^2-atoms for H^p).

Applying the same idea as in Corollary 6.6.7 to H^p, we obtain the following result.

Theorem 6.6.10. *Let $0 < p \le 1$. Given a distribution $f \in H^p(\mathbf{R}^n)$, there exists a sequence of L^2 atoms for H^p, $\{A_j\}_{j=1}^\infty$, and a sequence of scalars $\{\lambda_j\}_{j=1}^\infty$ such that*

$$\sum_{j=1}^{N} \lambda_j A_j \to f \qquad \text{in } H^p.$$

Moreover we have

$$\|f\|_{H^p} \approx \inf\bigg\{\Big(\sum_{j=1}^{\infty} |\lambda_j|^p\Big)^{\frac{1}{p}} : \ f = \lim_{N\to\infty} \sum_{j=1}^{N} \lambda_j A_j, \quad A_j \text{ are } L^2\text{-atoms for } H^p \text{ and the limit is taken in } H^p\bigg\}. \tag{6.6.15}$$

Proof. Let A_j be L^2-atoms for H^p and $\sum_{j=1}^\infty |\lambda_j|^p < \infty$. It follows from Theorem 6.6.9 that

$$\Big\|\sum_{j=1}^{N} \lambda_j A_j\Big\|_{H^p}^p \le C_{n,p}^p \sum_{j=1}^{N} |\lambda_j|^p.$$

Thus if the sequence $\sum_{j=1}^N \lambda_j A_j$ converges to f in H^p, then

$$\|f\|_{H^p} \le C_{n,p}\Big(\sum_{j=1}^{\infty} |\lambda_j|^p\Big)^{\frac{1}{p}},$$

which proves the direction $\le$ in (6.6.15). The gist of the theorem is contained in the converse statement.

Using Theorem 6.6.3 (with $L = [\frac{n}{p} - n]$), we can write every element f in $\dot{F}^p_{0,2} = H^p$ as a sum of the form $f = \sum_{Q \in \mathcal{D}} s_Q a_Q$, where each a_Q is a smooth L-atom for the cube Q and $s = \{s_Q\}_{Q \in \mathcal{D}}$ is a sequence in $\dot{f}^p_{0,2}$. We now use Theorem 6.6.5 to write the sequence $s = \{s_Q\}_Q$ as

$$s = \sum_{j=1}^{\infty} \lambda_j r_j$$

as a sum of ∞-atoms r_j for $\dot{f}^p_{0,2}$ such that

$$(\sum_{j=1}^{\infty} |\lambda_j|^p)^{\frac{1}{p}} \le C \|s\|_{\dot{f}^p_{0,2}} \le C \|f\|_{H^p} . \tag{6.6.16}$$

Then we have

$$f = \sum_{Q \in \mathcal{D}} s_Q a_Q = \sum_{Q \in \mathcal{D}} \sum_{j=1}^{\infty} \lambda_j \, r_{j,Q} \, a_Q = \sum_{j=1}^{\infty} \lambda_j A_j , \tag{6.6.17}$$

where we set

$$A_j = \sum_{Q \in \mathcal{D}} r_{j,Q} \, a_Q \tag{6.6.18}$$

and the series in (6.6.17) converges in $\mathcal{S}'(\mathbf{R}^n)$. Next we show that each A_j is a fixed multiple of an L^2-atom for H^p. Let us fix an index j. By the definition of the ∞-atom for $\dot{f}^p_{0,2}$, there exists a dyadic cube Q_0^j such that $r_{j,Q} = 0$ for all dyadic cubes Q not contained in Q_0^j. Then the support of each a_Q that appears in (6.6.18) is contained in $3Q$, hence in $3Q_0^j$. This implies that the function A_j is supported in $3Q_0^j$. The same is true for the function $g^{0,2}(r_j)$ defined in (6.6.1). Using this fact, we have

$$\begin{aligned}
\|A_j\|_{L^2} &\approx \|A_j\|_{\dot{F}^2_{0,2}} \\
&\le c \|r_j\|_{\dot{f}^2_{0,2}} \\
&= c \|g^{0,2}(r_j)\|_{L^2} \\
&\le c \|g^{0,2}(r_j)\|_{L^\infty} |3Q_0^j|^{\frac{1}{2}} \\
&\le c \, |3Q_0^j|^{-\frac{1}{p}+\frac{1}{2}} .
\end{aligned}$$

Since the series (6.6.18) defining A_j converges in L^2 and A_j is supported in some cube, this series also converges in L^1. It follows that the vanishing moment conditions of A_j are inherited from those of each a_Q. We conclude that each A_j is a fixed multiple of an L^2-atom for H^p.

Finally, we need to show that the series in (6.6.17) converges in $H^p(\mathbf{R}^n)$. But

$$\Big\| \sum_{j=N}^{M} \lambda_j A_j \Big\|_{H^p} \le C_{n,p} \Big(\sum_{j=N}^{M} |\lambda_j|^p \Big)^{\frac{1}{p}} \to 0$$

as $M, N \to \infty$ in view of the convergence of the series in (6.6.16). This implies that the series $\sum_{j=1}^{\infty} \lambda_j A_j$ is Cauchy in H^p and since it converges to f in $\mathcal{S}'(\mathbf{R}^n)$, it must converge to f in H^p. Combining this fact with (6.6.16) yields the direction $\ge$ in (6.6.15).

□

Remark 6.6.11. Property (c) in Definition 6.6.8 can be replaced by

$$\int x^\gamma A(x)\, dx = 0 \qquad \text{for all multiindices } \gamma \text{ with } |\gamma| \le L,$$

for any $L \ge [\frac{n}{p} - n]$ and the atomic decomposition of H^p holds unchanged. In fact, in the proof of Theorem 6.6.10 we may take $L \ge [\frac{n}{p} - n]$ instead of $L = [\frac{n}{p} - n]$ and then apply Theorem 6.6.3 for this L. Observe that Theorem 6.6.3 was valid for all $L \ge [\frac{n}{p} - n]$.

This observation can be very useful in concrete applications.

Exercises

6.6.1. (a) Prove that there exists a Schwartz function Θ supported in the unit ball $|x| \le 1$ such that $\int_{\mathbf{R}^n} x^\gamma \Theta(x)\, dx = 0$ for all multiindices γ with $|\gamma| \le N$ and such that $|\widehat{\Theta}| \ge \frac{1}{2}$ on the annulus $\frac{1}{2} \le |\xi| \le 2$.
(b) Prove there exists a Schwartz function Ψ whose Fourier transform is supported in the annulus $\frac{1}{2} \le |\xi| \le 2$ and is at least $c > 0$ in the smaller annulus $\frac{3}{5} \le |\xi| \le \frac{5}{3}$ such that we have

$$\sum_{j \in \mathbf{Z}} \widehat{\Psi}(2^{-j}\xi) \widehat{\Theta}(2^{-j}\xi) = 1$$

for all $\xi \in \mathbf{R}^n \setminus \{0\}$.
[*Hint:* Part (a): Let θ be a real-valued Schwartz function supported in the ball $|x| \le 1$ and such that $\widehat{\theta}(0) = 1$. Then for some $\varepsilon > 0$ we have $\widehat{\theta}(\xi) \ge \frac{1}{2}$ for all ξ satisfying $|\xi| < 2\varepsilon < 1$. Set $\Theta = (-\Delta)^N(\theta_\varepsilon)$. Part (b): Set $\widehat{\Psi(\xi)} = \widehat{\eta}(\xi) \big(\sum_{j \in \mathbf{Z}} \widehat{\eta}(2^{-j}\xi) \widehat{\Theta}(2^{-j}\xi) \big)^{-1}$ for a suitable η.]

6.6.2. Let $\alpha \in \mathbf{R}$, $0 < p \le 1$, $p \le q \le +\infty$.
(a) For all f, g in $\mathcal{S}'(\mathbf{R}^n)$ show that

$$\|f + g\|_{\dot{F}^p_{\alpha,q}}^p \le \|f\|_{\dot{F}^p_{\alpha,q}}^p + \|g\|_{\dot{F}^p_{\alpha,q}}^p .$$

(b) For all sequences $\{s_Q\}_{Q\in\mathcal{D}}$ and $\{t_Q\}_{Q\in\mathcal{D}}$ show that

$$\big\|\{s_Q\}_Q + \{t_Q\}_Q\big\|^p_{\dot f^p_{\alpha,q}} \le \big\|\{s_Q\}_Q\big\|^p_{\dot f^p_{\alpha,q}} + \big\|\{t_Q\}_Q\big\|^p_{\dot f^p_{\alpha,q}}\,.$$

[*Hint:* Use $|a+b|^p \le |a|^p + |b|^p$ and apply Minkowski's inequality on $L^{q/p}$ (or on $\ell^{q/p}$).]

6.6.3. Let Φ is a smooth function supported in the unit ball of $\mathbf{R}^n$. Use the same idea as in Theorem 6.6.9 to show directly (without appealing to any other theorem) that the smooth maximal function $M(\cdot,\Phi)$ of an L^2-atom for H^p lies in L^p when $p<1$. Recall $M(f,\Phi) = \sup\limits_{t>0} |\Phi_t * f|$.

6.6.4. Extend Theorem 6.6.9 to the case $1 < q \le \infty$. Precisely, prove that there is a constant $C_{n,p,q}$ such that every L^q-atom A for H^p satisfies

$$\big\|A\big\|_{H^p} \le C_{n,p,q}\,.$$

[*Hint:* If $1<q<2$, use the boundedness of the square function on L^q and for $2\le q\le\infty$ its boundedness on L^2.]

6.6.5. Show that the space H^p_F of all finite linear combinations of L^2-atoms for H^p is dense in H^p.
[*Hint:* Use Theorem 6.6.10.]

6.6.6. Show that for all $\mu, j\in\mathbf{Z}$, all $N,b>0$ satisfying $N>n/b$ and $b<1$, all scalars s_Q (indexed by dyadic cubes Q with centers c_Q), and all $x\in\mathbf{R}^n$ we have

$$\sum_{\substack{Q\in\mathcal{D}\\ \ell(Q)=2^{-\mu}}} \frac{|s_Q|}{(1+2^{\min(j,\mu)}|x-c_Q|)^N}$$

$$\le c(n,N,b)\, 2^{\max(\mu-j,0)\frac{n}{b}} \bigg\{ M\Big(\sum_{\substack{Q\in\mathcal{D}\\ \ell(Q)=2^{-\mu}}} |s_Q|^b \chi_Q\Big)(x)\bigg\}^{\frac{1}{b}},$$

where M is the Hardy-Littlewood maximal operator and $c(n,N,b)$ is a constant.
[*Hint:* Define $\mathcal{F}_0 = \big\{Q\in\mathcal{D}:\ \ell(Q)=2^{-\mu},\ |c_Q-x|\,2^{\min(j,\mu)}\le 1\big\}$ and $\mathcal{F}_k = \big\{Q\in\mathcal{D}:\ \ell(Q)=2^{-\mu},\ 2^{k-1}<|c_Q-x|\,2^{\min(j,\mu)}\le 2^k\big\}$ for $k\ge 1$. Break up the sum on the left as a sum over the families $\mathcal{F}_k$ and use that $\sum_{Q\in\mathcal{F}_k}|s_Q| \le \big(\sum_{Q\in\mathcal{F}_k}|s_Q|^b\big)^{1/b}$ and the fact that $\big|\bigcup_{Q\in\mathcal{F}_k} Q\big| \le c_n 2^{-\min(j,\mu)n+kn}$.]

6.6.7. Let A be an L^2-atom for $H^p(\mathbf{R}^n)$ for some $0 < p < 1$. Show that there is a constant C such that for all multiindices α with $|\alpha| \le k = [\frac{n}{p} - n]$ we have

$$\sup_{\xi \in \mathbf{R}^n} |\xi|^{|\alpha|-k-1} |(\partial^\alpha A)(\xi)| \le C \left\| A \right\|_{L^2(\mathbf{R}^n)}^{-\frac{2p}{2-p}(\frac{k+1}{n}+\frac{1}{2})-1} .$$

[*Hint:* Subtract the Taylor polynomial of degree $k - |\alpha|$ at 0 of the function $x \to e^{-2\pi i x \cdot \xi}$.]

6.6.8. Let A be an L^2-atom for $H^p(\mathbf{R}^n)$ for some $0 < p < 1$. Show that for all multiindices α and all $1 \le r \le \infty$ there is a constant C such that

$$\left\| |\partial^\alpha \widehat{A}|^2 \right\|_{L^{r'}(\mathbf{R}^n)} \le C \left\| A \right\|_{L^2(\mathbf{R}^n)}^{-\frac{2p}{2-p}(\frac{2|\alpha|}{n}+\frac{1}{r})-2} .$$

[*Hint:* In the case $r = 1$ use the $L^1 \to L^\infty$ boundedness of the Fourier transform and in the case $r = \infty$ use Plancherel's theorem. For general r use interpolation.]

6.6.9. Let f be in $H^p(\mathbf{R}^n)$ for some $0 < p \le 1$. Then the Fourier transform of f originally defined as a tempered distribution, is a continuous function that satisfies

$$|\widehat{f}(\xi)| \le C_{n,p} \left\| f \right\|_{H^p(\mathbf{R}^n)} |\xi|^{\frac{n}{p}-n}$$

for some constant $C_{n,p}$ independent of f.
[*Hint:* If f is an L^2 atom for H^p, combine the estimates of Exercises 6.6.7 and 6.6.8 with $\alpha = 0$ (and $r = 1$). In general, apply Theorem 6.6.10.]

6.6.10. Let A be an L^∞-atom for $H^p(\mathbf{R}^n)$ for some $0 < p < 1$ and let $\alpha = \frac{n}{p} - n$. Show that there is a constant $C_{n,p}$ such that for all g in $\dot{\Lambda}_\alpha(\mathbf{R}^n)$ we have

$$\left| \int_{\mathbf{R}^n} A(x) g(x)\, dx \right| \le C_{n,p} \left\| g \right\|_{\dot{\Lambda}_\alpha(\mathbf{R}^n)} .$$

[*Hint:* Suppose that A is supported in a cube Q of side length $2^{-\nu}$ and center c_Q. Write the previous integrand as $\sum_j \Delta_j(A)\Delta_j(g)$ for a suitable Littlewood-Paley operator Δ_j and apply the result of Appendix K.2 to obtain the estimate

$$\left| \Delta_j(A)(x) \right| \le C_N |Q|^{-\frac{1}{p}+1} \frac{2^{\min(j,\nu)n} 2^{-|j-\nu|D}}{\left(1 + 2^{\min(j,\nu)} |x - c_Q|\right)^N}$$

where $D = [\alpha] + 1$ when $\nu \ge j$ and $D = 0$ when $\nu < j$. Use Theorem 6.3.6.]

6.7. Singular Integrals on Function Spaces

Our final task in this chapter is to investigate the action of singular integrals on function spaces. The emphasis of our study focuses on Hardy spaces, although we point out that with no additional effort the action of singular integrals on other spaces can also be obtained.

6.7.a. Singular Integrals on the Hardy Space H^1

Before we state the main results in this subject, we review some basic background on singular integrals discussed in Chapter 4.

Let $K(x)$ be a function defined away from the origin on $\mathbf{R}^n$ that satisfies the size estimate

$$\sup_{0<R<\infty} \frac{1}{R}\int_{|x|\le R} |K(x)|\,|x|\,dx \le A_1\,, \tag{6.7.1}$$

Hörmander's condition

$$\sup_{y\in\mathbf{R}^n\setminus\{0\}} \int_{|x|\ge 2|y|} |K(x-y)-K(x)|\,dx \le A_2\,, \tag{6.7.2}$$

and the cancellation condition

$$\sup_{0<R_1<R_2<\infty} \left| \int_{R_1<|x|<R_2} K(x)\,dx \right| \le A_3\,, \tag{6.7.3}$$

and $A_1, A_2, A_3 < \infty$. Condition (6.7.3) implies that there exists a sequence $\varepsilon_j \downarrow 0$ as $j \to \infty$ such that the limit exists:

$$\lim_{j\to\infty} \int_{\varepsilon_j\le|x|\le 1} K(x)\,dx = L_0.$$

This gives that for a smooth and compactly supported function f on $\mathbf{R}^n$, the limit

$$\lim_{j\to\infty} \int_{|x-y|>\varepsilon_j} K(x-y)f(y)\,dy = T(f)(x) \tag{6.7.4}$$

exists and defines a linear operator T. This operator T is given by convolution with a tempered distribution W that coincides with the function K on $\mathbf{R}^n \setminus \{0\}$.

We know by the results of Chapter 4 that such a T, which is initially defined on $C_0^\infty(\mathbf{R}^n)$, admits an extension that is L^p bounded for all $1 < p < \infty$ and also of weak type $(1,1)$. All these norms are bounded above by dimensional multiples of the quantity $A_1 + A_2 + A_3$ (c.f. Theorem 4.4.1). Therefore, such a T is well defined on $L^1(\mathbf{R}^n)$ and in particular on $H^1(\mathbf{R}^n)$, which is contained in $L^1(\mathbf{R}^n)$. We begin with the following result.

Theorem 6.7.1. *Let K satisfy (6.7.1), (6.7.2), and (6.7.3), and let T be defined as in (6.7.4). Then there is a constant C_n such that for all f in $H^1(\mathbf{R}^n)$ we have*

$$\big\|T(f)\big\|_{L^1} \le C_n(A_1 + A_2 + A_3)\big\|f\big\|_{H^1}\,. \tag{6.7.5}$$

Proof. To prove this theorem we have a powerful tool at our disposal, the atomic decomposition of $H^1(\mathbf{R}^n)$. It is therefore natural to start by checking the validity of (6.7.5) whenever f is an L^2-atom for H^1.

Since T is a convolution operator (i.e., it commutes with translations), it suffices to take the atom f supported in a cube Q centered at the origin. Let $f = a$ be such an atom, supported in Q, and let $Q^* = 2\sqrt{n}\,Q$. We write

$$\int_{\mathbf{R}^n} |T(a)(x)|\,dx = \int_{Q^*} |T(a)(x)|\,dx + \int_{(Q^*)^c} |T(a)(x)|\,dx \tag{6.7.6}$$

and we estimate each term separately. We have

$$\begin{aligned}
\int_{Q^*} |T(a)(x)|\,dx \;&\le |Q^*|^{\frac{1}{2}}\bigg(\int_{Q^*} |T(a)(x)|^2\,dx\bigg)^{\frac{1}{2}} \\
&\le C_n(A_1 + A_2 + A_3)|Q^*|^{\frac{1}{2}}\bigg(\int_{Q} |a(x)|^2\,dx\bigg)^{\frac{1}{2}} \\
&\le C_n(A_1 + A_2 + A_3)|Q^*|^{\frac{1}{2}}|Q|^{\frac{1}{2}-\frac{1}{1}} = C_n'(A_1 + A_2 + A_3)\,,
\end{aligned}$$

where we used property (b) of atoms in Definition 6.6.8. Now note that if $x \notin Q^*$ and $y \in Q$, then $|x| \ge 2|y|$ and $x - y$ stays away from zero; thus $K(x - y)$ is well defined. Moreover, in this case $T(a)(x)$ can be expressed as a convergent integral of $a(y)$ against $K(x - y)$. We have

$$\begin{aligned}
\int_{(Q^*)^c} |T(a)(x)|\,dx \;&= \int_{(Q^*)^c} \bigg|\int_Q K(x-y)a(y)\,dy\bigg|\,dx \\
&= \int_{(Q^*)^c} \bigg|\int_Q \big(K(x-y) - K(x)\big)a(y)\,dy\bigg|\,dx \\
&\le \int_Q \int_{(Q^*)^c} \big|K(x-y) - K(x)\big|\,dx\,|a(y)|\,dy \\
&\le \int_Q \int_{|x|\ge 2|y|} \big|K(x-y) - K(x)\big|\,dx\,|a(y)|\,dy \\
&\le A_2 \int_Q |a(x)|\,dx \\
&\le A_2|Q|^{\frac{1}{2}}\bigg(\int_Q |a(x)|^2\,dx\bigg)^{\frac{1}{2}} \\
&\le A_2|Q|^{\frac{1}{2}}|Q|^{\frac{1}{2}-\frac{1}{1}} = A_2\,.
\end{aligned}$$

Combining this calculation with the previous one and inserting the final conclusions in (6.7.6) we obtain that for all L^2-atoms a for H^1 we have

$$\big\|T(a)\big\|_{L^1} \le (C_n'+1)(A_1+A_2+A_3)\,.$$

We now pass to general f in H^1. In view of Theorem 6.6.10 we can write a general $f\in H^1$ as

$$f=\sum_{j=1}^{\infty}\lambda_j a_j\,,$$

where the series converges in H^1, the a_j are L^2 atoms for H^1, and

$$\big\|f\big\|_{H^1}\approx\sum_{j=1}^{\infty}|\lambda_j|\,.$$

Since T maps L^1 into weak L^1 (Theorem 4.3.3), $T(f)$ is already a well-defined $L^{1,\infty}$ function. We plan to prove that

$$T(f)=\sum_{j=1}^{\infty}\lambda_j T(a_j)\qquad \text{a.e.} \tag{6.7.7}$$

Once (6.7.7) is established, the required conclusion (6.7.5) follows easily by taking L^1 norms. We observe that the series in (6.7.7) converges in L^1.

To prove (6.7.7), we will use the fact that T is of weak type $(1,1)$. For a given $\delta>0$ we have

$$\begin{aligned}
&\Big|\Big\{\Big|T(f)-\sum_{j=1}^{\infty}\lambda_jT(a_j)\Big|>\delta\Big\}\Big|\\
&\le\Big|\Big\{\Big|T(f)-\sum_{j=1}^{N}\lambda_jT(a_j)\Big|>\delta/2\Big\}\Big|+\Big|\Big\{\Big|\sum_{j=N+1}^{\infty}\lambda_jT(a_j)\Big|>\delta/2\Big\}\Big|\\
&\le\frac{2}{\delta}\big\|T\big\|_{L^1\to L^{1,\infty}}\Big\|f-\sum_{j=1}^{N}\lambda_ja_j\Big\|_{L^1}+\frac{2}{\delta}\Big\|\sum_{j=N+1}^{\infty}\lambda_jT(a_j)\Big\|_{L^1}\\
&\le\frac{2}{\delta}\big\|T\big\|_{L^1\to L^{1,\infty}}\Big\|f-\sum_{j=1}^{N}\lambda_ja_j\Big\|_{L^1}+\frac{2}{\delta}C(A_1+A_2+A_3)\sum_{j=N+1}^{\infty}|\lambda_j|\,.
\end{aligned}$$

Since $\sum_{j=1}^{N}\lambda_ja_j$ converges to f in H^1, it converges in L^1 and both terms in the sum converge to zero as $N\to\infty$. We conclude that

$$\Big|\Big\{\Big|T(f)-\sum_{j=1}^{\infty}\lambda_jT(a_j)\Big|>\delta\Big\}\Big|=0$$

for all $\delta>0$, which implies (6.7.7). $\square$

6.7.b. Singular Integrals on Besov-Lipschitz Spaces

We continue with a corollary concerning Besov-Lipschitz spaces.

Corollary 6.7.2. *Let K satisfy (6.7.1), (6.7.2), and (6.7.3), and let T be defined as in (6.7.4). Let $1 \le p \le \infty$, $0 < q \le \infty$ and $\alpha \in \mathbf{R}$. Then there is a constant $C_{n,p,q,\alpha}$ such that for all f in $\mathcal{S}(\mathbf{R}^n)$ we have*

$$\big\|T(f)\big\|_{\dot{B}^p_{\alpha,q}} \le C_n(A_1 + A_2 + A_3)\big\|f\big\|_{\dot{B}^p_{\alpha,q}}. \tag{6.7.8}$$

Therefore, T admits a bounded extension on all homogeneous Besov-Lipschitz spaces $\dot{B}^p_{\alpha,q}$ with $p \ge 1$; in particular, on all homogeneous Lipschitz spaces.

Proof. Let Ψ be a Schwartz function whose Fourier transform is supported in the annulus $\frac{1}{2} \le |\xi| \le 2$ and that satisfies

$$\sum_{j\in\mathbf{Z}} \widehat{\Psi}(2^{-j}\xi) = 1\,, \qquad \xi \ne 0\,.$$

Pick a Schwartz function ζ whose Fourier transform $\widehat{\zeta}$ is supported in the annulus $\frac{1}{4} < |\xi| < 8$ and that is equal to one on the support of $\widehat{\Psi}$. Let W be the tempered distribution that coincides with K on $\mathbf{R}^n \setminus \{0\}$ so that $T(f) = f * W$. Then we have $\zeta_{2^{-j}} * \Psi_{2^{-j}} = \Psi_{2^{-j}}$ for all j and hence

$$\begin{aligned}\big\|\Delta_j(T(f))\big\|_{L^p} &= \big\|\zeta_{2^{-j}} * \Psi_{2^{-j}} * W * f\big\|_{L^p} \\ &\le \big\|\zeta_{2^{-j}} * W\big\|_{L^1}\big\|\Delta_j(f)\big\|_{L^p},\end{aligned} \tag{6.7.9}$$

since $1 \le p \le \infty$. Note that the function $\zeta_{2^{-j}}$ is in L^1 and has mean value zero. Therefore, $\zeta_{2^{-j}}$ is in H^1. Using Theorem 6.7.1, we conclude that

$$\big\|T(\zeta_{2^{-j}})\big\|_{L^1} = \big\|\zeta_{2^{-j}} * W\big\|_{L^1} \le C\big\|\zeta_{2^{-j}}\big\|_{L^1} = C'\,.$$

Inserting this in (6.7.9), multiplying by $2^{j\alpha}$, and taking ℓ^q norms, we obtain the required conclusion.

$\square$

6.7.c. Singular Integrals on $H^p(\mathbf{R}^n)$

We are now interested in extending Theorem 6.7.1 to other H^p spaces for $p < 1$. It turns out that if some additional smoothness assumptions on K are imposed, then we may actually obtain boundedness of singular integrals from H^p into H^p for $p \le 1$.

Theorem 6.7.3. *Let $p \le 1$ and $N = [\frac{n}{p} - n]$. Let K be a C^N function on $\mathbf{R}^n \setminus \{0\}$ that satisfies*

$$|\partial^\beta K(x)| \le A\,|x|^{-n-|\beta|} \tag{6.7.10}$$

for all multiindices $|\beta| \le N$ and all $x \neq 0$. Let W be a tempered distribution that coincides with K on $\mathbf{R}^n \setminus \{0\}$ whose Fourier transform is bounded function that satisfies

$$|\widehat{W}(\xi)| \le B.$$

Then the operator $T(f) = f * W$ initially defined for f in the Schwartz class whose support vanishes in a neighborhood of thc origin admits an extension which is bounded on H^p, that is, it satisfies for all $f \in H^p$

$$\big\|T(f)\big\|_{H^p} \le C_{n,p}(A+B)\big\|f\big\|_{H^p}$$

for some constant $C_{n,p}$.

Proof. The proof of this theorem provides another classical application of the atomic decomposition of H^p. However, we only use the atomic decomposition for the domain H^p space, while it is best to use the maximal (or square function) characterization of H^p for the target H^p space.

We fix a smooth function Φ supported in the unit ball $B(0,1)$ in $\mathbf{R}^n$ whose mean value is not equal to zero. For $t>0$ we define the smooth functions

$$W^{(t)} = \Phi_t * W$$

and we observe that they satisfy

$$\sup_{t>0} \big|\widehat{W^{(t)}}(\xi)\big| \le \big\|\widehat{\Phi}\big\|_{L^\infty} B \tag{6.7.11}$$

and that

$$\sup_{t>0} |\partial^\beta W^{(t)}(x)| \le C_\Phi A\, |x|^{-n-|\beta|} \tag{6.7.12}$$

for all $|\beta| \le N$, where

$$C_\Phi = \sup_{|\gamma|\le N} \int_{\mathbf{R}^n} |\xi|^{|\gamma|} |\widehat{\Phi}(\xi)|\, d\xi\,.$$

Indeed, assertion (6.7.11) is easily verified while assertion (6.7.12) follows from the identity

$$W^{(t)}(x) = \big((\Phi_t * W)^\frown\big)^\vee(x) = \int_{\mathbf{R}^n} e^{2\pi i x\cdot\xi}\, \widehat{W}(\xi)\, \widehat{\Phi}(t\xi)\, d\xi$$

whenever $|x| \le 2t$ and from (6.7.10) and the fact that for $|x| \ge 2t$ we have the integral representation

$$\partial^\beta W^{(t)}(x) = \int_{|y|\le t} \partial^\beta K(x-y)\, \Phi_t(y)\, dy\,.$$

We now take $f = a$ to be an L^2-atom for H^p and without loss of generality we may assume that a is supported in a cube Q centered at the origin. We set Q^* be the cube with side length $2\sqrt{n}\,\ell(Q)$, where $\ell(Q)$ is the side length of Q. Recall the smooth maximal function $M(f;\Phi)$ from Section 6.4. Then $M(T(a);\Phi)$ is pointwise controlled by the Hardy-Littlewood maximal function of $T(a)$. Using an argument similar to that in Theorem 6.7.1, we have

$$\begin{aligned}
\Big(\int_{Q^*} |M(T(a);\Phi)(x)|^p\,dx\Big)^{\frac{1}{p}} &\le \|\Phi\|_{L^1}\Big(\int_{Q^*} |M(T(a))(x)|^p\,dx\Big)^{\frac{1}{p}} \\
&\le C|Q^*|^{\frac{1}{p}-\frac{1}{2}}\Big(\int_{Q^*} |M(T(a))(x)|^2\,dx\Big)^{\frac{1}{2}} \\
&\le C'|Q|^{\frac{1}{p}-\frac{1}{2}}\Big(\int_{\mathbf{R}^n} |T(a)(x)|^2\,dx\Big)^{\frac{1}{2}} \\
&\le C''B|Q|^{\frac{1}{p}-\frac{1}{2}}\Big(\int_{Q} |a(x)|^2\,dx\Big)^{\frac{1}{2}} \\
&\le C_nB|Q|^{\frac{1}{p}-\frac{1}{2}}|Q|^{\frac{1}{2}-\frac{1}{p}} \\
&= C_nB\,.
\end{aligned}$$

It therefore remains to estimate the contribution of $M(T(a);\Phi)$ on the complement of Q^*.

Now note that if $x\notin Q^*$ and $y\in Q$, then $|x|\ge 2|y|$ stays away from zero and $K(x-y)$ is well defined. We have

$$(T(a)*\Phi_t)(x) = (a*W^{(t)})(x) = \int_Q K^{(t)}(x-y)\,a(y)\,dy\,.$$

Recall $N=[\frac{n}{p}-n]$. Using the cancellation of atoms for H^p we deduce

$$\begin{aligned}
(T(a)*\Phi_t)(x) &= \int_Q a(y)K^{(t)}(x-y)\,dy \\
&= \int_Q a(y)\Big[K^{(t)}(x-y) - \sum_{|\beta|\le N-1}(\partial^\beta K^{(t)})(x)\frac{(y)^\beta}{\beta!}\Big]dy \\
&= \int_Q a(y)\Big[\sum_{|\beta|=N}(\partial^\beta K^{(t)})(x-\theta_y y)\frac{(y)^\beta}{\beta!}\Big]dy
\end{aligned}$$

for some $0\le\theta_y\le 1$. Using that $|x|\ge 2|y|$ and (6.7.12) we obtain the estimate

$$|(T(a)*\Phi_t)(x)| \le c_{n,N}\frac{A}{|x|^{N+n}}\int_Q |a(y)|\,|y|^{|\beta|}\,dy$$

from which it follows that for $x\notin Q^*$ we have

$$|(T(a)*\Phi_t)(x)| \le c_{n,p}\frac{A}{|x|^{N+n}}|Q|^{1+\frac{N}{n}-\frac{1}{p}}$$

via a calculation using properties of atoms (see the proof of Theorem 6.6.9). Taking the supremum over all $t > 0$ and integrating over $(Q^*)^c$, we obtain that

$$\Big(\int_{(Q^*)^c}\sup_{t>0}|(T(a)*\Phi_t)(x)|^p dx\Big)^{\frac{1}{p}} \le c_{n,p}|Q|^{1+\frac{N}{n}-\frac{1}{p}}\Big(\int_{(Q^*)^c}\frac{1}{|x|^{p(N+n)}}\,dx\Big)^{\frac{1}{p}}$$

and the latter is easily seen to be finite and controlled by a constant multiple of A. Combining this estimate with the previously obtained for the integral of $M(T(a);\Phi) = \sup_{t>0}|(T(a)*\Phi_t|$ over Q^* yields the conclusion of the theorem when $f = a$ is an atom.

We have now shown that there exists a constant $C_{n,p}$ such that

$$\big\|T(a)\big\|_{H^p} \le C_{n,p}(A+B)$$

whenever a is an L^2 atom for H^p. Let H^p_F be the set of all finite linear combination of L^2-atoms for H^p. Let f be an element of H^p_F and pick a representation of $f = \sum_{j=1}^N \lambda_j a_j$ such that

$$\big\|f\big\|_{H^p} \approx \Big(\sum_{j=1}^N |\lambda_j|^p\Big)^{\frac{1}{p}}.$$

Using the sublinearity of the quantity $\|\cdot\|^p_{H^p}$, it follows that

$$\big\|T(f)\big\|_{H^p} \le C'_{n,p}(A+B)\big\|f\big\|_{H^p}.$$

Since H^p_F is dense in H^p (see Exercise 6.6.5), it follows that T admits an extension that maps H^p into itself.

□

6.7.d. A Singular Integral Characterization of $H^1(\mathbf{R}^n)$

In this subsection we obtain a characterization of $H^1(\mathbf{R}^n)$ using the Riesz transforms.

Theorem 6.7.4. *(a) Let $n \ge 2$. A function lies in the Hardy space $H^1(\mathbf{R}^n)$ if and only if it lies in $L^1(\mathbf{R}^n)$ and its Riesz transforms are also in $L^1(\mathbf{R}^n)$. Precisely, there exist dimensional constants c_n, C_n such that for f in $L^1(\mathbf{R}^n)$ we have*

$$c_n\big\|f\big\|_{H^1} \le \big\|f\big\|_{L^1} + \sum_{k=1}^n \big\|R_k(f)\big\|_{L^1} \le C_n\big\|f\big\|_{H^1}. \tag{6.7.13}$$

When $n = 1$ the corresponding statement is

$$c_1\big\|f\big\|_{H^1} \le \big\|f\big\|_{L^1} + \big\|H(f)\big\|_{L^1} \le C_1\big\|f\big\|_{H^1}, \tag{6.7.14}$$

which says that a function f lies in the Hardy space $H^1(\mathbf{R})$ if and only if it lies in $L^1(\mathbf{R})$ and its Hilbert transform lies also in $L^1(\mathbf{R})$.
(b) Functions in $H^1(\mathbf{R}^n)$, $n \ge 1$, have integral zero.

Proof. The upper estimates in (6.7.13) and (6.7.14) are consequences of Theorem 6.7.1. To obtain the converse statements, we use the square function characterization of H^1. We fix a nice Schwartz function Ψ whose Fourier transform is supported in the annulus $\frac{1}{2} \le |\xi| \le 2$ and that satisfies (6.2.6). Then we have

$$\big\|f\big\|_{L^1} \le \big\|f\big\|_{H^1} \le C_n \Big\| \Big(\sum_j |\Delta_j(f)|^2 \Big)^{\frac{1}{2}} \Big\|_{L^1} .$$

Assuming that $R_k(f) \in L^1$, we can apply this estimate to $R_k(f)$. We obtain

$$\big\|R_k(f)\big\|_{L^1} \le C_n \Big\| \Big(\sum_j |\Delta_j(R_k(f))|^2 \Big)^{\frac{1}{2}} \Big\|_{L^1} = C_n \Big\| \Big(\sum_j |\Delta_j^{\theta_k}(f)|^2 \Big)^{\frac{1}{2}} \Big\|_{L^1} ,$$

where $\widehat{\theta_k}(\xi) = -i\widehat{\Psi}(\xi)\frac{\xi_k}{|\xi|}$. Then Corollary 6.5.7 implies that

$$\Big\| \Big(\sum_j |\Delta_j^{\theta_k}(f)|^2 \Big)^{\frac{1}{2}} \Big\|_{L^1} \le C_n' \Big\| \Big(\sum_j |\Delta_j(f)|^2 \Big)^{\frac{1}{2}} \Big\|_{L^1} ,$$

which combined with the previous estimate yields (6.7.13). The corresponding estimate (6.7.14) is obtained analogously.

Part (b) is a consequence of part (a). Indeed, if $f \in H^1(\mathbf{R}^n)$, we must have $R_1(f) \in L^1(\mathbf{R}^n)$, thus $\widehat{R_1(f)}$ is uniformly continuous. But since

$$\widehat{R_1(f)}(\xi) = -i\widehat{f}(\xi)\frac{\xi_1}{|\xi|} ,$$

it follows that $\widehat{R_1(f)}$ is continuous at zero if and only if $\widehat{f}(\xi) = 0$. But this happens exactly when f has integral zero.

□

We refer to Exercise 6.7.1 for an extension of this characterization to H^p for $p < 1$.

Exercises

6.7.1. Prove the following generalization of Theorem 6.7.4. Let φ be a fixed Schwartz function with nonvanishing integral and let $0 < p < 1$. Prove that a distribution lies in the Hardy space $H^p(\mathbf{R}^n)$ if and only if

$$c_n \big\|f\big\|_{H^p} \le \sup_{\varepsilon>0} \big\|\varphi_\varepsilon * f\big\|_{L^p} + \sum_{k=1}^{n} \sup_{\varepsilon>0} \big\|\varphi_\varepsilon * R_k(f)\big\|_{L^p} \le C_n \big\|f\big\|_{H^p}$$

when $n \ge 2$ and

$$c_1 \big\|f\big\|_{H^p} \le \sup_{\varepsilon>0} \big\|\varphi_\varepsilon * f\big\|_{L^p} + \sup_{\varepsilon>0} \big\|\varphi_\varepsilon * H(f)\big\|_{L^p} \le C_1 \big\|f\big\|_{H^p}$$

when $n = 1$.
[*Hint:* Note that $\sup_{\varepsilon>0} \big\|\varphi_\varepsilon * f\big\|_{L^p} \le \big\|\sup_{\varepsilon>0}|\varphi_\varepsilon * f|\,\big\|_{L^p}$, which is controlled by a multiple of $\big\|\big(\sum_j|\Delta_j(f)|^2\big)^{1/2}\big\|_{L^p}$. Apply this estimate to $R_k(f)$.]

6.7.2. (a) Let h be a function on $\mathbf{R}$ such that $h(x)$ and $xh(x)$ are in $L^2(\mathbf{R})$. Show that h is integrable over $\mathbf{R}$ and satisfies

$$\big\|h\big\|_{L^1}^2 \le 8\,\big\|h\big\|_{L^2}\big\|xh(x)\big\|_{L^2}.$$

(b) Suppose that g is an integrable function on $\mathbf{R}$ that satisfies $\int_{\mathbf{R}} g(x)\,dx = 0$ and $g(x)$ and $xg(x)$ are in $L^2(\mathbf{R})$. Show that g lies in $H^1(\mathbf{R})$ and that for some constant C we have

$$\big\|g\big\|_{H^1}^2 \le C\,\big\|g\big\|_{L^2}\big\|xg(x)\big\|_{L^2}.$$

[*Hint:* To prove part (a) split the integral of $|h(x)|$ over the regions $|x| \le R$ and $|x| > R$ and pick a suitable R. Part (b): Show that both $H(g)$ and $H(yg(y))$ lie in L^2. But since g has vanishing integral, we have $xH(g)(x) = H(yg(y))(x)$.]

6.7.3. (a) Let H be the Hilbert transform. Prove the identity

$$H(fg - H(f)H(g)) = fH(g) + gH(f)$$

for all f, g in $\bigcup_{1\le p<\infty} L^p(\mathbf{R})$.
(b) Show that the bilinear operators

$$(f,g) \to f\,H(g) + H(f)\,g$$
$$(f,g) \to f\,g - H(f)\,H(g)$$

map $L^p(\mathbf{R}) \times L^{p'}(\mathbf{R}) \to H^1(\mathbf{R})$ whenever $1 < p < \infty$.
[*Hint:* Part (a): Consider the boundary values of the product of the analytic extensions of $f + iH(f)$ and $g + iH(g)$ on the upper half-space. Part (b): Use part (a) and Theorem 6.7.4.]

6.7.4. Follow the steps given to prove the following interpolation result. Let $1 < p_1 \le \infty$ and let T be a subadditive operator that maps $H^1(\mathbf{R}^n) + L^{p_1}(\mathbf{R}^n)$ into measurable functions on $\mathbf{R}^n$. Suppose that there is $A_0 < \infty$ such that for all $f \in H^1(\mathbf{R}^n)$ we have

$$\sup_{\lambda>0} \lambda\big|\{x \in \mathbf{R}^n :\ |T(f)(x)| > \lambda\}\big| \le A_0\big\|f\big\|_{H^1}$$

and that it also maps $L^{p_1}(\mathbf{R}^n)$ into $L^{p_1,\infty}(\mathbf{R}^n)$ with norm at most A_1. Show that for any $1 < p < p_1$, T maps $L^p(\mathbf{R}^n)$ into itself with norm at most

$$C\,A_0^{\frac{\frac{1}{p}-\frac{1}{p_1}}{1-\frac{1}{p_1}}}\,A_1^{\frac{1-\frac{1}{p}}{1-\frac{1}{p_1}}},$$

where $C = C(n,p,p_1)$.
(a) Fix $1 < q < p < p_1 < \infty$ and f and let Q_j be the family of all maximal dyadic cubes such that $\lambda^q < |Q_j|^{-1}\int_{Q_j}|f|^q\,dx$. Write $E_\lambda = \bigcup Q_j$ and note that $E_\lambda \subseteq \big\{M(|f|^q)^{\frac{1}{q}} > \lambda\big\}$ and that $|f| \le \lambda$ a.e. on $(E_\lambda)^c$. Write f as the sum of the *good function*

$$g_\lambda = f\chi_{(E_\lambda)^c} + \sum_j (\operatorname*{Avg}_{Q_j} f)\,\chi_{Q_j}$$

and the *bad function*

$$b_\lambda = \sum_j b_\lambda^j, \qquad \text{where} \qquad b_\lambda^j = (f - \operatorname*{Avg}_{Q_j} f)\,\chi_{Q_j}.$$

(b) Show that g_λ lies in $L^{p_1}(\mathbf{R}^n) \cap L^\infty(\mathbf{R}^n)$ and that

$$\begin{aligned}\big\|g_\lambda\big\|_{L^\infty} &\le 2^{\frac{n}{q}}\lambda\\ \big\|g_\lambda\big\|_{L^{p_1}}^{p_1} &\le \int_{|f|\le\lambda}|f(x)|^{p_1}\,dx + 2^{\frac{np_1}{q}}\lambda^{p_1}|E_\lambda| < \infty.\end{aligned}$$

(c) Show that for $c = 2^{\frac{n}{q}+1}$, each $c^{-1}\lambda^{-1}|Q_j|^{-1}b_\lambda^j$ is an L^q-atom for H^1. Conclude that b_λ lies in $H^1(\mathbf{R}^n)$ and satisfies

$$\big\|b_\lambda\big\|_{H^1} \le c\,\lambda\sum_j |Q_j| \le c\,\lambda\,|E_\lambda| < \infty.$$

(d) Start with

$$\begin{aligned}\big\|T(f)\big\|_{L^p}^p \le p\,\gamma^p &\int_0^\infty \lambda^{p-1}\big|\{|T(g_\lambda)| > \tfrac{1}{2}\gamma\lambda\}\big|\,d\lambda\\ + p\,\gamma^p &\int_0^\infty \lambda^{p-1}\big|\{|T(b_\lambda)| > \tfrac{1}{2}\gamma\lambda\}\big|\,d\lambda\end{aligned}$$

and use the results in parts (c) and (d) to obtain that the preceding expression is at most $C(n,p,q,p_1)\max(A_1\gamma^{p-p_1}, \gamma^{p-1}A_0)$. Select $\gamma = A_1^{\frac{p_1}{p_1-1}}A_0^{-\frac{1}{p_1-1}}$ to obtain the required conclusion.
(e) In the case $p_1 = \infty$ we have $|T(g_\lambda)| \le A_1 2^{\frac{n}{q}}\lambda$ and pick $\gamma > 2A_1 2^{\frac{n}{q}}$ to make the integral involving g_λ vanishing.

6.7.5. Let f be a function on the line whose Fourier transform vanishes on the negative half line. Show that f lies in $H^1(\mathbf{R})$.

HISTORICAL NOTES

The strong type $L^p \to L^q$ estimates in Theorem 6.1.3 were obtained by Hardy and Littlewood [**236**] (see also [**238**]) when $n = 1$ and by Sobolev [**463**] for general n. The weak type estimate $L^1 \to L^{\frac{n}{n-s},\infty}$ first appeared in Zygmund [**570**]. The proof of Theorem 6.1.3 using estimate (6.1.11) is taken from Hedberg [**243**]. The best constants in this theorem when $p = \frac{2n}{n+s}$, $q = \frac{2n}{n-s}$, and $0 < s < n$, were precisely evaluated by Lieb [**334**]. A generalization of Theorem 6.1.3 for nonconvolution operators was obtained by Folland and Stein [**266**].

The Riesz potentials were systematically studied by Riesz [**428**] on $\mathbf{R}^n$ although their one-dimensional version appeared in earlier work of Weyl [**551**]. The Bessel potentials were introduced by Aronszajn and Smith [**13**] and also by Calderón [**60**], who was the first to observe that the potential space $\mathcal{L}^p_s$ (i.e., the Sobolev space L^p_s) coincides with the space L^p_k given in the classical Definition 6.2.1 when $s = k$ is an integer. Theorem 6.2.4 is due to Sobolev [**463**] when s is a positive integer. The case $p = 1$ of Sobolev's theorem (Exercise 6.2.9) was later obtained independently by Gagliardo [**201**] and Nirenberg [**394**]. We refer to the books of Adams [**2**], Lieb and Loss [**335**], and Maz'ya [**371**] for a more systematic study of Sobolev spaces and their use in analysis.

An early characterization of Lipschitz spaces using Littlewood-Paley type operators (built from the Poisson kernel) appears in the work of Hardy and Littlewood [**240**]. These and other characterizations were obtained and extensively studied in higher dimensions by Taibleson [**500**], [**501**], [**502**] in his extensive study. Lipschitz spaces can also be characterized via mean oscillation over cubes. This idea originated in the simultaneous but independent work of Campanato [**57**], [**58**] and Meyers [**374**] and led to duality theorems for these spaces. Incidentally, the predual of the space $\dot{\Lambda}_\alpha$ is the Hardy space H^p with $p = \frac{n}{n+\alpha}$, as shown by Duren, Romberg, and Shields [**171**] for the unit circle and by Walsh [**545**] for higher-dimensional spaces; see also Fefferman and Stein [**191**]. We refer to the book of García-Cuerva and Rubio de Francia [**204**] for a nice exposition of these results. An excellent expository reference on Lipschitz spaces is the article of Krantz [**316**].

Taibleson in his aforementioned work also studied the generalized Lipschitz spaces $\Lambda^{p,q}_\alpha$ called today Besov spaces. These spaces were named after Besov, who obtained a trace theorem and embeddings for them [**39**], [**40**]. The spaces $B^p_{\alpha,q}$, as defined in Section 6.5, were introduced by Peetre [**403**], although the case $p = q = 2$ was earlier considered by Hörmander [**254**]. The connection of Besov spaces with modern Littlewood-Paley theory was brought to the surface by Peetre [**403**]. The extension of the definition of Besov spaces to the case $p < 1$ is also due to Peetre [**404**] but there was a forerunner by Flett [**194**]. The spaces $F^p_{\alpha,q}$ with $1 < p, q < \infty$ were introduced by Triebel [**527**] and independently by Lizorkin [**343**]. The extension of the spaces $F^p_{\alpha,q}$ to the case $0 < p < \infty$ and $0 < q \le \infty$ first appeared in Peetre [**406**] who also obtained a maximal characterization for all of these spaces. Lemma 6.5.3 originated in Peetre [**406**]; the version given on the text is based on a refinement of Triebel [**529**]. The article of Lions, Lizorkin, and Nikol'skij [**337**] presents an account of the treatment of the spaces $F^p_{\alpha,q}$ introduced by Triebel and Lizorkin as well as the equivalent characterizations obtained by Lions, using interpolation between Banach spaces, and by Nikol'skij, using best approximation.

The theory of Hardy spaces is vast and complicated. In classical complex analysis, the Hardy spaces H^p were spaces of analytic functions and were introduced to characterize boundary values of analytic functions on the unit disc. Precisely, the space $H^p(\mathbb{D})$ was introduced by Hardy [**232**] to consist of all analytic functions F on the unit disc $\mathbb{D}$ with the property that $\sup_{0<r<1}\int_0^1 |F(re^{2\pi i\theta})|^p\,d\theta < \infty$, $0 < p < \infty$. When $1 < p < \infty$, this space coincides with the space of analytic functions whose real parts are Poisson integrals of functions in $L^p(\mathbf{T}^1)$. But for $0 < p \le 1$ this characterization fails and for several years a satisfactory characterization was missing. For a systematic treatment of these spaces we refer to the books of Duren [**170**] and Koosis [**313**].

With the illuminating work of Stein and Weiss [**486**] on systems of conjugate harmonic functions the road opened to higher dimensional extensions of Hardy spaces. Burkholder, Gundy, and Silverstein [**56**] proved the fundamental theorem that an analytic function F lies in $H^p(\mathbf{R}^2_+)$ [i.e., $\sup_{y>0}\int_{\mathbf{R}} |F(x+iy)|^p\,dx < \infty$] if and only if the nontangential maximal function of its real part lies in $L^p(\mathbf{R})$. This result was proved using Brownian motion, but later Koosis [**312**] obtained another proof using complex analysis. This theorem spurred the development of the modern theory of Hardy spaces as it provided the first characterization without the notion of conjugacy and it indicated that Hardy spaces are intrinsically defined. The pioneering article of Fefferman and Stein [**191**] furnished three new characterizations of Hardy spaces: using a maximal function associated with a general approximate identity, using the grand maximal function, and using the area function of Luzin. From this point on the role of the Poisson kernel faded into the background as it turned out that it was not essential in the study of Hardy spaces. A previous characterization of Hardy spaces using the g-function, a radial analogue of the Luzin area function, was obtained by Calderón [**62**]. Two alternative characterizations of Hardy spaces were obtained by Uchiyama in terms of the generalized Littlewood-Paley g-function [**532**] and in terms of Fourier multipliers [**533**]. Necessary and sufficient conditions for systems of singular integral operators to characterize $H^1(\mathbf{R}^n)$ were also obtained by Uchiyama [**531**]. The characterization of H^p using Littlewood-Paley theory was observed by Peetre [**405**]. The case $p = 1$ was later independently obtained by Rubio de Francia, Ruiz, and Torrea [**439**].

The one-dimensional atomic decomposition of Hardy spaces is due to Coifman [**106**] and its higher-dimensional extension to Latter [**327**]. A simplification of some of the technical details in Latter's proof was subsequently obtained by Latter and Uchiyama [**328**]. Using the atomic decomposition Coifman and Weiss [**122**] extended the definitions of Hardy spaces to more general spaces. The idea of obtaining the atomic decomposition from the reproducing formula (6.6.6) goes back to Calderón [**66**]. Another simple proof of the L^2-atomic decomposition for H^p (starting from the nontangential Poisson maximal function) was obtained by Wilson [**555**]. With only a little work, one can show that L^q-atoms for H^p can be written as sums of L^∞-atoms for H^p. We refer to the book of García-Cuerva and Rubio de Francia [**204**] for a proof of this fact. Atomic decompositions of general function spaces were obtained in the fundamental work of Frazier and Jawerth [**197**], [**198**]. The exposition in Section 6.6 is also based on the work of Frazier and Jawerth [**199**]. The work of these authors provided a solid manifestation that the atomic decomposition is intrinsically related to Littlewood-Paley theory and not wedded to a particular space. Littlewood-Paley theory therefore provides a comprehensive and unifying perspective on function spaces.

Main references on H^p spaces and their properties are the books of Baernstein and Sawyer [**23**], Folland and Stein [**267**] (in the context of homogeneous groups; this reference contains Lemma 6.4.5), Lu [**348**] (on which the proof of Theorem 6.4.4 is based), Strömberg and Torchinsky [**497**] (on weighted Hardy spaces), and Uchiyama [**534**]. The articles of

Calderón and Torchinsky [**67**], [**68**] develop and extend the theory of Hardy spaces to the nonisotropic setting. Hardy spaces can also be defined in terms of nonstandard convolutions, such as the "twisted convolution" on $\mathbf{R}^{2n}$. Characterizations of the space H^1 in this context have been obtained by Mauceri, Picardello, and Ricci [**369**]

The localized Hardy spaces h_p, $0 < p \le 1$ were introduced by Goldberg [**210**] as spaces of distributions for which the maximal operator $\sup_{0<t<1} |\Phi_t * f|$ lies in $L^p(\mathbf{R}^n)$ (here Φ is a Schwartz function with nonvanishing integral). These spaces can be characterized in ways analogous to those of the homogeneous Hardy spaces H^p; in particular, they admit an atomic decomposition. It was shown by Bui [**53**] that the space h^p coincides with the Triebel-Lizorkin space $F^p_{0,2}(\mathbf{R}^n)$; see also Meyer [**359**]. For the local theory of Hardy spaces we refer the reader to the articles of Dafni [**142**] and Chang, Krantz, and Stein [**89**].

Interpolation of operators between Hardy spaces was originally based on complex function theory; see the articles of Calderón and Zygmund [**74**] and Weiss [**547**]. The real interpolation approach discussed in Exercise 6.7.4 can be traced in the article of Igari [**269**]. Interpolation between Hardy spaces was further studied and extended by Riviere and Sagher [**430**] and Fefferman, Riviere, and Sagher [**189**].

The action of singular integrals on periodic spaces was studied by Calderón and Zygmund [**70**]. The preservation of Lipschitz spaces under singular integral operators is due to Taibleson [**499**]. The case $0 < \alpha < 1$ was earlier considered by Privalov [**418**] for the conjugate function on the circle. Fefferman and Stein [**191**] were the first to show that singular integrals map Hardy spaces into themselves. The boundedness of fractional integrals on H^p was obtained by Krantz [**315**]. The case $p = 1$ was earlier considered by Stein and Weiss [**486**]. The action of multilinear singular integrals on Hardy spaces was studied by Coifman and Grafakos [**109**] and Grafakos and Kalton [**215**]. An exposition on the subject of function spaces and the action of singular integrals on them was written by Frazier, Jawerth, and Weiss [**200**]. For a careful study of the action of singular integrals on function spaces, we refer the reader to the book of Torres [**526**].

CHAPTER 7

BMO and Carleson Measures

The space of functions of bounded mean oscillation, or BMO, naturally arises as the class of functions whose deviation from their means over cubes is bounded. L^∞ functions have this property, but there exist unbounded functions with bounded mean oscillation. Such functions are slowly growing and they typically have at most logarithmic blowup. The space BMO shares similar properties with the space L^∞ and it often serves as a substitute for it. For instance, classical singular integrals do not map L^∞ into L^∞ but L^∞ into BMO. And in many instances interpolation between L^p and BMO works just as well between L^p and L^∞. But the role of the space BMO is deeper and more far reaching than that. This space crucially arises in many situations in analysis, such as in the characterization of the L^2 boundedness of nonconvolution singular integral operators with standard kernels.

Carleson measures are among the most important tools in harmonic analysis. These measures capture essential orthogonality properties and exploit properties of extensions of functions on the upper half-space. There exists a natural and deep connection between Carleson measures and BMO functions as a certain type of measure is Carleson, if and only if the underlying function is in BMO. Carleson measures are especially crucial in the study of L^2 problems, where the Fourier transform cannot be used to provide boundedness via Plancherel's theorem. The power of the Carleson measure techniques will become apparent in the next chapter, where they play a crucial role in the proof of several important results.

7.1. Functions of Bounded Mean Oscillation

What exactly is bounded mean oscillation and what kind of functions have this property? The mean of a (locally integrable) function over a set is another word for its average over that set. The oscillation of a function over a set is the absolute value of the difference of the function from its mean over this set. Mean oscillation is therefore the average of this oscillation over a set. A function is said to be of bounded mean oscillation if its mean oscillation over all cubes is bounded. Precisely, given a locally integrable function f on $\mathbf{R}^n$ and a measurable set Q in $\mathbf{R}^n$, denote by

$$\operatorname*{Avg}_Q f = \frac{1}{|Q|}\int_Q f(x)\,dx$$

the mean (or average) of f over Q. Then the *oscillation* of f over Q is the function $|f - f_Q|$ and the *mean oscillation* of f over Q is

$$\frac{1}{|Q|}\int_Q |f(x) - \operatorname*{Avg}_Q f|\,dx.$$

7.1.a. Definition and Basic Properties of *BMO*

Definition 7.1.1. For f a complex-valued locally integrable function on $\mathbf{R}^n$, set

$$\|f\|_{BMO} = \sup_Q \frac{1}{|Q|}\int_Q |f(x) - \operatorname*{Avg}_Q f|\,dx,$$

where the supremum is taken over all cubes Q in $\mathbf{R}^n$. The function f is called of bounded mean oscillation if $\|f\|_{BMO} < \infty$ and $BMO(\mathbf{R}^n)$ is the set of all locally integrable functions f on $\mathbf{R}^n$ with $\|f\|_{BMO} < \infty$.

Several remarks are in order. First it is a simple fact that $BMO(\mathbf{R}^n)$ is a linear space, that is, if $f, g \in BMO(\mathbf{R}^n)$ and $\lambda \in \mathbf{C}$, then so $f + g$ and λf are also in $BMO(\mathbf{R}^n)$ and

$$\begin{aligned}\|f+g\|_{BMO} &\le \|f\|_{BMO} + \|g\|_{BMO}\\ \|\lambda f\|_{BMO} &= |\lambda|\,\|f\|_{BMO}.\end{aligned}$$

But $\|\ \|_{BMO}$ is not a norm. The problem is that if $\|f\|_{BMO} = 0$, this does not imply that $f = 0$ but that f is a constant. See Proposition 7.1.2. Moreover, every constant function c satisfies $\|c\|_{BMO} = 0$. One way to avoid this problem is to define BMO as the quotient space of all equivalent classes of functions whose difference is a constant. We are not going to do this, but we will keep in mind to identify two elements of BMO whose difference is a constant. Although $\|\ \|_{BMO}$ is only a seminorm, we will occasionally refer to it as a norm with no reason to cause confusion.

We begin with a list of basic properties of BMO.

Proposition 7.1.2. *The following properties of the space* $BMO(\mathbf{R}^n)$ *are valid:*

(1) *If* $\|f\|_{BMO} = 0$, *then* f *is a.e. equal to a constant.*

(2) $L^\infty(\mathbf{R}^n)$ *is contained in* $BMO(\mathbf{R}^n)$ *and* $\|f\|_{BMO} \le 2\|f\|_{L^\infty}$.

(3) *Suppose that there exists an* $A > 0$ *such that for all cubes* Q *in* $\mathbf{R}^n$ *there exists a constant* c_Q *such that*

$$\sup_Q \frac{1}{|Q|}\int_Q |f(x) - c_Q|\,dx \le A. \tag{7.1.1}$$

Then $f \in BMO(\mathbf{R}^n)$ *and* $\|f\|_{BMO} \le 2A$.

(4) *For all* f *locally integrable we have*

$$\frac{1}{2}\|f\|_{BMO} \le \sup_Q \frac{1}{|Q|} \inf_{c_Q} \int_Q |f(x) - c_Q|\, dx \le \|f\|_{BMO}.$$

(5) *If* $f \in BMO(\mathbf{R}^n)$ *and* $h \in \mathbf{R}^n$, *then the function* $\tau^h(f)$ *given by* $\tau^h(f)(x) = f(x-h)$ *is also in* $BMO(\mathbf{R}^n)$ *and*

$$\|\tau^h(f)\|_{BMO} = \|f\|_{BMO}.$$

(6) *If* $f \in BMO(\mathbf{R}^n)$ *and* $\lambda > 0$, *then the function* $\delta^\lambda(f)$ *defined by* $\delta^\lambda(f)(x) = f(\lambda x)$ *is also in* $BMO(\mathbf{R}^n)$ *and*

$$\|\delta^\lambda(f)\|_{BMO} = \|f\|_{BMO}.$$

(7) *If* $f \in BMO$ *then so is* $|f|$. *Similarly, if* f, g *are real-valued* BMO *functions, then so are* $\max(f,g)$, *and* $\min(f,g)$. *In other words,* BMO *is a lattice. Moreover,*

$$\begin{aligned} \||f|\|_{BMO} &\le 2\|f\|_{BMO} \\ \|\max(f,g)\|_{BMO} &\le \|f\|_{BMO} + \|g\|_{BMO}, \\ \|\min(f,g)\|_{BMO} &\le \|f\|_{BMO} + \|g\|_{BMO}. \end{aligned}$$

(8) *For locally integrable functions* f *define*

$$\|f\|_{BMO_{\text{balls}}} = \sup_B \frac{1}{|B|} \int_B |f(x) - \operatorname*{Avg}_B f\,|\, dx, \tag{7.1.2}$$

where the supremum is taken over all balls B *in* $\mathbf{R}^n$. *Then there are positive constants* c_n, C_n *such that*

$$c_n\|f\|_{BMO} \le \|f\|_{BMO_{\text{balls}}} \le C_n\|f\|_{BMO}.$$

Proof. To prove (1) note that f has to be a.e. equal to its average c_N over every cube $[-N,N]^n$. Since $[-N,N]^n$ is contained in $[-N-1,N+1]^n$, it follows that $c_N = c_{N+1}$ for all N. This implies the required conclusion. To prove (2) observe that

$$\operatorname*{Avg}_Q |f - \operatorname*{Avg}_Q f| \le 2 \operatorname*{Avg}_Q |f| \le 2\|f\|_{L^\infty}.$$

For part (3) note that

$$|f - \operatorname*{Avg}_Q f| \le |f - c_Q| + |\operatorname*{Avg}_Q f - c_Q| \le |f - c_Q| + \frac{1}{|Q|}\int_Q |f(t) - c_Q|\, dt\,.$$

Averaging over Q and using (7.1.1), we obtain that $\|f\|_{BMO} \le 2A$. The lower inequality in (4) follows from (3) while the upper one is trivial. Property (5) is immediate. For (6) note that $\operatorname*{Avg}_{Q} \delta^{\lambda}(f) = \operatorname*{Avg}_{\lambda Q} f$ and thus

$$\frac{1}{|Q|}\int_Q |f(\lambda x) - \operatorname*{Avg}_{Q} \delta^{\lambda}(f)|\,dx = \frac{1}{|\lambda Q|}\int_{\lambda Q} |f(x) - \operatorname*{Avg}_{\lambda Q} f|\,dx.$$

Property (7) is a consequence of the fact that $\big||f| - \operatorname*{Avg}_{Q}|f|\big| \le |f - \operatorname*{Avg}_{Q} f|$. Also, the maximum and the minimum of two functions can be expressed in terms of the absolute value of their difference. We now turn to (8). Given any cube Q in $\mathbf{R}^n$, we let B be the smallest ball that contains it. Then $|B|/|Q| = 2^{-n}v_n\sqrt{n^n}$, where v_n is the volume of the unit ball, and

$$\frac{1}{|Q|}\int_Q |f(x) - \operatorname*{Avg}_{B} f|\,dx \le \frac{|B|}{|Q|}\frac{1}{|B|}\int_B |f(x) - \operatorname*{Avg}_{B} f|\,dx \le \frac{v_n\sqrt{n^n}}{2^n}\big\|f\big\|_{BMO_{\text{balls}}}.$$

It follows from (3) that $\big\|f\big\|_{BMO} \le 2^{1-n}v_n\sqrt{n^n}\,\big\|f\big\|_{BMO_{\text{balls}}}$. To obtain the reverse conclusion given any ball B find the smallest cube Q that contains it and argue similarly using a version of (3) for the space BMO_{balls}.

$\square$

Example 7.1.3. We indicate why $L^\infty(\mathbf{R}^n)$ is a proper subspace of $BMO(\mathbf{R}^n)$. We claim that the function $\log|x|$ is in $BMO(\mathbf{R}^n)$ but not in $L^\infty(\mathbf{R}^n)$. Clearly, $\log|x|$ is not in L^∞. To prove that it is in $BMO(\mathbf{R}^n)$ for every $x_0 \in \mathbf{R}^n$ and $R > 0$, we need to find a constant $C_{x_0,R}$ so that the average of $f - C_{x_0,R}$ over the ball $|x - x_0| \le R$ is uniformly bounded. Since

$$\frac{1}{v_nR^n}\int_{|x-x_0|\le R} \big|\log|x| - C_{x_0,R}\big|\,dx = \frac{1}{v_n}\int_{|z-x_0/R|\le 1} \big|\log|z| - C_{x_0,R} + \log R\big|\,dz,$$

things reduce to the case where $R = 1$ and x_0 is arbitrary. If $R = 1$ and $|x_0| \le 2$, take $C_{x_0,1} = 0$ and observe that

$$\int_{|x-x_0|\le 1} \big|\log|x|\big|\,dx \le \int_{|x|\le 3} \big|\log|x|\big|\,dx = C.$$

When $R = 1$ and $|x_0| \ge 2$, take $C_{x_0,1} = \log|x_0|$. In this case notice that

$$\begin{aligned}\frac{1}{v_n}\int_{|z-x_0|\le 1} \big|\log|z| - \log|x_0|\big|\,dz &= \frac{1}{v_n}\int_{\left|\frac{z}{|x_0|} - \frac{x_0}{|x_0|}\right|\le 1} \left|\log\frac{|z|}{|x_0|}\right|\,dz \\ &\le \frac{\log\frac{3}{2}}{v_n}|x_0|^n \int_{|z|\le\frac{3}{2}|x_0|} dz \\ &= \left(\frac{3}{2}\right)^n \log\frac{3}{2}\,.\end{aligned}$$

This proves that $\log|x|$ is in BMO.

The function $\log |x|$ turns out to be a typical element of BMO, but we will make this statement a bit more precise later. It is interesting to observe that an abrupt cutoff of a BMO function may not give a function in the same space.

Example 7.1.4. The function $h(x) = \chi_{x>0} \log \frac{1}{x}$ is not in $BMO(\mathbf{R})$. Indeed, the problem is at the origin. Consider the intervals $(-\varepsilon, \varepsilon)$, where $0 < \varepsilon < \frac{1}{2}$. We have that

$$\operatorname*{Avg}_{(-\varepsilon,\varepsilon)} h = \frac{1}{2\varepsilon} \int_{-\varepsilon}^{+\varepsilon} h(x)\, dx = \frac{1}{2\varepsilon} \int_0^{\varepsilon} \log \frac{1}{x}\, dx = \frac{1 + \log \frac{1}{\varepsilon}}{2}.$$

But then

$$\frac{1}{2\varepsilon} \int_{-\varepsilon}^{+\varepsilon} |h(x) - \operatorname*{Avg}_{(-\varepsilon,\varepsilon)} h|\, dx \ge \frac{1}{2\varepsilon} \int_{-\varepsilon}^{0} |\operatorname*{Avg}_{(-\varepsilon,\varepsilon)} h|\, dx = \frac{1 + \log \frac{1}{\varepsilon}}{4}$$

and the latter is clearly unbounded as $\varepsilon \to 0$.

Let us now look at some basic properties of BMO functions. Observe that if a cube Q_1 is contained in a cube Q_2, then

$$\begin{aligned} |\operatorname*{Avg}_{Q_1} f - \operatorname*{Avg}_{Q_2} f| &\le \frac{1}{|Q_1|} \int_{Q_1} |f - \operatorname*{Avg}_{Q_2} f|\, dx \\ &\le \frac{1}{|Q_1|} \int_{Q_2} |f - \operatorname*{Avg}_{Q_2} f|\, dx \\ &\le \frac{|Q_2|}{|Q_1|} \|f\|_{BMO}. \end{aligned} \tag{7.1.3}$$

We make the point that the same estimate holds if the sets Q_1 and Q_2 are balls and $Q_1 \subseteq Q_2$.

A version of this inequality is the first statement in the following proposition. For simplicity, in (7.1.2) we denote by $\|f\|_{BMO}$ the expression given by $\|f\|_{BMO_{\text{balls}}}$, as these quantities are comparable. We also recall that for a ball B and $a > 0$, aB denotes the ball that is concentric with B and whose radius is a times the radius of B.

Proposition 7.1.5. *(i) Let f be in $BMO(\mathbf{R}^n)$. Given a ball B and a positive integer m, we have*

$$|\operatorname*{Avg}_{B} f - \operatorname*{Avg}_{2^m B} f| \le 2^n m \|f\|_{BMO}. \tag{7.1.4}$$

(ii) For any $\delta > 0$ there is a positive constant $C_{n,\delta}$ so that

$$\int_{\mathbf{R}^n} \frac{|f(x) - \operatorname*{Avg}_{B(0,1)} f|}{(1 + |x|)^{n+\delta}}\, dx \le C_{n,\delta} \|f\|_{BMO}. \tag{7.1.5}$$

(iii) There exists a constant C_n such that for all $f \in BMO(\mathbf{R}^n)$ we have

$$\sup_{y\in\mathbf{R}^n}\sup_{t>0}\int_{\mathbf{R}^n}|f(x)-(P_t*f)(y)|P_t(x-y)\,dx \le C_n\|f\|_{BMO}. \tag{7.1.6}$$

Here P_t denotes the Poisson kernel introduced in Chapter 2.
(iv) Conversely, there is a constant C'_n such that for all locally integrable functions f for which

$$\int_{\mathbf{R}^n}\frac{|f(x)|}{(1+|x|)^{n+1}}\,dx<\infty$$

we have

$$C'_n\|f\|_{BMO} \le \sup_{y\in\mathbf{R}^n}\sup_{t>0}\int_{\mathbf{R}^n}|f(x)-(P_t*f)(y)|P_t(x-y)\,dx\,. \tag{7.1.7}$$

Proof. (i) We have

$$\begin{aligned}|\operatorname*{Avg}_B f-\operatorname*{Avg}_{2B} f| &= \frac{1}{|B|}\left|\int_B\big(f(t)-\operatorname*{Avg}_{2B} f\big)\,dt\right|\\ &\le \frac{2^n}{|2B|}\int_{2B}\big|f(t)-\operatorname*{Avg}_{2B} f\big|\,dt\\ &\le 2^n\|f\|_{BMO}.\end{aligned}$$

Using this inequality, we derive (7.1.4) by adding and subtracting the terms

$$\operatorname*{Avg}_{2B} f\,,\quad \operatorname*{Avg}_{2^2B} f\,,\quad \dots,\quad \operatorname*{Avg}_{2^{m-1}B} f\,.$$

(ii) We have

$$\begin{aligned}&\int_{\mathbf{R}^n}\frac{\big|f(x)-\operatorname*{Avg}_B f\big|}{(1+|x|)^{n+\delta}}\,dx\\ &\le \int_B\frac{\big|f(x)-\operatorname*{Avg}_B f\big|}{(1+|x|)^{n+\delta}}\,dx+\sum_{k=0}^{\infty}\int_{2^{k+1}B\setminus 2^kB}\frac{\big|f(x)-\operatorname*{Avg}_{2^{k+1}B} f\big|+\big|\operatorname*{Avg}_{2^{k+1}B} f-\operatorname*{Avg}_B f\big|}{(1+|x|)^{n+\delta}}\,dx\\ &\le \int_B\big|f(x)-\operatorname*{Avg}_B f\big|\,dx\\ &\qquad+\sum_{k=0}^{\infty}2^{-k(n+\delta)}\int_{2^{k+1}B}\Big(\big|f(x)-\operatorname*{Avg}_{2^{k+1}B} f\big|+\big|\operatorname*{Avg}_{2^{k+1}B} f-\operatorname*{Avg}_B f\big|\Big)\,dx\\ &\le v_n\|f\|_{BMO}+\sum_{k=0}^{\infty}2^{-k(n+\delta)}\big(1+2^n(k+1)\big)(2^{k+1})^n v_n\|f\|_{BMO}\\ &= C'_{n,\delta}\|f\|_{BMO}.\end{aligned}$$

(iii) The proof of (7.1.6) is a reprise of the argument given in (ii). Set $B_t = B(y,t)$. We first prove a version of (7.1.6) in which the expression $(P_t * f)(y)$ is replaced by $\operatorname{Avg}_{B_t} f$. For fixed y, t we have

$$
\begin{aligned}
&\int_{\mathbf{R}^n} \frac{t\,\big|f(x) - \operatorname*{Avg}_{B_t} f\big|}{(t^2+|x-y|^2)^{\frac{n+1}{2}}}\,dx \\
&\le \int_{B_t} \frac{t\big|f(x) - \operatorname*{Avg}_{B_t} f\big|}{(t^2+|x-y|^2)^{\frac{n+1}{2}}}\,dx \\
&\quad + \sum_{k=0}^{\infty} \int_{2^{k+1}B_t \setminus 2^k B_t} \frac{t\big(\big|f(x) - \operatorname*{Avg}_{2^{k+1}B_t} f\big| + \big|\operatorname*{Avg}_{2^{k+1}B_t} f - \operatorname*{Avg}_{B_t} f\big|\big)}{(t^2+|x-y|^2)^{\frac{n+1}{2}}}\,dx \\
&\le \int_{B_t} \frac{\big|f(x) - \operatorname*{Avg}_{B_t} f\big|}{t^n}\,dx \\
&\quad + \sum_{k=0}^{\infty} \frac{2^{-k(n+1)}}{t^n} \int_{2^{k+1}B_t} \big(\big|f(x) - \operatorname*{Avg}_{2^{k+1}B} f\big| + \big|\operatorname*{Avg}_{2^{k+1}B_t} f - \operatorname*{Avg}_{B_t} f\big|\big)\,dx \\
&\le v_n\big\|f\big\|_{BMO} + \sum_{k=0}^{\infty} 2^{-k(n+1)}\big(1+2^n(k+1)\big)(2^{k+1})^n v_n\big\|f\big\|_{BMO} \\
&= C_n\big\|f\big\|_{BMO}.
\end{aligned}
\tag{7.1.8}
$$

Using the inequality just proved, we also obtain

$$
\begin{aligned}
&\int_{\mathbf{R}^n} \big|(P_t * f)(y) - \operatorname*{Avg}_{B_t} f\big| P_t(x-y)\,dx \\
&= \big|(P_t * f)(y) - \operatorname*{Avg}_{B_t} f\big| \\
&\le \int_{\mathbf{R}^n} P_t(x-y)\big|f(x) - \operatorname*{Avg}_{B_t} f\big|\,dx \\
&\le C_n\big\|f\big\|_{BMO},
\end{aligned}
$$

which, combined with the inequality in (7.1.8), yields (7.1.6) with constant $2C_n$.

(iv) The converse is easier. Let A be the expression on the right in (7.1.7). For $|x-y| \le t$ we have $P_t(x-y) \ge c_n t(2t^2)^{-\frac{n+1}{2}} = c_n' t^{-n}$, which gives

$$
A \ge \int_{\mathbf{R}^n} |f(x) - (P_t * f)(y)| P_t(x-y)\,dx \ge \frac{c_n'}{t^n} \int_{|x-y|\le t} |f(x) - (P_t * f)(y)|\,dx.
$$

Proposition 7.1.2 (3) now implies that $\big\|f\big\|_{BMO} \le 2A/(v_n c_n')$. □

7.1.b. The John-Nirenberg Theorem

Having set down some basic facts about BMO, we now turn to a deeper property of BMO functions: their exponential integrability. We begin with a preliminary remark. As we saw in Example 7.1.3, the function $g(x) = \log|x|$ is in $BMO(\mathbf{R})$. This function is exponentially integrable over any compact subset K of $\mathbf{R}$ in the sense that

$$\int_K e^{c|g(x)|}\, dx < \infty$$

for any $c < 1$. It turns out that this is a general property of BMO functions, and this is the main gist of the next theorem.

Theorem 7.1.6. *For all $f \in BMO(\mathbf{R}^n)$, for all cubes Q, and all $\alpha > 0$ we have*

$$|\{x \in Q : \ |f(x) - \operatorname*{Avg}_Q f| > \alpha\}| \le C|Q|e^{-A\alpha/\|f\|_{BMO}} \tag{7.1.9}$$

with $C = e^{2^n e}$ and $A = (2^n e)^{-1}$.

Proof. Since inequality (7.1.9) is scale invariant, it suffices to assume that $\|f\|_{BMO} = 1$. Let us now fix a closed cube Q and a constant $b > 1$ to be chosen later.

We apply the Calderón-Zygmund decomposition to the function $f - \operatorname{Avg}_Q f$ inside the cube Q. We introduce the following selection criterion for a cube R:

$$\frac{1}{|R|}\int_R |f(x) - \operatorname*{Avg}_Q f|\, dx > b. \tag{7.1.10}$$

Since

$$\frac{1}{|Q|}\int_{|Q|} |f(x) - \operatorname*{Avg}_Q f|\, dx \le \|f\|_{BMO} = 1 < b\,,$$

the cube Q does not satisfy the selection criterion (7.1.10). Set $Q^{(0)} = Q$ and subdivide $Q^{(0)}$ into 2^n equal closed subcubes of side length equal to half of the side length of Q. Select such a subcube R if it satisfies the selection criterion (7.1.10). Now subdivide all nonselected cubes into 2^n equal subcubes of half their side length by bisecting the sides and select among these subcubes those that satisfy (7.1.10). Continue this process indefinitely. We obtain a countable collection of cubes $\{Q_j^{(1)}\}_j$ satisfying the following properties:

(A-1) The interior of every $Q_j^{(1)}$ is contained in $Q^{(0)}$.

(B-1) $b < |Q_j^{(1)}|^{-1} \displaystyle\int_{Q_j^{(1)}} |f(x) - \operatorname*{Avg}_{Q^{(0)}} f|\, dx \le 2^n b$.

(C-1) $|\operatorname*{Avg}_{Q_j^{(1)}} f - \operatorname*{Avg}_{Q^{(0)}} f| \le 2^n b$.

(D-1) $\sum_j |Q_j^{(1)}| \le \frac{1}{b}\int_{Q_j^{(1)}} |f(x) - \operatorname*{Avg}_{Q^{(0)}} f|\,dx \le \frac{1}{b}|Q^{(0)}|$.

(E-1) $|f - \operatorname*{Avg}_{Q^{(0)}} f| \le b$ a.e. on the set $Q^{(0)} \setminus \bigcup_j Q_j^{(1)}$.

We call the cubes $Q_j^{(1)}$ of first generation. Note that in the second inequality in (D-1) we used both (B-1) and (C-1).

We now fix a selected first generation cube $Q_j^{(1)}$ and we introduce the following selection criterion for a cube R:

$$\frac{1}{|R|}\int_R |f(x) - \operatorname*{Avg}_{Q_j^{(1)}} f|\,dx > b. \tag{7.1.11}$$

Observe that $Q_j^{(1)}$ does not satisfy the selection criterion (7.1.11). We apply a similar Calderón-Zygmund decomposition to the function

$$f - \operatorname*{Avg}_{Q_j^{(1)}} f$$

inside the cube $Q_j^{(1)}$. Subdivide $Q_j^{(1)}$ into 2^n equal closed subcubes of side length equal to half of the side length of $Q_j^{(1)}$ by bisecting the sides and select such a subcube R if it satisfies the selection criterion (7.1.11). Continue this process indefinitely. Also repeat this process for any other cube $Q_j^{(1)}$ of the first generation. We obtain a collection of cubes $\{Q_l^{(2)}\}_l$ of second generation each contained in some $Q_j^{(1)}$ such that versions of (A-1)–(E-1) are satisfied, with the superscript (2) replacing (1) and the superscript (1) replacing (0). We use the superscript (k) to denote the generation of the selected cubes.

For a fixed selected cube $Q_l^{(2)}$ of second generation, introduce the selection criterion

$$\frac{1}{|R|}\int_R |f(x) - \operatorname*{Avg}_{Q_l^{(2)}} f|\,dx > b$$

and repeat the previous process to obtain a collection of cubes of third generation inside $Q_l^{(2)}$. Repeat this procedure for any other cube $Q_j^{(2)}$ of the second generation. Denote by $\{Q_s^{(3)}\}_s$ the thus obtained collection of all cubes of the third generation.

We iterate this procedure indefinitely to obtain a doubly indexed family of cubes $Q_j^{(k)}$ satisfying the following properties:

(A-k) The interior of every $Q_j^{(k)}$ is contained in a unique $Q_{j'}^{(k-1)}$.

(B-k) $b < |Q_j^{(k)}|^{-1}\int_{Q_j^{(k)}} |f(x) - \operatorname*{Avg}_{Q_{j'}^{(k-1)}} f|\,dx \le 2^n b$.

(C-k) $|\operatorname{Avg}_{Q_j^{(k)}} f - \operatorname{Avg}_{Q_{j'}^{(k-1)}} f| \le 2^n b$.

(D-k) $\sum_j |Q_j^{(k)}| \le \frac{1}{b} \sum_{j'} |Q_{j'}^{(k-1)}|$.

(E-k) $|f - \operatorname{Avg}_{Q_j^{(k-1)}} f| \le b$ a.e. on the set $Q_{j'}^{(k-1)} \setminus \bigcup_j Q_j^{(k)}$.

We prove (A-k)–(E-k). Note that (A-k) and the lower inequality in (B-k) are satisfied by construction. The upper inequality in (B-k) is a consequence of the fact that the unique cube $Q_{j_0}^{(k)}$ with double the side length of $Q_j^{(k)}$ that contains it was not selected in the process. Now (C-k) follows from the upper inequality in (B-k). (E-k) is a consequence of the Lebesgue differentiation theorem since for every point in $Q_{j'}^{(k-1)} \setminus \bigcup_j Q_j^{(k)}$ there is a sequence of cubes shrinking to that point and the averages of f over all these cubes is at most b. It remains to prove (D-k). We have

$$
\begin{aligned}
\sum_j |Q_j^{(k)}| &< \frac{1}{b} \sum_j \int_{Q_j^{(k)}} |f(x) - \operatorname{Avg}_{Q_{j'}^{(k-1)}} f|\, dx \\
&\le \frac{1}{b} \sum_{j'} \int_{Q_{j'}^{(k-1)}} |f(x) - \operatorname{Avg}_{Q_{j'}^{(k-1)}} f|\, dx \\
&\le \frac{1}{b} \sum_{j'} |Q_{j'}^{(k-1)}| \, \|f\|_{BMO} = \frac{1}{b} \sum_{j'} |Q_{j'}^{(k-1)}| .
\end{aligned}
$$

Having established (A-k)–(E-k) we turn to some consequences. Applying (D-k) successively $k-1$ times, we obtain

$$
\sum_j |Q_j^{(k)}| \le b^{-k} |Q^{(0)}|. \tag{7.1.12}
$$

For any fixed j we have that $|\operatorname{Avg}_{Q_j^{(1)}} f - \operatorname{Avg}_{Q^{(0)}} f| \le 2^n b$ and also $|f - \operatorname{Avg}_{Q_j^{(1)}} f| \le b$ a.e. on $Q_j^{(1)} \setminus \bigcup_l Q_l^{(2)}$. This gives

$$
|f - \operatorname{Avg}_{Q^{(0)}} f| \le 2^n b + b \qquad \text{a.e.} \quad \text{on} \quad Q_j^{(1)} \setminus \bigcup_l Q_l^{(2)},
$$

which, combined with (E-1), yields

$$
|f - \operatorname{Avg}_{Q^{(0)}} f| \le 2^n 2b \qquad \text{a.e.} \quad \text{on} \quad Q^{(0)} \setminus \bigcup_l Q_l^{(2)}. \tag{7.1.13}
$$

For every fixed l we also have that $|f - \operatorname{Avg}_{Q_l^{(2)}} f| \leq b$ a.e. on $Q_l^{(2)} - \bigcup_s Q_s^{(3)}$, which combined with $|\operatorname{Avg}_{Q_l^{(2)}} f - \operatorname{Avg}_{Q_{l'}^{(1)}} f| \leq 2^n b$ and $|\operatorname{Avg}_{Q_{l'}^{(1)}} f - \operatorname{Avg}_{Q^{(0)}} f| \leq 2^n b$, yields

$$|f - \operatorname{Avg}_{Q^{(0)}} f| \leq 2^n 3b \qquad \text{a.e.} \qquad \text{on} \qquad Q_l^{(2)} \setminus \bigcup_s Q_s^{(3)}.$$

In view of (7.1.13) the same estimate is valid on $Q_l^{(0)} \setminus \bigcup_s Q_s^{(3)}$. Continuing this reasoning, we obtain by induction that for all $k \geq 1$ we have

$$|f - \operatorname{Avg}_{Q^{(0)}} f| \leq 2^n kb \qquad \text{a.e.} \qquad \text{on} \qquad Q^{(0)} \setminus \bigcup_s Q_s^{(k)}. \tag{7.1.14}$$

This proves the almost everywhere inclusion

$$\{|f - \operatorname{Avg}_Q f| > 2^n kb\} \subseteq \bigcup_j Q_j^{(k)}.$$

We now use (7.1.12) and (7.1.14) to prove (7.1.9). We fix an $\alpha > 0$. If

$$2^n kb < \alpha \leq 2^n (k+1)b$$

for some $k \geq 1$, then

$$\begin{aligned} |\{x \in Q: \ |f - \operatorname{Avg}_Q f| > \alpha\}| &\leq |\{x \in Q: \ |f - \operatorname{Avg}_Q f| > 2^n kb\}| \\ &\leq \sum_j |Q_j^{(k)}| \leq \frac{1}{b^k}|Q^{(0)}| \\ &= |Q| e^{-k \log b} \\ &\leq |Q| b e^{-\alpha \log b/(2^n b)}, \end{aligned}$$

since $-k \leq 1 - \dfrac{\alpha}{2^n b}$. On the other hand, if $\alpha \leq 2^n b$, then use the trivial estimate

$$|\{x \in Q: \ |f - \operatorname{Avg}_Q f| > \alpha\}| \leq |Q| \leq |Q| e^{2^n b} e^{-\alpha}.$$

We have now proved (7.1.9) with constants $A = \log b/(2^n b)$ and $C = e^{2^n b}$. Recall that b here is any real number greater than 1. Observe that A becomes largest when $b = e$. This choice of b yields (7.1.9).

□

7.1.c. Consequences of Theorem 7.1.6

Having proved the important distribution inequality (7.1.9), we are now in a position to obtain a couple of corollaries.

Corollary 7.1.7. *Every BMO function is exponentially integrable over any cube. More precisely, for any $\gamma < 1/(2^n e)$ there exists a constant $C_{n,\gamma}$ so that for all $f \in BMO(\mathbf{R}^n)$ and all cubes Q we have*

$$\frac{1}{|Q|}\int_Q e^{\gamma|f(x)-\operatorname{Avg}_Q f|/\|f\|_{BMO}}\,dx \le C_{n,\gamma}.$$

Proof. Using identity (1.1.7), we obtain that

$$\frac{1}{|Q|}\int_Q e^{|f|}\,dx = \frac{1}{|Q|}\int_0^\infty e^\alpha |\{x\in Q:\ |f(x)|>\alpha\}|\,d\alpha$$

and inequality (7.1.9) to obtain for $\gamma < A = (2^n e)^{-1}$

$$\frac{1}{|Q|}\int_Q e^{\gamma|f(x)-\operatorname{Avg}_Q f|/\|f\|_{BMO}}\,dx \le \int_0^\infty e^\alpha C e^{-A(\frac{\alpha}{\gamma}\|f\|_{BMO})/\|f\|_{BMO}}\,d\alpha = C_{n,\gamma},$$

where C is as in (7.1.9).

□

Another important corollary of Theorem 7.1.6 is the following.

Corollary 7.1.8. *For all $0 < p < \infty$, there exists a finite constant B_p such that*

$$\sup_Q\left(\frac{1}{|Q|}\int_Q |f(x)-\operatorname{Avg}_Q f|^p\,dx\right)^{\frac{1}{p}} \le B_p\|f\|_{BMO}.$$

Proof. This result can be obtained from the one in the previous corollary or directly:

$$\begin{aligned}
&\frac{1}{|Q|}\int_Q |f(x)-\operatorname{Avg}_Q f|^p\,dx\\
&= \frac{p}{|Q|}\int_0^\infty \alpha^{p-1}|\{x\in Q:\ |f(x)-\operatorname{Avg}_Q f|>\alpha\}|\,d\alpha\\
&\le \frac{p}{|Q|}C|Q|\int_0^\infty \alpha^{p-1}e^{-A\alpha/\|f\|_{BMO}}\,d\alpha\\
&= p\,\Gamma(p)\frac{C}{A^p}\|f\|_{BMO}^p,
\end{aligned}$$

where $A = (2^n e)^{-1}$. Setting $B_p = (p\,\Gamma(p)\frac{C}{A^p})^{\frac{1}{p}}$, we conclude the proof of the corollary.

□

Since the inequality in Corollary 7.1.8 can be reversed when $p > 1$ via Hölder's inequality, we obtain the following interesting L^p characterization of the BMO norm.

Corollary 7.1.9. *For all $1 < p < \infty$ we have*

$$\sup_Q \left(\frac{1}{|Q|} \int_Q |f(x) - \operatorname*{Avg}_Q f|^p \, dx \right)^{\frac{1}{p}} \approx \|f\|_{BMO}.$$

Proof. Obvious. □

Exercises

7.1.1. Prove that BMO is a complete space, that is, every BMO-Cauchy sequence converges in BMO.
[*Hint:* Use Proposition 7.1.5 (ii) to show first that such a sequence is Cauchy in L^1 of every compact set.]

7.1.2. Find an example showing that the product of two BMO functions may not be in BMO.

7.1.3. Prove that

$$\| \, |f|^\alpha \, \|_{BMO} \le C_\alpha \|f\|_{BMO}^\alpha$$

whenever $0 < \alpha \le 1$.

7.1.4. Let f be a real-valued BMO function on $\mathbf{R}^n$. Prove that the functions

$$f_{KL}(x) = \begin{cases} K & \text{if } f(x) < K \\ f(x) & \text{if } K \le f(x) \le L \\ L & \text{if } f(x) > L \end{cases}$$

satisfy $\|f_{KL}\|_{BMO} \le \|f\|_{BMO}$.

7.1.5. Let B be a ball (or a cube) in $\mathbf{R}^n$ and let aB be a dilation of B with the same center. Show that there is a dimensional constant C_n such that for all f in BMO we have

$$|\operatorname*{Avg}_{aB} f - \operatorname*{Avg}_{B} f| \le C_n (\log a) \, \|f\|_{BMO} \, .$$

7.1.6. Let $a > 1$ and f be a BMO function on $\mathbf{R}^n$. Show that there exist dimensional constants C_n, C_n' such that
(a) for all balls B_1 and B_2 in $\mathbf{R}^n$ with radius R whose centers are at distance aR we have

$$|\operatorname*{Avg}_{aB_1} f - \operatorname*{Avg}_{B_2} f| \le C_n (\log a) \, \|f\|_{BMO};$$

(b) conclude that

$$|\operatorname*{Avg}_{B_1} f - \operatorname*{Avg}_{B_2} f| \le C_n' (\log a) \, \|f\|_{BMO} \, .$$

7.1.7. Let f be locally integrable on $\mathbf{R}^n$. Suppose that there exist positive constants m and b such that for all cubes Q in $\mathbf{R}^n$ and for all $0 < p < \infty$ we have

$$\alpha|\{x \in Q : \ |f(x) - \operatorname*{Avg}_Q f| > \alpha\}|^{\frac{1}{p}} \le b\, p^m \, |Q| \, .$$

Show that f satisfies the estimate

$$|\{x \in Q : \ |f(x) - \operatorname*{Avg}_Q f| > \alpha\}| \le |Q| \, e^{-c\alpha^{1/m}}$$

with $c = (2b)^{1/m} \log 2$.
[*Hint:* Try $p = (\alpha/2b)^{1/m}$.]

7.1.8. Prove that $|\log|x||^p$ is not in $BMO(\mathbf{R})$ when $1 < p < \infty$.
[*Hint:* Show that if $|\log|x||^p$ were in BMO, then estimate (7.1.9) would be violated for large α.]

7.1.9. Let $f \in BMO(\mathbf{R})$ have mean value equal to zero on the fixed interval I. Find a BMO function g on $\mathbf{R}$ such that
(1) $g = f$ on I.
(2) $g = 0$ on $\mathbf{R} \setminus \frac{4}{3} I$.
(3) $\|g\|_{BMO} \le C \|f\|_{BMO}$ for some constant C independent of f.
[*Hint:* Let I_0 be the middle third of I. Let I_1, I_2 the middle thirds of $I \setminus I_0$. Let $I_3, I_4, \dots, I_8$ the middle thirds of $I \setminus (I_0 \cup I_1 \cup I_2)$, etc. Also let J_k be the reflection of I_k with respect to the closest endpoint of I and set $g = \operatorname*{Avg}_{I_k} f$ on J_k for $k > 1$, $g = f$ on I, and zero otherwise.]

7.2. Duality between H^1 and BMO

The next main result we discuss about BMO is a certain remarkable duality relationship with the Hardy space H^1. We will show that BMO is the dual space of H^1. This means that every continuous linear functional on the Hardy space H^1 can be realized as integration against a fixed BMO function, where *integration* in this context is an abstract operation, not necessarily given by an absolutely convergent integral. Restricting our attention, however, on a dense subspace of H^1 such as the space of all finite sums of atoms, the use of the word *integration* is well justified. Indeed, for an L^2 atom for H^1 a and a BMO function b, the integral

$$\int |a(x)b(x)| \, dx < \infty$$

converges absolutely since $a(x)$ is compactly supported and bounded and $b(x)$ is locally (square) integrable.

Definition 7.2.1. Denote by $H^1_0(\mathbf{R}^n)$ the set of all finite linear combinations of L^2 atoms for $H^1(\mathbf{R}^n)$. For $b \in BMO(\mathbf{R}^n)$ we define a linear functional L_b on $H^1_0(\mathbf{R}^n)$ by setting

$$L_b(f) = \int_{\mathbf{R}^n} f(x)b(x)\,dx. \tag{7.2.1}$$

Note that (7.2.1) remains unchanged if b is replaced by $b + c$, where c is an additive constant. Therefore, this integral is unambiguously defined for $b \in BMO$.

Observe that L_b is a bounded linear functional on $H^1_0(\mathbf{R}^n)$. Indeed, if a_k are L^2 atoms for H^1 supported in cubes Q_k,

$$f = \sum_{k=1}^{N} \lambda_k a_k \in H^1_0(\mathbf{R}^n)\,,$$

and b is in $BMO(\mathbf{R}^n)$, then we have

$$\begin{aligned}
|L_b(f)| &= \left| \int_{\mathbf{R}^n} f(x)b(x)\,dx \right| \\
&= \left| \sum_{k=1}^{N} \lambda_k \int_{Q_k} a_k(x)(b(x) - \operatorname*{Avg}_{Q_k} b)\,dx \right| \\
&\le \sum_{k=1}^{N} |\lambda_k| \, \big\|a_{Q_k}\big\|_{L^2} |Q_k|^{\frac{1}{2}} \left(\frac{1}{|Q_k|} \int_{Q_k} |b(x) - \operatorname*{Avg}_{Q_k} b|^2\,dx \right)^{\frac{1}{2}} \\
&\le C_n \big\|f\big\|_{H^1} \big\|b\big\|_{BMO}\,,
\end{aligned}$$

where in the last step we used Corollary 7.1.8 and the fact that L^2 atoms for H^1 satisfy $\big\|a_{Q_k}\big\|_{L^2} \le |Q_k|^{-\frac{1}{2}}$.

Since $H^1_0(\mathbf{R}^n)$ is a dense subspace of $H^1(\mathbf{R}^n)$, the linear functional L_b has unique bounded extension on $H^1(\mathbf{R}^n)$. Precisely, given f in $H^1(\mathbf{R}^n)$, find a sequence $f_j \in H^1_0(\mathbf{R}^n)$ such that $f_j \to f$ in $H^1(\mathbf{R}^n)$. The estimate obtained in the preceding equations gives

$$|L_b(f_j) - L_b(f_k)| \le C_n \big\|b\big\|_{BMO} \big\|f_j - f_k\big\|_{H^1}\,,$$

which implies that the sequence $\{L_b(f_j)\}_j$ is Cauchy in $\mathbf{C}$. The same estimate gives that for any other sequence $f'_j \in H^1_0(\mathbf{R}^n)$ converging to f in $H^1(\mathbf{R}^n)$ we have

$$|L_b(f_j) - L_b(f'_j)| \le C_n \big\|b\big\|_{BMO} \big\|f_j - f'_j\big\|_{H^1} \to 0$$

as $j \to \infty$. If we set

$$L_b(f) = \lim_{j\to\infty} L_b(f_j)\,,$$

then L_b is well defined on $H^1(\mathbf{R}^n)$, coincides with L_b previously defined on $H_0^1(\mathbf{R}^n)$, and satisfies

$$|L_b(f)| \le C_n \|b\|_{BMO} \|f\|_{H^1}$$

for all $f \in H^1(\mathbf{R}^n)$. Thus L_b admits a bounded extension on all of $H^1(\mathbf{R}^n)$.

We have just proved that every BMO function b gives rise to a bounded linear functional L_b on $H^1(\mathbf{R}^n)$ satisfying

$$\|L_b\|_{H^1 \to \mathbf{C}} \le C_n \|b\|_{BMO}. \tag{7.2.2}$$

The fact that every bounded linear functional on H^1 arises this way is the main gist of the equivalence in the next theorem.

Theorem 7.2.2. *There exist dimensional constants C_n and C_n' such that the following statements are valid:*
(a) Given $b \in BMO(\mathbf{R}^n)$, the linear functional initially defined for f in H_0^1 by (7.2.1) has a unique bounded extension on $H^1(\mathbf{R}^n)$ with norm at most $C_n \|b\|_{BMO}$.
(b) For every bounded linear functional L on H^1 there exists a BMO function b such that for all $f \in H_0^1$ we have $L(f) = L_b(f)$ and also

$$\|b\|_{BMO} \le C_n' \|L_b\|_{H^1 \to \mathbf{C}}.$$

Proof. We have proved (a), and so it suffices to prove (b). Fix a bounded linear functional L on $H^1(\mathbf{R}^n)$ and also fix a cube Q. Consider the space $L^2(Q)$ of all square integrable functions supported in Q with norm

$$\|g\|_{L^2(Q)} = \left(\int_Q |g(x)|^2 \, dx \right)^{\frac{1}{2}}.$$

We denote by $L_0^2(Q)$ the closed subspace of $L^2(Q)$ consisting of all functions in $L^2(Q)$ with mean value zero. Next we show that every element in $L_0^2(Q)$ is in $H^1(\mathbf{R}^n)$ and we have the inequality

$$\|g\|_{H^1} \le c_n |Q|^{\frac{1}{2}} \|g\|_{L^2}. \tag{7.2.3}$$

To prove (7.2.3) we use the square function characterization of H^1. We fix a Schwartz function Ψ on $\mathbf{R}^n$ whose Fourier transform is supported in the annulus $\frac{1}{2} \le |\xi| \le 2$ and that satisfies (6.2.6) for all $\xi \neq 0$ and we let $\Delta_j(g) = \Psi_{2^{-j}} * g$. To estimate the L^1 norm of $\left(\sum_j |\Delta_j(g)|^2\right)^{1/2}$ over $\mathbf{R}^n$, consider the part of the integral over $3Q$ and the integral over $(3Q)^c$. First we use Hölder's inequality and an L^2 estimate to prove that

$$\int_{3Q} \Big(\sum_j |\Delta_j(g)(x)|^2 \Big)^{\frac{1}{2}} dx \le c_n |Q|^{\frac{1}{2}} \|g\|_{L^2}.$$

Now for $x \notin 3Q$ we use the mean value property of g to obtain

$$|\Delta_j(g)(x)| \le \frac{c_n \|g\|_{L^2} 2^{nj+j} |Q|^{\frac{2+n}{2n}}}{(1 + 2^j |x - c_Q|)^{n+2}}, \tag{7.2.4}$$

where c_Q is the center of Q. Estimate (7.2.4) is obtained in a way similar to that we obtained the corresponding estimate for one atom; see Theorem 6.6.9 for details. Now (7.2.4) implies that

$$\int_{(3Q)^c} \Big(\sum_j |\Delta_j(g)(x)|^2\Big)^{\frac{1}{2}} dx \le c_n |Q|^{\frac{1}{2}} \big\|g\big\|_{L^2},$$

which proves (7.2.3).

Let L be a bounded linear functional on $H^1(\mathbf{R}^n)$ with norm $\big\|L\big\|$. It follows from (7.2.3) that L is a bounded linear functional on $L^2_0(Q)$ with norm at most $c_n|Q|^{1/2}\big\|L\big\|$. By the Riesz representation theorem for the Hilbert space $L^2_0(Q)$, there is an element F^Q in $L^2_0(Q) = L^2(Q)/\{\text{constants}\}$ so that

$$L(g) = \int_Q F^Q(x)g(x)\,dx, \tag{7.2.5}$$

for all $g \in L^2_0(Q)$ and this F^Q has norm

$$\big\|F^Q\big\|_{L^2(Q)} \le c_n |Q|^{\frac{1}{2}} \big\|L\big\|. \tag{7.2.6}$$

Thus for any cube Q in $\mathbf{R}^n$, there is square integrable function F^Q supported in Q with mean value zero over Q such that (7.2.5) is satisfied. We observe that if a cube Q is contained in the cube Q', then F^Q differs from $F^{Q'}$ by a constant on Q. Indeed, F^Q and $F^{Q'}$ give rise to the same linear functional L on Q and thus

$$\int (F^{Q'}(x) - F^Q(x))g(x)\,dx = 0 \qquad \text{for all } g \in L^2_0(Q).$$

It follows that $F^{Q'} - F^Q$ is equal to a constant on Q by Exercise 7.2.2.

Next we claim that there is a locally integrable function b on $\mathbf{R}^n$ such that for any cube Q there is a constant c_Q such that

$$F^Q = b - c_Q \qquad \text{on } Q. \tag{7.2.7}$$

To prove this, define cubes $Q_m = [-m,m]^n$ for $m = 1, 2, \dots$. Observe that for all positive integers $l < m$ we have

$$F^{Q_m} - \frac{1}{|Q_l|}\int_{Q_l} F^{Q_m}(t)\,dt = F^{Q_l} - \frac{1}{|Q_l|}\int_{Q_l} F^{Q_l}(t)\,dt \qquad \text{on } Q_l, \tag{7.2.8}$$

since $F^{Q_m} - F^{Q_l}$ is a constant on Q_l and this constant is equal to the average of $F^{Q_m} - F^{Q_l}$ over Q_l. We now define a locally integrable function $b(x)$ on $\mathbf{R}^n$ by setting

$$b(x) = F^{Q_m}(x) - \frac{1}{|Q_1|}\int_{Q_1} F^{Q_m}(t)\,dt$$

whenever $x \in Q_m$. Identity (7.2.8) gives that $b(x)$ is a well-defined function on $\mathbf{R}^n$.

We prove (7.2.7). We first observe that (7.2.7) is valid by definition when $Q = Q_m$ for some $m = 1, 2, \dots$. Let Q be any other cube Q in $\mathbf{R}^n$. Pick an m so that $Q \subseteq Q_m$. Then $F^{Q_m} - F^Q$ is equal to a constant on Q. Thus

$$b - F^Q + \frac{1}{|Q_1|}\int_{Q_1} F^{Q_m}(t)\,dt$$

is a constant on Q, which proves (7.2.7).

We have now found a locally integrable function b such that for all cubes Q and all $g \in L^2_0(Q)$ we have

$$L(g) = \int_Q b(x)g(x)\,dx \tag{7.2.9}$$

as it follows from (7.2.5) and (7.2.7). We conclude the proof by showing that $b \in BMO(\mathbf{R}^n)$. By (7.2.7) and (7.2.6) we have

$$\begin{aligned}\sup_Q \frac{1}{|Q|}\int_Q |b(x) - c_Q|\,dx &= \sup_Q \frac{1}{|Q|}\int_Q |F^Q(x)|\,dx \\ &\le |Q|^{-\frac{1}{2}}\big\|F^Q\big\|_{L^2(Q)} \\ &\le c_n\big\|L\big\|,\end{aligned}$$

which shows that $\big\|b\big\|_{BMO} \le 2c_n\big\|L\big\|$ in view of Proposition 7.1.2 (3). Finally, (7.2.9) implies that

$$L(f) = \int b(x)f(x)\,dx$$

for all $f \in H^1_0(\mathbf{R}^n)$, thus proving the required conclusion.

□

Exercises

7.2.1. Prove that

$$\big\|f\big\|_{BMO} \approx \sup_{\|g\|_{H^1}\le 1} \big|\langle f, g\rangle\big|$$

$$\big\|f\big\|_{H^1} \approx \sup_{\|g\|_{BMO}\le 1} \big|\langle g, f\rangle\big|$$

7.2.2. Suppose that a locally integrable function u is supported in a cube Q in $\mathbf{R}^n$ and satisfies

$$\int_Q u(x)g(x)\,dx = 0$$

for all square integrable functions on Q with mean value zero. Show that u is equal to a constant.

7.2.3. Suppose that a locally integrable function u is supported in a cube Q in $\mathbf{R}^n$ and satisfies

$$\int_Q u(x)g(x)\,dx = 0$$

for all square integrable functions on Q with vanishing moments up to order N. Show that u is equal to a polynomial of degree at most N.

7.2.4. Define $BMO(\ell^2)$ and prove a duality theorem between $H^1(\ell^2)$ and $BMO(\ell^2)$.

7.3. Nontangential Maximal Functions and Carleson Measures

Many properties of functions defined on $\mathbf{R}^n$ are related to corresponding properties of associated functions defined on $\mathbf{R}^{n+1}_+$ in a natural way. A typical example of this situation is the relation between an $L^p(\mathbf{R}^n)$ function f and its Poisson integral $f * P_t$ or the more general $f * \Phi_t$, where $\{\Phi\}_{t>0}$ is an approximate identity. Here Φ is a Schwartz function on $\mathbf{R}^n$ with integral 1. A maximal operator associated to the approximate identity $\{f * \Phi_t\}_{t>0}$ is

$$f \to \sup_{t>0} |f * \Phi_t|\,,$$

which we know is pointwise controlled by a multiple of the Hardy-Littlewood maximal function $M(f)$. Another example of a maximal operator associated to the previous approximate identity is the nontangential maximal function

$$f \to M^*(f;\Phi)(x) = \sup_{t>0} \sup_{|y-x|<t} |(f * \Phi_t)(y)|.$$

To study nontangential behavior we will consider general functions F defined on $\mathbf{R}^{n+1}_+$ that are not necessarily given as an average of functions defined on $\mathbf{R}^n$. Throughout this section we will use capital letters to denote functions defined on $\mathbf{R}^{n+1}_+$. When we write $F(x,t)$ we mean that $x \in \mathbf{R}^n$ and $t > 0$.

7.3.a. Definition and Basic Properties of Carleson Measures

Definition 7.3.1. Let F be a measurable function on $\mathbf{R}^{n+1}_+$. For x in $\mathbf{R}^n$ let $\Gamma(x)$ be the cone with vertex x defined by

$$\Gamma(x) = \{(y,t) \in \mathbf{R}^n \times \mathbf{R}^+ :\ |y-x| < t\}.$$

A picture of this cone is shown in Figure 7.1. The *nontangential maximal function* of F is the function

$$F^*(x) = \sup_{(y,t)\in\Gamma(x)} |F(y,t)|$$

defined on $\mathbf{R}^n$. This function is obtained by taking the supremum of the values of F inside the cone $\Gamma(x)$.

We remark that if $F^*(x) = 0$ for almost all $x \in \mathbf{R}^n$, then F is identically equal to zero on $\mathbf{R}^{n+1}_+$.

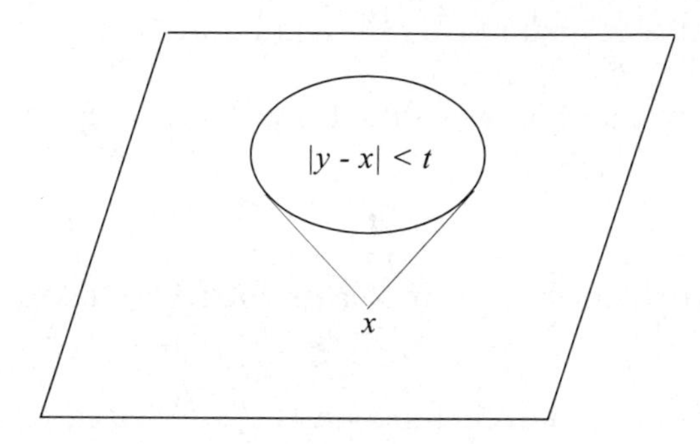

FIGURE 7.1. The cone $\Gamma(x)$ with vertex x in $\mathbf{R}^n$. At height t we have the disc $\{(y,t): \ |y-x|<t\}$.

Definition 7.3.2. Given a ball $B = B(x_0, r)$ in $\mathbf{R}^n$ we define the *tent* or *cylindrical tent* over B to be the "cylindrical set"

$$T(B) = \{(x,t) \in \mathbf{R}^{n+1}_+ : \ x \in B, \quad 0 < t \le r\}.$$

For a cube Q in $\mathbf{R}^n$ we define the tent over Q to be the cube

$$T(Q) = Q \times (0, \ell(Q)].$$

A tent over a ball and over a cube are shown in Figure 7.2. A positive measure μ on $\mathbf{R}^{n+1}_+$ is called a *Carleson measure* if

$$\big\|\mu\big\|_{\mathcal{C}} = \sup_Q \frac{1}{|Q|}\mu(T(Q)) < \infty, \tag{7.3.1}$$

where the supremum in (7.3.1) is taken over all cubes Q in $\mathbf{R}^n$. The *Carleson function* of the measure μ is defined as

$$\mathcal{C}(\mu)(x) = \sup_{Q \ni x} \frac{1}{|Q|}\mu(T(Q)), \tag{7.3.2}$$

where the supremum in (7.3.2) is taken over all cubes in $\mathbf{R}^n$ containing the point x. Observe that $\big\|\mathcal{C}(\mu)\big\|_{L^\infty} = \big\|\mu\big\|_{\mathcal{C}}$.

We also define

$$\big\|\mu\big\|_{\mathcal{C}}^{\text{cylinder}} = \sup_B \frac{1}{|B|}\mu(T(B)), \tag{7.3.3}$$

where the supremum is taken over all balls B in $\mathbf{R}^n$. The reader is asked to verify that there exist dimensional constants c_n and C_n such that

$$c_n\big\|\mu\big\|_{\mathcal{C}} \le \big\|\mu\big\|_{\mathcal{C}}^{\text{cylinder}} \le C_n\big\|\mu\big\|_{\mathcal{C}}$$

for all measures μ on $\mathbf{R}_+^{n+1}$, that is, a measure satisfies the Carleson condition (7.3.1) with respect to cubes if and only if it satisfies the analogous condition (7.3.3) with respect to balls. Likewise the Carleson function $\mathcal{C}(\mu)$ defined with respect to cubes is comparable to $\mathcal{C}^{\text{cylinder}}(\mu)$ defined with respect to cylinders over balls.

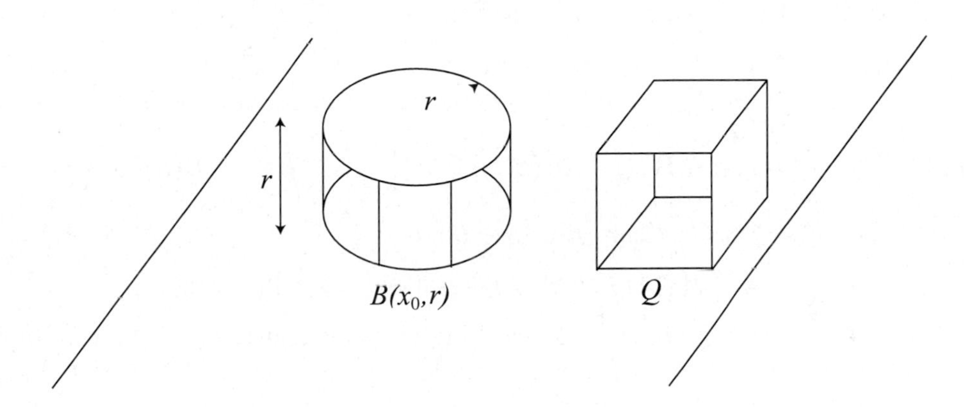

FIGURE 7.2. The tents over the ball $B(x_0, r)$ and over a cube Q in $\mathbf{R}^2$.

Examples 7.3.3. The Lebesgue measure on $\mathbf{R}_+^{n+1}$ is not a Carleson measure. Indeed, it is not difficult to see that condition (7.3.1) cannot hold for large balls.

Let L be a line in $\mathbf{R}^2$. For A measurable subsets of $\mathbf{R}_+^2$ define $\mu(A)$ to be the linear Lebesgue measure of the set $L \cap A$. Then μ is a Carleson measure on $\mathbf{R}_+^2$. Indeed, the linear measure of the part of a line inside the box $[x_0 - r, x_0 + r] \times (0, r]$ is at most equal to the diagonal of the box, that is, $\sqrt{5}r$.

Likewise, let P be an affine plane in $\mathbf{R}^{n+1}$ and define a measure ν by setting $\nu(A)$ to be the n-dimensional Lebesgue measure of the set $A \cap P$ for any $A \subseteq \mathbf{R}_+^{n+1}$. A similar idea shows that ν is a Carleson measure on $\mathbf{R}_+^{n+1}$.

We now turn to the study of some interesting boundedness properties of functions on $\mathbf{R}_+^{n+1}$ with respect to Carleson measures.

A useful tool in this study will be the *Whitney decomposition* of an open set in $\mathbf{R}^n$. This is a decomposition of general open set Ω in $\mathbf{R}^n$ as a union of disjoint cubes whose lengths are proportional to their distance from the boundary of the open set. For a given cube Q in $\mathbf{R}^n$, we will denote by $\ell(Q)$ its length.

Proposition 7.3.4. *(**Whitney decomposition**) Let Ω be an open nonempty proper subset of $\mathbf{R}^n$. Then there exists a family of closed cubes $\{Q_j\}_j$ such that*

(a) *$\bigcup_j Q_j = \Omega$ and the Q_j's have disjoint interiors.*

(b) *$\sqrt{n}\,\ell(Q_j) \leq dist\,(Q_j, \Omega^c) \leq 4\sqrt{n}\,\ell(Q_j)$.*

(c) *If the boundaries of two cubes* Q_j *and* Q_k *touch, then*

$$\frac{1}{4} \le \frac{\ell(Q_j)}{\ell(Q_k)} \le 4 .$$

(d) *For a given* Q_j *there exist at most* 12^n Q_k*'s that touch it.*

The proof of Proposition 7.3.4 is given in Appendix J.

Theorem 7.3.5. *There exists a dimensional constant* C_n *so that for all* $\alpha > 0$*, all measures* $\mu \ge 0$ *on* $\mathbf{R}^{n+1}_+$*, and all* μ*-measurable functions* F *on* $\mathbf{R}^{n+1}_+$*, we have*

$$\mu\big(\{(x,t) \in \mathbf{R}^{n+1}_+ : \,|F(x,t)| > \alpha\}\big) \le C_n \int_{\{F^* > \alpha\}} \mathcal{C}(\mu)(x)\,dx. \tag{7.3.4}$$

In particular, if μ *is a Carleson measure, then*

$$\mu\big(\{|F| > \alpha\}\big) \le C_n \big\|\mu\big\|_{\mathcal{C}} |\{F^* > \alpha\}|.$$

Proof. We prove this theorem by working with the equivalent definition of Carleson measures and Carleson functions using balls and cylinders over balls. As observed earlier, these quantities are comparable to the corresponding quantities using cubes.

We begin by observing that for any function F the set $\Omega = \{F^* > \alpha\}$ is open and in particular, F^* is Lebesgue measurable.

Let $\{Q_k\}$ be the Whitney decomposition of the set Ω. For each $x \in \Omega$, set $\delta_\alpha(x) = \text{dist}\,(x, \Omega^c)$. Then for $z \in Q_k$ we have

$$\delta_\alpha(z) \le \sqrt{n}\,\ell(Q_k) + \text{dist}\,(Q_k, \Omega^c) \le 5\sqrt{n}\,\ell(Q_k) \tag{7.3.5}$$

in view of Proposition 7.3.4 (b). For each Q_k, let B_k be the smallest ball that contains Q_k. Then the radius of B_k is $\sqrt{n}\,\ell(Q_k)/2$. Combine this observation with (7.3.5) to obtain that

$$z \in Q_k \quad \Longrightarrow \quad B(z, \delta_\alpha(z)) \subseteq 12\, B_k.$$

This implies that

$$\bigcup_{z\in\Omega} T\big(B(z, \delta_\alpha(z))\big) \subseteq \bigcup_k T(12\, B_k). \tag{7.3.6}$$

Next we claim that

$$\{|F| > \alpha\} \subseteq \bigcup_{z\in\Omega} T\big(B(z, \delta_\alpha(z))\big). \tag{7.3.7}$$

Indeed, let $(x,t) \in \mathbf{R}^{n+1}_+$ so that $|F(x,t)| > \alpha$. Then by the definition of F^* we have that $F^*(y) > \alpha$ for all $y \in \mathbf{R}^n$ satisfying $|x-y| < t$. Thus $B(x,t) \subseteq \Omega$ and so $\delta_\alpha(x) \ge t$. This gives that $(x,t) \in T\big(B(x, \delta_\alpha(x))\big)$, which proves (7.3.7).

Combining (7.3.6) and (7.3.7) we obtain

$$\{|F| > \alpha\} \subseteq \bigcup_k T(12\, B_k).$$

Applying the measure μ and using the definition of the Carleson function, we obtain

$$\begin{aligned}\mu(\{|F|>\alpha\}) &\le \sum_k \mu(T(12\,B_k))\\ &\le \sum_k |12\,B_k| \inf_{x\in Q_k} \mathcal{C}^{\text{cylinder}}(\mu)(x)\\ &\le 12^n \sum_k \frac{|B_k|}{|Q_k|}\int_{Q_k} \mathcal{C}^{\text{cylinder}}(\mu)(x)\,dx\\ &\le C_n \int_\Omega \mathcal{C}(\mu)(x)\,dx\end{aligned}$$

since $|B_k| = 2^{-n}n^{n/2}v_n|Q_k|$. This proves (7.3.4).

□

Corollary 7.3.6. *For any Carleson measure μ and every μ-measurable function F on $\mathbf{R}^{n+1}_+$ we have*

$$\int_{\mathbf{R}^{n+1}_+} |F(x,t)|^p\,d\mu(x,t) \le C_n\big\|\mu\big\|_{\mathcal{C}} \int_{\mathbf{R}^n} (F^*(x))^p\,dx$$

for all $0<p<\infty$.

Proof. Simply use Proposition 1.1.4 and the previous theorem.

□

It is noteworthy that the preceding corollary has a converse if $F(x,t)$ is the convolution of f with the Poisson kernel.

Theorem 7.3.7. *Let Φ be a function on $\mathbf{R}^n$ satisfying for some $0<C,\delta<\infty$*

$$|\Phi(x)| \le \frac{C}{(1+|x|)^{n+\delta}} \tag{7.3.8}$$

and for $t>0$ let $\Phi_t(x)=t^{-n}\Phi(t^{-1}x)$. Then a measure μ on $\mathbf{R}^{n+1}_+$ is Carleson if and only if for every p with $1<p<\infty$ there is a constant $C_{p,n}(\mu)$ such that for all $f\in L^p(\mathbf{R}^n)$ we have

$$\int_{\mathbf{R}^{n+1}_+} |(\Phi_t * f)(x)|^p\,d\mu(x,t) \le C_{p,n}(\mu)\int_{\mathbf{R}^n} |f(x)|^p\,dx \tag{7.3.9}$$

and the constant $C_{p,n}(\mu)$ is at most $C_{p,n}\big\|\mu\big\|_{\mathcal{C}}$, if μ is a Carleson measure. In particular (7.3.9) holds if $\Phi_t(x)$ is the Poisson kernel $P_t(x)$.

Proof. If μ is a Carleson measure, then we can obtain (7.3.9) as a consequence of Corollary 7.3.6. Indeed, for $F(x,t)=(\Phi_t * f)(x)$ we have

$$F^*(x) = \sup_{t>0}\ \sup_{\substack{y\in\mathbf{R}^n\\ |y-x|<t}} \big|(\Phi_t * f)(y)\big|\,.$$

Using (7.3.8) and Corollary 2.1.12, this is easily seen to be pointwise controlled by the Hardy-Littlewood maximal operator, which is L^p bounded. Conversely, if (7.3.9) holds, then we fix a ball $B = B(x_0, r)$ in $\mathbf{R}^n$ with center x_0 and radius $r > 0$. Then for (x,t) in $T(B)$ we have

$$|(\Phi_t * \chi_{2B})(x)| = \int_{x-2B} \Phi_t(y)\,dy \geq \int_{B(0,t)} \Phi_t(y)\,dy = \int_{B(0,1)} \Phi(y)\,dy = C_n\,,$$

since $B(0,t) \subseteq x - 2B(x_0, r)$ whenever $t \leq r$. Therefore, we have

$$\begin{aligned}\mu(T(B)) &\leq \frac{1}{C_n^p}\int_{\mathbf{R}^{n+1}_+} |(\Phi_t * \chi_{2B})(x)|^p\,d\mu(x,t)\\ &\leq \frac{C_p}{C_n^p}\int_{\mathbf{R}^n} |\chi_{2B}(x)|^p\,dx = \frac{2^{np}C_p}{C_n^p}|B|\end{aligned}$$

which proves that μ is a Carleson measure.

□

The reader may observe that the preceding proof also yields that μ is a Carleson measure if and only if (7.3.9) holds for some p in $(1,\infty)$.

7.3.b. BMO Functions and Carleson Measures

We now turn to an interesting connection between *BMO* functions and Carleson measures. We have the following.

Theorem 7.3.8. *Let b be a BMO function on $\mathbf{R}^n$ and let Ψ be an integrable function with mean value zero on $\mathbf{R}^n$ that satisfies*

$$|\Psi(x)| \leq A(1+|x|)^{-n-1}\,. \tag{7.3.10}$$

Consider the dilations $\Psi_t = t^{-n}\Psi(t^{-1}x)$ and define the Littlewood-Paley operators $\Delta_j(f) = f * \Psi_{2^{-j}}$.
(a) Suppose that

$$\sup_{\xi\in\mathbf{R}^n}\sum_{j\in\mathbf{Z}} |\widehat{\Psi}(2^{-j}\xi)|^2 \leq A^2 < \infty \tag{7.3.11}$$

and let $\delta_{2^{-j}}(t)$ be Dirac mass at the point $t = 2^{-j}$. Then there is a dimensional constant C_n so that

$$d\mu(x,t) = \sum_{j\in\mathbf{Z}} |(\Psi_{2^{-j}} * b)(x)|^2\,dx\,\delta_{2^{-j}}(t)$$

is a Carleson measure on $\mathbf{R}^{n+1}_+$ with norm at most $C_nA^2\|b\|^2_{BMO}$.
(b) Suppose that

$$\sup_{\xi\in\mathbf{R}^n}\int_0^\infty |\widehat{\Psi}(t\xi)|^2\frac{dt}{t} \leq A^2 < \infty\,. \tag{7.3.12}$$

Then the continuous version of $d\mu(x,t)$,

$$d\nu(x,t) = |(\Psi_t * b)(x)|^2\, dx\, \frac{dt}{t}\,,$$

is a Carleson measure on $\mathbf{R}^{n+1}_+$ with norm at most $C_n A^2 \big\|b\big\|^2_{BMO}$ for some dimensional constant C_n.
(c) Let $\delta, A > 0$. Suppose that $\{K_t\}_{t>0}$ are functions on $\mathbf{R}^n \times \mathbf{R}^n$ that satisfy

$$|K_t(x,y)| \le \frac{At}{(t+|x-y|)^{n+\delta}} \tag{7.3.13}$$

for all $t>0$ and all $x,y \in \mathbf{R}^n$. Let R_t be the linear operator

$$R_t(f)(x) = \int_{\mathbf{R}^n} K_t(x,y)\, f(y)\, dy$$

which is well defined for all $f \in \bigcup_{1\le p\le\infty} L^p(\mathbf{R}^n)$. Suppose that $R_t(1) = 0$ for all $t>0$ and that there is a constant $B>0$ such that

$$\int_0^\infty \int_{\mathbf{R}^n} \big|R_t(f)(x)\big|^2 \frac{dx\,dt}{t} \le B\big\|f\big\|^2_{L^2(\mathbf{R}^n)} \tag{7.3.14}$$

for all $f \in L^2(\mathbf{R}^n)$. Then for all b in BMO the measure

$$\big|R_t(b)\big|^2 \frac{dx\,dt}{t}$$

is Carleson with norm at most a dimensional multiple of $(A+B)^2\big\|b\big\|^2_{BMO}$.

Proof. We prove (a). The measure μ is defined so that for every μ-integrable function F on $\mathbf{R}^{n+1}_+$ we have

$$\int_{\mathbf{R}^{n+1}_+} F(x,t)\, d\mu(x,t) = \sum_{j\in\mathbf{Z}} \int_{\mathbf{R}^n} |(\Psi_{2^{-j}} * b)(x)|^2 F(x, 2^{-j})\, dx. \tag{7.3.15}$$

For a cube Q in $\mathbf{R}^n$ we let Q^* be the cube with the same center and orientation whose side length is $3\sqrt{n}\,\ell(Q)$, where $\ell(Q)$ is the side length of Q. Fix a cube Q in $\mathbf{R}^n$, take F to be the characteristic function of the tent of Q and split b as

$$b = \big(b - \operatorname*{Avg}_Q b\big)\chi_{Q^*} + \big(b - \operatorname*{Avg}_Q b\big)\chi_{(Q^*)^c} + \operatorname*{Avg}_Q b.$$

Since Ψ has mean value zero, the constant term $\operatorname*{Avg}_Q b$ has no contribution in (7.3.15). Then we have

$$\mu(T(Q)) = \sum_{2^{-j}\le\ell(Q)} \int_Q |\Delta_j(b)(x)|^2\, dx \le 2\Sigma_1 + 2\Sigma_2,$$

where

$$\Sigma_1 = \sum_{j\in\mathbf{Z}} \int_{\mathbf{R}^n} \left|\Delta_j\big((b - \operatorname*{Avg}_Q b)\chi_{Q^*}\big)(x)\right|^2 dx,$$

$$\Sigma_2 = \sum_{2^{-j}\le \ell(Q)} \int_Q \left|\Delta_j\big((b - \operatorname*{Avg}_Q b)\chi_{(Q^*)^c}\big)(x)\right|^2 dx.$$

Using Plancherel's theorem and (7.3.11), we obtain

$$\begin{aligned}\Sigma_1 &\le \sup_\xi \sum_{j\in\mathbf{Z}} |\widehat{\Psi}(2^{-j}\xi)|^2 \int_{\mathbf{R}^n} \left|\big((b - \operatorname*{Avg}_Q b)\chi_{Q^*}\big)\widehat{\ }(\xi)\right|^2 d\xi \\ &\le A^2 \int_{Q^*} |b(x) - \operatorname*{Avg}_Q b|^2\, dx \\ &\le C_n A^2 \|b\|_{BMO}^2 |Q|\end{aligned}$$

in view of Proposition 7.1.2 (3). To estimate Σ_2, we use the size estimate of the function Ψ. We obtain

$$\left|\big(\Psi_{2^{-j}} * (b - \operatorname*{Avg}_Q b)\chi_{(Q^*)^c}\big)(x)\right| \le \int_{(Q^*)^c} \frac{A\, 2^{-j}\left|b(y) - \operatorname*{Avg}_Q b\right|}{(2^{-j} + |x-y|)^{n+1}}\, dy \tag{7.3.16}$$

But note that if c_Q is the center of Q, then

$$\begin{aligned}2^{-j} + |x-y| &\ge |y-x| \\ &\ge |y - c_Q| - |c_Q - x| \\ &\ge \frac{1}{2}|c_Q - y| + \frac{3\sqrt{n}}{4}\ell(Q) - |c_Q - x| \\ &\ge \frac{1}{2}|c_Q - y| + \frac{3\sqrt{n}}{4}\ell(Q) - \frac{\sqrt{n}}{2} l(Q) \\ &= \frac{1}{2}\Big(|c_Q - y| + \frac{\sqrt{n}}{2}\ell(Q)\Big)\end{aligned}$$

when $y \in (Q^*)^c$ and $x \in Q$. Inserting this estimate in (7.3.16), integrating over Q, and summing over j with $2^{-j} \le \ell(Q)$, we obtain

$$\begin{aligned}\Sigma_2 &\le C_n \sum_{2^{-j}\le \ell(Q)} 2^{-2j} \int_Q \left(A \int_{\mathbf{R}^n} \frac{\left|b(y) - \operatorname*{Avg}_Q b\right|}{(\ell(Q) + |c_Q - y|)^{n+1}}\, dy\right)^2 dx \\ &\le C_n A^2 |Q| \left(\int_{\mathbf{R}^n} \frac{\ell(Q)\left|b(y) - \operatorname*{Avg}_Q b\right|}{(\ell(Q)^2 + |y - c_Q|^2)^{\frac{n+1}{2}}}\, dy\right)^2.\end{aligned}$$

But in (7.1.8) we showed that the last expression inside the parentheses in the preceding equation is controlled by $C_n\|b\|_{BMO}$. This proves that

$$\Sigma_1 + \Sigma_2 \le C_n A^2 |Q| \, \|b\|_{BMO}^2 ,$$

which implies that $\mu(T(Q)) \le C_n A^2 \|b\|_{BMO}^2 |Q|$.

We now observe that the proof of part (b) of the theorem is completely similar to that of part (a).

Finally, part (c) is a generalization of part (b) and is proved similarly. We quickly sketch its proof. Write

$$b = (b - \operatorname*{Avg}_Q b)\chi_{Q^*} + (b - \operatorname*{Avg}_Q b)\chi_{(Q^*)^c} + \operatorname*{Avg}_Q b$$

as in the proof of Theorem 7.3.8. Note that $R_t(\operatorname*{Avg}_Q b) = 0$ and apply an L^2 estimate on $(b - \operatorname*{Avg}_Q b)\chi_{Q^*}$ using condition (7.3.14) and an L^1 estimate on $(b - \operatorname*{Avg}_Q b)\chi_{(Q^*)^c}$ using condition (7.3.13).

□

Exercises

7.3.1. Let a_j, b_j be sequences of positive real numbers such that $\sum_j b_j < \infty$. Define a measure μ on $\mathbf{R}^{n+1}_+$ by setting

$$\mu(E) = \sum_j b_j |E \cap \{(x, a_j) : \; x \in \mathbf{R}^n\}| ,$$

where E is a subset of $\mathbf{R}^{n+1}_+$ and $|\;\;|$ denotes n-dimensional Lebesgue measure on the affine planes $t = a_j$. Show that μ is a Carleson measure with norm

$$\|\mu\|_{\mathcal{C}}^{\text{cylinder}} = \|\mu\|_{\mathcal{C}} = \sum_j b_j \, .$$

7.3.2. Let $x_0 \in \mathbf{R}^n$ and $\mu = \delta_{(x_0,1)}$ be the Dirac mass at the point $(x_0, 1)$. Show that μ is Carleson measure and compute $\|\mu\|_{\mathcal{C}}^{\text{cylinder}}$ and $\|\mu\|_{\mathcal{C}}$. Which of these norms is larger?

7.3.3. Define *conical* and *hemispherical* tents over balls in $\mathbf{R}^n$ as well as *pyramidal* tents over cubes in $\mathbf{R}^n$ and define the expressions $\|\mu\|_{\mathcal{C}}^{\text{cone}}$, $\|\mu\|_{\mathcal{C}}^{\text{hemisphere}}$, and $\|\mu\|_{\mathcal{C}}^{\text{pyramid}}$. Show that

$$\|\mu\|_{\mathcal{C}}^{\text{cone}} \approx \|\mu\|_{\mathcal{C}}^{\text{hemisphere}} \approx \|\mu\|_{\mathcal{C}}^{\text{pyramid}} \approx \|\mu\|_{\mathcal{C}} ,$$

where all the implicit constants in the previous estimates depend only on the dimension.

7.3.4. (a) Suppose that Φ has a radial, bounded, symmetrically decreasing integrable majorant. Set $F(x,t) = (f * \Phi_t)(x)$, where f is a locally integrable function on $\mathbf{R}^n$. Prove that

$$F^*(x) \le CM(f)(x)$$

where M is the Hardy-Littlewood maximal operator and C is a constant that only depends on the dimension.

(b) Use this result to obtain a sharpening of Theorem 7.3.7 for such functions Φ.

[*Hint:* If $\varphi(|x|)$ is the claimed majorant of $\Phi(x)$, then the function $\psi(|x|) = \varphi(0)$ for $|x| \le 1$ and $\psi(|x|) = \varphi(|x|-1)$ for $|x| \ge 1$ is a majorant for the function $\Psi(x) = \sup_{|u|\le 1} |\Phi(x-u)|$.]

7.3.5. Let F be a function on $\mathbf{R}^{n+1}_+$, let F^* be the nontangential maximal function derived from F, and let $\mu \ge 0$ be a measure on $\mathbf{R}^{n+1}_+$. Prove that

$$\big\|F\big\|_{L^r(\mathbf{R}^{n+1}_+,\mu)} \le C_n^{1/r}\bigg(\int_{\mathbf{R}^n} \mathcal{C}(\mu)(x)F^*(x)^r\,dx\bigg)^{1/r},$$

where C_n is the constant of Theorem 7.3.5 and $0 < r < \infty$.

7.3.6. (a) Given A a closed subset of $\mathbf{R}^n$ and $0 < \gamma < 1$, define

$$A^*_\gamma = \{x \in \mathbf{R}^n : \inf_{r>0} \frac{|A \cap B(x,r)|}{|B(x,r)|} \ge \gamma\}.$$

Show that A^* is a closed subset of A and that it satisfies

$$|(A^*_\gamma)^c| \le \frac{3^n}{1-\gamma}|A^c|.$$

[*Hint:* Consider the Hardy-Littlewood maximal function of χ_{A^c}.]

(b) For a function F on $\mathbf{R}^{n+1}_+$ and $0 < a < \infty$, set

$$F^*_a(x) = \sup_{t>0} \sup_{|y-x|<at} |F(y,t)|$$

Let $0 < a < b < \infty$ be given. Prove that for all $\lambda > 0$ we have

$$|\{F^*_a > \lambda\}| \le |\{F^*_b > \lambda\}| \le 3^n a^{-n}(a+b)^n |\{F^*_a > \lambda\}|.$$

7.3.7. Let μ be a Carleson measure on $\mathbf{R}^{n+1}_+$. Show that for any $z_0 \in \mathbf{R}^n$ and $t > 0$ we have

$$\iint\limits_{\mathbf{R}^n\times(0,t)} \frac{t}{(|z-z_0|^2+t^2+s^2)^{\frac{n+1}{2}}}\,d\mu(z,s) \le \big\|\mu\big\|_{\mathcal{C}}^{\text{cylinder}} \frac{\pi^{\frac{n+1}{2}}}{\Gamma(\frac{n+1}{2})}.$$

[*Hint:* Begin by writing

$$\frac{t}{(|z-z_0|^2+t^2+s^2)^{\frac{n+1}{2}}} = (n+1)t\int_Q^\infty \frac{dr}{r^{n+2}}\,,$$

where $Q = \sqrt{|z-z_0|^2+t^2+s^2}$. Apply Fubini's theorem to estimate the required expression by

$$\begin{aligned} t(n+1)\int_t^\infty \int_{T(B(z_0,\sqrt{r^2-t^2}))} d\mu(z,s)\,\frac{dr}{r^{n+2}} \\ \le t(n+1)v_n\big\|\mu\big\|_{\mathcal{C}}^{\text{cylinder}}\int_t^\infty (r^2-t^2)^{\frac{n}{2}}\,\frac{dr}{r^{n+2}} \end{aligned}$$

where v_n is the volume of the unit ball in $\mathbf{R}^n$. Reduce the last integral to a beta function.]

7.3.8. (*Verbitsky* [**540**]) Let μ be a Carleson measure on $\mathbf{R}^{n+1}_+$. Show that for all $p>2$ there exists a dimension-free constant C_p such that

$$\int_{\mathbf{R}^{n+1}_+} |(P_t*f)(x)|^p\,d\mu(x,t) \le C_p\big\|\mu\big\|_{\mathcal{C}}^{\text{cylinder}}\int_{\mathbf{R}^n}|f(x)|^p\,dx\,.$$

[*Hint:* It suffices to prove that the operator $f\to P_t*f$ maps $L^2(\mathbf{R}^n)$ into $L^{2,\infty}(\mathbf{R}^{n+1}_+,d\mu)$ with a dimension-free constant C since then the conclusion will follow by interpolation with the corresponding L^∞ estimate which holds with constant 1. By duality and Exercise 1.4.7 this is equivalent to showing that

$$\int_{\mathbf{R}^n}\left[\iint_E P_t(x-y)\,d\mu(y,t)\iint_E P_s(x-z)\,d\mu(z,s)\right]dx \le C\mu(E)$$

for any set E in $\mathbf{R}^{n+1}_+$ with $\mu(E)<\infty$. Apply Fubini's theorem, use the identity

$$\int_{\mathbf{R}^n} P_t(x-y)P_s(x-z)\,dx = P_{t+s}(y-z)\,,$$

and consider the cases $t\le s$ and $s\le t$.]

7.4. The Sharp Maximal Function

In Section 7.1 we defined BMO as the space of all locally integrable functions on $\mathbf{R}^n$ whose mean oscillation is at most a finite constant. In this section we will introduce a quantitative way to measure the mean oscillation of a function near any point.

7.4.a. Definition and Basic Properties of the Sharp Maximal Function

The "measurement" of this local behavior of the mean oscillation of a function will be achieved with the introduction of the sharp maximal function. This is a device that will enable us to relate integrability properties of a function to those of its mean oscillations.

Definition 7.4.1. Given a locally integrable function f on $\mathbf{R}^n$, we define its *sharp maximal function* $M^{\#}(f)$ as

$$M^{\#}(f)(x) = \sup_{Q \ni x} \frac{1}{|Q|} \int_Q |f(t) - \operatorname*{Avg}_Q f| \, dt,$$

where the supremum is taken over all cubes Q in $\mathbf{R}^n$ that contain the given point x.

The sharp maximal function is an analogue of the Hardy-Littlewood maximal function, but it has some advantages over it, especially when dealing with the endpoint space L^∞. The very definition of $M^{\#}(f)$ brings up a connection with BMO that will be crucial in interpolation. Precisely, we have

$$BMO(\mathbf{R}^n) = \{f \in L^1_{\text{loc}}(\mathbf{R}^n) : \; M^{\#}(f) \in L^\infty(\mathbf{R}^n)\},$$

and in this case

$$\|f\|_{BMO} = \|M^{\#}(f)\|_{L^\infty}.$$

We summarize some properties of the sharp maximal function.

Proposition 7.4.2. *Let f, g be a locally integrable functions on $\mathbf{R}^n$. Then*

(1) $M^{\#}(f) \le 2M_c(f)$, *where M_c is the Hardy-Littlewood maximal operator with respect to cubes in $\mathbf{R}^n$.*

(2) *For all cubes Q in $\mathbf{R}^n$ we have*

$$\frac{1}{2} M^{\#}(f)(x) \le \sup_{x \in Q} \inf_{a \in \mathbf{C}} \frac{1}{|Q|} \int_Q |f(y) - a| \, dy \le M^{\#}(f)(x).$$

(3) $M^{\#}(|f|) \le 2M^{\#}(f)$.

(4) *We have* $M^{\#}(f+g) \le M^{\#}(f) + M^{\#}(g)$.

Proof. The proof of (1) is trivial. To prove (2) we fix $\varepsilon > 0$ and for any cube Q we pick a constant a_Q such that

$$\frac{1}{|Q|} \int_Q |f(y) - a_Q| \, dy \le \inf_{a \in Q} \frac{1}{|Q|} \int_Q |f(y) - a| \, dy + \varepsilon \, .$$

Then

$$\begin{aligned}\frac{1}{|Q|}\int_Q |f(y)-\operatorname{Avg}_Q f|\,dy &\le \frac{1}{|Q|}\int_Q |f(y)-a_Q|\,dy+\frac{1}{|Q|}\int_Q |\operatorname{Avg}_Q f - a_Q|\,dy\\ &\le \frac{1}{|Q|}\int_Q |f(y)-a_Q|\,dy+\frac{1}{|Q|}\int_Q |f(y) - a_Q|\,dy\\ &\le 2\inf_{a\in Q}\frac{1}{|Q|}\int_Q |f(y)-a|\,dy+2\varepsilon\,.\end{aligned}$$

Taking the supremum over all cubes Q in $\mathbf{R}^n$, we obtain the first inequality in (2) since $\varepsilon > 0$ was arbitrary. The other inequality in (2) is simple. The proofs of (3) and (4) are immediate.

□

We saw that $M^{\#}(f) \le 2M_c(f)$, which implies that

$$\big\|M^{\#}(f)\big\|_{L^p} \le 2\big\|M_c(f)\big\|_{L^p} \le C_n p(p-1)^{-1}\big\|f\big\|_{L^p}$$

for $1 < p < \infty$. Thus the sharp function of an L^p function is also in L^p whenever $1 < p < \infty$. The fact that the converse inequality is also valid is one of the main results in this section. We will obtain this estimate via a distributional inequality for the sharp function called a *good lambda* inequality.

7.4.b. A Good Lambda Estimate for the Sharp Function

A crucial tool in obtaining this estimate will be the dyadic maximal function.

Definition 7.4.3. Given a locally integrable function f on $\mathbf{R}^n$, we define its *dyadic maximal function* $M_d(f)$ as

$$M_d(f)(x) = \sup_{\substack{Q\ni x\\ Q \text{ dyadic cube}}} \frac{1}{|Q|}\int_Q |f(t)|\,dt\,.$$

The supremum is taken over all dyadic cubes in $\mathbf{R}^n$ that contain a given point x. Recalling the expectation operators E_k from Section 5.4, we can write

$$M_d(f)(x) = \sup_{k\in\mathbf{Z}} E_k(f)(x)\,.$$

The operator M_d was first introduced in Exercise 2.1.12 and the reader was asked there to show that M_d is of weak type $(1,1)$ with norm at most 1. By interpolation it follows that M_d maps $L^p(\mathbf{R}^n)$ into itself when $1 < p < \infty$ with norm at most

$$\big\|M_d\big\|_{L^p(\mathbf{R}^n)\to L^p(\mathbf{R}^n)} \le \frac{p}{p-1}$$

in view of Exercise 1.3.3. Finally, note that M_d is comparable to M_c. To see this fact, we note that given any cube in $\mathbf{R}^n$, there exist at most 2^n dyadic cubes of comparable side length whose union contains it [see Exercise 5.4.1(c)]. Therefore, for some dimensional constants we have the estimates

$$M_d(f) \approx M_c(f) \approx M(f) .$$

The next theorem provides an example of a *good lambda distributional inequality.*

Theorem 7.4.4. *For all $\gamma > 0$, all $\lambda > 0$, and all locally integrable functions f on $\mathbf{R}^n$, we have the estimate*

$$\big|\{x \in \mathbf{R}^n : M_d(f)(x) > 2\lambda,\ M^{\#}(f)(x) \le \gamma\lambda\}\big| \le 2^n \gamma \big|\{x \in \mathbf{R}^n : M_d(f)(x) > \lambda\}\big|.$$

Proof. We may suppose that the set $\Omega_\lambda = \{x \in \mathbf{R}^n :\ M_d(f)(x) > \lambda\}$ has finite measure, otherwise there is nothing to prove. Then for each $x \in \Omega_\lambda$ there is a maximal dyadic cube Q^x that contains x such that

$$\frac{1}{|Q^x|}\int_{Q^x} |f(y)|\, dy > \lambda\,; \tag{7.4.1}$$

otherwise Ω_λ would have infinite measure. Let Q_j be the collection of all such maximal dyadic cubes containing all x in Ω_λ (i.e., $\{Q_j\}_j = \{Q^x :\ x \in \Omega_\lambda\}$). Maximal dyadic cubes are disjoint hence any two different Q_j's are disjoint; Moreover, it we note that if $x, y \in Q_j$, then $Q_j = Q^x = Q^y$. It follows that

$$\Omega_\lambda = \bigcup_j Q_j\,.$$

To prove the required estimate, it will suffice to prove that for all Q_j we have the estimate

$$\big|\{x \in Q_j :\ M_d(f)(x) > 2\lambda,\ M^{\#}(f)(x) \le \gamma\lambda\}\big| \le 2^n\gamma\,|Q_j|\,, \tag{7.4.2}$$

for, once (7.4.2) is established, the conclusion of the theorem follows by a simple summation on j.

We fix j and $x \in Q_j$ such that $M_d(f)(x) > 2\lambda$. Then the following supremum

$$M_d(f)(x) = \sup_{R \ni x} \frac{1}{|R|}\int_R |f(y)|\, dy \tag{7.4.3}$$

is taken over all dyadic cubes R that either contain Q_j or are contained in Q_j (since $Q_j \cap R \neq \emptyset$). If $R \supsetneqq Q_j$, the maximality of Q_j implies that (7.4.1) does not hold for R; thus the average of $|f|$ over R is at most λ. Thus if $M_d(f)(x) > 2\lambda$, then the supremum in (7.4.3) is attained for some dyadic cube R contained (not properly) in Q_j. Therefore, if $x \in Q_j$ and $M_d(f)(x) > 2\lambda$, then we can replace f by $f\chi_{Q_j}$ in (7.4.3) and we must have $M_d(f\chi_{Q_j})(x) > 2\lambda$. We let Q_j' be the unique dyadic cube of twice the side length of Q_j. Therefore, for $x \in Q_j$ we have

$$M_d\big((f - \operatorname*{Avg}_{Q_j'} f)\chi_{Q_j}\big)(x) \ge M_d(f\chi_{Q_j})(x) - |\operatorname*{Avg}_{Q_j'} f| > 2\lambda - \lambda = \lambda$$

since $|\operatorname{Avg}_{Q_j'} f| \le \operatorname{Avg}_{Q_j'} |f| \le \lambda$ because of the maximality of Q_j. We conclude that

$$\big|\{x \in Q_j : M_d(f)(x) > 2\lambda\}\big| \le \big|\{x \in Q_j : M_d\big((f - \operatorname{Avg}_{Q_j'} f)\chi_{Q_j}\big)(x) > \lambda\}\big| \tag{7.4.4}$$

and, using the fact that M_d is of weak type $(1,1)$ with constant 1, we can control the last expression in (7.4.4) by

$$\begin{aligned} \frac{1}{\lambda}\int_{Q_j} \big|f(y) - \operatorname{Avg}_{Q_j'} f\big|\,dy \;&\le \frac{2^n|Q_j|}{\lambda}\frac{1}{|Q_j'|}\int_{Q_j'} \big|f(y) - \operatorname{Avg}_{Q_j'} f\big|\,dy \\ &\le \frac{2^n|Q_j|}{\lambda} M^{\#}(f)(\xi_j) \end{aligned} \tag{7.4.5}$$

for all $\xi_j \in Q_j$. In proving (7.4.2) we may assume that for some $\xi_j \in Q_j$ we have $M^{\#}(f)(\xi_j) \le \gamma\lambda$; otherwise there is nothing to prove. For this ξ_j, using (7.4.4) and (7.4.5) we obtain (7.4.2).

□

Good lambda inequalities contain useful information concerning the L^p norms of the quantities in question. For example, we use Theorem 7.4.4 to obtain the equivalence of the L^p norms of $M_d(f)$ and $M^{\#}(f)$. Since $M^{\#}(f)$ is pointwise controlled by $2M_c(f)$ and hence by a multiple of $M_d(f)$, we have the estimate

$$\big\|M^{\#}(f)\big\|_{L^p(\mathbf{R}^n)} \le C_n\big\|M_d(f)\big\|_{L^p(\mathbf{R}^n)}$$

for some C_n and all f in $L^p(\mathbf{R}^n)$. The point of the next theorem is that the converse estimate is valid.

Theorem 7.4.5. *Let $0 < p_0 < \infty$. Then for any p with $p_0 \le p < \infty$ there is a constant $C_n(p)$ so that for all functions f with $M_d(f) \in L^{p_0}(\mathbf{R}^n)$ we have*

$$\big\|M_d(f)\big\|_{L^p(\mathbf{R}^n)} \le C_n(p)\big\|M^{\#}(f)\big\|_{L^p(\mathbf{R}^n)}\,. \tag{7.4.6}$$

Proof. For a positive real number N we set

$$I_N = \int_0^N p\lambda^{p-1}\big|\{x \in \mathbf{R}^n :\ M_d(f)(x) > \lambda\}\big|\,d\lambda\,.$$

We note that I_N is finite since it is bounded by

$$\frac{pN^{p-p_0}}{p_0}\int_0^N p_0\lambda^{p_0-1}\big|\{x \in \mathbf{R}^n : M_d(f)(x) > \lambda\}\big|\,d\lambda \le \frac{pN^{p-p_0}}{p_0}\big\|M_d(f)\big\|_{L^{p_0}}^{p_0} < \infty\,,$$

since $p \ge p_0$. We now write

$$I_N = 2^p\int_0^{\frac{N}{2}} p\lambda^{p-1}\big|\{x \in \mathbf{R}^n :\ M_d(f)(x) > 2\lambda\}\big|\,d\lambda$$

and we use Theorem 7.4.4 to obtain the following sequence of inequalities:

$$
\begin{aligned}
I_N &\le 2^p \int_0^{\frac{N}{2}} p\lambda^{p-1}\big|\{x\in \mathbf{R}^n:\ M_d(f)(x)>2\lambda,\ M^{\#}(f)(x)\le\gamma\lambda\}\big|\,d\lambda \\
&\qquad\qquad + 2^p \int_0^{\frac{N}{2}} p\lambda^{p-1}\big|\{x\in \mathbf{R}^n:\ M^{\#}(f)(x)>\gamma\lambda\}\big|\,d\lambda \\
&\le 2^p 2^n\gamma \int_0^{\frac{N}{2}} p\lambda^{p-1}\big|\{x\in \mathbf{R}^n:\ M_d(f)(x)>\lambda\}\big|\,d\lambda \\
&\qquad\qquad + 2^p \int_0^{\frac{N}{2}} p\lambda^{p-1}\big|\{x\in \mathbf{R}^n:\ M^{\#}(f)(x)>\gamma\lambda\}\big|\,d\lambda \\
&\le 2^p 2^n\gamma\, I_N + \frac{2^p}{\gamma^p}\int_0^{\frac{N\gamma}{2}} p\lambda^{p-1}\big|\{x\in \mathbf{R}^n:\ M^{\#}(f)(x)>\lambda\}\big|\,d\lambda\,.
\end{aligned}
$$

At this point we pick a γ such that $2^p2^n\gamma=\frac{1}{2}$. Since I_N is finite, we can subtract from both sides of the inequality the quantity $\frac{1}{2}I_N$ to obtain

$$
I_N \le 2^{p+1}2^{p(n+p+1)}\int_0^{\frac{N\gamma}{2}} p\lambda^{p-1}\big|\{x\in \mathbf{R}^n:\ M^{\#}(f)(x)>\lambda\}\big|\,d\lambda
$$

from which we obtain (7.4.6) letting $N\to\infty$ with $C_n(p)=2^{n+p+2+\frac{1}{p}}$.

□

Corollary 7.4.6. *Let $1\le p_0<\infty$. Then for any p with $p_0\le p<\infty$ there is a constant $C_n(p)$ so that for all functions f with $M(f)\in L^{p_0}(\mathbf{R}^n)$ we have*

$$
\big\|M(f)\big\|_{L^p(\mathbf{R}^n)} \le C_{n,p}\big\|M^{\#}(f)\big\|_{L^p(\mathbf{R}^n)}\,. \tag{7.4.7}
$$

Proof. The proof is immediate since, as observed earlier, we have that the functions $M_d(f)$ and $M(f)$ are pointwise comparable to each other whenever f is locally integrable.

□

7.4.c. Interpolation Using *BMO*

We continue this section by proving an interpolation result in which the space L^∞ is replaced by *BMO*. The sharp function plays a key role in the following theorem.

Theorem 7.4.7. *Let $1\le p_0<\infty$. Let T be a linear operator that maps $L^{p_0}(\mathbf{R}^n)$ into $L^{p_0}(\mathbf{R}^n)$ with bound A_0 and $L^\infty(\mathbf{R}^n)$ into $BMO(\mathbf{R}^n)$ with bound A_1. Then for all p with $p_0<p<\infty$ there is a constant $C_{n,p}$ such that for all $f\in L^p$ we have*

$$
\big\|T(f)\big\|_{L^p(\mathbf{R}^n)} \le C_{n,p,p_0}\big\|f\big\|_{L^p(\mathbf{R}^n)}\,. \tag{7.4.8}
$$

Remark 7.4.8. In some applications, T may not be a priori defined on all of $L^{p_0} + L^\infty$ but only in some subspace of them. We can then state the hypotheses and the conclusion of the preceding theorem to hold for a subspace of these spaces.

Proof. We consider the operator

$$S(f) = M^\#(T(f))$$

defined for $f \in L^{p_0} + L^\infty$. It is easy to see that S is a sublinear operator. We prove that S maps L^{p_0} into itself and L^∞ into itself:

$$\begin{aligned} \big\|S(f)\big\|_{L^{p_0}} &= \big\|M^\#(T(f))\big\|_{L^{p_0}} \le 2\big\|M_c(T(f))\big\|_{L^{p_0}} \\ &\le C_{n,p_0}\big\|T(f)\big\|_{L^{p_0}} \le C_{n,p_0}A_0\big\|f\big\|_{L^{p_0}}\,, \\ \big\|S(f)\big\|_{L^\infty} &= \big\|M^\#(T(f))\big\|_{L^\infty} = \big\|T(f)\big\|_{BMO} \le A_1\big\|f\big\|_{L^\infty}\,. \end{aligned}$$

Interpolating between these estimates using Theorem 1.3.2, we obtain the estimate

$$\big\|M^\#(T(f))\big\|_{L^p} = \big\|S(f)\big\|_{L^p} \le C'_{n,p,p_0}A_0^{\frac{p_0}{p}}A_1^{1-\frac{p_0}{p}}\big\|f\big\|_{L^p}$$

for all $p_0 < p < \infty$. Since $M(T(f)) \in L^{p_0}$, Corollary 7.4.6 is applicable and gives

$$\big\|M(T(f))\big\|_{L^p} \le C_{n,p,p_0}A_0^{\frac{p_0}{p}}A_1^{1-\frac{p_0}{p}}\big\|f\big\|_{L^p}$$

from which we obtain the required estimate

$$\big\|T(f)\big\|_{L^p} \le C_{n,p,p_0}A_0^{\frac{p_0}{p}}A_1^{1-\frac{p_0}{p}}\big\|f\big\|_{L^p}$$

for all $f \in L^p(\mathbf{R}^n)$.

□

7.4.4. Estimates for Singular Integrals Involving the Sharp Function

Next we use the sharp function to obtain pointwise estimates for singular integrals. These estimates will enable us to recover some estimates for singular integrals previously obtained, but we will also be able to obtain a new endpoint estimate from L^∞ into BMO.

Let us recall some facts from Chapter 4. Suppose that K is defined on $\mathbf{R}^n \setminus \{0\}$ and satisfies

$$|K(x)| \le A_1|x|^{-n} \tag{7.4.9}$$

$$|K(x-y) - K(x)| \le A_2|y|^\delta\,|x|^{-n-\delta} \qquad \text{whenever } |x| \ge 2|y| > 0 \tag{7.4.10}$$

$$\sup_{r<R<\infty}\left|\int_{r\le|x|\le R} K(x)\,dx\right| \le A_3 \tag{7.4.11}$$

Let W be a tempered distribution that coincides with K on $\mathbf{R}^n \setminus \{0\}$ and let T be the linear operator given by convolution with W.

Under these assumptions we have that T is L^2 bounded with norm at most a constant multiple of $A_1 + A_2 + A_3$ (Theorem 4.4.1) and hence it is also L^p bounded with a similar norm on L^p for $1 < p < \infty$ (Theorem 4.3.3). Furthermore, under the preceding conditions, the maximal singular integral $T^{(*)}$ is also bounded from $L^p(\mathbf{R}^n)$ into itself for $1 < p < \infty$ (Exercise 4.4.6).

Theorem 7.4.9. *Let T be given by convolution with a distribution W that coincides with a function K on $\mathbf{R}^n \setminus \{0\}$ that satisfies (7.4.10). Assume that T has an extension that is L^2 bounded with a norm B. Then there is a constant C_n such that for any $s > 1$ the estimate*

$$M^{\#}(T(f))(x) \le C_n(A_2 + B)\max(s,(s-1)^{-1})\, M(|f|^s)^{\frac{1}{s}}(x) \tag{7.4.12}$$

is valid for all f in $\bigcup\limits_{s\le p<\infty} L^p$ and almost all $x \in \mathbf{R}^n$.

Proof. In view of Proposition 7.4.2 (2), given any cube Q, it will suffice to find a constant a such that

$$\frac{1}{|Q|}\int_Q |T(f)(y) - a|\, dy \le C_{s,n}(A_2 + B)\, M(|f|^s)^{\frac{1}{s}}(x) \tag{7.4.13}$$

for all $x \in Q$. To prove this estimate we employ a theme that we have seen several times before. We write $f = f_Q^0 + f_Q^\infty$, where $f_Q^0 = f\chi_{6\sqrt{n}\,Q}$ and $f_Q^\infty = f\chi_{(6\sqrt{n}\,Q)^c}$. Here $6\sqrt{n}\,Q$ denotes the cube that is concentric with Q and has sides parallel to those of Q and that has side length $6\sqrt{n}\,\ell(Q)$, where $\ell(Q)$ is the side length of Q.

We now fix an f in $\bigcup\limits_{s\le p<\infty} L^p$ and we select $a = T(f_Q^\infty)(x)$. Then a is finite for almost all $x \in Q$. It follows that

$$\begin{aligned} &\frac{1}{|Q|}\int_Q |T(f)(y) - a|\, dy \\ &\le \frac{1}{|Q|}\int_Q |T(f_Q^0)(y)|\, dy + \frac{1}{|Q|}\int_Q |T(f_Q^\infty)(y) - T(f_Q^\infty)(x)|\, dy\,. \end{aligned} \tag{7.4.14}$$

In view of Theorem 4.3.3, T maps L^s into L^s with norm at most a dimensional multiple of $\max(s,(s-1)^{-1})(B + A_2)$. We can therefore control the first term in (7.4.14) by

$$\begin{aligned} &\left(\frac{1}{|Q|}\int_Q |T(f_Q^0)(y)|^s\, dy\right)^{\frac{1}{s}} \\ &\le C_n \max(s,(s-1)^{-1})(B + A_2)\left(\frac{1}{|Q|}\int_{\mathbf{R}^n} |f_Q^0(y)|^s\, dy\right)^{\frac{1}{s}} \\ &\le C_n' \max(s,(s-1)^{-1})(B + A_2) M(|f|^s)^{\frac{1}{s}}(x)\,. \end{aligned}$$

To estimate the second term in (7.4.14), we first note that

$$\int_Q |T(f_Q^\infty)(y) - T(f_Q^\infty)(x)|\, dy \le \int_Q \bigg| \int_{(6\sqrt{n}Q)^c} \big(K(y-z) - K(x-z)\big) f(z)\, dz \bigg|\, dy\,.$$

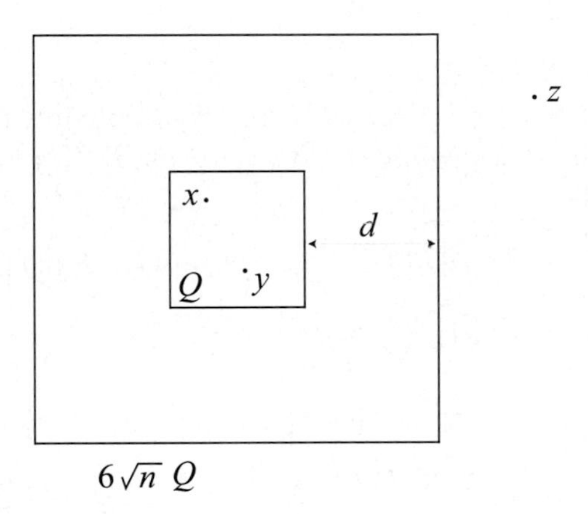

FIGURE 7.3. The cubes Q and $6\sqrt{n}\,Q$. The distance d is equal to $(3\sqrt{n}-\frac{1}{2})\,\ell(Q)$.

Let us make some geometric observations. Since both x and y are in Q we must have $|x-y| \le \sqrt{n}\,\ell(Q)$. Also since $z \notin 6\sqrt{n}\,Q$ and $x \in Q$ we must have $|x-z| \ge \operatorname{dist}\big(Q, (6\sqrt{n}\,Q)^c\big) \ge (3\sqrt{n}-\frac{1}{2})\,\ell(Q)$ and the latter is at least $2\sqrt{n}\,\ell(Q) \ge 2|x-y|$. See Figure 7.3. Therefore, we have $|x-z| \ge 2|x-y|$, which allows us to conclude that

$$\big|K(y-z) - K(x-z)\big| = \big|K((x-z)-(x-y)) - K(x-z)\big| \le A_2 \frac{|x-y|^\delta}{|x-z|^{n+\delta}}$$

using condition (7.4.10). Using these observations, we can bound the second term in (7.4.14) by

$$\begin{aligned}
&\frac{1}{|Q|}\int_Q \int_{(6\sqrt{n}\,Q)^c} \frac{A_2|x-y|^\delta}{|x-z|^{n+\delta}}\,|f(z)|\,dz\,dy \\
&\le C_n \frac{A_2}{|Q|}\int_{(6\sqrt{n}\,Q)^c} \frac{\ell(Q)^{n+\delta}}{|x-z|^{n+\delta}}\,|f(z)|\,dz \\
&\le C_n A_2 \int_{\mathbf{R}^n} \frac{\ell(Q)^\delta}{(\ell(Q)+|x-z|)^{n+\delta}}\,|f(z)|\,dz \\
&\le C_n A_2 M(f)(x) \\
&\le C_n A_2 (M(|f|^s)(x))^{\frac{1}{s}}\,,
\end{aligned}$$

where we used the fact that $|x-z|$ is at least $\ell(Q)$ and Theorem 2.1.10. This proves (7.4.13) and hence (7.4.12).

□

The inequality (7.4.12) in Theorem 7.4.9 is noteworthy as it provides a pointwise estimate for $T(f)$ in terms of a maximal function. This clearly strengthens the L^p boundedness of T. As a consequence of this estimate, we also obtain the following result.

Corollary 7.4.10. *Let T be given by convolution with a distribution W that coincides with a function K on $\mathbf{R}^n \setminus \{0\}$ that satisfies (7.4.10). Assume that T has an extension that is L^2 bounded with a norm B. Then there is a constant C_n such that the estimate*

$$\big\|T(f)\big\|_{BMO} \le C_n(A_2+B)\big\|f\big\|_{L^\infty} \tag{7.4.15}$$

is valid for all $f \in L^\infty \bigcap \big(\bigcup_{1\le p<\infty} L^p\big)$.

Proof. We take $s=2$ in Theorem 7.4.9 and we observe that

$$\big\|T(f)\big\|_{BMO} = \big\|M^\#(T(f))\big\|_{L^\infty} \le C_n(A_2+B)\big\|M(|f|^2)^{\frac{1}{2}}\big\|_{L^\infty},$$

and the last expression is easily controlled by $C_n(A_2+B)\big\|f\big\|_{L^\infty}$.

□

At this point we have not defined the action of T on f when f lies merely in L^∞; and for this reason we restricted f in Corollary 7.4.10 to be also in some L^p. However, there is a way to define T on L^∞ abstractly via duality. Theorem 6.7.1 gives that T and its adjoint T^* map H^1 into L^1. Then the adjoint operator of T^* (i.e., T) maps L^∞ into BMO and is therefore well defined on L^∞. In this way, however, $T(f)$ is not defined explicitly when f is in L^∞. Such an explicit definition will be given in the next chapter in a slightly more general setting.

We end with the following remark.

Remark 7.4.11. In the hypotheses of Theorem 7.4.9 we could have replaced the condition T maps L^2 into L^2 by the condition that T maps L^r into $L^{r,\infty}$ with norm B for some $1<r<\infty$.

Exercises

7.4.1. Suppose that $f \in L^q(\mathbf{R}^n)$ for some $0<q<\infty$. Prove that for every p with $q<p<\infty$ there is a constant $C_{n,p,q}$ such that

$$\big\|f\big\|_{L^p} \le C_{n,p,q}\big\|f\big\|_{L^q}^{1-\theta}\big\|f\big\|_{BMO}^{\theta},$$

where $\frac{1}{p} = \frac{1-\theta}{q}$.

7.4.2. Let μ be a positive Borel measure.
(a) Show that the maximal operator

$$M_\mu^d(f)(x) = \sup_{\substack{Q \ni x \\ Q \text{ dyadic cube}}} \frac{1}{\mu(Q)} \int_Q |f(t)|\, d\mu(t)$$

maps $L^1(\mathbf{R}^n, d\mu)$ into $L^{1,\infty}(\mathbf{R}^n, d\mu)$ with constant 1.
(b) For a μ-locally integrable function f, define the *sharp maximal function with respect to* μ

$$M_\mu^\#(f)(x) = \sup_{Q \ni x} \frac{1}{\mu(Q)} \int_Q \big|f(t) - \operatorname*{Avg}_{Q,\mu} f\big|\, d\mu(t),$$

where $\text{Avg}_{Q,\mu} f$ denotes the average of f over Q with respect to μ. Assume that μ is a doubling measure with doubling constant $C(\mu)$ [this means $\mu(3Q) \le C(\mu)\mu(Q)$ for all cubes Q]. Prove that for all $\gamma > 0$, all $\lambda > 0$, and all μ-locally integrable functions f on $\mathbf{R}^n$ we have the estimate

$$\begin{aligned} &\mu\big(\{x \in \mathbf{R}^n : M_\mu^d(f)(x) > 2\lambda,\ M_\mu^\#(f)(x) \le \gamma\lambda\}\big) \\ &\qquad\qquad \le C(\mu)\,\gamma\,\mu\big(\{x \in \mathbf{R}^n : \ M_\mu^d(f)(x) > \lambda\}\big)\,. \end{aligned}$$

[*Hint:* Part (a): For any x in the set $\{x :\ M_\mu^d(f)(x) > \lambda\}$, choose a maximal dyadic cube $Q = Q(x)$ such that $\int_Q |f(t)|\, d\mu(t) > \lambda\mu(Q)$. Part (b): Mimic the proof of Theorem 7.4.4.]

7.4.3. Let $0 < p_0 < \infty$ and let M_μ^d and $M_\mu^\#$ be as in the previous Exercise. Prove that for any p with $p_0 \le p < \infty$ there is a constant $C_n(p)$ so that for all functions f with $M_\mu^d(f) \in L^{p_0}(\mathbf{R}^n)$ we have

$$\big\|M_\mu^d(f)\big\|_{L^p(\mathbf{R}^n, d\mu)} \le C_n(p)\big\|M_\mu^\#(f)\big\|_{L^p(\mathbf{R}^n, d\mu)}\,.$$

7.4.4. We say that a function f on $\mathbf{R}^n$ is in BMO_d (or dyadic BMO) if and only if

$$\|f\|_{BMO_d} = \sup_{Q \text{ dyadic cube}} \frac{1}{|Q|} \int_Q \big|f(x) - \operatorname*{Avg}_Q f\big|\, dx < \infty$$

(a) Show that BMO is a proper subset of BMO_d.
(b) Suppose that f is in dyadic BMO. Then f is in BMO if and only if

$$\big|\operatorname*{Avg}_{Q_1} f - \operatorname*{Avg}_{Q_2} f\big| \le A$$

for all adjacent dyadic cubes of the same length. Precisely show that

$$\|f\|_{BMO} \approx A + \|f\|_{BMO_d}$$

with constants that depend only on the dimension.

7.4.5. Suppose that K is a function on $\mathbf{R}^n \setminus \{0\}$ that satisfies (7.4.9), (7.4.10), and (7.4.11). Let η be a smooth function that vanishes in a neighborhood of the origin and is equal to 1 in a neighborhood of infinity. For $\varepsilon > 0$ let $K_\eta^{(\varepsilon)}(x) = K(x)\eta(x/\varepsilon)$ and let $T_\eta^{(\varepsilon)}$ be the operator given by convolution with $K_\eta^{(\varepsilon)}$. Prove that for any $1 < s < \infty$ there is a constant $C_{n,s}$ such that for all p with $s < p < \infty$ and f in L^p we have

$$\Big\| \sup_{\varepsilon>0} M^{\#}(T_\eta^{(\varepsilon)}(f))\Big\|_{L^p(\mathbf{R}^n)} \le C_{n,s}(A_1 + A_2 + A_3)\big\|f\big\|_{L^p(\mathbf{R}^n)} .$$

[*Hint:* Observe that the kernels $K_\eta^{(\varepsilon)}$ satisfy (7.4.9), (7.4.10), and (7.4.11) uniformly in $\varepsilon > 0$ and use Theorems 4.4.1 and 7.4.9.]

7.4.6. Let $0 < p_0 < \infty$ and suppose that for some locally integrable function f we have that $M_d(f)$ lies in $L^{p_0,\infty}(\mathbf{R}^n)$. Show that for any p in (p_0, ∞) there exists a constant $C_n(p)$ such that

$$\big\|M_d(f)\big\|_{L^p(\mathbf{R}^n)} \le C_n(p)\big\|M^{\#}(f)\big\|_{L^p(\mathbf{R}^n)} ,$$

where $C_n(p)$ is independent of f.
[*Hint:* With the same notation as in the proof of Theorem 7.4.5, use the hypothesis $\big\|M_d(f)\big\|_{L^{p_0,\infty}} < \infty$ to prove that $I_N < \infty$ whenever $p > p_0$. Then the arguments in the proofs of Theorem 7.4.5 and Corollary 7.4.6 follow unchanged.]

7.4.7. Prove that expressions

$$\Sigma_N(x) = \sum_{k=1}^{N} \frac{\sin(2\pi k x)}{k}$$

are uniformly bounded in N and x. Then use Corollary 7.4.10 to prove that

$$\sup_{N\ge 1} \bigg\| \sum_{k=1}^{N} \frac{e^{2\pi i k x}}{k} \bigg\|_{BMO} \le C < \infty .$$

Deduce that the limit of $\Sigma_N(x)$ as $N \to \infty$ can be defined as an element of *BMO*.
[*Hint:* Use that the Hilbert transform of $\sin(2\pi k x)$ is $\cos(2\pi k x)$. Also note that the series $\sum_{k=1}^{\infty} \frac{\sin(2\pi k x)}{k}$ coincides with the periodic extension of the (bounded) function $= \pi(\frac{1}{2} - x)$ on $[0, 1)$.]

7.5. Commutators of Singular Integrals with BMO Functions*

The mean value zero property of $H^1(\mathbf{R}^n)$ is often manifested when its elements are paired with functions in BMO. It is therefore natural to expect that BMO can be utilized to express and quantify the cancellation of expressions in H^1. Let us be specific through an example. We saw in Exercise 6.7.3 that the bilinear operator

$$f\, H(g) + H(f)\, g$$

maps $L^2(\mathbf{R}^n) \times L^2(\mathbf{R}^n)$ into $H^1(\mathbf{R}^n)$; here H is the Hilbert transform. Pairing with a BMO function b and using that $H^t = -H$, we obtain that

$$\big\langle f\, H(g) + H(f)\, g\,,\, b\big\rangle = \big\langle f\,,\, H(g)\, b - H(g\, b)\big\rangle\,,$$

and hence the operator $g \to H(g)\, b - H(g\, b)$ should be L^2 bounded. This expression $H(g)\, b - H(g\, b)$ is called the *commutator* of H with the BMO function b. More generally, we give the following definition.

Definition 7.5.1. The *commutator* of a singular integral operator T with a function b is defined as

$$[b, T](f) = b\, T(f) - T(b\, f)\,.$$

If the function b is locally integrable and has at most polynomial growth at infinity, then the operation $[b, T]$ is well defined when acting on Schwartz functions f.

In view of the preceding remarks, the L^p boundedness of the commutator $[b, T]$ for b in BMO exactly captures the cancellation property of the bilinear expression

$$(f, g) \to T(f)\, g - f\, T^t(g)\,.$$

As in the case with the Hilbert transform, it is natural to expect that the commutator $[b, T]$ of a general singular integral T is L^p bounded for all $1 < p < \infty$. This fact will be proved in this section. As BMO functions are unbounded in general, the reader may surmise that the presence of the negative sign in the definition of the commutator plays a crucial cancellation role.

We now introduce some material to be used in our approach of the boundedness of the commutator.

7.5.a. An Orlicz-Type Maximal Function

We can express the L^p norm $(1 \le p < \infty)$ of a function f on a measure space X by

$$\big\|f\big\|_{L^p(X)} = \left(\int_X |f|^p\, d\mu\right)^{\frac{1}{p}} = \inf\left\{\lambda > 0 :\ \int_X \left|\frac{|f|}{\lambda}\right|^p d\mu \le 1\right\}.$$

Motivated by the second expression, we may replace the function t^p by a general increasing convex function $\Phi(t)$. We give the following definition.

Definition 7.5.2. A *Young's function* is a continuous increasing convex function Φ on $[0, \infty)$ that satisfies $\Phi(0) = 0$ and $\lim_{t\to\infty} \Phi(t) = \infty$. The *Orlicz norm* of measurable function f on a measure space (X, μ) with respect to a Young's function Φ is defined as

$$\big\|f\big\|_{\Phi(L)(X,\mu)} = \inf\Big\{\lambda > 0 : \int_X \Phi(|f|/\lambda)\, d\mu \le 1\Big\}.$$

The *Orlicz space* $\Phi(L)(X, \mu)$ is then defined as the space of all measurable functions f on X such that $\big\|f\big\|_{\Phi(L)(X,\mu)} < \infty$.

We will be concerned mostly with the case in which the measure space X is a cube in $\mathbf{R}^n$ with normalized Lebesgue measure $|Q|^{-1}dx$. For a measurable function f on a cube Q in $\mathbf{R}^n$, the Orlicz norm of f is therefore

$$\big\|f\big\|_{\Phi(L)(Q,\frac{dx}{|Q|})} = \inf\Big\{\lambda > 0 : \frac{1}{|Q|}\int_Q \Phi(|f|/\lambda)\, dx \le 1\Big\},$$

which will be simply denoted by $\big\|f\big\|_{\Phi(L)(Q)}$, as the measure will be understood to be normalized Lebesgue whenever the ambient space is a cube.

As we do not intend to study Orlicz spaces in detail but only use them as a tool, we refer to the Exercises for certain useful properties of these spaces. Since for $C > 1$ convexity gives $\Phi(t/C) \le \Phi(t)/C$ for all $t \ge 0$, it follows that

$$\big\|f\big\|_{C\,\Phi(Q)} \le C\,\big\|f\big\|_{\Phi(Q)}\,, \tag{7.5.1}$$

which implies that the norms with respect to Φ and $C\,\Phi$ are comparable.

A case of particular interest arises when $\Phi(t) = t\log(1 + t)$. We will need in the sequel a certain maximal operator defined in terms of the corresponding Orlicz norm.

Definition 7.5.3. We define the *Orlicz maximal operator*

$$M_{L\log(1+L)}(f)(x) = \sup_{Q\ni x} \big\|f\big\|_{L\log(1+L)(Q)}\,,$$

where the supremum is taken over all cubes Q with sides parallel to the axes that contain the given point x.

The boundedness properties of this maximal operator will be a consequence of the following lemma.

Lemma 7.5.4. *There is a positive a constant $c(n)$ such that for any cube Q in $\mathbf{R}^n$ and any measurable function w,*

$$\big\|w\big\|_{L\log(1+L)(Q)} \le \frac{c(n)}{|Q|}\int_Q M(w)\, dx\,, \tag{7.5.2}$$

where M is the Hardy-Littlewood maximal operator. Hence,

$$M_{L\log(1+L)}(w)(x) \le c\,M^2(w)(x), \tag{7.5.3}$$

where $M^2 = M \circ M$.

Proof. Fix a cube Q in $\mathbf{R}^n$ with sides parallel to the axes. We introduce a *maximal operator associated with* Q as follows:

$$M_c^Q(f)(x) = \sup_{\substack{R \ni x \\ R \subseteq Q}} \frac{1}{|R|} \int_R |f(y)|\, dy\,,$$

where all the sets R in the supremum are cubes in $\mathbf{R}^n$ with sides parallel to the axes. The key estimate follows from the following local version of the reverse weak type $(1,1)$ estimate of Exercise 2.1.4(b). For each nonnegative function f on $\mathbf{R}^n$ and $\alpha \ge \operatorname{Avg}_Q f$, we have

$$\frac{1}{\alpha}\int_{Q\cap\{f>\alpha\}} f\, dx \le 2^n \left|\{x\in Q: M_c^Q(f)(x) > \alpha\}\right|. \tag{7.5.4}$$

Indeed, to prove (7.5.4), we apply Corollary 2.1.21 to the function f and the number α. With the notation of that corollary, we have $Q\backslash(\bigcup_j Q_j) \subseteq \{f \le \alpha\}$, which implies that $Q\cap\{f>\alpha\} \subseteq \bigcup_j Q_j$ and the latter is contained in $\{x \in Q: \ M_c^Q(f)(x) > \alpha\}$. Multiplying both sides of (2.1.21) by $|Q_j|$, summing over all j, and using these observations, we obtain (7.5.4).

Using the definition of $M_{L\log(1+L)}$, (7.5.2) will follow from the fact that for some constant $c > 1$ independent of w we have

$$\frac{1}{|Q|}\int_Q \frac{w}{\lambda_Q}\log\left(1+\frac{w}{\lambda_Q}\right) d\mu \le 1, \tag{7.5.5}$$

where

$$\lambda_Q = \frac{c}{|Q|}\int_Q M_c(w)\, dx = c \operatorname*{Avg}_Q M_c(w).$$

We let $f = w/\lambda_Q$; by the Lebesgue differentiation theorem we have that $0 \le \operatorname{Avg}_Q f \le \frac{1}{c}$. Using identity (1.1.7), we obtain

$$\int_X \Phi(f)\, d\nu = \int_0^\infty \Phi'(t)\, \nu(\{x\in X: f(x) > t\})\, dt\,,$$

with $\Phi(t) = \log(1+t)$. Taking $d\nu = f\chi_Q\, dx$ we have

$$\frac{1}{|Q|}\int_Q f\log(1+f)\, dx = \frac{1}{|Q|}\int_0^\infty \frac{1}{1+t}\left(\int_{Q\cap\{f>t\}} f\, dx\right) dt = I_1 + I_2\,,$$

where

$$\begin{aligned} I_1 &= \frac{1}{|Q|}\int_0^{\operatorname{Avg}_Q f} \frac{1}{1+t}\left(\int_{Q\cap\{f>t\}} f\, dx\right) dt\,, \\ I_2 &= \frac{1}{|Q|}\int_{\operatorname{Avg}_Q f}^\infty \frac{1}{1+t}\left(\int_{Q\cap\{f>t\}} f\, dx\right) dt\,. \end{aligned}$$

Now we clearly have

$$I_1 \le (\operatorname*{Avg}_Q f)^2 \le \frac{1}{c^2}\,.$$

For I_2 we use estimate (7.5.4):

$$\begin{aligned} I_2 &= \frac{1}{|Q|}\int_{\mathrm{Avg}_Q f}^{\infty} \frac{1}{1+t}\left(\int_{Q\cap\{f>t\}} f\,dx\right) dt \\ &\le \frac{2^n}{|Q|}\int_{\mathrm{Avg}_Q f}^{\infty} \frac{t}{1+t}|\{x\in Q : M_c^Q(f)(x) > t\}|\,dt \\ &\le \frac{2^n}{|Q|}\int_0^{\infty} |\{x\in Q : M_c^Q(f)(x) > \lambda\}|\,d\lambda \\ &= \frac{2^n}{|Q|}\int_Q M_c^Q(f)\,dx \\ &= \frac{2^n}{|Q|}\int_Q M_c(w)\,dx\,\frac{1}{\lambda_Q} \le \frac{2^n}{c} \end{aligned}$$

using the definition of λ_Q. Therefore,

$$I_1 + I_2 \le \frac{1}{c^2} + \frac{2^n}{c} \le 1$$

provided c is large enough.

□

7.5.b. A Pointwise Estimate for the Commutator

For $\delta > 0$, $M_\delta^\#$ will denote the following modification of the sharp maximal operator introduced in Section 7.4:

$$M_\delta^\#(f) = M^\#(|f|^\delta)^{1/\delta}\,.$$

It will be often useful to work for the following characterization of $M^\#$ [see Proposition 7.4.2 (2)]:

$$M^\#(f)(x) \approx \sup_{Q\ni x}\inf_c \frac{1}{|Q|}\int_Q |f(y)-c|\,dy\,.$$

We will also need the following version of the Hardy-Littlewood maximal operator:

$$M_\varepsilon(f) = M(|f|^\varepsilon)^{1/\varepsilon}\,.$$

The next lemma expresses the fact that commutators of singular integral operators with *BMO* functions are pointwise controlled by the maximal function $M^2 = M\circ M$.

Lemma 7.5.5. *Let T be a linear operator given by convolution with a tempered distribution on $\mathbf{R}^n$ that coincides with a function $K(x)$ on $\mathbf{R}^n\setminus\{0\}$ satisfying (7.4.9), (7.4.10), and (7.4.11). Let b be in $BMO(\mathbf{R}^n)$, and let $0<\delta<\varepsilon$. Then there exists a positive constant $C=C_{\delta,\varepsilon,n}$ such that for any smooth function f with compact support we have*

$$M_\delta^\#([b,T](f)) \le C\|b\|_{BMO}\{M_\varepsilon(T(f)) + M^2(f)\}. \tag{7.5.6}$$

Proof. Fix a cube Q in $\mathbf{R}^n$ with sides parallel to the axes centered at the point x. Since for $0<\delta<1$ we have $\big||\alpha|^\delta-|\beta|^\delta\big|\le|\alpha-\beta|^\delta$ for $\alpha,\beta\in\mathbf{R}$, it is enough to show for some complex constant $c=c_Q$ that there exists $C=C_\delta>0$ such that

$$\left(\frac{1}{|Q|}\int_Q\big|[b,T](f)(y)-c\big|^\delta dy\right)^{\frac{1}{\delta}} \le C\,\|b\|_{BMO}\Big\{M_\varepsilon(T(f))(x)+M^2(f)(x)\Big\}. \tag{7.5.7}$$

Denote by Q^* the cube $5\sqrt{n}\,Q$ that has side length $5\sqrt{n}$ times the side length of Q and the same center x as Q. Let $f=f_1+f_2$, where $f_1=f\,\chi_{Q^*}$. For arbitrary constant a we can write

$$[b,T](f)=(b-a)T(f)-T((b-a)f_1)-T((b-a)f_2).$$

Selecting

$$c=\operatorname*{Avg}_{Q} T((b-a)f_2) \qquad\text{and}\qquad a=\operatorname*{Avg}_{Q^*} b,$$

we can estimate the left-hand side of (7.5.7) by a multiple of $L_1+L_2+L_3$, where

$$\begin{aligned}
L_1 &= \left(\frac{1}{|Q|}\int_Q\big|(b(y)-\operatorname*{Avg}_{Q^*} b)\,T(f)(y)\big|^\delta dy\right)^{\frac{1}{\delta}}\\
L_2 &= \left(\frac{1}{|Q|}\int_Q\big|T\big((b-\operatorname*{Avg}_{Q^*} b)f_1\big)(y)\big|^\delta dy\right)^{\frac{1}{\delta}}\\
L_3 &= \left(\frac{1}{|Q|}\int_Q\big|T\big((b-\operatorname*{Avg}_{Q^*} b)f_2\big)-\operatorname*{Avg}_{Q} T\big((b-\operatorname*{Avg}_{Q^*} b)f_2\big)\big|^\delta dy\right)^{\frac{1}{\delta}}.
\end{aligned}$$

To estimate L_1, we use Hölder's inequality with exponents r and r' for some $1<r<\varepsilon/\delta$:

$$\begin{aligned}
L_1 &\le \left(\frac{1}{|Q|}\int_Q\big|b(y)-\operatorname*{Avg}_{Q^*} b\big|^{\delta r'} dy\right)^{\frac{1}{\delta r'}}\left(\frac{1}{|Q|}\int_Q|T(f)(y)|^{\delta r} dy\right)^{\frac{1}{\delta r}}\\
&\le C\,\|b\|_{BMO}M_{\delta r}(T(f))(x)\\
&\le C\,\|b\|_{BMO}M_\varepsilon(T(f))(x),
\end{aligned}$$

where recall that x is the center of Q. Since $T : L^1(\mathbf{R}^n) \to L^{1,\infty}(\mathbf{R}^n)$ and $0 < \delta < 1$, Kolmogorov's inequality (Exercise 2.1.5) yields

$$\begin{aligned} L_2 &\le \frac{C}{|Q|}\int_Q \big|(b(y) - \operatorname*{Avg}_{Q^*} b) f_1(y)\big|\, dy \\ &= \frac{C'}{|Q^*|}\int_{Q^*} \big|(b(y) - \operatorname*{Avg}_{Q^*} b) f(y)\big|\, dy \\ &\le 2C' \big\|b - \operatorname*{Avg}_{Q^*} b\big\|_{(e^L-1)(Q^*)} \big\|f\big\|_{L\log(1+L)(Q^*)}, \end{aligned}$$

using Exercise 7.5.2(c).

For some $0 < \gamma < (2^n e)^{-1}$, let $C_{n,\gamma} > 2$ be a constant that works in Corollary 7.1.7. We set $c_0 = C_{n,\gamma} - 1 > 1$. We use (7.5.1) and we claim that

$$\big\|b - \operatorname*{Avg}_{Q^*} b\big\|_{(e^L-1)(Q^*)} \le c_0 \big\|b - \operatorname*{Avg}_{Q^*} b\big\|_{c_0^{-1}(e^L-1)(Q^*)} \le \frac{c_0}{\gamma} \big\|b\big\|_{BMO}. \tag{7.5.8}$$

Indeed, the last inequality is equivalent to

$$\frac{1}{|Q^*|}\int_{Q^*} c_0^{-1}\Big[e^{\gamma|b(y)-\operatorname{Avg}_{Q^*} b|/\|b\|_{BMO}} - 1\Big]\, dy \le 1,$$

which is a restatement of Corollary 7.1.7. We therefore conclude that

$$L_2 \le C \big\|b\big\|_{BMO} M_{L\log(1+L)}(f)(x).$$

Finally, we turn our attention to term L_3. Note that if $z, y \in Q$ and $w \notin Q^*$, then $|z - w| \ge 2|z - y|$. Using Fubini's theorem and property (7.4.10) succesively, we can control L_3 pointwise by

$$\begin{aligned} &\frac{1}{|Q|}\int_Q \big|T\big((b - \operatorname*{Avg}_{Q^*} b) f_2\big)(y) - \operatorname*{Avg}_{Q} T\big((b - \operatorname*{Avg}_{Q^*} b) f_2\big)\big|\, dy \\ &\le \frac{1}{|Q|^2}\int_Q\int_Q\int_{\mathbf{R}^n\setminus Q^*} |K(y-w) - K(z-w)|\big|(b(w) - \operatorname*{Avg}_{Q^*} b) f(w)\big|\, dw\, dz\, dy \\ &\le \frac{1}{|Q|^2}\int_Q\int_Q\sum_{j=0}^{\infty}\int_{2^{j+1}Q^*\setminus 2^jQ^*} \frac{A_2|y-z|^\delta}{|z-w|^{n+\delta}}\big|b(w) - \operatorname*{Avg}_{Q^*} b\big|\, |f(w)|\, dw\, dz\, dy \\ &\le C\,A_2 \sum_{j=0}^{\infty} \frac{\ell(Q)^\delta}{(2^j\ell(Q))^{n+\delta}} \int_{2^{j+1}Q^*} \big|b(w) - \operatorname*{Avg}_{Q^*} b\big|\, |f(w)|\, dw \\ &\le C\,A_2 \bigg(\sum_{j=0}^{\infty} \frac{2^{-j\delta}}{(2^j\ell(Q))^n} \int_{2^{j+1}Q^*} \big|b(w) - \operatorname*{Avg}_{2^{j+1}Q^*} b\big|\, |f(w)|\, dw \\ &\qquad + \sum_{j=0}^{\infty} 2^{-j\delta}\big|\operatorname*{Avg}_{2^{j+1}Q^*} b - \operatorname*{Avg}_{Q^*} b\big| \frac{1}{(2^j\ell(Q))^n} \int_{2^{j+1}Q^*} |f(w)|\, dw\bigg) \end{aligned}$$

$$
\begin{aligned}
&\le C'A_2 \sum_{j=0}^{\infty} 2^{-j\delta} \big\| b - \operatorname*{Avg}_{2^{j+1}Q^*} b \big\|_{(e^L-1)(2^{j+1}Q^*)} \big\| f \big\|_{L\log(1+L)(2^{j+1}Q^*)} \\
&\qquad\qquad + C'A_2 \big\| b \big\|_{BMO} \sum_{j=1}^{\infty} \frac{j}{2^{j\delta}} M(f)(x) \\
&\le C''A_2 \big\| b \big\|_{BMO} M_{L\log(1+L)}(f)(x) + C''A_2 \big\| b \big\|_{BMO} M(f)(x) \\
&\le C'''A_2 \big\| b \big\|_{BMO} M^2(f)(x),
\end{aligned}
$$

where we have used inequality (7.5.8), Lemma 7.5.4, and the simple estimate

$$
\big| \operatorname*{Avg}_{2^{j+1}Q^*} b - \operatorname*{Avg}_{Q^*} b \big| \le C_n j \big\| b \big\|_{BMO}
$$

of Exercise 7.1.5.

□

7.5.c. L^p Boundedness of the Commutator

We note that if f has compact support and b is in BMO, then bf lies in $L^q(\mathbf{R}^n)$ for all $q < \infty$ and therefore $T(bf)$ is well defined, whenever T is a singular integral operator. Likewise, $[b, T]$ is a well-defined operator on C_0^∞ for all b in BMO.

Having obtained the crucial Lemma 7.5.5, we can now pass to an important result concerning its L^p boundedness.

Theorem 7.5.6. *Let T be as in Lemma 7.5.5. Then for any $1 < p < \infty$ there exists a constant $C = C_{p,n}$ such that for all smooth functions with compact support f and all BMO functions b, the following estimate is valid:*

$$
\big\| [b, T](f) \big\|_{L^p(\mathbf{R}^n)} \le C \big\| b \big\|_{BMO} \big\| f \big\|_{L^p(\mathbf{R}^n)}. \tag{7.5.9}
$$

Consequently, the linear operator

$$
f \to [b, T](f)
$$

admits a bounded extension from $L^p(\mathbf{R}^n)$ into $L^p(\mathbf{R}^n)$ for all $1 < p < \infty$ with norm at most a multiple of $\big\| b \big\|_{BMO}$.

Proof. Using the inequality of Theorem 7.4.4, we obtain for functions g with $|g|^\delta$ locally integrable

$$
\big| \{ M_d(|g|^\delta)^{\frac{1}{\delta}} > 2^{\frac{1}{\delta}} \lambda \} \cap \{ M_\delta^\#(g) \le \gamma\lambda \} \big| \le 2^n \gamma^\delta \big| \{ M_d(|g|^\delta)^{\frac{1}{\delta}} > \lambda \} \big| \tag{7.5.10}
$$

for all $\lambda, \gamma, \delta > 0$. Then a repetition of the proof of Theorem 7.4.5 yields the second inequality:

$$
\big\| M(|g|^\delta)^{\frac{1}{\delta}} \big\|_{L^p} \le C_n \big\| M_d(|g|^\delta)^{\frac{1}{\delta}} \big\|_{L^p} \le C_n(p) \big\| M_\delta^\#(g) \big\|_{L^p} \tag{7.5.11}
$$

for all $p \in (p_0, \infty)$, provided $M_d(|g|^\delta)^{\frac{1}{\delta}} \in L^{p_0}(\mathbf{R}^n)$ for some $p_0 > 0$.

For the following argument, it will be convenient to replace b by the bounded function

$$b_k(x) = \begin{cases} k & \text{if } b(x) < k \\ b(x) & \text{if } -k \le b(x) \le k \\ -k & \text{if } b(x) > -k, \end{cases}$$

which satisfies $\|b_k\|_{BMO} \le \|b\|_{BMO}$ for any $k > 0$; see Exercise 7.1.4.

For given $1 < p < \infty$, select p_0 so that $1 < p_0 < p$. Given a smooth function with compact support f, we note that the function $b_k f$ lies in L^{p_0}; thus $T(b_k f)$ also lies in L^{p_0}. Likewise, $b_k T(f)$ also likes in L^{p_0}. As M_δ is bounded on L^{p_0} for $0 < \delta < 1$, we conclude that

$$\|M_\delta([b_k, T](f))\|_{L^{p_0}} \le C_\delta\big(\|M_\delta(b_k T(f))\|_{L^{p_0}} + \|M_\delta(T(b_k f))\|_{L^{p_0}}\big) < \infty .$$

This allows us to obtain (7.5.11) with $g = [b_k, T](f)$. We now turn to Lemma 7.5.5, in which we pick $0 < \delta < \varepsilon < 1$. Taking L^p norms on both sides of (7.5.6) and using (7.5.11) with $g = [b_k, T](f)$ and the boundedness of M_ε, T, and M^2 on $L^p(\mathbf{R}^n)$, we deduce the a priori estimate (7.5.9) for smooth functions with compact support f and the truncated *BMO* functions b_k.

The Lebesgue dominated convergence theorem gives that $b_k \to b$ in L^2 of every compact set and, in particular, in $L^2(\operatorname{supp} f)$. It follows that $b_k f \to bf$ in L^2 and therefore $T(b_k f) \to T(bf)$ in L^2 by the boundedness of T on L^2. We deduce that for some subsequence of integers k_j, $T(b_{k_j} f) \to T(bf)$ a.e. For this subsequence we have $[b_{k_j}, T](f) \to [b, T](f)$ a.e. Letting $j \to \infty$ and using Fatou's lemma, we deduce that (7.5.9) holds for all *BMO* functions b and smooth functions f with compact support.

Since smooth functions with compact support are dense in L^p, it follows that the commutator admits a bounded extension on L^p that satisfies (7.5.9).

□

We refer to Exercise 7.5.4 for an analogue of Theorem 7.5.6 for $p = 1$.

Exercises

7.5.1. Use Jensen's inequality to show that M is pointwise controlled by $M_{L\log(1+L)}$.

7.5.2. (*Young's inequality for Orlicz spaces*)

(a) Let φ be a continuous, real-valued, strictly increasing function defined on $[0,\infty)$ such that $\varphi(0) = 0$ and $\lim_{t\to\infty} \varphi(t) = \infty$. Let $\psi = \varphi^{-1}$ and for $x \in [0,\infty)$ define

$$\Phi(x) = \int_0^x \varphi(t)\,dt \qquad \Psi(x) = \int_0^x \psi(t)\,dt .$$

Show that for $s, t \in [0,\infty)$ we have

$$st \le \Phi(s) + \Psi(t) .$$

(b) (*cf. Exercise 4.2.3*) Choose a suitable function φ in part (a) to deduce for $s,t \in [0,\infty)$ the inequality.

$$st \le (t+1)\log(t+1) - t + e^s - s - 1 \le t\log(t+1) + e^s - 1\,.$$

(c) (*Hölder's inequality for Orlicz spaces*) Deduce the inequality

$$\big|\langle f,g\rangle\big| \le 2\big\|f\big\|_{\Phi(L)}\big\|g\big\|_{\Psi(L)}\,.$$

$\big[$*Hint:* Give a geometric proof distinguishing the cases $t > \varphi(s)$ and $t \le \varphi(s)$. To give an analytic proof, use the fact that for all $u \ge 0$ we have $\int_0^u \varphi(t)\,dt + \int_0^{\varphi(u)} \psi(s)\,ds = u\varphi(u)$. $\big]$

7.5.3. Let T be as in Lemma 7.5.5. Show that there is a constant $C_n < \infty$ such that for all $f \in L^p(\mathbf{R}^n)$ and $g \in L^{p'}(\mathbf{R}^n)$ we have

$$\big\|T(f)\,g - f\,T^t(g)\big\|_{H^1(\mathbf{R}^n)} \le C\big\|f\big\|_{L^p(\mathbf{R}^n)}\big\|g\big\|_{L^{p'}(\mathbf{R}^n)}\,.$$

In other words, show that the bilinear operator

$$(f,g) \to T(f)\,g - f\,T^t(g)$$

maps $L^p(\mathbf{R}^n) \times L^{p'}(\mathbf{R}^n)$ into $H^1(\mathbf{R}^n)$.

7.5.4. (*Pérez* [**411**]) Let $\Phi(t) = t\log(1+t)$. Then there exists a positive constant C, depending on the BMO constant of b, such that for any smooth function with compact support f the following is valid:

$$\sup_{\alpha>0} \frac{1}{\Phi(\frac{1}{\alpha})}\big|\{|[b,T](f)| > \alpha\}\big| \le C \sup_{\alpha>0} \frac{1}{\Phi(\frac{1}{\alpha})}\big|\{M^2(f) > \alpha\}\big|\,.$$

7.5.5. Let R_1, R_2 be the Riesz transforms in $\mathbf{R}^2$. Show that there is a constant $C < \infty$ such that for all g_1, g_2 square integrable functions on $\mathbf{R}^2$ the following is valid:

$$\big\|R_1(g_1)R_2(g_2) - R_1(g_2)R_2(g_1)\big\|_{H^1} \le C_p\big\|g_1\big\|_{L^2}\big\|g_2\big\|_{L^2}\,.$$

$\big[$*Hint:* Reduce matters to $R_1([b,R_2](g_2)) - R_1([b,R_1](g_1))$ pairing with a BMO function b.$\big]$

7.5.6. (*Coifman, Lions, Meyer, and Semmes* [**112**]) Use the previous Exercise to prove that the Jacobian J_f of a map $f = (f_1,f_2) : \mathbf{R}^2 \to \mathbf{R}^2$,

$$J_f = \det\begin{pmatrix} \partial_1 f_1 & \partial_2 f_1 \\ \partial_2 f_1 & \partial_2 f_2 \end{pmatrix},$$

lies in $H^1(\mathbf{R}^2)$ whenever $f_1, f_2 \in \dot{L}^2_1(\mathbf{R}^2)$.
$\big[$*Hint:* Set $g_j = \Delta^{1/2}(f_j)$.$\big]$

HISTORICAL NOTES

The space of functions of bounded mean oscillation first appeared in the work of John and Nirenberg [**279**] in the context of nonlinear partial differential equations that arise in the study of minimal surfaces. Theorem 7.1.6 was obtained by John and Nirenberg [**279**]. The relationship of *BMO* functions and Carleson measures is due to Fefferman and Stein [**191**]. For a variety of issues relating *BMO* to complex function theory we refer the reader to the book of Garnett [**206**]. The duality of H^1 and *BMO* (Theorem 7.2.2) was announced by Fefferman in [**184**] but its first proof appeared in the article of Fefferman and Stein [**191**]. This article actually contains two proofs of this result. Our proof of Theorem 7.2.2 is based on the atomic decomposition of H^1 which was obtained subsequently. An alternative proof of the duality between H^1 and *BMO* was given by Carleson [**86**].

Carleson measures first appeared in the work of Carleson [**82**] and [**83**]. Corollary 7.3.6 was first proved by Carleson, but the proof given here is due Stein. The characterization of Carleson measures in Theorem 7.3.8 was obtained by Carleson [**82**]. A theory of balayage for studying *BMO* was developed by Varopoulos [**537**]. The space *BMO* can also be characterized in terms Carleson measures via Theorem 7.3.8. The converse of Theorem 7.3.8 (see Fefferman and Stein [**191**]) states that if the function Ψ satisfies a nondegeneracy condition and $|f * \Psi_t|^2 \frac{dx\,dt}{t}$ is a Carleson measure, then f must be a *BMO* function. We refer the reader to Stein [**482**] (page 159) for a proof of this fact, which uses a duality idea related to tent spaces. The latter were introduced by Coifman, Meyer, and Stein [**117**] to systematically study the connection between square functions and Carleson measures.

The sharp maximal function was introduced by Fefferman and Stein [**191**], who first used it to prove Theorem 7.4.5 and also interpolation of analytic families of operators when one endpoint space is *BMO*. Theorem 7.4.7 provides the main idea why L^∞ can be replaced by *BMO* in this context. The fact that L^2-bounded singular integrals also map L^∞ into *BMO* was independently obtained by Peetre [**402**], Spanne [**467**], and Stein [**473**]. Peetre [**402**] also observed that translation invariant singular integrals (such as the ones in Corollary 7.4.10) actually map *BMO* into itself. Another interesting property of *BMO* is that it is preserved under the action of the Hardy-Littlewood maximal operator. This was proved by Bennett, DeVore, and Sharpley [**34**]; see also the almost simultaneous proof of Chiarenza and Frasca [**90**]. The decomposition of open sets given in Proposition 7.3.4 is due to Whitney [**553**].

An alternative characterization of *BMO* can be obtained in terms of commutators of singular integrals. Precisely, we have that the commutator $[b,T](f)$ is L^p bounded for $1 < p < \infty$ if and only if the function b is in *BMO*. The sufficiency of this result (Theorem 7.5.6) is due to Coifman, Rochberg, and Weiss [**119**], who used it to extend the classical theory of H^p spaces to higher dimensions. The necessity was obtained by Janson [**274**], who also obtained a simpler proof of the sufficiency. The exposition in Section 7.5 is based on the article of Pérez [**411**]. This approach is not the shortest available, but the information derived in Lemma 7.5.5 is often useful; for instance, it is used in the substitute of the weak type $(1,1)$ estimate of Exercise 7.5.4.

Orlicz spaces were introduced by Birbaum and Orlicz [**41**] and furher elaborated by Orlicz [**396**], [**397**]. For a modern treatment one may consult the book of Rao and Ren [**419**]. Bounded mean oscillation with Orlicz norms was considered by Strömberg [**495**].

The space of functions of vanishing mean oscillation (VMO) was introduced by Sarason [**444**] as the set of integrable functions f on $\mathbf{T}^1$ satisfying $\lim_{\delta\to 0}\sup_{I:\,|I|\le\delta} |I|^{-1}\int_I |f-\operatorname{Avg}_I f|\,dx = 0$. This space is the closure in the BMO norm of the subspace of $BMO(\mathbf{T}^1)$ consisting of all uniformly continuous functions on $\mathbf{T}^1$. An analogous $VMO(\mathbf{R}^n)$ can be defined as the set of functions on $\mathbf{R}^n$ that satisfy $\lim_{\delta\to 0}\sup_{Q:\,|Q|\le\delta} |Q|^{-1}\int_Q |f-\operatorname{Avg}_Q f|\,dx = 0$, $\lim_{N\to\infty}\sup_{Q:\,\ell(Q)\ge N} |Q|^{-1}\int_Q |f-\operatorname{Avg}_Q f|\,dx = 0$, and $\lim_{R\to\infty}\sup_{Q:\,Q\cap B(0,R)=\emptyset} |Q|^{-1}\int_Q |f-\operatorname{Avg}_Q f|\,dx = 0$; here I denotes intervals in $\mathbf{T}^1$ and Q cubes in $\mathbf{R}^n$. Then $VMO(\mathbf{R}^n)$ is the closure of the the space of continuous functions that vanish at infinity in the $BMO(\mathbf{R}^n)$ norm. One of the imporant features of $VMO(\mathbf{R}^n)$ is that it is the predual of $H^1(\mathbf{R}^n)$, as was shown by Coifman and Weiss [**122**]. As a companion to Corollary 7.4.10, singular integral operators can be shown to map the space of continuous functions that vanish at infinity into VMO. We refer to the article of Dafni [**143**] for a short and elegant exposition of these results as well as for a local version of the VMO-H^1 duality.

CHAPTER 8

Singular Integrals of Nonconvolution Type

Up to this point we have studied singular integrals given by convolution with some distribution. These operators commute with translations. We are now ready to broaden our perspective and study a class of more general singular integrals that are not necessarily translation invariant. This is the main topic of investigation in this chapter. These kinds of operators appear in many places in harmonic analysis and partial differential equations. For instance, a large class of pseudodifferential operators falls under the scope of this theory.

This broader point of view does not necessarily bring additional complications in the development of the theory except at one point, the study of L^2 boundedness. The problem here is the lack of Fourier transform techniques. The L^2 boundedness of convolution operators is easily understood by simple properties of the Fourier transform of the kernel. The situation in the nonconvolution setting is a bit more delicate and requires new tools in the study of L^2 boundedness. The main result of this chapter is the derivation of a set of necessary and sufficient conditions for nonconvolution singular integrals to be bounded on L^2. This result is referred to as the $T(1)$ theorem and owes its name to a condition expressed in terms of the action of the operator T on the function 1.

An extension of the $T(1)$ theorem, called the $T(b)$ theorem, is obtained in Section 8.6 and is used to deduce the L^2 boundedness of the Cauchy integral along Lipschitz curves. A variation of the $T(b)$ theorem is also used in the boundedness of the square root of a divergence form elliptic operator discussed in Section 8.7.

8.1. General Background and the Role of BMO

We begin by recalling the notion of the adjoint and transpose operator. We will be working with two kinds of inner products, the real and the complex. Precisely, for f, g complex-valued functions we denote the real inner product by

$$\langle f, g \rangle = \int_{\mathbf{R}^n} f(x) g(x)\, dx\,,$$

which is suitable when we think of f as a distribution. We also have the complex inner product

$$\langle f \,|\, g\rangle = \int_{\mathbf{R}^n} f(x)\overline{g(x)}\,dx\,,$$

which is suitable when we think of f and g as elements of a Hilbert space over the complex numbers. Now suppose that T is a linear operator bounded on L^p. Then the *adjoint* operator T^* of T is uniquely defined via the identity

$$\langle T(f)\,|\,g\rangle = \langle f\,|\,T^*(g)\rangle$$

for all f in L^p and g in $L^{p'}$. The *transpose* operator T^t of T is uniquely defined via the identity

$$\langle T(f), g\rangle = \langle f, T^t(g)\rangle = \langle T^t(g), f\rangle$$

for all functions f in L^p and g in $L^{p'}$. The name *transpose* comes from matrix theory, where, if A^t denotes the transpose of a complex $n\times n$ matrix A, then we have the identity

$$\langle Ax, y\rangle = \sum_{j=1}^{n}(Ax)_j\, y_j = Ax\cdot y = x\cdot A^t y = \sum_{j=1}^{n} x_j\,(A^t y)_j = \langle x, A^t y\rangle$$

for all $x=(x_1,\dots,x_n), y=(y_1,\dots,y_n)$ in $\mathbf{C}^n$. We can easily check the following relationship between the transpose and the adjoint of a linear operator T:

$$T^*(f) = \overline{T^t(\overline{f}\,)}$$

indicating that we could work with either one. However, in most cases, it will be more convenient for us to avoid complex conjugates and work with the transpose operator for simplicity. Observe that if a linear operator T has kernel $K(x,y)$, that is,

$$T(f)(x) = \int K(x,y)f(y)\,dy\,,$$

then the kernel of T^t is $K^t(x,y) = K(y,x)$ and of T^* is $K^*(x,y) = \overline{K(y,x)}$.

An operator is called *self-adjoint* if $T = T^*$ and *self-transpose* if $T = T^t$. For example, the operator iH, where H is the Hilbert transform, is self-adjoint but not self-transpose and the operator with kernel $i(x+y)^{-1}$ is self-transpose but not self-adjoint.

8.1.a. Standard Kernels

The singular integrals we will study in this chapter have kernels that satisfy size and regularity properties similar to those we encountered in Chapter 4 for convolution-type Calderón-Zygmund operators. Let us now be more specific and introduce the relevant background. We will consider functions $K(x,y)$ defined on $\mathbf{R}^n \times \mathbf{R}^n \setminus \{(x,x):\ x \in \mathbf{R}^n\}$ that satisfy for some $A > 0$ the size condition

$$|K(x,y)| \le \frac{A}{|x-y|^n} \tag{8.1.1}$$

and for some $\delta > 0$ the regularity conditions

$$|K(x,y) - K(x',y)| \le \frac{A\,|x-x'|^\delta}{(|x-y|+|x'-y|)^{n+\delta}}\,, \tag{8.1.2}$$

whenever $|x-x'| \le \frac{1}{2}\max\big(|x-y|,|x'-y|\big)$ and

$$|K(x,y) - K(x,y')| \le \frac{A\,|y-y'|^\delta}{(|x-y|+|x-y'|)^{n+\delta}}\,, \tag{8.1.3}$$

whenever $|y-y'| \le \frac{1}{2}\max\big(|x-y|,|x-y'|\big)$.

Remark 8.1.1. Observe that if

$$|x-x'| \le \frac{1}{2}\max\big(|x-y|,|x'-y|\big)\,,$$

then

$$\max\big(|x-y|,|x'-y|\big) \le \tfrac{3}{2}\min\big(|x-y|,|x'-y|\big)\,,$$

which says that the numbers $|x-y|$ and $|x'-y|$ are comparable. This fact will be useful in specific calculations.

Definition 8.1.2. Functions on $\mathbf{R}^n \times \mathbf{R}^n \setminus \{(x,x):\ x \in \mathbf{R}^n\}$ that satisfy (8.1.1), (8.1.2), and (8.1.3) are called *standard kernels* with constants δ, A. The class of all standard kernels with constants δ, A will be denoted by $SK(\delta,A)$. Given a kernel $K(x,y)$ in $SK(\delta,A)$, we observe that the functions $K(y,x)$ and $\overline{K(y,x)}$ are also in $SK(\delta,A)$. These functions have special names. The function

$$K^t(x,y) = K(y,x)$$

is called the *transpose kernel* of K and the function

$$K^*(x,y) = \overline{K(y,x)}$$

is called the *adjoint kernel* of K.

Example 8.1.3. The function $K(x,y) = |x-y|^{-n}$ defined away from the diagonal of $\mathbf{R}^n \times \mathbf{R}^n$ is in $SK(1, n\,4^{n+1})$. Indeed, for

$$|x-x'| \le \frac{1}{2}\max\big(|x-y|,|x'-y|\big)$$

the mean value theorem gives

$$\big|\,|x-y|^{-n} - |x'-y|^{-n}\big| \le \frac{n|x-x'|}{|\theta-y|^{n+1}}$$

for some θ that lies in the line segment joining x and x'. But then we have $|\theta - y| \ge \frac{1}{2}\max\big(|x-y|,|x'-y|\big)$, which gives (8.1.2) with $A = n\,4^{n+1}$.

Remark 8.1.4. The previous example can be modified to give that if $K(x,y)$ satisfies

$$|\nabla_x K(x,y)| \le A'|x-y|^{-n-1}$$

for all $x \neq y$ in $\mathbf{R}^n$, then $K(x,y)$ also satisfies (8.1.2) with $\delta = 1$ and A controlled by a constant multiple of A'. Likewise, if

$$|\nabla_y K(x,y)| \le A'|x-y|^{-n-1}$$

for all $x \neq y$ in $\mathbf{R}^n$, then $K(x,y)$ satisfies (8.1.3) with with $\delta = 1$ and A bounded by a multiple of A'.

We are interested in standard kernels K that can be extended to tempered distributions on $\mathbf{R}^n \times \mathbf{R}^n$. We begin by observing that given a standard kernel $K(x,y)$, there does not necessarily exist a tempered distribution W on $\mathbf{R}^n \times \mathbf{R}^n$ that coincides with the given $K(x,y)$ on $\mathbf{R}^n \times \mathbf{R}^n \setminus \{(x,x):\ x \in \mathbf{R}^n\}$. For example, in Exercise 8.1.2 the reader is asked to show that the function $K(x,y) = |x-y|^{-n}$ does not admit such an extension.

We will be mainly concerned with standard kernels $K(x,y)$ in $SK(\delta, A)$ for which there are tempered distributions W on $\mathbf{R}^n \times \mathbf{R}^n$ that coincide with K on $\mathbf{R}^n \times \mathbf{R}^n \setminus \{(x,x):\ x \in \mathbf{R}^n\}$. This means that we have

$$\langle W, F\rangle = \int_{\mathbf{R}^n}\int_{\mathbf{R}^n} K(x,y)F(x,y)\,dx\,dy \tag{8.1.4}$$

whenever the Schwartz function F on $\mathbf{R}^n \times \mathbf{R}^n$ is supported away from the diagonal $\{(x,x):\ x \in \mathbf{R}^n\}$. Note that the integral in (8.1.4) is well defined and absolutely convergent whenever F is a Schwartz function that vanishes in a neighborhood of the set $\{(x,x):\ x \in \mathbf{R}^n\}$. Also observe that there may be several distributions W coinciding with a fixed function $K(x,y)$. In fact, if W is such a distribution, then so is $W + \delta_{x=y}$, where $\delta_{x=y}$ denotes Lebesgue measure on the diagonal of $\mathbf{R}^{2n}$. (This is some sort of a Dirac distribution.)

We now consider continuous linear operators

$$T : \mathcal{S}(\mathbf{R}^n) \to \mathcal{S}'(\mathbf{R}^n)$$

from the space of Schwartz functions $\mathcal{S}(\mathbf{R}^n)$ into the space of all tempered distributions $\mathcal{S}'(\mathbf{R}^n)$. By the *Schwartz kernel theorem* (see Hörmander [**256**, p. 129]), for such an operator T there is a distribution W in $\mathcal{S}'(\mathbf{R}^{2n})$ that satisfies

$$\langle T(f), \varphi\rangle = \langle W, f\otimes\varphi\rangle \qquad \text{when} \quad f,\varphi \in \mathcal{S}(\mathbf{R}^n), \tag{8.1.5}$$

where $(f\otimes\varphi)(x,y) = f(x)\varphi(y)$. Furthermore, as a consequence of the Schwartz kernel theorem, there exist constants C, N, M so that for all $f, g \in \mathcal{S}(\mathbf{R}^n)$ we have

$$|\langle T(f), g\rangle| = |\langle W, f\otimes g\rangle| \le C\Big[\sum_{|\alpha|,|\beta|\le N} \rho_{\alpha,\beta}(f)\Big]\Big[\sum_{|\alpha|,|\beta|\le M} \rho_{\alpha,\beta}(g)\Big] \tag{8.1.6}$$

where $\rho_{\alpha,\beta}(\varphi) = \sup_{x\in\mathbf{R}^n} |\partial_x^\alpha(x^\beta\varphi)(x)|$ is the set of seminorms for the topology in $\mathcal{S}$. A distribution W that satisfies (8.1.5) and (8.1.6) is called a *Schwartz kernel*.

We will be interested in continuous linear operators $T : \mathcal{S}(\mathbf{R}^n) \to \mathcal{S}'(\mathbf{R}^n)$ whose Schwartz kernels coincide with some standard kernel $K(x,y)$ on $\mathbf{R}^n \times \mathbf{R}^n \setminus \{(x,x) : x \in \mathbf{R}^n\}$. This means that (8.1.5) takes the integral representation

$$\big\langle T(f), \varphi \big\rangle = \int_{\mathbf{R}^n} \int_{\mathbf{R}^n} K(x,y) f(x) \varphi(y)\, dx\, dy, \tag{8.1.7}$$

whenever f and φ are Schwartz functions whose supports do not intersect.

We make some remarks concerning duality in this context. Given a continuous linear operator $T : \mathcal{S}(\mathbf{R}^n) \to \mathcal{S}'(\mathbf{R}^n)$ with a Schwartz kernel W, we can define another distribution W^t as follows:

$$\big\langle W^t, F \big\rangle = \big\langle W, F^t \big\rangle,$$

where $F^t(x,y) = F(y,x)$. This means that for all $f, \varphi \in \mathcal{S}(\mathbf{R}^n)$ we have

$$\big\langle W, f \otimes \varphi \big\rangle = \big\langle W^t, \varphi \otimes f \big\rangle.$$

It is a simple fact that the transpose operator T^t of T, which satisfies

$$\big\langle T(\varphi), f \big\rangle = \big\langle T^t(f), \varphi \big\rangle \tag{8.1.8}$$

for all f, φ in $\mathcal{S}(\mathbf{R}^n)$, is the unique continuous linear operator from $\mathcal{S}(\mathbf{R}^n)$ into $\mathcal{S}'(\mathbf{R}^n)$ whose Schwartz kernel is the distribution W^t, that is, we have

$$\big\langle T^t(f), \varphi \big\rangle = \big\langle T(\varphi), f \big\rangle = \big\langle W, \varphi \otimes f \big\rangle = \big\langle W^t, f \otimes \varphi \big\rangle. \tag{8.1.9}$$

We now observe that a large class of standard kernels admit extensions to tempered distributions W on $\mathbf{R}^{2n}$.

Example 8.1.5. Suppose that $K(x,y)$ satisfies (8.1.1) and (8.1.2) and is *antisymmetric*, in the sense that

$$K(x,y) = -K(y,x)$$

for all $x \neq y$ in $\mathbf{R}^n$. Then K also satisfies (8.1.3) and moreover there is a distribution W on $\mathbf{R}^{2n}$ that extends K on $\mathbf{R}^n \times \mathbf{R}^n$.

Indeed, define

$$\big\langle W, F \big\rangle = \lim_{\varepsilon \to 0} \iint\limits_{|x-y|>\varepsilon} K(x,y) F(x,y)\, dy\, dx \tag{8.1.10}$$

for all F in the Schwartz class of $\mathbf{R}^{2n}$. Using antisymmetry, we can write

$$\iint\limits_{\varepsilon<|x-y|\le 1} K(x,y)F(x,y)\, dy\, dx = \frac{1}{2} \iint\limits_{\varepsilon<|x-y|\le 1} K(x,y)\big(F(x,y) - F(y,x)\big)\, dy\, dx\,.$$

Using the fact that $|F(x,y) - F(y,x)| \le 2\big\|\nabla F\big\|_{L^\infty} |x-y|$, it follows that the limit in (8.1.10) exists and gives a tempered distribution on $\mathbf{R}^{2n}$. We can therefore define an operator $T : \mathcal{S}(\mathbf{R}^n) \to \mathcal{S}'(\mathbf{R}^n)$ with kernel W as follows:

$$\langle T(f), \varphi \rangle = \lim_{\varepsilon \to 0} \iint_{|x-y|>\varepsilon} K(x,y) f(x) \varphi(y)\, dy\, dx.$$

Example 8.1.6. Let A be a Lipschitz function on $\mathbf{R}$. This means that it satisfies the estimate $|A(x) - A(y)| \le L|x-y|$ for some $L < \infty$ and all $x, y \in \mathbf{R}$. For $x, y \in \mathbf{R}$ we let

$$K(x,y) = \frac{1}{x - y + i(A(x) - A(y))} \tag{8.1.11}$$

and we observe that $K(x,y)$ is a standard kernel in $SK(1, 6+6L)$. The details are left to the reader. Note that the kernel K defined in (8.1.11) is antisymmetric.

Example 8.1.7. Let the function A be as in the previous example. For each integer $m \ge 1$ we set

$$K_m(x,y) = \left(\frac{A(x) - A(y)}{x - y} \right)^m \frac{1}{x-y}, \qquad x, y \in \mathbf{R}. \tag{8.1.12}$$

Clearly, K_m is an antisymmetric function. To see that each K_m is a standard kernel, we use the simple fact that

$$\left| \frac{A(x) - A(y)}{x - y} - \frac{A(x') - A(y)}{x' - y} \right| \le \frac{2L|x - x'|}{|x-y|},$$

the identity $a^m - b^m = (a-b)(a^{m-1} + a^{m-2}b + \cdots + b^{m-1})$, and the observation made in Remark 8.1.1. It follows that K_m lies in $SK(\delta, C)$ with $\delta = 1$ and C at most a dimensional multiple of mL^m. The linear operator in with kernel K_m is called the *mth Calderón commutator*.

8.1.b. Operators Associated with Standard Kernels

Having introduced standard kernels, we are ready to define linear operators associated with them.

Definition 8.1.8. Let $0 < \delta, A < \infty$ and K in $SK(\delta, A)$. A continuous linear operator T from $\mathcal{S}(\mathbf{R}^n)$ into $\mathcal{S}'(\mathbf{R}^n)$ that satisfies

$$T(f)(x) = \int_{\mathbf{R}^n} K(x,y) f(y)\, dy \tag{8.1.13}$$

for all $f \in C_0^\infty$ and x not in the support of f is said to be *associated with* K. If T is associated with K, then the Schwartz kernel W of T coincides with K on $\mathbf{R}^n \times \mathbf{R}^n \setminus \{(x,x) : x \in \mathbf{R}^n\}$.

If T is associated with K and it also admits a bounded extension on $L^2(\mathbf{R}^n)$, that is, it satisfies

$$\big\| T(f) \big\|_{L^2} \le B \big\| f \big\|_{L^2} \tag{8.1.14}$$

for all $f \in L^2(\mathbf{R}^n)$, then T is called a *Calderón-Zygmund operator* associated with the standard kernel K.

In the sequel we will denote by $CZO(\delta, A, B)$ the class of all Calderón-Zygmund operators associated with standard kernels in $SK(\delta, A)$ that are bounded on L^2 with norm at most B.

We make the point that there may be several Calderón-Zygmund operators associated with a given standard kernel K. For instance, we may check that the zero operator and the identity operator have the same kernel $K(x, y) = 0$. We will investigate connections between any two such operators in Proposition 8.1.11. Next we discuss the important fact that once an operator T admits an extension that is L^2 bounded, then (8.1.13) holds for all f that are bounded and compactly supported whenever x does not lie in its support.

Proposition 8.1.9. *Let T be an element of $CZO(\delta, A, B)$ associated with a standard kernel K. Then for all f in L^∞ with compact support and every $x \notin \operatorname{supp} f$ we have the integral representation*

$$T(f)(x) = \int_{\mathbf{R}^n} K(x, y) f(y)\, dy \tag{8.1.15}$$

as an absolutely convergent integral.

Proof. As a consequence of (8.1.1), the preceding integral converges absolutely on the support of f.

Identity (8.1.15) will be a consequence of the fact that whenever f and φ are bounded and compactly supported functions that satisfy

$$\operatorname{dist}(\operatorname{supp} \varphi, \operatorname{supp} f) \geq c > 0, \tag{8.1.16}$$

then we have the integral representation

$$\int_{\mathbf{R}^n} T(f)(x)\, \varphi(x)\, dx = \int_{\mathbf{R}^n} \int_{\mathbf{R}^n} K(x, y) f(y) \varphi(x)\, dy\, dx. \tag{8.1.17}$$

To see this, given f and φ as previously, select $f_j, \varphi_j \in C_0^\infty$ such that φ_j are uniformly bounded and supported in a small neighborhood of the support of φ, $\varphi_j \to \varphi$ in L^2 and almost everywhere, $f_j \to f$ in L^2 and almost everywhere, and

$$\operatorname{dist}(\operatorname{supp} \varphi_j, \operatorname{supp} f_j) \geq c/2 > 0$$

for all j. Because of (8.1.7), identity (8.1.17) is valid for the functions f_j and φ_j in place of f and φ. By the boundedness of T, it follows that $T(f_j)$ converges to $T(f)$ in L^2 and thus

$$\int_{\mathbf{R}^n} T(f_j)(x) \varphi_j(x)\, dx \to \int_{\mathbf{R}^n} T(f)(x) \varphi(x)\, dx.$$

Now write $f_j \varphi_j - f\varphi = (f_j - f)\varphi_j + f(\varphi_j - \varphi)$ and observe that

$$\int_{\mathbf{R}^n} \int_{\mathbf{R}^n} K(x, y) f(y) (\varphi_j(x) - \varphi(x))\, dy\, dx \to 0$$

since it is controlled by a multiple of $\big\|T(f)\big\|_{L^2}\displaystyle\int_{\mathbf{R}^n}|\varphi_j(x)-\varphi(x)|\,dx$, while

$$\int_{\mathbf{R}^n}\int_{\mathbf{R}^n}K(x,y)(f_j(y)-f(y))\varphi_j(x)\,dy\,dx\to 0$$

since it is controlled by a multiple of $\sup_j\big\|T^t(\varphi_j)\big\|_{L^2}\big\|f_j-f\big\|_{L^2}$. This gives that

$$\int_{\mathbf{R}^n}\int_{\mathbf{R}^n}K(x,y)f_j(y)\varphi_j(x)\,dy\,dx\to\int_{\mathbf{R}^n}\int_{\mathbf{R}^n}K(x,y)f(y)\varphi(x)\,dy\,dx$$

as $j\to\infty$, which proves the validity of (8.1.17).

□

We now define truncated kernels and operators.

Definition 8.1.10. Given a kernel K in $SK(\delta,A)$ and $\varepsilon>0$, we define the *truncated kernel*

$$K^{(\varepsilon)}(x,y)=K(x,y)\chi_{|x-y|>\varepsilon}\,.$$

Given a continuous linear operator T from $\mathcal{S}(\mathbf{R}^n)$ into $\mathcal{S}'(\mathbf{R}^n)$ and $\varepsilon>0$, we define the *truncated operator* $T^{(\varepsilon)}$ as

$$T^{(\varepsilon)}(f)(x)=\int_{\mathbf{R}^n}K^{(\varepsilon)}(x,y)\,f(y)\,dy$$

and the *maximal singular operator* associated with T as follows:

$$T^{(*)}(f)(x)=\sup_{\varepsilon>0}\big|T^{(\varepsilon)}(f)(x)\big|\,.$$

Note that both $T^{(\varepsilon)}$ and $T^{(*)}$ are well defined for f in $\displaystyle\bigcup_{1\le p<\infty}L^p(\mathbf{R}^n)$.

We would like to investigate connections between the boundedness of T and the boundedness of the family of the $T^{(\varepsilon)}$'s uniformly in $\varepsilon>0$. We have the following.

Proposition 8.1.11. *Let K be a standard kernel in $SK(\delta,A)$ and let T in $CZO(\delta,A,B)$ be associated with K. For $\varepsilon>0$, let $T^{(\varepsilon)}$ be the truncated operators obtained from T. Assume that there exists a constant $B'<\infty$ such that*

$$\sup_{\varepsilon>0}\big\|T^{(\varepsilon)}\big\|_{L^2\to L^2}\le B'. \tag{8.1.18}$$

Then there exists a linear operator T_0 defined on $L^2(\mathbf{R}^n)$ such that

(1) *The Schwartz kernel of T_0 coincides with K on*

$$\mathbf{R}^n\times\mathbf{R}^n\setminus\{(x,x):\ x\in\mathbf{R}^n\}.$$

(2) *For some subsequence $\varepsilon_j\downarrow 0$, we have*

$$\int_{\mathbf{R}^n}T^{(\varepsilon_j)}(f)(x)g(x)\,dx\to\int_{\mathbf{R}^n}(T_0f)(x)g(x)\,dx$$

as $j\to\infty$ for all f,g in $L^2(\mathbf{R}^n)$.

(3) *T_0 is bounded on $L^2(\mathbf{R}^n)$ with norm*

$$\big\|T_0\big\|_{L^2\to L^2} \le B'.$$

(4) *There exists a measurable function b on $\mathbf{R}^n$ with $\big\|b\big\|_{L^\infty} \le B + B'$ such that*

$$T(f) - T_0(f) = bf\,,$$

for all $f \in L^2(\mathbf{R}^n)$.

Proof. Consider the Banach space $X = \mathcal{B}(L^2, L^2)$ of all bounded linear operators from $L^2(\mathbf{R}^n)$ into itself. Then X is isomorphic to $\mathcal{B}((L^2)^*, (L^2)^*)^*$, which is a dual space. Since the unit ball of a dual space is weak* compact, and the operators $T^{(\varepsilon)}$ lie in a multiple of this unit ball, the Banach-Alaoglu theorem gives the existence of a sequence $\varepsilon_j \downarrow 0$ such that $T^{(\varepsilon_j)}$ converges to some T_0 in the weak* topology of $\mathcal{B}(L^2, L^2)$ as $j \to \infty$. This means that

$$\int_{\mathbf{R}^n} T^{(\varepsilon_j)}(f)(x) g(x)\,dx \to \int_{\mathbf{R}^n} T_0(f)(x) g(x)\,dx \tag{8.1.19}$$

for all f, g in $L^2(\mathbf{R}^n)$ as $j \to \infty$. This proves (2). The L^2 boundedness of T_0 is a consequence of (8.1.19), hypothesis (8.1.18), and duality, since

$$\big\|T_0(f)\big\|_{L^2} \le \sup_{\|g\|_{L^2}\le 1} \limsup_{j\to\infty} \left| \int_{\mathbf{R}^n} T^{(\varepsilon_j)}(f)(x) g(x)\,dx \right| \le B'\big\|f\big\|_{L^2}.$$

This proves (3). Finally, (1) is a consequence of the integral representation

$$\int_{\mathbf{R}^n} T^{(\varepsilon_j)}(f)(x) g(x)\,dx = \int_{\mathbf{R}^n}\int_{\mathbf{R}^n} K^{(\varepsilon_j)}(x,y) f(y)\,dy\, g(x)\,dx,$$

whenever f, g are Schwartz functions with disjoint supports, by letting $j \to \infty$.

We finally prove (4). We first observe that if g is a bounded function with compact support and Q is an open cube in $\mathbf{R}^n$, we have

$$(T^{(\varepsilon)} - T)(g\chi_Q)(x) = \chi_Q(x)\,(T^{(\varepsilon)} - T)(g)(x)\,, \tag{8.1.20}$$

whenever $x \notin \partial Q$ and ε is small enough. Indeed, take first $x \notin \overline{Q}$; then x is not in the support of $g\chi_Q$. Note that since $g\chi_Q$ is bounded and has compact support, we can use the integral representation formula (8.1.15) obtained in Proposition 8.1.9. Then we have that for $\varepsilon < \operatorname{dist}(x, \operatorname{supp} g\chi_Q)$, the left-hand side in (8.1.20) is zero. Moreover, for $x \in Q$, we have that x does not lie in the support of $g\chi_{Q^c}$ and again because of (8.1.15) we obtain $(T^{(\varepsilon)} - T)(g\chi_{Q^c})(x) = 0$ whenever $\varepsilon < \operatorname{dist}(x, \operatorname{supp} g\chi_{Q^c})$. This proves (8.1.20) for all x not on the boundary ∂Q of Q. Taking weak limits in (8.1.20) as $\varepsilon \to 0$, we obtain that

$$(T_0 - T)(g\chi_Q) = \chi_Q\,(T_0 - T)(g) \qquad \text{a.e.} \tag{8.1.21}$$

for all open cubes Q in $\mathbf{R}^n$. By linearity we extend (8.1.21) to simple functions. Using the fact that $T_0 - T$ is L^2 bounded and a simple density argument, we obtain

$$(T_0 - T)(gf) = f\,(T_0 - T)(g) \qquad \text{a.e.} \tag{8.1.22}$$

whenever f is in L^2 and g is bounded and has compact support. If $B(0,j)$ is the open ball with center 0 and radius j on $\mathbf{R}^n$, when $j \le j'$ we have

$$(T_0 - T)(\chi_{B(0,j)}) = (T_0 - T)(\chi_{B(0,j)}\chi_{B(0,j')}) = \chi_{B(0,j)}\,(T_0 - T)(\chi_{B(0,j')})\,.$$

Therefore, the sequence of functions $(T_0 - T)(\chi_{B(0,j)})$ satisfies the "consistency" property

$$(T_0 - T)(\chi_{B(0,j)}) = (T_0 - T)(\chi_{B(0,j')}) \quad \text{in } B(0,j)$$

when $j \le j'$. It follows that there exists a well-defined function b such that

$$b = (T_0 - T)(\chi_{B(0,j)}) \quad \text{a.e. in } B(0,j)\,.$$

Applying (8.1.22) with f supported in $B(0,j)$ and $g = \chi_{B(0,j)}$, we obtain

$$(T_0 - T)(f) = (T_0 - T)(f\chi_{B(0,j)}) = f\,(T_0 - T)(\chi_{B(0,j)}) = f\,b \qquad \text{a.e.}$$

from which it follows that $(T_0 - T)(f) = b\,f$ for all $f \in L^2$. Since the norm of $T - T_0$ on L^2 is at most $B + B'$, it follows that the norm of the linear map $f \to b\,f$ from L^2 into itself is at most $B + B'$. From this we obtain that $\big\|b\big\|_{L^\infty} \le B + B'$.

□

Remark 8.1.12. We will show in the next section that if a Calderón-Zygmund operator maps L^2 into L^2, then so does its corresponding maximal operator $T^{(*)}$. This implies that if T is L^2 bounded, then so are all of its truncations uniformly in $\varepsilon > 0$. Therefore, there exists a linear operator T_0 that is equal to T plus a bounded function times the identity operator and that has the form

$$T_0(f)(x) = \lim_{j\to\infty} \int_{|x-y|>\varepsilon_j} K(x,y) f(y)\, dy\,,$$

where the limit is taken in the weak topology of L^2.

We will give a special name to operators of this form.

Definition 8.1.13. Any operator in $CZO(\delta, A, B)$ that satisfies

$$\int_{\mathbf{R}^n} T(f)(x)\, g(x)\, dx = \lim_{j\to\infty} \iint_{|x-y|>\varepsilon_j} K(x,y) f(y) g(x)\, dx\, dy$$

for some sequence $\varepsilon_j \to 0$ and for all f, g in L^2 is called a *Calderón-Zygmund singular integral operator*. Thus Calderón-Zygmund singular integral operators are special kinds of Calderón-Zygmund operators.

In view of Proposition 8.1.11 and Remark 8.1.12, a Calderón-Zygmund operator is equal to a Calderón-Zygmund singular integral operator plus a bounded function times the identity operator. For this reason the study of Calderón-Zygmund operators is equivalent to the study of Calderón-Zygmund singular integral operators, and in most situations we restrict our attention to the study of the latter.

8.1.c. Calderón-Zygmund Operators Acting on Bounded Functions

We are now interested in defining the action of a Calderón-Zygmund operator T on bounded and smooth functions. To achieve this we will first need to define the space of special testing functions $\mathcal{D}_0$.

Definition 8.1.14. Recall the space $\mathcal{D}(\mathbf{R}^n) = C_0^\infty(\mathbf{R}^n)$ of all smooth functions with compact support on $\mathbf{R}^n$. We define $\mathcal{D}_0(\mathbf{R}^n)$ to be the space of all smooth functions with compact support and integral zero. We equip $\mathcal{D}_0(\mathbf{R}^n)$ with the same topology as the space $\mathcal{D}(\mathbf{R}^n)$. The dual space of $\mathcal{D}_0(\mathbf{R}^n)$ under this topology will be denoted by $\mathcal{D}_0'(\mathbf{R}^n)$. This is a space of distributions larger than $\mathcal{D}'(\mathbf{R}^n)$.

Example 8.1.15. BMO functions are examples of elements of $\mathcal{D}_0'(\mathbf{R}^n)$. Indeed, given $b \in BMO(\mathbf{R}^n)$, for any compact set K there is a constant $C_K = \|b\|_{L^1(K)}$ such that

$$\left| \int_{\mathbf{R}^n} b(x)\varphi(x)\,dx \right| \le C_K \|\varphi\|_{L^\infty}$$

for any $\varphi \in \mathcal{D}_0(\mathbf{R}^n)$. Moreover, observe that the preceding integral remains unchanged if the BMO function b is replaced by $b+c$, where c is a constant.

Definition 8.1.16. Let T be a continuous linear operator $T : \mathcal{S}(\mathbf{R}^n) \to \mathcal{S}'(\mathbf{R}^n)$ that satisfies (8.1.5) for some distribution W that coincides with a standard kernel $K(x,y)$ satisfying (8.1.1), (8.1.2), and (8.1.3). Given f bounded and smooth, we define an element $T(f)$ of $\mathcal{D}_0'(\mathbf{R}^n)$ as follows: Let φ be in $\mathcal{D}_0(\mathbf{R}^n)$ and pick a ball $B(x_0,R)$ that contains its support. Select η a smooth function with compact support that is equal to 1 on the ball $B(x_0,2R)$. Since T maps $\mathcal{S}$ to $\mathcal{S}'$, the expression $T(f\eta)$ is a tempered distribution and thus its action on φ is well defined. We define

$$\langle T(f),\varphi\rangle = \langle T(f\eta),\varphi\rangle + \int_{\mathbf{R}^n}\left[\int_{\mathbf{R}^n} K(x,y)\varphi(x)\,dx\right] f(y)(1-\eta(y))\,dy \tag{8.1.23}$$

and we will need to show that this definition is independent of η. But before we prove that this definition is independent of η, we show that the double integral in (8.1.23) is absolutely convergent. Since φ has mean value zero, we can write the integral inside the square brackets in (8.1.23) as

$$\int_{\mathbf{R}^n} \big(K(x,y) - K(x_0,y)\big)\varphi(x)\,dx.$$

But when $y \in \operatorname{supp}(1-\eta)f$, we have $y \notin B(x_0,2R)$; thus $2|x-x_0| \le 2R \le |x_0-y|$. Applying (8.1.2), we obtain that the double integral in (8.1.23) is controlled by

$$\iint_{|y-x_0|\ge 2|x-x_0|} |K(x,y)-K(x_0,y)|\,|\varphi(x)|\,|f(y)|\,dx\,dy \le c\,A\|f\|_{L^\infty}.$$

This gives the required absolute convergence.

Remark 8.1.17. We note that the definition of $T(f)$ is independent of the choice of the function η. Indeed, if ζ is another function that is equal to 1 in a neighborhood of the support of φ, then $f\,(\eta-\zeta)$ and φ have disjoint supports and by (8.1.7) we have the singular integral realization

$$\langle T(f(\eta-\zeta)),\varphi\rangle=\int_{\mathbf{R}^n}\int_{\mathbf{R}^n}K(x,y)f(y)(\eta-\zeta)(y)\,dy\,\varphi(x)\,dx\,.$$

It follows that (8.1.23) coincides with the corresponding expression obtained when η is replaced by ζ.

Finally, we make the comment that if f is a Schwartz function, then the tempered distribution $T(f)$ as given in Definition 8.1.16 coincides with the original $T(f)$ of Definition 8.1.8.

Remark 8.1.18. When T has an extension that maps L^2 into itself, then we can define $T(f)$ for all $f\in L^\infty(\mathbf{R}^n)$, not necessarily smooth. Simply observe that under this assumption, the expression $T(f\eta)$ is a well-defined L^2 function and thus

$$\langle T(f\eta),\varphi\rangle=\int_{\mathbf{R}^n}T(f\eta)(x)\varphi(x)\,dx$$

is given by an absolutely convergent integral for all $\varphi\in\mathcal{D}_0$.

We now have a definition for $T(f)$ when f is bounded and smooth. Moreover, when T is L^2 bounded, we have a definition for $T(f)$ when f is simply bounded. In both cases $T(f)$ is an element of $\mathcal{D}_0'$. The following theorem relates singular integrals to BMO.

Theorem 8.1.19. *Let T be in $CZO(\delta,A,B)$. Then for any bounded function f, the distribution $T(f)$ can be identified with a BMO function that satisfies*

$$\big\|T(f)\big\|_{BMO}\le C_n\big(\big\|T\big\|_{L^2\to L^2}+A_2\big)\big\|f\big\|_{L^\infty}\,, \tag{8.1.24}$$

where C_n is a fixed dimensional constant.

Proof. Let us recall the space $H_0^1(\mathbf{R}^n)$ of all finite sums of H^1 atoms. Fix a function $f\in L^\infty(\mathbf{R}^n)$ and define a linear functional L on $H_0^1(\mathbf{R}^n)$ by setting for all atoms a_Q

$$\begin{aligned}L(a_Q)=&\int_{\mathbf{R}^n}T(f\eta_Q)(x)a_Q(x)\,dx\\&+\int_{\mathbf{R}^n}\Big\{\int_{\mathbf{R}^n}K(x,y)a_Q(x)\,dx\Big\}f(y)(1-\eta_Q(y))\,dy,\end{aligned} \tag{8.1.25}$$

where a_Q is supported in the cube Q and η_Q is a smooth function with compact support that is equal to 1 on $2\sqrt{n}\,Q$ and supported in $4\sqrt{n}\,Q$. Observe that the expression inside the curly brackets in (8.1.25) is $T^t(a_Q)$ which is an L^2 function and since $f(1-\eta_Q)$ is also in L^2 the second integral in (8.1.25) converges absolutely. Also, the first integral in (8.1.25) converges absolutely. We indicate why this definition of L is independent of the specific choice of the function η_Q. Given another C_0^∞

function ζ_Q that is equal to 1 on $2\sqrt{n}\,Q$, the supports of $f(\eta_Q - \zeta_Q)$ and a_Q have a positive distance and in view of (8.1.17) we have the singular integral realization

$$\int_{\mathbf{R}^n} T(f(\eta_Q - \zeta_Q))(x) a_Q(x)\,dx = \int_{\mathbf{R}^n}\int_{\mathbf{R}^n} K(x,y)\,a_Q(x) f(y)(\eta_Q - \zeta_Q)(y)\,dx\,dy.$$

This proves that the expression inside the curly brackets in (8.1.25) is independent of the function η_Q as long as this function is equal to 1 on $2\sqrt{n}\,Q$.

We now show that for some dimensional constant C_n' we have

$$|L(a_Q)| \le C_n'(A + \|T\|_{L^2\to L^2})\|f\|_{L^\infty} \tag{8.1.26}$$

for all a_Q L^2-atoms for H^1. Indeed, using the Cauchy-Schwarz inequality, we obtain

$$\begin{aligned}\left|\int_{\mathbf{R}^n} T(f\eta_Q)(x) a_Q(x)\,dx\right| &\le \|T(f\eta_Q)\|_{L^2}\|a_Q\|_{L^2} \\ &\le \|T\|_{L^2\to L^2}\|f\|_{L^\infty}\|\eta_Q\|_{L^2}|Q|^{-\frac{1}{2}} \\ &\le (4\sqrt{n})^{\frac{n}{2}}\|T\|_{L^2\to L^2}\|f\|_{L^\infty},\end{aligned}$$

since η_Q is supported in $4\sqrt{n}\,Q$. To estimate the second integral in (8.1.25), we use that atoms have mean value zero to write

$$\begin{aligned}&\int_{\mathbf{R}^n}\int_{\mathbf{R}^n} K(x,y) a_Q(x)\,dx\,f(y)(1-\eta_Q(y))\,dy \\ =&\int_{\mathbf{R}^n}\int_{\mathbf{R}^n} \big[K(x,y) - K(c_Q,y)\big] a_Q(x)\,dx\,f(y)(1-\eta_Q(y))\,dy,\end{aligned}$$

where c_Q is the center of the cube Q. Since $|x - c_Q| \le \frac{1}{2}|y - c_Q|$, when $x \in Q$ and y in the support of $1-\eta$ [which is contained in $(2\sqrt{n}\,Q)^c$], using (8.1.2) we can control the last double integral by

$$\int_Q \left(\int_{|y-c_Q|\ge 2|x-c_Q|} \frac{A\,dy}{|y-c_Q|^{n+\delta}}\right)|a_Q(x)|\,|x-c_Q|^\delta dx\,\|f\|_{L^\infty} \le c\,A\|f\|_{L^\infty},$$

where the last estimate follows from the fact that L^2-atoms for H^1 have L^1 norm at most 1. This establishes the proof of (8.1.26).

We have now proved that L is a bounded linear functional on $H_0^1(\mathbf{R}^n)$, and in view of (8.1.26) we have that

$$\|L\| \le C_n'(A + \|T\|_{L^2\to L^2})\|f\|_{L^\infty}.$$

By Theorem 7.2.2 there exists a BMO function b with $\|b\|_{BMO} \le C_n''\|L\|$ so that

$$L(g) = \int_{\mathbf{R}^n} b(x) g(x)\,dx \tag{8.1.27}$$

for all g finite sums of L^2-atoms for H^1. In particular, (8.1.27) is valid for functions $g \in \mathcal{D}_0$, since elements of $\mathcal{D}_0$ are multiples of L^2-atoms for H^1. The definition of

$T(f)$ gives that

$$\langle T(f), \varphi \rangle = L(\varphi).$$

Therefore, for all $\varphi \in \mathcal{D}_0(\mathbf{R}^n)$ we have

$$\langle T(f), \varphi \rangle = L(\varphi) = \int_{\mathbf{R}^n} b(x)\varphi(x)\, dx.$$

This fact proves that $T(f)$ can be identified with the BMO function b. Since

$$\|b\|_{BMO} \le C_n''\|L\| \le C_n''C_n'\big(A + \|T\|_{L^2 \to L^2}\big)\|f\|_{L^\infty},$$

the last conclusion (8.1.24) follows.

□

There is a result analogous to that of Theorem 8.1.19.

Theorem 8.1.20. *Let T be an element of $CZO(\delta, A, B)$. Then T has an extension that maps $H^1(\mathbf{R}^n)$ to $L^1(\mathbf{R}^n)$. Precisely, there is a dimensional constant C_n such that*

$$\|T\|_{H^1 \to L^1} \le C_n\big(A + \|T\|_{L^2 \to L^2}\big).$$

Proof. This result is immediately obtained from Theorem 8.1.19 via duality (use Exercise 7.2.1). It can also be proved directly using the atomic decomposition of H^1. Then we will only need to check the action of T on atoms. This can be done in a way similar to that in the proof of Theorem 6.7.1 (and 6.7.3). We skip the details, which can be adapted easily from those proofs.

□

Exercises

8.1.1. Suppose that K is a function defined away from the diagonal on $\mathbf{R}^n \times \mathbf{R}^n$ that satisfies for some $\delta > 0$ the condition

$$|K(x,y) - K(x',y)| \le A' \frac{|x - x'|^\delta}{|x - y|^{n+\delta}}$$

whenever $|x - x'| \le \frac{1}{2}|x - y|$. Prove that K satisfies (8.1.2) with constant $A = (\frac{5}{2})^{n+\delta} A'$. Obtain an analogous statement for condition (8.1.3).

8.1.2. Prove that there does not exist a tempered distribution W on $\mathbf{R}^{2n}$ that extends the function $|x - y|^{-n}$ defined on $\mathbf{R}^{2n} \setminus \{(x,x) :\ x \in \mathbf{R}^n\}$.
[*Hint:* Apply such a distribution to a positive smooth bump that does not vanish at the origin.]

8.1.3. Let $\varphi(x)$ be a smooth radial function that is equal to 1 when $|x| \ge 1$ and vanishes when $|x| \le \frac{1}{2}$. Prove that if K lies in $SK(\delta, A)$, then all the smooth truncations $K_\varphi^{(\varepsilon)}(x,y) = K(x,y)\varphi(\frac{x-y}{\varepsilon})$ lie in $SK(\delta, c\,A)$ for some $c > 0$ independent of $\varepsilon > 0$.

8.1.4. Prove that the functions $K_m(x,y)$ in Example 8.1.7 are in $SK(1, c\, m\, L^m)$, where c is a constant.

8.1.5. Suppose that A is a Lipschitz map from $\mathbf{R}^n$ into $\mathbf{R}^m$. This means that there exists a constant L such that $|A(x) - A(y)| \le L|x-y|$ for all $x, y \in \mathbf{R}^n$. Suppose that F is a C^∞ odd function defined on $\mathbf{R}^m$. Show that the kernel

$$K(x,y) = \frac{1}{|x-y|^n} F\left(\frac{A(x) - A(y)}{|x-y|}\right)$$

is in $SK(1, C)$ for some $C > 0$.

8.1.6. Prove Theorem 8.1.20 using only Theorem 8.1.19 and duality.
[*Hint:* You will need Exercise 7.2.1.]

8.1.7. Extend the result of Proposition 8.1.11 to the case where the space L^2 is replaced by L^q for some $1 < q < \infty$.

8.1.8. Observe for an operator T as in Definition 8.1.16, the condition $T(1) = 0$ is equivalent to the statement that for all φ smooth with compact support and integral zero we have $\int_{\mathbf{R}^n} T^t(\varphi)(x)\, dx = 0$. A similar statement holds for T^t.

8.1.9. Suppose that $K(x,y)$ is continuous and nonnegative on $\mathbf{R}^n \times \mathbf{R}^n$, which satisfies $\int_{\mathbf{R}^n} K(x,y)\, dy = 1$ for all $x \in \mathbf{R}^n$. Define a linear operator T by setting $T(f)(x) = \int_{\mathbf{R}^n} K(x,y)\, f(y)\, dy$.
(a) Suppose that h is a continuous function on $\mathbf{R}^n$ that has a global minimum [i.e., there exists $x_0 \in \mathbf{R}^n$ such that $h(x_0) \le h(x)$ for all $x \in \mathbf{R}^n$]. If we have

$$T(h)(x) = h(x),$$

then prove that h is a constant function.
(b) Show that T preserves the set of functions that are bounded below by a fixed constant.
(c) Suppose that $T(T(f)) = f$ for some everywhere positive continuous function f on $\mathbf{R}^n$. Show that $T(f) = f$.
[*Hint:* Part (c): Let $L(x,y)$ be the kernel of $T \circ T$. Show that

$$\int_{\mathbf{R}^n} L(x,y) \frac{f(y)}{f(x)} \frac{T(f)(y)}{f(y)}\, dy = \frac{T(f)(x)}{f(x)}$$

and conclude by part (a) that $\frac{T(f)(y)}{f(y)}$ is a constant.]

8.2. Consequences of L^2 Boundedness

We saw in the previous section that any operator in $CZO(\delta, A, B)$ must map L^∞ to BMO and H^1 to L^1. Both of these results were simple consequences of the L^2 boundedness of elements of $CZO(\delta, A, B)$. In this section we will be concerned with other consequences of the L^2 boundedness of operators in this class. Throughout the entire discussion, all kernels satisfy the standard size and regularity conditions (8.1.1), (8.1.2), and (8.1.3). These conditions may be relaxed. The reader who may wish to do so is urged to work on the Exercises at the end of this section.

8.2.a. L^p Boundedness of Singular Integrals

We begin by proving that operators in $CZO(\delta, A, B)$ are bounded from L^1 into weak L^1. The reader should compare the following result with Theorem 4.3.3.

Theorem 8.2.1. *Assume that $K(x,y)$ is in $SK(\delta, A)$ and let T be an element of $CZO(\delta, A, B)$ that is associated with the kernel K. Then T has an extension that maps $L^1(\mathbf{R}^n)$ into $L^{1,\infty}(\mathbf{R}^n)$ with norm*

$$\big\|T\big\|_{L^1\to L^{1,\infty}} \le C_n(A+B),$$

and therefore T maps $L^p(\mathbf{R}^n)$ into itself for $1<p<\infty$ with norm

$$\big\|T\big\|_{L^p\to L^p} \le C_n \max(p,(p-1)^{-1})(A+B),$$

where C_n is a constant that depends on the dimension and not on p.

Proof. The proof of this theorem is a reprise of the argument of the proof of Theorem (4.3.3). Fix $\alpha>0$ and let f be in $L^1(\mathbf{R}^n)$. Since $T(f)$ may not be defined when f is a general integrable function, we take f to be a Schwartz class function. Once we obtain our weak type (1,1) estimate for Schwartz functions, it is only a matter of density to extend it to all f in L^1.

We apply the Calderón-Zygmund decomposition to f at height $\gamma\alpha$, where γ is a positive constant to be chosen later. Write the $f=g+b$, where $b=\sum_j b_j$ and conditions (1)–(6) of Theorem 4.3.1 are satisfied with the constant α replaced by $\gamma\alpha$. Since we are assuming that f is Schwartz function, it follows that each bad function b_j is bounded and compactly supported. Thus $T(b_j)$ is an L^2 function, and when x is not in the support of b_j we have the integral representation

$$T(b_j)(x) = \int_{Q_j} b_j(y)K(x,y)\,dy$$

in view of Proposition 8.1.9.

As usual, we denote by $\ell(Q)$ the side length of a cube Q. Let Q_j^* be the unique cube with sides parallel to the axes having the same center as Q_j and having side length $\ell(Q_j^*) = 2\sqrt{n}\,\ell(Q_j)$. We have

$$\begin{aligned}
&|\{x \in \mathbf{R}^n : \; |T(f)(x)| > \alpha\}| \\
&\le \Big|\Big\{x \in \mathbf{R}^n : \; |T(g)(x)| > \frac{\alpha}{2}\Big\}\Big| + \Big|\Big\{x \in \mathbf{R}^n : \; |T(b)(x)| > \frac{\alpha}{2}\Big\}\Big| \\
&\le \frac{2^2}{\alpha^2}\big\|T(g)\big\|_{L^2}^2 + \Big|\bigcup_j Q_j^*\Big| + \Big|\Big\{x \notin \bigcup_j Q_j^* : \; |T(b)(x)| > \frac{\alpha}{2}\Big\}\Big| \\
&\le \frac{2^2}{\alpha^2}B^2\big\|g\big\|_{L^2}^2 + \sum_j |Q_j^*| + \frac{2}{\alpha}\int_{(\bigcup_j Q_j^*)^c} |T(b)(x)|\,dx \\
&\le \frac{2^2}{\alpha^2}2^n B^2(\gamma\alpha)\big\|f\big\|_{L^1} + (2\sqrt{n})^n\frac{\big\|f\big\|_{L^1}}{\gamma\alpha} + \frac{2}{\alpha}\sum_j \int_{(Q_j^*)^c} |T(b_j)(x)|\,dx \\
&\le \left(\frac{(2^{n+1}B\gamma)^2}{2^n\gamma} + \frac{(2\sqrt{n})^n}{\gamma}\right)\frac{\big\|f\big\|_{L^1}}{\alpha} + \frac{2}{\alpha}\sum_j \int_{(Q_j^*)^c} |T(b_j)(x)|\,dx\,.
\end{aligned}$$

It suffices to show that the last sum is bounded by some constant multiple of $\big\|f\big\|_{L^1}$. Let y_j be the center of the cube Q_j. We control the last sum by

$$\begin{aligned}
\sum_j \int_{(Q_j^*)^c} |T(b_j)(x)|\,dx &= \sum_j \int_{(Q_j^*)^c} \left|\int_{Q_j} b_j(y)K(x,y)\,dy\right|\,dx \\
= \;& \sum_j \int_{(Q_j^*)^c} \left|\int_{Q_j} b_j(y)\big(K(x,y) - K(x,y_j)\big)\,dy\right|\,dx \\
\le \;& \sum_j \int_{Q_j} |b_j(y)| \int_{(Q_j^*)^c} |K(x,y) - K(x,y_j)|\,dx\,dy \\
\le \;& \sum_j \int_{Q_j} |b_j(y)| \int_{|x-y_j|\ge 2|y-y_j|} |K(x,y) - K(x,y_j)|\,dx\,dy \\
\le \;& A_2 \sum_j \int_{Q_j} |b_j(y)|\,dy = A_2 \sum_j \big\|b_j\big\|_{L^1} \le A_2 2^{n+1}\big\|f\big\|_{L^1}\,,
\end{aligned}$$

where we used the fact that if $x \in (Q_j^*)^c$, then $|x - y_j| \ge \frac{1}{2}\ell(Q_j^*) = \sqrt{n}\,\ell(Q_j)$. But if $y \in Q_j$ we have $|y - y_j| \le \sqrt{n}\,\ell(Q_j)/2$; thus $|x - y_j| \ge 2|y - y_j|$, since the diameter of a cube is equal to $\sqrt{n}$ times its side length.

Choosing $\gamma = B^{-1}$, we have proved a weak type $(1,1)$ estimate for $T(f)$ when f is in Schwartz. We obtain that T has a bounded extension from L^1 into $L^{1,\infty}$ with bound at most $C_n(A+B)$. The L^p result for $1 < p < 2$ follows by interpolation and Exercise 1.3.2. The result for $2 < p < \infty$ follows by duality.

□

8.2.b. Boundedness of Maximal Singular Integrals

We now ask whether there is an analogous theorem regarding the maximal singular integral operator $T^{(*)}$. We note that given f in $L^p(\mathbf{R}^n)$ for some $1 \le p < \infty$, the expression $T^{(*)}(f)(x)$ is well defined for all $x \in \mathbf{R}^n$. This is a simple consequence of estimate (8.1.1) and Hölder's inequality.

Theorem 8.2.2. *Let K be in $SK(\delta, A)$ and T in $CZO(\delta, A, B)$ be associated with K. Then there exist dimensional constants C_n, C_n' such that*

$$\big\|T^{(*)}(f)\big\|_{L^{1,\infty}(\mathbf{R}^n)} \le C_n'(A+B)\big\|f\big\|_{L^1(\mathbf{R}^n)} \tag{8.2.1}$$

$$\big\|T^{(*)}(f)\big\|_{L^p(\mathbf{R}^n)} \le C_n(A+B)\max(p,(p-1)^{-1})\big\|f\big\|_{L^p(\mathbf{R}^n)} \tag{8.2.2}$$

for all $1 < p < \infty$ and all f in the corresponding spaces.

Proof. Let us fix an $0 < r < 1$. Our proof will be based on the estimate

$$|T^{(*)}(f)(x)| \le C(n,r)\Big[M(|T(f)|^r)(x)^{\frac{1}{r}} + (A+B)\,M(f)(x)\Big], \tag{8.2.3}$$

where $C(n,r)$ only depends on r and n.

To prove (8.2.3), we fix $\varepsilon > 0$ and we let $B^\varepsilon = B(x, \frac{\varepsilon}{2})$. Then $2B^\varepsilon = B(x,\varepsilon)$. We set $f_0^\varepsilon = f\chi_{2B^\varepsilon}$ and $f_\infty^\varepsilon = f\chi_{(2B^\varepsilon)^c}$. Since $|x-y| \ge \varepsilon$ whenever $y \notin 2B^\varepsilon$, we have

$$T(f_\infty^\varepsilon)(x) = \int_{|x-y|\ge\varepsilon} K(x,y)\, f_\infty^\varepsilon(y)\, dy = \int_{|x-y|\ge\varepsilon} K(x,y)\, f(y)\, dy = T^{(\varepsilon)}(f)(x)\,.$$

Since K satisfies (8.1.2), for $z \in B^\varepsilon$ we have $|z-x| \le \frac{1}{2}|x-y|$ whenever $|x-y| \ge \varepsilon$ and thus

$$\begin{aligned}
|T(f_\infty^\varepsilon)(x) - T(f_\infty^\varepsilon)(z)| &= \left|\int_{|x-y|\ge\varepsilon} \big(K(z,y) - K(x,y)\big) f(y)\, dy\right| \\
&\le |z-x|^\delta \int_{|x-y|\ge\varepsilon} \frac{A\,|f(y)|}{(|x-y|+|y-z|)^{n+\delta}}\, dy \\
&\le \varepsilon^\delta \int_{|x-y|\ge\varepsilon} \frac{A\,|f(y)|}{|x-y|^{n+\delta}}\, dy \\
&\le C_{n,\delta}\, A\, M(f)(x)\,,
\end{aligned}$$

in view of Theorem 2.1.10. We conclude that for all $z \in B^\varepsilon$ we have

$$\begin{aligned}
|T^{(\varepsilon)}(f)(x)| &= |T(f_\infty^\varepsilon)(x)| \\
&\le |T(f_\infty^\varepsilon)(x) - T(f_\infty^\varepsilon)(z)| + |T(f_\infty^\varepsilon)(z)| \\
&\le C_{n,\delta}\, A\, M(f)(x) + |T(f_0^\varepsilon)(z)| + |T(f)(z)|\,.
\end{aligned} \tag{8.2.4}$$

For $0 < r < 1$ it follows from (8.2.4) that for $z \in B^\varepsilon$ we have

$$|T^{(\varepsilon)}(f)(x)|^r \le C_{n,\delta}^r\, A^r\, M(f)(x)^r + |T(f_0^\varepsilon)(z)|^r + |T(f)(z)|^r\,. \tag{8.2.5}$$

Integrating over $z \in B^\varepsilon$, dividing by $|B^\varepsilon|$, and raising to the power $\frac{1}{r}$, we obtain

$$|T^{(\varepsilon)}(f)(x)| \le 3^{\frac{1}{r}} \left[C_{n,\delta}^r\, A\, M(f)(x) + \left(\frac{1}{|B^\varepsilon|} \int_{B^\varepsilon} |T(f_0^\varepsilon)(z)|^r dz \right)^{\frac{1}{r}} + M(|T(f)|^r)(x)^{\frac{1}{r}} \right].$$

Using Exercise 2.1.5, we estimate the middle term on the right-hand side of the preceding equation by

$$\left(\frac{1}{|B^\varepsilon|} \frac{\|T\|_{L^1 \to L^{1,\infty}}^r}{1-r} |B^\varepsilon|^{1-r} \|f_0^\varepsilon\|_{L^1}^r \right)^{\frac{1}{r}} \le C_{n,r}(A+B) M(f)(x).$$

This proves (8.2.3).

We now use estimate (8.2.3) to show that T is L^p bounded and of weak type $(1,1)$. To obtain the weak type $(1,1)$ estimate for $T^{(*)}$ we will need to use that the Hardy-Littlewood maximal operator maps $L^{p,\infty}$ into $L^{p,\infty}$ for all $1 < p < \infty$. See Exercise 2.1.13. We will also use the trivial fact that for all $0 < p, q < \infty$ we have

$$\big\||f|^q\big\|_{L^{p,\infty}} = \big\|f\big\|_{L^{pq,\infty}}^q .$$

Take any $r < 1$ in (8.2.3). Then we have

$$\begin{aligned}
\big\|M(|T(f)|^r)^{\frac{1}{r}}\big\|_{L^{1,\infty}} &= \big\|M(|T(f)|^r)\big\|_{L^{\frac{1}{r},\infty}}^{\frac{1}{r}} \\
&\le C_{n,r} \big\||T(f)|^r\big\|_{L^{\frac{1}{r},\infty}}^{\frac{1}{r}} \\
&= C_{n,r} \big\|T(f)\big\|_{L^{1,\infty}} \\
&\le \widetilde{C}_{n,r}(A+B)\big\|f\big\|_{L^1},
\end{aligned}$$

where we used the weak type $(1,1)$ bound for T in the last estimate.

To obtain the L^p boundedness of $T^{(*)}$ for $1 < p < \infty$, we use the same argument as before. We fix $r = \frac{1}{2}$. Recall that the maximal function is bounded on L^{2p} with norm at most $3^{\frac{n}{2p}} \frac{2p}{2p-1} \le 2 \cdot 3^{\frac{n}{2}}$, [see (2.1.5)]. We have

$$\begin{aligned}
\big\|M(|T(f)|^{\frac{1}{2}})^2\big\|_{L^p} &= \big\|M(|T(f)|^{\frac{1}{2}})\big\|_{L^{2p}}^2 \\
&\le \big(3^{\frac{n}{2p}} \tfrac{2p}{2p-1}\big)^2 \big\||T(f)|^{\frac{1}{2}}\big\|_{L^{2p}}^2 \\
&\le 4 \cdot 3^n \big\|T(f)\big\|_{L^p} \\
&\le C_n \max(\tfrac{1}{p-1}, p)(A+B)\big\|f\big\|_{L^p},
\end{aligned}$$

where we used the L^p boundedness of T in the last estimate.

□

We end this section with two corollaries, the first of which confirms a fact mentioned in Remark 8.1.12.

Corollary 8.2.3. *Let K be in $SK(\delta, A)$ and T in $CZO(\delta, A, B)$ be associated with K. Then there exists a dimensional constant C_n such that*

$$\sup_{\varepsilon>0} \|T^{(\varepsilon)}\|_{L^2\to L^2} \le C_n\big(A + \|T\|_{L^2\to L^2}\big)\,.$$

Corollary 8.2.4. *Let K be in $SK(\delta, A)$ and let T in $CZO(\delta, A, B)$ be associated with K. Then for $1 \le p < \infty$ and all $f \in L^p(\mathbf{R}^n)$, the functions*

$$T^{(\varepsilon)}(f) \tag{8.2.6}$$

converge almost everywhere.

Proof. Using (8.1.1), (8.1.2), and (8.1.3), we see that the alleged convergence holds (everywhere) for smooth functions with compact support. The general case follows from Theorem 8.2.2 and Theorem 2.1.14.

□

We note that the limit of $T^{(\varepsilon)}(f)$ in (8.2.6) may not be $T(f)$, even for f in $\mathcal{S}$. See Proposition 8.1.11.

Exercises

8.2.1. Let $T : \mathcal{S}(\mathbf{R}^n) \to \mathcal{S}'(\mathbf{R}^n)$ be a continuous linear operator whose Schwartz kernel coincides with a function $K(x, y)$ on $\mathbf{R}^n \times \mathbf{R}^n$ minus its diagonal. Suppose that the function $K(x, y)$ satisfies

$$\sup_{R>0} \int_{R\le|x-y|\le 2R} |K(x, y)|\, dy \le A < \infty\,.$$

(a) Show that the previous condition is equivalent to

$$\sup_{R>0} \frac{1}{R} \int_{|x-y|\le R} |x-y|\, |K(x, y)|\, dy \le A' < \infty$$

by proving that $A' \le A \le 2A'$.
(b) For $\varepsilon > 0$, let $T^{(\varepsilon)}$ be the truncated linear operators with kernels $K^{(\varepsilon)}(x, y) = K(x, y)\chi_{|x-y|>\varepsilon}$. Show that the operation $f \to T^{(\varepsilon)}(f)$ is well defined for Schwartz functions.
[*Hint:* Consider the annuli $\varepsilon 2^j \le |x| \le \varepsilon 2^{j+1}$ for $j \ge 0$.]

8.2.2. Let T be as in the previous Exercise. Prove that the limit $T^{(\varepsilon)}(f)(x)$ exists for all f in the Schwartz class and for almost all $x \in \mathbf{R}^n$ as $\varepsilon \to 0$ if and only if the limit

$$\lim_{\varepsilon\to 0} \int_{\varepsilon<|x-y|<1} K(x, y)\, dy$$

exists for almost all $x \in \mathbf{R}^n$.

8.2.3. Let T be as in Exercise 8.2.1. We say that K satisfies *Hörmander's condition* if

$$\sup_{\substack{y,y'\in\mathbf{R}^n\\ y\neq y'}} \int_{|x-y|\geq 2|y-y'|} |K(x,y)-K(x,y')|\,dx \leq A < \infty\,.$$

Assuming that K satisfies Hörmander's condition show that all of its truncations $K^{(\varepsilon)}(x,y)$ also satisfy Hörmander's condition uniformly in $\varepsilon > 0$.

8.2.4. Let T be as in in Exercise 8.2.1 and assume that T maps $L^r(\mathbf{R}^n)$ into itself for some $1 < r \leq \infty$.
(a) Assume that $K(x,y)$ satisfies Hörmander's condition [this follows from the stronger condition $|\nabla_x K(x,y)| \leq A|x-y|^{-n-1}$ for all $x \neq y$]. Then T has an extension that maps $L^1(\mathbf{R}^n)$ into $L^{1,\infty}(\mathbf{R}^n)$ with norm

$$\big\|T\big\|_{L^1\to L^{1,\infty}} \leq C_n(A+B),$$

and therefore T maps $L^p(\mathbf{R}^n)$ into itself for $1 < p < r$ with norm

$$\big\|T\big\|_{L^p\to L^p} \leq C_n(p-1)^{-1}(A+B),$$

where C_n is a dimensional constant.
(b) Assuming that $K^t(x,y) = K(y,x)$ satisfies Hörmander's condition [this is a consequence of the condition $|\nabla_y K(x,y)| \leq A|x-y|^{-n-1}$ for all $x \neq y$], prove that T maps $L^p(\mathbf{R}^n)$ into itself for $r < p < \infty$ with norm

$$\big\|T\big\|_{L^p\to L^p} \leq C_n\,p\,(A+B),$$

where C_n is independent of p.

8.2.5. Carefully state and prove an extension of the previous Exercise for Banach-valued operators. For simplicity assume that the Banach spaces are reflexive, although this is not necessary. (For the last statement consult with Exercise 4.6.7.)

8.2.6. Show that estimate (8.2.3) also holds when $r = 1$.
$\big[$*Hint:* Estimate (8.2.5) also holds when $r = 1$. For a fixed $\varepsilon > 0$, take $0 < b < |T^{(\varepsilon)}(f)(x)|$ and define $B_1^\varepsilon = B^\varepsilon \cap \{|T(f)| > \frac{b}{3}\}$, $B_2^\varepsilon = B^\varepsilon \cap \{|T(f_0^\varepsilon)| > \frac{b}{3}\}$, and $B_3^\varepsilon = B^\varepsilon$ if $C_{n,\delta}M(f)(x) > \frac{b}{3}$ and empty otherwise. Then $|B^\varepsilon| \leq |B_1^\varepsilon| + |B_2^\varepsilon| + |B_3^\varepsilon|$. Use the weak type $(1,1)$ estimate for T to show $b \leq C(n)\big(M(|T(f)|)(x) + M(f)(x)\big)$, and take the supremum over all $b < |T^{(\varepsilon)}(f)(x)|$.$\big]$

8.2.7. Prove that if $|f|\log^+|f|$ is integrable over a ball, then $T^{(*)}(f)$ is integrable over the same ball.
$\big[$*Hint:* Use the behavior of the norm of $T^{(*)}$ on L^p as $p \to 1$ and use Exercise 1.3.7.$\big]$

8.3. The $T(1)$ Theorem

We now turn to one of the main results of this chapter, the so-called $T(1)$ theorem. This theorem gives necessary and sufficient conditions for linear operators T with standard kernels to be bounded on $L^2(\mathbf{R}^n)$. In this section we will obtain several such equivalent conditions. The name of theorem $T(1)$ is due to the fact that one of the conditions that we will derive is expressed in terms of properties of the distribution $T(1)$, which was introduced in Definition 8.1.16.

8.3.a. Preliminaries and Statement of the Theorem

We begin with some preliminary facts and definitions.

Definition 8.3.1. A *normalized bump* is a smooth function φ supported in the ball $B(0,10)$ that satisfies

$$|(\partial_x^\alpha \varphi)(x)| \le 1$$

for all multiindices $|\alpha| \le 2\left[\frac{n}{2}\right]+2$ ($[x]$ denotes here the integer part of x.)

Observe that every smooth function supported inside the ball $B(0,10)$ is a constant multiple of a normalized bump. Also note that if a normalized bump is supported in a compact subset of $B(0,10)$, then small translations of it are also normalized bumps.

Given a function f on $\mathbf{R}^n$, $R>0$, and $x_0 \in \mathbf{R}^n$, we will use the notation f_R to denote the function $f_R(x) = R^{-n} f(R^{-1}x)$ and $\tau^{x_0}(f)$ to denote the function $\tau^{x_0}(f)(x) = f(x-x_0)$. Thus

$$\tau^{x_0}(f_R)(y) = f_R(y-x_0) = R^{-n} f\big(R^{-1}(y-x_0)\big).$$

Set $N = \left[\frac{n}{2}\right]+1$. Using that all derivatives up to order $2N$ of normalized bumps are bounded by 1, we easily deduce that for all $x_0 \in \mathbf{R}^n$, all $R>0$, and all normalized bumps φ we have

$$\begin{aligned}
\int_{\mathbf{R}^n} \big|\widehat{\tau^{x_0}(\varphi_R)}(\xi)\big|\, d\xi &= \int_{\mathbf{R}^n} \big|\widehat{\varphi}(\xi)\big|\, d\xi \\
&= \int_{\mathbf{R}^n} \left| \int_{\mathbf{R}^n} \varphi(y) e^{-2\pi i y\cdot\xi}\, dy \right| d\xi \qquad (8.3.1)\\
&= \int_{\mathbf{R}^n} \left| \int_{\mathbf{R}^n} (I-\Delta)^N(\varphi)(y) e^{-2\pi i y\cdot\xi}\, dy \right| \frac{d\xi}{(1+4\pi^2|\xi|^2)^N} \le C_n,
\end{aligned}$$

where C_n is independent of the bump φ and

$$(I-\Delta)(\varphi) = \varphi + \sum_{j=1}^n \frac{\partial^2 \varphi}{\partial x_j^2}.$$

This fact will be useful to us later.

Definition 8.3.2. We say that a continuous linear operator

$$T:\ \mathcal{S}(\mathbf{R}^n)\to\mathcal{S}'(\mathbf{R}^n)$$

satisfies the *weak boundedness property* (WBP) if there is a constant C such that for all f and g normalized bumps and for all $x_0\in\mathbf{R}^n$ and $R>0$ we have

$$|\langle T(\tau^{x_0}(f_R)),\tau^{x_0}(g_R)\rangle|\le CR^{-n}. \tag{8.3.2}$$

The smallest constant C in (8.3.2) will be denoted by $\|T\|_{WB}$.

Note that $\|\tau^{x_0}(f_R)\|_{L^2}=\|f\|_{L^2}R^{-n/2}$ and thus if T has a bounded extension from $L^2(\mathbf{R}^n)$ into itself, then T satisfies the weak boundedness property with bound

$$\|T\|_{WB}\le 10^n v_n\|T\|_{L^2\to L^2},$$

where v_n is the volume of the unit ball in $\mathbf{R}^n$.

We now state one of the main theorems in this chapter.

Theorem 8.3.3. *Let T be a continuous linear operator from $\mathcal{S}(\mathbf{R}^n)$ into $\mathcal{S}'(\mathbf{R}^n)$ whose Schwartz kernel coincides with a function K on $\mathbf{R}^n\times\mathbf{R}^n\setminus\{(x,x):\ x\in\mathbf{R}^n\}$ that satisfies (8.1.1), (8.1.2), and (8.1.3) for some $0<\delta,A<\infty$. Let $K^{(\varepsilon)}$ and $T^{(\varepsilon)}$ be the usual truncated kernel and operator for $\varepsilon>0$. Assume that there exists a sequence $\varepsilon_j\downarrow 0$ such that for all $f,g\in\mathcal{S}(\mathbf{R}^n)$ we have*

$$\langle T^{(\varepsilon_j)}(f),g\rangle\to\langle T(f),g\rangle\,. \tag{8.3.3}$$

Consider the following statements:

(i) *The following quantity is finite, that is,*

$$\sup_{B}\sup_{\varepsilon>0}\left[\frac{\|T^{(\varepsilon)}(\chi_B)\|_{L^2}}{|B|^{\frac12}}+\frac{\|(T^{(\varepsilon)})^t(\chi_B)\|_{L^2}}{|B|^{\frac12}}\right]=B_1<\infty\,,$$

where the first supremum is taken over all balls B in $\mathbf{R}^n$.

(ii) *The following quantity is finite, that is,*

$$\sup_{\varepsilon,N,x_0}\left[\frac{1}{N^n}\int_{B(x_0,N)}\left|\int_{|x-y|<N}K^{(\varepsilon)}(x,y)\,dy\right|^2dx\right.$$

$$\left.+\frac{1}{N^n}\int_{B(x_0,N)}\left|\int_{|x-y|<N}K^{(\varepsilon)}(y,x)\,dy\right|^2dx\right]^{\frac12}=B_2<\infty\,,$$

where the supremum is taken over all $0<\varepsilon<N<\infty$ and all $x_0\in\mathbf{R}^n$.

(iii) *The following quantity is finite:*

$$\sup_{\varphi}\sup_{x_0\in\mathbf{R}^n}\sup_{R>0}R^{\frac{n}{2}}\Big[\|T(\tau^{x_0}(\varphi_R))\|_{L^2}+\|T^t(\tau^{x_0}(\varphi_R))\|_{L^2}\Big]=B_3<\infty\,,$$

where the first supremum is taken over all normalized bumps φ.

(iv) *The operator T satisfies the weak boundedness property and the distributions $T(1)$ and $T^t(1)$ coincide with BMO functions, that is, we have*

$$\|T(1)\|_{BMO} + \|T^t(1)\|_{BMO} + \|T\|_{WB} = B_4 < \infty .$$

(v) *For every $\xi \in \mathbf{R}^n$ the distributions $T(e^{2\pi i(\cdot)\cdot\xi})$ and $T^t(e^{2\pi i(\cdot)\cdot\xi})$ coincide with BMO functions that satisfy*

$$\sup_{\xi\in\mathbf{R}^n} \|T(e^{2\pi i(\cdot)\cdot\xi})\|_{BMO} + \sup_{\xi\in\mathbf{R}^n} \|T^t(e^{2\pi i(\cdot)\cdot\xi})\|_{BMO} = B_5 < \infty .$$

(vi) *The following quantity is finite:*

$$\sup_{\varphi}\sup_{x_0\in\mathbf{R}^n}\sup_{R>0} R^n\Big[\|T(\tau^{x_0}(\varphi_R))\|_{BMO} + \|T^t(\tau^{x_0}(\varphi_R))\|_{BMO}\Big] = B_6 < \infty ,$$

where the first supremum is taken over all normalized bumps φ.

Then (i)–(vi) *are all equivalent to each other and to the L^2 boundedness of T, and we have the following equivalence of the previous quantities:*

$$c_{n,\delta}(A + B_j) \le \|T\|_{L^2\to L^2} \le C_{n,\delta}(A + B_j),$$

for all $j \in \{1, 2, 3, 4, 5, 6\}$, for some constants $c_{n,\delta}, C_{n,\delta}$ that depend only on the dimension n and on the parameter $\delta > 0$.

Remark 8.3.4. Condition (8.3.3) is saying that the operator T is the weak limit of a sequence of its truncations. We already know that if T is bounded on L^2, then it must be equal to an operator that satisfies (8.3.3) plus a bounded function times the identity operator. (See Proposition 8.1.11.) Therefore, it is not a serious restriction to assume this. See Remark 8.3.6 for a version of Theorem 8.3.3 in which this assumption is not imposed. However, the reader should always keep in mind the following pathological situation: Let K be a function on $\mathbf{R}^n \times \mathbf{R}^n \setminus \{(x, x) : x \in \mathbf{R}^n\}$ that satisfies condition (ii) of the theorem. Then nothing prevents the Schwartz kernel W of T to have the form

$$W = K(x, y) + h(x)\delta_{x=y},$$

where $h(x)$ is an unbounded function and $\delta_{x=y}$ is Lebesgue measure on the subspace $x = y$. In this case, although the $T^{(\varepsilon)}$'s are uniformly bounded on L^2, T cannot be L^2 bounded, since h is not a bounded function.

Before we begin the lengthy proof of this theorem, we state one lemma that we will need later.

Lemma 8.3.5. *Under assumptions (8.1.1), (8.1.2), and (8.1.3), there is a constant C_n so that for all normalized bumps φ we have*

$$\sup_{x_0\in\mathbf{R}^n}\sup_{R>0} \int_{|x-x_0|\ge 20R} \left| \int_{\mathbf{R}^n} K(x, y)\tau^{x_0}(\varphi_R)(y)\, dy \right|^2 dx \le \frac{C_n A^2}{R^n}. \tag{8.3.4}$$

Proof. Note that the interior integral in (8.3.4) is absolutely convergent since $\tau^{x_0}(\varphi_R)$ is supported in the ball $B(x_0, 10R)$ and x lies in complement of the double of this ball. To prove (8.3.4), simply observe that since $K(x,y)| \le A|x-y|^{-n}$ we have that

$$|T(\tau^{x_0}(\varphi_R))(x)| \le \frac{C_n A}{|x-x_0|^n}$$

whenever $|x - x_0| \ge 20R$. The estimate follows easily.

□

8.3.b. The Proof of Theorem 8.3.3

This subsection is entirely dedicated to the proof of Theorem 8.3.3.

Proof. $\boxed{\text{(iii)} \implies \text{(iv)}}$

We begin the proof by showing that condition (iii) implies condition (iv). This part of the proof is similar to the argument of the proof of Theorem 8.1.19. Recall the space $H^1_0(\mathbf{R}^n)$ of all finite sums of H^1 atoms. Fix a bump η supported in the unit ball $B(0,1)$ that is equal to 1 on the ball $B(0,\frac{1}{2})$. For each cube Q in $\mathbf{R}^n$ with center c_Q and side length R, we define the function

$$\eta^Q(x) = \eta\left(\frac{x-c_Q}{2\sqrt{n}\,R}\right). \tag{8.3.5}$$

Then η^Q is equal to 1 on the ball $B(c_Q, \sqrt{n}\,R)$, which contains the double of the cube Q, and vanishes off the ball $B(c_Q, 2\sqrt{n}\,R)$.

Define a linear functional L on $H^1_0(\mathbf{R}^n)$ by setting for all atoms a_Q

$$\begin{aligned} L(a_Q) = &\int_{\mathbf{R}^n} T(\eta^Q)(x) a_Q(x)\,dx \\ &+ \int_{\mathbf{R}^n}\int_{\mathbf{R}^n} K(x,y) a_Q(x)\,dx\,(1-\eta^Q(y))\,dy\,, \end{aligned} \tag{8.3.6}$$

where a_Q is supported in the cube Q.

We first show that for some constant C_n we have

$$|L(a_Q)| \le C_n(A + B_3) \tag{8.3.7}$$

for all a_Q L^2-atoms for H^1. Indeed, we first observe that η is a multiple of a normalized bump and that $\eta^Q = (2\sqrt{n}\,R)^n \tau^{c_Q}(\eta_{2\sqrt{n}\,R})$. Therefore, hypothesis (iii) gives

$$\big\|T(\eta^Q)\big\|_{L^2} \le B_3\,(2\sqrt{n}\,R)^{\frac{n}{2}} = C_n B_3\,\ell(Q)^{\frac{n}{2}}\,.$$

Using this fact, the Cauchy-Schwarz inequality, and the fact that L^2-atoms for H^1 have norm at most $|Q|^{-\frac{1}{2}}$, we obtain

$$\left| \int_{\mathbf{R}^n} T(\eta^Q)(x) a_Q(x)\, dx \right| \le \big\|T(\eta^Q)\big\|_{L^2} \big\|a_Q\big\|_{L^2} \le C_n B_3.$$

Using that atoms have mean value zero, we write the double integral in (8.3.6) as

$$\int_{\mathbf{R}^n} \int_{\mathbf{R}^n} \big[K(x,y) - K(c_Q,y)\big] a_Q(x)\, dx\, (1-\eta^Q(y))\, dy, \tag{8.3.8}$$

where c_Q is the center of the cube Q. Since $|x-c_Q| \le \frac{1}{2}|y-c_Q|$, when $x \in Q$ and y in the support of $1-\eta^Q$, using (8.1.2) we can control the expression in (8.3.8) by

$$\int_Q \bigg(\int_{|y-c_Q|\ge 2|x-c_Q|} \frac{A\, dy}{|y-c_Q|^{n+\delta}} \bigg) |a_Q(x)|\, |x-c_Q|^\delta\, dx\,,$$

and this by a constant multiple of A since $\big\|a_Q\big\|_{L^1} \le 1$. This establishes (8.3.7).

We have now proved that L is a bounded linear functional on $H^1_0(\mathbf{R}^n)$ whose norm $\big\|L\big\|$ is controlled by $C_n(A+B_3)$. Theorem 7.2.2 yields the existence of a BMO function b with $\big\|b\big\|_{BMO} \le C_n \big\|L\big\|$ such that

$$L(g) = \int_{\mathbf{R}^n} b(x) g(x)\, dx \tag{8.3.9}$$

for all finite sums of L^2-atoms g for H^1. In particular, (8.3.9) is valid for functions $g \in \mathcal{D}_0$, since elements of $\mathcal{D}_0$ are multiples of L^2-atoms for H^1. The definition of $T(1)$ gives that

$$\big\langle T(1), \varphi \big\rangle = \big\langle T(\eta^{Q_0}), \varphi \big\rangle + \int_{\mathbf{R}^n} \bigg[\int_{\mathbf{R}^n} K(x,y)\varphi(x)\, dx \bigg] (1-\eta^{Q_0}(y))\, dy\,,$$

where η^{Q_0} is as in (8.3.5) and Q_0 contains the support of φ. Therefore, for all $\varphi \in \mathcal{D}_0(\mathbf{R}^n)$ we have

$$\big\langle T(1), \varphi \big\rangle = L(\varphi) = \int_{\mathbf{R}^n} b(x)\varphi(x)\, dx\,.$$

This fact proves that $T(1)$ can be identified with the BMO function b. Since

$$\big\|b\big\|_{BMO} \le C_n \big\|L\big\| \le C_n(A+B_3),$$

we have identified $T(1)$ with a BMO function whose norm is controlled by a multiple of $A+B_3$. Similarly, we can identify $T^t(1)$ with a BMO function that satisfies $\big\|T^t(1)\big\|_{BMO} \le C_{n,\delta}(A+B_3)$. To finish the proof of (iv), we need to prove that T satisfies the weak boundedness property. But this is trivial since for all normalized bumps φ and ψ and all $x \in \mathbf{R}^n$ and $R>0$ we have

$$\begin{aligned} \big|\big\langle T(\tau^x(\psi_R)), \tau^x(\varphi_R)\big\rangle\big| &\le \big\|T(\tau^x(\psi_R))\big\|_{L^2} \big\|\tau^x(\varphi_R)\big\|_{L^2} \\ &\le B_3 R^{-\frac{n}{2}} \big\|\tau^x(\varphi_R)\big\|_{L^2} \le C_n B_3 R^{-n}. \end{aligned}$$

This gives $\|T\|_{WB} \le C_n B_3$, which implies the estimate $B_4 \le C_{n,\delta}(A + B_3)$ and concludes the proof that condition (iii) implies (iv).

(iv) $\Longrightarrow$ boundedness of T on L^2

We now assume condition (iv) and we prove that T has an extension that maps $L^2(\mathbf{R}^n)$ into itself. Our basic assumption is that the distributions $T(1)$ and $T^t(1)$ coincide with BMO functions. These will allow us to construct Carleson measures that will provide the key tool in the boundedness of T. This is the most important step of the proof.

We pick a smooth radial function Φ with compact support that is supported in the ball $B(0, \frac{1}{2})$ and that satisfies $\int_{\mathbf{R}^n} \Phi(x)\, dx = 1$. We define

$$\Psi(x) = \Phi(x) - 2^{-n}\Phi(\tfrac{x}{2}). \tag{8.3.10}$$

We note that Ψ is also radial and has mean value zero. For $t > 0$ we define

$$\Phi_t(x) = t^{-n}\Phi(\tfrac{x}{t}) \qquad \text{and} \qquad \Psi_t(x) = t^{-n}\Psi(\tfrac{x}{t})$$

and we observe that both Φ and Ψ are supported in $B(0, 1)$ and they are multiples of normalized bumps. We also define self-transpose operators P_t and Q_t as follows:

$$Q_t(f) = f * \Psi_t \qquad\qquad P_t(f) = f * \Phi_t. \tag{8.3.11}$$

These are the continuous analogues of the operators Δ_j and $S_j = \sum_{k\le j} \Delta_k$ that we have encountered earlier, and they are self-transpose since Φ and Ψ are radial functions.

We now fix a smooth function f with compact support and we obtain an integral representation of $T(f)$. We begin with the observations that

$$\begin{aligned} T(f) &= \lim_{s\to 0} P_s^2 T P_s^2(f) \\ 0 &= \lim_{s\to\infty} P_s^2 T P_s^2(f), \end{aligned}$$

where the limits are interpreted in the topology of $\mathcal{S}'(\mathbf{R}^n)$. See Exercises 8.3.1 and 8.3.2. Using these facts, the fundamental theorem of calculus, and the product rule, we write

$$\begin{aligned} T(f) &= \lim_{s\to 0} P_s^2 T P_s^2(f) - \lim_{s\to\infty} P_s^2 T P_s^2(f) \\ &= -\lim_{\varepsilon\to 0} \int_{\varepsilon}^{\frac{1}{\varepsilon}} s \frac{d}{ds}(P_s^2 T P_s^2)(f)\, \frac{ds}{s} \\ &= -\lim_{\varepsilon\to 0} \int_{\varepsilon}^{\frac{1}{\varepsilon}} \left[s\Big(\frac{d}{ds} P_s^2\Big) T P_s^2(f) + P_s^2 \Big(T s \frac{d}{ds} P_s^2 \Big)(f) \right] \frac{ds}{s}. \end{aligned} \tag{8.3.12}$$

We have

$$\begin{aligned}\Big(s\frac{d}{ds}P_s^2(g)\Big)\widehat{\ }(\xi) &= \widehat{g}(\xi)\, s\,\frac{d}{ds}\widehat{\Phi}(s\xi)^2\\ &= \widehat{g}(\xi)\,\widehat{\Phi}(s\xi)\,\big(2s\xi\cdot\nabla\widehat{\Phi}(s\xi)\big)\\ &= \widehat{g}(\xi)\sum_{k=1}^{n}\widehat{\Psi_k}(s\xi)\widehat{\Theta_k}(s\xi)\\ &= \sum_{k=1}^{n}\Big(\widetilde{Q}_{k,s}Q_{k,s}(g)\Big)\widehat{\ }(\xi)=\sum_{k=1}^{n}\Big(Q_{k,s}\widetilde{Q}_{k,s}(g)\Big)\widehat{\ }(\xi)\,,\end{aligned}$$

where for $1\le k\le n$, $\widehat{\Psi_k}(\xi)=2\xi_k\widehat{\Phi}(\xi)$, $\widehat{\Theta_k}(\xi)=\partial_k\widehat{\Phi}(\xi)$ and $Q_{k,s}$, $\widetilde{Q}_{k,s}$ are operators defined by

$$Q_{k,s}(g)=g*(\Psi_k)_s \qquad\qquad \widetilde{Q}_{k,s}(g)=g*(\Theta_k)_s\,,$$

where $(\Theta_k)_s(x)=s^{-n}\Theta_k(s^{-1}x)$ and $(\Psi_k)_s$ are defined similarly. Observe that Ψ_k and Θ_k are smooth odd bumps supported in $B(0,\frac12)$ and have integral zero. Therefore, the operators $Q_{k,s}$ and $\widetilde{Q}_{k,s}$ have similar behavior with the operator Q_s defined earlier. However, since Ψ_k and Θ_k are odd, they are anti-self-transpose, the latter meaning that $Q_{k,s}^t=-Q_{k,s}$ and $\widetilde{Q}_{k,s}^t=-\widetilde{Q}_{k,s}$. We can now write the expression in (8.3.12) as

$$-\lim_{\varepsilon\to0}\sum_{k=1}^{n}\bigg[\int_{\varepsilon}^{\frac1\varepsilon}\widetilde{Q}_{k,s}Q_{k,s}TP_sP_s(f)\frac{ds}{s}+\int_{\varepsilon}^{\frac1\varepsilon}P_sP_sTQ_{k,s}\widetilde{Q}_{k,s}(f)\frac{ds}{s}\bigg]\,, \tag{8.3.13}$$

where the limit converges in $\mathcal{S}'(\mathbf{R}^n)$. We set

$$T_{k,s}=Q_{k,s}TP_s\,,$$

and we observe that the operator $P_sTQ_{k,s}$ is equal to $-((T^t)_{k,s})^t$, which is the negative transpose of $T_{k,s}$. Recall the translation operator $\tau^x(f)(z)=f(z-x)$. By a simple calculation we see that the operator $T_{k,s}$ has kernel

$$K_{k,s}(x,y)=-\big\langle\tau^x((\Psi_k)_s),T(\tau^y(\Phi_s))\big\rangle=-\big\langle T^t(\tau^x((\Psi_k)_s)),\tau^y(\Phi_s)\big\rangle \tag{8.3.14}$$

and the operator $-(T^t)_{k,s}^t$ has kernel

$$\big\langle\tau^y((\Psi_k)_s),T^t(\tau^x(\Phi_s))\big\rangle=\big\langle T(\tau^y((\Psi_k)_s)),\tau^x(\Phi_s)\big\rangle\,.$$

We will need the following facts regarding the kernels of these operators for all $1\le k\le n$:

$$\big|\big\langle T(\tau^x((\Psi_k)_s)),\tau^y(\Phi_s)\big\rangle\big|\le C_{n,\delta}\big(\big\|T\big\|_{WB}+A\big)\,p_s(x-y)\,, \tag{8.3.15}$$

$$\big|\big\langle T^t(\tau^x((\Psi_k)_s)),\tau^y(\Phi_s)\big\rangle\big|\le C_{n,\delta}\big(\big\|T\big\|_{WB}+A\big)\,p_s(x-y)\,, \tag{8.3.16}$$

where

$$p_t(u)=\frac{1}{t^n}\frac{1}{(1+|\frac{u}{t}|)^{n+\delta}}$$

is the L^1 dilation of the function $p(u) = (1+|u|)^{-n-\delta}$.

To prove (8.3.16), we consider the following two cases: If $|x-y| \le 5\,s$, then the weak boundedness property gives

$$\big|\big\langle T(\tau^y(\Phi_s)),\tau^x((\Psi_k)_s)\big\rangle\big| = \big|\big\langle T(\tau^x((\tau^{\frac{y-x}{s}}(\Phi))_s)),\tau^x((\Psi_k)_s)\big\rangle\big| \le \frac{C_n\|T\|_{WB}}{s^n}\,,$$

since both Ψ_k and $\tau^{\frac{y-x}{s}}(\Phi)$ are multiples of normalized bumps. Notice here that both of these functions are supported in $B(0,10)$ since $\frac{1}{s}|x-y| \le 5$. This estimate proves (8.3.16) when $|x-y| \le 5\,s$.

We now turn to the case $|x-y| \ge 5\,s$. Then the functions $\tau^y(\Phi_s)$ and $\tau^x((\Psi_k)_s)$ have disjoint supports and so we have the integral representation

$$\big\langle T^t(\tau^x((\Psi_k)_s)),\tau^y(\Phi_s)\big\rangle = \int_{\mathbf{R}^n}\int_{\mathbf{R}^n} \Phi_s(v-y)K(u,v)(\Psi_k)_s(u-x)\,du\,dv\,.$$

Using that Ψ_k has mean value zero, we can write the previous expression as

$$\int_{\mathbf{R}^n}\int_{\mathbf{R}^n} \Phi_s(v-y)\big(K(u,v)-K(x,v)\big)(\Psi_k)_s(u-x)\,du\,dv\,.$$

We observe that $|u-x| \le s$ and $|v-y| \le s$ in the preceding double integral. Since $|x-y| \ge 5\,s$, this makes $|u-v| \ge |x-y| - 2\,s \ge 3s$, which implies that $|u-x| \le \frac{1}{2}|u-v|$. Using (8.1.2), we obtain

$$|K(u,v)-K(x,v)| \le \frac{A|x-u|^\delta}{(|u-v|+|x-v|)^{n+\delta}} \le C_{n,\delta}A\frac{s^\delta}{|x-y|^{n+\delta}}\,,$$

where we used the fact that $|u-v| \approx |x-y|$. Inserting this estimate in the double integral, we obtain (8.3.16). Estimate (8.3.15) is proved similarly.

Let us see what we have achieved so far. At this point we will drop the indices k as we can concentrate on one term of the sum in (8.3.13). We have written $T(f)$ as a finite sum of operators of the form

$$\int_0^\infty \widetilde{Q}_s T_s P_s(f)\,\frac{ds}{s} \tag{8.3.17}$$

and of the form

$$\int_0^\infty P_s T_s \widetilde{Q}_s(f)\,\frac{ds}{s}\,, \tag{8.3.18}$$

where the preceding integrals converge in $\mathcal{S}'(\mathbf{R}^n)$ and the T_s's have kernels $K_s(x,y)$, which are pointwise dominated by a constant multiple of

$$(A+B_4)p_s(x-y)\,.$$

It suffices to obtain L^2 bounds for an operator of the form (8.3.17) with constant at most a multiple of $A+B_4$. Then by duality the same estimate will also hold for the operators of the form (8.3.18). We make one more observation. Using (8.3.14) (recall that we have dropped the indices k), we obtain

$$T_s(1)(x) = \int_{\mathbf{R}^n} K_s(x,y)\,dy = -\langle T^t(\tau^x(\Psi_s)), 1\rangle = -(\Psi_s * T(1))(x)\,,$$

where all integrals converge absolutely.

We can therefore concentrate on the L^2 boundedness of the operator in (8.3.17). The intuition here is as follows: T_s is an averaging operator at scale s and $P_s(f)$ is essentially constant on that scale. Therefore, the expression $T_sP_s(f)$ must look like $T_s(1)P_s(f)$. To be precise, we will introduce this term and try to estimate the error that occurs. We have

$$\begin{aligned}\int_0^\infty \widetilde{Q}_sT_sP_s(f)\frac{ds}{s} &= \int_0^\infty \widetilde{Q}_s\big[T_s(1)P_s(f)\big]\frac{ds}{s} + \int_0^\infty \widetilde{Q}_s\big[T_sP_s(f)-T_s(1)P_s(f)\big]\frac{ds}{s}\\ &= \mathrm{Main}(f)+\mathrm{Error}(f)\,,\end{aligned}$$

where we set

$$\begin{aligned}\mathrm{Main}(f) &= -\int_0^\infty \widetilde{Q}_s\big[(\Psi_s * T(1))\,P_s(f)\big]\frac{ds}{s}\,,\\ \mathrm{Error}(f) &= \int_0^\infty \widetilde{Q}_s\big[T_sP_s(f)-T_s(1)\,P_s(f)\big]\frac{ds}{s}\,.\end{aligned}$$

We estimate both terms using duality. Pairing with a Schwartz function g, we write

$$\Big\langle \int_0^\infty \widetilde{Q}_s\big[(\Psi_s * T(1))\,P_s(f)\big]\frac{ds}{s}\,,\,g\Big\rangle = \int_0^\infty \Big\langle \widetilde{Q}_s\big[(\Psi_s * T(1))\,P_s(f)\big]\,,\,g\Big\rangle\frac{ds}{s}\,,$$

since the integral converges in $\mathcal{S}'(\mathbf{R}^n)$, and this is equal to

$$-\int_0^\infty \Big\langle (\Psi_s * T(1))P_s(f), \widetilde{Q}_s(g)\Big\rangle\frac{ds}{s}$$

which is controlled by

$$\begin{aligned}&\Big(\int_0^\infty \big\|P_s(f)\,(\Psi_s * T(1))\big\|_{L^2}^2\frac{ds}{s}\Big)^{\frac12}\Big(\int_0^\infty \big\|\widetilde{Q}_s(g)\big\|_{L^2}^2\frac{ds}{s}\Big)^{\frac12}\\ \text{(8.3.19)}\qquad &= \bigg\|\Big(\int_0^\infty \big|P_s(f)\,(\Psi_s * T(1))\big|^2\frac{ds}{s}\Big)^{\frac12}\bigg\|_{L^2}\bigg\|\Big(\int_0^\infty \big|\widetilde{Q}_s(g)\big|^2\frac{ds}{s}\Big)^{\frac12}\bigg\|_{L^2}.\end{aligned}$$

Since $T(1)$ is a BMO function, $|(\Psi_s * T(1))(x)|^2dx\frac{ds}{s}$ is a Carleson measure on $\mathbf{R}^{n+1}_+$. Using Theorem 7.3.8 and the Littlewood-Paley theorem (Exercise 5.1.4), we obtain that (8.3.19) is controlled by

$$C_n\|T(1)\|_{BMO}\|f\|_{L^2}\|g\|_{L^2} \le C_nB_4\|f\|_{L^2}\|g\|_{L^2}\,.$$

This gives the L^2 boundedness of the term Main(f). For the term Error(f) we use duality again. We have

$$
\begin{aligned}
&\left|\left\langle \int_0^\infty \widetilde{Q}_s\big[T_sP_s(f) - T_s(1)\,P_s(f)\big]\,\frac{ds}{s}\,,\,g\right\rangle\right| \\
&= \left|\int_0^\infty\int_{\mathbf{R}^n} \widetilde{Q}_s(g)(x)\big[T_sP_s(f) - T_s(1)\,P_s(f)\big](x)\,dx\,\frac{ds}{s}\right| \\
&\le \left(\int_0^\infty\int_{\mathbf{R}^n} |\widetilde{Q}_s(g)(x)|^2\,dx\frac{ds}{s}\right)^{\frac12}\left(\int_0^\infty\int_{\mathbf{R}^n} |(T_sP_s(f) - T_s(1)P_s(f))(x)|^2\,dx\frac{ds}{s}\right)^{\frac12} \\
&\le C_n\big\|g\big\|_{L^2}\left(\int_0^\infty\int_{\mathbf{R}^n}\left|\int_{\mathbf{R}^n} K_s(x,y)[P_s(f)(y) - P_s(f)(x)]\,dy\right|^2 dx\frac{ds}{s}\right)^{\frac12} \\
&\le C_n(A + B_4)\big\|g\big\|_{L^2}\left(\int_0^\infty\int_{\mathbf{R}^n}\int_{\mathbf{R}^n} p_s(x-y)\big|P_s(f)(y) - P_s(f)(x)\big|^2\,dy\,dx\,\frac{ds}{s}\right)^{\frac12}
\end{aligned}
$$

where in the last estimate we used the fact that the measure $p_t(x-y)\,dy$ is a multiple of a probability measure. It suffices to estimate the last displayed square root. Changing variables $u = x - y$ and applying Plancherel's theorem, we can write this square root as

$$
\begin{aligned}
&\left(\int_0^\infty\int_{\mathbf{R}^n}\int_{\mathbf{R}^n} p_s(u)\big|P_s(f)(y) - P_s(f)(y+u)\big|^2\,du\,dy\,\frac{ds}{s}\right)^{\frac12} \\
&= \left(\int_0^\infty\int_{\mathbf{R}^n}\int_{\mathbf{R}^n} p_s(u)\big|\widehat{\Phi}(s\xi) - \widehat{\Phi}(s\xi)e^{2\pi i u\cdot\xi}\big|^2\big|\widehat{f}(\xi)\big|^2\,du\,d\xi\,\frac{ds}{s}\right)^{\frac12} \\
&\le \left(\int_0^\infty\int_{\mathbf{R}^n}\int_{\mathbf{R}^n} p_s(u)\big|\widehat{\Phi}(s\xi)\big|^2 2\pi^{\frac{\delta}{2}}|u|^{\frac{\delta}{2}}|\xi|^{\frac{\delta}{2}}\big|\widehat{f}(\xi)\big|^2\,du\,d\xi\,\frac{ds}{s}\right)^{\frac12} \\
&= 2^{\frac12}\pi^{\frac{\delta}{4}}\left(\int_{\mathbf{R}^n}\int_0^\infty\left(\int_{\mathbf{R}^n} p_s(u)\big|\tfrac{u}{s}\big|^{\frac{\delta}{2}}\,du\right)\big|\widehat{\Phi}(s\xi)\big|^2|s\xi|^{\frac{\delta}{2}}\frac{ds}{s}\,\big|\widehat{f}(\xi)\big|^2\,d\xi\right)^{\frac12}.
\end{aligned}
$$

We first bound the expression $\int_{\mathbf{R}^n} p_s(u)\big|\frac{u}{s}\big|^{\frac{\delta}{2}}\,du$ by a constant, and then we use the estimate

$$
\int_0^\infty \big|\widehat{\Phi}(s\xi)\big|^2|s\xi|^{\frac{\delta}{2}}\frac{ds}{s} = \int_0^\infty \big|\widehat{\Phi}(se_1)\big|^2 s^{\frac{\delta}{2}}\frac{ds}{s} \le C_{n,\delta} < \infty
$$

and Plancherel's theorem to obtain the required conclusion. [Here $e_1 = (1,0,\dots,0)$.] We have now finished the proof that (iv) implies the boundedness of T on L^2 and, moreover we have obtained the estimate $\big\|T\big\|_{L^2\to L^2} \le C_{n,\delta}(A + B_4)$.

Boundedness of T on L^2 $\Longrightarrow$ (iii)

The boundedness of T on L^2 easily implies that

$$
B_3 \le C_{n,\delta}\big(A + \big\|T\big\|_{L^2\to L^2}\big),
$$

and hence (iii).

We have now established the sequence of equivalent statements

$$\text{(iii)} \iff \text{(iv)} \iff (L^2 \text{ boundedness of } T).$$

We proceed by showing that these three statements are also equivalent to statements (v) and (vi).

Boundedness of T on $L^2 \implies$ (v)

If T has an extension that maps L^2 into itself, then by Theorem 8.1.19 we have

$$B_5 \le C_{n,\delta}\big(A + \|T\|_{L^2\to L^2}\big) < \infty.$$

Thus the boundedness of T on L^2 implies condition (v).

(v) $\implies$ (vi)

At a formal level the proof of this fact is clear since we can write a normalized bump as the inverse Fourier transform of its Fourier transform and interchange the integrations with the action of T to obtain

$$T(\tau^{x_0}(\varphi_R)) = \int_{\mathbf{R}^n} \widehat{\tau^{x_0}(\varphi_R)}(\xi) T(e^{2\pi i \xi\cdot(\,\cdot\,)})\, d\xi. \tag{8.3.20}$$

The conclusion will follow by taking BMO norms. To make identity (8.3.20) precise we provide the following argument.

Let us fix a normalized bump φ and a smooth and compactly supported function g with mean value zero. We pick a smooth function η with compact support that is equal to 1 on the double of a ball containing the support of g and vanishes off the triple of that ball. Define $\eta_k(\xi) = \eta(\xi/k)$ and note that η_k tend pointwise to 1 as $k \to \infty$. Observe that $\eta_k \tau^{x_0}(\varphi_R)$ converges to $\tau^{x_0}(\varphi_R)$ in $\mathcal{S}(\mathbf{R}^n)$ as $k \to \infty$ and by the continuity of T we obtain

$$\lim_{k\to\infty} \big\langle T(\eta_k \tau^{x_0}(\varphi_R)), g\big\rangle = \big\langle T(\tau^{x_0}(\varphi_R)), g\big\rangle.$$

The continuity and linearity of T also allow us to write

$$\big\langle T(\tau^{x_0}(\varphi_R)), g\big\rangle = \lim_{k\to\infty} \int_{\mathbf{R}^n} \widehat{\tau^{x_0}(\varphi_R)}(\xi) \, \big\langle T\big(\eta_k e^{2\pi i \xi\cdot(\,\cdot\,)}\big), g\big\rangle \, d\xi. \tag{8.3.21}$$

Let W be the Schwartz kernel of T. By (8.1.5) we have

$$\big\langle T(\eta_k e^{2\pi i \xi\cdot(\,\cdot\,)}), g\big\rangle = \big\langle W, g \otimes \eta_k e^{2\pi i \xi\cdot(\,\cdot\,)}\big\rangle. \tag{8.3.22}$$

Using (8.1.6), we obtain that the expression in (8.3.22) is controlled by a finite sum of L^∞ norms of derivatives of the function

$$g(x)\, \eta_k(y) e^{2\pi i \xi\cdot y}$$

on some compact set (that depends on g). Then for some $M > 0$ and some constant $C(g)$ depending on g, we have that this sum of L^∞ norms of derivatives is controlled by

$$C(g)\,(1+|\xi|)^M$$

uniformly in $k \ge 1$. Since $\widehat{\tau^{x_0}(\varphi_R)}$ is integrable, the Lebesgue dominated convergence theorem allows us to pass the limit inside the integrals in (8.3.21) to obtain

$$\big\langle T(\tau^{x_0}(\varphi_R)), g\big\rangle = \int_{\mathbf{R}^n} \widehat{\tau^{x_0}(\varphi_R)}(\xi)\, \big\langle T(e^{2\pi i\xi\cdot(\,\cdot\,)}), g\big\rangle\, d\xi.$$

We now use assumption (v). The distributions $T\big(e^{2\pi i\xi\cdot(\,\cdot\,)}\big)$ coincide with BMO functions whose norm is at most B_5. It follows that

$$\begin{aligned}
&\big|\big\langle T(\tau^{x_0}(\varphi_R)), g\big\rangle\big| \\
&\le \big\|\widehat{\tau^{x_0}(\varphi_R)}\big\|_{L^1} \sup_{\xi\in\mathbf{R}^n} \big\|T\big(e^{2\pi i\xi\cdot(\,\cdot\,)}\big)\big\|_{BMO}\big\|g\big\|_{H^1} \qquad (8.3.23)\\
&\le C_n B_5 R^{-n}\big\|g\big\|_{H^1},
\end{aligned}$$

where the constant C_n is independent of the normalized bump φ in view of (8.3.1). It follows from (8.3.23) that

$$g \to \big\langle T(\tau^{x_0}(\varphi_R)), g\big\rangle$$

is a bounded linear functional on BMO with norm at most a multiple of B_5R^{-n}. It follows from Theorem 7.2.2 that $T(\tau^{x_0}(\varphi_R))$ coincides with a BMO function that satisfies

$$R^n\big\|T(\tau^{x_0}(\varphi_R))\big\|_{BMO} \le C_n B_5.$$

The same argument is valid for T^t, and this shows that

$$B_6 \le C_{n,\delta}(A+B_5)$$

and concludes the proof that (v) implies (vi).

(vi) $\Longrightarrow$ (iii)

We fix $x_0\in\mathbf{R}^n$ and $R>0$. Pick z_0 in $\mathbf{R}^n$ such that $|x_0-z_0| = 40\,R$. Then if $|y-x_0|\le 10\,R$ and $|x-z_0|\le 20\,R$ we have

$$\begin{aligned}
10\,R &\le |z_0-x_0| - |x-z_0| - |y-x_0| \\
&\le |x-y| \\
&\le |x-z_0| + |z_0-x_0| + |x_0-y| \le 70\,R\,.
\end{aligned}$$

From this it follows that when $|x-z_0|\le 20\,R$ we have

$$\left|\int_{|y-x_0|\le 10R} K(x,y)\tau^{x_0}(\varphi_R)(y)\,dy\right| \le \int_{10R\le|x-y|\le 70R} |K(x,y)|\,\frac{dy}{R^n} \le \frac{C_{n,\delta}A}{R^n}$$

and thus

$$\Big| \operatorname*{Avg}_{B(z_0,20R)} T(\tau^{x_0}(\varphi_R))\Big| \le \frac{C_{n,\delta}A}{R^n}, \tag{8.3.24}$$

where $\operatorname*{Avg}_B g$ denotes the average of g over B. Because of assumption (vi), the BMO norm of the function $T(\tau^{x_0}(\varphi_R))$ is bounded by a multiple of B_6R^{-n}, a fact that will be used in the following sequence of implications. We have

$$\begin{aligned}
&\big\|T(\tau^{x_0}(\varphi_R))\big\|_{L^2(B(x_0,20R))}\\
&\le \big\|T(\tau^{x_0}(\varphi_R)) - \operatorname*{Avg}_{B(x_0,20R)} T(\tau^{x_0}(\varphi_R))\big\|_{L^2(B(x_0,20R))}\\
&\qquad + v_n^{\frac{1}{2}}(20R)^{\frac{n}{2}}\Big| \operatorname*{Avg}_{B(x_0,20R)} T(\tau^{x_0}(\varphi_R)) - \operatorname*{Avg}_{B(z_0,20R)} T(\tau^{x_0}(\varphi_R))\Big|\\
&\qquad + v_n^{\frac{1}{2}}(20R)^{\frac{n}{2}}\Big| \operatorname*{Avg}_{B(z_0,20R)} T(\tau^{x_0}(\varphi_R))\Big|\\
&\le C_{n,\delta}\big(R^{\frac{n}{2}}\big\|T(\tau^{x_0}(\varphi_R))\big\|_{BMO} + R^{\frac{n}{2}}\big\|T(\tau^{x_0}(\varphi_R))\big\|_{BMO} + R^{-\frac{n}{2}}A\big)\\
&\le C_{n,\delta}R^{-\frac{n}{2}}\big(B_6 + A\big),
\end{aligned}$$

where we used (8.3.24) and Exercise 7.1.6. Now we have that

$$\big\|T(\tau^{x_0}(\varphi_R))\big\|_{L^2(B(x_0,20R)^c)} \le C_{n,\delta}AR^{-\frac{n}{2}}$$

in view of Lemma 8.3.5. Since the same computations apply to T^t, it follows that

$$R^{\frac{n}{2}}\big(\big\|T(\tau^{x_0}(\varphi_R))\big\|_{L^2} + \big\|T^t(\tau^{x_0}(\varphi_R))\big\|_{L^2}\big) \le C_{n,\delta}(A + B_6), \tag{8.3.25}$$

which proves that $B_3 \le C_{n,\delta}(A + B_6)$ and hence (iii). This concludes the proof of the fact that (vi) implies (iii)

We have now completed the proof of the sequence of equivalent statements:

$$\big(L^2 \text{ boundedness of } T\big) \iff \text{(iii)} \iff \text{(iv)} \iff \text{(v)} \iff \text{(vi)}.$$

(i) $\iff$ (ii)

More precisely, we will show that the quantities $A+B_1$ and $A+B_2$ are controlled by constant multiples of each other. Let us set

$$I_{\varepsilon,N}(x) = \int_{\varepsilon<|x-y|<N} K(x,y)\,dy \qquad \text{and} \qquad I^t_{\varepsilon,N}(x) = \int_{\varepsilon<|x-y|<N} K^t(x,y)\,dy\,.$$

Observe that

$$I_{\varepsilon,N}(x) - T^{(\varepsilon)}(\chi_{B(x_0,3N)})(x)$$

$$(8.3.26) \qquad = \int\limits_{\varepsilon<|x-y|<N} K(x,y)\,dy - \int\limits_{\substack{\varepsilon<|x-y|\\ |x_0-y|<3N}} K(x,y)\,dy$$

$$= -\int_{S_{\varepsilon,N}(x,x_0)} K(x,y)\,dy,$$

where $S_{\varepsilon,N}(x,x_0)$ is the set of all $y \in \mathbf{R}^n$ that satisfy $\varepsilon < |x-y|$ and $|x_0-y| < 3N$ but do not satisfy $\varepsilon < |x-y| < N$. But observe that when $|x_0 - x| < N$, then

$$(8.3.27) \qquad S_{\varepsilon,N}(x,x_0) \subseteq \{y \in \mathbf{R}^n : \ N \le |x-y| < 4N\}.$$

Using (8.3.26), (8.3.27), and (8.1.2), we obtain

$$(8.3.28) \qquad \big|I_{\varepsilon,N}(x) - T^{(\varepsilon)}(\chi_{B(x_0,3N)})(x)\big| \le \int\limits_{N\le|x-y|\le 4N} |K(x,y)|\,dy \le 2A$$

whenever $|x_0 - x| < N$. It follows that

$$\big\|I_{\varepsilon,N} - T^{(\varepsilon)}(\chi_{B(x_0,3N)})\big\|_{L^2(B(x_0,N))} \le C_n A N^{\frac{n}{2}}$$

and, similarly, it follows that

$$\big\|I^t_{\varepsilon,N} - (T^{(\varepsilon)})^t(\chi_{B(x_0,3N)})\big\|_{L^2(B(x_0,N))} \le C_n A N^{\frac{n}{2}}.$$

These two estimates easily imply the equivalence of conditions (i) and (ii).

We now consider the following condition analogous to (iii):

$$\text{(iii)}' \qquad B_3' < \infty\,,$$

where

$$\sup_{\varphi}\ \sup_{x_0\in\mathbf{R}^n}\ \sup_{\substack{\varepsilon>0\\ R>0}} R^{\frac{n}{2}}\Big[\big\|T^{(\varepsilon)}(\tau^{x_0}(\varphi_R))\big\|_{L^2} + \big\|(T^{(\varepsilon)})^t(\tau^{x_0}(\varphi_R))\big\|_{L^2}\Big] = B_3'\,.$$

and the first supremum is taken over all normalized bumps φ. We continue by showing that this condition is a consequence of (ii).

(ii) $\Longrightarrow$ (iii)′

More precisely, we will prove that $B_3' \le C_{n,\delta}(A + B_2)$. To prove (iii)′, fix a normalized bump φ, a point $x_0 \in \mathbf{R}^n$ and $R > 0$. Also fix $x \in \mathbf{R}^n$ with $|x - x_0| \le 20R$. Then we have

$$T^{(\varepsilon)}(\tau^{x_0}(\varphi_R))(x) = \int_{\varepsilon<|x-y|\le 30R} K^{(\varepsilon)}(x,y)\tau^{x_0}(\varphi_R)(y)\,dx = U_1(x) + U_2(x),$$

where

$$U_1(x) = \int_{\varepsilon<|x-y|\le 30R} K(x,y)\big(\tau^{x_0}(\varphi_R)(y)-\tau^{x_0}(\varphi_R)(x)\big)\,dy,$$

$$U_2(x) = \tau^{x_0}(\varphi_R)(x)\int_{\varepsilon<|x-y|\le 30R} K(x,y)\,dy.$$

But we have that $|\tau^{x_0}(\varphi_R)(y)-\tau^{x_0}(\varphi_R)(x)|\le C_nR^{-1-n}|x-y|$; thus we obtain

$$|U_1(x)|\le C_nAR^{-n}$$

on $B(x_0,20R)$; hence $\big\|U_1\big\|_{L^2(B(x_0,20R))}\le C_nAR^{-\frac{n}{2}}$. Condition (ii) gives that

$$\big\|U_2\big\|_{L^2(B(x_0,20R))}\le R^{-n}\big\|I_{\varepsilon,30R}\big\|_{L^2(B(x_0,30R))}\le B_2(30\,R)^{\frac{n}{2}}R^{-n}.$$

Combining these two, we obtain

$$\big\|T^{(\varepsilon)}(\tau^{x_0}(\varphi_R))\big\|_{L^2(B(x_0,20R))}\le C_n(A+B_2)R^{-\frac{n}{2}} \tag{8.3.29}$$

and likewise for $(T^{(\varepsilon)})^t$. It follows from Lemma 8.3.5 that

$$\big\|T^{(\varepsilon)}(\tau^{x_0}(\varphi_R))\big\|_{L^2(B(x_0,20R)^c)}\le C_{n,\delta}AR^{-\frac{n}{2}},$$

which combined with (8.3.29) gives condition (iii)$'$ with constant

$$B_3'\le C_{n,\delta}(A+B_2).$$

This concludes the proof that condition (ii) implies (iii)$'$.

(iii)$'$ $\Longrightarrow$ The $T^{(\varepsilon)}$'s map $L^2\to L^2$ uniformly in $\varepsilon>0$

For $\varepsilon>0$ we introduce the smooth truncations $T_\zeta^{(\varepsilon)}$ of T by setting

$$T_\zeta^{(\varepsilon)}(f)(x)=\int_{\mathbf{R}^n}K(x,y)\zeta(\tfrac{x-y}{\varepsilon})\,f(y)\,dy,$$

where $\zeta(x)$ is a smooth function that is equal to 1 for $|x|\ge 1$ and vanishes for $|x|\le\frac12$. We observe that

$$\big|T_\zeta^{(\varepsilon)}(f)-T^{(\varepsilon)}(f)\big|\le C_nA\,M(f); \tag{8.3.30}$$

thus the uniform boundedness of $T^{(\varepsilon)}$ on L^2 is equivalent to those of $T_\zeta^{(\varepsilon)}$. In view of Exercise 8.1.3, the kernels of the operators $T_\zeta^{(\varepsilon)}$ lie in $SK(\delta,c\,A)$ uniformly in $\varepsilon>0$ (for some constant c). Moreover, because of (8.3.30), we see that the operators $T_\zeta^{(\varepsilon)}$ satisfy (iii)$'$ with constant C_nA+B_3'. The point to be noted here is that condition (iii) for T (with constant B_3) is identical to condition (iii)$'$ for the operators $T_\zeta^{(\varepsilon)}$ uniformly in $\varepsilon>0$ (with constant C_nA+B_3').

A careful examination of the proof of the implications

$$\text{(iii)}\Longrightarrow\text{(iv)}\Longrightarrow\big(L^2\text{ boundedness of }T\big)$$

reveals that all the estimates obtained only depended on the constants B_3, B_4, and A, but not on the specific operator T. Therefore, these estimates are valid for the operators $T_\zeta^{(\varepsilon)}$ that satisfy condition (iii)$'$. This gives the uniform boundedness of the $T_\zeta^{(\varepsilon)}$ on $L^2(\mathbf{R}^n)$ with bounds at most a constant multiple of $A + B_3'$. The same conclusion also holds for the operators $T^{(\varepsilon)}$.

The $T^{(\varepsilon)}$'s map $L^2 \to L^2$ uniformly in $\varepsilon > 0 \implies$ (i)

This is implication holds trivially. We have therefore established the equivalence

$$\text{(i)} \iff \text{(ii)} \iff \text{(iii)}' \iff \big[L^2 \text{ boundedness of } T^{(\varepsilon)} \text{ uniformly in } \varepsilon > 0\big].$$

(iii) $\iff$ (iii)$'$

Finally, we use (8.3.3) to link the two previous conditions. We first observe that in view of (8.3.3), we have that (iii)$'$ implies (iii). Indeed, using duality and (8.3.3), we obtain

$$\begin{aligned} \big\|T(\tau^{x_0}(\varphi_R))\big\|_{L^2} &= \sup_{\|h\|_{L^2}\le 1} \left| \int_{\mathbf{R}^n} T(\tau^{x_0}(\varphi_R))(x)\, h(x)\, dx \right| \\ &\le \sup_{\|h\|_{L^2}\le 1} \limsup_{j\to\infty} \left| \int_{\mathbf{R}^n} T^{(\varepsilon_j)}(\tau^{x_0}(\varphi_R))(x)\, h(x)\, dx \right| \\ &\le B_3' R^{-\frac{n}{2}}, \end{aligned}$$

which gives $B_3 \le B_3'$. Thus under assumption (8.3.3), (ii) implies (iii) and, as we have shown, (iii) implies the boundedness of T on L^2. But in view of Corollary 8.2.3, the boundedness of T on L^2 implies the boundedness of $T^{(\varepsilon)}$ on L^2 uniformly in $\varepsilon > 0$, which implies (iii)$'$.

This completes the proof of the equivalence of the six statements (i)–(vi) in such a way so that

$$\big\|T\big\|_{L^2\to L^2} \approx (A + B_j)$$

for all $j \in \{1, 2, 3, 4, 5, 6\}$. The proof of the theorem is now complete.

□

Remark 8.3.6. Suppose that T is as in the hypothesis of Theorem 8.3.3 but does not necessarily satisfy (8.3.3). Observe that the proof of Theorem 8.3.3 actually shows that (i) and (ii) are equivalent to each other and to the statement that the $T^{(\varepsilon)}$'s have bounded extensions on $L^2(\mathbf{R}^n)$ that satisfy

$$\sup_{\varepsilon>0} \big\|T^{(\varepsilon)}\big\|_{L^2\to L^2} < \infty .$$

Also note that without hypothesis (8.3.3), conditions (iii), (iv), (v), and (vi) are equivalent to each other and to the statement that T has an extension that maps $L^2(\mathbf{R}^n)$ to $L^2(\mathbf{R}^n)$.

8.3.c. An Application

We end this section with one application of the $T(1)$ theorem. We begin with the following observation.

Corollary 8.3.7. *Let K be a standard kernel that is antisymmetric, i.e., it satisfies $K(x, y) = -K(y, x)$ for all $x \neq y$. Then a linear continuous operator T associated with K is L^2 bounded if and only if $T(1)$ is in BMO.*

Proof. In view of Exercise 8.3.3, T automatically satisfies the weak boundedness property. Moreover, $T^t = -T$. Therefore, the three conditions of Theorem 8.3.3 (iv) reduce to the single condition $T(1) \in BMO$.

□

Example 8.3.8. Let us recall the kernels K_m in Example 8.1.7. These give rise to operators

$$\mathcal{C}_m(f)(x) = \lim_{\varepsilon \to 0} \int_{|x-y|>\varepsilon} \left(\frac{A(x) - A(y)}{x - y} \right)^m \frac{1}{x - y} f(y)\, dy\,, \tag{8.3.31}$$

which were called the Calderón commutators. The kernels K_m in Example 8.1.7 arise by expanding the kernel in Example 8.1.6 in geometric series

$$\frac{1}{x - y + i(A(x) - A(y))} = \frac{1}{x - y} \sum_{m=0}^{\infty} \left(i \frac{A(x) - A(y)}{x - y} \right)^m \tag{8.3.32}$$

when $L = \sup_{x \neq y} \frac{|A(x)-A(y)|}{|x-y|} < 1$. We will use the $T(1)$ theorem to show that the operators $\mathcal{C}_m$ are L^2 bounded.

We will show that there exists a constant $R > 0$ such that for all $m \geq 0$ we have

$$\big\|\mathcal{C}_m(1)\big\|_{BMO} \leq R^m L^m\,. \tag{8.3.33}$$

We prove (8.3.33) by induction. First we note that (8.3.33) is trivially true when $m = 0$ since $\mathcal{C}_0 = H$ is the Hilbert transform and $H(1) = 0$. (See Exercise 8.3.5.) The key observation is that

$$\mathcal{C}_{m+1}(1) = \mathcal{C}_m(A') \tag{8.3.34}$$

for which we refer the reader to Exercise 8.3.4. A' here denotes the derivative of A, which exists almost everywhere since Lipschitz functions are differentiable almost everywhere. Moreover, we have $\big\|A'\big\|_{L^\infty} \leq L$. Using (8.3.34), we obtain

$$\big\|\mathcal{C}_{m+1}(1)\big\|_{BMO} = \big\|\mathcal{C}_m(A')\big\|_{BMO} \leq C_1\big(c\, m\, L^m + \big\|\mathcal{C}_m\big\|_{L^2 \to L^2}\big)\big\|A'\big\|_{L^\infty}\,.$$

In the last estimate we used the fact that K_m lies in $SK(1, c\, m\, L^m)$, a fact discussed in Exercise 8.1.4. Also, in Theorem 8.3.3 we showed that

$$\big\|\mathcal{C}_m\big\|_{L^2 \to L^2} \leq C_2\Big[\big\|\mathcal{C}_m(1)\big\|_{BMO} + \big\|\mathcal{C}_m^t(1)\big\|_{BMO} + \big\|\mathcal{C}_m\big\|_{WB}\Big]\,,$$

and Exercise 8.3.3 gives that $\|\mathcal{C}_m\|_{WB} \le C_3 c\, m\, L^m$. Combining all these facts with the induction hypothesis (8.3.33) and with the fact that $\|A'\|_{L^\infty} \le L$, we obtain

$$\|\mathcal{C}_{m+1}(1)\|_{BMO} \le C_1\big(c\, m\, L^m + C_2(2R^m L^m + C_3 c\, m\, L^m)\big)L \le R^{m+1}L^{m+1},$$

provided that

$$R > \max\left(4C_1C_2, [2c\, m(C_1 + C_1C_2C_3)]^{\frac{1}{m+1}}\right) \tag{8.3.35}$$

for all $m \ge 0$. This completes the proof of (8.3.33) by induction. We conclude that the operator $\mathcal{C}_m$ is L^2 bounded with norm at most $(RL)^m$, where R is a fixed constant.

Exercises

8.3.1. Let T be a continuous linear operator from $\mathcal{S}(\mathbf{R}^n)$ into $\mathcal{S}'(\mathbf{R}^n)$ and let f be in $\mathcal{S}(\mathbf{R}^n)$. Let P_t be as in (8.3.11).
(a) Show that $P_t(f)$ converges to f in $\mathcal{S}(\mathbf{R}^n)$ as $t \to 0$.
(b) Conclude that $TP_t(f) \to T(f)$ in $\mathcal{S}'(\mathbf{R}^n)$ as $t \to 0$.
(c) Conclude that $P_tTP_t(f) \to T(f)$ in $\mathcal{S}'(\mathbf{R}^n)$ as $t \to 0$.
(d) Observe that (a)–(c) are also valid if P_t is replaced by P_t^2.
[*Hint:* Use the Fourier transform.]

8.3.2. Let T, f, and P_t be as in the previous Exercise.
(a) Show that $P_t(f)$ converges to 0 in $\mathcal{S}(\mathbf{R}^n)$ as $t \to \infty$.
(b) Conclude that $TP_t(f) \to 0$ in $\mathcal{S}'(\mathbf{R}^n)$ as $t \to \infty$.
(c) Conclude that $P_tTP_t(f) \to 0$ in $\mathcal{S}'(\mathbf{R}^n)$ as $t \to \infty$.
(d) Observe that (a)–(c) are also valid if P_t is replaced by P_t^2.
[*Hint:* Use the Fourier transform.]

8.3.3. Prove that every linear operator T from $\mathcal{S}(\mathbf{R}^n)$ into $\mathcal{S}'(\mathbf{R}^n)$ associated with an antisymmetric kernel in $SK(\delta, A)$ satisfies the weak boundedness property. Precisely, for some dimensional constant C_n we have

$$\|T\|_{WB} \le C_n A\,.$$

[*Hint:* Write $2\langle T(\tau^{x_0}(f_R)), \tau^{x_0}(g_R)\rangle$ as

$$\iint_{\mathbf{R}^n} K(x,y)\big(\tau^{x_0}(f_R)(y)\tau^{x_0}(g_R)(x) - \tau^{x_0}(f_R)(x)\tau^{x_0}(g_R)(y)\big)\, dy\, dx$$

and use the mean value theorem.]

8.3.4. Prove identity (8.3.34). Note that at the formal level it is obvious by an integration by parts, but to show it properly, you should interpret things in the sense of distributions.

8.3.5. Suppose that a standard kernel $K(x,y)$ has the form $k(x-y)$ for some function k on $\mathbf{R}^n \setminus \{0\}$. Suppose that k extends to a tempered distribution on $\mathbf{R}^n$ whose Fourier transform is a bounded function. Let T be a continuous linear operator from $\mathcal{S}(\mathbf{R}^n)$ into $\mathcal{S}'(\mathbf{R}^n)$ associated with K.
(a) Identify the functions $T(e^{2\pi i\xi\cdot()})$ and $T^t(e^{2\pi i\xi\cdot()})$ and restrict to $\xi=0$ to obtain $T(1)$ and $T^t(1)$.
(b) Use Theorem 8.3.3 to obtain the L^2 boundedness of T.
(c) What are $H(1)$ and $H^t(1)$ when H is the Hilbert transform?

8.3.6. (*A. Calderón*) Let A be a Lipschitz function on $\mathbf{R}$. Use expansion (8.3.32) and estimate (8.3.33) to show that the operator

$$\mathcal{C}_A(f)(x) = \lim_{\varepsilon\to 0}\int_{|x-y|>\varepsilon} \frac{f(y)\,dy}{x-y+i(A(x)-A(y))}$$

is bounded on $L^2(\mathbf{R})$ when $\big\|A'\big\|_{L^\infty} < R^{-1}$, where R satisfies (8.3.35).

8.3.7. Prove that condition (ii) of Theorem 8.3.3 is equivalent to the statement that

$$\sup_Q \sup_{\varepsilon>0}\left(\frac{\big\|T^{(\varepsilon)}(\chi_Q)\big\|_{L^2}}{|Q|^{\frac{1}{2}}} + \frac{\big\|(T^{(\varepsilon)})^t(\chi_Q)\big\|_{L^2}}{|Q|^{\frac{1}{2}}}\right) = B_1' < \infty\,,$$

where the first supremum is taken over all cubes Q in $\mathbf{R}^n$.

8.4. Paraproducts

In this section we study a useful class of operators called paraproducts. This name suggests a certain relation with products that will soon become precise. Paraproducts give us interesting examples of nonconvolution operators with standard kernels whose L^2 boundedness was the topic of discussion in the last section. They have found use in many situations, including the first proof of the main implication in Theorem 8.3.3. This is discussed in this section.

8.4.a. Introduction of Paraproducts

Throughout this section we fix a Schwartz radial function Ψ whose Fourier transform is supported in the annulus $\frac{1}{2} \le |\xi| \le 2$ and that satisfies

$$\sum_{j\in\mathbf{Z}} \widehat{\Psi}(2^{-j}\xi) = 1, \qquad \text{when} \quad \xi \in \mathbf{R}^n \setminus \{0\}. \tag{8.4.1}$$

Associated with this Ψ we define the Littlewood-Paley operator $\Delta_j(f) = f * \Psi_{2^{-j}}$, where $\Psi_t(x) = t^{-n}\Psi(t^{-1}x)$. Using (8.4.1), we easily obtain

$$\sum_{j\in\mathbf{Z}} \Delta_j = I, \tag{8.4.2}$$

where (8.4.2) is interpreted as an identity on Schwartz functions with mean value zero. See Exercise 8.4.1. Note that by construction, the function Ψ is radial and thus even. This makes the operator Δ_j equal to its transpose.

We now observe that in view of the properties of Ψ, the function

$$\xi \to \sum_{j\le 0} \widehat{\Psi}(2^{-j}\xi) \tag{8.4.3}$$

is supported in $|\xi| \le 2$, and is equal to 1 when $0 < |\xi| \le \frac{1}{2}$. But $\widehat{\Psi}(0) = 0$, which implies the function in (8.4.3) also vanishes at the origin. We can easily fix this discontinuity by introducing the Schwartz function whose Fourier transform is equal to

$$\widehat{\Phi}(\xi) = \begin{cases} \sum_{j\le 0} \widehat{\Psi}(2^{-j}\xi) & \text{when } \xi \neq 0 \\ 1 & \text{when } \xi = 0. \end{cases}$$

Definition 8.4.1. We define the *partial sum operator* S_j as

$$S_j = \sum_{k\le j} \Delta_k. \tag{8.4.4}$$

In view of the preceding discussion, S_j is given by convolution with $\Phi_{2^{-j}}$, that is,

$$S_j(f)(x) = (f * \Phi_{2^{-j}})(x), \tag{8.4.5}$$

and the expression in (8.4.5) is well defined for all f in $\bigcup_{1\le p\le\infty} L^p(\mathbf{R}^n)$. Since Φ is a radial function by construction, the operator S_j is self-transpose.

Similarly, $\Delta_j(g)$ is also well defined for all g in $\bigcup_{1\le p\le\infty} L^p(\mathbf{R}^n)$. Moreover, since Δ_j is given by convolution with function with mean value zero, it also follows that $\Delta_j(b)$ is well defined when $b \in BMO(\mathbf{R}^n)$. See Exercise 8.4.2 for details.

Definition 8.4.2. Given a function g on $\mathbf{R}^n$, we define the *paraproduct operator* P_g as follows:

$$P_g(f) = \sum_{j\in\mathbf{Z}} \Delta_j(g)\, S_{j-3}(f) = \sum_{j\in\mathbf{Z}} \sum_{k\le j-3} \Delta_j(g)\, \Delta_k(f), \tag{8.4.6}$$

for f in $L^1_{\text{loc}}(\mathbf{R}^n)$. At this point it is not clear for which functions g and in what sense the series in (8.4.6) converges even when f is a Schwartz function. The reader should note that the series in (8.4.6) converges absolutely almost everywhere when g is a Schwartz function with mean value zero; in this case by Exercise 8.4.1 the series $\sum_j \Delta_j(g)$ converges absolutely (everywhere) and $S_j(f)$ is uniformly bounded by the Hardy-Littlewood maximal function $M(f)$, which is finite almost everywhere.

One of the main goals of this section is to show that the series in (8.4.6) converges in L^2 when f is in $L^2(\mathbf{R}^n)$ and g is a BMO function.

The name *paraproduct* is derived from the fact that $P_g(f)$ is essentially 'half' the product of fg. Namely, in view of the identity in (8.4.2) the product fg can be written as

$$fg = \sum_j \sum_k \Delta_j(f)\,\Delta_k(g)\,.$$

Restricting the summation of the indices to $k < j$ defines an operator that corresponds to "half" the product of fg. It is only for technical reasons (which will become apparent later) that we take $k \le j-3$ in (8.4.6).

The main feature of the paraproduct operator P_g is that it is essentially a sum of orthogonal L^2 functions. Indeed, the Fourier transform of the function $\widehat{\Delta_j(g)}$ is supported in the set

$$\{\xi \in \mathbf{R}^n :\ 2^{j-1} \le |\xi| \le 2^{j+1}\}$$

while the Fourier transform of the function $\widehat{S_j(f)}$ is supported in the set

$$\bigcup_{k\le j-3} \{\xi \in \mathbf{R}^n :\ 2^{k-1} \le |\xi| \le 2^{k+1}\}\,.$$

This implies that the Fourier transform of the function $\Delta_j(g)\,S_{j-3}(f)$ is supported in the algebraic sum of the two sets

$$\{\xi \in \mathbf{R}^n :\ 2^{j-1} \le |\xi| \le 2^{j+1}\} + \{\xi \in \mathbf{R}^n :\ |\xi| \le 2^{j-2}\}\,.$$

But this sum is contained in the set

$$\{\xi \in \mathbf{R}^n :\ 2^{j-2} \le |\xi| \le 2^{j+2}\}\,, \tag{8.4.7}$$

and the family of sets in (8.4.7) are "almost disjoint" as j varies. This means that every point in $\mathbf{R}^n$ belongs to at most 4 annuli of the form (8.4.7). Therefore, the paraproduct $P_g(f)$ can be written as a sum of functions h_j so that the families $\{h_j :\ j \in 4\mathbf{Z} + r\}$ are mutually orthogonal in L^2, for all $r \in \{0,1,2,3\}$. This orthogonal decomposition of the paraproduct has as an immediate consequence its L^2 boundedness, when g is an element of BMO.

8.4.b. L^2 Boundedness of Paraproducts

The following theorem is the main result of this subsection.

Theorem 8.4.3. *For fixed $b \in BMO(\mathbf{R}^n)$ and $f \in L^2(\mathbf{R}^n)$ the series*

$$\sum_{|j|\le M} \Delta_j(b)\,S_{j-3}(f)$$

converges in L^2 as $M \to \infty$ to a function that we will denote by $P_b(f)$. The operator P_b thus defined is bounded on $L^2(\mathbf{R}^n)$, and there is a dimensional constant C_n such *that for all $b \in BMO(\mathbf{R}^n)$ we have*

$$\|P_b\|_{L^2\to L^2} \le C_n \|b\|_{BMO}.$$

Proof. The proof of this result follows by putting together some of the powerful sequence of ideas developed in Chapter 7. First we define a measure on $\mathbf{R}^{n+1}_+$ by setting

$$d\mu(x,t) = \sum_{j\in\mathbf{Z}} |\Delta_j(b)(x)|^2 \, dx \, \delta_{2^{-(j-3)}}(t)\,.$$

By Theorem 7.3.8 we have that μ is a Carleson measure on $\mathbf{R}^{n+1}_+$ whose norm is controlled by a constant multiple of $\|b\|^2_{BMO}$. Now fix $f \in L^2(\mathbf{R}^n)$ and recall that $\Phi(x) = \sum_{r\le 0} \Psi_{2^{-r}}(x)$. We define a function $F(x,t)$ on $\mathbf{R}^{n+1}_+$ by setting

$$F(x,t) = (\Phi_t * f)(x)\,.$$

Observe that $F(x, 2^{-k}) = S_k(f)(x)$ for all $k \in \mathbf{Z}$. We estimate the L^2 norm of a finite sum of terms of the form $\Delta_j(b)\, S_{j-3}(f)$. For $M, N \in \mathbf{Z}^+$ with $M \ge N$ we have

$$\begin{aligned} &\int_{\mathbf{R}^n} \Big| \sum_{N\le|j|\le M} \Delta_j(b)(x)\, S_{j-3}(f)(x)\Big|^2 dx \\ &= \int_{\mathbf{R}^n} \Big| \sum_{N\le|j|\le M} (\Delta_j(b)\, S_{j-3}(f))\widehat{\ }(\xi)\Big|^2 d\xi\,. \end{aligned} \tag{8.4.8}$$

It is a simple fact that every $\xi \in \mathbf{R}^n$ belongs to at most 4 annuli of the form (8.4.7). It follows that at most four terms in the last sum in (8.4.8) are nonzero. Thus

$$\begin{aligned} &\int_{\mathbf{R}^n} \Big| \sum_{N\le|j|\le M} (\Delta_j(b) S_{j-3}(f))\widehat{\ }(\xi)\Big|^2 d\xi \\ &\le 4 \sum_{N\le|j|\le M} \int_{\mathbf{R}^n} \big|(\Delta_j(b)\, S_{j-3}(f))\widehat{\ }(\xi)\big|^2 d\xi \\ &\le 4 \sum_{j\in\mathbf{Z}} \int_{\mathbf{R}^n} \big|\Delta_j(b)(x) S_{j-3}(f)(x)\big|^2 dx \\ &= 4 \int_{\mathbf{R}^n} |F(x,t)|^2 \, d\mu(x,t) \le C_n \|b\|^2_{BMO} \int_{\mathbf{R}^n} F^*(x)^2 \, dx, \end{aligned} \tag{8.4.9}$$

where we used Corollary 7.3.6 in the last inequality.

Next we note that the nontangential maximal function F^* of F is controlled by the Hardy-Littlewood maximal function of f. Indeed, since Φ is a Schwartz function we have

$$F^*(x) \le C_n \sup_{t>0} \sup_{|y-x|<t} \int_{\mathbf{R}^n} \frac{1}{t^n} \frac{|f(z)|}{(1+\frac{|z-y|}{t})^{n+1}} \, dz\,. \tag{8.4.10}$$

Now break the previous integral into the parts where $|z-y| \geq 3t$ and $|z-y| \leq 3t$. In the first case we have $|z-y| \geq |z-x| - t \geq \frac{1}{2}|z-x|$, and the last inequality is valid since $|z-x| \geq |z-y| - t \geq 2t$. Using this estimate together with Theorem 2.1.10 we obtain that this part of the integral is controlled by a constant multiple of $M(f)(x)$. The part of the integral in (8.4.10) where $|z-y| \leq 3t$ is controlled by the integral over the larger set $|z-x| \leq 4t$, and since the denominator in (8.4.10) is always bounded by 1, we also obtain that this part of the integral is controlled by a constant multiple of $M(f)(x)$. We conclude that

$$\int_{\mathbf{R}^n} F^*(x)^2\, dx \leq C_n \int_{\mathbf{R}^n} M(f)(x)^2\, dx \leq C_n \int_{\mathbf{R}^n} |f(x)|^2\, dx. \tag{8.4.11}$$

Combining (8.4.9) and (8.4.11), we obtain the estimate

$$4\sum_{j\in\mathbf{Z}} \int_{\mathbf{R}^n} |(\Delta_j(b)\, S_{j-3}(f))\widehat{\ }(\xi)|^2\, d\xi \leq C_n \|b\|_{BMO}^2 \|f\|_{L^2}^2 < \infty.$$

This implies that given $\varepsilon > 0$, we can find an $N_0 > 0$ such that for we have

$$M \geq N \geq N_0 \implies \sum_{N\leq|j|\leq M} \int_{\mathbf{R}^n} \left|(\Delta_j(b)\, S_{j-3}(f))\widehat{\ }(\xi)\right|^2 d\xi < \varepsilon.$$

But recall from (8.4.8) and (8.4.9) that

$$\int_{\mathbf{R}^n} \left| \sum_{N\leq|j|\leq M} \Delta_j(b)(x)\, S_{j-3}(f)(x)\right|^2 dx \leq 4 \sum_{N\leq|j|\leq M} \int_{\mathbf{R}^n} \left|(\Delta_j(b) S_{j-3}(f))\widehat{\ }(\xi)\right|^2 d\xi.$$

We conclude that the sequence

$$\Big\{ \sum_{|j|\leq M} \Delta_j(b) S_{j-3}(f) \Big\}_M$$

is Cauchy in $L^2(\mathbf{R}^n)$ and therefore it converges in L^2 to a function $P_b(f)$. The boundedness of P_b on L^2 follows from the sequence of inequalities already proved. □

8.4.c. Fundamental Properties of Paraproducts

Having established the L^2 boundedness of paraproducts, we turn to some properties that they possess. First we study their kernels. Paraproducts are not operators of convolution type but more general integral operators of the form discussed in Section 8.1. We will show that the kernel of P_b is a tempered distribution L_b that coincides with a standard kernel on $\mathbf{R}^n \times \mathbf{R}^n \setminus \{(x,x) :\ x \in \mathbf{R}^n\}$.

First we study the kernel of the operator $f \to \Delta_j(b) S_{j-3}(f)$ for any $j \in \mathbf{Z}$. We have that

$$\Delta_j(b)(x) S_{j-3}(f)(x) = \int_{\mathbf{R}^n} L_j(x,y) f(y)\, dy,$$

where L_j is the integrable function

$$L_j(x,y) = (b * \Psi_{2^{-j}})(x) 2^{(j-3)n} \Phi(2^{j-3}(x-y)).$$

Next we can easily verify the following size and regularity estimates for L_j:

$$|L_j(x,y)| \le C_n \big\|b\big\|_{BMO} \frac{2^{nj}}{(1+2^j|x-y|)^{n+1}} \tag{8.4.12}$$

$$|\partial_x^\alpha \partial_y^\beta L_j(x,y)| \le C_{n,\alpha,\beta,N} \big\|b\big\|_{BMO} \frac{2^{j(n+|\alpha|+|\beta|)}}{(1+2^j|x-y|)^{n+1+N}} \tag{8.4.13}$$

for all α and β multiindices and all $N \ge |\alpha| + |\beta|$.

It follows from (8.4.12) that when $x \neq y$ the series

$$\sum_{j\in\mathbf{Z}} L_j(x,y) \tag{8.4.14}$$

converges absolutely and is controlled in absolute value by

$$C_n \big\|b\big\|_{BMO} \sum_{j\in\mathbf{Z}} \frac{2^{nj}}{(1+2^j|x-y|)^{n+1}} \le \frac{C_n' \big\|b\big\|_{BMO}}{|x-y|^n}.$$

Similarly, by taking $N \ge |\alpha| + |\beta|$, it can be shown that the series

$$\sum_{j\in\mathbf{Z}} \partial_x^\alpha \partial_y^\beta L_j(x,y) \tag{8.4.15}$$

converges absolutely when $x \neq y$ and is controlled in absolute value by

$$C_{n,\alpha,\beta,N} \big\|b\big\|_{BMO} \sum_{j\in\mathbf{Z}} \frac{2^{j(n+|\alpha|+|\beta)}}{(1+2^j|x-y|)^{n+1+N}} \le \frac{C_{n,\alpha,\beta}' \big\|b\big\|_{BMO}}{|x-y|^{n+|\alpha|+|\beta|}}$$

for all multiindices α and β.

The Schwartz kernel of P_b is a distribution W_b on $\mathbf{R}^{2n}$. It follows from the preceding discussion that the distribution W_b coincides with the function

$$L_b(x,y) = \sum_{j\in\mathbf{Z}} L_j(x,y)$$

on $\mathbf{R}^n \times \mathbf{R}^n \setminus \{(x,x) : \ x \in \mathbf{R}^n\}$, and also that the function L_b satisfies the estimates

$$|\partial_x^\alpha \partial_y^\beta L_b(x,y)| \le \frac{C_{n,\alpha,\beta}' \big\|b\big\|_{BMO}}{|x-y|^{n+|\alpha|+|\beta|}} \tag{8.4.16}$$

away from the diagonal $x = y$.

We note that the transpose of the operator P_b is formally given by the identity

$$P_b^t(f) = \sum_{j\in\mathbf{Z}} S_{j-3}(f \Delta_j(b)).$$

As remarked in the previous section, the kernel of the operator P_b^t is a distribution W_b^t that coincides with the function

$$L_b^t(x,y) = L_b(y,x)$$

away from the diagonal of $\mathbf{R}^{2n}$. It is trivial to observe that L_b^t satisfies the same size and regularity estimates (8.4.16) as L_b. Moreover, it follows from Theorem 8.4.3 that the operator P_b^t is bounded on $L^2(\mathbf{R}^n)$ with norm at most a multiple of the BMO norm of b.

We now turn to two important properties of paraproducts. In view of Definition 8.1.16, we have a meaning for $P_b(1)$ and $P_b^t(1)$, where P_b is the paraproduct operator. The first property we prove is that $P_b(1) = b$. Observe that this statement is trivially valid at a formal level, since $S_j(1) = 1$ for all j and $\sum_j \Delta_j(b) = b$. The second property is that $P_b^t(1) = 0$. This is also trivially checked at a formal level since $S_{j-3}(\Delta_j(b)) = 0$ for all j, as a Fourier transform calculation shows. We make both of these statements precise in the following proposition.

Proposition 8.4.4. *Given $b \in BMO(\mathbf{R}^n)$, let P_b be the paraproduct operator defined as in (8.4.6). Then the distributions $P_b(1)$ and $P_b^t(1)$ coincide with elements of BMO. Precisely, we have*

$$P_b(1) = b \qquad \text{and} \qquad P_b^t(1) = 0. \tag{8.4.17}$$

Proof. Let φ be an element of $\mathcal{D}_0(\mathbf{R}^n)$. Find a uniformly bounded sequence of smooth functions with compact support $\{\eta_N\}_{N=1}^\infty$ that converges to the function 1 as $N \to \infty$. Without loss of generality assume that all the functions η_N are equal to 1 on the ball $B(y_0, 3R)$, where $B(y_0,R)$ is a ball that contains the support of φ. As we observed in Remark 8.1.17, the definition of $P_b(1)$ is independent of the choice of sequence η_N, so we have the following identity for all $N \ge 1$:

$$\begin{aligned} \langle P_b(1), \varphi\rangle = &\int_{\mathbf{R}^n} \sum_{j\in\mathbf{Z}} \Delta_j(b)(x)\, S_{j-3}(\eta_N)(x)\, \varphi(x)\, dx \\ &+ \int_{\mathbf{R}^n} \bigg[\int_{\mathbf{R}^n} L_b(x,y)\,\varphi(x)\, dx\bigg] (1-\eta_N(y))\, dy. \end{aligned} \tag{8.4.18}$$

Since φ has mean value zero, we can subtract the constant $L_b(y_0,y)$ from $L_b(x,y)$ in the integral inside the square brackets in (8.4.18). Then we can estimate the absolute value of the double integral in (8.4.18) by

$$\int_{|y-y_0|\ge 3R} \int_{|x-y_0|\le R} A \frac{|y_0-x|}{|y_0-y|^{n+1}} |1-\eta_N(y)|\, |\varphi(x)|\, dx\, dy\,,$$

which tends to zero as $N \to \infty$ by the Lebesgue dominated convergence theorem.

It suffices to prove that the first integral in (8.4.18) as $N \to \infty$ tends to $\int_{\mathbf{R}^n} b(x)\varphi(x)\,dx$. Let us make some preliminary observations. Since the Fourier transform of the product $\Delta_j(b) S_{j-3}(\eta_N)$ is supported in the annulus

$$\{\xi \in \mathbf{R}^n : \; 2^{j-2} \le |\xi| \le 2^{j+2}\}, \tag{8.4.19}$$

we can introduce a smooth and compactly supported function $\widehat{Z}(\xi)$ such that for all $j \in \mathbf{Z}$ the function $\widehat{Z}(2^{-j}\xi)$ is equal to 1 on the annulus (8.4.19) and vanishes outside the annulus $\{\xi \in \mathbf{R}^n : \ 2^{j-3} \le |\xi| \le 2^{j+3}\}$. Let us denote by Q_j the operator given by multiplication on the Fourier transform by the function $\widehat{Z}(2^{-j}\xi)$.

Note that $S_j(1)$ is well defined and equal to 1 for all j. This is because Φ has integral equal to 1. Also, the duality identity

$$\int f\, S_j(g)\, dx = \int g\, S_j(f)\, dx \tag{8.4.20}$$

holds for all $f \in L^1$ and $g \in L^\infty$. For φ in $\mathcal{D}_0(\mathbf{R}^n)$ we have

$$
\begin{aligned}
&\int_{\mathbf{R}^n} \sum_{j\in\mathbf{Z}} \Delta_j(b)\, S_{j-3}(\eta_N)\, \varphi\, dx \\
= &\sum_{j\in\mathbf{Z}} \int_{\mathbf{R}^n} \Delta_j(b)\, S_{j-3}(\eta_N)\, \varphi\, dx && \text{(series converges in } L^2 \text{ and } \varphi \in L^2) \\
= &\sum_{j\in\mathbf{Z}} \int_{\mathbf{R}^n} \Delta_j(b)\, S_{j-3}(\eta_N)\, Q_j(\varphi)\, dx && (\widehat{Q_j(\varphi)} = \widehat{\varphi} \text{ on the} \\
& && \text{support of } ((\Delta_j(b)\, S_{j-3}(\eta_N))\widehat{\ }\,)) \\
= &\sum_{j\in\mathbf{Z}} \int_{\mathbf{R}^n} \eta_N\ S_{j-3}\big(\Delta_j(b) Q_j(\varphi)\big)\, dx && \text{(duality)} \\
= &\int_{\mathbf{R}^n} \eta_N \sum_{j\in\mathbf{Z}} S_{j-3}\big(\Delta_j(b) Q_j(\varphi)\big)\, dx && \text{(series converges in } L^1 \text{ and } \eta_N \in L^\infty.)
\end{aligned}
$$

We now explain why the last series of the foregoing converges in L^1. Since φ is in $\mathcal{D}_0(\mathbf{R}^n)$, Exercise 8.4.1 gives that the series $\sum_{j\in\mathbf{Z}} Q_j(\varphi)$ converges in L^1. Since S_j preserves L^1 and

$$\sup_j \big\|\Delta_j(b)\big\|_{L^\infty} \le C_n \big\|b\big\|_{BMO}$$

by Exercise 8.4.2, it follows that the series $\sum_{j\in\mathbf{Z}} S_{j-3}\big(\Delta_j(b) Q_j(\varphi)\big)$ also converges in L^1.

We now use the Lebesgue dominated convergence theorem to obtain that the expression

$$\int_{\mathbf{R}^n} \eta_N \sum_{j\in\mathbf{Z}} S_{j-3}\big(\Delta_j(b) Q_j(\varphi)\big)\, dx$$

converges as $N \to \infty$ to

$$
\begin{aligned}
&\int_{\mathbf{R}^n} \sum_{j\in\mathbf{Z}} S_{j-3}\big(\Delta_j(b) Q_j(\varphi)\big)\,dx \\
= &\sum_{j\in\mathbf{Z}} \int_{\mathbf{R}^n} S_{j-3}\big(\Delta_j(b) Q_j(\varphi)\big)\,dx && \text{(series converges in } L^1) \\
= &\sum_{j\in\mathbf{Z}} \int_{\mathbf{R}^n} S_{j-3}(1)\,\Delta_j(b)\,Q_j(\varphi)\,dx && \text{(in view of (8.4.20))} \\
= &\sum_{j\in\mathbf{Z}} \int_{\mathbf{R}^n} \Delta_j(b)\,Q_j(\varphi)\,dx && \text{(since } S_{j-3}(1)=1) \\
= &\sum_{j\in\mathbf{Z}} \int_{\mathbf{R}^n} \Delta_j(b)\,\varphi\,dx && (\widehat{Q_j(\varphi)} = \widehat{\varphi} \text{ on support } \widehat{\Delta_j(b)}) \\
= &\sum_{j\in\mathbf{Z}} \langle b, \Delta_j(\varphi)\rangle && \text{(duality)} \\
= &\langle b, \sum_{j\in\mathbf{Z}} \Delta_j(\varphi)\rangle && \text{(series converges in } H^1,\ b\in BMO) \\
= &\langle b, \varphi\rangle && \text{[Exercise 8.4.1(a)].}
\end{aligned}
$$

For the fact that the series $\sum_j \Delta_j(\varphi)$ converges in H^1, we refer to Exercise 8.4.1. We now obtain that the first integral in (8.4.18) tends to $\langle b, \varphi\rangle$ as $N \to \infty$. We have therefore proved that

$$\langle P_b(1), \varphi\rangle = \langle b, \varphi\rangle$$

for all φ in $\mathcal{D}_0(\mathbf{R}^n)$. In other words, we have now identified $P_b(1)$ as an element of $\mathcal{D}_0'$ with the BMO function b.

For the transpose operator P_b^t we observe that we have the identity

$$
\begin{aligned}
\langle P_b^t(1), \varphi\rangle = &\int_{\mathbf{R}^n} \sum_{j\in\mathbf{Z}} S_{j-3}^t\big(\Delta_j(b)\,\eta_N\big)(x)\,\varphi(x)\,dx \\
&+ \int_{\mathbf{R}^n}\int_{\mathbf{R}^n} L_b^t(x,y)(1-\eta_N(y))\,\varphi(x)\,dy\,dx\,.
\end{aligned}
\tag{8.4.21}
$$

As before, we can use the Lebesgue dominated convergence theorem to show that the double integral in (8.4.21) tends to zero. As far as the first integral in (8.4.21), we have the identity

$$\int_{\mathbf{R}^n} P_b^t(\eta_N)\,\varphi\,dx = \int_{\mathbf{R}^n} \eta_N\, P_b(\varphi)\,dx\,.$$

Since φ is a multiple of an L^2-atom for H^1, Theorem 8.1.20 gives that $P_b(\varphi)$ is an L^1 function. The Lebesgue dominated convergence theorem now implies that

$$\int_{\mathbf{R}^n} \eta_N \, P_b(\varphi)\, dx \to \int_{\mathbf{R}^n} P_b(\varphi)\, dx = \int_{\mathbf{R}^n} \sum_{j\in\mathbf{Z}} \Delta_j(b)\, S_{j-3}(\varphi)\, dx$$

as $N \to \infty$. The required conclusion will follow if we can prove that the function $P_b(\varphi)$ has integral zero. Since $\Delta_j(b)$ and $S_{j-3}(\varphi)$ have disjoint Fourier transforms, it follows that

$$\int_{\mathbf{R}^n} \Delta_j(b)\, S_{j-3}(\varphi)\, dx = 0$$

for all j in $\mathbf{Z}$. But the series

$$\sum_{j\in\mathbf{Z}} \Delta_j(b)\, S_{j-3}(\varphi) \tag{8.4.22}$$

defining $P_b(\varphi)$ converges in L^2 and not necessarily in L^1 and for this reason we need to justify the interchange of the following integrals

$$\int_{\mathbf{R}^n} \sum_{j\in\mathbf{Z}} \Delta_j(b)\, S_{j-3}(\varphi)\, dx = \sum_{j\in\mathbf{Z}} \int_{\mathbf{R}^n} \Delta_j(b)\, S_{j-3}(\varphi)\, dx\,. \tag{8.4.23}$$

To complete the proof, it suffices to show that when φ is in $\mathcal{D}_0(\mathbf{R}^n)$, the series in (8.4.22) converges in L^1. To prove this, pick a ball B that contains the support of φ. The series in (8.4.22) converges in $L^2(3B)$ and hence converges in $L^1(3B)$. It remains to prove that it converges in $L^1((3B)^c)$. For a fixed $x \in (3B)^c$ and a finite subset F of $\mathbf{Z}$, we have

$$\sum_{j\in F} \int_{\mathbf{R}^n} L_j(x,y)\varphi(y)\, dy = \sum_{j\in F} \int_B \big(L_j(x,y) - L_j(x,y_0)\big)\varphi(y)\, dy\,, \tag{8.4.24}$$

where y_0 is the center of the ball B. Using estimates (8.4.13), we obtain that the expression in (8.4.24) is controlled by a constant multiple of

$$\int_B \sum_{j\in F} \frac{|y-y_0| 2^{nj} 2^j}{(1+2^j|x-y_0|)^{n+2}} |\varphi(y)|\, dy \le C \big\|\varphi\big\|_{L^1} \frac{|y-y_0|}{|x-y_0|^{n+1}} \chi_{|x-y_0|\ge 3R}\,,$$

where R is the radius of B. Integrating with respect to $x \in (3B)^c$, we obtain that

$$\sum_{j\in F} \big\|\Delta_j(b) S_{j-3}(\varphi)\big\|_{L^1((3B)^c)} \le C_\varphi < \infty$$

for all F finite subsets of $\mathbf{Z}$. This proves that the series in (8.4.22) converges in L^1.

We have now proved that $\langle P_b^t(1), \varphi\rangle = 0$ for all $\varphi \in \mathcal{D}_0(\mathbf{R}^n)$. This shows that the distribution $P_b^t(1)$ is a constant function, which is of course identified with zero if considered as an element of BMO.

□

Remark 8.4.5. Observe that the boundedness of P_b on L^2 is a consequence of Theorem 8.3.3 since hypothesis (iv) is satisfied. Indeed, $P_b(1) = b$, $P_b^t(1) = 0$ are both BMO functions and the weak boundedness property is seen easily to hold for P_b; see Exercise 8.4.4 for a sketch of proof of the estimate $\|P_b\|_{WB} \le C_n\|b\|_{BMO}$. This provides another proof of the fact that $\|P_b\|_{L^2\to L^2} \le C_n\|b\|_{BMO}$. We chose not to prove the boundedness of paraproducts in this way so that we can use this result to obtain a different proof of the main direction in the $T(1)$ theorem in the next section.

Exercises

8.4.1. Let $f \in \mathcal{S}(\mathbf{R}^n)$ have mean value zero, and consider the series

$$\sum_{j\in\mathbf{Z}} \Delta_j(f)\,.$$

(a) Show that this series converges to f absolutely everywhere.
(b) Show that this series converges in L^1.
(b) Show that this series converges in H^1.
[*Hint:* To obtain convergence in L^1 for $j \ge 0$ use the estimate $\|\Delta_j(f)\|_{L^1} \le 2^{-j}\int_{\mathbf{R}^n}\int_{\mathbf{R}^n} 2^{jn}|\Psi(2^j y)|\,|2^j y|\,|(\nabla f)(x-\theta y)|\,dy\,dx$ for some θ in $[0,1]$ and consider the cases $|x| \ge 2|y|$ and $|x| \le 2|y|$. When $j \le 0$ use the simple identity $f*\Psi_{2^{-j}} = (f_{2^j}*\Psi)_{2^{-j}}$ and reverse the roles of f and Ψ. To show convergence in H^1, use that $\|\Delta_j(\varphi)\|_{H^1} \approx \|(\sum_k |\Delta_k\Delta_j(\varphi)|^2)^{\frac{1}{2}}\|_{L^1}$ and that only at most three terms in the square function are nonzero.]

8.4.2. Without using the H^1-BMO duality theorem, prove that there is a dimensional constant C_n so that for all $b \in BMO(\mathbf{R}^n)$ we have

$$\sup_{j\in\mathbf{Z}} \|\Delta_j(b)\|_{L^\infty} \le C_n\|b\|_{BMO}.$$

8.4.3. (a) Prove that for all $1 < p,q,r < \infty$ with $\frac{1}{p}+\frac{1}{q} = \frac{1}{r}$ there is a constant C_{pqr} such that for all Schwartz functions f,g on $\mathbf{R}^n$ we have

$$\|P_g(f)\|_{L^r} \le C_{pqr}\|f\|_{L^p}\|g\|_{L^q}.$$

(b) Obtain the same conclusion for the bilinear operator

$$\widetilde{P}_g(f) = \sum_j\sum_{k\le j} \Delta_j(g)\,\Delta_k(f)\,.$$

[*Hint:* Part (a): Estimate the L^r norm using duality. Part (b): Use part (a).]

8.4.4. (a) Let f be a normalized bump (see Definition 8.3.1). Prove that

$$\big\|\Delta_j(f_R)\big\|_{L^\infty} \le C(n,\Psi)\min(2^{-j}R^{-(n+1)}, 2^{nj})$$

for all $R>0$. Then interpolate between L^1 and L^∞ to obtain

$$\big\|\Delta_j(f_R)\big\|_{L^2} \le C(n,\Psi)\min(2^{-\frac{j}{2}}R^{-\frac{n+1}{2}}, 2^{\frac{nj}{2}}).$$

(b) Observe that the same result is valid for the operators Q_j be as defined in Proposition 8.4.4. Conclude that for some constant C_n we have

$$\sum_{j\in\mathbf{Z}}\big\|Q_j(g_R)\big\|_{L^2} \le C_n R^{-\frac{n}{2}}.$$

(c) Show that there is a constant C_n such that for all normalized bumps f and g we have

$$\big|\big\langle P_b(\tau^{x_0}(f_R)), \tau^{x_0}(g_R)\big\rangle\big| \le C_n R^{-n}\big\|b\big\|_{BMO}.$$

[*Hint:* Part (a): Use the cancellation of the functions f and Ψ. Part (c): Write

$$\big\langle P_b(\tau^{x_0}(f_R)), \tau^{x_0}(g_R)\big\rangle = \sum_j \int_{\mathbf{R}^n} S_{j-3}[\Delta_j(\tau^{-x_0}(b))Q_j(g_R)]f_R\,dx$$

Apply the Cauchy-Schwarz inequality, and use the boundedness of S_{j-3} on L^2, Exercise 8.4.2, and part (b).]

8.4.5. (*Continuous paraproducts*)
(a) Let Φ and Ψ be Schwartz functions on $\mathbf{R}^n$ with $\int_{\mathbf{R}^n}\Phi(x)\,dx = 1$ and $\int_{\mathbf{R}^n}\Psi(x)\,dx = 0$. For $t>0$ define operators $P_t(f) = \Phi_t * f$ and $Q_t(f) = \Psi_t * f$. Let $b \in BMO(\mathbf{R}^n)$ and $f\in L^2(\mathbf{R}^n)$. Show that the limit

$$\lim_{\substack{\varepsilon\to 0\\ N\to\infty}} \int_\varepsilon^N Q_t\big(Q_t(b)\,P_t(f)\big)\,\frac{dt}{t}$$

converges in $L^2(\mathbf{R}^n)$ and defines an operator $\Pi_b(f)$ that satisfies

$$\big\|\Pi_b\big\|_{L^2\to L^2} \le C_n\big\|b\big\|_{BMO}$$

for some dimensional constant C_n.
(b) Under the additional assumption that

$$\lim_{\substack{\varepsilon\to 0\\ N\to\infty}} \int_\varepsilon^N Q_t^2\,\frac{dt}{t} = I,$$

identify $\Pi_b(1)$ and $\Pi_b(b)$.
[*Hint:* Suitably adapt the proofs of Theorem 8.4.3 and Proposition 8.4.4.]

8.5. An Almost Orthogonality Lemma and Applications

In this section we discuss an important lemma regarding orthogonality of operators and some of its applications.

It is often the case that a linear operator T is given as an infinite sum of other linear operators T_j such that the T_j's are uniformly bounded on L^2. This sole condition is not enough to imply that the sum of the T_j's is also L^2 bounded, although this is often the case. Let us consider, for instance, the linear operators $\{T_j\}_{j\in\mathbf{Z}}$ given by convolution with the smooth functions $e^{2\pi ijt}$ on the circle $\mathbf{T}^1$. Each T_j can be written as $T_j(f) = (\widehat{f} \otimes \delta_j)^\vee$, where $\widehat{f}$ is the sequence of Fourier coefficients of f, δ_j is the infinite sequence consisting of zeros everywhere except at the jth entry in which it has 1, and $\otimes$ denotes term-by-term multiplication of infinite sequences. It follows that each operator T_j is bounded on $L^2(\mathbf{T}^1)$ with norm 1. Moreover, the sum of the T_j's is the identity operator, which is also L^2 bounded with norm 1.

It is apparent from the preceding discussion that the crucial property of the T_j's that makes their sum to be bounded is their orthogonality. In the preceding example we have $T_jT_k = 0$ unless $j = k$. It turns out that this orthogonality condition is a bit too strong, and it can be improved significantly.

8.5.a. The Cotlar-Knapp-Stein Almost Orthogonality Lemma

The next result provides a sufficient orthogonality criterion for boundedness of sums of linear operators on a Hilbert space.

Lemma 8.5.1. *Let $\{T_j\}_{j\in\mathbf{Z}}$ be a family of operators mapping a Hilbert space H into itself. Assume that there is a a function $\gamma : \mathbf{Z} \to \mathbf{R}^+$ such that*

$$\big\|T_j^*T_k\big\|_{H\to H} + \big\|T_jT_k^*\big\|_{H\to H} \le \gamma(j-k) \tag{8.5.1}$$

for all j,k in $\mathbf{Z}$. Suppose that

$$A = \sum_{j\in\mathbf{Z}} \sqrt{\gamma(j)} < \infty\,.$$

Then the following three conclusions are valid:

(i) *For all finite subsets Λ of $\mathbf{Z}$ we have*

$$\Big\|\sum_{j\in\Lambda} T_j\Big\|_{H\to H} \le A.$$

(ii) *For all $x \in H$ we have*

$$\sum_{j\in\mathbf{Z}} \big\|T_j(x)\big\|_H^2 \le A^2\|x\|_H^2.$$

(iii) *For all $x \in H$ the sequence $\sum_{|j|\le N} T_j(x)$ converges to some $T(x)$ as $N \to \infty$ in the norm topology of H. The linear operator T defined this way is bounded from H to H with norm*

$$\|T\|_{H\to H} \le A.$$

Proof. As usual we denote by S^* the adjoint of a linear operator S. It is a simple fact that any bounded linear operator $S: H \to H$ satisfies

$$\|S\|_{H\to H}^2 = \|SS^*\|_{H\to H}. \tag{8.5.2}$$

See Exercise 8.5.1. By taking $j = k$ in (8.5.1) and using (8.5.2), we obtain

$$\|T_j\|_{H\to H} \le \sqrt{\gamma(0)} \tag{8.5.3}$$

for all $j \in \mathbf{Z}$. It also follows from (8.5.2) that if an operator S is self-adjoint, then $\|S\|_{H\to H}^2 = \|S^2\|_{H\to H}$, and more generally,

$$\|S\|_{H\to H}^m = \|S^m\|_{H\to H} \tag{8.5.4}$$

for m that are powers of 2. Now observe that the linear operator

$$\Big(\sum_{j\in\Lambda} T_j\Big)\Big(\sum_{j\in\Lambda} T_j^*\Big)$$

is self-adjoint. Applying (8.5.2) and (8.5.4) to this operator, we obtain

$$\Big\|\sum_{j\in\Lambda} T_j\Big\|_{H\to H}^2 = \Big\|\Big[\Big(\sum_{j\in\Lambda} T_j\Big)\Big(\sum_{j\in\Lambda} T_j^*\Big)\Big]^m\Big\|_{H\to H}^{\frac{1}{m}}, \tag{8.5.5}$$

where m is a power of 2. We now expand the mth power of the expression in (8.5.5). So we write the right side of the identity in (8.5.5) as

$$\Big\|\sum_{j_1,\dots,j_{2m}\in\Lambda} T_{j_1}T_{j_2}^*\dots T_{j_{2m-1}}T_{j_{2m}}^*\Big\|_{H\to H}^{\frac{1}{m}}, \tag{8.5.6}$$

which is controlled by

$$\Big(\sum_{j_1,\dots,j_{2m}\in\Lambda} \|T_{j_1}T_{j_2}^*\dots T_{j_{2m-1}}T_{j_{2m}}^*\|_{H\to H}\Big)^{\frac{1}{m}}. \tag{8.5.7}$$

We estimate the expression inside the sum in (8.5.7) in two different ways. First we group j_1 with j_2, j_3 with j_4, $\dots$, j_{2m-1} with j_{2m} and we apply (8.5.3) and (8.5.1) to control this expression by

$$\gamma(j_1-j_2)\gamma(j_3-j_4)\dots\gamma(j_{2m-1}-j_{2m}).$$

Grouping j_2 with j_3, j_4 with j_5, $\dots$, j_{2m-2} with j_{2m-1} and leaving j_1 and j_{2m} alone, we also control the expression inside the sum in (8.5.7) by

$$\sqrt{\gamma(0)}\gamma(j_2-j_3)\gamma(j_4-j_5)\dots\gamma(j_{2m-2}-j_{2m-1})\sqrt{\gamma(0)}.$$

Taking the geometric mean of these two estimates, we obtain the following bound for (8.5.7):

$$\Big(\sum_{j_1,\dots,j_{2m}\in\Lambda} \sqrt{\gamma(0)}\sqrt{\gamma(j_1-j_2)}\sqrt{\gamma(j_2-j_3)}\dots\sqrt{\gamma(j_{2m-1}-j_{2m})}\Big)^{\frac{1}{m}}.$$

Summing first over j_1, then over j_2, and finally over j_{2m-1}, we obtain the estimate

$$\gamma(0)^{\frac{1}{2m}}A^{\frac{2m-1}{m}}\Big(\sum_{j_{2m}\in\Lambda}1\Big)^{\frac{1}{m}}$$

for (8.5.7). Using (8.5.5), we conclude that

$$\|\sum_{j\in\Lambda}T_j\|_{H\to H}^2 \le \gamma(0)^{\frac{1}{2m}}A^{\frac{2m-1}{m}}|\Lambda|^{\frac{1}{m}},$$

and letting $m\to\infty$, we obtain conclusion (i) of the proposition.

To prove (ii) we use the Rademacher functions r_j of Appendix C.1. These functions are defined for j nonnegative integers, but we can reindex them so that they are the subscript j runs through the integers. The fundamental property of these functions is their orthogonality, that is, $\int_0^1 r_j(\omega)r_k(\omega)\,d\omega = 0$ when $j\neq k$. Using the fact that the norm $\|\cdot\|_H$ comes from an inner product, for every finite subset Λ of $\mathbf{Z}$ and x in H we obtain

$$\begin{aligned}&\int_0^1\Big\|\sum_{j\in\Lambda}r_j(\omega)T_j(x)\Big\|_H^2\,d\omega\\ &= \sum_{j\in\Lambda}\|T_j(x)\|_H^2 + \int_0^1\sum_{\substack{j,k\in\Lambda\\ j\neq k}} r_j(\omega)r_k(\omega)\left\langle T_j(x),T_k(x)\right\rangle_H d\omega \qquad (8.5.8)\\ &= \sum_{j\in\Lambda}\|T_j(x)\|_H^2.\end{aligned}$$

For any fixed $\omega\in[0,1]$ we now use conclusion (i) of the proposition for the operators $r_j(\omega)T_j$, which also satisfy assumption (8.5.1) since $r_j(\omega)=\pm1$. We obtain that

$$\Big\|\sum_{j\in\Lambda}r_j(\omega)T_j(x)\Big\|_H^2 \le A^2\|x\|_H^2,$$

which, combined with (8.5.8), gives conclusion (ii).

We now prove (iii). First we show that given $x\in H$ the sequence

$$\Big\{\sum_{j=-N}^{N}T_j(x)\Big\}_N$$

is Cauchy in H. Suppose that this is not the case. This means that there is some $\varepsilon > 0$ and a subsequence of integers $1 \le N_1 < N_2 < N_3 < \dots$ such that

$$\big\|\widetilde{T}_k(x)\big\|_H \ge \varepsilon\,, \tag{8.5.9}$$

where we set

$$\widetilde{T}_k(x) = \sum_{N_k \le |j| < N_{k+1}} T_j(x).$$

For any fixed $\omega \in [0,1]$, apply conclusion (i) to the operators $S_j = r_k(\omega)T_j$ whenever $N_k \le |j| < N_{k+1}$ since these operators clearly satisfy hypothesis (8.5.1). Taking $N_1 \le |j| \le N_{K+1}$, we obtain

$$\Big\|\sum_{k=1}^{K} r_k(\omega) \sum_{N_k \le |j| < N_{k+1}} T_j(x)\Big\|_H = \Big\|\sum_{k=1}^{K} r_k(\omega)\widetilde{T}_k(x)\Big\|_H \le A\,\big\|x\big\|_H\,.$$

Squaring and integrating this inequality with respect to ω in $[0,1]$, and using (8.5.8) with $\widetilde{T}_k$ in the place of T_k and $\{1,2,\dots,K\}$ in the place of Λ, we obtain

$$\sum_{k=1}^{K} \big\|\widetilde{T}_k(x)\big\|_H^2 \le A^2\big\|x\big\|_H^2.$$

But this clearly contradicts (8.5.9) as $K \to \infty$.

We conclude that every sequence $\big\{\sum_{j=-N}^{N} T_j(x)\big\}_N$ is Cauchy in H and thus it converges to Tx for some linear operator T. In view of conclusion (i), it follows that T is a bounded operator on H with norm at most A.

□

Remark 8.5.2. At first sight, it appears strange that the norm of the operator T is independent of the norm of every piece T_j and depends only on the quantity A in (8.5.1). But, as observed in the proof, if we take $j = k$ in (8.5.1), we obtain

$$\big\|T_j\big\|_{H\to H}^2 = \big\|T_jT_j^*\big\|_{H\to H} \le \gamma(0) \le A^2\,;$$

thus the norm of each individual T_j is also controlled by the constant A.

We also note that there wasn't anything special about the role of the index set $\mathbf{Z}$ in Lemma 8.5.1. Indeed, the set $\mathbf{Z}$ can be replaced by any countable group, such as $\mathbf{Z}^k$ for some k. For instance, see Theorem 8.5.7, in which the index set is $\mathbf{Z}^{2n}$. See also Exercises 8.5.7 and 8.5.8, in which versions of Lemma 8.5.1 are given with no assumed group structure in the set of indices.

8.5.b. An Application

We now discuss an application of the almost orthogonality lemma just proved concerning sums of nonconvolution operators on $L^2(\mathbf{R}^n)$. We begin with the following version Theorem 8.3.3, in which it is assumed that $T(1) = T^t(1) = 0$.

Proposition 8.5.3. *Suppose that $K_j(x,y)$ are functions on $\mathbf{R}^n \times \mathbf{R}^n$ indexed by $j \in \mathbf{Z}$ that satisfy*

$$|K_j(x,y)| \le \frac{A2^{nj}}{(1+2^j|x-y|)^{n+\delta}}\,, \tag{8.5.10}$$

$$|K_j(x,y) - K_j(x,y')| \le A2^{\gamma j}2^{nj}|y-y'|^{\gamma}\,, \tag{8.5.11}$$

$$|K_j(x,y) - K_j(x',y)| \le A2^{\gamma j}2^{nj}|x-x'|^{\gamma} \tag{8.5.12}$$

for some $0 < A, \gamma, \delta < \infty$ and all $x, y, x', y' \in \mathbf{R}^n$. Suppose also that

$$\int_{\mathbf{R}^n} K_j(z,y)\,dz = 0 = \int_{\mathbf{R}^n} K_j(x,z)\,dz\,, \tag{8.5.13}$$

for all $x, y \in \mathbf{R}^n$ and all $j \in \mathbf{Z}$. For $j \in \mathbf{Z}$ define integral operators

$$T_j(f)(x) = \int_{\mathbf{R}^n} K_j(x,y)\,f(y)\,dy$$

for $f \in L^2(\mathbf{R}^n)$. Then the series

$$\sum_{j\in\mathbf{Z}} T_j(f)$$

converges in L^2 to some $T(f)$ for all $f \in L^2(\mathbf{R}^n)$, and the linear operator T defined this way is L^2 bounded.

Proof. It is a consequence of (8.5.10) that the kernels K_j are in $L^1(dy)$ uniformly in $x \in \mathbf{R}^n$ and $j \in \mathbf{Z}$ and hence the operators T_j map $L^2(\mathbf{R}^n)$ to $L^2(\mathbf{R}^n)$ uniformly in j. Our goal is to show that the sum of the T_j's is also bounded on $L^2(\mathbf{R}^n)$. We will achieve this using the orthogonality Lemma 8.5.1. To be able to use Lemma 8.5.1, we need to prove (8.5.1). Indeed, we will show that for all $k, j \in \mathbf{Z}$ we have

$$\big\|T_jT_k^*\big\|_{L^2\to L^2} \le C\,A^2\,2^{-\frac{1}{4}\frac{\delta}{n+\delta}\min(\gamma,\delta)|j-k|}\,, \tag{8.5.14}$$

$$\big\|T_j^*T_k\big\|_{L^2\to L^2} \le C\,A^2\,2^{-\frac{1}{4}\frac{\delta}{n+\delta}\min(\gamma,\delta)|j-k|}\,, \tag{8.5.15}$$

for some $0 < C = C_{n,\gamma,\delta} < \infty$. We will only prove (8.5.15) since the proof of (8.5.14) is similar. In fact, since the kernels of T_j and T_j^* satisfy similar size, regularity, and cancellation estimates, (8.5.15) is directly obtained from (8.5.14) by replacing T_j by T_j^*.

It suffices to prove (8.5.15) under the extra assumption that $k \le j$. Once (8.5.15) is established under this assumption, taking $j \le k$ yields

$$\big\|T_j^*T_k\big\|_{L^2\to L^2} = \big\|(T_k^*T_j)^*\big\|_{L^2\to L^2} = \big\|T_k^*T_j\big\|_{L^2\to L^2} \le C\,A^2 2^{-\frac{1}{2}\min(\gamma,\delta)|j-k|},$$

thus proving (8.5.15) also under the assumption $j \le k$.

We therefore take $k \le j$ in the proof of (8.5.15). Note that the kernel of $T_j^*T_k$ is

$$L_{jk}(x,y) = \int_{\mathbf{R}^n} \overline{K_j(z,x)}K_k(z,y)\,dz.$$

We will prove that

$$\sup_{x\in\mathbf{R}^n}\int_{\mathbf{R}^n}|L_{kj}(x,y)|\,dy\;\le C\,A^2\,2^{-\frac14\frac{\delta}{n+\delta}\min(\gamma,\delta)|k-j|}\,, \tag{8.5.16}$$

$$\sup_{y\in\mathbf{R}^n}\int_{\mathbf{R}^n}|L_{kj}(x,y)|\,dx\;\le C\,A^2\,2^{-\frac14\frac{\delta}{n+\delta}\min(\gamma,\delta)|k-j|}\,. \tag{8.5.17}$$

Once (8.5.16) and (8.5.17) are established, (8.5.15) will follow directly from the lemma in Appendix I.1.

We will need to use the estimate valid for $k\le j$:

$$\int_{\mathbf{R}^n}\frac{2^{nj}\min(1,(2^k|u|)^\gamma)}{(1+2^j|u|)^{n+\delta}}\,du\le C_{n,\delta}2^{-\frac12\min(\gamma,\delta)(j-k)}. \tag{8.5.18}$$

Indeed, to prove (8.5.18), we observe that by changing variables we may assume that $j=0$ and $k\le 0$. Taking $r=k-j\le 0$, we have

$$\begin{aligned}
&\int_{\mathbf{R}^n}\frac{\min(1,(2^r|u|)^\gamma)}{(1+|u|)^{n+\delta}}\,du\\
\le&\int_{\mathbf{R}^n}\frac{\min\big(1,(2^r|u|)^{\frac12\min(\gamma,\delta)}\big)}{(1+|u|)^{n+\delta}}\,du\\
\le&\int_{|u|\le 2^{-r}}\frac{(2^r|u|)^{\frac12\min(\gamma,\delta)}}{(1+|u|)^{n+\delta}}\,du+\int_{|u|\ge 2^{-r}}\frac{1}{(1+|u|)^{n+\delta}}\,du\\
\le&\;2^{\frac12\min(\gamma,\delta)r}\int_{\mathbf{R}^n}\frac{1}{(1+|u|)^{n+\frac{\delta}{2}}}\,du+\int_{|u|\ge 2^{-r}}\frac{1}{|u|^{n+\delta}}\,du\\
\le&\;C_{n,\delta}\big[2^{\frac12\min(\gamma,\delta)r}+2^{\delta r}\big]\le C_{n,\delta}\,2^{nk}\,2^{-\frac12\min(\gamma,\delta)|r|}\,,
\end{aligned}$$

which establishes (8.5.18).

$k\le j$ and we obtain estimates for L_{jk}. Using (8.5.13), we write

$$\begin{aligned}
|L_{jk}(x,y)|&=\left|\int_{\mathbf{R}^n}K_k(z,y)\overline{K_j(z,x)}\,dz\right|\\
&=\left|\int_{\mathbf{R}^n}\big[K_k(z,y)-K_k(x,y)\big]\overline{K_j(z,x)}\,dz\right|\\
&\le A^2\int_{\mathbf{R}^n}2^{nk}\min(1,(2^k|x-z|)^\gamma)\frac{2^{nj}}{(1+2^j|z-x|)^{n+\delta}}\,dz\\
&\le C\,A^2\,2^{-\frac12\min(\gamma,\delta)(j-k)}
\end{aligned}$$

using estimate (8.5.18). Combining this estimate with

$$|L_{jk}(x,y)|\le\int_{\mathbf{R}^n}|K_j(z,x)|\,|K_k(z,y)|\,dz\le\frac{C\,A^2 2^{kn}}{(1+2^k|x-y|)^{n+\delta}}\,,$$

which follows from (8.5.10) and the result in Appendix K.1 (since $k \le j$), gives

$$|L_{jk}(x,y)| \le C_{n,\gamma,\delta}\, A^2\, 2^{-\frac{1}{2}\frac{\delta/2}{n+\delta}\min(\gamma,\delta)(j-k)} \frac{2^{kn}}{(1+2^k|x-y|)^{n+\frac{\delta}{2}}},$$

which easily implies (8.5.16) and (8.5.17) This concludes the proof of the proposition. □

8.5.c. Almost Orthogonality and the $T(1)$ Theorem

We now give an important application of the proposition just proved. We will reprove the difficult direction of the $T(1)$ theorem proved in Section 8.3. We have the following:

Theorem 8.5.4. *Let K be in $SK(\delta,A)$ and let T be a continuous linear operator from $\mathcal{S}(\mathbf{R}^n)$ into $\mathcal{S}'(\mathbf{R}^n)$ associated with K. Assume that*

$$\|T(1)\|_{BMO} + \|T^t(1)\|_{BMO} + \|T\|_{WB} = B_4 < \infty .$$

Then T extends to bounded linear operator on $L^2(\mathbf{R}^n)$ with norm at most a constant multiple of $A + B_4$.

Proof. Consider the paraproduct operators $P_{T(1)}$ and $P_{T^t(1)}$ introduced in the previous section. Then, as we showed in Proposition 8.4.4, we have

$$P_{T(1)}(1) = T(1) \qquad\qquad (P_{T(1)})^t(1) = 0$$
$$P_{T^t(1)}(1) = T^t(1) \qquad\qquad (P_{T^t(1)})^t(1) = 0.$$

Let us define an operator

$$L = T - P_{T(1)} - (P_{T^t(1)})^t.$$

Using Proposition 8.4.4, we obtain that

$$L(1) = L^t(1) = 0.$$

In view of (8.4.16), we have that L is an operator whose kernel satisfies the estimates (8.1.1), (8.1.2), and (8.1.3) with constants controlled by a dimensional multiple of

$$A + \|T(1)\|_{BMO} + \|T^t(1)\|_{BMO} .$$

Both of these numbers are controlled by $A + B_4$. We also have

$$\begin{aligned} \|L\|_{WB} &\le C_n(\|T\|_{WB} + \|P_{T(1)}\|_{L^2\to L^2} + \|(P_{T^t(1)})^t\|_{L^2\to L^2}) \\ &\le C_n(\|T\|_{WB} + \|T(1)\|_{BMO} + \|T^t(1)\|_{BMO}) \\ &\le C_n(A + B_4)\,, \end{aligned}$$

a fact that will be useful to us later.

We introduce operators Δ_j and S_j as follows. (Be careful; these are not the operators Δ_j and S_j introduced in Section 8.4 but rather discrete analogues of the ones introduced in the proof of Theorem 8.3.3.) We pick a smooth radial function

Φ with compact support that is supported in the unit ball $B(0,\frac{1}{2})$ and that satisfies $\int_{\mathbf{R}^n}\Phi(x)\,dx=1$ and we define

$$\Psi(x)=\Phi(x)-2^{-n}\Phi(\tfrac{x}{2}). \tag{8.5.19}$$

We note that Ψ has mean value zero. We define

$$\Phi_{2^{-j}}(x)=2^{nj}\Phi(2^j x) \qquad \text{and} \qquad \Psi_{2^{-j}}(x)=2^{nj}\Psi(2^j x)$$

and we observe that both Φ and Ψ are supported in $B(0,1)$ and they are multiples of normalized bumps. We then define Δ_j to be the operator given by convolution with the function $\Psi_{2^{-j}}$ S_j the operator given by convolution with the function $\Phi_{2^{-j}}$. In view of identity (8.5.19) we have that $\Delta_j=S_j-S_{j-1}$. Notice that

$$S_jLS_j=S_{j-1}LS_{j-1}+\Delta_jLS_j+S_{j-1}L\Delta_j\,,$$

which implies that for all integers $N<M$ we have

$$\begin{aligned} S_MLS_M-S_{N-1}LS_{N-1} &= \sum_{j=N}^{M}\big(S_jLS_j-S_{j-1}LS_{j-1}\big)\\ &= \sum_{j=N}^{M}\Delta_jLS_j-\sum_{j=N}^{M}S_{j-1}L\Delta_j. \end{aligned}$$

We plan to show that

$$\sup_{M\in\mathbf{Z}}\sup_{N<M}\big\|S_MLS_M-S_{N-1}LS_{N-1}\big\|_{L^2\to L^2}\le C_n(A_2+B_4) \tag{8.5.20}$$

and that the operators $S_MLS_M-S_{N-1}LS_{N-1}$ converge in the L^2 operator norm (as $M\to\infty$ and $N\to-\infty$) to an operator $\widetilde{L}$ that satisfies the same norm estimate. When these statements are proved, we can easily deduce that $L=\widetilde{L}$. To establish this, it suffices to prove that $S_MLS_M-S_{N-1}LS_{N-1}$ converges to L weakly in L^2. To see this, we introduce the space $\mathcal{S}_0(\mathbf{R}^n)$ of all Schwartz functions on $\mathbf{R}^n$ whose Fourier transform vanishes in a neighborhood of the origin, we let $\mathcal{S}_0'$ be its dual, and we fix $f,g\in\mathcal{S}_0$. The space $\mathcal{S}_0$ is dense in L^2. Then

$$\begin{aligned} &\big\langle S_MLS_M(f)-S_{N-1}LS_{N-1}(f),g\big\rangle-\big\langle L(f),g\big\rangle\\ &=\big\langle S_MLS_M(f),g\big\rangle-\big\langle L(f),g\big\rangle+\big\langle S_{N-1}LS_{N-1}(f),g\big\rangle. \end{aligned} \tag{8.5.21}$$

We first prove that the sum of the first two expressions in (8.5.21) tends to zero. Indeed, we have

$$\begin{aligned} &\big\langle S_MLS_M(f),g\big\rangle-\big\langle L(f),g\big\rangle\\ &=\big\langle LS_M(f),S_Mg\big\rangle-\big\langle L(f),g\big\rangle\\ &=\big\langle LS_M(f)-f,S_M(g)-g\big\rangle-\big\langle L(f),g\big\rangle+\big\langle L(f),S_M(g)-g\big\rangle\,, \end{aligned} \tag{8.5.22}$$

and the preceding converges to zero as $M\to\infty$ since $S_M(g)-g$ is a bounded sequence in $\mathcal{S}$ and the continuity of L on $\mathcal{S}$ gives that $L(S_M(f)-f)$ tends to zero in

$\mathcal{S}'$. For this part of the argument we do not need f, g to be in $\mathcal{S}_0$. See also Exercise 8.3.1.

We note that when f is in $\mathcal{S}_0$, then $S_N(f) \to 0$ in $\mathcal{S}$ as $N \to -\infty$. See also Exercise 8.3.2. Using this fact and an argument similar to (8.5.22), we obtain that the distributions $S_{N-1}LS_{N-1}(f)$ converge to zero in $\mathcal{S}_0'$ as $N \to -\infty$. This implies that $\langle S_{N-1}LS_{N-1}(f), g\rangle$ tends to zero as $N \to -\infty$. We conclude that $\widetilde{L} = L$, and it suffices to prove (8.5.20).

We now define

$$L_j = \Delta_j L S_j \qquad \text{and} \qquad L_j' = S_{j-1}L\Delta_j$$

for $j \in \mathbf{Z}$. We plan to use the almost orthogonality Lemma 8.5.1 to prove any finite sum of the L_j's and L_j''s has L^2 operator norm controlled by a constant multiple of $A_2 + B_4$. This will yield (8.5.20).

Using Definition 2.3.13, we see that whenever $f \in \mathcal{S}$ we have

$$(L(f * \Phi) * \Psi)(x) = \int_{\mathbf{R}^n} \big\langle L(\tau^y(\Phi)), \tau^x(\Psi)\big\rangle f(y)\, dy, \tag{8.5.23}$$

where $\tau^y(g)(u) = g(u-y)$. It follows from (8.5.23) that the kernel K_j of L_j satisfies

$$K_j(x,y) = \big\langle L(\tau^y(\Phi_{2^{-j}})), \tau^x(\Psi_{2^{-j+1}})\big\rangle .$$

We plan to prove that

$$|K_j(x,y)| + 2^{-j}|\nabla K_j(x,y)| \le C_n(A + B_4)2^{nj}(1 + 2^j|x-y|)^{-n-\delta}. \tag{8.5.24}$$

Once (8.5.24) is established, then Proposition 8.5.3, combined with Exercise 8.5.2, will yield the required conclusion.

To prove (8.5.24) we quickly repeat the corresponding argument from the proof of Theorem 8.3.3. We consider the following two cases: If $|x-y| \le 5 \cdot 2^{-j}$, then the weak boundedness property gives

$$\begin{aligned} \big|\langle L(\tau^y(\Phi_{2^{-j}})), \tau^x(\Psi_{2^{-j+1}})\rangle\big| &= \big|\langle L(\tau^x(\tau^{2^j(y-x)}(\Phi)_{2^{-j}})), \tau^x((\delta^{\frac{1}{2}}(\Psi))_{2^{-j}})\rangle\big| \\ &\le C_n \big\|L\big\|_{WB} 2^{jn}, \end{aligned}$$

since $\tau^{2^j(y-x)}(\Phi)$ [which is supported in $B(0,\frac{1}{2}) + B(0,5) \subseteq B(0,10)$] and $\delta^{\frac{1}{2}}(\Psi)(u) = \frac{1}{2}\Psi(\frac{1}{2}u)$ [which is supported in $B(0,2)$] are multiples of normalized bumps. This proves the first of the two estimates in (8.5.24) when $|x-y| \le 5 \cdot 2^{-j}$.

We now turn to the case $|x-y| \ge 5 \cdot 2^{-j}$. Then the functions $\tau^y(\Phi_{2^{-j}})$ and $\tau^x(\Psi_{2^{-j+1}})$ have disjoint supports and so we have the integral representation

$$K_j(x,y) = \int_{\mathbf{R}^n}\int_{\mathbf{R}^n} \Phi_{2^{-j}}(v-y)K(u,v)\Psi_{2^{-j+1}}(u-x)\, du\, dv .$$

Using that Ψ has mean value zero, we can write the previous expression as

$$\int_{\mathbf{R}^n}\int_{\mathbf{R}^n} \Phi_{2^{-j}}(v-y)\big(K(u,v) - K(x,v)\big)\Psi_{2^{-j+1}}(u-x)\, du\, dv .$$

We observe that $|u-x| \le 2^{-j+1}$ and $|v-y| \le 2^{-j}$ in the preceding integral. Since $|x-y| \ge 5 \cdot 2^{-j}$, this makes $|u-v| \ge |x-y| - 3\cdot 2^{-j} \ge 2 \cdot 2^{-j}$, which implies that $|u-x| \le \frac{1}{2}|u-v|$. Using the regularity condition (8.1.2), we deduce

$$|K(u,v) - K(x,v)| \le A \frac{|x-u|^\delta}{|u-v|^{n+\delta}} \le C_{n,\delta} A \frac{2^{-j\delta}}{|x-y|^{n+\delta}} .$$

Inserting this estimate in the preceding double integral we obtain the first estimate in (8.5.24). The second estimate is proved similarly. Since the same estimates can be obtained for L_j' by duality, the proof is complete.

□

8.5.d. Pseudodifferential Operators

We now turn to another elegant application of Lemma 8.5.1 regarding pseudodifferential operators. We first quickly introduce pseudodifferential operators.

Definition 8.5.5. Let $m \in \mathbf{R}$ and $0 < \rho, \delta \le 1$. A C^∞ function $\sigma(x,\xi)$ on $\mathbf{R}^n \times \mathbf{R}^n$ is called a *symbol of class* $S^m_{\rho,\delta}$ if for all multiindices α and β there is a constant $B_{\alpha,\beta}$ so that

$$|\partial_x^\alpha \partial_\xi^\beta \sigma(x,\xi)| \le B_{\alpha,\beta}(1+|\xi|)^{m-\rho|\beta|+\delta|\alpha|} . \tag{8.5.25}$$

For $\sigma \in S^m_{\rho,\delta}$, the linear operator

$$T_\sigma(f)(x) = \int_{\mathbf{R}^n} \sigma(x,\xi)\widehat{f}(\xi)e^{2\pi i x\cdot\xi}\, d\xi$$

initially defined for f in $\mathcal{S}(\mathbf{R}^n)$ is called a *pseudodifferential operator* with symbol $\sigma(x,\xi)$.

Example 8.5.6. The paraproduct P_b introduced in the previous section is a pseudodifferential operator with symbol

$$\sigma_b(x,\xi) = \sum_{j\in\mathbf{Z}} \Delta_j(b)(x)\widehat{\Psi}(2^{-j}\xi) . \tag{8.5.26}$$

It is not hard to see that the symbol σ_b satisfies

$$|\partial_x^\alpha \partial_\xi^\beta \sigma_b(x,\xi)| \le B_{\alpha,\beta}|\xi|^{-|\beta|+|\alpha|} \tag{8.5.27}$$

for all multiindices α and β. Indeed, every differentiation in x produces a factor of 2^j while every differentiation in ξ produces a factor of 2^{-j}. But since $\widehat{\Psi}$ is supported in $\frac{1}{2}\cdot 2^j \le |\xi| \le 2\cdot 2^j$, it follows that $|\xi| \approx 2^j$, which yields (8.5.27). It follows that σ_b is not in any of the classes $S^m_{\rho,\delta}$ introduced in Definition 8.5.5. However, if we restrict the indices of summation in (8.5.26) to $j \ge 0$, then $|\xi| \approx 1 + |\xi|$ and we obtain a symbol of class $S^0_{1,1}$. Note that not all symbols in $S^0_{1,1}$ give rise to bounded operators on L^2. See Exercise 8.5.6.

An example of a symbol in $S^m_{0,0}$ is $(1+|\xi|^2)^{\frac{1}{2}(m+it)}$ when $m,t \in \mathbf{R}$.

We do not plan to embark in a systematic study of pseudodifferential operators here, but we would like to study the L^2 boundedness of symbols of class $S^0_{0,0}$.

Theorem 8.5.7. *Suppose that a symbol σ belongs to the class $S^0_{0,0}$. Then the pseudodifferential operator T_σ with symbol σ, initially defined on $\mathcal{S}(\mathbf{R}^n)$, has a bounded extension on $L^2(\mathbf{R}^n)$.*

Proof. In view of Plancherel's theorem, it will suffice to obtain the L^2 boundedness of the linear operator

$$\widetilde{T_\sigma}(f)(x) = \int_{\mathbf{R}^n} \sigma(x,\xi) f(\xi) e^{2\pi i x\cdot\xi}\, d\xi\,. \tag{8.5.28}$$

We fix a nonnegative smooth function $\varphi(\xi)$ supported in a small multiple of the unit cube $Q_0 = [0,1]^n$ (say in $[-\frac{1}{9},\frac{10}{9}]^n$) that satisfies

$$\sum_{j\in\mathbf{Z}^n} \varphi(x-j) = 1\,, \qquad x\in\mathbf{R}^n\,. \tag{8.5.29}$$

For $j,k\in\mathbf{Z}^n$ we define symbols

$$\sigma_{j,k}(x,\xi) = \varphi(x-j)\sigma(x,\xi)\varphi(\xi-k)$$

and corresponding operators $T_{j,k}$ given by (8.5.28) in which $\sigma(x,\xi)$ is replaced by $\sigma_{j,k}(x,\xi)$. Using (8.5.29), we obtain that

$$\widetilde{T_\sigma} = \sum_{j,k\in\mathbf{Z}^n} T_{j,k}\,,$$

where the double sum is shown easily to converge in the topology of $\mathcal{S}(\mathbf{R}^n)$. Our goal will be to show that for all $N\in\mathbf{Z}^+$ we have

$$\big\|T^*_{j,k}T_{j',k'}\big\|_{L^2\to L^2} \le C_N(1+|j-j'|+|k-k'|)^{-2N}\,, \tag{8.5.30}$$

$$\big\|T_{j,k}T^*_{j',k'}\big\|_{L^2\to L^2} \le C_N(1+|j-j'|+|k-k'|)^{-2N}\,, \tag{8.5.31}$$

where C_N depends on N and n but is independent of j,j',k,k'.

We note that

$$T^*_{j,k}T_{j',k'}(f)(x) = \int_{\mathbf{R}^n} K_{j,k,j',k'}(x,y) f(y)\, dy\,,$$

where

$$K_{j,k,j',k'}(x,y) = \int_{\mathbf{R}^n} \overline{\sigma_{j,k}(z,x)}\sigma_{j',k'}(z,y) e^{2\pi i(y-x)\cdot z}\, dz\,. \tag{8.5.32}$$

We integrate by parts in (8.5.32) by using the identity

$$e^{2\pi i z\cdot(y-x)} = \frac{(I-\Delta_z)^N(e^{2\pi i z\cdot(y-x)})}{(1+4\pi^2|x-y|^2)^N}\,,$$

and we obtain the pointwise estimate

$$\frac{\varphi(x-k)\varphi(y-k')}{(1+4\pi^2|x-y|^2)^N}\big|(I-\Delta_z)^N(\varphi(z-j)\overline{\sigma(z,x)}\sigma(z,y)\varphi(z-j'))\big|$$

for the integrand in (8.5.32). The support property of φ forces $|j-j'| \le c_n$ for some dimensional constant c_n; indeed, $c_n = 2\sqrt{n}$ suffices. Moreover, all derivatives of σ and φ are controlled by constants, and φ is supported in a cube of finite measure. We also have $1+|x-y| \approx 1+|k-k'|$. It follows that

$$|K_{j,k,j',k'}(x,y)| \le \begin{cases} \dfrac{C_N\varphi(x-k)\varphi(y-k')}{(1+|k-k'|)^{2N}} & \text{when } |j-j'| \le c_n, \\ 0 & \text{otherwise.} \end{cases}$$

We can rewrite the preceding estimates in a more compact (and symmetric) form as

$$|K_{j,k,j',k'}(x,y)| \le \frac{C_{n,N}\varphi(x-k)\varphi(y-k')}{(1+|j-j'|+|k-k'|)^{2N}},$$

from which we easily obtain that

$$\sup_{x\in\mathbf{R}^n} \int_{\mathbf{R}^n} |K_{j,k,j',k'}(x,y)|\,dy \le \frac{C_{n,N}}{(1+|j-j'|+|k-k'|)^{2N}}, \tag{8.5.33}$$

$$\sup_{y\in\mathbf{R}^n} \int_{\mathbf{R}^n} |K_{j,k,j',k'}(x,y)|\,dx \le \frac{C_{n,N}}{(1+|j-j'|+|k-k'|)^{2N}}. \tag{8.5.34}$$

Using the classical Schur lemma in Appendix I.1, we obtain that

$$\big\|T_{j,k}^*T_{j',k'}\big\|_{L^2\to L^2} \le \frac{C_{n,N}}{(1+|j-j'|+|k-k'|)^{2N}},$$

which proves (8.5.30). Since $\rho=\delta=0$, the roles of the variables x and ξ are symmetric and (8.5.31) can be proved in exactly the same way as (8.5.30). The almost orthogonality Lemma 8.5.1 now applies since

$$\sum_{j,k\in\mathbf{Z}^n}\sqrt{\frac{1}{(1+|j|+|k|)^{2N}}} \le \sum_{j\in\mathbf{Z}^n}\sum_{k\in\mathbf{Z}^n}\frac{1}{(1+|j|)^{\frac{N}{2}}}\frac{1}{(1+|k|)^{\frac{N}{2}}} < \infty$$

for $N \ge 2n+2$, and the boundedness of $\widetilde{T}_\sigma$ on L^2 follows.

□

Remark 8.5.8. The reader may want to check that the argument in Theorem 8.5.7 is also valid for symbols of the kind $S^0_{\rho,\rho}$ whenever $0<\rho<1$.

Exercises

8.5.1. Prove that any bounded linear operator $S:\ H\to H$ satisfies

$$\|S\|_{H\to H}^2 = \|SS^*\|_{H\to H}.$$

8.5.2. Show that if a family of kernels K_j satisfy condition (8.5.10) and

$$|\nabla_x K_j(x,y)| + |\nabla_y K_j(x,y)| \le \frac{A2^{(n+1)j}}{(1+2^j|x-y|)^{n+\delta}}$$

for all $x, y \in \mathbf{R}^n$, then conditions (8.5.11) and (8.5.12) hold with $\gamma = 1$.

8.5.3. Prove the boundedness of the Hilbert transform using Lemma 8.5.1 and without using the Fourier transform.
$\big[$*Hint:* Pick an even smooth function η supported in $[1/2, 2]$ such that $\sum_{j\in\mathbf{Z}} \eta(2^{-j}x) = 1$ for $x \neq 0$ and set $K_j(x) = x^{-1}\eta(2^{-j}|x|)$ and $H_j(f) = f * K_j$. Note that $H_j^* = -H_j$. Estimate $\big\|H_k H_j\big\|_{L^2\to L^2}$ by $\big\|K_k * K_j\big\|_{L^1} \le \big\|K_k * K_j\big\|_{L^\infty} |\text{supp}\,(K_k * K_j)|$. When $j < k$, use the mean value property of K_j and that $\big\|K_k'\big\|_{L^\infty} \le C2^{-2k}$ to obtain that $\big\|K_k * K_j\big\|_{L^\infty} \le C2^{-2k+j}$. Conclude that $\big\|H_k H_j\big\|_{L^2\to L^2} \le C2^{-|j-k|}$.$\big]$

8.5.4. For a symbol $\sigma(x,\xi)$ in $S^0_{1,0}$, let $k(x,z)$ denote the inverse Fourier transform (evaluated at z) of the function $\sigma(x,\cdot)$ with x fixed. Show that for all $x \in \mathbf{R}^n$, the distribution $k(x,\cdot)$ coincides with a smooth function away from the origin in $\mathbf{R}^n$ that satisfies the following estimates:

$$|\partial_x^\alpha \partial_z^\beta k(x,z)| \le C_{\alpha,\beta}|z|^{-n-|\beta|}$$

and conclude that the kernels $K(x,y) = k(x, x-y)$ are well defined and smooth functions away from the diagonal in $\mathbf{R}^{2n}$ that belong to $SK(1,A)$ for some $A > 0$. Conclude that pseudodifferential operators with symbols in $S^0_{1,0}$ are associated with standard kernels.
$\big[$*Hint:* Consider the distribution $(\partial^\gamma\sigma(x,\cdot))^\vee = (-2\pi i z)^\gamma k(x,\cdot)$. Since $\partial_\xi^\gamma\sigma(x,\xi)$ is integrable in ξ when $|\gamma| \ge n+1$, it follows that $k(x,\cdot)$ coincides with a smooth function away on $\mathbf{R}^n \setminus \{0\}$. Next, set $\sigma_j(x,\xi) = \sigma(x,\xi)\widehat{\Psi}(2^{-j}\xi)$, where Ψ is as in Section 8.4 and k_j the inverse Fourier transform of σ_j in z. Use that

$$(-2\pi i z)^\gamma \partial_x^\alpha \partial_\xi^\beta k_j(x,z) = \int_{\mathbf{R}^n} \partial_\xi^\gamma\big((2\pi i\xi)^\beta \partial_x^\alpha \sigma_j(x,\xi)\big) 2^{2\pi i \xi\cdot z}\, d\xi$$

with $|\gamma| = M$ to obtain $\partial_x^\alpha \partial_\xi^\beta k_j(x,z) \le B_{M,\alpha,\beta} 2^{jn} 2^{j|\alpha|}(2^j n|z|)^{-M}$ and then sum over $j \in \mathbf{Z}$.$\big]$

8.5.5. Prove that pseudodifferential operators with symbols in $S^0_{1,0}$ that have compact support in x are elements of $CZO(1,A,B)$ for some $A, B > 0$.
$\big[$*Hint:* Write

$$T_\sigma(f)(x) = \int_{\mathbf{R}^n}\bigg(\int_{\mathbf{R}^n} \widehat{\sigma}(a,\xi)\widehat{f}(\xi)e^{2\pi i x\cdot\xi}\, d\xi\bigg)e^{2\pi i x\cdot a}\, da\,,$$

where $\widehat{\sigma}(a,\xi)$ denotes the Fourier transform of $\sigma(x,\xi)$ in the variable x. Use integration by parts to obtain $\sup_\xi |\widehat{\sigma}(a,\xi)| \le C_N(1+|a|)^{-N}$ and pass the L^2 norm inside the integral in a to obtain the required conclusion using the translation invariant case.$\big]$

8.5.6. Let $\widehat{\eta}(\xi)$ be a smooth bump on $\mathbf{R}$ that is supported in $2^{-\frac{1}{2}} \le |\xi| \le 2^{\frac{1}{2}}$ and is equal to 1 on $2^{-\frac{1}{4}} \le |\xi| \le 2^{\frac{1}{4}}$. Let

$$\sigma(x,\xi) = \sum_{k=1}^{\infty} e^{-2\pi i 2^k x} \widehat{\eta}(2^{-k}\xi) .$$

Show that σ is an element of $S^0_{1,1}$ on the line but the corresponding pseudo-differential operator T_σ is not L^2 bounded.
[*Hint:* To see the latter statement, consider the sequence of functions $f_N(x) = \sum_{k=5}^{\infty} \frac{1}{k} e^{2\pi i 2^k x} h(x)$, where $h(x)$ is a Schwartz function whose Fourier transform is supported in the set $|\xi| \le \frac{1}{4}$. Show that $\|f_N\|_{L^2} \le C\|h\|_{L^2}$ but $\|T_\sigma(f_N)\|_{L^2} \ge c \log N \|h\|_{L^2}$ for some positive constants c, C.]

8.5.7. Prove conclusions (i) and (ii) of Lemma 8.5.1 if hypothesis (8.5.1) is replaced by

$$\|T_j^* T_k\|_{H\to H} + \|T_j T_k^*\|_{H\to H} \le \Gamma(j,k) ,$$

where Γ is a nonnegative function on $\mathbf{Z} \times \mathbf{Z}$ such that

$$\sup_j \sum_{k\in\mathbf{Z}} \sqrt{\Gamma(j,k)} = A < \infty .$$

8.5.8. Let $\{T_t\}_{t\in\mathbf{R}^+}$ be a family of operators mapping a Hilbert space H into itself. Assume that there is a function $\gamma : \mathbf{R}^+ \times \mathbf{R}^+ \to \mathbf{R}^+ \cup \{0\}$ such that

$$\|T_t^* T_s\|_{H\to H} + \|T_t T_s^*\|_{H\to H} \le \gamma(t,s)$$

for all t, s in $\mathbf{R}^+$ and such that

$$A = \sup_{t>0} \int_0^\infty \sqrt{\gamma(t,s)}\, \frac{ds}{s} < \infty .$$

Then prove that for all $0 < \varepsilon < N$ we have

$$\Big\| \int_\varepsilon^N T_t \, \frac{dt}{t} \Big\|_{H\to H} \le A .$$

[We note that in most applications of this result we have

$$\gamma(t,s) = \min\Big(\frac{s}{t}, \frac{t}{s}\Big)^\varepsilon$$

for some $\varepsilon > 0$.]

8.6. The Cauchy Integral of Calderón and the $T(b)$ Theorem*

The Cauchy integral is almost as old as complex analysis itself. In the classical theory of complex analysis, if Γ is a curve in $\mathbf{C}$ and f is a function on the curve, the Cauchy integral of f is given by

$$\frac{1}{2\pi i}\int_\Gamma \frac{f(\zeta)}{z-\zeta}\,d\zeta\,.$$

One of the problems in which this operator appears is the following: If Γ is a closed simple curve (i.e., a Jordan curve), Ω_+ is the interior connected component of $\mathbf{C}\setminus\Gamma$, Ω_- is the exterior connected component of $\mathbf{C}\setminus\Gamma$, and f is a smooth complex function on Γ, is it possible to find analytic functions F_+ on Ω_+ and F_- on Ω_-, respectively, that have continuous extensions on Γ so that their difference $F_+ - F_-$ is equal to the given f on Γ?

This problem was solved by Plemelj, who showed that one may take

$$F_+(w) = \frac{1}{2\pi i}\int_\Gamma \frac{f(\zeta)}{\zeta-w}\,d\zeta\,, \qquad w\in\Omega_+$$

and

$$F_-(w) = \frac{1}{2\pi i}\int_\Gamma \frac{f(\zeta)}{\zeta-w}\,d\zeta\,, \qquad w\in\Omega_-\,.$$

We will be concerned with the case where the Jordan curve Γ passes through infinity, in particular, when it is the graph of a Lipschitz function on $\mathbf{R}$. In this case we will compute the boundary limits of F_+ and F_- and we will see that they give rise to a very interesting operator on the curve Γ.

To fix notation we let

$$A:\ \mathbf{R}\to\mathbf{R}$$

be a Lipschitz function. This means that there is a constant $L>0$ such that for all $x,y\in\mathbf{R}$ we have $|A(x)-A(y)|\le L|x-y|$. We define a curve

$$\gamma:\ \mathbf{R}\to\mathbf{C}$$

by setting

$$\gamma(x) = x + iA(x)$$

and we denote by

$$\Gamma = \{\gamma(x):\ x\in\mathbf{R}\} \tag{8.6.1}$$

the graph of γ. Given a smooth function f on Γ we set

$$F(w) = \frac{1}{2\pi i}\int_\Gamma \frac{f(\zeta)}{\zeta-w}\,d\zeta\,, \qquad w\in\mathbf{C}\setminus\Gamma\,. \tag{8.6.2}$$

We will now show that for $z\in\Gamma$ both $F(z+i\delta)$ and $F(z-i\delta)$ have limits as $\delta\downarrow 0$, and these limits give rise to an operator on the curve Γ that we would like to study. In particular, we obtain Plemelj's result in this case.

8.6.a. Introduction of the Cauchy Integral Operator along a Lipschitz Curve

For a smooth function f on the curve Γ and $z \in \Gamma$ we define the *Cauchy integral of f at z* as

$$\mathcal{C}_\Gamma(f)(z) = \lim_{\epsilon\to 0+} \frac{1}{\pi i} \int\limits_{\substack{\zeta\in\Gamma \\ |\operatorname{Re}\zeta - \operatorname{Re} z|>\epsilon}} \frac{f(\zeta)}{\zeta - z}\, d\zeta\,, \tag{8.6.3}$$

assuming that $f(\zeta)$ has some decay as $|\zeta| \to \infty$. The latter assumption makes the integral in (8.6.3) converge when $|\operatorname{Re}\zeta - \operatorname{Re} z| \geq 1$. The fact that the limit in (8.6.3) exists as $\epsilon \to 0$ for almost all $z \in \Gamma$ will be shown in the following proposition.

Proposition 8.6.1. *Let Γ be as in (8.6.1). Let $f(\zeta)$ be a smooth function on Γ that has decay as $|\zeta| \to \infty$. Given f, we define a function F as in (8.6.2) related to f. Then the limit in (8.6.3) exists as $\epsilon \to 0$ for almost all $z \in \Gamma$ and gives rise to a well-defined operator $\mathcal{C}_\Gamma(f)$ acting on such functions f. Moreover, for almost all $z \in \Gamma$ we have that*

$$\lim_{\delta\downarrow 0} F(z + i\delta) = \frac{1}{2}\mathcal{C}_\Gamma(f)(z) + \frac{1}{2} f(z) \tag{8.6.4}$$

$$\lim_{\delta\downarrow 0} F(z - i\delta) = \frac{1}{2}\mathcal{C}_\Gamma(f)(z) - \frac{1}{2} f(z)\,. \tag{8.6.5}$$

Proof. To show that the limit in (8.6.3) exists as $\epsilon \to 0$, for $z \in \Gamma$, we write

$$\begin{aligned} \frac{1}{\pi i} \int\limits_{\substack{\zeta\in\Gamma \\ |\operatorname{Re}\zeta - \operatorname{Re} z|>\epsilon}} \frac{f(\zeta)\, d\zeta}{\zeta - z} &= \frac{1}{\pi i} \int\limits_{\substack{\zeta\in\Gamma \\ |\operatorname{Re}\zeta - \operatorname{Re} z|>1}} \frac{f(\zeta)\, d\zeta}{\zeta - z} \\ &\quad + \frac{1}{\pi i} \int\limits_{\substack{\zeta\in\Gamma \\ \epsilon\leq|\operatorname{Re}\zeta - \operatorname{Re} z|\leq 1}} \frac{(f(\zeta) - f(z))\, d\zeta}{\zeta - z} \\ &\quad + \frac{f(z)}{\pi i} \int\limits_{\substack{\zeta\in\Gamma \\ \epsilon\leq|\operatorname{Re}\zeta - \operatorname{Re} z|\leq 1}} \frac{d\zeta}{\zeta - z}\,. \end{aligned} \tag{8.6.6}$$

Let $\tau = \operatorname{Re} z$; then $z = \gamma(\tau) = \tau + iA(\tau)$. Then the last summand in (8.6.6) can be computed easily and found to be equal to

$$\begin{aligned} &\frac{f(z)}{\pi i}\Big[\log\big(\gamma(1+\tau) - \gamma(\tau)\big) - \log\big(\gamma(\epsilon+\tau) - \gamma(\tau)\big) \\ &\quad - \log\big(\gamma(-1+\tau) - \gamma(\tau)\big) + \log\big(\gamma(-\epsilon+\tau) - \gamma(\tau)\big)\Big]\,, \end{aligned} \tag{8.6.7}$$

where $\log z = \log|z| + i\arg z$ denotes the branch of the complex logarithm in which the argument $\arg z$ lies in the open interval $(-\frac{\pi}{2}, \frac{3\pi}{2})$ (for example, positive real numbers have argument 0 and negative real numbers have argument π).

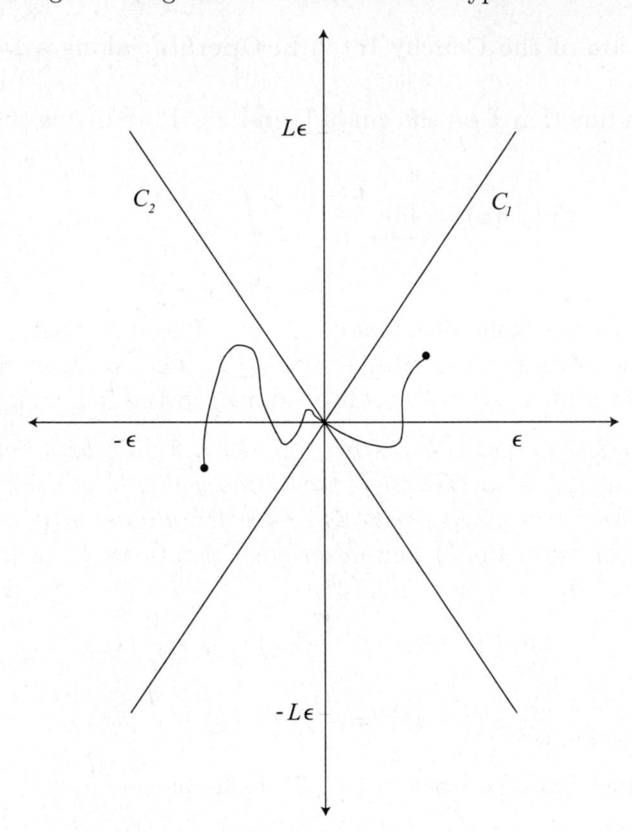

FIGURE 8.1. The cone C_1 is defined as the set of all complex numbers w such that $|\mathrm{Im}\, w| \le L\,\mathrm{Re}\, w$ while the cone C_2 is defined as the set of all complex numbers w such that $|\mathrm{Im}\, w| \le -L\,\mathrm{Re}\, w$.

Note that all the complex numbers inside the logarithms in (8.6.7) lie inside the cones C_1 and C_2 (see Figure 8.1) and the logarithm is a well-defined analytic function inside this region. We observe that the function $\gamma(\tau) = \tau + iA(\tau)$ is differentiable for almost all τ in $\mathbf{R}$ since A is a Lipschitz function and hence differentiable almost everywhere. Moreover, $\gamma'(\tau) \neq 0$ whenever γ is differentiable at τ. Fix a $\tau = \mathrm{Re}\, z$ at which γ is differentiable. To show that the expression in (8.6.7) has a limit as $\varepsilon \to 0$, write

$$\log\big(\gamma(-\epsilon+\tau)-\gamma(\tau)\big) - \log\big(\gamma(\epsilon+\tau)-\gamma(\tau)\big) = \log \frac{\dfrac{\gamma(-\epsilon+\tau)-\gamma(\tau)}{\varepsilon}}{\dfrac{\gamma(\epsilon+\tau)-\gamma(\tau)}{\varepsilon}}\,.$$

Letting $\varepsilon \to 0$, we obtain that the preceding expression converges to $\log(-1) = i\pi$. This establishes that the limit in (8.6.7) [thus in (8.6.6) and hence in (8.6.3)] exists as $\epsilon \to 0$ for almost all z on the curve. Thus $C_\Gamma(f)$ is a well-defined operator whenever f is a smooth function with decay at infinity.

We proceed with the proofs of (8.6.4) and (8.6.5). We only prove (8.6.4) since the proof of (8.6.5) is similar. For a fixed $\epsilon > 0$ we write

$$
\begin{aligned}
F(z+i\delta) &= \frac{1}{2\pi i}\int\limits_{\substack{\zeta\in\Gamma\\ |\mathrm{Re}\,\zeta-\mathrm{Re}\,z|>\epsilon}} \frac{f(\zeta)}{\zeta-z-i\delta}\,d\zeta \\
&\quad+\frac{1}{2\pi i}\int\limits_{\substack{\zeta\in\Gamma\\ |\mathrm{Re}\,\zeta-\mathrm{Re}\,z|\le\epsilon}} \frac{f(\zeta)-f(z)}{\zeta-z-i\delta}\,d\zeta \\
&\quad+f(z)\,\frac{1}{2\pi i}\int\limits_{\substack{\zeta\in\Gamma\\ |\mathrm{Re}\,\zeta-\mathrm{Re}\,z|\le\epsilon}} \frac{1}{\zeta-z-i\delta}\,d\zeta
\end{aligned} \tag{8.6.8}
$$

But the last expression can be computed easily and found to be equal to

$$
\frac{f(z)}{2\pi i}\log\frac{\gamma(\epsilon+\mathrm{Re}\,z)-z-i\delta}{\gamma(-\epsilon+\mathrm{Re}\,z)-z-i\delta},
$$

where $\log z = \log|z| + i\arg z$ denotes the same branch of the complex logarithm. So, taking limits as $\delta\downarrow 0$ in (8.6.8), we obtain that

$$
\begin{aligned}
\lim_{\delta\downarrow 0} F(z+i\delta) &= \frac{1}{2\pi i}\int\limits_{\substack{\zeta\in\Gamma\\ |\mathrm{Re}\,\zeta-\mathrm{Re}\,z|>\epsilon}} \frac{f(\zeta)}{\zeta-z}\,d\zeta \\
&\quad+\frac{1}{2\pi i}\int\limits_{\substack{\zeta\in\Gamma\\ |\mathrm{Re}\,\zeta-\mathrm{Re}\,z|\le\epsilon}} \frac{f(\zeta)-f(z)}{\zeta-z}\,d\zeta+\frac{f(z)}{2\pi i}\log\frac{\gamma(\tau+\epsilon)-\gamma(\tau)}{\gamma(\tau-\epsilon)-\gamma(\tau)},
\end{aligned} \tag{8.6.9}
$$

in which both integrals converge and we set $z=\gamma(\tau)=\tau+iA(\tau)$.

At this point $\epsilon>0$ was arbitrary and we can let it tend to zero. In doing so we first observe that the middle integral in (8.6.9) tends to zero because of the smoothness of f. But we previously showed that for almost all $\tau\in\mathbf{R}$ the limit of the logarithm in (8.6.9) is equal to πi as $\epsilon\to 0$. From this we conclude that for almost $z\in\Gamma$ we have

$$
\lim_{\delta\downarrow 0} F(z+i\delta) = \lim_{\epsilon\to 0}\frac{1}{2\pi i}\int\limits_{\substack{\zeta\in\Gamma\\ |\mathrm{Re}\,\zeta-\mathrm{Re}\,z|>\epsilon}} \frac{f(\zeta)}{\zeta-z}\,d\zeta+f(z)\,\frac{1}{2\pi i}(\pi i),
$$

which proves (8.6.4). The proof of (8.6.5) is similar.

□

Remark 8.6.2. If we let F_+ be the restriction of F on the region above the graph Γ and F_- be the restriction of F on the region below the graph Γ, we have that F_+ and F_- have continuous extensions on Γ and, moreover,

$$F_+ - F_- = f$$

where f is the given smooth function on the curve. We also note that argument given in Proposition 8.6.1 does not require f to be smoother than C^1.

8.6.b. Resolution of the Cauchy Integral and Reduction of its L^2 Boundedness to a Quadratic Estimate

Having now introduced the Cauchy integral $\mathcal{C}_\Gamma$ as an operator defined on smooth functions on the graph Γ of a Lipschitz function A, we turn to some of its properties. We are mostly interested in obtaining an a priori L^2 estimate for $\mathcal{C}_\Gamma$. Before we achieve this goal, we make some observations. First we can write $\mathcal{C}_\Gamma$ as

$$\mathcal{C}_\Gamma(H)(x+iA(x)) = \lim_{\varepsilon\to 0}\frac{1}{\pi i}\int\limits_{|x-y|>\varepsilon}\frac{H(y+iA(y))(1+iA'(y))}{y+iA(y)-x-iA(x)}\,dy\,, \tag{8.6.10}$$

where the integral is over the real line and H is a function on the curve Γ. (Recall that Lipschitz functions are differentiable almost everywhere.) To any function H on Γ we can associate a function h on the line $\mathbf{R}$ by setting

$$h(y) = H(y+iA(y))\,.$$

We have that

$$\int_{\mathbf{R}}|h(y)|^2\,dy \approx \int_{\mathbf{R}}|H(y)|^2\,dy$$

for some constants that depend on the Lipschitz constant L of A. See Exercise 8.6.1. Therefore, the boundedness of the operator in (8.6.10) is equivalent to that of the operator

$$C_\Gamma(h)(x) = \lim_{\varepsilon\to 0}\frac{1}{\pi i}\int\limits_{|x-y|>\varepsilon}\frac{h(y)(1+iA'(y))}{y-x+i(A(y)-A(x))}\,dy \tag{8.6.11}$$

acting on Schwartz functions h on the line. It is this operator that we will concentrate on in the remainder of this section. We recall that (see Example 8.1.6) the function

$$\frac{1}{y-x+i(A(y)-A(x))}$$

defined on $\mathbf{R}\times\mathbf{R}\setminus\{(x,x):\ x\in\mathbf{R}\}$ is a standard kernel in $SK(1,c\,L)$ for some $c>0$. We note that this is not the case with the kernel

$$\frac{1+iA'(y)}{y-x+i(A(y)-A(x))}\,, \tag{8.6.12}$$

for, conditions (8.1.2) and (8.1.3) fail for this kernel since the function $1+iA'$ does not possess any smoothness. [Condition (8.1.1) trivially holds for the function in

(8.6.12).] We note, however, that the L^p boundedness of the operator in (8.6.11) is equivalent to that of

$$\widetilde{C}_\Gamma(h)(x) = \lim_{\varepsilon\to 0} \frac{1}{\pi i} \int\limits_{|x-y|>\varepsilon} \frac{h(y)}{y-x+i(A(y)-A(x))}\, dy \tag{8.6.13}$$

since the function $1+iA'$ is bounded above and below and can be absorbed with h. Therefore, the L^2 boundedness of C_Γ is equivalent to that of $\widetilde{C}_\Gamma$, which has a kernel that satisfies standard estimates. This equivalence, however, will not be very useful to us in the approach we take in the sequel. We choose to work with the operator C_Γ, in which the appearance of the term $1+iA'(y)$ will play a crucial cancellation role.

In the proof of Theorem 8.3.3 we used a *resolution* of a given operator T that had the form

$$\int_0^\infty P_s T_s Q_s \frac{ds}{s},$$

where P_s and Q_s are nice averaging operators that approximate the identity and the zero operator, respectively. Our goal is to achieve a similar resolution for the operator C_Γ defined in (8.6.11). To achieve this, for every $s>0$ we introduce the auxiliary operator

$$C_\Gamma(h)(x;s) = \frac{1}{\pi i} \int\limits_{\mathbf{R}} \frac{h(y)(1+iA'(y))}{y-x+i(A(y)-A(x))+is}\, dy \tag{8.6.14}$$

defined for Schwartz functions h on the line. We make two preliminary observations regarding this operator: For almost all $x\in\mathbf{R}$ we have

$$\lim_{s\to\infty} C_\Gamma(h)(x;s) = 0, \tag{8.6.15}$$

$$\lim_{s\to 0} C_\Gamma(h)(x;s) = C_\Gamma(h)(x)+h(x). \tag{8.6.16}$$

Identity (8.6.15) is trivial. To obtain (8.6.16), for a fixed $\varepsilon>0$ we write

$$\begin{aligned} C_\Gamma(h)(x;s) &= \frac{1}{\pi i} \int\limits_{|x-y|>\varepsilon} \frac{h(y)(1+iA'(y))}{y-x+i(A(y)-A(x))+is}\, dy \\ &\quad + \frac{1}{\pi i} \int\limits_{|x-y|\le\varepsilon} \frac{(h(y)-h(x))(1+iA'(y))}{y-x+i(A(y)-A(x))+is}\, dy \\ &\quad + h(x)\frac{1}{\pi i}\log\frac{\varepsilon+i(A(x+\varepsilon)-A(x))+is}{-\varepsilon+i(A(x-\varepsilon)-A(x))+is}, \end{aligned} \tag{8.6.17}$$

where log denotes the same branch of the complex logarithm as in Proposition 8.6.1. We now take successive limits first as $s\to 0$ and then as $\varepsilon\to 0$ in (8.6.17). We obtain that

$$
\begin{aligned}
\lim_{s\to 0} C_\Gamma(h)(x;s) &= \lim_{\varepsilon\to 0}\frac{1}{\pi i}\int_{|x-y|>\varepsilon}\frac{h(y)(1+iA'(y))}{y-x+i(A(y)-A(x))}\,dy \\
&\qquad + h(x)\lim_{\varepsilon\to 0}\frac{1}{\pi i}\log\frac{\varepsilon+i(A(x+\varepsilon)-A(x))}{-\varepsilon+i(A(x-\varepsilon)-A(x))},
\end{aligned}
\tag{8.6.18}
$$

and the expression inside the logarithm in (8.6.18) is equal to that inside the logarithm in (8.6.9). But it was shown in Proposition 8.6.1 that this logarithm tends to πi a.e. as $\varepsilon \to 0$, and this concludes the proof of (8.6.16).

We now consider the second derivative in s of the auxiliary operator $C_\Gamma(h)(x;s)$. Applying integration by parts, we obtain

$$
\begin{aligned}
&\int_0^\infty s^2\frac{d^2}{ds^2}C_\Gamma(h)(x;s)\,\frac{ds}{s} \\
&= \int_0^\infty s\frac{d^2}{ds^2}C_\Gamma(h)(x;s)\,ds \\
&= \lim_{s\to\infty} s\frac{d}{ds}C_\Gamma(h)(x;s) - \lim_{s\to 0} s\frac{d}{ds}C_\Gamma(h)(x;s) - \int_0^\infty \frac{d}{ds}C_\Gamma(h)(x;s)\,ds \\
&= 0 - 0 + \lim_{s\to 0} C_\Gamma(h)(x;s) - \lim_{s\to\infty} C_\Gamma(h)(x;s) \\
&= C_\Gamma(h)(x) + h(x),
\end{aligned}
$$

where we used the facts that for almost all $x \in \mathbf{R}$ we have

$$
\lim_{s\to\infty} s\frac{d}{ds}C_\Gamma(h)(x;s) = \lim_{s\to 0} s\frac{d}{ds}C_\Gamma(h)(x;s) = 0
\tag{8.6.19}
$$

and identities (8.6.15) and (8.6.16) whenever h is a Schwartz function. We refer the reader to Exercise 8.6.2 for a proof of the identities in (8.6.19). So we have succeeded in writing the operator $C_\Gamma(h) + h$ as an average of smoother operators. Precisely, we have shown that for $h \in \mathcal{S}(\mathbf{R})$ we have

$$
C_\Gamma(h)(x) + h(x) = \int_0^\infty s^2\frac{d^2}{ds^2}C_\Gamma(h)(x;s)\,\frac{ds}{s},
\tag{8.6.20}
$$

and it remains to understand what the operator

$$
\frac{d^2}{ds^2}C_\Gamma(h)(x;s) = C_\Gamma(h)''(x;s)
$$

really is. Differentiating (8.6.14) twice, we obtain

$$\begin{aligned} C_\Gamma(h)(x)+h(x) &= \int_0^\infty s^2 C_\Gamma(h)''(x;s)\,\frac{ds}{s} \\ &= 4\int_0^\infty s^2 C_\Gamma(h)''(x;2s)\,\frac{ds}{s} \\ &= -\frac{8}{\pi i}\int_0^\infty\int_{\mathbf{R}} \frac{s^2 h(y)(1+iA'(y))}{(y-x+i(A(y)-A(x))+2is)^3}\,dy\,\frac{ds}{s} \\ &= -\frac{8}{\pi i}\int_0^\infty\int_\Gamma \frac{s^2 H(\zeta)}{(\zeta-z+2is)^3}\,d\zeta\,\frac{ds}{s}\,, \end{aligned}$$

where in the last step we set $z = x+iA(x)$, $H(z)=h(x)$, and we switched to complex integration over the curve Γ. We will now use the following simple fact from complex analysis. For $\zeta, z\in\Gamma$ we have

$$\frac{1}{(\zeta-z+2is)^3} = -\frac{1}{4\pi i}\int_\Gamma \frac{1}{(\zeta-w+is)^2}\,\frac{1}{(w-z+is)^2}\,dw\,, \tag{8.6.21}$$

for which we refer the reader to Exercise 8.6.3. Using this estimate in the preceding calculation, we obtain

$$C_\Gamma(h)(x)+h(x) = -\frac{2}{\pi^2}\int_0^\infty\left[\int_\Gamma \frac{s}{(w-z+is)^2}\left(\int_\Gamma \frac{s\,H(\zeta)}{(\zeta-w+is)^2}\,d\zeta\right)dw\right]\frac{ds}{s}\,,$$

where we recall that $z=x+iA(x)$. Introducing the linear operator

$$\Theta_s(h)(x) = \int_{\mathbf{R}} \theta_s(x,y)\,h(y)\,dy\,, \tag{8.6.22}$$

where

$$\theta_s(x,y) = \frac{s}{(y-x+i(A(y)-A(x))+is)^2}\,, \tag{8.6.23}$$

we can therefore write

$$C_\Gamma(h)(x)+h(x) = -\frac{2}{\pi^2}\int_0^\infty \Theta_s\big((1+iA')\Theta_s\big((1+iA')h\big)\big)(x)\,\frac{ds}{s}\,. \tag{8.6.24}$$

We also introduce the multiplication operator

$$M_b(h) = b\,h,$$

which enables us to write (8.6.24) in a more compact form as

$$C_\Gamma(h) = -h - \frac{2}{\pi^2}\int_0^\infty \Theta_s M_{1+iA'}\Theta_s M_{1+iA'}(h)\,\frac{ds}{s}\,. \tag{8.6.25}$$

This gives us the desired resolution of the operator C_Γ. It will suffice to obtain an L^2 estimate for the integral expression in (8.6.25). Using duality, we can write

$$\left\langle \int_0^\infty \Theta_s M_{1+iA'}\Theta_s M_{1+iA'}(h)\,\frac{ds}{s}, g\right\rangle = \int_0^\infty \big\langle M_{1+iA'}\Theta_s M_{1+iA'}(h), \Theta_s^t(g)\big\rangle\,\frac{ds}{s}\,,$$

which is easily controlled by

$$
\begin{aligned}
&\sqrt{1+L^2}\int_0^\infty \big\|\Theta_s M_{1+iA'}(h)\big\|_{L^2}\big\|\Theta_s^t(g)\big\|_{L^2}\frac{ds}{s} \\
\le\ &\sqrt{1+L^2}\bigg(\int_0^\infty \big\|\Theta_s M_{1+iA'}(h)\big\|_{L^2}^2\frac{ds}{s}\bigg)^{\frac12}\bigg(\int_0^\infty \big\|\Theta_s(g)\big\|_{L^2}^2\frac{ds}{s}\bigg)^{\frac12},
\end{aligned}
$$

and we have therefore reduced matters to the following estimate:

$$
\bigg(\int_0^\infty \big\|\Theta_s(h)\big\|_{L^2}^2\frac{ds}{s}\bigg)^{\frac12} \le C\big\|h\big\|_{L^2}. \tag{8.6.26}
$$

We will derive (8.6.26) as a consequence of Theorem 8.6.6 stated and proved later.

8.6.c. A Quadratic $T(1)$ Type Theorem

We review what we have achieved so far and we introduce definitions that place matters into a new framework.

For the purposes of the subsequent exposition we can switch to $\mathbf{R}^n$ as there are no differences from the one-dimensional argument. Suppose that for all $s>0$, there is a family of functions θ_s defined on $\mathbf{R}^n\times\mathbf{R}^n$ such that

$$
|\theta_s(x,y)| \le \frac{1}{s^n}\frac{A}{\big(1+\frac{|x-y|}{s}\big)^{n+\delta}} \tag{8.6.27}
$$

and

$$
|\theta_s(x,y)-\theta_s(x,y')| \le \frac{A}{s^n}\frac{|y-y'|^\gamma}{s^\gamma} \tag{8.6.28}
$$

for all $x,y,y'\in\mathbf{R}^n$ and some $0<\gamma,\delta,A<\infty$. Let Θ_s be the operator with kernel θ_s, that is,

$$
\Theta_s(h)(x) = \int_{\mathbf{R}^n}\theta_s(x,y)\,h(y)\,dy\,, \tag{8.6.29}
$$

which is well defined for all h in $\bigcup\limits_{1\le p\le\infty} L^p(\mathbf{R}^n)$ in view of (8.6.27).

At this point we observe that both (8.6.27) and (8.6.28) hold for the θ_s defined in (8.6.23) with $\gamma=\delta=1$ and A a constant multiple of L. We leave the details of this calculation to the reader but we note that (8.6.28) can be obtained quickly using the mean value theorem. Our goal will be to figure out under what additional conditions on Θ_s, the quadratic estimate (8.6.26) holds. If we can find such a condition that is easily verifiable for the Θ_s associated with the Cauchy integral, this will conclude the proof of its L^2 boundedness.

We first consider a simple condition that implies the quadratic estimate (8.6.26).

Theorem 8.6.3. *For $s>0$, let θ_s be a family of kernels satisfying (8.6.27) and (8.6.28) and let Θ_s be the linear operator whose kernel is θ_s. Suppose that for all $s>0$ we have*

$$\Theta_s(1)=0\,. \tag{8.6.30}$$

Then there is a constant $C_{n,\delta}$ such that for all $f\in L^2$ we have

$$\left(\int_0^\infty \big\|\Theta_s(f)\big\|_{L^2}^2\,\frac{ds}{s}\right)^{\frac12}\le C_{n,\delta}A\big\|f\big\|_{L^2}\,. \tag{8.6.31}$$

We note that condition (8.6.30) is not satisfied for the operators Θ_s associated with the Cauchy integral as defined in (8.6.22). However, Theorem 8.6.3 gives us an idea of what we are looking for, something like the action of Θ_s on a specific function. We also observe that condition (8.6.30) is "basically" saying that $\Theta(1)=0$, where

$$\Theta=\int_0^\infty \Theta_s\,\frac{ds}{s}\,.$$

Proof. We introduce Littlewood-Paley operators Q_s given by convolution with a smooth function $\Psi_s=\frac{1}{s^n}\Psi(\frac{\cdot}{s})$ whose Fourier transform is supported in the annulus $\frac{1}{2s}\le|\xi|\le 2s$ and that satisfies

$$\int_0^\infty Q_s^2\,\frac{ds}{s}=\lim_{\substack{\varepsilon\to0\\ N\to\infty}}\int_\varepsilon^N Q_s^2\,\frac{ds}{s}=I\,,$$

where the limit is taken in the sense of distributions and the identity holds in $\mathcal{S}'(\mathbf{R}^n)/\mathcal{P}$. This identity and properties of Θ_t imply the operator identity

$$\Theta_t=\Theta_t\int_0^\infty Q_s^2\,\frac{ds}{s}=\int_0^\infty \Theta_tQ_s^2\,\frac{ds}{s}\,.$$

The key fact is the following estimate:

$$\big\|\Theta_tQ_s\big\|_{L^2\to L^2}\le A\,C_{n,\Psi}\min\left(\frac{s}{t},\frac{t}{s}\right)^{\varepsilon}, \tag{8.6.32}$$

which holds for some $\varepsilon=\varepsilon(\gamma,\delta,n)>0$. [Recall that A, γ, and δ are as in (8.6.27) and (8.6.28).] Assuming momentarily estimate (8.6.32), we can quickly prove Theorem 8.6.3 using duality. Indeed, let us take a function $G(x,t)$ such that

$$\int_0^\infty\int_{\mathbf{R}^n}|G(x,t)|^2\,dx\,\frac{dt}{t}\le 1\,. \tag{8.6.33}$$

Then we have

$$
\begin{aligned}
&\int_0^\infty \int_{\mathbf{R}^n} G(x,t)\,\Theta_t(f)(x)\,dx\,\frac{dt}{t}\\
&= \int_0^\infty \int_{\mathbf{R}^n} G(x,t) \int_0^\infty \Theta_t Q_s^2(f)(x)\,\frac{ds}{s}\,dx\,\frac{dt}{t}\\
&= \int_0^\infty \int_0^\infty \int_{\mathbf{R}^n} G(x,t)\,\Theta_t Q_s^2(f)(x)\,dx\,\frac{dt}{t}\frac{ds}{s}\\
&\le \left(\int_0^\infty \int_0^\infty \int_{\mathbf{R}^n} |G(x,t)|^2 dx\,\min\Big(\frac{s}{t},\frac{t}{s}\Big)^{\varepsilon}\frac{dt}{t}\frac{ds}{s}\right)^{\frac{1}{2}}\\
&\qquad\left(\int_0^\infty \int_0^\infty \int_{\mathbf{R}^n} |\Theta_t Q_s(Q_s(f))(x)|^2\,dx\,\min\Big(\frac{s}{t},\frac{t}{s}\Big)^{-\varepsilon}\frac{dt}{t}\frac{ds}{s}\right)^{\frac{1}{2}}.
\end{aligned}
$$

But we have the estimate

$$
\sup_{s>0}\int_0^\infty \min\Big(\frac{s}{t},\frac{t}{s}\Big)^{\varepsilon}\frac{ds}{s} \le C_\varepsilon\,,
$$

which, combined with (8.6.33), yields that the first term in the product of the two preceding square functions is controlled by $\sqrt{C_\varepsilon}$. Using this fact and (8.6.32), we write

$$
\begin{aligned}
&\int_0^\infty \int_{\mathbf{R}^n} G(x,t)\,\Theta_t(f)(x)\,dx\,\frac{dt}{t}\\
&\le \sqrt{C_\varepsilon}\left(\int_0^\infty \int_0^\infty \int_{\mathbf{R}^n} |\Theta_t Q_s(Q_s(f))(x)|^2\,dx\,\min\Big(\frac{s}{t},\frac{t}{s}\Big)^{-\varepsilon}\frac{dt}{t}\frac{ds}{s}\right)^{\frac{1}{2}}\\
&\le A\sqrt{C_\varepsilon}\left(\int_0^\infty \int_0^\infty \int_{\mathbf{R}^n} |Q_s(f)(x)|^2\,dx\,\min\Big(\frac{s}{t},\frac{t}{s}\Big)^{2\varepsilon}\min\Big(\frac{s}{t},\frac{t}{s}\Big)^{-\varepsilon}\frac{dt}{t}\frac{ds}{s}\right)^{\frac{1}{2}}\\
&\le A\sqrt{C_\varepsilon}\left(\int_0^\infty \int_0^\infty \int_{\mathbf{R}^n} |Q_s(f)(x)|^2\,dx\,\min\Big(\frac{s}{t},\frac{t}{s}\Big)^{\varepsilon}\frac{dt}{t}\frac{ds}{s}\right)^{\frac{1}{2}}\\
&\le C_\varepsilon\,A\left(\int_0^\infty \int_{\mathbf{R}^n} |Q_s(f)(x)|^2\,dx\,\frac{ds}{s}\right)^{\frac{1}{2}} \le C_{n,\varepsilon}A\big\|f\big\|_{L^2}\,,
\end{aligned}
$$

where in the last step we used the continuous version of Theorem 5.1.2 (see Exercise 5.1.4). Taking the supremum over all functions $G(x,t)$ that satisfy (8.6.33) yields estimate (8.6.31).

It remains to prove (8.6.32). What it crucial here is that both Θ_t and Q_s satisfy the cancellation conditions $\Theta_t(1) = 0$ and $Q_s(1) = 0$. The proof of estimate (8.6.32) is similar to that of estimates (8.5.14) and (8.5.15) in Proposition 8.5.3. Using ideas from the proof of Proposition 8.5.3, we quickly dispose of the proof of (8.6.32).

The kernel of $\Theta_t Q_s$ is seen easily to be

$$
L_{t,s}(x,y) = \int_{\mathbf{R}^n} \theta_t(x,z)\Psi_s(z-y)\,dz\,.
$$

Notice that the function $(y,z) \to \Psi_s(z-y)$ satisfies (8.6.27) with $\delta = 1$ and $A = C_\Psi$ and satisfies

$$|\Psi_s(z-y) - \Psi_s(z'-y)| \le \frac{C_\Psi}{s^n}\frac{|z-z'|}{s}$$

for all $z, z', y \in \mathbf{R}^n$ for some $C_\Psi < \infty$. We will prove that

$$\sup_{x\in\mathbf{R}^n}\int_{\mathbf{R}^n}|L_{t,s}(x,y)|\,dy \le C_\Psi\, A \min\left(\frac{t}{s},\frac{s}{t}\right)^{\frac{1}{4}\frac{\min(\delta,1)}{n+\min(\delta,1)}\min(\gamma,\delta,1)}, \tag{8.6.34}$$

$$\sup_{y\in\mathbf{R}^n}\int_{\mathbf{R}^n}|L_{t,s}(x,y)|\,dx \le C_\Psi\, A \min\left(\frac{t}{s},\frac{s}{t}\right)^{\frac{1}{4}\frac{\min(\delta,1)}{n+\min(\delta,1)}\min(\gamma,\delta,1)}. \tag{8.6.35}$$

Once (8.6.34) and (8.6.35) are established, (8.6.32) will follow directly from the lemma in Appendix I.1 with $\varepsilon = \frac{1}{4}\frac{\min(\delta,1)}{n+\min(\delta,1)}\min(\gamma,\delta,1)$.

We begin by observing that when $t \le s$, we have the estimate

$$\int_{\mathbf{R}^n}\frac{t^{-n}\min(2,(s^{-1}|u|)^\gamma)}{(1+t^{-1}|u|)^{n+1}}\,du \le C_n\left(\frac{t}{s}\right)^{\frac{1}{2}\min(\gamma,1)}, \tag{8.6.36}$$

and when $s \le t$, we have

$$\int_{\mathbf{R}^n}\frac{s^{-n}\min(2,t^{-1}|u|)}{(1+s^{-1}|u|)^{n+\delta}}\,du \le C_n\left(\frac{s}{t}\right)^{\frac{1}{2}\min(\delta,1)}. \tag{8.6.37}$$

Both (8.6.36) and (8.6.37) are trivial reformulations of estimate (8.5.18) and can be proved in identical fashion. We leave the details to the reader.

We now take $t \le s$ and we use that $Q_s(1) = 0$ for all $s > 0$ to obtain

$$\begin{aligned}
|L_{t,s}(x,y)| &= \left|\int_{\mathbf{R}^n}\theta_t(x,z)\Psi_s(z-y)\,dz\right| \\
&= \left|\int_{\mathbf{R}^n}\big[\theta_t(x,z)-\theta_t(x,y)\big]\Psi_s(z-y)\,dz\right| \\
&\le C\,A\int_{\mathbf{R}^n}\frac{\min(2,(t^{-1}|z-y|)^\gamma)}{t^n}\frac{s^{-n}}{(1+s^{-1}|z-y|)^{n+1}}\,dz \\
&\le C\,A\frac{1}{s^n}\left(\frac{t}{s}\right)^{\frac{1}{2}\min(\gamma,1)} \le C\,A\min\left(\frac{1}{t},\frac{1}{s}\right)^n\left(\frac{t}{s}\right)^{\frac{1}{2}\min(\gamma,\delta,1)}
\end{aligned} \tag{8.6.38}$$

using estimate (8.6.36). Similarly, using (8.6.37) and the hypothesis that $\Theta_t(1) = 0$ for all $t > 0$, we obtain for $s \le t$

$$
\begin{aligned}
|L_{t,s}(x,y)| &= \left| \int_{\mathbf{R}^n} \theta_t(x,z)\Psi_s(z-y)\,dz \right| \\
&= \left| \int_{\mathbf{R}^n} \theta_t(x,z)\big[\Psi_s(z-y) - \Psi_s(x-y)\big]\,dz \right| \\
&\le C\,A \int_{\mathbf{R}^n} \frac{t^{-n}}{(1+t^{-1}|x-z|)^{n+\delta}} \frac{\min(2,(s^{-1}|x-z|)^{\gamma})}{s^n}\,dz \\
&\le C\,A \frac{1}{t^n}\Big(\frac{t}{s}\Big)^{\frac{1}{2}\min(\delta,1)} \le C\,A \min\Big(\frac{1}{t},\frac{1}{s}\Big)^n \Big(\frac{t}{s}\Big)^{\frac{1}{2}\min(\gamma,\delta,1)} .
\end{aligned}
\tag{8.6.39}
$$

Combining (8.6.38) and (8.6.39) with

$$
|L_{t,s}(x,y)| \le \int_{\mathbf{R}^n} |\theta_t(x,z)|\,|\Psi_s(z-y)|\,dz \le \frac{C\,A\min(\frac{1}{t},\frac{1}{s})^n}{\big(1+\min(\frac{1}{t},\frac{1}{s})|x-y|\big)^{n+\min(\delta,1)}} ,
$$

which follows from the result in Appendix K.1, gives

$$
|L_{t,s}(x,y)| \le \frac{C\,\min\Big(\frac{t}{s},\frac{s}{t}\Big)^{\frac{1}{2}\min(\gamma,\delta,1)(1-\beta)} A\min(\frac{1}{t},\frac{1}{s})^n}{\Big(\big(1+\min(\frac{1}{t},\frac{1}{s})|x-y|\big)^{n+\min(\delta,1)}\Big)^{\beta}}
$$

for any $0 < \beta < 1$. Choosing $\beta = (n + \frac{1}{2}\min(\delta,1))(n+\min(\delta,1))^{-1}$ yields (8.6.34) and (8.6.35) and concludes the proof of estimate (8.6.32).

□

We end this subsection with a small generalization of the previous theorem that follows by an examination of its proof. The simple details are left to the reader.

Corollary 8.6.4. *For $s > 0$ let Θ_s be linear operators that are uniformly bounded on $L^2(\mathbf{R}^n)$ by a constant B. Suppose that for some $C_{n,\Psi}, A, \varepsilon < \infty$, estimate (8.6.32) is satisfied for all $t, s > 0$. Then there is a constant $C_{n,\Psi,\varepsilon}$ such that estimate (8.6.31) holds for all $f \in L^2$, that is,*

$$
\Big(\int_0^\infty \|\Theta_s(f)\|_{L^2}^2 \frac{ds}{s}\Big)^{\frac{1}{2}} \le C_{n,\Psi,\varepsilon}(A+B)\|f\|_{L^2} .
$$

8.6.d. A $T(b)$ Theorem and the L^2 Boundedness of the Cauchy Integral

It turns out that the operators Θ_s defined in (8.6.22) and (8.6.23) that appear in the resolution of the Cauchy integral operator $\mathcal{C}_\Gamma$ do not satisfy the condition $\Theta_s(1) = 0$ of Theorem 8.6.3. It will turn out that a certain variation of this theorem will be needed for the purposes of the application we have in mind, the L^2 boundedness of the Cauchy integral operator. This variation is a quadratic type

$T(b)$ theorem and is discussed in this subsection. Before we state the main theorem, we will need a definition.

Definition 8.6.5. An L^∞ complex-valued function b on $\mathbf{R}^n$ is said to be *accretive* if there is a constant $c_0 > 0$ such that $\operatorname{Re} b(x) \geq c_0$ for almost all $x \in \mathbf{R}^n$.

The following theorem is the main result of this section.

Theorem 8.6.6. *Let θ_s be a complex-valued function on $\mathbf{R}^n \times \mathbf{R}^n$ that satisfies (8.6.27) and (8.6.28), and let Θ_s be the linear operator in (8.6.29) whose kernel is θ_s. If there is an accretive function b such that*

$$\Theta_s(b) = 0 \tag{8.6.40}$$

for all $s > 0$, then there is a constant $C_n(b)$ such that the estimate

$$\left(\int_0^\infty \|\Theta_s(f)\|_{L^2}^2 \frac{ds}{s} \right)^{\frac{1}{2}} \leq C_n(b) \|f\|_{L^2} \tag{8.6.41}$$

holds for all $f \in L^2$.

Corollary 8.6.7. *The Cauchy integral operator $\mathcal{C}_\Gamma$ maps $L^2(\mathbf{R})$ into itself.*

The corollary is a consequence of Theorem 8.6.6 and of the crucial and important cancellation property

$$\Theta_s(1 + iA') = 0 \tag{8.6.42}$$

satisfied by the accretive function $1 + iA'$, where Θ_s and θ_s here are as in (8.6.22) and (8.6.23). To prove (8.6.42) we simply note that

$$\begin{aligned} \Theta_s(1 + iA')(x) &= \int_{\mathbf{R}} \frac{s\,(1 + iA'(y))\,dy}{(y - x + i(A(y) - A(x)) + is)^2} \\ &= \lim_{y \to \infty} \left[\frac{-s}{y - x + i(A(y) - A(x)) + is} \right]_{y=-\infty}^{y=+\infty} = 0 - 0 = 0\,. \end{aligned}$$

This condition plays exactly the role of (8.6.30), which may fail in general. The necessary "internal cancellation" of the family of operators Θ_s is exactly captured by the single condition (8.6.42).

It remains to prove Theorem 8.6.6.

Proof. We fix an approximation of the identity operator, such as

$$P_s(f)(x) = \int_{\mathbf{R}^n} \Phi_s(x - y)\, f(y)\, dy\,,$$

where $\Phi_s(x) = s^{-n}\Phi(s^{-1}x)$, and Φ is a nonnegative Schwartz function with integral 1. Then P_s is a nice positive averaging operator that satisfies $P_s(1) = 1$ for all $s > 0$. The key idea is to decompose the operator Θ_s as

$$\Theta_s = (\Theta_s - M_{\Theta_s(1)} P_s) + M_{\Theta_s(1)} P_s\,, \tag{8.6.43}$$

where $M_{\Theta_s(1)}$ is the operator given by multiplication by $\Theta_s(1)$. We begin with the first term in (8.6.43), which is essentially an error term. We simply observe that

$$\big(\Theta_s - M_{\Theta_s(1)}P_s\big)(1) = \Theta_s(1) - \Theta_s(1)\,P_s(1) = \Theta_s(1) - \Theta_s(1) = 0\,.$$

Therefore, Theorem 8.6.3 is applicable once we check that the kernel of the operator $\Theta_s - M_{\Theta_s(1)}P_s$ satisfies (8.6.27) and (8.6.28). But these are verified easily since the kernels of both Θ_s and P_s satisfy these estimates and $\Theta_s(1)$ is a bounded function uniformly in s. The latter statement is a consequence of condition (8.6.27).

We now need to obtain the required quadratic estimate for the term $M_{\Theta_s(1)}P_s$. With the use of Theorem 7.3.7, this will follow once we prove that the measure

$$|\Theta_s(1)(x)|^2 \frac{dx\,ds}{s}$$

is Carleson. It is here where we use condition (8.6.40). Since $\Theta_s(b) = 0$ we have

$$P_s(b)\,\Theta_s(1) = \big(P_s(b)\,\Theta_s(1) - \Theta_s P_s(b)\big) + \big(\Theta_s P_s(b) - \Theta_s(b)\big)\,. \tag{8.6.44}$$

Suppose we could show that the measures

$$\big|\Theta_s(b)(x) - \Theta_s P_s(b)(x)\big|^2 \frac{dx\,ds}{s} \tag{8.6.45}$$

$$\big|\Theta_s P_s(b)(x) - P_s(b)(x)\,\Theta_s(1)(x)\big|^2 \frac{dxds}{s} \tag{8.6.46}$$

are Carleson. Then it would follow from (8.6.44) that the measure

$$\big|P_s(b)(x)\,\Theta_s(1)(x)\big|^2 \frac{dx\,ds}{s}$$

is also Carleson. Using the accretivity condition on b and the positivity of P_s we obtain

$$\big|P_s(b)\big| \ge \operatorname{Re} P_s(b) = P_s(\operatorname{Re} b) \ge P_s(c_0) = c_0,$$

from which it follows that $\big|\Theta_s(1)(x)\big|^2 \le \frac{1}{c_0^2}\big|P_s(b)(x)\,\Theta_s(1)(x)\big|^2$. Thus the measure $\big|\Theta_s(1)(x)\big|^2 \frac{dx\,ds}{s}$ must be Carleson.

Therefore, our proof will be complete if we can show that both measures (8.6.45) and (8.6.46) are Carleson. Theorem 7.3.8 will play a role here.

We begin with the measure in (8.6.45). First we observe that the kernel

$$L_s(x,y) = \int_{\mathbf{R}^n} \theta_s(x,z)\Phi_s(z-y)\,dz$$

of $\Theta_s P_s$ satisfies (8.6.27) and (8.6.28). The verification of (8.6.27) is a straightforward consequence of the result in Appendix K.1, while (8.6.28) follows easily from the mean value theorem. It follows that the kernel of

$$R_s = \Theta_s - \Theta_s P_s$$

satisfies the same estimates. Moreover, it is easy to see that $R_s(1) = 0$ and thus the quadratic estimate (8.6.31) holds for R_s in view of Theorem 8.6.3. Therefore, the hypotheses of Theorem 7.3.8(c) are satisfied, and this gives that the measure in (8.6.45) is Carleson.

We now continue with the measure in (8.6.46). Here we set

$$T_s(f)(x) = \Theta_s P_s(f)(x) - P_s(f)(x)\Theta_s(1)(x)\,.$$

The kernel of T_s is $L_s(x,y) - \Theta_s(1)(x)\Phi_s(x-y)$ which clearly satisfies (8.6.27) and (8.6.28) since $\Theta_s(1)(x)$ is a bounded function uniformly in $s>0$. We also observe that $T_s(1) = 0$. Using Theorem 8.6.3, we conclude that the quadratic estimate (8.6.31) holds for T_s. Therefore, the hypotheses of Theorem 7.3.8(c) are satisfied; hence the measure in (8.6.45) is Carleson.

□

We conclude by observing that if we attempt to replace Θ_s with $\widetilde{\Theta}_s = \Theta_s M_{1+iA'}$ in the resolution identity (8.6.25), then $\widetilde{\Theta}_s(1) = 0$ would hold, but the kernel of $\widetilde{\Theta}_s$ would not satisfy the regularity estimate (8.6.28). The whole purpose of Theorem 8.6.6 was to find a certain balance between regularity and cancellation.

Exercises

8.6.1. Given a function H on a Lipschitz graph Γ, we associate a function h on the line by setting $h(t) = H(t+iA(t))$. Prove that for all $0<p<\infty$ we have

$$\|h\|_{L^p(\mathbf{R})}^p \le \|H\|_{L^p(\Gamma)}^p \le \sqrt{1+L^2}\,\|h\|_{L^p(\mathbf{R})}^p$$

where L is the Lipschitz constant of the defining function A of the graph Γ.

8.6.2. Let $A:\mathbf{R}\to\mathbf{R}$ satisfy $|A(x)-A(y)|\le L|x-y|$ for all $x,y\in\mathbf{R}$ for some $L>0$. Also, let h be a Schwartz function on $\mathbf{R}$.
(a) Show that for all $s>0$ and $x,y\in\mathbf{R}$ we have

$$\frac{s^2+|x-y|^2}{|x-y|^2+|A(x)-A(y)+s|^2} \le 4L^2+2\,.$$

(b) Use the Lebesgue dominated convergence theorem to prove that

$$\int_{|x-y|>\sqrt{s}} \frac{s(1+iA'(y))h(y)}{(y-x+i(A(y)-A(x))+is)^2}\,dy \to 0$$

as $s\to 0$.
(c) Integrate directly to show that as $s\to 0$,

$$\int_{|x-y|\le\sqrt{s}} \frac{s(1+iA'(y))}{(y-x+i(A(y)-A(x))+is)^2}\,dy \to 0$$

for every point x at which A is differentiable.
(d) Use part (a) and the Lebesgue dominated convergence theorem to show

that as $s \to 0$,

$$\int_{|x-y|\le\sqrt{s}} \frac{s(1+iA'(y))(h(y)-h(x))}{(y-x+i(A(y)-A(x))+is)^2}\,dy \to 0\,.$$

(e) Use part (a) and the Lebesgue dominated convergence theorem to show that as $s \to \infty$,

$$\int_{\mathbf{R}} \frac{s(1+iA'(y))h(y)}{(y-x+i(A(y)-A(x))+is)^2}\,dy \to 0\,.$$

Conclude the validity of the statements in (8.6.19) for almost all $x \in \mathbf{R}$.

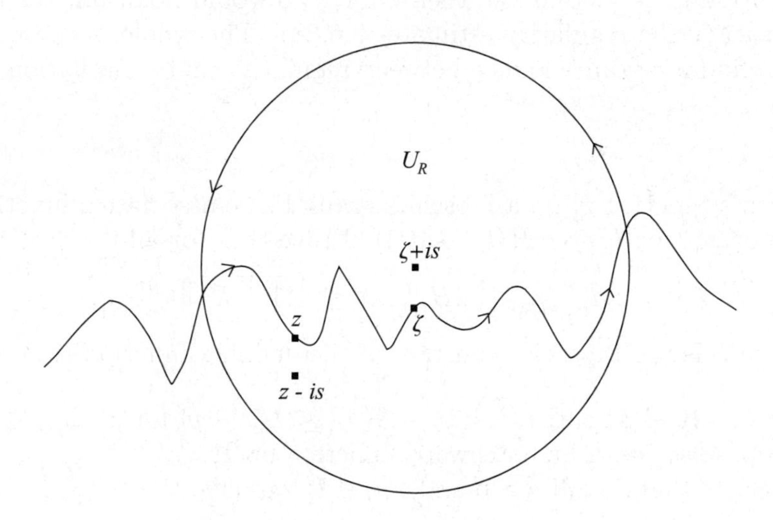

FIGURE 8.2. The region inside the circle and above the curve is U_R.

8.6.3. Prove identity (8.6.21).
[*Hint:* Write the identity in (8.6.21) as

$$\frac{-2}{((\zeta+is)-(z-is))^3} = \frac{1}{2\pi i}\int_\Gamma \frac{\frac{1}{(w-(z-is))^2}}{(w-(\zeta+is))^2}\,dw$$

and interpret it as Cauchy's integral formula for the derivative of the analytic function $w \to (w-(z-is))^{-2}$ defined on the region above Γ. If Γ were a closed curve containing $\zeta+is$ but not $z-is$, then the previous assertion would be immediate. In general, consider a circle of radius R centered at the point $\zeta+is$ and the region U_R inside this circle and above Γ. See Figure 8.2. Integrate over the boundary of U_R and let $R \to \infty$.]

8.6.4. Given an accretive function b, define an inner product

$$\langle f,g\rangle_b = \int_{\mathbf{R}^n} f(x)\,g(x)\,b(x)\,dx$$

on L^2. For an interval I, set $b_I = \int_I b(x)\,dx$. Let I_L denote the left half of a dyadic interval I and I_R denote its right half. For a complex number z, set $z^{\frac{1}{2}}$ be $e^{\frac{1}{2}\log z}$, where the logarithm is as in the proof of Proposition 8.6.1. Show that the family of functions

$$h_I = \frac{1}{b(I)^{\frac{1}{2}}}\left(\frac{b(I_L)^{\frac{1}{2}}}{b(I_R)^{\frac{1}{2}}}\chi_{I_L} - \frac{b(I_R)^{\frac{1}{2}}}{b(I_L)^{\frac{1}{2}}}\chi_{I_R}\right),$$

where I runs over all dyadic intervals, is an orthonormal family on $L^2(\mathbf{R})$ with respect to the preceding inner product. (This family of functions is called a *pseudo-Haar basis associated with* b.)

8.6.5. Let $I = (a,b)$ be a dyadic interval and let $3I$ be its triple. For a given $x \in \mathbf{R}$, let $d_I(x) = \min(|x-a|, |x-b|, |x-\frac{a+b}{2}|)$. Show that there exists a constant C such that

$$\big|C_\Gamma(h_I)(x)\big| \le C|I|^{-\frac{1}{2}} \log\frac{10|I|}{|x-d_I(x)|}$$

whenever $x \in 3I$ and also

$$\big|C_\Gamma(h_I)(x)\big| \le C|I|^{\frac{3}{2}}\frac{1}{|x-d_I(x)|^2}$$

for $x \notin 3I$. In the latter case $d_I(x)$ can be any of $a, b, \frac{a+b}{2}$.

8.6.6. Suppose that T is a linear operator associated with a standard kernel in $SK(\delta, A)$. Also suppose that $T^t(1) = 0$ and that $T(b) = 0$ for some accretive function b. Conclude that T is L^2 bounded.
[*Hint:* Pick a radial Schwartz function Ψ with integral zero such that $\int_0^\infty Q_s Q_s \frac{ds}{s}$ is a resolution of the identity operator, where $Q_s f = f * \Psi_s$, and $\Psi_s(x) = s^{-n}\Psi(s^{-1}x)$. Then write $T = \int_0^\infty Q_s\Theta_s \frac{ds}{s}$, where $\Theta_s = Q_s T$. Use Theorem 8.6.6 to obtain the required conclusion.]

8.6.7. (*Semmes* [**454**]) We say that a bounded function b is *para-accretive* if for all $s > 0$ there is a linear operator R_s with kernel satisfying (8.6.27) and (8.6.28) such that $|R_s(b)| \ge c_0$ for all $s > 0$. Let Θ_s and P_s be as in Theorem 8.6.6.
(a) Prove that

$$\big|R_s(b)(x) - R_s(1)(x)P_s(b)(x)\big|^2\,\frac{dx\,ds}{s}$$

is a Carleson measure.

(b) Use the result in part (a) and the fact that $\sup_{s>0}|R_s(1)| \le C$ to obtain that

$$\chi_\Omega(x,s)\,\frac{dx\,ds}{s}$$

is a Carleson measure, where

$$\Omega = \Big\{(x,s):\ |P_s(b)(x)| \le \frac{c_0}{2}\big(\sup_{s>0}|R_s(1)|\big)^{-1}\Big\}.$$

(c) Conclude that the measure

$$\big|\Theta_s(1)(x)\big|^2\,\frac{dx\,ds}{s}$$

is Carleson, thus obtaining a generalization of Theorem 8.6.6 for para-accretive functions.

8.6.8. Using the operator $\widetilde{C}_\gamma$ defined in (8.6.13), obtain that C_Γ is of weak type $(1,1)$ and bounded on $L^p(\mathbf{R})$ for all $1 < p < \infty$.

8.7. Square Roots of Elliptic Operators*

In this section we will prove an L^2 estimate for the square root of a divergence form second-order elliptic operator on $\mathbf{R}^n$. This estimate is based on an approach in the spirit of the $T(b)$ theorem discussed in the previous section. However, matters here are significantly more complicated because of two main reasons: the roughness of the variable coefficients of the aforementioned elliptic operator and the higher-dimensional nature of the problem.

8.7.a. Preliminaries and Statement of the Main Result

For $\xi = (\xi_1,\dots,\xi_n) \in \mathbf{C}^n$ we denote its complex conjugate $(\overline{\xi_1},\dots,\overline{\xi_n})$ by $\overline{\xi}$. Moreover, for $\xi,\zeta \in \mathbf{C}^n$ we use the inner product notation

$$\xi\cdot\zeta = \sum_{k=1}^n \xi_k\,\zeta_k.$$

Throughout this section $A = A(x)$ will be an $n\times n$ matrix of complex-valued L^∞ functions, defined on $\mathbf{R}^n$. We will assume everywhere in this section that the matrix A satisfies the *ellipticity* (or *accretivity*) conditions for some $0 < \lambda \le \Lambda < \infty$

$$\begin{aligned} &\lambda|\xi|^2 \le \operatorname{Re}\big(A(x)\,\xi\cdot\overline{\xi}\big) \\ &|A(x)\,\xi\cdot\overline{\zeta}| \le \Lambda|\xi|\,|\zeta| \end{aligned} \tag{8.7.1}$$

for all $x \in \mathbf{R}^n$ and $\xi, \zeta \in \mathbf{C}^n$. We will interpret an element ξ of $\mathbf{C}^n$ as a column vector in $\mathbf{C}^n$ when the matrix A acts on it.

Associated with such a matrix A, we define a second order *divergence form operator*

$$L(f) = -\operatorname{div}(A\nabla f) = \sum_{j=1}^{n} \partial_j\big((A\nabla f)_j\big), \tag{8.7.2}$$

which we interpret in the weak sense whenever f is a distribution.

The accretivity condition (8.7.1) enables us to define a square root operator $L^{1/2} = \sqrt{L}$ so that we have the operator identity $L = \sqrt{L}\sqrt{L}$. The *square root operator* can be written in several ways, one of which is

$$\sqrt{L}(f) = \frac{16}{\pi}\int_0^{+\infty} (I + t^2 L)^{-3} t^3 L^2(f)\,\frac{dt}{t}. \tag{8.7.3}$$

We refer the reader to Exercise 8.7.3 for the existence of the square root operator and the validity of identity (8.7.3).

An important problem in the subject is to determine whether the estimate

$$\big\|\sqrt{L}(f)\big\|_2 \le C_{n,\lambda,\Lambda}\big\|\nabla f\big\|_2 \tag{8.7.4}$$

holds for functions f in a dense subspace of the homogeneous Sobolev space $\dot{L}^2_1(\mathbf{R}^n)$, where $C_{n,\lambda,\Lambda}$ is a constant depending only on n, λ and Λ. Once (8.7.4) is known for a dense subspace of $\dot{L}^2_1(\mathbf{R}^n)$, then it can be extended to the entire space by density. The main purpose of this section is to discuss a detailed proof of the following result.

Theorem 8.7.1. *Let L be as in (8.7.2). Then there is a constant $C_{n,\lambda,\Lambda}$ such that for all smooth functions f with compact support estimate (8.7.4) is valid.*

The proof of this theorem requires some background and estimates concerning elliptic operators. These are presented in the next subsection; while the proof of the theorem follows in the remaining four subsections.

8.7.b. Estimates for Elliptic Operators on $\mathbf{R}^n$

The following lemma provides a quantitative expression for the mean decay of the resolvent kernel.

Lemma 8.7.2. *Let E and F be two closed sets of $\mathbf{R}^n$ and set*

$$d = \operatorname{dist}(E, F)$$

the distance between E and F. Then for all complex-valued functions f supported in E and all vector-valued functions $\vec{f}$ supported in E, we have

$$\int_F |(I + t^2 L)^{-1}(f)(x)|^2\,dx \le C e^{-c\frac{d}{t}} \int_E |f(x)|^2\,dx, \tag{8.7.5}$$

$$\int_F |t\nabla(I+t^2L)^{-1}(f)(x)|^2\,dx \le Ce^{-c\frac{d}{t}}\int_E |f(x)|^2\,dx\,, \tag{8.7.6}$$

$$\int_F |(I+t^2L)^{-1}(t\operatorname{div}\vec{f}\,)(x)|^2\,dx \le Ce^{-c\frac{d}{t}}\int_E |\vec{f}\,(x)|^2\,dx\,, \tag{8.7.7}$$

where $c = c(\lambda,\Lambda)$, $C = C(n,\lambda,\Lambda)$ are finite constants.

Proof. It suffices to obtain these inequalities whenever $d \ge t > 0$. Set $u_t = (I+t^2L)^{-1}(f)$. For all $v \in L^2_1(\mathbf{R}^n)$ we have

$$\int_{\mathbf{R}^n} u_t v\,dx + t^2\int_{\mathbf{R}^n} A\nabla u_t\cdot\nabla v\,dx = \int_{\mathbf{R}^n} fv\,dx\,.$$

Let η be a nonnegative smooth function with compact support that does not meet E and which satisfies with $\|\eta\|_\infty = 1$. Taking $v = \overline{u_t}\,\eta^2$ and using that f is supported in E, we obtain

$$\int_{\mathbf{R}^n} |u_t|^2\eta^2\,dx + t^2\int_{\mathbf{R}^n} A\nabla u_t\cdot\overline{\nabla u_t}\;\eta^2\,dx = -2t^2\int_{\mathbf{R}^n} A(\eta\nabla u_t)\cdot\overline{u_t\nabla\eta}\,dx\,.$$

Using (8.7.1) and the inequality $2ab \le \varepsilon|a|^2 + \varepsilon^{-1}|b|^2$, we obtain for all $\varepsilon > 0$

$$\begin{aligned}&\int_{\mathbf{R}^n} |u_t|^2\eta^2\,dx + \lambda t^2\int_{\mathbf{R}^n} |\nabla u_t|^2\;\eta^2\,dx\\&\qquad\le\;\Lambda\varepsilon t^2\int_{\mathbf{R}^n} |\nabla u_t|^2\;\eta^2\,dx + \Lambda\varepsilon^{-1}t^2\int_{\mathbf{R}^n} |u_t|^2|\nabla\eta|^2\,dx\,,\end{aligned}$$

and this reduces to

$$\int_{\mathbf{R}^n} |u_t|^2|\eta|^2\,dx \le \frac{\Lambda^2 t^2}{\lambda}\int_{\mathbf{R}^n} |u_t|^2|\nabla\eta|^2\,dx \tag{8.7.8}$$

by choosing $\varepsilon = \frac{\lambda}{\Lambda}$. Replacing η by $e^{k\eta}-1$ in (8.7.8), where $k = \frac{\sqrt{\lambda}}{2\Lambda t\|\nabla\eta\|_\infty}$, yields

$$\int_{\mathbf{R}^n} |u_t|^2|e^{k\eta}-1|^2\,dx \le \frac{1}{4}\int_{\mathbf{R}^n} |u_t|^2|e^{k\eta}|^2\,dx\,. \tag{8.7.9}$$

Using that $|e^{k\eta}-1|^2 \ge \frac{1}{2}|e^{k\eta}|^2 - 1$, we obtain

$$\int_{\mathbf{R}^n} |u_t|^2|e^{k\eta}|^2\,dx \le 4\int_{\mathbf{R}^n} |u_t|^2\,dx \le 4\,C\int_E |f|^2\,dx\,,$$

where in the last estimate we use the uniform boundedness of $(I+t^2L)^{-1}$ on $L^2(\mathbf{R}^n)$ (Exercise 8.7.2). If, in addition, we have $\eta = 1$ on F, then

$$|e^k|^2\int_F |u_t|^2\,dx \le \int_{\mathbf{R}^n} |u_t|^2|e^{k\eta}|^2\,dx\,,$$

and picking η so that $\|\nabla\eta\|_\infty \approx 1/d$ we conclude (8.7.5).

Next, choose $\varepsilon = \lambda/2\Lambda$ and η as before to obtain

$$\begin{aligned}\int_F |t\nabla u_t|^2\,dx &\le \int_{\mathbf{R}^n} |t\nabla u_t|^2\eta^2\,dx\\ &\le \frac{2\Lambda^2 t^2}{\lambda}\int_{\mathbf{R}^n} |u_t|^2|\nabla\eta|^2\,dx\\ &\le Ct^2 d^{-2} e^{-c\frac{d}{t}}\int_E |f|^2\,dx,\end{aligned}$$

which gives (8.7.6). Finally (8.7.7) is obtained by duality from (8.7.6) applied to $L^* = -\mathrm{div}\,(A^*\nabla)$ when the roles of E and F are interchanged.

□

Lemma 8.7.3. *Let M_f be the operator given by multiplication with a Lipschitz function f. Then there is a constant C that depends only on n, λ and Λ such that*

$$\big\|\,\big[(I+t^2L)^{-1}, M_f\big]\,\big\|_{L^2\to L^2} \le C\,t\,\big\|\nabla f\big\|_{\infty} \tag{8.7.10}$$

and

$$\big\|\,\nabla\big[(I+t^2L)^{-1}, M_f\big]\,\big\|_{L^2\to L^2} \le C\,\big\|\nabla f\big\|_{\infty} \tag{8.7.11}$$

for all $t>0$. Here $[T,S] = TS - ST$ is the commutator *of the operators T and S.*

Proof. Set $\vec{b} = A\nabla f$, $\vec{d} = A^t\nabla f$ and note that the operators given by pointwise multiplication with these vectors are L^2 bounded with norms at most a multiple of $C\big\|\nabla f\big\|_{\infty}$. Write

$$\begin{aligned}\big[(I+t^2L)^{-1}, M_f\big] &= -(I+t^2L)^{-1}\big[(I+t^2L), M_f\big](I+t^2L)^{-1}\\ &= -(I+t^2L)^{-1}t^2(\mathrm{div}\,\vec{b} + \vec{d}\cdot\nabla)(1+t^2L)^{-1}\,.\end{aligned}$$

The uniform L^2 boundedness of $(I+t^2L)^{-1}\,t\nabla(I+t^2L)^{-1}$ and $(I+t^2L)^{-1}t\,\mathrm{div}$ on L^2 (see Exercise 8.7.2) implies (8.7.10). Finally, using the L^2 boundedness of the operator $t^2\nabla(I+t^2L)^{-1}\mathrm{div}$ yields (8.7.11).

□

Next we have a technical lemma concerning the mean square deviation of f from $(I+t^2L)^{-1}$.

Lemma 8.7.4. *There exists a constant C depending only on n, λ, and Λ such that for all Q cubes in $\mathbf{R}^n$ with sides parallel to the axes, for all $t\le \ell(Q)$, and all Lipschitz functions f on $\mathbf{R}^n$ we have*

$$\frac{1}{|Q|}\int_Q |(I+t^2L)^{-1}(f) - f|^2\,dx \le C\,t^2\big\|\nabla f\big\|_{\infty}^2, \tag{8.7.12}$$

$$\frac{1}{|Q|}\int_Q |\nabla((I+t^2L)^{-1}(f) - f)|^2\,dx \le C\,\big\|\nabla f\big\|_{\infty}^2. \tag{8.7.13}$$

Proof. We begin by proving (8.7.12) while we omit the proof of (8.7.13) since it is similar. By a simple rescaling, we may assume that $\ell(Q) = 1$ and that $\|\nabla f\|_\infty = 1$. Set $Q_0 = 2Q$ (i.e., the cube with the same center as Q with twice its side length) and write $\mathbf{R}^n$ as a union of cubes Q_k of side length 2 with disjoint interiors and sides parallel to the axes. Lemma 8.7.2 implies that

$$(I + t^2 L)^{-1}(1) = 1$$

in the sense that

$$\lim_{R\to\infty} (I + t^2 L)^{-1}(\eta_R) = 1$$

in $L^2_{\text{loc}}(\mathbf{R}^n)$, where $\eta_R(x) = \eta(x/R)$ and η is a smooth bump function with $\eta \equiv 1$ near 0. Hence, we may write

$$(I + t^2 L)^{-1}(f)(x) - f(x) = \sum_{k\in\mathbf{Z}^n} (I + t^2 L)^{-1}((f - f(x))\chi_{Q_k})(x) = \sum_{k\in\mathbf{Z}^n} g_k(x).$$

The term for $k = 0$ in the sum is $[(I+t^2 L)^{-1}, M_f](\chi_{Q_0})(x)$. Hence, its $L^2(Q)$ norm is controlled by $Ct\|\chi_{Q_0}\|_2$ by (8.7.10). The terms for $k \neq 0$ are dealt with using the further decomposition

$$g_k(x) = (I + t^2 L)^{-1}((f - f(x_k))\chi_{Q_k})(x) + (f(x_k) - f(x))(I + t^2 L)^{-1}(\chi_{Q_k})(x)\,,$$

where x_k is the center of Q_k. Applying Lemma 8.7.2 for $(I + t^2 L)^{-1}$ on the sets $E = Q_k$ and $F = Q$ and using that f is Lipschitz, we obtain

$$\int_Q |g_k|^2\, dx \le Ct^2 e^{-c\frac{|x_k|}{t}} \|\chi_{Q_k}\|_2^2 = Ct^2 e^{-c\frac{|x_k|}{t}} 2^n |Q|.$$

The desired bound on the $L^2(Q)$ norm of $(I + t^2 L)^{-1}(f) - f$ follows from these estimates, Minkowski's inequality, and the fact that $t \le 1 = \ell(Q)$.

□

8.7.c. Reduction to a Quadratic Estimate

We are given a divergence form elliptic operator as in (8.7.2) with ellipticity constants λ and Λ in (8.7.1). Our goal is to obtain the a priori estimate (8.7.4) for functions f in some dense subspace of $\dot{L}^2_1(\mathbf{R}^n)$.

To obtain this estimate we will need to resolve the operator $\sqrt{L}$ as an average of simpler operators that are uniformly bounded from $\dot{L}^2_1(\mathbf{R}^n)$ into $L^2(\mathbf{R}^n)$. In the sequel we will use the following resolution of the square root:

$$\sqrt{L}(f) = \frac{16}{\pi}\int_0^\infty (I + t^2 L)^{-3} t^3 L^2(f)\,\frac{dt}{t},$$

in which the integral converges in $L^2(\mathbf{R}^n)$ for $f \in C_0^\infty(\mathbf{R}^n)$. Take $g \in C_0^\infty(\mathbf{R}^n)$ with $\|g\|_2 = 1$. Using duality and the Cauchy-Schwarz inequality, we can control the quantity $\left|\langle\sqrt{L}(f)\,|\,g\rangle\right|^2$ by

$$\frac{256}{\pi^2}\left(\int_0^\infty \big\|(I+t^2L)^{-1}tL(f)\big\|_2^2\,\frac{dt}{t}\right)\left(\int_0^\infty \big\|V_t(g)\big\|_2^2\,\frac{dt}{t}\right), \tag{8.7.14}$$

where we set

$$V_t = t^2L^*(I+t^2L^*)^{-2}\,.$$

Here L^* is the adjoint operator to L and note that the matrix corresponding to L^* is the conjugate-transpose matrix A^* of A (i.e., the transpose of the matrix whose entries are the complex conjugates of the matrix A). We explain why the estimate

$$\int_0^\infty \big\|V_t(g)\big\|_2^2\,\frac{dt}{t} \le C\big\|g\big\|_2^2 \tag{8.7.15}$$

is valid. Fix a real-valued function $\Psi \in C_0^\infty(\mathbf{R}^n)$ with mean value zero and normalized (by multiplying by a suitable constant) so that

$$\int_0^\infty \big\|Q_s(g)\big\|_{L^2}^2\,\frac{ds}{s} = \big\|g\big\|_{L^2}^2$$

for all L^2 functions g. Define $\Psi_s(x) = \frac{1}{s^n}\Psi(\frac{x}{s})$ and an operator Q_s by setting

$$Q_s(h) = h * \Psi_s \tag{8.7.16}$$

for some function Ψ to be chosen.

We will obtain estimate (8.7.15) as a consequence of Corollary 8.6.4. applied to the operators V_t that have uniform (in t) bounded extensions on $L^2(\mathbf{R}^n)$. To apply Corollary 8.6.4, we need to check that condition (8.6.32) holds for $\Theta_t = V_t$. Since

$$V_tQ_s = -(I+t^2L^*)^{-2}t^2\operatorname{div} A^*\nabla Q_s\,,$$

we have

$$\big\|V_tQ_s\big\|_{L^2\to L^2} \le \big\|(I+t^2L^*)^{-2}t^2\operatorname{div} A^*\big\|_{L^2\to L^2}\big\|\nabla Q_s\big\|_{L^2\to L^2} \le c\,\frac{t}{s}\,, \tag{8.7.17}$$

with C depending only on n, λ and Λ. Choose $\Psi = \Delta\varphi$ with $\varphi \in C_0^\infty(\mathbf{R}^n)$ radial so that, in particular, $\Psi = \operatorname{div}\vec{h}$. This yields $Q_s = s\operatorname{div}\vec{R}_s$ with $\vec{R}_s$ uniformly bounded; hence

$$\big\|V_tQ_s\big\|_{L^2\to L^2} \le \big\|t^2L^*(I+t^2L^*)^{-2}\operatorname{div}\big\|_{L^2\to L^2}\big\|s\vec{R}_s\big\|_{L^2\to L^2} \le c\,\frac{s}{t}\,, \tag{8.7.18}$$

with C depending only on n, λ, and Λ.

Combining (8.7.17) and (8.7.18) proves (8.6.32) with $\Theta_t = V_t$. Hence Corollary 8.6.4 is applicable and (8.7.15) follows.

Therefore, the second integral on the right-hand side of (8.7.14) is bounded and estimate (8.7.4) is reduced to proving

$$\int_0^\infty \big\|(I+t^2L)^{-1}tL(f)\big\|_2^2\,\frac{dt}{t} \le C\int_{\mathbf{R}^n}|\nabla f|^2\,dx \tag{8.7.19}$$

for all $f \in C_0^\infty(\mathbf{R}^n)$.

8.7.d. Reduction to a Carleson Measure Estimate

Our next goal will be to reduce matters to a Carleson measure estimate. We first introduce some notation that will be used throughout. For $\mathbf{C}^n$-valued functions $\vec{f} = (f_1, \dots, f_n)$ define

$$Z_t(\vec{f}) = -\sum_{k=1}^{n}\sum_{j=1}^{n}(I + t^2 L)^{-1} t\partial_j(a_{j,k} f_k).$$

In short, we write $Z_t = -(I + t^2 L)^{-1} t \operatorname{div} A$. With this notation, we can write (8.7.19) as

$$\int_0^\infty \|Z_t(\nabla f)\|_2^2 \frac{dt}{t} \le C \int_{\mathbf{R}^n} |\nabla f|^2 \, dx. \tag{8.7.20}$$

Also, define

$$\gamma_t(x) = Z_t(\mathbf{1})(x) = \Big(-\sum_{j=1}^{n}(I + t^2 L)^{-1} t\, \partial_j(a_{j,k})(x)\Big)_{1\le k\le n},$$

where $\mathbf{1}$ is the $n \times n$ identity matrix and the action of Z_t on $\mathbf{1}$ is columnwise.

The reduction to a Carleson measure estimate and to a $T(b)$ argument will require the following inequality:

$$\int_{\mathbf{R}^n}\int_0^\infty |\gamma_t(x) \cdot P_t^2(\nabla g)(x) - Z_t(\nabla g)(x)|^2 \frac{dx\, dt}{t} \le C \int_{\mathbf{R}^n} |\nabla g|^2 \, dx, \tag{8.7.21}$$

where C depends only on n, λ, and Λ. Here, P_t denotes the operator

$$P_t(h) = h * p_t\,, \tag{8.7.22}$$

where $t^{-n}p(t^{-1}x)$ where p is a smooth real-valued function supported in the unit ball of $\mathbf{R}^n$ with integral equal to 1. To prove this, we need to handle Littlewood-Paley theory in a setting a bit more general than the one encountered in the previous section.

Lemma 8.7.5. *For $t > 0$, let U_t be integral operators defined on $L^2(\mathbf{R}^n)$ with measurable kernels $L_t(x, y)$. Suppose that for some $m > n$ and for all $y \in \mathbf{R}^n$ and $t > 0$ we have*

$$\int_{\mathbf{R}^n} \left(1 + \frac{|x-y|}{t}\right)^{2m} |L_t(x, y)|^2 \, dx \le t^{-n}. \tag{8.7.23}$$

Assume that for any ball $B(y, t)$, U_t has a bounded extension from $L^\infty(\mathbf{R}^n)$ into $L^2(B(y, t))$ such that for all f in $L^\infty(\mathbf{R}^n)$ and $y \in \mathbf{R}^n$ we have

$$\frac{1}{t^n}\int_{B(y,t)} |U_t(f)(x)|^2 \, dx \le \|f\|_{L^\infty}^2. \tag{8.7.24}$$

Finally, assume that $U_t(1) = 0$ *in the sense that*

$$U_t(\chi_{B(0,R)}) \to 0 \quad \text{in} \quad L^2(B(y,t)) \tag{8.7.25}$$

as $R \to \infty$ *for all* $y \in \mathbf{R}^n$ *and* $t > 0$.

Let Q_s *and* P_t *be as in (8.7.16) and (8.7.22), respectively. Then for some* $\alpha > 0$ *and* C *depending on* n *and* m *we have*

$$\big\|U_t P_t Q_s\big\|_{L^2\to L^2} \le C \inf\left(\frac{t}{s}, \frac{s}{t}\right)^{\alpha}. \tag{8.7.26}$$

Proof. We begin by observing that $U_t^* U_t$ has a kernel $K_t(x,y)$ given by

$$K_t(x,y) = \int_{\mathbf{R}^n} \overline{L_t(z,x)} L_t(z,y)\, dz\,.$$

The simple inequality $(1+a+b) \le (1+a)(1+b)$ for $a, b > 0$ combined with the Cauchy-Schwarz inequality and (8.7.23) yield that $\left(1+\frac{|x-y|}{t}\right)^m |K_t(x,y)|$ is bounded by

$$\int_{\mathbf{R}^n} \left(1+\frac{|x-z|}{t}\right)^m |L_t(z,x)|\,|L_t(z,y)| \left(1+\frac{|z-y|}{t}\right)^m dy \le t^{-n}\,.$$

We conclude that

$$|K_t(x,y)| \le \frac{1}{t^n}\left(1+\frac{|x-y|}{t}\right)^{-m}.$$

Hence $U_t^* U_t$ is bounded on all L^p, $1 \le p \le +\infty$ and, in particular, for $p=2$. Since L^2 is a Hilbert space, it follows that U_t is bounded on $L^2(\mathbf{R}^n)$ uniformly in $t > 0$.

For $s \le t$ we use that $\big\|U_t\big\|_{L^2\to L^2} \le B < \infty$ and basic estimates to deduce that

$$\big\|U_t P_t Q_s\big\|_{L^2\to L^2} \le B \big\|P_t Q_s\big\|_{L^2\to L^2} \le C\,B\left(\frac{s}{t}\right)^{\alpha}.$$

Next, we consider the case $t \le s$. Since P_t has an integrable kernel, $W_t = U_t^* U_t P_t$ also has an L^1 kernel. If we prove that $W_t(1) = 0$, then we can deduce from standard arguments that

$$\big\|W_t Q_s\big\|_{L^2\to L^2} \le C\left(\frac{t}{s}\right)^{2\alpha}$$

for $0 < \alpha < m - n$. This would imply the required estimate (8.7.26) as

$$\big\|U_t P_t Q_s\big\|_{L^2\to L^2}^2 \le C \big\|U_t^* U_t P_t Q_s\big\|_{L^2\to L^2}\,.$$

We have that $W_t(1) = U_t^* U_t(1)$. Suppose that a function φ in $L^2(\mathbf{R}^n)$ is compactly supported. Then φ is integrable over $\mathbf{R}^n$ and we have

$$\langle U_t^* U_t(1)\,|\,\varphi\rangle = \lim_{R\to\infty} \langle U_t^* U_t(\chi_{B(0,R)})\,|\,\varphi\rangle = \lim_{R\to\infty} \langle U_t(\chi_{B(0,R)})\,|\,U_t(\varphi)\rangle.$$

We have

$$\langle U_t(\chi_{B(0,R)})\,|\,U_t(\varphi)\rangle = \int_{\mathbf{R}^n}\int_{\mathbf{R}^n} U_t(\chi_{B(0,R)})(x)\overline{U_t(x,y)\varphi(y)}\, dy\, dx$$

and this is in absolute value at most a constant multiple of

$$\left(t^{-n}\int_{\mathbf{R}^n}\int_{\mathbf{R}^n}\left(1+\frac{|x-y|}{t}\right)^{-2m}|U_t(\chi_{B(0,R)})(x)|^2|\varphi(y)|\,dy\,dx\right)^{\frac{1}{2}}\|\varphi\|_{L^1}^{\frac{1}{2}}$$

by (8.7.23) and the Cauchy-Schwarz inequality for the measure $|\varphi(y)|\,dy\,dx$. Using a covering in the x variable by a lattice of balls $B(y+ckt,t)$, $k\in\mathbf{Z}^n$, we deduce easily that the last displayed expression is at most

$$C_\varphi\left(\sum_{k\in\mathbf{Z}^n}\int_{\mathbf{R}^n}(1+|k|)^{-2m}c_R(y,k)|\varphi(y)|\,dy\right)^{\frac{1}{2}},$$

where C_φ is a constant that depends on φ and

$$c_R(y,k)=t^{-n}\int_{B(y+ckt,t)}|U_t(\chi_{B(0,R)})(x)|^2\,dx\,.$$

Applying the dominated convergence theorem and invoking (8.7.24) and (8.7.25) as $R\to\infty$, we conclude that $\langle U_t^*U_t(1)\,|\,\varphi\rangle=0$. The latter implies that $U_t^*U_t(1)=0$. The same conclusion follows for W_t since $P_t(1)=1$.

□

Lemma 8.7.6. *Let P_t be as in Lemma 8.7.5. Then the operator U_t defined by $U_t(\vec{f})(x)=\gamma_t(x)\cdot P_t(\vec{f})(x)-Z_tP_t(\vec{f})(x)$ satisfies*

$$\int_0^\infty\|U_tP_t(\vec{f})\|_2^2\,\frac{dt}{t}\le C\|\vec{f}\|_2^2,$$

where C depends only on n, λ, and Λ. Here the action of P_t on $\vec{f}$ is componentwise.

Proof. By the off-diagonal estimates of Lemma 8.7.2 for Z_t and the fact that p has support in the unit ball, it is simple to show that there is a constant C depending on n, λ, and Λ such that for all $y\in\mathbf{R}^n$

$$\frac{1}{t^n}\int_{B(y,t)}|\gamma_t(x)|^2\,dx\le C$$

and that the kernel of $C^{-1}U_t$ satisfies the hypotheses in Lemma 8.7.5. The conclusion follows from Corollary 8.6.4 applied to U_tP_t.

□

We now return to (8.7.21). We begin by writing

$$\gamma_t(x)\cdot P_t^2(\nabla g)(x)-Z_t(\nabla g)(x)=U_tP_t(\nabla g)(x)+Z_t(P_t^2-I)(\nabla g)(x)\,,$$

and we prove (8.7.21) for each term that appears on the right. For the first term we apply Lemma 8.7.6. As P_t commutes with partial derivatives, we may use that

$$\|Z_t\nabla\|_{L^2\to L^2}=\|(I+t^2L)^{-1}t\,L\|_{L^2\to L^2}\le Ct^{-1}\,,$$

and therefore we obtain for the second term

$$\int_{\mathbf{R}^n}\int_0^\infty |Z_t(P_t^2 - I)(\nabla g)(x)|^2 \,\frac{dx\,dt}{t} \le C^2 \int_{\mathbf{R}^n}\int_0^\infty |(P_t^2 - I)(g)(x)|^2 \,\frac{dt}{t^3}\,dx$$
$$\le C^2 c(p) \big\|\nabla g\big\|_2^2$$

by Plancherel's theorem, where C depends only on n, λ and Λ. This concludes the proof of (8.7.21).

Lemma 8.7.7. *The required estimate (8.7.4) follows from the Carleson measure estimate*

$$\sup_Q \frac{1}{|Q|}\int_Q\int_0^{\ell(Q)} |\gamma_t(x)|^2 \,\frac{dxdt}{t} < \infty, \tag{8.7.27}$$

where the supremum is taken over all cubes in $\mathbf{R}^n$ with sides parallel to the axes.

Proof. Indeed, (8.7.27) and Theorem 7.3.7 imply

$$\int_{\mathbf{R}^n}\int_0^\infty |P_t^2(\nabla g)(x)\cdot \gamma_t(x)|^2 \,\frac{dx\,dt}{t} \le C\int_{\mathbf{R}^n} |\nabla g|^2\,dx,$$

and together with (8.7.21) we deduce that (8.7.20) holds. □

Next we introduce an auxiliary averaging operator whose presence will simplify the analysis of the problem. Suppose that for any cube in $\mathbf{R}^n$ there is a mesh $\mathcal{C}_Q = \{Q'\}$ of dyadic cubes of $\mathbf{R}^n$ of all possible sizes that, in particular, contains Q. We define a dyadic averaging operator S_t^Q as follows:

$$S_t^Q(\vec{f}\,)(x) = \left(\frac{1}{|Q'|}\int_{Q'} \vec{f}(y)\,dy\right)\chi_{Q'}(x)\,,$$

where Q' is the unique dyadic cube in $\mathcal{C}_Q$ that contains x and satisfies $\frac{1}{2}\ell(Q') < t \le \ell(Q')$. Note that S_t^Q is a projection (i.e., $S_t^Q S_t^Q = S_t^Q$), since when applied it remains constant on each cube Q'. We have the following technical lemma concerning S_t^Q.

Lemma 8.7.8. *For some C depending only on n, λ, and Λ, we have*

$$\int_Q\int_0^{\ell(Q)} |\gamma_t(x)\cdot(S_t^Q - P_t^2)(\vec{f}\,)(x)|^2 \,\frac{dx\,dt}{t} \le C\int_{\mathbf{R}^n} |\vec{f}\,|^2\,dx. \tag{8.7.28}$$

Proof. We will actually obtain a stronger version of (8.7.28) in which the t-integration on the left is taken over $(0,+\infty)$. The proof of (8.7.28) follows by an adaptation of the proof of Lemma 8.7.5 provided we use a couple of new ingredients. First, the operator $U_t = (\gamma_t\cdot S_t^Q)$ is L^2 bounded; this follows from a standard $U_t^*U_t$ argument using condition (8.7.23). Second, as already observed, S_t^Q is an orthogonal projection (i.e., it satisfies $S_t^Q S_t^Q = S_t^Q$). Therefore, we have

$$\begin{aligned}\big\|(\gamma_t\cdot S_t^Q)Q_s\big\|_{L^2\to L^2} &\le \big\|(\gamma_t\cdot S_t^Q)S_t^Q Q_s\big\|_{L^2\to L^2}\\ &\le \big\|S_t^Q Q_s\big\|_{L^2\to L^2}\\ &\le \big\|S_t^Q\big\|_{L^2\to \dot L^2_\alpha}\big\|Q_s\big\|_{\dot L^2_\alpha\to L^2}\\ &\le C\Big(\frac{s}{t}\Big)^\alpha .\end{aligned}$$

The last inequality follows from the facts that, for any α in $(0,\frac12)$, Q_s maps the homogeneous Sobolev space $\dot L^2_\alpha$ into L^2 with norm at most a multiple of $C\,s^\alpha$ and that the dyadic averaging operator S_t^Q maps $L^2(\mathbf{R}^n)$ into $\dot L^2_\alpha(\mathbf{R}^n)$ with norm $C\,t^{-\alpha}$. The former of these statements is trivially verified by taking the Fourier transform, while the latter statement requires some explanation.

Let us fix an $\alpha\in(0,\frac12)$ and take $h,g\in L^2(\mathbf{R}^n)$. For $2^{-j-1}\le t<2^{-j}$ we have

$$\big\langle S_t(-\Delta)^{\frac{\alpha}{2}}h,g\big\rangle=\sum_{k\in\mathbf{Z}^n}\big\langle(-\Delta)^{\frac{\alpha}{2}}h,2^{nj}\chi_{J_{j,k}}(x)(\operatorname*{Avg}_{J_{j,k}}\overline{g})\big\rangle ,$$

where $J_{j,k}=\prod_{r=1}^n[2^{-j}k_r,2^{-j}(k_r+1))$ and $k=(k_1,\dots,k_n)$. It follows that

$$\begin{aligned}\big\langle S_t(-\Delta)^{\frac{\alpha}{2}}h,g\big\rangle &= \sum_{k\in\mathbf{Z}^n}\big\langle h,2^{nj}(\operatorname*{Avg}_{J_{j,k}}\overline{g})\,(-\Delta)^{\frac{\alpha}{2}}(\chi_{J_{j,k}})(x)\big\rangle\\ &=\Big\langle h,\sum_{k\in\mathbf{Z}^n}2^{(n+\alpha)j}(\operatorname*{Avg}_{J_{j,k}}\overline{g})\,(-\Delta)^{\frac{\alpha}{2}}(\chi_{[0,1)^n})(2^j(\cdot)-k)\Big\rangle .\end{aligned}$$

We now control the last expression using the Cauchy-Schwarz inequality. Set $\chi_\alpha=(-\Delta)^{\frac{\alpha}{2}}(\chi_{[0,1)^n})$, $c_{jk}=2^{(n+\alpha)j}(\operatorname{Avg}_{J_{j,k}}\overline{g})$. We note that

$$\int_{\mathbf{R}^n}\Big|\sum_{k\in\mathbf{Z}^n}c_{jk}\chi_\alpha(2^j-k)\Big|^2dx\le c_{n,\alpha}2^{-nj}\sum_{k\in\mathbf{Z}^n}|c_{jk}|^2$$

by taking the Fourier transform, where

$$c_{n,\alpha}=\sup_{\xi\in\mathbf{R}^n}\sum_{l\in\mathbf{Z}^n}|\widehat{\chi_\alpha}(\xi+2l\pi)|^2 .$$

Since

$$\widehat{\chi_\alpha}(\xi)=|\xi|^\alpha\prod_{j=1}^n\frac{1-e^{-2\pi i\xi_j}}{2\pi i\xi_j},$$

it follows that $c_{n,\alpha}<\infty$ when $\alpha<\frac12$. In this case we conclude that

$$\big\langle S_t(-\Delta)^{\frac{\alpha}{2}}h,g\big\rangle\le C\big\|h\big\|_{L^2}2^{j\alpha}\Big(2^{nj}\sum_{k\in\mathbf{Z}^n}\big|\operatorname*{Avg}_{J_{j,k}}\overline{g}\big|^2\Big)^{\frac12}\le C\big\|h\big\|_{L^2}t^{-\alpha}\big\|g\big\|_{L^2} .$$

□

8.7.e. The $T(b)$ Argument

To obtain (8.7.27), we adapt the $T(b)$ theorem of the previous section for square roots of divergence form elliptic operators. We fix a cube Q with center c_Q, an $\varepsilon \in (0,1)$, and a unit vector w in $\mathbf{C}^n$. We define a scalar-valued function

$$f_{Q,w}^{\varepsilon} = (1 + (\varepsilon\ell(Q))^2 L)^{-1}(\Phi_Q \cdot \overline{w})\,, \tag{8.7.29}$$

where

$$\Phi_Q(x) = x - c_Q\,.$$

We begin by observing that the following estimates are consequences of Lemma 8.7.4:

$$\int_{5Q} |f_{Q,w}^{\varepsilon} - \Phi_Q \cdot \overline{w}|^2\,dx \le C_1 \varepsilon^2 \ell(Q)^2 |Q| \tag{8.7.30}$$

and

$$\int_{5Q} |\nabla(f_{Q,w}^{\varepsilon} - \Phi_Q \cdot \overline{w})|^2\,dx \le C_2 |Q|\,, \tag{8.7.31}$$

where C_1, C_2 depend on n, λ, Λ and not on ε, Q, and w. It is important to observe that the constants C_1, C_2 are independent of ε.

The proof of (8.7.27) follows by combining the next two lemmata. The rest of this section is devoted to their proofs.

Lemma 8.7.9. *There exists an $\varepsilon > 0$ depending on n, λ, Λ, and a finite set $\mathcal{F}$ of unit vectors in $\mathbf{C}^n$ whose cardinality depends on ε and n, such that*

$$\sup_Q \frac{1}{|Q|}\int_Q\int_0^{\ell(Q)} |\gamma_t(x)|^2 \frac{dxdt}{t} \le$$
$$C \sum_{w\in\mathcal{F}} \sup_Q \frac{1}{|Q|}\int_Q\int_0^{\ell(Q)} |\gamma_t(x)\cdot(S_t^Q \nabla f_{Q,w}^{\varepsilon})(x)|^2 \frac{dxdt}{t}$$

where C depends only on ε, n, λ, and Λ. The suprema are taken over all cubes Q in $\mathbf{R}^n$ with sides parallel to the axes.

Lemma 8.7.10. *For C depending only on n, λ, Λ and $\varepsilon > 0$, we have*

$$\int_Q\int_0^{\ell(Q)} \big|\gamma_t(x)\cdot(S_t^Q \nabla f_{Q,w}^{\varepsilon})(x)\big|^2 \frac{dxdt}{t} \le C|Q|. \tag{8.7.32}$$

We begin with the proof of Lemma 8.7.10 which is the easiest of the two.

Proof of Lemma 8.7.10. Pick a smooth bump function $\mathcal{X}_Q$ localized on $4Q$ and equal to 1 on $2Q$ with $\|\mathcal{X}_Q\|_\infty + \ell(Q)\|\nabla\mathcal{X}_Q\|_\infty \le c_n$. By Lemma 8.7.5 and estimate

(8.7.21), the left-hand side of (8.7.32) is bounded by

$$C\int_{\mathbf{R}^n}\big|\nabla(\mathcal{X}_Q f^{\varepsilon}_{Q,w})\big|^2\,dx + 2\int_Q\int_0^{\ell(Q)}\big|\gamma_t(x)\cdot(P_t^2\nabla(\mathcal{X}_Q f^{\varepsilon}_{Q,w}))(x)\big|^2\,\frac{dx\,dt}{t}$$
$$\leq C\int_{\mathbf{R}^n}\big|\nabla(\mathcal{X}_Q f^{\varepsilon}_{Q,w})\big|^2\,dx + 4\int_Q\int_0^{\ell(Q)}\big|(Z_t\nabla(\mathcal{X}_Q f^{\varepsilon}_{Q,w}))(x)\big|^2\,\frac{dx\,dt}{t}.$$

It remains to control the last displayed expression by $C|Q|$.

First, it follows easily from (8.7.30) and (8.7.31) that

$$\int_{\mathbf{R}^n}|\nabla(\mathcal{X}_Q f^{\varepsilon}_{Q,w})|^2\,dx\leq C|Q|\,,$$

where C is independent of Q and w (but it may depend on ε). Next, we write

$$Z_t\nabla(\mathcal{X}_Q f^{\varepsilon}_{Q,w}) = W_t^1 + W_t^2 + W_t^3\,,$$

where

$$\begin{aligned}
W_t^1 &= (I+t^2L)^{-1}t\left(\mathcal{X}_Q L(f^{\varepsilon}_{Q,w})\right),\\
W_t^2 &= -\,(I+t^2L)^{-1}t\left(\operatorname{div}\left(Af^{\varepsilon}_{Q,w}\nabla\mathcal{X}_Q\right)\right),\\
W_t^3 &= -\,(I+t^2L)^{-1}t\left(A\nabla f^{\varepsilon}_{Q,w}\cdot\nabla\mathcal{X}_Q\right),
\end{aligned}$$

and we will use different arguments to treat each term W_t^j.

To handle W_t^1, observe that

$$L(f^{\varepsilon}_{Q,w}) = \frac{f^{\varepsilon}_{Q,w}-\Phi_Q\cdot\overline{w}}{\varepsilon^2\ell(Q)^2}\,,$$

and therefore it follows from (8.7.30) that

$$\int_{\mathbf{R}^n}|\mathcal{X}_Q L(f^{\varepsilon}_{Q,w})|^2\leq C|Q|(\varepsilon\ell(Q))^{-2}\,,$$

where C is independent of Q and w. Using the (uniform in t) boundedness of $(I+t^2L)^{-1}$ on $L^2(\mathbf{R}^n)$, we obtain

$$\int_Q\int_0^{\ell(Q)}\big|W_t^1(x)\big|^2\,\frac{dx\,dt}{t}\leq\int_0^{\ell(Q)}\frac{C|Q|\,t^2}{(\varepsilon\ell(Q))^2}\,\frac{dt}{t}\leq\frac{C|Q|}{\varepsilon^2}\,,$$

which establishes the required quadratic estimate for W_t^1.

To obtain a similar quadratic estimate for W_t^2, we apply Lemma 8.7.2 for the operator $(I+t^2L)^{-1}t\operatorname{div}$ with sets $F=Q$ and $E=\operatorname{supp}(f^{\varepsilon}_{Q,w}\nabla\mathcal{X}_Q)\subseteq 4Q\setminus 2Q$. We obtain that

$$\int_Q\int_0^{\ell(Q)}|W_t^2(x)|^2\,\frac{dx\,dt}{t}\leq C\int_0^{\ell(Q)}e^{-\frac{\ell(Q)}{ct}}\,\frac{dt}{t}\int_{4Q\setminus 2Q}|Af^{\varepsilon}_{Q,w}\nabla\mathcal{X}_Q|^2\,dx\,.$$

The first integral on the right provides at most a constant factor, while we handle the second integral by writing

$$f^{\varepsilon}_{Q,w} = (f^{\varepsilon}_{Q,w} - \Phi_Q \cdot \overline{w}) + \Phi_Q \cdot \overline{w}\,.$$

Using (8.7.30) and the facts that $\big\|\nabla \mathcal{X}_Q\big\|_\infty \le c_n \ell(Q)^{-1}$ and that $|\Phi_Q| \le c_n \ell(Q)$ on the support of $\mathcal{X}_Q$, we obtain that

$$\int_{4Q\setminus 2Q} |A f^{\varepsilon}_{Q,w} \nabla \mathcal{X}_Q|^2\,dx \le C\,|Q|\,,$$

where C depends only on n, λ, and Λ. This yields the required result for W_t^2.

To obtain a similar estimate for W_t^3, we use the (uniform in t) boundedness of $(I+t^2L)^{-1}$ on $L^2(\mathbf{R}^n)$ (Exercise 8.7.2) to obtain that

$$\int_Q\int_0^{\ell(Q)} |W_t^3(x)|^2\,\frac{dx\,dt}{t} \le C\int_0^{\ell(Q)} t^2\,\frac{dt}{t}\int_{4Q\setminus 2Q} |A\nabla f^{\varepsilon}_{Q,w}\cdot \nabla \mathcal{X}_Q|^2\,dx\,.$$

But the last integral is shown easily to be bounded by $C|Q|$ by writing $f^{\varepsilon}_{Q,w}$, as in the previous case, and using (8.7.31) and the properties of $\mathcal{X}_Q$ and Φ_Q. Note that C here depends only on n, λ, and Λ. This concludes the proof of Lemma 8.7.10.

□

It remains to prove Lemma 8.7.9. This will be achieved in the next (and final) subsection of this section.

8.7.f. The Proof of Lemma 8.7.9

The main ingredient in the proof of Lemma 8.7.9 is the following proposition, which we state and prove first.

Proposition 8.7.11. *There exists an $\varepsilon > 0$ depending on n, λ, and Λ, and $\eta = \eta(\varepsilon) > 0$ such that for each unit vector w in $\mathbf{C}^n$ and each cube Q with sides parallel to the axes, there exists a collection $\mathcal{S}'_w = \{Q'\}$ of nonoverlapping dyadic subcubes of Q such that*

$$\Big|\bigcup_{Q'\in\mathcal{S}'_w}\Big| \le (1-\eta)|Q| \tag{8.7.33}$$

and, moreover, if $\mathcal{S}''_w$ is the collection of all dyadic subcubes of Q not contained in any $Q'\in\mathcal{S}'_w$, then for any $Q''\in\mathcal{S}''_w$ we have

$$\frac{1}{|Q''|}\int_{Q''} \mathrm{Re}\,(\nabla f^{\varepsilon}_{Q,w}(y)\cdot w)\,dy \ge \frac{3}{4} \tag{8.7.34}$$

and

$$\frac{1}{|Q''|}\int_{Q''} |\nabla f^{\varepsilon}_{Q,w}(y)|^2\,dy \le (4\varepsilon)^{-2}. \tag{8.7.35}$$

Proof. We begin by proving the following crucial estimate:

$$\left| \int_Q (1 - \nabla f^{\varepsilon}_{Q,w}(x) \cdot w) \, dx \right| \le C \varepsilon^{\frac{1}{2}} |Q|, \tag{8.7.36}$$

where C depends on n, λ, and Λ, but not on ε, Q and w. Indeed, we observe that

$$\nabla(\Phi_Q \cdot \overline{w})(x) \cdot w = |w|^2 = 1,$$

so that

$$1 - \nabla f^{\varepsilon}_{Q,w}(x) \cdot w = \nabla g^{\varepsilon}_{Q,w}(x) \cdot w,$$

where we set

$$g^{\varepsilon}_{Q,w}(x) = \Phi_Q(x) \cdot \overline{w} - f^{\varepsilon}_{Q,w}(x) \, .$$

Next we state another lemma that will be of use to us. Its proof is postponed until the end of this section.

Lemma 8.7.12. *There exists a constant $C = C_n$ such that for all $h \in \dot{L}^2_1$ we have*

$$\left| \int_Q \nabla h(x) \, dx \right| \le C \ell(Q)^{\frac{n-1}{2}} \left(\int_Q |h(x)|^2 \, dx \right)^{\frac{1}{4}} \left(\int_Q |\nabla h(x)|^2 \, dx \right)^{\frac{1}{4}} .$$

Applying Lemma 8.7.12 to the function $g^{\varepsilon}_{Q,w}$, we deduce (8.7.36) as a consequence of (8.7.30) and (8.7.31).

We now proceed with the proof of Proposition 8.7.11. First we deduce from (8.7.36) that

$$\frac{1}{|Q|} \int_Q \operatorname{Re} (\nabla f^{\varepsilon}_{Q,w}(x) \cdot w) \, dx \ge \frac{7}{8}$$

provided that ε is small enough. We also observe that as a consequence of (8.7.31) we have

$$\frac{1}{|Q|} \int_Q |\nabla f^{\varepsilon}_{Q,w}(x)|^2 \, dx \le C_3,$$

where C_3 is independent of ε. Now, we perform a stopping-time decomposition to select a collection $\mathcal{S}'_w$ of dyadic subcubes of Q that are maximal with respect to either one of the following conditions:

$$\frac{1}{|Q'|} \int_{Q'} \operatorname{Re} (\nabla f^{\varepsilon}_{Q,w}(x) \cdot w) \, dx \le \frac{3}{4} \tag{8.7.37}$$

$$\frac{1}{|Q'|} \int_{Q'} |\nabla f^{\varepsilon}_{Q,w}(x)|^2 \, dx \ge (4\varepsilon)^{-2} . \tag{8.7.38}$$

This is achieved by subdividing Q dyadically and by selecting those cubes Q' for which either (8.7.37) or (8.7.38) holds, subdividing all the nonselected cubes, and repeating the procedure. The validity of (8.7.34) and (8.7.35) now follows from the construction and (8.7.37) and (8.7.38).

It remains to establish (8.7.33). Let B_1 be the union of the cubes in $\mathcal{S}'_w$ for which (8.7.37) holds. Also, let B_2 be the union of those cubes in $\mathcal{S}'_w$ for which (8.7.38) holds. We then have

$$\Big|\bigcup_{Q'\in\mathcal{S}'_w} Q'\Big| \le |B_1| + |B_2|\,.$$

The fact that the cubes in $\mathcal{S}'_w$ do not overlap yields

$$|B_2| \le (4\varepsilon)^2 \int_Q |\nabla f^{\varepsilon}_{Q,w}(x)|^2\,dx \le (4\varepsilon)^2 C_3|Q|.$$

Setting $b^{\varepsilon}_{Q,w}(x) = 1 - \mathrm{Re}\,(\nabla f^{\varepsilon}_{Q,w}(x)\cdot w)$, we also have

$$|B_1| \le 4\sum \int_{Q'} b^{\varepsilon}_{Q,w}\,dx = 4\int_Q b^{\varepsilon}_{Q,w}\,dx - 4\int_{Q\setminus B_1} b^{\varepsilon}_{Q,w}\,dx\,, \tag{8.7.39}$$

where the sum is taken over all cubes Q' that comprise B_1. The first term on the right in (8.7.39) is bounded above by $C\varepsilon^{\frac{1}{2}}|Q|$ in view of (8.7.36). The second term on the right in (8.7.39) is controlled in absolute value by

$$4|Q\setminus B_1| + 4|Q\setminus B_1|^{\frac{1}{2}}(C_3|Q|)^{\frac{1}{2}} \le 4|Q\setminus B_1| + 4C_3\varepsilon^{\frac{1}{2}}|Q| + \varepsilon^{-\frac{1}{2}}|Q\setminus B_1|.$$

Since $|Q\setminus B_1| = |Q| - |B_1|$, we obtain

$$(5+\varepsilon^{-\frac{1}{2}})|B_1| \le (4 + C\varepsilon^{\frac{1}{2}} + \varepsilon^{-\frac{1}{2}})|Q|\,,$$

which yields $|B_1| \le (1-\varepsilon^{\frac{1}{2}} + o(\varepsilon^{\frac{1}{2}}))|Q|$ if ε is small enough. Hence

$$|B| \le (1-\eta(\varepsilon))|Q|$$

with $\eta(\varepsilon)\sim\varepsilon^{\frac{1}{2}}$ for small ε. This concludes the proof of Proposition 8.7.11.

□

Next, we will need the following simple geometric fact.

Lemma 8.7.13. *Let w,u,v be in $\mathbf{C}^n$ such that $|w|=1$ and $0<\varepsilon\le 1$ be such that*

$$|u-(u\cdot\overline{w})w| \le \varepsilon\,|u\cdot\overline{w}| \tag{8.7.40}$$

$$\mathrm{Re}\,(v\cdot w) \ge \frac{3}{4} \tag{8.7.41}$$

$$|v| \le (4\varepsilon)^{-1}\,. \tag{8.7.42}$$

Then we have $|u|\le 4\,|u\cdot v|$.

Proof. It follows from (8.7.41) that

$$\tfrac{3}{4}\,|u\cdot\overline{w}| \le |(u\cdot\overline{w})(v\cdot w)|\,. \tag{8.7.43}$$

Moreover, (8.7.40) and the triangle inequality imply that

$$|u| \le (1+\varepsilon)|u\cdot\overline{w}| \le 2\,|u\cdot\overline{w}|\,. \tag{8.7.44}$$

Also, as a consequence of (8.7.40) and (8.7.42), we obtain

$$|(u-(u\cdot\overline{w})w)\cdot v| \le \tfrac{1}{4}\,|u\cdot\overline{w}|\,. \tag{8.7.45}$$

Finally, using (8.7.43) and (8.7.45) together with the triangle inequality, we deduce that

$$|u\cdot v| \ge |(u\cdot\overline{w})(v\cdot w)| - |(u-(u\cdot\overline{w})w)\cdot v| \ge (\tfrac{3}{4}-\tfrac{1}{4})\,|u\cdot\overline{w}| \ge \tfrac{1}{4}\,|u|\,,$$

where in the last inequality we used (8.7.44).

□

We may now proceed with the proof of Lemma 8.7.9.

We fix an $\varepsilon>0$ to be chosen later and we choose a finite number of cones $\mathcal{C}_w$ indexed by a finite set $\mathcal{F}$ of unit vectors w in $\mathbf{C}^n$ defined by

$$\mathcal{C}_w = \big\{u\in\mathbf{C}^n:\ |u-(u\cdot\overline{w})w| \le \varepsilon\,|u\cdot\overline{w}|\big\} \tag{8.7.46}$$

so that

$$\mathbf{C}^n = \bigcup_{w\in\mathcal{F}} \mathcal{C}_w\,.$$

Note that the size of the set $\mathcal{F}$ can be chosen to depend only on ε and the dimension n.

It will suffice to show for each fixed $w\in\mathcal{F}$ we have a Carleson measure estimate for $\gamma_{t,w}(x)\equiv\chi_{\mathcal{C}_w}(\gamma_t(x))\gamma_t(x)$, where $\chi_{\mathcal{C}_w}$ denotes the characteristic function of $\mathcal{C}_w$. To achieve this we define

$$A_w \equiv \sup_Q \frac{1}{|Q|}\int_Q\int_0^{\ell(Q)} |\gamma_{t,w}(x)|^2\,,\frac{dx\,dt}{t} \tag{8.7.47}$$

where the supremum is taken over all cubes Q in $\mathbf{R}^n$ with sides parallel to the axes. By truncating $\gamma_{t,w}(x)$ for t small and t large, we may assume that this quantity is finite. Once an a priori bound independent of these truncations is obtained, we can pass to the limit by monotone convergence to deduce the same bound for $\gamma_{t,w}(x)$.

We now fix a cube Q and let $\mathcal{S}_w''$ be as in Proposition 8.7.11. We pick Q'' in $\mathcal{S}_w''$ and we set

$$v = \frac{1}{|Q''|}\int_{Q''} \nabla f_{Q,w}^{\varepsilon}(y)\,dy \in \mathbf{C}^n.$$

It is obvious that statements (8.7.34) and (8.7.35) in Proposition 8.7.11 yield conditions (8.7.41) and (8.7.42) of Lemma 8.7.13. Set $u=\gamma_{t,w}(x)$ and note that if $x\in Q''$ and $\frac{1}{2}\ell(Q'')<t\le\ell(Q'')$, then $v=S_t^Q(\nabla f_{Q,w}^{\varepsilon})(x)$; hence

$$\big|\gamma_{t,w}(x)\big| \le 4\,\big|\gamma_{t,w}(x)\cdot S_t^Q(\nabla f_{Q,w}^{\varepsilon})(x)\big| \le 4\,\big|\gamma_t(x)\cdot S_t^Q(\nabla f_{Q,w}^{\varepsilon})(x)\big| \tag{8.7.48}$$

from Lemma 8.7.13 and the definition of $\gamma_{t,w}(x)$.

We next observe that we may partition the Carleson box $Q \times (0, \ell(Q)]$ as a union Carleson boxes $Q' \times (0, \ell(Q')]$ for Q' in $\mathcal{S}'_w$ and Whitney rectangles $Q'' \times (\frac{1}{2}\ell(Q''), \ell(Q'')]$ for Q'' in $\mathcal{S}''_w$. This allows us to write

$$\int_Q \int_0^{\ell(Q)} |\gamma_{t,w}(x)|^2 \frac{dxdt}{t} = \sum_{Q' \in \mathcal{S}'_w} \int_{Q'} \int_0^{\ell(Q')} |\gamma_{t,w}(x)|^2 \frac{dxdt}{t} + \sum_{Q'' \in \mathcal{S}''_w} \int_{Q''} \int_{\frac{1}{2}\ell(Q'')}^{\ell(Q'')} |\gamma_{t,w}(x)|^2 \frac{dxdt}{t}.$$

First observe that

$$\sum_{Q' \in \mathcal{S}'_w} \int_{Q'} \int_0^{\ell(Q')} |\gamma_{t,w}(x)|^2 \frac{dxdt}{t} \le \sum_{Q' \in \mathcal{S}'_w} A_w |Q'| \le A_w(1-\eta)|Q|.$$

Second, using (8.7.48), we obtain

$$\begin{aligned} &\sum_{Q'' \in \mathcal{S}''_w} \int_{Q''} \int_{\frac{1}{2}\ell(Q'')}^{\ell(Q'')} |\gamma_{t,w}(x)|^2 \frac{dxdt}{t} \\ &\le 16 \sum_{Q'' \in \mathcal{S}''_w} \int_{Q''} \int_{\frac{1}{2}\ell(Q'')}^{\ell(Q'')} |\gamma_t(x) \cdot S_t^Q(\nabla f^\varepsilon_{Q,w})(x)|^2 \frac{dxdt}{t} \\ &\le 16 \int_Q \int_0^{\ell(Q)} |\gamma_t(x) \cdot S_t^Q(\nabla f^\varepsilon_{Q,w})(x)|^2 \frac{dxdt}{t}. \end{aligned}$$

Altogether, we obtain the bound

$$\begin{aligned} &\int_Q \int_0^{\ell(Q)} |\gamma_{t,w}(x)|^2 \frac{dxdt}{t} \\ &\le A_w(1-\eta)|Q| + 16 \int_Q \int_0^{\ell(Q)} |\gamma_t(x) \cdot S_t^Q(\nabla f^\varepsilon_{Q,w})(x)|^2 \frac{dxdt}{t}. \end{aligned}$$

We divide by $|Q|$, we take the supremum over all cubes Q with sides parallel to the axes, and we use the definition and the finiteness of A_w to obtain the required estimate:

$$A_w \le 16\,\eta^{-1} \sup_Q \frac{1}{|Q|} \int_Q \int_0^{\ell(Q)} |\gamma_t(x) \cdot S_t^Q(\nabla f^\varepsilon_{Q,w})(x)|^2 \frac{dxdt}{t}.$$

□

We end by verifying the validity of Lemma 8.7.12 used earlier.

Proof of Lemma 8.7.12. For simplicity we may take Q to be the cube $[-1,1]^n$. Once this case is established, the case of a general cube will follow by translation and rescaling. Set

$$M = \left(\int_Q |h(x)|^2\,dx\right)^{\frac{1}{2}} \qquad M' = \left(\int_Q |\nabla h(x)|^2\,dx\right)^{\frac{1}{2}}$$

If $M \geq M'$, there is nothing to prove, so we may assume that $M < M'$. Take $t \in (0,1)$ and $\varphi \in C_0^\infty(Q)$ with $\varphi(x) = 1$ when $\operatorname{dist}(x, \partial Q) \geq t$ and $0 \leq \varphi \leq 1$, $\|\nabla\varphi\|_\infty \leq C/t$, $C = C(n)$. Then

$$\int_Q \nabla h(x)\,dx = \int_Q (1 - \varphi(x))\nabla h(x)\,dx - \int_Q h(x)\nabla\varphi(x)\,dx$$

and the Cauchy-Schwarz inequality yields

$$\left|\int_Q \nabla h(x)\,dx\right| \leq C(M'\,t^{\frac{1}{2}} + M\,t^{-\frac{1}{2}}).$$

Choosing $t = M/M'$, we conclude the proof. □

The proof of Theorem 8.7.1 is now complete.

Exercises

8.7.1. Let A and L be as in the statement of Theorem 8.7.1.

(a) Consider the generalized heat equation

$$\frac{\partial u}{\partial t} - \operatorname{div}(A\nabla u) = 0$$

on $\mathbf{R}_+^{n+1}$ with initial condition $u(0,x) = u_0$. Assume a uniqueness theorem for solutions of these equations to obtain that the solution of the equation in part (a) is

$$u(t,x) = e^{-tL}(u_0).$$

(b) Take $u_0 = 1$ to deduce the identity

$$e^{-tL}(1) = 1$$

for all $t > 0$. Conclude that the family of $\{e^{-tL}\}_{t>0}$ is an approximate identity, in the sense that

$$\lim_{t\to 0} e^{-tL} = I.$$

8.7.2. Let L be as in (8.7.2). Show that the operators

$$\begin{aligned} L_1 &= (I + t^2L)^{-1} \\ L_2 &= t\nabla(I + t^2L)^{-1} \\ L_3 &= (I + t^2L)^{-1}t\operatorname{div} \end{aligned}$$

are bounded on $L^2(\mathbf{R}^n)$ uniformly in t with bounds depending only on n, λ, and Λ.

[*Hint:* Note that the L^2 boundedness of L_3 follows from that of L_2 using duality and integration by parts. To prove the L^2 boundedness of L_1 and L_2, let $u_t = (I+t^2L)^{-1}(f)$. Then $u_t + t^2L(u_t) = f$, which implies $\int_{\mathbf{R}^n} |u_t|^2\,dx + t^2\int_{\mathbf{R}^n} \overline{u_t}\,L(u_t)\,dx = \int_{\mathbf{R}^n} \overline{u_t}\,f\,dx$. The definition of L and integration by parts yield $\int_{\mathbf{R}^n} |u_t|^2\,dx + t^2\int_{\mathbf{R}^n} A\nabla u_t\,\nabla\overline{u_t}\,dx = \int_{\mathbf{R}^n} \overline{u_t}\,f\,dx$. Apply the ellipticity condition to bound the left side of this identity from below by $\int_{\mathbf{R}^n} |u_t|^2\,dx + \lambda\int_{\mathbf{R}^n} |t\nabla u_t|^2\,dx$. Also $\int_{\mathbf{R}^n} \overline{u_t}\,f\,dx$ is at most $\varepsilon^{-1}\int_{\mathbf{R}^n} |f|^2\,dx + \varepsilon\int_{\mathbf{R}^n} |u_t|^2\,dx$ by the Cauchy-Schwarz inequality. Choose ε small enough to complete the proof when $\big\|u_t\big\|_{L^2} < \infty$. In the case $\big\|u_t\big\|_{L^2} = \infty$, multiply the identity $u_t + t^2L(u_t) = f$ by $\overline{u_t}\eta_R$, where η_R is a suitable cutoff localized in a ball $B(0,R)$, and use the idea of Lemma 8.7.2. Then let $R\to\infty$.]

8.7.3. Let L be as in the proof of Theorem 8.7.1.
(a) Show that for all $t>0$ we have

$$(I+t^2L^2)^{-2} = \int_0^\infty e^{-u(I+t^2L)}u\,du$$

by checking the identities

$$\int_0^\infty (I+t^2L)^2 e^{-u(I+t^2L)}u\,du = \int_0^\infty e^{-u(I+t^2L)}(I+t^2L)^2u\,du = I\,.$$

(b) Prove that the operator

$$T = \frac{4}{\pi}\int_0^\infty L(I+t^2L)^{-2}\,dt$$

satisfies $TT = L$.
(c) Conclude that the operator

$$S = \frac{16}{\pi}\int_0^{+\infty} t^3L^2(I+t^2L)^{-3}\,\frac{dt}{t}$$

satisfies $SS = L$, that is S is the square root of L. Moreover, all the integrals converge in $L^2(\mathbf{R}^n)$ whenever restricted to functions in $f\in C_0^\infty(\mathbf{R}^n)$.
[*Hint:* Part (a): Write $(I+t^2L)e^{-u(I+t^2L)} = -\frac{d}{du}(e^{-u(I+t^2L)})$, apply integration by parts twice and use Exercise 8.7.1. Part (b): Write the integrand as in part (a) and use the identity

$$\int_0^\infty\int_0^\infty e^{-(ut^2+vs^2)L}L^2\,dt\,ds = \frac{\pi}{4}(uv)^{-\frac12}\int_0^\infty e^{-r^2L}L^2\,2r\,dr.$$

Set $\rho = r^2$ and use $e^{-\rho L}L = \frac{d}{d\rho}(e^{-\rho L})$. Part (c): Show that $T=S$ using an integration by parts starting with the identity $L = \frac{d}{dt}(tL)$.]

8.7.4. Suppose that μ is a measure on $\mathbf{R}^{n+1}_+$. For a cube Q in $\mathbf{R}^n$ we define the tent $T(Q)$ of Q as the set $Q\times(0,\ell(Q))$. Suppose that there exist two positive constants $\alpha < 1$ and β such that for all cubes Q in $\mathbf{R}^n$ there exist subcubes Q_j of Q with disjoint interiors such that

(a) $\left|Q \setminus \bigcup_j Q_j\right| > \alpha\,|Q|$,

(b) $\mu\Big(T(Q) \setminus \bigcup_j T(Q_j)\Big) \le \beta\,|Q|$.

Then μ is a Carleson measure with constant

$$\big\|\mu\big\|_{\mathcal{C}} \le \frac{\beta}{\alpha}.$$

[*Hint:* We have

$$\begin{aligned}\mu(T(Q)) &\le \mu\Big(T(Q) \setminus \bigcup_j T(Q_j)\Big) + \sum_j \mu(T(Q_j)) \\ &\le \beta|Q| + \big\|\mu\big\|_{\mathcal{C}} \sum_j |Q_j|\end{aligned}$$

and the last expression is at most $(\beta + (1-\alpha)\big\|\mu\big\|_{\mathcal{C}})|Q|$. Assuming that $\big\|\mu\big\|_{\mathcal{C}} < \infty$, we obtain the required conclusion. In general, approximate the measure by a sequence of truncated measures.]

HISTORICAL NOTES

Most of the material in Sections 8.1 and 8.2 has been in the literature since the 1960s. Theorem 8.1.19 was independently obtained by Peetre [**402**], Spanne [**467**], and Stein [**473**].

The original proof of the $T(1)$ theorem obtained by David and Journé [**149**] stated that if $T(1)$, $T^t(1)$ are in BMO and T satisfies the weak boundedness property, then T is L^2 bounded. This proof is based on the boundedness of paraproducts and is given in Theorem 8.5.4. Paraproducts were first exploited by Bony [**46**] and Coifman and Meyer [**115**]. The proof of L^2 boundedness using condition (iv) given in the proof of Theorem 8.3.3 was later obtained by Coifman and Meyer [**116**]. The equivalent conditions (ii), (iii), and (vi) first appeared in Stein [**482**], while condition (iv) is also due to David and Journé [**149**]. Condition (i) appears in the article of Nazarov, Volberg, and Treil [**389**] in the context of nondoubling measures. The same authors [**390**] obtained a proof of Theorems 8.2.1 and 8.2.2 for Calderón-Zygmund operators on nonhomogeneous spaces. Multilinear versions of the $T(1)$ theorem were obtained by Christ and Journé [**102**] and Grafakos and Torres [**227**]. The article [**102**] also contains a proof of the quadratic $T(1)$ type Theorem 8.6.3.

The orthogonality Lemma 8.5.1 was first proved by Cotlar [**132**] for self-adjoint and mutually commuting operators T_j. The case of general noncommuting operators operators was obtained by Knapp and Stein [**305**]. Theorem 8.5.7 is due to Calderón and Vaillancourt [**75**] and is also valid for symbols of class $S^0_{\rho,\rho}$ when $0 \le \rho < 1$. For additional topics on pseudodifferential operators we refer to the books of Coifman and Meyer [**115**] Journé [**282**], Stein [**482**], Taylor [**514**], Torres [**526**], and the references therein. The latter reference presents a careful study of the action of linear operators with standard kernels on general function spaces. The continuous version of the orthogonality Lemma 8.5.1 given in Exercise

8.5.8 is due to Calderón and Vaillancourt [**75**]. Conclusion (iii) in the orthogonality Lemma 8.5.1 follows from a general principle saying that if $\sum x_j$ is a series in a Hilbert space so that $\| \sum_{j\in F} x_j\| \le M$ for all finite sets F, then the series $\sum x_j$ converges in norm. This is a consequence of the Orlicz-Pettis theorem which states that in any Banach space, if $\sum x_{n_j}$ converges weakly for every subsequence of integers n_j, then $\sum x_j$ converges in norm.

A nice exposition on the Cauchy integral that presents several historical aspects of its study is the book of Muskhelishvili [**387**]. See also the book of Journé [**282**]. Proposition 8.6.1 is due to Plemelj [**414**] when Γ is a closed Jordan curve. The L^2 boundedness of the first commutator $\mathcal{C}_1$ in Example 8.3.8 is due to Calderón [**62**]. The L^2 boundedness of the remaining commutators $\mathcal{C}_m$, $m \ge 2$ is due to Coifman and Meyer [**114**], but with bounds of the order $m!\,\big\|A'\big\|_{L^\infty}^m$. These bounds are not as good as those obtained in Example 8.3.8 and do not suffice in obtaining the boundedness of the Cauchy integral by summing the series of commutators. The L^2 boundedness of the Cauchy integral when $\big\|A'\big\|_{L^\infty}$ is small enough is due to Calderón [**65**]. The first proof of the boundedness of the Cauchy integral with arbitrary $\big\|A'\big\|_{L^\infty}$ was obtained by Coifman, McIntosh, and Meyer [**113**]. This proof is based on an improved operator norm for the commutators $\big\|\mathcal{C}_m\big\|_{L^2\to L^2} \le C_0 m^4 \big\|A'\big\|_{L^\infty}^m$. The quantity m^4 was improved by Christ and Journé [**102**] to $m^{1+\delta}$ for any $\delta > 0$; it was announced in Verdera [**541**] that Mateu and Verdera improved this result by taking $\delta = 0$. Another proof of the L^2 boundedness of the Cauchy integral was given by David [**148**] by employing the following bootstrapping argument: If the Cauchy integral is L^2 bounded whenever $\big\|A'\big\|_{L^\infty} \le \varepsilon$, then it is also L^2 bounded whenever $\big\|A'\big\|_{L^\infty} \le \frac{10}{9}\varepsilon$. A refinement of this bootstrapping technique was independently obtained by Murai [**385**], who was also able to obtain the best possible bound for the operator norm $\big\|\widetilde{C}_\Gamma\big\|_{L^2\to L^2} \le C\big(1+\big\|A'\big\|_{L^\infty}\big)^{\frac{1}{2}}$ in terms of $\big\|A'\big\|_{L^\infty}$. Here $\widetilde{C}_\Gamma$ is the operator defined in (8.6.13). Note that the corresponding estimate for C_Γ involves the power $\frac{3}{2}$ instead of $\frac{1}{2}$. See the book of Murai [**386**] for this result and a variety of topics related to the commutators and the Cauchy integral. Two elementary proofs of the L^2 boundedness of the Cauchy integral were given by Coifman, Jones, and Semmes [**111**]. The first of these proofs uses complex variables and the second a pseudo-Haar basis of L^2 adapted to the accretive function $1+iA'$. A geometric proof was given by Melnikov and Verdera [**373**]. Other proofs were obtained by Verdera [**541**] and Tchamitchian [**516**]. The proof of boundedness of the Cauchy integral given in Section 8.6 is taken from Semmes [**454**]. The book of Christ [**98**] contains an insightful exposition of many of the preceding results and discusses connections between the Cauchy integral and analytic capacity. The book of David and Semmes [**151**] presents several extensions of the results in this chapter to singular integrals along higher-dimensional surfaces.

The $T(1)$ theorem is applicable to many problems only after a considerable amount of work; see, for instance, Christ [**98**] for the case of the Cauchy integral. A more direct approach to many problems was given by McIntosh and Meyer [**358**], who replaced the function 1 by an accretive function b and showed that any operator T with standard kernel that satisfies $T(b) = T^t(b) = 0$ and $\big\|M_bTM_b\big\|_{WB} < \infty$ must be L^2 bounded. (M_b here is the operator given by multiplication with b.) This theorem easily implies the boundedness of the Cauchy integral. David, Journé, and Semmes [**150**] generalized this theorem even further as follows: If b_1 and b_2 are para-accretive functions such that T maps $b_1C_0^\infty \to (b_2C_0^\infty)'$ and is associated with a standard kernel, then T is L^2 bounded if and only if $T(b_1) \in BMO$, $T^t(b_2) \in BMO$, and $\big\|M_{b_1}TM_{b_2}\big\|_{WB} < \infty$. This is called the $T(b)$ theorem. The article of

Semmes [**454**] contains a different proof of this theorem in the special case $T(b) = 0$ and $T^t(1) = 0$ (Exercise 8.6.7). Our proof of Theorem 8.6.6 is based on ideas from [**454**]. An alternative proof of the $T(b)$ theorem was given by Fabes, Mitrea, and Mitrea [**178**] based on a lemma due to Krein [**318**]. Another version of the $T(b)$ theorem that is applicable to spaces with no euclidean structure was obtained by Christ [**97**].

Theorem 8.7.1 was posed as a problem by Kato [**291**] for maximal accretive operators and reformulated by M^c^Intosh [**356**], [**357**] for square roots of elliptic operators. The reformulation was motivated by counterexamples found to Kato's original abstract formulation, first by Lions [**336**] for maximal accretive operators, and later by M^c^Intosh [**354**] for regularly accretive ones. The one-dimensional Kato problem and the boundeness of the Cauchy integral along Lipschitz curves are equivalent problems as shown by Kenig and Meyer [**301**]. See also Auscher, M^c^Intosh, and Nahmod [**18**]. Coifman, Deng, and Meyer [**107**] and independently Fabes, Jerison, and Kenig [**176**], [**177**] solved the square root problem for small peturbations of the identity matrix. This method used multilinear expansions and can be extended to operators with smooth coefficients. M^c^Intosh [**355**] considered coefficients in Sobolev spaces, Escauriaza in VMO (unpublished), and Alexopoulos [**3**] real Hölder coefficients using homogenization techniques. Peturbations of real symmetric matrices with L^∞ coefficients were treated in Auscher, Hofmann, Lewis, and Tchamitchian [**19**]. The solution of the two-dimensional Kato problem was obtained by Hofmann and M^c^Intosh [**250**] using a previously derived $T(b)$ type reduction due to Auscher and Tchamitchian [**17**]. Hofmann, Lacey, and M^c^Intosh [**251**] extended this theorem to the case in which the heat kernel of e^{-tL} satisfies Gaussian bounds. Theorem 8.7.1 was obtained by Auscher, Hofmann, Lacey, M^c^Intosh, and Tchamitchian [**20**]; the exposition in the text is based on this reference. Combining Theorem 8.7.1 with a theorem of Lions [**336**], it follows that the domain of $\sqrt{L}$ is $\dot{L}^2_1(\mathbf{R}^n)$ and that for functions f in this space the equivalence of norms $\big\|\sqrt{L}(f)\big\|_{L^2} \approx \big\|\nabla f\big\|_{L^2}$ is valid.

CHAPTER 9

Weighted Inequalities

Weighted inequalities arise naturally in Fourier analysis, but their use is best justified by the variety of applications in which they appear. For example, the theory of weights plays an important role in the study of boundary value problems for Laplace's equation on Lipschitz domains. Other applications of weighted inequalities include extrapolation of operators, vector-valued inequalities, and applications to certain classes of nonlinear partial differential and integral equations.

The theory of weighted inequalities is a natural development of the principles and methods we have acquainted ourselves with in earlier chapters. Although a variety of ideas related to weighted inequalities appeared almost simultaneously with the birth of singular integrals, it was only in the 1970s that a better understanding of the subject was obtained. This was spurred by Muckenhoupt's characterization of positive functions w for which the Hardy-Littlewood maximal operator M maps $L^p(\mathbf{R}^n, w(x)\,dx)$ into itself. This characterization led to the introduction of the class A_p and the development of weighted inequalities. We will pursue exactly this approach in the next section to motivate our introduction of the A_p classes.

9.1. The A_p Condition

A *weight* is a locally integrable function on $\mathbf{R}^n$ that takes values in $(0, \infty)$ almost everywhere. Therefore, weights are allowed to be zero or infinite only on a set of Lebesgue measure zero. Hence, if w is a weight and $1/w$ is locally integrable, then $1/w$ is also a weight.

Given a weight w and a measurable set E, we will use the notation

$$w(E) = \int_E w(x)\,dx$$

to denote the w-measure of the set E. Since weights are locally integrable functions, $w(E) < \infty$ if E is a bounded set. The weighted L^p spaces will be denoted by $L^p(\mathbf{R}^n, w)$ or simply $L^p(w)$. Also recall the definition of the uncentered Hardy-Littlewood maximal operators on $\mathbf{R}^n$ over balls

$$M(f)(x) = \sup_{B \ni x} \operatorname*{Avg}_B |f| = \sup_{B \ni x} \frac{1}{|B|} \int_B |f(y)|\,dy$$

and over cubes

$$M_c(f)(x) = \sup_{Q \ni x} \operatorname*{Avg}_{Q} |f| = \sup_{Q \ni x} \frac{1}{|Q|} \int_Q |f(y)|\, dy\,,$$

where the suprema are taken over all balls B and cubes Q (with sides parallel to the axes) that contain the given point x. It is a classical result proved in Section 2.1 that for all $1 < p < \infty$ there is a constant $C_p(n) > 0$ such that

$$\int_{\mathbf{R}^n} M(f)(x)^p\, dx \le C_p(n)^p \int_{\mathbf{R}^n} |f(x)|^p\, dx \tag{9.1.1}$$

for all functions $f \in L^p(\mathbf{R}^n)$. In Exercise 2.1.1 we saw that a variant of (9.1.1) holds is Lebesgue measure dx is replaced by a doubling measure $d\mu$ on $\mathbf{R}^n$. We will be concerned with the situation in which the measure $d\mu$ is absolutely continuous with respect to Lebesgue measure [i.e., it has the form $d\mu(x) = w(x)\, dx$ for some weight $w(x)$].

9.1.a. Motivation for the A_p Condition

The question we raise is whether there is a characterization of all weights $w(x)$ such that the strong type (p,p) inequality

$$\int_{\mathbf{R}^n} M(f)(x)^p\, w(x)\, dx \le C_p^p \int_{\mathbf{R}^n} |f(x)|^p\, w(x)\, dx \tag{9.1.2}$$

is valid for all $f \in L^p(w)$.

Suppose that (9.1.2) is valid for some weight w and all $f \in L^p(w)$ for some $1 < p < \infty$. Apply (9.1.2) to the function $f\chi_B$ supported in a ball B and use that $\operatorname*{Avg}_{B} |f| \le M(f\chi_B)(x)$ for all $x \in B$ to obtain

$$w(B)(\operatorname*{Avg}_{B} |f|)^p \le \int_B M(f\chi_B)^p\, w\, dx \le C_p^p \int_B |f|^p\, w\, dx\,. \tag{9.1.3}$$

It follows that

$$\left(\frac{1}{|B|} \int_B |f(t)|\, dt \right)^p \le \frac{C_p^p}{w(B)} \int_B |f(x)|^p\, w(x)\, dx \tag{9.1.4}$$

for all balls B and all functions f. At this point, it is tempting to choose a function so that the two integrands are equal. We do so by setting $f = w^{-p'/p}$, which gives $f^p w = w^{-p'/p}$. Assuming for a moment that $\int_B (w(x))^{-p'/p}\, dx < \infty$, it would follow from (9.1.4) that

$$\sup_{B \text{ balls}} \left(\frac{1}{|B|} \int_B w(x)\, dx \right) \left(\frac{1}{|B|} \int_B w(x)^{-\frac{1}{p-1}}\, dx \right)^{p-1} \le C_p^p\,. \tag{9.1.5}$$

If w had a zero in B, then we take $f = (w+\varepsilon)^{-p'/p}$ to get

$$\left(\frac{1}{|B|}\int_B w(x)\,dx\right)\left(\frac{1}{|B|}\int_B (w(x)+\varepsilon)^{-p'}w(x)\,dx\right)^{p-1} \le C_p^p \tag{9.1.6}$$

for all $\varepsilon > 0$, from which we can still deduce (9.1.5) by letting $\varepsilon \to 0$. We have now obtained that every weight w that satisfies (9.1.2) must also satisfy the rather strange-looking condition (9.1.5), which we will be refer to in the sequel as the A_p condition. It is a remarkable fact, which we will prove in this chapter, that the implication obtained can be reversed, that is, (9.1.2) is a consequence of (9.1.5). This will be the first significant achievement of the theory of weights [i.e., a characterization of all functions w for which (9.1.2) holds]. This characterization is based on some deep principles that will be discussed in the next section and provides a solid motivation for the introduction and careful examination of condition (9.1.5).

Before we study the converse statements, we will consider for a moment the case $p=1$. Assume that for some weight w the weak type $(1,1)$ inequality

$$w\big(\{x\in\mathbf{R}^n:\ M(f)(x)>\alpha\}\big) \le \frac{C_1}{\alpha}\int_{\mathbf{R}^n} |f(x)|w(x)\,dx \tag{9.1.7}$$

holds for all functions $f \in L^1(\mathbf{R}^n)$. Since $M(f)(x) \ge \operatorname{Avg}_B |f|$ for all $x \in B$, it follows from (9.1.7) that for all $\alpha < \operatorname{Avg}_B |f|$ we have

$$w(B) \le w\big(\{x\in\mathbf{R}^n:\ M(f)(x)>\alpha\}\big) \le \frac{C_1}{\alpha}\int_{\mathbf{R}^n} |f(x)|w(x)\,dx\,. \tag{9.1.8}$$

Taking $f\chi_B$ instead of f in (9.1.8), we deduce that

$$\operatorname*{Avg}_B |f| = \frac{1}{|B|}\int_B |f(t)|\,dt \le \frac{C_1}{w(B)}\int_B |f(x)|\,w(x)\,dx \tag{9.1.9}$$

for all functions f and balls B. Taking $f = \chi_S$, we obtain

$$\frac{|S|}{|B|} \le C_1\frac{w(S)}{w(B)}, \tag{9.1.10}$$

where S is any measurable subset of the ball B.

Recall that the *essential infimum* of a function w over a set E is defined as

$$\operatorname*{essinf}_E(w) = \inf\big\{b>0:\ |\{x\in E:\ w(x)<b\}|>0\big\}.$$

Then for every $a > \operatorname{essinf}_B(w)$ there exists a subset S_a of B with positive measure such that $w(x) < a$ for all $x \in S_a$. Applying (9.1.10) to the set S_a, we obtain

$$\frac{1}{|B|}\int_B w(t)\,dt \le \frac{C_1}{|S_a|}\int_{S_a} w(t)\,dt \le C_1 a\,, \tag{9.1.11}$$

which implies

$$\frac{1}{|B|}\int_B w(t)\,dt \le C_1 w(x) \qquad \text{for all balls } B \text{ and almost all } x\in B. \tag{9.1.12}$$

It remains to understand what condition (9.1.12) really means. For every ball B, there exists a null set $N(B)$ such that (9.1.12) holds for all x in $B\setminus N(B)$. Let N be the union of all the null sets $N(B)$ for all B balls with centers in $\mathbf{Q}^n$ and rational radii. Then N is a null set and for every x in $\mathbf{R}^n \setminus N$, (9.1.12) holds for all balls B with centers in $\mathbf{Q}^n$ and rational radii. By density (9.1.12) must also hold for all balls B that contain a fixed x in $\mathbf{R}^n \setminus N$. It follows that for $x \in \mathbf{R}^n \setminus N$ we have

$$M(w)(x) = \sup_{B \ni x} \frac{1}{|B|} \int_B w(t)\,dt \le C_1 w(x)\,. \tag{9.1.13}$$

Therefore, assuming (9.1.7), we have arrived at the condition

$$M(w)(x) \le C_1 w(x) \qquad \text{for almost all } x \in \mathbf{R}^n, \tag{9.1.14}$$

where C_1 is the same constant as in (9.1.12). As in the case $p > 1$, this deduction can be reversed and we can obtain (9.1.7) as a consequence of (9.1.14). This motivates a careful study of condition (9.1.14) which we will refer to as the A_1 condition.

Definition 9.1.1. A function $w(x) \ge 0$ is called an A_1 *weight* if

$$M(w)(x) \le C_1 w(x) \qquad \text{for almost all } x \in \mathbf{R}^n \tag{9.1.15}$$

for some constant C_1. If w is an A_1 weight, then the (finite) quantity

$$[w]_{A_1} = \sup_{Q \text{ cubes in } \mathbf{R}^n} \left(\frac{1}{|Q|} \int_Q w(t)\,dt \right) \big\| w^{-1} \big\|_{L^\infty(Q)} \tag{9.1.16}$$

will be called the A_1 *Muckenhoupt characteristic constant*, of w or simply the A_1 *characteristic constant* of w. Note that A_1 weights w satisfy

$$\frac{1}{|Q|} \int_Q w(t)\,dt \le [w]_{A_1} \inf_{y \in Q} w(y) \tag{9.1.17}$$

for all cubes Q in $\mathbf{R}^n$.

Remark 9.1.2. We also define

$$[w]_{A_1}^{\text{balls}} = \sup_{B \text{ balls in } \mathbf{R}^n} \left(\frac{1}{|B|} \int_B w(t)\,dt \right) \big\| w^{-1} \big\|_{L^\infty(B)}\,. \tag{9.1.18}$$

Using (9.1.12), we see that the smallest constant C_1 that appears in (9.1.15) is equal to the A_1 characteristic constant of w as defined in (9.1.18). This is also equal to the smallest constant that appears in (9.1.12). All these constants are bounded above and below by dimensional multiples of $[w]_{A_1}$.

We now recall condition (9.1.5), which motivates the following definition of A_p weights for $1 < p < \infty$.

Definition 9.1.3. Let $1 < p < \infty$. A weight w is said to be *of class* A_p if

$$\sup_{Q \text{ cubes in } \mathbf{R}^n} \left(\frac{1}{|Q|} \int_Q w(x)\,dx \right) \left(\frac{1}{|Q|} \int_Q w(x)^{-\frac{1}{p-1}}\,dx \right)^{p-1} < \infty\,. \tag{9.1.19}$$

The expression in (9.1.19) is called the A_p Muckenhoupt characteristic constant of w (or simply the A_p characteristic constant of w) and will be denoted by $[w]_{A_p}$.

Remark 9.1.4. Note that Definitions 9.1.1 and 9.1.3 could have been given with the set of all cubes in $\mathbf{R}^n$ replaced by the set of all balls in $\mathbf{R}^n$. Defining $[w]_{A_p}^{\text{balls}}$ as in (9.1.19) except that cubes are replaced by balls, we see that

$$(v_n 2^{-n})^p \le \frac{[w]_{A_p}}{[w]_{A_p}^{\text{balls}}} \le \left(n^{n/2} v_n 2^{-n}\right)^p . \tag{9.1.20}$$

9.1.b. Properties of A_p Weights

It is straightforward that translations, isotropic dilations, and scalar multiples of A_p weights are also A_p weights with the same A_p characteristic. We summarize some basic properties of A_p weights in the following proposition.

Proposition 9.1.5. *Let $w \in A_p$ for some $1 \le p < \infty$. Then*

(1) $[\delta^\lambda(w)]_{A_p} = [w]_{A_p}$, *where* $\delta^\lambda(w)(x) = w(\lambda x_1, \dots, \lambda x_n)$.

(3) $[\tau^z(w)]_{A_p} = [w]_{A_p}$, *where* $\tau^z(w)(x) = w(x - z)$, $z \in \mathbf{R}^n$.

(3) $[\lambda w]_{A_p} = [w]_{A_p}$ *for all* $\lambda > 0$.

(4) *When* $1 < p < \infty$, *the function* $w^{-\frac{1}{p-1}}$ *is in* $A_{p'}$ *with characteristic constant*

$$\left[w^{-\frac{1}{p-1}}\right]_{A_{p'}} = [w]_{A_p}^{\frac{1}{p-1}} .$$

Therefore, $w \in A_2$ *if and only if* $w^{-1} \in A_2$ *and both weights have the same* A_2 *characteristic constant.*

(5) $[w]_{A_p} \ge 1$ *for all* $w \in A_p$. *Equality holds if and only if* w *is a constant.*

(6) *The classes* A_p *are increasing as* p *increases; precisely, for* $1 \le p < q < \infty$ *we have*

$$[w]_{A_q} \le [w]_{A_p} .$$

(7) $\lim_{q \to 1+} [w]_{A_q} = [w]_{A_1}$ *if* $w \in A_1$.

(8) *The following is an equivalent characterization of the* A_p *characteristic constant of* w*:*

$$[w]_{A_p} = \sup_{\substack{Q \text{ cubes} \\ \text{in } \mathbf{R}^n}} \sup_{\substack{f \text{ in } L^p(Q,\, w\,dx) \\ |Q \cap \{|f|=0\}|=0}} \left\{ \frac{\left(\frac{1}{|Q|}\int_Q |f(t)|\,dt\right)^p}{\frac{1}{w(Q)}\int_Q |f(t)|^p w(t)\,dt} \right\}.$$

(9) *The measure* $w(x)\,dx$ *is doubling: precisely, for all* $\lambda > 1$ *and all cubes* Q *we have*

$$w(\lambda Q) \le \lambda^{np} [w]_{A_p} w(Q) .$$

(λQ denotes the cube with the same center as Q and side length λ times the side length of Q.)

Proof. The simple proofs of (1), (2), and (3) are left as an exercise. Property (4) is also easy to check and plays the role of duality in this context. To prove (5) we use Hölder's inequality with exponents p and p' to obtain

$$1 = \frac{1}{|Q|}\int_Q dx = \frac{1}{|Q|}\int_Q w(x)^{\frac{1}{p}} w(x)^{-\frac{1}{p}}\,dx \le [w]_{A_p}^{\frac{1}{p}},$$

with equality holding only when $w(x)^{\frac{1}{p}} = c\,w(x)^{-\frac{1}{p}}$ for some $c > 0$ (i.e., when w is a constant). To prove (6), observe that $0 < q' - 1 < p' - 1 \le \infty$ and that the statement $[w]_{A_q} \le [w]_{A_p}$ is equivalent to the fact

$$\left\|w^{-1}\right\|_{L^{q'-1}(Q,\frac{dx}{|Q|})} \le \left\|w^{-1}\right\|_{L^{p'-1}(Q,\frac{dx}{|Q|})}.$$

Property (7) is a consequence of part (a) of Exercise 1.1.3.

To prove (8), apply Hölder's inequality with exponents p and p' to get

$$\begin{aligned}
(\operatorname*{Avg}_Q |f|)^p &= \left(\frac{1}{|Q|}\int_Q |f(x)|\,dx\right)^p \\
&= \left(\frac{1}{|Q|}\int_Q |f(x)| w(x)^{\frac{1}{p}} w(x)^{-\frac{1}{p}}\,dx\right)^p \\
&\le \frac{1}{|Q|^p}\left(\int_Q |f(x)|^p w(x)\,dx\right)\left(\int_Q w(x)^{-\frac{p'}{p}}\,dx\right)^{\frac{p}{p'}} \\
&= \left(\frac{1}{w(Q)}\int_Q |f(x)|^p w(x)\,dx\right)\left(\frac{1}{|Q|}\int_Q w(x)\,dx\right)\left(\frac{1}{|Q|}\int_Q w(x)^{-\frac{1}{p-1}}\,dx\right)^{p-1} \\
&\le [w]_{A_p}\left(\frac{1}{w(Q)}\int_Q |f(x)|^p w(x)\,dx\right).
\end{aligned}$$

This argument proves the inequality $\ge$ in (8) when $p > 1$. In the case $p = 1$ the obvious modification yields the same inequality. The reverse inequality follows by taking $f = (w+\varepsilon)^{-p'/p}$ as in (9.1.6) and letting $\varepsilon \to 0$.

Applying (8) to the function $f = \chi_Q$ and putting λQ in the place of Q in (8), we obtain

$$w(\lambda Q) \le \lambda^{np}[w]_{A_p} w(Q),$$

which says that $w(x)\,dx$ is a doubling measure. This proves (9).

□

Example 9.1.6. We investigate for which real numbers a the power function $|x|^a$ is an A_p weight on $\mathbf{R}^n$. For $1 < p < \infty$, we will need to examine for which a the following expression is finite:

$$\sup_{B \text{ balls}} \left(\frac{1}{|B|}\int_B |x|^a\,dx\right)\left(\frac{1}{|B|}\int_B |x|^{-a\frac{p'}{p}}\,dx\right)^{\frac{p}{p'}}. \tag{9.1.21}$$

Let us divide the set of all balls in $\mathbf{R}^n$ into two categories: The balls of type I, whose radius is less than three times their distance from the origin, and the remaining ones, which we call of type II. If $B = B(x_0, R)$ is of type I, then the presence of the origin is not affecting the behavior of neither integral in (9.1.21), and we see that the expression inside the supremum in (9.1.21) is at most a multiple of

$$|x_0|^a \big(|x_0|^{-a\frac{p'}{p}}\big)^{\frac{p}{p'}} = 1.$$

If $B = B(x_0, R)$ is a ball of type II, then $B(0, 5R)$ has size comparable to B and contains it. It suffices therefore to estimate (9.1.21), in which we have replaced B by $B(0, 5R)$. But this is

$$\begin{aligned} &\left(\frac{1}{v_n(5R)^n}\int_{B(0,5R)} |x|^a\,dx\right)\left(\frac{1}{v_n(5R)^n}\int_{B(0,5R)} |x|^{-a\frac{p'}{p}}\,dx\right)^{\frac{p}{p'}} \\ &= \left(\frac{n}{(5R)^n}\int_0^{5R} r^{a+n-1}dr\right)\left(\frac{n}{(5R)^n}\int_0^{5R} r^{-a\frac{p'}{p}+n-1}dr\right)^{\frac{p}{p'}}, \end{aligned}$$

which is seen easily to be bounded above by a constant $C(n,p)$ only when $-n < a < n\frac{p}{p'}$. We conclude that $|x|^a$ is an A_p weight only when $-n < a < n(p-1)$. This is exactly the range of a's for which the integral in (9.1.21) is finite.

The previous proof can be suitably modified to also include the case $p = 1$. In this case we obtain that $|x|^a$ is an A_1 weight if and only if $-n < a \le 0$. Finally, we observe that the measure $|x|^a dx$ is doubling on the larger range $-n < a < \infty$. Thus for $a > n(p-1)$ we have an absolutely continuous doubling measure $|x|^a\,dx$ for which the Radon-Nikodym derivative is not in A_p.

We now return to a point alluded to earlier, that the A_p condition implies the boundedness of the Hardy-Littlewood maximal function M on the space $L^p(w)$. This is not immediate and requires tools that will be developed in the next section. Nevertheless, with our present knowledge, we can prove a weaker version of this statement, namely that the A_p condition implies the weak type (p,p) boundedness of M with respect to w. This is a straightfoward issue but requires the verification of some details. For simplicity we state the following proposition for the uncentered Hardy-Littlewood maximal function M_c with respect to cubes in $\mathbf{R}^n$ (with sides parallel to the axes). The analogous statement for M follows from the doubling property of A_p weights [Proposition 9.1.5 (9)].

Proposition 9.1.7. *Let $1 \le p < \infty$ and $w \in A_p$. Then M_c maps $L^p(w)$ into $L^{p,\infty}(w)$ with norm at most $3^n [w]_{A_p}^{\frac{2}{p}}$. Conversely, if M_c maps $L^p(w)$ into $L^{p,\infty}(w)$ with norm C_p, then w is in A_p with characteristic constant at most C_p^p.*

Proof. We define

$$M_c^{(w)}(f)(x) = \sup_{Q \ni x} \frac{1}{w(Q)} \int_Q |f(t)| w(t)\,dt\,, \tag{9.1.22}$$

the supremum taken over all cubes Q in $\mathbf{R}^n$ that contain x. It follows from Hölder's inequality that for all functions f we have

$$M_c(f) \le [w]_{A_p}^{\frac{1}{p}} M_c^{(w)}(|f|^p)^{\frac{1}{p}} . \tag{9.1.23}$$

Since $\mu = w\,dx$ is a doubling measure and $\mu(3Q) \le 3^{np}[w]_{A_p}\mu(Q)$, using Proposition 9.1.5 (9) and Exercise 2.1.1 we obtain that $M_c^{(w)}$ maps $L^1(w)$ into $L^{1,\infty}(w)$ with norm at most $3^{np}[w]_{A_p}$. This implies that the operator $f \to M_c^{(w)}(|f|^p)^{\frac{1}{p}}$ maps $L^p(w)$ into $L^{p,\infty}(w)$ with norm at most $3^n[w]_{A_p}^{\frac{1}{p}}$ and thus, in view of (9.1.23), M_c maps $L^p(w)$ into $L^{p,\infty}(w)$ with norm at most $3^n[w]_{A_p}^{\frac{2}{p}}$.

The observation needed to prove the converse statement has already been discussed in the introduction of this section. Assume that the weak type (p,p) inequality

$$w\big(\{x\in\mathbf{R}^n :\ M_c(f)(x) > \alpha\}\big) \le \frac{C_p^p}{\alpha^p}\int_{\mathbf{R}^n} |f(x)|^p w(x)\,dx \tag{9.1.24}$$

holds for all functions $f \in L^p(\mathbf{R}^n, w\,dx)$. Since $M_c(f)(x) \ge \operatorname{Avg}_Q |f|$ for all $x \in Q$, it follows from (9.1.24) that for all $\alpha < \operatorname{Avg}_Q |f|$ we have

$$w(Q) \le w\big(\{x\in\mathbf{R}^n :\ M_c(f)(x) > \alpha\}\big) \le \frac{C_p^p}{\alpha^p}\int_{\mathbf{R}^n} |f(x)|^p w(x)\,dx$$

and, using property (8) of Proposition 9.1.5, we obtain that $[w]_{A_p} \le C_p^p$.

□

Example 9.1.8. On $\mathbf{R}^n$ the function

$$u(x) = \begin{cases} \log\frac{1}{|x|} & \text{when } |x| < \frac{1}{e} \\ 1 & \text{otherwise} \end{cases}$$

is an A_1 weight. Indeed, to check condition (9.1.16) it suffices to consider balls of type I and type II as defined in Example 9.1.6. In either case the required estimate follows easily.

Exercises

9.1.1. Let k be a nonnegative measurable function such that k, k^{-1} are in $L^\infty(\mathbf{R}^n)$. Prove that if w is an A_p weight, then so is kw.

9.1.2. Let w_1, w_2 be two A_1 weights. Prove that $w_1 w_2^{1-p}$ is an A_p weight by showing that

$$[w_1 w_2^{1-p}]_{A_p} \le [w_1]_{A_1} [w_2]_{A_1}^{p-1} .$$

9.1.3. Suppose that $w \in A_p$ and $0 < \delta < 1$. Prove that $w^\delta \in A_q$, where $q = \delta p + 1 - \delta$, by showing that

$$[w^\delta]_{A_q} \le [w]_{A_p}^\delta .$$

9.1.4. Let $w \in A_1$ and $1 < p < \infty$. Prove that the maximal operator M_c maps $L^p(w)$ into itself with norm at most $2(3^n p)^{\frac{1}{p}}(p-1)^{-\frac{1}{p}}[w]_{A_1}^{\frac{1}{p}}$. Conclude that a similar estimate holds for M.
[*Hint:* Use that w is doubling and Exercise 2.1.1.]

9.1.5. Let $w_0 \in A_{p_0}$ and $w_1 \in A_{p_1}$ for some $1 \le p_0, p_1 < \infty$. Let $0 \le \theta \le 1$ and define

$$\frac{1}{p} = \frac{1-\theta}{p_0} + \frac{\theta}{p_1} \qquad \text{and} \qquad w^{\frac{1}{p}} = w_0^{\frac{1-\theta}{p_0}} w_1^{\frac{\theta}{p_1}} .$$

Prove that

$$[w]_{A_p} \le [w_0]_{A_{p_0}}^{(1-\theta)\frac{p}{p_0}} [w_1]_{A_{p_1}}^{\theta\frac{p}{p_1}} ;$$

thus w is in A_p.

9.1.6. Let $1 < p < \infty$. A pair of weights (u, w) that satisfies

$$[u, w]_{(A_p,A_p)} = \sup_{\substack{Q \text{ cubes} \\ \text{in } \mathbf{R}^n}} \left(\frac{1}{|Q|}\int_Q u\, dx\right)\left(\frac{1}{|Q|}\int_Q w^{-\frac{1}{p-1}}\, dx\right)^{p-1} < \infty$$

is said to be of class (A_p, A_p). The quantity $[u, w]_{(A_p,A_p)}$ is called the (A_p, A_p) characteristic constant of the pair.
(a) Show that for any locally integrable function f, $0 < f < \infty$ a.e., the pair $(f, M(f))$ is of *class* (A_p, A_p) for every $1 < p < \infty$ with characteristic constant independent of f.
(b) If (u, w) is of class (A_p, A_p), then the Hardy-Littlewood maximal operator M may not map $L^p(w)$ into $L^p(u)$.
[*Hint:* Try the pair $\big(M(g)^{1-p}, |g|^{1-p}\big)$ for a suitable g.]

9.1.7. In contrast with part (b) of the previous Exercise, show that if the pair of weights (u, w) is of class (A_p, A_p) for some $1 < p < \infty$, then M must map $L^p(w)$ into $L^{p,\infty}(u)$ with norm at most $C(n,p)[u, w]_{(A_p,A_p)}^{\frac{1}{p}}$.
[*Hint:* Show first using Hölder's inequality that for all functions f and all cubes Q' we have

$$\left(\frac{1}{|Q'|}\int_{Q'} |f|\, dx\right)^p u(Q') \le [u, w]_{(A_p,A_p)} \int_{Q'} |f|^p w\, dx .$$

Replacing f by $f\chi_Q$, where $Q \subseteq Q'$ obtain that

$$u(Q') \le [u,w]_{(A_p,A_p)} |Q'|^p \frac{\int_Q |f|^p w\,dx}{\left(\int_Q |f|\,dx\right)^p}.$$

Then use Exercise 4.3.9 to find disjoint cubes Q_j so that the set $E_\alpha = \{x \in \mathbf{R}^n : M_c(f)(x) > \alpha\}$ is contained in the union of $3Q_j$ and $\frac{\alpha}{4^n} < \frac{1}{|Q_j|}\int_{Q_j} |f(t)|\,dt \le \frac{\alpha}{2^n}$. Then $u(E_\alpha) \le \sum_j u(3Q_j)$, and bound each $u(3Q_j)$ by taking $Q' = 3Q_j$ and $Q = Q_j$ in the preceding estimate.]

9.1.8. Use the previous Exercise to prove that for all $1 < q < \infty$ there is a constant $C_q < \infty$ such that for all $f, g \ge 0$ locally integrable functions on $\mathbf{R}^n$ we have

$$\int_{\mathbf{R}^n} M(f)(x)^q\, g(x)\,dx \le C_q \int_{\mathbf{R}^n} f(x)^q\, M(g)(x)\,dx.$$

[*Hint:* Take $1 < p < q$ and interpolate between L^p and L^∞.]

9.1.9. Let $w \in A_p$ for some $1 \le p < \infty$ and $k \ge 1$. Show that $\min(w,k)$ is in A_p and satisfies

$$[\min(w,k)]_{A_p} \le c_p\left([w]_{A_p} + 1\right),$$

where $c_p = 1$ when $p \le 2$ and $c_p = 2^{p-2}$ when $p > 2$.
[*Hint:* Use that $\frac{1}{|Q|}\int_Q \min(w,k)^{-\frac{1}{p-1}}\,dx \le \frac{1}{|Q|}\int_Q w^{-\frac{1}{p-1}}\,dx + k^{-\frac{1}{p-1}}$, raise to the power $p-1$, and multiply by $\min\left(k, \frac{1}{|Q|}\int_Q w\,dx\right)$.]

9.1.10. Use Jensen's inequality to show that for all A_1 weights w and all cubes Q in $\mathbf{R}^n$ we have

$$\operatorname*{Avg}_Q w - \inf_Q \log w \le \log [w]_{A_1}.$$

9.1.11. Suppose that $w_j \in A_{p_j}$ with $1 \le j \le m$ for some $1 \le p_1, \dots, p_m < \infty$ and let $0 < \theta_1, \dots, \theta_m < 1$ be such that $\theta_1 + \cdots + \theta_m = 1$. Show that

$$w_1^{\theta_1} \dots w_m^{\theta_m} \in A_{\max\{p_1,\dots,p_m\}}.$$

[*Hint:* First note that each weight w_j lies in $A_{\max\{p_1,\dots,p_m\}}$ and then apply Hölder's inequality.]

9.1.12. Let $w_1 \in A_{p_1}$ and $w_2 \in A_{p_2}$ for some $1 \le p_1, p_2 < \infty$. Prove that

$$[w_1 + w_2]_{A_p} \le [w_1]_{A_{p_1}} + [w_2]_{A_{p_2}},$$

where $p = \min(p_1, p_2)$.
[*Hint:* Use the definition of the A_p condition.]

9.1.13. Show that if the A_p characteristic constants of a weight w are uniformly bounded for all $p > 1$, then $w \in A_1$. Is it true that

$$A_1 = \bigcap_{p>1} A_p \, ?$$

9.2. Reverse Hölder Inequality for A_p Weights and Consequences

An essential property of A_p weights is that they assign to subsets of balls mass proportional to the percentage of the Lebesgue measure of the subset within the ball. The following lemma provides a way to quantify this statement.

Lemma 9.2.1. *Let $w \in A_p$ for some $1 \le p < \infty$ and let $0 < \alpha < 1$. Then there exists $\beta < 1$ such that whenever S is a measurable subset of a cube Q that satisfies $|S| \le \alpha|Q|$, we have $w(S) \le \beta\, w(Q)$.*

Proof. Taking $f = \chi_A$ in property (8) of Proposition 9.1.5, we obtain

$$\left(\frac{|A|}{|Q|}\right)^p \le [w]_{A_p} \frac{w(A)}{w(Q)} \, . \tag{9.2.1}$$

We write $S = Q \setminus A$ to get

$$\left(1 - \frac{|S|}{|Q|}\right)^p \le [w]_{A_p} \left(1 - \frac{w(S)}{w(Q)}\right) . \tag{9.2.2}$$

Given $0 < \alpha < 1$, set

$$\beta = 1 - \frac{(1-\alpha)^p}{[w]_{A_p}} \tag{9.2.3}$$

and use (9.2.2) to obtain the required conclusion. □

9.2.a. The Reverse Hölder Property of A_p Weights

We are now ready to state and prove one of the main results of the theory of weights, the reverse Hölder inequality for A_p weights.

Theorem 9.2.2. *Let $w \in A_p$ for some $1 \le p < \infty$. Then there exist constants C and $\gamma > 0$ that depend only on the dimension n, on p, and on $[w]_{A_p}$ such that for every cube Q we have*

$$\left(\frac{1}{|Q|}\int_Q w(t)^{1+\gamma}\, dt\right)^{\frac{1}{1+\gamma}} \le \frac{C}{|Q|}\int_Q w(t)\, dt \, . \tag{9.2.4}$$

Proof. Let us fix a cube Q and set

$$\alpha_0 = \frac{1}{|Q|} \int_Q w(x)\,dx\,.$$

We also fix an $0 < \alpha < 1$. We define an increasing sequence of scalars

$$\alpha_0 < \alpha_1 < \alpha_2 < \cdots < \alpha_k < \ldots$$

for $k \geq 0$ by setting

$$\alpha_{k+1} = 2^n \alpha^{-1} \alpha_k \qquad \text{or} \qquad \alpha_k = (2^n \alpha^{-1})^k \alpha_0,$$

and for each $s \geq 1$ we divide the cube Q into a mesh of 2^{ns} subcubes of side length equal to $2^{-s}\ell(Q)$. Among all these cubes, we select those with the property that the average of w over them is strictly greater than α_k and we isolate all maximal cubes with this property. In this way we obtain a collection $\{Q_{k,j}\}_j$ of subcubes of Q so that the following are satisfied:

(1) $\alpha_k < \frac{1}{|Q_{k,j}|} \displaystyle\int_{Q_{k,j}} w(t)\,dt \leq 2^n \alpha_k$.

(2) For almost all $x \notin U_k$ we have $w(x) \leq \alpha_k$, where $U_k = \displaystyle\bigcup_j Q_{k,j}$.

(3) Each $Q_{k+1,j}$ is contained in some $Q_{k,l}$.

Property (1) is satisfied since the unique dyadic parent of $Q_{k,j}$ was not chosen in the selection procedure, while (2) follows from the Lebesgue differentiation theorem using the fact that for almost all $x \notin U_k$ there exists a sequence of nonselected cubes of decreasing lengths whose intersection is $\{x\}$. Property (3) is satisfied since each $Q_{k,j}$ is the maximal subcube of Q with the property that the average of w over it is bigger than α_k. And since the average of w over $Q_{k+1,j}$ is also bigger than α_k, it follows that $Q_{k+1,j}$ must be contained in some maximal cube that possesses this property.

We will now compute the portion of $Q_{k,l}$ that is covered by cubes of the form $Q_{k+1,j}$ for some j. We have

$$\begin{aligned} 2^n \alpha_k &\geq \frac{1}{|Q_{k,l}|} \int_{Q_{k,l} \cap U_{k+1}} w(t)\,dt \\ &= \frac{1}{|Q_{k,l}|} \sum_{j:\, Q_{k+1,j} \subseteq Q_{k,l}} |Q_{k+1,j}| \frac{1}{|Q_{k+1,j}|} \int_{Q_{k+1,j}} w(t)\,dt \\ &> \frac{\big|Q_{k,l} \cap U_{k+1}\big|}{|Q_{k,l}|} \alpha_{k+1} = \frac{\big|Q_{k,l} \cap U_{k+1}\big|}{|Q_{k,l}|} 2^n \alpha^{-1} \alpha_k\,. \end{aligned}$$

It follows that $\big|Q_{k,l} \cap U_{k+1}\big| \leq \alpha |Q_{k,l}|$; thus, applying Lemma 9.2.1, we obtain

$$\frac{w(Q_{k,l} \cap U_{k+1})}{w(Q_{k,l})} < \beta = 1 - \frac{(1-\alpha)^p}{[w]_{A_p}}$$

from which, summing over all l, we obtain

$$w(U_{k+1}) \le \beta w(U_k)\,.$$

The latter gives $w(U_k) \le \beta^k w(U_0)$. We also have $|U_{k+1}| \le \alpha |U_k|$; hence $|U_k| \to 0$ as $k \to \infty$. Therefore, the intersection of the U_k's is a null set. We can therefore write

$$Q = (Q \setminus U_0) \bigcup \Big(\bigcup_{k=0}^{\infty} U_k \setminus U_{k+1} \Big)$$

modulo a set of Lebesgue measure zero. Let us now find a $\gamma > 0$ so that the reverse Hölder inequality (9.2.4) holds. We have $w(x) \le \alpha_k$ for almost all x in $Q \setminus U_k$ and therefore

$$\begin{aligned}
\int_Q w(t)^{1+\gamma}\,dt &= \int_{Q\setminus U_0} w(t)^\gamma w(t)\,dt + \sum_{k=0}^{\infty} \int_{U_k \setminus U_{k+1}} w(t)^\gamma w(t)\,dt \\
&\le \alpha_0^\gamma w(Q \setminus U_0) + \sum_{k=0}^{\infty} \alpha_{k+1}^\gamma w(U_k) \\
&\le \alpha_0^\gamma w(Q \setminus U_0) + \sum_{k=0}^{\infty} ((2^n\alpha^{-1})^{k+1}\alpha_0)^\gamma \beta^k w(U_0) \\
&\le \alpha_0^\gamma \big(1 + (2^n\alpha^{-1})^\gamma \sum_{k=0}^{\infty} (2^n\alpha^{-1})^{\gamma k}\beta^k\big)\big(w(Q \setminus U_0) + w(U_0)\big) \\
&= \Big(\frac{1}{|Q|}\int_Q w(t)\,dt\Big)^\gamma \Big(1 + \frac{(2^n\alpha^{-1})^\gamma}{1 - (2^n\alpha^{-1})^\gamma\beta}\Big) \int_Q w(t)\,dt\,,
\end{aligned}$$

provided $\gamma > 0$ is chosen small enough so that $(2^n\alpha^{-1})^\gamma\beta < 1$. Keeping track of the constants, we conclude the proof of the theorem with

$$\gamma < \frac{-\log\beta}{\log 2^n - \log\alpha} = \frac{\log\big([w]_{A_p}\big) - \log\big([w]_{A_p} - (1-\alpha)^p\big)}{\log 2^n - \log\alpha} \tag{9.2.5}$$

and

$$C = 1 + \frac{(2^n\alpha^{-1})^\gamma}{1 - (2^n\alpha^{-1})^\gamma\beta} = 1 + \frac{(2^n\alpha^{-1})^\gamma}{1 - (2^n\alpha^{-1})^\gamma\big(1 - \frac{(1-\alpha)^p}{[w]_{A_p}}\big)}\,. \tag{9.2.6}$$

Note that up to now α was an arbitrary number in $(0,1)$ and it may be chosen to maximize (9.2.5).

□

Remark 9.2.3. It is worth observing that for any fixed $0 < \alpha < 1$, the constant in (9.2.5) decreases as $[w]_{A_p}$ increases while the constant in (9.2.6) increases as $[w]_{A_p}$ increases. This allows us to obtain the following stronger version of Theorem 9.2.2:

For any $1 \le p < \infty$ and $B > 1$, there exist positive constants $C = C(n,p,B)$ and $\gamma = \gamma(n,p,B)$ such that for all $w \in A_p$ satisfying $[w]_{A_p} \le B$ the reverse Hölder condition (9.2.4) holds for every cube Q. See Exercise 9.2.4(a) for details.

Observe that in the proof of Theorem 9.2.2 it was crucial to know that for some $0 < \alpha, \beta < 1$ we have

$$|S| \le \alpha\,|Q| \implies w(S) \le \beta\,w(Q) \tag{9.2.7}$$

whenever S is a subset of the cube Q. No special property of Lebesgue measure was used in the proof of Theorem 9.2.2 except for its doubling property. Therefore, it is reasonable to ask whether Lebesgue measure in (9.2.7) can be replaced by a general measure μ satisfying the doubling property

$$\mu(2Q) \le C_n\,\mu(Q) < \infty \tag{9.2.8}$$

for all cubes Q in $\mathbf{R}^n$. A straightforward adjustment of the proof of the previous theorem indicates that this is indeed the case.

Corollary 9.2.4. *Let w be a weight and let μ be a measure on $\mathbf{R}^n$ satisfying (9.2.8). Suppose that there exist $0 < \alpha, \beta < 1$, such that*

$$\mu(S) \le \alpha\,\mu(Q) \implies \int_S w(t)\,d\mu(t) \le \beta \int_Q w(t)\,d\mu(t)$$

whenever S is a μ-measurable subset of a cube Q. Then there exist $0 < C, \gamma < \infty$ [which depend only on the dimension n, the constant C_n in (9.2.8), α, and β] such that for every cube Q in $\mathbf{R}^n$ we have

$$\left(\frac{1}{\mu(Q)}\int_Q w(t)^{1+\gamma}\,d\mu(t)\right)^{\frac{1}{1+\gamma}} \le \frac{C}{\mu(Q)}\int_Q w(t)\,d\mu(t). \tag{9.2.9}$$

Proof. The proof of the corollary can be obtained almost verbatim from that of Theorem 9.2.2 by replacing Lebesgue measure with the doubling measure $d\mu$ and the constant 2^n by C_n.

Precisely, we define $\alpha_k = (C_n\alpha^{-1})^k\alpha_0$, where α_0 is the μ-average of w over Q; then properties (1), (2), (3) concerning the selected cubes $\{Q_{k,j}\}_j$ are replaced by the following properties:

(1_μ) $\alpha_k < \frac{1}{\mu(Q_{k,j})}\displaystyle\int_{Q_{k,j}} w(t)\,d\mu(t) \le C_n\,\alpha_k$.

(2_μ) On $Q \setminus U_k$ we have $w \le \alpha_k$ μ-almost everywhere, where $U_k = \bigcup_j Q_{k,j}$.

(3_μ) Each $Q_{k+1,j}$ is contained in some $Q_{k,l}$.

To prove (2_μ) we need a differentiation theorem for doubling measures, analogous to that in Corollary 2.1.16. This can be found in Exercise 2.1.1. The remaining details of the proof are trivially adapted to the new setting. The conclusion is that for

$$0 < \gamma < \frac{-\log\beta}{\log C_n - \log\alpha} \tag{9.2.10}$$

and

$$C = 1 + \frac{(2\alpha^{-1})^\gamma}{1 - (C_n\alpha^{-1})^\gamma \beta}, \tag{9.2.11}$$

(9.2.9) is satisfied.

□

9.2.b. Consequences of the Reverse Hölder Property

Having established the crucial reverse Hölder inequality for A_p weights, we now pass to some very important applications. Among them, the first result of this section yields that an A_p weight that lies a priori in $L^1_{\text{loc}}(\mathbf{R}^n)$ must actually lie in the better space $L^{1+\sigma}_{\text{loc}}(\mathbf{R}^n)$ for some $\sigma > 0$ depending on the weight.

Theorem 9.2.5. *If $w \in A_p$ for some $1 \le p < \infty$, then there exists a number $\gamma > 0$ (which depends on w, p, and n) such that $w^{1+\gamma} \in A_p$.*

Proof. The proof is simple. When $p = 1$, we apply the reverse Hölder inequality of Theorem 9.2.2 to the weight w to obtain

$$\frac{1}{|Q|}\int_Q w(t)^{1+\gamma}\,dt \le \left(\frac{C}{|Q|}\int_Q w(t)\,dt\right)^{1+\gamma} \le C^{1+\gamma}[w]_{A_1}^{1+\gamma} w(x)$$

for almost all x in the cube Q. Therefore, $w^{1+\gamma}$ is an A_1 weight with characteristic constant at most $C^{1+\gamma}[w]_{A_1}^{1+\gamma}$. ($C$ is here the constant of Theorem 9.2.2.) When $p > 1$, there exist $\gamma_1, \gamma_2 > 0$ and $C_1, C_2 > 0$ so that the reverse Hölder inequality of Theorem 9.2.2 holds for the weights $w \in A_p$ and $w^{-\frac{1}{p-1}} \in A_{p'}$, that is,

$$\begin{aligned}
\frac{1}{|Q|}\int_Q w(t)^{1+\gamma_1}\,dt \;&\le\; \left(\frac{C_1}{|Q|}\int_Q w(t)\,dt\right)^{1+\gamma_1}, \\
\frac{1}{|Q|}\int_Q w(t)^{-(1+\gamma_2)\frac{1}{p-1}}\,dt \;&\le\; \left(\frac{C_2}{|Q|}\int_Q w(t)^{-\frac{1}{p-1}}\,dt\right)^{1+\gamma_2}.
\end{aligned}$$

Taking $\gamma = \min(\gamma_1, \gamma_2)$, both inequalities are satisfied with γ in the place of γ_1, γ_2. It follows that $w^{1+\gamma}$ is in A_p and satisfies

$$[w^{1+\gamma}]_{A_p} \le (C_1 C_2^{p-1})^{1+\gamma}[w]_{A_p}^{1+\gamma}. \tag{9.2.12}$$

□

Corollary 9.2.6. *For any $1 < p < \infty$ and for every $w \in A_p$ there is a $q = q([w]_{A_p}, p, n)$ with $q < p$ such that $w \in A_q$. In other words, we have*

$$A_p = \bigcup_{q\in(1,p)} A_q.$$

Proof. Given $w \in A_p$, let γ, C_1, C_2 be as in the proof of Theorem 9.2.5. In view of the result in Exercise 9.1.3, if $w^{1+\gamma} \in A_p$ and

$$q = p\frac{1}{1+\gamma} + 1 - \frac{1}{1+\gamma} = \frac{p+\gamma}{1+\gamma},$$

then $w \in A_q$ and

$$[w]_{A_q} \leq \big[w^{1+\gamma}\big]_{A_p}^{\frac{1}{1+\gamma}} \leq C_1 C_2^{p-1} [w]_{A_p},$$

where the last estimate comes from (9.2.12). Since $1 < q = \frac{p+\gamma}{1+\gamma} < p$, the required conclusion follows. Observe that the constants $C_1 C_2^{p-1}$, q, and $\frac{1}{\gamma}$ increase as $[w]_{A_p}$ increases.

□

We can now return to a point promised earlier, the improvement of Proposition 9.1.7

Corollary 9.2.7. *Let $1 < p < \infty$. Then the following statements are equivalent:*
(a) $w \in A_p$.
(b) The Hardy-Littlewood maximal operator M maps $L^p(w)$ into itself.

Proof. We already know that (b) implies (a). Conversely, if (a) holds, then w lies in $A_{p-\varepsilon}$ for some $\varepsilon > 0$. Therefore, by Proposition 9.1.7, M maps $L^{p-\varepsilon}(w)$ into $L^{p-\varepsilon,\infty}(w)$. Interpolating between this estimate and the trivial $L^\infty(w)$ estimate, we obtain that M maps $L^p(w)$ into itself.

□

Another powerful consequence of the reverse Hölder property of A_p weights is the following characterization of all A_1 weights.

Theorem 9.2.8. *Let w be an A_1 weight. Then there exist $0 < \varepsilon < 1$, a nonnegative function k such that $k, k^{-1} \in L^\infty$, and a nonnegative locally integrable function f that satisfies $M(f) < \infty$ a.e., such that*

$$w(x) = k(x)\, M(f)(x)^{\varepsilon}. \tag{9.2.13}$$

Conversely, every weight w of the form (9.2.13) for some k, f as previously is in A_1 with

$$[w]_{A_1} \leq \frac{C_n}{1-\varepsilon} \big\|k\big\|_{L^\infty} \big\|k^{-1}\big\|_{L^\infty},$$

where C_n is a universal dimensional constant.

Proof. In view of Theorem 9.2.2, there exist $0 < \gamma, C < \infty$ such that the reverse Hölder condition

$$\left(\frac{1}{|Q|}\int_Q w(t)^{1+\gamma}\, dt\right)^{\frac{1}{1+\gamma}} \leq \frac{C}{|Q|}\int_Q w(t)\, dt \leq C\, [w]_{A_1} w(x) \tag{9.2.14}$$

holds for all cubes Q for almost all x in Q. We set

$$\varepsilon = \frac{1}{1+\gamma} \qquad \text{and} \qquad f(x) = w(x)^{1+\gamma} = w(x)^{\frac{1}{\varepsilon}}\,.$$

It follows from (9.2.14) that the uncentered Hardy-Littlewood maximal function $M_c(f)$ with respect to cubes is at most $C^{1+\gamma}[w]_{A_1}^{1+\gamma} f$, which implies that $M(f) \le C_n C^{1+\gamma}[w]_{A_1}^{1+\gamma} f$ for some constant C_n that depends only on the dimension. We now set

$$k(x) = \frac{f(x)^\varepsilon}{M(f)(x)^\varepsilon}\,,$$

and we observe that $C^{-1}C_n^{-\varepsilon}[w]_{A_1}^{-1} \le k \le 1$ a.e.

It remains to prove the converse. Given a weight $w = kM(f)^\varepsilon$ in the form (9.2.13) and a cube Q, we observe that it suffices to show that

$$\frac{1}{|Q|}\int_Q M(f)(t)^\varepsilon\,dt \le \frac{C_n}{1-\varepsilon} M(f)^\varepsilon(x) \qquad \text{for almost all } x \in Q, \tag{9.2.15}$$

since the corresponding statement for $kM(f)^\varepsilon$ will follow trivially from (9.2.15) using that $k, k^{-1} \in L^\infty$. To show (9.2.15), we write

$$f = f\chi_{3Q} + f\chi_{(3Q)^c}\,.$$

Then

$$\frac{1}{|Q|}\int_Q M(f\chi_{3Q})(t)^\varepsilon\,dt \le \frac{C_n'}{1-\varepsilon}\left(\frac{1}{|Q|}\int_{\mathbf{R}^n} (f\chi_{3Q})(t)\,dt\right)^\varepsilon \tag{9.2.16}$$

in view of Kolmogorov's inequality (Exercise 2.1.5). But the last expression in (9.2.16) is at most a dimensional multiple of $M(f)(x)^\varepsilon$ for almost all $x \in Q$, which proves (9.2.15) when f is replaced by $f\chi_{3Q}$ on the left-hand side of the inequality. And for $f\chi_{(3Q)^c}$ we only need to notice that

$$M(f\chi_{(3Q)^c})(t) \le 2^n \mathcal{M}(f\chi_{(3Q)^c})(t) \le 2^n n^{\frac{n}{2}} M(f)(x)$$

for all x, t in Q since any ball B centered at t that gives a nonzero average for $f\chi_{(3Q)^c}$, must have radius at least the side length of Q, and thus $\sqrt{n}\,B$ must also contain x. (Here $\mathcal{M}$ is the centered Hardy-Littlewood maximal operator introduced in Definition 2.1.1.) Hence (9.2.15) also holds when f is replaced by $f\chi_{(3Q)^c}$ on the left-hand side. Combining these two estimates and using the subbaditivity property $M(f+g)^\varepsilon \le M(f)^\varepsilon + M(g)^\varepsilon$, we obtain (9.2.15).

□

Exercises

9.2.1. Let $w \in A_p$ for some $1 < p < \infty$ and let $1 \le q < \infty$. Prove that the sublinear operator

$$S(f) = \big(M(|f|^q w)w^{-1}\big)^{\frac{1}{q}}$$

is bounded on $L^{p'q}(w)$.

9.2.2. Let v be a real-valued locally integrable function on $\mathbf{R}^n$ and let $1 < p < \infty$. Prove that e^v is an A_p weight if and only if the following two conditions are satisfied for some constant $C < \infty$:

$$\sup_{Q \text{ cubes}} \frac{1}{|Q|} \int_Q e^{v(t)-v_Q}\, dt \le C\,,$$

$$\sup_{Q \text{ cubes}} \frac{1}{|Q|} \int_Q e^{-(v(t)-v_Q)\frac{1}{p-1}}\, dt \le C\,.$$

[*Hint:* If $e^v \in A_p$, use that

$$\frac{1}{|Q|} \int_Q e^{v(t)-v_Q}\, dt \le \big(\operatorname{Avg}_Q e^{-\frac{v}{p-1}}\big)^{p-1} \big(\operatorname{Avg}_Q e^{v}\big)$$

and obtain a similar estimate for the second quantity.]

9.2.3. Let v be a real-valued locally integrable function on $\mathbf{R}^n$ and let $1 < p < \infty$.
(a) Use the result of the previous Exercise to show that e^v is in A_2 if and only if for some constant $C < \infty$, we have

$$\sup_{Q \text{ cubes}} \frac{1}{|Q|} \int_Q e^{|v(t)-v_Q|}\, dt \le C\,.$$

Conclude that $\big\|\log\varphi\big\|_{BMO} \le [\varphi]_{A_2}$; thus if $\varphi \in A_2$, then $\log\varphi \in BMO$.
(b) Use part (a) and Theorem 7.1.6 to prove the converse, namely that every BMO function is equal to a constant multiple of the logarithm of an A_2 weight.
(c) Prove that if φ is in A_p for some $1 < p < \infty$, then $\log\varphi$ is in BMO by showing that

$$\big\|\log\varphi\big\|_{BMO} \le \begin{cases} [\varphi]_{A_p} & \text{when } 1 < p \le 2 \\ (p-1)[\varphi]_{A_p}^{\frac{1}{p-1}} & \text{when } 2 < p < \infty\,. \end{cases}$$

[*Hint:* Use that $\varphi^{-\frac{1}{p-1}} \in A_{p'}$ when $p > 2$.]

9.2.4. Prove the following quantitative versions of Theorem 9.2.2 and Corollaries 9.2.6 and 9.2.7.
(a) For any $1 \le p < \infty$ and $B > 1$, there exist positive constants $C = C_1(n,p,B)$ and $\gamma = \gamma(n,p,B)$ such that for all $w \in A_p$ satisfying $[w]_{A_p} \le B$ we have that (9.2.4) holds for every cube Q.
(b) Given any $1 < p < \infty$ and $B > 1$ there exist constants $C = C_2(n,p,B)$ and $\delta = \delta(n,p,B)$ such that for all $w \in A_p$ we have

$$[w]_{A_p} \le B \implies [w]_{A_{p-\delta}} \le C\,.$$

(c) Given any $B > 1$ and $1 < p < \infty$ there exists $C_3(n,p,B)$ such that for all $w \in A_p$ with $[w]_{A_p} \le B$, M maps $L^p(w)$ into itself with norm at most $C_3(n,p,B)$.
[*Hint:* Use estimate (1.3.11).]

9.2.5. Given a positive doubling measure μ on $\mathbf{R}^n$ [i.e., a measure that satisfies (9.2.8)], define the characteristic constant $[w]_{A_p(\mu)}$ and the class $A_p(\mu)$ for $1 < p < \infty$.
(a) Show that statement (8) of Proposition 9.1.5 remains valid if Lebesgue measure is replaced by μ.
(b) Obtain as a consequence that if $w \in A_p(\mu)$, then for all cubes Q and all μ-measurable subsets A of Q we have

$$\left(\frac{\mu(A)}{\mu(Q)}\right)^p \le [w]_{A_p(\mu)} \frac{w(A)}{w(Q)}.$$

Conclude that if Lebesgue measure is replaced by μ in Lemma 9.2.1, then the lemma is valid for $w \in A_p(\mu)$.
(c) Use Corollary 9.2.4 to obtain that weights in $A_p(\mu)$ satisfy a reverse Hölder condition.
(d) Prove that given a weight $w \in A_p(\mu)$, there exists $1 < q < p$ [which depends on $[w]_{A_p(\mu)}$] such that $w \in A_q(\mu)$.

9.2.6. Let $1 < q < \infty$ and μ a positive measure on $\mathbf{R}^n$. We say that a positive function K on $\mathbf{R}^n$ satisfies a *reverse Hölder condition* of order q with respect to μ [symbolically $K \in RH_q(\mu)$] if

$$[K]_{RH_q(\mu)} = \sup_{Q \text{ cubes in } \mathbf{R}^n} \frac{\left(\frac{1}{\mu(Q)}\int_Q K^q\, d\mu\right)^{\frac{1}{q}}}{\frac{1}{\mu(Q)}\int_Q K\, d\mu} < \infty.$$

For positive functions u, v on $\mathbf{R}^n$ and $1 < p < \infty$, show that

$$[vu^{-1}]_{RH_{p'}(u\,dx)} = [uv^{-1}]^{\frac{1}{p}}_{A_p(v\,dx)},$$

that is, vu^{-1} satisfies a reverse Hölder condition of order p' with respect to $u\,dx$ if and only if uv^{-1} is in $A_p(v\,dx)$. Conclude that

$$\begin{aligned} w \in RH_{p'}(dx) &\iff w^{-1} \in A_p(w\,dx) \\ w \in A_p(dx) &\iff w^{-1} \in RH_{p'}(w\,dx). \end{aligned}$$

9.2.7. (*Gehring* [**207**]) Suppose that a positive function K on $\mathbf{R}^n$ lies in $RH_p(dx)$ for some $1 < p < \infty$. Show that there exists a $\delta > 0$ such that K lies in $RH_{p+\delta}(dx)$.
[*Hint:* By the previous Exercise $K \in RH_p(dx)$ is equivalent to the fact that $K \in A_{p'}(K^{-1}\,dx)$ and the index p' can be improved by Exercise 9.2.5(d).]

9.2.8. (a) Show that for any $w \in A_1$ and any cube Q in $\mathbf{R}^n$ and $a > 1$ we have

$$\inf_Q w \le a^n [w]_{A_1} \inf_{aQ} w\,.$$

(b) Prove that there is a constant C_n such that for all locally integrable functions f on $\mathbf{R}^n$ and all cubes Q in $\mathbf{R}^n$ we have

$$\inf_Q M(f) \le C_n \inf_{3Q} M(f)$$

and an analogous statement is valid for M_c.
[*Hint:* Part (a): Use (9.1.17). Part (b): Apply part (a) to $M(f)^{\frac{1}{2}}$, which is an A_1 weight in view of Theorem 9.2.8.]

9.2.9. Define the class $RH_\infty(dx)$ as the class of all weights v satisfying

$$\sup_Q \big(\sup_Q v\big) \left(\frac{1}{|Q|}\int_Q v\,dx\right)^{-1} = [v]_{RH_\infty} < \infty\,,$$

where the first supremum is taken over all cubes Q in $\mathbf{R}^n$ with sides parallel to the axes. Show that for all $v \in A_1$ we have

$$v \in RH_\infty(dx) \implies u \in A_1(v\,dx)\,.$$

9.3. The A_∞ condition*

In this section we will examine more closely the class of all A_p weights. It will turn out that A_p weights possess properties that are p-independent but delicate enough to characterize them without reference to a specific value of p. The A_p classes increase as p increases, and it is only natural to consider their limit as $p \to \infty$. Not surprisingly, a condition obtained as a limit of the A_p conditions as $p \to \infty$, provides some unexpected but insightful characterizations of the class of all A_p weights.

9.3.a. The Class of A_∞ Weights

Let us start by recalling a simple consequence of Jensen's inequality:

$$\left(\int_X |h(t)|^q\,d\mu(t)\right)^{\frac{1}{q}} \ge \exp\left(\int_X \log|h(t)|\,d\mu(t)\right), \tag{9.3.1}$$

which holds for all measurable functions h on a probability space (X,μ) and all $0 < q < \infty$. See Exercise 1.1.3(b). Moreover, part (c) of the same Exercise says that the limit of the expressions on the left in (9.3.1) as $q \to 0$ is equal to the expression on the right in (9.3.1).

Let us apply (9.3.1) to the A_p characteristic constant of a weight w. We obtain

$$\frac{w(Q)}{|Q|}\left(\frac{1}{|Q|}\int_Q w(t)^{-\frac{1}{p-1}}\,dt\right)^{p-1} \geq \frac{w(Q)}{|Q|}\exp\left(\frac{1}{|Q|}\int_Q \log w(t)^{-1}\,dt\right), \tag{9.3.2}$$

and the limit of the expressions on the left in (9.3.2) as $p \to \infty$ is equal to the expression on the right in (9.3.2). This observation provides the motivation for the following definition.

Definition 9.3.1. A weight w is called an A_∞ weight if

$$[w]_{A_\infty} = \sup_{Q \text{ cubes in } \mathbf{R}^n} \left\{\left(\frac{1}{|Q|}\int_Q w(t)\,dt\right)\exp\left(\frac{1}{|Q|}\int_Q \log w(t)^{-1}\,dt\right)\right\} < \infty .$$

The quantity $[w]_{A_\infty}$ is called the A_∞ *characteristic constant* of w.

It follows from the previous definition and (9.3.2) that for all $1 \leq p < \infty$ we have

$$[w]_{A_\infty} \leq [w]_{A_p} .$$

This means that

$$\bigcup_{1\leq p<\infty} A_p \subseteq A_\infty , \tag{9.3.3}$$

but the remarkable thing is that equality actually holds in (9.3.3), a deep property that requires a good deal of work.

Before we examine this and other characterizations of A_∞ weights, we discuss some of their elementary properties.

Proposition 9.3.2. *Let $w \in A_\infty$. Then*

(1) $[\delta^\lambda(w)]_{A_\infty} = [w]_{A_\infty}$, *where $\delta^\lambda(w)(x) = w(\lambda x_1, \dots, \lambda x_n)$ and $\lambda > 0$.*

(3) $[\tau^z(w)]_{A_\infty} = [w]_{A_\infty}$, *where $\tau^z(w)(x) = w(x - z)$, $z \in \mathbf{R}^n$.*

(3) $[\lambda w]_{A_\infty} = [w]_{A_\infty}$ *for all $\lambda > 0$.*

(4) $\lim_{p\to\infty} [w]_{A_p} = [w]_{A_\infty}$ *and $[w]_{A_\infty} \geq 1$.*

(5) *The following is an equivalent characterization of the A_∞ characteristic constant of w:*

$$[w]_{A_\infty} = \sup_{\substack{Q \text{ cubes} \\ \text{in } \mathbf{R}^n}} \sup_{\substack{\log|f| \in L^1(Q) \\ |Q\cap\{|f|=0\}|=0}} \left\{\frac{w(Q)}{\int_Q |f(t)|w(t)\,dt}\exp\left(\frac{1}{|Q|}\int_Q \log|f(t)|\,dt\right)\right\}.$$

(6) *The measure $w(x)\,dx$ is doubling; precisely, for all $\lambda > 1$ and all cubes Q we have*

$$w(\lambda Q) \leq 2^{\lambda^n}[w]_{A_\infty}^{\lambda^n} w(Q) .$$

As usual, λQ here denotes the cube with the same center as Q and side length λ times that of Q.

We note that estimate (6) is not as good as $\lambda \to \infty$ but it can be substantially improved using the case $\lambda = 2$. We refer to Exercise 9.3.1 for an improvement.

Proof. Properties (1)–(3) are elementary, while property (4) is a consequence of part (c) of Exercise 1.1.3. To show (5), first observe that by taking $f = w^{-1}$, the expression on the right in (5) is at least as big as $[w]_{A_\infty}$. Conversely, (9.3.1) gives

$$\exp\left(\frac{1}{|Q|}\int_Q \log\big(|f(t)|w(t)\big)\,dt\right) \le \frac{1}{|Q|}\int_Q |f(t)|w(t)\,dt\,,$$

which, after a simple algebraic manipulation, can be written as

$$\frac{w(Q)}{\int_Q |f|w\,dt}\exp\left(\frac{1}{|Q|}\int_Q \log|f|\,dt\right) \le \frac{w(Q)}{|Q|}\exp\left(-\frac{1}{|Q|}\int_Q \log|w|\,dt\right),$$

whenever f does not vanish almost everywhere on Q. Taking the supremum over all such f and all cubes Q in $\mathbf{R}^n$, we obtain that the expression on the right in (5) is at most $[w]_{A_\infty}$.

To prove the doubling property for A_∞ weights, we fix $\lambda > 1$ and we apply property (5) to the cube λQ in place of Q and to the function

$$f = \begin{cases} c & \text{on } Q \\ 1 & \text{on } \mathbf{R}^n \setminus Q, \end{cases} \tag{9.3.4}$$

where c is chosen so that $c^{1/\lambda^n} = 2[w]_{A_\infty}$. We obtain

$$\frac{w(\lambda Q)}{w(\lambda Q \setminus Q) + c\,w(Q)}\exp\left(\frac{\log c}{\lambda^n}\right) \le [w]_{A_\infty}\,,$$

which implies (6) if we take into account the chosen value of c. $\square$

9.3.b. Characterizations of A_∞ Weights

Having established some elementary properties of A_∞ weights, we now turn to some of their deeper properties, one of which is that every A_∞ weight lies in some A_p for $p < \infty$. It will also turn out that A_∞ weights are characterized by the reverse Hölder property, which as we saw is a fundamental property of A_p weights. The following is the main theorem of this section.

Theorem 9.3.3. *Suppose that w is a weight. Then w is in A_∞ if and only if any one of the following conditions holds:*
(a) There exist $0 < \gamma, \delta < 1$ such that for all cubes Q in $\mathbf{R}^n$ we have

$$\big|\{x \in Q :\ w(x) \le \gamma \operatorname*{Avg}_Q w\}\big| \le \delta\,|Q|\,.$$

(b) There exist $0 < \alpha, \beta < 1$ *such that for all cubes* Q *and all measurable subsets* A *of* Q *we have*

$$|A| \le \alpha\,|Q| \implies w(A) \le \beta\, w(Q)\,.$$

(c) The reverse Hölder condition holds for w, *that is, there exist* $0 < C_1, \varepsilon < \infty$ *such that for all cubes* Q *we have*

$$\left(\frac{1}{|Q|}\int_Q w(t)^{1+\varepsilon}\,dt\right)^{\frac{1}{1+\varepsilon}} \le \frac{C_1}{|Q|}\int_Q w(t)\,dt\,.$$

(d) There exist $0 < C_2, \varepsilon_0 < \infty$ *such that for all cubes* Q *and all measurable subsets* A *of* Q *we have*

$$\frac{w(A)}{w(Q)} \le C_2\left(\frac{|A|}{|Q|}\right)^{\varepsilon_0}.$$

(e) There exist $0 < \alpha', \beta' < 1$ *such that for all cubes* Q *and all measurable subsets* A *of* Q *we have*

$$w(A) < \alpha'\, w(Q) \implies |A| < \beta'\,|Q|\,.$$

(f) There exist $p, C_3 < \infty$ *such that* $[w]_{A_p} \le C_3$, *that is,* w *lies in* A_p.

All the constants $C_1, C_2, C_3, \alpha, \beta, \gamma, \delta, \alpha', \beta', \varepsilon, \varepsilon_0$, *and* p *in (a)–(f) only depend on the dimension* n *and on* $[w]_{A_\infty}$. *Moreover, if any of the statements in (a)–(f) is valid, then so is any other statement in (a)–(f) with constants that depend only on the dimension* n *and the constants that appear in the assumed statement.*

Proof. The proof will follow from the sequence of implications:

$$w \in A_\infty \implies (a) \implies (b) \implies (c) \implies (d) \implies (e) \implies (f) \implies w \in A_\infty\,.$$

At each step we will keep track of the way the constants depend on the constants of the previous step. This is needed to validate the last assertion of the theorem. We begin with the first implication.

$w \in A_\infty \implies (a)$

Fix a cube Q. Since multiplication of an A_∞ weight with a positive scalar does not alter its A_∞ characteristic, we may assume that $\int_Q \log w(t)\,dt = 0$. This implies that $\operatorname*{Avg}_Q w \le [w]_{A_\infty}$. Then we have

$$\begin{aligned}
&\big|\{x\in Q:\ w(x)\le \gamma \operatorname*{Avg}_Q w\}\big| \\
&\le \big|\{x\in Q:\ w(x)\le \gamma [w]_{A_\infty}\}\big| \\
&= \big|\{x\in Q:\ \log(1+w(x)^{-1})\ge \log(1+(\gamma[w]_{A_\infty})^{-1})\}\big| \\
&\le \frac{1}{\log(1+(\gamma[w]_{A_\infty})^{-1})}\int_Q \log\frac{1+w(t)}{w(t)}\,dt \\
&= \frac{1}{\log(1+(\gamma[w]_{A_\infty})^{-1})}\int_Q \log(1+w(t))\,dt \\
&\le \frac{1}{\log(1+(\gamma[w]_{A_\infty})^{-1})}\int_Q w(t)\,dt \\
&\le \frac{[w]_{A_\infty}|Q|}{\log(1+(\gamma[w]_{A_\infty})^{-1})} = \frac{1}{2}|Q|\,,
\end{aligned}$$

which proves (a) with $\gamma = [w]_{A_\infty}^{-1}(e^{2[w]_{A_\infty}}-1)^{-1}$ and $\delta=\frac{1}{2}$.

$\boxed{(a) \implies (b)}$

Let Q be fixed and let A be a subset of Q with $w(A) > \beta w(Q)$ for some β to be chosen later. Setting $S = Q\setminus A$, we have $w(S) < (1-\beta)w(Q)$. We write $S = S_1\cup S_2$, where

$$S_1 = \{x\in S:\ w(x) > \gamma \operatorname*{Avg}_Q w\} \quad\text{and}\quad S_2 = \{x\in S:\ w(x)\le \gamma \operatorname*{Avg}_Q w\}\,.$$

For S_2 we have $|S_2|\le \delta|Q|$ by assumption (a). For S_1 we use Chebychev's inequality to obtain

$$|S_1| \le \frac{1}{\gamma \operatorname*{Avg}_Q w}\int_S w(t)\,dt \le \frac{|Q|}{\gamma}\frac{w(S)}{w(Q)} \le \frac{1-\beta}{\gamma}|Q|\,.$$

Adding the estimates for $|S_1|$ and $|S_2|$, we obtain

$$|S| \le |S_1| + |S_2| \le \frac{1-\beta}{\gamma}|Q| + \delta|Q| = \Big(\delta + \frac{1-\beta}{\gamma}\Big)|Q|\,.$$

Choosing numbers α,β in $(0,1)$ such that $\delta + \frac{1-\beta}{\gamma} = 1-\alpha$ [for example $\alpha = \frac{1-\delta}{2}$ and $\beta = 1-\frac{(1-\delta)\gamma}{2}$], we obtain $|S|\le (1-\alpha)|Q|$, that is, $|A| > \alpha|Q|$.

$\boxed{(b) \implies (c)}$

This was proved in Corollary 9.2.4. To keep track of the constants, we note that the choices

$$\varepsilon = \frac{-\frac{1}{2}\log\beta}{\log 2^n - \log\alpha} \quad\text{and}\quad C_1 = 1 + \frac{(2^n\alpha^{-1})^{\varepsilon}}{1-(2^n\alpha^{-1})^{\varepsilon}\beta}$$

as given in (9.2.5) and (9.2.6) serve our purposes.

$\boxed{(c) \implies (d)}$

We apply first Hölder's inequality with exponents $1+\varepsilon$ and $(1+\varepsilon)/\varepsilon$ and then the reverse Hölder estimate in

$$\begin{aligned}\int_A w(x)\,dx &\le \Big(\int_A w(x)^{1+\varepsilon}\,dx\Big)^{\frac{1}{1+\varepsilon}} |A|^{\frac{\varepsilon}{1+\varepsilon}}\\ &= \Big(\frac{1}{|Q|}\int_A w(x)^{1+\varepsilon}\,dx\Big)^{\frac{1}{1+\varepsilon}} |Q|^{\frac{1}{1+\varepsilon}}\,|A|^{\frac{\varepsilon}{1+\varepsilon}}\\ &\le \frac{C_1}{|Q|}\int_Q w(x)\,dx\,|Q|^{\frac{1}{1+\varepsilon}}\,|A|^{\frac{\varepsilon}{1+\varepsilon}}\,,\end{aligned}$$

which gives

$$\frac{w(A)}{w(Q)} \le C_1\Big(\frac{|A|}{|Q|}\Big)^{\frac{\varepsilon}{1+\varepsilon}}.$$

This proves (d) with $\varepsilon_0 = \frac{\varepsilon}{1+\varepsilon}$ and $C_2 = C_1$.

$\boxed{(d) \implies (e)}$

Pick an $0 < \alpha'' < 1$ small enough so that $\beta'' = C_2(\alpha'')^{\varepsilon_0} < 1$. It follows from (d) that

$$|A| < \alpha''|Q| \implies w(A) < \beta'' w(Q) \tag{9.3.5}$$

for all cubes Q and all A measurable subsets of Q. Replacing A by $Q \setminus A$, the implication in (9.3.5) can be equivalently written as

$$|A| \ge (1-\alpha'')|Q| \implies w(A) \ge (1-\beta'')w(Q)\,.$$

In other words, for measurable subsets A of Q we have

$$w(A) < (1-\beta'')w(Q) \implies |A| < (1-\alpha'')|Q|\,, \tag{9.3.6}$$

which is the statement in (e) if we set $\alpha' = (1-\beta'')$ and $\beta' = 1-\alpha''$. Note that (9.3.5) and (9.3.6) are indeed equivalent.

$\boxed{(e) \implies (f)}$

This is the deepest of the implications but we already have available all the tools we need to obtain it. We begin by examining condition (e), which can be written as

$$\int_A w(t)\,dt \le \alpha' \int_Q w(t)\,dt \implies \int_A w(t)^{-1}w(t)\,dt \le \beta' \int_Q w(t)^{-1}w(t)\,dt\,,$$

or, equivalently, as

$$\mu(A) \le \alpha'\mu(Q) \implies \int_A w(t)^{-1}\,d\mu(t) \le \beta' \int_Q w(t)^{-1}\,d\mu(t)$$

after defining the measure $d\mu(t) = w(t)\,dt$. Observe that in the proof of (d) $\Longrightarrow$ (e) we established the equivalence of (9.3.5) and (9.3.6). Using Exercise 9.3.2 we deduce that the measure μ is doubling [i.e., it satisfies property (9.2.8) for some constant $C_n = C_n(\alpha', \beta')$]. Therefore, the hypotheses of Corollary 9.2.4 are satisfied. We conclude that the weight w^{-1} satisfies a reverse Hölder estimate with respect to the measure μ, that is, if γ, C are defined by (9.2.10) and (9.2.11) [in which α is replaced by α', β by β', and C_n is the doubling constant of $w(x)\,dx$], then we have

$$\left(\frac{1}{\mu(Q)}\int_Q w(t)^{-1-\gamma}\,d\mu(t)\right)^{\frac{1}{1+\gamma}} \le \frac{C}{\mu(Q)}\int_Q w(t)^{-1}\,d\mu(t) \tag{9.3.7}$$

for all cubes Q in $\mathbf{R}^n$. Setting $p = 1+\frac{1}{\gamma}$ and raising to the pth power, we can rewrite (9.3.7) as the A_p condition for w. We can therefore take $C_3 = C^p$ to conclude the proof of (f).

$(f) \Longrightarrow w \in A_\infty$

This is trivial since $[w]_{A_\infty} \le [w]_{A_p}$. □

An immediate consequence of the result just proved is the following.

Corollary 9.3.4. *The following equality is valid:*

$$A_\infty = \bigcup_{1\le p<\infty} A_p.$$

Exercises

9.3.1. (a) Show that property (6) in Proposition 9.3.2 can be improved to

$$w(\lambda Q) \le \min_{\varepsilon>0} \frac{(1+\varepsilon)^{\lambda^n}[w]_{A_\infty}^{\lambda^n} - 1}{\varepsilon}\, w(Q)\,.$$

(b) Take $\lambda = 2$ in property (6) of Proposition 9.3.2 and iterate the estimate obtained to deduce that

$$w(\lambda Q) \le (2\lambda)^{2^n(1+\log_2[w]_{A_\infty})} w(Q)\,.$$

[*Hint:* Part (a): Take c in (9.3.4) such that $c^{1/\lambda^n} = (1+\varepsilon)[w]_{A_\infty}$.]

9.3.2. Suppose that w is a weight on $\mathbf{R}^n$ with the property that for all cubes Q and all measurable subsets A of Q we have

$$|A| < \alpha|Q| \Longrightarrow \mu(A) < \beta\mu(Q)$$

for some fixed $0 < \alpha, \beta < 1$. Show that μ is doubling [i.e., it satisfies (9.2.8)]. [*Hint:* Find a $\lambda > 0$ such that $R = \lambda Q \setminus Q$ has Lebesgue measure $< \alpha|Q|$. Write R as the disjoint union of 2^n cubes Q_j (touching only its vertices), union $2n$ parallelepipeds P_k (adjacent to its faces). Since each P_k is contained in a small shift of Q, we obtain $\mu(P_k) < \beta\mu(Q) + \beta\mu(P_k)$. Argue similarly for the Q_j to conclude that $\mu(\lambda Q) \le (1 + \frac{(2^n+2n)\beta}{1-\beta} + \frac{n2^{n+1}\beta^2}{(1-\beta)^2})\mu(Q)$. Iterate to obtain an estimate for $\mu(3Q)$.]

9.3.3. Prove that a weight w is in A_p if and only if both w and $w^{-\frac{1}{p-1}}$ are in A_∞.
[*Hint:* You may want to use the result of Exercise 9.2.2.]

9.3.4. (*Stein* [**480**]) Prove that if $P(x)$ is a polynomial of degree k in $\mathbf{R}^n$, then $\log|P(x)|$ is in BMO with norm depending only on k and n and not on the coefficients of the polynomial.
[*Hint:* Use that all norms on the finite-dimensional space of polynomials of degree at most k are equivalent to show that $|P(x)|$ satisfies a reverse Cauchy-Schwarz inequality. Therefore, $|P(x)|$ is an A_∞ weight and thus Exercise 9.2.3 is applicable.]

9.3.5. Show that the product of two A_1 weights may not be an A_∞ weight.

9.3.6. (*D. Cruz-Uribe*) Let $u \in A_1$ and $v \ge 0$. Prove that

$$uv \in A_\infty \iff v \in A_\infty(u)\,.$$

[*Hint:* Use the following characterization of the property $v \in A_\infty(u)$. There are $C_1, \delta > 0$ such that for all cubes Q and all subsets E of Q we have $\frac{uv(E)}{uv(Q)} \le C_1\big(\frac{u(E)}{u(Q)}\big)^\delta$. Use Hölder's inequality to estimate $\frac{|E|}{|Q|}$ by $[u]_{A_2}^{\frac12}\big(\frac{u(E)}{u(Q)}\big)^{\frac12}\big(\frac{|Q|}{u(Q)}\big)^{\frac12}$.]

9.3.7. (*Pérez* [**412**]) Show that a weight w lies in A_∞ if and only if there exist $\gamma, C > 0$ such that for all $\lambda > 0$ we have

$$w\big(\{x \in Q:\ w(x) > \lambda\}\big) \le C\,\lambda\big|\{x \in Q:\ w(x) > \gamma\lambda\}\big|$$

for all $\lambda > \operatorname{Avg}_Q w$.
[*Hint:* The displayed condition easily implies that

$$\frac{1}{|Q|}\int_Q w_k^{1+\varepsilon}\,dx \le \Big(\frac{w(Q)}{|Q|}\Big)^{\varepsilon+1} + \frac{C'\delta}{\gamma^{1+\varepsilon}}\frac{1}{|Q|}\int_Q w_k^{1+\varepsilon}\,dx\,,$$

where $w_k = \min(w,k)$ and k is a positive constant. Take $\varepsilon > 0$ small enough to obtain the reverse Hölder condition (c) in Theorem 9.3.3 for w_k. Let $k \to \infty$ to obtain the same conclusion for w. Conversely, find constants $\gamma, \delta \in (0,1)$ as in condition (a) of Theorem 9.3.3 and for $\lambda > \operatorname{Avg}_Q w$ write the set $\{w > \lambda\} \cap Q$ as a union of maximal dyadic cubes Q_j such that $\lambda < \operatorname{Avg}_{Q_j} w \le 2^n\lambda$ for all j. Then $w(Q_j) \le 2^n\lambda|Q_j| \le \frac{2^n\lambda}{1-\delta}|Q_j \cap \{w > \gamma\lambda\}|$ and the required conclusion follows by summing on j.]

9.4. Weighted Norm Inequalities for Singular Integrals

We now address a topic of great interest in the theory of singular integrals, their boundedness properties on weighted L^p spaces. It will turn out that a certain amount of regularity must be imposed on the kernels of these operators to obtain the aforementioned weighted estimates.

9.4.a. A Review of Singular Integrals

We begin by recalling some definitions from Chapter 8.

Definition 9.4.1. Let $0 < \delta, A < \infty$. A function $K(x,y)$ on $\mathbf{R}^n \times \mathbf{R}^n \setminus \{0\}$ is called a standard kernel with constants δ, A if

$$|K(x,y)| \le \frac{A}{|x-y|^n} \tag{9.4.1}$$

for all $x \neq y$ in $\mathbf{R}^n$ and

$$|K(x,y) - K(x',y)| \le \frac{A|x-x'|^\delta}{(|x-y|+|x'-y|)^{n+\delta}} \tag{9.4.2}$$

and that

$$|K(x,y) - K(x,y')| \le \frac{A|y-y'|^\delta}{(|x-y|+|x-y'|)^{n+\delta}} \tag{9.4.3}$$

for all $x \neq y$ and $x' \neq y$. The class of all kernels that satisfy (9.4.1), (9.4.2), and (9.4.3) will be denoted by $SK(\delta, A)$.

Definition 9.4.2. Let $0 < \delta, A < \infty$ and K in $SK(\delta, A)$. A Calderón-Zygmund operator associated with K is a linear operator T defined on $\mathcal{S}(\mathbf{R}^n)$ that admits a bounded extension on $L^2(\mathbf{R}^n)$

$$\big\|T(f)\big\|_{L^2} \le B\big\|f\big\|_{L^2} \tag{9.4.4}$$

and that satisfies

$$T(f)(x) = \int_{\mathbf{R}^n} K(x,y) f(y)\, dy \tag{9.4.5}$$

for all $f \in C_0^\infty$ and x not in the support of f. The class of all Calderón-Zygmund operators associated with kernels in $SK(\delta, A)$ that are bounded on L^2 with norm at most B will be denoted by $CZO(\delta, A, B)$. Given a Calderón-Zygmund operator T in $CZO(\delta, A, B)$, we define the truncated operator $T^{(\varepsilon)}$ as

$$T^{(\varepsilon)}(f)(x) = \int_{|x-y|>\varepsilon} K(x,y)\, f(y)\, dy$$

and the maximal operator associated with T as follows:

$$T^{(*)}(f)(x) = \sup_{\varepsilon>0} \left|T^{(\varepsilon)}(f)(x)\right|.$$

We note that if T is in $CZO(\delta, A, B)$, then $T^{(\varepsilon)} f$ and $T^{(*)}(f)$ is well defined for all f in $\bigcup_{1\le p<\infty} L^p(\mathbf{R}^n)$.

9.4.b. A Good Lambda Estimate for Singular Integrals*

The following theorem is the main result of this section.

Theorem 9.4.3. *Let $w \in A_\infty$ and T in $CZO(\delta, A, B)$. Then there exists a positive constant $C_0 = C_0(n, [w]_{A_\infty})$, a positive constant $\varepsilon_0 = \varepsilon_0(n, [w]_{A_\infty})$, and a positive constant $\gamma_0 = \gamma_0(n, \delta, A)$ such that for all f in $\bigcup_{1\le p<\infty} L^p(\mathbf{R}^n)$ and for all $0 < \gamma < \gamma_0$ we have*

$$w\big(\{T^{(*)}(f) > 3\lambda\} \cap \{M(f) \le \gamma\lambda\}\big) \le C_0\gamma^{\varepsilon_0}(A+B)^{\varepsilon_0} w\big(\{T^{(*)}(f) > \lambda\}\big), \tag{9.4.6}$$

where M denotes the Hardy-Littlewood maximal operator.

Proof. We write the open set

$$\Omega = \{T^{(*)}(f) > \lambda\} = \bigcup_j Q_j,$$

where Q_j are the Whitney cubes of Proposition 7.3.4. We set

$$Q_j^* = 10\sqrt{n}\, Q_j,$$
$$Q_j^{**} = 10\sqrt{n}\, Q_j^*,$$

where $a\,Q$ denotes the cube with the same center as Q whose side length is $a\,\ell(Q)$, where $\ell(Q)$ is the side length of Q. We note that in view of Proposition 7.3.4, the distance from Q_j to Ω^c is at least $2\sqrt{n}\,\ell(Q_j)$ and at most $4\sqrt{n}\,\ell(Q_j)$. But the distance from Q_j to the boundary of Q_j^* is $(5\sqrt{n} - \frac{1}{2})\,\ell(Q_j)$ which is bigger than $4\sqrt{n}\,\ell(Q_j)$. Therefore, Q_j^* must meet Ω^c and for every cube Q_j we fix a point $y_j \in \Omega^c \cap Q_j^*$. See Figure 9.1.

We also fix f in $\bigcup_{1\le p<\infty} L^p(\mathbf{R}^n)$ and for each j we write

$$f = f_0^j + f_\infty^j,$$

where f_0^j is the part of f near Q_j and f_∞^j is the part of f away from Q_j defined as follows:

$$f_0^j = f\chi_{Q_j^{**}},$$
$$f_\infty^j = f\chi_{(Q_j^{**})^c}.$$

We now claim that the following estimate is true:

$$\big|Q_j \cap \{T^{(*)}(f) > 3\lambda\} \cap \{M(f) \le \gamma\lambda\}\big| \le C_n \gamma (A+B)\big|Q_j\big|. \tag{9.4.7}$$

Once the validity of (9.4.7) is established, we apply Theorem 9.3.3 part (d) to obtain constants $\varepsilon_0, C_2 > 0$ (which depend on $[w]_{A_\infty}$ and the dimension n) such that

$$w\big(Q_j \cap \{T^{(*)}(f) > 3\lambda\} \cap \{M(f) \le \gamma\lambda\}\big) \le C_2 (C_n)^{\varepsilon_0} \gamma^{\varepsilon_0} (A+B)^{\varepsilon_0} w(Q_j).$$

Then a simple summation on j gives (9.4.6) with $C_0 = C_2(C_n)^{\varepsilon_0}$, and recall that C_2 and ε_0 depend on n and $[w]_{A_\infty}$.

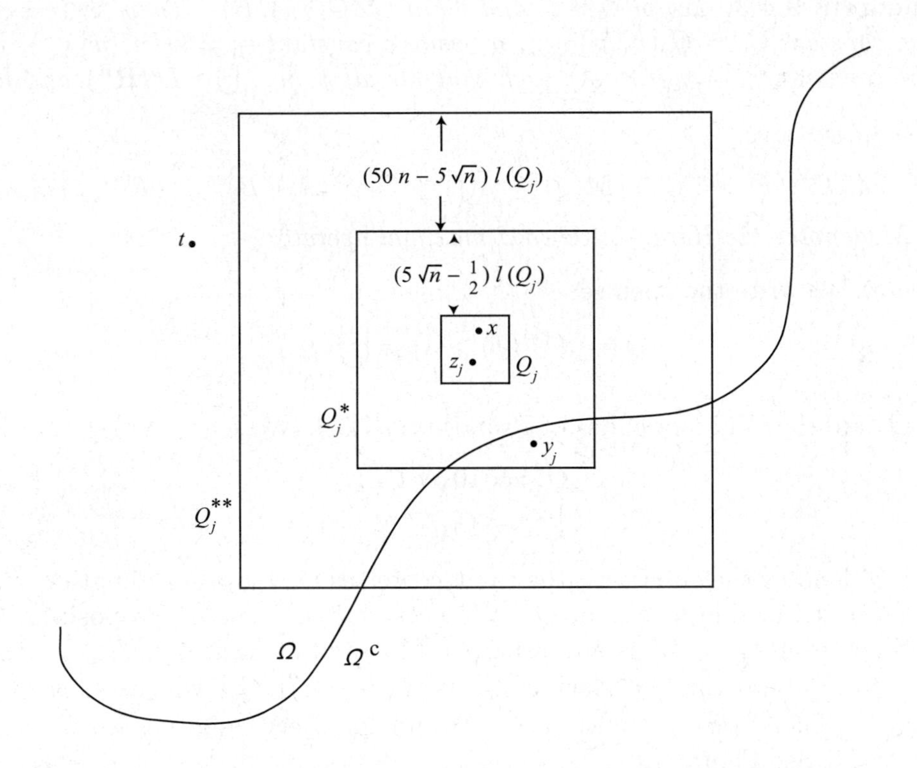

FIGURE 9.1. A picture of the proof.

In proving estimate (9.4.7), we may assume that for each cube Q_j there exists a $z_j \in Q_j$ such that $M(f)(z_j) \le \gamma\lambda$; otherwise the set on the left in (9.4.7) is empty.

We will invoke Theorem 8.2.2, which gives that $T^{(*)}$ maps $L^1(\mathbf{R}^n)$ into $L^{1,\infty}(\mathbf{R}^n)$ with norm at most $C(n)(A+B)$. We have the estimate

$$\big|Q_j \cap \{T^{(*)}(f) > 3\lambda\} \cap \{M(f) \le \gamma\lambda\}\big| \le I_0^\lambda + I_\infty^\lambda, \tag{9.4.8}$$

where

$$\begin{aligned} I_0^\lambda &= \big|Q_j \cap \{T^{(*)}(f_0^j) > \lambda\} \cap \{M(f) \le \gamma\lambda\}\big|\,, \\ I_\infty^\lambda &= \big|Q_j \cap \{T^{(*)}(f_\infty^j) > 2\lambda\} \cap \{M(f) \le \gamma\lambda\}\big|\,. \end{aligned}$$

To control I_0^λ we argue as follows:

$$\begin{aligned} I_0^\lambda &\le \big|\{T^{(*)}(f_0^j) > \lambda\}\big| \\ &\le \frac{\big\|T^{(*)}\big\|_{L^1\to L^{1,\infty}}}{\lambda}\int_{\mathbf{R}^n} |f_0(x)|\,dx \\ &\le C(n)\,(A+B)\,\frac{|Q_j^{**}|}{\lambda}\frac{1}{|Q_j^{**}|}\int_{Q_j^{**}} |f(x)|\,dx \\ &\le C(n)\,(A+B)\,\frac{|Q_j^{**}|}{\lambda} M_c(f)(z_j) \\ &\le \widetilde{C}(n)\,(A+B)\,\frac{|Q_j^{**}|}{\lambda} M(f)(z_j) \\ &\le \widetilde{C}(n)\,(A+B)\,\frac{|Q_j^{**}|}{\lambda}\,\lambda\gamma \\ &= C_n\,(A+B)\,\gamma\,|Q_j|\,. \end{aligned} \tag{9.4.9}$$

Next we claim that $I_\infty^\lambda = 0$ if we take γ sufficiently small. We first show that for all $x \in Q_j$ we have

$$\sup_{\varepsilon>0}\big|T^{(\varepsilon)}(f_\infty^j)(x) - T^{(\varepsilon)}(f_\infty^j)(y_j)\big| \le C_{n,\delta}^{(1)}\,A\,M(f)(z_j)\,. \tag{9.4.10}$$

Indeed, let us fix an $\varepsilon > 0$. We have

$$\begin{aligned} &\big|T^{(\varepsilon)}(f_\infty^j)(x) - T^{(\varepsilon)}(f_\infty^j)(y_j)\big| \\ &= \Bigg|\int_{|t-x|>\varepsilon} K(x,t) f_\infty^j(t)\,dt - \int_{|t-y_j|>\varepsilon} K(y_j,t) f_\infty^j(t)\,dt\Bigg| \\ &\le L_1 + L_2 + L_3\,, \end{aligned}$$

where

$$
\begin{aligned}
L_1 &= \Big| \int_{|t-y_j|>\varepsilon} \big[K(x,t) - K(y_j,t)\big] f_\infty^j(t)\, dt \Big| \\
L_2 &= \Big| \int_{\substack{|t-x|>\varepsilon \\ |t-y_j|\le\varepsilon}} K(x,t) f_\infty^j(t)\, dt \Big| \\
L_3 &= \Big| \int_{\substack{|t-x|\le\varepsilon \\ |t-y_j|>\varepsilon}} K(x,t) f_\infty^j(t)\, dt \Big|
\end{aligned}
$$

in view of identity (4.4.6). Note that in all three integrals we can multiply the integrands by the characteristic function of the set

$$|t - z_j| \ge 49\, n\, \ell(Q_j)\,,$$

since the distance from $t \in (Q_j^{**})^c$ to z_j is at least $(50\,n - \frac{1}{2})\ell(Q_j)$.

We now make a couple of observations. For $t \notin Q_j^{**}$, $x, z_j \in Q_j$ and $y_j \in Q_j^*$ we have

$$
\frac{3}{4} \le \frac{|t-x|}{|t-y_j|} \le \frac{5}{4}\,, \qquad\qquad \frac{40}{41} \le \frac{|t-z_j|}{|t-x|} \le \frac{40}{39}\,. \tag{9.4.11}
$$

Indeed,

$$|t - y_j| \ge (50\, n - 5\sqrt{n})\, \ell(Q_j) \ge 40\, n\, \ell(Q_j)$$

and

$$|x - y_j| \le \sqrt{n}\, 10\sqrt{n}\, \ell(Q_j) \le \frac{1}{4}|t - y_j|\,.$$

Using this estimate and the inequalities

$$|t - y_j| - |x - y_j| \le |t - x| \le |t - y_j| + |x - y_j|\,,$$

we obtain the first estimate in (9.4.11). The second estimate is proved similarly. Now use that

$$|K(x,t) - K(y_j,t)| \le \frac{A|x-y_j|^\delta}{(|t-x| + |t-y_j|)^{n+\delta}} \le C'_{n,\delta} A\, \frac{\ell(Q_j)^\delta}{|t-z_j|^{n+\delta}}$$

to obtain

$$L_1 \le \int_{|t-z_j|\ge 49n\,\ell(Q_j)} C'_{n,\delta} A\, \frac{\ell(Q_j)^\delta}{|t-z_j|^{n+\delta}} |f(t)|\, dt \le C''_{n,\delta} A\, M(f)(z_j)$$

using Theorem 2.1.10. Using (9.4.11), we then deduce

$$L_2 \le \int_{\frac{50}{39}\varepsilon \ge |t-z_j| \ge \frac{40}{41}\varepsilon} \frac{A}{|x-t|^n} |f_\infty^j(t)|\, dt \le C'_n A\, M(f)(z_j)\,.$$

Again using (9.4.11), we obtain

$$L_3 \le \int\limits_{\frac{30}{41}\varepsilon \le |t-z_j| \le \frac{40}{39}\varepsilon} \frac{A}{|x-t|^n} |f_\infty^j(t)|\, dt \le C_n'' A\, M(f)(z_j)\,.$$

This proves (9.4.10) with constant $C_{n,\delta}^{(1)} = C_{n,\delta}'' + C_n' + C_n''$.

Having established (9.4.10), we next claim that

$$\sup_{\varepsilon>0} \left|T^{(\varepsilon)}(f_\infty^j)(y_j)\right| \le T^{(*)}(f)(y_j) + C_n^{(2)} A\, M(f)(z_j)\,. \tag{9.4.12}$$

To prove (9.4.12) we fix a cube Q_j and $\varepsilon > 0$. We let R_2^j be the smallest number and R_1^j be the largest number so that

$$B(y_j, R_1^j) \subseteq Q_j^{**} \subseteq B(y_j, R_2^j)\,.$$

See Figure 9.2. We consider the following three cases.

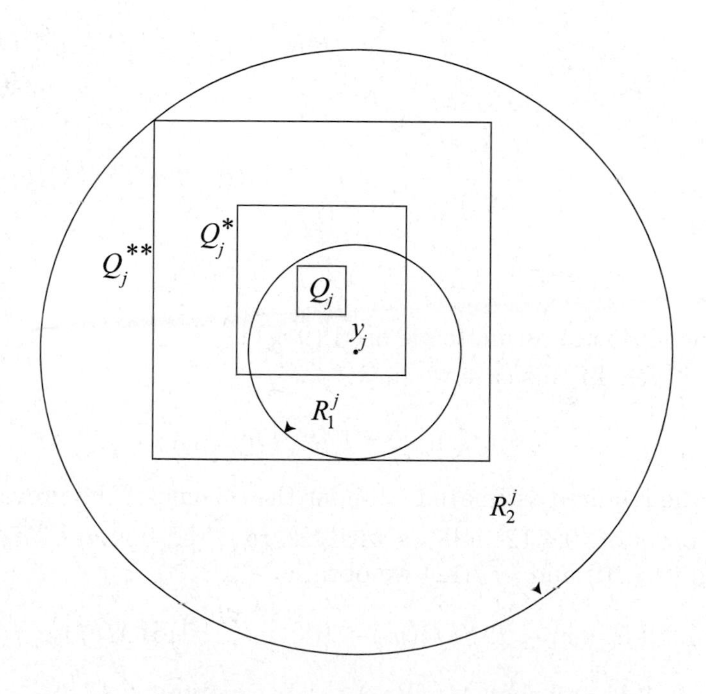

FIGURE 9.2. The balls $B(y_j, R_1^j)$ and $B(y_j, R_2^j)$.

Case (1): $\varepsilon \ge R_2^j$. Since $Q_j^{**} \subseteq B(y_j, \varepsilon)$, we have $B(y_j,\varepsilon)^c \subseteq (Q_j^{**})^c$ and therefore

$$T^{(\varepsilon)}(f_\infty^j)(y_j) = T^{(\varepsilon)}(f)(y_j)\,,$$

so (9.4.12) holds easily in this case.

Case (2): $R_1^j \le \varepsilon < R_2^j$. Note that if $t \in (Q_j^{**})^c$, then $|t - y_j| \ge 40\, n\, \ell(Q_j)$. On the other hand $R_2^j \le \operatorname{diam}(Q_j^{**}) = 100\, n^{\frac{3}{2}}\, \ell(Q_j)$. This implies that

$$R_2^j \le \tfrac{5\sqrt{n}}{2}|t - y_j|\,, \qquad \text{when} \quad t \in (Q_j^{**})^c\,.$$

Notice also that in this case we have $B(y_j, R_2^j)^c \subseteq (Q_j^{**})^c$, hence

$$T^{(R_2^j)}(f_\infty^j)(y_j) = T^{(R_2^j)}(f)(y_j)\,.$$

Therefore, we have

$$\begin{aligned}
\big|T^{(\varepsilon)}(f_\infty^j)(y_j)\big| &= \big|T^{(\varepsilon)}(f_\infty^j)(y_j) - T^{(R_2^j)}(f_\infty^j)(y_j) + T^{(R_2^j)}(f)(y_j)\big| \\
&\le \int\limits_{\varepsilon \le |y_j - t| \le R_2^j} |K(y_j,t)|\, |f_\infty^j(t)|\, dt + T^{(*)}(f)(y_j) \\
&\le \int\limits_{\frac{2}{5\sqrt{n}} R_2^j \le |y_j - t| \le R_2^j} |K(y_j,t)||f_\infty^j(t)|\, dt + T^{(*)}(f)(y_j) \\
&\le \frac{A(\frac{2}{5\sqrt{n}})^{-n}}{(R_2^j)^n} \int\limits_{|z_j - t| \le \frac{50}{39} R_2^j} |f(t)|\, dt + T^{(*)}(f)(y_j) \\
&\le C_n^{(2)} A\, M(f)(z_j) + T^{(*)}(f)(y_j)\,,
\end{aligned}$$

where in the penultimate estimate we used (9.4.11).
Case (3): $\varepsilon \le R_1^j$. In this case we have

$$T^{(\varepsilon)}(f_\infty^j)(y_j) = T^{(R_1^j)}(f_\infty^j)(y_j)\,,$$

and therefore the required estimate fits under the scheme of the previous case (with $\varepsilon = R_1^j$). The proof of (9.4.12) follows with the required bound $C_n^{(2)} A$.

Combining (9.4.10) and (9.4.12) we obtain

$$T^{(*)}(f_\infty^j)(x) \le T^{(*)}(f)(y_j) + \big(C_{n,\delta}^{(1)} + C_n^{(2)}\big)\, A\, M(f)(z_j)\,.$$

Recalling that $y_j \notin \Omega$ and that $M(f)(z_j) \le \gamma\lambda$, we obtain

$$T^{(*)}(f_\infty^j)(x) \le \lambda + \big(C_{n,\delta}^{(1)} + C_n^{(2)}\big)\, A\, \gamma\lambda\,.$$

Setting $\gamma_0 = \big(C_{n,\delta}^{(1)} + C_n^{(2)}\big)^{-1} A^{-1}$, for $0 < \gamma < \gamma_0$, we have that the set

$$Q_j \cap \{T^{(*)}(f_\infty^j) > 2\lambda\} \cap \{M(f) \le \gamma\lambda\}$$

is empty. This shows that the quantity $I_\infty^\gamma = 0$ if γ is smaller than γ_0. Returning to (9.4.8) and using the estimate (9.4.9) proved earlier, we conclude the proof of (9.4.7), which, as indicated earlier, implies the theorem.

□

Remark 9.4.4. We observe that for any $\delta > 0$ estimate (9.4.6) also holds for the operator

$$T_\delta^{(*)}(f)(x) = \sup_{\varepsilon \geq \delta} |T^{(\varepsilon)}(f)(x)|$$

with the same constant (which is independent of δ).

Indeed, to see the validity of the statement made in this remark, we consider the following three situations that may arise:
(a) $\delta < R_1^j < R_2^j$. Here the proof of estimate (9.4.6) remains unchanged as the three cases are not affected.
(b) $R_1^j \leq \delta < R_2^j$. Here only cases (1) and (2) in the proof of estimate (9.4.6) appear.
(c) $R_1^j < R_2^j \leq \delta$. Only case (1) in the proof of estimate (9.4.6) appears.
Moreover, all estimates previously obtained in cases (1), (2), and (3) equally apply in cases (a), (b), and (c) and are therefore independent of δ.

9.4.c. Consequences of the Good Lambda Estimate

Having obtained the important good lambda weighted estimate for singular integrals, we now pass to some of its consequences.

Theorem 9.4.5. *Let T be a $CZO(\delta, A, B)$. Then for all $1 \leq p < \infty$ and for every weight $w \in A_p$ there is a constant $C_p = C_p(n, [w]_{A_\infty})$ such that*

$$\big\|T^{(*)}(f)\big\|_{L^{1,\infty}(w)} \leq C_1 \big\|f\big\|_{L^1(w)} \tag{9.4.13}$$

for all $f \in L^1(w)$ and

$$\big\|T^{(*)}(f)\big\|_{L^p(w)} \leq C_p \big\|f\big\|_{L^p(w)} \tag{9.4.14}$$

for all $f \in L^p(w)$.

Proof. This theorem is a consequence of the estimate proved in the previous theorem. For technical reasons, it will be useful to us to fix a $\delta > 0$ and work with the auxiliary maximal operator $T_\delta^{(*)}$ instead of $T^{(*)}$. We begin by taking $1 < p < \infty$ and $f \in L^p(w)$ for some $w \in A_p$. We write

$$\begin{aligned}\|T_\delta^{(*)}(f)\|_{L^p(w)}^p &= \int_0^\infty p\lambda^{p-1} w(\{T_\delta^{(*)}(f) > \lambda\})\, d\lambda \\ &= 3^p \int_0^\infty p\lambda^{p-1} w(\{T_\delta^{(*)}(f) > 3\lambda\})\, d\lambda\end{aligned}$$

which we control by

$$3^p \int_0^\infty p\lambda^{p-1} w\big(\{T_\delta^{(*)}(f) > 3\lambda\} \cap \{M(f) \le \gamma\lambda\}\big)\, d\lambda + 3^p \int_0^\infty p\lambda^{p-1} w\big(\{M(f) > \gamma\lambda\}\big)\, d\lambda .$$

Using Theorem 9.4.3 (or rather Remark 9.4.4), we estimate the last terms by

$$3^p C_0 \gamma^{\varepsilon_0} (A+B)^{\varepsilon_0} \int_0^\infty p\lambda^{p-1} w\big(\{T_\delta^{(*)}(f) > \lambda\}\big)\, d\lambda + \frac{3^p}{\gamma^p} \int_0^\infty p\lambda^{p-1} w\big(\{M(f) > \lambda\}\big)\, d\lambda ,$$

which is equal to

$$3^p C_0 \gamma^{\varepsilon_0} (A+B)^{\varepsilon_0} \big\|T_\delta^{(*)}(f)\big\|_{L^p(w)}^p + \frac{3^p}{\gamma^p} \big\|M(f)\big\|_{L^p(w)}^p .$$

Taking $\gamma = \min\big(\frac{1}{2}\gamma_0, \frac{1}{2}(2C_0 3^p)^{-\frac{1}{\varepsilon_0}} (A+B)^{-1}\big)$, we conclude that

$$\begin{aligned} &\big\|T_\delta^{(*)}(f)\big\|_{L^p(w)}^p \\ &\le \frac{1}{2} \big\|T_\delta^{(*)}(f)\big\|_{L^p(w)}^p + \widetilde{C}_p(n, \delta, A+B, [w]_{A_\infty}) \big\|M(f)\big\|_{L^p(w)}^p . \end{aligned} \tag{9.4.15}$$

We now prove a similar estimate when $p = 1$. For $f \in L^1(w)$ and $w \in A_1$ we have

$$\begin{aligned} &3\lambda w\big(\{T_\delta^{(*)}(f) > 3\lambda\}\big) \\ &\le 3\lambda w\big(\{T_\delta^{(*)}(f) > 3\lambda\} \cap \{M(f) \le \gamma\lambda\big) + 3\lambda w\big(\{M(f) > \gamma\lambda\}\big) \end{aligned}$$

and this expression is controlled by

$$3\lambda C_0 \gamma^{\varepsilon_0} (A+B)^{\varepsilon_0} w\big(\{T_\delta^{(*)}(f) > \lambda\}\big) + \frac{3}{\gamma} \big\|M(f)\big\|_{L^{1,\infty}(w)} .$$

It follows that

$$\begin{aligned} &\big\|T_\delta^{(*)}(f)\big\|_{L^{1,\infty}(w)} \\ &\le \frac{1}{2} \big\|T_\delta^{(*)}(f)\big\|_{L^{1,\infty}(w)} + \widetilde{C}_1(n, \delta, A+B, [w]_{A_\infty}) \big\|M(f)\big\|_{L^{1,\infty}(w)} . \end{aligned} \tag{9.4.16}$$

At this point the proof of estimates (9.4.14) and (9.4.13) would follow from (9.4.15) and (9.4.16), respectively, if we knew that $\big\|T_\delta^{(*)}(f)\big\|_{L^p(w)} < \infty$ whenever $1 \le p < \infty$ and $f \in L^p(w)$. However, at this point we don't know that $\big\|T_\delta^{(*)}(f)\big\|_{L^p(w)} < \infty$ when $f \in L^p(w)$, and a certain amount of work is needed to overcome this difficulty.

To deal with this problem we will momentarily restrict our attention to a special class of functions on $\mathbf{R}^n$, the class of bounded functions with compact support. Note that in view of Exercise 9.4.1 such functions are dense in $L^p(w)$ when $w \in A_\infty$ and $1 \le p < \infty$. Let f be a bounded function with compact support on $\mathbf{R}^n$. Then

$T_\delta^{(*)}(f) \le \delta^{-n} C_1(f)$ and $T_\delta^{(*)}(f)(x) \le C_2(f)|x|^{-n}$ for x away from the support of f. It follows that

$$T_\delta^{(*)}(f)(x) \le C_3(f,\delta)(1+|x|)^{-n}$$

for all $x \in \mathbf{R}^n$. Furthermore, if f is nonzero, then

$$M(f)(x) \ge \frac{C_4(f)}{(1+|x|)^n}$$

and therefore

$$\big\|T_\delta^{(*)}(f)\big\|_{L^{1,\infty}(w\,dx)} \le C_f \big\|M(f)\big\|_{L^{1,\infty}(w\,dx)} < \infty$$

while for $1 < p < \infty$

$$\int_{\mathbf{R}^n} (T_\delta^{(*)}(f))^p(x) w(x)\,dx \le \frac{C_3(f,\delta)^p}{C_4(f)^p} \int_{\mathbf{R}^n} M(f)(x)^p w(x)\,dx < \infty$$

in view of Corollary 9.2.7. Using this corollary, (9.4.15), and (9.4.16), we now conclude that for all $\delta > 0$ we have

$$\begin{aligned} \big\|T_\delta^{(*)}(f)\big\|_{L^p(w)}^p &\le 2\widetilde{C}_p \big\|M(f)\big\|_{L^p(w)}^p \le C_p^p \big\|f\big\|_{L^p(w)}^p \\ \big\|T_\delta^{(*)}(f)\big\|_{L^{1,\infty}(w)} &\le 2\widetilde{C}_1 \big\|M(f)\big\|_{L^{1,\infty}(w)} \le C_1 \big\|f\big\|_{L^1(w)} \end{aligned} \tag{9.4.17}$$

whenever f a bounded function with compact support, $1 < p < \infty$, and $w \in A_p$. Both constants $\widetilde{C}_p$ and $\widetilde{C}_1$ depend only on the parameters $A+B$, n, δ, and $[w]_{A_\infty}$. At this point we introduce another auxiliary maximal operator for $0 < \delta < L < \infty$:

$$T_{\delta,L}^{(*)}(f)(x) = \sup_{L \ge \varepsilon \ge \delta} |T^{(\varepsilon)}(f)(x)|\,.$$

Since $T_{\delta,L}^{(*)}(f) \le T_\delta^{(*)}(f)$, it follows from (9.4.17) that for all f bounded functions on $\mathbf{R}^n$ with compact support we have the estimate

$$\big\|T_{\delta,L}^{(*)}(f)\big\|_{L^p(w)}^p \le C_p^p \big\|f\big\|_{L^p(w)}^p\,, \tag{9.4.18}$$

where $1 < p < \infty$ and C_p is independent of f, δ, and L.

We now extend estimate (9.4.18) to all functions in $L^p(\mathbf{R}^n, w\,dx)$. Given a weight $w \in A_p$ and $f \in L^p(w)$, let

$$f_N(x) = f(x)\chi_{|f| \le N}\chi_{|x| \le N}\,.$$

Then f_N is a bounded function with compact support that converges to f in $L^p(w)$ by the Lebesgue dominated convergence theorem. But observe that

$$|T_{\delta,L}^{(*)}(f_N) - T_{\delta,L}^{(*)}(f)| \le T_{\delta,L}^{(*)}(f - f_N)$$

and the latter satisfies

$$T_{\delta,L}^{(*)}(f - f_N)(x) \le A\,\delta^{-n} \int_{\delta \le |x-y| \le L} |f(y) - f_N(y)|\,dy\,. \tag{9.4.19}$$

Using Hölder's inequality, we can estimate the expression on the right in (9.4.19) by

$$A\,\delta^{-n}\big\|f-f_N\big\|_{L^p(w)}\bigg(\int_{B(x,L)} w^{-\frac{p'}{p}}\bigg)^{\frac{1}{p'}}$$

when $1<p<\infty$ and by

$$A\,\delta^{-n}\big\|f-f_N\big\|_{L^1(w)}\big\|w^{-1}\big\|_{L^\infty(B(x,L))}$$

when $p=1$. Both of the preceding expressions tend to zero for almost every $x\in\mathbf{R}^n$ as $N\to\infty$ whenever δ, L are fixed. It follows from (9.4.19) and the Lebesgue dominated convergence theorem that $T^{(*)}_{\delta,L}(f-f_N)$ converges to zero in $L^p(w)$ as $N\to\infty$ and therefore so does $T^{(*)}_{\delta,L}(f_N)-T^{(*)}_{\delta,L}(f)$. This proves (9.4.18) for all $f\in L^p(w)$.

It remains to prove (9.4.14) and (9.4.13) for all $f\in L^p(w)$. But this is a consequence of Fatou's lemma since

$$\lim_{\substack{\delta\to 0\\ L\to\infty}} T^{(*)}_{\delta,L}(f)(x)=T^{(*)}(f)(x)$$

for all $f\in L^p(w)$ and $x\in\mathbf{R}^n$.

□

Corollary 9.4.6. *Let T be a $CZO(\delta,A,B)$. Then for all $1\le p<\infty$ and for every weight $w\in A_p$ there is a constant $C_p=C_p(n,[w]_{A_\infty},\delta,A,B)$ such that*

$$\big\|T(f)\big\|_{L^{1,\infty}(w)}\le C_1\big\|f\big\|_{L^1(w)}$$

for all $f\in L^1(w)$ and

$$\big\|T(f)\big\|_{L^p(w)}\le C_p\big\|f\big\|_{L^p(w)}$$

for all $f\in L^p(w)$.

Proof. We use the fact that any element of $CZO(\delta,A,B)$ is a weak limit of a sequence of its truncations plus a bounded function times the identity operator, that is, $T=T_0+aI$, where $\big\|a\big\|_{L^\infty}\le C_n(A+B)$ (cf. Proposition 8.1.11). Then $T^{(\varepsilon_j)}(f)\to T_0(f)$ weakly for some sequence $\varepsilon_j\to 0+$ and we have $|T_0(f)|\le T^{(*)}(f)$. Therefore, $|T(f)|\le T^{(*)}(f)+C_n(A+B)|f|$, and this estimate implies the required result by using the previous theorem.

□

9.4.d. Necessity of the A_p Condition

We have established the main theorems relating Calderón-Zygmund operators and A_p weights, namely that such operators are bounded on $L^p(w)$ whenever w lies in A_p. It is natural to ask whether the A_p condition is necessary for the boundedness of singular integrals on L^p. We end this section by indicating the necessity of the A_p condition for the boundedness of the Riesz transforms on weighted L^p spaces.

Theorem 9.4.7. *Let w be a weight in $\mathbf{R}^n$ and let $1 \le p < \infty$. Suppose that each of the Riesz transforms R_j is of weak type (p,p) with respect to w. Then w must be an A_p weight. Similarly, if w be a weight in $\mathbf{R}$ and the Hilbert transform H is of weak type (p,p) with respect to w, then w must be an A_p weight.*

Proof. We prove the nth dimensional case when $n \ge 2$. The one-dimensional case is obtained by a simple adaptation of the following argument.

Let Q be a cube and let f be a nonnegative function on $\mathbf{R}^n$ supported in Q that satisfies $\operatorname{Avg}_Q f > 0$. Let Q' be the cube that shares a corner with Q, which has the same length as Q, and which satisfies $x_j \ge y_j$ for all $1 \le j \le n$ whenever $x \in Q'$ and $y \in Q$. Then for $x \in Q'$ we have

$$\sum_{j=1}^n R_j(f)(x) = \frac{\Gamma(\frac{n+1}{2})}{\pi^{\frac{n+1}{2}}} \sum_{j=1}^n \int_Q \frac{x_j - y_j}{|x-y|^{n+1}} f(y)\, dy \ge \frac{\Gamma(\frac{n+1}{2})}{\pi^{\frac{n+1}{2}}} \int_Q \frac{f(y)}{|x-y|^n}\, dy\,.$$

But if $x \in Q'$ and $y \in Q$ we must have that $|x-y| \le 2\sqrt{n}\,\ell(Q)$, which implies that $|x-y|^{-n} \ge (2\sqrt{n})^{-n}|Q|^{-1}$. Let $C_n = \Gamma(\frac{n+1}{2})(2\sqrt{n})^{-n}\pi^{-\frac{n+1}{2}}$. It follows that for all $0 < \alpha < C_n \operatorname*{Avg}_Q f$ we have

$$Q' \subseteq \{x \in \mathbf{R}^n : \big|\sum_{j=1}^n R_j(f)(x)\big| > \alpha\}\,.$$

Since the operator $\sum_{j=1}^n R_j$ is of weak type (p,p) with respect to w (with constant C), we must have

$$w(Q') \le \frac{C^p}{\alpha^p} \int_Q f(x)^p w(x)\, dx$$

for all $\alpha < C_n \operatorname*{Avg}_Q f$, which implies that

$$(\operatorname*{Avg}_Q f)^p \le \frac{C_n^{-p} C^p}{w(Q')} \int_Q f(x)^p w(x)\, dx\,. \tag{9.4.20}$$

We observe that we can reverse the roles of Q and Q' and obtain

$$(\operatorname*{Avg}_{Q'} g)^p \le \frac{C_n^{-p} C^p}{w(Q)} \int_{Q'} g(x)^p w(x)\, dx \tag{9.4.21}$$

for all g supported in Q'. In particular taking $g = \chi_{Q'}$ in (9.4.21) gives that $w(Q) \le C_n^{-p} C^p w(Q')$. Using this estimate and (9.4.20), we obtain

$$(\operatorname*{Avg}_Q f)^p \le \frac{(C_n^{-p} C^p)^2}{w(Q)} \int_Q f(x)^p w(x)\, dx\,. \tag{9.4.22}$$

Using the characterization of the A_p characteristic constant in Proposition 9.1.5 (8), it follows that $[w]_{A_p} \le (C_n^{-p}C^p)^2 < \infty$; hence $w \in A_p$.

□

Exercises

9.4.1. Show that C_0^∞ is dense in $L^p(w)$ for all $w \in A_\infty$.

9.4.2. (*Córdoba and Fefferman* [**129**]) Let T be in $CZO(\delta, A, B)$. Show that for all $\varepsilon > 0$ and all $1 < p < \infty$ there exists a constant $C_{\varepsilon,p,n}$ such that for all functions u and f on $\mathbf{R}^n$ we have

$$\int_{\mathbf{R}^n} |T^{(*)}(f)|^p \, u \, dx \le C_{\varepsilon,p} \int_{\mathbf{R}^n} |f|^p M(u^{1+\varepsilon})^{\frac{1}{1+\varepsilon}} \, dx$$

whenever the right-hand side is finite.
[*Hint:* Obtain this result as a consequence of Theorem 9.4.5.]

9.4.3. Use the idea of the proof of Theorem 9.4.5 to prove the following result. Suppose that the nonnegative μ-measurable functions f and $T(f)$ satisfy the distributional inequality

$$\mu(\{T(f) > \alpha\} \cap \{f \le c\alpha\}) \le A\,\mu(\{T(f) > B\alpha\})$$

for some fixed $A, B > 0$ and all $\alpha > 0$. Show that if $0 < p < \infty$ and $A < B^p$, then if $\big\|T(f)\big\|_{L^p(\mu)} < \infty$, it must satisfy

$$\big\|T(f)\big\|_{L^p(\mu)} \le C(c, p, A, B)\big\|f\big\|_{L^p(\mu)}$$

for some constant $C(c, p, A, B)$ that depends only on the indicated parameters.

9.4.4. Let f be in $L^1(\mathbf{R}^n, w)$, where $w \in A_1$. Apply the Calderón-Zygmund decomposition to f at height $\alpha > 0$ to write $f = g + b$ as in Theorem 4.3.1. Prove that there exists a constant $C = C(n, [w]_{A_1})$ such that

$$\big\|g\big\|_{L^1(w)} + \big\|b\big\|_{L^1(w)} \le C\big\|f\big\|_{L^1(w)}\,.$$

9.4.5. Assume that T is in $CZO(\delta, A, B)$ and suppose that T maps $L^2(w)$ into $L^2(w)$ for every $w \in A_1$. Prove that T maps $L^1(w)$ into $L^{1,\infty}(w)$ for every $w \in A_1$.
[*Hint:* Use Theorem 4.3.1 to write $f = g + b$, where $b = \sum_j b_j$ and each b_j is supported in a cube Q_j with center c_j. To estimate $T(g)$ use an $L^2(w)$ estimate and the previous Exercise. To estimate $T(b)$ use the mean value

property, the fact that

$$\int_{\mathbf{R}^n \setminus Q_j^*} \frac{|y-c_j|^\delta}{|x-c_j|^{n+\delta}}\, w(x)\, dx \le C_{\delta,n} M(w)(y) \le C'_{\delta,n} w(y)\,,$$

and the previous Exercise to obtain the required estimate.]

9.4.6. Recall that the transpose of a linear operator T is defined by

$$\langle T(f), g\rangle = \langle f, T^t(g)\rangle$$

for all suitable f and g. Suppose that T is a linear operator that maps $L^p(v)$ into itself for some $1 < p < \infty$ and all $v \in A_p$. Show that the transpose operator T^t of T maps $L^{p'}(w)$ into $L^{p'}(w)$ for all $w \in A_{p'}$.

9.4.7. Suppose that T is a linear operator that maps $L^2(v)$ into itself for all v such that $v^{-1} \in A_1$. Show that the transpose operator T^t of T maps $L^2(w)$ into $L^2(w)$ for all $w \in A_1$.

9.4.8. Let $1 < p < \infty$. Suppose that T is a linear operator that maps $L^p(v)$ into itself for all v satisfying $v^{-1} \in A_p$. Show that the transpose operator T^t of T maps $L^{p'}(w)$ into itself for all w satisfying $w^{-1} \in A_{p'}$.

9.4.9. Let $w \in A_\infty$ and assume that for some locally integrable function f we have $M(f) \in L^{p_0}(w)$ for some $0 < p_0 < \infty$. Show that for all p with $p_0 \le p < \infty$ there is a constant $C(p, n, [w]_{A_\infty})$ such that

$$\big\|M_d(f)\big\|_{L^p(w)} \le C(p, n, [w]_{A_\infty})\big\|M^\#(f)\big\|_{L^p(w)}\,,$$

where M_d is the dyadic maximal operator given in Definition 7.4.3. Conclude the same estimate for M.
[*Hint:* Combine estimate (7.4.2) with property (d) of Theorem 9.3.3

$$w\big(Q_j \cap \{M_d(f) > 2\lambda,\ M^\#(f) \le \gamma\lambda\}\big) \le C_2(2^n\gamma)^{\varepsilon_0}\, w(Q_j)\,,$$

where both C_2 and ε_0 depend on the dimension n and $[w]_{A_\infty}$. Obtain the result of Theorem 7.4.4 in which the Lebesgue measure is replaced by w in A_∞ and the quantity $2^n\gamma$ is replaced by $C_2(2^n\gamma)^{\varepsilon_0}$. Finally, observe that Theorem 7.4.5 can be can adapted to a general weight w in A_∞.]

9.5. Further Properties of A_p Weights*

In this section we will discuss other properties of A_p weights. Many of these properties indicate some deep connections with other branches of analysis. We focus our attention on three such properties: factorization, extrapolation, and relations of weighted inequalities with vector-valued inequalities.

9.5.a. Factorization of Weights

Recall the simple fact that if w_1, w_2 are A_1 weights, then $w = w_1\, w_2^{1-p}$ is an A_p weight (Exercise 9.1.2). The factorization theorem for weights says that the converse of this statement is true. This provides a surprising and striking representation of A_p weights.

Theorem 9.5.1. *Suppose that w is an A_p weight for some $1 < p < \infty$. Then there exist A_1 weights w_1 and w_2 such that*

$$w = w_1\, w_2^{1-p}.$$

Proof. Let us fix a $p \geq 2$ and $w \in A_p$. We define an operator T as follows:

$$T(f) = \big(w^{-\frac{1}{p}} M(f^{p-1} w^{\frac{1}{p}})\big)^{\frac{1}{p-1}} + w^{\frac{1}{p}} M(f w^{-\frac{1}{p}}),$$

where M is the Hardy-Littlewood maximal operator. We observe that T is well defined and bounded on $L^p(\mathbf{R}^n)$. This is a consequence of the facts that $w^{-\frac{1}{p-1}}$ is an $A_{p'}$ weight and that M maps $L^{p'}(w^{-\frac{1}{p-1}})$ into itself and also $L^p(w)$ into itself. Thus the norm of T on L^p only depends on the A_p characteristic constant of w. Let $B(w) = \big\|T\big\|_{L^p \to L^p}$, the norm of T on L^p. Next, we observe that for $f, g \geq 0$ in $L^p(\mathbf{R}^n)$ and $\lambda \geq 0$ we have

$$T(f+g) \leq T(f) + T(g), \quad T(\lambda f) = \lambda T(f). \tag{9.5.1}$$

To see the first assertion, we only need to note that for every ball B, the operator

$$f \to \left(\frac{1}{|B|}\int_B |f|^{p-1} w^{\frac{1}{p}}\, dx\right)^{\frac{1}{p-1}}$$

is sublinear as a consequence of Minkowski's integral inequality, since $p - 1 \geq 1$.

We now fix an L^p function f_0 with $\big\|f_0\big\|_{L^p} = 1$ and we define a function φ in $L^p(\mathbf{R}^n)$ as the sum of the L^p convergent series

$$\varphi = \sum_{j=1}^{\infty} (2B(w))^{-j} T^j(f_0). \tag{9.5.2}$$

We define

$$w_1 = w^{\frac{1}{p}} \varphi^{p-1}, \qquad w_2 = w^{-\frac{1}{p}} \varphi,$$

so that $w = w_1\, w_2^{1-p}$. It remains to show that w_1, w_2 are A_1 weights. Applying T and using (9.5.1), we obtain

$$\begin{aligned} T(\varphi) &\le 2B(w)\sum_{j=1}^{\infty}(2B(w))^{-j-1}T^{j+1}(f_0) \\ &= 2B(w)\Big(\varphi - \frac{T(f_0)}{2B(w)}\Big) \\ &\le 2B(w)\,\varphi\,, \end{aligned}$$

that is,

$$\big(w^{-\frac{1}{p}}M(\varphi^{p-1}w^{\frac{1}{p}})\big)^{\frac{1}{p-1}} + w^{\frac{1}{p}}M(\varphi w^{-\frac{1}{p}}) \le 2B(w)\,\varphi\,.$$

Using that $\varphi = (w^{-\frac{1}{p}}w_1)^{\frac{1}{p-1}} = w^{\frac{1}{p}}w_2$, we obtain

$$M(w_1) \le (2B(w))^{p-1}w_1\,, \qquad \text{and} \qquad M(w_2) \le 2B(w)w_2\,.$$

These show that w_1 and w_2 are A_1 weights whose characteristic constants depend on $[w]_{A_p}$ (and also the dimension n and p). This concludes the case $p \ge 2$.

We now turn to the case $p < 2$. Given a weight $w \in A_p$ for $1 < p < 2$, we consider the weight $w^{-1/(p-1)}$ which is in $A_{p'}$. Since $p' > 2$, we can apply the factorization result just proved to write $w^{-1/(p-1)} = v_1 v_2^{1-p'}$, where v_1, v_2 are A_1 weights. It follows that $w = v_1^{1-p}v_2$, and the asserted factorized is completed.

□

Combining the result just obtained with Theorem 9.2.8, we obtain the following description of A_p weights.

Corollary 9.5.2. *Let w be an A_p weight for some $1 < p < \infty$. Then there exist locally integrable functions f_1 and f_2 with*

$$M(f_1) + M(f_2) < \infty \qquad \text{a.e.},$$

constants $0 < \varepsilon_1, \varepsilon_2 < 1$, and a nonnegative function k satisfying $k, k^{-1} \in L^\infty$, such that

$$w = k\, M(f_1)^{\varepsilon_1} M(f_2)^{\varepsilon_2(1-p)}\,. \tag{9.5.3}$$

9.5.b. Extrapolation from Weighted Estimates on a Single L^r

Our next topic concerns a striking application of weighted norm inequalities. This says that an estimate on a weighted L^r for a single r and all A_r weights implies a corresponding L^p estimate for all p. This property is referred to as extrapolation.

Surprisingly the operator T is not needed to be linear or sublinear in the following extrapolation theorem. The only condition needed is that T is well defined on $\bigcup\limits_{1\le q<\infty}\bigcup\limits_{w\in A_q} L^q(w)$. If T happens to be a linear operator, this condition can be relaxed to T being well defined on $C_0^\infty(\mathbf{R}^n)$.

Theorem 9.5.3. *Suppose that T is defined on $\bigcup_{1\le q<\infty}\bigcup_{w\in A_q} L^q(w)$ and takes values in the space of all measurable complex-valued functions. Fix $1 \le r < \infty$ and suppose that for all $B > 1$ there exists a constant $N_r(B) > 0$ such that for all weights v in A_r with $[v]_{A_r} \le B$ we have*

$$\big\|T\big\|_{L^r(v)\to L^r(v)} \le N_r(B)\,. \tag{9.5.4}$$

Then for any $1 < p < \infty$ and all $B > 1$, there exists $N_p(B) > 0$ such that for all weights w in A_p satisfying $[w]_{A_p} \le B$ we have

$$\big\|T\big\|_{L^p(w)\to L^p(w)} \le N_p(B)\,. \tag{9.5.5}$$

The following lemma provides the crucial idea in the proof of the extrapolation Theorem 9.5.3.

Lemma 9.5.4. *(a) Let $1 \le r < p < \infty$ and $w \in A_p$. Then there exists a constant $C_1 = C_1(n, r, p, [w]_{A_p})$ such that for every nonnegative function g in $L^{(p/r)'}(w)$, there is a function $G(g)$ such that*

$$g \le G(g) \tag{9.5.6}$$

$$\big\|G(g)\big\|_{L^{\frac{p}{p-r}}(w)} \le 2\,\big\|g\big\|_{L^{\frac{p}{p-r}}(w)}\,, \tag{9.5.7}$$

$$\big[G(g)w\big]_{A_r} \le C_1\,. \tag{9.5.8}$$

(b) Let $1 < p < r < \infty$ and $w \in A_p$. Then there exists a constant $C_2 = C_2(n, r, p, [w]_{A_p})$ such that for every nonnegative function h in $L^{\frac{p}{r-p}}(w)$, there is a function $H(h)$ with the properties

$$h \le H(h) \tag{9.5.9}$$

$$\big\|H(h)\big\|_{L^{\frac{p}{r-p}}(w)} \le 2^{r-1}\big\|h\big\|_{L^{\frac{p}{r-p}}(w)}\,, \tag{9.5.10}$$

$$\big[H(h)^{-1}w\big]_{A_r} \le C_2\,. \tag{9.5.11}$$

Moreover, both constants $C_1(n, r, p, B)$ and $C_2(n, r, p, B)$ increase as B increases.

Proof. (a) Given $1 \le r < p$, let $t = \frac{p'}{(p/r)'} = \frac{p-r}{p-1}$, which satisfies $0 < t \le 1$. Define a sublinear map S as follows:

$$S(g) = \big(M_c(g^{\frac{1}{t}}w)w^{-1}\big)^t\,.$$

It is simple to see that S maps $L^{\frac{p}{p-r}}(w)$ into itself with norm

$$\big\|S\big\| \le \big\|M_c\big\|^t_{L^{p'}(w^{1-p'})\to L^{p'}(w^{1-p'})}\,.$$

For a function $g \ge 0$ define

$$G(g) = \sum_{k=0}^{\infty} \frac{S^k(g)}{2^k\big\|S\big\|^k}\,.$$

Then we have that $g \le G(g)$ and

$$\big\|G(g)\big\|_{L^{\frac{p}{p-r}}(w)} \le \sum_{k=0}^{\infty} \frac{\big\|S\big\|^k \|g\|_{L^{\frac{p}{p-r}}(w)}}{2^k \big\|S\big\|^k} = 2\big\|g\big\|_{L^{\frac{p}{p-r}}(w)}.$$

It remains to see that $G(g)w$ is an A_r weight. Note that

$$S(G(g)) \le 2\big\|S\big\| \sum_{k=0}^{\infty} \frac{S^{k+1}(g)}{2^{k+1}\big\|S\big\|^{k+1}} \le 2\big\|S\big\|\, G(g)\,.$$

That is,

$$\big(M_c(G(g)^{\frac{1}{t}} w)w^{-1}\big)^t \le 2\big\|S\big\|G(g)\,,$$

which implies that $G(g)^{\frac{1}{t}} w$ is an A_1 weight with characteristic constant

$$\big[G(g)^{\frac{1}{t}} w\big]_{A_1} \le (2\big\|S\big\|)^{\frac{1}{t}} = 2^{\frac{1}{t}} \big\|M_c\big\|_{L^{p'}(w^{1-p'})\to L^{p'}(w^{1-p'})}\,.$$

It follows that for all cubes Q we have

$$\frac{1}{|Q|}\int_Q G(g)^{\frac{1}{t}} w\,dx \le (2\big\|S\big\|)^{\frac{1}{t}} G(g)^{\frac{1}{t}} w$$

or, equivalently,

$$G(g) \ge (2\big\|S\big\|)^{-1}\Big(\frac{1}{|Q|}\int_Q G(g)^{\frac{1}{t}} w\,dx\Big)^t w^{-t}\,. \tag{9.5.12}$$

We now use Hölder and (9.5.12) to obtain

$$\begin{aligned}
&\Big(\frac{1}{|Q|}\int_Q G(g)w\,dx\Big)\Big(\frac{1}{|Q|}\int_Q G(g)^{-\frac{1}{r-1}} w^{-\frac{1}{r-1}}\,dx\Big)^{r-1} \\
&\le \Big(\frac{1}{|Q|}\int_Q G(g)^{\frac{1}{t}} w\,dx\Big)^t \Big(\frac{1}{|Q|}\int_Q w\,dx\Big)^{1-t} \\
&\qquad \Big(\frac{1}{|Q|}\int_Q (2\big\|S\big\|)^{\frac{1}{r-1}} \Big(\frac{1}{|Q|}\int_Q G(g)^{\frac{1}{t}} w\,dy\Big)^{-\frac{t}{r-1}} w^{\frac{t}{r-1}} w^{-\frac{1}{r-1}}\,dx\Big)^{r-1} \\
&= 2\,\big\|S\big\| \Big\{\Big(\frac{1}{|Q|}\int_Q w\,dx\Big)\Big(\frac{1}{|Q|}\int_Q w^{-\frac{1}{p-1}}\,dx\Big)^{p-1}\Big\}^{1-t} \le 2\,\big\|S\big\|\,[w]_{A_p}^{1-t}\,,
\end{aligned}$$

where we used the fact that $\frac{t-1}{r-1} = -\frac{1}{p-1}$. We conclude that $G(g)w$ is an A_r weight with characteristic constant

$$\big[G(g)w\big]_{A_r} \le 2\big\|M_c\big\|^t_{L^{p'}(w^{1-p'})\to L^{p'}(w^{1-p'})}[w]_{A_p}^{1-t}\,.$$

(b) The proof of part (b) is similar to that given in part (a). We set $s = \frac{r-p}{r-1}$ and we observe that $0 < s < 1$. We define an operator

$$R(h) = M_c(h^{\frac{1}{s}})^s\,,$$

and we observe that R maps $L^{\frac{p}{s}}(w)$ into itself, say with norm $\|R\|$. Note that $\|R\|$ depends on $[w]_{A_p}$ and increases as $[w]_{A_p}$ increases. Then we set

$$H(h)=\bigg(\sum_{k=0}^{\infty}\frac{R^k\big(h^{\frac{1}{r-1}}\big)}{2^k\|R\|^k}\bigg)^{r-1}$$

and we observe that $H(h)\ge h$ by comparing with the term $k=0$. Using that R maps $L^{\frac{p}{s}}(w)$ into itself, we observe that

$$\begin{aligned}\|H(h)\|_{L^{\frac{p}{r-p}}(w)} &= \bigg(\int_{\mathbf{R}^n}\bigg(\sum_{k=0}^{\infty}\frac{R^k\big(h^{\frac{1}{r-1}}\big)}{2^k\|R\|^k}\bigg)^{\frac{p}{s}} w\,dx\bigg)^{\frac{r-p}{p}}\\ &\le \bigg\|\sum_{k=0}^{\infty}\frac{R^k\big(h^{\frac{1}{r-1}}\big)}{2^k\|R\|^k}\bigg\|_{L^{\frac{p}{s}}(w)}^{\frac{p}{s}\frac{r-p}{p}}\\ &\le 2^{r-1}\big\|h^{\frac{1}{r-1}}\big\|_{L^{\frac{p}{s}}(w)}^{r-1} = 2^{r-1}\|h\|_{L^{\frac{p}{r-p}}(w)},\end{aligned}$$

which proves (9.5.10). Now notice that

$$R\big(H(h)^{\frac{1}{r-1}}\big)\le\sum_{k=0}^{\infty}\frac{R^{k+1}\big(h^{\frac{1}{r-1}}\big)}{2^k\|R\|^k}\le 2\,\|R\|\,H(h)^{\frac{1}{r-1}},$$

from which it follows that

$$M_c\big(H(h)^{\frac{1}{r-p}}\big)^{r-p}\le(2\,\|R\|)^{r-1}H(h).$$

We conclude that for any cube Q with sides parallel to the axes we have

$$\bigg(\frac{1}{|Q|}\int_Q H(h)^{\frac{1}{r-p}}\,dx\bigg)^{r-p}\le(2\,\|R\|)^{r-1}H(h)(x)$$

for almost every $x\in Q$. It follows that

$$\bigg(\frac{1}{|Q|}\int_Q H(h)^{-1}w\,dx\bigg)\bigg(\frac{1}{|Q|}\int_Q H(h)^{\frac{1}{r-p}}\,dx\bigg)^{r-p}\le(2\,\|R\|)^{r-1}\frac{1}{|Q|}\int_Q w\,dx.$$

Multiply both sides of the inequality by $\big(\frac{1}{|Q|}\int_Q w^{-\frac{1}{p-1}}\,dx\big)^{p-1}$ and use the following consequence of Hölder's inequality with exponents $\frac{1}{s}$ and $(\frac{1}{s})'$:

$$\bigg(\frac{1}{|Q|}\int_Q (H(h)^{-1}w)^{-\frac{1}{r-1}}\,dx\bigg)^{r-1}\le\bigg(\frac{1}{|Q|}\int_Q w^{-\frac{1}{p-1}}\,dx\bigg)^{p-1}\bigg(\frac{1}{|Q|}\int_Q H(h)^{\frac{1}{r-p}}\,dx\bigg)^{r-p}$$

to obtain the required estimate

$$\big[H(h)^{-1}w\big]_{A_r}\le(2\,\|R\|)^{r-1}[w]_{A_p}.$$

Setting $C_2=(2\,\|R\|)^{r-1}[w]_{A_p}$ yields the asserted conclusion.

□

Having completed the proof of Lemma 9.5.4, we can now easily prove Theorem 9.5.3.

Proof. Let $N_r(B)$ denote the smallest constant that satisfies (9.5.4). Then $N_r(B)$ increases as B increases, a property that we will use.

We first consider the case $1 \le r < p < \infty$. For a function $|g| \in L^{(p/r)'}(w)$ we let $G(|g|)$ be as in Lemma 9.5.4 (a). Using the consequence of this lemma, the boundedness of T on $L^p(w)$, and Hölder's inequality, we obtain

$$
\begin{aligned}
&\bigg(\int_{\mathbf{R}^n} |T(f)|^p w\,dx\bigg)^{\frac{1}{p}} \\
&= \bigg(\int_{\mathbf{R}^n} (|T(f)|^r)^{\frac{p}{r}} w\,dx\bigg)^{\frac{r}{p}\frac{1}{r}} \\
&= \sup_{\|g\|_{L^{(p/r)'}(w)}\le 1} \bigg(\int_{\mathbf{R}^n} |T(f)|^r |g| w\,dx\bigg)^{\frac{1}{r}} \\
&\le \sup_{\|g\|_{L^{(p/r)'}(w)}\le 1} \bigg(\int_{\mathbf{R}^n} |T(f)|^r G(|g|) w\,dx\bigg)^{\frac{1}{r}} \\
&\le \sup_{\|g\|_{L^{(p/r)'}(w)}\le 1} N_r([G(|g|)w]_{A_r})\bigg(\int_{\mathbf{R}^n} |f|^r G(|g|) w\,dx\bigg)^{\frac{1}{r}} \\
&\le \sup_{\|g\|_{L^{(p/r)'}(w)}\le 1} N_r(C_1(n,p,r,[w]_{A_p}))\, \big\||f|^r\big\|_{L^{p/r}(w)}^{\frac{1}{r}} \big\|G(|g|)\big\|_{L^{(p/r)'}(w)}^{\frac{1}{r}} \\
&\le N_r(C_1(n,p,r,[w]_{A_p}))2^{\frac{1}{r}} \big\|f\big\|_{L^p(w)}\,.
\end{aligned}
$$

This concludes the case $1 \le r < p < \infty$.

The argument for the case $1 < p < r < \infty$ is similar. Given $w \in A_p$ and $f \in L^p(w)$, we observe that

$$
|f|^{r-p} \in L^{\frac{p}{r-p}}(w)\,.
$$

Applying Lemma 9.5.4 (b), we have

$$
H(|f|^{r-p}) \ge |f|^{r-p}
$$

and

$$
[H(|f|^{r-p})^{-1} w]_{A_r} \le C_2 = C_2(n,p,r,[w]_{A_p})\,.
$$

The boundedness of T on $L^p(w)$ and Hölder's inequality now give

$$\begin{aligned}
\Big(\int_{\mathbf{R}^n}|T(f)|^p w\,dx\Big)^{\frac{1}{p}} &= \big\||T(f)|^r\big\|_{L^{\frac{p}{r}}(w)}^{\frac{1}{r}}\\
&= \big\||T(f)|^r H(|f|^{r-p})^{-1}H(|f|^{r-p})\big\|_{L^{\frac{p}{r}}(w)}^{\frac{1}{r}}\\
&\le \big\||T(f)|^r H(|f|^{r-p})^{-1}\big\|_{L^1(w)}^{\frac{1}{r}}\big\|H(|f|^{r-p})\big\|_{L^{\frac{p}{r-p}}(w)}^{\frac{1}{r}}\\
&\le \Big(\int_{\mathbf{R}^n}|T(f)|^r H(|f|^{r-p})^{-1}w\,dx\Big)^{\frac{1}{r}} 2^{\frac{r-1}{r}}\big\||f|^{r-p}\big\|_{L^{\frac{p}{r-p}}(w)}^{\frac{1}{r}}\\
&\le 2^{\frac{r-1}{r}} N_r([H(|f|^{r-p})^{-1}w]_{A_r})\Big(\int_{\mathbf{R}^n}|f|^r H(|f|^{r-p})^{-1}w\,dx\Big)^{\frac{1}{r}}\big\|f\big\|_{L^p}^{\frac{r-p}{r}}\\
&\le 2^{\frac{r-1}{r}} N_r(C_2(n,p,r,[w]_{A_p}))\Big(\int_{\mathbf{R}^n}|f|^r|f|^{p-r}w\,dx\Big)^{\frac{1}{r}}\big\|f\big\|_{L^p}^{\frac{r-p}{r}}\\
&= 2^{\frac{r-1}{r}} N_r(C_2(n,p,r,[w]_{A_p}))\big\|f\big\|_{L^p}\,.
\end{aligned}$$

The theorem is proved with $N_p(B) = a_r\, N_r(C_j(n,p,r,[w]_{A_p}))$, where $a_r = 2^{1/r}$ when $p > r$ and $a_r = 2^{(r-1)/r}$ when $p < r$.

$\square$

Remark 9.5.5. It is interesting to observe that in the proof of the preceding theorem, it was not necessary to use that T is a linear operator, not even sublinear. The only thing needed was that T is well defined on $L^q(w)$ for all $w \in A_q$ and all $1 \le q < \infty$.

If the operator T happened to be linear, then we could weaken the hypotheses of Theorem 9.5.3 by assuming that T is only initially defined on $C_0^\infty(\mathbf{R}^n)$ since this space is dense in $L^p(w)$ for all $1 \le p < \infty$ and all $w \in A_p$. Unfortunately, this simplification cannot be made when T is sublinear.

We now observe that our extrapolation theorem 9.5.3 can be improved by weakening the initial strong type assumption to a weak type estimate. However, to obtain a strong type estimate as a conclusion, we need to assume that the operator in question is sublinear (or quasi-sublinear). We have the following.

Theorem 9.5.6. *Suppose that a sublinear operator T is defined on the union $\bigcup\limits_{1<q<\infty}\bigcup\limits_{w\in A_q} L^q(w)$ and takes values in the space of all measurable complex-valued functions. Fix $1 \le r < \infty$ and suppose that for all $B > 1$ there exists a constant $N_r(B) > 0$ such that for all weights v in A_r with $[v]_{A_r} \le B$ we have*

$$\big\|T\big\|_{L^r(v)\to L^{r,\infty}(v)} \le N_r(B)\,. \tag{9.5.13}$$

Then for any $1 < p < \infty$ and all $B > 1$, there exists $N_p(B) > 0$ such that for all weights w in A_p satisfying $[w]_{A_p} \le B$ we have

$$\big\|T\big\|_{L^p(w)\to L^p(w)} \le N_p(B)\,. \tag{9.5.14}$$

Proof. The proof of this theorem is a direct consequence of the proof of Theorem 9.5.3. Indeed, let us fix a $\lambda > 0$ and define an operator

$$T_\lambda(f) = \lambda \chi_{|T(f)|>\lambda}\,.$$

Note that T_λ is not a linear operator, but this is not an obstacle in using Theorem 9.5.3. We show that T_λ maps $L^r(w)$ into $L^r(w)$ for every $w \in A_r$ with $[w]_{A_r} \le B$. We have

$$\begin{aligned}\bigg(\int_{\mathbf{R}^n} |T_\lambda(f)|^r\, w\, dx\bigg)^{\frac{1}{r}} &= \bigg(\int_{\mathbf{R}^n} \lambda^r \chi_{|T(f)|>\lambda}\, w\, dx\bigg)^{\frac{1}{r}} \\ &= \Big(\lambda^r w(\{|T(f)|>\lambda\})\Big)^{\frac{1}{r}} \\ &\le N_r(B)\big\|f\big\|_{L^r(w)}\end{aligned}$$

using the hypothesis on T. Applying Theorem 9.5.3, we obtain that T_λ maps $L^p(w)$ into itself for all $1 < p < \infty$ and all $w \in A_p$ with a constant independent of λ. Precisely, for any $B > 0$ there exists a constant $N_p(B) > 0$ such that if $[w]_{A_p} \le B$ we have

$$\big\|T_\lambda(f)\big\|_{L^p(w)} \le N_p(B)\big\|f\big\|_{L^p(w)}\,.$$

Since we have

$$\big\|T(f)\big\|_{L^{p,\infty}(w)} = \sup_{\lambda>0}\big\|T_\lambda(f)\big\|_{L^p(w)}\,,$$

it follows that T maps $L^p(w)$ into $L^{p,\infty}(w)$ for any $1<p<\infty$ and $w \in A_p$.

At this point we will use the hypothesis that T is sublinear. We use the Marcinkiewicz interpolation theorem to strengthen the weak type estimate and obtain boundedness into $L^p(w)$. Let $w \in A_p$ with $[w]_{A_p} \le B$. Then there is a B' and an $\varepsilon > 0$ such that

$$[w]_{A_p} \le B \implies [w]_{A_{p-\varepsilon}} \le B'\,.$$

Also $w \in A_{p+\varepsilon}$ with $[w]_{A_{p+\varepsilon}} \le [w]_{A_p} \le B$. We have proved that the operator T maps $L^{p-\varepsilon}(w)$ into $L^{p-\varepsilon,\infty}(w)$ with bound at most a constant $C_{1,\infty}(B')$ and also $L^{p+\varepsilon}(w)$ into $L^{p+\varepsilon,\infty}(w)$ with bound at most a constant $C_{2,\infty}(B)$. It follows from Theorem 1.3.2 that T maps $L^p(w)$ into $L^p(w)$ with constant at most $2(2p/\varepsilon)^{\frac{1}{p}} C_{1,\infty}(B')^{\frac{p-\varepsilon}{2p}} C_{2,\infty}(B)^{\frac{p+\varepsilon}{2p}}$. A careful examination of the proof of Corollary 9.2.6 gives that

$$\frac{1}{\varepsilon} = \Big(1+\frac{1}{\gamma}\Big)\frac{1}{p-1}, \qquad \text{where} \qquad \gamma = \frac{\log\big([w]_{A_p}\big) - \log\big([w]_{A_p} - (\frac{1}{2})^p\big)}{\log 2^n - \log\frac{1}{2}}\,,$$

and we see that if $[w]_{A_p} \le B$, then $1/\varepsilon$ is bounded by some constant depending on B. This completes the proof of the theorem.

□

We end this subsection by observing that the conclusion of the extrapolation Theorem 9.5.3 can be strengthened to yield vector-valued estimates.

Corollary 9.5.7. *Suppose that T is defined on $\bigcup_{1\le q<\infty}\bigcup_{w\in A_q} L^q(w)$ and takes values in the space of all measurable complex-valued functions. Fix $1\le r<\infty$ and suppose that for all $B>1$ there exists a constant $N_r(B)>0$ such that for all weights v in A_r with $[v]_{A_r}\le B$ we have*

$$\big\|T\big\|_{L^r(v)\to L^r(v)} \le N_r(B)\,.$$

Then for every $1<p<\infty$ and every $B>1$, there exists a constant $N_p(B,r)$ such that for all weights $w\in A_p$ with $[w]_{A_p}\le N_p(B,r)$ we have

$$\Big\|\Big(\sum_j |T(f_j)|^r\Big)^{\frac{1}{r}}\Big\|_{L^p(w)} \le N_p(B,r)\Big\|\Big(\sum_j |f_j|^r\Big)^{\frac{1}{r}}\Big\|_{L^p(w)}$$

for all sequences of functions f_j in $L^p(w)$.

Proof. Note that in the proof of Theorem 9.5.3, the function f may be replaced by $(\sum_j |f_j|^r)^{\frac{1}{r}}$ and $T(f)$ may be replaced by $(\sum_j |T(f_j)|^r)^{\frac{1}{r}}$.

□

9.5.c. Weighted Inequalities Versus Vector-Valued Inequalities

We now turn to the last topic we are going to address in relation to A_p weights, connections between weighted inequalities and vector-valued inequalities. The next result provides strong evidence that there is a nontrivial connection of this sort. The following is a general theorem saying that any vector-valued inequality is equivalent to some weighted inequality. The proof of the theorem is based on a minimax lemma whose precise formulation and proof can be found in Appendix H.

Theorem 9.5.8. *(a) Let $0<p<q,r<\infty$. Let $\{T_j\}_j$ be a sequence of sublinear operators that map $L^q(\mu)$ into $L^r(\nu)$ where μ and ν are arbitrary measures. Then the vector-valued inequality*

$$\Big\|\Big(\sum_j |T_j(f_j)|^p\Big)^{\frac{1}{p}}\Big\|_{L^r} \le C\Big\|\Big(\sum_j |f_j|^p\Big)^{\frac{1}{p}}\Big\|_{L^q} \tag{9.5.15}$$

holds for all $f_j\in L^q(\mu)$ if and only if for every u in $L^{\frac{r}{r-p}}(\nu)$ there exists U in $L^{\frac{q}{q-p}}(\mu)$ with

$$\begin{gathered}\big\|U\big\|_{L^{\frac{q}{q-p}}} \le \big\|u\big\|_{L^{\frac{r}{r-p}}}\\ \sup_j \int |T_j(f)|^p\,u\,d\nu \le C^p\int |f|^p\,U\,d\mu\,.\end{gathered} \tag{9.5.16}$$

(b) Let $0 < q, r < p < \infty$. *Let* $\{T_j\}_j$ *be as before. Then the vector-valued inequality (9.5.15) holds for all* $f_j \in L^q(\mu)$ *if and only if for every* u *in* $L^{\frac{q}{p-q}}(\mu)$ *there exists* U *in* $L^{\frac{r}{p-r}}(\nu)$ *with*

$$\begin{aligned} &\big\|U\big\|_{L^{\frac{r}{p-r}}} \le \big\|u\big\|_{L^{\frac{q}{p-q}}} \\ &\sup_j \int |T_j(f)|^p \, U^{-1} \, d\nu \le C^p \int |f|^p \, u^{-1} \, d\mu \,. \end{aligned} \tag{9.5.17}$$

Proof. We begin with part (a). Given $f_j \in L^q(\mathbf{R}^n, \mu)$, we use (9.5.16) to obtain

$$\begin{aligned} \Big\| \Big(\sum_j |T_j(f_j)|^p \Big)^{\frac{1}{p}} \Big\|_{L^r(\nu)} &= \Big\| \sum_j |T_j(f_j)|^p \Big\|_{L^{\frac{r}{p}}(\nu)}^{\frac{1}{p}} \\ &= \sup_{\|u\|_{L^{\frac{r}{r-p}}} \le 1} \Big(\int_{\mathbf{R}^n} \sum_j |T_j(f_j)|^p \, u \, d\nu \Big)^{\frac{1}{p}} \\ &\le \sup_{\|u\|_{L^{\frac{r}{r-p}}} \le 1} C \Big(\int_{\mathbf{R}^n} \sum_j |f_j|^p \, U \, d\mu \Big)^{\frac{1}{p}} \\ &\le \sup_{\|u\|_{L^{\frac{r}{r-p}}} \le 1} C \Big\| \sum_j |f_j|^p \Big\|_{L^{\frac{q}{p}}(\mu)}^{\frac{1}{p}} \big\|U\big\|_{L^{\frac{q}{q-p}}}^{\frac{1}{p}} \\ &\le C \Big\| \Big(\sum_j |f_j|^p \Big)^{\frac{1}{p}} \Big\|_{L^q(\mu)} , \end{aligned}$$

which proves (9.5.15) with the same constant C as in (9.5.16). To prove the converse we define

$$A = \Big\{ a = (a_0, a_1) : \; a_0 = \sum_j |f_j|^p, \quad a_1 = \sum_j |T_j(f_j)|^p, \quad f_j \in L^q(\mu) \Big\}$$

and

$$B = \Big\{ b \in L^{\frac{q}{q-p}}(\mu) : \; b \ge 0 \,, \quad \big\|b\big\|_{L^{\frac{q}{q-p}}} \le 1 \Big\} .$$

Notice that A and B are convex sets and B is weakly compact. (The sublinearity of each T_j is used here.) Given a fixed u in $L^{\frac{r}{r-p}}(\nu)$ with $\big\|u\big\|_{L^{\frac{r}{r-p}}} \le 1$, we define the function Φ on $A \times B$ by setting

$$\Phi(a, b) = \int a_1 u \, d\mu - C^p \int a_0 b \, d\nu = \sum_j \Big(\int |T_j(f_j)|^p u \, d\nu - C^p \int |f_j|^p b \, d\mu \Big) .$$

Then Φ is concave on A and weakly continuous and convex on B. Thus the *minimax lemma* in Appendix H is applicable. This gives

$$\min_{b \in B} \sup_{a \in A} \Phi(a, b) = \sup_{a \in A} \min_{b \in B} \Phi(a, b) \,. \tag{9.5.18}$$

At this point observe that for a fixed $a = \big(\sum_j |f_j|^p, \sum_j |T_j(f_j)|^p\big)$ in A we have

$$\begin{aligned}\min_{b\in B} \Phi(a,b) &\le \Big\|\sum_j |T_j(f_j)|^p\Big\|_{L^{\frac{r}{p}}(\nu)} \big\|u\big\|_{L^{\frac{r}{r-p}}} - C^p \max_{b\in B} \int \sum_j |f_j|^p \, b \, d\mu \\ &\le \Big\|\sum_j |T_j(f_j)|^p\Big\|_{L^{\frac{r}{p}}(\nu)} - C^p \Big\|\sum_j |f_j|^p\Big\|_{L^{\frac{q}{p}}(\mu)} \le 0\end{aligned}$$

using the hypothesis (9.5.15). It follows that $\sup_{a\in A} \min_{b\in B} \Phi(a,b) \le 0$ and hence (9.5.18) yields $\min_{b\in B} \sup_{a\in A} \Phi(a,b) \le 0$. Thus there exists a $U \in B$ such that $\Phi(a,U) \le 0$ for every $a \in A$. This completes the proof of part (a).

The proof of part (b) is similar. Using the result of Exercise 9.5.1 and (9.5.17), given $f_j \in L^q(\mathbf{R}^n, \mu)$ we have

$$\begin{aligned}\Big\|\big(\sum_j |f_j|^p\big)^{\frac{1}{p}}\Big\|_{L^q(\mu)} &= \Big\|\sum_j |f_j|^p\Big\|_{L^{\frac{q}{p}}(\mu)}^{\frac{1}{p}} \\ &= \inf_{\|u\|_{L^{\frac{q}{p-q}}} \le 1} \Big(\int_{\mathbf{R}^n} \sum_j |f_j|^p \, u^{-1} \, d\mu\Big)^{\frac{1}{p}} \\ &\ge \frac{1}{C} \inf_{\|U\|_{L^{\frac{r}{p-r}}} \le 1} \Big(\int_{\mathbf{R}^n} \sum_j |T_j(f_j)|^p \, U^{-1} \, d\nu\Big)^{\frac{1}{p}} \\ &= \frac{1}{C}\Big\|\sum_j |T_j(f_j)|^p\Big\|_{L^{\frac{r}{p}}(\nu)}^{\frac{1}{p}} = \frac{1}{C}\Big\|\big(\sum_j |T_j(f_j)|^p\big)^{\frac{1}{p}}\Big\|_{L^r(\nu)} .\end{aligned}$$

To prove the converse direction in part (b), we define A as in part (a) and

$$B = \big\{b \in L^{\frac{p}{p-r}}(\nu) : \ b \ge 0, \quad \big\|b\big\|_{L^{\frac{p}{p-r}}} \le 1\big\} .$$

Given a fixed u in $L^{\frac{q}{p-q}}(\mu)$ with $\big\|u\big\|_{L^{\frac{q}{p-q}}} \le 1$, we define the function Φ on $A \times B$ by setting

$$\begin{aligned}\Phi(a,b) &= \int a_1 b^{-1} d\nu - C^p \int a_0 u^{-1} \, d\mu \\ &= \sum_j \Big(\int |T_j(f_j)|^p b^{-1} \, d\nu - C^p \int |f_j|^p u^{-1} \, d\mu\Big) .\end{aligned}$$

Then Φ is concave on A and weakly continuous and convex on B. Also, using Exercise 9.5.1, for any $a = \big(\sum_j |f_j|^p, \sum_j |T_j(f_j)|^p\big)$ in A, we have

$$\min_{b\in B} \Phi(a,b) \le \Big\|\sum_j |T_j(f_j)|^p\Big\|_{L^{\frac{r}{p}}(\nu)} - C^p \Big\|\sum_j |f_j|^p\Big\|_{L^{\frac{q}{p}}(\mu)} \le 0 .$$

Thus $\sup_{a\in A}\min_{b\in B}\Phi(a,b)\le 0$. Using (9.5.18), yields $\min_{b\in B}\sup_{a\in A}\Phi(a,b)\le 0$, and the latter implies the existence of a U in B such that $\Phi(a,U)\le 0$ for all $a\in A$. This proves (9.5.17).

□

Example 9.5.9. We use the previous theorem to obtain another proof of the vector-valued Hardy-Littlewood maximal inequality in Corollary 4.6.5. We take $T_j = M$ for all j. For given $1<p<q<\infty$ and u in $L^{\frac{q}{q-p}}$ we set $s=\frac{q}{q-p}$ and $U=\big\|M\big\|_{L^s\to L^s}^{-1}M(u)$. In view of Exercise 9.1.8 we have

$$\big\|U\big\|_{L^s}\le\big\|u\big\|_{L^s}\qquad\text{and}\qquad\int_{\mathbf{R}^n}M(f)^p\,u\,dx\le C^p\int_{\mathbf{R}^n}|f|^p\,U\,dx\,.$$

Using Theorem 9.5.8, we obtain

$$\Big\|\Big(\sum_j|M(f_j)|^p\Big)^{\frac{1}{p}}\Big\|_{L^q}\le C_{n,p,q}\Big\|\Big(\sum_j|f_j|^p\Big)^{\frac{1}{p}}\Big\|_{L^q}\tag{9.5.19}$$

whenever $1<p<q<\infty$, an inequality obtained earlier in (4.6.17).

It turns out that no specific properties of the Hardy-Littlewood maximal function are used in the preceding inequality, and we can obtain a general result along these lines. For simplicity we take the operators T_j in the next theorem to be linear.

Theorem 9.5.10. *Let $1<r<\infty$ be fixed. Let T_j be a sequence of linear operators on $L^p(\mathbf{R}^n,w\,dx)$ with the property that for each $B>1$, there is a constant $C_r(B)>0$ such that for all w in A_r with $[w]_{A_r}\le B$ we have*

$$\int_{\mathbf{R}^n}|T_j(f)(x)|^r w(x)\,dx\le (C_r(B))^r\int_{\mathbf{R}^n}|f(x)|^r w(x)\,dx\,.\tag{9.5.20}$$

Then the vector-valued inequality (9.5.15) is valid for all $1<p\neq q<\infty$ for some constant $C(n,p,q)$ and all $f_j\in L^q(\mathbf{R}^n,w\,dx)$.

Proof. In view of Theorem 9.5.3, (9.5.20) also holds if r is replaced by any other $1<p<\infty$ [with some constant $N_p(B)$ replacing $C_r(B)$]. Therefore, it suffices to prove the vector-valued inequality (9.5.15) when $p=r$. To achieve this we consider two cases, $r<q$ and $r>q$.

We first consider the case $1<r<q$. Given $r<q$, we set $s=\frac{q}{q-r}$ and $\sigma=\frac{s+1}{2}$. For a given $u\ge 0$ in $L^s(w)$, we set $U=\big\|M\big\|_{L^s\to L^s}^{-1}M(u^\sigma)^{1/\sigma}$. Then U satisfies $\big\|U\big\|_{L^s}\le\big\|u\big\|_{L^s}$ [hence the first assumption in (9.5.16) holds] and U pointwise controls a constant multiple of u [hence the second assumption in (9.5.16) holds]. Applying Theorem 9.5.8, we obtain (9.5.15).

Suppose now that $1<q<r<\infty$. We denote by T^t the transpose of a linear operator T defined by the identity

$$\langle T(f),g\rangle=\langle f,T^t(g)\rangle$$

for all suitable f and g. Consider the vector-valued operators

$$\vec{T}(\{f_j\}_j) = \{T_j(f_j)\}_j, \qquad \vec{T}^t(\{f_j\}_j) = \{T_j^t(f_j)\}_j .$$

Duality considerations give that $\vec{T}$ maps $L^q(\ell^r)$ into itself if and only if $\vec{T}^t$ maps $L^{q'}(\ell^{r'})$ into itself and the norms of these operators are the same. Using the result of Exercise 9.4.6, we obtain that

$$\int_{\mathbf{R}^n} |T_j^t(f)(x)|^{r'} v(x)\, dx \le (C_r(B))^r \int_{\mathbf{R}^n} |f(x)|^{r'} v(x)\, dx$$

for every $v \in A_{r'}$ with $[v]_{A_{r'}} \le B$. But when $r > q$ we have $r' < q'$, and applying the result in the previous case we obtain the required conclusion in this case as well. This concludes the proof of the theorem.

□

Exercises

9.5.1. Let $0 < s < 1$ and f be in $L^s(X, \mu)$. Show that

$$\big\|f\big\|_{L^s} = \inf\left\{ \int_X |f| u^{-1}\, d\mu : \ \big\|u\big\|_{L^{\frac{s}{1-s}}} \le 1 \right\}$$

and that the infimum is attained.
[*Hint:* Try $u = c\,|f|^{1-s}$ for a suitable constant c.]

9.5.2. Use the same idea of the proof of Theorem 9.5.1 to prove the following general result: Let μ be a positive measure on a measure space X and let T be a bounded sublinear operator on $L^p(X, \mu)$ for some $1 \le p < \infty$. Suppose that $T(f) \ge 0$ for all f in $L^p(X, \mu)$. Prove that for all $f_0 \in L^p(X, \mu)$, there exists an $f \in L^p(X, \mu)$ such that
(a) $f(x) \le f_0(x)$ for μ-almost all $x \in X$.
(b) $\big\|f_0\big\|_{L^p(X)} \le 2\,\big\|f\big\|_{L^p(X)}$.
(c) $T(f)(x) \le 2\,\big\|T\big\|_{L^p \to L^p} f(x)$ for μ-almost all $x \in X$.
[*Hint:* Try the expression in (9.5.2) starting the sum at $j = 0$.]

9.5.3. Prove the following result without using Theorem 9.5.3. If T satisfies the hypotheses of Theorem 9.5.3, then for $1 < q < r$ and all $v \in A_1$, T maps $L^q(v)$ into $L^q(v)$ with constant depending on q, r, and $[v]_{A_1}$.
[*Hint:* Hölder's inequality gives that

$$\big\|T(f)\big\|_{L^q(v)} \le \left(\int_{\mathbf{R}^n} |T(f)(x)|^r M(f)(x)^{-(r-q)}\, v(x)\, dx \right)^{\frac{1}{r}} \left(\int_{\mathbf{R}^n} M(f)(x)^q v(x)\, dx \right)^{\frac{r-q}{rq}} .$$

Then use the fact that the weight $M(f)^{\frac{r-q}{r-1}}$ is in A_1 and Exercise 9.1.2.]

9.5.4. (*Duoandikoetxea* [**167**]) Use the result of the previous Exercise to give another proof of Theorem 9.5.3.
[*Hint:* Given $1 < p < \infty$ and $B > 1$, use Exercise 9.2.6(b) to find numbers $\delta = \delta(n,p,B)$ and $B' = B'(n,p,B)$ such that $[\omega]_{A_p} \le B \implies [\omega]_{A_{p-\delta}} \le B'$. Let $\delta > 0$ so that $1 < \frac{p}{p-\delta} < r$ and $B'' = B''(n,(p-\delta)',B')$, $\theta < (p-\delta)'-1$, $\theta = \theta(n,(p-\delta)',B')$, such that $[\omega]_{A_{(p-\delta)'}} \le B' \implies [\omega]_{A_{(p-\delta)'-\theta}} \le B''$. Use the previous Exercise with $v = M\big(|T(f)|^{\frac{p(1+\eta)}{(p-\delta)'}} w^{1+\eta}\big)(x)^{\frac{1}{1+\eta}}$, where $1+\eta = \frac{(p-\delta)'}{(p-\delta)'-\theta}$, to control $\big\|T(f)\big\|_{L^p(w)}$ by a constant multiple of

$$\left(\int_{\mathbf{R}^n} |f|^{\frac{p}{p-\delta}} w^{\frac{1}{p-\delta}} M\big(|T(f)|^{\frac{p(1+\eta)}{(p-\delta)'}} w^{1+\eta}\big)^{\frac{1}{1+\eta}} w^{-\frac{1}{p-\delta}}\, dx\right)^{\frac{1}{p}}.$$

Finally, estimate the latter applying Hölder's inequality.]

9.5.5. Let T be a sublinear operator defined on $\bigcup_{2\le q<\infty} L^q$. Suppose that for all functions f and u we have

$$\int_{\mathbf{R}^n} |T(f)|^2 u\, dx \le \int_{\mathbf{R}^n} |f|^2 M(u)\, dx\,.$$

Prove that T maps $L^p(\mathbf{R}^n)$ into itself for all $2 < p < \infty$.
[*Hint:* $\big\|T(f)\big\|_{L^p} = \sup\limits_{\|u\|_{L^{(p/2)'}}\le 1} \left(\int_{\mathbf{R}^n} |T(f)|^2 u\, dx\right)^{\frac{1}{2}}$. Apply the hypothesis and Hölder's inequality.]

9.5.6. (*X. C. Li*) Let T be a sublinear operator defined on $\bigcup_{1<q\le 2}\bigcup_{w\in A_q} L^q(w)$. Suppose that for all $B > 1$ there exists a $C(B)$ such that T maps $L^2(w)$ into $L^2(w)$ with norm at most $C(B)$ for all w that satisfy $w^{-1} \in A_1$. Prove that T maps L^p into itself for all $1 < p < 2$.
[*Hint:* We have

$$\big\|T(f)\big\|_{L^p} \le \left(\int_{\mathbf{R}^n} |T(f)|^2 M(f)^{-(2-p)}\, dx\right)^{\frac{1}{2}} \left(\int_{\mathbf{R}^n} M(f)^p\, dx\right)^{\frac{2-p}{2p}}$$

by Hölder's inequality. Apply the hypothesis to the first term of the product.]

9.5.7. Prove the following extension of Theorem 9.5.10. Let $1 < r < \infty$ be fixed. Let T_j be a sequence of linear operators on $L^p(\mathbf{R}^n)$ with the property that for each $B > 1$, there is a constant $C_r(B) > 0$ such that for all w in A_r with

$[w]_{A_r} \le B$ we have

$$\sup_j \big\|T_j(f)\big\|_{L^{r,\infty}(w)} \le C_r(B)\big\|f\big\|_{L^r(w)}\,.$$

Then the vector-valued inequality (9.5.15) holds for all $1 < p, q < \infty$.

9.5.8. (*García-Cuerva* [**203**]) Prove part (b) of Lemma 9.5.4 using only the conclusion in part (a).
[*Hint:* Let $1 < p < r < \infty$ and $w \in A_p$. Then we have $1 < r' < p' < \infty$ and $u = w^{-\frac{1}{p-1}} \in A_{p'}$. Applying part (a), we conclude that for every $g \in L^{(p'/r')'}(u)$ there exists an $G(g) \ge g$ such that $\big\|G(g)\big\|_{L^{(p'/r')'}(u)} \le C_1\big\|g\big\|_{L^{(p'/r')'}(u)}$ and $[G(g)u]_{A_{r'}} \le C_1$. Note that $(p'/r')' = (r-1)p(r-p)^{-1}$ and that

$$g \in L^{(p'/r')'}(u) \iff g^{r-1}w^{-\frac{r-p}{p-1}} \in L^{\frac{p}{r-p}}(w)\,.$$

We also have

$$G(g)u \in A_{r'} \iff (G(g)u)^{-\frac{1}{r'-1}} = \Big(\big(G(g)^{r-1}w^{-\frac{r-p}{p-1}}\big)^{-1}\Big)^{-1} w \in A_r\,.$$

Thus if we are given $h \ge 0$ in $L^{\frac{p}{r-p}}(w)$, we just write $h = g^{r-1}w^{-\frac{r-p}{p-1}}$ with $g \in L^{(p'/r')'}(u)$ and we obtain the corresponding $G(g)$. Defining $H(h) = G(g)^{r-1}w^{-\frac{r-p}{p-1}}$, we observe that (9.5.10) and (9.5.11) are satisfied.]

HISTORICAL NOTES

Weighted inequalities can be probably traced back to the beginning of integration, but the A_p condition first appeared in a paper of Rosenblum [**432**] in a somewhat different form. The characterization of A_p when $n = 1$ in terms of the boundedness of Hardy-Littlewood maximal operator was obtained by Muckenhoupt [**381**]. The fact that A_∞ is the union of the A_p spaces was independently obtained by Muckenhoupt [**382**] and Coifman and Fefferman [**108**]. The latter paper also contains a proof that A_p weights satisfy the crucial reverse Hölder condition. This condition first appeared in the work of Gehring [**207**] in the following context: If F is a quasi-conformal homeomorphism from $\mathbf{R}^n$ into itself, then $|\det(\nabla F)|$ satisfies a reverse Hölder inequality. The boundedness of the Hardy-Littlewood maximal operator on $L^p(w)$ for $w \in A_p$ can also be proved without the reverse Hölder condition as shown by Christ and Fefferman [**100**]. The characterization of A_1 weights is due to Coifman and Rochberg [**118**]. The fact that $M(f)^\delta$ is in A_∞ when $\delta < 1$ was previously obtained by Córdoba and Fefferman [**129**]. The different characterizations of A_∞ (Theorem 9.3.3) are implicit in [**381**] and [**108**]. Our definition of A_∞ using the reverse Jensen inequality was obtained as an equivalent characterization of that space by García-Cuerva and Rubio de Francia [**204**] (p. 405) and independently by Hruščev [**258**]. The reverse Hölder condition was extensively studied by Cruz-Uribe and Neugebauer [**136**].

Weighted inequalities with weights of the form $|x|^a$ for the Hilbert transform were first obtained by Hardy and Littlewood [**239**] and later by Stein [**469**] for other singular integrals. The necessity and sufficiency of the A_p condition for the boundedness of the Hilbert transform on weighted L^p spaces was obtained by Hunt, Muckenhoupt, and Wheeden [**264**]. Historically, the first result relating A_p weights and the Hilbert transform is the Helson-Szegö theorem [**244**] which says that the Hilbert transform is bounded on $L^2(w)$ if and only if $\log w = u + Hv$, where $u, v \in L^\infty(\mathbf{R})$ and $\|v\|_{L^\infty} < \frac{\pi}{2}$. The Helson-Szegö condition easily implies the A_2 condition but the only known direct proof for the converse gives $\|v\|_{L^\infty} < \pi$; see Coifman, Jones, and Rubio de Francia [**110**]. A related result in higher dimensions was obtained by Garnett and Jones [**205**]. Weighted L^p estimates controlling Calderón-Zygmund operators by the Hardy-Littlewood maximal operator were obtained by Coifman [**105**]. Coifman and Fefferman [**108**] extended one-dimensional weighted norm inequalities to higher dimensions and also obtained good lambda inequalities for A_∞ weights for more general singular integrals and maximal singular integrals (Theorem 9.4.3). Bagby and Kurtz [**24**], and later Alvarez and Pérez [**5**], gave a sharper version of Theorem 9.4.3, by replacing the good lambda inequality by a rearrangement inequality. See also the related work of Lerner [**332**].

The factorization of A_p weights was conjectured by Muckenhoupt and proved by Jones [**280**]. The simple proof given in the text can be found in [**110**]. Extrapolation of operators (Theorem 9.5.3) is due to Rubio de Francia [**434**]. The simpler proof of this theorem given in the text is inspired by that given in García-Cuerva [**203**]. Our treatment presents some differences in the proof of the crucial Lemma 9.5.4 which was communicated to the author by J. M. Martell. The simple proof of Theorem 9.5.6 was conceived by J. M. Martell and first appeared in the treatment of extrapolation of operators of many variables in Grafakos and Martell [**218**]. The equivalence between vector-valued inequalities and weighted norm inequalities of Theorem 9.5.8 is also due to Rubio de Francia [**437**]. The difficult direction in this equivalence is obtained by using a minimax principle (see Fan [**181**]). Alternatively, one can use the factorization theory of Maurey [**370**], which brings a connection with Banach space theory. The operators T_j in Theorem 9.5.10 can also be taken to be linearizable, which means $|T_j(f)| = \|U_j(f)\|_B$, where U_j are linear operators taking values in a Banach space B, such as maximal operators. For details we refer the reader to the book of García-Cuerva and Rubio de Francia [**204**], which provides an excellent reference on the topic of weighted norm inequalities.

A primordial double-weighted norm inequality is the observation of Fefferman and Stein [**190**] that the maximal function maps $L^p(M(w))$ into $L^p(w)$ for nonnegative measurable functions w (Exercise 9.1.8). Sawyer [**446**] obtained that the single condition $\sup_Q \big(\int_Q v^{1-p'}\,dx\big)^{-1}\int_Q M(v^{1-p'}\chi_Q)^p w\,dx < \infty$ provides a characterization of all pairs of weights (v,w) for which the Hardy-Littlewood maximal operator M maps $L^p(v)$ into $L^p(w)$. Simpler proofs of this result were obtained by Cruz-Uribe [**135**] and Verbitsky [**539**]. The fact that Sawyer's condition reduces to the usual A_p condition when $v = w$ was shown by Hunt, Kurtz, and Neugebauer [**263**]. The two-weight problem for singular integrals is more delicate since they are not necessarily bounded from $L^p(M(w))$ into $L^p(w)$. Known results in this direction are that singular integrals map $L^p(M^{[p]+1}(w))$ into $L^p(w)$ where M^r denotes the rth-iterate of the maximal operator. See Wilson [**556**] (for $1 < p < 2$) and Pérez [**410**] for the remaining p's. A necessary condition for the boundedness of the Hilbert transform from $L^p(v)$ into $L^p(w)$ was obtained by Muckenhoupt and Wheeden [**383**]. A necessary and sufficient such condition is yet to be found. For an approach to two weighted inequalities using Bellman functions, we refer to the article of Nazarov, Treil, and Volberg [**391**].

The theory of A_p weights presented in this chapter carries through without any complications to the case when Lebesgue measure is replaced by a general doubling measure. The theory also has a substantial analogue when the underlying measure is nondoubling but satisfies $\mu(\partial Q) = 0$ for all cubes Q in $\mathbf{R}^n$ with sides parallel to the axes; see Orobitg and Pérez [**398**].

CHAPTER 10

Boundedness and Convergence of Fourier Integrals

In this chapter we return to fundamental questions in Fourier analysis related to convergence of Fourier series and Fourier integrals. Our main goal is to understand in which sense the inversion property of the Fourier transform holds

$$f(x) = \int_{\mathbf{R}^n} \widehat{f}(\xi) e^{2\pi i x\cdot\xi}\, d\xi$$

when f is a function on $\mathbf{R}^n$. This question is equivalent to the corresponding question for the Fourier series

$$f(x) = \sum_{m\in\mathbf{Z}^n} \widehat{f}(m) e^{2\pi i x\cdot m}$$

when f is a function on $\mathbf{T}^n$. The main problem is that the function (or sequence) $\widehat{f}$ may not be integrable and the convergence of the preceding integral (or series) needs to be suitably interpreted. To overcome this difficulty a summability method should be used. This is achieved by introducing a localizing factor $\Phi(\xi/R)$ and leads to the study of the convergence of the expressions

$$\int_{\mathbf{R}^n} \Phi(\xi/R)\widehat{f}(\xi) e^{2\pi i x\cdot\xi}\, d\xi$$

as $R \to \infty$. Here Φ is a function on $\mathbf{R}^n$ that decays sufficiently rapidly at infinity and satisfies $\Phi(0) = 1$. For instance, we may take $\Phi = \chi_{B(0,1)}$. Analogous summability methods can be introduced on the torus.

A very interesting case arises when $\Phi(\xi) = (1-|\xi|^2)_+^\lambda$ for some $\lambda \geq 0$ in which we obtain the Bochner-Riesz means introduced by Riesz when $n = 1$ and $\lambda = 0$ and Bochner for $n \geq 2$ and general $\lambda > 0$. The main question of interest here is whether the Bochner-Riesz means

$$\sum_{m_1^2+\cdots+m_n^2 \leq R^2} \left(1 - \frac{m_1^2+\cdots+m_n^2}{R^2}\right)^\lambda \widehat{f}(m_1,\dots,m_n)\, e^{2\pi i(m_1x_1+\cdots+m_nx_n)}$$

converge in L^p and almost everywhere to f as $R\to\infty$. As we have seen in Chapter 3, the convergence of the previous series in $L^p(\mathbf{T}^n)$ as $R\to\infty$ is equivalent to the fact that the function $(1-|\xi|^2)_+^\lambda$ is an L^p multiplier on $\mathbf{R}^n$. In this chapter we will show that when $\lambda = 0$, the function $\chi_{B(0,1)}$ is not an L^p multiplier on $\mathbf{R}^n$ for $n \geq 2$ unless $p = 2$. This settles the question of norm convergence for $\lambda = 0$. The analogous almost everywhere problem as well as the corresponding questions for other $\lambda > 0$ will be investigated in this chapter.

10.1. The Multiplier Problem for the Ball

In this section we will show that the characteristic function of the unit disc in $\mathbf{R}^2$ is not an L^p multiplier when $p \neq 2$. We observe that this suffices to establish the same conclusion in dimensions $n \geq 3$ since by Theorem 2.5.16 we have that if $\chi_{B(0,1)} \notin \mathcal{M}_p(\mathbf{R}^2)$, then $\chi_{B(0,1)} \notin \mathcal{M}_p(\mathbf{R}^n)$ for any $n \geq 3$.

We begin with a certain geometric construction that at first sight has no apparent relationship with the multiplier problem for the ball in $\mathbf{R}^n$. Given a triangle ABC with base $b = AB$ and height h_0 we let M be the midpoint of AB. We construct two other triangles AMF and BME from ABC as follows. We fix a height $h_1 > h_0$ and we extend the sides AC and BC in the direction away from its base until they reach a certain height h_1. We let E be the unique point on the line passing through the points B and C such that the triangle EMB has height h_1. Similarly, F is uniquely chosen on the line through A and C so that the triangle AMF has height h_1.

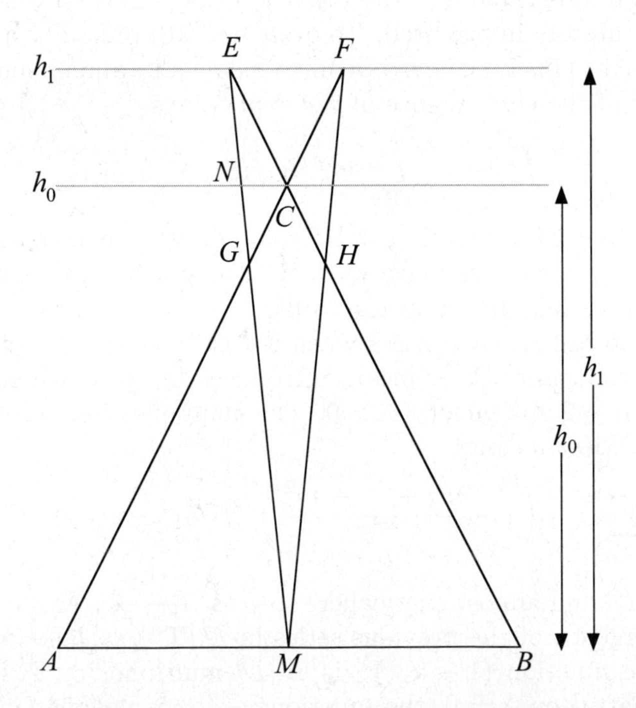

FIGURE 10.1. The sprouting of the triangle ABC.

The triangle ABC now gives rise to two triangles AMF and BME called the *sprouts* of ABC. The union of the two sprouts AMF and BME is called the *sprouted figure* obtained from ABC and is denoted by $\mathrm{Spr}(ABC)$. Clearly $\mathrm{Spr}(ABC)$ contains ABC. We call the difference

$$\mathrm{Spr}(ABC) \setminus ABC$$

the arms of the sprouted figure. The sprouted figure $\mathrm{Spr}(ABC)$ has two arms of equal area, the triangles EGC and FCH as shown in Figure 10.1, and we can precisely compute the area of each arm. We have the fact that

$$\text{Area (each arm of } \mathrm{Spr}(ABC)) = \frac{b}{2}\frac{(h_1 - h_0)^2}{2h_1 - h_0}, \tag{10.1.1}$$

where $b = AB$. The proof of this fact can be shown easily using similar triangles and is left to the reader. See Exercise 10.1.1.

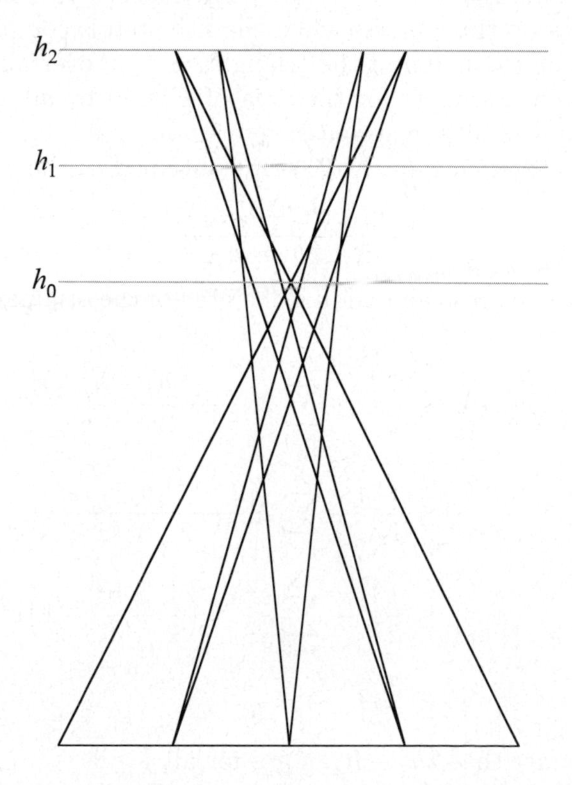

FIGURE 10.2. The second step of the construction.

We now start with an isosceles triangle $\Lambda = ABC$ in $\mathbf{R}^2$ with base AB of side length $b_0 = \varepsilon$ and height $MC = h_0 = \varepsilon$, where as before M is the midpoint of AB.

We define the sequence of heights

$$\begin{aligned} h_1 &= (1+\frac{1}{2})\varepsilon \\ h_2 &= (1+\frac{1}{2}+\frac{1}{3})\varepsilon \\ &\dots \\ h_j &= (1+\frac{1}{2}+\dots+\frac{1}{j+1})\varepsilon. \end{aligned}$$

We apply the previously described sprouting procedure to Λ to obtain two sprouts $\Lambda_1 = AMF$ and $\Lambda_2 = EMB$, as in Figure 10.1, each with height h_1 and base length $b_0/2$. We now apply the same procedure to the triangles Λ_1 and Λ_2. We then obtain two sprouts Λ_{11} and Λ_{12} from Λ_1 and two sprouts Λ_{21} and Λ_{22} from Λ_2, a total of four sprouts with height h_2. See Figure 10.2. We continue this process obtaining at the jth step 2^j sprouts $\Lambda_{r_1\dots r_j}$, $r_1,\dots,r_j \in \{1,2\}$ each with base length $b_j = 2^{-j}b_0$ and height h_j. We stop this process when the kth step is completed.

We let $E(\varepsilon,k)$ be the union of the triangles $\Lambda_{r_1\dots r_k}$ over all sequences r_j of 1's and 2's. We obtain an estimate for the area of $E(\varepsilon,k)$ by adding to the area of Λ the areas of the arms of all the sprouted figures obtained during the construction. By (10.1.1) we have that each of the 2^j arms obtained at the jth step has area

$$\frac{b_{j-1}}{2}\frac{(h_j-h_{j-1})^2}{2h_j-h_{j-1}}.$$

Summing over all these areas and adding the area of the original triangle, we obtain the estimate

$$\begin{aligned} |E(\varepsilon,k)| &\le \frac{1}{2}\varepsilon^2 + \sum_{j=1}^{k} 2^j \frac{b_{j-1}}{2}\frac{(h_j-h_{j-1})^2}{2h_j-h_{j-1}} \\ &\le \frac{1}{2}\varepsilon^2 + \sum_{j=1}^{k} 2^j \frac{2^{-(j-1)}b_0}{2}\frac{\varepsilon^2}{(j+1)^2\varepsilon} \\ &\le \frac{1}{2}\varepsilon^2 + \sum_{j=2}^{\infty} \frac{\varepsilon^2}{j^2} \le \Big(\frac{1}{2}+\frac{\pi^2}{6}-1\Big)\varepsilon^2 \\ &\le \frac{3}{2}\varepsilon^2, \end{aligned}$$

where we used the fact that $2h_j - h_{j-1} \ge \varepsilon$ for all $j \ge 1$.

Having completed the construction of the set $E(\varepsilon,k)$, we are now in a position to indicate some of the ideas that appear in the solution of the Kakeya problem. We first observe that no matter what k is, the measure of the set $E(\varepsilon,k)$ can be made as small as we wish if we take ε small enough. Our purpose is to make a needle of infinitesimal width and unit length move continuously from one side of this angle

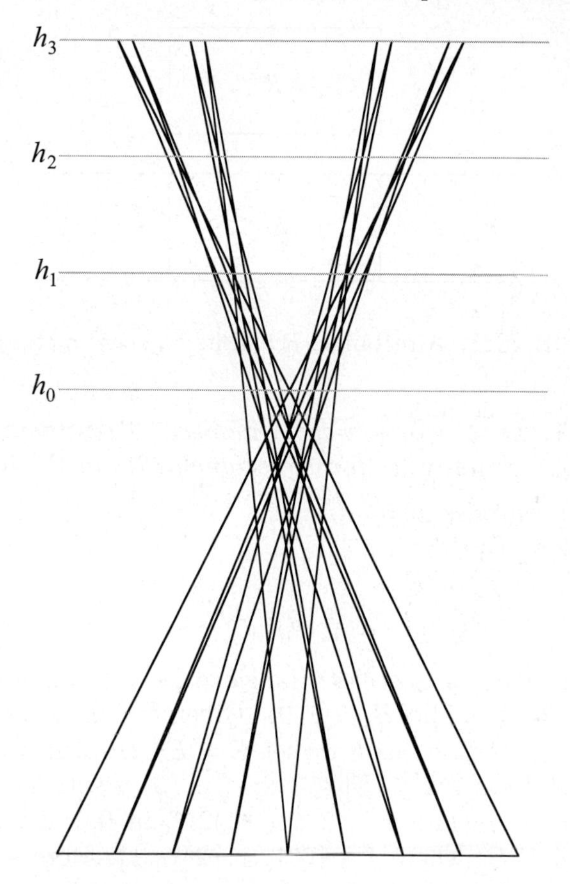

FIGURE 10.3. The third step of the construction.

to the other by utilizing each sprouted triangle in succession. To achieve this, we need to apply a similar construction to any of the 2^k triangles that make up the set $E(\varepsilon, k)$ and repeat the sprouting procedure a large enough number of times. We refer to [**137**] for details. An elaborate construction of this sort will only yield a set within which the needle can be turned through a fixed angle. But adjoining a few such sets together we can rotate a needle through a half-turn within a set that still has arbitrarily small area. This is the idea used to solve the aforementioned needle problem.

We now return to the multiplier problem for the ball, which has an interesting connection with the Kakeya needle problem.

In the discussion that follows we will need the following notation. Given a rectangle R in $\mathbf{R}^2$, we let R' be two copies of R adjacent to R along its shortest side so that $R \cup R'$ has the same width as R but three times its length. See Figure 10.4.

We will need the following lemma.

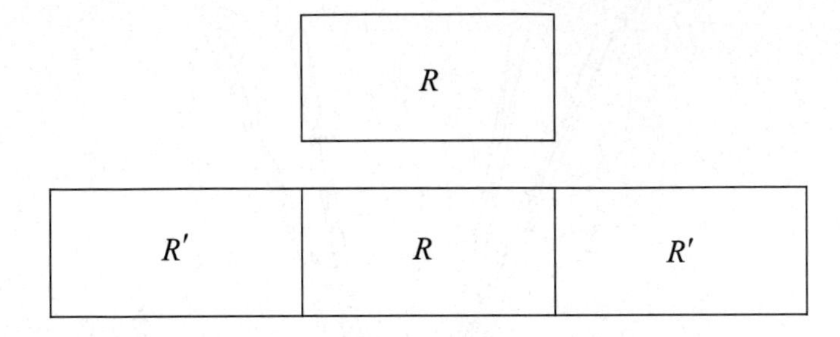

FIGURE 10.4. A rectangle R and its adjacent rectangles R'.

Lemma 10.1.1. *Let $\delta > 0$ be a given number. Then there exists a measurable subset E of $\mathbf{R}^2$ and a finite collection of rectangles R_j in $\mathbf{R}^2$ in such that*

(1) *The R_j's are pairwise disjoint.*
(2) *We have $1/2 \le |E| \le 3/2$.*
(3) *We have $|E| \le \delta \sum_j |R_j|$.*
(4) *For all j we have $|R_j' \cap E| \ge \frac{1}{12}|R_j|$.*

Proof. We start with an isosceles triangle ABC in the plane with height 1 and base AB, where $A = (0,0)$ and $B = (1,0)$. Given $\delta > 0$, we find a positive integer k such that $k+2 > e^{1/\delta}$. For this k we set $E = E(1,k)$, the set constructed earlier with $\varepsilon = 1$. We then have $1/2 \le |E| \le 3/2$; thus (3) is satisfied.

Recall that each dyadic interval $[j2^{-k}, (j{+}1)2^{-k}]$ in $[0,1]$ is the base of exactly one sprouted triangle $A_jB_jC_j$ where $j \in \{0, 1, \dots, 2^k - 1\}$. Here we set $A_j = (j2^{-k}, 0)$ and $B_j = ((j+1)2^{-k}, 0)$ and C_j the other vertex of the sprouted rectangle. We define a rectangle R_j inside the angle $\angle A_jC_jB_j$ as in Figure 10.6. R_j is defined so that one of its vertices is either A_j or B_j and the length of its longest side is $3\log(k+2)$.

We now make some calculations. First we observe that the longest possible length that either A_jC_j or B_jC_j can achieve is $\sqrt{5}h_k/2$. By symmetry we may assume that the length of A_jC_j is larger than that of B_jC_j as in Figures 10.5 and 10.6. We now have that

$$\frac{\sqrt{5}}{2}h_k < \frac{3}{2}\Big(1 + \frac{1}{2} + \cdots + \frac{1}{k+1}\Big) < \frac{3}{2}(1 + \log(k+1)) < 3\log(k+2),$$

since $k \ge 1$ and $e < 3$. Hence R_j' contains the triangle $A_jB_jC_j$. We also have that

$$h_k = 1 + \frac{1}{2} + \cdots + \frac{1}{k+1} > \log(k+2).$$

Using these two facts, we obtain

$$|R_j' \cap E| \ge \text{Area}(A_jB_jC_j) = \frac{1}{2}2^{-k}h_k > 2^{-k-1}\log(k+2). \tag{10.1.2}$$

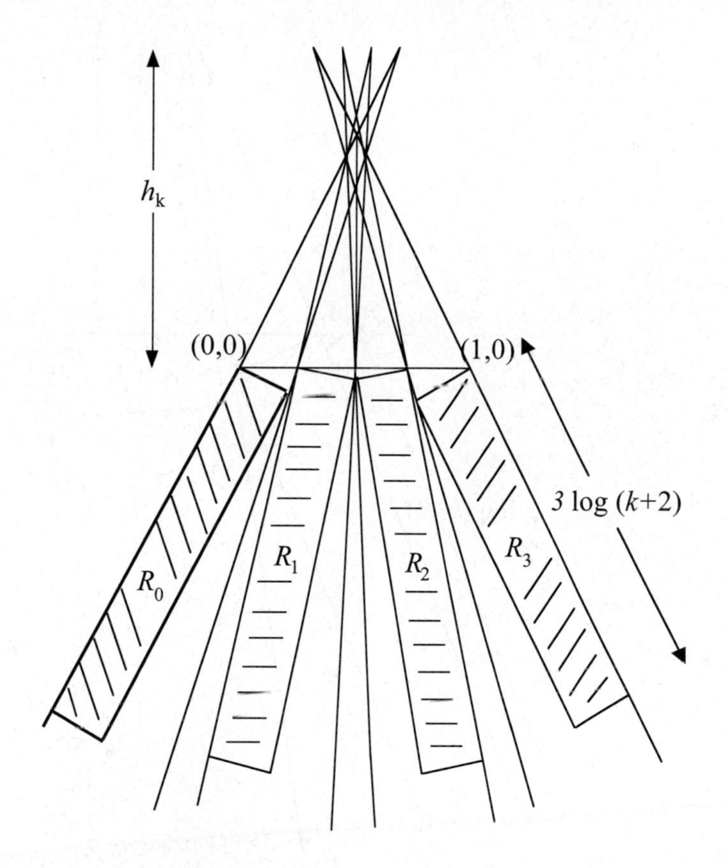

FIGURE 10.5. The rectangles R_j.

Denote by $|XY|$ the length of the line segment through the points X and Y. The law of sines applied to the triangle $A_jB_jD_j$ gives

$$|A_jD_j| = 2^{-k}\frac{\sin(\angle A_jB_jD_j)}{\sin(\angle A_jD_jB_j)} \le \frac{2^{-k}}{\cos(\angle A_jC_jB_j)}\,. \tag{10.1.3}$$

But the law of cosines applied to the triangle $A_jB_jC_j$ combined with the estimates $h_k \le |A_jC_j|, |B_jC_j| \le \sqrt{5}h_k/2$ give that

$$\cos(\angle A_jC_jB_j) \ge \frac{h_k^2 + h_k^2 - (2^{-k})^2}{2\,\frac{5}{4}\,h_k^2} \ge \frac{4}{5} - \frac{2}{5}\cdot\frac{1}{4} \ge \frac{1}{2}\,. \tag{10.1.4}$$

Combining (10.1.3) and (10.1.4), we obtain

$$|A_jD_j| \le 2^{-k+1} = 2|A_jB_j|\,.$$

Using this fact and (10.1.2), we deduce

$$|R'_j \cap E| \ge 2^{-k-1}\log(k+2) = \frac{1}{12}2^{-k+1}3\log(k+2) \ge \frac{1}{12}|R_j|\,,$$

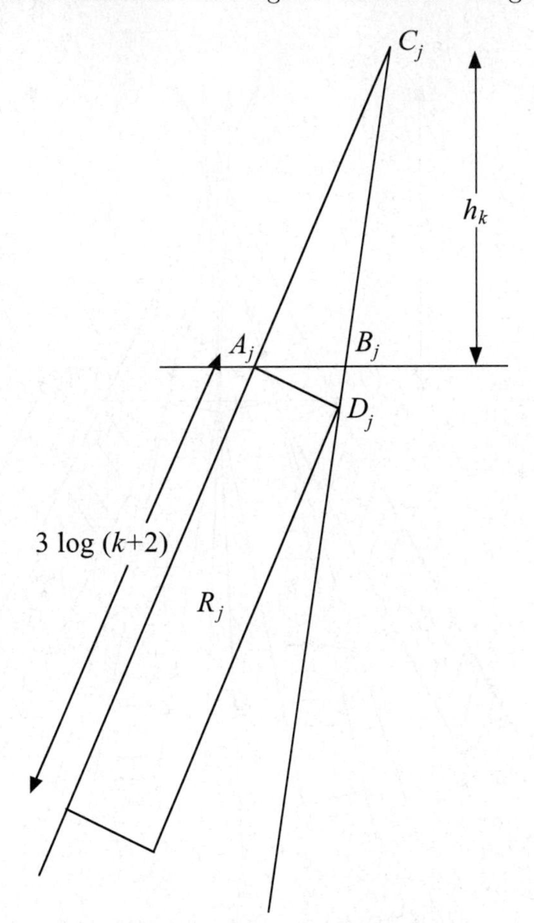

FIGURE 10.6. A closer look at R_j.

which proves the required conclusion (4).

Conclusion (1) in Lemma 10.1.1 follows from the fact that the regions inside the angles $A_jC_jB_j$ under the triangles ABC are pairwise disjoint. This is shown in Figure 10.6. This can be proved rigorously by a careful examination of the construction of the sprouted triangles $A_jC_jB_j$, but the details are left to the reader.

It remains to prove (3). To achieve this we first estimate the length of the line segment A_jD_j from below. The law of sines gives

$$\frac{|A_jD_j|}{\sin(\angle A_jB_jD_j)} = \frac{2^{-k}}{\sin(\angle A_jD_jB_j)},$$

from which we obtain that $|A_jD_j| \geq 2^{-k}\sin(\angle A_jB_jD_j) \geq 2^{-k-1}\angle A_jB_jD_j \geq 2^{-k-1}\angle B_jA_jC_j$. (All angles are measured in radians.) But the smallest possible value of the angle $\angle B_jA_jC_j$ is attained when $j = 0$, in which case $\angle B_0A_0C_0 =$

$\arctan 2 > 1$. This gives that

$$|A_jD_j| \geq 2^{-k-1}.$$

It follows that each R_j has area at least $2^{-k-1}3\log(k+2)$. Therefore,

$$\left|\bigcup_{j=0}^{2^k-1} R_j\right| = \sum_{j=0}^{2^k-1} |R_j| = 2^k 2^{-k-1} 3\log(k+2) \geq |E|\log(k+2) \geq \frac{|E|}{\delta}$$

since k was chosen so that $k+2 > e^{1/\delta}$ and $|E| \leq 3/2$.

□

Next we have a calculation involving the Fourier transforms of characteristic functions of rectangles.

Proposition 10.1.2. *Let R be a rectangle whose center is the origin in $\mathbf{R}^2$ and let v be a unit vector parallel to its longest side. Consider the half-plane*

$$\mathcal{H} = \{x \in \mathbf{R}^2 : \ x \cdot v \geq 0\}$$

and the multiplier operator

$$S_{\mathcal{H}}(f) = (\widehat{f}\chi_{\mathcal{H}})^{\vee}.$$

Then we have $|S_{\mathcal{H}}(\chi_R)| \geq \frac{1}{10}\chi_{R'}$.

Remark 10.1.3. Applying a translation, we see that the same conclusion is valid for any rectangle in $\mathbf{R}^2$ if $\mathcal{H}$ is defined as one of the two half-planes formed by the line passing through its center and parallel to its shortest side.

Proof. Applying a rotation, we reduce the problem to the case where $R = [-a,a] \times [-b,b]$, where $0 < a \leq b < \infty$, and $v = e_2 = (0,1)$. Since the Fourier transform acts in each variable independently, we have the identity

$$\begin{aligned} S_{\mathcal{H}}(\chi_R)(x_1,x_2) &= \chi_{[-a,a]}(x_1)\big(\widehat{\chi_{[-b,b]}}\chi_{[0,\infty)}\big)^{\vee}(x_2) \\ &= \chi_{[-a,a]}(x_1)\,\frac{I+iH}{2}(\chi_{[-b,b]})(x_2). \end{aligned}$$

It follows that

$$\begin{aligned} |S_{\mathcal{H}}(\chi_R)(x_1,x_2)| &\geq \frac{1}{2}\chi_{[-a,a]}(x_1)|H(\chi_{[-b,b]})(x_2)| \\ &= \frac{1}{2\pi}\chi_{[-a,a]}(x_1)\left|\log\left|\frac{x_2+b}{x_2-b}\right|\right|. \end{aligned}$$

But for $(x_1,x_2) \in R'$, $\chi_{[-a,a]}(x_1) = 1$, and $b < |x_2| < 3b$. So we have two cases, $b < x_2 < 3b$ and $-3b < x_2 < -b$. When $b < x_2 < 3b$ we see that

$$\left|\frac{x_2+b}{x_2-b}\right| = \frac{x_2+b}{x_2-b} > 2$$

and, similarly, when $-3b < x_2 < -b$ we have

$$\left|\frac{x_2 - b}{x_2 + b}\right| = \frac{b - x_2}{-b - x_2} > 2.$$

It follows that for $(x_1, x_2) \in R'$ the lower estimate is valid:

$$|S_{\mathcal{H}}(\chi_R)(x_1, x_2)| \geq \frac{\log 2}{2\pi} \geq \frac{1}{10}.$$

□

Next we have a lemma regarding vector-valued inequalities of half-plane multipliers.

Lemma 10.1.4. *Let $v_1, v_2, \dots, v_j, \dots$ be a sequence of unit vectors in $\mathbf{R}^2$. Define the half-planes*

$$\mathcal{H}_j = \{x \in \mathbf{R}^2 : \ x \cdot v_j \geq 0\} \tag{10.1.5}$$

and linear operators

$$S_{\mathcal{H}_j}(f) = (\widehat{f}\, \chi_{\mathcal{H}_j})^\vee.$$

Assume that the disc multiplier operator

$$T(f) = (\widehat{f}\, \chi_{B(0,1)})^\vee$$

maps $L^p(\mathbf{R}^2)$ into itself with norm $B_p < \infty$. Then we have the inequality

$$\Big\| \Big(\sum_j |S_{\mathcal{H}_j}(f_j)|^2 \Big)^{\frac{1}{2}} \Big\|_{L^p} \leq B_p \Big\| \Big(\sum_j |f_j|^2 \Big)^{\frac{1}{2}} \Big\|_{L^p} \tag{10.1.6}$$

for all bounded and compactly supported functions f_j.

Proof. We prove the lemma for Schwartz functions f_j and we obtain the general case by a simple limiting argument. We let $D_{j,R}$ be the discs $\{x \in \mathbf{R}^2 : \ |x - Rv_j| \leq R\}$. We also let

$$T_{j,R}(f) = (\widehat{f}\, \chi_{D_{j,R}})^\vee$$

be the multiplier operator associated to the disc $D_{j,R}$. We observe that $\chi_{D_{j,R}} \to \chi_{\mathcal{H}_j}$ pointwise as $R \to \infty$, as shown in Figure 10.7. For $f \in \mathcal{S}(\mathbf{R}^2)$ and every $x \in \mathbf{R}^2$ we have

$$\lim_{R\to\infty} T_{j,R}(f)(x) = S_{\mathcal{H}_j}(f)(x)$$

by passing the limit inside the convergent integral. Fatou's lemma now yields

$$\Big\| \Big(\sum_j |S_{\mathcal{H}_j}(f_j)|^2 \Big)^{\frac{1}{2}} \Big\|_{L^p} \leq \liminf_{R\to\infty} \Big\| \Big(\sum_j |T_{j,R}(f_j)|^2 \Big)^{\frac{1}{2}} \Big\|_{L^p}. \tag{10.1.7}$$

Next we observe that the following identity is valid:

$$T_{j,R}(f)(x) = e^{2\pi i R v_j \cdot x} T_R(e^{-2\pi i R v_j \cdot (\cdot)} f)(x), \tag{10.1.8}$$

where T_R is the multiplier operator

$$T_R(f) = (\widehat{f}\chi_{B(0,R)})^\vee.$$

Setting $g_j = e^{-2\pi i R v_j \cdot (\cdot)} f_j$ and using (10.1.7) and (10.1.8), we deduce

$$\Big\|\Big(\sum_j |S_{\mathcal{H}_j}(f_j)|^2\Big)^{\frac{1}{2}}\Big\|_{L^p} \le \liminf_{R\to\infty} \Big\|\Big(\sum_j |T_R(g_j)|^2\Big)^{\frac{1}{2}}\Big\|_{L^p}. \tag{10.1.9}$$

Observe that the operator T_R is L^p bounded with the same norm B_p as T in view of identity (2.5.15). Applying the Theorem 4.5.1 we obtain that the last term in (10.1.9) is bounded by

$$\liminf_{R\to\infty} \|T_R\|_{L^p\to L^p} \Big\|\Big(\sum_j |g_j|^2\Big)^{\frac{1}{2}}\Big\|_{L^p} = B_p \Big\|\Big(\sum_j |f_j|^2\Big)^{\frac{1}{2}}\Big\|_{L^p}.$$

Combining this inequality with (10.1.9), we obtain (10.1.6).

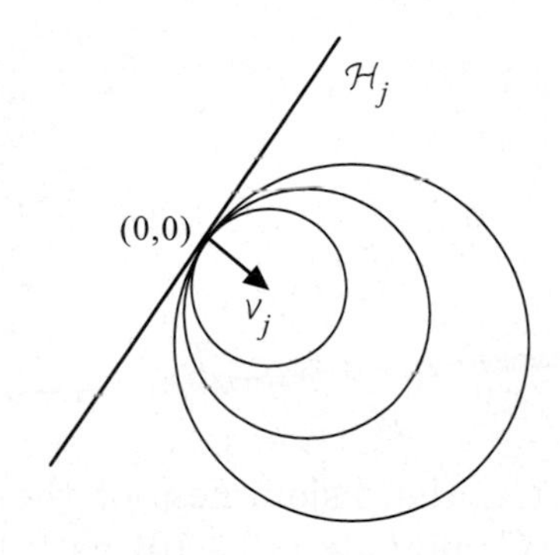

FIGURE 10.7. A sequence of discs converging to a half-plane.

□

We have now completed all the preliminary material we need to prove that the characteristic function of the unit disc in $\mathbf{R}^2$ is not an L^p multiplier if $p \neq 2$.

Theorem 10.1.5. *The characteristic function of the unit ball in $\mathbf{R}^n$ is not an L^p multiplier when $1 < p \neq 2 < \infty$.*

Proof. As mentioned earlier, in view of Theorem 2.5.16, it suffices to prove the result in dimension $n = 2$. By duality is suffices to prove the result when $p > 2$. Suppose that $\chi_{B(0,1)} \in \mathcal{M}_p(\mathbf{R}^2)$ for some $p > 2$, say with norm $B_p < \infty$.

Suppose that $\delta > 0$ is given. Let E and R_j be as in Lemma 10.1.1. We let $f_j = \chi_{R_j}$. Let v_j be the unit vector parallel to the long side of R_j and H_j be the half-plane defined as in (10.1.5). Using Proposition 10.1.2, we obtain

$$
\begin{aligned}
\int_E \sum_j |S_{\mathcal{H}_j}(f_j)(x)|^2\,dx &= \sum_j \int_E |S_{\mathcal{H}_j}(f_j)(x)|^2\,dx \\
&\geq \sum_j \int_E \frac{1}{10^2}\chi_{R'_j}(x)\,dx \\
&= \frac{1}{100}\sum_j |E\cap R'_j| \\
&\geq \frac{1}{1200}\sum_j |R_j|,
\end{aligned} \tag{10.1.10}
$$

where we used condition (4) of Lemma 10.1.1 in the last inequality. Hölder's inequality with exponents $p/2$ and $(p/2)' = p/(p-2)$ gives

$$
\begin{aligned}
\int_E \sum_j |S_{\mathcal{H}_j}(f_j)(x)|^2\,dx &\leq |E|^{\frac{p-2}{p}} \Big\| \Big(\sum_j |S_{\mathcal{H}_j}(f_j)|^2\Big)^{\frac12}\Big\|_{L^p}^2 \\
&\leq B_p|E|^{\frac{p-2}{p}} \Big\| \Big(\sum_j |f_j|^2\Big)^{\frac12}\Big\|_{L^p}^2 \\
&\leq B_p|E|^{\frac{p-2}{p}} \Big(\sum_j |R_j|\Big)^{\frac{2}{p}} \\
&\leq B_p\delta^{\frac{p-2}{p}} \sum_j |R_j|,
\end{aligned} \tag{10.1.11}
$$

where we used Lemma 10.1.4, the disjointness of the R_j's, and condition (3) of Lemma 10.1.1 successively. Combining (10.1.10) with (10.1.11), we obtain the inequality

$$
\sum_j |R_j| \leq 1200\, B_p\, \delta^{\frac{p-2}{p}} \sum_j |R_j|,
$$

which provides a contradiction when δ is very small. □

Exercises

10.1.1. Prove identity (10.1.1).
[*Hint:* With the notation of Figure 10.1, first prove

$$
\frac{h_1 - h_0}{h_1} = \frac{NC}{b/2}, \qquad \frac{\text{height}\,(NGC)}{h_0} = \frac{NC}{NC + b/2}
$$

using similar triangles.]

10.1.2. Given a rectangle R, let R'' denote any of the two parts that make up R'. Prove that for any $k \in \mathbf{Z}^+$ and any $\delta > 0$, there exist rectangles S_j in $\mathbf{R}^2$, $0 \le j < 2^k$, such that each S_j has dimensions $2^{-k} \times 1$,

$$\left| \bigcup_{j=0}^{2^k-1} S_j \right| < \delta$$

and for some choice of S_j'', the S_j'''s are disjoint.
[*Hint:* Consider the 2^k triangles that make up the set $E(\varepsilon, k)$ and choose each rectangle S_j inside a corresponding triangle. Then the parts of the S_j''s that point downward are disjoint. Choose ε depending on δ.]

10.1.3. Is the characteristic function of the cylinder

$$\{(\xi_1, \xi_2, \xi_3) \in \mathbf{R}^3 : \ \xi_1^2 + \xi_2^2 < 1\}$$

a Fourier multiplier on $L^p(\mathbf{R}^3)$ for any $1 < p < \infty$?

10.1.4. Modify the ideas of the proof of Theorem 10.1.5 to show that the characteristic function of the set

$$\{(\xi, \eta) \in \mathbf{R}^2 : \ \xi > \eta^2\}$$

is not in $\mathcal{M}_p(\mathbf{R}^2)$ when $p \neq 2$.

10.1.5. Let $a_1, \dots, a_n > 0$. Show that the characteristic function of the ellipsoids

$$\left\{(\xi, \dots, \xi_n) \in \mathbf{R}^n : \ \frac{\xi_1^2}{a_1^2} + \cdots + \frac{\xi_n^2}{a_n^2} < 1\right\}$$

is not in $\mathcal{M}_p(\mathbf{R}^n)$ when $p \neq 2$.
[*Hint:* Think about dilations.]

10.2. Bochner-Riesz Means and the Carleson-Sjölin Theorem

We now address the problem of norm convergence for the Bochner-Riesz means. In this section we will provide a satisfactory answer in dimension $n = 2$, although a key ingredient required in the proof will be left for the next section.

Definition 10.2.1. For a function f on $\mathbf{R}^n$ we define its *Bochner-Riesz means* of order $\lambda > 0$ to be the family of operators

$$B_R^\lambda(f)(x) = \int_{\mathbf{R}^n} (1 - |\xi/R|^2)_+^\lambda \widehat{f}(\xi) e^{2\pi i x \cdot \xi} \, d\xi\,, \quad R > 0.$$

We will be interested in the convergence properties of the operators $B_R^\lambda(f)$ as $R \to \infty$. Observe that when $R \to \infty$ and f is a Schwartz function, the sequence $B_R^\lambda(f)$ converges pointwise to f. Does it also converge in norm? This is the main question addressed here. Using Exercise 10.2.1, this is equivalent to showing that the function $(1-|\xi|^2)_+^\lambda$ is an L^p multiplier [it lies in $\mathcal{M}_p(\mathbf{R}^n)$], that is, the linear operator

$$B^\lambda(f)(x) = \int_{\mathbf{R}^n} (1-|\xi|^2)^\lambda \widehat{f}(\xi) e^{2\pi i x\cdot\xi}\, d\xi$$

maps $L^p(\mathbf{R}^n)$ into itself. The precise range of λ and p for which this is the case will be investigated in this section.

Recall that the analogous question for the operators B_R^λ introduced in Definition 3.4.1 is also equivalent to the fact that the function $(1-|\xi|^2)_+^\lambda$ is a Fourier multiplier in $\mathcal{M}_p(\mathbf{R}^n)$. This was shown in Corollary 3.6.10. Therefore the Bochner-Riesz problem for the torus $\mathbf{T}^n$ and the Euclidean space $\mathbf{R}^n$ are completely equivalent. We start our investigation of the Bochner-Riesz problem on $\mathbf{R}^n$ by studying the kernel of the operator B^λ.

10.2.a. The Bochner-Riesz Kernel and Simple Estimates

In view of the last identity in Appendix B.5, B^λ is a convolution operator with kernel

$$K_\lambda(x) = \frac{\Gamma(\lambda+1)}{\pi^\lambda} \frac{J_{\frac{n}{2}+\lambda}(2\pi|x|)}{|x|^{\frac{n}{2}+\lambda}}. \tag{10.2.1}$$

Using the asymptotics for Bessel functions in Appendix B.6, we see that K_λ is a smooth function that is equal to

$$\sqrt{2}\,\frac{\Gamma(\lambda+1)}{\pi^{\lambda+\frac{1}{2}}} \frac{\cos(2\pi|x| - \frac{\pi(n+1)}{4} - \frac{\pi\lambda}{2})}{|x|^{\frac{n+1}{2}+\lambda}} + O(|x|^{-\frac{n+3}{2}-\lambda}) \tag{10.2.2}$$

as $|x| \to \infty$. Thus, when $\lambda > \frac{n-3}{2}$, the error term in (10.2.2) is contributing an integrable term on $\mathbf{R}^n$, and we need to focus on understanding the contribution of the main term in (10.2.2). This can be written as

$$c(n,\lambda)\frac{e^{2\pi i|x|}}{|x|^{\frac{n+1}{2}+\lambda}} + \overline{c(n,\lambda)}\frac{e^{-2\pi i|x|}}{|x|^{\frac{n+1}{2}+\lambda}},$$

where $c(n,\lambda)$ is a complex constant. When $\lambda > \frac{n-1}{2}$, then K_λ is a smooth integrable function on $\mathbf{R}^n$ and it easily follows that B^λ is a bounded operator on the L^p spaces, $1 \le p \le \infty$. We state this result.

Proposition 10.2.2. *For all $1 \le p \le \infty$ and $\lambda > \frac{n-1}{2}$, B^λ is a bounded operator on $L^p(\mathbf{R}^n)$.*

Proof. Obvious. □

We refer the reader to Exercise 10.2.7 for an analogous result for the maximal Bochner-Riesz operator.

It is natural to examine whether the operators B^λ are bounded on certain L^p spaces by testing them on specific functions. This sort of test will give some indication for which range of p's these operators are expected to be bounded.

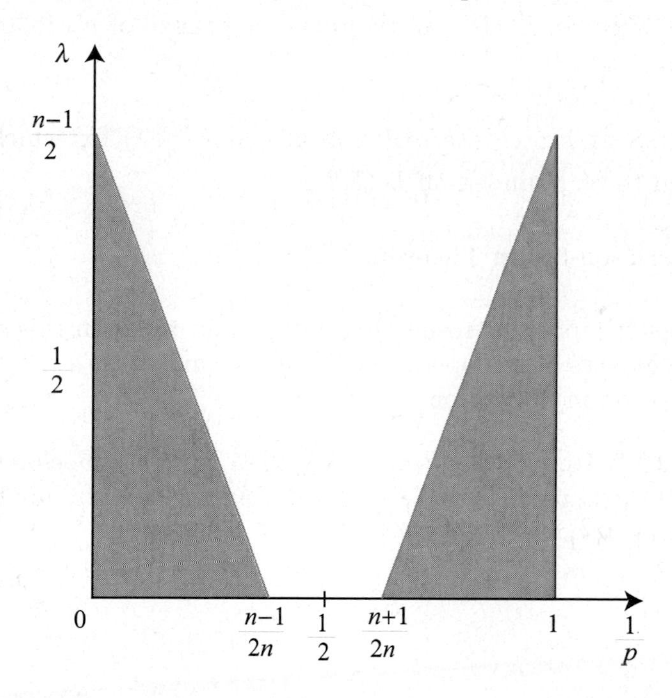

FIGURE 10.8. The operators B^λ are unbounded on $L^p(\mathbf{R}^n)$ when the pair $(1/p, \lambda)$ lies in the shaded region.

Proposition 10.2.3. *When $\lambda > 0$ and $p \le \frac{2n}{n+1+2\lambda}$ or $p \ge \frac{2n}{n-1+2\lambda}$, the operators B^λ are not bounded on $L^p(\mathbf{R}^n)$.*

Proof. Let h be a Schwartz function whose Fourier transform is equal to 1 on the ball $B(0,2)$ and vanishes off the ball $B(0,3)$. Then

$$B^\lambda(h)(x) = \int_{|\xi|\le 1} (1-|\xi|^2)^\lambda e^{2\pi i \xi\cdot x}\, dx = K_\lambda(x)$$

and it suffices to show that K_λ is not in $L^p(\mathbf{R}^n)$ for the specified range of $p's$. But we have

$$|K_\lambda(x)|^p \approx |x|^{-p\frac{n+1}{2}-p\lambda} + O(|x|^{-p\frac{n+3}{2}-p\lambda}) \tag{10.2.3}$$

for all $|x|$ satisfying

$$k + \frac{n+\lambda}{8} \le |x| \le k + \frac{n+\lambda}{8} + \frac{1}{4}$$

for positive large integers k. Indeed, in this range of x's the cosine in (10.2.2) stays away from zero and (10.2.3) follows easily. Now observe that the error term in (10.2.3) is of lower order than the main term, which implies that it has no influence in the behavior of $|K_\lambda(x)|^p$ at infinity. Putting these observations together we conclude that $|K_\lambda(x)|^p$ is not integrable when $-p\frac{n+1}{2}-p\lambda \geq -n$ (i.e., when $p \leq \frac{2n}{n+1+2\lambda}$). The unboundedness of B^λ on $L^p(\mathbf{R}^n)$ in the remaining range of p's follows by duality.

□

In Figure 10.8 we sketch the region of all pairs $(\frac{1}{p}, \lambda)$ for which the operators B^λ are expected to be bounded on $L^p(\mathbf{R}^n)$.

10.2.b. The Carleson-Sjölin Theorem

We now pass to a positive result that is the main result in this section. We will prove the boundedness of the operators B^λ in the range of p's not excluded by the previous proposition in dimension $n = 2$.

Theorem 10.2.4. *Suppose that* $0 < \lambda \leq 1/2$. *Then the Bochner-Riesz operator* B^λ *maps* $L^p(\mathbf{R}^2)$ *into itself when* $\frac{4}{3+2\lambda} < p < \frac{4}{1-2\lambda}$. *Moreover, for this range of p's and for all* $f \in L^p(\mathbf{R}^2)$ *we have that*

$$B_R^\lambda(f) \to f$$

in $L^p(\mathbf{R}^2)$ *as* $R \to \infty$.

Proof. Once the first assertion of the theorem is established, the second assertion will be a direct consequence of it and of the fact that the means $B_R^\lambda(h)$ converge to h in L^p for h in a dense subclass of L^p. Such a dense class is, for instance, the class of all Schwartz functions h whose Fourier transforms are compactly supported (Exercise 5.1.7). For a function h in this class, we can see easily that $B_R^\lambda(h) \to h$ pointwise. But if $\widehat{h}$ is supported in $|\xi| \leq c$, then for $R \geq 2c$, integration by parts gives that the functions $B_R^\lambda(h)(x)$ are pointwise controlled by the function $(1+|x|)^{-N}$ with N large; then the Lebesgue dominated convergence theorem gives that $B_R^\lambda(h)$ converge to h in L^p. Finally, a standard $\varepsilon/3$ argument [using that $\sup_{R>0} \big\|B_R^\lambda\big\|_{L^p\to L^p} = \big\|(1-|\xi|^2)_+^\lambda\big\|_{\mathcal{M}_p} < \infty$] yields $B_R^\lambda(f) \to f$ in L^p for general L^p functions f.

It suffices therefore to focus our attention on the first part of the theorem. Although the number λ in the statement of the theorem is real, for purposes of complex interpolation, we will work with a complex number λ with positive real part. So, until the end of the proof of the theorem, λ will be a complex number and all the estimates will grow at most exponentially in $|\mathrm{Im}\,\lambda|^2$. [This is because of the presence of the gamma function in the kernel K_λ in (10.2.1).]

We start by picking a smooth function φ supported in $[-\frac{1}{2}, \frac{1}{2}]$ and a smooth function ψ supported in $[\frac{1}{8}, \frac{5}{8}]$ that satisfy

$$\varphi(t) + \sum_{k=0}^{\infty} \psi\Big(\frac{1-t}{2^{-k}}\Big) = 1$$

for all $t \in [0,1]$. We now decompose the multiplier $(1-|\xi|^2)_+^\lambda$ as

$$(1-|\xi|^2)_+^\lambda = m_{00}(\xi) + \sum_{k=0}^{\infty} 2^{-k\lambda} m_k(\xi)\,, \tag{10.2.4}$$

where $m_{00}(\xi) = \varphi(|\xi|)(1-|\xi|^2)^\lambda$ and for $k \ge 0$, m_k is defined by

$$m_k(\xi) = \Big(\frac{1-|\xi|}{2^{-k}}\Big)^\lambda \psi\Big(\frac{1-|\xi|}{2^{-k}}\Big)(1+|\xi|)^\lambda\,.$$

Note that m_{00} is a smooth function with compact support; hence the multiplier m_{00} lies in $\mathcal{M}_p(\mathbf{R}^2)$ for all $1 \le p \le \infty$. Each function m_k is also smooth, radial, and supported in the small annulus

$$1 - \tfrac{5}{8}\, 2^{-k} \le |\xi| \le 1 - \tfrac{1}{8}\, 2^{-k}$$

and therefore also lies in $\mathcal{M}_p$; nevertheless the $\mathcal{M}_p$ norms of the m_k's grow as k increases and it will be crucial for us to determine how this growth depends on k so that we can sum the series in (10.2.4). We observe that

$$|\partial^\gamma m_k(\xi)| \le C_\gamma 2^{k|\gamma|}$$

for all $\xi \in \mathbf{R}^2$ and all multiindices γ.

We will show that the multiplier norm of each m_k on $L^4(\mathbf{R}^2)$ grows at most polynomially in k as $k \to \infty$. By a simple summation on k, this implies that when $\operatorname{Re}\lambda > 0$, B^λ maps $L^4(\mathbf{R}^2)$ into itself. Next, we would like to use complex interpolation between the estimates

$$\begin{aligned}
B^\lambda &:\ L^4(\mathbf{R}^2) \to L^4(\mathbf{R}^2) && \text{when } \operatorname{Re}\lambda > 0,\\
B^\lambda &:\ L^1(\mathbf{R}^2) \to L^1(\mathbf{R}^2) && \text{when } \operatorname{Re}\lambda > \tfrac{1}{2},\\
B^\lambda &:\ L^\infty(\mathbf{R}^2) \to L^\infty(\mathbf{R}^2) && \text{when } \operatorname{Re}\lambda > \tfrac{1}{2}.
\end{aligned}$$

A careful examination of the constants gives that the norms of B^λ on the previous spaces grow at most exponentially in $|\operatorname{Im}\lambda|^2$ as $|\operatorname{Im}\lambda| \to \infty$. We can now apply the interpolation Theorem 1.3.7 to conclude the proof of the first (and main) statement of the theorem. We refer to Figure 10.8 for a pictorial representation of the set of indices p obtained via interpolation.

To obtain the polynomial growth of the norm of each m_k in $\mathcal{M}_4(\mathbf{R}^2)$, we will need an additional decomposition of the operator m_k. This decomposition should take into account the circular nature of m_k. For each $k \ge 0$ we define the sectorial arcs (parts of a sector between two arcs)

$$\Gamma_{k,\ell} = \left\{ re^{2\pi i\theta} \in \mathbf{R}^2 : \ |\theta - \ell\, 2^{-\frac{k}{2}}| < 2^{-\frac{k}{2}}, \quad 1 - \tfrac{5}{8}\, 2^{-k} \le r \le 1 - \tfrac{1}{8}\, 2^{-k} \right\}$$

for all $\ell \in \{0,1,2,\dots,[2^{k/2}]-1\}$. We now introduce a smooth function ω supported in $[-1,1]$ and equal to 1 on a small neighborhood of the origin such that for all $x \in \mathbf{R}$ we have

$$\sum_{\ell \in \mathbf{Z}} \omega(x-\ell) = 1 .$$

Then we define $m_{k,\ell}(re^{2\pi i\theta}) = m_k(re^{2\pi i\theta})\omega(2^{k/2}\theta - \ell)$ for integers ℓ in the set $\{0,1,2,\dots,[2^{k/2}]-1\}$. We suitably adjust the function $m_{k,[2^{k/2}]-1}$, by extending its support, so that we have

$$m_k(\xi) = \sum_{\ell=0}^{[2^{k/2}]-1} m_{k,\ell}(\xi)$$

for all ξ in $\mathbf{R}^2$. This provides the circular (angular) decomposition of m_k. Observe that for all positive integers α and β there exist constants $C_{\alpha,\beta}$ such that

$$|\partial_r^\alpha \partial_\theta^\beta m_{k,\ell}(re^{2\pi i\theta})| \le C_{\alpha,\beta} 2^{k\alpha} 2^{\frac{k}{2}\beta}$$

and that each $m_{k,\ell}$ is a smooth function supported in the sectorial arcs $\Gamma_{k,\ell}$.

We proceed by estimating the $L^4(\mathbf{R}^2)$ norm of the operator with multiplier m_k. At this point we fix a $k \ge 0$ and we write each $m_{k,\ell}$ as a sum

$$m_{k,\ell}^1 + m_{k,\ell}^2 + m_{k,\ell}^3 + m_{k,\ell}^4 + m_{k,\ell}^5 ,$$

where each $m_{k,\ell}^j$ is properly supported in a sector centered at the origin of angle at most $\pi/2$. It suffices to consider only the sectorial arcs $\Gamma_{k,\ell}$ contained entirely in one of these sectors. By rotational invariance we may assume that this sector is contained in the quarter-plane

$$Q = \{(x,y) \in \mathbf{R}^2 : \ x > 0 \ |y| < |x|\} .$$

Let I be the set of all indices ℓ in $\{0,1,2,\dots,[2^{k/2}]-1\}$ such that the sectorial arcs $\Gamma_{k,\ell}$ are contained in the quarter-plane Q. Let $T_{k,\ell}$ be the operator given on the Fourier transform by multiplication with the function $m_{k,\ell}$. We have

$$\begin{aligned} \Big\| \sum_{\ell \in I} T_{k,\ell}(f) \Big\|_{L^4}^4 &= \int_{\mathbf{R}^2} \Big| \sum_{\ell \in I} T_{k,\ell}(f) \Big|^4 dx \\ &= \int_{\mathbf{R}^2} \Big| \sum_{\ell \in I} \sum_{\ell' \in I} T_{k,\ell}(f) T_{k,\ell'}(f) \Big|^2 dx \qquad (10.2.5) \\ &= \int_{\mathbf{R}^2} \Big| \sum_{\ell \in I} \sum_{\ell' \in I} \widehat{T_{k,\ell}(f)} * \widehat{T_{k,\ell'}(f)} \Big|^2 d\xi , \end{aligned}$$

where we used Plancherel's identity in the last equality. Each function $\widehat{T_{k,\ell}(f)}$ is supported in the sectorial arc $\Gamma_{k,\ell}$. Therefore, $\widehat{T_{k,\ell}(f)} * \widehat{T_{k,\ell'}(f)}$ is supported in $\Gamma_{k,\ell} + \Gamma_{k,\ell'}$ and we can write the last integral as

$$\int_{\mathbf{R}^2} \Big| \sum_{\ell\in I} \sum_{\ell'\in I} \big(\widehat{T_{k,\ell}(f)} * \widehat{T_{k,\ell'}(f)}\big) \chi_{\Gamma_{k,\ell}+\Gamma_{k,\ell'}} \Big|^2 d\xi .$$

In view of the Cauchy-Schwarz inequality, the last expression is controlled by

$$\int_{\mathbf{R}^2} \Big(\sum_{\ell\in I} \sum_{\ell'\in I} |\widehat{T_{k,\ell}(f)} * \widehat{T_{k,\ell'}(f)}|^2 \Big) \Big(\sum_{\ell\in I} \sum_{\ell'\in I} |\chi_{\Gamma_{k,\ell}+\Gamma_{k,\ell'}}|^2 \Big) d\xi . \tag{10.2.6}$$

At this point we will make use of the following lemma, in which the curvature of the circle makes its presence felt.

Lemma 10.2.5. *There exists a constant C_0 such that for all $k \ge 0$ the following estimate holds:*

$$\sum_{\ell\in I} \sum_{\ell'\in I} \chi_{\Gamma_{k,\ell}+\Gamma_{k,\ell'}} \le C_0 .$$

We postpone the proof of this lemma until the end of this section. Using Lemma 10.2.5, we control the expression in (10.2.6) by

$$C_0 \int_{\mathbf{R}^2} \sum_{\ell\in I} \sum_{\ell'\in I} \big|\widehat{T_{k,\ell}(f)} * \widehat{T_{k,\ell'}(f)}\big|^2 d\xi = C_0 \Big\| \Big(\sum_{\ell\in I} |T_{k,\ell}(f)|^2 \Big)^{\frac{1}{2}} \Big\|_{L^4}^4 . \tag{10.2.7}$$

We look at each $T_{k,\ell}$ a bit more carefully. When $\ell = 0$ we have that the $m_{k,0}$ is supported in a rectangle with sides parallel to the axes and dimensions 2^{-k} (along the x-axis) and $2^{-\frac{k}{2}+1}$ (along the y-axis). Moreover, $m_{k,0}$ is a smooth function that satisfies

$$|\partial_{\xi_1}^{\alpha} \partial_{\xi_2}^{\beta} m_{k,0}(\xi_1, \xi_2)| \le C_{\alpha,\beta} 2^{k\alpha} 2^{\frac{k}{2}\beta}$$

for all positive integers α and β. This estimate can also be written as

$$|\partial_{\xi_1}^{\alpha} \partial_{\xi_2}^{\beta} m_{k,0}(2^k \xi_1, 2^{\frac{k}{2}} \xi_2)| \le C_{\alpha,\beta}$$

and easily implies that

$$2^{-\frac{3}{2}k} |\partial_{x_1}^{\alpha} \partial_{x_2}^{\beta} m_{k,0}^{\vee}(2^{-k} x_1, 2^{-\frac{k}{2}} x_2)| \le C_{\alpha,\beta,M} (1 + |x|)^{-M}$$

for all positive integers M. Here $m_{k,0}^{\vee}$ denotes the inverse Fourier transform of $m_{k,0}$. Therefore, the kernel $m_{k,0}^{\vee}$ is integrable over $\mathbf{R}^2$ uniformly in k. Applying a rotation, we obtain that the kernels $m_{k,\ell}^{\vee}$ satisfy

$$\big|m_{k,\ell}^{\vee}(x_1, x_2)\big| \le C_M 3^{\frac{3k}{2}} (1 + |2^k x_1| + |2^{\frac{k}{2}} x_2|)^{-M} \tag{10.2.8}$$

for all $M > 0$ and hence

$$\sup_{k\ge 0} \sup_{\ell\in I} \big\|m_{k,\ell}^{\vee}\big\|_{L^1} \le C . \tag{10.2.9}$$

The crucial fact is that the constant C in (10.2.9) is independent of ℓ and k.

At this point, for each fixed $k \geq 0$ and $\ell \in I$ we let $J_{k,\ell}$ be the ξ_2-projection of the support of $m_{k,\ell}$. Based on the earlier definition of $m_{k,\ell}$, we can see easily that when $\ell > 0$

$$J_{k,\ell} = \left[(1 - \tfrac{5}{8}2^{-k})\sin(2\pi 2^{-\frac{k}{2}}(\ell - 1)), (1 - \tfrac{1}{8}2^{-k})\sin(2\pi 2^{-\frac{k}{2}}(\ell + 1))\right].$$

A similar formula holds for $\ell < 0$ in I. The crucial observation is that for any fixed $k \geq 0$ the sets $J_{k,\ell}$ are "almost disjoint" for different $\ell \in I$. Indeed, the sets $J_{k,\ell}$ are contained in the intervals

$$\widetilde{J}_{k,\ell} = \left[(1 - \tfrac{3}{8}2^{-k})\sin(2\pi 2^{-\frac{k}{2}}\ell) - 10 \cdot 2^{-\frac{k}{2}}, (1 - \tfrac{3}{8}2^{-k})\sin(2\pi 2^{-\frac{k}{2}}\ell) + 10 \cdot 2^{-\frac{k}{2}}\right],$$

which have length $20 \cdot 2^{-\frac{k}{2}}$ and are centered at the points

$$(1 - \tfrac{3}{8}2^{-k})\sin(2\pi 2^{-\frac{k}{2}}\ell).$$

For $\sigma \in \mathbf{Z}$ and $\tau \in \{0, 1, \dots, 39\}$ we define the strips

$$S_{k,\sigma,\tau} = \left\{(\xi_1, \xi_2) : \ \xi_2 \in \left[40\sigma 2^{-\frac{k}{2}} + \tau, 40(\sigma + 1)2^{-\frac{k}{2}} + \tau\right)\right\}.$$

These strips have length $40 \cdot 2^{-\frac{k}{2}}$ and have the property that each $\widetilde{J}_{k,\ell}$ is contained in one of them; say $\widetilde{J}_{k,\ell}$ is contained in some $S_{k,\sigma_\ell,\tau_\ell}$, which we call $B_{k,\ell}$. Then we have

$$T_{k,\ell}(f) = T_{k,\ell}(f_{k,\ell}),$$

where we set

$$f_{k,\ell} = \left(\chi_{B_{k,\ell}} \widehat{f}\,\right)^{\vee} = \chi_{B_{k,\ell}}^{\vee} * f.$$

As a consequence of the Cauchy-Schwarz inequality (with respect to the measure $|m_{k,\ell}^{\vee}|\,dx$), we obtain

$$\begin{aligned} |T_{k,\ell}(f_{k,\ell})|^2 &\leq \big\|m_{k,\ell}^{\vee}\big\|_{L^1} \left(|m_{k,\ell}^{\vee}| * |f_{k,\ell}|^2\right) \\ &\leq C \left(|m_{k,\ell}^{\vee}| * |f_{k,\ell}|^2\right) \end{aligned}$$

in view of (10.2.9). We now return to (10.2.7), which controls (10.2.6) and hence (10.2.5). Using this estimate, we bound the term in (10.2.7) by

$$\begin{aligned}
\Big\|\Big(\sum_{\ell\in I}|T_{k,\ell}(f)|^2\Big)^{\frac12}\Big\|_{L^4}^4 &= \Big\|\sum_{\ell\in I}|T_{k,\ell}(f_{k,\ell})|^2\Big\|_{L^2}^2 \\
&\le C\Big\|\sum_{\ell\in I}|m_{k,\ell}^\vee| * |f_{k,\ell}|^2\Big\|_{L^2}^2 \\
&= C\Big(\int_{\mathbf{R}^2}\sum_{\ell\in I}(|m_{k,\ell}^\vee| * |f_{k,\ell}|^2)\, w\, dx\Big)^2 \\
&= C\Big(\sum_{\ell\in I}\int_{\mathbf{R}^2}(|m_{k,\ell}^\vee| * w)\, |f_{k,\ell}|^2\, dx\Big)^2 \\
&\le C\Big(\int_{\mathbf{R}^2}\sup_{\ell\in I}\big(|m_{k,\ell}^\vee| * w|\big)\sum_{\ell\in I}|f_{k,\ell}|^2\, dx\Big)^2 \\
&\le C\Big\|\sup_{\ell\in I}\big(|m_{k,\ell}^\vee| * w\big)\Big\|_{L^2}^2\Big\|\Big(\sum_{\ell\in I}|f_{k,\ell}|^2\Big)^{\frac12}\Big\|_{L^4}^4,
\end{aligned}$$

where w is an appropriate nonnegative function in $L^2(\mathbf{R}^2)$ of norm 1.

The proof of the polynomial growth estimate

$$\|m_k\|_{\mathcal{M}_p} \le C(1+k)^{\frac12} \tag{10.2.10}$$

will be completed once we establish the validity of the following estimates:

$$\Big\|\sup_{\ell\in I}\big(|m_{k,\ell}^\vee| * w\big)\Big\|_{L^2} \le C(1+k)\|w\|_{L^2} \tag{10.2.11}$$

and also

$$\Big\|\Big(\sum_{\ell\in I}|f_{k,\ell}|^2\Big)^{\frac12}\Big\|_{L^4} \le C\|f\|_{L^4}. \tag{10.2.12}$$

Note that (10.2.10) allows us to sum the series in (10.2.4).

Estimates (10.2.11) and (10.2.12) will be discussed in the next two subsections.

□

10.2.c. The Kakeya Maximal Function

We showed in the previous subsection that $m_{k,0}^\vee$ is integrable over $\mathbf{R}^2$ and satisfies the estimate

$$2^{-\frac32 k}|m_{k,0}^\vee(2^{-k}x_1, 2^{\frac{-k}{2}}x_2)| \le \frac{C}{(1+|x|)^3}.$$

Since the latter is at most a constant multiple of

$$\sum_{s=0}^{\infty} \frac{2^{-s}}{2^{2s}} \chi_{[-2^s,2^s]\times[-2^s,2^s]}(x)\,,$$

it follows that

$$|m_{k,0}^{\vee}(x)| \le C' \sum_{s=0}^{\infty} 2^{-s} \frac{1}{|R_s|} \chi_{R_s}(x)\,,$$

where $R_s = [-2^s 2^{-k}, 2^s 2^{-k}] \times [-2^s 2^{-\frac{k}{2}}, 2^s 2^{-\frac{k}{2}}]$. Since a general $m_{k,\ell}^{\vee}$ can be obtained from $m_{k,0}^{\vee}$ via a rotation, a similar estimate holds for it. Precisely, we have

$$|m_{k,\ell}^{\vee}(x)| \le C' \sum_{s=0}^{\infty} 2^{-s} \frac{1}{|R_{s,\ell}|} \chi_{R_{s,\ell}}(x)\,, \tag{10.2.13}$$

where $R_{s,\ell}$ is a rectangle with main axes along the directions $e^{2\pi i \ell \delta^{1/2}}$ and $ie^{2\pi i \ell \delta^{1/2}}$ and side lengths $2^s 2^{-k}$ and $2^s 2^{-\frac{k}{2}}$, respectively. Using (10.2.13), we obtain the following pointwise estimate for the maximal function in (10.2.11):

$$\sup_{\ell \in I} \big(|m_{k,\ell}^{\vee}| * w\big) \;\le\; C' \sum_{s=0}^{\infty} 2^{-s} \sup_{\ell \in I} \frac{1}{|R_{s,\ell}|} \int_{R_{s,\ell}} w(x-y)\, dy\,, \tag{10.2.14}$$

where $R_{s,\ell}$ are rectangles with dimensions $2^s 2^{-k}$ and $2^s 2^{-\frac{k}{2}}$.

Motivated by (10.2.14), for fixed $N \ge 10$ and $a > 0$, we introduce the *Kakeya maximal operator without dilations*

$$\mathcal{K}_N^a(w)(x) = \sup_{R \ni x} \frac{1}{|R|} \int_R |w(y)|\, dy\,, \tag{10.2.15}$$

acting on functions $w \in L^1_{\text{loc}}$, where the supremum is taken over all rectangles R in $\mathbf{R}^2$ of dimensions a and aN and arbitrary orientation. What makes this maximal operator interesting is that the rectangles R that appear in the supremum in (10.2.16) are allowed to be rotated arbitrarily. We also define the *Kakeya maximal operator* $\mathcal{K}_N$ as

$$\mathcal{K}_N(w)(x) = \sup_{a>0} \mathcal{K}_N^a(w)\,, \tag{10.2.16}$$

for w locally integrable. The maximal quantity $\mathcal{K}_N(w)(x)$ is therefore obtained as the supremum of the averages of a function over all rectangles in $\mathbf{R}^2$ that contain the point x and have arbitrary orientation but fixed eccentricity equal to N. (The eccentricity of a rectangle is the ratio of its longest side over its shortest side.)

We see that $\mathcal{K}_N(f)$ is pointwise controlled by a $c\,N\,M(f)$, where M is the Hardy-Littlewood maximal operator M. This implies that $\mathcal{K}_N$ is of weak type $(1,1)$ with bound at most a multiple of N. Since $\mathcal{K}_N$ is bounded on L^∞ with norm 1, it follows that $\mathcal{K}_N$ maps $L^p(\mathbf{R}^2)$ into itself with norm at most a multiple of $N^{1/p}$. However, we will show in the next section that this estimate is very rough and can be improved significantly. In fact, we will obtain an L^p estimate for $\mathcal{K}_N$ with norm that grows logarithmically in N (when $p \geq 2$), and this will be very crucial to us since $N = 2^{k/2}$ in the following application.

Using this new terminology, we can write the estimate in (10.2.14) as

$$\sup_{\ell \in I} \left(|m_{k,\ell}^{\vee}| * w \right) \leq C' \sum_{s=0}^{\infty} 2^{-s} \mathcal{K}_{2^{k/2}}^{2^{s-k}}(w) . \tag{10.2.17}$$

The required estimate (10.2.11) will be a consequence of (10.2.17) and of the following theorem, whose proof will be a consequence of the results obtained in the next section.

Theorem 10.2.6. *There exists a constant C such that for all $N \geq 10$ and all $f \in L^2(\mathbf{R}^2)$ the following norm inequality is valid:*

$$\sup_{a>0} \left\| \mathcal{K}_N^a(f) \right\|_{L^2(\mathbf{R}^2)} \leq C \, (\log N) \left\| f \right\|_{L^2(\mathbf{R}^2)} .$$

Theorem 10.2.6 will be a consequence of Theorem 10.3.5, in which the preceding estimate is proved for a more general maximal operator $\mathfrak{M}_{\Sigma_N}$, which in particular, controls $\mathcal{K}_N$ and hence $\mathcal{K}_N^a$ for all $a > 0$. This maximal operator will be introduced in the next section.

10.2.d. Boundedness of a Square Function

We now turn to the proof of estimate (10.2.12). This will be a consequence of the following result, which is a version of the Littlewood-Paley theorem for intervals of equal length.

Theorem 10.2.7. *For $j \in \mathbf{Z}$, let I_j be intervals of equal length and disjoint interior whose union is $\mathbf{R}$. We define operators P_j with multipliers χ_{I_j}. Then for any $2 \leq p < \infty$, there is a constant C_p such that for all $f \in L^p(\mathbf{R})$ we have*

$$\left\| \Big(\sum_j |P_j(f)|^2 \Big)^{\frac{1}{2}} \right\|_{L^p(\mathbf{R})} \leq C_p \left\| f \right\|_{L^p(\mathbf{R})} . \tag{10.2.18}$$

In particular, the same estimate holds if the intervals I_j have disjoint interiors and equal length but do not necessarily cover $\mathbf{R}$.

Proof. Multiplying the function f by a suitable exponential, we may assume that the intervals I_j have the form $\big((j-\frac{1}{2})a, (j+\frac{1}{2})a\big)$ for some $a > 0$. Applying a dilation to f reduces matters to the case $a = 1$. We conclude that the constant C_p is independent of the common size of the intervals I_j and it suffices to obtain estimate (10.2.18) in the case $a = 1$.

We assume therefore that $I_j = (j - \frac{1}{2}, j + \frac{1}{2})$ for all $j \in \mathbf{Z}$. Our next goal will be to replace the operators P_j by smoother analogues of them. To achieve this we introduce a smooth function ψ with compact support that is identically equal to 1 on the interval $[-\frac{1}{2}, \frac{1}{2}]$ and vanishes off the interval $[-\frac{3}{4}, \frac{3}{4}]$. We introduce operators S_j by setting

$$\widehat{S_j(f)}(\xi) = \widehat{f}(\xi)\psi(\xi - j)$$

and we note that the identity

$$P_j = P_j S_j \tag{10.2.19}$$

is valid for all $j \in \mathbf{Z}$. For $t \in \mathbf{R}$ we define multipliers m_t as follows:

$$m_t(\xi) = \sum_{j\in\mathbf{Z}} e^{2\pi i j t}\psi(\xi - j)$$

and we set $k_t = m_t^\vee$. Then we have

$$\begin{aligned} \int_{I_0} |(k_t * f)(x)|^2\,dt &= \int_{I_0} \Big|\sum_{j\in\mathbf{Z}} e^{2\pi i j t} S_j(f)(x)\Big|^2\,dt \\ &= \sum_{j\in\mathbf{Z}} |S_j(f)(x)|^2\,, \end{aligned} \tag{10.2.20}$$

where the last equality is just Plancherel's identity on $I_0 = [-\frac{1}{2}, \frac{1}{2}]$. In view of the last identity, it will suffice to analyze the operator given by convolution with the family of kernels k_t. By the Poisson summation formula (Theorem 3.1.17) applied to the function $x \to \psi(-x)e^{2\pi i x t}$, we obtain

$$\begin{aligned} m_t(\xi) &= e^{2\pi i \xi t}\sum_{j\in\mathbf{Z}} \psi(-\xi + j)e^{2\pi i(-\xi+j)t} \\ &= \sum_{j\in\mathbf{Z}} \big(\psi(-(\cdot))e^{2\pi i(\cdot)t}\big)^\wedge(j)\, e^{2\pi i j(-\xi)} e^{2\pi i \xi t} \\ &= \sum_{j\in\mathbf{Z}} e^{2\pi i(j+t)\xi}\widehat{\psi}(t+j)\,. \end{aligned}$$

Taking inverse Fourier transforms, we obtain

$$k_t = \sum_{j\in\mathbf{Z}} \widehat{\psi}(t+j)\delta_{-j-t}\,,$$

where δ_b denotes Dirac mass at the point b. Therefore, k_t is a sum of Dirac masses with rapidly decaying coefficients. Since each Dirac mass has Borel norm at most 1, we conclude that

$$\|k_t\|_{\mathcal{M}} \le \sum_{j\in\mathbf{Z}} |\widehat{\psi}(t+j)| \le \sum_{j\in\mathbf{Z}} (1 + |t+j|)^{-10} \le 10\,, \tag{10.2.21}$$

which is independent of t. This is saying that the measures k_t have uniformly bounded norms. Take now $f \in L^p(\mathbf{R})$ and $p \ge 2$. Using identity (10.2.19), we obtain

$$\begin{aligned}\int_{\mathbf{R}}\Big(\sum_{j\in\mathbf{Z}}|P_j(f)(x)|^2\Big)^{\frac{p}{2}}dx &= \int_{\mathbf{R}}\Big(\sum_{j\in\mathbf{Z}}|P_jS_j(f)(x)|^2\Big)^{\frac{p}{2}}dx\\ &\le c_p\int_{\mathbf{R}}\Big(\sum_{j\in\mathbf{Z}}|S_j(f)(x)|^2\Big)^{\frac{p}{2}}dx\,,\end{aligned}$$

and the last inequality follows from Exercise 4.6.1(a). The constant c_p only depends on p. Recalling identity (10.2.20), we can write

$$\begin{aligned}c_p\int_{\mathbf{R}}\Big(\sum_{j\in\mathbf{Z}}|S_j(f)(x)|^2\Big)^{\frac{p}{2}}dx &\le c_p\int_{\mathbf{R}}\bigg(\int_{I_0}|(k_t*f)(x)|^2\,dt\bigg)^{\frac{p}{2}}dx\\ &\le c_p\int_{\mathbf{R}}\bigg(\int_{I_0}|(k_t*f)(x)|^p dt\bigg)^{\frac{p}{p}}dx\\ &= c_p\int_{I_0}\int_{\mathbf{R}}|(k_t*f)(x)|^p\,dx\,dt\\ &\le 10c_p\int_{I_0}\int_{\mathbf{R}}|f(x)|^p\,dx\,dt = 10c_p\big\|f\big\|_{L^p}^p\,,\end{aligned}$$

where we used Hölder's inequality on the interval I_0 (together with the fact that $p\ge 2$) and (10.2.21). The proof of theorem is complete with constant $C_p=(10c_p)^{1/p}$. Note that C_p behaves like p as $p\to\infty$.

□

We now return to estimate (10.2.12). First recall the strips

$$S_{k,\sigma,\tau}=\big\{(\xi_1,\xi_2):\ \xi_2\in[40\sigma2^{-\frac{k}{2}}+\tau,40(\sigma+1)2^{-\frac{k}{2}}+\tau)\big\}$$

defined for $\sigma\in\mathbf{Z}$ and $\tau\in\{0,1,\dots,39\}$. These strips have length $40\cdot2^{-\frac{k}{2}}$, and each $\widetilde{J}_{k,\ell}$ is contained in one of them, which we called $S_{k,\sigma_\ell,\tau_\ell}=B_{k,\ell}$.

The family $\{B_{k,\ell}\}_{\ell\in I}$ does not consist of disjoint sets, but we split it into 40 subfamilies by placing $B_{k,\ell}$ in different subfamilies if the indices τ_ℓ and $\tau_{\ell'}$ are different. We can now write the set I as

$$I=I^1\cup I^2\cup\cdots\cup I^{40}\,,$$

where for each $\ell,\ell'\in I^j$ the sets $B_{k,\ell}$ and $B_{k,\ell'}$ are disjoint.

We can now use Theorem 10.2.7 to obtain the required quadratic estimate (10.2.12). Things now are relatively simple. We observe that the multiplier operators $f\to(\chi_{B_{k,\ell}}\widehat{f})^\vee$ on $\mathbf{R}^2$ obey the estimates (10.2.18), in which $L^p(\mathbf{R})$ is replaced by $L^p(\mathbf{R}^2)$ since they are the identity operators in the x_1-variable.

We conclude that

$$\Big\|\Big(\sum_{\ell\in I^j}|T_{k,\ell}(f)|^2\Big)^{\frac{1}{2}}\Big\|_{L^p(\mathbf{R}^2)}\le C_p\big\|f\big\|_{L^p(\mathbf{R}^2)} \tag{10.2.22}$$

holds for all $p \geq 2$ and, in particular, for $p = 4$. This proves (10.2.12) for a single I^j, and the same conclusion follows for I with a constant 40 times as big.

10.2.e. The Proof of Lemma 10.2.5

We finally discuss the proof of Lemma 10.2.5.

Proof. If $k = 0, 1, \dots, k_0$ up to a fixed integer k_0, then there exist only finitely many pairs of sets $\Gamma_\ell + \Gamma_{\ell'}$ depending on k_0 and the lemma is trivially true. We may therefore assume that k is a large integer; in particular we may take $\delta = 2^{-k} \leq 2400^{-2}$. In the sequel, for simplicity we replace 2^{-k} by δ and we denote the set $\Gamma_{k,\ell}$ by Γ_ℓ. In the proof that follows we will be working with a fixed $\delta \in [0, 2400^{-2}]$. Elements of the set $\Gamma_\ell + \Gamma_{\ell'}$ have the form

$$re^{2\pi i(\ell+\alpha)\delta^{1/2}} + r' e^{2\pi i(\ell'+\alpha')\delta^{1/2}}, \tag{10.2.23}$$

where α, α' range in the interval $[-1, 1]$ and r, r' range in $[1 - \frac{5}{8}\delta, 1 - \frac{1}{8}\delta]$. We set

$$w(\ell, \ell') = e^{2\pi i \ell \delta^{1/2}} + e^{2\pi i \ell' \delta^{1/2}} = 2\cos(\pi|\ell - \ell'|\delta^{\frac{1}{2}}) e^{\pi i(\ell+\ell')\delta^{1/2}}, \tag{10.2.24}$$

where the last equality is a consequence of a trigonometric identity that can be found in Appendix E. Using similar identities (see Appendix E) and performing algebraic manipulations, the reader can verify that the general element (10.2.23) of the set $\Gamma_\ell + \Gamma_{\ell'}$ can be written as

$$\begin{aligned} w(\ell, \ell') &+ \Big\{ r\, \frac{\cos(2\pi\alpha\delta^{\frac{1}{2}}) + \cos(2\pi\alpha'\delta^{\frac{1}{2}}) - 2}{2} \Big\}\, w(\ell, \ell') \\ &+ \Big\{ r\, \frac{\sin(2\pi\alpha\delta^{\frac{1}{2}}) + \sin(2\pi\alpha'\delta^{\frac{1}{2}})}{2} \Big\}\, i\, w(\ell, \ell') \\ &+ \mathrm{E}(r, \ell, \ell', \alpha, \alpha', \delta), \end{aligned}$$

where

$$\begin{aligned} \mathrm{E}(r, \ell, \ell', \alpha, \alpha', \delta) &= (r-1)\Big(e^{2\pi i \ell \delta^{1/2}} + e^{2\pi i \ell' \delta^{1/2}}\Big) \\ &+ (r' - r) e^{2\pi i(\ell'+\alpha')\delta^{1/2}} \\ &+ r\Big(e^{2\pi i \ell \delta^{1/2}} - e^{2\pi i \ell' \delta^{1/2}}\Big)\Big(\frac{\cos(2\pi\alpha\delta^{\frac{1}{2}}) - \cos(2\pi\alpha'\delta^{\frac{1}{2}})}{2}\Big) \\ &+ r\Big(e^{2\pi i \ell \delta^{1/2}} - e^{2\pi i \ell' \delta^{1/2}}\Big)\Big(\frac{\sin(2\pi\alpha\delta^{\frac{1}{2}}) - \sin(2\pi\alpha'\delta^{\frac{1}{2}})}{2}\Big). \end{aligned}$$

The coefficients in the curly brackets are real and $\mathrm{E}(r, \ell, \ell', \alpha, \alpha', \delta)$ is an error of magnitude at most $2\delta + 8\pi^2|\ell - \ell'|\delta$. These observations and the facts $|\sin x| \leq |x|$ and $|1 - \cos x| \leq |x|^2/2$ (see Appendix E) imply that the set $\Gamma_\ell + \Gamma_{\ell'}$ is contained in the rectangle $R(\ell, \ell')$ centered at the point $w(\ell, \ell')$ with half-width

$$2\pi^2\delta + (2\delta + 8\pi^2|\ell - \ell'|\delta) \leq 80\,(1 + |\ell - \ell'|)\delta$$

in the direction along $w(\ell,\ell')$ and half-length

$$2\pi\delta^{\frac{1}{2}} + (2\delta + 8\pi^2|\ell-\ell'|\delta) \le 30\,\delta^{\frac{1}{2}}$$

in the direction along $iw(\ell,\ell')$ [which is perpendicular to that along $w(\ell,\ell')$]. Since $2\pi|\ell-\ell'|\delta^{\frac{1}{2}} < \frac{\pi}{2}$, this rectangle is contained in a disc of radius $105\,\delta^{\frac{1}{2}}$ centered at the point $w(\ell,\ell')$.

We immediately deduce that if $|w(\ell,\ell') - w(m,m')|$ is bigger than $210\,\delta^{\frac{1}{2}}$, then the sets $\Gamma_\ell + \Gamma_{\ell'}$ and $\Gamma_m + \Gamma_{m'}$ do not intersect. Therefore, if these sets intersected, we should have

$$|w(\ell,\ell') - w(m,m')| \le 210\,\delta^{\frac{1}{2}}\,.$$

In view of Exercise 10.2.2, the left hand side of the last expression is at least

$$\tfrac{2}{\pi}\cos(\tfrac{\pi}{4})|\pi(\ell+\ell') - \pi(m+m')|\delta^{\frac{1}{2}}$$

(here we use the hypothesis that $|2\pi\ell\delta^{\frac{1}{2}}| < \frac{\pi}{4}$ twice.) We conclude that if the sets $\Gamma_\ell + \Gamma_{\ell'}$ and $\Gamma_m + \Gamma_{m'}$ intersected, then

$$|(\ell+\ell') - (m+m')| \le 210/\sqrt{2} \le 150\,. \tag{10.2.25}$$

In this case the angle between the vectors $w(\ell,\ell')$ and $w(m,m')$ is

$$\varphi_{\ell,\ell',m,m'} = \pi|(\ell+\ell') - (m+m')|\delta^{\frac{1}{2}}\,,$$

which is smaller than $\pi/16$ provided (10.2.25) holds and $\delta < 2400^{-2}$. This is saying that in this case the rectangles $R(\ell,\ell')$ and $R(m,m')$ are essentially parallel to each other (the angle between them is smaller than $\pi/16$).

Let us fix a rectangle $R(\ell,\ell')$, and for another rectangle $R(m,m')$ we denote by $\widetilde{R}(m,m')$ the smallest rectangle containing $R(m,m')$ with sides parallel to the corresponding sides of $R(\ell,\ell')$. Easy trigonometry shows that $\widetilde{R}(m,m')$ has the same center as $R(m,m')$ and has half-sides at most

$$30\delta^{\frac{1}{2}}\cos(\varphi_{\ell,\ell',m,m'}) + 80(1+|\ell-\ell'|)\delta\sin(\varphi_{\ell,\ell',m,m'})$$
$$80(1+|\ell-\ell'|)\delta\cos(\varphi_{\ell,\ell',m,m'}) + 30\delta^{\frac{1}{2}}\sin(\varphi_{\ell,\ell',m,m'})$$

in view of Exercise 10.2.3. Then $\widetilde{R}(m,m')$ has half-sides at most $60{,}000\delta^{\frac{1}{2}}$ and $18{,}000(1+|\ell-\ell'|)\delta$ and is therefore contained in a fixed multiple of $R(m,m')$. If $\Gamma_\ell + \Gamma_{\ell'}$ and $\Gamma_m + \Gamma_{m'}$ intersect, then so would $\widetilde{R}(m,m')$ and $R(\ell,\ell')$ and both of these rectangles have sides parallel to the vectors $w(\ell,\ell')$ and $iw(\ell,\ell')$. But in the direction of $w(\ell,\ell')$, these rectangles have sides with half lengths at most $80(1+|\ell-\ell'|)\delta$ and $16{,}000(1+|m-m'|)\delta$. Note that the distance of the lines parallel to the direction $i\,w(\ell,\ell')$ and passing through the centers of the rectangles $\widetilde{R}(m,m')$ and $R(\ell,\ell')$ is

$$2\big|\cos(\pi|\ell-\ell'|\delta^{\frac{1}{2}}) - \cos(\pi|m-m'|\delta^{\frac{1}{2}})\big|\,,$$

as we can see easily using (10.2.24). If these rectangles intersect, we must have

$$2\big|\cos(\pi|\ell-\ell'|\delta^{\frac{1}{2}})-\cos(\pi|m-m'|\delta^{\frac{1}{2}})\big|\le 16{,}080\,(2+|\ell-\ell'|+|m-m'|)\delta\,.$$

We conclude that if the sets $R(m,m')$ and $R(\ell,\ell')$ intersect and $(\ell,\ell')\neq(m,m')$, then

$$\big|\cos(\pi|\ell-\ell'|\delta^{\frac{1}{2}})-\cos(\pi|m-m'|\delta^{\frac{1}{2}})\big|\le 50{,}000\,(|\ell-\ell'|+|m-m'|)\delta\,.$$

But the expression on the left is equal to

$$2\big|\sin(\pi\tfrac{|\ell-\ell'|-|m-m'|}{2}\delta^{\frac{1}{2}})\sin(\pi\tfrac{|\ell-\ell'|+|m-m'|}{2}\delta^{\frac{1}{2}})\big|\,,$$

which is at least

$$2\big||\ell-\ell'|-|m-m'|\big|\,(|\ell-\ell'|+|m-m'|)\delta$$

in view of the simple estimate $|\sin t|\ge\frac{2}{\pi}|t|$ for $|t|<\frac{\pi}{2}$. We conclude that if the sets $R(m,m')$ and $R(\ell,\ell')$ intersect and $(\ell,\ell')\neq(m,m')$, then

$$\big||\ell-\ell'|-|m-m'|\big|\le 25{,}000\,. \tag{10.2.26}$$

Combining (10.2.25) with (10.2.26), it follows that for each fixed pair (ℓ,ℓ') the set $\Gamma_m+\Gamma_{m'}$ intersects the set $\Gamma_\ell+\Gamma_{\ell'}$ for only finitely many pairs (m,m'). This concludes the proof of the lemma.

□

Exercises

10.2.1. For $\lambda\ge 0$ show that for all $f\in L^p(\mathbf{R}^n)$ the Bochner-Riesz operators

$$B_R^\lambda(f)(x)=\int_{\mathbf{R}^n}(1-|\xi/R|^2)^\lambda\widehat{f}(\xi)e^{2\pi i x\cdot\xi}\,d\xi$$

converge to f in $L^p(\mathbf{R}^n)$ if and only if the function $(1-|\xi|^2)_+^\lambda$ lies in $\mathcal{M}_p(\mathbf{R}^n)$. [*Hint:* In the beginning of the proof of Theorem 10.2.4 it was shown that if $(1-|\xi|^2)_+^\lambda$ lies in $\mathcal{M}_p(\mathbf{R}^n)$, then $B_R^\lambda(f)$ converge to f in $L^p(\mathbf{R}^n)$. Conversely, if $B_R^\lambda(f)$ converge to f in L^p as $R\to\infty$, then for every f in $L^p(\mathbf{R}^n)$ there is a constant C_f such that $\sup_{R>0}\big\|B_R^\lambda(f)\big\|_{L^p}$ is at most $C_f<\infty$. It follows that $\sup_{R>0}\big\|B_R^\lambda\big\|_{L^p\to L^p}<\infty$ by the uniform boundedness principle; hence $\big\|B^\lambda\big\|_{L^p\to L^p}<\infty$.]

10.2.2. Let $|\theta_1|,|\theta_2|<\frac{\pi}{4}$ be two angles. Show geometrically that

$$|r_1e^{i\theta_1}-r_2e^{i\theta_2}|\ge\min(r_1,r_2)\sin|\theta_1-\theta_2|$$

and use the estimate $|\sin t|\ge\frac{2|t|}{\pi}$ for $|t|<\frac{\pi}{2}$ to obtain a lower bound for the second expression in terms of $|\theta_1-\theta_2|$.

10.2.3. Let R be a rectangle in $\mathbf{R}^2$ having length $b > 0$ along a direction $\vec{v} = (\xi_1, \xi_2)$ and length $a > 0$ along the perpendicular direction $\vec{v}^\perp = (-\xi_2, \xi_1)$. Let $\vec{w}$ be another vector that forms an angle $\varphi < \frac{\pi}{2}$ with $\vec{v}$. Show that the smallest rectangle R' that contains R and having sides parallel to $\vec{w}$ and $\vec{w}^\perp$ has side lengths $a\sin(\varphi) + b\cos(\varphi)$ along the direction $\vec{w}$ and $a\cos(\varphi) + b\sin(\varphi)$ along the direction $\vec{w}^\perp$.

10.2.4. Prove that Theorem 10.2.7 does not hold when $p < 2$.
[*Hint:* Try the intervals $I_j = [j, j+1]$ and $\widehat{f} = \chi_{[0,N]}$ as $N \to \infty$.]

10.2.5. Let $\{I_k\}_k$ be a family of intervals in the real line with $|I_k| = |I_{k'}|$ and $I_k \cap I_{k'} = \emptyset$ for all $k \neq k'$. Define the sets

$$S_k = \big\{(\xi_1, \dots, \xi_n) \in \mathbf{R}^n : \ \xi_1 \in I_k\big\}.$$

Prove that for all $p \ge 2$ and all $f \in L^p(\mathbf{R}^n)$ we have

$$\Big\| \Big(\sum_k |(\widehat{f}\chi_{S_k})^\vee|^2 \Big)^{\frac{1}{2}} \Big\|_{L^p(\mathbf{R}^n)} \le C_p \big\|f\big\|_{L^p(\mathbf{R}^n)},$$

where C_p is the constant of Theorem 10.2.7.

10.2.6. (a) Let $\{I_k\}_k$, $\{J_\ell\}_\ell$ be two families of intervals in the real line with $|I_k| = |I_{k'}|$, $I_k \cap I_{k'} = \emptyset$ for all $k \neq k'$, and $|J_\ell| = |J_{\ell'}|$, $J_\ell \cap J_{\ell'} = \emptyset$ for all ℓ, ℓ'. Prove that for all $p \ge 2$ there is a constant C_p such that

$$\Big\| \Big(\sum_k \sum_\ell |(\widehat{f}\,\chi_{I_k \times J_\ell})^\vee|^2 \Big)^{\frac{1}{2}} \Big\|_{L^p(\mathbf{R}^2)} \le C_p^2 \big\|f\big\|_{L^p(\mathbf{R}^2)},$$

for all $f \in L^p(\mathbf{R}^2)$.
(b) State and prove an analogous result on $\mathbf{R}^n$.
[*Hint:* Use the Rademacher functions and apply Theorem 10.2.7 twice.]

10.2.7. (*Rubio de Francia* [**433**]) On $\mathbf{R}^n$ consider the points $x_\ell = \ell\sqrt{\delta}$, where $\ell \in \mathbf{Z}^n$. Fix a Schwartz function h whose Fourier transform is supported in the unit ball in $\mathbf{R}^n$. Given a function f on $\mathbf{R}^n$, define $\widehat{f_\ell}(\xi) = \widehat{f}(\xi)\widehat{h}(\delta^{-\frac{1}{2}}(\xi - x_\ell))$. Prove that for some constant C (which depends only on h and n) the estimate

$$\Big(\sum_{\ell \in \mathbf{Z}^n} |f_\ell|^2 \Big)^{\frac{1}{2}} \le C M(|f|^2)^{\frac{1}{2}}$$

holds for all functions f. Deduce the $L^p(\mathbf{R}^n)$ boundedness of the preceding square function for all $p \ge 2$.
[*Hint:* For a sequence λ_ℓ with $\sum_\ell |\lambda_\ell|^2 = 1$, set

$$G(f)(x) = \sum_{\ell \in \mathbf{Z}^n} \lambda_\ell f_\ell(x) = \int_{\mathbf{R}^n} \Big[\sum_{\ell \in \mathbf{Z}^n} \lambda_\ell \, e^{2\pi i \frac{x_\ell \cdot y}{\sqrt{\delta}}} \Big] f(x - \tfrac{y}{\sqrt{\delta}}) h^\vee(y)\, dy.$$

Split $\mathbf{R}^n$ as the union of $Q_0 = [-\frac{1}{2}, \frac{1}{2}]^n$ and $2^{j+1}Q_0 \setminus 2^j Q_0$ for $j \ge 0$ and control the integral on each such set using the decay of $h^\vee$ and the $L^2(2^{j+1}Q_0)$ norms of the other two terms. Finally, exploit the orthogonality of the functions $e^{2\pi i \ell \cdot y}$ to estimate the $L^2(2^{j+1}Q_0)$ norm of the expression inside the square brackets by $C2^{nj/2}$. Sum in $j \ge 0$ to obtain the required conclusion.]

10.2.8. For $\lambda > 0$ define the *maximal Bochner-Riesz operator*

$$B_*^\lambda(f)(x) = \sup_{R>0} \left| \int_{\mathbf{R}^n} (1 - |\xi/R|^2)^\lambda \widehat{f}(\xi) e^{2\pi i x\cdot\xi} \, d\xi \right|.$$

Prove that B_*^λ maps $L^p(\mathbf{R}^n)$ into itself when $\lambda > \frac{n-1}{2}$ for $1 \le p \le \infty$. [*Hint:* Use (10.2.2) and Corollary 2.1.12.]

10.3. Kakeya Maximal Operators

We recall the Hardy-Littlewood maximal operator with respect to cubes on $\mathbf{R}^n$ defined as

$$M_c(f)(x) = \sup_{\substack{Q \in \mathcal{F} \\ Q \ni x}} \frac{1}{|Q|} \int_Q |f(y)| \, dy \,, \tag{10.3.1}$$

where $\mathcal{F}$ is the set of all closed cubes in $\mathbf{R}^n$ (with sides not necessarily parallel to the axes). The operator M_c is equivalent (bounded above and below by constants) with the corresponding maximal operator M_c' in which the family $\mathcal{F}$ is replaced by the more restrictive family $\mathcal{F}'$ of cubes in $\mathbf{R}^n$ with sides parallel to the coordinate axes.

It is interesting to observe that if the family of all cubes $\mathcal{F}$ in (10.3.1) is replaced by the family of all rectangles (or parallelepipeds) $\mathcal{R}$ in $\mathbf{R}^n$, then we obtain an operator M_0 that is unbounded on $L^p(\mathbf{R}^n)$; see also Exercise 2.1.9. If we substitute the family of all parallelepipeds $\mathcal{R}$, however, with the more restrictive family $\mathcal{R}'$ of all parallelepipeds with sides parallel to the coordinate axes, then we obtain the so-called strong maximal function

$$M_s(f)(x) = \sup_{\substack{R \in \mathcal{R}' \\ R \ni x}} \frac{1}{|R|} \int_R |f(y)| \, dy \,, \tag{10.3.2}$$

which was introduced in Exercise 2.1.6. The operator M_s is bounded on $L^p(\mathbf{R}^n)$ for $1 < p < \infty$ but it is not of weak type $(1,1)$. See Exercise 10.3.1.

Theses examples indicate that averaging over long and skinny rectangles may have a different outcome than averaging over squares. In general, the direction and the dimensions of the averaging rectangles play a significant role in the boundedness properties of the maximal functions. In this section we will investigate this topic in some detail.

10.3.a. Maximal Functions Associated with a Set of Directions

Definition 10.3.1. Let Σ be a set of unit vectors in $\mathbf{R}^2$, i.e., a subset of the unit circle $\mathbf{S}^1$. Associated with Σ we define $\mathcal{R}_\Sigma$ to be the set of all closed rectangles in $\mathbf{R}^2$ whose longest side is parallel to some vector in Σ. We also define a maximal operator $\mathfrak{M}_\Sigma$ associated with Σ as follows:

$$\mathfrak{M}_\Sigma(f)(x) = \sup_{\substack{R\in\mathcal{R}_\Sigma \\ R\ni x}} \frac{1}{|R|}\int_R |f(y)|\,dy\,,$$

where f is a locally integrable function on $\mathbf{R}^2$.

We also recall the definition given in (10.2.16) of the *Kakeya maximal operator*

$$\mathcal{K}_N(w)(x) = \sup_{R\ni x} \frac{1}{|R|}\int_R |w(y)|\,dy\,, \tag{10.3.3}$$

where the supremum is taken over all rectangles R in $\mathbf{R}^2$ of dimensions a and aN where $a>0$ is arbitrary. N here is a fixed real number that is at least 10.

Example 10.3.2. Let $\Sigma = \{v\}$ consist of only one vector $v = (a,b)$. Then

$$\mathfrak{M}_\Sigma(f)(x) = \sup_{0<r\le 1}\sup_{N>0} \frac{1}{rN^2}\int_{-N}^{N}\int_{-rN}^{rN} |f(x - t(a,b) - s(-b,a))|\,ds\,dt\,.$$

If $\Sigma = \{(1,0),(0,1)\}$ consists of the two unit vectors along the axes, then

$$\mathfrak{M}_\Sigma = M_s\,,$$

where M_s is the strong maximal function defined in (10.3.2).

It is obvious that for each $\Sigma \subseteq \mathbf{S}^1$, the maximal function $\mathfrak{M}_\Sigma$ maps $L^\infty(\mathbf{R}^2)$ into itself with constant 1. But $\mathfrak{M}_\Sigma$ may not always be of weak type $(1,1)$ as the example M_s indicates; see Exercise 10.3.1. The boundedness of $\mathfrak{M}_\Sigma$ on $L^p(\mathbf{R}^2)$ in general depends on the set Σ.

An interesting case arises in the following example as well.

Example 10.3.3. For $N \in \mathbf{Z}^+$, let

$$\Sigma = \Sigma_N = \big\{\big(\cos(\tfrac{2\pi j}{N}), \sin(\tfrac{2\pi j}{N})\big) \,:\, j = 0,1,2,\dots,N-1\big\}$$

be the set of N uniformly spread directions on the circle. Then we expect $\mathfrak{M}_{\Sigma_N}$ to be L^p bounded with constant depending on N. There is a connection between the operator $\mathfrak{M}_{\Sigma_N}$ previously defined and the Kakeya maximal operator $\mathcal{K}_N$ defined in (10.2.16). In fact, in Exercise 10.3.3 the reader is asked to show that

$$\mathcal{K}_N(f) \le 20\,\mathfrak{M}_{\Sigma_N}(f) \tag{10.3.4}$$

for all locally integrable functions f on $\mathbf{R}^2$.

We now indicate why the norms of $\mathcal{K}_N$ and $\mathfrak{M}_{\Sigma_N}$ on $L^2(\mathbf{R}^2)$ grow as $N \to \infty$. We refer to Exercises 10.3.4 and 10.3.7 for the corresponding result for $p \neq 2$.

Proposition 10.3.4. *There is a constant c such that for any $N \geq 10$ we have*

$$\big\|\mathcal{K}_N\big\|_{L^2(\mathbf{R}^2)\to L^2(\mathbf{R}^2)} \geq c\,(\log N) \tag{10.3.5}$$

and

$$\big\|\mathcal{K}_N\big\|_{L^2(\mathbf{R}^2)\to L^{2,\infty}(\mathbf{R}^2)} \geq c\,(\log N)^{\frac{1}{2}}\,. \tag{10.3.6}$$

Therefore, a similar conclusion follows for $\mathfrak{M}_{\Sigma_N}$.

Proof. We consider the family of functions $f_N(x) = \frac{1}{|x|}\chi_{3\leq|x|\leq N}$ defined on $\mathbf{R}^2$ for $N \geq 10$. Then we have

$$\big\|f_N\big\|_{L^2(\mathbf{R}^2)} \leq c_1(\log N)^{\frac{1}{2}}\,. \tag{10.3.7}$$

On the other hand, for every x in the annulus $6 < |x| < N$, we consider the rectangle R_x of dimensions $|x| - 3$ and $\frac{|x|-3}{N}$, one of whose shortest sides touches the circle $|y| = 3$ and the other has midpoint x. Then

$$\mathcal{K}_N(f_N)(x) \geq \frac{1}{|R_x|}\int_{R_x} |f_N(y)|\,dy \geq \frac{c_2 N}{(|x|-3)^2}\iint\limits_{\substack{6\leq y_1\leq|x|\\ |y_2|\leq\frac{|x|-3}{N}}} \frac{dy_1dy_2}{y_1} \geq c_3\,\frac{\log|x|}{|x|}\,.$$

It follows that

$$\big\|\mathcal{K}_N(f_N)\big\|_{L^2(\mathbf{R}^2)} \geq c_3\bigg(\int\limits_{6\leq|x|\leq N}\Big(\frac{\log|x|}{|x|}\Big)^2 dx\bigg)^{\frac{1}{2}} \geq c_4\,(\log N)^{\frac{3}{2}}\,. \tag{10.3.8}$$

Combining (10.3.7) with (10.3.8) we obtain (10.3.5) with $c = c_4/c_1$.

We now turn to estimate (10.3.6). Since for all $6 < |x| < N$ we have

$$\mathcal{K}_N(f_N)(x) \geq c_3\,\frac{\log|x|}{|x|} > c_3\frac{\log N}{N}\,,$$

it follows that $\big|\{\mathcal{K}_N(f_N) > c_3\frac{\log N}{N}\}\big| \geq \pi(N^2 - 6^2) \geq c_5N^2$ and hence

$$\begin{aligned}
\frac{\big\|\mathcal{K}_N(f_N)\big\|_{L^{2,\infty}}}{\big\|f_N\big\|_{L^2}} &\geq \frac{\sup_{\lambda>0}\lambda\big|\{\mathcal{K}_N(f_N)>\lambda\}\big|^{\frac{1}{2}}}{c_1(\log N)^{\frac{1}{2}}}\\
&\geq c_3\frac{\log N}{N}\,\frac{\big|\{\mathcal{K}_N(f_N) > c_3\frac{\log N}{N}\}\big|^{\frac{1}{2}}}{c_1(\log N)^{\frac{1}{2}}}\\
&\geq \frac{c_3\sqrt{c_5}}{c_1}\,(\log N)^{\frac{1}{2}}\,.
\end{aligned}$$

□

10.3.b. The Boundedness of $\mathfrak{M}_{\Sigma_N}$ on $L^p(\mathbf{R}^2)$

It is rather remarkable that both estimates of Proposition 10.3.4 are sharp in terms of their behavior as $N \to \infty$, as the following result indicates.

Theorem 10.3.5. *There exist constants $0 < B, C < \infty$ such that for every $N \geq 10$ and all $f \in L^2(\mathbf{R}^2)$ we have*

$$\big\|\mathfrak{M}_{\Sigma_N}(f)\big\|_{L^{2,\infty}(\mathbf{R}^2)} \leq B\,(\log N)^{\frac{1}{2}}\big\|f\big\|_{L^2(\mathbf{R}^2)} \tag{10.3.9}$$

and

$$\big\|\mathfrak{M}_{\Sigma_N}(f)\big\|_{L^2(\mathbf{R}^2)} \leq C\,(\log N)\big\|f\big\|_{L^2(\mathbf{R}^2)}\,. \tag{10.3.10}$$

In view of (10.3.4), similar estimates also hold for $\mathcal{K}_N$.

Proof. We will deduce (10.3.10) from the weak type estimate (10.3.9), which we rewrite as

$$\big|\{x \in \mathbf{R}^2 :\ \mathfrak{M}_{\Sigma_N}(f)(x) > \lambda\}\big| \leq B^2\,(\log N)\,\frac{\big\|f\big\|_{L^2}^2}{\lambda^2}\,. \tag{10.3.11}$$

We will prove this estimate for some constant $B > 0$ independent of N. But prior to doing this we indicate why (10.3.11) implies (10.3.10).

Using Exercise 10.3.2, we have that $\mathfrak{M}_{\Sigma_N}$ maps $L^p(\mathbf{R}^2)$ into $L^p(\mathbf{R}^2)$ (and hence into $L^{p,\infty}$) with constant at most a multiple of $N^{1/p}$ for all $1 < p < \infty$. Using this with $p = 3/2$, we have

$$\big\|\mathfrak{M}_{\Sigma_N}\big\|_{L^{\frac{3}{2}}\to L^{\frac{3}{2},\infty}} \leq \big\|\mathfrak{M}_{\Sigma_N}\big\|_{L^{\frac{3}{2}}\to L^{\frac{3}{2}}} \leq A\,N^{\frac{2}{3}} \tag{10.3.12}$$

for some constant $A > 0$. Now split f as the sum

$$f = f_1 + f_2 + f_3\,,$$

where

$$\begin{aligned} f_1 &= f\,\chi_{|f|\leq\frac{1}{4}\lambda}\\ f_2 &= f\,\chi_{\frac{1}{4}\lambda<|f|\leq N^2\lambda}\\ f_3 &= f\,\chi_{N^2\lambda<|f|}\,. \end{aligned}$$

It follows that

$$\big|\{\mathfrak{M}_{\Sigma_N}(f) > \lambda\}\big| \leq \big|\{\mathfrak{M}_{\Sigma_N}(f_2) > \tfrac{\lambda}{3}\}\big| + \big|\{\mathfrak{M}_{\Sigma_N}(f_3) > \tfrac{\lambda}{3}\}\big| \tag{10.3.13}$$

since the set $\{\mathfrak{M}_{\Sigma_N}(f_1) > \frac{\lambda}{3}\}$ is empty. To obtain the required result we will use the $L^{2,\infty}$ estimate (10.3.11) for f_2 and the $L^{\frac{3}{2},\infty}$ estimate (10.3.12) for f_3. We have

$$\begin{aligned}
&\big\|\mathfrak{M}_{\Sigma_N}(f)\big\|_{L^2}^2\\
&= 2\int_0^\infty \lambda\big|\{\mathfrak{M}_{\Sigma_N}(f) > \lambda\}\big|\,d\lambda\\
&\le \int_0^\infty 2\lambda\big|\{\mathfrak{M}_{\Sigma_N}(f_2) > \tfrac{\lambda}{3}\}\big|\,d\lambda + \int_0^\infty 2\lambda\big|\{\mathfrak{M}_{\Sigma_N}(f_3) > \tfrac{\lambda}{3}\}\big|\,d\lambda\\
&\le \int_0^\infty \frac{2\lambda B^2\,(\log N)}{\lambda^2}\int_{\frac{1}{4}\lambda<|f|\le N^2\lambda} |f|^2 dx\,d\lambda + \int_0^\infty \frac{2\lambda A^{\frac{3}{2}}\,N}{\lambda^{\frac{3}{2}}}\int_{|f|>N^2\lambda} |f|^{\frac{3}{2}}\,dx\,d\lambda\\
&\le 2B^2(\log N)\int_{\mathbf{R}^2}|f(x)|^2\int_{\frac{|f(x)|}{N^2}}^{4|f(x)|}\frac{d\lambda}{\lambda}\,dx + 2A^{\frac{3}{2}}\,N\int_{\mathbf{R}^2}|f(x)|^{\frac{3}{2}}\int_0^{\frac{|f(x)|}{N^2}}\frac{d\lambda}{\lambda^{\frac{1}{2}}}\,dx\\
&= \big(4B^2(\log 2N)(\log N) + 4A^{\frac{3}{2}}\big)\big\|f\big\|_{L^2}^2\\
&\le C(\log N)^2\big\|f\big\|_{L^2}^2
\end{aligned}$$

using Fubini's theorem for integrals. This proves (10.3.10).

To avoid problems with antipodal points, it will be convenient to split Σ_N as the union of eight sets, in each of which the angle between any two vectors does not exceed $2\pi/8$. It will suffice therefore to obtain (10.3.11) for each such subset of Σ_N. Let us fix one such subset of Σ_N, which we call Σ_N^1. To prove (10.3.10), we fix a $\lambda > 0$ and we start with a compact subset K of the set $\{x \in \mathbf{R}^2 :\ \mathfrak{M}_{\Sigma_N^1}(f)(x) > \lambda\}$. Then for every $x \in K$, there exists an open rectangle R_x that contains x and whose longest side is parallel to a vector in Σ_N^1. By compactness of K, there exists a finite subfamily $\{R_\alpha\}_{\alpha\in\mathcal{A}}$ of the family $\{R_x\}_{x\in K}$ such that

$$\int_{R_\alpha} |f(y)|\,dy > \lambda\,|R_\alpha|$$

for all $\alpha \in \mathcal{A}$.

We claim that there is a constant C such that for any finite family $\{R_\alpha\}_{\alpha\in\mathcal{A}}$ of rectangles whose longest side is parallel to a vector in Σ_N^1 there is a subset $\mathcal{B}$ of $\mathcal{A}$ such that

$$\int_{\mathbf{R}^2}\Big(\sum_{\beta\in\mathcal{B}}\chi_{R_\beta}(x)\Big)^2\,dx \le C\Big|\bigcup_{\beta\in\mathcal{B}} R_\beta\Big| \tag{10.3.14}$$

and that

$$\Big|\bigcup_{\alpha\in\mathcal{A}} R_\alpha\Big| \le C(\log N)\Big|\bigcup_{\beta\in\mathcal{B}} R_\beta\Big|. \tag{10.3.15}$$

Assuming (10.3.14) and (10.3.15), we can easily deduce (10.3.11). Indeed,

$$
\begin{aligned}
\Big|\bigcup_{\beta\in\mathcal{B}} R_\beta\Big| &\le \sum_{\beta\in\mathcal{B}} |R_\beta| \\
&< \frac{1}{\lambda}\sum_{\beta\in\mathcal{B}} \int_{R_\beta} |f(y)|\,dy \\
&= \frac{1}{\lambda}\int_{\mathbf{R}^2} \Big(\sum_{\beta\in\mathcal{B}} \chi_{R_\beta}\Big)\,|f(y)|\,dy \\
&\le \frac{1}{\lambda}\bigg(\int_{\mathbf{R}^2} \Big(\sum_{\beta\in\mathcal{B}} \chi_{R_\beta}\Big)^2 dx\bigg)^{\frac{1}{2}} \|f\|_{L^2} \\
&\le \frac{C^{\frac{1}{2}}}{\lambda}\Big|\bigcup_{\beta\in\mathcal{B}} R_\beta\Big|^{\frac{1}{2}} \|f\|_{L^2}\,,
\end{aligned}
$$

from which it follows that

$$
\Big|\bigcup_{\beta\in\mathcal{B}} R_\beta\Big| \le \frac{C}{\lambda^2}\,\|f\|_{L^2}^2\,.
$$

Then, using (10.3.15), we obtain

$$
|K| \le \Big|\bigcup_{\alpha\in\mathcal{A}} R_\alpha\Big| \le C\,(\log N)\Big|\bigcup_{\beta\in\mathcal{B}} R_\beta\Big| \le \frac{C^2}{\lambda^2}(\log N)\big\|f\big\|_{L^2}^2
$$

and since K was an arbitrary compact subset of $\{x:\ \mathfrak{M}_{\Sigma_N^1}(f)(x)>\lambda\}$, the same conclusion holds for the latter set.

We now turn to the selection of the subfamily $\{R_\beta\}_{\beta\in\mathcal{B}}$ and the proof of (10.3.14) and (10.3.15).

Let R_{β_1} be the rectangle in $\{R_\alpha\}_{\alpha\in\mathcal{A}}$ with the longest side. Suppose we have chosen $R_{\beta_1}, R_{\beta_2}, \dots, R_{\beta_{j-1}}$. Then among all rectangles R_α that satisfy

$$
\sum_{k=1}^{j-1} |R_{\beta_k}\cap R_\alpha| \le \frac{1}{2}|R_\alpha|\,, \tag{10.3.16}
$$

we choose a rectangle R_{β_j} such that its longest side is as large as possible. Since the collection $\{R_\alpha\}_{\alpha\in\mathcal{A}}$ is finite, this selection stops after m steps, for some positive integer m. Let $\mathcal{B}=\{\beta_1,\beta_2,\dots,\beta_m\}$.

Using (10.3.16), we obtain

$$
\begin{aligned}
\int_{\mathbf{R}^2}\Big(\sum_{\beta\in\mathcal{B}}\chi_{R_\beta}\Big)^2dx &\le 2\sum_{j=1}^{m}\sum_{k=1}^{j}|R_{\beta_k}\cap R_{\beta_j}| \\
&= 2\sum_{j=1}^{m}\Big[\Big(\sum_{k=1}^{j-1}|R_{\beta_k}\cap R_{\beta_j}|\Big)+|R_{\beta_j}|\Big] \\
&\le 2\sum_{j=1}^{m}\Big[\frac{1}{2}|R_{\beta_j}|+|R_{\beta_j}|\Big] \\
&= 3\sum_{j=1}^{m}|R_{\beta_j}|\,.
\end{aligned}
\tag{10.3.17}
$$

A consequence of this fact is that

$$
\begin{aligned}
\sum_{j=1}^{m}|R_{\beta_j}| &= \int_{\bigcup_{j=1}^m R_{\beta_j}}\Big(\sum_{j=1}^{m}\chi_{R_{\beta_j}}\Big)\,dx \\
&\le \Big|\bigcup_{j=1}^{m}R_{\beta_j}\Big|^{\frac12}\Big(\int_{\mathbf{R}^n}\Big(\sum_{\beta\in\mathcal{B}}\chi_{R_\beta}\Big)^2dx\Big)^{\frac12} \\
&\le \Big|\bigcup_{j=1}^{m}R_{\beta_j}\Big|^{\frac12}\sqrt{3}\Big(\sum_{j=1}^{m}|R_{\beta_j}|\Big)^{\frac12},
\end{aligned}
$$

which implies

$$
\sum_{j=1}^{m}|R_{\beta_j}|\le 3\Big|\bigcup_{j=1}^{m}R_{\beta_j}\Big|.
\tag{10.3.18}
$$

Using (10.3.18) in conjunction with the last estimate in (10.3.17), we obtain (10.3.14) with $C=9$.

We now turn to the proof of (10.3.15). Let M_c be the usual Hardy-Littlewood maximal operator with respect to cubes in $\mathbf{R}^n$ (or squares in $\mathbf{R}^2$; recall $n=2$). Since M_c is of weak type $(1,1)$, (10.3.15) will be a consequence of the estimate

$$
\bigcup_{\alpha\in\mathcal{A}\setminus\mathcal{B}}R_\alpha\subseteq\Big\{x\in\mathbf{R}^2:\ M_c\big(\sum_{\beta\in\mathcal{B}}\chi_{(R_\beta)^*}\big)(x)>c\,(\log N)^{-1}\Big\}
\tag{10.3.19}
$$

for some absolute constant c, where $(R_\beta)^*$ is the rectangle R_β expanded 5 times in both directions. Indeed, if (10.3.19) holds, then

$$\begin{aligned}\Big|\bigcup_{\alpha\in\mathcal{A}} R_\alpha\Big| &\le \Big|\bigcup_{\beta\in\mathcal{B}} R_\beta\Big| + \Big|\bigcup_{\alpha\in\mathcal{A}\setminus\mathcal{B}} R_\alpha\Big| \\ &\le \Big|\bigcup_{\beta\in\mathcal{B}} R_\beta\Big| + \frac{10}{c}(\log N)\sum_{\beta\in\mathcal{B}}|(R_\beta)^*| \\ &\le \Big|\bigcup_{\beta\in\mathcal{B}} R_\beta\Big| + \frac{250}{c}(\log N)\sum_{\beta\in\mathcal{B}}|R_\beta| \\ &\le C\,(\log N)\Big|\bigcup_{\beta\in\mathcal{B}} R_\beta\Big|,\end{aligned}$$

where we just used (10.3.18) and the fact that N is large.

It remains to prove (10.3.19). At this point we will need the following lemma. In the sequel we will denote by θ_α the angle between the x axis and the vector pointing in the longest direction of R_α for any $\alpha\in\mathcal{A}$. We will also denote by l_α the shortest side of R_α and by L_α the longest side of R_α for any $\alpha\in\mathcal{A}$. Finally, we set

$$\omega_k = \frac{2\pi 2^k}{N}$$

for $k\in\mathbf{Z}^+$ and $\omega_0 = 0$.

Lemma 10.3.6. *Let R_α be a rectangle in the family $\{R_\alpha\}_{\alpha\in\mathcal{A}}$ and let $0\le k < \big[\frac{\log(N/8)}{\log 2}\big]$. Suppose that $\beta\in\mathcal{B}$ is such that*

$$\omega_k \le |\theta_\alpha - \theta_\beta| < \omega_{k+1}$$

and such that $L_\beta \ge L_\alpha$. Let $s_\alpha = 8\max(l_\alpha, \omega_k L_\alpha)$. For an arbitrary $x\in R_\alpha$, let Q be a square centered at x with sides of length s_α parallel to the sides of R_α. Then we have

$$\frac{|R_\beta\cap R_\alpha|}{|R_\alpha|} \le 16\,\frac{|(R_\beta)^*\cap Q|}{|Q|}. \tag{10.3.20}$$

Assuming Lemma 10.3.6, we conclude the proof of (10.3.19). Suppose that $\alpha\in\mathcal{A}\setminus\mathcal{B}$. Then the rectangle R_α was not selected in the selection procedure. This means that for all $l\in\{1,2,\dots,m\}$ we have exactly one of the following: either

$$\sum_{j=1}^{l-1}|R_{\beta_j}\cap R_\alpha| > \frac{1}{2}|R_\alpha| \tag{10.3.21}$$

or

$$\sum_{j=1}^{l-1}|R_{\beta_j}\cap R_\alpha| \le \frac{1}{2}|R_\alpha| \qquad\text{and}\qquad L_\alpha \le L_{\beta_l}. \tag{10.3.22}$$

Since (10.3.22) holds for $l=1$, we let $\mu \le m$ be the largest integer so that (10.3.22) holds for all $l \le \mu$. Then (10.3.21) fails for $l=\mu$ but holds for $l=\mu+1$; thus

$$\frac{1}{2}|R_\alpha| < \sum_{j=1}^{\mu} |R_{\beta_j} \cap R_\alpha| \le \sum_{\substack{\beta\in\mathcal{B}\\ L_\beta \ge L_\alpha}} |R_\beta \cap R_\alpha|\,. \tag{10.3.23}$$

[It should be noted that even in the special case $\mu = m$, (10.3.21) also holds for $l = m+1$; otherwise for this $\alpha \in \mathcal{A}\setminus\mathcal{B}$ we would have

$$\sum_{j=1}^{m} |R_{\beta_j} \cap R_\alpha| \le \frac{1}{2}|R_\alpha| \tag{10.3.24}$$

and thus there would exist an R_α with maximum longest length satisfying (10.3.24), which would have been defined as $R_{\beta_{m+1}}$, contradicting that $\mathcal{B}$ has m elements.]

Now there exists at least one k with $0 \le k < \left[\frac{\log(N/8)}{\log 2}\right]$ such that

$$\frac{\log 2}{2\log(N/8)}\,|R_\alpha| \;<\; \sum_{\substack{\beta\in\mathcal{B}\\ L_\beta\ge L_\alpha\\ \omega_k \le |\theta_\beta-\theta_\alpha|<\omega_{k+1}}} |R_\beta \cap R_\alpha|\,. \tag{10.3.25}$$

By Lemma 10.3.6, for any $x \in R_\alpha$ there is a square Q such that (10.3.20) holds for any R_β with $\beta\in\mathcal{B}$ satisfying $L_\beta \ge L_\alpha$ and $\omega_k \le |\theta_\beta - \theta_\alpha| < \omega_{k+1}$. It follows that

$$\frac{\log 2}{2\log(N/8)} < 2 \sum_{\substack{\beta\in\mathcal{B}\\ L_\beta\ge L_\alpha\\ \omega_k \le |\theta_\beta-\theta_\alpha|<\omega_{k+1}}} \frac{|(R_\beta)^*\cap Q|}{|Q|}\,,$$

which implies

$$\frac{c}{\log N} < \frac{\log 2}{4\log(N/8)} < \frac{1}{|Q|}\int_Q \sum_{\beta\in\mathcal{B}} \chi_{(R_\beta)^*}\,dx\,.$$

This proves (10.3.19) since for $\alpha\in\mathcal{A}\setminus\mathcal{B}$, any $x\in R_\alpha$ must be an element of the set $\big\{x\in\mathbf{R}^2:\ M_c\big(\sum_{\beta\in\mathcal{B}}\chi_{(R_\beta)^*}\big)(x) > c\,(\log N)^{-1}\big\}$.

□

It remains to prove Lemma 10.3.6, which we do now. The proof will be based on a few geometric observations.

Proof. Let τ be the angle between the directions of the rectangles R_α and R_β, that is,

$$\tau = |\theta_\alpha - \theta_\beta|\,.$$

By assumption we have $\tau < \omega_{k+1} \le \frac{\pi}{4}$ since $k+1 \le \left[\frac{\log(N/8)}{\log 2}\right] \le \frac{\log(N/8)}{\log 2}$.

We first indicate why the expanded rectangle $(R_\beta)^*$ meets the square Q. We may assume that $\overline{R_\alpha}\cap\overline{R_\beta} \ne \emptyset$; otherwise there is nothing to prove. The worst possible case appears on the left in Figure 10.9 in which the upper left corner of R_β

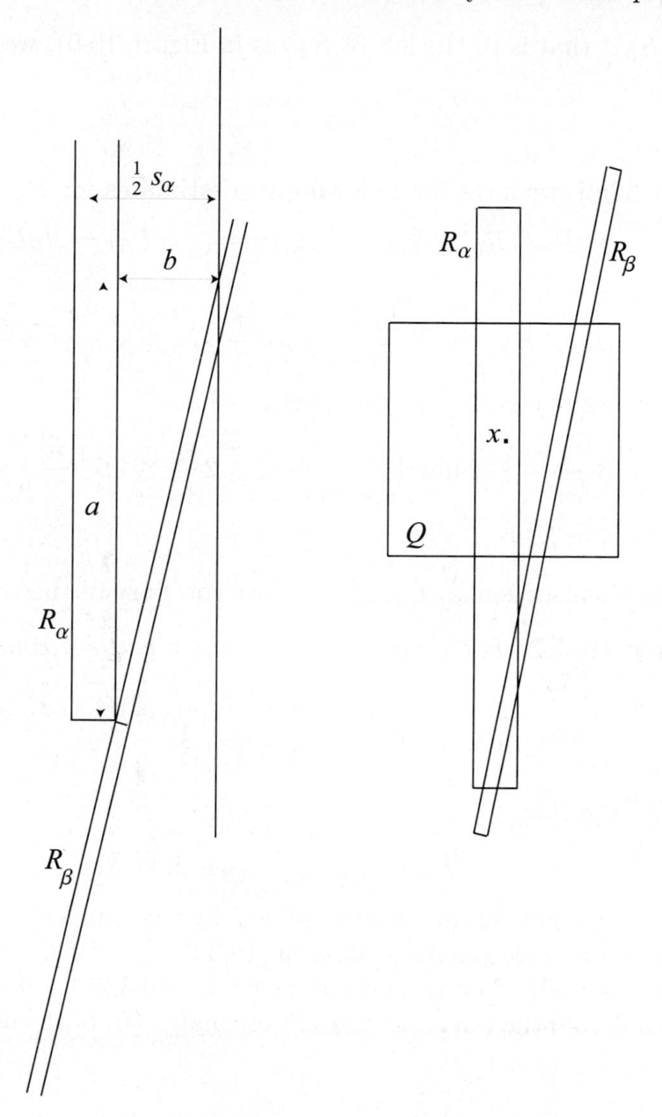

FIGURE 10.9. The rectangles R_α, R_β and the square Q.

barely touches the lower right corner of R_α. Using the notation in Figure 10.9, we have that

$$\frac{3}{2}\, 2\,\omega_k > \frac{3}{2}\,\tau > \tan(\tau) = \frac{b}{a} \geq \frac{\frac{1}{2}\, s_\alpha - l_\alpha}{a} \geq \frac{\frac{3}{8}\, s_\alpha}{a} \geq \frac{3\,\omega_k\, L_\alpha}{a}\,,$$

from which it follows that $a > L_\alpha$. This implies that the rectangle $(R_\beta)^*$ meets the square Q. Next we examine the size of the intersection $(R_\beta)^* \cap Q$. By considering the part of $(R_\beta)^*$ that is to the left of R_β (as in Figure 10.9), we can see easily that

$$\frac{|(R_\beta)^*\cap Q|}{|Q|} \ge \frac{l_\beta\, s_\alpha \frac{1}{\cos(\tau)}}{s_\alpha^2} \ge \frac{l_\beta}{s_\alpha}\,.$$

On the other hand, we have the following two estimates for $R_\alpha\cap R_\beta$:

$$|R_\alpha\cap R_\beta| \le 2\min(l_\alpha,l_\beta)\max(L_\alpha,L_\beta) = 2l_\beta L_\alpha$$

and

$$|R_\alpha\cap R_\beta| \le l_\alpha\frac{l_\beta}{\sin(\tau)} \le l_\alpha l_\beta \frac{\pi}{2\tau} \le l_\alpha l_\beta\frac{\pi}{2\omega_k} \le 2\frac{l_\alpha l_\beta}{\omega_k}\,,$$

as a geometric argument shows. We conclude that

$$\frac{|R_\alpha\cap R_\beta|}{|R_\alpha|} \le 2\min\Big(\frac{l_\beta}{l_\alpha},\frac{l_\beta}{\omega_k L_\alpha}\Big) \le 2\frac{l_\beta}{s_\alpha} \le 16\,\frac{|(R_\beta)^*\cap Q|}{|Q|}\,.$$

□

We end this subsection with an immediate corollary of the theorem just proved.

Corollary 10.3.7. *For every $1<p<\infty$ there exists a constant c_p such that*

$$\big\|\mathcal{K}_N\big\|_{L^p(\mathbf{R}^2)\to L^p(\mathbf{R}^2)} \le c_p \begin{cases} N^{\frac{2}{p}-1}(\log N)^{\frac{1}{p'}} & \text{when } 1<p<2,\\ (\log N)^{\frac{1}{p}} & \text{when } 2<p<\infty.\end{cases} \tag{10.3.26}$$

Proof. We see that

$$\big\|\mathcal{K}_N\big\|_{L^1(\mathbf{R}^2)\to L^{1,\infty}(\mathbf{R}^2)} \le CN \tag{10.3.27}$$

by replacing a rectangle of dimensions $a\times aN$ by the smallest square of side length aN that contains it. Interpolating between (10.3.9) and (10.3.27), we obtain the first statement in (10.3.26). The second statement in (10.3.26) follows by interpolation between (10.3.9) and the trivial $L^\infty\to L^\infty$ estimate. (In both cases we use Theorem 1.3.2.) □

10.3.c. The Higher-Dimensional Kakeya Maximal Operator

The reader may recall that only the boundedness of the Kakeya maximal operator without dilations $\mathcal{K}_N^a$ on $L^2(\mathbf{R}^2)$ (with logarithmic bound) was needed in obtaining the boundedness of the Bochner-Riesz operator B^λ on $L^4(\mathbf{R}^2)$. An analogous operator can be defined on $\mathbf{R}^n$.

Definition 10.3.8. Given fixed $a>0$ and $N\ge 10$, we introduce the *Kakeya maximal operator without dilations* on $\mathbf{R}^n$ as

$$\mathcal{K}_N^a(f)(x) = \sup_R \frac{1}{|R|}\int_R |f(y)|\,dy\,,$$

where the supremum is taken over all rectangular parallelepipeds (boxes) of arbitrary orientation in $\mathbf{R}^n$ that contain the point x and have dimensions

$$\underbrace{a \times a \times \cdots \times a}_{n-1 \text{ times}} \times aN\,.$$

We also define the higher-dimensional analogue of the Kakeya maximal operator $\mathcal{K}_N$ introduced in (10.3.3).

Definition 10.3.9. Let $N \geq 10$. We denote by $\mathcal{R}(N)$ the set of all rectangular parallelepipeds (boxes) in $\mathbf{R}^n$ with arbitrary orientation and dimensions

$$\underbrace{a \times a \times \cdots \times a}_{n-1 \text{ times}} \times aN$$

with arbitrary $a > 0$. Given a locally integrable function f on $\mathbf{R}^n$, we define

$$\mathcal{K}_N(f)(x) = \sup_{\substack{R \in \mathcal{R}(N) \\ R \ni x}} \frac{1}{|R|} \int_R |f(y)|\, dy\,.$$

The operator $\mathcal{K}_N$ is called the *nth-dimensional Kakeya maximal function.*

For convenience we will call rectangular parallelepipeds [i.e., elements of $\mathcal{R}(N)$] higher-dimensional rectangles, or simply rectangles. We clearly have

$$\sup_{a>0} \mathcal{K}_N^a = \mathcal{K}_N$$

hence the boundedness of $\mathcal{K}_N^a$ can be deduced from that of $\mathcal{K}_N$; however, this deduction can essentially be reversed with only logarithmic loss in N. The reader may consult the notes at the end of this chapter for references. In the sequel we will restrict our attention to the operator $\mathcal{K}_N^a$, whose study already presents all the essential difficulties and requires a novel set of ideas in its analysis. We may restrict our attention to a specific value of a, as a simple dilation argument yields that the norms of $\mathcal{K}_N^a$ and $\mathcal{K}_N^b$ on any $L^p(\mathbf{R}^n)$ are equal for all $a, b > 0$.

It would be desirable to know the following estimate for $\mathcal{K}_N^1$:

$$\big\|\mathcal{K}_N^1\big\|_{L^n(\mathbf{R}^n) \to L^{n,\infty}(\mathbf{R}^n)} \leq c_n' (\log N)^{\frac{n-1}{n}} \tag{10.3.28}$$

for some dimensional constant c_n'. It would then follow that

$$\big\|\mathcal{K}_N^1\big\|_{L^n(\mathbf{R}^n) \to L^n(\mathbf{R}^n)} \leq c_n \log N \tag{10.3.29}$$

for some other dimensional constant c_n; see Exercise 10.3.8(b). Moreover, we have the estimate

$$\big\|\mathcal{K}_N^1\big\|_{L^1(\mathbf{R}^n) \to L^{1,\infty}(\mathbf{R}^n)} \leq c_n'' N^{n-1} \tag{10.3.30}$$

by replacing a rectangle of dimensions $\overbrace{1 \times 1 \times \cdots \times 1}^{n-1 \text{ times}} \times N$ by the smallest cube of side length N that contains it. (This estimate is sharp; see Exercise 10.3.7.)

If estimate (10.3.28) were true, then interpolating between (10.3.30) and (10.3.28) would yield the bound

$$\big\|\mathcal{K}_N^1\big\|_{L^p(\mathbf{R}^n)\to L^p(\mathbf{R}^n)} \le c_{n,p} N^{\frac{n}{p}-1}(\log N)^{\frac{1}{p'}}\,, \qquad 1<p<n\,. \tag{10.3.31}$$

It is estimate (10.3.31) that we would like to concentrate on. We have the following result for a certain range of p's in $(1,n)$.

Theorem 10.3.10. *For every* $1<p\le\frac{n+1}{2}$ *there exists a constant* $C_{n,p}$ *such that for all* $N\ge 10$ *we have*

$$\big\|\mathcal{K}_N^1\big\|_{L^p(\mathbf{R}^n)\to L^p(\mathbf{R}^n)} \le C_{n,p} N^{\frac{n}{p}-1}(\log N)^{\frac{1}{p'}}\,. \tag{10.3.32}$$

Proof. Let $p_n=\frac{n+1}{2}$. It suffices to prove (10.3.32) for $p=p_n$; indeed, if we know the claimed result for $p=p_n$, the case $1<p<p_n$ will follow by interpolation between the case $p=p_n$ and the case $p=1$, in which a weak type estimate holds with constant C_nN^{n-1}. We therefore concentrate on estimate (10.3.32) when $p=p_n$. In view of Exercise 10.3.8(a) (with $q=1$, $p=p_n$), it will suffice to show that

$$\big\|\mathcal{K}_N^1\big\|_{L^{p_n,1}(\mathbf{R}^n)\to L^{p_n,\infty}(\mathbf{R}^n)} \le C_{n,p} N^{\frac{n}{p}-1}\,. \tag{10.3.33}$$

At this point, in order to make the geometric idea of the proof a bit more transparent, we set $\delta=\frac{1}{N}\le\frac{1}{10}$ and we choose to work with the equivalent operator $\mathcal{K}_N^\delta$ (whose norm is the same as that of $\mathcal{K}_N^1$). This is only a change of scale. Instead of averaging over rectangles of dimensions $1\times1\times\cdots\times1\times N$, we average over rectangles of dimensions $\delta\times\delta\times\cdots\times\delta\times1$.

Since the operators in question are positive, we also choose to work with positive functions. Our next reduction involves a linearization of the operator $\mathcal{K}_N^\delta$. Let us call a rectangle of dimensions $\delta\times\delta\times\cdots\times\delta\times1$ a δ-tube or simply a *tube*. The idea here is for every x in $\mathbf{R}^n$ to select (in some measurable way) a tube $\tau(x)$ that contains x so that the average of f over $\tau(x)$ satisfies

$$\mathcal{K}_N^\delta(f)(x)\le\frac{2}{|\tau(x)|}\int_{\tau(x)}f(y)\,dy\,.$$

Then we observe that if we have a cubical grid in $\mathbf{R}^n$ of side length δ, then for every z in the cube that contains the given x we have

$$\mathcal{K}_N^\delta(f)(z)\le c_n\mathcal{K}_N^\delta(f)(x)\le\frac{2c_n}{|\tau(x)|}\int_{\tau(x)}f(y)\,dy \tag{10.3.34}$$

for some dimensional constant. This is saying that $\tau(x)$ can be chosen as the tube for which the maximum is attained for all points in the cube of the grid the contains the point x. This observation motivates us to introduce a grid of width δ in $\mathbf{R}^n$ and for every cube Q_j in the grid to associate with it a tube τ_j so that

$$\tau_j\cap Q_j\neq\emptyset.$$

Then we define a linear operator

$$L^\delta(f) = \left(\int_{\tau_j} f(y)\,dy\right)\chi_{Q_j}$$

and we claim that the $L^{p_n,1} \to L^{p_n,\infty}$ boundedness of L^δ with constant $C_n(\delta^{-1})^{\frac{n}{p}-1}$ (which is independent of the choice of cubes Q_j and the choice of τ_j) implies the boundedness of $\mathcal{K}_N^\delta$ with a slightly worse dimensional constant. The reason for this is that given a function f, we can control $\big\|\mathcal{K}_N^\delta(f)\big\|_{L^{p_n,\infty}}$ by $c_n\big\|L^\delta(f)\big\|_{L^{p_n,\infty}}$ for the choice of grid and tubes that satisfy (10.3.34) for the given function f. Then we obtain the required estimate for $\mathcal{K}_N^\delta$.

Our next reduction will be to take f to be the characteristic function of a set. The space $L^{p_n,\infty}$ is normable [i.e., it has an equivalent norm under which it is a Banach space (Exercise 1.1.12)]; hence by Exercise 1.4.7, the boundedness of L^δ from $L^{p_n,1}$ into $L^{p_n,\infty}$ will be a consequence of the restricted weak type estimate

$$\sup_{\lambda>0} \lambda\big|\{L^\delta(\chi_A) > \lambda\}\big|^{\frac{1}{p_n}} \le C_n'(\delta^{-1})^{\frac{n}{p_n}-1}|A|^{\frac{1}{p_n}}\,, \tag{10.3.35}$$

for some dimensional constant C_n and all sets A of finite measure. This estimate can be written as

$$\lambda^{\frac{n+1}{2}}\delta^{\frac{n-1}{2}}|E_\lambda| \le C_n|A| \tag{10.3.36}$$

where

$$E_\lambda = \{x \in \mathbf{R}^n : \ L^\delta(\chi_A)(x) > \lambda\} = \{L^\delta(\chi_A) > \lambda\}\,.$$

Our final reduction stems from the observation that the operator L^δ is "local." This means that if f is supported in a cube Q, then $L^\delta(f)$ is supported in a multiple of Q, in our case in the cube $10\sqrt{n}\,Q$. [It is simple to verify that if $x \notin 10\sqrt{n}\,Q$, then $L^\delta(f)(x) = 0$ if f is supported in Q, as no tube containing x can reach Q.] For "local" operators, it suffices to prove their boundedness for functions supported in cubes; see Exercise 10.3.9. We may therefore work with a measurable set A contained in a cube in $\mathbf{R}^n$.

Having completed all the required reductions we proceed by showing the restricted weak type estimate (10.3.35) for sets A supported in a cube of side length 1. In proving (10.3.35) we may take $\lambda \le 1$; otherwise there is nothing to prove. We may also take $\delta \le \frac{\lambda}{100\sqrt{n}}$ as the remaining case is simple. Indeed, if $\delta > \frac{\lambda}{100\sqrt{n}}$, then

$$|E_\lambda| \le C_n^1(1/\delta)^{n-1}\,\frac{|A|}{\lambda} \tag{10.3.37}$$

by the weak type $(1,1)$ boundedness of L^δ (or $\mathcal{M}_N^\delta$) with constant $C_n^1\delta^{1-n}$. It follows from (10.3.37) that

$$C_n^1|A| \ge |E_\lambda|\delta^{n-1}\lambda \ge (100\sqrt{n})^{-\frac{n-1}{2}}\,|E_\lambda|\lambda^{\frac{n+1}{2}}\delta^{\frac{n-1}{2}}\,,$$

which proves (10.3.36) in this case.

We now consider the main case $100\sqrt{n}\,\delta \le \lambda \le 1$. Since $L(\chi_A)$ is constant on each Q_j, we have that each Q_j is either entirely contained in the set E_λ or it is disjoint from it. Denote

$$\mathcal{E} = \{j:\ Q_j \subseteq E_\lambda\}.$$

Then

$$|\mathcal{E}| = \#\{j \in \mathcal{E}\} = \frac{|E_\lambda|}{\delta^n}$$

and for all $j \in \mathcal{E}$ we have $|R_j \cap A| > \lambda |R_j| = \lambda\,\delta^{n-1}$. Then we have

$$\int_A \sum_{j\in\mathcal{E}} \chi_{\tau_j}\,dy = \sum_{j\in\mathcal{E}} |\tau_j \cap A| > \lambda\,\delta^{n-1}|\mathcal{E}| = \lambda\,\delta^{n-1}\frac{|E_\lambda|}{\delta^n} = \frac{\lambda\,|E_\lambda|}{\delta}.$$

Therefore, there exists an x_0 in A such that

$$\#\{j \in \mathcal{E}:\ x_0 \in \tau_j\} > \frac{\lambda\,|E_\lambda|}{\delta|A|}.$$

Let us call a set F of tubes ε-separated, if for every τ and τ' in F we have that

$$\text{angle}(\tau,\tau') \ge \varepsilon$$

for some fixed $\varepsilon > 0$.

Let $S(x_0,\frac{1}{2})$ be a sphere of radius $\frac{1}{2}$ centered at the point x_0. We can find on this sphere a family of spherical caps S_k of diameter δ (and area $\omega_{n-1}\delta^{n-1}$) that are at distance at least $10\sqrt{n}\,\delta$ from each other. We see that given such a cap S_k, there exist at most $c_n\delta^{-1}$ tubes (from the initial family) that contain the point x_0 and intersect S_k. Since we have at least $\frac{\lambda|E_\lambda|}{\delta|A|}$ tubes that contain the given point x_0, it follows that at least $\frac{\lambda|E_\lambda|}{c_n|A|}$ tubes have to pass through different such caps S_k. But tubes that intersect different caps S_k and contain x_0 are δ-separated. We have therefore extracted at least $\frac{\lambda|E_\lambda|}{c_n|A|}$ δ-separated tubes from the original family that contain the point x_0. Next, we extract a subfamily of these δ-separated tubes that is $\frac{30\sqrt{n}\,\delta}{\lambda}$-separated. This is straightforward. We can "thin out" this family by removing the $\frac{c_n}{\lambda^{n-1}}$ tubes that are the closest neighbors of a given tube, thus making the angle of separation at least $\frac{30\sqrt{n}\,\delta}{\lambda}$. In this way we can extract at least $c_n'\frac{\lambda|E_\lambda|}{|A|}\lambda^{n-1} = c_n'\frac{\lambda^n|E_\lambda|}{|A|}$ tubes that are $\frac{30\sqrt{n}\,\delta}{\lambda}$-separated and contain the point x_0. We denote these tubes by $\{\tau_j:\ j \in \mathcal{F}\}$. We have therefore found a subset $\mathcal{F}$ of $\mathcal{E}$ such that

$$x_0 \in \tau_j \quad \text{for all} \quad j \in \mathcal{F}, \tag{10.3.38}$$

$$\text{angle}(\tau_k,\tau_j) \ge 30\sqrt{n}\,\frac{\delta}{\lambda} \quad \text{if} \quad j,k \in \mathcal{F},\ j \ne k, \tag{10.3.39}$$

$$|\mathcal{F}| \ge c_n'\frac{|E_\lambda|\lambda^n}{|A|}. \tag{10.3.40}$$

Recall that for any $j \in \mathcal{E}$ (and hence also in $\mathcal{F}$) we have $|A \cap \tau_j| > \lambda\delta^{n-1}$. But we see that

$$|A \cap \tau_j \cap B(x_0, \tfrac{\lambda}{3})| \le \frac{2}{3}\lambda\delta^{n-1};$$

hence it follows that

$$|A \cap \tau_j \cap B(x_0, \tfrac{\lambda}{3})^c| > \frac{1}{3}\lambda\delta^{n-1}.$$

Moreover, it is crucial to note that since the tubes $\{\tau_j\}_{j\in\mathcal{F}}$ are $\frac{30\sqrt{n}\,\delta}{\lambda}$-separated, the sets

$$A \cap \tau_j \cap B(x_0, \tfrac{\lambda}{3})^c, \qquad j \in \mathcal{F}$$

are pairwise disjoint. This is because the tips of two vectors of length $\frac{\lambda}{3}$ and of angle at least $30\sqrt{n}\,\frac{\delta}{\lambda}$ that start from the same point x_0 must have distance at least $10\sqrt{n}\,\delta$ from each other. On the other hand, any tubes that point in the directions of these vectors and contain x_0 must be at most $\sqrt{n}$ away from these vectors. We can now conclude the proof of the theorem as follows:

$$\begin{aligned}
|A| &\ge \Big|A \cap \bigcup_{j\in\mathcal{F}} (\tau_j \cap B(x_0, \tfrac{\lambda}{3})^c)\Big| \\
&= \sum_{j\in\mathcal{F}} |A \cap \tau_j \cap B(x_0, \tfrac{\lambda}{3})^c| \\
&\ge \sum_{j\in\mathcal{F}} \frac{\lambda\delta^{n-1}}{3} \\
&= |\mathcal{F}|\,\frac{\lambda\delta^{n-1}}{3} \\
&\ge c_n' \frac{|E_\lambda|\lambda^n}{|A|}\frac{\lambda\delta^{n-1}}{3}.
\end{aligned}$$

We conclude that

$$|A|^2 \ge \tfrac{c_n'}{3}\lambda^{n+1}\delta^{n-1}|E_\lambda| \ge c_n''\lambda^{n+1}\delta^{n-1}|E_\lambda|^2$$

since the set E_λ is contained in a cube of side length at most $10\sqrt{n}$ (recall that A is contained in a cube of side length 1). Taking square roots, we obtain (10.3.36). This proves (10.3.35) and hence (10.3.32).

□

Exercises

10.3.1. Let h be the characteristic function of the square $[0,1]^2$ in $\mathbf{R}^2$. Prove that for any $0 < \lambda < 1$ we have

$$\big|\{x \in \mathbf{R}^2 : M_s(h)(x) > \lambda\}\big| \ge \frac{1}{\lambda}\log\frac{1}{\lambda}.$$

Use this to show that M_s is not of weak type $(1,1)$. Compare this result with that of Exercise 2.1.6.

10.3.2. (a) Given a unit vector v in $\mathbf{R}^2$ define the *directional maximal function along* $\vec{v}$ by

$$M_{\vec{v}}(f)(x) = \sup_{\varepsilon>0} \frac{1}{2\varepsilon}\int_{-\varepsilon}^{+\varepsilon} |f(x - t\vec{v})|\,dt$$

wherever f is locally integrable over $\mathbf{R}^2$. Prove that for such f, $M_{\vec{v}}(f)(x)$ is well defined for almost all x contained in any line not parallel to $\vec{v}$.
(b) Use the method of rotations to show that $M_{\vec{v}}$ is bounded on $L^p(\mathbf{R}^2)$ for all $1 < p < \infty$ with the same constant as the norm of the centered Hardy-Littlewood maximal operator M on $L^p(\mathbf{R})$.
(c) Let Σ be a finite set of directions. Prove that for all $1 \le p \le \infty$, there is a constant $C_p > 0$ such that

$$\big\|\mathfrak{M}_\Sigma(f)\big\|_{L^p(\mathbf{R}^2)} \le C_p\,|\Sigma|^{\frac{1}{p}}\,\big\|f\big\|_{L^p(\mathbf{R}^2)}$$

for all f in $L^p(\mathbf{R}^2)$.
[*Hint:* Control $\mathfrak{M}_\Sigma(f)^p$ pointwise by $\sum_{\vec{v}\in\Sigma}[M_{\vec{v}}M_{\vec{v}^\perp}(f)]^p$.]

10.3.3. Show that

$$\mathcal{K}_N \le 20\,\mathfrak{M}_{\Sigma_N}\,,$$

where Σ_N is a set of N uniformly distributed vectors in $\mathbf{S}^1$.
[*Hint:* Use Exercise 10.2.3.]

10.3.4. This exercise indicates a connection between the Besicovitch construction in Section 10.1 and the Kakeya maximal function. Recall the set E of Lemma 10.1.1, which satisfies $\frac{1}{2} \le |E| \le \frac{3}{2}$.
(a) Show that there is a positive constant c such that for all $N \ge 10$ we have

$$\big|\big\{x \in \mathbf{R}^2 :\ \mathcal{K}_N(\chi_E)(x) > \tfrac{1}{144}\big\}\big| \ge c\log\log N\,.$$

(b) Conclude that for all $2 < p < \infty$ there is a constant c_p such that

$$\big\|\mathcal{K}_N\big\|_{L^p(\mathbf{R}^2)\to L^p(\mathbf{R}^2)} \ge c_p(\log\log N)^{\frac{1}{p}}\,.$$

[*Hint:* Using the notation of Lemma 10.1.1, first show that

$$\big|\big\{x \in \mathbf{R}^2 :\ \mathcal{K}_{3\cdot 2^k\log(k+3)}(\chi_E)(x) > \tfrac{1}{36}\big\}\big| \ge \log(k+3)\,,$$

by showing that the previous set contains all the disjoint rectangles R_j for $j = 1, 2, \dots, 2^k$; here k is a large positive integer. To show this, for x in $\bigcup_{j=1}^{2^k} R_j$ consider the unique rectangle R_{j_x} that contains x union $(R_{j_x})'$ and set $R_x = R_{j_x} \cup (R_{j_x})'$. Then $|R_x| = 3|R_{j_x}| = 3 \cdot 2^{-k}\log(k+3)$ and we have

$$\frac{1}{|R_x|}\int_{R_x} |\chi_E(y)|\,dy = \frac{|E\cap R_x|}{|R_x|} \geq \frac{|E\cap (R_{j_x})'|}{3|R_{j_x}|} \geq \frac{1}{36}$$

in view of conclusion (4) in Lemma 10.1.1. Part (b): Express the L^p norm of $\mathcal{K}_N(\chi_E)$ in terms of its distribution function.]

10.3.5. Show that $\mathfrak{M}_{\mathbf{S}^1}$ is unbounded on $L^p(\mathbf{R}^2)$ for any $p<\infty$.
[*Hint:* You may use Proposition 10.3.4 when $p \leq 2$. When $p > 2$ you will need the previous Exercise.]

10.3.6. Consider the nth-dimensional Kakeya maximal operator $\mathcal{K}_N$. Show that there exist dimensional constants c_n and c_n' such that for all $N \geq 10$ we have

$$\begin{aligned} \big\|\mathcal{K}_N\big\|_{L^n(\mathbf{R}^n)\to L^n(\mathbf{R}^n)} &\geq c_n\,(\log N) \\ \big\|\mathcal{K}_N\big\|_{L^n(\mathbf{R}^n)\to L^{n,\infty}(\mathbf{R}^n)} &\geq c_n'\,(\log N)^{\frac{n-1}{n}}\,. \end{aligned}$$

[*Hint:* Consider the functions $f_N(x) = \frac{1}{|x|}\chi_{3\leq|x|\leq N}$ and adapt the argument in Proposition 10.3.4 to an nth-dimensional setting.]

10.3.7. For all $1 \leq p < n$ show that there exist constants $c_{n,p}$ such that the nth-dimensional Kakeya maximal operator $\mathcal{K}_N$ satisfies

$$\big\|\mathcal{K}_N\big\|_{L^p(\mathbf{R}^n)\to L^p(\mathbf{R}^n)} \geq \big\|\mathcal{K}_N\big\|_{L^p(\mathbf{R}^n)\to L^{p,\infty}(\mathbf{R}^n)} \geq c_{n,p}\,N^{\frac{n}{p}-1}\,.$$

[*Hint:* Consider the functions $h_N(x) = |x|^{-\frac{n+1}{p}}\chi_{3\leq|x|\leq N}$ and show that $\mathcal{K}_N(h_N)(x) > c/|x|$ for all x in the annulus $6 < |x| < N$.]

10.3.8. (*Carbery, Hernández, and Soria* [**80**]) Let T be a sublinear operator defined on $L^1(\mathbf{R}^n) + L^\infty(\mathbf{R}^n)$ and taking values in a set of measurable functions. Let $10 \leq N < \infty$ and $1 \leq q < p < \infty$.
(a) Suppose that for some $0 < A, B < \infty$ we have

$$\begin{aligned} &(A) \qquad && \|T\|_{L^q\to L^{q,\infty}} \leq A\,N\,, \\ &(B) \qquad && \|T\|_{L^{p,1}\to L^{p,\infty}} \leq B\,, \\ &(C) \qquad && \|T\|_{L^\infty\to L^\infty} \leq 1\,. \end{aligned}$$

Show that there is a constant $D = D(A,B,p,q)$ such that

$$(D) \qquad \|T\|_{L^p\to L^{p,\infty}} \leq D\,(\log N)^{\frac{1}{p'}}\,.$$

(b) Suppose that T satisfies (A), (B), and (D) for some $A, B, D < \infty$. Show that there is a constant $C = C(A, B, D, p, q)$ such that

$$\|T\|_{L^p \to L^p} \le C \log N \, .$$

$\big[$*Hint:* Part (a): Split $f = f_1 + f_2 + f_3$, where $f_1 = f\chi_{|f| \le \frac{\lambda}{4}}$, $f_2 = f\chi_{\frac{\lambda}{4} < |f| \le L\lambda}$, and $f_3 = f\chi_{|f| > L\lambda}$, where $L = N^{q/(p-q)}$. Apply (A) on f_3 and (B) on f_2. You will need the auxiliary result

$$\big\|f\chi_{a \le |f| \le b}\big\|_{L^{p,1}} \le a\, d_f(a)^{\frac{1}{p}} + \int_a^b d_f(t)^{\frac{1}{p}}\, dt \le C_p(1 + \log \tfrac{b}{a})^{\frac{1}{p'}} \big\|f\big\|_{L^p} \, ,$$

which can be obtained easily using Hölder's inequality (d_f is the distribution function of f). Part (b): Use the same splitting and the idea employed in the proof of Theorem 10.3.5.$\big]$

10.3.9. Suppose that T is a linear operator defined on a subspace of measurable functions on $\mathbf{R}^n$ has the property that whenever f is supported in a cube Q of side length s, then $T(f)$ is supported in aQ for some $a > 1$. Show the following.
(a) If T is defined on $L^p(\mathbf{R}^n)$ for some $0 < p < \infty$ and

$$\big\|T(f)\big\|_{L^p} \le B\big\|f\big\|_{L^p}$$

for all f supported in a cube of side length s, then the same estimate holds (with a larger constant) for all functions in $L^p(\mathbf{R}^n)$.
(b) If T satisfies for some $0 < p < \infty$

$$\big\|T(\chi_A)\big\|_{L^{p,\infty}} \le B|A|^{\frac{1}{p}}$$

for all measurable sets A contained in a cube of side length s, then the same estimate holds (with a larger constant) for all all measurable sets A in $\mathbf{R}^n$.

10.4. Fourier Transform Restriction and Bochner-Riesz Means

If g is a continuous function on $\mathbf{R}^n$, its restriction to a hypersurface $S \subseteq \mathbf{R}^n$ is a well-defined function. By a hypersurface we mean a submanifold of $\mathbf{R}^n$ of dimension $n-1$. So, if f is an integrable function on $\mathbf{R}^n$, its Fourier transform $\widehat{f}$ is continuous and hence its restriction $\widehat{f}\,\big|_S$ on S is well defined.

Definition 10.4.1. Let $1 \le p, q \le \infty$. We say that a hypersurface S in $\mathbf{R}^n$ satisfies a (p, q) *restriction theorem* if for all compact subsets S_0 of S, the restriction operator

$$f \to \widehat{f}\,\big|_{S_0} \, ,$$

which is initially defined on $L^1(\mathbf{R}^n) \cap L^p(\mathbf{R}^n)$, has an extension that maps $L^p(\mathbf{R}^n)$ boundedly into $L^q(S_0)$. (The norm of this extension may depend on S_0 in addition to p, q, n.) If S satisfies a (p,q) restriction theorem, then we will write that property $R_{p\to q}(S)$ holds.

If S is compact, we will say that property $R_{p\to q}(S)$ holds with constant C, if for all $f \in L^1(\mathbf{R}^n) \cap L^p(\mathbf{R}^n)$ we have

$$\|\widehat{f}\,\|_{L^q(S)} \le C\|f\|_{L^p(\mathbf{R}^n)}\,.$$

Example 10.4.2. Property $R_{1\to\infty}(S)$ holds for any hypersurface S.

Let us denote by $\mathcal{R}(f) = \widehat{f}\,\big|_{\mathbf{S}^{n-1}}$ (i.e., the restriction of the Fourier transform on a hypersurface S). Let $d\sigma$ be the canonically induced surface measure on S. Then for a function φ defined on S we have

$$\int_{\mathbf{S}^{n-1}} \widehat{f}\,\varphi\,d\sigma = \int_{\mathbf{R}^n} \widehat{f}\,(\widehat{\varphi\,d\sigma})^\vee\,d\xi = \int_{\mathbf{R}^n} f\,\widehat{\varphi\,d\sigma}\,dx$$

saying that the transpose of the linear operator $\mathcal{R}$ is the linear operator

$$\mathcal{R}^t(\varphi) = \widehat{\varphi\,d\sigma}\,. \tag{10.4.1}$$

Using duality considerations, we can see easily that a (p,q) restriction theorem for a hypersurface S is equivalent to the following (q',p') *extension theorem*:

$$\mathcal{R}^t : \; L^{q'}(S_0) \to L^{p'}(\mathbf{R}^n)$$

for all compact subsets S_0 of S.

Our objective will be to determine all pairs of indices (p,q) for which the sphere $\mathbf{S}^{n-1}$ satisfies a (p,q) restriction theorem. As we will shortly see in this section, this problem will be relevant in the understanding of the norm convergence of the Bochner-Riesz means. We will devote the rest of this section to a partial study of the restriction theorem for the sphere.

10.4.a. Necessary Conditions for $R_{p\to q}(\mathbf{S}^{n-1})$ to Hold

We look at basic examples, which tell us that a restriction theorem for the sphere cannot be satisfied for all pairs of indices (p,q).

Example 10.4.3. Let $d\sigma$ be surface measure on the unit sphere $\mathbf{S}^{n-1}$. In view of the result in Appendix B.4, we have

$$\widehat{d\sigma}(\xi) = \frac{2\pi}{|\xi|^{\frac{n-2}{2}}} J_{\frac{n-2}{2}}(2\pi|\xi|)\,.$$

Using the asymptotics in Appendix B.6, the last expression is equal to

$$\frac{2\sqrt{2\pi}}{|\xi|^{\frac{n-1}{2}}}\cos(2\pi|\xi| - \tfrac{\pi(n-1)}{4}) + O(|\xi|^{-\frac{n+1}{2}})$$

as $|\xi| \to \infty$. It follows that $\widehat{d\sigma}(\xi)$ [which is the operator in (10.4.1) applied to the function 1] does not lie in $L^{p'}(\mathbf{R}^n)$ if $\frac{n-1}{2}p' \le n$. Therefore, a necessary condition for $R_{p\to q}(\mathbf{S}^{n-1})$ to hold is that

$$1 \le p < \frac{2n}{n+1}. \tag{10.4.2}$$

In addition to this condition, there is another necessary condition on the indices p, q for $R_{p\to q}(\mathbf{S}^{n-1})$ to hold. This is a consequence of the following revealing example.

Example 10.4.4. Let φ be a Schwartz function on $\mathbf{R}^n$ such that $\widehat{\varphi} \ge 0$ and $\widehat{\varphi}(\xi) \ge 1$ for all ξ in the closed ball $|\xi| \le 2$. For $N \ge 1$ define functions

$$f_N(x_1, x_2, \dots, x_{n-1}, x_n) = \varphi\Big(\frac{x_1}{N}, \frac{x_2}{N}, \dots, \frac{x_{n-1}}{N}, \frac{x_n}{N^2}\Big).$$

In view of the result in Exercise 10.4.2(a), to test property $R_{p\to q}(\mathbf{S}^{n-1})$, instead of working with $\mathbf{S}^{n-1}$, we may work with the translated sphere $S = \mathbf{S}^{n-1} + (0, 0, \dots, 0, 1)$ in $\mathbf{R}^n$. We have

$$\widehat{f_N}(\xi) = N^{n+1}\widehat{\varphi}(N\xi_1, N\xi_2, \dots, N\xi_{n-1}, N^2\xi_n).$$

We note that for all $\xi = (\xi_1, \dots, \xi_n)$ in the spherical cap

$$S' = S \cap \{x \in \mathbf{R}^n : \ \xi_1^2 + \dots + \xi_{n-1}^2 \le N^{-2} \quad \text{and} \quad \xi_n < 1\}, \tag{10.4.3}$$

we have $\xi_n \le 1 - (1 - \frac{1}{N})^{\frac{1}{2}} \le \frac{1}{N}$ and therefore $|\xi| \le 2$. This implies that for all ξ in S' we have $\widehat{f_N}(\xi) \ge N^{n+1}$. But the spherical cap S' in (10.4.3) has surface measure $c(N^{-1})^{n-1}$. Thus we obtain

$$\big\|\widehat{f_N}\big\|_{L^q(S)} \ge \big\|\widehat{f_N}\big\|_{L^q(S')} \ge c^{\frac{1}{q}} N^{n+1} N^{\frac{1-n}{q}}.$$

On the other hand, $\|f_N\|_{L^p(\mathbf{R}^n)} = \|\varphi\|_{L^p(\mathbf{R}^n)} N^{\frac{n+1}{p}}$. Therefore, if $R_{p\to q}(\mathbf{S}^{n-1})$ holds we must have

$$\|\varphi\|_{L^p(\mathbf{R}^n)} N^{\frac{n+1}{p}} \ge C c^{\frac{1}{q}} N^{n+1} N^{\frac{1-n}{q}},$$

and letting $N \to \infty$, we obtain the following necessary condition on p and q for $R_{p\to q}(\mathbf{S}^{n-1})$ to hold:

$$\frac{1}{q} \ge \frac{n+1}{n-1}\frac{1}{p'}. \tag{10.4.4}$$

We shade in Figure 10.10 the region for which we have shown that $R_{p\to q}(\mathbf{S}^{n-1})$ fails. In this section we will try to understand whether property $R_{p\to q}(\mathbf{S}^{n-1})$ holds for $(\frac{1}{p}, \frac{1}{q})$ in the unshaded region of Figure 10.10.

We note that if $R_{p\to q}(\mathbf{S}^{n-1})$ holds, then so does $R_{p\to r}(\mathbf{S}^{n-1})$ for any $r \le q$ by Hölder's inequality (since $\mathbf{S}^{n-1}$ is a compact set). Therefore, the problem is to

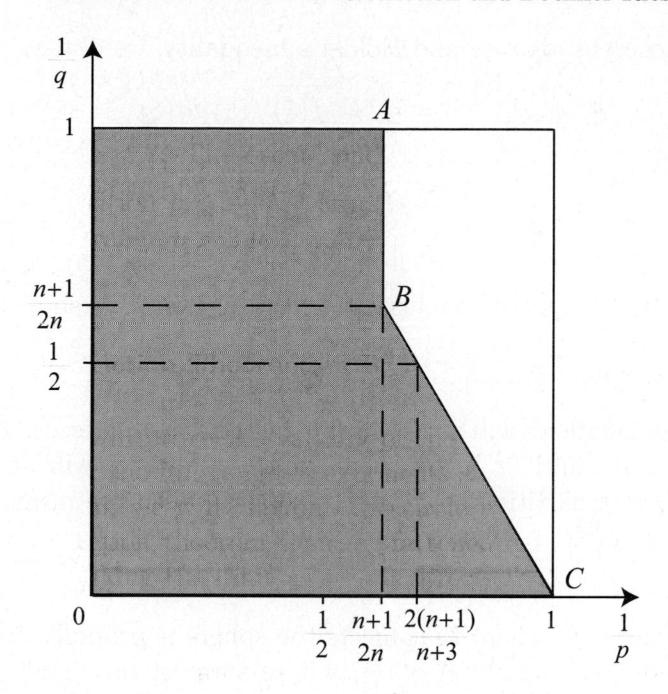

FIGURE 10.10. The set of points $(\frac{1}{p}, \frac{1}{q})$ with $1 \le p, q \le \infty$. The restriction property $R_{p\to q}(\mathbf{S}^{n-1})$ fails in the shaded region and on the closed line segment AB above but may hold on the line segment BC and inside the unshaded region.

establish the restriction property $R_{p\to q}(\mathbf{S}^{n-1})$ for the points

$$\left\{(p,q) : \frac{1}{q} = \frac{n+1}{n-1}\frac{1}{p'} \qquad 1 \le p < \frac{2n}{n+1}\right\}. \tag{10.4.5}$$

10.4.b. A Restriction Theorem for the Fourier Transform

In this subsection we establish the following restriction theorem for the Fourier transform.

Theorem 10.4.5. *Property $R_{p\to q}(\mathbf{S}^{n-1})$ holds for the set:*

$$\left\{(p,q) : \frac{1}{q} = \frac{n+1}{n-1}\frac{1}{p'} \qquad 1 \le p \le \frac{2(n+1)}{n+3}\right\}. \tag{10.4.6}$$

Proof. The case $p = 1$ and $q = \infty$ is trivial. Therefore, we only need to establish the case $p = \frac{2(n+1)}{n+3}$ and $q = 2$, since the remaining cases follow by interpolation.

Using Plancherel's identity and Hölder's inequality, we obtain

$$\begin{aligned}\|\widehat{f}\|_{L^2(\mathbf{S}^{n-1})}^2 &= \int_{\mathbf{S}^{n-1}} \overline{\widehat{f}(\xi)}\,\widehat{f}(\xi)\,d\sigma(\xi)\\ &= \int_{\mathbf{S}^{n-1}} \overline{f(x)}\,(f * d\sigma^\vee)(x)\,dx\\ &\le \|f\|_{L^p(\mathbf{R}^n)}\|f * d\sigma^\vee\|_{L^{p'}(\mathbf{R}^n)}.\end{aligned}$$

To establish the required conclusion it will be enough to show that

$$\|f * d\sigma^\vee\|_{L^{p'}(\mathbf{R}^n)} \le C_n\|f\|_{L^p(\mathbf{R}^n)} \qquad \text{when} \quad p = \frac{2(n+1)}{n+3}. \tag{10.4.7}$$

To obtain this estimate we will need to split the sphere into pieces. Each hyperplane $\xi_k = 0$ cuts the sphere $\mathbf{S}^{n-1}$ in two hemispheres which we will denote by H_k^1 and H_k^2. We introduce a partition of unity $\{\varphi_j\}_j$ of $\mathbf{R}^n$ with the property that for any j there exist k and $l \in \{1,2\}$ such that

$$(\text{support}\,\varphi_j) \cap \mathbf{S}^{n-1} \subsetneqq H_k^l\,;$$

that is, the support of each φ_j that meets the sphere is properly contained in some hemisphere. Then the family of all φ_j whose support meets $\mathbf{S}^{n-1}$ forms a finite partition of unity of the sphere when restricted on it. We can therefore write

$$d\sigma = \sum_{j\in F} \varphi_j\,d\sigma\,,$$

where F is a finite set. If we obtain (10.4.7) for each measure $\varphi_j d\sigma$ instead of $d\sigma$, then (10.4.7) will follow by a simple (finite) summation on j. We fix such a measure $\varphi_j\,d\sigma$, which without loss of generality we assume is properly supported in the top hemisphere H_n^1. In the sequel we will write $\mathbf{R}^n = \mathbf{R}^{n-1}\times\mathbf{R}$ [i.e., elements of $\mathbf{R}^n$ will be written as $x = (x',t)$, where $x' \in \mathbf{R}^{n-1}$ and $t \in \mathbf{R}$]. Then for $x \in \mathbf{R}^n$ we have

$$(\varphi_j\,d\sigma)^\vee(x) = \int_{\mathbf{S}^{n-1}} \varphi_j(\xi)e^{2\pi i x\cdot\xi}\,d\sigma(\xi) = \int_{\mathbf{R}^{n-1}} e^{2\pi i x\cdot\xi}\,\frac{\varphi_j(\xi',\sqrt{1-|\xi'|^2}\,)\,d\xi'}{\sqrt{1-|\xi'|^2}}\,,$$

where $\xi = (\xi',\xi_n)$; for the last identity we refer to Appendix D.5. Writing $x = (x',t) \in \mathbf{R}^n$, we have

$$\begin{aligned}(\varphi_j\,d\sigma)^\vee(x',t) &= \int_{\mathbf{R}^{n-1}} e^{2\pi i x'\cdot\xi'}\,e^{2\pi i t\sqrt{1-|\xi'|^2}}\,\frac{\varphi_j(\xi',\sqrt{1-|\xi'|^2}\,)}{\sqrt{1-|\xi'|^2}}\,d\xi'\\ &= \left(e^{2\pi i t\sqrt{1-|\xi'|^2}}\,\frac{\varphi_j(\xi',\sqrt{1-|\xi'|^2}\,)}{\sqrt{1-|\xi'|^2}}\right)^{\triangledown}(x')\,,\end{aligned} \tag{10.4.8}$$

where $^\triangledown$ indicates the inverse Fourier transform in the ξ' variable. We now introduce a function on $\mathbf{R}^{n-1}$ by setting

$$K_t(x') = (\varphi_j\,d\sigma)^\vee(x',t)\,.$$

We observe that identity (10.4.8) and the fact that $1-|\xi'|^2 > c > 0$ on the support of φ_j imply that

$$\sup_{t\in\mathbf{R}}\sup_{\xi'\in\mathbf{R}^{n-1}} |(K_t)^{\triangle}(\xi')| \le C_n < \infty\,, \tag{10.4.9}$$

where $^{\triangle}$ indicates the Fourier transform on $\mathbf{R}^{n-1}$. We also have that

$$K_t(x') = (\varphi_j\, d\sigma)^{\vee}(x',t) = \big(\varphi_j^{\vee} * d\sigma^{\vee}\big)(x',t)\,.$$

Since $\varphi_j^{\vee}$ is a Schwartz function on $\mathbf{R}^n$ and the function $|d\sigma^{\vee}(x',t)|$ is bounded by $(1+|(x',t)|)^{-\frac{n-1}{2}}$ (see Appendices B.4 and B.6) , if follows from Exercise 2.2.4 that

$$|K_t(x')| \le C(1+|(x',t)|)^{-\frac{n-1}{2}} \le C(1+|t|)^{-\frac{n-1}{2}} \tag{10.4.10}$$

for all $x'\in\mathbf{R}^{n-1}$. Estimate (10.4.9) says that the operator given by convolution with K_t maps $L^2(\mathbf{R}^{n-1})$ into itself with norm at most a constant, while estimate (10.4.10) says that the same operator maps $L^1(\mathbf{R}^{n-1})$ into $L^\infty(\mathbf{R}^{n-1})$ with norm at most a constant multiple of $(1+|t|)^{-\frac{n-1}{2}}$. Interpolation between these two estimates yields

$$\big\|K_t * g\big\|_{L^{p'}(\mathbf{R}^{n-1})} \le C_{p,n}|t|^{-(n-1)(\frac{1}{p}-\frac{1}{2})}\big\|g\big\|_{L^p(\mathbf{R}^{n-1})}$$

for all $1\le p\le 2$.

We now return to the proof of the required estimate (10.4.7) in which $d\sigma^{\vee}$ is replaced by $(\varphi_j\, d\sigma)^{\vee}$. Let $f(x) = f(x',t)$ be a function on $\mathbf{R}^n$. Denoting by $\star$ convolution on $\mathbf{R}^{n-1}$ (and by $*$ convolution on $\mathbf{R}^n$), we have

$$\begin{aligned}
\big\|f*(\varphi_j\, d\sigma)^{\vee}\big\|_{L^{p'}(\mathbf{R}^n)} &= \Big\|\Big\|\int_{\mathbf{R}} f(\,\cdot\,,\tau)\star K_{t-\tau}\,d\tau\Big\|_{L^{p'}(\mathbf{R}^{n-1})}\Big\|_{L^{p'}(\mathbf{R})}\\
&\le \Big\|\int_{\mathbf{R}}\big\|f(\,\cdot\,,\tau)\star K_{t-\tau}\big\|_{L^{p'}(\mathbf{R}^{n-1})}\,d\tau\Big\|_{L^{p'}(\mathbf{R})}\\
&\le C_{p,n}\Big\|\int_{\mathbf{R}}\frac{\big\|f(\,\cdot\,,\tau)\big\|_{L^p(\mathbf{R}^{n-1})}}{|t-\tau|^{(n-1)(\frac{1}{p}-\frac{1}{2})}}\,d\tau\Big\|_{L^{p'}(\mathbf{R})}\\
&= C_{p,n}\Big\|I_\beta\Big(\big\|f(\,\cdot\,,t)\big\|_{L^p(\mathbf{R}^{n-1})}\Big)\Big\|_{L^{p'}(\mathbf{R},dt)}\,,
\end{aligned}$$

where $\beta = 1-(n-1)(\frac{1}{p}-\frac{1}{2})$ and I_β is the Riesz potential (or fractional integral) given in Definition 6.1.1. Using Theorem 6.1.3 with $s=\beta$, $n=1$, and $q=p'$, we obtain that the last displayed equation is bounded by a constant multiple of

$$\Big\|\big\|f(\,\cdot\,,t)\big\|_{L^p(\mathbf{R}^{n-1})}\Big\|_{L^p(\mathbf{R},dt)} = \big\|f\big\|_{L^p(\mathbf{R}^n)}\,.$$

Note that the condition on the indices

$$\frac{1}{p}-\frac{1}{q}=\frac{s}{n} \iff \frac{1}{p}-\frac{1}{p'}=\frac{\beta}{1}$$

assumed in Theorem 6.1.3 translates exactly to $p = \frac{n+3}{2(n+1)}$. This concludes the proof of estimate (10.4.7) in which the measure $\sigma^\vee$ is replaced by $(\varphi_j\, d\sigma)^\vee$. Estimates for the remaining $(\varphi_j\, d\sigma)^\vee$ follows by a similar argument in which the role of the last coordinate is played by some other coordinate. The final estimate (10.4.7) follows by a simple summation on j over the finite set F. The proof of the theorem is now completed.

□

10.4.c. Applications to Bochner-Riesz Multipliers

We will now apply the restriction theorem obtained in the previous subsection to the Bochner-Riesz problem. In this subsection we will prove the following result.

Theorem 10.4.6. *For $\lambda > \frac{n-1}{2(n+1)}$, the Bochner-Riesz operator B^λ is bounded on $L^p(\mathbf{R}^n)$ for p in the optimal range*

$$\frac{2n}{n+1+2\lambda} < p < \frac{2n}{n-1-2\lambda}\,.$$

Proof. We will base our proof on a decomposition of the kernel instead of a multiplier. But first we isolate the smooth part of the multiplier (near the origin) and we only consider the part of it near the boundary of the unit disc. Precisely, we start with a Schwartz function $0 \le \eta \le 1$ whose Fourier transform is supported in the ball $B(0, \frac{3}{4})$ and is equal to 1 on the smaller ball $B(0, \frac{1}{2})$. Then we write

$$m_\lambda(\xi) = (1-|\xi|^2)_+^\lambda = (1-|\xi|^2)_+^\lambda \eta(\xi) + (1-|\xi|^2)_+^\lambda (1-\eta(\xi))\,.$$

Since the function $(1-|\xi|^2)_+^\lambda \eta(\xi)$ is smooth and compactly supported, it is a Fourier multiplier [element of $\mathcal{M}_p(\mathbf{R}^n)$] for all $1 \le p \le \infty$ and we need to concentrate our attention on the nonsmooth piece of the multiplier $(1-|\xi|^2)_+^\lambda(1-\eta(\xi))$, which is supported in $B(0, \frac{1}{2})^c$.

Next we pick a smooth *radial* function φ with support inside the ball $B(0,2)$ which is equal to 1 on the closed unit ball $\overline{B(0,1)}$. Let

$$K^\lambda(x) = \Big((1-|\xi|^2)_+^\lambda(1-\eta(\xi))\Big)^\vee(x)$$

be the kernel of the nonsmooth piece of the multiplier. For $j = 1, 2, \dots$ we introduce functions

$$\psi_j(x) = \varphi(2^{-j}x) - \varphi(2^{-j+1}x)$$

supported in the annuli $2^{j-1} \le |x| \le 2^{j+1}$. Then we write the operator given by convolution with K_λ as the sum

$$K^\lambda * f = T_0^\lambda(f) + \sum_{j=1}^{\infty} T_j^\lambda(f)\,, \tag{10.4.11}$$

where T_0^λ is given by convolution with $K^\lambda\varphi$ and each T_j^λ is given by convolution with $K^\lambda\psi_j$.

We begin by examining the kernel $K^\lambda\varphi$. Introducing a compactly supported function ζ that is equal to 1 on $B(0,\frac{3}{2})$, we can write

$$\begin{aligned}\Big((1-|\cdot|^2)_+^\lambda(1-\eta)\Big)^\vee &= \Big((1-|\cdot|^2)_+^\lambda(1-\eta)\zeta\Big)^\vee \\ &= \big((1-|\cdot|^2)_+^\lambda\big)^\vee * \big((1-\eta)\zeta\big)^\vee \\ &= \frac{\Gamma(\lambda+1)\,J_{\frac{n}{2}+\lambda}(2\pi|\cdot|)}{\pi^\lambda|\cdot|^{\frac{n}{2}+\lambda}} * \big((1-\eta)\zeta\big)^\vee\end{aligned} \tag{10.4.12}$$

in view of the identity in Appendix B.5. Using the asymptotic properties of Bessel functions (Appendix B.6), we see that the displayed fraction in (10.4.12) is a bounded function and therefore its convolution with a Schwartz function is also bounded. We conclude that the function K^λ is bounded and thus the operator T_0^λ given by convolution with the kernel $K^\lambda\varphi$ is bounded on all the L^p spaces.

Next we study the boundedness of the operators T_j^λ in which the dependence on the index j will play a role. Fix $p<2$ as in the statement of the theorem. Our goal will be to show that there exist positive constants C,δ such that for all functions f in $L^p(\mathbf{R}^n)$ we have the estimate

$$\big\|T_j^\lambda(f)\big\|_{L^p(\mathbf{R}^n)} \le C\,2^{-j\delta}\big\|f\big\|_{L^p(\mathbf{R}^n)}\,. \tag{10.4.13}$$

Once (10.4.13) is established, then the L^p boundedness of the operator $f\to K^\lambda * f$ will follow by a simple summation of the series in (10.4.11).

A localization argument (Exercise 10.4.4) allows us to reduce estimate (10.4.13) to functions f that are supported in a cube of side length 2^j. Let us assume, henceforth, that f is supported in some cube Q of side length 2^j.

Using the asymptotics in Appendix B.6, we note that the last displayed function in identity (10.4.12) is the convolution of a Schwartz function with a function bounded by a constant multiple of $(1+|x|)^{-\frac{n+1}{2}-\lambda}$. Using Exercise 2.2.4, we deduce that this convolution is also bounded by a constant multiple of $(1+|x|)^{-\frac{n+1}{2}-\lambda}$. It follows that

$$|K_j^\lambda(x)| \le C(1+|x|)^{-\frac{n+1}{2}-\lambda}|\psi_j(x)| \le C'2^{-(\frac{n+1}{2}+\lambda)j} \tag{10.4.14}$$

since $\psi_j(x)=\psi(2^{-j}x)$ and ψ is supported in the annulus $\frac{1}{2}\le|x|\le 2$. (The constants C and C' only depend on the dimension and λ.) Since K_j^λ is supported in a ball of radius 2^{j+1}, using Hölder's inequality and (10.4.14), we deduce the estimate

$$\|\widehat{K_j^\lambda}\|_{L^2}^2 = \|K_j^\lambda\|_{L^2}^2 \le C''2^{-(n+1+2\lambda)j}2^{nj} = C''2^{-(1+2\lambda)j}\,. \tag{10.4.15}$$

We now obtain an estimate on $\widehat{K_j^\lambda}$. We claim that for all $M \ge n+1$ there is a constant C_M such that

$$\int_{|\xi|\le\frac{1}{8}} |\widehat{K_j^\lambda}(\xi)|^2|\xi|^{-\beta}\,d\xi \le C_{M,n,\beta}\,2^{-2j(M-n)}, \quad \beta<n. \tag{10.4.16}$$

Indeed, since $\widehat{K^\lambda}(\xi)$ is supported in $|\xi|\ge\frac{1}{2}$ [recall the function η was chosen equal to 1 on $B(0,\frac{1}{2})$], we have

$$|\widehat{K_j^\lambda}(\xi)| = |(\widehat{K^\lambda}*\widehat{\psi_j})(\xi)| \le 2^{jn}\int_{\frac{1}{2}\le|\xi-\omega|\le 1}(1-|\xi-\omega|^2)_+^\lambda|\widehat{\psi}(2^j\omega)|\,d\omega\,.$$

Suppose that $|\xi|\le\frac{1}{8}$. Since $|\xi-\omega|\ge\frac{1}{2}$, we must have $|\omega|\ge\frac{3}{8}$. Then

$$|\widehat{\psi}(2^j\omega)| \le C_M(2^j|\omega|)^{-M} \le (8/3)^M C_M 2^{-jM},$$

from which it follows easily that

$$\sup_{|\xi|\le\frac{1}{8}}|\widehat{K_j^\lambda}(\xi)| \le C_M' 2^{-j(M-n)}\,. \tag{10.4.17}$$

Then (10.4.16) is a consequence of (10.4.17) and of the fact that the function $|\xi|^{-\beta}$ is integrable near the origin.

We now return to estimate (10.4.13) which we need to establish for a function f supported in a cube Q of side length 2^j. Observe that K_j^λ is supported in $5Q$ and we have for $1\le p<2$ by Hölder's inequality

$$\begin{aligned}\big\|T_j^\lambda(f)\big\|_{L^p(5Q)}^2 &\le |5Q|^{2(\frac{1}{p}-\frac{1}{2})}\big\|T_j^\lambda(f)\big\|_{L^2(5Q)}^2\\ &\le C_n 2^{(\frac{1}{p}-\frac{1}{2})2nj}\big\|\widehat{K_j^\lambda}\widehat{f}\,\big\|_{L^2}^2\end{aligned} \tag{10.4.18}$$

Having returned to L^2, we will be able to use the $L^p\to L^2$ restriction theorem obtained in the previous subsection. To this end we use polar coordinates and the fact that K_j^λ is a radial function to write

$$\big\|\widehat{K_j^\lambda}\widehat{f}\,\big\|_{L^2}^2 = \int_0^\infty|\widehat{K_j^\lambda}(re_1)|^2\int_{\mathbf{S}^{n-1}}|\widehat{f}(r\theta)|^2\,d\theta\,r^{n-1}dr\,, \tag{10.4.19}$$

where $e_1=(1,0,\dots,0)\in\mathbf{S}^{n-1}$. Using Exercise 10.4.2(b) we obtain

$$\int_{\mathbf{S}^{n-1}}|\widehat{f}(r\theta)|^2\,d\theta = \frac{1}{r^{n-1}}\int_{r\mathbf{S}^{n-1}}|\widehat{f}(\theta)|^2\,d\theta \le \frac{(C_{pn}r^{\frac{n-1}{2}-\frac{n}{p'}})^2}{r^{n-1}}\big\|f\big\|_{L^p}^2 \tag{10.4.20}$$

where C_{pn} is the constant of the restriction Theorem 10.4.5 that holds whenever $p\le\frac{2(n+1)}{n+3}$. So, under the assumption $p\le\frac{2(n+1)}{n+3}$, combining (10.4.19) and (10.4.20) yields

$$\begin{aligned}\|\widehat{K_j^\lambda f}\|_{L^2}^2 &\le C_{pn}^2\|f\|_{L^p}^2\int_0^\infty|\widehat{K_j^\lambda}(re_1)|^2 r^{n-1-\frac{2n}{p'}}\,dr\\ &\le \frac{C_{pn}^2}{\omega_{n-1}}\|f\|_{L^p}^2\int_{\mathbf{R}^n}|\widehat{K_j^\lambda}(\xi)|^2|\xi|^{-\frac{2n}{p'}}\,d\xi\,,\end{aligned}\tag{10.4.21}$$

where $\omega_{n-1}=|\mathbf{S}^{n-1}|$. Appealing to estimate (10.4.16) for $|\xi|\le\frac18$ with $\beta=\frac{2n}{p'}<n$ (since $p<2$) and to estimate (10.4.15) for $|\xi|\ge\frac18$, we obtain

$$\|\widehat{K_j^\lambda f}\|_{L^2}^2\le C\,2^{-(1+2\lambda)j}\|f\|_{L^p}^2\,.$$

Combining this inequality with the one previously obtained in (10.4.18) yields (10.4.13) with

$$\delta=\frac{n+1}{2}+\lambda-\frac{n}{p}\,.$$

This number is positive exactly when $\frac{2n}{n+1+2\lambda}<p$, which was the condition assumed by the theorem when $p<2$. The other condition $\lambda>\frac{n-1}{2(n+1)}$ is naturally imposed by the restriction condition $p\le\frac{2(n+1)}{n+3}$ used earlier. Note that the result claimed by the theorem in the range $p>2$ follows by duality.

□

10.4.d. The Full Restriction Theorem on $\mathbf{R}^2$

In this section we will prove the validity of the restriction condition $R_{p\to q}(\mathbf{S}^1)$ in dimension $n=2$, for the full range of exponents suggested in Figure 10.10.

To achieve this goal, we will "fatten" the circle by a small amount 2δ. Then we will obtain a restriction theorem for the "fattened circle" and then obtain the required estimate by taking the limit as $\delta\to0$. Precisely, we will use the following simple identity:

$$\int_{\mathbf{S}^1}|\widehat{f}(\omega)|^q\,d\omega=\lim_{\delta\to0}\frac{1}{2\delta}\int_{1-\delta}^{1+\delta}\int_{\mathbf{S}^1}|\widehat{f}(r\theta)|^q\,d\theta\,r\,dr\tag{10.4.22}$$

to recover the restriction theorem for the circle from a restriction theorem for annuli of width 2δ.

Throughout this subsection, δ will be a number satisfying $0<\delta<\frac{1}{1000}$ and for simplicity we will use the notation

$$\chi^\delta(\xi)=\chi_{(1-\delta,1+\delta)}(|\xi|)\,,\qquad\qquad \xi\in\mathbf{R}^2\,.$$

We note that, in view of identity (10.4.22), the restriction property $R_{p\to q}(\mathbf{S}^1)$ will be a trivial consequence of the estimate

$$\frac{1}{2\delta}\int_0^\infty\int_{\mathbf{S}^1}|\chi^\delta(r\theta)\widehat{f}(r\theta)|^q\,d\theta\,r\,dr\le C^q\|f\|_{L^p}^q\tag{10.4.23}$$

or, equivalently, of

$$\|\chi^\delta \widehat{f}\|_{L^q(\mathbf{R}^2)} \le (2\delta)^{\frac{1}{q}} C \|f\|_{L^p(\mathbf{R}^2)}. \tag{10.4.24}$$

We have the following result.

Theorem 10.4.7. *(a) Given $1 \le p < \frac{4}{3}$, set $q = \frac{p'}{3}$. Then there is a constant C_p such that for all L^p functions f on $\mathbf{R}^2$ and all small positive δ we have*

$$\|\chi^\delta \widehat{f}\|_{L^q(\mathbf{R}^2)} \le C_p \delta^{\frac{1}{q}} \|f\|_{L^p(\mathbf{R}^2)}. \tag{10.4.25}$$

(b) When $p = q = 4/3$, there is a constant C such that for all $L^{4/3}$ functions f on $\mathbf{R}^2$ and all small $\delta > 0$ we have

$$\|\chi^\delta \widehat{f}\|_{L^{\frac{4}{3}}(\mathbf{R}^2)} \le C \delta^{\frac{3}{4}} (\log \tfrac{1}{\delta})^{\frac{1}{4}} \|f\|_{L^{\frac{4}{3}}(\mathbf{R}^2)}. \tag{10.4.26}$$

Proof. To prove this theorem, we will work with the *extension operator*

$$E^\delta(g) = \widehat{\chi^\delta g} = \widehat{\chi^\delta} * \widehat{g},$$

which is dual (i.e., transpose) to $f \to \chi^\delta \widehat{f}$, and we will need to show that

$$\|E^\delta(f)\|_{L^{p'}(\mathbf{R}^2)} \le C \delta^{\frac{1}{q}} (\log \tfrac{1}{\delta})^\beta \|f\|_{L^{q'}(\mathbf{R}^2)}, \tag{10.4.27}$$

where $\beta = \frac{1}{4}$ when $p = \frac{4}{3}$ and $\beta = 0$ when $p < \frac{4}{3}$.

We will employ a splitting similar to that used in Theorem 10.2.4, with the only difference that our partition of unity will be nonsmooth and hence simpler. We define functions

$$\chi_\ell^\delta(\xi) = \chi^\delta(\xi) \chi_{2\pi\ell\delta^{1/2} \le \mathrm{Arg}\, \xi < 2\pi(\ell+1)\delta^{1/2}}$$

for $\ell \in \{0, 1, \dots, [\delta^{-1/2}]\}$. We suitably adjust the support of the function $\chi^\delta_{[\delta^{-1/2}]}$ so that the sum of all these functions equals χ^δ. We now split the indices that appear in the set $\{0, 1, \dots, [\delta^{-1/2}]\}$ into nine different subsets so that the supports of the functions indexed by them are properly contained in some sector centered at the origin of amplitude $\pi/4$. We can therefore write E^δ as a sum of nine pieces, each properly supported in a sector of amplitude $\pi/4$. Let I be the set of indices that correspond to one of these nine sectors and let

$$E_I^\delta(f) = \sum_{\ell \in I} \widehat{\chi_\ell^\delta f}.$$

It will suffice therefore to obtain (10.4.27) for each E_I^δ in lieu of E^δ. Let us fix such an index set I and by rotational symmetry let us assume that

$$I = \{0, 1, \dots, [\tfrac{1}{8}\delta^{-1/2}]\}.$$

Since the theorem is trivial when $p = 1$, to prove part (a) we fix a p with $1 < p < \frac{4}{3}$. We set

$$r = (p'/2)'$$

and we observe that this r satisfies $\frac{1}{r}=\frac{1}{p'}+\frac{1}{q'}$. We note that $1<r<2$ and we can apply the Hausdorff-Young inequality $\|h\|_{L^{r'}}\le\|h^\vee\|_{L^r}$. We have

$$\begin{aligned}\|E_I^\delta(f)\|_{L^{p'}(\mathbf{R}^2)}^{p'} &= \int_{\mathbf{R}^2}|E_I^\delta(f)^2|^{r'}\,dx\\ &\le \Big(\int_{\mathbf{R}^2}|(E_I^\delta(f)^2)^\vee|^r\,dx\Big)^{\frac{r'}{r}} \qquad (10.4.28)\\ &= \Big(\int_{\mathbf{R}^2}\Big|\sum_{\ell\in I}\sum_{\ell'\in I}(\chi_\ell^\delta f)*(\chi_{\ell'}^\delta f)\Big|^r\,dx\Big)^{\frac{r'}{r}}.\end{aligned}$$

We will obtain the estimate

$$\Big(\int_{\mathbf{R}^2}\Big|\sum_{\ell\in I}\sum_{\ell'\in I}(\chi_\ell^\delta f)*(\chi_{\ell'}^\delta f)\Big|^r\,dx\Big)^{\frac{r'}{r}}\le C\delta^{\frac{p'}{q}}\|f\|_{L^{q'}(\mathbf{R}^2)}^{p'}, \tag{10.4.29}$$

which will suffice to prove the theorem.

Denote by $S_{\delta,\ell,\ell'}$ the support of $\chi_\ell^\delta+\chi_{\ell'}^\delta$. Then we can write the left-hand side of (10.4.29) as

$$\int_{\mathbf{R}^2}\Big|\sum_{\ell\in I}\sum_{\ell'\in I}((\chi_\ell^\delta f)*(\chi_{\ell'}^\delta f))\chi_{S_{\delta,\ell,\ell'}}\Big|^r\,dx, \tag{10.4.30}$$

which, via Hölder's inequality, is controlled by

$$\int_{\mathbf{R}^2}\Big(\sum_{\ell\in I}\sum_{\ell'\in I}|(\chi_\ell^\delta f)*(\chi_{\ell'}^\delta f)|^r\Big)^{\frac{r}{r}}\Big(\sum_{\ell\in I}\sum_{\ell'\in I}|\chi_{S_{\delta,\ell,\ell'}}|^{r'}\Big)^{\frac{r}{r'}}\,dx. \tag{10.4.31}$$

We now recall Lemma 10.2.5, in which the curvature of the circle was crucial. In view of that lemma, the second factor of the integrand in (10.4.31) is bounded by a constant independent of δ. We have therefore obtained the estimate

$$\|E_I^\delta(f)\|_{L^{p'}}^{p'}\le C\Big(\sum_{\ell\in I}\sum_{\ell'\in I}\int_{\mathbf{R}^2}|(\chi_\ell^\delta f)*(\chi_{\ell'}^\delta f)|^r\,dx\Big)^{\frac{r'}{r}}. \tag{10.4.32}$$

We will prove at the end of this section the following auxiliary result.

Lemma 10.4.8. *With the same notation as in the proof of Theorem 10.4.7, for any $1<r<\infty$, there is a constant C (independent of δ and f) such that*

$$\|(\chi_\ell^\delta f)*(\chi_{\ell'}^\delta f)\|_{L^r}\le C\Big(\frac{\delta^{\frac{3}{2}}}{|\ell-\ell'|+1}\Big)^{\frac{1}{r'}}\|\chi_\ell^\delta f\|_{L^r}\|\chi_{\ell'}^\delta f\|_{L^r} \tag{10.4.33}$$

for all $\ell,\ell'\in I=\{0,1,\dots,[\frac{1}{8}\delta^{-1/2}]\}$.

Assuming Lemma 10.4.8, we estimate the term on the right in (10.4.32) by

$$
\begin{aligned}
&C\delta^{\frac{3}{2}}\Big[\sum_{\ell\in I}\|\chi_\ell^\delta f\|_{L^r}^r\Big(\sum_{\ell'\in I}\frac{\|\chi_{\ell'}^\delta f\|_{L^r}^r}{(|\ell-\ell'|+1)^{\frac{r}{r'}}}\Big)\Big]^{\frac{r'}{r}}\\
&\le C\delta^{\frac{3}{2}}\Big[\sum_{\ell\in I}\|\chi_\ell^\delta f\|_{L^r}^{rs}\Big]^{\frac{r'}{rs}}\Big[\sum_{\ell\in I}\Big(\sum_{\ell'\in I}\frac{\|\chi_{\ell'}^\delta f\|_{L^r}^r}{(|\ell-\ell'|+1)^{\frac{r}{r'}}}\Big)^{s'}\Big]^{\frac{r'}{rs'}},
\end{aligned}
\tag{10.4.34}
$$

where we used Hölder's inequality for some $1 < s < \infty$. We now recall the discrete fractional integral operator

$$
\{a_j\}_j \to \Big\{\sum_{j'}\frac{a_{j'}}{(|j-j'|+1)^{1-\alpha}}\Big\}_j ,
$$

which maps $\ell^s(\mathbf{Z}) \to \ell^{s'}(\mathbf{Z})$ when

$$
\frac{1}{s}-\frac{1}{s'}=\alpha\,,\qquad 0<\alpha<1\,.
\tag{10.4.35}
$$

(See Exercise 6.1.10.) When $1 < p < \frac{4}{3}$, we have $1 < r < 2$ and choosing $\alpha = 2-r = 1-\frac{r}{r'}$, we estimate the last expression in (10.4.34) (and thus $\|E_I^\delta(f)\|_{L^{p'}}^{p'}$) by

$$
C'\delta^{\frac{3}{2}}\Big[\sum_{\ell\in I}\|\chi_\ell^\delta f\|_{L^r}^{rs}\Big]^{\frac{r'}{rs}}\Big[\sum_{\ell\in I}\|\chi_\ell^\delta f\|_{L^r}^{rs}\Big]^{\frac{r'}{rs}} = C'\delta^{\frac{3}{2}}\Big[\sum_{\ell\in I}\|\chi_\ell^\delta f\|_{L^r}^{rs}\Big]^{\frac{2r'}{rs}}.
\tag{10.4.36}
$$

The unique s that solves equation (10.4.35) is seen easily to be $s = q'/r$. Moreover, since $q = p'/3$, we have $1 < s < 2$. We use again Hölder's inequality to pass from $\|\chi_\ell^\delta f\|_{L^r}$ to $\|\chi_\ell^\delta f\|_{L^{q'}}$. Indeed, recalling that the support of χ_ℓ^δ has measure $\approx \delta^{\frac{3}{2}}$, we have

$$
\|\chi_\ell^\delta f\|_{L^r} \le C(\delta^{\frac{3}{2}})^{\frac{1}{r}-\frac{1}{q'}}\|\chi_\ell^\delta f\|_{L^{q'}}\,.
$$

Inserting this in (10.4.36) provides us with

$$
\begin{aligned}
\|E_I^\delta(f)\|_{L^{p'}}^{p'} &\le C\delta^{\frac{3}{2}}\Big[\sum_{\ell\in I}\Big(C(\delta^{\frac{3}{2}})^{\frac{1}{r}-\frac{1}{q'}}\|\chi_\ell^\delta f\|_{L^{q'}}\Big)^{rs}\Big]^{\frac{2r'}{rs}}\\
&= C'\delta^{\frac{3}{2}}(\delta^{\frac{3}{2}})^{2r'(\frac{1}{r}-\frac{1}{q'})}\Big[\sum_{\ell\in I}\|\chi_\ell^\delta f\|_{L^{q'}}^{q'}\Big]^{\frac{2r'}{q'}}\\
&\le C\delta^{3}\|f\|_{L^{q'}}^{p'} = C\delta^{\frac{p'}{q}}\|f\|_{L^{q'}}^{p'},
\end{aligned}
$$

which is the required estimate since $\frac{1}{r}=\frac{1}{p'}+\frac{1}{q'}$ and $p' = 2r'$. In the last inequality we used the fact that the supports of the functions χ_ℓ^δ are disjoint and that these add up to a function that is at most 1.

To prove part (b) of the theorem, we need to adjust the previous argument to obtain the case $p = \frac{4}{3}$. Here we repeat part of the preceding argument taking $r = r' = s = s' = 2$.

Using Lemma 10.4.8, we estimate the term on the right in (10.4.32) by

$$
\begin{aligned}
& C\delta^{\frac{3}{2}}\Big[\sum_{\ell\in I}\|\chi_\ell^\delta f\|_{L^2}^2\Big(\sum_{\ell'\in I}\frac{\|\chi_{\ell'}^\delta f\|_{L^2}^2}{|\ell-\ell'|+1}\Big)\Big] \\
\le\; & C\delta^{\frac{3}{2}}\Big[\sum_{\ell\in I}\|\chi_\ell^\delta f\|_{L^2}^4\Big]^{\frac{1}{2}}\Big[\sum_{\ell\in I}\Big(\sum_{\ell'\in I}\frac{\|\chi_{\ell'}^\delta f\|_{L^2}^2}{|\ell-\ell'|+1}\Big)^2\Big]^{\frac{1}{2}} \\
\le\; & C\delta^{\frac{3}{2}}\Big[\sum_{\ell\in I}\|\chi_\ell^\delta f\|_{L^2}^4\Big]^{\frac{1}{2}}\Big[\sum_{\ell\in I}\|\chi_\ell^\delta f\|_{L^2}^4\Big]^{\frac{1}{2}}\Big[\sum_{\ell\in I}\frac{1}{|\ell|+1}\Big] \\
\le\; & C\delta^{\frac{3}{2}}\Big[\sum_{\ell\in I}\|\chi_\ell^\delta f\|_{L^2}^4\Big]\log(\delta^{-\frac{1}{2}}) \\
\le\; & C\delta^{\frac{3}{2}}(\delta^{\frac{3}{2}})^{(\frac{1}{2}-\frac{1}{4})4}\Big[\sum_{\ell\in I}\|\chi_\ell^\delta f\|_{L^4}^4\Big]\log\tfrac{1}{\delta} \\
\le\; & C\delta^3\big(\log\tfrac{1}{\delta}\big)\|f\|_{L^4}^4 .
\end{aligned}
$$

□

It remains to prove Lemma 10.4.8. This is our next goal in this section.

Proof. We can easily prove estimate using interpolation. For fixed $\ell, \ell' \in I$ we define the bilinear operator

$$T_{\ell,\ell'}(g,h) = (g\chi_\ell^\delta) * (h\chi_{\ell'}^\delta) .$$

As we have already observed, as a matter of simple geometry, we have that the support of χ_ℓ^δ is contained in a rectangle of side length $\approx \delta$ in the direction $e^{2\pi i\delta^{1/2}\ell}$ and of side length $\approx \delta^{\frac{1}{2}}$ in the direction $ie^{2\pi i\delta^{1/2}\ell}$. Any two rectangles with the preceding dimensions in the aforementioned directions have an intersection that depends on the angle between them. Indeed, if $\delta^{\frac{1}{2}}|\ell-\ell'| \le \frac{1}{10}$, then the measure of their intersection is at most $\delta^{\frac{3}{2}}$, while if $\delta^{\frac{1}{2}}|\ell-\ell'| \ge \frac{1}{10}$, the measure of their intersection is seen easily to be at most a constant multiple of

$$\delta\cdot\frac{\delta}{\sin(2\pi\delta^{\frac{1}{2}}|\ell-\ell'|)}$$

via a simple geometric argument. We conclude that the measure of such an intersection is at most

$$C\,\delta^{\frac{3}{2}}(1+|\ell-\ell'|)^{-1} .$$

It follows that

$$\big\|\chi_\ell^\delta * \chi_{\ell'}^\delta\big\|_{L^\infty} = \sup_{z\in\mathbf{R}^2} |(z - \operatorname{supp}(\chi_\ell^\delta)) \cap \operatorname{supp}(\chi_{\ell'}^\delta)| \le \frac{C\delta^{\frac{3}{2}}}{1+|\ell-\ell'|},$$

which implies the estimate

$$\begin{aligned} \big\|T_{\ell,\ell'}(g,h)\big\|_{L^\infty} &\le \big\|\chi_\ell^\delta * \chi_{\ell'}^\delta\big\|_{L^\infty}\big\|g\big\|_{L^\infty}\big\|h\big\|_{L^\infty} \\ &\le \frac{C\delta^{\frac{3}{2}}}{1+|\ell-\ell'|}\big\|g\big\|_{L^\infty}\big\|h\big\|_{L^\infty}. \end{aligned} \tag{10.4.37}$$

Also, the estimate

$$\begin{aligned} \big\|T_{\ell,\ell'}(g,h)\big\|_{L^1} &\le \big\|g\chi_\ell^\delta\big\|_{L^1}\big\|h\chi_{\ell'}^\delta\big\|_{L^1} \\ &\le \big\|g\big\|_{L^1}\big\|h\big\|_{L^1} \end{aligned} \tag{10.4.38}$$

holds trivially. Interpolating between (10.4.37) and (10.4.38) yields the required estimate (10.4.33). Here we used bilinear interpolation (Exercise 1.4.16).

□

Example 10.4.9. The presence of the logarithmic factor in estimate (10.4.26) is necessary. In fact, this estimate is sharp. We can prove this, by showing that the corresponding estimate for the "dual" extension operator E^δ is sharp. Let I be the set of indices we worked with in Theorem 10.4.7 (i.e. $I = \{0, 1, \dots, [\frac{1}{8}\delta^{-1/2}]\}$.) Let

$$f^\delta = \sum_{\ell\in I} \chi_\ell^\delta .$$

Then

$$\big\|f^\delta\big\|_{L^4} \approx \delta^{\frac{1}{4}} .$$

However,

$$E^\delta(f^\delta) = \sum_{\ell\in I} \widehat{\chi_\ell^\delta}$$

and we have

$$\begin{aligned} \big\|E^\delta(f^\delta)\big\|_{L^4} &= \bigg(\int_{\mathbf{R}^2} \Big|\sum_{\ell\in I}\sum_{\ell'\in I} \widehat{\chi_\ell^\delta}\widehat{\chi_{\ell'}^\delta}\Big|^2 \,d\xi\bigg)^{\frac{1}{4}} \\ &= \bigg(\int_{\mathbf{R}^2} \Big|\sum_{\ell\in I}\sum_{\ell'\in I} \chi_\ell^\delta * \chi_{\ell'}^\delta\Big|^2 \,dx\bigg)^{\frac{1}{4}}. \\ &\ge \bigg(\sum_{\ell\in I}\sum_{\ell'\in I} \int_{\mathbf{R}^2} |\chi_\ell^\delta * \chi_{\ell'}^\delta|^2 \,dx\bigg)^{\frac{1}{4}}. \end{aligned}$$

At this point observe that the function $\chi_\ell^\delta * \chi_{\ell'}^\delta$ is at least a constant multiple of $\delta^{\frac{3}{2}}(|\ell-\ell'|+1)^{-1}$ on a set of measure $c\delta^{\frac{3}{2}}(|\ell-\ell'|+1)$. (See Exercise 10.4.5.) Using this fact and the previous estimates, we deduce easily that

$$\big\|E^\delta(f^\delta)\big\|_{L^4} \ge c\Big(\sum_{\ell\in I}\sum_{\ell'\in I}\frac{\delta^3}{(|\ell-\ell'|+1)^2}\delta^{\frac{3}{2}}(|\ell-\ell'|+1)\Big)^{\frac14} \approx \delta(\log\tfrac{1}{\delta})^{\frac14}$$

since $|I|\approx\delta^{-\frac12}$. It follows that

$$\frac{\big\|E^\delta(f^\delta)\big\|_{L^4}}{\big\|f^\delta\big\|_{L^4}} \ge c\delta^{\frac34}(\log\tfrac{1}{\delta})^{\frac14},$$

which justifies the sharpness of estimate (10.4.26).

Exercises

10.4.1. Let S be a hypersurface in $\mathbf{R}^n$ and let $d\sigma$ be surface measure on it. Suppose that for some $0<b<n$ we have

$$|\widehat{d\sigma}(\xi)| \le C\,(1+|\xi|)^{-b}$$

for all $\xi\in\mathbf{R}^n$. Prove that $R_{p\to q}(S)$ does not hold for any $1\le q\le\infty$ when $p\ge\frac{n}{n-b}$.

10.4.2. Let S be a hypersurface and let $1\le p,q\le\infty$.
(a) Suppose that $R_{p\to q}(S)$ holds for S. Show that $R_{p\to q}(\tau+S)$ holds for the translated hypersurface $\tau+S$.
(b) Suppose that the hypersurface S is compact and its interior contains the origin. For $r>0$ let $rS=\{r\xi:\ \xi\in S\}$. Suppose that $R_{p\to q}(\mathbf{S}^{n-1})$ holds with constant C_{pqn}. Show that $R_{p\to q}(r\mathbf{S}^{n-1})$ holds with constant $C_{pqn}r^{\frac{n-1}{q}-\frac{n}{p'}}$.

10.4.3. Obtain a different proof of estimate (10.4.7) (and hence of Theorem 10.4.5) by following the sequence of steps outlined here:
(a) Consider the analytic family of functions

$$(K_z)^\vee(\xi) = 2\pi^{1-z}\frac{J_{\frac{n-2}{2}+z}(2\pi|\xi|)}{|\xi|^{\frac{n-2}{2}+z}}$$

and observe that in view of the identity in Appendix B.4, $(K_z)^\vee(\xi)$ reduces to $d\sigma^\vee(\xi)$ when $z=0$, where $d\sigma$ is surface measure on $\mathbf{S}^{n-1}$.
(b) Use for free that the Bessel function $J_{-\frac12+i\theta}$, $\theta\in\mathbf{R}$ satisfies

$$|J_{-\frac12+i\theta}(x)| \le C_\theta|x|^{-\frac12},$$

where C_θ grows at most exponentially in $|\theta|$, to obtain that the family of operators given by convolution with $(K_z)^\vee$ map $L^1(\mathbf{R}^n)$ into $L^\infty(\mathbf{R}^n)$ when $z=-\frac{n-1}{2}+i\theta$.

(c) Appeal to the result in Appendix B.5 to obtain that for z not equal to $0, -1, -2, \dots$ we have

$$K_z(x) = \frac{2}{\Gamma(z)}(1-|x|^2)^{z-1}.$$

Use this identity to deduce that for $z = 1 + i\theta$ the family of operators given by convolution with $(K_z)^\vee$ map $L^2(\mathbf{R}^n)$ into itself with constants that grow at most exponentially in $|\theta|$.
(d) Use Exercise 1.3.4 to obtain that for $z = 0$ the operator given by convolution with $d\sigma^\vee$ maps $L^p(\mathbf{R}^n)$ into $L^{p'}(\mathbf{R}^n)$ when $p = \frac{2(n+1)}{n+3}$.

10.4.4. Suppose that T is a linear operator given by convolution with a kernel K that is supported in the ball $B(0, 2R)$. Assume that there is a constant C such that for all functions f supported in a cube of side length R we have

$$\big\|T(f)\big\|_{L^p} \le B\big\|f\big\|_{L^p}$$

for some $1 \le p < \infty$. Show that this estimate also holds for all L^p functions f with constant $5^n B$.
[*Hint:* Write $f = \sum_j f\chi_{Q_j}$, where each cube Q_j has side length R.]

10.4.5. Using the notation of Theorem 10.4.7, show that there exist constants c, c' such that the function $\chi_\ell^\delta * \chi_{\ell'}^\delta$ is at least $c'\delta^{\frac{3}{2}}(|\ell - \ell'| + 1)^{-1}$ on a set of measure $c\delta^{\frac{3}{2}}(|\ell - \ell'| + 1)$.
[*Hint:* Show the required conclusion for characteristic functions of rectangles with the same orientation and comparable dimensions. Then use that the support of each χ_ℓ^δ contains such a rectangle.]

10.5. Almost Everywhere Convergence of Fourier Integrals*

In this section we will discuss in detail the proof of one of the most celebrated theorems in Fourier analysis, Carleson's theorem on the almost everywhere convergence of Fourier series. This theorem states that the Fourier series of a square integrable function on the circle converges (back to the function) almost everywhere. The same result is also valid for functions f on the line, if the Fourier series is replaced by the Fourier integrals defined by

$$\int_{|\xi|\le N} \widehat{f}(\xi)e^{2\pi i x\xi}\,d\xi.$$

The equivalence of these assertions follows from the transference methods discussed in Chapter 3.

For square-integrable functions f on the line, define the *Carleson operator*

$$\mathcal{C}(f)(x) = \sup_{N>0} \left| \int_{|\xi|\le N} \widehat{f}(\xi) e^{2\pi i x \xi}\, d\xi \right| .$$

We note that $\mathcal{C}(f)$ is well defined for f in $L^2(\mathbf{R})$ since so are the operators $(\widehat{f}\chi_{[a,b]})^\vee$ for all $-\infty \le a < b \le \infty$. We have the following result.

Theorem 10.5.1. *There is a constant $C > 0$ such that for all square-integrable functions f on the line the following estimate is valid:*

$$\big\|\mathcal{C}(f)\big\|_{L^{2,\infty}} \le C\big\|f\big\|_{L^2} .$$

It follows that for all f in $L^2(\mathbf{R})$ we have

$$\lim_{N\to\infty} \int_{|\xi|\le N} \widehat{f}(\xi) e^{2\pi i x \xi}\, d\xi = f(x) \tag{10.5.1}$$

for almost all $x \in \mathbf{R}$.

Proof. Because of the simple identity

$$\int_{|\xi|\le N} \widehat{f}(\xi) e^{2\pi i x \xi}\, d\xi = \int_{-\infty}^{N} \widehat{f}(\xi) e^{2\pi i x \xi}\, d\xi - \int_{-\infty}^{-N} \widehat{f}(\xi) e^{2\pi i x \xi}\, d\xi ,$$

it suffices to obtain $L^2 \to L^{2,\infty}$ bounds for the *one-sided maximal operators*

$$\mathcal{C}_1(f)(x) = \sup_{N>0} \left| \int_{-\infty}^{N} \widehat{f}(\xi) e^{2\pi i x \xi}\, d\xi \right| ,$$

$$\mathcal{C}_2(f)(x) = \sup_{N>0} \left| \int_{-\infty}^{-N} \widehat{f}(\xi) e^{2\pi i x \xi}\, d\xi \right| .$$

Once these bounds are obtained, we can use the simple fact that (10.5.1) holds for Schwartz functions and Theorem 2.1.14 to obtain (10.5.1) for all square integrable functions f on the line. Note that $\widetilde{\mathcal{C}_2(f)} = \mathcal{C}_1(\widetilde{f})$, where $\widetilde{f}(x) = f(-x)$ is the usual reflection operator. Therefore, it suffices to obtain bounds only for $\mathcal{C}_1$. [Note that just as is the case with $\mathcal{C}$, the operators $\mathcal{C}_1$ and $\mathcal{C}_2$ are well defined on $L^2(\mathbf{R})$.]

Next we recall the translation operator τ^y introduced in Section 2.2. For $a > 0$ and $y \in \mathbf{R}$ we introduce the modulation operator M^a and dilation operator D^a as follows:

$$\begin{aligned}
\tau^y(f)(x) &= f(x-y)\\
D^a(f)(x) &= a^{-\frac{1}{2}} f(a^{-1}x)\\
M^y(f)(x) &= f(x)e^{2\pi i y x} .
\end{aligned}$$

These operators are isometries on $L^2(\mathbf{R})$; for this reason we used the factor $a^{-\frac{1}{2}}$ in the definition of the dilation operator.

We will break down the proof of Theorem 10.5.1 into five main steps. We begin with the first step:

10.5.a. Preliminaries

We fix a Schwartz function φ such that $\widehat{\varphi}$ is real, nonnegative, and supported in the interval $[-1/10, 1/10]$. We will denote rectangles of area 1 in the (x, ξ) plane by s, t, u, etc. All rectangles considered in the sequel will have sides parallel to the axes. We will think of x as the time coordinate and of ξ as the frequency coordinate. For this reason we will refer to the (x, ξ) coordinate plane as the time-frequency plane. The projection of a rectangle s on the time axis will be denoted by I_s while its projection on the frequency axis by ω_s. Thus a rectangle s is just $s = I_s \times \omega_s$.

The center of an interval I will be denoted by $c(I)$. Also for $a > 0$, aI will denote an interval with the same center as I whose length is $a|I|$. Given a rectangle s, we will denote by $s(1)$ its bottom half and by $s(2)$ its upper half defined by

$$s(1) = I_s \times \big(\omega_s \cap (-\infty, c(\omega_s)\big), \qquad s(2) = I_s \times \big(\omega_s \cap [c(\omega_s), +\infty)\big) .$$

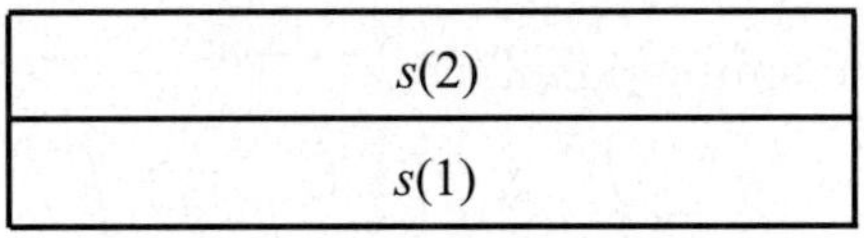

s(2)

s(1)

FIGURE 10.11. The lower and the upper parts of a tile s.

A dyadic interval is an interval of the form $[m2^k, (m+1)2^k)$, where k and m are integers. We will denote by $\mathbf{D}$ the set of all rectangles $I \times \omega$ with I, ω dyadic intervals and $|I|\,|\omega| = 1$. Such rectangles will be called *tiles* and $\mathbf{D}$ will be the set of all tiles. For each tile $s \in \mathbf{D}$, we introduce a function φ_s as follows:

$$\varphi_s(x) = |I_s|^{-\frac{1}{2}}\varphi\Big(\frac{x - c(I_s)}{|I_s|}\Big)e^{2\pi i c(\omega_{s(1)})x} . \tag{10.5.2}$$

This function localized in frequency near $c(\omega_{s(1)})$. Using our previous notation, we have $\varphi_s = M^{c(\omega_{s(1)})}\tau^{c(I_s)}D^{|I_s|}(\varphi)$. Observe that

$$\widehat{\varphi_s}(\xi) = |\omega_s|^{-\frac{1}{2}}\widehat{\varphi}\Big(\frac{\xi - c(\omega_{s(1)})}{|\omega_s|}\Big)e^{2\pi i (c(\omega_{s(1)})-\xi)c(I_s)} , \tag{10.5.3}$$

from which it follows that $\widehat{\varphi_s}$ is supported in $\frac{1}{5}\omega_{s(1)}$. Also observe that the functions φ_s have the same $L^2(\mathbf{R})$ norm.

Recall our complex inner product notation for $f, g \in L^2(\mathbf{R})$:

$$\langle f \,|\, g\rangle = \int_{\mathbf{R}} f(x)\overline{g(x)}\,dx . \tag{10.5.4}$$

Given a real number ξ, we introduce an operator

$$A_\xi(f) = \sum_{s \in \mathbf{D}} \chi_{\omega_{s(2)}}(\xi)\langle f \,|\, \varphi_s\rangle\, \varphi_s \tag{10.5.5}$$

initially defined for f in the Schwartz class. We will show in the next subsection that the series in (10.5.5) converges absolutely for f in the Schwartz class and thus A_ξ is well defined on this class. Moreover, we show in Lemma 10.5.2 that A_ξ admits an extension that is L^2 bounded, and therefore it can thought as well defined on $L^2(\mathbf{R})$.

For every integer m, let us denote by $\mathbf{D}_m$ the set of all tiles $s \in \mathbf{D}$ such that $|I_s| = 2^m$. We will call these tiles *of scale* m. Then

$$A_\xi(f) = \sum_{m\in\mathbf{Z}} A_\xi^m(f),$$

where

$$A_\xi^m(f) = \sum_{s\in\mathbf{D}_m} \chi_{\omega_{s(2)}}(\xi)\,\langle f\,|\,\varphi_s\rangle\,\varphi_s\,, \tag{10.5.6}$$

and observe that for each scale m, the second sum above ranges over all dyadic rectangles of a fixed scale whose tops contain the line perpendicular to the frequency axis at height ξ. As we will later see, the operators A_ξ^m are discretized versions of the multiplier operator $f \to \big(\widehat{f}\chi_{(-\infty,\xi]}\big)^\vee$. Indeed, the Fourier transform of $A_\xi^m(f)$ is supported in the frequency projection of the lower part $s(1)$ of all the tiles s that appear in (10.5.6). But the sum in (10.5.6) is taken over all tiles s whose frequency projection of the upper part $s(2)$ contains ξ. So the Fourier transform of $A_\xi^m(f)$ is supported in $(-\infty,\xi]$. On the other hand, in view of Exercise 10.5.8, summing over all s in (10.5.6) yields essentially the identity operator. Therefore, A_ξ^m can be viewed as the "part" of the identity operator whose frequency multiplier consists of the function $\chi_{(-\infty,\xi]}$ instead of the function 1. As m becomes larger, we obtain a better and better approximation to this multiplier. This heuristic explanation provides our motivation for the introduction of the operators A_ξ^m and A_ξ. In the sequel we will focus our study on their properties.

Let us make some preliminary observations regarding the operators A_ξ^m. We have the following

Lemma 10.5.2. *For any fixed ξ, the operators A_ξ^m are bounded from $L^2(\mathbf{R})$ into itself uniformly in m and ξ and moreover the operator A_ξ is also L^2 bounded uniformly in ξ.*

Proof. Let us make a few observations about the operators A_ξ^m. First recall that the adjoint of an operator T is uniquely defined by the identity

$$\langle T(f)\,|\,g\rangle = \langle f\,|\,T^*(g)\rangle$$

for all f and g. Now observe that A_ξ^m are self-adjoint operators, in the sense that $(A_\xi^m)^* = A_\xi^m$. Moreover, we claim that if $m \neq m'$, then $A_\xi^{m'}(A_\xi^m)^* = (A_\xi^{m'})^* A_\xi^m = 0$.

Indeed, given f and g we have

$$\begin{aligned} &\left\langle (A_\xi^{m'})^* A_\xi^m(f)\,\middle|\, g\right\rangle = \left\langle A_\xi^m f \,\middle|\, A_\xi^{m'} g\right\rangle \\ &= \sum_{s\in\mathbf{D}_m}\sum_{s'\in\mathbf{D}_{m'}} \langle f\,|\,\varphi_s\rangle\overline{\langle g\,|\,\varphi_{s'}\rangle}\langle\varphi_s\,|\,\varphi_{s'}\rangle\chi_{\omega_{s(2)}}(\xi)\chi_{\omega_{s'(2)}}(\xi)\,. \end{aligned} \tag{10.5.7}$$

Suppose that $\langle\varphi_s\,|\,\varphi_{s'}\rangle\chi_{\omega_{s(2)}}(\xi)\chi_{\omega_{s'(2)}}(\xi)$ is nonzero. Then $\langle\varphi_s\,|\,\varphi_{s'}\rangle$ is also nonzero, which implies that $\omega_{s(1)}$ and $\omega_{s'(1)}$ intersect. Also, the function $\chi_{\omega_{s(2)}}(\xi)\chi_{\omega_{s'(2)}}(\xi)$ is nonzero; hence $\omega_{s(2)}$ and $\omega_{s'(2)}$ must intersect. It follows that if

$$\langle\varphi_s\,|\,\varphi_{s'}\rangle\chi_{\omega_{s(2)}}(\xi)\chi_{\omega_{s'(2)}}(\xi)$$

is nonzero, then the dyadic intervals ω_s and $\omega_{s'}$ are not disjoint and hence one contains the other. If ω_s were properly contained in $\omega_{s'}$, then it would follow that ω_s is contained in $\omega_{s'(1)}$ or in $\omega_{s'(2)}$. But then either $\omega_{s(1)}\cap\omega_{s'(1)}$ or $\omega_{s(2)}\cap\omega_{s'(2)}$ would have to be empty, which does not happen as observed. It follows that if $\langle\varphi_s\,|\,\varphi_{s'}\rangle\chi_{\omega_{s(2)}}(\xi)\chi_{\omega_{s'(2)}}(\xi)$ is nonzero, then $\omega_s=\omega_{s'}$, which is impossible if $m\neq m'$. Thus the expression in (10.5.7) has to be zero. We conclude that the operators $\{A_\xi^m\}_{m\in\mathbf{Z}}$ are self-adjoint and have orthogonal ranges. By a simple orthogonality argument (we may also use Lemma 8.5.1), the uniform boundedness of each A_ξ^m implies that of their sum.

We now discuss the boundedness of each operator A_ξ^m. We have

$$\begin{aligned} \big\|A_\xi^m(f)\big\|_{L^2}^2 &= \sum_{s\in\mathbf{D}_m}\sum_{s'\in\mathbf{D}_m} \langle f\,|\,\varphi_s\rangle\overline{\langle f\,|\,\varphi_{s'}\rangle}\langle\varphi_s\,|\,\varphi_{s'}\rangle\chi_{\omega_{s(2)}}(\xi)\chi_{\omega_{s'(2)}}(\xi) \\ &= \sum_{s\in\mathbf{D}_m}\sum_{\substack{s'\in\mathbf{D}_m\\ \omega_{s'}=\omega_s}} \langle f\,|\,\varphi_s\rangle\overline{\langle f\,|\,\varphi_{s'}\rangle}\langle\varphi_s\,|\,\varphi_{s'}\rangle\chi_{\omega_{s(2)}}(\xi)\chi_{\omega_{s'(2)}}(\xi) \\ &\le \sum_{s\in\mathbf{D}_m}\sum_{\substack{s'\in\mathbf{D}_m\\ \omega_{s'}=\omega_s}} \big|\langle f\,|\,\varphi_s\rangle\big|^2\chi_{\omega_{s(2)}}(\xi)\big|\langle\varphi_s\,|\,\varphi_{s'}\rangle\big| \\ &\le C'\sum_{s\in\mathbf{D}_m}\big|\langle f\,|\,\varphi_s\rangle\big|^2\chi_{\omega_{s(2)}}(\xi)\,, \end{aligned} \tag{10.5.8}$$

where we used our earlier observation about s and s', the Cauchy-Schwarz inequality, and the fact that

$$\sum_{\substack{s'\in\mathbf{D}_m\\ \omega_{s'}=\omega_s}}\big|\langle\varphi_s\,|\,\varphi_{s'}\rangle\big|\le C\sum_{\substack{s'\in\mathbf{D}_m\\ \omega_{s'}=\omega_s}}\left(1+\frac{\operatorname{dist}\,(I_s,I_{s'})}{2^m}\right)^{-10}\le C'\,,$$

which follows from the result in Appendix K. To estimate (10.5.8), we use that

$$\begin{aligned}
|\langle f \,|\, \varphi_s\rangle| &\le C \int_{\mathbf{R}} |f(y)|\,|I_s|^{-\frac{1}{2}} \left(1+\frac{|y-c(I_s)|}{|I_s|}\right)^{-10} dy \\
&\le C'\,|I_s|^{\frac{1}{2}} \int_{\mathbf{R}} |f(y)| \left(1+\frac{|y-z|}{|I_s|}\right)^{-10} \frac{dy}{|I_s|} \\
&\le C''\,|I_s|^{\frac{1}{2}} M(f)(z),
\end{aligned}$$

for all $z \in I_s$, in view of Theorem 2.1.10. Since the preceding estimate holds for all $z \in I_s$, it follows that

$$|\langle f \,|\, \varphi_s\rangle|^2 \le (C'')^2 |I_s| \inf_{z\in I_s} M(f)(z)^2 \le C \int_{I_s} M(f)(x)^2\,dx. \tag{10.5.9}$$

Next we observe that the rectangles $s \in \mathbf{D}_m$ with the property that $\xi \in \omega_{s(2)}$ are all disjoint. This implies that the corresponding time intervals I_s are also disjoint. Thus, summing (10.5.9) over all $s \in \mathbf{D}_m$ with $\xi \in \omega_{s(2)}$, we obtain that

$$\begin{aligned}
\sum_{s\in\mathbf{D}_m} |\langle f \,|\, \varphi_s\rangle|^2 \chi_{\omega_{s(2)}}(\xi) &\le C \sum_{s\in\mathbf{D}_m} \chi_{\omega_{s(2)}}(\xi) \int_{I_s} M(f)(x)^2\,dx \\
&\le C \int_{\mathbf{R}} M(f)(x)^2\,dx,
\end{aligned}$$

which establishes the required claim using the boundedness of the Hardy-Littlewood maximal operator M on $L^2(\mathbf{R})$.

□

10.5.b. Discretization of the Carleson Operator

We let $h \in \mathcal{S}(\mathbf{R})$, $\xi \in \mathbf{R}$, and for each $m \in \mathbf{Z}$, $y, \eta \in \mathbf{R}$ and $\lambda \in [0,1]$ we introduce the operators

$$B^m_{\xi,y,\eta,\lambda}(h) = \sum_{s\in\mathbf{D}_m} \chi_{\omega_{s(2)}}(2^{-\lambda}(\xi+\eta)) \,\langle D^{2^\lambda}\tau^y M^\eta(h) \,|\, \varphi_s\rangle M^{-\eta}\tau^{-y} D^{2^{-\lambda}}(\varphi_s)\,.$$

It is not hard to see that for all $x \in \mathbf{R}$ and $\lambda \in [0,1]$ we have

$$B^m_{\xi,y,\eta,\lambda}(h)(x) = B^m_{\xi,y+2^{m-\lambda},\eta+2^{-m+\lambda},\lambda}(h)(x)\,;$$

in other words, the function $(y,\eta) \to B^m_{\xi,y,\eta,\lambda}(h)(x)$ is periodic in $\mathbf{R}^2$ with period $(2^{m-\lambda}, 2^{-m+\lambda})$. See Exercise 10.5.1.

Using Exercise 10.5.2, we obtain that for $|m|$ large (with respect to ξ) we have

$$\begin{aligned}
&\Big|\sum_{s\in\mathbf{D}_m}\chi_{\omega_{s(2)}}(2^{-\lambda}(\xi+\eta))\,\langle D^{2^\lambda}\tau^y M^\eta(h)\,|\,\varphi_s\rangle\, M^{-\eta}\tau^{-y}D^{2^{-\lambda}}(\varphi_s)(x)\Big| \\
\le\;& C_h\min(2^m,1)\sum_{s\in\mathbf{D}_m}\chi_{\omega_{s(2)}}(2^{-\lambda}(\xi+\eta))2^{-m/2}\Big|\varphi\Big(\frac{x+y-c(I_s)}{2^{m-\lambda}}\Big)\Big| \\
\le\;& C_h\min(2^{m/2},2^{-m/2})\sum_{k\in\mathbf{Z}}\Big|\varphi\Big(\frac{x+y-k2^m}{2^{m-\lambda}}\Big)\Big|\le C_h\min(2^{m/2},2^{-m/2}),
\end{aligned}$$

since the last sum is seen easily to converge to some quantity that remains bounded in x, y, η, and λ. It follows that for $h\in\mathcal{S}(\mathbf{R})$ we have

$$\sup_{x\in\mathbf{R}}\sup_{y\in\mathbf{R}}\sup_{\eta\in\mathbf{R}}\sup_{0\le\lambda\le1}\big|B^m_{\xi,y,\eta,\lambda}(h)(x)\big|\le C_h\min(2^{m/2},2^{-m/2}). \tag{10.5.10}$$

Using Exercise 10.5.3 and the periodicity of the functions $B^m_{\xi,y,\eta,\lambda}(h)$, we conclude that the averages

$$\frac{1}{4KL}\int_{-L}^{L}\int_{-K}^{K}\int_0^1 B^m_{\xi,y,\eta,\lambda}(h)\,d\lambda\,dy\,d\eta$$

converge pointwise to some $\Pi^m_\xi(h)$ as $K,L\to\infty$. Estimate (10.5.10) implies the uniform convergence for the series $\sum_{m\in\mathbf{Z}}B^m_{\xi,y,\eta,\lambda}(h)$ and therefore

$$\begin{aligned}
&\lim_{\substack{K\to\infty\\L\to\infty}}\frac{1}{4KL}\int_{-L}^{L}\int_{-K}^{K}\int_0^1 M^{-\eta}\tau^{-y}D^{2^{-\lambda}}A_{\frac{\xi+\eta}{2^\lambda}}D^{2^\lambda}\tau^y M^\eta(h)\,d\lambda\,dy\,d\eta \\
=\;&\lim_{\substack{K\to\infty\\L\to\infty}}\frac{1}{4KL}\int_{-L}^{L}\int_{-K}^{K}\int_0^1\sum_{m\in\mathbf{Z}}B^m_{\xi,y,\eta,\lambda}(h)\,d\lambda\,dy\,d\eta \\
=\;&\sum_{m\in\mathbf{Z}}\lim_{\substack{K\to\infty\\L\to\infty}}\frac{1}{4KL}\int_{-L}^{L}\int_{-K}^{K}\int_0^1 B^m_{\xi,y,\eta,\lambda}(h)\,d\lambda\,dy\,d\eta \\
=\;&\sum_{m\in\mathbf{Z}}\Pi^m_\xi(h)=\Pi_\xi(h).
\end{aligned}\tag{10.5.11}$$

We now make a few observations about the operator Π_ξ defined on $\mathcal{S}(\mathbf{R})$, which is also given by the expression in (10.5.11). First we observe that in view of Lemma 10.5.2 and Fatou's lemma, we have that Π_ξ is bounded on L^2 uniformly in ξ. Next we observe that Π_ξ commutes with all translations τ^z for $z\in\mathbf{R}$. To see this, we use the fact that $\tau^{-z}M^{-\eta} = e^{-2\pi i\eta z}M^{-\eta}\tau^{-z}$ to obtain

$$\begin{aligned}
&\sum_{s\in\mathbf{D}_m} \chi_{\omega_{s(2)}}(2^{-\lambda}(\xi+\eta))\big\langle D^{2^\lambda}\tau^y M^\eta \tau^z(h)\,|\,\varphi_s\big\rangle\,\tau^{-z}M^{-\eta}\tau^{-y}D^{2^{-\lambda}}(\varphi_s)\\
&= \sum_{s\in\mathbf{D}_m} \chi_{\omega_{s(2)}}(2^{-\lambda}(\xi+\eta))\big\langle h\,|\,\tau^{-z}M^{-\eta}\tau^{-y}D^{2^{-\lambda}}(\varphi_s)\big\rangle\,\tau^{-z}M^{-\eta}\tau^{-y}D^{2^{-\lambda}}(\varphi_s)\\
&= \sum_{s\in\mathbf{D}_m} \chi_{\omega_{s(2)}}(2^{-\lambda}(\xi+\eta))\big\langle h\,|\,M^{-\eta}\tau^{-y-z}D^{2^{-\lambda}}(\varphi_s)\big\rangle\,M^{-\eta}\tau^{-y-z}D^{2^{-\lambda}}(\varphi_s)\,.
\end{aligned}$$

Recall that $\tau^{-z}\Pi_\xi^m\tau^z(h)$ is equal to the limit of the averages of the preceding expressions over all $(y,\eta,\lambda)\in[-K,K]\times[-L,L]\times[0,1]$. But in view of the previous identity, this is equal to the limit of the averages of the expressions

$$\sum_{s\in\mathbf{D}_m} \chi_{\omega_{s(2)}}(2^{-\lambda}(\xi+\eta))\big\langle D^{2^\lambda}\tau^{y'}M^\eta(h)\,|\,\varphi_s\big\rangle M^{-\eta}\tau^{-y'}D^{2^{-\lambda}}(\varphi_s) \tag{10.5.12}$$

over all $(y',\eta,\lambda)\in[-K+z,K+z]\times[-L,L]\times[0,1]$. Since (10.5.12) is periodic in (y',η), it follows that its average over the set $[-K+z,K+z]\times[-L,L]\times[0,1]$ is equal to its average over the set $[-K,K]\times[-L,L]\times[0,1]$. Taking limits as $K,L\to\infty$, we obtain the identity $\tau^{-z}\Pi_\xi^m\tau^z(h)=\Pi_\xi^m(h)$. Summing over all $m\in\mathbf{Z}$ it follows that $\tau^{-z}\Pi_\xi\tau^z(h)=\Pi_\xi(h)$.

A similar argument using averages over shifted rectangles of the form $[-K,K]\times[-L+\theta,L+\theta]$ yields the identity

$$M^{-\theta}\Pi_{\xi+\theta}M^\theta = \Pi_\xi \tag{10.5.13}$$

for all $\xi,\theta\in\mathbf{R}$. The details are left to the reader. Next, we claim that the operator $M^{-\xi}\Pi_\xi M^\xi$ commutes with dilations D^{2^a}, $a\in\mathbf{R}$. First we observe that for all integers k we have

$$A_\xi(h) = D^{2^{-k}}A_{2^{-k}\xi}D^{2^k}(h)\,, \tag{10.5.14}$$

which is simply saying that A_ξ is well-behaved under change of scale. This identity is left as an exercise to the reader. Identity (10.5.14) may not hold for noninteger k, and this is exactly we have averaged over all dilations 2^λ, $0\le\lambda\le1$ in (10.5.11).

Let us denote by $[a]$ the integer part of a real number a. Using the identities $D^b M^\eta = M^{\eta/b} D^b$ and $D^b \tau^z = \tau^{bz} D^b$, we obtain

$$
\begin{aligned}
(10.5.15) \qquad & D^{2^{-a}} M^{-(\xi+\eta)} \tau^{-y} D^{2^{-\lambda}} A_{\frac{\xi+\eta}{2^\lambda}} D^{2^\lambda} \tau^y M^{\xi+\eta} D^{2^a} \\
&= M^{-2^a(\xi+\eta)} \tau^{-2^{-a}y} D^{2^{-(a+\lambda)}} A_{\frac{\xi+\eta}{2^\lambda}} D^{2^{a+\lambda}} \tau^{2^{-a}y} M^{2^a(\xi+\eta)} \\
&= M^{-2^a(\xi+\eta)} \tau^{-y'} D^{2^{-\lambda'}} D^{2^{-[a+\lambda]}} A_{\frac{2^a(\xi+\eta)}{2^{\lambda'} 2^{[a+\lambda]}}} D^{2^{[a+\lambda]}} D^{2^{\lambda'}} \tau^{y'} M^{2^a(\xi+\eta)} \\
&= M^{-2^a\xi} M^{-\eta'} \tau^{-y'} D^{2^{-\lambda'}} A_{\frac{2^a\xi+2^a\eta}{2^{\lambda'}}} D^{2^{\lambda'}} \tau^{y'} M^{\eta'} M^{2^a\xi} \\
(10.5.16) \qquad &= M^{-\xi} M^{-\theta} \big(M^{-\eta'} \tau^{-y'} D^{2^{-\lambda'}} A_{\frac{\xi+\theta+\eta'}{2^{\lambda'}}} D^{2^{\lambda'}} \tau^{y'} M^{\eta'}\big) M^\theta M^\xi ,
\end{aligned}
$$

where we set $y' = 2^{-a}y$, $\eta' = 2^a\eta$, $\lambda' = a + \lambda - [a+\lambda]$, and $\theta = (2^a - 1)\xi$. The average of (10.5.15) over all (y, η, λ) in $[-K, K] \times [-L, L] \times [0, 1]$ converges to the operator $D^{2^{-a}} M^{-\xi} \Pi_\xi M^\xi D^{2^a}$ as $K, L \to \infty$. But this limit is equal to the limit of the averages of the expression in (10.5.16) over all (y', η', λ') in $[-2^{-a}K, 2^{-a}K] \times [-2^a L, 2^a L] \times [0, 1]$, which is

$$
M^{-\xi} M^{-\theta} \Pi_{\xi+\theta} M^\theta M^\xi .
$$

Using the identity (10.5.13), we obtain that

$$
D^{2^{-a}} M^{-\xi} \Pi_\xi M^\xi D^{2^a} = M^{-\xi} \Pi_\xi M^\xi ,
$$

saying that the operator $M^{-\xi} \Pi_\xi M^\xi$ commutes with dilations.

Next we observe that if $\widehat{h}$ is supported in $[0, \infty)$, then $M^{-\xi} \Pi_\xi M^\xi(h) = 0$. This is a consequence of the fact that the inner products

$$
\big\langle D^{2^\lambda} \tau^y M^\eta M^\xi(h) \,|\, \varphi_s \big\rangle = \big\langle M^\xi(h) \,|\, M^{-\eta} \tau^{-y} D^{2^{-\lambda}}(\varphi_s) \big\rangle
$$

vanish, since the Fourier transform of $\tau^{-z} M^{-\eta} \tau^{-y} D^{2^{-\lambda}} \varphi_s$ is supported in the set $(-\infty, 2^\lambda c(\omega_{s(1)}) - \eta + \frac{2^\lambda}{10}|\omega_s|)$ which is disjoint from $(\xi, +\infty)$ whenever $2^{-\lambda}(\xi+\eta) \in \omega_{s(2)}$. Finally, we observe that Π_ξ is a positive semidefinite operator, that is, it satisfies

$$
\langle \Pi_\xi(h) \,|\, h \rangle \geq 0 . \qquad (10.5.17)
$$

This follows easily from the fact that the inner product in (10.5.17) is equal to

$$
\lim_{\substack{K\to\infty \\ L\to\infty}} \frac{1}{4KL} \int_{-L}^{L} \int_{-K}^{K} \int_0^1 \sum_{s\in\mathbf{D}} \chi_{\omega_{s(2)}}\big(\tfrac{\xi+\eta}{2^\lambda}\big) \big|\big\langle D^{2^\lambda} \tau^y M^\eta(h) \,|\, \varphi_s \big\rangle\big|^2 \, d\lambda \, dy \, d\eta.
$$

This expression also implies that Π_ξ is not the zero operator, since we can always find a Schwartz function h and a tile s such that $\langle D^{2^\lambda} \tau^y M^\eta(h) \,|\, \varphi_s \rangle$ is not zero for (y, η, λ) near $(0, 0, 0)$. It follows that the operators and $M^{-\xi} \Pi_\xi M^\xi$ are nonzero for every ξ.

Let us summarize what we have already proved: The operator $M^{-\xi}\Pi_\xi M^\xi$ is nonzero, is bounded on $L^2(\mathbf{R})$, it commutes with translations and dilations, and vanishes when applied to functions whose Fourier transform is supported in the positive semi-axis $[0,\infty)$. In view of Exercise 4.1.11, it follows that for some constant $c_\xi \neq 0$ we have

$$M^{-\xi}\Pi_\xi M^\xi(h)(x) = c_\xi \int_{-\infty}^{0} \widehat{h}(\eta) e^{2\pi i x\eta}\, d\eta\,,$$

which identifies Π_ξ with the convolution operator whose multiplier is the function $c_\xi \chi_{(-\infty,\xi]}$. Using the identity (10.5.13), we obtain

$$c_{\xi+\theta} = c_\xi$$

for all ξ and θ, saying that c_ξ does not depend on ξ. We have therefore proved that for all Schwartz functions h the following identity is valid:

$$\Pi_\xi(h) = c\,\big(\widehat{h}\,\chi_{(-\infty,\,\xi]}\big)^\vee \tag{10.5.18}$$

for some fixed nonzero constant c. This completely identifies the operator Π_ξ. By density it follows that

$$\mathcal{C}_1(f) = \frac{1}{c}\sup_\xi |\Pi_\xi(f)| \tag{10.5.19}$$

for all $f \in \bigcup_{1\le p<\infty} L^p(\mathbf{R})$.

10.5.c. Linearization of the Maximal Dyadic Sum Operator $\sup_\xi |A_\xi(f)|$

Recall the operators A_ξ, Π_ξ and $\mathcal{C}_1$ introduced thus far that were defined on $L^2(\mathbf{R})$. Our goal will be to show that there exists a constant $C>0$ so that for all $f\in L^2(\mathbf{R})$ we have

$$\big\| \sup_{\xi\in\mathbf{R}} |A_\xi(f)|\, \big\|_{L^{2,\infty}(\mathbf{R})} \le C\|f\|_{L^2(\mathbf{R})}\,. \tag{10.5.20}$$

Once (10.5.20) is established, averaging yields the same conclusion for the operator $f\to \sup_\xi |\Pi_\xi(f)|$, thus establishing the required bound for $\mathcal{C}_1$. Let us describe this averaging argument. Identity (10.5.11) gives

$$\Pi_\xi(f) = \lim_{\substack{K\to\infty\\ L\to\infty}} \frac{1}{4KL}\int_{-L}^{L}\int_{-K}^{K}\int_0^1 G_{\xi,y,\eta,\lambda}(f)\, d\lambda\, dy\, d\eta\,,$$

where

$$G_{\xi,y,\eta,\lambda}(f) = M^{-\eta}\tau^{-y}D^{2^{-\lambda}}A_{\frac{\xi+\eta}{2^\lambda}}D^{2^\lambda}\tau^y M^\eta(f)\,.$$

This, in turn, implies

$$\sup_{\xi\in\mathbf{R}} |\Pi_\xi(f)| \le \liminf_{\substack{K\to\infty\\ L\to\infty}} \frac{1}{4KL}\int_{-L}^{L}\int_{-K}^{K}\int_0^1 \sup_{\xi\in\mathbf{R}} |G_{\xi,y,\eta,\lambda}(f)|\, d\lambda\, dy\, d\eta\,. \tag{10.5.21}$$

We now take the $L^{2,\infty}$ quasi-norms of both sides and we use Fatou's lemma for weak L^2 [Exercise 1.1.12(d)]. We thus reduce the estimate for the operator $\sup_{\xi\in\mathbf{R}}|\Pi_\xi(f)|$ to the corresponding estimate for $\sup_{\xi\in\mathbf{R}}|A_\xi(f)|$. In this way we obtain the $L^{2,\infty}$ boundedness of $\sup_{\xi\in\mathbf{R}}|\Pi_\xi(f)|$ and therefore that of $\mathcal{C}_1$ in view of identity (10.5.19).

Matters are now reduced to the study of the discretized maximal operator $\sup_{\xi\in\mathbf{R}}|A_\xi(f)|$ and, in particular, to the proof of estimate (10.5.20). It will be convenient to study the maximal operator $\sup_\xi|A_\xi(f)|$ via a linearization and a simplification. Here is how we will achieve this. For a measurable real-valued function $x\to N(x)$ on the line we define a linear operator $\mathfrak{D}_N$ by setting for $f\in L^2(\mathbf{R})$

$$\mathfrak{D}_N(f)(x) = A_{N(x)}(f)(x) = \sum_{s\in\mathbf{D}}(\chi_{\omega_{s(2)}}\circ N)(x)\,\langle f\,|\,\varphi_s\rangle\,\varphi_s(x)\,, \tag{10.5.22}$$

where the sum on the right converges in $L^2(\mathbf{R})$ [and also uniformly for $f\in\mathcal{S}(\mathbf{R})$].

Using Exercise 10.5.4, to prove (10.5.20) it will suffice to show that there exists a $C>0$ such that for all $f\in L^2(\mathbf{R})$, and all real-valued measurable functions N on the line we have

$$\big\|\mathfrak{D}_N(f)\big\|_{L^{2,\infty}} \le C\big\|f\big\|_{L^2}\,. \tag{10.5.23}$$

To justify some algebraic manipulations we fix a finite subset $\mathbf{P}$ of $\mathbf{D}$ and we define

$$\mathfrak{D}_{N,\mathbf{P}}(f)(x) = \sum_{s\in\mathbf{P}}(\chi_{\omega_{s(2)}}\circ N)(x)\,\langle f\,|\,\varphi_s\rangle\,\varphi_s(x)\,. \tag{10.5.24}$$

To prove (10.5.23) it will suffice to show that there exists a $C>0$ such that for all $f\in L^2(\mathbf{R})$, all finite subsets $\mathbf{P}$ of $\mathbf{D}$, and all real-valued measurable functions N on the line we have

$$\big\|\mathfrak{D}_{N,\mathbf{P}}(f)\big\|_{L^{2,\infty}} \le C\big\|f\big\|_{L^2}\,. \tag{10.5.25}$$

The important point is that the constant C in (10.5.25) is independent of f, $\mathbf{P}$, and N.

Using Exercise 1.4.14, to prove (10.5.25), it will suffice to prove that for every measurable set E of finite measure, there exists a measurable subset E' of E with $|E'|\ge\frac{1}{2}|E|$ such that

$$\left|\int_{E'}\mathfrak{D}_{N,\mathbf{P}}(f)\,dx\right| = \left|\sum_{s\in\mathbf{P}}\big\langle(\chi_{\omega_{s(2)}}\circ N)\varphi_s,\chi_{E'}\big\rangle\big\langle\varphi_s\,|\,f\big\rangle\right| \le C|E|^{\frac{1}{2}}\big\|f\big\|_{L^2} \tag{10.5.26}$$

Taking $E'=E$, we will obtain estimate (10.5.26) as a consequence of

$$\sum_{s\in\mathbf{P}}\big|\big\langle(\chi_{\omega_{s(2)}}\circ N)\varphi_s,\chi_E\big\rangle\big\langle f\,|\,\varphi_s\big\rangle\big| \le C|E|^{\frac{1}{2}}\big\|f\big\|_{L^2} \tag{10.5.27}$$

for all L^2 functions f, all measurable functions N, all measurable sets E, and all finite subsets $\mathbf{P}$ of $\mathbf{D}$.

Finally, we use a dilation argument to simplify matters a bit. It is straightforward to check that for all integers j we have the identity

$$\sum_{s\in\mathbf{P}} \big|\big\langle(\chi_{\omega_{s(2)}}\circ N)\varphi_s,\chi_E\big\rangle\big\langle f\,|\,\varphi_s\big\rangle\big| = 2^{-\frac{j}{2}}\sum_{u\in\mathbf{P}(j)}\big|\big\langle(\chi_{\omega_{u(2)}}\circ N_j)\varphi_u,\chi_{2^j\otimes E}\big\rangle\big\langle D^{2^j}(f)\,|\,\varphi_u\big\rangle\big|,$$

where for any set A we set $2^j\otimes A=\{2^j y:\ y\in A\}$,

$$N_j(x)=2^{-j}N(2^{-j}x)$$

and

$$\mathbf{P}(j)=\{(2^j\otimes I_s)\times(2^{-j}\otimes\omega_s):\ s\in\mathbf{P}\}\,.$$

By picking an integer j such that $\frac{1}{2}\le 2^j|E|\le 1$, we obtain that (10.5.27) will be a consequence of the estimate

$$\sum_{s\in\mathbf{P}}\big|\big\langle\chi_{E\cap N^{-1}[\omega_{s(2)}]},\varphi_s\big\rangle\big\langle f\,|\,\varphi_s\big\rangle\big| = \sum_{s\in\mathbf{P}}\big|\big\langle(\chi_{\omega_{s(2)}}\circ N)\varphi_s,\chi_E\big\rangle\big\langle f\,|\,\varphi_s\big\rangle\big|\le C\big\|f\big\|_{L^2} \tag{10.5.28}$$

for all functions f in $L^2(\mathbf{R})$, all measurable functions N, all measurable sets E with $|E|\le 1$, and all finite subsets $\mathbf{P}$ of $\mathbf{D}$. (Here we set $N^{-1}[\omega_s]=\{x:\ N(x)\in\omega_s\}$.) It is estimate (10.5.28) that we will concentrate on.

10.5.d. Iterative Selection of Sets of Tiles with Large Mass and Energy

Throughout the rest of this section, we will fix a square integrable function f on the line, a measurable functions N, and a measurable subset E of $\mathbf{R}$ with $|E|\le 1$. We introduce a partial order in the set of all tiles that will help us organize them in our study.

Definition 10.5.3. We define a *partial order* $<$ in the set of all tiles $\mathbf{D}$ by setting

$$s<s'\iff I_s\subseteq I_{s'}\quad\text{and}\quad \omega_{s'}\subseteq\omega_s\,.$$

Tiles can be written as a union of two *semitiles* $I_s\times\omega_{s(1)}$ and $I_s\times\omega_{s(2)}$. Since tiles have area 1, semitiles have area $\frac{1}{2}$.

If two tiles $s,'\in\mathbf{D}$ intersect, then we must have either $s<s'$ or $s'<s$. Indeed, both the time and frequency components of the tiles must intersect; then either $I_s\subseteq I_{s'}$ or $I_{s'}\subseteq I_s$. In the first case we must have $|\omega_s|\ge|\omega_{s'}|$, thus $\omega_{s'}\subseteq\omega_s$, which gives $s<s'$, while in the second case a similar argument gives $s'<s$. As a consequence of this observation, if $\mathbf{R}_0$ is a finite set of tiles, then all maximal elements of $\mathbf{R}_0$ under $<$ must be disjoint sets.

Definition 10.5.4. A finite set of tiles $\mathbf{P}$ is called a *tree* if there exists a tile $t \in \mathbf{P}$ such that all $s \in \mathbf{P}$ satisfy $s < t$. We call t the top of $\mathbf{P}$ and we denote it by $t = \mathrm{top}(\mathbf{P})$. Observe that the top of a tree is unique.

We will denote trees by $\mathbf{T}$, $\mathbf{T}'$, $\mathbf{T}_1$, $\mathbf{T}_2$, and so on.

We observe that every finite set of tiles $\mathbf{P}$ can be written as a union of trees. Indeed, consider all maximal elements of $\mathbf{P}$ under the partial order $<$. Then every nonmaximal element s of $\mathbf{P}$ satisfies $s < t$ for some maximal element $t \in \mathbf{P}$ and thus it belongs to a tree with top t.

Definition 10.5.5. A tree $\mathbf{T}$ is called a 1-tree if

$$\omega_{\mathrm{top}(\mathbf{T})(1)} \subseteq \omega_{s(1)}$$

all $s \in \mathbf{T}$. A tree $\mathbf{T}'$ is called a 2-tree if for all $s \in \mathbf{T}'$ we have

$$\omega_{\mathrm{top}(\mathbf{T}')(2)} \subseteq \omega_{s(2)} \,.$$

We make a few observations about 1-trees and 2-trees. First note that every tree can be written as the union of a 1-tree and of a 2-tree whose intersection is exactly the top of the tree. Also, if $\mathbf{T}$ is an 1-tree, then the intervals $\omega_{\mathrm{top}(\mathbf{T})(2)}$ and $\omega_{s(2)}$ are disjoint for all $s \in \mathbf{T}$, and similarly for 2-trees. See Figure 10.12.

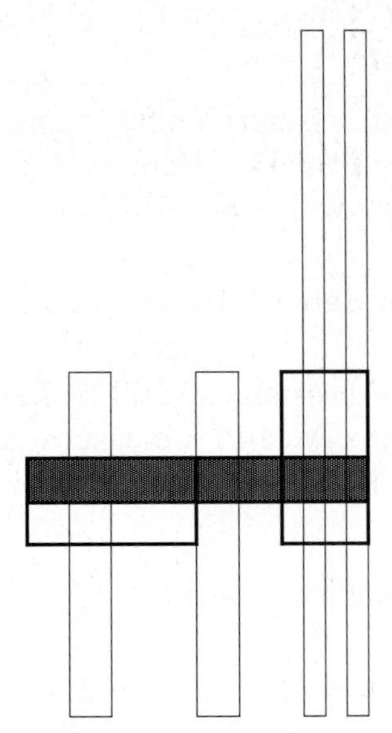

FIGURE 10.12. A tree of seven tiles including the darkened top. The three tiles on the right and the top form a 1-tree while the three tiles on the left and the top form a 2-tree.

Definition 10.5.6. Let $s \in \mathbf{D}$. Then we introduce the quantity

$$\mathcal{M}(\{s\}) = \sup_{\substack{u \in \mathbf{D} \\ s<u}} \int_{E \cap N^{-1}[\omega_u]} \frac{|I_u|^{-1}\, dx}{(1 + \frac{|x - c(I_u)|}{|I_u|})^{10}} \,,$$

and we call $\mathcal{M}(\{s\})$ the *mass* of $\{s\}$. Given a subset $\mathbf{P}$ of $\mathbf{D}$, we define the mass of $\mathbf{P}$ as

$$\mathcal{M}(\mathbf{P}) = \sup_{s\in\mathbf{P}} \mathcal{M}(\{s\}).$$

We observe that the mass of any set of tiles is at most

$$\int_{-\infty}^{+\infty} \frac{dx}{(1+|x|)^{10}} \leq 1.$$

Definition 10.5.7. Given a finite subset $\mathbf{P}$ of $\mathbf{D}$ and a function f in $L^2(\mathbf{R})$, we introduce the quantity

$$\mathcal{E}(f;\mathbf{P}) = \frac{1}{\|f\|_{L^2}} \sup_{\mathbf{T}} \left(\frac{1}{|I_{\mathrm{top}(\mathbf{T})}|} \sum_{s\in\mathbf{T}} |\langle f \,|\, \varphi_s\rangle|^2 \right)^{\frac{1}{2}},$$

where the supremum is taken over all 2-trees $\mathbf{T}$ contained in $\mathbf{P}$. We call $\mathcal{E}(f;\mathbf{P})$ the *energy* of the function f with respect to the set of tiles $\mathbf{P}$.

We now state the following three important lemmata which we will prove in the remaining three subsections.

Lemma 10.5.8. *There exists a constant C_1 such that for any measurable function $N: \mathbf{R}\to\mathbf{R}$ and for any finite set of tiles $\mathbf{P}$, there is a subset $\mathbf{P}'$ of $\mathbf{P}$ such that*

$$\mathcal{M}(\mathbf{P}\setminus\mathbf{P}') \leq \frac{1}{4}\mathcal{M}(\mathbf{P})$$

and $\mathbf{P}'$ is a union of trees $\mathbf{T}_j$ satisfying

$$\sum_j |I_{\mathrm{top}(\mathbf{T}_j)}| \leq \frac{C_1}{\mathcal{M}(\mathbf{P})}. \tag{10.5.29}$$

Lemma 10.5.9. *There exists a constant C_2 such that for any finite set of tiles $\mathbf{P}$ and for all functions f in $L^2(\mathbf{R})$, there is a subset $\mathbf{P}''$ of $\mathbf{P}$ such that*

$$\mathcal{E}(f;\mathbf{P}\setminus\mathbf{P}'') \leq \frac{1}{2}\mathcal{E}(f;\mathbf{P})$$

and $\mathbf{P}''$ is a union of trees $\mathbf{T}_j$ satisfying

$$\sum_j |I_{\mathrm{top}(\mathbf{T}_j)}| \leq \frac{C_2}{\mathcal{E}(f;\mathbf{P})^2}. \tag{10.5.30}$$

Lemma 10.5.10. *(The basic estimate) There is a finite constant C_3 such that for all trees $\mathbf{T}$, all functions f in $L^2(\mathbf{R})$, for any measurable function $N: \mathbf{R}\to\mathbf{R}$, and for all measurable sets E with $|E|\leq 1$ we have*

$$\sum_{s\in\mathbf{T}} |\langle f \,|\, \varphi_s\rangle\langle \chi_{E\cap N^{-1}[\omega_{s(2)}]} \,|\, \varphi_s\rangle| \leq C_3\, |I_{\mathrm{top}(\mathbf{T})}|\, \mathcal{E}(f;\mathbf{T})\mathcal{M}(\mathbf{T})\|f\|_{L^2}. \tag{10.5.31}$$

In the remaining of this subsection, we conclude the proof of Theorem 10.5.1 assuming Lemmata 10.5.8, 10.5.9, and 10.5.10.

In the next argument set

$$C_0 = C_1 + C_2 \,,$$

where C_1 and C_2 are the constants that appear in Lemmata 10.5.8 and 10.5.9, respectively. Given a finite set of tiles $\mathbf{P}$ and a function f in $L^2(\mathbf{R})$, find a very large integer n_0 such that $\mathcal{E}(f;\mathbf{P}) \le 2^{n_0}$ and $\mathcal{M}(\mathbf{P}) \le 2^{2n_0}$. We construct by decreasing induction a sequence of pairwise disjoint sets $\mathbf{P}_{n_0}$, $\mathbf{P}_{n_0-1}$, $\mathbf{P}_{n_0-2}$, $\mathbf{P}_{n_0-3}$, ... such that

$$\bigcup_{j=-\infty}^{n_0} \mathbf{P}_j = \mathbf{P} \tag{10.5.32}$$

and such that the following properties are satisfied:

(1) $\mathcal{E}(f;\mathbf{P}_j) \le 2^{j+1}$ for all $j \le n_0$.
(2) $\mathcal{M}(\mathbf{P}_j) \le 2^{2j+2}$ for all $j \le n_0$.
(3) $\mathcal{E}\big(f;\mathbf{P} \setminus (\mathbf{P}_{n_0} \cup \cdots \cup \mathbf{P}_j)\big) \le 2^j$ for all $j \le n_0$.
(4) $\mathcal{M}\big(\mathbf{P} \setminus (\mathbf{P}_{n_0} \cup \cdots \cup \mathbf{P}_j)\big) \le 2^{2j}$ for all $j \le n_0$.
(5) $\mathbf{P}_j$ is a union of trees $\mathbf{T}_{jk}$ such that $\sum_k |I_{\mathrm{top}(\mathbf{T}_{jk})}| \le C_0 2^{-2j}$ for all $j \le n_0$.

Assuming momentarily that we have constructed a sequence $\mathbf{P}_j$ with the described properties. Then to obtain estimate (10.5.28) we use (1), (2), (5), the observation that the mass is always bounded by 1, and Lemma 10.5.10 to obtain

$$\begin{aligned}
&\sum_{s\in\mathbf{P}} \big|\langle f \,|\, \varphi_s\rangle\langle \chi_{E\cap N^{-1}[\omega_{s(2)}]} \,|\, \varphi_s\rangle\big| \\
=\;& \sum_j \sum_{s\in\mathbf{P}_j} \big|\langle f \,|\, \varphi_s\rangle\langle \chi_{E\cap N^{-1}[\omega_{s(2)}]} \,|\, \varphi_s\rangle\big| \\
\le\;& \sum_j \sum_k \sum_{s\in\mathbf{T}_{jk}} \big|\langle f \,|\, \varphi_s\rangle\langle \chi_{E\cap N^{-1}[\omega_{s(2)}]} \,|\, \varphi_s\rangle\big| \\
\le\;& C_3 \sum_j \sum_k |I_{\mathrm{top}(\mathbf{T}_{jk})}|\, \mathcal{E}(f;\mathbf{T}_{jk})\, \mathcal{M}(\mathbf{T}_{jk}) \big\|f\big\|_{L^2} \\
\le\;& C_3 \sum_j \sum_k |I_{\mathrm{top}(\mathbf{T}_{jk})}|\, 2^{j+1} \min(1, 2^{2j+2}) \big\|f\big\|_{L^2} \\
\le\;& C_3 \sum_j C_0 2^{-2j} 2^{j+1} \min(1, 2^{2j+2}) \big\|f\big\|_{L^2} \\
\le\;& 8C_0C_3 \sum_j \min(2^j, 2^{-j}) \big\|f\big\|_{L^2} \\
\le\;& C \big\|f\big\|_{L^2} \,.
\end{aligned}$$

This proves estimate (10.5.28).

It remains to construct a sequence of disjoint sets $\mathbf{P}_j$ satisfying (1)–(5). We start our induction at $j = n_0$ by setting $\mathbf{P}_{n_0} = \emptyset$. Then (1), (2), and (5) are clearly satisfied, while

$$\mathcal{E}(f; \mathbf{P} \setminus \mathbf{P}_{n_0}) = \mathcal{E}(f; \mathbf{P}) \leq 2^{n_0}$$
$$\mathcal{M}(\mathbf{P} \setminus \mathbf{P}_{n_0}) = \mathcal{M}(\mathbf{P}) \leq 2^{2n_0};$$

hence (3) and (4) are also satisfied for $\mathbf{P}_{n_0}$.

Suppose we have selected pairwise disjoint sets $\mathbf{P}_{n_0}, \mathbf{P}_{n_0-1}, \dots, \mathbf{P}_n$ for some $n < n_0$ such that (1)–(5) are satisfied for all $j \in \{n_0, n_0 - 1, \dots, n\}$. We will construct a set of tiles $\mathbf{P}_{n-1}$ disjoint from all the sets already constructed such that (1)–(5) are satisfied for all $j = n - 1$. This procedure is given by decreasing induction. We will need to consider the following four cases.

Case 1: $\mathcal{E}\big(f; \mathbf{P} \setminus (\mathbf{P}_{n_0} \cup \cdots \cup \mathbf{P}_n)\big) \leq 2^{n-1}$, $\mathcal{M}\big(\mathbf{P} \setminus (\mathbf{P}_{n_0} \cup \cdots \cup \mathbf{P}_n)\big) \leq 2^{2(n-1)}$

In this case set $\mathbf{P}_{n-1} = \emptyset$ and observe that (1)-(5) trivially hold.

Case 2: $\mathcal{E}\big(f; \mathbf{P} \setminus (\mathbf{P}_{n_0} \cup \cdots \cup \mathbf{P}_n)\big) > 2^{n-1}$, $\mathcal{M}\big(\mathbf{P} \setminus (\mathbf{P}_{n_0} \cup \cdots \cup \mathbf{P}_n)\big) \leq 2^{2(n-1)}$

Use Lemma 10.5.9 to find a subset $\mathbf{P}_{n-1}$ of $\mathbf{P} \setminus (\mathbf{P}_{n_0} \cup \cdots \cup \mathbf{P}_n)$ such that

$$\begin{aligned} &\mathcal{E}\big(f; \mathbf{P} \setminus (\mathbf{P}_{n_0} \cup \cdots \cup \mathbf{P}_n \cup \mathbf{P}_{n-1})\big) \\ &\qquad \leq \frac{1}{2} \mathcal{E}\big(f; \mathbf{P} \setminus (\mathbf{P}_{n_0} \cup \cdots \cup \mathbf{P}_n)\big) \leq \frac{1}{2} 2^n \end{aligned} \tag{10.5.33}$$

and $\mathbf{P}_{n-1}$ is a union of trees (whose set of tops we denote by $\mathbf{P}^*_{n-1}$) such that

$$\sum_{t \in \mathbf{P}^*_{n-1}} |I_t| \leq C_2\, \mathcal{E}\big(f; \mathbf{P} \setminus (\mathbf{P}_{n_0} \cup \cdots \cup \mathbf{P}_n)\big)^{-2} \leq C_2 2^{-2(n-1)}. \tag{10.5.34}$$

Then (10.5.33) gives (3) and (10.5.34) gives (5) for $j = n - 1$. Since

$$\mathcal{E}(f; \mathbf{P}_{n-1}) \leq \mathcal{E}\big(f; \mathbf{P} \setminus (\mathbf{P}_{n_0} \cup \cdots \cup \mathbf{P}_n)\big) \leq 2^n = 2^{(n-1)+1},$$

estimate (1) is satisfied for $j = n - 1$. Also, by our induction hypothesis we have

$$\mathcal{M}\big(\mathbf{P} \setminus (\mathbf{P}_{n_0} \cup \cdots \cup \mathbf{P}_n \cup \mathbf{P}_{n-1})\big) \leq \mathcal{M}\big(\mathbf{P} \setminus (\mathbf{P}_{n_0} \cup \cdots \cup \mathbf{P}_n)\big) \leq 2^{2(n-1)};$$

hence (4) is satisfied for $j = n-1$. Finally $\mathbf{P}_{n-1}$ is contained in $\mathbf{P} \setminus (\mathbf{P}_{n_0} \cup \cdots \cup \mathbf{P}_n)$ and hence its mass is at most the mass of the latter which is trivially bounded by $2^{2(n-1)+2}$; thus (2) is also satisfied for $j = n - 1$.

Case 3: $\mathcal{E}\big(f; \mathbf{P} \setminus (\mathbf{P}_{n_0} \cup \cdots \cup \mathbf{P}_n)\big) \leq 2^{n-1}$, $\mathcal{M}\big(\mathbf{P} \setminus (\mathbf{P}_{n_0} \cup \cdots \cup \mathbf{P}_n)\big) > 2^{2(n-1)}$

In this case we repeat the argument in case 2 with the roles of the mass and energy reversed. Precisely, use Lemma 10.5.8 to find a subset $\mathbf{P}_{n-1}$ of the set

$\mathbf{P}\setminus(\mathbf{P}_{n_0}\cup\cdots\cup\mathbf{P}_n)$ such that

$$\mathcal{M}\big(\mathbf{P}\setminus(\mathbf{P}_{n_0}\cup\cdots\cup\mathbf{P}_n\cup\mathbf{P}_{n-1})\big)\le\frac{1}{4}\mathcal{M}\big(\mathbf{P}\setminus(\mathbf{P}_{n_0}\cup\cdots\cup\mathbf{P}_n)\big)\le\frac{1}{4}\,2^{2n} \tag{10.5.35}$$

and $\mathbf{P}_{n-1}$ is a union of trees (whose set of tops we denote by $\mathbf{P}^*_{n-1}$) such that

$$\sum_{t\in\mathbf{P}^*_{n-1}}|I_t|\le C_1\mathcal{M}\big(\mathbf{P}\setminus(\mathbf{P}_{n_0}\cup\cdots\cup\mathbf{P}_n)\big)^{-1}\le C_1 2^{-2(n-1)}. \tag{10.5.36}$$

Then (10.5.35) gives (4) and (10.5.36) gives (5) for $j=n-1$. By induction we have

$$\mathcal{M}(\mathbf{P}_{n-1})\le\mathcal{M}\big(\mathbf{P}\setminus(\mathbf{P}_{n_0}\cup\cdots\cup\mathbf{P}_n)\big)\le 2^{2n}=2^{2(n-1)+2};$$

thus (2) is satisfied for $j=n-1$. Finally, (1) and (3) follow from the inclusion $\mathbf{P}_{n-1}\subseteq\mathbf{P}\setminus(\mathbf{P}_{n_0}\cup\cdots\cup\mathbf{P}_n)$ and the assumption $\mathcal{E}\big(f;\mathbf{P}\setminus(\mathbf{P}_{n_0}\cup\cdots\cup\mathbf{P}_n)\big)\le 2^{n-1}$. This concludes the proof of (1)–(5) for $j=n-1$.

Case 4: $\mathcal{E}\big(f;\mathbf{P}\setminus(\mathbf{P}_{n_0}\cup\cdots\cup\mathbf{P}_n)\big)>2^{n-1}$, $\mathcal{M}\big(\mathbf{P}\setminus(\mathbf{P}_{n_0}\cup\cdots\cup\mathbf{P}_n)\big)>2^{2(n-1)}$

This is the most involved case since it requires elements from both of the previous cases. We use Lemma 10.5.8 to find a subset $\mathbf{P}'_{n-1}$ of $\mathbf{P}\setminus(\mathbf{P}_{n_0}\cup\cdots\cup\mathbf{P}_n)$ such that

$$\mathcal{M}\big(\mathbf{P}\setminus(\mathbf{P}_{n_0}\cup\cdots\cup\mathbf{P}_n\cup\mathbf{P}'_{n-1})\big)\le\frac{1}{4}\,\mathcal{M}\big(\mathbf{P}\setminus(\mathbf{P}_{n_0}\cup\cdots\cup\mathbf{P}_n)\big)\le\frac{1}{4}\,2^{2n} \tag{10.5.37}$$

and $\mathbf{P}'_{n-1}$ is a union of trees (whose set of tops we denote by $(\mathbf{P}'_{n-1})^*$) such that

$$\sum_{t\in(\mathbf{P}'_{n-1})^*}|I_t|\le C_1\mathcal{M}\big(\mathbf{P}\setminus(\mathbf{P}_{n_0}\cup\cdots\cup\mathbf{P}_n)\big)^{-1}\le C_1 2^{-2(n-1)}. \tag{10.5.38}$$

We now consider the following two subcases of case 4.

Subcase 4(a): $\mathcal{E}\big(f;\mathbf{P}\setminus(\mathbf{P}_{n_0}\cup\cdots\cup\mathbf{P}_n\cup\mathbf{P}'_{n-1})\big)\le 2^{n-1}$

In this subcase, we set $\mathbf{P}_{n-1}=\mathbf{P}'_{n-1}$. Then (3) is automatically satisfied for $j=n-1$ and also (5) is satisfied in view of (10.5.38). By the inductive hypothesis we have $\mathcal{E}\big(f;\mathbf{P}\setminus(\mathbf{P}_{n_0}\cup\cdots\cup\mathbf{P}_n)\big)\le 2^n=2^{(n-1)+1}$ and also $\mathcal{M}\big(\mathbf{P}\setminus(\mathbf{P}_{n_0}\cup\cdots\cup\mathbf{P}_n)\big)\le 2^{2n}=2^{2(n-1)+2}$. Since $\mathbf{P}_{n-1}$ is contained in $\mathbf{P}\setminus(\mathbf{P}_{n_0}\cup\cdots\cup\mathbf{P}_n)$, the same estimates hold for $\mathcal{E}(f;\mathbf{P}_{n-1})$ and $\mathcal{M}(\mathbf{P}_{n-1})$; thus (1) and (2) also hold for $j=n-1$. Finally, (4) for $j=n-1$ follows from (10.5.37) since $\mathbf{P}'_{n-1}=\mathbf{P}_{n-1}$.

Subcase 4(b): $\mathcal{E}\big(f;\mathbf{P}\setminus(\mathbf{P}_{n_0}\cup\cdots\cup\mathbf{P}_n\cup\mathbf{P}'_{n-1})\big)>2^{n-1}$

Here we use Lemma 10.5.9 one more time to find a subset $\mathbf{P}''_{n-1}$ of the set $\mathbf{P}\setminus(\mathbf{P}_{n_0}\cup\cdots\cup\mathbf{P}_n\cup\mathbf{P}'_{n-1})$ such that

$$\begin{aligned}&\mathcal{E}\big(f;\mathbf{P}\setminus(\mathbf{P}_{n_0}\cup\cdots\cup\mathbf{P}_n\cup\mathbf{P}'_{n-1}\cup\mathbf{P}''_{n-1})\big)\\&\qquad\le\frac{1}{2}\,\mathcal{E}\big(f;\mathbf{P}\setminus(\mathbf{P}_{n_0}\cup\cdots\cup\mathbf{P}_n\cup\mathbf{P}'_{n-1})\big)\end{aligned} \tag{10.5.39}$$

and $\mathbf{P}''_{n-1}$ is a union of trees [whose set of tops we denote by $(\mathbf{P}''_{n-1})^*$] such that

$$\sum_{t\in(\mathbf{P}''_{n-1})^*} |I_t| \le \frac{C_2}{\mathcal{E}\big(f;\mathbf{P}\setminus(\mathbf{P}_{n_0}\cup\cdots\cup\mathbf{P}_n\cup\mathbf{P}'_{n-1})\big)^2} \le \frac{C_2}{2^{2(n-1)}}\,. \tag{10.5.40}$$

We set $\mathbf{P}_{n-1} = \mathbf{P}'_{n-1}\cup\mathbf{P}''_{n-1}$ and we observe that $\mathbf{P}_{n-1}$ is disjoint from all the previously selected $\mathbf{P}_j$'s. Since by the induction hypothesis the last term in (10.5.39) is bounded by $\frac{1}{2}\mathcal{E}\big(f;\mathbf{P}\setminus(\mathbf{P}_{n_0}\cup\cdots\cup\mathbf{P}_n)\big)\le\frac{1}{2}2^n$, the first term in (10.5.39) is also bounded by 2^{n-1}, thus (3) holds for $j=n-1$. Likewise, since

$$\mathcal{E}(f;\mathbf{P}_{n-1})\le\mathcal{E}\big(f;\mathbf{P}\setminus(\mathbf{P}_{n_0}\cup\cdots\cup\mathbf{P}_n)\big)\le 2^n = 2^{(n-1)+1}$$

$$\mathcal{M}(\mathbf{P}_{n-1})\le\mathcal{M}\big(\mathbf{P}\setminus(\mathbf{P}_{n_0}\cup\cdots\cup\mathbf{P}_n)\big)\le 2^{2n} = 2^{2(n-1)+2}\,,$$

(1) and (2) are satisfied for $j=n-1$. Since

$$\mathcal{M}\big(\mathbf{P}\setminus(\mathbf{P}_{n_0}\cup\cdots\cup\mathbf{P}_n\cup\mathbf{P}_{n-1})\big)\le\mathcal{M}\big(\mathbf{P}\setminus(\mathbf{P}_{n_0}\cup\cdots\cup\mathbf{P}_n\cup\mathbf{P}'_{n-1})\big)\,,$$

(10.5.37) implies that (4) is satisfied for $j=n-1$. Now each of $\mathbf{P}'_{n-1}$ and $\mathbf{P}''_{n-1}$ is given as a union of trees; thus the same is true for $\mathbf{P}_{n-1}$. The set of tops of all of these trees, call it $(\mathbf{P}_{n-1})^*$, is contained in the union of the set of tops of the trees in $\mathbf{P}'_{n-1}$ and the trees in $\mathbf{P}''_{n-1}$ [i.e., in $(\mathbf{P}'_{n-1})^*\cup(\mathbf{P}''_{n-1})^*$]. This implies that

$$\sum_{t\in(\mathbf{P}_{n-1})^*}|I_t|\le\sum_{t\in(\mathbf{P}'_{n-1})^*}|I_t|+\sum_{t\in(\mathbf{P}''_{n-1})^*}|I_t|\le(C_1+C_2)2^{-2(n-1)}=C_0 2^{-2(n-1)}$$

in view of (10.5.38) and (10.5.40). This proves (5) for $j=n-1$ and concludes the inductive step $j=n-1$.

Pick $j\in\mathbf{Z}$ with $0<2^{2j}<\min_{s\in\mathbf{P}}\mathcal{M}(\{s\})$. Then

$$\mathcal{M}\big(\mathbf{P}\setminus(\mathbf{P}_{n_0}\cup\cdots\cup\mathbf{P}_j)\big)=0\,,$$

and since the only set of tiles with zero mass is the empty set, we conclude that (10.5.32) holds. It also follows that there exists a n_1 such that for all $n\le n_1$, $\mathbf{P}_n=\emptyset$. The construction of the $\mathbf{P}_j$'s is now complete.

10.5.e. The Proof of Lemma 10.5.8

Given a finite set of tiles $\mathbf{P}$, we set $\mu=\mathcal{M}(\mathbf{P})$ to be the mass of $\mathbf{P}$. We define

$$\mathbf{P}'=\{s\in\mathbf{P}:\ \mathcal{M}(\{s\})>\tfrac{1}{4}\mu\}$$

and we observe that $\mathcal{M}(\mathbf{P}\setminus\mathbf{P}')\le\frac{1}{4}\mu$. We will now show that $\mathbf{P}'$ is a union of trees whose tops satisfy (10.5.29).

It follows from the definition of mass that for each $s \in \mathbf{P}'$, there is a tile $u(s) \in \mathbf{D}$ such that $u(s) > s$ and

$$\int_{E\cap N^{-1}[\omega_{u(s)}]} \frac{|I_{u(s)}|^{-1}\,dx}{(1+\frac{|x-c(I_{u(s)})|}{|I_{u(s)}|})^{10}} > \frac{\mu}{4}\,. \tag{10.5.41}$$

Let $\mathbf{U} = \{u(s) :\ s \in \mathbf{P}'\}$. Also, let $\mathbf{U}_{\max}$ be the subset of $\mathbf{U}$ containing all maximal elements of $\mathbf{U}$ under the partial order of tiles $<$. Likewise define $\mathbf{P}'_{\max}$ as the set of all maximal elements in $\mathbf{P}'$. Notice that tiles in $\mathbf{P}'$ can be grouped into trees $\mathbf{T}_j = \{s \in \mathbf{P}' :\ s < t_j\}$ with tops $t_j \in \mathbf{P}'_{\max}$. Next, observe that

$$\sum_j |I_{t_j}| \le \sum_{u\in\mathbf{U}_{\max}} |I_u|\,,$$

since for each top t_j, the corresponding $u(t_j) \in \mathbf{U}$ [which satisfies $t_j < u(t_j)$] either belongs to $\mathbf{U}_{\max}$ or satisfies $u(t_j) < u$ for some $u \in \mathbf{U}_{\max}$. Therefore, estimate (10.5.29) will be a consequence of

$$\sum_{u\in\mathbf{U}_{\max}} |I_u| \le C\mu^{-1}\,. \tag{10.5.42}$$

For $u \in \mathbf{U}_{\max}$ we rewrite (10.5.41) as

$$\sum_{k=0}^{\infty} \int_{E\cap N^{-1}[\omega_u]\cap\left(2^kI_u\setminus 2^{k-1}I_u\right)} \frac{|I_u|^{-1}\,dx}{\left(1+\frac{|x-c(I_u)|}{|I_u|}\right)^{10}} > \frac{\mu}{8}\sum_{k=0}^{\infty} 2^{-k}$$

with the interpretation that $2^{-1}I_u = \emptyset$. It follows that for all u in $\mathbf{U}_{\max}$ there exists an integer $k \ge 0$ such that

$$\frac{\mu}{8}\,|I_u|\,2^{-k} < \int_{E\cap N^{-1}[\omega_u]\cap\left(2^kI_u\setminus 2^{k-1}I_u\right)} \frac{dx}{(1+\frac{|x-c(I_u)|}{|I_u|})^{10}} \le \frac{|E\cap N^{-1}[\omega_u]\cap 2^kI_u|}{(\frac{4}{5})^{10}(1+2^{k-2})^{10}}\,.$$

We therefore conclude that

$$\mathbf{U}_{\max} = \bigcup_{k=0}^{\infty} \mathbf{U}_k\,,$$

where

$$\mathbf{U}_k = \{u \in \mathbf{U}_{\max} :\ \tfrac{5^{-10}}{8}\,\mu\,|I_u|\,2^{9k} \le |E\cap N^{-1}[\omega_u]\cap 2^kI_u|\}\,.$$

The required estimate (10.5.42) will be a consequence of the sequence of estimates

$$\sum_{u\in\mathbf{U}_k} |I_u| \le C2^{-8k}\mu^{-1}\,, \qquad k \ge 0\,. \tag{10.5.43}$$

We now fix a $k \ge 0$ and we concentrate on (10.5.43). Select an element $v_0 \in \mathbf{U}_k$ such that $|I_{v_0}|$ is the largest possible among elements of $\mathbf{U}_k$. Then select an element $v_1 \in \mathbf{U}_k \setminus \{u_0\}$ such that the enlarged rectangle $(2^kI_{v_1}) \times \omega_{v_1}$ is disjoint from the enlarged rectangle $(2^kI_{v_0}) \times \omega_{v_0}$ and $|I_{v_1}|$ is the largest possible. Continue this process by induction. At the jth step select an element of $\mathbf{U}_k \setminus \{v_0, \dots, v_{j-1}\}$ such that the enlarged rectangle $(2^k I_{v_j}) \times \omega_{v_j}$ is disjoint from all the enlarged rectangles of the previously selected tiles and the length $|I_{v_j}|$ is the largest possible. This process will terminate after a finite number of steps. We denote by $\mathbf{V}_k$ the set of all selected tiles in $\mathbf{U}_k$.

We make a few observations. Recall that all elements of $\mathbf{U}_k$ are maximal rectangles in $\mathbf{U}$ and therefore disjoint. For any $u \in \mathbf{U}_k$ there exists a selected $v \in \mathbf{V}_k$ with $|I_u| \le |I_v|$ such that the enlarged rectangles corresponding to u and v intersect. Let us associate this u to the selected v. Observe that if u and u' are associated to the same selected v, then since u and u' are disjoint and since both ω_u and $\omega_{u'}$ contain ω_v, the intervals I_u and $I_{u'}$ must be disjoint. Moreover, all tiles $u \in \mathbf{U}_k$ associated with a fixed $v \in \mathbf{V}_k$ satisfy

$$I_u \subseteq 2^{k+2} I_v \,.$$

Putting these observations together, we obtain

$$\begin{aligned}
\sum_{u \in \mathbf{U}_k} |I_u| &\le \sum_{v \in \mathbf{V}_k} \sum_{\substack{u \in \mathbf{U}_k \\ u \text{ associated with } v}} |I_u| \\
&= \sum_{v \in \mathbf{V}_k} \Bigg| \bigcup_{\substack{u \in \mathbf{U}_k \\ u \text{ associated with } v}} I_u \Bigg| \\
&\le \sum_{v \in \mathbf{V}_k} |2^{k+2} I_v| = 2^{k+2} \sum_{v \in \mathbf{V}_k} |I_v| \\
&\le 2^{k+5} 5^{10} \, \mu^{-1} \, 2^{-9k} \sum_{v \in \mathbf{V}_k} |E \cap N^{-1}[\omega_v] \cap 2^k I_v| \\
&\le 32 \cdot 5^{10} \, \mu^{-1} \, 2^{-8k} |E| \le 32 \cdot 5^{10} \, \mu^{-1} \, 2^{-8k} \,,
\end{aligned}$$

since the enlarged rectangles $2^k I_v \times \omega_v$ of the selected tiles v are disjoint and therefore so are the subsets $E \cap N^{-1}[\omega_v] \cap 2^k I_v$ of E. This concludes the proof of estimate (10.5.43) and therefore of Lemma 10.5.8.

10.5.f. The Proof of Lemma 10.5.9

As before, we fix a finite set of tiles $\mathbf{P}$. For a 2-tree $\mathbf{T}'$, let us denote by

$$\Delta(f; \mathbf{T}') = \frac{1}{\big\|f\big\|_{L^2}} \left\{ \frac{1}{|I_{\mathrm{top}(\mathbf{T}')}|} \sum_{s \in \mathbf{T}'} |\langle f \,|\, \varphi_s \rangle|^2 \right\}^{\frac{1}{2}}$$

the quantity associated with $\mathbf{T}'$ appearing in the definition of the energy. Consider the set of all 2-trees $\mathbf{T}'$ contained in $\mathbf{P}$ that satisfy

$$\Delta(f; \mathbf{T}') \ge \frac{1}{2} \mathcal{E}(f; \mathbf{P}) \tag{10.5.44}$$

and among them select a 2-tree $\mathbf{T}_1'$ with $c(\omega_{\mathrm{top}(\mathbf{T}_1')})$ as small as possible. We let $\mathbf{T}_1$ be the set of $s \in \mathbf{P}$ satisfying $s < \mathrm{top}(\mathbf{T}_1')$. Then $\mathbf{T}_1$ is the largest tree in $\mathbf{P}$ whose top is $\mathrm{top}(\mathbf{T}_1')$. We now repeat this procedure to the set $\mathbf{P} \setminus \mathbf{T}_1$. Among all 2-trees contained in $\mathbf{P} \setminus \mathbf{T}_1$ that satisfy (10.5.44) we pick a 2-tree $\mathbf{T}_2'$ with $c(\omega_{\mathrm{top}(\mathbf{T}_2')})$ as small as possible. Then we let $\mathbf{T}_2$ be the of $s \in \mathbf{P}\setminus\mathbf{T}_1$ satisfying $s < \mathrm{top}(\mathbf{T}_2')$. Then $\mathbf{T}_2$ is the largest tree in $\mathbf{P}\setminus\mathbf{T}_1$ whose top is $\mathrm{top}(\mathbf{T}_2')$. We continue this procedure by induction until there is no 2-tree left in $\mathbf{P}$ that satisfies (10.5.44). We have therefore constructed a finite sequence of pairwise disjoint 2-trees

$$\mathbf{T}_1', \mathbf{T}_2', \mathbf{T}_3', \dots, \mathbf{T}_q',$$

and a finite sequence of pairwise disjoint trees

$$\mathbf{T}_1, \mathbf{T}_2, \mathbf{T}_3, \dots, \mathbf{T}_q,$$

such that $\mathbf{T}_j' \subseteq \mathbf{T}_j$, $\mathrm{top}(\mathbf{T}_j) = \mathrm{top}(\mathbf{T}_j')$, and the $\mathbf{T}_j'$ satisfy (10.5.44). We now let

$$\mathbf{P}'' = \bigcup_j \mathbf{T}_j,$$

and observe that this selection of trees ensures that

$$\mathcal{E}(f;\mathbf{P}\setminus\mathbf{P}'') \le \frac{1}{2}\,\mathcal{E}(f;\mathbf{P}).$$

It remains to prove (10.5.30). Using (10.5.44), we obtain that

$$\begin{aligned}
&\frac{1}{4}\,\mathcal{E}(f;\mathbf{P})^2 \sum_j |I_{\mathrm{top}(\mathbf{T}_j)}| \\
&\le \frac{1}{\|f\|_{L^2}^2} \sum_j \sum_{s\in\mathbf{T}_j'} |\langle f \,|\, \varphi_s\rangle|^2 \\
&= \frac{1}{\|f\|_{L^2}^2} \sum_j \sum_{s\in\mathbf{T}_j'} \langle f \,|\, \varphi_s\rangle \overline{\langle f \,|\, \varphi_s\rangle} \qquad (10.5.45)\\
&= \frac{1}{\|f\|_{L^2}^2} \Big\langle f \,\Big|\, \sum_j \sum_{s\in\mathbf{T}_j'} \langle f \,|\, \varphi_s\rangle \varphi_s \Big\rangle \\
&\le \frac{1}{\|f\|_{L^2}} \Big\| \sum_j \sum_{s\in\mathbf{T}_j'} \langle \varphi_s \,|\, f\rangle \varphi_s \Big\|_{L^2},
\end{aligned}$$

and we will use this estimate to obtain (10.5.30). Let us set $\mathbf{U} = \bigcup_j \mathbf{T}_j'$. We will prove that

$$\frac{1}{\|f\|_{L^2}} \Big\| \sum_{s\in\mathbf{U}} \langle \varphi_s \,|\, f\rangle \varphi_s \Big\|_{L^2} \le C\Big(\mathcal{E}(f;\mathbf{P})^2 \sum_j |I_{\mathrm{top}(\mathbf{T}_j)}| \Big)^{\frac{1}{2}}. \qquad (10.5.46)$$

Once this estimate is established, then (10.5.45) combined with (10.5.46) will yield (10.5.30). (Here we are using the fact that all the involved quantities are a priori finite, a consequence of the fact that $\mathbf{P}$ is a finite set of tiles.)

We estimate the square of the left-hand side in (10.5.46) by

$$\sum_{\substack{s,u\in\mathbf{U}\\ \omega_s=\omega_u}} \big|\langle\varphi_s\,|\,f\rangle\langle f\,|\,\varphi_u\rangle\langle\varphi_s\,|\,\varphi_u\rangle\big| + 2\sum_{\substack{s,u\in\mathbf{U}\\ \omega_s\subsetneqq\omega_u}} \big|\langle\varphi_s\,|\,f\rangle\langle f\,|\,\varphi_u\rangle\langle\varphi_s\,|\,\varphi_u\rangle\big| \tag{10.5.47}$$

since $\langle\varphi_s\,|\,\varphi_u\rangle = 0$ unless ω_s contains ω_u or vice versa. We now estimate the quantities $\big|\langle\varphi_s\,|\,f\rangle\big|$ and $\big|\langle\varphi_u\,|\,f\rangle\big|$ by the larger one and we use Exercise 10.5.5 to obtain the following bound for the first term in (10.5.47):

$$\begin{aligned}
&\sum_{s\in\mathbf{U}} \big|\langle f\,|\,\varphi_s\rangle\big|^2 \sum_{\substack{u\in\mathbf{U}\\ \omega_u=\omega_s}} \big|\langle\varphi_s\,|\,\varphi_u\rangle\big| \\
&\le \sum_{s\in\mathbf{U}} \big|\langle f\,|\,\varphi_s\rangle\big|^2 \sum_{\substack{u\in\mathbf{U}\\ \omega_u=\omega_s}} C' \int_{I_u} \frac{1}{|I_s|}\left(1+\frac{|x-c(I_s)|}{|I_s|}\right)^{-100} dx \\
&\le C'' \sum_{s\in\mathbf{U}} \big|\langle f\,|\,\varphi_s\rangle\big|^2 \\
&= C'' \sum_j \sum_{s\in\mathbf{T}'_j} \big|\langle f\,|\,\varphi_s\rangle\big|^2 \\
&\le C'' \sum_j |I_{\mathrm{top}(\mathbf{T}_j)}|\,|I_{\mathrm{top}(\mathbf{T}_j)}|^{-1} \sum_{s\in\mathbf{T}'_j} \big|\langle f\,|\,\varphi_s\rangle\big|^2 \\
&\le C'' \sum_j |I_{\mathrm{top}(\mathbf{T}_j)}|\,\mathcal{E}(f;\mathbf{P})^2 \|f\|_{L^2}^2 ,
\end{aligned} \tag{10.5.48}$$

where in the derivation of the second inequality we used the fact that for fixed $s\in\mathbf{U}$, the intervals I_u with $\omega_u=\omega_s$ are pairwise disjoint.

Our next goal is to obtain a similar estimate for the second term in (10.5.47). That is, we need to prove that

$$\sum_{\substack{s,u\in\mathbf{U}\\ \omega_s\subsetneqq\omega_u}} \big|\langle\varphi_s\,|\,f\rangle\langle f\,|\,\varphi_u\rangle\langle\varphi_s\,|\,\varphi_u\rangle\big| \le C\,\mathcal{E}(f;\mathbf{P})^2 \|f\|_{L^2}^2 \sum_j |I_{\mathrm{top}(\mathbf{T}_j)}| . \tag{10.5.49}$$

Then the required estimate (10.5.46) will follow by combining (10.5.48) and (10.5.49). To prove (10.5.49), we argue as follows:

$$
\begin{aligned}
&\sum_{\substack{s,u\in\mathbf{U}\\ \omega_s\subsetneq\omega_u}} \big|\langle\varphi_s\,|\,f\rangle\langle f\,|\,\varphi_u\rangle\langle\varphi_s\,|\,\varphi_u\rangle\big| \\
&= \sum_j\sum_{s\in\mathbf{T}'_j} \big|\langle f\,|\,\varphi_s\rangle\big| \sum_{\substack{u\in\mathbf{U}\\ \omega_s\subsetneq\omega_u}} \big|\langle f\,|\,\varphi_u\rangle\langle\varphi_s\,|\,\varphi_u\rangle\big| \\
&\le \sum_j |I_{\mathrm{top}(\mathbf{T}_j)}|^{\frac12}\Delta(f;\mathbf{T}'_j)\big\|f\big\|_{L^2}\bigg\{\sum_{s\in\mathbf{T}'_j}\bigg(\sum_{\substack{u\in\mathbf{U}\\ \omega_s\subsetneq\omega_u}} \big|\langle f\,|\,\varphi_u\rangle\langle\varphi_s\,|\,\varphi_u\rangle\big|\bigg)^2\bigg\}^{\frac12} \\
&\le \mathcal{E}(f;\mathbf{P})\big\|f\big\|_{L^2}\sum_j |I_{\mathrm{top}(\mathbf{T}_j)}|^{\frac12}\bigg\{\sum_{s\in\mathbf{T}'_j}\bigg(\sum_{\substack{u\in\mathbf{U}\\ \omega_s\subseteq\omega_{u(1)}}} \big|\langle f\,|\,\varphi_u\rangle\langle\varphi_s\,|\,\varphi_u\rangle\big|\bigg)^2\bigg\}^{\frac12},
\end{aligned}
$$

where we used the Cauchy-Schwarz inequality and the fact that if $\omega_s \subseteq \omega_u$ and $\langle\varphi_s\,|\,\varphi_u\rangle \neq 0$, then $\omega_s \subseteq \omega_{u(1)}$ The proof of (10.5.49) will be complete if we can show that the expression inside the curly brackets is at most a multiple of $\mathcal{E}(f;\mathbf{P})^2\big\|f\big\|_{L^2}^2|I_{\mathrm{top}(\mathbf{T}_j)}|$. Since any singleton $\{s\}\subseteq\mathbf{P}$ is a 2-tree, we have

$$
\mathcal{E}(f;\{u\}) = \frac{1}{\big\|f\big\|_{L^2}}\bigg(\frac{\big|\langle f\,|\,\varphi_u\rangle\big|^2}{|I_u|}\bigg)^{\frac12} = \frac{1}{\big\|f\big\|_{L^2}}\frac{\big|\langle f\,|\,\varphi_u\rangle\big|}{|I_u|^{\frac12}} \le \mathcal{E}(f;\mathbf{P});
$$

hence

$$
\big|\langle f\,|\,\varphi_u\rangle\big| \le \big\|f\big\|_{L^2}|I_u|^{\frac12}\mathcal{E}(f;\mathbf{P})
$$

and it follows that

$$
\sum_{s\in\mathbf{T}'_j}\bigg[\sum_{\substack{u\in\mathbf{U}\\ \omega_s\subseteq\omega_{u(1)}}} \big|\langle f\,|\,\varphi_u\rangle\langle\varphi_s\,|\,\varphi_u\rangle\big|\bigg]^2 \le \mathcal{E}(f;\mathbf{P})^2\big\|f\big\|_{L^2}^2\sum_{s\in\mathbf{T}'_j}\bigg[\sum_{\substack{u\in\mathbf{U}\\ \omega_s\subseteq\omega_{u(1)}}} |I_u|^{\frac12}\big|\langle\varphi_s\,|\,\varphi_u\rangle\big|\bigg]^2.
$$

Thus (10.5.49) will be proved if we can establish that

$$
\sum_{s\in\mathbf{T}'_j}\bigg(\sum_{\substack{u\in\mathbf{U}\\ \omega_s\subseteq\omega_{u(1)}}} |I_u|^{\frac12}\big|\langle\varphi_s\,|\,\varphi_u\rangle\big|\bigg)^2 \le C|I_{\mathrm{top}(\mathbf{T}_j)}|. \tag{10.5.50}
$$

We will need the following crucial lemma.

Lemma 10.5.11. *Let* $\mathbf{T}_j$, $\mathbf{T}'_j$ *be as previously. Let* $s\in\mathbf{T}'_j$ *and* $u\in\mathbf{T}'_k$. *Then if* $\omega_s\subseteq\omega_{u(1)}$, *we have* $I_u\cap I_{\mathrm{top}(\mathbf{T}_j)}=\emptyset$. *Moreover, if* $u\in\mathbf{T}'_k$ *and* $v\in\mathbf{T}'_l$ *are different tiles and satisfy* $\omega_s\subseteq\omega_{u(1)}$ *and* $\omega_s\subseteq\omega_{v(1)}$ *for some fixed* $s\in\mathbf{T}'_j$, *then* $I_u\cap I_v=\emptyset$.

Proof. We observe that if $s \in \mathbf{T}'_j$, $u \in \mathbf{T}'_k$ and $\omega_s \subseteq \omega_{u(1)}$, then the 2-trees $\mathbf{T}'_j$ and $\mathbf{T}'_k$ have different tops and therefore they cannot be the same tree, thus $j \neq k$.

Next we observe that the center of $\omega_{\mathrm{top}(\mathbf{T}'_j)}$ is contained in ω_s which is contained in $\omega_{u(1)}$. Therefore, the center of $\omega_{\mathrm{top}(\mathbf{T}'_j)}$ is contained in $\omega_{u(1)}$, and therefore it must be smaller than the center of $\omega_{\mathrm{top}(\mathbf{T}'_k)}$ since $\mathbf{T}'_k$ is a 2-tree. This means that the 2-tree $\mathbf{T}'_j$ was selected before $\mathbf{T}'_k$, that is, we must have $j < k$. If I_u had a nonempty intersection with $I_{\mathrm{top}(\mathbf{T}_j)} = I_{\mathrm{top}(\mathbf{T}'_j)}$, then since

$$|I_{\mathrm{top}(\mathbf{T}'_j)}| = \frac{1}{|\omega_{\mathrm{top}(\mathbf{T}'_j)}|} \geq \frac{1}{|\omega_s|} \geq \frac{1}{|\omega_{u(1)}|} = \frac{2}{|\omega_u|} = 2|I_u|\,,$$

I_u would have to be contained in $I_{\mathrm{top}(\mathbf{T}'_j)}$. Since also $\omega_{\mathrm{top}(\mathbf{T}'_j)} \subseteq \omega_s \subseteq \omega_u$, it follows that $u < \mathrm{top}(\mathbf{T}'_j)$; thus u would belong to the tree $\mathbf{T}_j$ [which is the largest tree with top $\mathrm{top}(\mathbf{T}'_j)$] since this tree was selected first. But if u belonged to $\mathbf{T}_j$, then it could not belong to $\mathbf{T}'_k$ which is disjoint from $\mathbf{T}_j$; hence we get contradiction. We conclude that $I_u \cap I_{\mathrm{top}(\mathbf{T}_j)} = \emptyset$.

Next assume that $u \in \mathbf{T}'_k$, $v \in \mathbf{T}'_l$, $u \neq v$ and that $\omega_s \subseteq \omega_{u(1)} \cap \omega_{v(1)}$ for some fixed $s \in \mathbf{T}'_j$. Since the left halves of two dyadic intervals ω_u and ω_v intersect, three things can happen: (a) $\omega_u \subseteq \omega_{v(1)}$, in which case I_v is disjoint from the top of $\mathbf{T}'_k$ and thus from I_u; (b) $\omega_v \subseteq \omega_{u(1)}$, in which case I_u is disjoint from the top of $\mathbf{T}'_l$ and thus from I_v; and (c) $\omega_u = \omega_v$, in which case $|I_u| = |I_v|$, thus I_u and I_v are either disjoint or they coincide. Since $u \neq v$, it follows that I_u and I_v cannot coincide, thus $I_u \cap I_v = \emptyset$. This finishes the proof of the lemma.

□

We now return to (10.5.50). In view of Lemma 10.5.11, different $u \in \mathbf{U}$ that appear in the interior sum in (10.5.50) have disjoint intervals I_u and all of these are contained in $(I_{\mathrm{top}(\mathbf{T}_j)})^c$. Using Exercise 10.5.5, we obtain

$$\begin{aligned}
&\sum_{s \in \mathbf{T}'_j} \Bigg(\sum_{\substack{u \in \mathbf{U} \\ \omega_s \subseteq \omega_{u(1)}}} |I_u|^{\frac{1}{2}} \big|\langle \varphi_s \,|\, \varphi_u \rangle\big| \Bigg)^2 \\
&\leq C \sum_{s \in \mathbf{T}'_j} \Bigg(\sum_{\substack{u \in \mathbf{U} \\ \omega_s \subseteq \omega_{u(1)}}} |I_u|^{\frac{1}{2}} \bigg(\frac{|I_s|}{|I_u|}\bigg)^{\frac{1}{2}} \int_{I_u} \frac{|I_s|^{-1}\, dx}{\big(1 + \frac{|x - c(I_s)|}{|I_s|}\big)^{20}} \Bigg)^2 \\
&\leq C \sum_{s \in \mathbf{T}'_j} |I_s| \Bigg(\sum_{\substack{u \in \mathbf{U} \\ \omega_s \subseteq \omega_{u(1)}}} \int_{I_u} \frac{|I_s|^{-1}\, dx}{\big(1 + \frac{|x - c(I_s)|}{|I_s|}\big)^{20}} \Bigg)^2 \\
&\leq C \sum_{s \in \mathbf{T}'_j} |I_s| \Bigg(\int_{(I_{\mathrm{top}(\mathbf{T}_j)})^c} \frac{|I_s|^{-1}\, dx}{\big(1 + \frac{|x - c(I_s)|}{|I_s|}\big)^{20}} \Bigg)^2 ,
\end{aligned}$$

and the latter is at most

$$C \sum_{s\in \mathbf{T}_j'} |I_s| \int_{(I_{\mathrm{top}(\mathbf{T}_j)})^c} \frac{|I_s|^{-1}\, dx}{\big(1+\frac{|x-c(I_s)|}{|I_s|}\big)^{20}}$$

since $\int_{\mathbf{R}} (1+|x|)^{-20}\, dx \le 1$. Set $t_j = \mathrm{top}(\mathbf{T}_j)$. For each scale $k \ge 0$ the tiles s that appear in the tree $\mathbf{T}_j$ are disjoint; therefore, we must have

$$\begin{aligned}
&\sum_{s\in \mathbf{T}_j'} |I_s| \int_{(I_{t_j})^c} \frac{|I_s|^{-1}\, dx}{\big(1+\frac{|x-c(I_s)|}{|I_s|}\big)^{20}} \\
&\le \sum_{k=0}^{\infty} (2^{-k}|I_{t_j}|)^{-1} \sum_{\substack{s\in \mathbf{T}_j' \\ |I_s|=2^{-k}|I_{t_j}|}} |I_s| \int_{(I_{t_j})^c} \frac{dx}{\big(1+\frac{|x-c(I_s)|}{2^{-k}|I_{t_j}|}\big)^{20}} \\
&\le C \sum_{k=0}^{\infty} (2^{-k}|I_{t_j}|)^{-1} \int_{I_{t_j}} \int_{(I_{t_j})^c} \frac{1}{\big(1+\frac{|x-y|}{2^{-k}|I_{t_j}|}\big)^{20}}\, dx\, dy \\
&\le C' \sum_{k=0}^{\infty} (2^{-k}|I_{t_j}|)^{-1} (2^{-k}|I_{t_j}|)^2 \\
&= C''|I_{t_j}| = C''|I_{\mathrm{top}(\mathbf{T}_j)}|\,,
\end{aligned}$$

in view of Exercise 10.5.6. This completes the proof of (10.5.50) and thus of Lemma 10.5.9.

10.5.g. The Proof of Lemma 10.5.10 (the Basic Estimate)

In proving the main estimate we may assume that $\|f\|_{L^2} = 1$, for, we can always replace f by $f/\|f\|_{L^2}$. So throughout this subsection fix a square integrable function with L^2 norm 1. Let $\mathcal{J}'$ be the set of all dyadic intervals J such that $3J$ does not contain any I_s with $s \in \mathbf{P}$. It is not hard to see that any point in $\mathbf{R}$ belongs to a set in $\mathcal{J}'$. Let $\mathcal{J}$ be the set of all maximal (under inclusion) elements of $\mathcal{J}'$. Then $\mathcal{J}$ consists of disjoint sets that cover $\mathbf{R}$, thus it forms a partition of $\mathbf{R}$. This partition of $\mathbf{R}$ is shown in Figure 10.13 when the tree consists of two tiles.

Let us fix a square-integrable function f with $\|f\|_{L^2} = 1$, a tree $\mathbf{T}$, and a measurable set E with $|E| \le 1$. For each $s \in \mathbf{T}$ pick an $\varepsilon_s \in \mathbf{C}$ with $|\varepsilon_s| = 1$ such that

$$|\langle f \,|\, \varphi_s\rangle\langle \chi_{E\cap N^{-1}[\omega_{s(2)}]} \,|\, \varphi_s\rangle| = \varepsilon_s \langle f \,|\, \varphi_s\rangle\langle \varphi_s \,|\, \chi_{E\cap N^{-1}[\omega_{s(2)}]}\rangle\,.$$

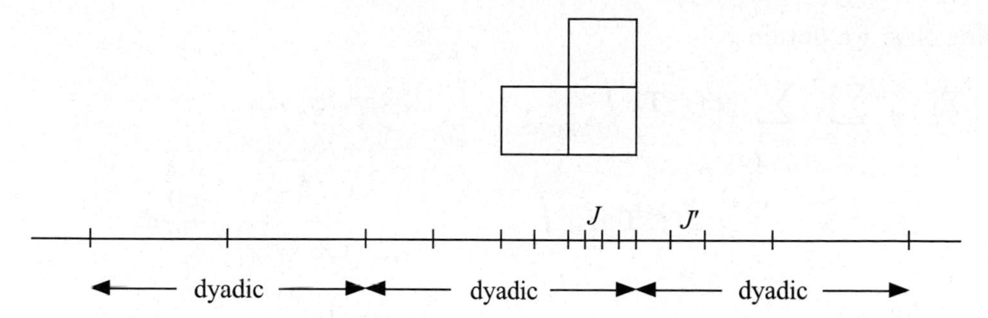

FIGURE 10.13. A tree of two tiles. The corresponding partition $\mathcal{J}$ of $\mathbf{R}$ is shown on the bottom.

We can now write the left-hand side of (10.5.31) as

$$\begin{aligned}
&\sum_{s\in\mathbf{T}} \varepsilon_s \langle f \,|\, \varphi_s \rangle \langle \varphi_s \,|\, \chi_{E\cap N^{-1}[\omega_{s(2)}]} \rangle \\
&\le \Big\| \sum_{s\in\mathbf{T}} \varepsilon_s \langle f \,|\, \varphi_s \rangle \chi_{E\cap N^{-1}[\omega_{s(2)}]} \varphi_s \Big\|_{L^1(\mathbf{R})} \\
&= \sum_{J\in\mathcal{J}} \Big\| \sum_{s\in\mathbf{T}} \varepsilon_s \langle f \,|\, \varphi_s \rangle \chi_{E\cap N^{-1}[\omega_{s(2)}]} \varphi_s \Big\|_{L^1(J)} \\
&\le \Sigma_1 + \Sigma_2 \,,
\end{aligned}$$

where

$$\Sigma_1 = \sum_{J\in\mathcal{J}} \Big\| \sum_{\substack{s\in\mathbf{T} \\ |I_s|\le 2|J|}} \varepsilon_s \langle f \,|\, \varphi_s \rangle \chi_{E\cap N^{-1}[\omega_{s(2)}]} \varphi_s \Big\|_{L^1(J)} , \tag{10.5.51}$$

$$\Sigma_2 = \sum_{J\in\mathcal{J}} \Big\| \sum_{\substack{s\in\mathbf{T} \\ |I_s|> 2|J|}} \varepsilon_s \langle f \,|\, \varphi_s \rangle \chi_{E\cap N^{-1}[\omega_{s(2)}]} \varphi_s \Big\|_{L^1(J)} . \tag{10.5.52}$$

We start with Σ_1. Observe that for every $s \in \mathbf{T}$, the singleton $\{s\}$ is a 2-tree contained in $\mathbf{T}$ and we therefore have the estimate

$$\big|\langle f \,|\, \varphi_s \rangle\big| \le |I_s|^{\frac{1}{2}} \mathcal{E}(f;\mathbf{T}) \,. \tag{10.5.53}$$

Using this, we obtain

$$
\begin{aligned}
\Sigma_1 &\le \sum_{J\in\mathcal{J}} \sum_{\substack{s\in\mathbf{T}\\ |I_s|\le 2|J|}} \mathcal{E}(f;\mathbf{T}) \int_{J\cap E\cap N^{-1}[\omega_{s(2)}]} |I_s|^{\frac12} |\varphi_s(x)|\,dx \\
&\le C \sum_{J\in\mathcal{J}} \sum_{\substack{s\in\mathbf{T}\\ |I_s|\le 2|J|}} \mathcal{E}(f;\mathbf{T}) |I_s| \int_{J\cap E\cap N^{-1}[\omega_{s(2)}]} \frac{|I_s|^{-1}}{\big(1+\frac{|x-c(I_s)|}{|I_s|}\big)^{20}}\,dx \\
&\le C \sum_{J\in\mathcal{J}} \sum_{\substack{s\in\mathbf{T}\\ |I_s|\le 2|J|}} \mathcal{E}(f;\mathbf{T})\mathcal{M}(\mathbf{T}) |I_s| \sup_{x\in J} \frac{1}{\big(1+\frac{|x-c(I_s)|}{|I_s|}\big)^{10}} \\
&\le C\mathcal{E}(f;\mathbf{T})\mathcal{M}(\mathbf{T}) \sum_{J\in\mathcal{J}} \sum_{k=-\infty}^{\log_2 2|J|} 2^k \sum_{\substack{s\in\mathbf{T}\\ |I_s|=2^k}} \frac{1}{\big(1+\frac{\mathrm{dist}\,(J,I_s)}{2^k}\big)^5} \frac{1}{\big(1+\frac{\mathrm{dist}\,(J,I_s)}{2^k}\big)^5}\,.
\end{aligned}
$$

But note that all I_s with $s\in\mathbf{T}$ and $|I_s|=2^k$ are pairwise disjoint and contained in $I_{\mathrm{top}(\mathbf{T})}$. Therefore, $2^{-k}\mathrm{dist}\,(J,I_s) \ge |I_{\mathrm{top}(\mathbf{T})}|^{-1}\mathrm{dist}\,(J,I_{\mathrm{top}(\mathbf{T})})$ and we can estimate $\big(1+\frac{\mathrm{dist}\,(J,I_s)}{2^k}\big)^{-5}$ by $\big(1+\frac{\mathrm{dist}\,(J,I_{\mathrm{top}(\mathbf{T})})}{|I_{\mathrm{top}(\mathbf{T})}|}\big)^{-5}$. Moreover, the sum

$$
\sum_{\substack{s\in\mathbf{T}\\ |I_s|=2^k}} \frac{1}{\big(1+\frac{\mathrm{dist}\,(J,I_s)}{2^k}\big)^5} \tag{10.5.54}
$$

is controlled by a finite number since all the I_s that appear in (10.5.54) are not contained in $3J$ and for every positive integer m there exist at most 4 tiles s with $|I_s|=2^k$ such that $m2^k \le \mathrm{dist}\,(J,I_s) < (m+1)2^k$. Therefore, we obtain

$$
\begin{aligned}
\Sigma_1 &\le C\mathcal{E}(f;\mathbf{T})\mathcal{M}(\mathbf{T}) \sum_{J\in\mathcal{J}} \sum_{k=-\infty}^{\log_2 2|J|} \frac{2^k}{\big(1+\frac{\mathrm{dist}\,(J,I_{\mathrm{top}(\mathbf{T})})}{|I_{\mathrm{top}(\mathbf{T})}|}\big)^5} \\
&\le C\mathcal{E}(f;\mathbf{T})\mathcal{M}(\mathbf{T}) \sum_{J\in\mathcal{J}} \frac{|J|}{\big(1+\frac{\mathrm{dist}\,(J,I_{\mathrm{top}(\mathbf{T})})}{|I_{\mathrm{top}(\mathbf{T})}|}\big)^5} \\
&\le C\mathcal{E}(f;\mathbf{T})\mathcal{M}(\mathbf{T}) \sum_{J\in\mathcal{J}} \int_J \frac{1}{\big(1+\frac{|x-c(I_{\mathrm{top}(\mathbf{T})})|}{|I_{\mathrm{top}(\mathbf{T})}|}\big)^5}\,dx \\
&\le C\,|I_{\mathrm{top}(\mathbf{T})}|\,\mathcal{E}(f;\mathbf{T})\,\mathcal{M}(\mathbf{T})\,,
\end{aligned} \tag{10.5.55}
$$

since $\mathcal{J}$ forms a partition of $\mathbf{R}$. We need to justify, however, the penultimate inequality in (10.5.55). Since J and $I_{\mathrm{top}(\mathbf{T})}$ are dyadic intervals, there are only two possibilities: (a) $J\cap I_{\mathrm{top}(\mathbf{T})}=\emptyset$ and (b) $J\subseteq I_{\mathrm{top}(\mathbf{T})}$. [The third possibility $I_{\mathrm{top}(\mathbf{T})}\subseteq J$ is excluded since $3J$ does not contain $I_{\mathrm{top}(\mathbf{T})}$.] In case (a) we have $|J|\le \mathrm{dist}\,(J,I_{\mathrm{top}(\mathbf{T})})+|I_{\mathrm{top}(\mathbf{T})}|$ since $3J$ does not contain $I_{\mathrm{top}(\mathbf{T})}$. Also, for any

$x \in J$ we have

$$\begin{aligned} |x - c(I_{\mathrm{top}(\mathbf{T})})| &\le |J| + \mathrm{dist}\,(J, I_{\mathrm{top}(\mathbf{T})}) + \frac{1}{2}|I_{\mathrm{top}(\mathbf{T})}| \\ &\le 2\,\mathrm{dist}\,(J, I_{\mathrm{top}(\mathbf{T})}) + \frac{3}{2}|I_{\mathrm{top}(\mathbf{T})}|\,. \end{aligned}$$

Therefore, it follows that

$$\int_J \frac{dx}{\big(1+\frac{|x-c(I_{\mathrm{top}(\mathbf{T})})|}{|I_{\mathrm{top}(\mathbf{T})}|}\big)^5} \ge \frac{|J|}{\big(\frac{5}{2}+\frac{2\mathrm{dist}\,(J,I_{\mathrm{top}(\mathbf{T})})}{|I_{\mathrm{top}(\mathbf{T})}|}\big)^5} \ge \frac{\big(\frac{2}{5}\big)^5|J|}{\big(1+\frac{\mathrm{dist}\,(J,I_{\mathrm{top}(\mathbf{T})})}{|I_{\mathrm{top}(\mathbf{T})}|}\big)^5}\,.$$

In case (b) we have $J \subseteq I_{\mathrm{top}(\mathbf{T})}$ and therefore any x in J lies in $I_{\mathrm{top}(\mathbf{T})}$; thus $|x - c(I_{\mathrm{top}(\mathbf{T})})| \le \frac{1}{2}|I_{\mathrm{top}(\mathbf{T})}|$. We conclude that

$$\int_J \frac{dx}{\big(1+\frac{|x-c(I_{\mathrm{top}(\mathbf{T})})|}{|I_{\mathrm{top}(\mathbf{T})}|}\big)^5} \ge \frac{|J|}{(3/2)^5} = \Big(\frac{2}{3}\Big)^5 \frac{|J|}{\big(1+\frac{\mathrm{dist}\,(J,I_{\mathrm{top}(\mathbf{T})})}{|I_{\mathrm{top}(\mathbf{T})}|}\big)^5}\,.$$

These observations justify the second to last inequality in (10.5.55) and complete the required estimate for Σ_1.

We now turn our attention to Σ_2. We may assume that for all J appearing in the sum in (10.5.52), the set of s in $\mathbf{T}$ with $2|J| < |I_s|$ is nonempty. Thus, if J appears in the sum in (10.5.52), we have $2|J| < |I_{\mathrm{top}(\mathbf{T})}|$ and it is easy to see that J is contained in $3I_{\mathrm{top}(\mathbf{T})}$. [The intervals J in $\mathcal{J}$ which are not contained in $3I_{\mathrm{top}(\mathbf{T})}$ have size larger than $|I_{\mathrm{top}(\mathbf{T})}|$.]

We let $\mathbf{T}_2$ be the 2-tree of all s in $\mathbf{T}$ such that $\omega_{\mathrm{top}(\mathbf{T})(2)} \subseteq \omega_{s(2)}$ and we also let $\mathbf{T}_1 = \mathbf{T} \setminus \mathbf{T}_2$. Then $\mathbf{T}_1$ is a 1-tree minus its top. We set

$$\begin{aligned} F_{1J} &= \sum_{\substack{s\in\mathbf{T}_1\\ |I_s|>2|J|}} \varepsilon_s \langle f \,|\, \varphi_s\rangle \varphi_s \,\chi_{E\cap N^{-1}[\omega_{s(2)}]}\,, \\ F_{2J} &= \sum_{\substack{s\in\mathbf{T}_2\\ |I_s|>2|J|}} \varepsilon_s \langle f \,|\, \varphi_s\rangle \varphi_s \,\chi_{E\cap N^{-1}[\omega_{s(2)}]}\,. \end{aligned}$$

Clearly

$$\Sigma_2 \le \sum_{J\in\mathcal{J}} \big\|F_{1J}\big\|_{L^1(J)} + \sum_{J\in\mathcal{J}} \big\|F_{2J}\big\|_{L^1(J)} = \Sigma_{21} + \Sigma_{22}\,,$$

and we need to estimate both sums. We start by estimating F_{1J}. If the tiles s and s' that appear in the definition of F_{1J} have different scales, then the sets $\omega_{s(2)}$ and $\omega_{s'(2)}$ are disjoint and thus so are the sets $E \cap N^{-1}[\omega_{s(2)}]$ and $E \cap N^{-1}[\omega_{s'(2)}]$. Let us set

$$G_J = J \cap \bigcup_{\substack{s\in\mathbf{T}\\ |I_s|>2|J|}} E \cap N^{-1}[\omega_{s(2)}]\,.$$

Then $F_{1J}\chi_J$ is supported in the set G_J and we have

$$\begin{aligned}
\big\|F_{1J}\big\|_{L^1(J)} &\le \big\|F_{1J}\big\|_{L^\infty(J)}|G_J| \\
&= \Big\| \sum_{k>\log_2 2|J|} \sum_{\substack{s\in\mathbf{T}_1\\ |I_s|=2^k}} \varepsilon_s \left\langle f \,|\, \varphi_s\right\rangle \varphi_s \, \chi_{E\cap N^{-1}[\omega_{s(2)}]}\Big\|_{L^\infty(J)} |G_J| \\
&\le \sup_{k>\log_2 2|J|} \Big\| \sum_{\substack{s\in\mathbf{T}_1\\ |I_s|=2^k}} \varepsilon_s \left\langle f \,|\, \varphi_s\right\rangle \varphi_s \, \chi_{E\cap N^{-1}[\omega_{s(2)}]}\Big\|_{L^\infty(J)} |G_J| \\
&\le \sup_{k>\log_2 2|J|} \sup_{x\in J} \sum_{\substack{s\in\mathbf{T}_1\\ |I_s|=2^k}} \mathcal{E}(f;\mathbf{T}) 2^{k/2} \frac{2^{-k/2}}{\big(1+\frac{|x-c(I_s)|}{2^k}\big)^{10}} |G_J| \\
&\le C\,\mathcal{E}(f;\mathbf{T})|G_J|\,,
\end{aligned}$$

using (10.5.53) and the fact that all the I_s that appear in the sum are disjoint. We now claim that for all $J \in \mathcal{J}$ we have

$$|G_J| \le C\mathcal{M}(\mathbf{T})|J|\,. \tag{10.5.56}$$

Once (10.5.56) is established, summing over all the intervals J that appear in the definition of F_{1J} and keeping in mind that all of these intervals are pairwise disjoint and contained in $3I_{\text{top}(\mathbf{T})}$, we obtain the desired estimate for Σ_{21}.

To prove (10.5.56), we consider the unique dyadic interval $\widetilde{J}$ of length $2|J|$ that contains J. Then by the maximality of $\mathcal{J}$, $3\widetilde{J}$ contains the time interval I_{s_J} of a tile s_J in $\mathbf{T}$. We consider the following two cases: (a) If I_{s_J} is either $\big(\widetilde{J}-|\widetilde{J}|\big)\cup\widetilde{J}$ or $\widetilde{J}\cup\big(\widetilde{J}+|\widetilde{J}|\big)$, we let $u_J = s_J$; in this case $|I_{u_J}| = 2|\widetilde{J}|$. (This is the case for the interval J in Figure 10.13.) Otherwise we have case (b) in which I_{s_J} is contained in one of the two dyadic intervals $\widetilde{J}-|\widetilde{J}|$, $\widetilde{J}+|\widetilde{J}|$. (This is the case for the interval J' in Figure 10.13.) Whichever of these two dyadic intervals contains I_{s_J} is also contained in $I_{\text{top}(\mathbf{T})}$ since it intersects it and has smaller length than it. In case (b) there exists a tile $u_J \in \mathbf{D}$ with $|I_{u_J}| = |\widetilde{J}|$ such that $I_{s_J} \subseteq I_{u_J} \subseteq I_{\text{top}(\mathbf{T})}$ and $\omega_{\text{top}(\mathbf{T})} \subseteq \omega_{u_J} \subseteq \omega_{s_J}$. In both cases we have a tile u_J satisfying $s_J < u_J < \text{top}(\mathbf{T})$ with $|\omega_{u_J}|$ being either $\frac{1}{4}|J|^{-1}$ or $\frac{1}{2}|J|^{-1}$.

Then for any $s \in \mathbf{T}$ with $|I_s| > 2|J|$ we have $|\omega_s| \le |\omega_{u_J}|$. But since both ω_s and ω_{u_J} contain $\omega_{\text{top}(\mathbf{T})}$, they must intersect, and thus $\omega_s \subseteq \omega_{u_J}$. We conclude that any $s \in \mathbf{T}$ with $|I_s| > 2|J|$ must satisfy $N^{-1}[\omega_s] \subseteq N^{-1}[\omega_{u_J}]$. It follows that

$$G_J \subseteq J \cap E \cap N^{-1}[\omega_{u_J}] \tag{10.5.57}$$

and therefore we have

$$
\begin{aligned}
\mathcal{M}(\mathbf{T}) &= \sup_{s\in\mathbf{T}} \sup_{\substack{u\in\mathbf{D}\\ s<u}} \int_{E\cap N^{-1}[\omega_u]} \frac{|I_u|^{-1}}{\left(1+\frac{|x-c(I_u)|}{|I_u|}\right)^{10}}\,dx \\
&\ge \int_{J\cap E\cap N^{-1}[\omega_{u_J}]} \frac{|I_{u_J}|^{-1}}{\left(1+\frac{|x-c(I_{u_J})|}{|I_{u_J}|}\right)^{10}}\,dx \\
&\ge c\,|I_{u_J}|^{-1}\,|J\cap E\cap N^{-1}[\omega_{u_J}]|\,, \\
&\ge c\,|I_{u_J}|^{-1}\,|G_J|\,,
\end{aligned}
$$

using (10.5.57) and the fact that for $x\in J$ we have $|x-c(I_{u_J})|\le 4|J| = 2|I_{u_J}|$. It follows that

$$
|G_J| \le \frac{1}{c}\mathcal{M}(\mathbf{T})|I_{u_J}| = \frac{2}{c}\mathcal{M}(\mathbf{T})|J|\,,
$$

and this is exactly (10.5.56) what we wanted to prove.

We now turn to the estimate for $\Sigma_{22} = \sum_{J\in\mathcal{J}} \big\|F_{2J}\big\|_{L^1(J)}$. All the intervals $\omega_{s(2)}$ with $s\in\mathbf{T}_2$ are nested since $\mathbf{T}_2$ is a 2-tree. Therefore, for each $x\in J$ for which $F_{2J}(x)$ is nonzero, there exists a largest dyadic interval ω_{u_x} and a smallest dyadic interval ω_{v_x} (for some $u_x, v_x \in \mathbf{T}_2\cap\{s:\ |I_s|\ge 4|J|\}$) such that for $s\in\mathbf{T}_2\cap\{s:\ |I_s|\ge 4|J|\}$ we have $N(x)\in\omega_{s(2)}$ if and only if $\omega_{v_x}\subseteq\omega_s\subseteq\omega_{u_x}$. Then we have

$$
\begin{aligned}
F_{2J}(x) &= \sum_{\substack{s\in\mathbf{T}_2\\ |I_s|\ge 4|J|}} \varepsilon_s\langle f\,|\,\varphi_s\rangle(\varphi_s\chi_{E\cap N^{-1}[\omega_{s(2)}]})(x) \\
&= \chi_E(x) \sum_{\substack{s\in\mathbf{T}_2\\ |\omega_{v_x}|\le|\omega_s|\le|\omega_{u_x}|}} \varepsilon_s\langle f\,|\,\varphi_s\rangle\,\varphi_s(x)\,.
\end{aligned}
$$

Pick a Schwartz function ψ whose Fourier transform $\widehat{\psi}(t)$ is supported in $|t|\le 1+\frac{1}{100}$ and that is equal to 1 on $|t|\le 1$. We can easily check that for all $z\in\mathbf{R}$ if $|\omega_{v_x}|\le|\omega_s|\le|\omega_{u_x}|$, then

$$
\left(\varphi_s * \left\{\frac{M^{c(\omega_{u_x})}D^{|\omega_{u_x}|^{-1}}(\psi)}{|\omega_{u_x}|^{-\frac{1}{2}}} - \frac{M^{c(\omega_{v_x(2)})}D^{|\omega_{v_x(2)}|^{-1}}(\psi)}{|\omega_{v_x(2)}|^{-\frac{1}{2}}}\right\}\right)(z) = \varphi_s(z) \tag{10.5.58}
$$

by a simple examination of the Fourier transforms. Basically, the Fourier transform (in z) of the function inside the curly brackets is equal to

$$
\widehat{\psi}\Big(\tfrac{\xi-c(\omega_{u_x})}{|\omega_{u_x}|}\Big) - \widehat{\psi}\Big(\tfrac{\xi-c(\omega_{v_x(2)})}{|\omega_{v_x(2)}|}\Big)\,,
$$

which is equal to 1 on the support of $\widehat{\varphi_s}$ for all s in $\mathbf{T}_2$ that satisfy $|\omega_{v_x}| \le |\omega_s| \le |\omega_{u_x}|$ but vanishes on $\omega_{v_x(2)}$. Taking $z = x$ in (10.5.58) yields

$$\begin{aligned} F_{2J}(x) &= \sum_{\substack{s\in\mathbf{T}_2\\ |\omega_{v_x}|\le|\omega_s|\le|\omega_{u_x}|}} \varepsilon_s\langle f\,|\,\varphi_s\rangle\,\varphi_s(x)\chi_E(x)\\ &= \Big[\sum_{s\in\mathbf{T}_2}\varepsilon_s\langle f\,|\,\varphi_s\rangle\,\varphi_s\Big] * \Big\{\frac{M^{c(\omega_{u_x})}D^{|\omega_{u_x}|^{-1}}(\psi)}{|\omega_{u_x}|^{-\frac12}} - \frac{M^{c(\omega_{v_x(2)})}D^{|\omega_{v_x(2)}|^{-1}}(\psi)}{|\omega_{v_x(2)}|^{-\frac12}}\Big\}(x)\chi_E(x)\,. \end{aligned}$$

Since all s that appear in the definition of F_{2J} satisfy $|\omega_s| \le (4|J|)^{-1}$, it follows that we have the estimate

$$\begin{aligned} |F_{2J}(x)| &\le 2\chi_E(x) \sup_{\delta>|\omega_{u_x}|^{-1}} \int_{\mathbf{R}} \Big|\sum_{s\in\mathbf{T}_2}\varepsilon_s\langle f\,|\,\varphi_s\rangle\varphi_s(z)\Big|\,\tfrac{1}{\delta}\big|\psi\big(\tfrac{x-z}{\delta}\big)\big|\,dz\\ &\le C \sup_{\delta>4|J|}\frac{1}{2\delta}\int_{x-\delta}^{x+\delta}\Big|\sum_{s\in\mathbf{T}_2}\varepsilon_s\langle f\,|\,\varphi_s\rangle\varphi_s(z)\Big|\,dz\,. \end{aligned} \tag{10.5.59}$$

(The last inequality follows from Exercise 2.1.14.) Observe that the maximal function in (10.5.59) satisfies the property

$$\sup_{x\in J}\sup_{\delta>4|J|}\frac{1}{2\delta}\int_{x-\delta}^{x+\delta}|h(t)|\,dt \le 2\inf_{x\in J}\sup_{\delta>4|J|}\frac{1}{2\delta}\int_{x-\delta}^{x+\delta}|h(t)|\,dt\,.$$

Using this property, we obtain

$$\begin{aligned} \Sigma_{22} &\le \sum_{J\in\mathcal{J}}\big\|F_{2J}\big\|_{L^1(J)} \le \sum_{J\in\mathcal{J}}\big\|F_{2J}\big\|_{L^\infty(J)}|G_J|\\ &\le C\sum_{\substack{J\in\mathcal{J}\\ J\subseteq 3I_{\mathrm{top}(\mathbf{T})}}} \mathcal{M}(\mathbf{T})|J|\sup_{x\in J}\sup_{\delta>4|J|}\frac{1}{2\delta}\int_{x-\delta}^{x+\delta}\Big|\sum_{s\in\mathbf{T}_2}\varepsilon_s\langle f\,|\,\varphi_s\rangle\varphi_s(z)\Big|\,dz\\ &\le 2C\mathcal{M}(\mathbf{T})\sum_{\substack{J\in\mathcal{J}\\ J\subseteq 3I_{\mathrm{top}(\mathbf{T})}}}\int_J\sup_{\delta>4|J|}\frac{1}{2\delta}\int_{x-\delta}^{x+\delta}\Big|\sum_{s\in\mathbf{T}_2}\varepsilon_s\langle f\,|\,\varphi_s\rangle\varphi_s(z)\Big|\,dz\,dx\\ &\le C\mathcal{M}(\mathbf{T})\Big\|M\Big(\sum_{s\in\mathbf{T}_2}\varepsilon_s\langle f\,|\,\varphi_s\rangle\varphi_s\Big)\Big\|_{L^1(3I_{\mathrm{top}(\mathbf{T})})}, \end{aligned}$$

where M is the Hardy-Littlewood maximal operator. Using the Cauchy-Schwarz inequality and the boundedness of M on $L^2(\mathbf{R})$, we obtain the following estimate:

$$\Sigma_{22} \le C\mathcal{M}(\mathbf{T})\,|I_{\mathrm{top}(\mathbf{T})}|^{\frac12}\Big\|\sum_{s\in\mathbf{T}_2}\varepsilon_s\langle f\,|\,\varphi_s\rangle\varphi_s\Big\|_{L^2}.$$

Appealing to the result of Exercise 10.5.7(a) we deduce

$$\Big\| \sum_{s\in \mathbf{T}_2} \varepsilon_s \langle f \,|\, \varphi_s\rangle \varphi_s \Big\|_{L^2} \le C\Big(\sum_{s\in \mathbf{T}_2} |\varepsilon_s \langle f \,|\, \varphi_s\rangle|^2 \Big)^{\frac{1}{2}} \le C' |I_{\text{top}(\mathbf{T})}|^{\frac{1}{2}}\, \mathcal{E}(f;\mathbf{T})\,.$$

The first estimate was also shown in (10.5.46); the same argument applies here and the presence of the ε_s's presents no differences. We conclude that

$$\Sigma_{22} \le C\mathcal{M}(\mathbf{T}) |I_{\text{top}(\mathbf{T})}|\, \mathcal{E}(f;\mathbf{T})\,,$$

which is what we needed to prove. The proof of the theorem is now complete. □

Exercises

10.5.1. Show that for every f in the Schwartz class, $x,\xi \in \mathbf{R}$, and $\lambda \in [0,1]$, the function $(y,\eta) \to B^m_{\xi,y,\eta,\lambda}(f)(x)$ is periodic in $\mathbf{R}^2$ with period $(2^{m-\lambda}, 2^{-m+\lambda})$.

10.5.2. Fix a function h in the Schwartz class, $\xi, y, \eta \in \mathbf{R}$, $s \in \mathbf{D}_m$, and $\lambda \in [0,1]$. Suppose that $2^{-\lambda}(\xi+\eta) \in \omega_{s(2)}$.
(a) Assume that $m \le 0$ and that $2^{-m} \ge 40|\xi|$. Show that for some C that does not depend on y, η, and λ we have

$$\begin{aligned} \big|\langle D^{2^\lambda} \tau^y M^\eta (h) \,|\, \varphi_s\rangle\big| &= \big|\langle h \,|\, M^{-\eta}\tau^{-y} D^{2^{-\lambda}}(\varphi_s)\rangle\big| \\ &\le C 2^{\frac{m}{2}} \big\|\widehat{h}\,\big\|_{L^1((-\infty,-\frac{1}{40\cdot 2^m})\cup(\frac{1}{40\cdot 2^m},\infty))}\,. \end{aligned}$$

[*Hint:* Use Plancherel's theorem noting that $\eta \ge 2^\lambda c(\omega_{s(1)}) + \frac{9}{40}2^{-m}$.]
(b) Using the trivial fact that $\big|\langle D^{2^\lambda}\tau^y M^\eta(h) \,|\, \varphi_s\rangle\big| \le C\big\|h\big\|_{L^2}$, conclude that whenever $|m|$ is large with respect to ξ we have

$$\chi_{\omega_{s(2)}}(2^{-\lambda}(\xi+\eta)) |\langle D^{2^\lambda}\tau^y M^\eta(h) \,|\, \varphi_s\rangle| \le C_h \min(1, 2^m)\,,$$

where C_h may depend on h but is independent of y, η, and λ.

10.5.3. (a) Let g be a bounded periodic function on $\mathbf{R}$ with period κ. Show that

$$\lim_{K\to\infty} \frac{1}{2K}\int_{-K}^{K} g(t)\,dt \to \frac{1}{\kappa}\int_0^\kappa g(t)\,dt$$

(b) Let g be a bounded periodic function on $\mathbf{R}^n$ that is periodic with period $(\kappa_1,\dots,\kappa_n)$. Show that

$$\lim_{K_1,\dots,K_n\to\infty} \frac{2^{-n}}{K_1\dots K_n}\int_{-K_1}^{K_1}\cdots\int_{-K_n}^{K_n} g(t)\,dt = \frac{1}{\kappa_1\dots\kappa_n}\int_0^{\kappa_1}\cdots\int_0^{\kappa_n} g(t)\,dt$$

10.5.4. Let (X,μ) be a measure space and let $0 < p, B < \infty$. Let $\{T_\varepsilon\}_{\varepsilon>0}$ be linear operators mapping $L^p(X,d\mu)$ into itself. Define maximal operators $T_*(f) = \sup_{\varepsilon>0}|T_\varepsilon(f)|$ and $T_*^B(f) = \sup_{0<\varepsilon\le B}|T_\varepsilon(f)|$ for f in $L^p(X)$. Also, for a measurable function $N : X \to \mathbf{R}^+$ define a linear operator $\mathfrak{T}_N$ on $L^p(X)$ by setting for all $x \in X$

$$\mathfrak{T}_N(f)(x) = T_{N(x)}(f)(x)\,.$$

Show that

$$\sup_{N\in\mathcal{M}_B} \big\|\mathfrak{T}_N\big\|_{L^p\to L^p} = \big\|T_*^B\big\|_{L^p\to L^p}$$

and that

$$\sup_{N\in\mathcal{M}} \big\|\mathfrak{T}_N\big\|_{L^p\to L^p} = \big\|T_*\big\|_{L^p\to L^p}\,,$$

where $\mathcal{M}$ is the set of all measurable functions $N : X \to \mathbf{R}^+$ and $\mathcal{M}_B$ is the set of all measurable functions $N : X \to (0,B]$.
[*Hint:* The second identity is a consequence of the first. To prove that the left-hand side of the second identity is at least as big as its right hand-side, given $\varepsilon > 0$, for any f in $L^p(X)$ choose N_f in $\mathcal{M}_B$ such that $|T_{N_f(x)}(f)(x)| \ge (1-\varepsilon)T_*^B(f)(x)$. The converse inequality is trivial.]

10.5.5. Use the result in Appendix K.1 to obtain the size estimate

$$\big|\langle \varphi_s \,|\, \varphi_u\rangle\big| \le C_M \frac{\min\left(\dfrac{|I_s|}{|I_u|}, \dfrac{|I_u|}{|I_s|}\right)^{\frac{1}{2}}}{\left(1 + \dfrac{|c(I_s)-c(I_u)|}{\max(|I_s|,|I_u|)}\right)^M}$$

for every $M > 0$. Conclude that if $|I_u| \le |I_s|$, then

$$\big|\langle \varphi_s \,|\, \varphi_u\rangle\big| \le C_M'\left(\frac{|I_s|}{|I_u|}\right)^{\frac{1}{2}} \int_{I_u} \frac{|I_s|^{-1}\,dx}{(1+\frac{|x-c(I_s)|}{|I_s|})^M}\,.$$

[*Hint:* Use that

$$\left|\frac{|x-c(I_s)|}{|I_s|} - \frac{|c(I_u)-c(I_s)|}{|I_s|}\right| \le \frac{1}{2}$$

for all $x \in I_u$.]

10.5.6. Prove that there is a constant $C > 0$ such that for any interval J and any $b > 0$

$$\int_J\int_{J^c} \frac{1}{(1+\frac{|x-y|}{b|J|})^{20}}\,dx\,dy \le Cb^2|J|^2\,.$$

[*Hint:* Translate J to the interval $[-\frac{1}{2}|J|, \frac{1}{2}|J|]$ and change variables. The resulting integral can be computed explicitly.]

10.5.7. Let $\mathbf{T}_2$ be a 2-tree and $f \in L^2(\mathbf{R})$.

(a) Show that there is a constant C such that for all sequences of complex scalars $\{\lambda_s\}_{s\in\mathbf{T}_2}$ we have

$$\Big\|\sum_{s\in\mathbf{T}_2} \lambda_s\, \varphi_s\Big\|_{L^2(\mathbf{R})} \le C\Big(\sum_{s\in\mathbf{T}_2} |\lambda_s|^2\Big)^{\frac{1}{2}}.$$

(b) Use duality to conclude that

$$\sum_{s\in\mathbf{T}_2} \big|\langle f\,|\,\varphi_s\rangle\big|^2 \le C^2\|f\|_{L^2}^2 .$$

[*Hint:* To prove part (a) define $\mathcal{G}_m = \{s \in \mathbf{T}_2 : |I_s| = 2^m\}$. Then for $s \in \mathcal{G}_m$ and $s' \in \mathcal{G}_{m'}$, the functions φ_s and $\varphi_{s'}$ are orthogonal to each other and it suffices to obtain the corresponding estimate when the summation is restricted in a given $\mathcal{G}_m$. But for s in $\mathcal{G}_m$, the intervals I_s are disjoint and we may use the idea of the proof of Lemma 10.5.2. Use that $\sum\limits_{u:\ \omega_u=\omega_s} \big|\langle \varphi_s\,|\,\varphi_u\rangle\big| \le C$ for every fixed s.]

10.5.8. Fix $A \ge 1$. Let $\mathbf{S}$ be a finite collection of tiles such that for all s_1, s_2 in $\mathbf{S}$ we have either $\omega_{s_1} \cap \omega_{s_2} = \emptyset$ or $AI_{s_1} \cap AI_{s_2} = \emptyset$. Let $N_{\mathbf{S}}$ be the *counting function* of $\mathbf{S}$, defined by

$$N_{\mathbf{S}} = \sup_{x\in\mathbf{R}} \#\{I_s : \text{ with } \quad s \in \mathbf{S} \quad \text{and} \quad x \in I_s\}.$$

(a) Show that for any $M > 0$ there exists a $C_M > 0$ such that for all $f \in L^2(\mathbf{R})$ we have

$$\sum_{s\in\mathbf{S}} \Big|\Big\langle f, |I_s|^{-\frac{1}{2}}\Big(1 + \frac{\operatorname{dist}(\cdot, I_s)}{|I_s|}\Big)^{-\frac{M}{2}}\Big\rangle\Big|^2 \le C_M N_{\mathbf{S}}\|f\|_{L^2}^2 .$$

(b) Show that for any $M > 0$ there exists a $C_M > 0$ such that for all finite sequences of scalars $\{a_s\}_{s\in\mathbf{S}}$ we have

$$\Big\|\sum_{s\in\mathbf{S}} a_s\varphi_s\Big\|_{L^2}^2 \le C_M(1 + A^{-M}N_{\mathbf{S}})\sum_{s\in\mathbf{S}} |a_s|^2 .$$

(c) Conclude that for any $M > 0$ there exists a $C_M > 0$ such that for all $f \in L^2(\mathbf{R})$ we have

$$\sum_{s\in\mathbf{S}} \big|\langle f, \varphi_s\rangle\big|^2 \le C_M(1 + A^{-M}N_{\mathbf{S}})\|f\|_{L^2}^2.$$

[*Hint:* Use the idea of Lemma 10.5.2 to prove part (a) when $N_{\mathbf{S}} = 1$. Suppose now that $N_{\mathbf{S}} > 1$. Call an element $s \in \mathbf{S}$ *h*-maximal if the region in $\mathbf{R}^2$ that is directly horizontally above the tile s does not intersect no other tile $s' \in \mathbf{S}$. Let $\mathbf{S}_1$ be the set of all *h*-maximal tiles in $\mathbf{S}$. Then $N_{\mathbf{S}_1} = 1$, otherwise some $x \in \mathbf{R}$ would belong to both I_s and $I_{s'}$ for $s \neq s' \in \mathbf{S}_1$ and thus the horizontal regions directly above s and s' would have to intersect, contradicting the h-maximality of $\mathbf{S}_1$. Now define $\mathbf{S}_2$ to be the set of all h-maximal tiles in $\mathbf{S}\setminus\mathbf{S}_1$. As before we have $N_{\mathbf{S}_2}=1$. Continue this way and write $\mathbf{S}$ as a union of at most $N_{\mathbf{S}}$ families of tiles $\mathbf{S}_j$, each of which has the property $N_{\mathbf{S}_j}=1$. Apply the result to each $\mathbf{S}_j$ and then sum over j. Next, note that part (c) follows from part (b) by duality. To show part (b), observe that whenever $s_1,s_2\in\mathbf{S}$ and $s_1\neq s_2$ we must have either $\langle\varphi_{s_1},\varphi_{s_2}\rangle=0$ or $\operatorname{dist}(I_{s_1},I_{s_2})\geq(A-1)\max(|I_{s_1}|,|I_{s_2}|)$, which implies

$$\Big(1+\frac{\operatorname{dist}(I_{s_1},I_{s_2})}{\max(|I_{s_1}|,|I_{s_2}|)}\Big)^{-M}\leq A^{-\frac{M}{2}}\Big(1+\frac{\operatorname{dist}(I_{s_1},I_{s_2})}{\max(|I_{s_1}|,|I_{s_2}|)}\Big)^{-\frac{M}{2}}.$$

Use this estimate to obtain

$$\Big\|\sum_{s\in\mathbf{S}}a_s\varphi_s\Big\|_{L^2}^2\leq\sum_{s\in\mathbf{S}}|a_s|^2+\frac{C_M}{A^{\frac{M}{2}}}\bigg\|\sum_{s\in\mathbf{S}}\frac{|a_s|}{|I_s|^{\frac12}}\Big(1+\frac{\operatorname{dist}(x,I_s)}{|I_s|}\Big)^{-\frac{M}{2}}\bigg\|_{L^2}^2$$

by expanding the square on the left. The required estimate follows from the dual statement to part (a).]

10.5.9. (*M. Christ*) For $M\in\mathbf{Z}^+$ and for each $1\leq j\leq M$ and $0\leq k\leq 2^j-1$, let s_{jk} be the tile $[2^j,2^{j+1}]\times[k2^{-j},(k+1)2^{-j})$.
(a) Show that for every M there is a sequence $\varepsilon_{jk}\in\{-1,+1\}$, $1\leq j\leq M$, $0\leq k\leq 2^j-1$ such that

$$\Big\|\sum_{j=1}^{M}\sum_{k=1}^{2^j-1}\varepsilon_{jk}2^{-k}\widehat{\varphi_{s_{jk}}}\Big\|_{L^2}^2\geq c\,\Big\|\sum_{k=1}^{2^j-1}\varepsilon_{jk}2^{-k}\widehat{\varphi_{s_{jk}}}\Big\|_{L^2}^2$$

for some $c>0$ independent of M.
(b) Take $\mathbf{S}=\{s_{jk}:\ 1\leq j\leq M \text{ and } 0\leq k\leq 2^j\}$ to show that the expressions

$$\sup_{\{a_s\}}\frac{\Big\|\sum_{s\in\mathbf{S}}a_s\varphi_s\Big\|_{L^2}^2}{\big\|\{a_s\}\big\|_{\ell^2}^2}$$

cannot be bounded from above by a constant independent of $\mathbf{S}$.
[*Hint:* Part (a): Try $\varepsilon_{jk}=r_{2^j+k}(t)$ for some $t\in[0,1)$, where r_j are the Rademacher functions (Appendix C.1). Use the orthogonality of the $\varphi_{s_{jk}}$ for $k\neq k'$ to show that

$$\Big\|\sum_{j=1}^{M}\sum_{k=1}^{2^j-1}\varepsilon_{jk}2^{-k}\widehat{\varphi_{s_{jk}}}\Big\|_{L^2}^2\geq c'M\,.$$

Part (b): Observe that $2^M=N_{\mathbf{S}}$.]

10.5.10. Fix φ a Schwartz function whose Fourier transform is supported in the interval $[-\frac{3}{8}, \frac{3}{8}]$ and that satisfies

$$\sum_{l\in\mathbf{Z}} |\widehat{\varphi}(t+\tfrac{l}{2})|^2 = c_0$$

for all real numbers t. Define functions φ_s as follows. Fix an integer m and set

$$\varphi_s(x) = 2^{-\frac{m}{2}}\varphi(2^{-m}x - k)e^{2\pi i 2^{-m}x\frac{l}{2}},$$

whenever $s = [k2^m, (k+1)2^m) \times [l2^{-m}, (l+1)2^{-m})$ is a tile in $\mathbf{D}$. Prove that for all Schwartz functions f we have

$$\sum_{s\in\mathbf{D}_m} \langle f \,|\, \varphi_s\rangle\varphi_s = c_0 f\,.$$

Observe that m does not appear on the right of this identity.
[*Hint:* First prove that

$$\sum_{s\in\mathbf{D}_m} \varphi_s(x)\overline{\widehat{\varphi_s}(y)} = c_0 e^{2\pi i x y}.$$

using the Poisson summation formula.]

10.5.11. This is a continuous version of the previous Exercise. Fix a Schwartz function φ on $\mathbf{R}^n$ and define a *wavepacket*

$$\varphi_{y,\xi}(x) = \varphi(x-y)e^{2\pi i\xi\cdot x}\,.$$

Prove that for all f Schwartz functions on $\mathbf{R}^n$, the following identity is valid:

$$\|\varphi\|_{L^2}^2 f(x) = \int_{\mathbf{R}^n}\int_{\mathbf{R}^n} \varphi_{y,\xi}(x)\langle f \,|\, \varphi_{y,\xi}\rangle\, dy\, d\xi\,.$$

[*Hint:* Prove first that $\displaystyle\int_{\mathbf{R}^n}\int_{\mathbf{R}^n} \varphi_{y,\xi}(x)\overline{\widehat{\varphi_{y,\xi}}(z)}\, dy d\xi = \|\varphi\|_{L^2}^2 e^{2\pi i x\cdot z}$.]

10.6. L^p Boundedness of the Carleson Operator*

In this section we will prove Hunt's extension of Carleson's theorem to L^p. As in the case of the proof of Theorem 10.5.1, the approach we will follow is not based on the original proof of this theorem but on the time-frequency analysis developed in the previous section. Our goal will be to obtain a restricted weak type (p,p) estimate for the Carleson operator for all $1 < p < \infty$. This reduction will allow us to work with the characteristic function of a measurable set of finite measure and will be crucial in obtaining a sharper energy estimate for this function, which will be the key of the proof of the theorem. This sharper energy estimate will also be a consequence of a further reduction on the tiles that appear in the set $\mathbf{P}$. Later in this section we will obtain weighted estimates for the Carleson operator $\mathcal{C}$. These estimates will be reminiscent of the corresponding estimates for maximal singular integrals we encountered in the previous chapter.

10.6.a. The Main Theorem and Preliminary Reductions

We begin by stating the main theorem of this section, the boundedness of the Carleson operator $\mathcal{C}$ on $L^p(\mathbf{R})$. Recall that this operator is well defined on $\bigcup_{1\le p<\infty} L^p(\mathbf{R})$ since so are $(\widehat{f}\chi_{[a,b]})^\vee$ for all $-\infty \le a < b \le \infty$ in view of the boundedness of the Hilbert transform on these spaces.

Theorem 10.6.1. *For any $1 < p < \infty$ there is a constant $C_p > 0$ such that for all f in $L^p(\mathbf{R})$ we have the estimate*

$$\big\|\mathcal{C}(f)\big\|_{L^p(\mathbf{R})} \le C_p \big\|f\big\|_{L^p(\mathbf{R})}\,. \tag{10.6.1}$$

The proof of this theorem will follow by a modification of the proof of Theorem 10.5.1. We will use in the sequel the notation introduced in Section 10.5. Recall that in (10.5.19) we had identified the one-sided Carleson operator $\mathcal{C}_1(f)$ as

$$\mathcal{C}_1(f)(x) = \sup_N \left| \int_{-\infty}^{N} \widehat{f}(\eta) e^{2\pi i x\cdot\eta}\, d\eta \right| = \frac{1}{c} \sup_\xi |\Pi_\xi(f)|\,, \tag{10.6.2}$$

where $0 < c < \infty$ and Π_ξ was given by

$$\Pi_\xi(f) = \lim_{\substack{K\to\infty\\ L\to\infty}} \frac{1}{4KL} \int_{-L}^{L}\int_{-K}^{K}\int_0^1 G_{\xi,y,\eta,\lambda}(f)\, d\lambda\, dy\, d\eta\,. \tag{10.6.3}$$

Also recall that $G_{\xi,y,\eta,\lambda}(f)$ was

$$G_{\xi,y,\eta,\lambda}(f) = M^{-\eta}\tau^{-y} D^{2^{-\lambda}} A_{\frac{\xi+\eta}{2^\lambda}} D^{2^\lambda} \tau^y M^\eta(f) \tag{10.6.4}$$

where A_ξ was defined in (10.5.5).

Our goal will be to show that for all $1 < p < \infty$ there exists a constant $C_p > 0$ so that for all $f \in L^p(\mathbf{R})$ we have

$$\big\| \sup_{\xi\in\mathbf{R}} |A_\xi(f)|\, \big\|_{L^p(\mathbf{R})} \le C_p \big\|f\big\|_{L^p(\mathbf{R})} \tag{10.6.5}$$

or, equivalently (see Exercise 10.5.4),

$$\sup_{N\colon \mathbf{R}\to\mathbf{R}} \big\|\mathfrak{D}_N(f)\big\|_{L^p(\mathbf{R})} \le C_p \big\|f\big\|_{L^p(\mathbf{R})}\,. \tag{10.6.6}$$

There is a small issue that we need to address here. At this point, the operators A_ξ and $\mathfrak{D}_N$ are well defined on $\mathcal{S}(\mathbf{R})$ and also on $L^2(\mathbf{R})$ but we don't know that they have a well-defined extension on $L^p(\mathbf{R})$. To deal with this small inconvenience we fix a finite subset $\mathbf{P}$ of $\mathbf{D}$ and a real-valued measurable function $x \to N(x)$ on the line and we define a linear operator $\mathfrak{D}_{N,\mathbf{P}}$ by setting

$$\mathfrak{D}_{N,\mathbf{P}}(f)(x) = \sum_{s\in\mathbf{P}} (\chi_{\omega_{s(2)}} \circ N)(x) \left\langle f \,|\, \varphi_s \right\rangle \varphi_s(x) . \tag{10.6.7}$$

Note that the operator $\mathfrak{D}_{N,\mathbf{P}}$ is well defined on $L^p(\mathbf{R})$ for all $1 \le p \le \infty$. To prove (10.6.6) it will suffice to show that for all $1 < p < \infty$ there exists a $C_p > 0$ such that for all $f \in L^p(\mathbf{R})$, all measurable functions $N : \mathbf{R} \to \mathbf{R}$, and all finite subsets $\mathbf{P}$ of $\mathbf{D}$ we have

$$\big\|\mathfrak{D}_{N,\mathbf{P}}(f)\big\|_{L^p} \le C_p \big\|f\big\|_{L^p} . \tag{10.6.8}$$

Once (10.6.8) is established, a density argument gives that the operator $\mathfrak{D}_N$, originally defined on $\mathcal{S}(\mathbf{R})$, admits an extension on $L^p(\mathbf{R})$ that satisfies (10.6.6). This extension on $L^p(\mathbf{R})$ will still denoted by $\mathfrak{D}_N$. We have therefore reduced (10.6.6) [or, equivalently, (10.6.5)] to (10.6.8).

Once (10.6.5) is established, an averaging argument will yield

$$\frac{1}{c}\big\|\mathcal{C}(f)\big\|_{L^p} = \Big\| \sup_{\xi\in\mathbf{R}} |\Pi_\xi(f)| \,\Big\|_{L^p(\mathbf{R})} \le C_p \big\|f\big\|_{L^p(\mathbf{R})} , \tag{10.6.9}$$

thus establishing the required L^p bound for $\mathcal{C}_1$ with the aid of (10.6.2). We quickly explain why (10.6.9) is a consequence of (10.6.5). Using (10.6.3), we obtain

$$\sup_{\xi\in\mathbf{R}} |\Pi_\xi(f)| \le \liminf_{\substack{K\to\infty \\ L\to\infty}} \frac{1}{4KL} \int_{-L}^{L} \int_{-K}^{K} \int_0^1 \sup_{\xi\in\mathbf{R}} |G_{\xi,y,\eta,\lambda}(f)| \, d\lambda \, dy \, d\eta \tag{10.6.10}$$

and by Fatou's lemma we can pass the L^p norm inside the integral and reduce (10.6.9) to (10.6.5). Here we use that translations τ^y and modulations M^η do not alter L^p norms while the dilation D^{2^λ} produces a factor that cancels precisely the factor produced by the dilation $D^{2^{-\lambda}}$ once (10.6.5) is applied.

Using the interpolation Theorem 1.4.19, estimate (10.6.8) will be a consequence of the restricted weak type (p,p) estimate (for some constant C)

$$\big\|\mathfrak{D}_{N,\mathbf{P}}(\chi_F)\big\|_{L^{p,\infty}} \le C \, \frac{p^2}{p-1} |F|^{\frac{1}{p}} \tag{10.6.11}$$

for all sets F of finite measure. In view of Exercise 1.4.14, to prove (10.6.11) it will suffice to prove that for every measurable set E of finite measure, there is a subset E' of E that satisfies $|E'| \ge \frac{1}{2}|E|$ such that

$$\left| \int_{E'} \mathfrak{D}_{N,\mathbf{P}}(\chi_F)(x) \, dx \right| \le C' \, \frac{p^2}{p-1} |E|^{\frac{p-1}{p}} |F|^{\frac{1}{p}} , \tag{10.6.12}$$

where C' is a constant independent of all the parameters involved. We will concentrate therefore on estimate (10.6.12) that implies (10.6.11).

We now introduce a set

$$\Omega_{E,F} = \big\{ M(\chi_F) > 4 \min\big(1, \tfrac{|F|}{|E|}\big) \big\} . \tag{10.6.13}$$

It follows that $|\Omega_{E,F}| \le \frac{1}{2}|E|$ since the Hardy-Littlewood maximal operator is of weak type $(1,1)$ with constant at most 2. (Note that in the case $|F| \ge |E|$ the set $\Omega_{E,F}$ is empty.) We conclude that the set

$$E' = E \setminus \Omega_{E,F}$$

satisfies $|E'| \ge \frac{1}{2}|E|$.

Thus (10.6.11) will be a consequence of the following two estimates:

$$\left| \int_{E'} \sum_{\substack{s\in\mathbf{P} \\ I_s\subseteq\Omega_{E,F}}} \langle \chi_F \,|\, \varphi_s\rangle \chi_{\omega_{s(2)}}(N(x))\varphi_s(x)\,dx \right| \le C|E|^{\frac{p-1}{p}}|F|^{\frac{1}{p}} \tag{10.6.14}$$

and

$$\left| \int_{E'} \sum_{\substack{s\in\mathbf{P} \\ I_s\not\subseteq\Omega_{E,F}}} \langle \chi_F \,|\, \varphi_s\rangle \chi_{\omega_{s(2)}}(N(x))\varphi_s(x)\,dx \right| \le \frac{C\,p^2}{p-1}|E|^{\frac{p-1}{p}}|F|^{\frac{1}{p}}, \tag{10.6.15}$$

where the constant C is independent of the sets E, F, of the measurable function N, of the finite set of tiles $\mathbf{P}$, and of the number $p \in (1,\infty)$.

In proving (10.6.14), we may assume that $|F| \le |E|$, otherwise the set $\Omega_{E,F}$ is empty and there is nothing to prove.

We denote by $\mathcal{I}(\mathbf{P})$ the dyadic grid that consists of all the time projections I_s of tiles s in $\mathbf{P}$. For a fixed interval J in $\mathcal{I}(\mathbf{P})$ we define

$$\mathbf{P}(J) = \{s \in \mathbf{P} : \ I_s = J\}$$

and a function

$$\psi_J(x) = |J|^{-\frac{1}{2}}\left(1 + \frac{|x - c(J)|}{|J|}\right)^{-M},$$

where M is a large integer to be chosen momentarily. We note that for each $s \in \mathbf{P}(J)$ we have $|\varphi_s(x)| \le C_M\,\psi_J(x)$.

For each $k = 0, 1, 2, \dots$ we introduce families

$$\mathcal{F}_k = \big\{J \in \mathcal{I}(\mathbf{P}) : \ 2^k J \subseteq \Omega_{E,F},\ 2^{k+1}J \not\subseteq \Omega_{E,F}\big\}.$$

We begin by controlling the left-hand side of (10.6.14) by

$$\begin{aligned}
&\sum_{\substack{J\in\mathcal{I}(\mathbf{P}) \\ J\subseteq\Omega_{E,F}}} \left| \sum_{s\in\mathbf{P}(J)} \int_{E'} \langle \chi_F \,|\, \varphi_s\rangle\, \chi_{\omega_{s(2)}}(N(x))\varphi_s(x)\,dx \right| \\
&\le \sum_{k=0}^{\infty} \sum_{\substack{J\in\mathcal{I}(\mathbf{P}) \\ J\in\mathcal{F}_k}} \left| \int_{E'} \sum_{s\in\mathbf{P}(J)} \langle \chi_F \,|\, \varphi_s\rangle\, \chi_{\omega_{s(2)}}(N(x))\varphi_s(x)\,dx, \right|.
\end{aligned} \tag{10.6.16}$$

Using Exercise 9.2.8(b) we obtain the existence of a constant $C_0 < \infty$ such that for each $k = 0, 1, \dots$ and $J \in \mathcal{F}_k$ we have

$$\begin{aligned} \langle \chi_F, \psi_J \rangle &\le |J|^{\frac{1}{2}} \inf_J M(\chi_F) \\ &\le |J|^{\frac{1}{2}} C_0^k \inf_{2^{k+1}J} M(\chi_F) \\ &\le 4\, C_0^k |J|^{\frac{1}{2}} \frac{|F|}{|E|} \end{aligned} \tag{10.6.17}$$

since $2^{k+1}J$ meets the complement of $\Omega_{E,F}$.

For $J \in \mathcal{F}_k$ we also have that $E' \cap 2^k J = \emptyset$ and hence

$$\int_{E'} \psi_J(y)\, dy \le \int_{(2^k J)^c} \psi_J(y)\, dy \le |J|^{\frac{1}{2}} C_M 2^{-kM} \,. \tag{10.6.18}$$

Next we note that for each $J \in \mathcal{I}(\mathbf{P})$ and $x \in \mathbf{R}$ there is at most one $s = s_x \in \mathbf{P}(J)$ such that $N(x) \in \omega_{s_x(2)}$. Using this observation along with (10.6.17) and (10.6.18), we can therefore estimate the expression on the right in (10.6.16) as follows:

$$\begin{aligned} &\sum_{k=0}^{\infty} \sum_{\substack{J \in \mathcal{I}(\mathbf{P}) \\ J \in \mathcal{F}_k}} \left| \int_{E'} \langle \chi_F \,|\, \varphi_{s_x} \rangle\, \chi_{\omega_{s_x(2)}}(N(x)) \varphi_{s_x}(x)\, dx \right| \\ &\le C_M^2 \sum_{k=0}^{\infty} \sum_{\substack{J \in \mathcal{I}(\mathbf{P}) \\ J \in \mathcal{F}_k}} \int_{E'} \langle \chi_F, \psi_J \rangle \psi_J(x)\, dx \\ &\le C_M^2\, 4 \frac{|F|}{|E|} \sum_{k=0}^{\infty} C_0^k \sum_{J \in \mathcal{F}_k} |J|^{\frac{1}{2}} \int_{E'} \psi_J(x)\, dx \\ &\le 4\, C_M^3 \frac{|F|}{|E|} \sum_{k=0}^{\infty} (C_0 2^{-M})^k \sum_{J \in \mathcal{F}_k} |J| \end{aligned} \tag{10.6.19}$$

and we pick $M > \log C_0 / \log 2$. It remains to control

$$\sum_{J \in \mathcal{F}_k} |J|$$

for each nonnegative integer k. In doing this we let $\mathcal{F}_k^*$ be all elements of $\mathcal{F}_k$ that are maximal under inclusion. Then we observe that if $J \in \mathcal{F}_k^*$ and $J' \in \mathcal{F}_k$ satisfy $J' \subseteq J$ then $\operatorname{dist}(J', J^c) = 0$ (otherwise $2J'$ would be contained in J and thus $2^{k+1}J' \subseteq 2^k J \subseteq \Omega_{E,F}$). Therefore, for any J in $\mathcal{F}_k^*$ and any scale m there are at most 2 intervals J' from $\mathcal{F}_k$ contained in J with $|J'| = 2^m$. Summing over all possible scales, we obtain a bound of at most 4 times the length of J. We conclude that

$$\sum_{J\in\mathcal{F}_k}|J| = \sum_{J\in\mathcal{F}_k^*}\sum_{\substack{J'\in\mathcal{F}_k\\ J'\subseteq J}}|J'| \le \sum_{J\in\mathcal{F}_k^*}4\,|J| \le 4\,|\Omega_{E,F}|$$

since elements of $\mathcal{F}_k^*$ are disjoint and contained in $\Omega_{E,F}$. Inserting this estimate in (10.6.19), we obtain the required bound

$$C'_M\frac{|F|}{|E|}\,|\Omega_{E,F}| \le C''_M\,|F| \le C''_M\,|E|^{\frac{p-1}{p}}\,|F|^{\frac{1}{p}}$$

for the expression on the right in (10.6.16).

This concludes the proof of (10.6.14). For the purposes of Exercise 10.6.5 note that the expression on the right in (10.6.14) was shown to be bounded by $C\,\min(|F|,|E|)$.

It therefore remains to establish estimate (10.6.15); this requires a considerable amount of work and will be achieved in the next two subsections.

10.6.b. The Proof of Estimate (10.6.15)

To simplify matters a little we will use a scaling argument to reduce the proof of (10.6.15) to the special case in which $\frac{1}{2}\le|E|\le 1$. We briefly discuss this scaling argument.

Let us fix an integer j. We recall the notation introduced in the previous section: For any set A we set $2^j\otimes A=\{2^jy:\ y\in A\}$. Given a measurable function N and a finite set of tiles $\mathbf{P}$, we define

$$N_j(x)=2^{-j}N(2^{-j}x)$$

and

$$\mathbf{P}(j)=\{(2^j\otimes I_s)\times(2^{-j}\otimes\omega_s):\ s\in\mathbf{P}\}\,.$$

Recall the sets

$$\Omega_{E,F}=\{M(\chi_F)>4\left(1,\tfrac{|F|}{|E|}\right)\}\qquad\text{and}\qquad E'=E\setminus\Omega_{E,F}\,.$$

We note that

$$2^j\otimes\Omega_{E,F}=\left\{M(\chi_{2^j\otimes F})>4\,\min\left(1,\frac{|2^j\otimes F|}{|2^j\otimes E|}\right)\right\}=\Omega_{2^j\otimes E,2^j\otimes F}$$

and, defining $(2^j\otimes E)'=(2^j\otimes E)\setminus\Omega_{2^j\otimes E,2^j\otimes F}$, we have that

$$(2^j\otimes E)'=2^j\otimes E'\,.$$

It is a matter of simple algebra to check that for all integers j we have the identity

$$\begin{aligned}&\sum_{\substack{s\in\mathbf{P}\\ I_s\not\subseteq\Omega_{E,F}}}\big|\langle\chi_{E'},(\chi_{\omega_{s(2)}}\circ N)\varphi_s\rangle\langle\chi_F,\varphi_s\rangle\big|\\ &\quad=2^{-\frac{j}{2}}\sum_{\substack{u\in\mathbf{P}(j)\\ I_u\not\subseteq\Omega_{2^j\otimes E,2^j\otimes F}}}\big|\langle\chi_{(2^j\otimes E)'},(\chi_{\omega_{u(2)}}\circ N_j)\varphi_u\rangle\langle 2^{-\frac{j}{2}}\chi_{2^j\otimes F},\varphi_u\rangle\big|\,.\end{aligned}\tag{10.6.20}$$

Next we claim that (10.6.15) will be a consequence of the estimate

$$\sum_{\substack{s\in\mathbf{P}\\ I_s\not\subseteq\Omega_{E,F}}}\big|\langle\chi_{E'},(\chi_{\omega_{s(2)}}\circ N)\varphi_s\rangle\langle\chi_F,\varphi_s\rangle\big|\le\frac{C\,p^2}{p-1}|F|^{\frac{1}{p}}\tag{10.6.21}$$

for all measurable functions N, all measurable sets E with $\frac{1}{2}\le|E|\le 1$, all subsets E' of E with $|E'|\ge\frac{1}{2}|E|$, and all finite subsets $\mathbf{P}$ of $\mathbf{D}$. Indeed, suppose that (10.6.21) holds. Then for general E we pick an integer j such that $\frac{1}{2}\le 2^j|E|\le 1$ and we apply (10.6.21) to the sets $2^j\otimes E$, $2^j\otimes F$, the function N_j, and the sets $\mathbf{P}(j)$ and $\Omega_{2^j\otimes E,2^j\otimes F}$ [note that $|2^j\otimes E|\le 1$ hence (10.6.21) is applicable]. Using (10.6.20), we obtain the estimate

$$\sum_{\substack{s\in\mathbf{P}\\ I_s\not\subseteq\Omega_{E,F}}}\big|\langle\chi_{E'},(\chi_{\omega_{s(2)}}\circ N)\varphi_s\rangle\langle\chi_F,\varphi_s\rangle\big|\le\frac{C\,p^2}{p-1}2^{-j}|2^j\otimes F|^{\frac{1}{p}}\le\frac{C'\,p^2}{p-1}|E|^{1-\frac{1}{p}}|F|^{\frac{1}{p}}$$

for the left hand side of the quantity in (10.6.15), recalling that $|E|\approx 2^{-j}$.

Setting $N^{-1}[\omega_s]=\{x:\ N(x)\in\omega_s\}$, we can also write (10.6.21) as

$$\sum_{\substack{s\in\mathbf{P}\\ I_s\not\subseteq\Omega_{E,F}}}\big|\langle\chi_F,\varphi_s\rangle\langle\chi_{E'\cap N^{-1}[\omega_{s(2)}]},\varphi_s\rangle\big|\le\frac{C\,p^2}{p-1}|F|^{\frac{1}{p}}\,.\tag{10.6.22}$$

Note that when $\frac{1}{2}\le|E|\le 1$ we have

$$I_s\not\subseteq\{M(\chi_F)>4\min\big(1,\tfrac{|F|}{|E|}\big)\}\implies I_s\not\subseteq\{M(\chi_F)>8\min(1,|F|)\}\,,$$

and we may therefore assume that the set $\Omega_{E,F}$ is replaced by the slightly smaller set

$$\Omega_F=\{M(\chi_F)>8\min(1,|F|)\}\,.\tag{10.6.23}$$

To simplify notation, in the sequel we set

$$\mathbf{P}_F=\{s\in\mathbf{P}:\ I_s\not\subseteq\Omega_F\}\,.$$

The following lemma is the main ingredient of the proof and will be proved in the next section.

Lemma 10.6.2. *There is a constant C such that for all measurable sets F of finite measure we have*

$$\mathcal{E}(\chi_F;\mathbf{P}_F) \le C \min\left(|F|^{\frac{1}{2}}, |F|^{-\frac{1}{2}}\right). \tag{10.6.24}$$

Assuming Lemma 10.6.2, we argue as follows to prove (10.6.15). Given the finite set of tiles $\mathbf{P}_F$, we write it as the union

$$\mathbf{P}_F = \bigcup_{j=-\infty}^{n_0} \mathbf{P}_j\,,$$

where the sets $\mathbf{P}_j$ satisfy properties (1)–(5) of page 810.

Given the sequence of sets $\mathbf{P}_j$, we use properties (1), (2), (5) on page 810, the observation that the mass is always bounded by 1, and Lemmata 10.6.2 and 10.5.10 to obtain the following bound for the expression on the left in (10.6.15):

$$\begin{aligned}
&\sum_{s\in\mathbf{P}_F} \left|\langle \chi_F \,|\, \varphi_s\rangle\right| \left|\langle \chi_{E'\cap N^{-1}[\omega_{s(2)}]}, \varphi_s\rangle\right| \\
&= \sum_{j\in\mathbf{Z}} \sum_{s\in\mathbf{P}_j} \left|\langle \chi_F \,|\, \varphi_s\rangle\right| \left|\langle \chi_{E'\cap N^{-1}[\omega_{s(2)}]}, \varphi_s\rangle\right| \\
&\le \sum_{j\in\mathbf{Z}} \sum_{k} \sum_{s\in\mathbf{T}_{jk}} \left|\langle \chi_F \,|\, \varphi_s\rangle\right| \left|\langle \chi_{E'\cap N^{-1}[\omega_{s(2)}]}, \varphi_s\rangle\right| \\
&\le C_3 \sum_{j} \sum_{k} |I_{\mathrm{top}(\mathbf{T}_{jk})}|\, \mathcal{E}(f;\mathbf{T}_{jk})\, \mathcal{M}(\mathbf{T}_{jk}) |F|^{\frac{1}{2}} \\
&\le C_3 |F|^{\frac{1}{2}} \sum_{j\in\mathbf{Z}} \sum_{k} |I_{\mathrm{top}(\mathbf{T}_{jk})}| \min\left(2^{j+1}, C\,|F|^{\frac{1}{2}}, C\,|F|^{-\frac{1}{2}}\right) \min(1, 2^{2j+2}) \\
&\le C_3' |F|^{\frac{1}{2}} \sum_{j\in\mathbf{Z}} 2^{-2j} \min\left(2^{j}, |F|^{\frac{1}{2}}, |F|^{-\frac{1}{2}}\right) \min(1, 2^{2j}) \\
&\le \begin{cases} C_4 |F| \left(1+\log|F|^{-1}\right) & \text{when } |F| \le |E| \le 1, \\ C_5 \left(1+\log|F|\right) & \text{when } |F| > |E| \ge \frac{1}{2}. \end{cases}
\end{aligned}$$

The last estimate follows by a simple calculation. Note that in both cases we obtain the bound $\frac{C\,p^2}{p-1}|F|^{\frac{1}{p}}$ for all $1 < p < \infty$ which establishes (10.6.22) and therefore (10.6.21).

Finally, we note that removing the scaling assumption $\frac{1}{2} \le |E| \le 1$, we have actually obtained the stronger estimate

$$C \min(|E|, |F|) \left(1 + \left|\log \frac{|E|}{|F|}\right|\right)$$

for the expression on the left in (10.6.15). (See Exercise 10.6.5.)

10.6.c. The Proof of Lemma 10.6.2

It remains to prove Lemma 10.6.2. Fix a 2-tree $\mathbf{T}$ contained in $\mathbf{P}_F$ and let $t = \mathrm{top}(\mathbf{T})$ denote its top. We will show that

$$\frac{1}{|I_t|}\sum_{s\in\mathbf{T}} \big|\langle \chi_F \,|\, \varphi_s\rangle\big|^2 \le C\,|F|^{2\gamma} \tag{10.6.25}$$

for some constant C independent of F and $\mathbf{T}$. Then (10.6.24) will follow from (10.6.25) by taking the supremum over all 2-trees $\mathbf{T}$ contained in $\mathbf{P}_F$.

We write the function χ_F as $\chi_{F\cap 3I_t} + \chi_{F\cap(3I_t)^c}$. We begin by observing that for s in $\mathbf{P}_F$ we have

$$\big|\langle \chi_{F\cap(3I_t)^c} \,|\, \varphi_s\rangle\big| \le \frac{C_M|I_s|^{\frac12}\inf\limits_{I_s} M(\chi_F)}{\Big(1+\dfrac{\mathrm{dist}((3I_t)^c, c(I_s))}{|I_s|}\Big)^M} \le C_M|I_s|^{\frac12}\min(1,|F|)\left(\frac{|I_s|}{|I_t|}\right)^M$$

since I_s meets the complement of Ω_F [defined in (10.6.23)] for every $s\in\mathbf{P}_F$. Square this inequality and sum over all s in $\mathbf{T}$ to obtain

$$\sum_{s\in\mathbf{T}} |\langle \chi_{F\cap(3I_t)^c} \,|\, \varphi_s\rangle|^2 \le C\,|I_t|\,\min(1,|F|)^2\,,$$

where the last estimate follows by Exercise 10.6.1.

We now turn to the corresponding estimate for the function $\chi_{F\cap 3I_t}$. At this point it will be convenient to distinguish the simple case $|F|>1$ from the difficult case $|F|\le 1$. In the first case (in which the set Ω_F is empty) Exercise 10.5.7(b) yields

$$\sum_{s\in\mathbf{T}} \big|\langle \chi_{F\cap 3I_t} \,|\, \varphi_s\rangle\big|^2 \le C\,\big\|\chi_{F\cap 3I_t}\big\|_{L^2}^2 \le C\,|I_t| \le C\,|I_t|\,\min(1,|F|)^2\,,$$

since $|F|>1$.

We can therefore concentrate on the case $|F|\le 1$. In proving (10.6.24) we may assume that there exists a point $x_0\in I_t$ such that

$$M(\chi_F)(x_0)\le c\,|F|\,;$$

otherwise there is nothing to prove.

We write the set $\Omega_F = \{M(\chi_F) > 8\,|F|\}$ as a disjoint union of dyadic intervals J'_ℓ such that the dyadic parent $\widetilde{J'_\ell}$ of J'_ℓ is not contained in Ω_F and therefore

$$|F\cap J'_\ell| \le |F\cap \widetilde{J'_\ell}| \le 16\,|F|\,|J'_\ell|\,.$$

Now some of these dyadic intervals may have size larger than or equal to $|I_t|$. Let J'_ℓ be such an interval. Then we split J'_ℓ in $\frac{|J'_\ell|}{|I_t|}$ intervals $J'_{\ell,m}$ each of size exactly

$|I_t|$. Since there is an $x_0 \in I_t$ with $M(\chi_F)(x_0) \le c|F|$, if K is the smallest interval that contains x_0 and $J'_{\ell,m}$, then

$$\frac{1}{|K|}\int_K \chi_F\,dx \le 8\,|F| \implies |F\cap J'_{\ell,m}| \le 8\,|F|\,|I_t|\,\frac{|K|}{|I_t|}\,.$$

We conclude that

$$|F\cap J'_{\ell,m}| \le c\,|F|\,|I_t|\left(1+\frac{\operatorname{dist}(I_t,J'_{\ell,m})}{|I_t|}\right). \tag{10.6.26}$$

We now have a new collection of dyadic intervals $\{J_k\}_k$ contained in Ω_F consisting of all the previous J'_ℓ when $|J'_\ell| < |I_t|$ and the $J'_{\ell,m}$'s when $|J'_{\ell,m}| \ge |I_t|$. In view of the construction we have

$$|F\cap J_k| \le \begin{cases} 2\,c\,|F|\,|J_k| & \text{when } |J_k|<|I_t| \\ 2\,c\,|F|\,|J_k|\left(1+\dfrac{\operatorname{dist}(I_t,J_k)}{|I_t|}\right) & \text{when } |J_k|=|I_t| \end{cases} \tag{10.6.27}$$

for all k. We now define the "bad functions"

$$b_k(x) = \left(e^{-2\pi i c(\omega_t)x}\chi_{F\cap 3I_t}(x) - \frac{1}{|J_k|}\int_{J_k} e^{-2\pi i c(\omega_t)y}\chi_{F\cap 3I_t}(y)\,dy\right)\chi_{J_k}(x)$$

which are supported in J_k, have mean value zero, and satisfy

$$\big\|b_k\big\|_{L^1} \le 2\,c\,|F|\,|J_k|\left(1+\frac{\operatorname{dist}(I_t,J_k)}{|I_t|}\right).$$

We also set

$$g(x) = e^{-2\pi i c(\omega_t)x}\chi_{F\cap 3I_t}(x) - \sum_k b_k(x)$$

the "good function" of the Calderón-Zygmund decomposition. We have therefore decomposed the function $\chi_{F\cap 3I_t}$ as follows:

$$\chi_{F\cap 3I_t}(x) = g(x)e^{2\pi i c(\omega_t)x} + \sum_k b_k(x)e^{2\pi i c(\omega_t)x}\,. \tag{10.6.28}$$

We check that $\big\|g\big\|_{L^\infty} \le C|F|$. Indeed, for x in J_k we have

$$g(x) = \frac{1}{|J_k|}\int_{J_k} e^{-2\pi i c(\omega_t)y}\chi_{F\cap 3I_t}(y)\,dy\,,$$

which implies

$$|g(x)| \le \frac{|F\cap 3I_t\cap J_k|}{|J_k|} \le \begin{cases} \dfrac{|F\cap J_k|}{|J_k|} & \text{when } |J_k|<|I_t| \\[2ex] \dfrac{|F\cap 3I_t|}{|I_t|} & \text{when } |J_k|=|I_t| \end{cases}$$

and both of the preceding are at most a multiple of $|F|$; the latter is because there is an $x_0 \in I_t$ with $M(\chi_F)(x_0) \le 8\,|F|$. Also, for $x \in (\cup_k J_k)^c = (\Omega_F)^c$ we have

$$|g(x)| = \chi_{F\cap 3I_t}(x) \le M(\chi_F)(x) \le c\,|F|\,.$$

We conclude that $\big\|g\big\|_{L^\infty} \le C\,|F|$. Moreover,

$$\big\|g\big\|_{L^1} \le \sum_k \int_{J_k} \frac{|F\cap 3I_t \cap J_k|}{|J_k|}\,dx + \big\|\chi_{F\cap 3I_t}\big\|_{L^1} \le C\,|F\cap 3I_t| \le C\,|F|\,|I_t|$$

since the J_k are disjoint. It follows that

$$\big\|g\big\|_{L^2} \le C\,|F|^{\frac{1}{2}}|F|^{\frac{1}{2}}\,|I_t|^{\frac{1}{2}} = C|F|\,|I_t|^{\frac{1}{2}}\,.$$

Using Exercise 10.5.7, we have

$$\sum_{s\in\mathbf{T}} \big|\big\langle g e^{2\pi i c(\omega_t)(\cdot)}\,\big|\,\varphi_s\big\rangle\big|^2 \le C\,\big\|g\big\|_{L^2}^2\,,$$

from which we obtain the required conclusion for the first function in the decomposition (10.6.28).

Next we turn to the corresponding estimate for the second function

$$\sum_k b_k e^{2\pi i c(\omega_t)(\cdot)}$$

in the decomposition (10.6.28), which requires some further analysis. We have the following two estimates for all s and k:

$$\big|\big\langle b_k e^{2\pi i c(\omega_t)(\cdot)}\,\big|\,\varphi_s\big\rangle\big| \le \frac{C_M\,|F|\,|J_k|^2\,|I_s|^{-\frac{3}{2}}}{(1+\frac{\operatorname{dist}(J_k,I_s)}{|I_s|})^M}\,, \tag{10.6.29}$$

$$\big|\big\langle b_k e^{2\pi i c(\omega_t)(\cdot)}\,\big|\,\varphi_s\big\rangle\big| \le \frac{C_M\,|F|\,|I_s|^{\frac{1}{2}}}{(1+\frac{\operatorname{dist}(J_k,I_s)}{|I_s|})^M} \tag{10.6.30}$$

for all $M>0$, where C_M depends only on M.

To prove (10.6.29) we use the mean value theorem together with the fact that b_k has vanishing integral to write for some ξ_y

$$
\begin{aligned}
&\big|\langle b_k e^{2\pi i c(\omega_t)(\cdot)} \,|\, \varphi_s\rangle\big| \\
&= \Big|\int_{J_k} b_k(y) e^{2\pi i c(\omega_t) y}\overline{\varphi_s(y)}\,dy\Big| \\
&= \Big|\int_{J_k} b_k(y)\big(e^{2\pi i c(\omega_t) y}\overline{\varphi_s(y)} - e^{2\pi i c(\omega_t) c(J_k)}\overline{\varphi_s(c(J_k))}\,dy\Big| \\
&\le |J_k| \int_{J_k} |b_k(y)| \Big[2\pi \tfrac{|c(\omega_s)-c(\omega_t)|}{|I_s|^{\frac12}} \big|\varphi\big(\tfrac{\xi_y - c(I_s)}{|I_s|}\big)\big| + |I_s|^{-\frac32}\big|\varphi'\big(\tfrac{\xi_y - c(I_s)}{|I_s|}\big)\big|\Big]\,dy \\
&\le \big\|b_k\big\|_{L^1}|J_k| \sup_{\xi\in J_k} \frac{C_M |I_s|^{-\frac32}}{(1+\frac{|\xi - c(I_s)|}{|I_s|})^{M+1}} \\
&\le C_M\,|F|\,|J_k|\Big(1+\frac{\operatorname{dist}(J_k, I_t)}{|I_t|}\Big)\frac{|J_k|\,|I_s|^{-\frac32}}{(1+\frac{\operatorname{dist}(J_k,I_s)}{|I_s|})^{M+1}} \\
&\le \frac{C_M\,|F|\,|J_k|^2\,|I_s|^{-\frac32}}{(1+\frac{\operatorname{dist}(J_k,I_s)}{|I_s|})^{M}},
\end{aligned}
$$

where we used the fact that $1+\frac{\operatorname{dist}(J_k,I_t)}{|I_t|} \le 1+\frac{\operatorname{dist}(J_k,I_s)}{|I_s|}$. To prove estimate (10.6.30) we note that

$$
\big|\langle b_k e^{2\pi i c(\omega_t)(\cdot)} \,|\, \varphi_s\rangle\big| \le \frac{C_M\,|I_s|^{\frac12} \inf_{I_s} M(b_k)}{(1+\frac{\operatorname{dist}(J_k,I_s)}{|I_s|})^{M}}
$$

and that

$$
M(b_k) \le M(\chi_F) + \frac{|F\cap 3I_t \cap J_k|}{|J_k|} M(\chi_{J_k}),
$$

and since $I_s \not\subseteq \Omega_F$ we have $\inf_{I_s} M(\chi_F) \le 8\,|F|$ while the second term in the sum was observed earlier to be at most $C\,|F|$.

Finally, we have the estimate

$$
\big|\langle b_k e^{2\pi i c(\omega_t)(\cdot)} \,|\, \varphi_s\rangle\big| \le \frac{C_M\,|F|\,|J_k|\,|I_s|^{-\frac12}}{(1+\frac{\operatorname{dist}(J_k,I_s)}{|I_s|})^{M}}, \tag{10.6.31}
$$

which follows by taking the geometric mean of (10.6.29) and (10.6.30).

Now for a fixed $s\in \mathbf{P}_F$ we may have either $J_k\subseteq I_s$ or $J_k\cap I_s = \emptyset$ (since I_s is not contained in Ω_F). Therefore, for fixed $s\in\mathbf{P}_F$ there are only three possibilities for J_k:

(a) $J_k \subseteq 3I_s$

(b) $J_k \cap 3I_s = \emptyset$

(c) $J_k \cap I_s = \emptyset$, $J_k \cap 3I_s \neq \emptyset$, and $J_k \not\subseteq 3I_s$

Observe that case (c) is equivalent to the following statement:

(c) $J_k \cap I_s = \emptyset$, $\operatorname{dist}(J_k, I_s) = 0$, and $|J_k| \geq 2|I_s|$

Note that in case (c) for each I_s there exists exactly one $J_k = J_{k(s)}$ with the previous properties; but for a given J_k there may be a sequence of I_s's that lie on the left of J_k such that $|J_k| \geq 2|I_s|$ and $\operatorname{dist}(J_k, I_s) = 0$ and another sequence with similar properties on the right of J_k. The I_s's that lie on either side of J_k must be nested and their lengths add up to $|I^L_{s_k}| + |I^R_{s_k}|$ where $I^L_{s_k}$ is the largest one among them on the left of J_k and $I^R_{s_k}$ is the largest one among them on the right of J_k. Using (10.6.30), we obtain

$$\begin{aligned}\sum_{s\in\mathbf{T}}\Big|\sum_{\substack{k:\ J_k\cap I_s=\emptyset\\ \operatorname{dist}(J_k,I_s)=0\\ |J_k|\geq 2|I_s|}}\big\langle b_k e^{2\pi i c(\omega_t)(\cdot)}\,\big|\,\varphi_s\big\rangle\Big|^2 &= \sum_{s\in\mathbf{T}}\Big|\big\langle b_{k(s)} e^{2\pi i c(\omega_t)(\cdot)}\,\big|\,\varphi_s\big\rangle\Big|^2\\ &\leq C\,|F|^2 \sum_{\substack{s\in\mathbf{T}:\ J_k\cap I_s=\emptyset\\ \operatorname{dist}(J_k,I_s)=0\\ |J_k|\geq 2|I_s|}} |I_s|\\ &\leq C\,|F|^2\sum_k |I^L_{s_k}| + |I^R_{s_k}|\,.\end{aligned}$$

But note that $I^L_{s_k} \subseteq 2J_k$ and since $I^L_{s_k} \cap J_k = \emptyset$ we must have $I^L_{s_k} \subseteq 2J_k \setminus J_k$ (and likewise for $I^R_{s_k}$). We define $I^{L+}_{s_k} = I^L_{s_k} + \frac{1}{2}|J_k|$ and $I^{R-}_{s_k} = I^R_{s_k} - \frac{1}{2}|J_k|$. We have $I^{L+}_{s_k} \cup I^{R-}_{s_k} \subseteq J_k$ and hence the sets $I^{L+}_{s_k}$ are pairwise disjoint for different k and the same is true for the $I^{R-}_{s_k}$. Moreover, since $\frac{1}{2}|J_k| \leq \frac{1}{2}|I_t|$ for all k, all the shifted sets $I^{L+}_{s_k}$, $I^{R-}_{s_k}$ are contained in $3I_t$. We conclude that

$$\sum_k |I^L_{s_k}| + \sum_k |I^R_{s_k}| = \sum_k |I^{L+}_{s_k}| + |I^{R-}_{s_k}| \leq \Big|\bigcup_k I^{L+}_{s_k}\Big| + \Big|\bigcup_k I^{R-}_{s_k}\Big| \leq 2\,|3I_t|\,,$$

which combined with the previously obtained estimate yields the required result in case (c).

We now consider case (a). Using (10.6.29), we can write

$$\bigg(\sum_{s\in\mathbf{T}}\Big|\sum_{k:J_k\subseteq 3I_s}\big\langle b_k e^{2\pi i c(\omega_t)(\cdot)}\,\big|\,\varphi_s\big\rangle\Big|^2\bigg)^{\frac{1}{2}} \leq C_M\,|F|^2\bigg(\sum_{s\in T}\Big|\sum_{k:\ J_k\subseteq 3I_s}|J_k|^{\frac{1}{2}}\frac{|J_k|^{\frac{3}{2}}}{|I_s|^{\frac{3}{2}}}\Big|^2\bigg)^{\frac{1}{2}}$$

and we control the second expression by

$$
\begin{aligned}
& C_M\,|F| \bigg\{ \sum_{s\in\mathbf{T}} \Big(\sum_{k:\ J_k\subseteq 3I_s} |J_k| \Big) \Big(\sum_{k:\ J_k\subseteq 3I_s} \frac{|J_k|^3}{|I_s|^3} \Big) \bigg\}^{\frac{1}{2}} \\
\le\ & C_M\,|F| \bigg\{ \sum_{k:\ J_k\subseteq 3I_t} |J_k|^3 \sum_{\substack{s\in\mathbf{T}\\ J_k\subseteq 3I_s}} \frac{1}{|I_s|^2} \bigg\}^{\frac{1}{2}},
\end{aligned}
$$

where we used that the dyadic intervals J_k are disjoint and the Cauchy-Schwarz inequality. We note that the last sum is equal to at most $C|J_k|^{-2}$ since for every dyadic interval J_k there exist at most 3 dyadic intervals of a given length whose triples contain it. The required estimate $C\,|F|\,|I_t|^{\frac{1}{2}}$ now follows in case (a).

Finally, we deal with case (b) which is the most difficult case. We split the set of k into two subsets, those for which $J_k \subseteq 3I_t$ and those for which $J_k \nsubseteq 3I_t$, (recall $|J_k| \le |I_t|$.) Whenever $J_k \nsubseteq 3I_t$, we have $\operatorname{dist}(J_k, I_s) \approx \operatorname{dist}(J_k, I_t)$. In this case we use Minkowski's inequality and estimate (10.6.31) to deduce

$$
\begin{aligned}
& \Big(\sum_{s\in\mathbf{T}} \Big| \sum_{k:\ J_k\nsubseteq 3I_t} \big\langle b_k e^{2\pi i c(\omega_t)(\cdot)} \,|\, \varphi_s \big\rangle \Big|^2 \Big)^{\frac{1}{2}} \\
\le\ & \sum_{k:\ J_k\nsubseteq 3I_t} \Big(\sum_{s\in\mathbf{T}} \big| \big\langle b_k e^{2\pi i c(\omega_t)(\cdot)} \,|\, \varphi_s \big\rangle \big|^2 \Big)^{\frac{1}{2}} \\
\le\ & C_M|F| \sum_{k:\ J_k\nsubseteq 3I_t} |J_k| \Big(\sum_{s\in\mathbf{T}} \frac{|I_s|^{2M-1}}{\operatorname{dist}(J_k, I_s)^{2M}} \Big)^{\frac{1}{2}} \\
\le\ & C_M|F| \sum_{k:\ J_k\nsubseteq 3I_t} \frac{|J_k|}{\operatorname{dist}(J_k, I_t)^M} \Big(\sum_{s\in\mathbf{T}} |I_s|^{2M-1} \Big)^{\frac{1}{2}} \\
\le\ & C_M|F|\,|I_t|^{M-\frac{1}{2}} \sum_{k:\ J_k\nsubseteq 3I_t} \frac{|J_k|}{\operatorname{dist}(J_k, I_t)^M} \\
\le\ & C_M|F|\,|I_t|^{M-\frac{1}{2}} \sum_{l=1}^{\infty} \sum_{\substack{k:\\ \operatorname{dist}(J_k,I_t)\approx 2^l|I_t|}} \frac{|J_k|}{(2^l|I_t|)^M}\,.
\end{aligned}
$$

But note that all the J_k with $\operatorname{dist}(J_k, I_t) \approx 2^l|I_t|$ are contained in $2^{l+2}I_t$ and since they are disjoint we estimate the last sum by $C2^l|I_t|(2^l|I_t|)^{-M}$. The required estimate $C_M|F|\,|I_t|^{\frac{1}{2}}$ follows.

Next we consider the case $J_k \subseteq 3I_t$, $J_k \cap 3I_s = \emptyset$, and $|J_k| \le |I_s|$ in which we use estimate (10.6.29). We have

$$\Big(\sum_{s\in\mathbf{T}}\Big|\sum_{\substack{k:\ J_k\subseteq 3I_t\\ J_k\cap 3I_s=\emptyset\\ |J_k|\le|I_s|}}\langle b_k e^{2\pi i c(\omega_t)(\cdot)}\,|\,\varphi_s\rangle\Big|^2\Big)^{\frac12}$$

$$\le C_M\,|F|\Big(\sum_{s\in\mathbf{T}}\Big|\sum_{\substack{k:\ J_k\subseteq 3I_t\\ J_k\cap 3I_s=\emptyset\\ |J_k|\le|I_s|}}|J_k|^2|I_s|^{-\frac32}\frac{|I_s|^M}{\operatorname{dist}(J_k,I_s)^M}\Big|^2\Big)^{\frac12}$$

$$\le C_M\,|F|\Big\{\sum_{s\in\mathbf{T}}\Big[\sum_{\substack{k:\ J_k\subseteq 3I_t\\ J_k\cap 3I_s=\emptyset\\ |J_k|\le|I_s|}}\frac{|J_k|^3}{|I_s|^2}\Big(\frac{|I_s|}{\operatorname{dist}(J_k,I_s)}\Big)^M\Big]$$

$$\Big[\sum_{\substack{k:\ J_k\subseteq 3I_t\\ J_k\cap 3I_s=\emptyset\\ |J_k|\le|I_s|}}\frac{|J_k|}{|I_s|}\Big(\frac{\operatorname{dist}(J_k,I_s)}{|I_s|}\Big)^{-M}\Big]\Big\}^{\frac12}$$

$$\le C_M\,|F|\Big\{\sum_{s\in\mathbf{T}}\Big[\sum_{\substack{k:\ J_k\subseteq 3I_t\\ J_k\cap 3I_s=\emptyset\\ |J_k|\le|I_s|}}\frac{|J_k|^3}{|I_s|^2}\Big(\frac{|I_s|}{\operatorname{dist}(J_k,I_s)}\Big)^M\Big]$$

$$\Big[\sum_{\substack{k:\ J_k\subseteq 3I_t\\ J_k\cap 3I_s=\emptyset\\ |J_k|\le|I_s|}}\int_{J_k}\Big(\frac{|x-c(I_s)|}{|I_s|}\Big)^{-M}\frac{dx}{|I_s|}\Big]\Big\}^{\frac12}$$

$$\le C_M\,|F|\Big\{\sum_{s\in\mathbf{T}}\Big[\sum_{\substack{k:\ J_k\subseteq 3I_t\\ J_k\cap 3I_s=\emptyset\\ |J_k|\le|I_s|}}\frac{|J_k|^3}{|I_s|^2}\Big(\frac{|I_s|}{\operatorname{dist}(J_k,I_s)}\Big)^M\Big]$$

$$\Big[\int_{(3I_s)^c}\Big(\frac{|x-c(I_s)|}{|I_s|}\Big)^{-M}\frac{dx}{|I_s|}\Big]\Big\}^{\frac12}$$

$$\le C_M\,|F|\Big\{\sum_{s\in\mathbf{T}}\sum_{\substack{k:\ J_k\subseteq 3I_t\\ J_k\cap 3I_s=\emptyset\\ |J_k|\le|I_s|}}|J_k|^3|I_s|^{-2}\Big(\frac{|I_s|}{\operatorname{dist}(J_k,I_s)}\Big)^M\Big\}^{\frac12}.$$

But since the last integral contributes at most a constant factor, we can estimate the last displayed expression by

$$
\begin{aligned}
&\leq C_M\,|F| \Bigg\{ \sum_{\substack{k:\ J_k \subseteq 3I_t \\ J_k \cap 3I_s = \emptyset \\ |J_k| \leq |I_s|}} |J_k|^3 \sum_{m \geq \log |J_k|} 2^{-2m} \sum_{\substack{s \in \mathbf{T} \\ |I_s| = 2^m}} \Big(\frac{\operatorname{dist}(J_k, I_s)}{2^m} \Big)^{-M} \Bigg\}^{\frac{1}{2}} \\
&\leq C_M\,|F| \Bigg\{ \sum_{\substack{k:\ J_k \subseteq 3I_t \\ J_k \cap 3I_s = \emptyset \\ |J_k| \leq |I_s|}} |J_k|^3 \sum_{m \geq \log |J_k|} 2^{-2m} \Bigg\}^{\frac{1}{2}} \\
&\leq C_M\,|F| \Bigg\{ \sum_{\substack{k:\ J_k \subseteq 3I_t \\ J_k \cap 3I_s = \emptyset \\ |J_k| \leq |I_s|}} |J_k|^3 |J_k|^{-2} \Bigg\}^{\frac{1}{2}} \\
&\leq C_M\,|F|\,|I_t|^{\frac{1}{2}}\,.
\end{aligned}
$$

There is also the subcase of case (b) in which $|J_k| > |I_s|$. Here we have the two special subcases: $I_s \cap 3J_k = \emptyset$ and $I_s \subseteq 3J_k = \emptyset$. We begin with the first of these special subcases in which we use estimate (10.6.30). We have

$$
\begin{aligned}
&\Bigg(\sum_{s \in \mathbf{T}} \Bigg| \sum_{\substack{k:\ J_k \subseteq 3I_t \\ J_k \cap 3I_s = \emptyset \\ |J_k| > |I_s| \\ I_s \cap 3J_k = \emptyset}} \langle b_k e^{2\pi i c(\omega_t)(\cdot)} \mid \varphi_s \rangle \Bigg|^2 \Bigg)^{\frac{1}{2}} \\
&\leq C_M\,|F| \Bigg(\sum_{s \in T} \Bigg| \sum_{\substack{k:\ J_k \subseteq 3I_t \\ J_k \cap 3I_s = \emptyset \\ |J_k| > |I_s| \\ I_s \cap 3J_k = \emptyset}} |I_s|^{\frac{1}{2}} \frac{|I_s|^M}{\operatorname{dist}(J_k, I_s)^M} \Bigg|^2 \Bigg)^{\frac{1}{2}} \\
&\leq C_M\,|F| \Bigg\{ \sum_{s \in \mathbf{T}} \Bigg[\sum_{\substack{k:\ J_k \subseteq 3I_t \\ J_k \cap 3I_s = \emptyset \\ |J_k| > |I_s| \\ I_s \cap 3J_k = \emptyset}} \frac{|I_s|^2}{|J_k|} \frac{|I_s|^M}{\operatorname{dist}(J_k, I_s)^M} \Bigg] \Bigg[\sum_{\substack{k:\ J_k \subseteq 3I_t \\ J_k \cap 3I_s = \emptyset \\ |J_k| > |I_s| \\ I_s \cap 3J_k = \emptyset}} \frac{|J_k|}{|I_s|} \frac{|I_s|^M}{\operatorname{dist}(J_k, I_s)^M} \Bigg] \Bigg\}^{\frac{1}{2}}.
\end{aligned}
$$

Since $I_s \cap 3J_k = \emptyset$ we have that $\operatorname{dist}(J_k, I_s) \approx |x - c(I_s)|$ for every $x \in J_k$ and therefore the second term inside square brackets satisfies

$$\sum_{\substack{k:\ J_k \subseteq 3I_t \\ J_k \cap 3I_s = \emptyset \\ |J_k| > |I_s| \\ I_s \cap 3J_k = \emptyset}} \frac{|J_k|}{|I_s|} \frac{|I_s|^M}{\operatorname{dist}(J_k, I_s)^M} \le \sum_k \int_{J_k} \Big(\frac{|x - c(I_s)|}{|I_s|}\Big)^{-M} \frac{dx}{|I_s|} \le C_M .$$

Using this estimate, we obtain

$$\begin{aligned}
& C_M\,|F| \Bigg\{ \sum_{s \in \mathbf{T}} \Bigg[\sum_{\substack{k:\ J_k \subseteq 3I_t \\ J_k \cap 3I_s = \emptyset \\ |J_k| > |I_s| \\ I_s \cap 3J_k = \emptyset}} \frac{|I_s|^2}{|J_k|} \frac{|I_s|^M}{\operatorname{dist}(J_k, I_s)^M} \Bigg] \Bigg[\sum_{\substack{k:\ J_k \subseteq 3I_t \\ J_k \cap 3I_s = \emptyset \\ |J_k| > |I_s| \\ I_s \cap 3J_k = \emptyset}} \frac{|J_k|}{|I_s|} \frac{|I_s|^M}{\operatorname{dist}(J_k, I_s)^M} \Bigg] \Bigg\}^{\frac{1}{2}} \\
& \le C_M\,|F| \Bigg\{ \sum_{s \in \mathbf{T}} \Bigg[\sum_{\substack{k:\ J_k \subseteq 3I_t \\ J_k \cap 3I_s = \emptyset \\ |J_k| > |I_s| \\ I_s \cap 3J_k = \emptyset}} \frac{|I_s|^2}{|J_k|} \frac{|I_s|^M}{\operatorname{dist}(J_k, I_s)^M} \Bigg] \Bigg\}^{\frac{1}{2}} \\
& = C_M\,|F| \Bigg\{ \sum_{k:\ J_k \subseteq 3I_t} \frac{1}{|J_k|} \sum_{\substack{s \in \mathbf{T} \\ J_k \cap 3I_s = \emptyset \\ |J_k| > |I_s| \\ I_s \cap 3J_k = \emptyset}} |I_s|^2 \frac{|I_s|^M}{\operatorname{dist}(J_k, I_s)^M} \Bigg\}^{\frac{1}{2}} \\
& \le C_M\,|F| \Bigg\{ \sum_{k:\ J_k \subseteq 3I_t} \frac{1}{|J_k|} \sum_{m=-\infty}^{\log_2 |J_k|} 2^{2m} \sum_{\substack{s \in \mathbf{T}:\ |I_s| = 2^m \\ J_k \cap 3I_s = \emptyset \\ |J_k| > |I_s| \\ I_s \cap 3J_k = \emptyset}} \frac{|I_s|^M}{\operatorname{dist}(J_k, I_s)^M} \Bigg\}^{\frac{1}{2}} \\
& \le C_M\,|F| \Bigg\{ \sum_{k:\ J_k \subseteq 3I_t} \frac{1}{|J_k|} \sum_{m=-\infty}^{\log_2 |J_k|} 2^{2m} \Bigg\}^{\frac{1}{2}} \\
& \le C_M \sum_{k:\ J_k \subseteq 3I_t} \frac{1}{|J_k|} |J_k|^2 \\
& \le C_M\,|F|\,|I_t|^{\frac{1}{2}} .
\end{aligned}$$

Finally there is the subcase of case (b) in which $|J_k| \ge |I_s|$ and $I_s \subseteq 3J_k = \emptyset$. Here again we use estimate (10.6.30). We have

$$\Big\{\sum_{s\in\mathbf{T}}\Big|\sum_{\substack{k:\ J_k\subseteq 3I_t\\ J_k\cap 3I_s=\emptyset\\ |J_k|>|I_s|\\ I_s\subseteq 3J_k}}\langle b_k e^{2\pi i c(\omega_t)(\cdot)}\,|\,\varphi_s\rangle\Big|^2\Big\}^{\frac12}$$

(10.6.32)

$$\le C_M\,|F|\Big\{\sum_{s\in\mathbf{T}}|I_s|\Big|\sum_{\substack{k:\ J_k\subseteq 3I_t\\ J_k\cap 3I_s=\emptyset\\ |J_k|>|I_s|\\ I_s\subseteq 3J_k}}\frac{|I_s|^M}{\operatorname{dist}(J_k,I_s)^M}\Big|^2\Big\}^{\frac12}.$$

Let us make some observations. For a fixed s there exist at most finitely many J_k's contained in $3I_t$ with size at least $|I_s|$. Let $J_L^1(s)$ be the interval that lies to the left of I_s and is closest to I_s among all J_k that satisfy the conditions in the preceding sum. Then $|J_L^1(s)|>|I_s|$ and

$$\operatorname{dist}(J_L^1(s),I_s)\ge|I_s|\,.$$

Let $J_L^2(s)$ be the interval to the left of $J_L^1(s)$ that is closest to $J_L^1(s)$ and that satisfies the conditions of the sum. Since $3J_L^2(s)$ contains I_s, it follows that $|J_L^2(s)|>2|I_s|$ and

$$\operatorname{dist}(J_L^2(s),I_s)\ge 2|I_s|\,.$$

Continuing in this way, we can find a finite number of intervals $J_L^r(s)$ that lie to the left of I_s and inside $3I_t$, satisfy $|J_L^r(s)|>2^r|I_s|$ and $\operatorname{dist}(J_L^r(s),I_s)\ge 2^r|I_s|$, and whose triples contain I_s. Likewise we find a finite collection of intervals $J_R^1(s), J_R^2(s)\dots$ that lie to the right of I_s and satisfy similar conditions. Then, using the Cauchy-Schwarz inequality, we obtain

$$\Big|\sum_{\substack{k:\ J_k\subseteq 3I_t\\ J_k\cap 3I_s=\emptyset\\ |J_k|>|I_s|\\ I_s\subseteq 3J_k}}\frac{|I_s|^M}{\operatorname{dist}(J_k,I_s)^M}\Big|^2$$

$$\le 2\Big|\sum_{r=1}^{\infty}\frac{|I_s|^{\frac M2}}{\operatorname{dist}(J_L^r(s),I_s)^{\frac M2}}\frac{1}{2^{\frac{rM}{2}}}\Big|^2+2\Big|\sum_{r=1}^{\infty}\frac{|I_s|^{\frac M2}}{\operatorname{dist}(J_R^r(s),I_s)^{\frac M2}}\frac{1}{2^{\frac{rM}{2}}}\Big|^2$$

$$\le C_M\sum_{r=1}^{\infty}\frac{|I_s|^M}{\operatorname{dist}(J_L^r(s),I_s)^M}+C_M\sum_{r=1}^{\infty}\frac{|I_s|^M}{\operatorname{dist}(J_R^r(s),I_s)^M}$$

$$\le C_M\sum_{\substack{k:\ J_k\subseteq 3I_t\\ J_k\cap 3I_s=\emptyset\\ |J_k|>|I_s|\\ I_s\subseteq 3J_k}}\frac{|I_s|^M}{\operatorname{dist}(J_k,I_s)^M}\,.$$

We use this estimate to control the expression on the left in (10.6.32) by

$$C_M|F|\Big\{\sum_{s\in\mathbf{T}}|I_s|\sum_{\substack{k:\ J_k\subseteq 3I_t\\ J_k\cap 3I_s=\emptyset\\ |J_k|>|I_s|\\ I_s\subseteq 3J_k}}\frac{|I_s|^M}{\operatorname{dist}(J_k,I_s)^M}\Big\}^{\frac12}$$

$$\le C_M|F|\Big\{\sum_{k:J_k\subseteq 3I_t}|J_k|\sum_{m=0}^{\infty}2^{-m}\sum_{\substack{s:\ I_s\subseteq 3J_k\\ J_k\cap 3I_s=\emptyset\\ |I_s|=2^{-m}|J_k|}}\frac{|I_s|^M}{\operatorname{dist}(J_k,I_s)^M}\Big\}^{\frac12}.$$

Since the last sum is at most a constant, it follows that the term on the left in (10.6.32) also satisfies the estimate $C_M\,|F|\,|I_t|^{\frac12}$. This concludes the proof of Lemma 10.6.2.

10.6.d. The Maximal Carleson Operator and Weighted Estimates

Recall the one-sided Carleson operator $\mathcal{C}_1$ defined in the previous section;

$$\mathcal{C}_1(f)(x)=\sup_{N>0}\left|\int_{-\infty}^{N}\widehat{f}(\xi)e^{2\pi i x\xi}\,d\xi\right|.$$

Recall also the modulation operator $M^a(g)(x)=g(x)e^{2\pi i a x}$. We begin by observing that the following identity is valid:

$$(\widehat{f}\chi_{(-\infty,b]})^{\vee}=M^b\frac{I-iH}{2}M^{-b}(f)=\frac12 f-\frac{i}{2}M^bHM^{-b}(f), \tag{10.6.33}$$

where H is the Hilbert transform. It follows from (10.6.33) that

$$\mathcal{C}_1(f)\le\frac12|f|+\frac12\sup_{\xi\in\mathbf{R}}|H(M^{\xi}(f))|$$

and that

$$\sup_{\xi\in\mathbf{R}}|H(M^{\xi}(f))|\le|f|+2\,\mathcal{C}_1(f).$$

We conclude that the L^p boundedness of the sublinear operator $f\to\mathcal{C}_1(f)$ is equivalent to that of the sublinear operator

$$f\to\sup_{\xi\in\mathbf{R}}|H(M^{\xi}(f))|.$$

Definition 10.6.3. The *maximal Carleson operator* is defined by

$$\begin{aligned}\mathcal{C}_*(f)(x)&=\sup_{\varepsilon>0}\sup_{\xi\in\mathbf{R}}\left|\int_{|x-y|>\varepsilon}f(y)e^{2\pi i\xi y}\frac{dy}{x-y}\right|\\&=\sup_{\xi\in\mathbf{R}}\left|H^{(*)}(M^{\xi}(f))(x)\right|,\end{aligned}\tag{10.6.34}$$

where $H^{(*)}$ is the maximal Hilbert transform. Observe that $\mathcal{C}_*(f)$ is well defined for all f in $\bigcup_{1\le p<\infty} L^p(\mathbf{R})$ and that $\mathcal{C}_*(f)$ controls the Carleson operator $\mathcal{C}(f)$ pointwise.

We begin with the following pointwise estimate, which reduces the boundedness of $\mathcal{C}_*$ to that of $\mathcal{C}$:

Lemma 10.6.4. *There is a positive constant $c > 0$ such that for all functions f in $\bigcup_{1\le p<\infty} L^p(\mathbf{R})$ we have*

$$\mathcal{C}_*(f) \le c\,M(f) + M(\mathcal{C}(f))\,, \tag{10.6.35}$$

where M is the Hardy-Littlewood maximal function.

Proof. The proof of (10.6.35) is based on the classical inequality

$$H^{(*)}(g) \le c\,M(g) + M(H(g)).$$

obtained in (4.1.32). Applying this to the functions $M^\xi(f)$ and taking supremum over $\xi \in \mathbf{R}$, we obtain

$$\mathcal{C}_*(f) \le c\,M(f) + \sup_{\xi\in\mathbf{R}} M\big(H(M^\xi(f))\big)$$

from which (10.6.35) easily follows by passing the supremum inside the maximal function.

□

It will be convenient to work with a power of the Hardy-Littlewood maximal operator M. For $0 < r < \infty$ we define

$$M_r(f) = M(|f|^r)^{\frac{1}{r}}$$

for f such that $|f|^r$ is locally integrable on $\mathbf{R}$. Note that $M(f) \le M_r(f)$ for any $r \in (1,\infty)$. Our next goal will be to obtain the boundedness of the Carleson operator on weighted L^p spaces.

Theorem 10.6.5. *For every $p \in (1,\infty)$ and every $w \in A_p$ there is a constant $C(p,[w]_{A_p})$ such that for all $f \in L^p(\mathbf{R})$ we have*

$$\big\|\mathcal{C}(f)\big\|_{L^p(w)} \le C(p,[w]_{A_p})\big\|f\big\|_{L^p(w)}\,, \tag{10.6.36}$$

$$\big\|\mathcal{C}_*(f)\big\|_{L^p(w)} \le C(p,[w]_{A_p})\big\|f\big\|_{L^p(w)}\,. \tag{10.6.37}$$

Proof. Fix a $1 < p < \infty$ and pick an $r \in (1,p)$ so that $w \in A_r$. We will show that for all $f \in L^p(w)$ we have the estimate

$$\int_{\mathbf{R}} \mathcal{C}(f)(x)^p\, w(x)\,dx \le C_p([w]_{A_p}) \int_{\mathbf{R}} M_r(f)(x)^p w(x)\,dx\,. \tag{10.6.38}$$

Then the boundedness of $\mathcal{C}$ on $L^p(w)$ will be a consequence of the boundedness of the Hardy-Littlewood maximal operator on $L^{\frac{p}{r}}(w)$.

If we can show that for any $w \in A_p$ there is a constant $C_p([w]_{A_p})$ such that

$$\int_{\mathbf{R}} M(\mathcal{C}(f))^p \, w \, dx \le C_p([w]_{A_p}) \int_{\mathbf{R}} M_r(f)^p \, w \, dx \,, \tag{10.6.39}$$

then the trivial fact $\mathcal{C}(f) \le M(\mathcal{C}(f))$, inserted in (10.6.39), yields (10.6.38).

Estimate (10.6.39) will be a consequence of the following two important observations:

$$M^{\#}(\mathcal{C}(f)) \le C_r \, M_r(f) \qquad \text{a.e.} \tag{10.6.40}$$

and

$$\big\|M(\mathcal{C}(f))\big\|_{L^p(w)} \le c_p([w]_{A_p}) \big\|M^{\#}(\mathcal{C}(f))\big\|_{L^p(w)} \,, \tag{10.6.41}$$

where $c_p([w]_{A_p})$ depends on $[w]_{A_p}$ and C_r depends only on r.

We begin with estimate (10.6.40), which was obtained in Theorem 7.4.9 for singular integral operators. Here this estimate is extended to maximally modulated singular integrals. To prove (10.6.40) we will use the result in Proposition 7.4.2 (2). We fix $x \in \mathbf{R}$ and we pick an interval I that contains x. We write $f = f_0 + f_\infty$, where $f_0 = \chi_{3I}$ and $f_\infty = f\chi_{(3I)^c}$. We set $a_I = \mathcal{C}(f_\infty)(c_I)$, where c_I is the center of I. Then we have

$$\begin{aligned}\frac{1}{|I|}\int_I |\mathcal{C}(f)(y) - a_I| \, dx \; &\le \frac{1}{|I|}\int_I \sup_{\xi\in\mathbf{R}} \big|H(M^\xi(f))(y) - H(M^\xi(f_\infty))(c_I)\big| \, dy \\ &\le B_1 + B_2\end{aligned}$$

where

$$\begin{aligned}B_1 &= \frac{1}{|I|}\int_I \sup_{\xi\in\mathbf{R}} \big|H(M^\xi(f_0))(y)\big| \, dy \\ B_2 &= \frac{1}{|I|}\int_I \sup_{\xi\in\mathbf{R}} \big|H(M^\xi(f_\infty))(y) - H(M^\xi(f_\infty))(c_I)\big| \, dy \,.\end{aligned}$$

But

$$\begin{aligned}B_1 \; &\le \frac{1}{|I|}\int_I \mathcal{C}(f_0)(y) \, dy \\ &\le \frac{1}{|I|} \big\|\mathcal{C}(f_0)\big\|_{L^r} \big\|\chi_I\big\|_{L^{r'}} \\ &\le \frac{\big\|\mathcal{C}\big\|_{L^r \to L^r}}{|I|} \big\|f_0\big\|_{L^r} |I|^{\frac{1}{r'}} \\ &\le C_r M_r(f)(x) \,,\end{aligned}$$

where we used the boundedness of the Carleson operator $\mathcal{C}$ from L^r into L^r and Theorem 1.4.17 (v).

We turn to the corresponding estimate for B_2. We have

$$\begin{aligned}
B_2 &\le \frac{1}{|I|}\int_I\int_{\mathbf{R}^n} |f_\infty(z)| \left|\frac{1}{y-z} - \frac{1}{c_I - z}\right| dz\, dy \\
&= \frac{1}{|I|}\int_I\int_{(3I)^c} |f(z)| \left|\frac{y-c_I}{(y-z)(c_I-z)}\right| dz\, dy \\
&\le \int_I \left(\int_{(3I)^c} |f(z)| \frac{C\,|I|}{(|c_I - z| + |I|)^2}\, dz\right) dy \\
&\le \int_I \frac{C}{|I|} M(f)(x)\, dy \\
&\le C M(f)(x) \\
&\le C M_r(f)(x)\,.
\end{aligned}$$

This completes the proof of estimate (10.6.40) and we now turn to the proof of estimate (10.6.41). We can derive (10.6.41) as a consequence of Exercise 9.4.9 provided we have that

$$\big\| M(\mathcal{C}(f))\big\|_{L^r(w)} < \infty\,. \tag{10.6.42}$$

Unfortunately, the finiteness estimate (10.6.42) for general functions f in $L^p(w)$ cannot be deduced easily without knowledge of the sought estimate (10.6.36) for $p = r$. However, we can show the validity of (10.6.42) for functions f with compact support and weights $w \in A_p$ that are bounded. This argument requires a few technicalities, which we now present. For a fixed constant B we introduce a truncated Carleson operator

$$\mathcal{C}^B(f) = \sup_{|\xi|\le B} |H(M^\xi(f))|\,.$$

Next we work with a weight w in A_p that is bounded. In fact, we will work with $w_k = \min(w, k)$, which satisfies $[w_k]_{A_p} \le (1 + 2^{p-2})\big(1 + [w]_{A_p}\big)$ for all $k \ge 1$ (see Exercise 9.1.9). Finally, we take $f = h$ to be a smooth function with support contained in an interval $[-R, R]$. Then for $|\xi| \le B$ we have

$$|H(M^\xi(h))(x)| \le 2R\big\|(M^\xi(h))'\big\|_{L^\infty}\chi_{|x|\le 2R} + \frac{\big\|h\big\|_{L^1}}{|x| + R}\chi_{|x|>2R} \le \frac{B\, C_h\, R}{|x| + R}\,,$$

where C_h is a constant that depends on h. This implies that the last estimate also holds for $\mathcal{C}^B(h)$. Using Example 2.1.8, we now obtain

$$M(\mathcal{C}^B(h))(x) \le B\, C_h\, \frac{\log\big(1 + \frac{|x|}{R}\big)}{1 + \frac{|x|}{R}}\,.$$

It follows that $M(\mathcal{C}^B(h))$ lies in $L^r(w_k)$ since $r > 1$ and $w_k \le k$. Therefore, $\big\|M(\mathcal{C}^B(f))\big\|_{L^r(w_k)} < \infty$ and thus (10.6.42) holds in this setting. Applying the previous argument to $\mathcal{C}^B(h)$ and the weight w_k [in lieu of $\mathcal{C}(f)$ and w], we obtain

(10.6.39) and thus (10.6.36) for $M(\mathcal{C}^B(h))$ and the weight w_k. This establishes the estimate

$$\big\|\mathcal{C}^B(h)\big\|_{L^p(w_k)} \le C(p,[w]_{A_p})\big\|h\big\|_{L^p(w_k)} \tag{10.6.43}$$

for some constant $C(p,[w]_{A_p})$ that is independent of B and k, for functions h that are smooth and compactly supported. Letting $k\to\infty$ in (10.6.43) and applying Fatou's lemma, we obtain (10.6.36) for smooth functions h with compact support. To deduce the validity of (10.6.36) for general functions f in $L^p(w)$, we use Exercise 10.5.4 to write this estimate as an equivalent norm inequality for a linearized operator. Then by density we obtain that (10.6.36) holds for all functions in $L^p(w)$ since smooth functions with compact support are dense in $L^p(w)$.

Finally, to obtain (10.6.37) for general $f\in L^p(w)$, we raise (10.6.35) to the power p, use the inequality $(a+b)^p\le 2^p(a^p+b^p)$, and integrate over $\mathbf{R}$ with the respect to the measure $w\,dx$ to obtain

$$\int_{\mathbf{R}} \mathcal{C}_*(f)^p w\,dx \le 2^p c\int_{\mathbf{R}} M(f)^p w\,dx + 2^p\int_{\mathbf{R}} M(\mathcal{C}(f))^p w\,dx\,. \tag{10.6.44}$$

Then we use estimate (10.6.36) and the boundedness of the Hardy-Littlewood maximal operator on $L^p(w)$ to obtain the required conclusion.

□

Exercises

10.6.1. Let $\mathbf{T}$ be a 2-tree with top I_t and let $M>1$ and L such that $2^L<|I_t|$. Show that there exists a constant $C_M>0$ such that

$$\sum_{s\in\mathbf{T}}|I_s|^M\le C_M|I_t|^M$$

$$\sum_{\substack{s\in\mathbf{T}\\ |I_s|\ge 2^L}}|I_s|^{-M}\le C_M\frac{|I_t|}{(2^L)^{M+1}}$$

$$\sum_{\substack{s\in\mathbf{T}\\ |I_s|\le 2^L}}|I_s|^{M}\le C_M|I_t|(2^L)^{M-1}\,.$$

[*Hint:* Group the s that appear in each sum in families $\mathcal{G}_m$ such that $|I_s|=2^{-m}|I_t|$ for each $s\in\mathcal{G}_m$.]

10.6.3. Show that the operator

$$g\to\sup_{-\infty<a<b<\infty}\big|(\widehat{g}\,\chi_{[a,b]})^\vee\big|$$

defined on the line is L^p bounded for all $1<p<\infty$.

10.6.4. On $\mathbf{R}^n$ fix a unit vector b and consider the maximal operator

$$T(g)(x) = \sup_{N>0} \left| \int_{|b\cdot\xi|\le N} \widehat{g}(\xi) e^{2\pi i x\cdot\xi}\, d\xi \right|$$

Show that T maps $L^p(\mathbf{R}^n)$ into $L^p(\mathbf{R}^n)$ for all $1 < p < \infty$.
[*Hint:* Apply a rotation.]

10.6.5. (a) Show that Theorem 10.6.1 actually gives the stronger estimate that for every set E of finite measure, there is a subset E' of E with $|E'| \ge \frac{1}{2}|E|$ such that

$$\left| \int_{E'} \mathfrak{D}_{N,\mathbf{P}}(\chi_F)\, dx \right| \le C \min(|E|,|F|)\Big(1 + \Big|\log\frac{|E|}{|F|}\Big|\Big),$$

where C is a constant independent of E, F, N, and $\mathbf{P}$. Deduce that the same conclusion also holds for $\mathfrak{D}_N$.
(b) Average to obtain the conclusion in part (a) for the operator

$$\mathfrak{C}_N(f)(x) = \int_{-\infty}^{N(x)} \widehat{f}(\xi) e^{2\pi i x\xi}\, d\xi$$

with constant independent of the measurable function $N : \mathbf{R} \to \mathbf{R}$.
(c) (*Sjölin* [**459**]) Prove that for some $C, c > 0$ and all $\lambda > 0$ we have

$$\big|\{|\mathfrak{C}_N(\chi_F)| > \lambda\}\big| \le C \begin{cases} \frac{1}{\lambda}\log(\frac{1}{\lambda})|F| & \text{when } \lambda < \frac{1}{2} \\ e^{-c\lambda}|F| & \text{when } \lambda \ge \frac{1}{2}. \end{cases}$$

[*Hint:* For $\lambda > 0$ define

$$E^1_\lambda = \{\operatorname{Re}\mathfrak{C}_N(\chi_F) > \lambda\} \qquad E^2_\lambda = \{\operatorname{Re}\mathfrak{C}_N(\chi_F) < -\lambda\}$$
$$E^3_\lambda = \{\operatorname{Im}\mathfrak{C}_N(\chi_F) > \lambda\} \qquad E^4_\lambda = \{\operatorname{Im}\mathfrak{C}_N(\chi_F) < -\lambda\}$$

and apply part (b) to each set E^j_λ. Use that if $\frac{A}{\log A} < \frac{1}{\lambda}$ for some $A > 1$, then $A \le \frac{2}{\lambda}\log\frac{1}{\lambda}$.]

10.6.6. Let $N : \mathbf{R} \to \mathbf{R}$ be a real-valued measurable function. Show that the operators

$$\mathfrak{G}_{N,\varepsilon}(g)(x) = \int_{|t-x|>\varepsilon} e^{2\pi i N(t)x} \frac{g(t)}{x-t}\, dt$$

are uniformly (in $\varepsilon > 0$) bounded on $L^p(\mathbf{R})$ for all $1 < p < \infty$.

10.6.7. Define the *directional Carleson operators* on $\mathbf{R}^n$ as follows:

$$\mathcal{C}^\theta(f)(x) = \sup_{a\in\mathbf{R}} \left| \lim_{\varepsilon\to 0} \int_{\varepsilon<|t|<\varepsilon^{-1}} e^{2\pi i a t} f(x - t\theta)\, \frac{dt}{t} \right|,$$

where θ is a vector in $\mathbf{S}^{n-1}$.
(a) Show that $\mathcal{C}^\theta(f)$ is well defined for f in $L^p(\mathbf{R}^n)$ and is also bounded on these spaces for all $1 < p < \infty$.
(b) Let Ω be an odd integrable function on $\mathbf{S}^{n-1}$. Define the operator

$$\mathcal{C}_\Omega(f)(x) = \sup_{\xi \in \mathbf{R}^n} \left| \lim_{\varepsilon \to 0} \int_{\varepsilon < |y| < \varepsilon^{-1}} e^{2\pi i \xi \cdot y} f(x-y) \frac{\Omega(\frac{y}{|y|})}{|y|^n} \, dy \right| .$$

Show that $\mathcal{C}_\Omega(f)$ is bounded on $L^p(\mathbf{R}^n)$ for $1 < p < \infty$.

10.6.8. For a fixed $\lambda > 0$ write

$$\mathcal{O} = \{x \in \mathbf{R} : \, \mathcal{C}_*(f)(x) > \lambda\} = \bigcup_j I_j \, ,$$

where $I_j = (\alpha_j, \alpha_j + \delta_j)$ are open disjoint intervals. Let $1 < r < \infty$. Show that there exists a $\gamma_0 > 0$ such that for every $0 < \gamma < \gamma_0$ there exists a constant $C_\gamma > 0$ such that $\lim_{\gamma \to 0} C_\gamma = 0$ and

$$\big|\{x \in I_j : \mathcal{C}_*(f)(x) > 3\lambda, \; M_r(f)(x) \le \gamma\,\lambda\}\big| \le C_\gamma \big|I_j\big|$$

for all f for which $\mathcal{C}_*(f)$ is defined.
[*Hint:* Note that we must have $\mathcal{C}_*(f)(\alpha_j) \le \lambda$ and $\mathcal{C}_*(f)(\alpha_j + \delta_j) \le \lambda$ for all j. Set $I_j^* = (\alpha_j - 5\delta_j, \alpha_j + 6\delta_j)$, $f_1(x) = f(x)$ for $x \in I_j^*$, $f_1(x) = 0$ for $x \notin I_j^*$, and $f_2(x) = f(x) - f_1(x)$. We may assume that for all j there exists a z_j in I_j such that $M_r(f)(z_j) \le \gamma\lambda$. For fixed $x \in I_j$ estimate the difference $|H^{(\varepsilon)}(f_2)(x) - H^{(\varepsilon)}(f_2)(\alpha_j)|$ by the sum of the three expressions

$$\left| \int_{|\alpha_j - t| > \varepsilon} f_2(t) e^{2\pi i \xi t} \left(\frac{2}{\alpha_j - t} - \frac{2}{x - t} \right) dt \right|$$

$$\left| \int_{|x - t| > \varepsilon \ge |\alpha_j - t|} f_2(t) e^{2\pi i \xi t} \frac{1}{x - t} \, dt \right|$$

$$\left| \int_{|\alpha_j - t| > \varepsilon \ge |x - t|} f_2(t) e^{2\pi i \xi t} \frac{1}{\alpha_j - t} \, dt \right| .$$

which can be shown easily to be controlled by $c_0 \, M(f)(z_j)$ for constant c_0. Thus $\mathcal{C}_*(f_2)(x) \le \mathcal{C}_*(f_2)(\alpha_j) + c_0 \, M(f)(z_j) \le \lambda + c_0\,\gamma\,\lambda$. Select γ_0 such that $c_0\,\gamma_0 < \frac{1}{2}$. Then $\lambda + c_0\,\gamma\,\lambda < \frac{3}{2}\lambda$ for $\gamma < \gamma_0$; hence for $x \in I_j$ we have $\mathcal{C}_*(f)(x) \le \mathcal{C}_*(f_1)(x) + \frac{3}{2}\lambda$ and thus $I_j \cap \{\mathcal{C}_*(f) > 3\,\lambda\} \subseteq \{\mathcal{C}_*(f_1) > \lambda\}$. Using the boundedness of $\mathcal{C}_*$ on L^r and the fact that $M_r(f)(z_j) \le \gamma\,\lambda$, we can easily obtain that the last set has measure at most a constant multiple of $\gamma^r |I_j|$.]

10.6.9. (*Hunt and Young* [**265**]) Show that for every w in A_∞ there is a finite constant $\gamma_0 > 0$ such that for all $0 < \gamma < \gamma_0$ and all $1 < r < \infty$ there is a constant B_γ such that

$$\omega\big(\{\mathcal{C}_*(f) > 3\lambda\} \cap \{M_r(f) \le \gamma\lambda\}\big) \le B_\gamma w\big(\{\mathcal{C}_*(f) > \lambda\}\big)$$

for all f for which $C_*(f)$ is finite. Moreover, the constants B_γ satisfy $B_\gamma \to 0$ as $\gamma \to 0$.

[*Hint:* Start with positive constants C_0 and δ such that for all intervals I and any measurable set E we have $|E \cap I| \le \varepsilon |I| \implies w(E \cap I) \le C_0 \, \varepsilon^\delta w(I)$. Use the previous estimate with $I = I_j$ and sum over j to obtain the required estimate with $B_\gamma = C_0 \, (C_\gamma)^\delta$.]

10.6.10. Use the result in Corollary 9.5.7 to prove the following vector-valued version of Theorem 10.6.1:

$$\Big\| \Big(\sum_j |\mathcal{C}(f_j)|^r \Big)^{\frac{1}{r}} \Big\|_{L^p(w)} \le C_{p,r}(w) \Big\| \Big(\sum_j |f_j|^r \Big)^{\frac{1}{r}} \Big\|_{L^p(w)}$$

for all $1 < p, r < \infty$, all weights $w \in A_p$, and all sequences of functions f_j in $L^p(w)$.

HISTORICAL NOTES

The geometric construction in Section 10.1 is based on ideas of Besicovitch, who used a similar construction to answer the following question posed in 1917 by the Japanese mathematician S. Kakeya: What is the smallest possible area of the trace of ink left on a piece of paper by an ink-covered needle of unit length when the positions of its two ends are reversed? This problem puzzled mathematicians for several decades until Besicovitch [**37**] showed that for any $\varepsilon > 0$ there is a way to move the needle so that the total area of the blot of ink left on the paper is smaller than ε. Fefferman [**185**] borrowed ideas from the construction of Besicovitch to provide the negative answer to the multiplier problem to the ball for $p \neq 2$ (Theorem 10.1.5). Prior to Fefferman's work, the fact that the characteristic function of the unit ball is not a multiplier on $L^p(\mathbf{R}^n)$ for $|\frac{1}{p} - \frac{1}{2}| \ge \frac{1}{2n}$ was pointed out by Herz [**246**] who also showed that this limitation is not necessary when this operator is restricted to radial L^p functions. The crucial Lemma 10.1.4 in Fefferman's proof is due to Y. Meyer.

The study of Bochner-Riesz means originated in the article of Bochner [**43**], who obtained their L^p boundedness for $\lambda > \frac{n-1}{2}$. Stein [**468**] improved this result to $\lambda > \frac{n-1}{2}|\frac{1}{p} - \frac{1}{2}|$ using interpolation for analytic families of operators. Theorem 10.2.4 was first proved by Carleson and Sjölin [**87**]. A second proof of this theorem was given by Fefferman [**187**]. A third proof was given by Hörmander [**255**]. The proof of Theorem 10.2.4 given in the text is due Córdoba [**127**]. This proof elaborated the use of the Kakeya maximal function in the study of spherical summation multipliers, which was implicitly pioneered in Fefferman [**187**]. The boundedness of the Kakeya maximal function $\mathcal{K}_N$ on $L^2(\mathbf{R}^2)$ with norm $C(\log N)^2$ was first obtained by Córdoba [**126**]. The sharp estimate $C \log N$ was later obtained by Strömberg [**494**]. The proof of Theorem 10.3.5 is taken from this article of Strömberg. Another proof of the boundedness of the Kakeya maximal function without dilations on $L^2(\mathbf{R}^2)$ was obtained by Müller [**384**]. Barrionuevo [**27**] showed that for any subset Σ of $\mathbf{S}^1$ with N elements the maximal operator $\mathfrak{M}_\Sigma$ maps $L^2(\mathbf{R}^2)$ into itself with norm $CN^{2(\log N)^{-1/2}}$ for some absolute constant C. Note that this bound is $O(N^\varepsilon)$ for any $\varepsilon > 0$. Katz [**294**] improved this bound to $C \log N$ for some absolute constant C; see also Katz [**295**]. The latter is a sharp bound as indicated in Proposition 10.3.4. Katz [**293**] also showed that the maximal operator $\mathfrak{M}_K$ associated with a set of unit vectors pointing along a Cantor set K of directions is unbounded on $L^2(\mathbf{R}^2)$. If Σ is an infinite set of vectors in $\mathbf{S}^1$ pointing in lacunary directions, then $\mathfrak{M}_\Sigma$ was studied by Strömberg [**493**], Córdoba and Fefferman [**131**], and Nagel, Stein, and Wainger [**388**]. The latter authors obtained its L^p boundedness for all $1 < p < \infty$. Theorem 10.2.7 was first proved by Carleson [**85**]. For a short account on extensions of this theorem, the reader may consult the historical notes at the end of Chapter 5.

The idea of restriction theorems for the Fourier transform originated in the work of E. M. Stein around 1967. Stein's original restriction result was published in the article of Fefferman [**182**], which was the first to point out connections between restriction theorems and boundedness of the Bochner-Riesz means. The full restriction theorem for the circle (Theorem 10.4.7 for $p < \frac{4}{3}$) is due to Fefferman and Stein and was published in the aforementioned article of Fefferman [**182**]. See also the related article of Zygmund [**573**]. The present proof of Theorem 10.4.7 is based in that of Córdoba [**128**]. This proof was further elaborated by Tomas [**524**], who pointed out the logarithmic blowup when $p = \frac{4}{3}$ for the corresponding restriction problem for annuli. The result in Example 10.4.4 is also due to Fefferman and Stein and was initially proved using arguments from spherical harmonics. The simple proof presented here was observed by A. W. Knapp. The restriction property in Theorem 10.4.5 for $p < \frac{2(n+1)}{n+3}$ is due to Tomas [**523**] while the case $p = \frac{2(n+1)}{n+3}$ is due to Stein [**480**]. Theorem 10.4.6 was first proved by Fefferman [**182**] for the smaller range of $\lambda > \frac{n-1}{4}$ using the restriction property $R_{p\to 2}(\mathbf{S}^{n-1})$ for $p < \frac{4n}{3n+1}$. The fact that the $R_{p\to 2}(\mathbf{S}^{n-1})$ restriction property (for $p < 2$) implies the boundedness of the Bochner-Riesz operator B^λ on $L^p(\mathbf{R}^n)$ is contained in the work of Fefferman [**182**]. A simpler proof of this fact, obtained later by E. M. Stein, appeared in the subsequent article of Fefferman [**187**]. This proof is given in Theorem 10.4.6, incorporating the Tomas-Stein restriction property $R_{p\to 2}(\mathbf{S}^{n-1})$ for $p \le \frac{2(n+1)}{n+3}$. It should be noted that the case $n = 3$ of this theorem was first obtained in unpublished work of Sjölin. For a short exposition and history of this material consult the book of Davis and Chang [**154**]. Much of the material in Sections 10.2, 10.3 and 10.4, is based on the notes of Vargas [**536**].

There is an extensive literature on restriction theorems for submanifolds of $\mathbf{R}^n$. It is noteworthy to mention (in chronological order) the results of Strichartz [**492**], Prestini [**417**], Greenleaf [**228**], Christ [**92**], Drury [**162**], Barceló [**25**], [**26**], Drury and Marshall [**164**], [**165**], Beckner, Carbery, Semmes, and Soria [**30**], Drury and Guo [**163**], De Carli and Iosevich [**155**], [**156**], Sjölin and Soria [**461**], Oberlin [**395**], Wolff [**561**], and Tao [**508**].

The boundedness of the Bochner-Riesz operators on the range excluded by Proposition 10.2.3 implies that the restriction property $R_{p\to q}(\mathbf{S}^{n-1})$ is valid when $\frac{1}{q} = \frac{n+1}{n-1}\frac{1}{p'}$ and $1 \le p < \frac{2n}{n+1}$, as shown by Tao [**506**]; in this article a hierarchy of conjectures in harmonic analysis and interrelations among them is discussed. In particular, the aforementioned restriction property would imply estimate (10.3.31) for the Kakeya maximal operator $\mathcal{K}_N$ on $\mathbf{R}^n$, which would in turn imply that Besicovitch sets have Minkowski dimension n. (A

Besicovitch set is defined as a subset of $\mathbf{R}^n$ that contains a unit line segment in every direction.) Katz, Laba, and Tao [**296**] have obtained good estimates on the Minkowski dimension of such sets in $\mathbf{R}^3$.

A general sieve argument obtained by Córdoba [**126**] reduces the boundedness of the Kakeya maximal operator $\mathcal{K}_N$ to the one without dilations $\mathcal{K}_N^a$. For applications to the Bochner-Riesz multiplier problem, only the latter is needed. Carbery, Hernández, and Soria [**80**] have proved estimate (10.3.28) for radial functions in all dimensions. Igari [**271**] proved estimate (10.3.29) for products of one-variable functions of each coordinate. The norm estimates in Corollary 10.3.7 can be reversed, as shown by Keich [**300**] for $p > 2$. The corresponding estimate for $1 < p < 2$ in the same corollary can be improved to $N^{\frac{2}{p}-1}$. Córdoba [**127**] proved the partial case $p \le 2$ of Theorem 10.3.10 on $\mathbf{R}^n$. This range was extended by Drury [**161**] to $p \le \frac{n+1}{n-1}$ using estimates for the x-ray transform. Theorem 10.3.10 (i.e., the further extension to $p \le \frac{n+1}{2}$) is due to Christ, Duoandikoetxea, and Rubio de Francia [**99**], and its original proof also used estimates for the x-ray transform; the proof of Theorem 10.3.10 given in the text is derived from that in Bourgain [**49**]. This article brought a breakthrough in many of the previous topics. In particular, Bourgain [**49**] showed that the Kakeya maximal operator $\mathcal{K}_N$ maps $L^p(\mathbf{R}^n)$ into itself with bound $C_\varepsilon N^{\frac{n}{p}-1+\varepsilon}$ for for all $\varepsilon > 0$ and some $p_n > \frac{n+1}{2}$. He also showed that the range of p's in Theorem 10.4.5 is not sharp as there exist indices $p = p(n) > \frac{2(n+1)}{n+3}$ for which property $R_{p\to q}(\mathbf{S}^{n-1})$ holds and that Theorem 10.4.6 is not sharp as there exist indices $\lambda_n < \frac{n-1}{2(n+1)}$ for which the Bochner-Riesz operators are bounded on $L^p(\mathbf{R}^n)$ in the optimal range of p's when $\lambda \ge \lambda_n$. Improvements on these indices were subsequently obtained by Bourgain [**50**], [**51**]. Some of Bourgain's results in $\mathbf{R}^3$ were reproved by Schlag [**449**] using different geometric methods. Wolff [**559**] showed that the Kakeya maximal operator $\mathcal{K}_N$ maps $L^p(\mathbf{R}^n)$ into itself with bound $C_\varepsilon N^{\frac{n}{p}-1+\varepsilon}$ for any $\varepsilon > 0$ whenever $p \le \frac{n+2}{2}$. In higher dimensions, this range of p's was later extended by Bourgain [**52**] to $p \le (1+\varepsilon)\frac{n}{2}$ for some dimension-free positive constant ε. When $n = 3$ further improvements on the restriction and the Kakeya conjectures were obtained by Tao, Vargas, and Vega [**512**]. For further historical advances in the subject the reader is referred to the survey articles of Wolff [**560**] and Katz and Tao [**297**].

Regarding the almost everywhere convergence of the Bochner-Riesz means the following are among the known resuts: Carbery [**77**] has shown that the maximal operator $B_*^\lambda(f) = \sup_{R>0} |B_R^\lambda(f)|$ is bounded on $L^p(\mathbf{R}^2)$ when $\lambda > 0$ and $2 \le p < \frac{4}{1-2\lambda}$, obtaining the convergence $B_R^\lambda(f) \to f$ almost everywhere for $f \in L^p(\mathbf{R}^2)$. For $n \ge 3$, $2 \le p < \frac{2n}{n-1-2\lambda}$, and $\lambda \ge \frac{n-1}{2(n+1)}$ the same result was obtained by Christ [**94**]. Although the boundedness of the maximal operator B_*^λ is not known on the full range of $\lambda > 0$, the almost everywhere convergence of $B_R^\lambda(f) \to f$ for f in $L^p(\mathbf{R}^n)$, $2 \le p < \frac{2n}{n-1-2\lambda}$ and $\lambda > 0$, was obtained by Carbery, Rubio de Francia, and Vega [**81**]. Their proof relied on embedding $L^p(\mathbf{R}^n)$ into $L^2(\mathbf{R}^n) + L^2(\mathbf{R}^n, |x|^{-\beta}dx)$ and on obtaining the boundedness of B_*^λ on both of the aforementioned spaces. Tao [**505**] obtained a counterexample to show that the maximal Bochner-Riesz operator B_*^λ is not L^p bounded on the range on which B^λ is known to be bounded whenever $1 < p < 2$ and $n = 2$. Tao [**510**] also obtained boundedness for the maximal Bochner-Riesz operators B_*^λ on $L^p(\mathbf{R}^2)$ whenever $1 < p < 2$ for an open range of pairs $(\frac{1}{p}, \lambda)$ that lie below the line $\lambda = \frac{1}{2}(\frac{1}{p} - \frac{1}{2})$.

On the critical line $\lambda = \frac{n}{p} - \frac{n+1}{2}$ boundedness into weak L^p for the Bochner-Riesz operators is possible in the range $1 \le p \le \frac{2n}{n+1}$. Christ [**96**], [**95**] first obtained such results for $1 \le p < \frac{2(n+1)}{n+3}$ in all dimensions. The point $p = \frac{2(n+1)}{n+3}$ was later included by Tao [**504**]. In two dimensions weak boundedness for the full range of indices was shown by Seeger [**453**]; in all dimensions the same conclusion was obtained by Colzani, Travaglini, and Vignati [**124**] for radial functions. Tao [**505**] has obtained a general argument that yields weak endpoint bounds for B^λ whenever strong type bounds are known above the critical line.

A version of Theorem 10.5.1 concerning the almost everywhere convergence of Fourier series of square integrable functions on the circle was first proved by Carleson [**84**]. The extension of this theorem to L^p functions for $1 < p < \infty$ (Theorem 10.6.1) was obtained by Hunt [**262**]. The reader is referred to the books of Jørsboe and Melbro [**281**], Mozzochi [**379**], and Arias de Reyna [**11**] for detailed presentation of these results. An alternative proof of Carleson's theorem was provided by Fefferman [**186**], pioneering a set of ideas called time-frequency analysis. A third proof of this theorem was obtained by Lacey and Thiele [**324**]. The proof of Theorem 10.5.1 follows closely that in Lacey and Thiele [**324**], which improves in some ways that of Fefferman's [**186**], by which it was inspired. The reader may also consult the expository article of Thiele [**519**]. The proof of Lacey and Thiele was a byproduct of their work [**322**], [**323**] on the boundedness of the bilinear Hilbert transforms $H_\alpha(f_1, f_2)(x) = \frac{1}{\pi}\text{p.v.}\int_{\mathbf{R}} f_1(x-t)f_2(x-\alpha t)\,\frac{dt}{t}$. This family of operators arose in early attempts of A. Calderón to show that the first commutator (Example 8.3.8, $m = 1$) is bounded on L^2 when A' is in L^∞, an approach completed only using the uniform boundedness of H_α obtained by Thiele [**518**], Grafakos and Li [**217**], and Li [**333**]. The proof of Theorem 10.6.1 given in the text is based on that in Grafakos, Tao, and Terwilleger [**224**]. This article also investigates higher-dimensional analogues of the theory that were initiated in Pramanik and Terwilleger [**416**]. Theorem 10.6.5 was first obtained by Hunt and Young [**265**] using a good lambda inequality for the Carleson operator. The particular proof of Theorem 10.6.5 given in the text is based on the approach of Rubio de Francia, Ruiz, and Torrea [**439**].

The subject of Fourier analysis is currently enjoying a surge of activity. Emerging connections with analytic number theory, combinatorics, geometric measure theory, partial differential equations, and multilinear analysis introduce new dynamics and present promising developments. These connections are also creating new research directions that extend beyond the scope of this book.

APPENDIX A

Gamma and Beta Functions

A.1. A Useful Formula

The following formula is valid:

$$\int_{\mathbf{R}^n} e^{-|x|^2}\,dx = \left(\sqrt{\pi}\right)^n .$$

This is an immediate consequence of the corresponding one-dimensional formula

$$\int_{-\infty}^{+\infty} e^{-x^2}\,dx = \sqrt{\pi}\,,$$

which is usually proved from its two dimensional version by switching to polar coordinates

$$I^2 = \int_{-\infty}^{+\infty}\int_{-\infty}^{+\infty} e^{-x^2}e^{-y^2}\,dydx = 2\pi\int_0^\infty re^{-r^2}\,dr = \pi.$$

A.2. Definitions of $\Gamma(z)$ and $B(z, w)$

For a complex number z with $\operatorname{Re} z > 0$ define

$$\Gamma(z) = \int_0^\infty t^{z-1}e^{-t}dt.$$

$\Gamma(z)$ is called the gamma function. It follows from its definition that $\Gamma(z)$ is analytic on the right half-plane $\operatorname{Re} z > 0$.

Two fundamental properties of the gamma function are that

$$\Gamma(z+1) = z\Gamma(z) \qquad \text{and} \qquad \Gamma(n) = (n-1)!$$

when $n \in \mathbf{Z}^+$ and z has positive real part. Indeed, integration by parts yields

$$\Gamma(z) = \int_0^\infty t^{z-1}e^{-t}\,dt = \left[\frac{t^z e^{-t}}{z}\right]_0^\infty + \frac{1}{z}\int_0^\infty t^z e^{-t}\,dt = \frac{1}{z}\Gamma(z+1).$$

Since $\Gamma(0) = 1$, the property $\Gamma(n) = (n-1)!$ follows by induction. Another important fact is that

$$\Gamma\left(\tfrac{1}{2}\right) = \sqrt{\pi}\,.$$

This follows easily from the identity

$$\Gamma(\tfrac{1}{2}) = \int_0^\infty t^{-\frac{1}{2}} e^{-t}\, dt = 2\int_0^\infty e^{-u^2}\, du = \sqrt{\pi}\,.$$

Next we define the beta function. Fix z and w complex numbers with positive real parts. We define

$$B(z,w) = \int_0^1 t^{z-1}(1-t)^{w-1}\, dt = \int_0^1 t^{w-1}(1-t)^{z-1}\, dt.$$

We have the following relationship between the gamma and the beta functions:

$$B(z,w) = \frac{\Gamma(z)\Gamma(w)}{\Gamma(z+w)},$$

when z and w have positive real parts.

The proof of this fact is as follows:

$$\begin{aligned}
&\Gamma(z+w)B(z,w)\\
&= \Gamma(z+w)\int_0^1 t^{w-1}(1-t)^{z-1}\, dt\\
&= \Gamma(z+w)\int_0^\infty u^{w-1}\left(\frac{1}{1+u}\right)^{z+w} du \qquad\qquad t = u/(1+u)\\
&= \int_0^\infty\int_0^\infty u^{w-1}\left(\frac{1}{1+u}\right)^{z+w} v^{z+w-1}e^{-v}\, dv\, du\\
&= \int_0^\infty\int_0^\infty u^{w-1}s^{z+w-1}e^{-s(u+1)}\, ds\, du \qquad\qquad s = v/(1+u)\\
&= \int_0^\infty s^z e^{-s}\int_0^\infty (us)^{w-1}e^{-su}\, du\, ds\\
&= \int_0^\infty s^{z-1}e^{-s}\Gamma(w)ds = \Gamma(z)\Gamma(w)\,.
\end{aligned}$$

A.3. Volume of the Unit Ball and Surface of the Unit Sphere

Let v_n be the volume of the unit ball in $\mathbf{R}^n$ and ω_{n-1} be the surface of the unit sphere S^{n-1}. We have the following:

$$v_n = \frac{2\pi^{\frac{n}{2}}}{n\Gamma(\frac{n}{2})} = \frac{\pi^{\frac{n}{2}}}{\Gamma(\frac{n}{2}+1)}$$

and

$$\omega_{n-1} = n v_n = \frac{2\pi^{\frac{n}{2}}}{\Gamma(\frac{n}{2})}\,.$$

The easy proofs are based on the formula in Appendix A.1. We have

$$\left(\sqrt{\pi}\right)^n = \int_{\mathbf{R}^n} e^{-|x|^2}\,dx = \omega_{n-1}\int_0^\infty e^{-r^2} r^{n-1}\,dr\,,$$

by switching to polar coordinates. Now change variables $t = r^2$ to obtain that

$$\pi^{\frac{n}{2}} = \tfrac{\omega_{n-1}}{2}\int_0^\infty e^{-t} t^{\frac{n}{2}-1}\,dt = \tfrac{\omega_{n-1}}{2}\Gamma\big(\tfrac{n}{2}\big)\,.$$

This proves the formula for the surface area of the unit sphere in $\mathbf{R}^n$.

The formula for the volume follows from the following simple argument. Let $B(0,R)$ be the ball in $\mathbf{R}^n$ of radius $R > 0$ centered at the origin. Then the volume of the shell $B(0,R+h)\setminus B(0,R)$ divided by h tends to the surface area of $B(0,R)$ as $h \to 0$. In other words, the derivative of the volume of $B(0,R)$ with respect to the radius R is equal to the surface area of $B(0,R)$. Since the volume of $B(0,R)$ is $v_n R^n$, it follows that the surface area of $B(0,R)$ is $n v_n R^{n-1}$. Taking $R = 1$, we obtain $\omega_{n-1} = n v_n$.

A.4. A Useful Integral

The following is a valid formula:

$$\int_0^{\pi/2} (\sin\varphi)^a(\cos\varphi)^b\,d\varphi = \frac{1}{2}B\big(\tfrac{a+1}{2},\tfrac{b+1}{2}\big) = \frac{1}{2}\frac{\Gamma(\frac{a+1}{2})\Gamma(\frac{b+1}{2})}{\Gamma(\frac{a+b+2}{2})}\,,$$

whenever $a > -1$ and $b > -1$.

The proof the first identity change variables by setting $u = (\sin\varphi)^2$; then $du = 2(\sin\varphi)(\cos\varphi)d\varphi$ and the preceding integral becomes

$$\frac{1}{2}\int_0^1 u^{\frac{a-1}{2}}(1-u)^{\frac{b-1}{2}}\,du = \frac{1}{2}B\big(\tfrac{a+1}{2},\tfrac{b+1}{2}\big).$$

A.5. Meromorphic Extensions of $B(z,w)$ and $\Gamma(z)$

Using the identity $\Gamma(z+1) = z\Gamma(z)$, we can easily define a meromorphic extension of the gamma function on the whole complex plane starting from its known values on the right half plane. We give an explicit description of the meromorphic extension of $\Gamma(z)$ on the whole plane. First write

$$\Gamma(z) = \int_0^1 t^{z-1}e^{-t}dt + \int_1^\infty t^{z-1}e^{-t}dt$$

and observe that the second integral is an analytic function of z for all $z \in \mathbf{C}$. Write the first integral as

$$\int_0^1 t^{z-1} \left\{ e^{-t} - \sum_{j=0}^{N} \frac{(-t)^j}{j!} \right\} dt + \sum_{j=0}^{N} \frac{(-1)^j/j!}{z+j} .$$

The last integral converges when $\operatorname{Re} z > -N-1$ since the expression inside the curly brackets is $O(t^{N+1})$ as $t \to 0$. It follows that the gamma function can be defined to be a analytic function on $\operatorname{Re} z > -N-1$ except at the points $z = -j$, $j = 0, 1, \ldots, N$ at which it has simple poles with residues $\frac{(-1)^j}{j!}$. Since N was arbitrary, it follows that the gamma function has a meromorphic extension on the whole plane.

In view of the identity

$$B(z, w) = \frac{\Gamma(z)\Gamma(w)}{\Gamma(z+w)} ,$$

the definition of $B(z, w)$ can be extended to $\mathbf{C} \times \mathbf{C}$. It follows that $B(z, w)$ is a meromorphic function in each argument.

A.6. Asymptotics of $\Gamma(x)$ as $x \to \infty$

We will now derive *Stirling's formula*:

$$\lim_{x \to \infty} \frac{\Gamma(x+1)}{\left(\frac{x}{e}\right)^x \sqrt{2\pi x}} = 1 .$$

First change variables $t = x + sx\sqrt{\frac{2}{x}}$ to obtain

$$\Gamma(x+1) = \int_0^\infty e^{-t} t^x \, dt = \left(\frac{x}{e}\right)^x \sqrt{2x} \int_{-\sqrt{x/2}}^{+\infty} \frac{\left(1 + s\sqrt{\frac{2}{x}}\right)^x}{e^{2s\sqrt{x/2}}} ds .$$

Setting $y = \sqrt{\dfrac{x}{2}}$, we obtain

$$\frac{\Gamma(x+1)}{\left(\frac{x}{e}\right)^x \sqrt{2x}} = \int_{-\infty}^{+\infty} \left(\frac{\left(1 + \frac{s}{y}\right)^y}{e^s} \right)^{2y} \chi_{(-y,\infty)}(s) \, ds.$$

To show that the last integral converges to $\sqrt{\pi}$ as $y \to \infty$, we need the following:

(1) The fact that $\lim_{y \to \infty} \left(\frac{(1 + s/y)^y}{e^s} \right)^{2y} \to e^{-s^2}$, which follows easily by taking logarithms and applying L' Hôpital's rule twice.

(2) The estimate, valid for $y \geq 1$:

$$\left(\frac{\left(1+\frac{s}{y}\right)^y}{e^s}\right)^{2y} \leq \begin{cases} \dfrac{(1+s)^2}{e^s} & \text{when } s \geq 0, \\ e^{-s^2} & \text{when } -y < s < 0, \end{cases}$$

which can be easily checked using calculus. Using these facts, the Lebesgue dominated convergence theorem, the trivial fact that $\chi_{-y<s<\infty} \to 1$ as $y \to \infty$, and the identity in Appendix A.1, we obtain that

$$\begin{aligned} \lim_{x\to\infty} \frac{\Gamma(x+1)}{\left(\frac{x}{e}\right)^x \sqrt{2x}} &= \lim_{y\to\infty} \int_{-\infty}^{+\infty} \left(\frac{\left(1+\frac{s}{y}\right)^y}{e^s}\right)^{2y} \chi_{(-y,\infty)}(s)\, ds \\ &= \int_{-\infty}^{+\infty} e^{-s^2}\, ds \\ &= \sqrt{\pi}. \end{aligned}$$

A.7. The Duplication Formula for the Gamma Function

This formula relates $\Gamma(2z)$ and $\Gamma(z)$ as follows:

$$\Gamma(2z) = \pi^{-\frac{1}{2}} 2^{2z-1} \Gamma(z)\Gamma(z+\tfrac{1}{2}).$$

To prove it, we first assume that $\operatorname{Re} z > 0$. Taking $z = w$ in the identity $\Gamma(z)\Gamma(w) = B(z,w)\Gamma(z+w)$, we obtain

$$\frac{\Gamma(z)^2}{\Gamma(2z)} = B(z,z) = \int_0^1 \big(t(1-t)\big)^{z-1}\, dt = 2\int_0^{\frac{1}{2}} \big(t(1-t)\big)^{z-1}\, dt\,,$$

where in the last step we used that the function $t(1-t)$ is symmetric about the point $t = \frac{1}{2}$. We use the substitution $t = \frac{1}{2}(1 - s^{\frac{1}{2}})$ [$dt = -\frac{1}{4}s^{-\frac{1}{2}}ds$, $t(1-t) = \frac{1}{4}(1-s)$], which yields

$$2\int_0^{\frac{1}{2}} \big(t(1-t)\big)^{z-1}\, dt = 2^{1-2z} \int_0^1 s^{-\frac{1}{2}}(1-s)^{z-1} ds = 2^{1-2z} \frac{\Gamma(\frac{1}{2})\Gamma(z)}{\Gamma(z+\frac{1}{2})}.$$

Since $\Gamma(\frac{1}{2}) = \sqrt{\pi}$ and the preceding is equal to

$$\frac{\Gamma(z)^2}{\Gamma(2z)},$$

the claimed duplication formula follows for $\operatorname{Re} z > 0$. The case $\operatorname{Re} z < 0$ is obtained by a change of variables.

APPENDIX B

Bessel Functions

B.1. Definition

We shall only consider Bessel functions J_k of real order $k > -1/2$ (although some of the results can be extended easily to complex numbers k with real part bigger than $-1/2$).

We will define the Bessel function J_k of order k by its *Poisson representation formula*

$$J_k(z) = \frac{\left(\frac{z}{2}\right)^k}{\Gamma(k+\frac{1}{2})\Gamma(\frac{1}{2})} \int_{-1}^{+1} e^{izs}(1-s^2)^k \frac{ds}{\sqrt{1-s^2}},$$

where $k > -1/2$ and $z \in \mathbf{C}$. Among all equivalent definitions of Bessel functions, the preceding definition will be the most useful to us. Observe that for t real, $J_k(t)$ is also a real number.

B.2. Some Basic Properties

Let us summarize a few properties of Bessel functions.

(1) We have the following recurrence formula:

$$\frac{d}{dz}\big(z^{-k}J_k(z)\big) = -z^{-k}J_{k+1}(z), \qquad k > -1/2.$$

(2) We also have the companion recurrence formula:

$$\frac{d}{dz}\big(z^{k}J_k(z)\big) = z^{k}J_{k-1}(z), \qquad k > 1/2.$$

(3) J_k satisfies the differential equation

$$z^2 f''(z) + zf'(z) + (z^2 - k^2)f(z) = 0.$$

(4) If $k \in \mathbf{Z}^+$, then J_k can be written in the form

$$J_k(z) = \frac{1}{2\pi}\int_0^{2\pi} e^{iz\sin\theta} e^{-ik\theta}\, d\theta = \frac{1}{2\pi}\int_0^{2\pi} \cos(z\sin\theta - k\theta)\, d\theta.$$

This was taken by Bessel as the definition of these functions for k integer.
(5) For $k > -1/2$ and t real we have the following identity:

$$J_k(t) = \frac{1}{\Gamma(\frac{1}{2})}\left(\frac{t}{2}\right)^k \sum_{j=0}^{\infty}(-1)^j \frac{\Gamma(j+\frac{1}{2})}{\Gamma(j+k+1)}\frac{t^{2j}}{(2j)!}.$$

We first verify property (1). We have

$$\begin{aligned}
\frac{d}{dz}(z^{-k}J_k(z)) &= \frac{i}{2^k\Gamma(k+\frac{1}{2})\Gamma(\frac{1}{2})}\int_{-1}^{1} se^{izs}(1-s^2)^{k-\frac{1}{2}}\,ds \\
&= \frac{i}{2^k\Gamma(k+\frac{1}{2})\Gamma(\frac{1}{2})}\int_{-1}^{1} \frac{iz}{2}e^{izs}\frac{(1-s^2)^{k+\frac{1}{2}}}{k+\frac{1}{2}}\,ds \\
&= -\,z^{-k}J_{k+1}(z),
\end{aligned}$$

where we integrated by parts and we we used the fact that $\Gamma(x+1) = x\Gamma(x)$. Property (2) can be proved similarly.

Property (3) follows from a direct calculation. A calculation using the the definition of the Bessel function gives that the left-hand side of (3) is equal to

$$\frac{2^{-k}z^{k+1}}{\Gamma(k+\frac{1}{2})\Gamma(\frac{1}{2})}\int_{-1}^{+1} e^{isz}\Big((1-s^2)z + 2is(k+\tfrac{1}{2})\Big)(1-s^2)^{k-\frac{1}{2}}\,ds\,,$$

which in turn is equal to

$$-i\int_{-1}^{+1}\frac{d}{ds}\big(e^{isz}(1-s^2)^{k+\frac{1}{2}}\big)\,ds = 0\,.$$

Property (4) can be derived directly from (1). Let

$$G_k(z) = \frac{1}{2\pi}\int_0^{2\pi} e^{iz\sin\theta}e^{-ik\theta}\,d\theta,$$

for $k > -1/2$ and $z \in \mathbf{C}$. We can show easily that $G_0 = J_0$. If we had

$$\frac{d}{dz}(z^{-k}G_k(z)) = -z^{-k}G_{k+1}(z), \qquad z \in \mathbf{C},$$

for $k \geq 0$ we would immediately conclude that $G_k = J_k$ for $k \in \mathbf{Z}^+$. We have

$$\begin{aligned}
&\frac{d}{dz}(z^{-k}G_k(z)) \\
&= -z^{-k}\Big(\frac{k}{z}G_k(z) - \frac{dG_k}{dz}(z)\Big) \\
&= -z^{-k}\int_0^{2\pi} \frac{k}{2\pi z}e^{iz\sin\theta}e^{-ik\theta} - \frac{1}{2\pi}\Big(\frac{d}{dz}e^{iz\sin\theta}\Big)e^{-ik\theta}\,d\theta \\
&= -\frac{z^{-k}}{2\pi}\int_0^{2\pi} i\frac{d}{d\theta}\Big(\frac{e^{iz\sin\theta - ik\theta}}{z}\Big) + (\cos\theta - i\sin\theta)e^{iz\sin\theta}e^{-ik\theta}\,d\theta \\
&= -\frac{z^{-k}}{2\pi}\int_0^{2\pi} e^{iz\sin\theta}e^{-i(k+1)\theta}\,d\theta = -z^{-k}G_{k+1}(z)\,.
\end{aligned}$$

Finally, the identity in (5) can be derived by inserting the expression

$$\sum_{j=0}^{\infty}(-1)^j\frac{(ts)^{2j}}{(2j)!} + i\sin(ts)$$

for e^{its} in the definition of the Bessel function $J_k(t)$ in Appendix B.1. Carrying out the algebra gives

$$\begin{aligned}
J_k(t) &= \frac{(t/2)^k}{\Gamma(\frac{1}{2})}\sum_{j=0}^{\infty}(-1)^j\frac{1}{\Gamma(k+\frac{1}{2})}\frac{t^{2j}}{(2j)!}2\int_0^1 s^{2j-1}(1-s^2)^{k-\frac{1}{2}}s\,ds \\
&= \frac{(t/2)^k}{\Gamma(\frac{1}{2})}\sum_{j=0}^{\infty}(-1)^j\frac{1}{\Gamma(k+\frac{1}{2})}\frac{t^{2j}}{(2j)!}\frac{\Gamma(j+\frac{1}{2})\Gamma(k+\frac{1}{2})}{\Gamma(j+k+1)} \\
&= \frac{(t/2)^k}{\Gamma(\frac{1}{2})}\sum_{j=0}^{\infty}(-1)^j\frac{\Gamma(j+\frac{1}{2})}{\Gamma(j+k+1)}\frac{t^{2j}}{(2j)!}\,.
\end{aligned}$$

For further results on Bessel functions, consult Watson's monograph [**546**].

B.3. An Interesting Identity

Let $\mu > -\frac{1}{2}$, $\nu > -1$ and $t > 0$. Then the following identity is valid:

$$\int_0^1 J_\mu(ts)s^{\mu+1}(1-s^2)^\nu\,ds = \frac{\Gamma(\nu+1)2^\nu}{t^{\nu+1}}J_{\mu+\nu+1}(t)\,.$$

To prove this identity we use formula (5) in Appendix B.2. We have

$$\begin{aligned}
&\int_0^1 J_\mu(ts)s^{\mu+1}(1-s^2)^\nu\,ds \\
&= \frac{(\frac{t}{2})^\mu}{\Gamma(\frac{1}{2})}\int_0^1 \sum_{j=0}^{\infty} \frac{(-1)^j\Gamma(j+\frac{1}{2})\,t^{2j}}{\Gamma(j+\mu+1)(2j)!} s^{2j+\mu+\mu}(1-s^2)^\nu s\,ds \\
&= \frac{1}{2}\frac{(\frac{t}{2})^\mu}{\Gamma(\frac{1}{2})}\sum_{j=0}^{\infty} \frac{(-1)^j\Gamma(j+\frac{1}{2})\,t^{2j}}{\Gamma(j+\mu+1)(2j)!}\int_0^1 u^{j+\mu}(1-u)^\nu\,du \\
&= \frac{1}{2}\frac{(\frac{t}{2})^\mu}{\Gamma(\frac{1}{2})}\sum_{j=0}^{\infty} \frac{(-1)^j\Gamma(j+\frac{1}{2})\,t^{2j}}{\Gamma(j+\mu+1)(2j)!}\frac{\Gamma(\mu+j+1)\Gamma(\nu+1)}{\Gamma(\mu+\nu+j+2)} \\
&= \frac{2^\nu\Gamma(\nu+1)}{t^{\nu+1}}\frac{(\frac{t}{2})^{\mu+\nu+1}}{\Gamma(\frac{1}{2})}\sum_{j=0}^{\infty}\frac{(-1)^j\Gamma(j+\frac{1}{2})\,t^{2j}}{\Gamma(j+\mu+\nu+2)(2j)!} \\
&= \frac{\Gamma(\nu+1)2^\nu}{t^{\nu+1}}J_{\mu+\nu+1}(t)\,.
\end{aligned}$$

B.4. The Fourier Transform of Surface Measure on $\mathbf{S}^{n-1}$

Let $d\sigma$ denote surface measure on $\mathbf{S}^{n-1}$ for $n \geq 2$. Then the following is true:

$$\widehat{d\sigma}(\xi) = \int_{\mathbf{S}^{n-1}} e^{-2\pi i\xi\cdot\theta}d\theta = \frac{2\pi}{|\xi|^{\frac{n-2}{2}}}J_{\frac{n-2}{2}}(2\pi|\xi|).$$

To see this use the result in Appendix D.3 to write

$$\begin{aligned}
\widehat{d\sigma}(\xi) &= \int_{\mathbf{S}^{n-1}} e^{-2\pi i\xi\cdot\theta}\,d\theta \\
&= \frac{2\pi^{\frac{n-1}{2}}}{\Gamma(\frac{n-1}{2})}\int_{-1}^{+1} e^{-2\pi i|\xi|s}(1-s^2)^{\frac{n-2}{2}}\frac{ds}{\sqrt{1-s^2}} \\
&= \frac{2\pi^{\frac{n-1}{2}}}{\Gamma(\frac{n-1}{2})}\frac{\Gamma(\frac{n-2}{2}+\frac{1}{2})\Gamma(\frac{1}{2})}{(\pi|\xi|)^{\frac{n-2}{2}}}J_{\frac{n-2}{2}}(2\pi|\xi|) \\
&= \frac{2\pi}{|\xi|^{\frac{n-2}{2}}}J_{\frac{n-2}{2}}(2\pi|\xi|)\,.
\end{aligned}$$

B.5. The Fourier Transform of a Radial Function on $\mathbf{R}^n$

Let $f(x) = f_0(|x|)$ be a radial function defined on $\mathbf{R}^n$, where f_0 is defined on $[0, \infty)$. Then the Fourier transform of f is given by the formula

$$\widehat{f}(\xi) = \frac{2\pi}{|\xi|^{\frac{n-2}{2}}} \int_0^\infty f_0(r) J_{\frac{n}{2}-1}(2\pi r|\xi|) r^{\frac{n}{2}}\, dr\,.$$

To obtain this formula, use polar coordinates to write

$$\begin{aligned}
\widehat{f}(\xi) &= \int_{\mathbf{R}^n} f(x) e^{-2\pi i \xi\cdot x}\, dx \\
&= \int_0^\infty \int_{\mathbf{S}^{n-1}} f_0(r) e^{-2\pi i \xi\cdot r\theta} d\theta\, r^{n-1} dr \\
&= \int_0^\infty f_0(r)\, \widehat{d\sigma}(r\xi) r^{n-1} dr \\
&= \int_0^\infty \frac{2\pi}{(r|\xi|)^{\frac{n-2}{2}}} J_{\frac{n-2}{2}}(2\pi r|\xi|) r^{n-1} dr \\
&= \frac{2\pi}{|\xi|^{\frac{n-2}{2}}} \int_0^\infty f_0(r) J_{\frac{n}{2}-1}(2\pi r|\xi|) r^{\frac{n}{2}}\, dr\,.
\end{aligned}$$

As an application we take $f(x) = \chi_{B(0,1)}$, where $B(0,1)$ is the unit ball in $\mathbf{R}^n$. We obtain

$$\widehat{\chi_{B(0,1)}}(\xi) = \frac{2\pi}{|\xi|^{\frac{n-2}{2}}} \int_0^1 J_{\frac{n}{2}-1}(2\pi|\xi|r) r^{\frac{n}{2}}\, dr = \frac{J_{\frac{n}{2}}(2\pi|\xi|)}{|\xi|^{\frac{n}{2}}}\,,$$

in view of the result in Appendix B.3. More generally, for $\lambda > -1$, let

$$m_\lambda(x) = \begin{cases} (1-|x|^2)^\lambda & \text{for } |x| \le 1, \\ 0 & \text{for } |x| > 1. \end{cases}$$

Then

$$\widehat{m_\lambda}(\xi) = \frac{2\pi}{|\xi|^{\frac{n-2}{2}}} \int_0^1 J_{\frac{n}{2}-1}(2\pi|\xi|r) r^{\frac{n}{2}} (1-r^2)^\lambda\, dr = \frac{\Gamma(\lambda+1)}{\pi^\lambda} \frac{J_{\frac{n}{2}+\lambda}(2\pi|\xi|)}{|\xi|^{\frac{n}{2}+\lambda}}\,,$$

also in view of the identity in Appendix B.3.

B.6. Asymptotics of Bessel Functions

Here we take $z = r$ a positive real number and we seek the asymptotic behavior of $J_k(r)$ as $r \to 0$ and as $r \to \infty$. Let us fix $k > -1/2$. The following is true:

$$J_k(r) = \begin{cases} \dfrac{r^k}{2^k\Gamma(k+1)} + O(r^{k+1}) & \text{as } r \to 0, \\ \sqrt{\frac{2}{\pi r}}\cos(r - \frac{\pi k}{2} - \frac{\pi}{4}) + O(r^{-3/2}) & \text{as } r \to \infty. \end{cases}$$

The asymptotic behavior of $J_k(r)$ as $r \to 0$ is rather trivial. We simply need to note that

$$\begin{aligned} \int_{-1}^{+1} e^{irt}(1-t^2)^{k-\frac{1}{2}}\,dt &= \int_{-1}^{+1} (1-t^2)^{k-\frac{1}{2}}\,dt + O(r) \\ &= \int_0^{\pi} (\sin^2\phi)^{k-\frac{1}{2}}(\sin\phi)\,d\phi + O(r) \\ &= \frac{\Gamma(k+\frac{1}{2})\Gamma(\frac{1}{2})}{\Gamma(k+1)} + O(r) \end{aligned}$$

as $r \to 0$. To evaluate the last integral, we used the result in Appendix A.4.

The asymptotic behavior of $J_k(r)$ as $r \to \infty$ is more delicate. Consider the region in the complex plane obtained by excluding the rays $(-\infty, -1)$ and $(1, \infty)$. We choose an analytic branch of $(1-z^2)^{k-\frac{1}{2}}$ in this region that is real valued and nonnegative on the interval $[-1, 1]$. We integrate the analytic function

$$(1-z^2)^{k-\frac{1}{2}}e^{irz}$$

over the boundary of the rectangle whose lower side is $[-1, 1]$ and whose height is $R > 0$. We obtain

$$i\int_0^R e^{ir(1+it)}(t^2-2it)^{k-\frac{1}{2}}\,dt + \int_{-1}^{+1} e^{irt}(1-t^2)^{k-\frac{1}{2}}\,dt$$
$$+\, i\int_R^0 e^{ir(-1+it)}(t^2+2it)^{k-\frac{1}{2}}\,dt + \varepsilon(R) = 0\,,$$

where $\varepsilon(R) \to 0$ as $R \to \infty$. It follows that

$$I = \int_{-1}^{+1} e^{irt}(1-t^2)^{k-\frac{1}{2}}\,dt = I_+ + I_-,$$

where

$$I_+ = +ie^{-ir}\int_0^{\infty} e^{-rt}(t^2+2it)^{k-\frac{1}{2}}dt\,, \qquad \text{and}$$

$$I_- = -ie^{ir}\int_0^{\infty} e^{-rt}(t^2-2it)^{k-\frac{1}{2}}\,dt\,.$$

Next we observe that

$$\begin{aligned} (t^2+2it)^{k-\frac{1}{2}} &= -\,i(2t)^{k-\frac{1}{2}}e^{i(\frac{k\pi}{2}+\frac{\pi}{4})} + \phi_+(t) \\ (t^2-2it)^{k-\frac{1}{2}} &= +\,i(2t)^{k-\frac{1}{2}}e^{-i(\frac{k\pi}{2}+\frac{\pi}{4})} + \phi_-(t)\,, \end{aligned}$$

where $|\phi_+(t)| + |\phi_-(t)| \le Ct^{k+\frac{1}{2}}$. Note that the Laplace transform

$$\mathcal{L}(f)(r) = \int_0^\infty f(t)e^{-tr}\, dt$$

of the function t^b is $r^{-b-1}\Gamma(b+1)$ when $b > -1/2$, and that the functions ϕ_+ and ϕ_- have Laplace transforms bounded by a constant multiple of $r^{-k-3/2}$. Therefore, we obtain

$$\begin{aligned} I_+ &= (-i)(+i)e^{-ir}e^{i(\frac{k\pi}{2}+\frac{\pi}{4})}r^{-k-\frac{1}{2}}2^{k-\frac{1}{2}}\Gamma(k+\tfrac{1}{2}) + O(r^{-k-3/2}) \\ I_- &= (+i)(-i)e^{ir}e^{-i(\frac{k\pi}{2}+\frac{\pi}{4})}r^{-k-\frac{1}{2}}2^{k-\frac{1}{2}}\Gamma(k+\tfrac{1}{2}) + O(r^{-k-3/2}) \end{aligned}$$

as $r \to \infty$. Adding these two last inequalities and multiplying by the missing factor $(r/2)^k/\Gamma(k+\frac{1}{2})\Gamma(\frac{1}{2})$, we obtain the equality

$$J_k(r) = \sqrt{\frac{2}{\pi r}}\cos(r - \frac{\pi k}{2} - \frac{\pi}{4}) + O(r^{-3/2})$$

as $r \to \infty$.

APPENDIX C

Rademacher Functions

C.1. Definition of the Rademacher Functions

The Rademacher functions are functions defined on $[0,1]$ as follows: $r_0(t) = 1$; $r_1(t) = 1$ for $0 \le t \le 1/2$ and $r_1(t) = -1$ for $1/2 < t \le 1$; $r_2(t) = 1$ for $0 \le t \le 1/4$, $r_2(t) = -1$ for $1/4 < t \le 1/2$, $r_2(t) = 1$ for $1/2 < t \le 3/4$, and $r_2(t) = -1$ for $3/4 < t \le 1$; and so on. According to this definition, we have that $r_j(t) = \operatorname{sgn}(\sin(2^j \pi t))$ for $j = 0, 1, 2, \dots$. It is easy to check that the r_j's are mutually independent random variables on $[0,1]$. This means that for all functions f_j we have

$$\int_0^1 \prod_{j=0}^n f_j(r_j(t))\, dt = \prod_{j=0}^n \int_0^1 f_j(r_j(t))\, dt\,.$$

To see the validity of this identity, we write its right-hand side as

$$\begin{aligned} f_0(1) \prod_{j=1}^n \int_0^1 f_j(r_j(t))\, dt &= f_0(1) \prod_{j=1}^n \frac{f_j(1) + f_j(-1)}{2} \\ &= \frac{f_0(1)}{2^n} \sum_{S \subset \{1,2,\dots,n\}} \prod_{j \in S} f_j(1) \prod_{j \notin S} f_j(-1) \end{aligned}$$

and we observe that there is a one-to-one and onto correspondence between subsets S of $\{1, 2, \dots, n\}$ and intervals $I_k = [\frac{k}{2^n}, \frac{k+1}{2^n}]$, $k = 0, 1, \dots, 2^n - 1$ such that the restriction of the function $\prod_{j=1}^n f_j(r_j(t))$ on I_k is equal to

$$\prod_{j \in S} f_j(1) \prod_{j \notin S} f_j(-1)\,.$$

It follows that the last of the three equal displayed expressions is

$$f_0(1) \sum_{k=0}^{2^n - 1} \int_{I_k} \prod_{j=1}^n f_j(r_j(t))\, dt = \int_0^1 \prod_{j=0}^n f_j(r_j(t))\, dt\,.$$

C.2. Khintchine's Inequalities

The following property of the Rademacher functions is of fundamental importance and with far-reaching consequences in analysis:

For any $0 < p < \infty$ and for any real-valued square summable sequences $\{a_j\}$ and $\{b_j\}$ we have

$$B_p\Big(\sum_j |a_j + ib_j|^2\Big)^{\frac{1}{2}} \le \Big\|\sum_j (a_j + ib_j)r_j\Big\|_{L^p([0,1])} \le A_p\Big(\sum_j |a_j + ib_j|^2\Big)^{\frac{1}{2}}$$

for some constants $0 < A_p, B_p < \infty$ that only depend on p.

These inequalities reflect the orthogonality of the Rademacher functions in L^p (especially when $p \neq 2$). Khintchine [**304**] was the first to prove a special form of this inequality, and he used it to estimate the asymptotic behavior of certain random walks. Later this inequality was systematically studied almost simultaneously by Littlewood [**339**] and by Paley and Zygmund [**400**], who proved the more general form stated previously. The foregoing inequalities are usually referred to by Khintchine's name.

C.3. Derivation of Khintchine's Inequalities

Both assertions in Appendix C.2 can be derived from an exponentially decaying distributional inequality for the function

$$F(t) = \sum_j (a_j + ib_j)r_j(t)\,, \qquad t \in [0,1],$$

when a_j, b_j are square summable real numbers.

We first obtain a distributional inequality for the above function F under the following three assumptions:

(a) The sequence $\{b_j\}$ is identically zero.
(b) All but finitely many terms of the sequence $\{a_j\}$ are zero.
(c) The sequence $\{a_j\}$ satisfies $(\sum_j |a_j|^2)^{1/2} = 1$.

Let $\rho > 0$. Under assumptions (a), (b), and (c), independence gives

$$\begin{aligned}\int_0^1 e^{\rho \sum a_j r_j(t)}\,dt &= \prod_j \int_0^1 e^{\rho a_j r_j(t)}\,dt \\ &= \prod_j \frac{e^{\rho a_j} + e^{-\rho a_j}}{2} \\ &\le \prod_j e^{\frac{1}{2}\rho^2 a_j^2} = e^{\frac{1}{2}\rho^2 \sum a_j^2} = e^{\frac{1}{2}\rho^2}\,,\end{aligned}$$

where we used the inequality $\frac{1}{2}(e^x + e^{-x}) \le e^{\frac{1}{2}x^2}$ for all real x, which can be checked using power series expansions. Since the same argument is also valid for $-\sum a_j r_j(t)$, we obtain that

$$\int_0^1 e^{\rho|F(t)|}\,dt \le 2e^{\frac{1}{2}\rho^2}.$$

From this it follows that

$$e^{\rho\alpha}|\{t\in[0,1]:\ |F(t)|>\alpha\}| \le \int_0^1 e^{\rho|F(t)|}\,dt \le 2e^{\frac{1}{2}\rho^2}$$

and hence we obtain the distributional inequality

$$d_F(\alpha) = |\{t\in[0,1]:\ |F(t)|>\alpha\}| \le 2e^{\frac{1}{2}\rho^2-\rho\alpha} = 2e^{-\frac{1}{2}\alpha^2},$$

by picking $\rho = \alpha$. The L^p norm of F can now be computed easily. Formula (1.1.6) gives

$$\big\|F\big\|_{L^p}^p = \int_0^\infty p\alpha^{p-1} d_F(\alpha)\,d\alpha \le \int_0^\infty p\alpha^{p-1} 2e^{-\frac{\alpha^2}{2}}\,d\alpha = 2^{\frac{p}{2}} p\,\Gamma(p/2).$$

We have now proved that

$$\big\|F\big\|_{L^p} \le \sqrt{2}\,(p\,\Gamma(p/2))^{\frac{1}{p}} \big\|F\big\|_{L^2}$$

under assumptions (a), (b), and (c).

We now dispose of assumptions (a), (b), and (c). Assumption (b) can be easily eliminated by a limiting argument and (c) by a scaling argument. To dispose of assumption (a), let a_j and b_j be real numbers. We have

$$\begin{aligned}
\Big\|\sum_j (a_j+ib_j)r_j\Big\|_{L^p} &\le \Big\|\big|\sum_j a_j r_j\big| + \big|\sum_j b_j r_j\big|\Big\|_{L^p} \\
&\le \Big\|\sum_j a_j r_j\Big\|_{L^p} + \Big\|\sum_j b_j r_j\Big\|_{L^p} \\
&\le \sqrt{2}\,(p\Gamma(p/2))^{\frac{1}{p}}\Big(\big(\sum_j |a_j|^2\big)^{\frac{1}{2}} + \big(\sum_j |b_j|^2\big)^{\frac{1}{2}}\Big) \\
&\le \sqrt{2}\,(p\Gamma(p/2))^{\frac{1}{p}}\sqrt{2}\,\Big(\sum_j |a_j+ib_j|^2\Big)^{\frac{1}{2}}.
\end{aligned}$$

Let us now set $A_p = 2\big(p\Gamma(p/2)\big)^{1/p}$ when $p>2$. Since we have the trivial estimate $\big\|F\big\|_{L^p} \le \big\|F\big\|_{L^2}$ when $0<p\le 2$, we obtain the required inequality $\big\|F\big\|_{L^p} \le A_p\big\|F\big\|_{L^2}$ with

$$A_p = \begin{cases} 1 & \text{when } 0<p\le 2, \\ 2\,p^{\frac{1}{p}}\,\Gamma(p/2)^{\frac{1}{p}} & \text{when } 2<p<\infty. \end{cases}$$

Using Sterling's formula in Appendix A.6, we see that A_p is asymptotic to $\sqrt{p}$ as $p \to \infty$.

We now discuss the converse inequality $B_p\big\|F\big\|_{L^2} \le \big\|F\big\|_{L^p}$. It is clear that $\big\|F\big\|_{L^2} \le \big\|F\big\|_{L^p}$ when $p \ge 2$ and we may therefore take $B_p = 1$ for $p \ge 2$. Let us now consider the case $0 < p < 2$. Pick an s such that $2 < s < \infty$. Find a $0 < \theta < 1$ such that

$$\frac{1}{2} = \frac{1-\theta}{p} + \frac{\theta}{s}.$$

Then

$$\big\|F\big\|_{L^2} \le \big\|F\big\|_{L^p}^{1-\theta}\big\|F\big\|_{L^s}^{\theta} \le \big\|F\big\|_{L^p}^{1-\theta} A_s^{\theta}\big\|F\big\|_{L^2}^{\theta}.$$

It follows that

$$\big\|F\big\|_{L^2} \le A_s^{\frac{\theta}{1-\theta}}\big\|F\big\|_{L^p}.$$

We have now proved the inequality $B_p\big\|F\big\|_{L^2} \le \big\|F\big\|_{L^p}$ with

$$B_p = \begin{cases} 1 & \text{when } 2 \le p < \infty, \\ \sup\limits_{s>2} A_s^{-\frac{\frac{1}{p}-\frac{1}{2}}{\frac{1}{2}-\frac{1}{s}}} & \text{when } 0 < p < 2. \end{cases}$$

Observe that the function $s \to A_s^{-(\frac{1}{p}-\frac{1}{2})/(\frac{1}{2}-\frac{1}{s})}$ tends to zero as $s \to 2+$ and as $s \to \infty$. Hence it must attain its maximum for some $s = s(p)$ in the interval $(2, \infty)$. We see that $B_p \ge 16 \cdot 256^{-1/p}$ when $p < 2$ by taking $s = 4$.

It is worthwhile to mention that the best possible values of the constants A_p and B_p in Khintchine's inequality are known exactly when the b_j's are zero. In this case Szarek [**498**] showed that the best possible value of B_1 is $1/\sqrt{2}$, and later Haagerup [**230**] found the best possible values of A_p and B_p for all $0 < p < \infty$. The best possible values of A_p and B_p when all $b_j = 0$ are the numbers

$$A_p = \begin{cases} 1 & \text{when } 0 < p \le 2, \\ 2^{\frac{1}{2}}\pi^{-\frac{1}{2p}}\Gamma(\frac{p+1}{2}) & \text{when } 2 < p < \infty, \end{cases}$$

and

$$B_p = \begin{cases} 2^{\frac{1}{2}-\frac{1}{p}} & \text{when } 0 < p \le p_0, \\ 2^{\frac{1}{2}}\pi^{-\frac{1}{2p}}\Gamma(\frac{p+1}{2}) & \text{when } p_0 < p < 2, \\ 1 & \text{when } 2 < p < \infty, \end{cases}$$

where $p_0 = 1.84742\ldots$ is the unique solution of the equation $2\Gamma(\frac{p+1}{2}) = \sqrt{\pi}$ in the interval $(1, 2)$.

C.4. Khintchine's Inequalities for Weak Type Spaces

We note that the following weak type estimates are valid:

$$4^{-\frac{1}{p}} B_{\frac{p}{2}} \Big(\sum_j |a_j + i b_j|^2 \Big)^{\frac{1}{2}} \le \Big\| \sum_j (a_j + i b_j) r_j \Big\|_{L^{p,\infty}} \le A_p \Big(\sum_j |a_j + i b_j|^2 \Big)^{\frac{1}{2}}$$

for all $0 < p < \infty$.

Indeed, the upper estimate is a simple consequence of the fact that L^p is a subspace of $L^{p,\infty}$. For the converse inequality we use the fact that $L^{p,\infty}([0,1])$ is contained in $L^{p/2}([0,1])$ and we have (see Exercise 1.1.11)

$$\big\| F \big\|_{L^{p/2}} \le 4^{\frac{1}{p}} \big\| F \big\|_{L^{p,\infty}} .$$

Since any Lorentz space $L^{p,q}([0,1])$ can be sandwiched between $L^{2p}([0,1])$ and $L^{p/2}([0,1])$, similar inequalities hold for all Lorentz spaces $L^{p,q}([0,1])$, $0 < p < \infty$, $0 < q \le \infty$.

C.5. Extension to Several Variables

We first extend the inequality on the right in Appendix C.2 to several variables. For a positive integer n we let

$$F_n(t_1, \dots, t_n) = \sum_{j_1} \cdots \sum_{j_n} c_{j_1,\dots,j_n} r_{j_1}(t_1) \dots r_{j_n}(t_n),$$

for $t_j \in [0,1]$, where $c_{j_1,\dots,j_n}$ is a sequence of complex numbers. F_n is a function defined on $[0,1]^n$.

For any $0 < p < \infty$ and for any complex-valued square summable sequence of n variables $\{c_{j_1,\dots,j_n}\}_{j_1,\dots,j_n}$, we have the following inequalities for F_n:

$$B_p^n \Big(\sum_{j_1} \cdots \sum_{j_n} |c_{j_1,\dots,j_n}|^2 \Big)^{\frac{1}{2}} \le \big\| F_n \big\|_{L^p} \le A_p^n \Big(\sum_{j_1} \cdots \sum_{j_n} |c_{j_1,\dots,j_n}|^2 \Big)^{\frac{1}{2}},$$

where A_p, B_p are the constants in Appendix C.2. The norms are over $[0,1]^n$.

The case $n = 2$ is indicative of the general case. For $p \geq 2$ we have

$$\begin{aligned}
&\int_0^1 \int_0^1 |F_2(t_1, t_2)|^p \, dt_1 \, dt_2 \\
&\leq A_p^p \int_0^1 \Big(\sum_{j_1} \Big| \sum_{j_2} c_{j_1, j_2} r_{j_2}(t_2) \Big|^2 \Big)^{\frac{p}{2}} \, dt_2 \\
&\leq A_p^p \Big(\sum_{j_1} \Big(\int_0^1 \Big| \sum_{j_2} c_{j_1, j_2} r_{j_2}(t_2) \Big|^p \, dt_2 \Big)^{\frac{2}{p}} \Big)^{\frac{p}{2}} \\
&\leq A_p^{2p} \Big(\sum_{j_1} \sum_{j_2} |c_{j_1, j_n}|^2 \Big)^{\frac{p}{2}} ,
\end{aligned}$$

where we used Minkowski's integral inequality (with exponent $p/2 \geq 1$) in the second inequality and the result in the case $n = 1$ twice.

The case $p < 2$ follows trivially from Hölder's inequality with constant $A_p = 1$. The reverse inequalities follow exactly as in the case of one variable. Replacing A_p by A_p^n in the argument, giving the reverse inequality in the case $n = 1$, we obtain the constant B_p^n.

Likewise we can extend the weak type inequalities of Appendix C.3 in several variables.

APPENDIX D

Spherical Coordinates

D.1. Spherical Coordinate Formula

The following formula switches integration from spherical coordinates to Cartesian:

$$\int_{R\mathbf{S}^{n-1}} f(x)\,d\sigma(x) = \int_{\varphi_1=0}^{\pi} \cdots \int_{\varphi_{n-2}=0}^{\pi} \int_{\varphi_{n-1}=0}^{2\pi} f(x(\varphi))J(n,R,\varphi)\,d\varphi_{n-1}\dots d\varphi_1,$$

where

$$\begin{aligned}
x_1 &= R\cos\varphi_1\\
x_2 &= R\sin\varphi_1\,\cos\varphi_2\\
x_3 &= R\sin\varphi_1\,\sin\varphi_2\,\cos\varphi_3\\
&\dots\\
x_{n-1} &= R\sin\varphi_1\,\sin\varphi_2\,\sin\varphi_3\dots\sin\varphi_{n-2}\,\cos\varphi_{n-1}\\
x_n &= R\sin\varphi_1\,\sin\varphi_2\,\sin\varphi_3\dots\sin\varphi_{n-2}\,\sin\varphi_{n-1}\,,
\end{aligned}$$

and $0 \le \varphi_1, \dots \varphi_{n-2} \le \pi$, $0 \le \varphi_{n-1} = \theta \le 2\pi$,

$$x(\varphi) = (x_1(\varphi_1,\dots,\varphi_{n-1}),\dots,x_n(\varphi_1,\dots,\varphi_{n-1}))\,,$$

and

$$J(n,R,\varphi) = R^{n-1}(\sin\varphi_1)^{n-2}\dots(\sin\varphi_{n-3})^2(\sin\varphi_{n-2})$$

is the Jacobian of the transformation. We denote by φ' the vector $(\varphi_2,\dots,\varphi_{n-1})$ so that $\varphi = (\varphi_1,\varphi')$.

D.2. A Useful Change of Variables Formula

The following formula is useful in computing integrals over the sphere $\mathbf{S}^{n-1}$ when $n \ge 2$. Let f be a function of n variables. Then

$$\int_{R\,\mathbf{S}^{n-1}} f(x)\,d\sigma(x) = \int_{-R}^{+R}\left\{\int_{|\theta|=\sqrt{R^2-s^2}} f(s,\theta)\,d\theta\right\}\frac{R\,ds}{\sqrt{R^2-s^2}}\,,$$

where the inner integral is over the sphere $\sqrt{R^2-s^2}\,\mathbf{S}^{n-2}$. To prove this formula, let $x'=(x_2,\dots,x_n)$. Then the change of variables $s=R\cos\varphi_1$ (which gives $ds=-\sin\varphi_1\,d\varphi_1$ and $\sqrt{R^2-s^2}=R\sin\varphi_1$) transforms the last identity to

$$\int_{R\mathbf{S}^{n-1}} f(x)\,d\sigma(x)=\int_0^{\pi}\left\{\int_{|\theta|=R\cos\varphi_1} f(R\cos\varphi_1,\theta)\,d\theta\right\}d\varphi_1 .$$

The validity of the last identity follows by applying the formula in Appendix D.1 on the left- and right-hand sides and using that

$$x(\varphi)=\big(R\cos\varphi_1,(R\cos\varphi_1)x'(\varphi')\big).$$

D.3. Computation of an Integral over the Sphere

Let K be a function on the line. We will use the result in the Appendix D.2 to show that for $n\ge 2$ we have

$$\int_{\mathbf{S}^{n-1}} K(x\cdot\theta)\,d\theta=\frac{2\pi^{\frac{n-1}{2}}}{\Gamma\big(\frac{n-1}{2}\big)}\int_{-1}^{+1} K(s|x|)(\sqrt{1-s^2}\,)^{n-3}\,ds$$

when $x\in\mathbf{R}^n\setminus\{0\}$. Let $x'=x/|x|$ and pick a matrix $A\in O(n)$ such that $Ae_1=x'$, where $e_1=(1,0,\dots,0)$. We have

$$\begin{aligned}
&\int_{\mathbf{S}^{n-1}} K(x\cdot\theta)\,d\theta=\int_{\mathbf{S}^{n-1}} K(|x|(x'\cdot\theta))\,d\theta\\
&=\int_{\mathbf{S}^{n-1}} K(|x|(Ae_1\cdot\theta))\,d\theta=\int_{\mathbf{S}^{n-1}} K(|x|(e_1\cdot A^{-1}\theta))\,d\theta\\
&=\int_{\mathbf{S}^{n-1}} K(|x|\theta_1)\,d\theta=\int_{-1}^{+1} K(|x|s)\omega_{n-2}(\sqrt{1-s^2}\,)^{n-2}\frac{ds}{\sqrt{1-s^2}}\\
&=\omega_{n-2}\int_{-1}^{+1} K(s|x|)(\sqrt{1-s^2}\,)^{n-3}\,ds\,,
\end{aligned}$$

where $\omega_{n-2}=2\pi^{\frac{n-1}{2}}\Gamma\big(\frac{n-1}{2}\big)^{-1}$ is the surface area of $\mathbf{S}^{n-2}$.

For example, we have

$$\int_{\mathbf{S}^{n-1}}\frac{d\theta}{|\xi\cdot\theta|^{\alpha}}=\omega_{n-2}\int_{-1}^{+1}\frac{1}{|s|^{\alpha}|\xi|^{\alpha}}(1-s^2)^{\frac{n-3}{2}}\,ds=\frac{1}{|\xi|^{\alpha}}\frac{2\pi^{\frac{n-1}{2}}\Gamma\big(\frac{1-\alpha}{2}\big)}{\Gamma\big(\frac{n-\alpha}{2}\big)},$$

and the integral converges only when $\operatorname{Re}\alpha<1$.

D.4. The Computation of Another Integral over the Sphere

We compute the following integral for $n \geq 2$:

$$\int_{\mathbf{S}^{n-1}} \frac{d\theta}{|\theta - e_1|^\alpha},$$

where $e_1 = (1, 0, \dots, 0)$. Applying the formula in Appendix D.2, we obtain

$$\begin{aligned}
\int_{\mathbf{S}^{n-1}} \frac{d\theta}{|\theta - e_1|^\alpha} &= \int_{-1}^{+1} \int_{\theta \in \sqrt{1-s^2}\,\mathbf{S}^{n-2}} \frac{d\theta}{(|s-1|^2 + |\theta|^2)^{\frac{\alpha}{2}}} \frac{ds}{\sqrt{1-s^2}} \\
&= \int_{-1}^{+1} \omega_{n-2} \frac{(1-s^2)^{\frac{n-2}{2}}}{((1-s)^2 + 1 - s^2)^{\frac{\alpha}{2}}} \frac{ds}{\sqrt{1-s^2}} \\
&= \frac{\omega_{n-2}}{2^{\frac{\alpha}{2}}} \int_{-1}^{+1} \frac{(1-s^2)^{\frac{n-3}{2}}}{(1-s)^{\frac{\alpha}{2}}} ds \\
&= \frac{\omega_{n-2}}{2^{\frac{\alpha}{2}}} \int_{-1}^{+1} (1-s)^{\frac{n-3-\alpha}{2}} (1+s)^{\frac{n-3}{2}} ds,
\end{aligned}$$

which converges exactly when $\operatorname{Re}\alpha < n - 1$.

D.5. Integration over a General Surface

Suppose that S is a hypersurface in $\mathbf{R}^n$ of the form

$$S = \{(u, \Phi(u)) : \; u \in D\},$$

where D is an open subset of $\mathbf{R}^{n-1}$ and Φ is a continuously differentiable mapping from D into $\mathbf{R}$. Let σ be the canonical surface measure on S. If g is a function on S, then we have

$$\int_S g(y)\, d\sigma(y) = \int_D g(x, \Phi(x)) \Big(1 + \sum_{j=1}^n |\partial_j \Phi(x)|^2\Big)^{\frac{1}{2}} dx.$$

Specializing to the sphere we obtain

$$\int_{\mathbf{S}^{n-1}} g(\theta)\, d\theta = \int_{\substack{\xi' \in \mathbf{R}^{n-1} \\ |\xi'| < 1}} \Big[g(\xi', \sqrt{1 - |\xi'|^2}) + g(\xi', -\sqrt{1 - |\xi'|^2})\Big] \frac{d\xi'}{\sqrt{1 - |\xi'|^2}}.$$

D.6. The Stereographic Projection

Define a map $\Pi : \mathbf{R}^n \to \mathbf{S}^n$ by the formula

$$\Pi(x_1, \dots, x_n) = \left(\frac{2x_1}{1+|x|^2}, \dots, \frac{2x_n}{1+|x|^2}, \frac{|x|^2-1}{1+|x|^2}\right).$$

It is easy to see that Π is a one-to-one map from $\mathbf{R}^n$ onto the sphere $\mathbf{S}^n$ minus the north pole $e_{n+1} = (0, \dots, 0, 1)$. Its inverse is given by the formula

$$\Pi^{-1}(\theta_1, \dots, \theta_{n+1}) = \left(\frac{\theta_1}{1-\theta_{n+1}}, \dots, \frac{\theta_n}{1-\theta_{n+1}}\right).$$

The Jacobian of the map is verified to be

$$J_\Pi(x) = \left(\frac{2}{1+|x|^2}\right)^n,$$

and the following change of variables formulas are valid:

$$\int_{\mathbf{S}^n} F(\theta)\, d\theta = \int_{\mathbf{R}^n} F(\Pi(x)) J_\Pi(x)\, dx$$

and

$$\int_{\mathbf{R}^n} F(x)\, dx = \int_{\mathbf{S}^n} F(\Pi^{-1}(\theta))\, J_{\Pi^{-1}}(\theta)\, d\theta\,,$$

where

$$J_{\Pi^{-1}}(\theta) = \frac{1}{J_\Pi(\Pi^{-1}(\theta))} = \left(\frac{|\theta_1|^2 + \dots + |\theta_n|^2 + |1-\theta_{n+1}|^2}{2|1-\theta_{n+1}|^2}\right)^n.$$

Another interesting formula about the stereographic projection Π is

$$|\Pi(x) - \Pi(y)| = 2|x-y|(1+|x|^2)^{-1/2}(1+|y|^2)^{-1/2},$$

for all x, y in $\mathbf{R}^n$.

APPENDIX E

Some Trigonometric Identities and Inequalities

The following inequalities are valid for t real:

$$\begin{aligned}
0 < t < \frac{\pi}{2} &\implies \sin(t) < t < \tan(t) \\
0 < |t| < \frac{\pi}{2} &\implies \frac{2}{\pi} < \frac{\sin(t)}{t} < 1 \\
-\infty < t < +\infty &\implies |\sin(t)| \le |t| \\
-\infty < t < +\infty &\implies |1 - \cos(t)| \le \frac{|t|^2}{2} \\
-\infty < t < +\infty &\implies |1 - e^{it}| \le |t| \\
|t| \le \frac{\pi}{2} &\implies |\sin(t)| \ge \frac{2|t|}{\pi} \\
|t| \le \pi &\implies |1 - \cos(t)| \ge \frac{2|t|^2}{\pi^2} \\
|t| \le \pi &\implies |1 - e^{it}| \ge \frac{2|t|}{\pi}.
\end{aligned}$$

The following sum to product formulas are valid:

$$\begin{aligned}
\sin(a) + \sin(b) &= 2\,\sin\left(\frac{a+b}{2}\right)\,\cos\left(\frac{a-b}{2}\right) \\
\sin(a) - \sin(b) &= 2\,\cos\left(\frac{a+b}{2}\right)\,\sin\left(\frac{a-b}{2}\right) \\
\cos(a) + \cos(b) &= 2\,\cos\left(\frac{a+b}{2}\right)\,\cos\left(\frac{a-b}{2}\right) \\
\cos(a) - \cos(b) &= -\,2\,\sin\left(\frac{a+b}{2}\right)\,\sin\left(\frac{a-b}{2}\right).
\end{aligned}$$

The following identities are also easily proved

$$\begin{aligned}
\sum_{k=1}^{N} \cos(kx) &= -\frac{1}{2} + \frac{\sin((N+\frac{1}{2})\,x)}{2\sin(\frac{x}{2})} \\
\sum_{k=1}^{N} \sin(kx) &= \frac{\cos(\frac{x}{2}) - \cos((N+\frac{1}{2})\,x)}{2\sin(\frac{x}{2})}.
\end{aligned}$$

APPENDIX F

Summation by Parts

Let $\{a_k\}_{k=0}^{\infty}$, $\{b_k\}_{k=0}^{\infty}$ be two sequences of complex numbers. Then for $N \geq 1$ we have

$$\sum_{k=0}^{N} a_k b_k = A_N b_N - \sum_{k=0}^{N-1} A_k (b_{k+1} - b_k),$$

where

$$A_k = \sum_{j=0}^{k} a_j \, .$$

More generally we have

$$\sum_{k=M}^{N} a_k b_k = A_N b_N - A_{M-1} b_M - \sum_{k=M}^{N-1} A_k (b_{k+1} - b_k) \, ,$$

whenever $0 \leq M \leq N$, where $A_{-1} = 0$ and

$$A_k = \sum_{j=0}^{k} a_j$$

for $k \geq 0$.

APPENDIX G

Basic Functional Analysis

A quasi-norm is a nonnegative functional $\|\cdot\|$ on a vector space X that satisfies $\|x+y\| \le K(\|x\|+\|y\|)$ for some $K \ge 0$ and all $x, y \in X$ and also $\|\lambda x\| = |\lambda|\|x\|$ for all scalars λ. When $K = 1$, the quasi-norm is called a norm. A quasi-Banach space is a quasi-normed space that is complete with respect to the topology generated by the quasi-norm.

Open Mapping Theorem. Suppose that X and Y are quasi-normed spaces and T is a bounded surjective linear map from X into Y. Then there exists a constant $K < \infty$ such that

$$\|x\| \le K\|T(x)\|$$

for all $x \in X$.

Closed Graph Theorem. Suppose that X and Y are quasi-normed spaces and T is a linear map from X into Y that satisfies the following: whenever $x_k, x \in X$ and $x_k \to x$ in X then $(x_k, T(x_k)) \to (x, T(x))$ in $X \times Y$. Then T is bounded.

Uniform Boundedness Principle. Suppose that X and Y are quasi-normed spaces and $(T_\alpha)_{\alpha \in I}$ is a family of linear maps from X into Y such that for all $x \in X$ there exists a $C_x < \infty$ so that

$$\sup_{\alpha \in I} \|T_\alpha(x)\| \le C_x .$$

Then there exists a constant $K < \infty$ such that

$$\sup_{\alpha \in I} \|T_\alpha\|_{X \to Y} \le K .$$

The Hahn-Banach Theorem. Let X be a normed space and X_0 a subspace. Every bounded linear functional Λ_0 on X_0 has a bounded extension on X with the same norm. (Note: This theorem may fail for quasi-normed vector spaces.)

APPENDIX H

The Minimax Lemma

Minimax type results are used in the theory of games and have their origin in the work of Von Neumann [**543**]. Much of the theory in this subject is based on convex analysis techniques. For instance, this is the case with the next proposition, which is needed in the "difficult" inequality in the proof of the minimax lemma. We refer the reader to Fan [**181**] for a general account of minimax results. The following exposition is based on the simple presentation in Appendix A2 of [**204**].

Minimax Lemma. *Let A, B be convex subsets of certain vector spaces. Assume that a topology is defined in B for which it is a compact Hausdorff space and assume that there is a function $\Phi : A \times B \to \mathbf{R} \cup \{+\infty\}$ that satisfies the following:*

(a) $\Phi(\,.\,, b)$ is a concave function on A for each $b \in B$,
(b) $\Phi(a, \,.\,)$ is a convex function on B for each $a \in A$,
(c) $\Phi(a, \,.\,)$ is lower semicontinuous on B for each $a \in A$.

Then the following identity holds:

$$\min_{b \in B} \sup_{a \in A} \Phi(a, b) = \sup_{a \in A} \min_{b \in B} \Phi(a, b)\,.$$

To prove the lemma we will need the following proposition:

Proposition. *Let B be a convex compact subset of a vector space and suppose that $g_j : B \to \mathbf{R} \cup \{+\infty\}$, $j = 1, 2, \ldots, n$ are convex and lower semicontinuous. If*

$$\max_{1 \le j \le n} g_j(b) > 0 \quad \textit{for all} \quad b \in B\,,$$

then there exist nonnegative numbers $\lambda_1, \lambda_2, \ldots, \lambda_n$ such that

$$\lambda_1 g_1(b) + \lambda_2 g_2(b) + \cdots + \lambda_n g_n(b) > 0 \quad \textit{for all} \quad b \in B\,.$$

Proof. We first consider the case $n = 2$. Define subsets of B

$$B_1 = \{b \in B : \; g_1(b) \le 0\}, \quad B_2 = \{b \in B : \; g_2(b) \le 0\}\,.$$

If $B_1 = \emptyset$, we take $\lambda_1 = 1$ and $\lambda_2 = 0$ and we similarly deal with the case $B_2 = \emptyset$. If B_1 and B_2 are nonempty, then they are closed and thus compact. The hypothesis of the proposition implies that $g_2(b) > 0 \ge g_1(b)$ for all $b \in B_1$. Therefore, the function

$-g_1(b)/g_2(b)$ is well defined and upper semicontinuous on B_1 and thus attains its maximum. The same is true for $-g_2(b)/g_1(b)$ defined on B_2. We set

$$\mu_1 = \max_{b \in B_1} \frac{-g_1(b)}{g_2(b)} \geq 0\,, \qquad \mu_2 = \max_{b \in B_2} \frac{-g_2(b)}{g_1(b)} \geq 0\,.$$

We need to find $\lambda > 0$ such that $\lambda g_1(b) + g_2(b) > 0$ for all $b \in B$. This is clearly satisfied if $b \notin B_1 \cup B_2$, while for $b_1 \in B_1$ and $b_2 \in B_2$ we have

$$\begin{aligned} \lambda g_1(b_1) + g_2(b_1) &\geq (1 - \lambda\mu_1)g_2(b_1) \\ \lambda g_1(b_2) + g_2(b_2) &\geq (\lambda - \mu_2)g_1(b_2) \end{aligned}$$

Therefore, it suffices to find a $\lambda > 0$ such that $1 - \lambda\mu_1 > 0$ and $\lambda - \mu_2 > 0$. Such a λ exists if and only if $\mu_1\mu_2 < 1$. To prove that $\mu_1\mu_2 < 1$, we can assume that $\mu_1 \neq 0$ and $\mu_2 \neq 0$. Then we take $b_1 \in B_1$ and $b_2 \in B_2$ for which the maxima μ_1 and μ_2 are attained, respectively. Then we have

$$\begin{aligned} g_1(b_1) + \mu_1 g_2(b_1) &= 0 \\ g_1(b_2) + \frac{1}{\mu_2} g_2(b_2) &= 0\,. \end{aligned}$$

But $g_1(b_1) < 0 < g_1(b_2)$; thus taking $b_\theta = \theta b_1 + (1-\theta)b_2$ for some θ in $(0,1)$, we have

$$g_1(b_\theta) \leq \theta g_1(b_1) + (1-\theta)g_1(b_2) = 0\,.$$

Considering the same convex combination of last displayed equations and using this identity, we obtain that

$$\mu_1\mu_2\theta g_2(b_1) + (1-\theta)g_2(b_2) = 0\,.$$

The hypothesis of the proposition implies that $g_2(b_\theta) > 0$ and the convexity of g_2:

$$\theta g_2(b_1) + (1-\theta)g_2(b_2) > 0\,.$$

Since $g_2(b_1) > 0$, we must have $\mu_1\mu_2 g_2(b_1) < g_2(b_1)$, which gives $\mu_1\mu_2 < 1$. This proves the required claim and completes the case $n = 2$.

We will now use induction to prove the proposition for arbitrary n. Assume that the result has been proved for $n-1$ functions. Consider the subset of B:

$$B_n = \{b \in B : \; g_n(b) \leq 0\}\,.$$

If $B_n = \emptyset$, we choose $\lambda_1 = \lambda_2 = \ldots = \lambda_{n-1} = 0$ and $\lambda_n = 1$. If B_n is not empty, then it is compact and convex and we can restrict $g_1, g_2, \ldots, g_{n-1}$ to B_n. Using the induction hypothesis, we can find $\lambda_1, \lambda_2, \ldots, \lambda_{n-1} \geq 0$ such that

$$g_0(b) = \lambda_1 g_1(b) + \lambda_2 g_2(b) + \ldots + \lambda_{n-1} g_{n-1}(b) > 0$$

for all $b \in B_n$. Then g_0 and g_n are convex lower semicontinuous functions on B, and $\max(g_0(b), g_n(b)) > 0$ for all $b \in B$. Using the case $n = 2$, which was first proved, we can find $\lambda_0, \lambda_n \geq 0$ such that for all $b \in B$ we have

$$\begin{aligned} 0 &< \lambda_0 g_0(b) + \lambda_n g_n(b) \\ &= \lambda_0\lambda_1 g_1(b) + \lambda_0\lambda_2 g_2(b) + \ldots + \lambda_0\lambda_{n-1} g_{n-1}(b) + \lambda_n g_n(b). \end{aligned}$$

This establishes the case of n functions and concludes the proof of the induction and hence of the proposition.

□

We now turn to the proof of the minimax lemma.

Proof. The fact that the left-hand side in the required conclusion of the minimax lemma is at least as big as the right-hand side is obvious. We can therefore concentrate on the converse inequality. In doing this we may assume that the right-hand side is finite. Without loss of generality we can subtract a finite constant from $\Phi(a, b)$, and so we can also assume that

$$\sup_{a \in A} \min_{b \in B} \Phi(a, b) = 0\,.$$

Then, by hypothesis *(c)* of the minimax lemma, the subsets

$$B_a = \{b \in B : \ \Phi(a, b) \leq 0\}, \qquad\qquad a \in A$$

of B are closed and nonempty, and we will see that they verify the finite intersection property. Indeed, suppose that

$$B_{a_1} \cap B_{a_2} \cap \ldots \cap B_{a_n} = \emptyset$$

for some $a_1, a_2, \ldots, a_n \in A$. We write $g_j(b) = \Phi(a_j, b)$, $j = 1, 2, \ldots, n$, and we observe that the conditions of the previous proposition are satisfied. Therefore we can find $\lambda_1, \lambda_2, \ldots, \lambda_n \geq 0$ such that for all $b \in B$ we have

$$\lambda_1\Phi(a_1, b) + \lambda_2\Phi(a_2, b) + \ldots + \lambda_n\Phi(a_n, b) > 0\,.$$

For simplicity we normalize the λ_j's by setting $\lambda_1 + \lambda_2 + \ldots + \lambda_n = 1$. If we set $a_0 = \lambda_1 a_1 + \lambda_2 a_2 + \ldots + \lambda_n a_n$, the concavity hypothesis (a) gives

$$\Phi(a_0, b) > 0$$

for all $b \in B$ contradicting the fact that $\sup_{a \in A} \min_{b \in B} \Phi(a, b) = 0$. Therefore, the family of closed subsets $\{B_a\}_{a \in A}$ of B satisfies the finite intersection property. The compactness of B now implies $\bigcap_{a \in A} B_a \neq \emptyset$. Take $b_0 \in \bigcap_{a \in A} B_a$. Then $\Phi(a, b_0) \leq 0$ for every $a \in A$, and therefore

$$\min_{b \in B} \sup_{a \in A} \Phi(a, b) \leq \sup_{a \in A} \Phi(a, b_0) \leq 0$$

as required.

□

APPENDIX I

The Schur Lemma

Schur's lemma provides sufficient conditions for linear operators to be bounded on L^p. Moreover, for positive operators it provides necessary and sufficient such conditions. We discuss these situations.

I.1. The Classical Schur Lemma

We begin with an easy situation. Suppose that $K(x, y)$ is a locally integrable function on a product of two σ-finite measure spaces $(X, \mu) \times (Y, \nu)$, and let T be a linear operator given by

$$T(f)(x) = \int_Y K(x, y) f(y) \, d\nu(y)$$

when f is bounded and compactly supported. It is a simple consequence of Fubini's theorem that for almost all $x \in X$ the integral defining T converges absolutely. The following lemma provides a sufficient criterion for the L^p boundedness of T.

Lemma. *Suppose that a locally integrable function $K(x, y)$ satisfies*

$$\sup_{x \in X} \int_Y |K(x, y)| \, d\nu(y) = A < \infty$$

$$\sup_{y \in Y} \int_X |K(x, y)| \, d\mu(x) = B < \infty.$$

Then the operator T extends to a bounded operator from $L^p(Y)$ into $L^p(X)$ with norm $A^{1-\frac{1}{p}} B^{\frac{1}{p}}$ for $1 \le p \le \infty$.

Proof. The second condition gives that T maps L^1 into L^1 with bound B while the first condition gives that T maps L^∞ into L^∞ with bound A. It follows by the Riesz-Thorin interpolation theorem that T maps L^p into L^p with bound $A^{1-\frac{1}{p}} B^{\frac{1}{p}}$. □

This lemma can be improved significantly when the operators are assumed to be positive.

I.2. Schur's Lemma for Positive Operators

We have the following necessary and sufficient condition for the L^p boundedness of positive operators.

Lemma. *Let (X,μ) and (Y,ν) be two σ-finite measure spaces, where μ and ν are positive measures, and suppose that $K(x,y)$ is a nonnegative measurable function on $X\times Y$. Let $1<p<\infty$ and $0<A<\infty$. Let T be the linear operator*

$$T(f)(x)=\int_Y K(x,y)f(y)\,d\nu(y)$$

and T^t be its transpose operator

$$T^t(g)(y)=\int_X K(x,y)g(x)\,d\mu(x)\,.$$

To avoid trivialities, we assume that there is a compactly supported, bounded, and positive ν-a.e. function h_1 on Y such that $T(h_1)>0$ μ-a.e. Then the following are equivalent:

(i) T maps $L^p(Y)$ into $L^p(X)$ with norm at most A.
(ii) For all $B>A$ there is a measurable function h on Y that satisfies $0<h<\infty$ ν-a.e., $0<T(h)<\infty$ μ-a.e., and such that

$$T^t\big(T(h)^{\frac{p}{p'}}\big)\le B^p\,h^{\frac{p}{p'}}.$$

(iii) For all $B>A$ there are measurable functions u on X and v on Y such that $0<u<\infty$ μ-a.e., $0<v<\infty$ ν-a.e., and such that

$$T(u^{p'})\;\le\;B\,v^{p'},$$
$$T^t(v^p)\;\le\;B\,u^p.$$

Proof. First we assume (ii) and we prove (iii). Define u,v by the equations $v^{p'}=B\,T(h)$ and $u^{p'}=Bh$ and observe that (iii) holds for this choice of u and v. Moreover, observe that $0<u,v<\infty$ a.e. with respect to the measures μ and ν, respectively.

Next we assume (iii) and we prove (i). For g in $L^{p'}(X)$ we have

$$\int_X T(f)(x)\,g(x)\,d\nu(x)=\int_X\int_Y K(x,y)f(y)g(x)\frac{v(x)}{u(y)}\frac{u(y)}{v(x)}\,d\nu(x)\,d\mu(y).$$

We now apply Hölder's inequality with exponents p and p' to the functions

$$f(y)\frac{v(x)}{u(y)}\qquad\text{and}\qquad g(x)\frac{u(y)}{v(x)}$$

with the respect to the measure $K(x,y)\,d\nu(x)\,d\mu(y)$ on $X\times Y$. Since

$$\left(\int_X\int_Y f(y)^p\frac{v(x)^p}{u(y)^p}K(x,y)\,d\nu(x)\,d\mu(y)\right)^{\frac{1}{p}}\le B^{\frac{1}{p}}\big\|f\big\|_{L^p(Y)}$$

and

$$\left(\int_X \int_Y g(x)^{p'} \frac{u(y)^{p'}}{v(x)^{p'}} K(x,y)\, d\mu(y)\, d\nu(x)\right)^{\frac{1}{p'}} \leq B^{\frac{1}{p'}} \|g\|_{L^{p'}(X)},$$

we conclude that

$$\left|\int_X T(f)(x) g(x)\, d\nu(x)\right| \leq B^{\frac{1}{p}+\frac{1}{p'}} \|f\|_{L^p(Y)} \|g\|_{L^{p'}(X)}.$$

Taking the supremum over all g with $L^{p'}(X)$ norm 1, we obtain

$$\|T(f)\|_{L^p(X)} \leq B\|f\|_{L^p(Y)}.$$

Since B was any number greater than A, we conclude that

$$\|T\|_{L^p(Y)\to L^p(X)} \leq A$$

which proves (i).

We finally assume (i) and we prove (ii). Without loss of generality, take here $A = 1$ and $B > 1$. Define a map $S: L^p(Y) \to L^p(Y)$ by setting

$$S(f)(y) = \left(T^t(T(f)^{\frac{p}{p'}})\right)^{\frac{p'}{p}}(y).$$

We observe two things. First, $f_1 \leq f_2$ implies $S(f_1) \leq S(f_2)$, which is an easy consequence of the fact that the same monotonicity is valid for T. Next, we observe that $\|f\|_{L^p} \leq 1$ implies that $\|S(f)\|_{L^p} \leq 1$ as a consequence of the boundedness of T on L^p (with norm at most 1).

Construct a sequence h_n, $n = 1, 2, \ldots$ by induction as follows. Pick $h_1 > 0$ on Y as in the hypothesis of the theorem such that $T(h_1) > 0$ μ-a.e. and such that $\|h_1\|_{L^p} \leq B^{-p'}(B^{p'} - 1)$. (The last condition can be obtained by multiplying h_1 by a small constant.) Assuming that h_n has been defined, we define

$$h_{n+1} = h_1 + \frac{1}{B^{p'}} S(h_n).$$

We check easily by induction that we have the monotonicity property $h_n \leq h_{n+1}$ and the fact that $\|h_n\|_{L^p} \leq 1$. We now define

$$h(x) = \sup_n h_n(x) = \lim_{n\to\infty} h_n(x).$$

Fatou's lemma gives that $\|h\|_{L^p} \leq 1$, from which it follows that $h < \infty$ ν-a.e. Since $h \geq h_1 > 0$ ν-a.e., we also obtain that $h > 0$ ν-a.e.

Next we use the Lebesgue dominated convergence theorem to obtain that $h_n \to h$ in $L^p(Y)$. Since T is bounded on L^p, it follows that $T(h_n) \to T(h)$ in $L^p(X)$. It follows that $T(h_n)^{\frac{p}{p'}} \to T(h)^{\frac{p}{p'}}$ in $L^{p'}(X)$. Our hypothesis gives that T^t maps $L^{p'}(X)$ into $L^{p'}(Y)$ with norm at most 1. It follows $T^t(T(h_n)^{\frac{p}{p'}}) \to T^t(T(h)^{\frac{p}{p'}})$ in $L^{p'}(Y)$. Raising to the power $\frac{p'}{p}$, we obtain that $S(h_n) \to S(h)$ in $L^p(Y)$.

It follows that for some subsequence n_k of the integers we have $S(h_{n_k}) \to S(h)$ a.e. in Y. Since the sequence $S(h_n)$ is increasing we conclude that the entire sequence $S(h_n)$ converges almost everywhere to $S(h)$. We use this information in conjunction with $h_{n+1} = h_1 + \frac{1}{B^{p'}} S(h_n)$. Indeed, letting $n \to \infty$ in this identity, we obtain

$$h = h_1 + \frac{1}{B^{p'}} S(h)\,.$$

Since $h_1 > 0$ ν-a.e. it follows that $S(h) \le B^{p'} h$ ν-a.e., which proves the required estimate in (ii).

It remains to prove that $0 < T(h) < \infty$ μ-a.e. Since $\|h\|_{L^p} \le 1$ and T is L^p bounded, it follows that $\|T(h)\|_{L^p} \le 1$, which implies that $T(h) < \infty$ μ-a.e. We also have $T(h) \ge T(h_1) > 0$ μ-a.e.

□

I.3. An Example

Consider the Hilbert operator

$$T(f)(x) = \int_0^\infty \frac{f(y)}{x+y}\, dy\,,$$

where $x \in (0, \infty)$. T takes measurable functions on $(0, \infty)$ to measurable functions on $(0, \infty)$. We claim that T maps $L^p(0, \infty)$ into itself for $1 < p < \infty$; precisely, we have the estimate

$$\int_0^\infty T(f)(x)\, g(x)\, dx \le \frac{\pi}{\sin(\pi/p)} \|f\|_{L^p(0,\infty)} \|g\|_{L^{p'}(0,\infty)}\,.$$

To see this we use Schur's lemma. We take

$$u(x) = v(x) = x^{-\frac{1}{pp'}}\,.$$

We have that

$$T(u^{p'})(x) = \int_0^\infty \frac{y^{-\frac{1}{p}}}{x+y}\, dy = x^{-\frac{1}{p}} \int_0^\infty \frac{t^{-\frac{1}{p}}}{1+t}\, dt = B(\tfrac{1}{p'}, \tfrac{1}{p})\, v(x)^{\frac{1}{p'}}\,,$$

where B is the usual beta function and the last identity follows from the change of variables $s = (1+t)^{-1}$. Now an easy calculation yields

$$B(\tfrac{1}{p'}, \tfrac{1}{p}) = \frac{\pi}{\sin(\pi/p)}\,,$$

so the lemma in Appendix H.2 gives that $\|T\|_{L^p \to L^p} \le \frac{\pi}{\sin(\pi/p)}$. The sharpness of this constant follows by considering the sequence of functions

$$h_\varepsilon(x) = \begin{cases} x^{-\frac{1}{p}+\varepsilon} & \text{when } x < 1, \\ x^{-\frac{1}{p}-\varepsilon} & \text{when } x \ge 1, \end{cases}$$

which satisfies

$$\lim_{\varepsilon\to 0} \frac{\big\|T(h_\varepsilon)\big\|_{L^p(0,\infty)}}{\big\|h_\varepsilon\big\|_{L^p(0,\infty)}} = \frac{\pi}{\sin(\pi/p)}\,.$$

We make some comments related to the history of Schur's lemma. Schur [**450**] first proved a matrix version of the lemma in Appendix H.1 when $p = 2$. Precisely, Schur's original version was the following: If $K(x, y)$ is a positive decreasing function in both variables and satisfies

$$\sup_m \sum_n K(m,n) + \sup_n \sum_m K(m,n) < \infty\,,$$

then

$$\sum_m \sum_n a_{mn} K(m,n) b_{mn} \le C\|a\|_{\ell^2}\|b\|_{\ell^2}\,.$$

Hardy-Littlewood and Pólya [**241**] extended this result to L^p for $1 < p < \infty$ and disposed of the condition that K is a decreasing function. Aronszajn, Mulla, and Szeptycki [**12**] proved that (iii) implies (i) in the lemma of Appendix H.2. Gagliardo in [**202**] proved the converse direction that (i) implies (iii) in the same lemma. The case $p = 2$ was previously obtained by Karlin [**290**]. Condition (ii) was introduced by Howard and Schep [**257**], who showed that it is equivalent to (i) and (iii). A multilinear analogue of the lemma in Appendix H.2 was obtained by Grafakos and Torres [**225**]; the easy direction (iii) implies (i) was independently observed by Bekollé, Bonami, Peloso, and Ricci [**31**]. See also Cwikel and Kerman [**141**] for an alternative multilinear formulation of the Schur lemma.

The case $p = p' = 2$ of the application in Appendix H.3 is a continuous version of Hilbert's double series theorem. The discrete version was first proved by Hilbert in his lectures on integral equations (published by Weyl [**550**]) without the determination of the exact constant. This exact constant turns out to be π as discovered by Schur [**450**]. The extension to other p's (with sharp constants) is due to Hardy and M. Riesz and published by Hardy [**235**].

APPENDIX J

The Whitney Decomposition of Open Sets in $\mathbf{R}^n$

An arbitrary open set in $\mathbf{R}^n$ can be decomposed as as a union of disjoint cubes whose lengths are proportional to their distance from the boundary of the open set. See for instance Figue J.1 when the open set is the unit disc in $\mathbf{R}^2$. For a given cube Q in $\mathbf{R}^n$, we will denote by $\ell(Q)$ its length.

Proposition. *Let Ω be an open nonempty proper subset of $\mathbf{R}^n$. Then there exists a family of closed cubes $\{Q_j\}_j$ such that*
(a) $\bigcup_j Q_j = \Omega$ and the Q_j's have disjoint interiors.
(b) $\sqrt{n}\,\ell(Q_j) \le \operatorname{dist}(Q_j, \Omega^c) \le 4\sqrt{n}\,\ell(Q_j)$.
(c) If the boundaries of two cubes Q_j and Q_k touch, then

$$\frac{1}{4} \le \frac{\ell(Q_j)}{\ell(Q_k)} \le 4 .$$

(d) For a given Q_j there exist at most 12^n Q_k's that touch it.

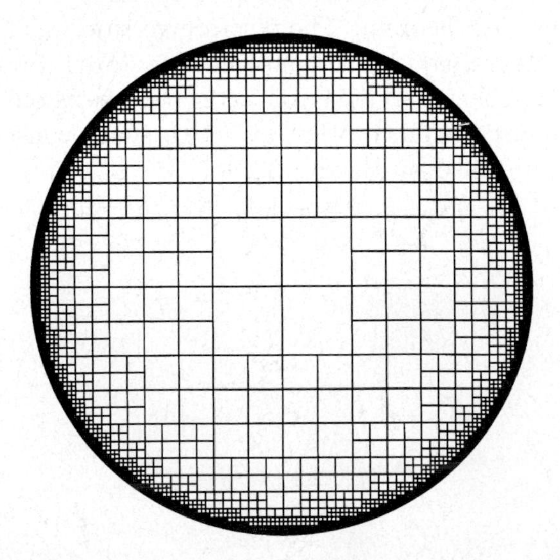

FIGURE J.1. The Whitney decomposition of the unit disc.

Proof. Let $\mathcal{D}_k$ be the collection of all dyadic cubes of the form

$$\{(x_1,\dots,x_n)\in\mathbf{R}^n:\ m_j2^{-k}\le x_j\le(m_j+1)2^{-k}\},$$

where $m_j\in\mathbf{Z}$. Observe that each cube in $\mathcal{D}_k$ gives rise to 2^n cubes in $\mathcal{D}_{k+1}$ by bisecting each side.

Write the set Ω as the union of the sets

$$\Omega_k=\{x\in\Omega:\ 2\sqrt{n}\,2^{-k}<\operatorname{dist}(x,\Omega^c)\le 4\sqrt{n}\,2^{-k}\}$$

over all $k\in\mathbf{Z}$. Let $\mathcal{F}'$ be the set of all cubes Q in $\mathcal{D}_k$ for some $k\in\mathbf{Z}$ such that $Q\cap\Omega_k\neq\emptyset$. We will show that the collection $\mathcal{F}'$ satisfies property (b). Let $Q\in\mathcal{F}'$ and pick $x\in\Omega_k\cap Q$ for some $k\in\mathbf{Z}$. Observe that

$$\sqrt{n}\,2^{-k}\le\operatorname{dist}(x,\Omega^c)-\sqrt{n}\,\ell(Q)\le\operatorname{dist}(Q,\Omega^c)\le\operatorname{dist}(x,\Omega^c)\le 4\sqrt{n}\,2^{-k},$$

which proves (b).

Next we observe that

$$\bigcup_{Q\in\mathcal{F}'}Q=\Omega.$$

Indeed every Q in $\mathcal{F}'$ is contained in Ω (since it has positive distance from its complement) and every $x\in\Omega$ lies in some Ω_k and in some dyadic cube in $\mathcal{D}_k$.

The problem is that the cubes in the collection $\mathcal{F}'$ may not be disjoint. We have to refine the collection $\mathcal{F}'$ by eliminating those cubes that are contained in some other cubes in the collection. Recall that two dyadic cubes either have disjoint interiors or one contains the other. For every cube Q in $\mathcal{F}'$ we can therefore consider the unique *maximal* cube $Q^{\max}$ in $\mathcal{F}'$ that contains it. Two different such maximal cubes must have disjoint interiors by maximality. Now set $\mathcal{F}=\{Q^{\max}:\ Q\in\mathcal{F}'\}$.

The collection of cubes $\{Q_j\}_j=\mathcal{F}$ clearly satisfies (a) and (b), and we now turn our attention to the proof of (c). Observe that for if Q_j and Q_k in $\mathcal{F}$, touch then

$$\sqrt{n}\,\ell(Q_j)\le\operatorname{dist}(Q_j,\Omega^c)\le\operatorname{dist}(Q_j,Q_k)+\operatorname{dist}(Q_k,\Omega^c)\le 0+4\sqrt{n}\,\ell(Q_k),$$

which proves (c). To prove (d), observe that any cube Q in $\mathcal{D}_k$ is touched by exactly 3^n-1 other cubes in $\mathcal{D}_k$. But each cube Q in $\mathcal{D}_k$ can contain at most 4^n cubes of $\mathcal{F}$ of length at least one-quarter of the length of Q. This fact combined with (c) yields (d).

□

The following observation is a consequence of the result just proved:

Let $\mathcal{F}=\{Q_j\}_j$ be the Whitney decomposition of a proper open subset Ω of $\mathbf{R}^n$. Fix $0<\varepsilon<1/4$ and denote by Q_k^* the cube with the same center as Q_k but with side length $(1+\varepsilon)$ times that of Q_k. Then Q_k and Q_j touch if and only if Q_k^* and Q_j intersect. Consequently, every point in Ω is contained in at most 12^n cubes Q_k^*.

APPENDIX K

Smoothness and Vanishing Moments

K.1. The Case of No Cancellation

Let $a, b \in \mathbf{R}^n$, $\mu, \nu \in \mathbf{R}$, and $M, N > n$. Set

$$I(a,\mu,M;b,\nu,N) = \int_{\mathbf{R}^n} \frac{2^{\mu n}}{(1+2^{\mu}|x-a|)^M} \frac{2^{\nu n}}{(1+2^{\nu}|x-b|)^N}\, dx\,.$$

Then we have

$$I(a,\mu,M;b,\nu,N) \le C_0 \frac{2^{\min(\mu,\nu)n}}{\big(1+2^{\min(\mu,\nu)}|a-b|\big)^{\min(M,N)}}\,,$$

where

$$C_0 = v_n\left(\frac{M4^N}{M-n} + \frac{N4^M}{N-n}\right)$$

and v_n is the volume of the unit ball in $\mathbf{R}^n$.

To prove this estimate, first observe that

$$\int_{\mathbf{R}^n} \frac{dx}{(1+|x|)^M} \le \frac{v_n M}{M-n}\,.$$

Without loss of generality, assume that $\nu \le \mu$. Consider the cases $2^{\nu}|a-b| \le 1$ and $2^{\nu}|a-b| \ge 1$. In the case $2^{\nu}|a-b| \le 1$ we use the estimate

$$\frac{2^{\nu n}}{(1+2^{\nu}|x-a|)^N} \le 2^{\nu n} \le \frac{2^{\nu n}2^{\min(M,N)}}{(1+2^{\nu}|a-b|)^{\min(M,N)}}$$

and the result follows by integrating the other term in $I(a,\mu,M;b,\nu,N)$. In the case $2^{\nu}|a-b| \ge 1$ let H_a and H_b be the two half-spaces, containing the points a and b, respectively, formed by the hyperplane perpendicular to the line segment $[a,b]$ at its midpoint. Split the integral over $\mathbf{R}^n$ as the integral over H_a and the integral over H_b. For $x \in H_a$ use that $|x-b| \ge \frac{1}{2}|a-b|$. For $x \in H_b$ use a similar inequality and the fact that $2^{\nu}|a-b| \ge 1$ to obtain

$$\frac{2^{\mu n}}{(1+2^{\mu}|x-a|)^M} \le \frac{2^{\mu n}}{(2^{\mu}\frac{1}{2}|a-b|)^M} \le \frac{4^M 2^{(\nu-\mu)(M-n)}2^{\nu n}}{(1+2^{\nu}|a-b|)^M}\,.$$

The required estimate follows.

K.2. The Case of Cancellation

Let $a, b \in \mathbf{R}^n$, $M, N > 0$, and L be a nonnegative integer. Suppose that ϕ_μ and ϕ_ν are two functions on $\mathbf{R}^n$ that satisfy

$$|(\partial_x^\alpha \phi_\mu)(x)| \le \frac{A_\alpha\, 2^{\mu n}\, 2^{\mu L}}{(1+2^\mu|x-x_\mu|)^M} \qquad \text{for all } |\alpha| = L,$$

$$|\phi_\nu(x)| \le \frac{B\, 2^{\nu n}}{(1+2^\nu|x-x_\nu|)^N}\,,$$

for some A_α and B positive, and

$$\int_{\mathbf{R}^n} \phi_\nu(x)\, x^\beta\, dx = 0 \qquad \text{for all } |\beta| \le L-1,$$

where the last condition is supposed to be vacuous when $L = 0$. Suppose that $N > M + L + n$ and that $\nu \ge \mu$. Then we have

$$\left| \int_{\mathbf{R}^n} \phi_\mu(x)\phi_\nu(x)\, dx \right| \le C_{00}\, \frac{2^{\mu n} 2^{-(\nu-\mu)L}}{(1+2^\mu|x_\mu - x_\nu|)^M}\,,$$

where

$$C_{00} = v_n\, \frac{N-L-M}{N-L-M-n}\, B \sum_{|\alpha|=L} \frac{A_\alpha}{\alpha!}\,.$$

To prove this statement, we subtract the Taylor polynomial of order $L-1$ of ϕ_μ at the point x_ν from the function $\phi_\mu(x)$ and use the remainder theorem to control the required integral by

$$B \sum_{|\alpha|=L} \frac{A_\alpha}{\alpha!} \int_{\mathbf{R}^n} \frac{|x-x_\nu|^L 2^{\mu n} 2^{\mu L}}{(1+2^\mu|\xi_x - x_\mu|)^M}\, \frac{2^{\nu n}}{(1+2^\nu|x-x_\nu|)^N}\, dx\,,$$

for some ξ_x on the segment joining x_ν to x. Using $\nu \ge \mu$ and the triangle inequality, we obtain

$$\frac{1}{1+2^\mu|\xi_x - x_\mu|} \le \frac{1+2^\nu|x-x_\nu|}{1+2^\mu|x_\mu - x_\nu|}\,.$$

We insert this estimate in the last integral and we use that $N > L + M + n$ to deduce the required conclusion.

K.3. The Case of Three Factors

Given three numbers a, b, c we denote by $\operatorname{med}(a,b,c)$ the number with the property $\min(a,b,c) \le \operatorname{med}(a,b,c) \le \max(a,b,c)$.

Let $x_\nu, x_\mu, x_\lambda \in \mathbf{R}^n$. Suppose that ψ_ν, ψ_μ, ψ_λ are functions defined on $\mathbf{R}^n$ such that for all $N > n$ sufficiently large there exist constants $A_\nu, A_\mu, A_\lambda < \infty$ such that

$$\begin{aligned}
|\psi_\nu(x)| &\le A_\nu \frac{2^{\nu n/2}}{(1+2^\nu|x-x_\nu|)^N}, \\
|\psi_\mu(x)| &\le A_\mu \frac{2^{\mu n/2}}{(1+2^\mu|x-x_\mu|)^N}, \\
|\psi_\lambda(x)| &\le A_\lambda \frac{2^{\lambda n/2}}{(1+2^\lambda|x-x_\lambda|)^N},
\end{aligned}$$

for all $x \in \mathbf{R}^n$. Then the following estimate is valid:

$$\int_{\mathbf{R}^n} |\psi_\nu(x)|\,|\psi_\mu(x)|\,|\psi_\lambda(x)|\,dx \le$$

$$\frac{C_{N,n} A_\nu A_\mu A_\lambda\, 2^{-\max(\mu,\nu,\lambda)n/2}\, 2^{\mathrm{med}\,(\mu,\nu,\lambda)n/2}\, 2^{\min(\mu,\nu,\lambda)n/2}}{((1+2^{\min(\nu,\mu)}|x_\nu - x_\mu|)(1+2^{\min(\mu,\lambda)}|x_\mu - x_\lambda|)(1+2^{\min(\lambda,\nu)}|x_\lambda - x_\nu|))^N}$$

for some constant $C_{N,n} > 0$ independent of the remaining parameters.

Analogous statements hold if some of these factors is assumed to have cancellation and the others vanishing moments. The reader is referred to Grafakos and Torres [**226**] for precise statements and proofs of these results.

Bibliography

[1] D. Adams, *A note on Riesz potentials*, Duke Math. J. **42** (1975), 765–778.

[2] R. A. Adams, *Sobolev Spaces*, Academic Press, New York, 1975.

[3] G. Alexopoulos, *La conjecture de Kato pour les opérateurs différentiels elliptiques à coefficients périodiques*, C. R. Acad. Sci. Paris **312** (1991), 263–266.

[4] S. A. Alimov, R. R. Ashurov, and A. K. Pulatov, *Multiplc Fourier Series and Fourier Integrals*, Commutative Harmonic Analysis, Vol. IV, pp. 1–95, V. P. Khavin and N. K. Nikol'skiĭ (eds.), Encyclopedia of Mathematical Sciences, Vol. 42, Springer–Verlag, Berlin–Heidelberg–New York, 1992.

[5] J. Alvarez and C. Pérez, *Estimates with A_∞ weights for various singular integral operators*, Boll. Unione Mat. Ital. **8-A** (1994), 123–133.

[6] N. Y. Antonov, *Convergence of Fourier series*, Proceedings of the XXth Workshop on Function Theory (Moscow, 1995), pp. 187–196, East J. Approx. **2** (1996).

[7] N. Y. Antonov, *The behavior of partial sums of trigonometric Fourier series*, PhD thesis, Institute of Mathematics and Mechanics, Ural Branch of the Russian Academy of Sciences, Ekaterinburg, 1998.

[8] T. Aoki, *Locally bounded spaces*, Proc. Imp. Acad. Tokyo **18** (1942) No. 10.

[9] G. I. Arhipov, A. A. Karachuba, and V. N. Čubarikov, *Trigonometric integrals*, Math USSR Izvestija **15** (1980), 211–239.

[10] J. Arias de Reyna, *Pointwise Convergence of Fourier Series*, J. London Math. Soc. **65** (2002), 139–153.

[11] J. Arias de Reyna, *Pointwise Convergence of Fourier Series*, Lect. Notes in Math. 1785, Springer, Berlin–Heidelberg–New York, 2002.

[12] N. Aronszajn, F. Mulla, and P. Szeptycki, *On spaces of potentials connected with L^p-spaces*, Ann. Inst. Fourier (Grenoble) **12** (1963), 211–306.

[13] N. Aronszajn and K. T. Smith, *Theory of Bessel potentials I*, Ann. Inst. Fourier (Grenoble) **11** (1961), 385–475.

[14] J. M. Ash, *Multiple trigonometric series*, Studies in harmonic analysis, pp. 76–96, MAA Stud. Math., Vol. 13, Math. Assoc. Amer., Washington, DC, 1976.

[15] N. Asmar, E. Berkson, and T. A. Gillespie, *Summability methods for transferring Fourier multipliers and transference of maximal inequalities*, Analysis and Partial Differential Equations, pp. 1–34, A collection of papers dedicated to Mischa Cotlar, C. Sadosky (ed.), Marcel Dekker Inc, New York and Basel, 1990.

[16] N. Asmar, E. Berkson, and T. A. Gillespie, *On Jodeit's multiplier extension theorems*, Journal d' Analyse Math. **64** (1994), 337–345.

[17] P. Auscher, A. McIntosh, and A. Nahmod, *Holomorphic functional calculi of operators, quadratic estimates, and interpolation*, Indiana Univ. Math. J. **46** (1997), 375–403.

[18] P. Auscher and P. Tchamitchian, *Square root problem for divergence operators and related topics*, Astérisque No. 249, Societé Mathematique de France, 1998.

[19] P. Auscher, S. Hofmann, J. L. Lewis, and P. Tchamitchian, *Extrapolation of Carleson measures and the analyticity of Kato's square-root operators*, Acta Math. **187** (2001), 161–190.

[20] P. Auscher, A. McIntosh, S. Hofmann, M. Lacey, and P. Tchamitchian, *The solution of the Kato Square Root Problem for Second Order Elliptic Operators on* $\mathbb{R}^n$, Ann. of Math. **156** (2002), 633–654.

[21] K. I. Babenko, *An inequality in the theory of Fourier integrals* [in Russian], Izv. Akad. Nauk SSSR Ser. Mat. **25** (1961), 531–542.

[22] A. Baernstein II, *Some sharp inequalities for conjugate functions*, Indiana Univ. Math. J. **27** (1978), 833–852.

[23] A. Baernstein II and E. T. Sawyer, *Embedding and multiplier theorems for* $H^p(\mathbf{R}^n)$, Mem. Amer. Math. Soc., No. 318, 1985.

[24] R. Bagby and D. Kurtz, *A rearranged good-*λ *inequality*, Trans. Amer. Math. Soc. **293** (1986), 71–81.

[25] B. Barceló, *On the restriction of the Fourier transform to a conical surface*, Trans. Amer. Math. Soc. **292** (1985), 321–333.

[26] B. Barceló, *The restriction of the Fourier transform to some curves and surfaces*, Studia Math. **84** (1986), 39–69.

[27] J. Barrionuevo, *A note on the Kakeya maximal operator*, Math. Res. Lett. **3** (1995), 61–65.

[28] N. Bary, *A Treatise on Trigonometric Series*, Pergamon Press, New York, 1964.

[29] W. Beckner, *Inequalities in Fourier analysis*, Ann. of Math. **102** (1975), 159–182.

[30] W. Beckner, A. Carbery, S. Semmes, and F. Soria, *A note on restriction of the Fourier transform to spheres*, Bull. London Math. Soc. **21** (1989), 394–398.

[31] D. Bekollé, A. Bonami, M. Peloso, and F. Ricci, *Boundedness of Bergman projections on tube domains over light cones*, Math. Zeit. **237** (2001), 31–59.

[32] A. Benedek, A. Calderón. and R. Panzone, *Convolution operators on Banach-space valued functions*, Proc. Nat. Acad. Sci. USA **48** (1962), 356–365.

[33] J. Benedetto and M. Frazier (eds.), *Wavelets, Mathematics and Applications*, CRC Press, Boca Raton, FL, 1994.

[34] C. Bennet, R. A. DeVore, and R. Sharpley, *Weak* L^∞ *and BMO*, Ann. of Math. **113** (1981), 601–611.

[35] C. Bennet and R. Sharpley, *Interpolation of Operators*, Academic Press, Series: Pure and Applied Math. 129, Orlando, FL, 1988.

[36] J. Berg and J. Löfström, *Interpolation Spaces, An Introduction*, Springer-Verlag, New York, 1976.

[37] A. Besicovitch, *On Kakeya's problem and a similar one*, Math. Z. **27** (1928), 312–320.

[38] A. Besicovitch, *A general form of the covering principle and relative differentiation of additive functions*, Proc. of Cambridge Phil. Soc. **41** (1945), 103–110.

[39] O. V. Besov, *On a family of function spaces. Embedding theorems and applications* [in Russian], Dokl. Akad. Nauk SSSR **126** (1959), 1163–1165.

[40] O. V. Besov, *On a family of function spaces in connection with embeddings and extensions* [in Russian], Trudy Mat. Inst. Steklov **60** (1961), 42–81.

[41] Z. Birnbaum and M. W. Orlicz, *Über die Verallgemeinerung des Begriffes der Zueinander konjugierten Potenzen*, Studia Math. **3** (1931), 1–67; reprinted in W. Orlicz, "Collected Papers," pp. 133–199, PWN, Warsaw, 1988.

[42] R. P. Boas and S. Bochner, *On a theorem of M. Riesz for Fourier series*, J. London Math. Soc. **14** (1939), 62–73.

[43] S. Bochner, *Summation of multiple Fourier series by spherical means*, Trans. Amer. Math. Soc. **40** (1936), 175–207.

[44] S. Bochner, *Lectures on Fourier integrals, (with an author's supplement on monotonic functions, Stieltjes integrals, and harmonic analysis)*, Annals of Math. Studies 42, Princeton Univ. Press, Princeton, NJ, 1959.

[45] S. Bochner, *Harmonic Analysis and the Theory of Probability*, University of California Press, Berkeley, CA, 1955.

[46] J. M. Bony, *Calcul symbolique et propagation des singularités pour les équations aux dérivées partielles non linéaires*, Ann. Sci. École Norm. Sup. **14** (1981), 209–246.

[47] J. Bourgain, *On square functions on the trigonometric system*, Bull. Soc. Math. Belg. Sér B **37** (1985), 20–26.

[48] J. Bourgain, *Averages in the plane over convex curves and maximal operators*, J. Analyse Math. **47** (1986), 69–85.

[49] J. Bourgain, *Besicovitch type maximal operators and applications to Fourier analysis*, Geom. Funct. Anal. **1** (1991), 147–187.

[50] J. Bourgain, *On the restriction and multiplier problems in* $\mathbb{R}^3$, Geometric Aspects of functional analysis (1989–90), pp. 179–191, Lect. Notes in Math. 1469, Springer, Berlin, 1991.

[51] J. Bourgain, *Some new estimates on oscillatory integrals*, Essays on Fourier Analysis in Honor of E. M. Stein, pp. 83-112, C. Fefferman, R. Fefferman, and S. Wainger (eds.), Princeton Univ. Press, Princeton, NJ, 1995.

[52] J. Bourgain, *On the dimension of Kakeya sets and related maximal inequalities*, Geom. Funct. Anal. **9** (1999), 256–282.

[53] H. Q. Bui, *Some aspects of weighted and non-weighted Hardy spaces*, Kôkyûroku Res. Inst. Math. Sci. **383** (1980), 38–56.

[54] D. L. Burkholder, *Martingale transforms*, Ann. of Math. Stat. **37** (1971), 1494–1505.

[55] D. L. Burkholder and R. F. Gundy, *Extrapolation and interpolation of quasilinear operators on martingales*, Acta Math. **124** (1970), 249–304.

[56] D. L. Burkholder, R. F. Gundy, and M. L. Silverstein, *A maximal characterization of the class* H^p, Trans. Amer. Math. Soc. **157** (1971), 137–153.

[57] S. Campanato, *Proprietà di hölderianità di alcune classi di funzioni*, Ann. Scuola Norm. Sup. Pisa **17** (1963), 175–188.

[58] S. Campanato, *Proprietà di una famiglia di spazi funzionali*, Ann. Scuola Norm. Sup. Pisa **18** (1964), 137–160.

[59] A. P. Calderón, *On the theorems of M. Riesz and Zygmund*, Proc. Amer. Math. Soc. **1** (1950), 533–535.

[60] A. P. Calderón, *Lebesgue spaces of differentiable functions and distributions*, Proc. Symp. Pure Math. **4** (1961), 33–49.

[61] A. P. Calderón, *Intermediate spaces and interpolation, the complex method*, Studia Math. **24** (1964), 113–190.

[62] A. P. Calderón, *Commutators of singular integral operators*, Proc. Nat. Acad. Sci. USA **53** (1965), 1092–1099.

[63] A. P. Calderón, *Singular integrals*, Bull. Amer. Math. Soc. **72** (1966), 427–465.

[64] A. P. Calderón, *Algebras of singular integral operators*, Singular Integrals, Proc. Sympos. Pure Math. (Chicago, Ill., 1966), pp. 18–55, Amer. Math. Soc., Providence, RI, 1967.

[65] A. P. Calderón, *Commutators of singular integral operators*, Proc. Nat. Acad. Sci. USA **74** (1977), 1324–1327.

[66] A. P. Calderón, *An atomic decomposition of distributions in parabolic* H^p *spaces*, Adv. in Math. **25** (1977), 216–225.

[67] A. P. Calderón and A. Torchinsky, *Parabolic maximal functions associated with a distribution*, Adv. in Math. **16** (1975), 1–63.

[68] A. P. Calderón and A. Torchinsky, *Parabolic maximal functions associated with a distribution*, Adv. in Math. **24** (1977), 101–171.

[69] A. P. Calderón and A. Zygmund, *On the existence of certain singular integrals*, Acta Math. **88** (1952), 85–139.

[70] A. P. Calderón and A. Zygmund, *Singular integrals and periodic functions*, Studia Math. **14** (1954), 249–271.

[71] A. P. Calderón and A. Zygmund, *A note on interpolation of sublinear operators*, Amer. J. Math. **78** (1956), 282–288.

[72] A. P. Calderón and A. Zygmund, *On singular integrals*, Amer. J. Math. **78** (1956), 289–309.

[73] A. P. Calderón and A. Zygmund, *Algebras of certain singular integral operators*, Amer. J. Math. **78** (1956), 310–320.

[74] A. P. Calderón and A. Zygmund, *Commutators, singular integrals on Lipschitz curves and applications*, Proceedings of the International Congress of Mathematicians (Helsinki, 1978), pp. 85–96, Acad. Sci. Fennica, Helsinki, 1980.

[75] A. P. Calderón and R. Vaillancourt, *A class of bounded pseudo-differential operators*, Proc. Nat. Acad. Sci. USA **69** (1972), 1185–1187.

[76] C. Calderón, *Lacunary spherical means*, Ill. J. Math. **23** (1979), 476–484.

[77] A. Carbery, *The boundedness of the maximal Bochner-Riesz operator on* $L^4(\mathbb{R}^2)$, Duke Math. J. **50** (1983), 409–416.

[78] A. Carbery, *Variants of the Calderón-Zygmund theory for* L^p*-spaces*, Rev. Mat. Iber. **2** (1986), 381–396.

[79] A. Carbery, M. Christ, and J. Wright, *Multidimensional van der Corput and sublevel set estimates*, J. Amer. Math. Soc. **12** (1999), 981–1015.

[80] A. Carbery, E. Hernández, and F. Soria, *Estimates for the Kakeya maximal operator on radial functions in* $\mathbf{R}^n$, Harmonic Analysis, ICM 1990 Satellite Conference Proceedings, pp. 41–50, S. Igari (ed.), Springer-Verlag, Tokyo, 1991.

[81] A. Carbery, J.-L. Rubio de Francia, and L. Vega, *Almost everywhere summability of Fourier integrals*, J. London Math. Soc. **38** (1988), 513–524.

[82] L. Carleson, *An interpolation problem for bounded analytic functions*, Amer. J. Math. **80** (1958), 921–930.

[83] L. Carleson, *Interpolation by bounded analytic functions and the corona problem*, Ann. of Math. **76** (1962), 547–559.

[84] L. Carleson, *On convergence and growth of partial sums of Fourier series*, Acta Math. **116** (1966), 135–157.

[85] L. Carleson, *On the Littlewood-Paley Theorem*, Mittag-Leffler Institute Report, Djursholm, Sweden 1967.

[86] L. Carleson, *Two remarks on* H^1 *and B.M.O*, Adv. in Math. **22** (1976), 269–277.

[87] L. Carleson and P. Sjölin, *Oscillatory integrals and a multiplier problem for the disc*, Studia Math. **44** (1972), 287–299.

[88] M. J. Carro, *New extrapolation estimates*, J. Funct. Anal. **174** (2000), 155–166.

[89] D.-C. Chang, S. G. Krantz, and E. M. Stein, H^p *theory on a smooth domain in* R^N *and elliptic boundary value problems*, J. Funct. Anal. **114** (1993), 286–347.

[90] F. Chiarenza and M. Frasca, *Morrey spaces and Hardy-Littlewood maximal function*, Rend. Mat. Appl. Series 7, **7** (1981), 273–279.

[91] M. Christ, *Estimates for the k-plane transform*, Indiana Univ. Math. J. **33** (1984), 891–910.

[92] M. Christ, *On the restriction of the Fourier transform to curves: endpoint results and the degenerate case*, Trans. Amer. Math. Soc. **287** (1985), 223–238.

[93] M. Christ, *A convolution inequality for Cantor-Lebesgue measures*, Rev. Mat. Iber. **1** (1985), 79–83.

[94] M. Christ, *On almost everywhere convergence of Bochner-Riesz means in higher dimensions*, Proc. Amer. Math. Soc. **95** (1985), 16–20.

[95] M. Christ, *Weak type endpoint bounds for Bochner-Riesz multipliers*, Rev. Mat. Iber. **3** (1987), 25–31.

[96] M. Christ, *Weak type* $(1,1)$ *bounds for rough operators I*, Ann. of Math. **128** (1988), 19–42.

[97] M. Christ, *A* $T(b)$ *theorem with remarks on analytic capacity and the Cauchy integral*, Colloq. Math. **60/61** (1990), 601–628.

[98] M. Christ, *Lectures on singular integral operators*, CBMS Regional Conference Series in Mathematics, Vol. 77, Amer. Math. Soc., Providence, RI, 1990.

[99] M. Christ, J. Duoandikoetxea and J.-L. Rubio de Francia, *Maximal operators related to the Radon transform and the Calderón-Zygmund method of rotations*, Duke Math. J. **53** (1986), 189–209.

[100] M. Christ and R. Fefferman, *A note on weighted norm inequalities for the Hardy-Littlewood maximal operator*, Proc. Amer. Math. Soc. **87** (1983), 447–448.

[101] M. Christ and L. Grafakos, *Best constants for two nonconvolution inequalities*, Proc. Amer. Math. Soc. **123** (1995), 1687–1693.

[102] M. Christ and J.-L. Journé, *Polynomial growth estimates for multilinear singular integral operators*, Acta Math. **159** (1987), 51–80.

[103] M. Christ and J.-L. Rubio de Francia, *Weak type* $(1,1)$ *bounds for rough operators II*, Invent. Math. **93** (1988), 225–237.

[104] C. Chui (ed.), *Wavelets: A Tutorial in Theory and Applications*, Academic Press, San Diego, CA, 1992.

[105] R. R. Coifman, *Distribution function inequalities for singular integrals*, Proc. Nat. Acad. Sci. USA **69** (1972), 2838–2839.

[106] R. R. Coifman, *A real variable characterization of* H^p, Studia Math. **51** (1974), 269–274.

[107] R. R. Coifman, D. G. Deng, and Y. Meyer, *Domaine de la racine carée de certaines opérateurs différentiels acrétifs*, Ann. Inst. Fourier **33** (1983), 123–134.

[108] R. R. Coifman and C. Fefferman, *Weighted norm inequalities for maximal functions and singular integrals*, Studia Math. **51** (1974), 241–250.

[109] R. R. Coifman and L. Grafakos, *Hardy space estimates for multilinear operators, I*, Rev. Mat. Iber. **8** (1992), 45–67.

[110] R. R. Coifman, P. Jones, and J. L. Rubio de Francia, *Constructive decomposition of BMO functions and factorization of* A_p *weights*, Proc. Amer. Math. Soc. **87** (1983), 675–676.

[111] R. R. Coifman, P. Jones, and S. Semmes, *Two elementary proofs of the* L^2 *boundedness of Cauchy integrals on Lipschitz curves*, J. Amer. Math. Soc. **2** (1989), 553–564.

[112] R. R. Coifman, P. L. Lions, Y. Meyer, and S. Semmes, *Compensated compactness and Hardy spaces*, J. Math. Pures Appl. **72** (1993), 247–286.

[113] R. R. Coifman, A. McIntosh, Y. Meyer, *L' intégrale de Cauchy définit un opérateur borné sur* L^2 *pour les courbes lipschitziennes*, Ann. of Math. **116** (1982), 361–387.

[114] R. R. Coifman and Y. Meyer, *Commutateurs d' intégrales singulières et opérateurs multilinéaires*, Ann. Inst. Fourier (Grenoble) **28** (1978), 177–202.

[115] R. R. Coifman and Y. Meyer, *Au Au-délà des opérateurs pseudo-différentiels*, Astérisque No. 57, Societé Mathematique de France, 1979.

[116] R. R. Coifman and Y. Meyer, *A simple proof of a theorem by G. David and J.-L. Journé on singular integral operators*, Probability Theory and Harmonic Analysis, pp. 61–65, J. Chao and W. Woyczyński (eds.), Marcel Dekker, New York, 1986.

[117] R. R. Coifman, Y. Meyer, and E. M. Stein, *Some new function spaces and their applications to harmonic analysis*, J. Funct. Anal. **62** (1985), 304–335.

[118] R. R. Coifman and R. Rochberg, *Another characterization of BMO*, Proc. Amer. Math. Soc. **79** (1980), 249–254.

[119] R. R. Coifman, R. Rochberg, and G. Weiss, *Factorization theorems for Hardy spaces in several variables*, Ann. of Math. **103** (1976), 611–635.

[120] R. R. Coifman, J.-L. Rubio de Francia, and S. Semmes, *Multiplicateurs de Fourier de $L^p(\mathbb{R})$ et estimations quadratiques*, C. R. Acad. Sci. Paris **306** (1988), 351–354.

[121] R. R. Coifman and G. Weiss, *Transference Methods in Analysis*, C.B.M.S. Regional Conference Series in Math. No. 31, Amer. Math. Soc., Providence, RI, 1976.

[122] R. R. Coifman and G. Weiss, *Extensions of Hardy spaces and their use in analysis*, Bull Amer. Math. Soc. **83** (1977), 569–645.

[123] R. R. Coifman and G. Weiss, *Book review of "Littlewood-Paley and multiplier theory" by R. E. Edwards and G. I. Gaudry*, Bull. Amer. Math. Soc. **84** (1978), 242–250.

[124] L. Colzani, G. Travaglini, and M. Vignati, *Bochner-Riesz means of functions in weak-L^p*, Monatsh. Math. **115** (1993), 35–45.

[125] W. C. Connett, *Singular integrals near L^1*, Harmonic analysis in Euclidean spaces, Proc. Sympos. Pure Math. (Williams Coll., Williamstown, Mass., 1978), pp. 163–165, Amer. Math. Soc., Providence, RI, 1979.

[126] A. Córdoba, *The Kakeya maximal function and the spherical summation multipliers*, Amer. J. Math. **99** (1977), 1–22.

[127] A. Córdoba, *A note on Bochner-Riesz operators*, Duke Math. J. **46** (1979), 505–511.

[128] A. Córdoba, *Multipliers of $F(L^p)$*, Euclidean harmonic analysis (Proc. Sem., Univ. Maryland, College Park, Md., 1979), pp. 162–177, Lecture Notes in Math. 779, Springer, Berlin, 1980.

[129] A. Córdoba and C. Fefferman, *A weighted norm inequality for singular integrals*, Studia Math. **57** (1976), 97–101.

[130] A. Córdoba and R. Fefferman, *A geometric proof of the strong maximal theorem*, Ann. of Math. **102** (1975), 95–100.

[131] A. Córdoba and R. Fefferman, *On differentiation of integrals*, Proc. Nat. Acad. Sci. USA **74** (1977), 2211–2213.

[132] M. Cotlar, *A combinatorial inequality and its applications to L^2 spaces*, Rev. Mat. Cuyana, **1** (1955), 41–55.

[133] M. Cotlar, *A unified theory of Hilbert transforms and ergodic theorems*, Rev. Mat. Cuyana, **1** (1955), 105–167.

[134] M. Cowling and G. Mauceri, *On maximal functions*, Rend. Sem. Mat. Fis. Milano **49** (1979), 79–87.

[135] D. Cruz-Uribe, *New proofs of two-weight norm inequalities for the maximal operator*, Georgian Math. J. **7** (2000), 33–42.

[136] D. Cruz-Uribe SFO and C. J. Neugebauer, *The structure of the reverse Hölder classes*, Trans. Amer. Math. Soc. **347** (1995), 2941–2960.

[137] F. Cunningham, *The Kakeya problem for simply connected and for star-shaped sets*, Amer. Math. Monthly **78** (1971), 114–129.

[138] M. Cwikel, *The dual of weak L^p*, Ann. Inst. Fourier (Grenoble) **25** (1975), 81–126.

[139] M. Cwikel and C. Fefferman, *Maximal seminorms on Weak L^1*, Studia Math. **69** (1980), 149–154.

[140] M. Cwikel and C. Fefferman, *The canonical seminorm on Weak L^1*, Studia Math. **78** (1984), 275–278.

[141] M. Cwikel and R. Kerman, *Positive multilinear operators acting on weighted L^p spaces*, J. Funct. Anal. **106** (1992), 130–144.

[142] G. Dafni, *Hardy spaces on some pseudoconvex domains*, J. Geom. Anal. **4** (1994), 273–316.

[143] G. Dafni, *Local VMO and weak convergence in h^1*, Can. Math. Bull. **45** (2002), 46–59.

[144] J. E. Daly, *A necessary condition for Calderón-Zygmund singular integral operators*, J. Four. Anal. and Appl. **5** (1999), 303–308.

[145] J. E. Daly and K. Phillips, *On the classification of homogeneous multipliers bounded on* $H^1(\mathbf{R}^2)$, Proc. Amer. Math. Soc. **106** (1989), 685–696.

[146] I. Daubechies, *Orthonormal bases of compactly supported wavelets*, Comm. Pure Appl. Math. **61** (1988), 909–996.

[147] I. Daubechies, *Ten Lectures on Wavelets*, Society for Industrial and Applied Mathematics, Philadelphia, PA, 1992.

[148] G. David, *Opérateurs intégraux singuliers sur certains courbes du plan complexe*, Ann. Sci. Ecole Norm. Sup. **17** (1984), 157–189.

[149] G. David and J.-L. Journé, *A boundedness criterion for generalized Calderón-Zygmund operators*, Ann. of Math. **120** (1984), 371–397.

[150] G. David, J.-L. Journé, and S. Semmes, *Opérateurs de Calderón-Zygmund, fonctions para-accrétives et interpolation*, Rev. Math. Iber. **1** (1985), 1–56.

[151] G. David and S. Semmes, *Singular integrals and rectifiable sets in* $\mathbf{R}^n$*: Beyond Lipschitz graphs*, Astérisque No. 193, Societé Mathematique de France, 1991.

[152] B. Davis, *On the integrability of the martingale square function*, Israel J. Math. **8** (1970), 187–190.

[153] B. Davis, *On the weak type* $(1,1)$ *inequality for conjugate functions*, Proc. Amer. Math. Soc. **44** (1974), 307–311.

[154] K. M. Davis and Y. C. Chang, *Lectures on Bochner-Riesz Means*, London Math. Soc. Lect. Notes **114**, Cambridge Univ. Press, Cambridge, UK, 1987.

[155] L. De Carli and A. Iosevich, *A restriction theorem for flat manifolds of codimension two*, Ill. J. Math. **39** (1995), 576–585.

[156] L. De Carli and A. Iosevich, *Some sharp restriction theorems for homogeneous manifolds*, J. Fourier Anal. Appl. **4** (1998), 105–128.

[157] M. de Guzmán, *Real-Variable Methods in Fourier Analysis*, North-Holland Math. Studies **46**, North-Holland, Amsterdam 1981.

[158] K. de Leeuw, *On* L_p *multipliers*, Ann. of Math. **81** (1965), 364–379.

[159] J. Diestel and J. J. Uhl Jr, *Vector Measures*, Amer. Math. Soc, Providence, RI, 1977.

[160] J. L. Doob, *Stochastic Processes*, Wiley, New York, 1953.

[161] S. W. Drury, L^p *estimates for the x-ray transform*, Ill. J. Math. **27** (1983), 125–129.

[162] S. W. Drury, *Restrictions of Fourier transforms to curves*, Ann. Inst. Fourier (Grenoble) **35** (1985), 117–123.

[163] S. W. Drury and K. Guo, *Some remarks on the restriction of the Fourier transform to surfaces*, Math. Proc. Cambridge Philos. Soc. **113** (1993), 153–159.

[164] S. W. Drury and B. P. Marshall, *Fourier restriction theorems for curves with affine and Euclidean arclengths*, Math. Proc. Cambridge Philos. Soc. **97** (1985), 111–125.

[165] S. W. Drury and B. P. Marshall, *Fourier restriction theorems for degenerate curves*, Math. Proc. Cambridge Philos. Soc. **101** (1987), 541–553.

[166] N. Dunford and J. T. Schwartz, *Linear Operators, I*, General Theory, Interscience, New York, 1959.

[167] J. Duoandikoetxea, *Fourier Analysis*, Grad. Studies in Math. **29**, Amer. Math. Soc., Providence, RI, 2000.

[168] J. Duoandikoetxea and J.-L. Rubio de Francia, *Maximal and singular integral operators via Fourier transform estimates*, Invent. Math. **84** (1986), 541–561.

[169] J. Duoandikoetxea and L. Vega, *Spherical means and weighted inequalities*, J. London Math. Soc. **53** (1996), 343–353.

[170] P. L. Duren, *Theory of* H^p *Spaces*, Dover Publications Inc., New York, 2000.

[171] P. L. Duren, B. W. Romberg, and A. L. Shields, *Linear functionals on H^p spaces with* $0 < p < 1$, J. Reine Angew. Math. **238** (1969), 32–60.

[172] H. Dym and H. P. McKean, *Fourier Series and Integrals*, Academic Press, New York, 1972.

[173] R. E. Edwards, *Fourier Series: A Modern Introduction*, 2nd ed., Springer-Verlag, New York, 1979.

[174] A. Erdélyi, *Asymptotic Expansions*, Dover Publications Inc., New York, 1956.

[175] M. Essén, *A superharmonic proof of the M. Riesz conjugate function theorem*, Arkiv f. Math. **22** (1984), 241–249.

[176] E. Fabes, D. Jerison, and C. Kenig, *Multilinear Littlewood-Paley estimates with applications to partial differential equations*, Proc. Nat. Acad. Sci. USA **79** (1982), 5746–5750.

[177] E. Fabes, D. Jerison, and C. Kenig, *Multilinear square functions and partial differential equations*, Amer. J. Math. **107** (1985), 1325–1367.

[178] E. Fabes, I. Mitrea, and M. Mitrea, *On the boundedness of singular integrals*, Pac. J. Math. **189** (1999), 21–29.

[179] D. Fan, K. Guo, and Y. Pan, *A note of a rough singular integral operator*, Math. Ineq. and Appl. **2** (1999), 73–81.

[180] D. Fan and Y. Pan, *Singular integral operators with rough kernels supported by subvarieties*, Amer J. Math. **119** (1997), 799–839.

[181] K. Fan, *Minimax theorems*, Proc. Nat. Acad. Sci. USA **39** (1953), 42–47.

[182] C. Fefferman, *Inequalities for strongly singular convolution operators*, Acta Math. **124** (1970), 9–36.

[183] C. Fefferman, *On the convergence of Fourier series*, Bull. Amer. Math. Soc. **77** (1971), 744–745.

[184] C. Fefferman, *Characterizations of bounded mean oscillation*, Bull. Amer. Math. Soc. **77** (1971), 587–588.

[185] C. Fefferman, *The multiplier problem for the ball*, Ann. of Math. **94** (1971), 330–336.

[186] C. Fefferman, *Pointwise convergence of Fourier series*, Ann. of Math. **98** (1973), 551–571.

[187] C. Fefferman, *A note on spherical summation multipliers*, Israel J. Math. **15** (1973), 44–52.

[188] C. Fefferman, *The uncertainty principle*, Bull. Amer. Math. Soc. **9** (1983), 129–206.

[189] C. Fefferman, N. Riviere, and Y. Sagher, *Interpolation between H^p spaces: The real method*, Trans. Amer. Math. Soc. **191** (1974), 75–81.

[190] C. Fefferman and E. M. Stein, *Some maximal inequalities*, Amer. J. Math. **93** (1971), 107–115.

[191] C. Fefferman and E. M. Stein, *H^p spaces of several variables*, Acta Math. **129** (1972), 137–193.

[192] R. Fefferman, *A theory of entropy in Fourier analysis*, Adv. in Math. **30** (1978), 171–201.

[193] R. Fefferman, *Strong differentiation with respect to measures*, Amer J. Math. **103** (1981), 33–40.

[194] T. M. Flett, *Lipschitz spaces of functions on the circle and the disc*, J. Math. Anal. Appl. **39** (1972), 125–158.

[195] J. Fourier, *Théorie Analytique de la Chaleur*, Institut de France, Paris, 1822.

[196] M. Frazier, *An Introduction to Wavelets Through Linear Algebra*, Springer-Verlag, New York, NY, 1999.

[197] M. Frazier and B. Jawerth, *Decomposition of Besov spaces*, Indiana Univ. Math. J. **34** (1985), 777–799.

[198] M. Frazier and B. Jawerth, *A discrete transform and decompositions of distribution spaces*, J. Funct. Anal. **93** (1990), 34–170.

[199] M. Frazier and B. Jawerth, *Applications of the φ and wavelet transforms to the theory of function spaces*, Wavelets and Their Applications, pp. 377–417, Jones and Bartlett (eds.), Boston, MA, 1992.

[200] M. Frazier, B. Jawerth and W. Weiss, *Littlewood-Paley Theory and the Study of Function Spaces*, CBMS Regional Conference Series in Mathematics, 79, Amer. Math. Soc. Providence, RI, 1991.

[201] E. Gagliardo, *Proprietà di alcune classi di funzioni in più variabili*, Ricerche di Mat. Napoli **7** (1958), 102–137.

[202] E. Gagliardo, *On integral transformations with positive kernel*, Proc. Amer. Math. Soc. **16** (1965), 429–434.

[203] J. García-Cuerva, *An extrapolation theorem in the theory of A_p weights*, Proc. Amer. Math. Soc. **87** (1983), 422–426.

[204] J. García-Cuerva and J.-L. Rubio de Francia, *Weighted Norm Inequalities and Related Topics*, North-Holland Math. Studies **116**, North-Holland, Amsterdam, 1985.

[205] J. Garnett and P. Jones, *The distance in BMO to L^∞*, Ann. of Math. **108** (1978), 373–393.

[206] J. Garnett, *Bounded Analytic Functions*, Academic Press, New York, 1981.

[207] F. W. Gehring, *The L^p-integrability of the partial derivatives of a quasiconformal mapping*, Acta Math. **130** (1973), 265–277.

[208] I. M. Gelfand and G. E. Šilov, *Generalized Functions, Vol. 1: Properties and Operations*, Academic Press, New York, London, 1964.

[209] I. M. Gelfand and G. E. Šilov, *Generalized Functions, Vol. 2: Spaces of Fundamental and Generalized Functions*, Academic Press, New York, London, 1968.

[210] D. Goldberg, *A local version of real Hardy spaces*, Duke Math. J. **46** (1979), 27–42

[211] I. Gohberg and N. Krupnik, *Norm of the Hilbert transformation in the L_p space*, Funct. Anal. Appl. **2** (1968), 180–181.

[212] L. Grafakos, *Best bounds for the Hilbert transform on $L^p(\mathbb{R}^1)$*, Math. Res. Lett. **4** (1997), 469–471.

[213] L. Grafakos and N. Kalton, *Some remarks on multilinear maps and interpolation*, Math. Ann. **319** (2001), 151–180.

[214] L. Grafakos and N. Kalton, *The Marcinkiewicz multiplier condition for bilinear operators*, Studia Math. **146** (2001), 115–156.

[215] L. Grafakos and N. Kalton, *Multilinear Calderón-Zygmund operators on Hardy spaces*, Collect. Math. **52** (2001), 169–179.

[216] L. Grafakos and J. Kinnunen, *Sharp inequalities for maximal functions associated to general measures*, Proc. Royal Soc. Edinb. **128** (1998), 717–723.

[217] L. Grafakos and X. Li, *Uniform bounds for the bilinear Hilbert transforms I*, Ann. of Math., to appear.

[218] L. Grafakos and J. M. Martell, *Extrapolation of operators of many variables*, to appear.

[219] L. Grafakos and C. Morpurgo, *A Selberg integral formula and applications*, Pac. J. Math. **191** (1999), 85–94.

[220] L. Grafakos and S. Montgomery-Smith, *Best constants for uncentred maximal functions*, Bull. London Math. Soc. **29** (1997), 60–64.

[221] L. Grafakos and A. Stefanov, *Convolution Calderón-Zygmund singular integral operators with rough kernels*, Analysis of Divergence: Control and Management of Divergent Processes, pp. 119–143, William Bray, Časlav V. Stanojević (eds.), Birkhäuser, Boston–Basel–Berlin, 1999.

[222] L. Grafakos and A. Stefanov, *L^p bounds for singular integrals and maximal singular integrals with rough kernels*, Indiana Univ. Math. J. **47** (1998), 455–469.

[223] L. Grafakos and T. Tao, *Multilinear interpolation between adjoint operators*, J. Funct. Anal. **199** (2003), 379–385.

[224] L. Grafakos, T. Tao, and E. Terwilleger, *L^p bounds for a maximal dyadic sum operator*, to appear.

[225] L. Grafakos and R. Torres, *A multilinear Schur test and multiplier operators*, J. Funct. Anal. **187** (2001), 1–24.

[226] L. Grafakos and R. Torres, *Discrete decompositions for bilinear operators and almost diagonal conditions*, Trans. Amer. Math. Soc. **354** (2002), 1153–1176.

[227] L. Grafakos and R. Torres, *Multilinear Calderón-Zygmund theory*, Adv. in Math. **165** (2002), 124–164.

[228] A. Greenleaf, *Principal curvature and harmonic analysis*, Indiana Univ. Math. J. **30** (1981), 519–537.

[229] K. Gröhening, *Foundations of Time-Frequency Analysis*, Birkhäuser, Boston, 2001.

[230] U. Haagerup, *The best constants in Khintchine's inequality*, Studia Math. **70** (1982), 231–283.

[231] L.-S. Hahn, *On multipliers of p-integrable functions*, Trans. Amer. Math. Soc. **128** (1967), 321–335.

[232] G. H. Hardy, *The mean value of the modulus of an analytic function*, Proc. London Math. Soc. **14** (1914), 269–277.

[233] G. H. Hardy, *Note on a theorem of Hilbert*, Math. Zeit. **6** (1920), 314–317.

[234] G. H. Hardy, *Note on some points in the integral calculus*, Messenger Math. **57** (1928), 12–16.

[235] G. H. Hardy, *Note on a theorem of Hilbert concerning series of positive terms*, Proc. London Math. Soc. **23** (1925), Records of Proc. XLV–XLVI.

[236] G. H. Hardy and J. E. Littlewood, *Some properties of fractional integrals I*, Math. Zeit. **27** (1927), 565–606.

[237] G. H. Hardy and J. E. Littlewood, *A maximal theorem with function-theoretic applications*, Acta Math. **54** (1930), 81–116.

[238] G. H. Hardy and J. E. Littlewood, *Some properties of fractional integrals II*, Math. Zeit. **34** (1932), 403–439.

[239] G. H. Hardy and J. E. Littlewood, *Some theorems on Fourier series and Fourier power series*, Duke Math. J. **2** (1936), 354–381.

[240] G. H. Hardy and J. E. Littlewood, *Generalizations of a theorem of Paley*, Quarterly Jour. **8** (1937), 161–171.

[241] G. H. Hardy, J. E. Littlewood, and G. Polya, *The maximum of a certain bilinear form*, Proc. London Math. Soc. **25** (1926), 265–282.

[242] G. H. Hardy, J. E. Littlewood, and G. Pólya, *Inequalities*, 2nd ed., Cambridge University Press, Cambridge, UK, 1952.

[243] L. Hedberg, *On certain convolution inequalities*, Proc. Amer. Math. Soc. **36** (1972), 505–510.

[244] H. Helson and G. Szegö, *A problem in prediction theory*, Ann. Math. Pura Appl. **51** (1960), 107–138.

[245] E. Hérnandez and G. Weiss, *A First Course on Wavelets*, CRC Press, Boca Raton, FL, 1996.

[246] C. Hertz, *On the mean inversion of Fourier and Hankel transforms*, Proc. Nat. Acad. Sci. USA **40** (1954), 996–999.

[247] E. Hewitt and K. Ross, *Abstract Harmonic Analysis I,* 2nd ed., Grundlehren der mathematischen Wissenschaften 115, Springer-Verlag, Berlin–Heidelberg–New York, 1979.

[248] I. I. Hirschman, *A convexity theorem for certain groups of transformations*, Jour. d' Analyse Math. **2** (1952), 209–218.

[249] S. Hofmann, *Weak* $(1,1)$ *boundedness of singular integrals with nonsmooth kernels*, *Proc. Amer. Math. Soc.* **103** (1988), 260–264.

[250] S. Hofmann and A. McIntosh, *The solution of the Kato problem in two dimensions*, Proceedings of the 6th International Conference on Harmonic Analysis and Partial Differential Equations (El Escorial, Spain, 2000), pp. 143–160, Publ. Mat. Extra Volume, 2002.

[251] S. Hofmann, M. Lacey, and A. McIntosh, *The solution of the Kato problem for divergence form elliptic operators with Gaussian heat kernel bounds*, Ann. of Math. **156** (2002), 623–631.

[252] B. Hollenbeck and I. E. Verbitsky, *Best constants for the Riesz projection*, J. Funct. Anal. **175** (2000), 370–392.

[253] L. Hörmander, *Estimates for translation invariant operators in* L^p *spaces*, Acta Math. **104** (1960), 93–140.

[254] L. Hörmander, *Linear Partial Differential Operators*, Springer-Verlag, Berlin–Göttingen–Heidelberg, 1963.

[255] L. Hörmander, *Oscillatory integrals and multipliers on* FL^p, Arkiv f. Mat. **11** (1973), 1–11.

[256] L. Hörmander, *The Analysis of Linear Partial Differential Operators I*, 2nd ed., Springer-Verlag, Berlin–Heidelberg–New York, 1990.

[257] R. Howard and A. R. Schep, *Norms of positive operators on* L^p *spaces*, Proc. Amer Math. Soc. **109** (1990), 135–146.

[258] S. V. Hruščev, *A description of weights satisfying the* A_∞ *condition of Muckenhoupt*, Proc. Amer. Math. Soc. **90** (1984), 253–257.

[259] S. Hudson, *A covering lemma for maximal operators with unbounded kernels*, Michigan Math. J. **34** (1987), 147–151.

[260] R. Hunt, *An extension of the Marcinkiewicz interpolation theorem to Lorentz spaces*, Bull. Amer. Math. Soc. **70** (1964), 803–807.

[261] R. Hunt, *On* $L(p,q)$ *spaces*, L' Einseignement Math. **12** (1966), 249–276.

[262] R. Hunt, *On the convergence of Fourier series*, Orthogonal Expansions and Their Continuous Analogues (Edwardsville, Ill., 1967), pp. 235–255, D. T. Haimo (ed.), Southern Illinois Univ. Press, Carbondale IL, 1968.

[263] R. Hunt, D. Kurtz, and C. J. Neugebauer, *A note on the equivalence of* A_p *and Sawyer's condition*, Conference on Harmonic Analysis in honor of Antoni Zygmund, Vol. 1, pp. 156–158, W. Beckner et al. (eds.), Wadsworth, Belmont, 1983.

[264] R. Hunt, B. Muckenhoupt, and R. Wheeden, *Weighted norm inequalities for the conjugate function and the Hilbert transform*, Trans. Amer. Math. Soc. **176** (1973), 227–251.

[265] R. Hunt and W.-S. Young, *A weighted norm inequality for Fourier series*, Bull. Amer. Math. Soc. **80** (1974), 274–277.

[266] G. B. Folland and E. M. Stein, *Estimates for the* $\overline{\partial}_b$ *complex and analysis on the Heisenberg group*, Comm. Pure and Appl. Math. **27** (1974), 429–522.

[267] G. B. Folland and E. M. Stein, *Hardy Spaces on Homogeneous Groups*, Mathematical Notes 28, Princeton Univ. Press, Princeton, NJ, 1982.

[268] G. B. Folland and A. Sitaram, *The uncertainty principle: a mathematical survey*, J. Fourier Anal. Appl. **3** (1997), 207–238.

[269] S. Igari, *An extension of the interpolation theorem of Marcinkiewicz II*, Tôhoku Math. J. **15** (1963), 343–358.

[270] S. Igari, *Lectures on Fourier Series of Several Variables*, Univ. Wisconsin Lecture Notes, Madison, WI, 1968.

[271] S. Igari, *On Kakeya's maximal function*, Proc. Japan Acad. Ser. A Math. Sci. **62** (1986), 292–293.

[272] R. S. Ismagilov, *On the Pauli problem*, Func. Anal. Appl. **30** (1996), 138–140.

[273] T. Iwaniec and G. Martin, *Riesz transforms and related singular integrals*, J. Reine Angew. Math. **473** (1996), 25–57.

[274] S. Janson, *Mean oscillation and commutators of singular integral operators*, Ark. Math. **16** (1978), 263–270.

[275] S. Janson, *On interpolation of multilinear operators*, Function Spaces and Applications (Lund, 1986), pp. 290–302, Lect. Notes in Math. 1302, Springer, Berlin–New York, 1988.

[276] B. Jessen, J. Marcinkiewicz, and A. Zygmund, *Note on the differentiability of multiple integrals*, Fund. Math. **25** (1935), 217–234.

[277] M. Jodeit, *Restrictions and extensions of Fourier multipliers*, Studia Math. **34** (1970), 215–226.

[278] M. Jodeit, *A note on Fourier multipliers*, Proc. Amer. Math. Soc. **27** (1971), 423–424.

[279] F. John and L. Nirenberg, *On functions of bounded mean oscillation*, Comm. Pure and Appl. Math. **14** (1961), 415–426.

[280] P. Jones, *Factorization of A_p weights*, Ann. of Math. **111** (1980), 511–530.

[281] O. G. Jørsboe and L. Melbro, *The Carleson-Hunt Theorem on Fourier Series*, Lect. Notes in Math. 911, Springer-Verlag, Berlin, 1982.

[282] J.-L. Journé *Calderón-Zygmund Operators, Pseudo-Differential Operators and the Cauchy Integral of Calderón*, Lect. Notes in Math. 994, Springer-Verlag, Berlin, 1983.

[283] J.- L. Journé *Calderón-Zygmund operators on product spaces*, Rev. Mat. Iber. **1** (1985), 55–91.

[284] J. P. Kahane and Y. Katznelson, *Sur les ensembles de divergence des séries trigonometriques*, Studia Math. **26** (1966), 305–306.

[285] G. Kaiser, *A Friendly Guide to Wavelets*, Birkhäuser, Boston–Basel–Berlin, 1994.

[286] N. J. Kalton, *Convexity, type, and the three space problem*, Studia Math. **69** (1981), 247–287.

[287] N. J. Kalton, *Plurisubharmonic functions on quasi-Banach spaces*, Studia Math. **84** (1986), 297–323.

[288] N. J. Kalton, *Analytic functions in non-locally convex spaces and applications*, Studia Math. **83** (1986), 275–303.

[289] N. J. Kalton, N. T. Peck, and J. W. Roberts, *An F-space sampler*, Lon. Math. Soc. Lect. Notes 89, Cambridge Univ. Press, Cambridge, UK, 1984.

[290] S. Karlin, *Positive operators*, J. Math. Mech. **6** (1959), 907–937.

[291] T. Kato, *Fractional powers of dissipative operators*, J. Math. Soc. Japan **13** (1961), 246–274.

[292] T. Kato, *Peturbation theory for linear operators*, 2nd ed., Springer-Verlag, Berlin–Heidelberg–New York, 1980.

[293] N. Katz, *A counterexample for maximal operators over a Cantor set of directions*, Math. Res. Lett. **3** (1996), 527–536.

[294] N. Katz, *Maximal operators over arbitrary sets of directions*, Duke Math. J. **97** (1999), 67–79.

[295] N. Katz, *Remarks on maximal operators over arbitrary sets of directions*, Bull. London Math. Soc. **31** (1999), 700–710.

[296] N. Katz, I. Laba, and T. Tao, *An improved bound on the Minkowski dimension of Besicovitch sets in $\mathbb{R}^3$*, Ann. of Math. **152** (2000), 383–446.

[297] N. Katz and T. Tao, *Recent progress on the Kakeya conjecture*, Proceedings of the 6th International Conference on Harmonic Analysis and Partial Differential Equations (El Escorial, Spain, 2000), pp. 161–179, Publ. Mat. Extra Volume, 2002.

[298] Y. Katznelson, *Sur les ensembles de divergence des séries trigonometriques*, Studia Math. **26** (1966), 301–304.

[299] Y. Katznelson, *An Introduction to Harmonic Analysis*, 2nd ed., Dover Publications, Inc., New York, 1976.

[300] U. Keich, *On L^p bounds for Kakeya maximal functions and the Minkowski dimension in $\mathbb{R}^2$*, Bull. London Math. Soc. **31** (1999), 213–221.

[301] C. Kenig and Y. Meyer, *Kato's square roots of accretive operators and Cauchy kernels on Lipschitz curves are the same*, Recent progress in Fourier analysis (El Escorial, Spain, 1983), pp. 123–143, North-Holland Math. Stud., 111, North-Holland, Amsterdam, 1985.

[302] C. Kenig and E. M. Stein, *Multilinear estimates and fractional integration*, Math. Res. Lett. **6** (1999), 1–15.

[303] C. Kenig and P. Tomas, *Maximal operators defined by Fourier multipliers*, Studia Math. **68** (1980), 79–83.

[304] A. Khintchine, *Über dyadische Brüche*, Math. Zeit. **18** (1923), 109–116.

[305] A. Knapp and E. M. Stein, *Intertwining operators for semisimple groups*, Ann. of Math. **93** (1971), 489–578.

[306] A. N. Kolmogorov, *Une série de Fourier-Lebesgue divergente presque partout*, Fund. Math. **4** (1923), 324–328.

[307] A. N. Kolmogorov, *Une contribution à l' étude de la convergence des séries de Fourier*, Fund. Math. **5** (1924), 96–97.

[308] A. N. Kolmogorov, *Sur les fonctions harmoniques conjuguées et les séries de Fourier*, Fund. Math. **7** (1925), 23–28.

[309] A. N. Kolmogorov, *Une série de Fourier-Lebesgue divergente partout*, C. R. Acad. Sci. Paris **183** (1926), 1327–1328.

[310] A. N. Kolmogorov, *Zur Normierbarkeit eines topologischen Raumes*, Studia Math. **5** (1934), 29–33.

[311] S. V. Konyagin, *On everywhere divergence of trigonometric Fourier series,* Sbornik: Mathematics **191** (2000), 97–120.

[312] P. Koosis, *Sommabilité de la fonction maximale et appartenance à H_1*, C. R. Acad. Sci. Paris **28** (1978), 1041–1043.

[313] P. Koosis, *Introduction to H_p Spaces*, 2nd ed., Cambridge Tracts in Math. 115, Cambridge Univ. Press, Cambridge, UK, 1998.

[314] T. Körner, *Fourier Analysis*, Cambridge Univ. Press, Cambridge, UK, 1988.

[315] S. G. Krantz, *Fractional integration on Hardy spaces*, Studia Math. **63** (1982), 87–94.

[316] S. G. Krantz, *Lipschitz spaces, smoothness of functions, and approximation theory*, Exposition. Math. **1** (1983), 193–260.

[317] S. G. Krantz, *A panorama of Harmonic Analysis*, Carus Math. Monographs # 27, Mathematical Association of America, Washington, DC, 1999.

[318] M. G. Krein, *On linear continuous operators in functional spaces with two norms*, Trudy Inst. Mat. Akad. Nauk Ukrain. SSRS **9** (1947), 104–129.

[319] J. L. Krivine, *Théorèmes de factorisation dans les éspaces reticulés*, Sém. Maurey-Schwartz 1973/74, exp. XXII-XXIII, Palaiseau, France.

[320] D. Kurtz, *Littlewood-Paley and multiplier theorems on weighted L^p spaces*, Trans. Amer. Math. Soc. **259** (1980), 235–254.

[321] D. Kurtz and R. Wheeden, *Results on weighted norm inequalities for multipliers*, Trans. Amer. Math. Soc. **255** (1979), 343–362.

[322] M. T. Lacey and C. M. Thiele, *L^p bounds for the bilinear Hilbert transform, $p > 2$*, Ann. of Math. **146** (1997), 693–724.

[323] M. Lacey and C. Thiele, *On Calderón's conjecture*, Ann. of Math. **149** (1999), 475–496.

[324] M. Lacey and C. Thiele, *A proof of boundedness of the Carleson operator*, Math. Res. Lett. **7** (2000), 361–370.

[325] E. Landau, *Zur analytischen Zahlentheorie der definiten quadratischen Formen (Über die Gitterpunkte in einem mehrdimensionalen Ellipsoid),* Berl. Sitzungsber. **31** (1915), 458–476; reprinted in E. Landau "Collected Works," Vol. 6, pp. 200–218, Thales-Verlag, Essen, 1986.

[326] S. Lang, *Real Analysis*, 2nd ed., Addison-Wesley, Reading, MA, 1983.

[327] R. H. Latter, *A decomposition of* $H^p(\mathbf{R}^n)$ *in terms of atoms*, Studia Math. **62** (1977), 92–101.

[328] R. H. Latter and A. Uchiyama, *The atomic decomposition for parabolic* H^p *spaces*, Trans. Amer. Math. Soc. **253** (1979), 391–398.

[329] H. Lebesgue, *Intégrale, longeur, aire*, Annali Mat. Pura Appl. **7** (1902), 231–359.

[330] H. Lebesgue, *Oeuvres Scientifiques*, Vol. I, L' Enseignement Math., pp. 201–331, Geneva, 1972.

[331] P. Lemarié and Y. Meyer, *Ondelettes et bases hilbertiennes*, Rev. Mat. Iber. **2** (1986), 1–18.

[332] A. K. Lerner, *On pointwise estimates for maximal and singular integral operators*, Studia Math. **138** (2000), 285–291.

[333] X. Li, *Uniform bounds for the bilinear Hilbert transforms II*, to appear.

[334] E. H. Lieb, *Sharp constants in the Hardy-Littlewood-Sobolev and related inequalities*, Ann. of Math. **18** (1983), 349–374.

[335] E. H. Lieb and M. Loss, *Analysis*, Grad. Studies in Math. **14**, Amer. Math. Soc., Providence, RI, 1997.

[336] J.-L. Lions, *Espaces d' interpolation and domaines de puissances fractionnaires*, J. Math. Soc. Japan **14** (1962), 233–241.

[337] J.-L. Lions, P. I. Lizorkin, and S. M. Nikol'skij, *Integral representation and isomorphic properties of some classes of functions*, Ann. Sc. Norm. Sup. Pisa **19** (1965), 127–178.

[338] J.-L. Lions and J. Peetre, *Sur une classe d' éspaces d' interpolation*, Inst. Hautes Etudes Sci. Publ. Math. No. **19** (1964), 5–68.

[339] J. E. Littlewood, *On a certain bilinear form*, Quart. J. Math. Oxford Ser. **1** (1930), 164–174.

[340] J. E. Littlewood and R. E. A. C. Paley, *Theorems on Fourier series and power series (I)*, J. London Math. Soc. **6** (1931), 230–233.

[341] J. E. Littlewood and R. E. A. C. Paley, *Theorems on Fourier series and power series (II)*, Proc. London Math. Soc. **42** (1936), 52–89.

[342] J. E. Littlewood and R. E. A. C. Paley, *Theorems on Fourier series and power series (III)*, Proc. London Math. Soc. **43** (1937), 105–126.

[343] P. I. Lizorkin, *Properties of functions of the spaces* $\Lambda^r_{p\Theta}$ [in Russian], Trudy Mat. Inst. Steklov **131** (1974), 158–181.

[344] L. H. Loomis, *A note on Hilbert's transform*, Bull. Amer. Math. Soc. **52** (1946), 1082–1086.

[345] L. H. Loomis and H. Whitney, *An inequality related to the isoperimetric inequality*, Bull. Amer. Math. Soc. **55** (1949), 961–962.

[346] G. Lorentz, *Some new function spaces*, Ann. of Math. **51** (1950), 37–55.

[347] G. Lorentz, *On the theory of spaces* Λ, Pacific. J. Math. **1** (1951), 411–429.

[348] S.-Z. Lu, *Four Lectures on Real* H^p *Spaces*, World Scientific, Singapore, 1995.

[349] N. Luzin, *Sur la convergence des séries trigonométriques de Fourier*, C. R. Acad. Sci. Paris **156** (1913), 1655–1658.

[350] A. Lyapunov, *Sur les fonctions-vecteurs complètement additives* [in Russian], Izv. Akad. Nauk SSSR Ser. Mat. **4** (1940), 465–478.

[351] A. Magyar, E. M. Stein, and S. Wainger, *Discrete analogues in harmonic analysis: spherical averages*, Ann. of Math. **155** (2002), 189–208.

[352] S. Mallat, *Multiresolution approximation and wavelets*, Trans. Amer. Math. Soc. **315** (1989), 69–88.

[353] S. Mallat, *A Wavelet Tour of Signal Processing*, Academic Press, San Diego, CA, 1998.

[354] A. McIntosh, *On the comparability of* $A^{1/2}$ *and* $A^{*1/2}$, *Proc. Amer. Math. Soc.* **32** (1972), 430–434.

[355] A. McIntosh, *Square roots of elliptic operators*, J. Funct. Anal. **61** (1985), 307–327.

[356] A. McIntosh, *On representing closed accretive sesquilinear forms as* $(A^{1/2}u, A^{*1/2}v)$, Collège de France Seminar, Vol. III, pp. 252–267, H. Brezis and J.-L. Lions (eds.), Research Notes in Mathematics, No. 70, Pitman, 1982.

[357] A. McIntosh, *Square roots of operators and applications to hyperbolic PDE*, Proceedings of the Miniconference on Operator Theory and PDE, CMA, The Australian National University, Canberra, 1983.

[358] A. McIntosh and Y. Meyer, *Algèbres d' opérateurs définits par des intégrales singulières* C. R. Acad. Sci. Paris **301** (1985), 395–397.

[359] Y. Meyer, *Régularité des solutions des équations aux dérivées partielles non linéaires* [d'aprés J.-M. Bony], Séminaire Bourbaki, 1979/80, No. 560.

[360] Y. Meyer, *Principe d' incertitude, bases hilbertiennes et algèbres d' opérateurs*, Séminaire Bourbaki, 1985/86, No. 662.

[361] Y. Meyer, *Ondelettes et Opérateurs*, Vol. I, Hermann, Paris, 1990.

[362] Y. Meyer, *Ondelettes et Opérateurs*, Vol. II, Hermann, Paris, 1990.

[363] Y. Meyer and R. R. Coifman, *Ondelettes et Opérateurs*, Vol. III, Hermann, Paris, 1991.

[364] Y. Meyer, *Wavelets, Vibrations, and Scalings*, CRM Monograph Series Vol. 9, American Mathematical Society, Providence, RI, 1998.

[365] J. Marcinkiewicz, *Sur l'interpolation d'operations*, C. R. Acad. Sci. Paris **208** (1939), 1272–1273.

[366] J. Marcinkiewicz, *Sur les multiplicateurs des séries de Fourier*, Studia Math. **8** (1939), 78–91.

[367] J. Marcinkiewicz and A. Zygmund, *Quelques inégalités pour les opérations linéaires*, Fund. Math. **32** (1939), 112–121.

[368] J. Marcinkiewicz and A. Zygmund, *On the summability of double Fourier series*, Fund. Math. **32** (1939), 122–132.

[369] G. Mauceri, M. Picardello, and F. Ricci, *A Hardy space associated with twisted convolution*, Adv. in Math. **39** (1981), 270–288.

[370] B. Maurey, *Théoremes de factorization pour les opérateurs linéaires à valeurs dans un éspace* L^p, Astérisque No. 11, Societé Mathematique de France, 1974.

[371] V. G. Maz'ya, *Sobolev Spaces*, Springer-Verlag, Berlin–New York, 1985.

[372] A. Melas, *The best constant for the centered Hardy-Littlewood maximal inequality*, Ann. of Math. **157** (2003), 647–688.

[373] M. Melnikov and J. Verdera, *A geometric proof of the* L^2 *boundedness of the Cauchy integral on Lipschitz graphs*, Internat. Math. Res. Notices **7** (1995), 325–331.

[374] N. G. Meyers, *Mean oscillation over cubes and Hölder continuity*, Proc. Amer. Math. Soc. **15** (1964), 717–721.

[375] S. G. Mihlin, *On the multipliers of Fourier integrals* [in Russian], Dokl. Akad. Nauk. **109** (1956), 701–703.

[376] G. Mockenhaupt, A. Seeger, and C. Sogge, *Wave front sets, local smoothing and Bourgain's circular maximal theorem*, Ann. of Math. **136** (1992), 207–218.

[377] K. H. Moon, *On restricted weak type* $(1,1)$, Proc. Amer. Math. Soc. **42** (1974), 148–152.

[378] A. P. Morse, *Perfect blankets*, Trans. Amer. Math. Soc. **69** (1947), 418–442.

[379] C. J. Mozzochi, *On the pointwise convergence of Fourier Series*, Lect. Notes in Math. 199, Springer, Berlin 1971.

[380] B. Muckenhoupt, *On certain singular integrals*, Pacific J. Math. **10** (1960), 239–261.

[381] B. Muckenhoupt, *Weighted norm inequalities for the Hardy maximal function*, Trans. Amer. Math. Soc. **165** (1972), 207–226.

[382] B. Muckenhoupt, *The equivalence of two conditions for weight functions*, Studia Math. **49** (1974), 101–106.

[383] B. Muckenhoupt and R. Wheeden, *Two weighted function norm inequalities for the Hardy-Littlewood maximal function and the Hilbert transform*, Studia Math. **60** (1976), 279–294.

[384] D. Müller, *A note on the Kakeya maximal function*, Arch. Math. (Basel) **49** (1987), 66–71.

[385] T. Murai, *Boundedness of singular integral integral operators of Calderón type*, Proc. Japan Acad. Ser. A Math. Sci. **59** (1983), 364–367.

[386] T. Murai, *A real variable method for the Cauchy transform and analytic capacity*, Lect. Notes in Math. 1307, Springer-Verlag, Berlin, 1988.

[387] N. I. Muskhelishvili, *Singular Integral Equations*, Wolters-Noordhoff Publishing, Groningen, the Netherlands, 1958.

[388] A. Nagel, E. M. Stein, and S. Wainger, *Differentiation in lacunary directions*, Proc. Nat. Acad. Sci. USA **75** (1978), 1060–1062.

[389] F. Nazarov, S. Treil, and A. Volberg, *Cauchy integral and Calderón-Zygmund operators on nonhomogeneous spaces*, Internat. Math. Res. Notices **15** (1997), 703–726.

[390] F. Nazarov, S. Treil, and A. Volberg, *Weak type estimates and Cotlar inequalities for Calderón-Zygmund operators on nonhomogeneous spaces*, Internat. Math. Res. Notices **9** (1998), 463–487.

[391] F. Nazarov, S. Treil, and A. Volberg, *The Bellman functions and two-weight inequalities for Haar multipliers*, J. Amer. Math. Soc. **12** (1999), 909–928.

[392] U. Neri, *Singular Integrals*, Lect. Notes in Math. 200, Springer-Verlag, Berlin, 1971.

[393] R. O' Neil, *Convolution operators and $L(p,q)$ spaces*, Duke Math. J. **30** (1963), 129–142.

[394] L. Nirenberg, *On elliptic partial differential equations*, Ann. di Pisa **13** (1959), 116–162.

[395] D. Oberlin, *Fourier restriction for affine arclength measures in the plane*, Proc. Amer. Math. Soc. **129** (2001), 3303–3305.

[396] M. W. Orlicz, *Über eine gewisse Klasse von Räumen vom Typus B*, Bull. Int. Acad. Pol. de Science, Ser A (1932), 207–220; reprinted in W. Orlicz "Collected Papers," pp. 217–230, PWN, Warsaw, 1988.

[397] M. W. Orlicz, *Über Räume* (L^M), Bull. Int. Acad. Pol. de Science, Ser A (1936), 93–107; reprinted in W. Orlicz "Collected Papers," pp. 345–359, PWN, Warsaw, 1988.

[398] J. Orobitg and C. Pérez, *A_p weights for nondoubling measures in R^n and applications*, Trans. Amer. Math. Soc. **354** (2002), 2013–2033.

[399] R. E. A. C. Paley, *A remarkable series of orthogonal functions*, Proc. London Math. Soc. **34** (1932), 241–264.

[400] R. E. A. C. Paley and A. Zygmund, *On some series of functions*, Proc. Cambridge Phil. Soc. **26** (1930), 337–357.

[401] J. Peetre, *Nouvelles propriétés d' éspaces d' interpolation*, C. R. Acad. Sci. Paris **256** (1963), 1424–1426.

[402] J. Peetre, *On convolution operators leaving $L^{p,\lambda}$ spaces invariant*, Ann. Mat. Pura Appl. **72** (1966), 295–304.

[403] J. Peetre, *Sur les éspaces de Besov*, C. R. Acad. Sci. Paris **264** (1967), 281–283.

[404] J. Peetre, *Remarques sur les éspaces de Besov. Le cas* $0 < p < 1$, C. R. Acad. Sci. Paris **277** (1973), 947–950.

[405] J. Peetre, *H_p Spaces*, Lecture Notes, University of Lund and Lund Institute of Technology, Lund, Sweden 1974.

[406] J. Peetre, *On spaces of Triebel-Lizorkin type*, Ark. Math. **13** (1975), 123–130.

[407] J. Peetre, *Correction to the paper "On spaces of Triebel-Lizorkin type,"* Ark. Math. **14** (1975), 299.

[408] J. Peetre, *New Thoughts on Besov Spaces*, Duke University Math. Series 1, Durham, 1976.

[409] M. C. Pereyra, *Lecture Notes on Dyadic Harmonic Analysis*, in "Second Summer School in Analysis and Mathematical Physics," pp. 1–60, S. Pérez-Esteva and C. Villegas-Blas (eds.), Contemp. Math. AMS, Vol. 289, Providence, RI, 2001.

[410] C. Pérez, *Weighted norm inequalities for singular integral operators*, J. London Math. Soc. **49** (1994), 296–308.

[411] C. Pérez, *Endpoint estimates for commutators of singular integral operators*, J. Funct. Anal. **128** (1995), 163–185.

[412] C. Pérez, *Banach function spaces and the two-weight problem for maximal functions*, Function Spaces, Differential Operators and Nonlinear Analysis (Paseki and Jizerou, 1995), pp. 141–158, Prometheus, Prague, 1996.

[413] S. Pichorides, *On the best values of the constants in the theorems of M. Riesz, Zygmund and Kolmogorov*, Studia Math. **44** (1972), 165–179.

[414] J. Plemelj, *Ein Ergänzungssatz zur Cauchyschen Integraldarstellung analytischer Functionen, Randwerte betreffend*, Monatsh. Math. Phys. **19** (1908), 205–210.

[415] M. Pinsky, N. Stanton, and P. Trapa, *Fourier series of radial functions in several variables*, J. Funct. Anal. **116** (1993), 111–132.

[416] M. Pramanik and E. Terwilleger, *A weak L^2 estimate for a maximal dyadic sum operator on $\mathbf{R}^n$*, Ill. J. Math., to appear.

[417] E. Prestini, *A restriction theorem for space curves*, Proc. Amer. Math. Soc. **70** (1978), 8–10.

[418] J. Privalov, *Sur les fonctions conjuguées*, Bull. Soc. Math. France **44** (1916), 100–103.

[419] M. M. Rao and Z. D. Ren, *Theory of Orlicz spaces*, Pure and Applied Mathematics, Marcel Dekker, New York–Basel–Hong Kong, 1991.

[420] M. Reed and B. Simon, *Methods of Mathematical Physics, Vols. I, II*, Academic Press, New York, 1975.

[421] M. Reed and B. Simon, *Methods of Mathematical Physics, Vols. III, IV*, Academic Press, New York, 1978.

[422] F. Ricci and G. Weiss, *A characterization of $H^1(\Sigma_{n-1})$*, Harmonic analysis in Euclidean spaces, Proc. Sympos. Pure Math., (Williams Coll., Williamstown, Mass., 1978), pp. 289–294, Amer. Math. Soc., Providence, RI, 1979.

[423] F. Riesz, *Untersuchungen über Systeme integrierbarer Funktionen*, Math. Ann. **69** (1910), 449–497.

[424] F. Riesz, *Sur un théorème du maximum de MM. Hardy et Littlewood*, J. London Math. Soc. **7** (1932), 10–13.

[425] M. Riesz, *Les fonctions conjuguées et les séries de Fourier*, C. R. Acad. Sci. Paris **178** (1924), 1464–1467.

[426] M. Riesz, *Sur les maxima des formes bilinéaires et sur les fonctionnelles linéaires*, Acta Math. **49** (1927), 465–497.

[427] M. Riesz, *Sur les fonctions conjuguées*, Math. Zeit. **27** (1927), 218–244.

[428] M. Riesz, *L' intégrale de Riemann-Liouville et le problème de Cauchy*, Acta Math. **81** (1949), 1–223.

[429] N. Riviere, *Singular integrals and multiplier operators*, Arkiv f. Math. **9** (1971), 243–278.

[430] N. Riviere and Y. Sagher, *Interpolation between L^∞ and H^1, the real method*, J. Funct. Anal. **14** (1973), 401–409.

[431] S. Rolewicz, *Metric Linear Spaces*, 2nd ed., Mathematics and Its Applications (East European Series), 20. D. Reidel Publishing Co., Dordrecht-Boston, MA; PWN, Warsaw, 1985.

[432] M. Rosenblum, *Summability of Fourier series in $L^p(\mu)$*, Trans. Amer. Math. Soc. **105** (1962), 32–42.

[433] J.-L. Rubio de Francia, *Estimates for some square functions of Littlewood-Paley type*, Publ. Mat. **27** (1983), 81–108.

[434] J.-L. Rubio de Francia, *Factorization theory and A_p weights*, Amer. J. Math. **106** (1984), 533–547.

[435] J.-L. Rubio de Francia, *A Littlewood-Paley inequality for arbitrary intervals*, Rev. Mat. Iber. **1** (1985), 1–14.

[436] J.-L. Rubio de Francia, *Maximal functions and Fourier transforms*, Duke Math. J. **53** (1986), 395–404.

[437] J.-L. Rubio de Francia, *Weighted norm inequalities and vector-valued inequalities*, Harmonic Analysis, (Minneapolis, Minn., 1981), pp. 86–101, Lect. Notes in Math. 908, Springer-Verlag, Berlin–Heidelberg–New York, 1982.

[438] J.-L. Rubio de Francia and J.-L. Torrea, *Vector extensions of operators in L^p spaces*, Pacific J. Math. **105** (1983), 227–235.

[439] J.-L. Rubio de Francia, F. J. Ruiz, and J. L. Torrea, *Calderón-Zygmund theory for operator-valued kernels*, Adv. in Math. **62** (1986), 7–48.

[440] W. Rudin, *Real and Complex Analysis*, 2nd ed., Tata McGraw-Hill Publishing, New Delhi, 1974.

[441] D. Ryabogin and B. Rubin, *Singular integrals generated by zonal measures*, Proc. Amer. Math. Soc. **130** (2002), 745–751.

[442] C. Sadosky, *Interpolation of Operators and Singular Integrals*, Marcel Dekker Inc., 1976.

[443] S. Saks, *Theory of the Integral*, Hafner Publ. Co, New York, 1938.

[444] D. Sarason, *Functions of bounded mean oscillation*, Trans. Amer. Math. Soc. **207** (1975), 391–405.

[445] S. Sato, *Note on a Littlewood-Paley operator in higher dimensions*, J. London Math. Soc. **42** (1990), 527–534.

[446] E. Sawyer, *A characterization of a two-weight norm inequality for maximal operators*, Studia Math. **75** (1982), 1–11.

[447] L. Schwartz, *Théorie de Distributions, I, II*, Hermann, Paris, 1950-51.

[448] W. Schlag, *A geometric proof of the circular maximal theorem*, Duke Math. J. **93** (1998), 505–533.

[449] W. Schlag, *A geometric inequality with applications to the Kakeya problem in three dimensions*, Geom. Funct. Anal. **8** (1998), 606–625.

[450] I. Schur, *Bemerkungen zur Theorie der beschränkten Bilinearformen mit unendlich vielen Veränderlichen*, Journal f. Math. **140** (1911), 1–28.

[451] A. Seeger, *Some inequalities for singular convolution operators in L^p-spaces*, Trans. Amer. Math. Soc. **308** (1988), 259–272.

[452] A. Seeger, *Singular integral operators with rough convolution kernels*, J. Amer. Math. Soc. **9** (1996), 95–105.

[453] A. Seeger, *Endpoint inequalities for Bochner-Riesz multipliers in the plane*, Pacific J. Math. **174** (1996), 543–553.

[454] S. Semmes, *Square function estimates and the $T(b)$ theorem*, Proc. Amer. Math. Soc. **110** (1990), 721–726.

[455] V. L. Shapiro, *Fourier series in several variables*, Bull. Amer. Math. Soc. **70** (1964), 48–93.

[456] R. Sharpley, *Interpolation of n pairs and counterexamples employing indices*, J. Approx. Theory **13** (1975), 117–127.

[457] R. Sharpley, *Multilinear weak type interpolation of mn-typles with applications*, Studia Math. **60** (1977), 179–194.

[458] P. Sjögren and F. Soria, *Rough maximal operators and rough singular integral operators applied to integrable radial functions*, Rev. Math. Iber. **13** (1997), 1–18.

[459] P. Sjölin, *On the convergence almost everywhere of certain singular integrals and multiple Fourier series*, Arkiv f. Math. **9** (1971), 65–90.

[460] P. Sjölin, *A note on Littlewood-Paley decompositions with arbitrary intervals*, J. Approx. Theory **48** (1986), 328–334.

[461] P. Sjölin and F. Soria, *Some remarks on restriction of the Fourier transform for general measures*, Publ. Mat. **43** (1999), 655–664.

[462] P. Sjölin and F. Soria, *Remarks on a theorem by N. Y. Antonov*, to appear.

[463] S. L. Sobolev, *On a theorem in functional analysis* [in Russian], Mat. Sob. **46** (1938), 471–497.

[464] C. Sogge, *Fourier Integrals in Classical Analysis*, Cambridge Tracts in Math. 105, Cambridge Univ. Press, Cambridge, UK, 1993.

[465] F. Soria, *A note on a Littlewood-Paley inequality for arbitrary intervals in* $\mathbb{R}^2$, J. London Math. Soc. **36** (1987), 137–142.

[466] F. Soria, *On an extrapolation theorem of Carleson Sjölin with applications to a.e. convergence of Fourier series*, Studia Math. **94** (1989), 235–244.

[467] S. Spanne, *Sur l' interpolation entre les éspaces* $\mathcal{L}_k^{p,\Phi}$, Ann. Scuola Norm. Sup. Pisa **20** (1966), 625–648.

[468] E. M. Stein, *Interpolation of linear operators*, Trans. Amer. Math. Soc. **83** (1956), 482–492.

[469] E. M. Stein, *Note on singular integrals*, Proc. Amer. Math. Soc. **8** (1957), 250–254.

[470] E. M. Stein, *Localization and summability of multiple Fourier series*, Acta Math. **100** (1958), 93–147.

[471] E. M. Stein, *On the functions of Littlewood-Paley, Lusin, and Marcinkiewicz*, Trans. Amer. Math. Soc. **88** (1958), 430–466.

[472] E. M. Stein, *On limits of sequences of operators*, Ann. of Math. **74** (1961), 140–170.

[473] E. M. Stein, *Singular integrals, harmonic functions, and differentiability properties of functions of several variables*, in Singular Integrals, Proc. Sympos. Pure Math., (Chicago, Ill., 1966), pp. 316–335, Amer. Math. Soc., Providence, RI, 1967.

[474] E. M. Stein, *Note on the class* $L \log L$, Studia Math. **32** (1969), 305–310.

[475] E. M. Stein, *Topics in Harmonic Analysis Related to the Littlewood-Paley Theory*, Annals of Math. Studies 63, Princeton Univ. Press, Princeton, NJ, 1970.

[476] E. M. Stein, *Maximal functions: Spherical means*, Proc. Nat. Acad. Sci. USA **73** (1976), 2174–2175.

[477] E. M. Stein, *The development of square functions in the work of A. Zygmund*, Bull. Amer. Math. Soc. **7** (1982), 359–376.

[478] E. M. Stein, *Some results in harmonic analysis in* $\mathbf{R}^n$, *for* $n \to \infty$, Bull. Amer. Math. Soc. **9** (1983), 71–73.

[479] E. M. Stein, *Boundary behavior of harmonic functions on symmetric spaces: Maximal estimates for Poisson integrals*, Invent. Math. **74** (1983), 63–83.

[480] E. M. Stein, *Oscillatory integrals in Fourier analysis*, Beijing Lectures in Harmonic Analysis, pp. 307–355, E. M. Stein (ed.), Annals of Math. Studies 112, Princeton Univ. Press, Princeton, NJ, 1986.

[481] E. M. Stein, *Singular Integrals and Differentiability Properties of Functions*, Princeton Univ. Press, Princeton, NJ, 1970.

[482] E. M. Stein, *Harmonic Analysis, Real Variable Methods, Orthogonality, and Oscillatory Integrals*, Princeton Univ. Press, Princeton, NJ, 1993.

[483] E. M. Stein and J. O. Strömberg, *Behavior of maximal functions in* $\mathbf{R}^n$ *for large* n, Arkiv f. Math. **21** (1983), 259–269.

[484] E. M. Stein and G. Weiss, *Interpolation of operators with change of measures*, Trans. Amer. Math. Soc. **87** (1958), 159–172.

[485] E. M. Stein and G. Weiss, *An extension of theorem of Marcinkiewicz and some of its applications*, J. Math. Mech. **8** (1959), 263–284.

[486] E. M. Stein and G. Weiss, *On the theory of harmonic functions of several variables, I: The theory of H^p spaces*, Acta Math. **103** (1960), 25–62.

[487] E. M. Stein and G. Weiss, *Introduction to Fourier Analysis on Euclidean Spaces*, Princeton Univ. Press, Princeton, NJ, 1971.

[488] E. M. Stein and N. J. Weiss, *On the convergence of Poisson integrals*, Trans. Amer. Math. Soc. **140** (1969), 34–54.

[489] P. Stein, *On a theorem of M. Riesz*, J. London Math. Soc. **8** (1933), 242–247.

[490] V. D. Stepanov, *On convolution integral operators*, Soviet Math. Dokl. **19** (1978), 1334–1337.

[491] R. Strichartz, *A multilinear version of the Marcinkiewicz interpolation theorem*, Proc. Amer. Math. Soc. **21** (1969), 441–444.

[492] R. Strichartz, *Restrictions of Fourier transforms to quadratic surfaces and decay of solutions of wave equations*, Duke Math. J. **44** (1977), 705–713.

[493] J.-O. Strömberg, *Maximal functions for rectangles with given directions*, Doctoral dissertation, Mittag-Leffler Institute, Djursholm, Sweden, 1976.

[494] J.-O. Strömberg, *Maximal functions associated to rectangles with uniformly distributed directions*, Ann. of Math. **107** (1978), 399–402.

[495] J.-O. Strömberg, *Bounded mean oscillation with Orlicz norms and duality of Hardy spaces*, Indiana Univ. Math. J. **28** (1979), 511–544.

[496] J.-O. Strömberg, *A modified Franklin system and higher-order spline systems on $\mathbb{R}^n$ as unconditional bases for Hardy spaces*, Conference on harmonic analysis in honor of Antoni Zygmund, Vol. I, II (Chicago, Ill., 1981), pp. 475–494, Wadsworth Math. Ser., Wadsworth, Belmont, CA, 1983.

[497] J.-O. Strömberg and A. Torchinsky, *Weighted Hardy spaces*, Lect. Notes in Mathematics 1381, Springer-Verlag, Berlin–New York, 1989.

[498] S. J. Szarek, *On the best constant in the Khintchine inequality*, Studia Math. **58** (1978), 197–208.

[499] M. Taibleson, *The preservation of Lipschitz spaces under singular integral operators*, Studia Math. **24** (1963), 105–111.

[500] M. Taibleson, *On the theory of Lipschitz spaces of distributions on Euclidean n-space, I*, J. Math. Mech. **13** (1964), 407–480.

[501] M. Taibleson, *On the theory of Lipschitz spaces of distributions on Euclidean n-space, II*, J. Math. Mech. **14** (1965), 821–840.

[502] M. Taibleson, *On the theory of Lipschitz spaces of distributions on Euclidean n-space, III*, J. Math. Mech. **15** (1966), 973–981.

[503] J. D. Tamarkin and A. Zygmund, *Proof of a theorem of Thorin*, Bull. Amer. Math. Soc. **50** (1944), 279–282.

[504] T. Tao, *Weak type endpoint bounds for Riesz means*, Proc. Math. Amer. Soc. **124** (1996), 2797–2805.

[505] T. Tao, *The weak-type endpoint Bochner-Riesz conjecture and related topics*, Indiana Univ. Math. J. **47** (1998), 1097–1124.

[506] T. Tao, *The Bochner-Riesz conjecture implies the restriction conjecture*, Duke Math. J. **96** (1999), 363–375.

[507] T. Tao, *The weak type* $(1,1)$ *of* $L\log L$ *homogeneous convolution operators*, Indiana Univ. Math. J. **48** (1999), 1547–1584.

[508] T. Tao, *Endpoint bilinear restriction theorems for the cone, and some sharp null form estimates*, Math. Zeit. **238** (2001), 215–268.

[509] T. Tao, *A converse extrapolation theorem for translation-invariant operators*, J. Funct. Anal. **180** (2001), 1–10.

[510] T. Tao, *On the Maximal Bochner-Riesz conjecture in the plane, for* $p < 2$, Trans. Amer. Math. Soc. **354** (2002), 1947–1959.

[511] T. Tao and A. Seeger, *Sharp Lorentz estimates for rough operators*, Math. Ann. **320** (2001), 381–415.

[512] T. Tao, A. Vargas, and L. Vega, *A bilinear approach to the restriction and Kakeya conjectures*, J. Amer. Math. Soc. **11** (1998), 967–1000.

[513] T. Tao and J. Wright, *Endpoint multiplier theorems of Marcinkiewicz type*, Rev. Mat. Iber. **17** (2001), 521–558.

[514] M. Taylor, *Pseudodifferential Operators and Nonlinear PDE*, Progress in mathematics 100, Birkhäuser, Boston, 1991.

[515] M. Taylor, *Partial Differential Equations, Basic Theory*, texts in applied mathematics 23, Springer-Verlag, New York, 1996.

[516] P. Tchamitchian, *Ondelettes et intégrale de Cauchy sur les courbes lipschitziennes*, Ann. of Math. **129** (1989), 641–649.

[517] N. R. Tevzadze, *On the convergence of double Fourier series of quadratic summable functions*, Soobšč. Akad. Nauk Gruzin. **5** (1970), 277-279.

[518] C. Thiele, *A uniform estimate*, Ann. of Math. **157** (2002), 1–45.

[519] C. Thiele, *Multilinear singular integrals*, Proceedings of the 6th International Conference on Harmonic Analysis and Partial Differential Equations (El Escorial, Spain, 2000), pp. 229–274, Publ. Mat. Extra Volume, 2002.

[520] G. O. Thorin, *An extension of a convexity theorem due to M. Riesz*, Fys. Säellsk. Förh. **8** (1938), No. 14.

[521] G. O. Thorin, *Convexity theorems generalizing those of M. Riesz and Hadamard with some applications*, Comm. Sem. Math. Univ. Lund [Medd. Lunds Univ. Mat. Sem.] **9** (1948), 1–58.

[522] E. C. Titchmarsh, *The Theory of the Riemann Zeta Function*, Oxford at the Clarendon Press, 1951.

[523] P. A. Tomas, *A restriction theorem for the Fourier transform*, Bull. Amer. Math. Soc. **81** (1975), 477–478.

[524] P. A. Tomas, *A note on restriction*, Indiana Univ. Math. J. **29** (1980), 287–292.

[525] A. Torchinsky, *Real-Variable Methods in Harmonic Analysis*, Academic Press, New York, 1986.

[526] R. Torres, *Boundedness results for operators with singular kernels on distribution spaces*, Mem. Amer. Math. Soc., No. 442, 1991.

[527] H. Triebel, *Spaces of distributions of Besov type on Euclidean n-space. Duality, interpolation*, Ark. Math. **11** (1973), 13–64.

[528] H. Triebel, *Spaces of Besov-Hardy-Sobolev type*, Teubner-Texte zur Mathematik, Leipzig, 1978.

[529] H. Triebel, *Theory of function spaces*, Monographs in Math. Vol. 78, Birkhäuser-Verlag, Basel–Boston–Stuttgart, 1983.

[530] H. Triebel, *Theory of function spaces II*, Monographs in Math. Vol. 84, Birkhäuser-Verlag, Basel–Boston–Stuttgart, 1992.

[531] A. Uchiyama, *A constructive proof of the Fefferman-Stein decomposition of* $BMO(R^n)$. Acta Math. **148** (1982), 215–241.

[532] A. Uchiyama, *Characterization of* $H^p(\mathbf{R}^n)$ *in terms of generalized Littlewood-Paley g-function*, Studia Math. **81** (1985), 135–158.

[533] A. Uchiyama, *On the characterization of $H^p(\mathbf{R}^n)$ in terms of Fourier multipliers*, Proc. Amer. Math. Soc. **109** (1990), 117–123.

[534] A. Uchiyama, *Hardy Spaces on the Euclidean Space*, Springer Monographs in Mathematics, Springer-Verlag, Tokyo, 2001.

[535] J. G. van der Corput, *Zahlentheoretische Abschätzungen*, Math. Ann. **84** (1921), 53–79.

[536] A. Vargas, *Bochner-Riesz multipliers, Maximal operators, Restriction theorems in $\mathbf{R}^n$*, Lecture Notes given at MSRI, August 1997.

[537] N. Varopoulos, *BMO functions and the $\overline{\partial}$-equation*, Pacific J. Math. **71** (1977), 221–273.

[538] I. E. Verbitsky, *Estimate of the norm of a function in a Hardy space in terms of the norms of its real and imaginary part*, Amer. Math. Soc. Transl. **124** (1984), 11–15.

[539] I. E. Verbitsky, *Weighted norm inequalities for maximal operators and Pisier's theorem on factorization through $L^{p\infty}$*, Integral Equations Operator Theory **15** (1992), 124–153.

[540] I. E. Verbitsky, *A dimension-free Carleson measure inequality*, Operator Theory: Advances and Applications, Vol. 113, pp. 393–398, Birkhäuser-Verlag, Basel, Switzerland, 2000.

[541] J. Verdera, *L^2 boundedness of the Cauchy Integral and Menger curvature, Contemp. Math.* **277** (2001), 139–158.

[542] G. Vitali, *Sui gruppi di punti e sulle funzioni di variabili reali*, Atti Accad. Sci. Torino **43** (1908), 75–92.

[543] J. Von Neuman, *Zur Theorie des Gesellschaftsspiele*, Math. Ann. **100** (1928), 295–320.

[544] S. Wainger, *Special trigonometric series in k dimensions*, Mem. Amer. Math. Soc., No. 59, 1965.

[545] T. Walsh, *The dual of $H^p(\mathbb{R}^{n+1})$ for $p < 1$*, Can. J. Math. **25** (1973), 567–577.

[546] G. N. Watson, *A Treatise on the Theory of Bessel Functions*, Cambridge University Press, Cambridge, UK, 1952.

[547] G. Weiss, *An interpolation theorem for sublinear operators on H^p spaces*, Proc. Amer. Math. Soc. **8** (1957), 92–99.

[548] M. Weiss and A. Zygmund, *An example in the theory of singular integrals*, Studia Math. **26** (1965), 101–111.

[549] G. V. Welland, *Weighted norm inequalities for fractional integrals*, Proc. Amer. Math. Soc. **51** (1975), 143–148.

[550] H. Weyl, *Singuläre Integralgleichungen mit besonderer Berücksichtigung des Fourierschen Integraltheorems*, Inaugural-Dissertation, Gottingen, 1908.

[551] H. Weyl, *Bemerkungen zum Begriff der Differentialquotienten gebrochener Ordnung*, Vie. Natur. Gesellschaft Zürich **62** (1917), 296–302.

[552] N. Wiener, *The ergodic theorem*, Duke Math. J. **5** (1939), 1–18.

[553] H. Whitney, *Analytic extensions of differentiable functions defined in closed sets*, Trans. Amer. Math. Soc. **36** (1934), 63–89.

[554] M. V. Wickerhauser, *Adapted Wavelet Analysis from Theory to Software*, A. K. Peters, Wellesley, 1994.

[555] J. M. Wilson, *On the atomic decomposition for Hardy spaces*, Pacific J. Math. **116** (1985), 201–207.

[556] J. M. Wilson, *Weighted norm inequalities for the continuous square function*, Trans. Amer. Math. Soc. **314** (1989), 661–692.

[557] P. Wojtaszczyk, *A Mathematical Introduction to Wavelets*, London Math. Soc. Student Texts 37, Cambridge University Press, Cambridge, UK, 1997.

[558] T. H. Wolff, *A note on interpolation spaces*, Harmonic Analysis (Minneapolis, Minn., 1981), pp. 199–204, Lect. Notes in Math. 908, Springer, Berlin–New York, 1982.

[559] T. H. Wolff, *An improved bound for Kakeya type maximal functions*, Rev. Mat. Iber. **11** (1995), 651–674.

[560] T. H. Wolff, *Recent work connected with the Kakeya problem*, Prospects in Mathematics, pp. 129–162, H. Rossi (ed.), Amer. Math. Soc., Providence, RI, 1998.

[561] T. H. Wolff, *A sharp bilinear cone restriction estimate*, Ann. of Math. **153** (2001), 661–698.

[562] S. Yano, *An extrapolation theorem*, J. Math. Soc. Japan **3** (1951), 296–305.

[563] A. I. Yanushauskas, *Multiple trigonometric series* [in Russian], Nauka, Novosibirsk 1989.

[564] K. Yosida, *Functional Analysis*, Springer-Verlag, Berlin, 1968.

[565] M. Zafran, *A multilinear interpolation theorem*, Studia Math. **62** (1978), 107–124.

[566] L. V. Zhizhiashvili, *Some problems in the theory of simple and multiple trigonometric and orthogonal series*, Russian Math. Surveys, **28** (1973), 65–127.

[567] L. V. Zhizhiashvili, *Trigonometric Fourier series and their Conjugates*, Mathematics and its Applications, Kluwer Academic Publishers, Vol. 372, Dordrecht, Boston, London, 1996.

[568] W. P. Ziemer, *Weakly Differentiable Functions*, Graduate Texts in Mathematics, Springer-Verlag, New York, 1989.

[569] F. Zo, *A note on approximation of the identity*, Studia Math. **55** (1976), 111–122.

[570] A. Zygmund, *On a theorem of Marcinkiewicz concerning interpolation of operators*, Jour. de Math. Pures et Appliquées **35** (1956), 223–248.

[571] A. Zygmund, *Trigonometric Series*, Vol. I, 2nd ed., Cambridge University Press, Cambridge, UK, 1959.

[572] A. Zygmund, *Trigonometric Series*, Vol. II, 2nd ed., Cambridge University Press, Cambridge, UK, 1959.

[573] A. Zygmund, *On Fourier coefficients and transforms of functions of two variables*, Studia Math. **50** (1974), 198–201.

[574] A. Zygmund, *Notes on the history of Fourier series,* Studies in harmonic analysis, 1–19, MAA Stud. Math., Vol. 13, Math. Assoc. Amer., Washington, DC, 1976.

Index of Notation

$A \subseteq B$	A is a subset of B (not necessarily proper subset)
$A \subsetneqq B$	A is a proper subset of B
A^c	complement of a set A
χ_E	characteristic function of the set E
d_f	the distribution function of a function f
f^*	the decreasing rearrangement of a function f
$f_n \uparrow f$	f_n increases monotonically to a function f
$\mathbf{Z}$	set of all integers
$\mathbf{Z}^+$	$\{1, 2, 3, \dots\}$ the set of all positive integers
$\mathbf{Z}^n$	n-fold product of the integers
$\mathbf{R}$	the set of real numbers
$\mathbf{R}^+$	the set of positive real numbers
$\mathbf{R}^n$	Euclidean n-space
$\mathbf{C}$	the set of complex numbers
$\mathbf{C}^n$	the n-fold product of complex numbers
$\mathbf{T}$	the unit circle idenitfied with the interval $[0, 1]$
$\mathbf{T}^n$	$[0, 1]^n$, the n-dimensional torus
$\lvert x \rvert$	$\sqrt{\lvert x_1\rvert^2 + \cdots + \lvert x_n\rvert^2}$ when $x = (x_1, \dots, x_n) \in \mathbf{R}^n$
$\mathbf{S}^{n-1}$	unit sphere $\{x \in \mathbf{R}^n : \lvert x\rvert = 1\}$
e_j	the vector $(0, \dots, 0, 1, 0, \dots, 0)$ in $\mathbf{R}^n$ with 1 in the jth entry
$\log t$	logarithm with base e of $t > 0$
$\log_a t$	logarithm with base a of $t > 0$, $(1 \neq a > 0)$
$\log^+ t$	$\log t$ when $t \geq 1$, 0 when $t < 1$
$[t]$	the integer part of the real number t
$x \cdot y$	$\sum_{j=1}^{n} x_j y_j$ when $x = (x_1, \dots, x_n)$ and $y = (y_1, \dots, y_n)$
$B(x, R)$	ball of radius R centered at x in $\mathbf{R}^n$
ω_{n-1}	surface area of the unit sphere $\mathbf{S}^{n-1}$
v_n	volume of the unit ball $\{x \in \mathbf{R}^n : \lvert x\rvert < 1\}$
$\lvert A\rvert$	Lebesgue measure of the set $A \subseteq \mathbf{R}^n$

dx	Lebesgue measure
$\operatorname{Avg}_B f$	average $\frac{1}{\|B\|}\int_B f(x)\,dx$ of f over the set B
$\langle f, g\rangle$	real inner product $\int_{\mathbf{R}^n} f(x)g(x)\,dx$
$\langle f \mid g\rangle$	complex inner product $\int_{\mathbf{R}^n} f(x)\overline{g(x)}\,dx$
$\langle u, f\rangle$	action of a distribution u on a function f
p'	$p/(p-1)$, whenever $0 < p \neq 1 < \infty$
$1'$	∞
∞'	1
$f \underset{x\to x_0}{=} O(g)$	$\|f(x)\| \le M\|g(x)\|$ for some M when x lies in a neighorhood of x_0
$f \underset{x\to x_0}{=} o(g)$	$\|f(x)\|\,\|g(x)\|^{-1} \to 0$ as $x \to x_0$
A^t	transpose of the matrix A
A^{-1}	inverse of the matrix A
$O(n)$	space of real matrices satisfying $A^{-1} = A^t$
$\\|T\\|_{X\to Y}$	norm of the (bounded) operator $T : X \to Y$
$A \approx B$	there exists a $c > 0$ such that $c^{-1} \le \frac{B}{A} \le c$
α	a multiindex $(\alpha_1, \dots, \alpha_n)$ with $\alpha_j \in \mathbf{Z}^+ \cup \{0\}$
$\|\alpha\|$	$\|\alpha_1\| + \cdots + \|\alpha_n\|$, the size of a multiindex α
$\partial_j^m f$	the mth partial derivative of $f(x_1, \dots, x_n)$ with respect to x_j
$\partial^\alpha f$	$\partial_1^{\alpha_1} \dots \partial_n^{\alpha_n} f$
C^k	space of functions f with $\partial^\alpha f$ continuous for all $\|\alpha\| \le k$
C_0	space of continous functions with compact support
C_{00}	space of continuous functions that vanish at infinity
$C_0^\infty = \mathcal{D}$	space of smooth functions with compact support
$\mathcal{S}$	space of Schwartz functions
C^∞	$\bigcup_{k=1}^{\infty} C^k$ the space of smooth functions
$\mathcal{D}'(\mathbf{R}^n)$	space of distributions on $\mathbf{R}^n$
$\mathcal{S}'(\mathbf{R}^n)$	space of tempered distributions on $\mathbf{R}^n$
$\mathcal{E}'(\mathbf{R}^n)$	space of distributions with compact support on $\mathbf{R}^n$

$\mathcal{P}$	the set of all complex-valued polynomials of n real variables
$\mathcal{S}'(\mathbf{R}^n)/\mathcal{P}$	the space of tempered distributions on $\mathbf{R}^n$ modulo polynomials
$\ell(Q)$	the side length of a cube Q in $\mathbf{R}^n$
$L^p(X,\mu)$	Lebesgue space over the measure space (X,μ)
$L^p(\mathbf{R}^n)$	$L^p(\mathbf{R}^n, \lvert\cdot\rvert)$
$L^{p,q}(X,\mu)$	Lorentz space over the measure space (X,μ)
$L^p_{\text{loc}}(\mathbf{R}^n)$	space of functions that lie in $L^p(K)$ for any compact set K in $\mathbf{R}^n$
$\lvert d\mu\rvert$	total variation of a finite Borel measure μ on $\mathbf{R}^n$
$\mathcal{M}(\mathbf{R}^n)$	space of all finite Borel measures on $\mathbf{R}^n$
$\mathcal{M}_p(\mathbf{R}^n)$	space of L^p Fourier multipliers, $1 \le p \le \infty$
$\mathcal{M}^{p,q}(\mathbf{R}^n)$	space of translation invariant operators that map $L^p(\mathbf{R}^n)$ into $L^q(\mathbf{R}^n)$
$\lVert\mu\rVert_{\mathcal{M}}$	$\int_{\mathbf{R}^n} \lvert d\mu\rvert$ the norm of a finite Borel measure μ on $\mathbf{R}^n$
$\mathfrak{M}$	centered Hardy-Littlewood maximal operator with respect to balls
M	uncentered Hardy-Littlewood maximal operator with respect to balls
$\mathfrak{M}_c$	centered Hardy-Littlewood maximal operator with respect to cubes
M_c	uncentered Hardy-Littlewood maximal operator with respect to cubes
$\mathfrak{M}_\mu$	centered maximal operator with respect to a measure μ
M_μ	uncentered maximal operator with respect to a measure μ
M_s	strong maximal operator
M_d	dyadic maximal operator
$\mathcal{M}$	grand maximal operator
$L^p_s(\mathbf{R}^n)$	inhomogeneous L^p Sobolev space
$\dot{L}^p_s(\mathbf{R}^n)$	homogeneous L^p Sobolev space
$\Lambda_\alpha(\mathbf{R}^n)$	inhomogeneous Lipschitz space
$\dot{\Lambda}_\alpha(\mathbf{R}^n)$	homogeneous Lipschitz space
$H^p(\mathbf{R}^n)$	real Hardy space on $\mathbf{R}^n$
$B^p_{s,q}(\mathbf{R}^n)$	inhomogeneous Besov space on $\mathbf{R}^n$
$\dot{B}^p_{s,q}(\mathbf{R}^n)$	homogeneous Besov space on $\mathbf{R}^n$
$F^p_{s,q}(\mathbf{R}^n)$	inhomogeneous Triebel-Lizorkin space on $\mathbf{R}^n$
$\dot{F}^p_{s,q}(\mathbf{R}^n)$	homogeneous Triebel-Lizorkin space on $\mathbf{R}^n$
$BMO(\mathbf{R}^n)$	the space of functions of bounded mean oscillation on $\mathbf{R}^n$

Index